Lecture Notes in Computer Science 4516

Commenced Publication in 1973
Founding and Former Series Editors:
Gerhard Goos, Juris Hartmanis, and Jan van Leeuwen

Lorne Mason   Tadeusz Drwiega
James Yan (Eds.)

Managing
Traffic Performance
in Converged Networks

20th International Teletraffic Congress, ITC20 2007
Ottawa, Canada, June 17-21, 2007
Proceedings

 Springer

Volume Editors

Lorne Mason
McGill University
Department of Electrical and Computer Engineering
3480 University Street, Montreal, Quebec, H3A 2A7, Canada
E-mail: lorne.mason@mcgill.ca

Tadeusz Drwiega
Nortel Networks
PO Box 3511, Station C, Ottawa, K1Y 4H7, ON, Canada
E-mail: drwiega@nortel.com

James Yan
Carleton University
Department of Systems and Computer Engineering
1125 Colonel By Drive, Ottawa, K1S 5B6, Ontario, Canada
E-mail: jim.yan@sympatico.ca

Library of Congress Control Number: 2007928129

CR Subject Classification (1998): C.2, C.4, H.4, D.2, J.1, K.6, K.4

LNCS Sublibrary: SL 5 – Computer Communication Networks
and Telecommunications

ISSN   0302-9743
ISBN-10   3-540-72989-5 Springer Berlin Heidelberg New York
ISBN-13   978-3-540-72989-1 Springer Berlin Heidelberg New York

Springer is a part of Springer Science+Business Media

springer.com

© Springer-Verlag Berlin Heidelberg 2007
Printed in Germany

Typesetting: Camera-ready by author, data conversion by Scientific Publishing Services, Chennai, India
Printed on acid-free paper   SPIN: 12074860   06/3180   5 4 3 2 1 0

Preface

Globally, major network operators have incorporated network convergence into their strategy to grow service revenues and reduce capital and operating costs. Convergence occurs in applications (integrated messaging, voice over IP), in network control (portable numbers, SIP, mobile IP), in the transport layer, as well as in the access network. This convergence of networks means that various types of traffic flows, which have been carried by separate specialized networks, now share the resources of a single core IP-based network. In the access, the trends are towards fixed/wireless convergence as well as convergence of various wireless access technologies.

Network convergence will be successful only if the quality of the individual services is maintained in the new network environment without undue increased costs. The quality of service delivery is critically dependent on how network performance and availability, as experienced by the traffic flows, are managed. Managing traffic performance is a critical enabler for success. Reaching the desired performance levels requires adapting processes such as network planning, resource engineering, and network monitoring to the converged network milieu.

ITC-20, with a conference theme *"Managing Traffic Performance in Converged Networks – The Interplay between Convergent and Divergent Forces,"* explored the role of teletraffic engineering in the balancing act between pursuing the vision of network convergence and still managing the performance of individual services and legacy-heterogeneous networks. Meeting customer needs in an ever-increasing service portfolio that permits personalization of these services adds to the complexity and cost of building and operating converged networks. While it has been argued that deregulation of the telecom industry has resulted in improved customer satisfaction through wider choice and competitive pricing, it is clear that this same competitive environment significantly increases the challenge of building a unified converged infrastructure. A consequence of this decentralized ownership and planning structure is greater volatility and uncertainty in the composition, volume and distribution of the traffic supported as well as absence of global network topology knowledge.

These challenges were addressed by the invited speakers. R. Srikant's invited speech entitled "Game-theoretic Models of ISP-ISP and ISP-Customer Interactions" explicitly accounted for the multi-player aspects of the current multi–autonomous system Internet. Alberto Leon-Garcia's invited speech on "Towards Autonomic Networking" provided a proposal for coordinated network management of resources across a diverse set of local or regional management systems. Debasis Mitra's invited talk entitled "Topics in Networking Research" considered the imminent problem of providing communications support for virtual communities by dynamic virtual networks spanning wireless and optical segments of the global Internet. Finally, Uma S. Jha's invited talk "Path Towards Network Convergence and Challenges" described the current directions in wireless networks. Taken together these four invited presentations illuminated a number of challenges and potential solutions to the problems faced in the interplay of convergent and divergent forces as converged networks are deployed.

Converged networks pose new challenges – novel architectures and technologies, varied modes of sharing network resources, new quality targets for services and applications, more flexible signaling and control protocols. The impact of multi-operator competition on traffic demand forecasting, end-to-end QoS provisioning, traffic engineering, and pricing policies needs to be accounted for in a converged network architecture. The sessions on peer-to-peer networks, service overlay networks, and mesh networks discussed the challenges of new architectures. How to manage the traffic performance of new applications and converged networks was covered in the sessions on IPTV, end-to-end delay in converged networks, QoS in converged networks, performance and traffic management of wireless networks, and network performance. Finally, the impact of network convergence on fundamental network functions and traffic engineering practices was discussed in the sessions on server/switch performance, routing, scheduling, traffic engineering, network design for capacity and performance, traffic source and aggregate models, traffic measurements, characterization and demand forecast, and queuing models.

By no means is the answer to the ITC-20 theme final; it is still a work in progress. To explore future paths, a panel session entitled "Alternative Resource Management Architectures for the NGI" wrapped up the technical part of the conference with a lively exchange of views among the panelists chosen for their distinct vision of the future. The panelists were: Jim Roberts (France Telecom R&D, France); Gerald Ash (ATT Labs, USA); Alberto Leon-Garcia (University of Toronto, Canada); and Andrew Odlyzko (University of Minnesota, USA).

The comprehensive and informative program documented in these proceedings would not be possible without the contributions of many people from around the world. First, we would like to thank the authors of all the submitted papers for their interest in ITC-20. Unfortunately, because of time and space limitations, we could not accommodate more papers. We are very grateful to the members of the Technical Program Committee and other reviewers for volunteering so many hours to ensure that high-quality papers were selected. Our gratitude goes to the invited speakers and panelists for their time and effort to share their thought-provoking ideas. To our sponsors and partners, we want to say a big "thank you" for their financial contribution and in-kind support. Finally, we would like to thank the members of the Organizing Committee for their steadfast commitment and contributions since the planning of ITC-20 started in March 2004.

June 2007

Lorne Mason
Tadeusz Drwiega
James Yan

Organization

Organizing Committee Executive Chair

David Hudson   Nortel, Canada

Organizing Committee Chair

James Yan   ITC IAC Member for Canada and Carleton
                             University, Canada

Organizing Committee

François Blouin   Nortel, Canada
Tadeusz Drwiega   Nortel, Canada
Paul Kühn   University of Stuttgart, Germany, ITC IAC
                             Liaison
Lorne Mason   McGill University, Canada
William McCrum   Industry Canada, Canada
Jim O'Shaughnessy   Honorary ITC IAC Member, Canada
David Plant   McGill University, Canada
Philip Richards   NewStep Networks and Honorary IAC Member,
                             Canada
Sergio Tilli   Bell Canada, Canada

Technical Program Committee Co-chairs

Lorne Mason,   McGill University, Canada
Tadeusz Drwiega,   Nortel, Canada

Technical Program Committee

International Members

Ron Addie   University of Southern Queensland, Australia
Sara Alouf,   INRIA, Sophia Antipolis, France
Ake Arvidsson   Ericsson Network Core Products, Soft Center
                             VII, Sweden

Daniel Kofman   ENST Telecom Paris, France
Ulf Korner   Lund Institute of Technology, Sweden
Paul Kühn   University of Stuttgart, Germany
Wai Sum Lai   AT&T Labs Research, USA
Guy Leduc   University of Liège, Belgium
Ralf Lehnert   Telecommunications Technology University
                                   Dresden, Germany
Will Leland   Telcordia, USA
Yonatan Levy   AT&T Labs Research, USA
Xiong-Jian Liang   BUPT, China
 Michel Mandjes   University of Amsterdam, The Netherlands
Deep Medhi   University of Missouri–Kansas City, USA
Debasis Mitra   Bell Laboratories, Alcatel-Lucent, USA
Sándor Molnár   Budapest University of Technology and
                                   Economics, Hungary
Hajime Nakamura   KDDI R&D Lab, Japan
José Neuman de Souza   Universidade Federal do Ceará - UFC, Brazil
Ilkka Norros   VTT Technical Research Centre, Finland
Andrew Odlyzko   University of Minnesota, USA
Achille Pattavina   Politecnico di Milano, Italy
Harry Perros   North Carolina State University, USA
Michal Pióro   Warsaw University of Technology, Poland
Guy Pujolle   Université Pierre et Marie Curie, France
James Roberts   France Telecom, France
Matthew Roughan   University of Adelaide, Australia
Hiroshi Saito   NTT, Japan
Tadao Saito   Toyota Info Technology Center, Japan
Iraj Saniee   Bell Laboratories, Alcatel-Lucent, USA
Khoshrow Sohraby   University of Missouri-Kansas City, USA
Maciej Stasiak   Poznan University of Technology, Poland
Yutaka Takahashi   Kyoto University, Japan
Peter Taylor   University of Melbourne, Australia
Phuoc Tran-Gia   University of Wurzburg, Germany
Danny H.K.Tsang   Hong Kong University of Science and
                                   Technology, Hong Kong
Ivan Tsitovich   Institute for Problems of Information
                                   Transmission, Russia
Kurt Tutschku   University of Wurzburg, Germany
Hans van den Berg   TNO Telecom/University of Twente,
                                   The Netherlands
Rob van der Mei   CWI, The Netherlands
J.P. Vasseur   Cisco Systems, USA
Manuel Villen-Altamirano   Telefónica I+D, Spain
Jorma Virtamo   Helsinki University of Technology, Finland
Jean Walrand   University of California, Berkeley, USA

Zhanhong Xin   BUPT, China
Qin Yang   Nanyang Technological University, Singapore
Yuanyuan Yang   State University of New York at Stony Brook,
                                    USA
Wuyi Yue   Konan University, Japan
Moshe Zukerman   University of Melbourne, Australia

Canadian National Members

Osama Aboul-Magd   Nortel, Ottawa
Attahiru Alfa   University of Manitoba, Winnipeg
Ron Armolavicius   Nortel, Ottawa
Maged Beshai   telecom Consultant, Ottawa
Gregor von Bochmann   University of Ottawa
Azzedine Boukerche   University of Ottawa
Raouf Boutaba   University of Waterloo
Mark Coates   McGill University, Montreal
Zbigniew Dziong   École de Technologie Supérieure, Montreal
Abraham Fapojuwo   University of Calgary
Sudhakar Ganti   University of Victoria
André Girard   INRS-EMT, Montreal
Wayne Grover   TRLabs, Edmonton
Hossam Hassanein   Queen's University, Kingston
Changcheng Huang   Carleton University, Ottawa
Rainer Iraschko   Telus, Calgary
Alberto Leon-Garcia   University of Toronto
Cyril Leung   University of British Columbia, Vancouver
Victor Leung   University of British Columbia, Vancouver
Jorg Liebeherr   University of Toronto
Mike MacGregor   TRLabs, Edmonton
Peter Marbach   University of Toronto
Mustafa Mehmet Ali   Concordia University, Montreal
Hussein Mouftah   University of Ottawa
Biswajit Nandy   Solana Networks, Ottawa
Jean Régnier   OZ Communications, Montreal
Brunilde Sansò   École Polytechnique de Montréal
Anand Srinivasan   EION, Ottawa
Ljiljana Trajkovic   Simon Fraser University, Vancouver
Tom Vilmansen   CRTC, Ottawa
Carey Williamson   University of Calgary
Vincent Wong   University of British Columbia, Vancouver
James Yan   Carleton University, Ottawa
Oliver Yang   University of Ottawa

List of Reviewers

Aalto Samuli
Abadpour Arash
Aboul-Magd Osama
Abu Ali Najah
Addie Ronald
Aida Masaki
Al-Manthari Bader
Alouf Sara
Alvarez Calvo Miguel
Anantharam Venkat
Aracil Javier
Armolavicius Ron
Arvidsson Åke
Ash Gerald
Banchs Albert
Bannazadeh Hadi
Basu Kalyan
Berberana Ignacio
Beshai Maged
Binzenhoefer Andreas
Blondia Chris
Bocci Matthew
Bochmann von Gregor
Boel Rene
Borgonovo Flaminio
Borst Sem
Bose Sanjay
Boukerche Azzedine
Boutaba Raouf
Braga Reinaldo
Bregni Stefano
Bruneel Herwig
Bziuk Wolfgang
Bæk Iversen Villy
Campista Miguel Elias
Cano Tinoco Alejandra
Casaca Augusto
Cesana Matteo
Cetinkaya Coskun
Charzinski Joachim
Chaturvedi Ajit
Chaudet Claude
Chemouil Prosper
Chen Min
Chen Thomas

Chen Yue
Cheng Tee Hiang
Choudhury Gagan
Clérot Fabrice
Coates Mark
Costeux Jean-Laurent
Courcoubetis Costas
Cunha Daniel
Cuthbert Laurie
de Souza José Neuman
Devetsikiotis Michael
Doshi Bharat
Doverspike Robert
Drwiega Tadeusz
Dziong Zbigniew
El-Khatib Khalil
Elbiaze Halima
Elkashlan Maged
Elteto Tamas
Emmert Barbara
Emstad Peder
Eun Do Young
Everitt David
Fallahi Afshin
Fapojuwo Abraham
Farago Andras
Farha Ramy
Fattah Hossam
Fawaz Wissam
Fernandes Natalia
Fernandez Marcial
Fernández-Palacios Juan
Fiedler Markus
Fitzpatrick Paul
Foh Chuan Heng
Fratta Luigi
Gale Charles
Ganti Sudhakar
García-Osma María Luisa
Gauthier Vincent
Görg Carmelita
Gefferth Andras
Ghosh Indradip
Gibbens Richard
Girard André

Gonzalez Soto Oscar
Granelli Fabrizio
Gravey Annie
Gregoire Jean-Charles
Grover Wayne
Guan Yong
Guillemin Fabrice
Guo Song
Gurtov Andrei
Gusak Oleg
Guyard Frederic
Harris Richard
Hassanein Hossam
He Peng
Henjes Robert
Herzberg Meir
Holanda Raimir
Hoßfeld Tobias
Huang Changcheng
Hwang Mintae
Hyytiä Esa
Ibrahim Ali
Imre Sándor
Iraschko Rainer
Ishibashi Brent
Ivanovich Milosh
Jamalipour Abbas
Jiang Libin
Jin Mushi
Jin Youngmi
Jourdan Guy-Vincent
Kamal Ahmed
Kasahara Shoji
Kawahara Ryoichi
Kawano Hiroyuki
Kawashima Konosuke
Kawata Takehiro
Kesidis George
Key Peter
Khavari Khashayar
Killat Ulrich
Kinsner Witold
Klaoudatos Gerasimos
Klopfenstein Olivier
Ko King-Tim

Körner Ulf
Kuehn Paul J.
Kwan Raymond
L. S. da Fonseca Nelson
Lai Waisum
Lambadaris Ioannis
Lambert Joke
Lan Tom
Lancieri Luigi
Langar Rami
Lassila Pasi
Lau Wing Cheong
Le Long
Le Grand Gwendal
Leduc Guy
Lee Ian
Lehnert Ralf
Leland Will
Levy Yoni
Li Jie
Liang Xiong Jian
Liebeherr Jorg
Lin Haitao
Liu Xiao-Gao
Lung Chung-Horng
Ma Chi
Ma Ming
MacGregor Mike
Maeder Andreas
Magedanz Thomas
Majumdar Shikharesh
Mandjes Michel
Mang Xiaowen
Manjunath D.
Marbach Peter
Martin Ruediger
Martinez-Bauset Jorge
Martins Philippe
Mason Lorne
Masuyama Hiroyuki
Matrawy Ashraf
Medhi Deep
Mehmet Ali Mustafa
Menth Michael
Michailidis George
Mitra Debasis
Molina Maurizio

Montuno Delfin
Moraes Igor
Mountrouidou Xenia
Musacchio John
Nakamura Hajime
Nandy Biswajit
Neame Timothy
Neglia Giovanni
Neishaboori Azin
Németh Felicián
Nikolova Dessislava
Niyato Dusit
Noel Eric
Nordström Ernst
Norros Ilkka
Northcote Bruce
Odlyzko Andrew
Oechsner Simon
Ohzahata Satoshi
Olivarrieta Ivonne
Olivier Philippe
Pan Deng
Pang Qixiang(Kevin)
Patangar Pushkar
Peng Cheng
Penttinen Aleksi
Perros Harry
Pióro Michal
Pitts Jonathan
Pla Vicent
Popescu Adrian
Pries Rastin
Pujolle Guy
Qian Haiyang
Qiao Ying
Qin Yang
Rabbat Michael
Radwan Ayman
Rahman Md. Mostafizur
Ramaswami Vaidyanathan
Regnier Jean
Ridoux Julien
Roberts James
Rodríguez Sánchez Jorge
Roijers Frank
Roodi Meysam
Rossi Dario

Roughan Matthew
Rubino Gerardo
Rumsewicz Michael
Saito Hiroshi
Saniee Iraj
Sanso Brunilde
Santos Aldri
Schlosser Daniel
Schormans John
Scoglio Caterina
Seddigh Nabil
Sharma Shad
Sheltami Tarek
Shioda Shigeo
Shu Feng
Shu Yantai
Shuaib Khaled
Singh Yatindra
Sofia Rute
Sohraby Khosrow
Song Joo-Han
Spaey Kathleen
Srinivasan Anand
Stasiak Maciej
Sulyman Ahmed Iyanda
Tachibana Takuji
Takahashi Yutaka
Talim Jerome
Taveira Danilo
Taylor Peter
Telikepalli Radha
Tian Wenhong
Tinnirello Ilenia
Tirana Plarent
Trajkovic Ljiljana
Trinh Tuan
Tsang Danny H. K.
Tsitovich Ivan
Tutschku Kurt
Valaee Shahrokh
Van den Berg Hans
Van den Wijngaert Nik
van der Mei Rob
Vaton Sandrine
Velloso Pedro
Verticale Giacomo
Villen-Altamirano Manuel

Vilmansen Tom
Virtamo Jorma
Voorhaen Michael
Walrand Jean
Wang Sheng
Wang Wenye
Whiting Phil
Williamson Carey
Wong Vincent
Woodside Murray
Wosinska Lena
Xiao Gaoxi
Xiao Jin
Xiong Kaiqi

Yan James
Yang Min
Yang Oliver
Yao Na
Yavuz Emre
Yue Wuyi
Zhang Baoxian
Zhang Yide
Zhang Yiming
Zhang Yingjun
Zhao Miao
Zhou Peifang
Zukerman Moshe
Østerbø Olav

20th International Teletraffic Congress
Ottawa, Canada   June 17-21, 2007

Charter Sponsors

Sponsors

Partners

Agile All-Photonic Networks (AAPN)   **Department of**
Réseaux agiles tout-photoniques(RATP)   **Systems and Computer Eng**

Table of Contents

Routing

Server/Switch Performance

Service Overlay Networks

Traffic Source and Aggregate Models

Mesh Networks - Performance Optimization - I

QoS in Converged Networks

Traffic Engineering

Mesh Networks - Performance Optimization - II

End-to-End Delay in Converged Networks

Queuing Models - I

Performance of Peer-to-Peer Networks

Loss/Blocking Probability

Traffic Management in Wireless Networks

Traffic Measurements and Characterization

Network Design for Capacity and Performance

Performance of Wireless Networks

Scheduling

Plenary Session - Contributed Papers

Game-Theoretic Models of ISP-ISP and ISP-Customer Interactions

R. Srikant

University of Illinois at Urbana-Champaign, Urbana, IL 61801, USA
rsrikant@uiuc.edu
http://www.ifp.uiuc.edu/~srikant

The talk will consist of two parts. In the first part of the talk, we will examine how transit and customer prices and quality of service are set in a network consisting of multiple ISPs. Some ISPs may face an identical set of circumstances in terms of potential customer pool and running costs. We will examine the existence of equilibrium strategies in this situation and show how positive profit can be achieved using threat strategies with multiple qualities of service. It will be shown that if the number of ISPs competing for the same customers is large then it can lead to price wars. ISPs that are not co-located may not directly compete for users, but may be nevertheless involved in a non-cooperative game of setting access and transit prices for each other. They are linked economically through a sequence of providers forming a hierarchy, and we study their interaction by considering a multi-stage game. We will also consider the economics of private exchange points and show that they could become far more wide spread then they currently are. This is joint work with Srinivas Shakkottai (UIUC) (see [1]).

In the second part of the work, we consider interactions between an ISP and its customers. To maximize its revenue, an ISP should set access charges such that it extracts the full utility from the users of the network. However, such a pricing scheme would have to be based on usage and network conditions. In reality, most networks charge a fixed periodic access price. We will present an algorithm for setting such a price which we call the marginal user principle. We will then characterize the price of simplicity, i.e., the loss in revenue to the ISP due to the use of such a simple fixed-price scheme and discuss strategies for increasing the revenue by using two-tier pricing schemes. This is joint work with Srinivas Shakkottai (UIUC), Daron Acemoglu (MIT) and Asuman Ozdaglar (MIT) (see [2]).

References

1. S. Shakkottai and R. Srikant. Economics of Pricing with Multiple ISPs. IEEE/ACM Transactions on Networking (2006) 1233-1245
2. S. Shakkottai, D. Acemoglu, A. Ozdaglar and R. Srikant. The Price of Simplicity. Working Paper (2007)

L. Mason, T. Drwiega, and J. Yan (Eds.): ITC 2007, LNCS 4516, p. 1, 2007.
© Springer-Verlag Berlin Heidelberg 2007

Towards Autonomic Communications

Alberto Leon-Garcia

University of Toronto, Canada
leongarcia@comm.utoronto.ca

Abstract. In this talk we provide an overview of our work on autonomic service management systems. The purpose of these systems is to ensure delivery of a multiplicity of network-based applications and content, according to service metrics and in the face of variable demand and equipment failure. We begin by describing our vision of the application-oriented infrastructure that must emerge from the convergence of computing and communications technologies. We then describe an autonomic service architecture that builds on this highly flexible and controllable infrastructure to support the delivery of applications at very low cost and to enable the introduction of new services at very rapid pace.

We describe approaches for creating resource abstractions that accommodate communications and computing capabilities so that service management can be carried out by a single common control framework. We also describe how an application can be supported by the allocation of virtual resources that are managed by autonomic managers that ensure that service metrics are met even in the presence of changing demand. We then present recent results for autonomic resource management algorithms for: 1. the management of network flows; 2. the management of application flows; 3. the management of radio resources in cognitive radio settings.

L. Mason, T. Drwiega, and J. Yan (Eds.): ITC 2007, LNCS 4516, p. 2, 2007.
© Springer-Verlag Berlin Heidelberg 2007

Topics in Networking Research*

Debasis Mitra

Bell Labs, Alcatel-Lucent,
Murray Hill, NJ 07974, USA
mitra@research.bell-labs.com

Abstract. What are the big movements in networking that researchers should heed? A standout is the global spread of communities of interest (the networking analogue of the flat world) and their need for "dynamic virtual networks" that support rich applications requiring resources from several domains. The imperative for inter-networking, i.e., the enablement of coordinated sharing of resources across multiple domains, is certain. This challenge has many facets, ranging from the organizational, e.g., different, possibly competing, owners to the technical, e.g., different technologies. Yet another key characteristic of the emerging networking environment is that the service provider is required to handle ever-increasing uncertainty in demand, both in volume and time. On the other hand there are new instruments available to handle the challenge. Thus, inter-networking and uncertainty management are important challenges of emerging networking that deserve attention from the research community.

We describe research that touch on both topics. First, we consider a model of data-optical inter-networking, where routes connecting end-points in data domains are concatenation of segments in the data and optical domains. The optical domain in effect acts as a carrier's carrier for multiple data domains. The challenge to inter-networking stems from the limited view that the data and optical domains have of each other. Coordination has to be enabled through parsimonious and qualitatively restrictive information exchange across domains. Yet the overall optimization objective, which is to maximize end-to-end carried traffic with minimum lightpath provisioning cost, enmeshes data and optical domains. This example of inter-networking also involves two technologies. A mathematical reflection of the latter fact is the integrality of some of the decision variables due to wavelengths being the bandwidth unit in optical transmission. Through an application of Generalized Bender's Decomposition the problem of optimizing provisioning and routing is decomposed into sub-problems, which are solved by the different domains and the results exchanged in iterations that provably converge to the global optimum.

In turning to uncertainty management we begin by presenting a framework for stochastic traffic management. Traffic demands are uncertain and given by probability distributions. While there are alternative perspectives (and metrics) to resource usage, such as social welfare and network revenue, we adopt the latter, which is aligned with the service provider's interests. Uncertainty

---

* Joint work with Qiong Wang and Anwar Walid, Bell Labs, Murray Hill.

L. Mason, T. Drwiega, and J. Yan (Eds.): ITC 2007, LNCS 4516, pp. 3–4, 2007.

introduces the risk of misallocation of resources. What is the right measure of risk in networking? We examine various definitions of risk, some taken from modern portfolio theory, and suggest a balanced solution. Next we consider the optimization of an objective which is a risk-adjusted measure of network revenue. We obtain conditions under which the optimization problem is an instance of convex programming. Studies of the properties of the solution show that it asymptotically meets the stochastic efficiency criterion. Service providers' risk mitigation policies are suggested. For instance, by selecting the appropriate mix of long-term contracts and opportunistic servicing of random demand, the service provider can optimize its risk-adjusted revenue. The "efficient frontier", which is the set of Pareto optimal pairs of mean revenue and revenue risk, is useful to the service provider in selecting its operating point.

Path Towards Network Convergence and Challenges

Uma S. Jha[*]

Qualcomm CDMA Technologies
ujha@qualcomm.com

Abstract. The emerging trends in the market place are driving towards convergence at device, network, and application levels. Today consumers are looking for a single device capable of fulfilling their daily communication, computing, and personal productivity application needs. Businesses are working tirelessly to offer enhanced quality of service and application portability to consumers irrespective of their locality, mobility, and system preferences. These user preferences squarely challenge existence of stovepipe platforms, islands of interoperable systems, centralized command and control structure, and fragmented connectivity landscape. The new paradigm is all about being technology, platform, operating system, and network agnostic. There are clear indications, which confirms this transformation - e.g. connectivity migrating form circuit switched to packet switched, system resources and application transitioning from centralized to distributed (e.g. thin client, P2P), applications becoming platform and operating systems agnostic (e.g. Java applets), fixed and mobile convergence (IP centric - IMS), and aggregation of voice, data, and video (triple play).

This paradigm shift brings challenges and opportunities to system developers and raises the threshold of entry. To become a successful player in this new world one needs to understand the full value chain of the new ecosystem. The existing base of deployed system, equipments, and hierarchies can't be overhauled overnight due to business, political, and regulatory reasons and one need to work in the confine of these constraints. Coexistence and collaboration across the heterogeneous systems and platforms need to be evaluated and commonality across must be exploited. We have to take advantage of the recent advancements in VLSI, system integration, and signal processing technologies. The mobile industry has been in the forefront and enabler of this transformation. They have taken full advantage of Moore's law in integrating myriad of complementary applications such as WLAN, BT, 3D gaming, GPS, camera, audio/video streaming, and personal productivity gadgets to name a few. And this all has been done under the constraints of the device size and battery power. It is no coincidence that today's mobile device deliver the capabilities of PC of yester years.

There has been a lot of progress already made towards the global convergence but more needs to be done. If current trend continues then there is a safe bet to say that today's mobile evolution would be able to reduce the "digital divide" for the human kind globally.

---

[*] Sr. Member IEEE.

L. Mason, T. Drwiega, and J. Yan (Eds.): ITC 2007, LNCS 4516, p. 5, 2007.
© Springer-Verlag Berlin Heidelberg 2007

Dimensioning Multicast-Enabled Networks for IP-Transported TV Channels

Zlatka Avramova[1], Danny De Vleeschauwer[1,2],
Sabine Wittevrongel[1], and Herwig Bruneel[1]

[1] SMACS Research Group, TELIN, Ghent University
Sint-Pietersnieuwstraat 41, B-9000 Gent, Belgium
{kayzlat,sw,hb}@telin.ugent.be
[2] Network Strategy Group, Alcatel-Lucent
Copernicuslaan 50, B-2018 Antwerp, Belgium
danny.de_vleeschauwer@alcatel-lucent.be

Abstract. The required capacity for an IP TV distribution network decreases if the right TV channels are multicast, while the others rely on unicast. In [1] we proposed mathematical models to determine this required capacity in two network scenarios: the "static" and "dynamic" scenarios. As the exact calculations turned out to be too tedious, we developed approximative models, assuming the variables have a Gaussian distribution and proved that the approximations yield results close to the exact results. In this paper, we build further on this approximation. First, we determine the optimal parameters in the static scenario. Second, we compare the static case resource demand function (with optimal settings) to the dynamic case function and we prove the superiority of the dynamic dimensioning approach. Then, the applicability regions of both scenarios (where gain is achieved) are explored. Finally, we illustrate the methodology by calculating the required capacity in some realistic examples.

Keywords: unicast, multicast, capacity planning, IPTV, mobile TV.

1 Introduction

Less than a decade ago communication operators acted only in their own sector specialised in delivering certain communication services and occupying their own well-defined market (e.g., PSTN (Public Switched Telephone Network) services, TV distribution, data transport). Currently, we are witnessing a changing network paradigm, i.e., a vertical and horizontal convergence of network platforms delivering mixed services. The ubiquitous IP network protocol, with proved flexibility and efficiency, is at the very center of this convergence. Mobile network operators migrate from purely voice and narrow-band data services to enhanced multimedia telecommunication services, such as video telephony or (IP-based) mobile TV. Similar trends exist in the fixed broadband networks, where we witness competition between traditional telephone operators willing to offer TV services and cable TV

L. Mason, T. Drwiega, and J. Yan (Eds.): ITC 2007, LNCS 4516, pp. 6–17, 2007.

network operators willing to offer voice and data services. As far as TV services offerings are concerned, all evolve to the distribution of digital video over an IP-based network. IPTV is offered over fixed networks (either DSL (Digital Subscriber Line) or Cable) by many operators already worldwide.

To offer a large enough set of TV channels to the users at reasonable capacity demand, the network has to be carefully dimensioned. For example, some mobile operators offer streaming TV/radio services setting up a new unicast channel for every new active user, which will lead to a capacity shortage when the service gets more popular. Recently, there have been some trials with multicast technologies for UMTS (Universal Mobile Telecommunication Services) networks (i.e., MBMS (Multimedia Broadcast Multicast Service)) aiming to alleviate this problem. Other service providers offer pure multicast/broadcast TV transport to an unlimited audience (e.g., in a DVB-H/DVB-T (Digital Video Broadcasting – Handheld/Terrestrial) network) but the return path (the interactivity feedback channel) is typically organised through a cellular network of a mobile operator. A mixed network architecture, referred to as a hybrid architecture, allows for a TV channel to be either unicast or multicast/broadcast. Such an architecture combines the advantages of both transmitting techniques and its potential to offload traffic from the unicast part of the network to the multicast-enabled part has been explored in [10], [11]. Similarly, in broadcasting TV over a fixed network, the channels to multicast must be elected such as to minimise the required capacity. The dimensioning schemes we put forward in [1] and further study in the current paper, addressing the described capacity planning problems, apply to hybrid as well as to all-unicast or all-multicast networks.

This paper is organised as follows. In Section 2 we present a channel popularity model, that gives for each channel the probability it is requested to be watched. Section 3 presents previous work and in Section 4 we explore the optimal parameters in the static case. Section 5 compares both dimensioning approaches and gives the regions of capacity gain in comparison to an all-multicast network. In Section 6 we illustrate the methodology by realistic examples. We also verify our theoretical results against simulation results for dimensioning multicast-enabled networks in Section 7. The last Section 8 summarises the contribution of our work and its perspectives.

2 Channel Popularity Distribution

For proper dimensioning of the network, user behaviour should be modeled. Some studies (e.g., [4], [13]) show that the channel popularity distribution (probability of a channel being watched by an active user) has a power-law form (Zipf distribution [3]). We will widely use this assumption throughout this article, although our models also hold for other types of channel popularity distributions.

We assume that the probability π_k that a channel is watched is given by the formula:

$$\pi_k = dk^{-\alpha}, \quad \text{for } k=1, 2, ..., K, \tag{1}$$

where k is the channel's (popularity) rank, d is a normalisation constant, and K is the number of channels. As α increases, the median of the Zipf distribution shifts to smaller k (see Figure 1a) and the distribution becomes less heavy-tailed.

We monitored the fifteen most popular channels that a French DSL IPTV operator [12] announces real-time on its website. In Figure 1b we show experimental results: eight snapshots of the fraction of the user population watching a channel versus the rank of the channel are shown (four snapshots taken between 2pm and 3pm and four between 6pm and 7pm). It can be seen that all of them are well approximated by Zipf law (the thick black line in the figure) with α=0.57 in this case. Remark that at other hours the shape of the popularity (i.e., the parameter α) can be different.

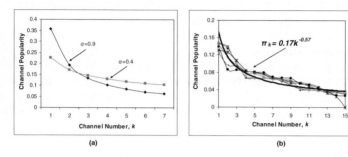

Fig. 1. Zipf distribution: (a) influence of the parameter α on the shape of the Zipf distribution; (b) eight snapshots of channel popularity distribution approximated by Zipf law with α=0.57

3 Previous Work

In [1] we studied the (instantaneous) capacity demand for two specific scenarios (referred to as "static" and "dynamic"), with as aim to dimension the network given the necessary network dimensioning parameters: the number of subscribers N, the activity grade a (i.e., the probability of a user actively watching a TV channel) and the channel popularity distribution, under a certain blocking probability P_{block}.

A network operator willing to offer a set (so-called "bouquet") of (IP-transported) TV channels, can elect for each channel to either multicast (broadcast) it or to unicast it to its active users. In certain technologies (e.g. MBMS), it may take more bandwidth to multicast a channel than to unicast it: for instance in the case of a mobile network, a multicast channel requires more power, hence more capacity, as it needs to reach also the outskirts of the cell. We denote this bandwidth ratio of a multicast towards a unicast channel by β.

The operator has still a choice between a static scenario and a dynamic scenario as will be explained below. In order to save on bandwidth in (the distribution part of) the network, the operator may choose once and for all (because providing a multicast tree for a particular channel may be slow) to multicast the M most popular channels (i.e., $M \le K$) and to unicast the other ones only on request. This scenario we call the *"static scenario"*. We note that it is not efficient to multicast a channel as long as there are less than β users tuned into it. Therefore, (for cases when the switch to multicast can happen without considerable delay) we also explore the *"dynamic scenario"* where a particular channel is either unicast or multicast depending on the number of viewers requesting it. If more than β users watch it, it is (instantaneously) multicast.

For both scenarios we derived closed-form expressions to estimate the capacity demand R, i.e., R_S in the static scenario and R_D in the dynamic scenario. As we ran into computational problems, we assumed that the variables of interest follow a Gaussian distribution [6] and we showed that this approximation yields accurate results (under certain assumptions which are easily met for the systems under consideration). In the present paper we will use only the Gaussian approximation formulae derived in [1] to estimate the resource demand R (R_S and R_D respectively). The approximate formula to estimate R_S is:

$$R_S = \beta M + N a P_u + erfc^{-1}(P_{block})\sqrt{N a P_u (1 - a P_u)} \quad , \tag{2}$$

where P_u is the probability that an active user watches a unicast channel and is actually the sum of the probabilities π_k for the channels from the bouquet that are set aside for unicast on demand (i.e., $M < k \le K$):

$$P_u = d \sum_{i=M+1}^{K} i^{-\alpha} . \tag{3}$$

The function $erfc^{-1}(X)$ is the inverse of the tail distribution function (TDF) of the normal distribution with zero mean and unit variance, i.e., it gives the value which a zero mean unit variance Gaussian variable exceeds with probability X.

The capacity R_D is calculated by more complex formulae. For the exact formulae for the average and standard deviations of the number of unicast channels n_u, the number of multicast channels n_m and the capacity demand R_D, we refer to [1].

4 Optimal Parameters for the Static Case

For the static scenario described in [1], we assumed that the number of multicast channels M is predetermined. In what follows we will develop a methodology for choosing the optimal M.

If all K channels are multicast, then the necessary bandwidth (expressed in unicast channel units) is simply equal to βK. In this case, there is no flexibility and all the channels are broadcast all the time regardless of whether or not they are watched. On the other hand, if all channels are unicast, then in the worst case the maximum bandwidth needed (in unicast channel units again) is equal to the number of users N in the system. Normally the number N of users is much larger than the number K of channels, even larger than βK (because β assumes small values). There is obviously a trade-off between flexibility and reserving only reasonable amount of network resources, which we try to address by our dimensioning methods.

A typical dimensioning curve is shown in Figure 2a. It gives an estimate of the necessary number of channels, i.e., the bandwidth (in unicast channel units) that is sufficient except for a small fraction P_{block} of the time. In other words, P_{block} is the probability that the actual capacity demand exceeds the calculated capacity R_S, i.e., that the available network resources are not enough.

Clearly, the resource demand R_S (given a blocking probability P_{block}) must be less than βK (in order to achieve resource gain) and is by nature greater than or equal to

βM. Otherwise, if $R_S \geq \beta K$, from an economical point of view it would be more beneficial to broadcast all channels (i.e., $M=K$) leading to a capacity of exactly βK.

Figure 2b shows that for a certain set of parameters (and a fixed P_{block}), the function $R_S=f(M)$ according to eq. (2) may have a minimum within the interval 0 to K (or may be monotonously decreasing as M increases and hence have an extremum for $M=K$). That is to say that in certain cases there could be an optimal M at which the target P_{block} is achieved with smallest network capacity R_S.

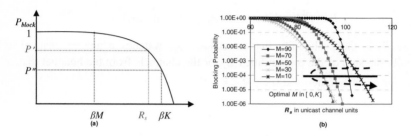

Fig. 2. (a) Tail distribution function of r_S and (b) its variation as a function of M

The minimum of the function $R_S=f(M)$ can be found either by evaluating it in the range $M=0..K$ or by taking its derivative with respect to M and finding the M at which it changes its sign. To this end, we approximate the sum P_u by an integral, which can be justified if the number of channels K is large:

$$P_u = d \int_{M+1}^{K+1} x^{-\alpha} dx = d \frac{(K+1)^{1-\alpha} - (M+1)^{1-\alpha}}{(1-\alpha)} . \qquad (4)$$

In what follows, we will elaborate on the choice of an optimal M. More specifically, we will consider the impact of the parameters α, N (a) and K.

If α is large, there are a few channels with large popularity, which are worth broadcasting all the time. Only in rare cases a channel outside the multicast set will be requested via unicast thus contributing only to a limited extent to the capacity demand. Figure 3a illustrates how the $R_S=f(M)$ function changes with increasing α from a monotonous one to a function with concave point (minimum) within the interval 0 to K. The cases with α_1 and α_2 do not lead to a capacity gain, because the capacity demand is (unless $M=K$) always larger than βK, implying that a switched broadcast technique is useless. On the contrary, in the case with α_3, if M is set to its optimal value (multicasting channels up to around channel 60), a capacity gain of approximately 25% is achieved. Therefore, the network operator is said to have extra programming capacity, called "virtual programming capacity" in [4]. This example proves that if α assumes large values, a capacity decrease is possible.

The absolute value of the optimal R_S increases with increasing number of offered channels K, but less than linearly to K. Hence, the relative resource gain (comparing to βK) is thus increasing. This is a very nice result since a network operator is more attractive when capable to offer a larger bouquet of channels to the audience.

Finding the minimum of the function $R_S=f(M)$ (if available in the interval 0 to K) means finding an M that minimises the resource demand for a given set of network parameters. However, we must also note that the function $R_S=f(M)$ varies more for small values of M (see Figure 3a and 3b). Since the two parameters studied above (i.e., α and K) are more or less characteristic of the system and not subject to very frequent changes, this variance is not so great importance, but the situation is different as far as the number of users N and their activity grade a are concerned.

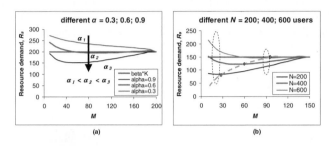

(a)   (b)

Fig. 3. (a) Impact of α and (b) impact of N on the extremum of $R_S =f(M)$

When dimensioning, it has to be kept in mind that the number of users can change (e.g., mobile user entering and leaving the cell). This type of changes can occur quite often and unpredictably and it can be misleading and risky to opt for the M that is chosen solely on the basis of minimising the function $R_S=f(M)$. It is safer to take M larger rather than smaller than its optimal value. We will prove this by considering the derivative of the function for R_S to N. It is given by the following expression:

$$\frac{\partial R_s}{\partial N} = aP_u + \frac{erfc^{-1}(P_{block})\sqrt{aP_u(1-aP_u)}}{2\sqrt{N}} \quad . \tag{5}$$

The variance of the function for R_S by N diminishes as M increases because then P_u assumes smaller values and the variation of the resource demand function decreases. The three graphs in Figure 3b correspond to a number of users N respectively of 200, 400, and 600. If the number of multicast channels is chosen (comparatively) low, fixed to e.g. $M=30$ (hence $P_u=0.344$ in the example), when the number of users grows (or respectively, when a grows), it is possible that R_S soon surpasses the limit βK channels (equal to 150 in this example as $\beta=1$ and $K=150$). On the other hand, if M is chosen equal to e.g. 130 (hence $P_u=0.033$ in the example), then the network engineers are still "on the safe side" in occurrence of great surge in any of both parameters (N or a). The derivative (eq. (5)) is 0.2 in the first example case and it is 0.009 in the latter case, which shows a lower sensitivity by a factor of more than 20 in the second case.

The impact of increasing the number N of users is studied in [4] as well. They discover that for a given increase in the number N of users, the number K of channels viewed does not change proportionally, but rather in a logarithmic fashion and this leads to (relative) saving on network resources (through switched broadcast). Here we observe the same behaviour in Figure 3b: the optimum of the three curves lies on an increasing, convex function (the thick dashed line in the graph).

5   Comparison Between Static and Dynamic Scenario

Above we presented considerations on how to choose an optimal threshold M of multicast channels with largest popularity in the static scenario.

In that scenario, the telecommunication operator selects the channels to multicast a priori and does so regardless of the fact whether they are watched or not. In a dynamic scenario, though, the IPTV provider will monitor the number of users tuned in to a certain channel and multicast the channel only if the number of users is greater than β. We will first illustrate the superiority of the dynamic scenario approach over the static one (with calculations executed with the approximate formulae from [1] for R_D and eq. (2) for R_S).

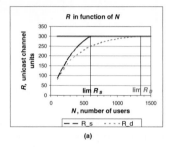
(a)

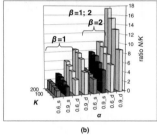
(b)

Fig. 4. Illustration of the superiority of the dynamic dimensioning scheme: (a) resource demand R_S or R_D as a function of the number of users N and (b) limit ratio N/K in both dimensioning schemes as a function of K and α (suffix _S and _D for static and dynamic cases respectively)

In Figure 4a, a comparison is made of R_S and R_D in function of N (with the same network settings). The network parameters are as follows: $K=150$, $a=0.7$, $\alpha=0.6$, $\beta=2$, $P_{block}=10^{-4}$, and M is taken to minimise the function $R_S=f(M)$. We demonstrate how the capacity demand R increases logically with increasing number of users. In both scenarios we consider that as long as R_S or R_D is less than βK (300 in this example), it is feasible to employ them and thus achieve capacity gain; otherwise, when R_S or $R_D \geq \beta K$, there is no benefit in deploying such dimensioning schemes but rather multicast (broadcast) all channels. We see that in the static case R reaches the limit earlier (at smaller value of N, i.e., around 600) than in the dynamic case (N is around 1350). This is an illustration of the superiority of the dynamic scheme.

To prove that the capacity R_S is always larger than (or equal to) R_D, we use the exact definitions of r_S and r_D from [1]:

$$r_S = \beta M + \sum_{k=M+1}^{K} n_k, \tag{6}$$

$$r_D = \beta \sum_{k=1}^{K} 1_{\{n_k > \beta\}} + \sum_{k=1}^{K} n_k 1_{\{n_k \leq \beta\}}, \tag{7}$$

where n_k is the number of users tuned in to channel k and $1_{\{\}}$ is the indicator function. We rewrite the equations in the following form, where both r_S and r_D have one term per channel and hence in total K terms:

$$r_S = \sum_{k=1}^{M} \beta + \sum_{k=M+1}^{K} n_k \, , \tag{8}$$

$$r_D = \sum_{k=1}^{K} \left(\beta 1_{\{n_k > \beta\}} + n_k 1_{\{n_k \leq \beta\}} \right) \, . \tag{9}$$

We compare the terms one by one. For the first M terms for any value of n_k holds:

$$\beta 1_{\{n_k > \beta\}} + n_k 1_{\{n_k \leq \beta\}} \leq \beta \, . \tag{10}$$

For the last K-M terms, it can be readily seen (for any value of n_k) that:

$$\beta 1_{\{n_k > \beta\}} + n_k 1_{\{n_k \leq \beta\}} \leq n_k \, . \tag{11}$$

So, each of the terms of r_S is larger than or equal to the corresponding term of r_D, which makes R_S always larger than or equal to R_D. For small values of N though, there is hardly a difference in the values of R_S or R_D, because there is hardly any need to multicast and both dimensioning approaches yield similar results (the variable "number of unicast channels" is dominant over "number of multicast channels").

Another way to show the superiority of the dynamic dimensioning scheme is by exploring the regions where both approaches lead to capacity gain. We explored it in terms of the limit ratio N/K in which these two bandwidth-saving techniques actually lead to a resource gain (and hence are worth deploying). For a given set of network parameters we find the limit number of users N for which applying the concrete scheme is justified (see Figure 4b). We juxtapose results for different numbers of offered channels K (for sizes of the bouquet of channels ranging from 50 to 200 with step of 50 channels) and for two values of α (α=0.6; 0.9) and two values of β (β=1; 2). For β=2, if all channels are to be multicast, the necessary bandwidth will be twice the size of the bouquet, i.e., a capacity demand of $2K$ unicast channel units. In both scenarios (static and dynamic one) we obtain the limit N/K ratio by taking the number of users N leading to the capacity demand βK (for P_{block}=10^{-4}). For the static case this is straightforward because it is not economically justified to employ bandwidth (expressed in unicast channels) more than βK. For the dynamic case, if all channels are multicast, then the necessary bandwidth is βK, while if all channels are unicast, then the required bandwidth is maximum N unicast channel units. Usually $\beta K < N$ (as β does not assume large values) and therefore we decided (also for uniformity reasons) to choose βK as reference value in the dynamic scenario. For all comparisons M is taken such as to minimise the resource demand function R_S=$f(M)$.

From both graphs it is evident that bigger values of α lead to better results both for the static and the dynamic cases. We already explained the nature of this phenomenon in Section 4. It also comes as no surprise that the larger the bouquet of channels K is, the greater the resource gain is (if the popularity distribution function is not changed). The superiority of the dynamic scenario is clearly seen for both values of β. At the same network settings, more users can be allowed (bigger N/K ratio) on a network with implemented dynamic case strategy compared to the static one. Also, we see that for a bigger value of β, the resource gain is higher. This is due to the fact that the limit βK is thus greater and the margin before reaching this limit value is larger.

6 Application Examples of the Resource Provisioning Techniques

We will give three examples to demonstrate the applicability of our results.

Example 1: Suppose an IPTV network operator would like to offer a pool of 150 channels (i.e., $K=150$). The channel popularity distribution is known experimentally and the parameter $\alpha=0.6$ (approximately as in Figure 1b). We assume the activity grade of users to be high, e.g. $a=0.7$. It takes the same bandwidth no matter if the channel is unicast or multicast (i.e., $\beta=1$). In the static dimensioning strategy, with optimal M, the number of users for which still resources can be gained (at $P_{block}=10^{-4}$) is up to about 270 subscribers, while for the dynamic dimensioning approach, it is about 800. As long as N is less than 270, both dimensioning schemes lead to capacity saving; when N is bigger than 270 but still smaller than 800, the dynamic scenario provides some gain; if N is greater than 800, it is best to multicast/broadcast all channels (because otherwise R exceeds 150 channels). Allowing, for instance, no more than 200 subscribers to the node, leads to capacity saving of 14% in the static and 35% in the dynamic case compared to deciding just to broadcast all the channels all the time ($R_S=129$ and $R_D=97$ physical channels for $P_{block}=10^{-4}$). In comparison with the scenario when all channels rely on unicast, (then R is 164 unicast channels for $P_{block}=10^{-4}$), the gain is greater, namely 21% of saved bandwidth in the static and 41% in the dynamic scenario. An example architecture is shown in Figure 5.

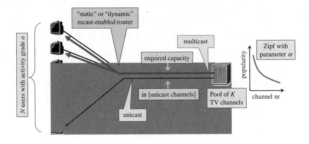

Fig. 5. Multicast-enabled network architecture

Example 2: As a second example for resource gain in the context of IPTV distribution network, we will reconsider the so-called "Trial B" case from [4]. The authors consider a network architecture called "MPEG Local Switch" according to [7] and achieve reduction in the number of necessary edge QAM modulators.

The architecture of the network is shown in Figure 6b. The parameters are as follows: $K=180$, $N=915$, $\alpha=0.64$, $\beta=1$, dynamic dimensioning scenario. As for the parameter a, we can guess that its peak value is $a=0.16$. The P_{block}, assumed by the authors, is not known either, therefore we calculate a graph for the capacity required for a range of P_{block} in Figure 7. The value stated by the authors for R is 72 channels, which is smaller than our estimates of R for P_{block} from 10^{-5} to 10^{-1}. The explanation for that is that we dimension for the peak value of a, and therefore our estimates are more conservative. If P_{block} is chosen to be 10^{-4}, it can be provided by 107 channels rounded to 11 QAMs in their scenario, resulting in a capacity gain of 40% (because 107/180=0.6). Compared to the all-broadcast case, this gain is less than the saving factor of 58% pointed out in [4] but still a pretty good result for an IPTV operator.

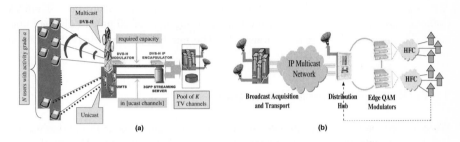

Fig. 6. (a) Hybrid mobile TV architecture and (b) IPTV Cable architecture

Example 3: In a hybrid mobile architecture (as for example the one depicted in Figure 6a), a mobile network operator can decide to offload the most popular channels to a DVB-H multiplex, for example, and broadcast (multicast) them to all users. Rarely watched channels from the bouquet are delivered individually (unicast) only on request through the UMTS spectrum. The case does not need necessarily to be a hybrid network (e.g., UMTS and DVB-H cells overlapping). An UMTS architecture enhanced with MBMS functionality to the cellular network is similar. Such approaches have been proposed and explored in [10], [11]. Mobile operators still cannot offer a large bouquet of channels because of technology constraints and most of them offer no more than 20 channels at present [8]. We will assume however K=40. In [11] α is assumed 2, but we set it to 0.9 staying conservative again as in the other examples. a=0.3, P_{block}=10^{-4} and the other parameters are case-specific.

Suppose a mobile operator has 4 UMTS nodes B with HSDPA (High Speed Download Packet Access) implemented, operating in unicast mode and thus every one with a capacity of 10 channels by 256 kbps (as in the trial described in [9] although for a different network technology). This means that R is 40 unicast channels. The number of users that can be served is calculated to be maximum N=80.

As the number of subscribers grows, the network operator must increase respectively the network capacity incrementally. The network operator however decides to deploy MBMS in order to save on resources. He adds 3 MBMS-enabled antennae each with an equivalent capacity 10 unicast channels at 256 kbps. Actually, propagating a multicast/broadcast MBMS radio channel requires more power than a unicast channel and because of interference considerations, this translates therefore in bandwidth ratio β more than 1 but typically less than 3. We assume here an intermediary value β=2. Thus, an MBMS antenna can be loaded with 5 multicast channels. This is the static scenario according to our terminology, with M set to 15, i.e. 15 channels out of 40 are constantly multicast/broadcast. In total, however, R is 70 unicast channel units. The number of clients that can be served by this capacity is about N=290. If this number of subscribers had to be met by incrementally adding simply unicast capacity, this would have led to tripling the capacity. This justifies the deployment of MBMS, which leads to resource savings despite its spectral inefficiency.

Suppose the mobile network operator decides to offload its cellular traffic by implementing a DVB-H cell overlapping with the UMTS one instead of deploying extra MBMS antennae. The throughput rates cited for a dedicated DVB-H multiplexer are usually in the range of 5 to 11 Mbps (see e.g. [8]). We assume that 30 channels at 256 kbps can be delivered through it. This capacity complements the 40 unicast channels through the UMTS spectrum, thus resulting in R of 70 unicast channel units again (in this case β=1). M is obviously 30. The number of subscribers that such an architecture can support is already 930!

7 Verification Against Simulations

We developed a C-based event-driven simulator program, similar to the one used in [14] to verify our mathematical models. The traffic load from a user is modelled as a Markov chain, with K+1 possible states (because every user can either be tuned to any of the K channels or can be inactive). State changes are governed by a continuous-time Markov model with a given transition matrix, determined by the (Zipf) channel popularity distribution and with exponential sojourn times.

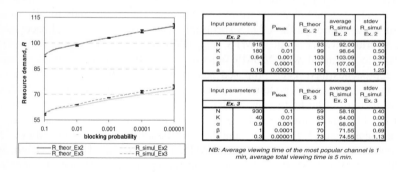

Fig. 7. Comparison of theoretical and simulation results for Example 2 and Example 3

In Figure 7 we compare theoretical and simulation results for Example 2 and Example 3 above (P_{block} ranging from 10^{-5} to 10^{-1}). The input parameters correspond to the described examples above, hence Example 2 is simulated as a dynamic case, while Example 3 as a static case (with β=1 and M=30). Visually (in the graph to the left) and numerically (in the corresponding table to the right), a good match of the results from both approaches estimating resource demand can be observed (zero variance appears because the TDF is calculated from the simulations in the form of integer value quantiles). The variance in the simulations with respect to the average is displayed in the graph of Figure 7 by error bars. Longer simulation runs lead to smaller standard deviations for smaller values of P_{block}. The difference in average values obtained from simulations and theoretically is less than 2 unicast channel units.

8 Conclusion and Future Work

This paper built on our previous work presented in [1]. In the current paper we used the normal distribution approximation formulae developed and proven to yield accurate results quickly in order to estimate the capacity demand in both static and dynamic capacity provisioning strategies.

We explored the optimality of the resource demand in the static case in function of the threshold of the number of multicast channels under different system parameters. Then, we compared the dynamic capacity planning approach to the static one with optimal settings. We proved that the dynamic scenario is superior at the same network settings and therefore we recommend it if the switch to multicast is instantaneous. Then, we studied which cases (i.e., sets of parameters) lead to capacity gains for both scenarios and illustrated this with realistic examples. We gained further confidence in our theoretical network dimensioning models by endorsing them with simulations: simulation results and theoretical predictions match very well for all cases studies.

In this work we assumed a constant channel popularity distribution over time. In our future work, we plan to explore its dynamicity, to extend the models for cases with evolving channel popularity distributions.

References

[1] Avramova, Z., De Vleeschauwer, D., Wittevrongel, S., Bruneel, H.: Models to Estimate the Unicast and Multicast Resource Demand for a Bouquet of IP-Transported TV Channels. In: Proceedings of Towards the Quality of Service Internet. Workshop, Networking Conference, Coimbra, Portugal, pp. 137–148 (May 15-19, 2006)

[2] 3GPP TS standards of the series 2x.x46

[3] Zipf, J.K.: Selective Studies and the Principle of Relative Frequency in Language (1932)

[4] Sinha, N., Oz, R., Vasudevan, S.V.: The Statistics of Switched Broadcast, SCTE 2005 Conference on Emerging Technologies, Tampa, USA (2005)

[5] RFC 1112 Host Extensions for IP Multicasting, Steve Deering (August 1989)

[6] Yates, R., Goodman, D.: Probability and stochastic processes, pp. 247–255. John Wiley & Sons, USA (1999)

[7] http://www.cable360.net/ct/strategy/emergingtech/15264.html

[8] http://www.dvb-h-online.org/services.htm

[9] T-Systems Media&Broadcast, Parameters of DVB-H Transmission, DVB-H Trial Berlin, pp. 3, www.dvb-h.org/PDF/Berlin-T-Systems-Parameters.pdf

[10] Heuck, C.: Benefits and Limitations of Hybrid (broadcast/mobile) Networks, 14th IST Mobile & Wireless Communications Summit, Dresden, (19–23 July, 2005)

[11] Unger, P., Kürner, Th.: Radio Network Planning of DVB-H/UMTS Hybrid Mobile Communication Networks, 14th IST Mobile & Wireless Communications Summit, Dresden, (19–23 July, 2005)

[12] audience.free.fr

[13] Aaltonen, J.: Content Distribution Using Wireless Broadcast and Multicast Communication Networks, Tampere University of Technology, Ph.D. thesis (2003)

[14] Avramova, Z., De Vleeschauwer, D., Wittevrongel, S., Bruneel, H.: Multiplexing Gain of Capped VBR Video, Workshop on QoS and Traffic Control, Paris (France), (7–9 December, 2005)

Video-on-Demand Server Selection and Placement

Frederic Thouin and Mark Coates

McGill University
Department Electrical and Computer Engineering
3480 University, Montreal, Quebec, Canada H3A 2A7
{frederic.thouin,mark.coates}mail.mcgill.ca

Abstract. Large-scale Video-on-Demand (VoD) systems with high storage and high bandwidth requirements need a substantial amount of resources to store, distribute and transport all of the content and deliver it to the clients. We define an extension to the *VoD equipment allocation problem* as determining the number and model of VoD servers to install at each potential replica location to minimize deployment costs for a given set of distributed demand and available VoD server models. We propose three novel heuristics that generate near-optimal solutions and show that the number of replica sites for networks where the load is unevenly distributed is low ($35 - 45\%$ of potential locations), but that the hit ratios at deployed replicas are high ($> 85\%$).

1   Introduction

As the number of available titles and usage of video-on-demand services is expected to grow dramatically in the next years, many providers are planning the deployment of large-scale video-on-demand (VoD) systems. These systems require significant resources (bandwidth and storage) to store the videos, distribute them to caches, and deliver them to clients. An important and complicated task part of the network planning phase is resource allocation. It consists of determining the location and number of resources to deploy such that user demand is satisfied, cost is minimized, and any quality of experience (QoE) constraints (delay, packet loss, frame loss, or packet jitter) are respected. The main challenge is to build sufficiently accurate models for all of the factors involved: the available infrastructure, the network topology, the peak/average usage of the system, the popularity of each title, and bandwidth and storage requirements.

In the case of a distributed video-on-demand network deployment, the resources to consider are the equipment required at the origin and proxy video servers and for the actual transport between each location. We assume an existing topology with a high bandwidth capacity and focus on the equipment required at each location to store and stream the content. A video server consists of storage devices to cache the desired content and streaming devices to deliver the videos to the users. In [1], we defined the *VoD equipment allocation problem* that consists of determining the number of streaming and storage

L. Mason, T. Drwiega, and J. Yan (Eds.): ITC 2007, LNCS 4516, pp. 18–29, 2007.

devices at each location in the topology such that the demand is satisfied and the deployment cost is minimized. We showed that the nature of the equipment installed at each location has a major impact on the design and on whether it is beneficial to even cache content. A natural extension of the problem thus involves identifying the best type of equipment to install at each location when many models are available and there is flexibility for variation from site to site. Therefore, in this paper, we address the problem of determining not only the number, but also the model of the VoD servers at each potential replica location.

This paper is organized as follows. In Sect. 2, we formulate the VoD equipment allocation problem such that the solution includes both the number and model of the VoD servers. In Sect. 3, we discuss the network cost model we used to solve this problem. In Sect. 4, we present two simple algorithms (Full Search and Centralized or Fully Distributed Heuristic) and three novel heuristics (Greedy Search, Integer Relaxation Heuristic and Improved Greedy Search) that can be employed to generate solutions. In Sect. 5, we show and discuss the results of simulation experiments performed on randomly generated topologies. Finally, in Sect. 6, we present our conclusions and suggest future extensions to our work.

1.1   Related Work and Contribution

Researchers have tackled the problem of generating cost-efficient VoD network designs using different optimization techniques: placement of replica servers, video objects or allocation of available resources to minimize cost. Solving the replica placement [2, 3] or video placement [4] problems independently of the resource allocation problem usually leads to suboptimal solutions because the location of the replicas has a direct impact on the amount of resources required. Laoutaris et al. defined the *storage capacity allocation problem* as determining the location of each object from a set to achieve minimal cost whilst enforcing a capacity constraint [5]. Although they determine the actual storage requirements at each node with their solution, the authors do not explicitly determine the equipment required. In [6], Wauters et al. define an Integer Linear Programming (ILP) model built on viewing behavior, grooming strategies, statistical multiplexing and Erlang modeling to specify the equipment required for transport (the number of ports, multiplexers and switch ports) at each of the candidate network nodes [6].

In [1], we defined the *VoD equipment allocation problem* as the task of determining the number of VoD servers (which include both a storage and streaming device) to deploy at each potential location in a network topology such that the total demand is satisfied and the deployment cost is minimized. Solving the VoD equipment allocation problem determines the location of the replicas, the amount of storage available to cache content, the streaming capacity available to serve clients and the explicit specification of the equipment installed at each location. However, our approach in [1] assumed that a fixed, single and predetermined type of VoD server was available at each location. This constraint rarely holds in practice and enforcing it leads to suboptimal designs if the nature of the equipment is not a good fit to the streaming (user demand) and storage (library size) requirements.

This optimization problem also has some similarities with the classical facility location problem which has been studied thoroughly (many algorithms and exact heuristics have already been developed to solve it). However, the presence of an origin server that gathers traffic from all other locations and the non-linearity in some constraints make our problem substantially different even from the generalized form of the facility location problem proposed in recent work [7] and thus unsolvable using available heuristics.

In this paper, we re-formulate the VoD equipment allocation problem to determine both the number and model of the servers to install at each location. Instead of fixing the streaming and storage capacity per VoD server at each site (the approach used in [1]), we require the pre-specification of a set of available VoD servers and select the model at each location that minimizes total network cost. This leads to the faster generation of lower-cost solutions because the network designer does not need to manually try all models for each potential site. To solve the problem, we develop a network cost model solely in terms of the numbers and models of servers and propose three novel heuristics: the Integer Relaxation Heuristic and two greedy-search based algorithms (GS and IGS).

2   Problem Statement

We consider a metropolitan area network with one origin server and N potential replica locations such as the one depicted in Fig. 1. Each cluster of clients has worst-case demand M_i (peak usage demand) and is assigned to a potential replica location with hit ratio h_i that represents an estimate of the fraction of M_i served at that replica, the other portion is served directly by the origin server.

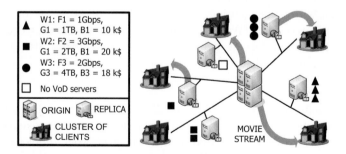

Fig. 1. Video-on-Demand equipment allocation problem. Logical connectivity between origin, $N = 5$ potential replica server locations and clients. Clients' requests (shown as movie stream arrows) are served by replicas (if content is available) or by origin. Key shows the specifications of $W = 3$ different VoD server models. We show the number and type of VoD servers installed at each potential location. The optimal solution can include locations with no equipment (empty square).

We address the *VoD equipment allocation problem* of determining not only the number, but also the model of the VoD servers at each potential replica location. To solve this problem, we require the specification of a set of available VoD server models $\mathcal{W} = \{w_j : j = 1, \ldots, W\}$ where w_j is a VoD server with streaming capacity F_j Gbps, storage capacity G_j TB and unit cost B_j k\$. We define the sets $\mathcal{N} = \{n_i : i = 1, \ldots, N\}$ and $\mathcal{V} = \{v_o, v_i \in \mathcal{W} : i = 1, \ldots, N\}$ where n_i is the number and v_i is the model of the servers installed at location i. The optimization problem is expressed as follows:

$$\{\mathcal{N}^*, \mathcal{V}^*\} = \arg \min_{\mathcal{N}, \mathcal{V}} C_{\text{TOT}_{\mathcal{V}}}(\mathcal{N}) \tag{1}$$

where $C_{\text{TOT}_{\mathcal{V}}}(\mathcal{N})$ is the total cost of the network C_{TOT} for a fixed set $\mathcal{V}$.

$$n_o \cdot G_o \geq Y \cdot \text{file size} \tag{2}$$

$$n_i \cdot F_i \geq h_i \cdot M_i, \quad \forall i \in \{1 \ldots N\} \tag{3}$$

$$n_o \cdot F_o \geq \sum_{i=1}^{N} (1 - h_i) M_i \tag{4}$$

$$\widehat{H}(n_i \cdot G_i) \geq h_i, \quad \forall i \in \{1 \ldots N\} \tag{5}$$

The first constraint states that the storage capacity at the origin must be large enough to host the entire initial library where Y is the number of files in the library. The constraints in (3) and (4) ensure that the streaming capacity at each replica and the origin is large enough. For each replica sites, the streaming capacity required is at least the fraction of the user demand coming from its associated cluster of clients. For the origin, the total demand is equal to the sum of all residuals fractions of the demand that are not handled by the replicas. In (5), we introduce $\widehat{H}$, an estimate of the hit ratio as a function of the storage capacity (we give more details about the form of $\widehat{H}$ in Sect. 3). This constraint states that the amount of storage at every location should be large enough such that the estimated hit ratio is greater than the actual hit ratio h_i; that a fraction of the requests equal to h_i are for files stored at the replica.

3   Network Cost Model

In order to perform a direct optimization, we derive an expression for the deployment cost solely in terms of $\mathcal{N}$ and $\mathcal{V}$. We express the total cost C_{TOT} as the sum of the cost of infrastructure, C_T, and the cost of transport, C_S:

$$C_T = f_1(n_o, v_o) + \sum_{i=1}^{N} f_1(n_i, v_i) \qquad C_S = \sum_{i=1}^{N} f_2(h_i, M_i)$$

$$C_{\text{TOT}} = f_1(n_o, v_o) + \sum_{i=1}^{N} f_1(n_i, v_i) + f_2(h_i, M_i) \tag{6}$$

The cost of infrastructure at each potential location and the origin includes a start-up cost for installation and software (A_i) and increases linearly with the number of VoD servers installed (n_i). Note that f_1 is also a function of v_i which defines B_i, F_i and G_i.

$$f_1(n_i, v_i) = A_i + B_i n_i \qquad (7)$$

The cost of transport for each location is inspired from the model proposed by Mason et al. is divided in two components: transport from the replica to the clients ($C_{S_{RC_i}}$) and from the origin to the replica ($C_{S_{OR_i}}$) [8]. Transport cost from the replica to the clients includes the cost of network interfaces (C_{IF}) and fiber (C_f). The number of network interfaces (n_{RC_i}) required is proportional to the demand M_i and the fiber capacity (c). For transport from the origin to the replica location, we add the cost of DWDM with w_{max}-ports multiplexers (C_{DWDM}) and line amplifiers (C_{LA}). The number of network interfaces (n_{OR_i}) required is a function of the demand M_i and the hit ratio h_i: the amount of traffic on this link is equal to the fraction of the demand un-served by the replica. For more details on the cost functions f_1 and f_2, the reader is referred to [1].

$$C_{S_{RC_i}} = n_{RC_i} \cdot (2 \cdot C_{IF} + d_{RC_i} \cdot C_f)$$

$$C_{S_{OR_i}} = n_{OR_i}(2 \cdot C_{IF}) + \frac{n_{OR_i}}{w_{max}} \left[2C_{DWDM} + d_{OR_i} \cdot C_f + \left(\frac{d_{OR_i}}{max_{amp}} \right) \cdot C_{LA} \right]$$

$$n_{OR_i} = \frac{(1 - h_i) \cdot M_i}{c} \qquad n_{RC_i} = \frac{M_i}{c}$$

$$f_2(h_i, M_i) = C_{S_{OR_i}} + C_{S_{RC_i}} \qquad (8)$$

To derive an expression for C_{TOT} solely in terms of n_i, we resolve the hit ratio h_i and number of servers at the origin n_o as functions of n_i. To estimate the hit ratio $\widehat{H}$, we use (9), a function of the cache size ratio X_i (number of files in the cache / number of total files in the library), the library size Y and the number of files added to the library every week Z. We designed the function and determined best-fit constants K_1 to K_8 using a discrete-time simulator based on the file access model proposed by Gummadi et al. in [9] (refer to [1] for more details).

$$\widehat{H} = A(Y, Z) + B(Y, Z) \cdot \log(X) \qquad (9)$$

$$A = K_1 + K_2 Z + K_3 \log(Y) + K_4 Z \log(Y) \qquad B = K_5 + K_6 Z + K_7 Y + K_8 ZY$$

The hit ratio at a location is limited by either the streaming or the storage capacity represented by constraints shown in (3) and (5). We isolate h_i in both expressions and define $f_3(n_i, v_i)$ as the minimum (worst-case) hit ratio:

$$f_3(n_i, v_i) = \min \left[\frac{n_i \cdot F_i}{M_i}, \widehat{H} \left(\frac{n_i \cdot G_i}{Y \cdot \text{file size}}, Y, Z \right) \right] \qquad (10)$$

The number of servers required at the origin, n_o, is also constrained by either streaming or storage (shown in (2) and (4)). In (11), we define n_o as $f_4(\mathcal{N}, \mathcal{V})$ by substituting h_i with the expression in (10).

$$f_4(\mathcal{N}, \mathcal{V}) = \max \left[\frac{\sum_{i=1}^{N}(1 - h_i) \cdot M_i}{F_o}, \frac{Y \cdot \text{file size}}{G_o} \right] \tag{11}$$

We replace the equations for n_o and h_i in (6):

$$C_{\text{TOT}} = f_1(f_4(\mathcal{N}, \mathcal{V})) + \sum_{i=1}^{N} f_1(n_i) + f_2(f_3(n_i, v_i)) \tag{12}$$

4   Description of Heuristics

4.1   Full Search (FS)

The Full Search is a very straightforward approach that consists of trying all the possible points in the solution space. We reduce this space by calculating the maximum number of servers it is worth installing at a given location using (13). We define $\mathbf{ub} = \{ub_i : i = 1, \ldots, N\}$ where ub_i is the number of servers required to store the entire library and handle 100% of the requests ($h_i = 1.0$).

$$ub_i = \max \left(\frac{M_i}{F_i}, \frac{Y \cdot \text{file size}}{G_i} \right) \tag{13}$$

For a given $\mathcal{V}$, the boundaries of the solution space are $\mathcal{N} = \mathbf{0}$ to $\mathbf{ub}$ where $\mathbf{0} = \{n_i = 0 : i = 1, \ldots, N\}$. To complete the full search, all the possible combinations of $\mathcal{V}$ must also be tried. Although this procedure is guaranteed to find the optimal solution, it is very computationally expensive and the amount of time to search the entire space grows exponentially with the size of the network.

4.2   Central or Fully Distributed Heuristic (CoFDH)

The Central or Fully Distributed Heuristic simply calculates the cost of a centralized design ($\forall i : n_i = 0$) and a fully distributed design ($\forall i : n_i = ub_i$) for each available VoD server model in $\mathcal{W}$ and picks the cheapest design. This heuristic is straight-forward and highly suboptimal, but it provides an upper-bound that can be used as a comparison base for other approaches.

4.3   Greedy Search (GS)

We define a search topology in the discrete solution space where each solution is connected to its neighbouring solutions. In this case, a neighbour consists of adding one server at one of the locations. Greedy Search is a searching heuristic that explores all neighbouring nodes and selects the one that yields the best solution at every iteration without considering the subsequent steps [10].

4.4   Improved Greedy Search (IGS)

The Improved Greedy Search is divided into two steps inspired by GS. The difference is that a neighbour solution is obtained by adding or removing ub_i servers at location i. The motivation behind this is that the hit ratio at replica location is often very high which leads to n_i close to ub_i. During the first step of the heuristic, we iteratively add servers in a greedy-fashion starting from a centralized design by setting $n_i = ub_i$ at the location that achieves the lowest cost. We complete the first step and determine an initial integer solution by repeating this procedure for each VoD server model.

The second step is an exploration procedure in the neighbourhood of the initial solution. In a greedy-type approach, we add or remove, at iteration k, one server to the initial solution at the location that minimizes the cost C_k. We stop the search when $C_j \geq C_{j-1} \; \forall j \in k - I + 1 \ldots k$ or when $C_j \geq C_{IGS}$ $\forall j \in k - 2I + 1 \ldots k$ (minimum cost has not decreased for 2I iterations).

4.5   Integer Relaxation Heuristic (IRH)

The first step of the Integer Relaxation Heuristic is to find a non-integer solution for each server model using a constrained nonlinear optimization algorithm based on sequential quadratic programming [11]. Then, we calculate the cost associated with each replica $(C_{T_i} + C_{S_{OR_i}})$ and determine the model that minimizes this cost for each location. We complete the initial solution by determining the best server model to install at the origin. In the second step, we perform two different searches to find a near-optimal integer solution: we iteratively (i) set $n_i = 0$ at each location to make sure it is profitable to setup a replica and then (ii) try to remove or add up to two servers at each location until we find a local minimum.

5   Simulation Experiments

Our simulation results were obtained by applying our heuristics to different networks (simulations were executed on a AMD Athlon 3000+ with 1 GB of OCZ Premier Series 400 MHz Dual Channel memory), each defined by the constant variables in Table 1 and choosing values for the other network parameters from uniform distributions with the ranges specified in Table 2 (these values were obtained from discussions with industrial partners [12]).

5.1   Complexity

Table 3 presents the size of the solution space for different network topologies (generated using the values displayed in Table 1 and Table 2) which consist of all possible number of servers ($n_i = 0$ to $n_i = ub_i$) and server models at each location. From other experiments, we measured that the machines we used for simulations explore 4000 solutions per second on average, which allows us to estimate to time it would take to explore the entire solution space in order to determine the global optimal solution. From our estimates, it is clear that

Table 1. Values of constants

Variable	Value
C_{IF}	10 k\$
C_{DWDM}	25 k\$
C_{LA}	10 k\$
C_f	0.006 k\$/km
w_{max}	16
c	10 Gbps
max_{amp}	75 km
bit rate	3.75 Mbps
duration	5400 s
file size	2.53 GB

Table 2. Range of the variables

Variable	Min	Max
d_{OR} (km)	0	50
d_{RC} (km)	0	5
Y (files)	1000	10000
Z (files/week)	0	100
priceGbps (k\$/Gbps)	0	4
priceTB (k\$/TB)	0	3
A (k\$)	6	36
F (Gbps)	1	5
G (TB)	1	11
M (Gbps)	1	20

Table 3. Number of possible solutions for topologies of N locations with K possible VoD server models and estimate of time taken to find the global optimal solution based on an observed average rate of 4000 solutions per second. Values obtained from 50 different topologies for each pair of (N,K).

N	K	Number of solutions			Estimated time (days)		
		min	median	max	min	median	max
10	1	7.8×10^5	4.5×10^8	9.5×10^{13}	2.2×10^{-3}	1.3×10^0	2.8×10^5
25	1	5.6×10^{14}	7.1×10^{20}	2.4×10^{35}	1.6×10^6	2.1×10^{12}	6.9×10^{26}
50	1	2.0×10^{28}	1.4×10^{40}	1.2×10^{68}	5.9×10^{19}	4.1×10^{31}	3.5×10^{59}
100	1	1.9×10^{60}	5.0×10^{79}	1.3×10^{130}	5.6×10^{51}	1.4×10^{71}	3.7×10^{121}
15	2	9.1×10^{13}	2.0×10^{18}	6.7×10^{23}	2.6×10^5	5.9×10^9	1.9×10^{15}
10	3	5.4×10^{10}	1.0×10^{14}	4.3×10^{17}	1.6×10^2	2.9×10^5	1.3×10^9

performing a full search is infeasible as even the smallest problems ($N = 10, K = 1$) can take up to thousands of days to solve depending on the demand and the specifications of the equipment. This shows that heuristics are essential to solve the VoD Equipment Allocation Problem.

5.2 Performance

In this section, we evaluate the performance of our heuristics. In our first set of tests, we analyse small networks to allow comparison with the full search, which cannot produce a solution for larger networks within a reasonable time frame. In Fig. 2, we show the performance of our heuristics by dividing the cost of the solution by the optimal solution provided by the full search. For these small networks, Integer Relaxation Heuristic and Improved Greedy Search perform within 4% of the optimal solution. For all values of N and W, both IRH and IGS perform better than the Greedy Search.

In Fig. 3, we show values of the ratio between the cost of Integer Relaxation Heuristic, Improved Greedy Search and Greedy Search and the cost of CoFDH. Whereas Greedy Search is actually very close to the cost produced by CoFDH,

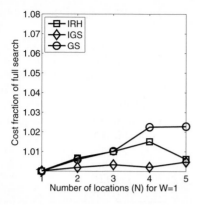
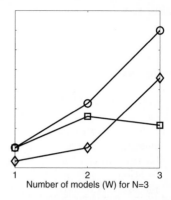

Fig. 2. Cost ratio between heuristics and the full search averaged over 30 runs for networks with the number of locations $N \in \{1,\ldots,5\}$ and the number server model $W = 1$ (LEFT) and another series with $N = 3$ and $W \in \{1,2,3\}$ (RIGHT)

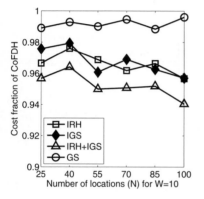
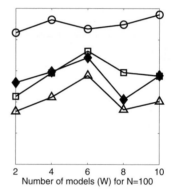

Fig. 3. Cost ratio between the heuristics solution and CoFDH averaged over 25 runs. IRH+IGS is the average of the minimum value between IRH and IGS for all runs.

the other two heuristics generate solutions that cost 2-5% less. It is not clear from those plots whether Integer Relaxation Heuristic or Improved Greedy Search performs better. By combining both (choosing the best solution of the two), we obtain a better heuristic (IRH+IGS), which achieves a 4-6% cost reduction compared to CoFDH. In the left panel, we notice the downward trend of the cost fraction as the number of locations in the network increases because more modifications to the CoFDH design can be made to reduce cost. For the same set of tests, we also observed the computational time in seconds of each heuristic. The Integer Relaxation Heuristic was the slowest of the tested heuristics because of the constrained optimizations using SQP, but it still converges in a reasonable amount of time (< 250 seconds for $N = 100$ and $W = 10$). Since the computation time of Improved Greedy Search is so low (< 10 seconds), we can combine IRH and IGS and obtain a solution in a timely fashion.

5.3   Analysis

Finally, we focus on the networks with six server models (similar behaviour was observed for other values of W) to analyze the ratio of locations with replicas and average demand at replica locations. The left panel of Fig. 4 shows that for networks of any size, where demand is not uniformly distributed among all locations (the demand at each location is different), the percentage of locations where a replica will be deployed is below 40% for both heuristics. Although a case where the demand load is evenly shared among all the locations is more plausible, this result indicates that it is not always advantageous to cache content. If the demand is too low then it can be more cost-effective to assume the entire load from a group of clients directly at the origin. An impact of this low percentage is the number of servers installed at the origin. Because the fraction of locations where replicas are installed remains approximately constant for any value of N, the total number of sites for which the origin must assume the demand grows as the network becomes larger.

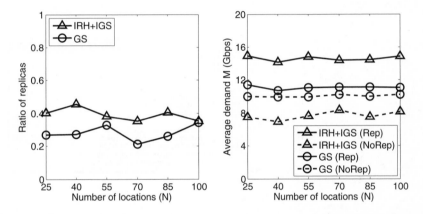

Fig. 4. LEFT: Ratio between the number of replicas (location with cached content) and potential locations. RIGHT: Average load on the locations where replicas are installed (Rep) and where no replicas are installed (NoRep). The values shown are averages of 25 runs with $W = 6$.

In the right panel of Fig. 4, we depict the difference between the average demand at replica locations and that at sites where no caching is performed. Whereas there is only a marginal difference in the GS case, the average demand at replica sites in the IRH+IGS solutions is almost twice the average demand of the other locations. The solutions generated by combining Integer Relaxation Heuristic and Improved Greedy Search have a much lower total cost than the GS solutions, indicating that it is more cost-efficient to install replicas at locations where demand is high and transport the entire load of locations with low demand to the origin.

6   Concluding Remarks

In this paper, we defined an extension of the *VoD equipment allocation problem* described in [1]. Instead of considering fixed and pre-determined streaming and storage capacity at each location, we require the specification of a set of available VoD servers models. The optimization problem consists of choosing the number and type of VoD servers to install at each potential location in the network such that cost is minimized. For most topologies, we showed that it is infeasible to obtain the global optimal solution. We described three heuristics to find a near-optimal solution including two greedy-type approaches (GS and IGS) and an integer relaxation method (IRH) that we implemented in an interactive design tool shown in Fig. 5. We combined IRH and IGS by choosing the best of the two to obtain a better solution while maintaining low computational time.

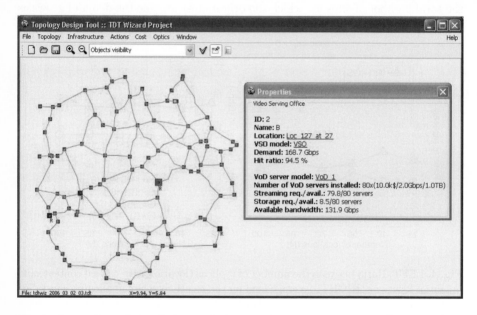

Fig. 5. Screenshot of the design tool that implements our heuristic (IRH+IGS) to solve the VoD equipment allocation problem. A sample topology of potential replica locations and the properties window of a selected replica location is shown in the figure.

We saw that a way to obtain a cost-efficient solution is to use equipment (VoD server model) that satisfies the streaming and storage requirements of most of the locations in the topology. Alternatively, the network designer could strive to divide the demand evenly among all locations such that it is optimal to deploy replicas at most locations using the same model of equipment. A sensible extension to the resource allocation problem we addressed in this paper is the problem of jointly designing the VoD network and the logical topology: choosing a topology that allows an allocation of resources which minimizes the deployment cost of the network. We also observed that limiting ourselves to a single VoD server model per location, solutions produced with our heuristics often deploy the same model at all replica sites. Our preliminary results show that relaxing this constraint allows finer tuning and the possibility to match more closely the demand in storage and streaming at each location and leads to cost savings.

Other future research avenues include the scenario where the service provider owns network equipment or infrastructures prior to the deployment. However, even if there is no installation cost, there are still fees incurred by the usage and maintenance of the equipment and the resources, which have to be considered when generating solutions for this scenario. Also, we focused on large-scale deployments, but there is also the issue of scalability of such deployments. As the library reaches tens of thousands of movies, the access model we assumed changes as a larger portion of requests are located in the heavy tail of the popularity distribution ('long-tail' of content).

References

[1] Thouin, F., Coates, M., Goodwill, D.: Video-on-demand equipment allocation. In: Proc. IEEE Int. Conf. Network Computing Applications (NCA), Cambridge, MA (July 2006)

[2] Krishnan, P., Raz, D., Shavitt, Y.: The cache location problem. IEEE/ACM Trans. Networking 8, 568–582 (2000)

[3] Almeida, J.M., Eager, D.L., Vernon, M.K., Wright, S.: Minimizing delivery cost in scalable streaming content distribution systems. IEEE Trans. Multimedia 6, 356–365 (2004)

[4] Tang, W., Wong, E., Chan, S., Ko, K.: Optimal video placement scheme for batching vod services. IEEE Trans. on Broad 50, 16–25 (2004)

[5] Laoutaris, N., Zissimopoulos, V., Stavrakakis, I.: On the optimization of storage capacity allocation for content distribution. Computer Networks Journal 47, 409–428 (2005)

[6] Wauters, T., Colle, D., Pickavet, M., Dhoedt, B., Demeester, P.: Optical network design for video on demand services. In: Proc. Conf. Optical Network Design and Modelling, Milan, Italy, (February 2005)

[7] Wu, L.-Y., Zhang, X.-S., Zhang, J.-L.: Capacitated facility location problem with general setup cost. Comp. Oper. Research 33, 1226–1241 (2006)

[8] Mason, L., Vinokurov, A., Zhao, N., Plant, D.: Topological design and dimensioning of agile all photonic networks. Computer Networks, Special issue on Optical Networking 50, 268–287 (2006)

[9] Gummadi, K.P., Dunn, R.J., Saroiu, S., Gribble, S.D., Levy, H.M., Zahorjan, J.: Measurement, modeling, and analysis of a peer-to-peer file-sharing workload. In: Proc. ACM Symp. OS Principles (SOSP), Bolton Landing, NY (October 2003)

[10] Cormen, T.H., Leiserson, C.E., Rivest, R.L.: Introduction to Algorithms. The MIT Press, Cambridge, MA (1990)

[11] Fletcher, R.: Practical Methods of Optimization. John Wiley and Sons, New York (1987)

[12] Goodwill, D.: private communication, Nortel Networks (2005)

Analysis of the Influence of Video-Aware Traffic Management on the Performance of the Dejittering Mechanism*

Kathleen Spaey and Chris Blondia

University of Antwerp - IBBT
Department of Mathematics and Computer Science
PATS Research Group - www.pats.ua.ac.be
Middelheimlaan 1, B-2020 Antwerpen, Belgium
{kathleen.spaey,chris.blondia}@ua.ac.be

Abstract. IPTV over DSL access networks is becoming a reality. This paper studies the impact of the presence of a video-aware traffic management module in the access network on the performance of the video dejittering mechanism at the receiver. The presented numerical results illustrate that with the traffic management module, the packet loss probability at the dejittering buffer decreases. Since the traffic management module will only drop the least important enhancement layer packets of the video in case of congestion, especially the important base layer packets of the video gain from the use of the traffic management module. All results in this paper are obtained using a mathematical model.

1 Introduction

Today, there are several drivers for the delivery of IPTV (Internet Protocol Television) over DSL (Digital Subscriber Line) access networks. These are both market drivers such as broadband service providers facing ongoing threats from cable operators, as well as technological enablers like improvements in video compression. The delivery of television quality video services over a DSL access network is however more demanding than that of traditional high-speed data-only Internet services: the video streams are more bandwidth intensive and also their sensitivity to packet loss and jitter (variation in the delay) is greater.

Several access network enhancements for the delivery of video services are possible [1]. One of them is video-aware traffic management that during congestion periods selectively discards the information that will least affect the picture quality. More and more, also the endpoints are asked to contribute to a performance increase of the applications. This contribution can take several forms, like the encoding of the information that will be transmitted such that the important information can be separated from the less important information, as in

---

* This work was carried out within the framework of the IWT (Flemish Institute for the promotion of Scientific and Technological Research in the Industry) project CHAMP.

L. Mason, T. Drwiega, and J. Yan (Eds.): ITC 2007, LNCS 4516, pp. 30–41, 2007.
© Springer-Verlag Berlin Heidelberg 2007

the forthcoming scalable extensions of H.264/AVC (Advanced Video Coding), known as SVC (Scalable Video Coding) [2]. SVC generates a base layer and one or more enhancement layers, where each additional layer contributes to enhancing the video quality. Other examples are letting the sources react to feedback obtained about the status of the network, as in DCCP (Datagram Congestion Control Protocol) [3], absorbing the jitter introduced in the network, etc.

Figure 1 shows the network scenario that will be considered in this paper. Video traffic will be sent in packets from a video server towards N video receivers of a DSL customer via the core network, aggregation network and DSL access network. At the receivers, there is a dejittering buffer before the decoders. Packets that arrive before their playout instant are temporarily stored in this buffer, packets that arrive too late are useless and considered lost. A simultaneous delivery of N video streams to one customer is necessary to allow different televisions of the customer to display different content. The actual number of simultaneous video streams sent to one customer is usually very limited due to bandwidth limitations. Assuming a well engineered core and aggregation network, and considering typical MPEG2 (Motion Pictures Expert Group 2) streams around 3.5 Mbit/s for SDTV (Standard Definition Television), or H.264 HDTV (High Definition Television) streams around 8 Mbit/s, it becomes clear that the bottleneck of the network scenario is the downstream DSL rate. This rate is typically 12 Mbit/s for ADSL2 (Asymmetric Digital Subscriber Line 2) or 24 Mbit/s for ADSL2+, but these speeds drop quickly with growing distance from the DSLAM (Digital Subscriber Line Access Multiplexer) [4].

The video traffic is presumed to be compressed in two layers: a base layer and an enhancement layer. The base layer packets are of high importance for the video decoding process at the receiver, while the enhancement layer packets of course contribute to the quality of the video, but can be discarded without error propagation. Further, it is assumed that at the bottleneck point a video-aware traffic management module is present, which during congestion periods prevents the enhancement layer packets to enter the bottleneck buffer in which the traffic of the N video streams is multiplexed. A threshold Th on the content of the bottleneck buffer is used to detect congestion, i.e., while the bottleneck buffer

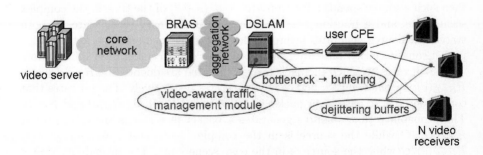

Fig. 1. Network scenario

content exceeds Th, the traffic management module discards all enhancement layer packets.

The aim of this paper is to understand which impact the presence of the video-aware traffic management module in the access network has on the performance of the dejittering mechanism. A mathematical multiplexing model will be used to analyze the discussed scenario. This model and the performance measures that are derived from it are discussed in Section 2. Section 3 applies the model to numerical examples and discusses the obtained results. The conclusions derived from this work can be found in Section 4.

2   Model Description and Performance Measures

The network scenario discussed in Section 1 will be analyzed using a mathematical model. It is assumed that time is divided in units of constant length, called slots, where the length of a slot corresponds to the time needed to place a packet of length b on the bottleneck link of capacity C. At the bottleneck, the traffic of the N video streams that is not discarded by the traffic management module, is multiplexed in a finite output buffer of capacity B packets. When this buffer overflows, packets get lost. All packets that were not discarded by the traffic management module and that did not get lost due to buffer overflow at the bottleneck buffer, reach the dejittering buffer of their receiver. But only the packets that reach this buffer on time will also effectively reach the receiver. Packets that arrive after their playout time are useless and considered lost.

2.1   Video Source Model

As video source model, consider a two-state Markov modulated Bernouilli source. A typical video contains scenes of different complexity. With constant-quality encoding, more bandwidth is used for complex scenes than for easy scenes. So one state of the Markov source is used to model the easy scenes, the other represents the complex scenes. If α, respectively β, denotes the probability that the source makes a transition to the easy, respectively complex scene state in the next slot, given that it is currently in the complex, respectively easy scene state, then such a source spends a fraction $\pi(c) = \beta/(\alpha+\beta)$ of the time in the complex scene state, and a fraction $\pi(e) = \alpha/(\alpha + \beta)$ in the easy scene state. In both states, the source generates layered video traffic in fixed-sized packets. Such a packet contains either base layer information, or enhancement layer information.

The bit rates at which the packets are generated are denoted by $R_{B+E}(c)$ and $R_{B+E}(e)$, for the complex and easy scene states respectively. The bit rates that correspond to the base layer packets only are symbolized by $R_B(c)$ and $R_B(e)$. The probability that a source generates a packet in a slot is then $p_{B+E}(c) = R_{B+E}(c)/C$ while the source is in the complex scene state, and $p_{B+E}(e) = R_{B+E}(e)/C$ while the source is in the easy scene state. The probability that a generated packet is a base layer packet is given by $R_B(c)/R_{B+E}(c)$ or $R_B(e)/R_{B+E}(e)$, depending on the state the source is in when it generates the packet.

2.2 Dejittering Principle

When the video packets traverse the network, they experience a delay that may vary from packet to packet. Consequently, the cadence with which the packets arrive at the receiver's side differs from that with which the video source generated the packets. The task of reducing this jitter is typically performed by placing a dejittering buffer before the decoder. In this buffer, the first packet of the video stream is retained for some time (the *dejittering delay*) before it is offered to the decoder. From then on, packets are played out with the same cadence as with which the encoder generated the video packet stream. Packets that arrive before their playout instant are temporarily stored in the dejittering buffer, packets that arrive too late are useless and considered lost. For each packet that arrives too late, the dejittering buffer underflows (i.e., runs empty) at the moment such a packet is supposed to be played out.

Choosing the dejittering delay T_{jit} always involves a trade-off between packets being lost to the video application due to dejittering buffer underflow and their delay. An application must decide about the maximum dejittering delay it can accept, and this in turn determines the fraction of packets that arrive in time to be played out [5]. Namely, if the first packet of a video stream is sent into the network at time 0, and the packet experiences a delay δ_1 before it arrives at the dejittering buffer, then this packet is played out at time $\delta_1 + T_{jit}$. If packet n experiences a delay δ_n in the network, and c_n denotes the interpacket time at the source between packet 1 and packet n, then packet n arrives at the dejittering buffer at time $c_n + \delta_n$. The target playout time for packet n is then c_n time after the first packet, or thus at time $\delta_1 + T_{jit} + c_n$. So packet n arrives too late at the dejittering buffer if $\delta_n > \delta_1 + T_{jit}$. This means that the dejittering buffer underflow probability depends on the delay distribution of the packets that arrive at that buffer. Remark however that only the variable component of this distribution matters, which for the considered network scenario is caused by the variable queueing delay at the bottleneck buffer. So if d^* denotes the queueing delay experienced by the first packet at the bottleneck buffer, and if $P_{late}(T_{jit}, d^*)$ denotes the dejittering buffer underflow probability for a given T_{jit} and d^*, then

$$P_{late}(T_{jit}, d^*) = P\{D > T_{jit} + d^*\} \ , \tag{1}$$

where D is the random variable that denotes the queueing delay of a packet at the bottleneck buffer. Depending on if D denotes the queueing delay of a random packet, a base layer packet or an enhancement layer packet, (1) gives the probability that respectively a random packet, a base layer packet or an enhancement layer packet arrives too late at the dejittering buffer.

2.3 Delay Distributions at the Bottleneck Buffer

From Section 2.2, it became clear that to calculate the dejittering buffer underflow probability for a random packet, a base layer packet or an enhancement layer packet, the probability distributions of the queueing delay experienced at

the bottleneck buffer by a random packet, a base layer packet or an enhancement layer packet, need to be known. We calculate these distributions using
the matrix-analytical approach. In this paper, the details of the calculations are
omitted, but a brief and informal outline of the reasoning behind them will be
given now.

The bottleneck buffer is modelled as a discrete-time single-server queueing
system with a finite capacity of B packets, whose input process is the superposition of the traffic generated by N video sources as defined in Section 2.1, minus
the traffic that is discarded at the traffic management module. This queueing
system is a level-dependent D-BMAP/D/1/B queueing system where the service times equal one slot. The packets that enter the system are served in FIFO
(First In First Out) order. There is a threshold on the content of the system,
and depending on whether the content of the system exceeds this threshold or
not, a different D-BMAP (Discrete-time Batch Markovian Arrival Process) describes the input process to the queueing system. Namely, a first D-BMAP that
describes the superimposed traffic of the N sources consisting of both the base
and the enhancement layer packets is used as input process if the system content
does not exceed the threshold at the beginning of a slot. In the other case, a
second D-BMAP that describes the superimposed traffic consisting of only the
base layer packets of the N sources, is used.

If $L(n)$ denotes the number of packets in the system at slot n, and $J(n)$
symbolizes the number of sources that are in the complex scene state in that
slot, then $\{(L(n), J(n)), n \geq 0\}$ is a two-dimensional Markov chain with state
space $\{(l, j)|0 \leq l \leq B, 0 \leq j \leq N\}$ and a $(B + 1)(N + 1)$-state transition
probability matrix $\mathbf{Q}$. First the invariant probability vector of $\mathbf{Q}$, denoted by
$\mathbf{x} = (\mathbf{x}_0 \ldots \mathbf{x}_B)$ with $\mathbf{x}_i = (x_{i,0} \ldots x_{i,N})$, is calculated. Element $x_{i,j}$ of vector $\mathbf{x}$
represents the stationary joint probability that in an arbitrary time slot there
are i packets in the system and j of the N sources are in the complex scene
state.

Once this invariant probability vector is known, the delay distributions at
the bottleneck buffer can be calculated. First, a formula for $y_m^{(A)}$, $0 \leq m \leq B$,
which is the probability that a packet that arrives at the bottleneck buffer sees
m packets in that buffer upon arrival, is worked out. Remark that since the
bottleneck buffer has a finite capacity of B packets, packets that see B packets
present in the bottleneck buffer upon their arrival cannot enter that buffer, and
will be lost. So $y_B^{(A)}$ also equals the packet loss probability of a random packet
at the bottleneck buffer. Analogously, also for the probabilities $v_m^{(A)}$ and $w_m^{(A)}$,
$0 \leq m \leq B$, defined as the probabilities that an enhancement layer packet,
respectively a base layer packet, sees m packets in the bottleneck buffer upon
arrival, an expression is constructed. Again $v_B^{(A)}$, respectively $w_B^{(A)}$, equals the
probability that an enhancement layer packet, respectively base layer packet,
gets lost at the bottleneck buffer due to buffer overflow. The probability d_m that
the queueing delay of a random packet is m slots, is then given by

$$d_m = \begin{cases} \frac{y_m^{(A)}}{1-y_B^{(A)}} & \text{if } m \in \{0, \dots, B-1\} \ , \\ 0 & \text{otherwise} \ , \end{cases} \tag{2}$$

where the denominator $1 - y_B^{(A)}$ accounts for the fact that the queueing delay of a packet at the bottleneck buffer only exists for packets that can effectively enter the bottleneck buffer. In a similar way, also the probabilities d_m^E and d_m^B that the queueing delay at the bottleneck buffer of an enhancement, respectively a base layer packet, equals m slots, are calculated.

From the invariant probability vector $\mathbf{x}$ of the queueing system, also the fractions of the time $P_{B+E}(c), P_B(c), P_{B+E}(e)$ and $P_B(e)$ that the video traffic arrives at the bottleneck buffer at the rate $R_{B+E}(c), R_B(c), R_{B+E}(e)$ and $R_B(e)$ respectively, can be computed.

2.4   Packet Loss Probabilities

In Section 2.3, the focus was on the calculation of the delay distributions at the bottleneck buffer, since these distributions determine the underflow probability of the dejittering buffer. However, not all originally generated packets reach the dejittering buffer. In the modeled network scenario, packet loss can occur at three places: at the traffic management module, at the bottleneck buffer, and at the dejittering buffer.

If there is congestion at the bottleneck, the enhancement layer packets of the video streams are already dropped at the traffic management module. The fraction of all packets that are discarded at that point is given by

$$\frac{P_B(c)(R_{B+E}(c) - R_B(c)) + P_B(e)(R_{B+E}(e) - R_B(e))}{\pi(c)R_{B+E}(c) + \pi(e)R_{B+E}(e)} \ , \tag{3}$$

and the fraction of all enhancement layer packets that are discarded at the traffic management module is then

$$\frac{P_B(c)(R_{B+E}(c) - R_B(c)) + P_B(e)(R_{B+E}(e) - R_B(e))}{\pi(c)(R_{B+E}(c) - R_B(c)) + \pi(e)(R_{B+E}(e) - R_B(e))} \ . \tag{4}$$

Base layer packets are never dropped by the traffic management module.

At the bottleneck buffer, packets will be lost if the bottleneck buffer overflows. The probabilities that a random packet, a base layer packet and an enhancement layer packet respectively, that arrives at the bottleneck buffer cannot enter this buffer, are given by the values $y_B^{(A)}, w_B^{(A)}$ and $v_B^{(A)}$ respectively, which were defined in Section 2.3.

At the dejittering buffer, packets get lost due to dejittering buffer underflow. From (1) it is known that the dejittering buffer underflow probabilities, given T_{jit} and d^*, are calculated as

$$\sum_{m=T_{jit}+d^*+1}^{B-1} d_m \ , \qquad \sum_{m=T_{jit}+d^*+1}^{B-1} d_m^B \ , \quad \text{and} \quad \sum_{m=T_{jit}+d^*+1}^{B-1} d_m^E \ , \tag{5}$$

for a random packet, a base layer packet and an enhancement layer packet respectively.

From all these loss probabilities, the fraction of all generated traffic of a certain type that reaches its receiver can be calculated as the product of the fractions of the traffic of that type that are not lost at the traffic management module, at the bottleneck buffer, and at the dejittering buffer.

3   Numerical Results and Discussion

Consider a system with the following characteristics:

- the capacity of the bottleneck DSL link is $C = 12$ Mbit/s,
- packets have a size of $b = 1500$ bytes, so 1 slot corresponds to 1 ms,
- the bottleneck buffer capacity B is 120 packets (or thus 120 ms),
- the sources switch on average every 30 seconds to the other scene type,
- the rates at which the sources generate the traffic are $R_{B+E}(c) = 3.5$ Mbit/s and $R_{B+E}(e) = 2.5$ Mbit/s, and
- the rates that correspond to the base layer packets only are $R_B(c) = 2.4$ Mbit/s and $R_B(e) = 1.5$ Mbit/s.

So one source generates on average 3 Mbit/s. This means that in case the traffic management module is not present, one source puts a load of 0.25 on the bottleneck buffer. A setup with 4 subscribers will be considered.

The Figures 2 – 4 show for three values of the threshold Th the probabilities that a random packet, a base layer packet and an enhancement layer packet arrive too late at the dejittering buffer, as a function of $T_{jit} + d^*$. Note the different scale of the y-axis in the three figures. A setting of $Th = B = 120$ corresponds to the situation where no traffic management module is present at the bottleneck, since irrespective of the bottleneck buffer content, no packets are prevented from entering this buffer. The curves illustrate that for all choices of $T_{jit} + d^* < 119$ slots, the underflow probabilities are always smaller when the traffic management module is present than when it is not. The smaller the threshold value that is used, the smaller the underflow probabilities become. Remark from the Figures 3 – 4 that once $T_{jit} + d^*$ becomes larger than the threshold value used, the underflow probability goes down more steeply with increasing $T_{jit} + d^*$ than before. For $T_{jit} + d^*$ equal to or larger than 119 slots, the underflow probability becomes zero in all cases, because a packet that arrives at the dejittering buffer will never have experienced a queueing delay larger than 119 slots in the bottleneck buffer.

Curves as shown in the Figures 2 – 4 can be used to dimension the parameter T_{jit} of the dejittering process. This parameter should be set as small as possible to reduce the extra delay introduced by the dejittering process, but large enough to keep the packet loss due to dejittering buffer underflow below some target value. For a certain target underflow probability, the corresponding value for $T_{jit} + d^*$ is read from the curve. Remark that in general the delay d^* experienced by the first packet of the video stream is not known at the moment a value for

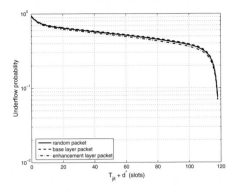

Fig. 2. Underflow probability given $T_{jit} + d^*$, for the scenario with $Th = 120$ (no traffic management module)

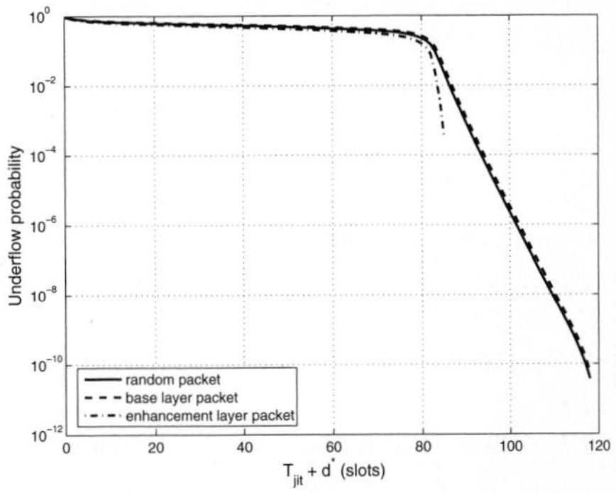

Fig. 3. Underflow probability given $T_{jit} + d^*$, for the scenario with $Th = 84$

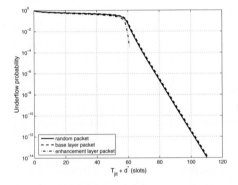

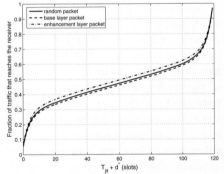

Fig. 4. Underflow probability given $T_{jit} + d^*$, for the scenario with $Th = 60$

Fig. 5. Fraction of generated packets that reach the receiver. Scenario without traffic management module ($Th = 120$).

T_{jit} should be set. So under the worst case assumption that the first packet of the video stream experiences a queueing delay of 0 slots in the bottleneck buffer, the worst case value for T_{jit} is thus found.

From the Figures 2 – 4 it is seen that when the traffic management module is not present ($Th = 120$), there is not much difference between the loss probability due to dejittering buffer underflow for a random packet and for a base layer packet or an enhancement layer packet. When the traffic management module is used, base layer packets have a higher probability than enhancement layer packets to get lost due to dejittering buffer underflow. This is because the underflow probability of the dejittering buffer depends on the delay distribution at the bottleneck buffer, and enhancement layer packets will never experience a delay much larger than the threshold value Th used in the scenario, since at moments the bottleneck buffer content exceeds the threshold, no enhancement layer packets can enter the bottleneck buffer.

Table 1. Packet loss probabilities at the traffic management module and at the bottleneck buffer

	At traffic management module		At bottleneck buffer		
	random	enhancement	random	base	enhancement
$Th = 120$	0	0	0.0314	0.0325	0.0294
$Th = 108$	0.0316	0.0904	8.41e-06	1.25e-05	0
$Th = 84$	0.0322	0.0920	9.88e-12	1.47e-11	0
$Th = 72$	0.0326	0.0931	1.15e-14	1.72e-14	0
$Th = 60$	0.0331	0.0944	$< 1e-16$	$< 1e-16$	0
$Th = 48$	0.0337	0.0963	$< 1e-16$	$< 1e-16$	0
$Th = 36$	0.0347	0.0993	$< 1e-16$	$< 1e-16$	0
$Th = 24$	0.0368	0.1051	$< 1e-16$	$< 1e-16$	0
$Th = 12$	0.0441	0.1259	$< 1e-16$	$< 1e-16$	0
$Th = 0$	0.35	1	$< 1e-16$	$< 1e-16$	0

Table 1 shows for several values of the threshold the packet loss probabilities at the traffic management module and at the bottleneck buffer. Remark that when the traffic management module is not used ($Th = 120$), obviously no packets get lost at it. When it is used, it discards the enhancement layer packets in case of congestion, but never the base layer packets. Therefore, the packet loss probability at the traffic management module of a random packet is 35% of the packet loss probability of an enhancement layer packet, since from the parameters of the source it is seen that 65% of all generated packets are base layer packets, and 35% of the packets are enhancement layer packets. Remark also the extreme case $Th = 0$, which corresponds to the scenario where all enhancement layer packets are discarded.

About the packet loss at the bottleneck buffer, it can be read from Table 1 that when the traffic management module is used, only base layer packets get lost due to bottleneck buffer overflow. This is because the bottleneck buffer will only overflow in case of congestion, and during congestion periods the enhancement layer packets are already dropped at the traffic management module (only in examples where the bottleneck buffer space behind the threshold is smaller than the number of sources, it might happen that also enhancement layer packets get lost at the bottleneck buffer). For the scenario with $Th = 120$, both types of packets are lost, and it is noticed that much more base layer packets get lost now than when the traffic management module is used.

Based on the values in Table 1, the amount of traffic (random packets, base layer packets or enhancement layer packets) that arrives at the dejittering buffer can be calculated. Namely, this are all packets that were not discarded at the traffic management module and that did not get lost due to buffer overflow at the bottleneck buffer. From these numbers it is then seen that with the traffic management module, all but a very small amount (corresponding to the packet loss at the bottleneck buffer) of the base layer packets reach the dejittering buffer. The smaller the threshold used in the scenario, the more base layer packets reach the dejittering buffer. When the traffic management module is not used, in total

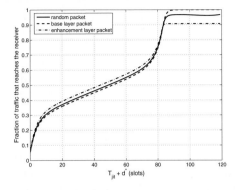
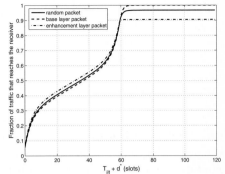

Fig. 6. Fraction of generated packets that reach the receiver. Scenario with $Th = 84$.

Fig. 7. Fraction of generated packets that reach the receiver. Scenario with $Th = 60$.

more packets reach the dejittering buffer, but considerably less of the important base layer packets do.

In case the delay budget is large enough to allow $T_{jit} + d^* \geq 119$ slots, all traffic that reaches the dejittering buffer will also reach the receiver, since the dejittering buffer never underflows. If $T_{jit} + d^* < 119$ slots, part of the packets that arrive at the dejittering buffer will get lost at that buffer, because they arrive there after their target playout time. The Figures 5 – 7 show for the three examples for which the dejittering buffer underflow probability was shown before, the fractions of the generated traffic that reach the receiver. It is seen that for $T_{jit} + d^* < 119$ slots, more video packets (base plus enhancement layer packets) reach the receiver when the traffic management module is used compared to when it is not used. The reason is that in case of congestion, the traffic management module already discards packets, so less traffic is sent into the bottleneck buffer at *critical* moments. This has as effect that the packet loss at the bottleneck buffer is reduced, and also that the delay the packets experience at the bottleneck decreases, which has a positive effect on the dejittering buffer underflow probability. Further, the figures show that in spite of the fact that base layer packets have a higher probability than enhancement layer packets to get lost at the bottleneck buffer and at the dejittering buffer, end-to-end the base layer packets reach the receiver with a higher probability than the enhancement layer packets when $T_{jit} + d^*$ is larger than the threshold value used in the scenario. The explanation for this is that when $T_{jit} + d^*$ is larger than the threshold, the underflow probability of the base layer packets at the dejittering buffer becomes small compared to the loss probability of an enhancement layer packet at the traffic management module, the latter which is independent of $T_{jit} + d^*$.

So if the goal is to let as much base layer packets as possible reach the receiver, then putting $Th = 0$ is obviously the best choice. The lower the threshold however, the more enhancement layer packets are discarded at the traffic management module, and the magnitude of this packet loss increases faster when the threshold position becomes smaller, as can be read from Table 1. And although the enhancement layer packets are the least important packets, they still have some value when they reach the receiver, and most likely reducing the base layer packet loss further from for example 1e-9 to much smaller is not worth anymore the loss of several percents of enhancement layer packets. A balance needs thus to be found between where to put the threshold such that at critical moments the bottleneck buffer is discharged because temporarily no enhancement layer packets are sent into it, and the superfluous dropping of the enhancement layer packets.

As becomes clear from the presented results, the choice of the position of the threshold Th in the bottleneck buffer and of the dejittering delay T_{jit} at the dejittering buffer cannot be uncoupled. Once one of them is fixed, there are limitations on how to choose the other. More concrete, the threshold Th should always be chosen low enough such that the packet loss of the base layer packets at the bottleneck buffer is below a required target, but preferable somewhere in the higher part of the buffer because otherwise there might be a large part of the enhancement layer packets that will never reach their receiver simply because they could not pass the traffic management module. If it is not possible to find a setting for Th such that both requirements are fulfilled, then typically too many connections are accepted into the network. Once the threshold position is chosen, the dejittering delay T_{jit} should be determined such that $T_{jit} + d^*$ is larger than Th, because then the packet loss due to dejittering buffer underflow lies in the region where the underflow probability goes down more steeply with increasing $T_{jit} + d^*$. As mentioned earlier, if the queueing delay d^* experienced at the bottleneck buffer by the first packet of the video stream is not known at the moment a value for T_{jit} should be set, it is best to start from the worst case assumption that the first packet of the video stream will experience a queueing delay of 0 slots.

4   Conclusions and Future Work

In this paper, the influence of the presence of a video-aware traffic management module in the access network on the performance of the dejittering mechanism, was studied. All results were derived by applying a mathematical multiplexing model to layered video sources, whose traffic first had to pass a video-aware traffic management module that discards the enhancement layer packets in case of congestion. A threshold Th on the content of the bottleneck buffer is used to detect the congestion.

From the numerical examples, it was concluded that with the presence of the traffic management module, both the packet loss probability at the bottleneck buffer and the dejittering buffer underflow probability (which depends on the delay distribution at the bottleneck buffer) decreases. The reason is that with the traffic management module, part of the packets are already discarded before they enter the bottleneck buffer, so less traffic is sent into the bottleneck buffer at *critical* moments. Since the traffic management module discards only enhancement layer packets, especially the highly important base layer packets gain from the presence of the traffic management module. Although in the scenarios where the traffic managment module is used it is observed that the base layer packets experience a higher packet loss probability at the bottleneck buffer and at the dejittering buffer than the enhancement layer packets, it appears that if the sum of $T_{jit} + d^*$, of the dejittering delay T_{jit} and of the queueing delay experienced at the bottleneck buffer by the first packet of the video stream d^*, is larger than the threshold Th used in the scenario, then a base layer packet has a higher probability of reaching the receiver than an enhancement layer packet.

So the choice of the position of the threshold in the bottleneck buffer and of the dejittering delay at the dejittering buffer cannot be uncoupled. A general guideline is that Th should always be chosen low enough such that the packet loss of the base layer packets at the bottleneck buffer is below a required target, but preferable somewhere in the higher part of that buffer because otherwise there will be a large part of the enhancement layer packets that will never reach their receiver because they are unnecessarily dropped by the traffic management module. Once the threshold position is chosen, the dejittering delay T_{jit} should be chosen such that $T_{jit} + d^*$ is larger than Th, because then the packet loss due to dejittering buffer underflow lies in the region where the underflow probability goes down more steeply with increasing $T_{jit} + d^*$.

Remark that with the network scenario considered in this paper, where the traffic management module is present immediately before the bottleneck buffer, the packets that are discarded by the traffic management module first travel (unnecessary) from the video source through the core and aggregation network until the access network. Another option would be to make the video sources responsive, as with DCCP, such that the sources themselves discard the enhancement layer packets in case of congestion, based on feedback signals they receive from the receiver about the congestion status of the network. If this feedback arrives with a negligible delay, the model and the results presented in this paper stay valid. When the feedback signals arrive however at the sources with a non-negligible delay, the results will change. The construction of a model that can assess the performance of such a network scenario, and a comparison of its performance with that of the network scenario considered in the current paper, will be the topic of some future work.

References

1. Sharpe, R., Zriny, D., De Vleeschauwer, D.: Access network enhancements for the delivery of video services. Alcatel Telecom Review (ATR), pp. 134–140 (2005)
2. Schwarz, H., Marpe, D., Wiegand, T.: Overview of the scalable H.264/MPEG4-AVC extension. In: Proceedings of the IEEE International Conference on Image Processing (ICIP'06) (2006)
3. Kohler, E., Handley, M., Floyd, S.: Datagram congestion control protocol (DCCP). RFC 4340 (2006)
4. Anderson, N.: An introduction to IPTV (2006), `http://arstechnica.com/guides/other/iptv.ars/`
5. Perkins, C.: RTP - Audio and Video for the Internet. Addison-Wesley, New York (2003)

Modeling of H.264 High Definition Video Traffic Using Discrete-Time Semi-Markov Processes

Sebastian Kempken and Wolfram Luther

Universität Duisburg-Essen, 47048 Duisburg, Germany,
{kempken, luther}@inf.uni-due.de
http://wrcg.inf.uni-due.de

Abstract. Semi-Markov processes (SMPs) are widely used to model various types of data traffic in communication networks. Also, efficient and reliable analysis techniques are available. In this paper, we consider several present methods of deriving the parameters of a discrete-time semi-Markov process from given H.264 video traces in order to model the original traffic adequately. We take the distribution of frame sizes and the autocorrelation of both the original trace and the resulting SMP model into account as key quality indicators. We propose a new evolutionary optimization approach using genetic programming, which is able to significantly improve the accuracy of semi-Markov models of video traces and, at the same time, requires a smaller number of states.

1 Introduction

High definition (HD) video content is gaining more and more popularity as the number of available distribution channels increases. HD video is delivered to the end user not just via satellite or cable links, but also via IP networks (IPTV). Hence, today's internet traffic consists of an increasing amount of HD video traffic encoded according to the H.264 standard, which needs to be reflected in the techniques used for traffic modeling in service-integrated communication networks.

Routers and switches in these networks receive data over a number of input links and distribute it over multiple output links according to their terminal destination. A particular output line is chosen by the router for each packet according to forwarding rules defined by routing protocols – like OSPF or BGP – standardized by the Internet engineering task force (IETF).

Moreover, buffers are also included in the routing process. In general, one buffer is available for each output link. In this way, the router is able to store all corresponding data packets and to forward them after some delay in cases of transmission collisions or temporary overload situations.

Hence, the usual model of such a router considers

1. an output link with a forwarding capacity C, which can be measured in data per time (e.g. Gb/sec),
2. a buffer of size B measured, for instance, in Mb or Gb,
3. the input process of the data arriving at the router.

L. Mason, T. Drwiega, and J. Yan (Eds.): ITC 2007, LNCS 4516, pp. 42–53, 2007.

In order to efficiently model and analyze such a router, we consider a time-slotted system. This approach is often taken in performance analysis of telecommunication systems [1,2,3]. In this case, the time is divided into time-slots of equal length Δ and the amount of data arriving at or leaving the router is traced per slot.

Usually, the output capacity is constant, so that a constant amount of data $C\Delta$ can be transmitted in each time slot. In contrast, the amount of arriving data is often highly variable and also shows a high level of autocorrelation.

For this reason, we consider semi-Markov processes to model the arrival process. This way, we can reflect the autocorrelation of the process [4]. An analysis of such models is possible via matrix-analytical [5] or factorization techniques [1,3,6,7,8].

Furthermore, several modeling techniques for video traffic using Markovian approaches are known [9,10,11,12]. We compare some methods to derive the parameters of a semi-Markov process. As key characteristics for the quality of such a model, we consider the accurate representation of both the distribution and the autocorrelation of the arrival process.

In terms of the latter criterion, video traffic shows a high level of long-term correlation [9]. This is difficult to include in semi-Markov models, as the autocorrelation of a SMP shows exponentially decaying behavior [13]. Present semi-Markov modeling techniques require a large state space to reflect the long-term behavior. However, the autocorrelation of a SMP depends on its parameters. We find that by choosing them carefully, the quality of the SMP model can be significantly improved. We implemented an evolutionary optimization technique to find suitable parameters. This results in a more accurate representation of the long-term autocorrelation, even in a much smaller state space.

In the following section, we describe the typical characteristics of H.264 video streams. Next, we give an overview on some present techniques for modeling these characteristics using semi-Markov processes. We also propose a new method that addresses the problems mentioned above. Finally, in Section 4, we give some examples and compare the results of the modeling approaches.

2   H.264 Video Traces

The H.264 standard stipulated by the International Telecommunication Union (ITU) defines a coding technique for video streams with compression [14]. It is intended for a broad range of applications from video conferences and portable media with relatively low bandwidth requirements to premium high definition content with a correspondingly high need for bandwidth capacity.

The compression is achieved by reducing both the spatial and temporal redundancies. The former refers to the compression of a particular frame using entropy coding and transforms; the latter considers changes on a frame-to-frame level. The compression here is done by predicting future frames using motion vectors. Hence, we may distinguish three types of frames:

1. I-frames: These use only intra-frame coding. The compression is achieved through such methods as the discrete cosine transform and entropy coding.
2. P-frames: Additional to the intra-frame coding techniques, these frames are predicted using motion compensation referring to the previous I- or P-frame.
3. B-frames: These bidirectional frames code the differences between the actual picture and an interpolation between the previous and next P- or I-frames.

The usual case is that B-frames have the lowest bandwidth requirements, and I-frames are the largest frames in a sequence. The video coding standard requires the frames to be in a deterministic periodic order, for example "IBBBPBBB" or "IBBPBBPBBPBB". Such a sequence is called group of pictures (GoP).

Hence, a video traffic stream can be considered at several levels (cf. [9]):

– Sequence layer, which is the whole video sequence (minutes to hours)
– Scene layer, which is given in intervals where the content of the pictures is almost the same (seconds to minutes)
– GoP layer, which is defined by a group of pictures (tenths of a second)
– Frame layer, which considers the period of a single frame (hundredths of a second)

Since we intend to provide a reliable model in terms of distribution and autocorrelation, we consider a low level of the video stream as a basis for the modeling approach. On the frame level, the periodic coding scheme together with the typical size relations of the frame types leads to an autocorrelation behavior typical of MPEG video streams. However, this behavior is very difficult to reflect in a semi-Markovian model and can only be achieved with a high number of states of the SMP [9]. On the GoP level, long-term dependency is typical of video streams. The slow decay of the autocorrelation function is difficult to model using a semi-Markov-based approach since this leads to an exponentially decaying autocorrelation function [4].

In the modeling of MPEG-based video streams, we therefore face two challenges: On the frame level, we must be able to reproduce the typical periodic autocorrelation behavior, while on the GoP level, we have to find a way to model the occurring long-term dependencies using an exponentially decaying autocorrelation.

3   Modeling Techniques

We introduce a discrete-time homogeneous semi-Markov process (SMP) to describe the workload behavior of a given queueing system (cf. [15]). A family of random variables $\{(R_t, \sigma_t)|t \in \mathbb{N}_0\}$ denotes such a process if

$$P\left(R_{t+1} = a, \sigma_{t+1} = j | R_k = a_k, \sigma_k = i_k, 1 \leq k \leq t\right) =$$

$$P\left(R_{t+1} = a, \sigma_{t+1} = j | \sigma_t = i_t\right)$$

holds for all $t \in \mathbb{N}$ and $\{\sigma_t\}$ is a Markov chain. In our case, R_t denotes the t-th workload change. In the following, we restrict ourselves to the case of a finite Markov chain. If the underlying chain $\{\sigma_t\}$ has M states, the semi-Markov process is called SMP(M).

To define a semi-Markov process completely, the transition matrix $P := (p_{ij})$ of the underlying chain $\{\sigma_t\}$ with M^2 distribution functions $f_{ij}(t)$, $i, j = 1, \ldots, M$ must be known.

A simplification of the transition is achieved using a special case of a semi-Markov process (SSMP). The SSMP is a SMP whose sojourn time distribution in a given state depends only on the actual state $\sigma_t = i_t$, but not on the next visited state σ_{t+1}. Please note that every SMP with M states can always be described as an SSMP with M^2 states. To do so, we identify each transition from state i to state j of the SMP with a state (i, j) of the SSMP (cf. [2]). Using special case SMPs, we are able to adjust the autocorrelation and the distribution of the process independently: The autocorrelation relies only on the state-dependent mean values (cf. Equation 8).

Furthermore, to make efficient analysis of the SMP feasible, we require the embedded Markov chain to be aperiodic and irreducible and the transition matrix to be diagonalizable. In this case, we can denote the autocorrelation function of the SMP as an exponential sum for all $n \in \mathbb{N}_0$:

$$\mathcal{A}_R(n) = \sum_{j=2}^{M} \alpha_j \lambda_j^n \tag{1}$$

with λ_j being the eigenvalues of the transition matrix (cf. [4]). The first eigenvalue λ_1 of a stochastic transition matrix is always 1. The coefficients α_j can be derived, for instance, by solving a Vandermonde-type equation system using a number of given autocorrelation values.

As already mentioned, the accurate modeling of a sequence of frame sizes is difficult because of its periodic structure. However, we are able to derive the size of particular frames from the size of one group of pictures (GoP) (cf. [9]). First, we determine the mean size of a GoP and the mean size of each frame inside a GoP from a given video trace. If, for instance, the GoP consists of 12 frames, we yield the same number of scaling factors by dividing the mean frame size by the mean size of a GoP. These scaling factors can now be used to reconstruct the sizes of particular frames from the size of a GoP, as determined by a semi-Markovian model. According to works by Rose [9], the loss of information concerning the correlation between frames within a group of pictures has almost no influence on the results. Hence, we can use this approach as an efficient approximation technique to model a sequence of frame sizes from a semi-Markov model based on GoP sizes.

The coding scheme requires the frames to be altered in order. Since bidirectional frames require information from both preceding and subsequent frames, the latter have to be transmitted before the particular B-frame. For example, the transmission order for the sequence "$IBBP\ldots$" is "$IPBB\ldots$". The groups of pictures, however, are transmitted in the same order as they are needed for video playback. Therefore, we focus on the GoP level.

In the following, we describe the modeling techniques considered. We denote the original data as G_j describing the size of the GoP in bytes with j being the index of a particular GoP.

3.1 Simple Approach

We now describe how the parameters of a special case semi-Markov process with M states SSMP(M) can be determined according to a simple size-oriented approach. The number of states M can be set according to (cf. [9]):

$$M = \left\lfloor \frac{\max_{j\in\{1,\dots,j_{\max}\}}(G_j)}{\sigma_G} \right\rfloor \tag{2}$$

with σ_G denoting the standard deviation of the values G_j. This equation defines just one possibility to obtain a value for M. In general, one has to find a tradeoff between accuracy of the model and the complexity of the model.

We split the range of values from $\min(G_j)$ to $\max(G_j)$ into M disjoint intervals of the same diameter. The values G_j are assigned to a state according to the interval containing the corresponding value. Since a sequence of states is now defined, we may approximate the transition probabilities:

$$p_{ij} = \frac{t_{ij}}{\sum_{k=1}^{M} t_{ik}} \tag{3}$$

Here, t_{ij} denotes the number of transitions from state i to state j. The stationary probabilities, however, cannot be approximated this way, as this approximation does not fit exactly with the values for the transition probabilities. Hence, we have to determine the stationary probabilities $p_i, i \in Z := \{1, \dots, M\}$ by solving the following system of equations:

$$\sum_{k\in Z} p_k p_{kj} = p_j \text{ for all } j \in Z, \sum_{j\in Z} p_j = 1 \tag{4}$$

The state-dependent discrete probability function $\rho_i(x)$ is determined by counting. If the lower and upper bounds of the value interval for state i are given by $\underline{g}_i$ and $\overline{g}_i$ respectively, then the probabilities can be computed in this way:

$$\rho_i(x) = P(G = x | \sigma = i) = \frac{|\{j | \underline{g}_i \leq G_j \leq \overline{g}_i \wedge G_j = x\}|}{|\{k | \underline{g}_i \leq G_k \leq \overline{g}_i\}|} \tag{5}$$

3.2 Scene-Oriented Approach

In the second approach for modeling video traffic using SMPs, the sequence is split into one or more scenes. The term "scene" denotes a part of the video sequence with limited variance, that is, similar GoP sizes. This definition may coincide with the common definition of a movie scene, but in our case refers to statistical characteristics of a GoP sequence.

Rose [9] suggests the following algorithm to find the boundaries of scenes : Let G_i denote the size of the GoP i and n the GoP number. Then the boundaries b_{left}, b_{right} are computed thus:

1. Set $n = 1$. Set current left scene boundary $b_{\text{left}} = 1$.
2. Increment n by 1. Compute the variation coefficient c_{new} of the sequence $G_{b_{\text{left}}} \ldots G_n$ as quotient of the standard deviation and the mean value of the considered sequence

$$c_{\text{new}} = \frac{\sigma(G_{b_{\text{left}}} \ldots G_n)}{E(G_{b_{\text{left}}} \ldots G_n)}.$$

3. Increment n by 1. Set $c_{\text{old}} = c_{\text{new}}$. Compute the coefficient c_{new} of the sequence $G_{b_{\text{left}}} \ldots G_n$.
   (a) If $|c_{\text{new}} - c_{\text{old}}|(n - b_{\text{left}} + 1) > \epsilon$, set the right scene boundary $b_{\text{right}} = n - 1$ and the left scene boundary of the new scene $b_{\text{left}} = n$. Go to step 2.
   (b) Otherwise, go to step 3.

The amount of variation allowed within a sequence can be adjusted using the parameter ϵ. Rose has yielded best results for a value of $\epsilon = 0.9$.

Next, we create a scene-transition SSMP. We calculate the mean values for the size of the GoPs per scene and determine the parameters of the SSMP as described in Section 3.1. The transitions between groups of pictures within a particular scene are also taken into account to compute the transition probabilities. If, for instance, a scene contains five GoPs, then four transitions within the scene have to be considered.

For each state of the scene-transition SSMP, we construct another SSMP which describes the behavior within all scenes of the same class. To do so, all scenes of the same class are combined while preserving their order in the original video stream. We determine the parameters of a SSMP describing the scene-specific behavior analogously to the size-oriented approach (Section 3.1). In the following, the SSMP for a scene of class k will be denoted by SSMP_k, the transition SSMP by SSMP_T.

The SSMPs determined are combined to one SSMP with more states: Let M_k denote the number of states of SSMP_k and M_T denote the number of states of the transition process. We need $\overline{M} = \sum_{k=1}^{M_T} M_k$ states to describe the combined SSMP. Each possible combination of a state of the transition SSMP and a state of the corresponding SSMP_k is reflected in a state of the combined SSMP. We denote these states as (k, i) with $1 \leq k \leq M_T$ and $1 \leq i \leq M_k$ and (l, j) respectively. Then the transition probabilities of the combined SSMP $\overline{p}$ can be computed as:

$$\overline{p}_{(k,i),(l,j)} = \begin{cases} p_{k,k} p_{i,j}^{(k)} & \text{if } l = k \\ p_{k,l} p_j^{(l)} & \text{if } l \neq k \end{cases} \tag{6}$$

$p_{k,l}$ denotes the transition probabilities of $SSMP_T$ and $p_{i,j}^{(k)}$ and $p_i^{(k)}$ the transition and steady state probabilities of $SSMP_k$ respectively.

3.3   Genetic Programming Approach

As already mentioned, video traffic shows a high level of long-term autocorrelation. The methods presented so far require a high number of states to accurately reflect this in the resulting model. We present a new approach employing genetic programming that offers a more accurate representation of the long-term correlation. At the same time, a much smaller number of states is required for the Markov process, making the subsequent analysis of these models easier in terms of computational complexity.

This approach does not derive the parameters of a SSMP model from given data in a deterministic way but instead efficiently tests new parameters and checks which fit best. To do so, a *population* is defined. Each member of this population (*individual*) represents a possible parameter configuration, and *fitness* values are computed for each one. The members with a high fitness level are reproduced, *mutated* and combined with each other (*crossover*). They form the members of the next generation of the population. This depicts a simple but efficient optimization technique for a broad range of problems. In the following, we describe how we adapt this idea to our problem of finding the parameters of an SSMP model. Further details about genetic programming techniques can be found, for instance, in [16] or [17].

The main problem in deriving SSMP parameters from a given empirical data series is the assignment of values to states of a Markov chain. If such an assignment is given, the transition probabilities, the stationary probabilities and the state-dependent distributions can be computed from the given data G_j by counting: The transition and steady state probabilities are derived according to Equations 3 and 4 respectively. To model the state-dependent distributions, we look at the distributions of the actual values for each state according to the particular GoP-state assignment.

If N denotes the number of groups of pictures and M the number of states of a SSMP model, then there are M^N possible GoP-state assignments. For a typical data series as examined in the following section, we have $N \approx 4300$ and $M = 4$. Hence, it is infeasible to test all combinations. However, we attempt to search this huge parameter space using genetic algorithms.

The total number of states M of the SSMP model to be determined is a parameter for the optimization process. In our experiments (cf. Section 4), we found that a even a smaller number of states than given by Equation 2 is sufficient to give a good approximation of the autocorrelation behavior of the empirical data set.

We code each possible assignment alternative as a sequence of the corresponding state indices. In the example mentioned, one scheme is therefore defined by 4300 values between 1 and 4. The idea is to optimize the assignment scheme so that the autocorrelation of an SSMP derived from it and the autocorrelation of the empirical data series proved the best match. The autocorrelation function of a data series $G_j, j \in \{1, \ldots, N\}, n \in \{0, \ldots, N-1\}$ is estimated by:

$$\mathcal{A}_G(n) = \frac{1}{\sigma_G^2 (N-n)} \sum_{j=1}^{N-n} (G_j - E(G)) (G_{j+n} - E(G)) \tag{7}$$

The corresponding values for the SSMP model can be derived from its parameters (cf. [4]) or by using:

$$\mathcal{A}_S(n) = \frac{1}{\sigma_S^2} \left(\sum_{i=1}^{M} \sum_{j=1}^{M} p_i E_i(S) p_{ij}^{(n)} E_j(S) - E^2(S) \right), \mathcal{A}_S(0) = 1 \tag{8}$$

Here, $E(S)$ denotes the overall expectation value of the semi-Markov process S and $E_i(S)$ the expectation value if S is in state i, which is the state dependent mean value. The n-step transition and the stationary probabilities are given by $p_{ij}^{(n)}$ and p_i respectively.

To evaluate the quality of an assignment scheme and the corresponding SSMP model S of the empirical data series G, we use the following function:

$$g_G(S) := \frac{1}{\sum_{i=1}^{N} (\mathcal{A}_G(i) - \mathcal{A}_S(i))^2} \tag{9}$$

Since all video traces examined in our experiments show a non-monotonic auto-correlation behavior, a perfect approximation cannot be achieved by SMPs. A division by zero can therefore be ignored. In terms of genetic programming, the intention is to find an individual (parameter configuration) that maximizes the above fitness function. Please note that it is also possible to introduce a scaling factor depending on i in the equation above. This allows to put more emphasis on the first values of the autocorrelation representing smaller lags, if an accurate representation of the short-time correlation is considered more important than long-time effects. We initialize a population of individuals by randomly taking values from the range of the state indices. These random assignment schemes are to be evaluated and optimized iteratively. Each iteration consists of several steps:

1. For each individual, construct an SSMP model with M states according to the given assignment scheme.
2. Compute the fitness of each individual according to Equation 9.
3. Build up a new generation of individuals.
   (a) The best individual – that is, the one with the highest fitness value – is taken into the next generation unchanged.
   (b) A *mating pool* is constructed from a selection of the population. Each individual has a probability proportional to its fitness value to be selected for the mating pool.
   (c) With given probabilities, some mutators are applied to the members of the mating pool (see below).
   (d) Genetic material is exchanged between random members of the mating pool (crossover, see below).

(e) The changed and unchanged members of the mating pool form the next generation of the population.

Since we cannot decide when the optimal configuration has been reached, we could extend the optimization process infinitely. However, we abort the optimization when no further improvement considering the maximum fitness value of the population has been achieved for a given number of generations. The parameters corresponding to the individual with the highest fitness value are returned as result of the genetic approach.

The genetic operators (mutation and crossover) and their probability are important to the optimization process. In our case, we used four types of mutation operators, each with a probability of 0.2:

- Swap. Two random values in the sequence are exchanged.
- Reverse. A randomly chosen, continuous part of the sequence is reversed.
- Shuffle. A randomly chosen, continuous part of the sequence is shuffled.
- Block. A randomly chosen, continuous part of the sequence is replaced by a block of identical random values.

In our experiments, we use the so-called 2-point crossover: A randomly chosen continuous part of the genetic sequence is replaced by a sequence taken from another individual with the same length and relative position.

An interesting enhancement of this approach is to start with a modified initial population: Instead of creating all individuals randomly from scratch, we may include the assignments according to the other two approaches mentioned. Because they are included in the sequence population, and the best sequence is kept in every iteration step, it is guaranteed that the results yielded by the evolutionary optimization process are at least of the same quality. However, this is only feasible if the number of states of the approaches match.

The state-dependent distributions of the of the SSMP model are approximated from the empirical data trace by counting. The mean value and the variance of the original data are therefore conserved up to a negligible approximation error.

4   Examples and Comparison

Reisslein et al. [18,19] provide three H.264 high definition video traces for network evaluation purposes[1]. We consider the two traces with a duration of 30 minutes to evaluate the approaches presented and also to give some illustrating examples. The first sequence to be modeled is taken from the documentary "From Mars to China" (A) and has a duration of 51,715 frames.The other trace we considered is taken from the "Horizon" talk show (B) and includes 49,177 frames.The GoP order is "IBBPBBPBBPBB" at a rate of 30 frames per second in both cases. We applied all three of the methods presented in this paper. Since the genetic approach is random-driven, we perform 20 runs with a different random seed and consider the best result. The autocorrelation functions from all approaches

---

[1] http://trace.eas.asu.edu/hd

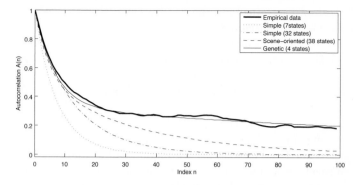

Fig. 1. Autocorrelation of sequence A and simple, scene and genetic models

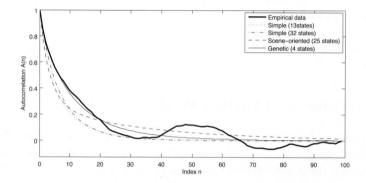

Fig. 2. Autocorrelation of sequence B and simple, scene and genetic models

are depicted in Figures 1 and 2. The sum of the remaining square error between the functions of the original trace $\mathcal{A}_G(n)$ and the models $\mathcal{A}_S(n)$ up to $n = 100$, which is the reciprocal of the fitness value, is given in the following table:

Sum of square error	Sequence A				Sequence B			
Simple (7 / 13)	5.4975				0.6756			
Simple (32)	3.6419				0.5665			
Scene (38 / 25)	1.6778				0.5560			
	min	max	mean	dev.	min	max	mean	dev.
Genetic (4)	0.0168	0.1808	0.0581	0.0400	0.2394	0.2501	0.2440	0.0031

For the genetic approach, we report the minimum, maximum and mean error along with the standard deviation of the fitness values from all 20 runs. We see that the simple size-oriented approach is somewhat inadequate for accurate modeling of the autocorrelation. We applied the approach with a higher number of states (32) in order to increase the quality, but doing so yields only a small improvement. The scene-oriented approach shows better results, as could also be expected from the higher number of states included in the model. Considering the two examples, the genetic approach proposed in this paper yields best results for the modeling of the autocorrelation of a given video trace: The autocorrelation of this model matches the empirical trace closely. Moreover, only four states are required to model the autocorrelation behavior.

The computation time of the approaches is very different. During our experiments, we experienced the following typical time requirements on the same test machine: While the simple approach requires a few milliseconds, the scene-oriented approach terminates after about one second. The genetic algorithm completes in about a minute. The huge increase in computation time has to be seen in connection with the low number of states that is required for an accurate representation of the autocorrelation behavior: In subsequent workload analysis of a resulting SSMP queue model, the computation time depends on the number of states. Furthermore, a verification of the workload results, which is a main issue in our related work, is also possible only if the number of states is kept sufficiently small.

5   Conclusion and Further Work

In this paper, we have considered two known approaches to model high-definition video traffic using discrete-time semi-Markov processes and also proposed one new technique using genetic programming. We have applied all three methods to modeling two empirical video traces provided for network performance analysis. We see that an improvement in accuracy of the model can be achieved on the one hand by considering a Markov chain with more states. On the other hand, the accuracy can be improved significantly while using a smaller number of states if the parameters of the semi-Markov process are chosen appropriately. The genetic programming approach proposed in this paper yields the best results considering the accurate modeling of the autocorrelation behavior in a small state space.

As further work, we intend to apply additional techniques (e.g [20]) to the considered modeling problem and compare them with the methods presented in this paper. A statistical evaluation of the techniques requires a large set of sample traces, therefore we work on a software tool to extract those traces from given video files. Furthermore, the genetic programming approach proposed in this work may be applied to other types of traffic as well.

Acknowledgements

This research was carried out in an ongoing project funded by the German Research Council (DFG). The authors also want to thank Gerhard Haßlinger for his valuable comments and suggestions.

References

1. Li, S.Q.: A general solution technique for discrete queueing analysis of multi-media traffic on ATM. IEEE Transactions on Communication 39, 1115–1132 (1991)
2. Haßlinger, G.: Semi-Markovian modelling and performance analysis of variable bit rate traffic in ATM networks. Telecommunication Systems 7, 281–298 (1997)
3. Haßlinger, G.: Waiting times, busy periods and output models of a server analyzed via Wiener-Hopf factorization. Performance Evaluation 40, 3–26 (2000)
4. Kempken, S., Luther, W.: Verified methods in stochastic traffic modeling. In: Reliable Implementation of Real Number Algorithms: Theory and Practice. LNCS, Springer (to appear) (2007)
5. Latouche, G., Ramasvami, V.: Introduction to matrix analytic methods in stochastic modeling. ASA-SIAM (1999)
6. Luther, W., Traczinski, D.: Reliable computation of workload distributions using semi-Markov processes. In: Proceedings of the 13th International Conference on Analytical and Stochastic Modelling Techniques and Applications, pp.111–116 (2006)
7. Kempken, S., Luther, W., Haßlinger, G.: A tool for verified analysis of transient and steady states of queues. In: Proceedings of the First International Conference on Performance Evaluation Methodologies and Tools (VALUETOOLS '06), electronic resource (2006)
8. Traczinski, D., Luther, W., Haßlinger, G.: Polynomial factorization for servers with semi-Markovian workload: Performance and numerical aspects of a verified solution technique. Stochastic Models 21, 643–668 (2005)
9. Rose, O.: Simple and efficient models for variable bit rate MPEG video traffic. Performance Evaluation 30, 69–85 (1997)
10. Haßlinger, G., Takes, P.: Real time video traffic characteristics and dimensioning regarding QoS demands. In: Proceedings of the 18th International Teletraffic Congress, pp. 1211–1220 (2003)
11. Krunz, M.M.: The correlation structure for a class of scene-based video models. IEEE Transactions on Multimedia 2(1), 27–36 (2000)
12. Salvador, P., Valadas, R., Pacheco, A.: Multiscale fitting procedure using Markov modulated poisson processes. Telecommunications Systems 23(1-2), 123–148 (2003)
13. Feldmann, A., Whitt, W.: Fitting mixtures of exponentials to long-tail distributions to analyze network performance models. Performance Evaluation 31, 245–279 (1998)
14. Wiegand, T., Sullivan, G., Bjntegaard, G., Luthra, A.: Overview of the H.264/AVC video coding standard. IEEE Transactions on Circuits and Systems for Video Technology 13(7), 560–576 (2003)
15. Kleinrock, L.: Queueing Systems. vol.1(2). Wiley, Chichester (1975)
16. Goldberg, D.E.: Genetic Algorithms in Search, Optimization and Machine Learning. Kluwer Academic Publishers, Boston, MA (1989)
17. Whitley, D.: A genetic algorithm tutorial. Statistics and Computing 4(2), 65–85 (1994)
18. Seeling, P., Reisslein, M., Kulapala, B.: Network performance evaluation with frame size and quality traces of single-layer and two-layer video: A tutorial. IEEE Communications Surveys and Tutorials 6(3), 58–78 (2004)
19. van der Auwera, G., David, P., Reisslein, M.: Bit rate-variability of H.264/AVC FRExt. Technical report, Arizona State University (2006)
20. Rose, O.: A memory Markov chain for VBR traffic with strong positive correlations. In: Proceedings of the 16th International Teletraffic Congress, pp. 827–836 (1999)

Initial Simulation Results That Analyze SIP Based VoIP Networks Under Overload

Eric C. Noel and Carolyn R. Johnson

AT&T Labs
{eric.noel, carolyn.johnson}@att.com

Abstract. This paper presents the results of a simulation study that assessed SIP based VOIP networks under overload. This work addresses the issue of network level congestion controls in SIP based telephony networks. The simulation network consists of Media Gateways and Call Controllers, each with internal overload detection and control mechanisms. The simulation includes SIP timers tuned to operate gracefully with PSTN interfacing protocols. The traffic model is SIP based VoIP calls not involving application server or media server interactions. This work demonstrates that combining external overload controls with internal overload controls reduces blocking and increases goodput across the overload levels that ranged from 1 to 4 times the engineered load.

Keywords: SIP signaling congestion control, internal overload controls, external overload controls, throughput, goodput, simulation.

1 Introduction

The growth of internet telephony is increasing dramatically as a result of wide spread offers of lower cost telecommunications solutions for both consumer and business services. Particularly for business applications, the opportunity to combine voice and data on the same customer facilities offers even greater economic and engineering benefits. Internet telephony also brings opportunities for advanced feature functionality not present in current TDM based services.

Internet telephony applications use the Session Initiation Protocol (SIP) to establish sessions between end users. SIP is a standardized protocol described in RFC3261 [1] of the IETF. SIP is an application layer protocol used between SIP servers. SIP is used to establish bearer path (voice path) sessions between media servers via control signals between the media servers and control servers. In this paper we focus on the signaling between SIP media servers and control servers. This work is the basis for more complex SIP based VoIP architectures and call types that will be addressed in later papers.

As the use of Voice over IP (VoIP) communications increase, the need for effective methods of controlling traffic during periods of congestion is essential. All SIP servers should provide adequate internal overload controls. These controls are needed to ensure servers can detect their overload condition, and shed load gracefully when the server is in overload. However, when internal controls are insufficient to

L. Mason, T. Drwiega, and J. Yan (Eds.): ITC 2007, LNCS 4516, pp. 54–64, 2007.

manage the traffic demands, traffic should be throttled as close to the source as possible. SIP provides limited congestion control with the use of the 503 service unavailable message. When a server detects it is unable to process a SIP message, it will respond with a 503 to the originating server. However, the 503 response will only stop the current session request. Subsequent attempts to establish sessions will continue to be forwarded to the congested server. Consequently, the overloaded server will continue to receive service requests, and will continue to spend resources replying with 503s, adding to its overloaded state. While the 503 has an option to stop new requests for a specified time, this approach is suboptimal for managing network congestion because it can over or under control, as described in [2] and [3].

This paper describes the initial results of simulation used to quantify the impact of overloads on a reference SIP-based VoIP network. The network consists of Media Gateways and Call Controllers. Each element is modeled with an internal overload detection and control mechanism that recognizes its input queue length and limiting process utilization. Internal overload control on and off thresholds are used to mitigate oscillations. When load increases beyond engineered limits, particularly during prolonged periods, it is more efficient to throttle requests at the source rather than at the overloaded element. We introduce external overload controls to limit offered load to an overloaded server. The external overload control uses flow rate commands to throttle requests to the overloaded server. Again, thresholds are used to mitigate oscillations. While internal overload detection and control is necessary to ensure element protection, it is not sufficient to efficiently manage overloads across a network. External overload controls have both the benefit of shedding load closer to the source, and relieving load shedding resources at the overloaded server. This paper shows initial results of simulations that demonstrate the superior performance of combining external overload controls with internal controls, as indicated by increased goodput and lower blocking probability.

Overload controls have been widely studied and deployed in PSTN networks. Typical methods include Automatic Congestion Control (ACC), percent blocking, and Automatic Call or Code Gapping (ACG). These Rate based controls are described in [4] and [5]. A comparison of call gapping and percent blocking is provided in [6]. Over time, the PSTN overload controls have been finely tuned to maximize network efficiency when the network is subjected to overloads. However, SIP has only limited capabilities to control network overloads with the use of a 503 Retry message, which indicates that a network element is unable to process requests. Work is underway in the IETF to develop standard overload control methods. In this paper we use the percent blocking control as a basis for evaluating overload controls in a SIP network, and compare the network throughput and blocking with the case of no controls and the 503 Retry method. In a subsequent paper, we will describe and compare our work on more sophisticated control methods for SIP networks.

The remainder of this paper covers the model and simulation results. In section 2 we describe the basic SIP call flow and timers used in our model. Section 3 covers the element models and associated parameters used in the simulation, and a description of the overload controls. Section 4 describes the network model and presents the results of the simulation. We use goodput and blocking as metrics for comparing the control options. Section 5 summarizes the conclusions from this study.

2 SIP Signaling

SIP is an application-layer control (signaling) protocol for creating, modifying, and terminating sessions with one or more participants. These sessions include Internet telephone calls, multimedia distribution, and multimedia conferences. The latest version of the specification is RFC 3261 from the IETF SIP Working Group [1]. It is widely used as a signaling protocol for VoIP.

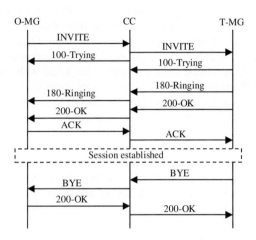

Fig. 1. Simple call flow (O-MG: Originating Media Gateway, CC: Call Controller, T-MG: Terminating Media Gateway)

We limit our analysis to *simple calls* (i.e., no advanced call features) as depicted in Figure 1 where the VoIP network consists only of Media Gateways (SIP proxies that control media) and Call Controllers (SIP proxies).

When messages associated with explicit acknowledgements (e.g., INVITE/100-Trying, 200-OK/ACK, BYE/200-ACK) are lost due to queue overflow or transmission packet loss, the SIP retransmission method for unreliable transport is triggered. SIP has two types of retransmission procedure, one for INVITE messages (INVITE request with 100-Trying response) and the other for non-INVITE messages (200-OK request with ACK response and BYE request with 200-OK response). The SIP protocol specifies timers for expected response to specific messages. See RFC 3261 [1] for the complete list. The key timers for this study are Timer A and B for INVITE messages, and Timer E and F for non-INVITE messages. Upon expiration of the short timer (Timer A for INVITE requests, Timer E for non-INVITE requests) a new transmission attempt is sent with a timeout set to twice the current short timer. Once the short timer exceeds the long timer (Timer B for INVITE, Timer F for non-INVITE) the associated call fails.

In RFC 3261, the short timer is initialized to Timer T1 (500 msec) and the long timers are set to 64 times T1. To allow for graceful failure detection when interworking with PSTN protocols, and consistency with typical PSTN Post Dial Delay (PDD), we reduced RFC 3261 Timer B (INVITE transaction timeout timer)

Table 1. SIP Timers values

T1	500 msec
Timer B	2 sec
Timer F	4 sec

and Timer F (non-INVITE transaction timeout timer) to the values shown in Table 1. The primary PSTN protocol timers of concern are timers T7 (20-30 sec) and T11 (15-20 sec) for ISUP as defined in ANSI T1.113 [7] and timers T303 (4 sec) and T310 (10 sec) for ISDN as defined in ITU-T Q.931 [8].

SIP provides limited congestion control with the use of the 503 service unavailable message. When a server detects it is unable to process a SIP message, it will respond with a 503 to the originating server. However, the 503 response will only stop the current session request. With the optional retry-after field set in the 503 message, the originating server will drop all new requests over a duration specified by in the retry-after field. Note RFC 3261 does not provide any guidelines for the 503 retry-after duration.

3   Simulation Model

In this paper, we report on the version of our simulation model limited to *simple calls* (i.e., no advanced call features) where the VoIP network consists of Media Gateways (MG) and Call Controllers (CC). Our enhanced version simulation model allows for *complex calls* (i.e., calls with advanced features requiring multiple server processing) and includes elements such as Application Servers and Media Servers. Future versions will support the 3rd Generation Partnership Project (3GPP) IP Multimedia Subsystem (IMS) [9].

To minimize run time and allow for call model flexibility, we implemented our VoIP simulation model in the C programming language. The simulation model is discrete event based and includes a heap sort scheduler. It is approximately 8,500 lines of code, with typical run times of a few minutes for the network described here (10 Media Gateways and 1 Call Controller).

3.1   Element Model

As shown in Figure 2, each network element is modeled as an input queue and a bottleneck process followed by a delay line that emulates the element processing delay.

The bottleneck process limits the element call processing rate and incurs a penalty for rejected call requests so that with increasing overload level, the element goodput (carried volume) reduces as shown in Figure 3. As offered load increases, the proportion of time the bottleneck process spends rejecting new calls increases so that the carried load decreases.

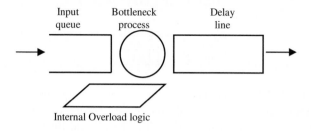

Fig. 2. Node model

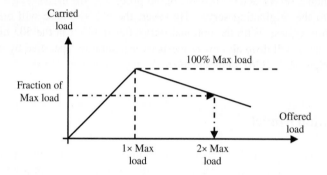

Fig. 3. Network element hypothetical goodput as a function of offered load

In addition, each element has a limit on the number of simultaneous calls it can sustain. In our simulation, we assume the Media Gateway limit is related to its input port capacity (LAN for IP side or trunks for TDM side), and for the Call Controller we assume the limit is related to memory. Of course the actual limitations can vary depending on vendor implementation.

Table 2. Parameters used for the simulated elements

Media Gateway	Queue length	5,000 SIP messages
	Bottleneck process	800K BHCA
	Mean delay line	50 msec for SIP INVITE
		25 msec for other SIP messages
	Max Simultaneous calls	15,000
	Rejected calls overhead	20%
Call Controller	Queue length	5,000 SIP messages
	Bottleneck process	1M BHCA
	Mean delay line	100 msec for SIP INVITE
		50 msec for other SIP messages
	Max Simultaneous calls	60,000
	Rejected calls overhead	20%

In Table 2 we capture all our element simulation parameters. There we deliberately engineer MGs and CCs so that CCs are always the bottleneck over the overload range under study. Note all processing delays are generated from a triangular distribution so that the produced delays are bounded, as is expected with simulated elements. We arbitrarily assumed the mean processing delay to be twice the distribution minimum and half the distribution maximum.

3.2 Overload Control

We modeled both internal and external overload controls. For the internal overload control, each network element monitors its input queue instantaneous length and its 1-sec average bottleneck process utilization. When either metric exceeds predefined *On* thresholds, new calls are rejected while messages for existing calls continue to be processed, and the network element is said to be in internal overload. Exiting the internal overload state requires both metrics to be below predefined *Off* thresholds. Simulation model values used for both thresholds are summarized in Table 3.

Table 3. Internal overload control parameters values

On Threshold	Queue length utilization	40%
	Bottleneck process utilization	95%
Off Threshold	Queue length utilization	20%
	Bottleneck process utilization	85%

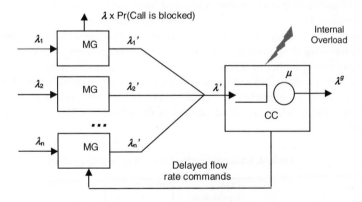

Fig. 4. External overload control block diagram (*MG: Media Gateway, CC: Call Controller*)

For external overload controls we implemented the optional retry-after 503 throttle mechanism (referred to as the IETF 503 Retry control algorithm) we previously described. We set the retry-after period to 32 sec independent of the overload level.

We also implemented a PSTN like Rate based throttling mechanism (referred to as the Rate based overload control algorithm). In the Rate based overload control algorithm, we rely on a *hypothetical*[1] control communication between the Call

---

[1] SIP based external overload controls are being defined in IETF [2].

Controller and the Media Gateways as depicted in Figure 4. Media Gateways reject calls on the behalf of the overloaded Call Controller, hence avoiding the capacity penalty of having the Call Controller reject calls.

Every measurement intervals of duration T, the Call Controller estimates the controlled load λ' and the number of active sources A_t. So every control interval C, where $T \leq C$, the Call Controller notifies the Media Gateways of the new controlled load. Then the Media Gateways execute the Percent Blocking overload control algorithm[2] set to the specified controlled load.

The controlled load estimate was taken from [10], and attempts to bound the bottleneck queueing delay by a predefined target queueing delay d_e. Its expression is captured below:

$$\lambda' = \mu_t \times (1 - (d_t - d_e) / C),\tag{1}$$

Where μ_t is the estimated bottleneck process service rate (during the last T interval) and d_t is the estimated bottleneck process queueing delay at the end of the last T interval (the number of queued jobs at the end of T divided by the estimated mean service rate). So the ratio $(d_t - d_e)/C$ is equivalent to the queue backlog.

While the active source estimator is captured as follows:

$$A_{t+1} = w \times A_t + (1 - w) \times A_t \times N_t / (\lambda' \times T).\tag{2}$$

Where A_t is the estimated number of active sources in the measurement previous interval (of duration T), w is a smoothing weight and N_t is the number of new call attempts during the last T interval. The ratio $N_t / (\lambda' \times T)$ can be interpreted as the amount of traffic that exceeds the controlled load. And scaling A_t by that quantity in the expression $A_t \times N_t / (\lambda' \times T)$ is our estimate for the number of active sources in the next measurement interval. So the expression in (2) represents a weighted average between the estimated number of active sources in the previous measurement interval and the expected number of active sources in the next measurement interval.

The Call Controller activates the external overload control when the estimated bottleneck process queueing delay d_t exceeds the pre-specified On threshold. Conversely, the external overload control is de-activated when d_t is below the Off threshold.

Table 4. External overload control parameters values

On Threshold	90%
Off Threshold	10%
w	0.8
d_e	1 sec
T	100 msec
C	2 sec

---

[2] Percent blocking is known to be a sub-optimal algorithm. In a future paper we will report on simulator implementations that rely on better algorithms (i.e. one that throttles to a limit independent of offered load level).

Simulation model values used for the Rate based overload control parameters are summarized in Table 4. Note the Off threshold was deliberately chosen to be low because the goodput penalty for maintaining the overload control algorithm active is negligible when the offered load is below the overload limit. This is because in expression (1), when the queueing delay is below the pre-defined threshold (i.e., offered load less than the overload limit) the controlled load would actually increase.

4 Simulation Results

4.1 Network Model

Our network model is depicted in Figure 5, where each Media Gateway hub consists of 2 Media Gateways. It also includes a single Call Controller, which is not collocated with Media Gateways. This implies that packet loss impacts SIP messages between MGs and the CC. Traffic sources generate Poisson distributed call attempts with exponentially distributed holding times of mean 3 minutes.

Blocked calls generate retries where a call retry is attempted with a 70% chance and a maximum of 3 times. Inter-retry delay distribution is exponential with mean of 10 seconds. When a SIP INVITE is rejected due to overload or transmission errors, the message originator (Media Gateway or Call Controller) will make a second attempt to a different element (another Media Gateway in the same hub or another Call Controller, if any). Lastly, packet loss associated with transmission impairments was accounted for. We arbitrarily set the packet loss probability to 1%.

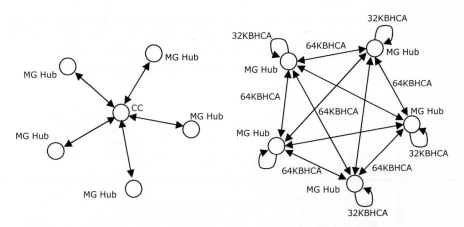

Fig. 5. Network under test that shows the signaling network on the left hand side and the bearer path network on the right hand side. Intra-hub paths are 32K BHCA loaded, and inter-hub paths are 64K BHCA loaded.

4.2 Results

We simulated network wide overload by scaling the offered load (set to CC engineered capacity, or 80% its maximum capacity) by an overload factor ranging from 1 to 4. Only the network-wide call blocking probability, Call Controller normalized throughput (ratio of goodput to maximum capacity) and retry factor (average number of attempts per calls) are reported.

Figure 6, Figure 7, and Figure 8 capture the simulation results with all metrics averaged over 10 independent runs. Each simulation runs consisted of 20 million call attempts. The corresponding 95% confidence interval was estimated to be approximately 10% of the reported mean value.

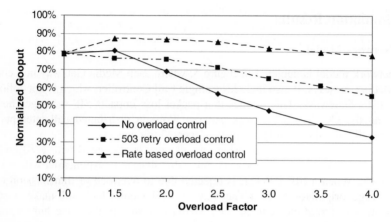

Fig. 6. Call Controller normalized goodput as a function of overload factor and overload control algorithm

From Fig. 6 we observe that with increasing overload factor, the Call Controller goodput (ratio of CC carrier load to CC maximum call rate capacity) reduces due to both SIP message retries (timer expiration) and the overhead incurred from rejecting call attempts and re-attempts. That degradation is mitigated by the external overload

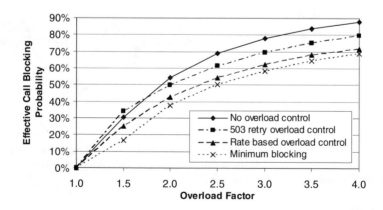

Fig. 7. Network wide blocking probability as a function of overload factor and overload control algorithm

controls. The Rate based overload control significantly improves the Call Controller goodput (at 4 times the overload factor, CC goodput increases from ~32% to ~78%).

In Figure 7, in addition to reporting measurements from the simulator, we included an estimate of the *minimum blocking probability* for our network, computed as the ratio of the offered load minus CC capacity and divided by the offered load (minimum because there are no penalties for rejected calls). In Figure 7, we note the blocking probability increases with the overload factor. We also note improvements in the blocking probability due to the overload control algorithms. It turns out our 503 retry overload control over controls at 1.5 times the engineered load (32 sec retry-after duration is too long) and our Rate based overload control is nearly optimal.

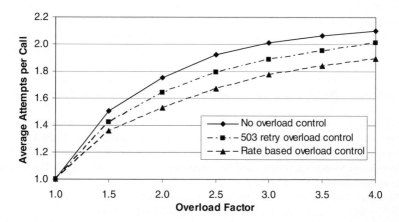

Fig. 8. Average number of attempts per call as a function of overload factor and overload control algorithm

Lastly, in Fig. 8 we show that the average number of attempts per call (bounded by 3) increases as the overload factor increases. We note the external overload control effectively reduces the number of call retries. This is directly a result of increased goodput and lower blocking probability associated with external overload controls.

5   Conclusion

This paper shows the initial results of simulations designed to assess SIP based VOIP networks under overload. The simulated network consists of Media Gateways and Call Controllers, each with internal overload detection and control mechanisms. The internal overload controls monitor both queue length utilization and a bottleneck process utilization. The simulation included SIP timers tuned to operate gracefully with PSTN interfacing protocols. The offered traffic consists of simple SIP based VoIP calls. The offered traffic ranged from 1 to 4 times the engineered load. External overload controls are evaluated based on flow rate controls. Both internal and external overload controls have On and Off thresholds to mitigate oscillations in control.

The simulation compares the results of using internal controls only with results from external overload controls combined with internal overload controls. Results are shown in terms of blocking probability and normalized goodput. Combining external overload controls with internal overload controls increases goodput and reduces blocking across the overload levels. In addition, call retries are reduced with external overload controls. Subsequent results to be reported will include more sophisticated external controls, as well as more complex call types.

References

1. Rosenberg, J., et al.: SIP: Session Initiation Protocol, RFC3261 (June 2002), http://www.ietf.org/rfc/rfc3261.txt
2. Rosenberg, J. et al.: Requirements for Management of Overload in the Session Initiation Protocol, Internet Draft (November 2006), http://www.ietf.org/internet-drafts/draft-ietf-sipping-overload-reqs-00.txt
3. Ejzak, R.P., Florkey, C.K., Hemmeter, R.W.: Network Overload and Congestion: A comparison of ISUP and SIP. Bell Labs Technical Journal Lucent Technologies, Inc. 9(3), 173–182 (2004)
4. Telcordia Technologies Generic Requirements, GR-477-CORE, Issue 4 (2000)
5. Telcordia Technologies Generic Requirements, GR-1298-CORE, Issue 7 (2001)
6. Berger, A.: Comparison of Call Gapping and Percent Blocking for Overload Control in Distributed Switching Systems and Telecommunications Networks. IEEE Transactions on Communications 39(4), 574–580 (1991)
7. American National Standards Institute: Telecommunications Signaling System No. 7 (SS7)– Integrated Services Digital Network (ISDN) User Part (ISUP)
8. International Telecommunication Union Telecommunication Standardization Sector: Q.931 Digital subscriber Signalling System No.1 (DSS 1) - ISDN user-network interface layer 3 specification for basic call control
9. http://www.3gpp.org/specs/specs.htm
10. Hosein, P.A.: United States Patent US 6,442,139 B1: Adaptive Rate Control Based on Estimation of Message Queueing Delay (2002)

Recovery, Routing and Load Balancing Strategy for an IP/MPLS Network

Eligijus Kubilinskas[1], Faisal Aslam[2], Mateusz Dzida[3], and Michał Pióro[1,3]

[1] Dept. of Communication Systems, Lund University, Sweden
[2] Computer Science Dept., Lahore University of Management Sciences, Pakistan
[3] Institute of Telecommunications, Warsaw University of Technology, Poland
eligik@telecom.lth.se, mdzida@tele.pw.edu.pl, mpp@telecom.lth.se

Abstract. The paper considers a problem of routing, protection and load balancing in the IP/MPLS network. A network design problem combining all these aspects is presented. Proportionally fair distribution of residual bandwidths on links is used for load balancing, and protection is achieved with failure–dependent backup paths. The efficiency of the proposed approach is tested by combining optimization and simulation tools. Numerical experiments show that using the proposed load balancing and protection mechanisms decreases the number of disrupted LSPs in case of failures, as compared to other recovery options considered.

1 Introduction

A number of factors influence network performance and efficiency of its resource utilization. Two such factors are recovery mechanisms and routing strategies. By using a capacity-efficient recovery mechanism some of the capacity, previously dedicated for protection, can be saved and used for transporting the revenue generating traffic. On the other hand, an efficient routing strategy can ensure a good trade-off between a well balanced load and the consumed capacity. A capacity-efficient recovery mechanism integrated with a routing strategy is central to this study.

In this paper we focus on IP/MPLS (Internet Protocol/Multi Protocol Label Switching) network. Traffic demands are assumed to have priority levels. High priority traffic can preempt low priority traffic. We assume that the high priority (HP) label switched paths (LSPs) are long-lived, e.g., they are used for virtual private networks (VPNs) or as virtual leased lines and thus are not reconfigured often, and also that they constitute a significant amount of network traffic. Bandwidth reserved for backup LSPs of HP demands can be used by low priority LSPs in the normal state. Since the high priority LSPs require 100% survivability, it is important how the network resources are used for their protection because this implies how many low priority LSPs are discarded from the network when backup paths for HP demands are activated. The demands in the network are divided into two categories–high priority demands (with priority= 0) and low priority demands with priorities ranging from 1 to 4; thus, there are 5 priority classes in total. We assume that only HP demands require 100% survivability

L. Mason, T. Drwiega, and J. Yan (Eds.): ITC 2007, LNCS 4516, pp. 65–76, 2007.

and other demands can be recovered only if resources are available, in the order of priority. Based on this differentiation of the demands we propose a network protection and routing strategy, which is as follows. Failure-dependent backup path protection mechanism with sharing of backup capacity is used for assuring availability of the HP demands. The selection of primary and failure-dependent backup paths for HP connections is performed by an off-line network design problem (NDP) defined as a mixed integer program (MIP). It is valid to assume that the set of HP demands is known at this design stage, since such a global optimization can be performed not only during the greenfield network design stage (using estimated demand volumes), but also periodically (however, with a long period, e.g., half a year) during the network operation. For routing the HP demands we propose a strategy combining shortest path routing and load balancing on the links. We propose a load balancing function using proportional fairness rule applied to residual bandwidths. The low priority connections are routed using on-line constrained shortest path first (CSPF) algorithm.

The proposed protection and routing strategy has been tested both by numerical experiments and simulations. The results show that for the considered network examples less LSPs are disrupted on the average due to failures using the proposed protection mechanism and routing strategy, as compared with the other recovery strategies. Thus, with the given link capacities the network is able to retain considerably more LSPs in the case of failures.

The rest of the paper is organized as follows. First, the related work is reviewed and an IP/MPLS network model is presented. Then the mathematical problem formulation for finding nominal and backup paths for HP demands is presented. Next, the proposed routing and load balancing criterion is discussed, leading to the objective function for the presented network design problem. Finally, some numerical experiments are presented, followed by final conclusions.

2   Related Work

There are a number of novel recovery mechanisms and routing strategies proposed in the recent literature, conserving network resources used for assuring network resilience. Most of such recovery mechanisms assume a kind of backup capacity multiplexing. Several of the proposed protection mechanisms are reviewed below.

In [1] a demand-wise shared protection (DSP) is presented. It combines advantages of dedicated and shared path protection. In DSP, a capacity is reserved for each demand and is not shared with other demands, but only among the paths of the same demand. The paths of a demand are not explicitly divided into nominal (working) and backup, and in general the whole demand volume is assumed to be split among several working paths.

In [2] and [3] two schemes, called original quasi-path restoration (OQPR) and improved quasi path restoration (IQPR), are presented. In QPR schemes, a failing path is divided into three parts: a subpath from source node to a critical node on the same side of the failing link, the failing link, and the subpath from the critical node on the other side of the failing link to the destination node. Then restoration implies rerouting only one of the subpaths, while the other subpath is left intact. Traffic affected by the failure is rerouted on single/multiple paths using spare capacity in the network.

In [4] a problem of lightpath routing and protection in wavelength division multiplexing (WDM) networks with dynamic traffic demands is considered. The authors propose a new multiplexing technique called primary-backup multiplexing and shows that it increases network resource utilization and improves the network blocking performance. The technique allows to share one wavelength channel among one primary and one or more backup lightpaths.

In [5] an $L + 1$ fault tolerance routing strategy is studied. The main idea is to transform the original network with L links into a collection of $L + 1$ subnetworks, where one of the links in the original (base) network is removed, this way modeling each single-link failure scenario. A connection is accepted to the network only if it can be routed in all of the subgraphs. When a failure occurs the connectivity is restored by putting the network to the state of the particular (failure-dependent) subgraph. The major drawback is that this strategy can potentially require a full network reconfiguration upon a single link failure (a.k.a., unrestricted reconfiguration).

In [6] a dynamic failure-dependent path protection scheme is developed. The study assumes a WDM network with heterogeneous grooming architectures. A method is developed to assign primary and (failure-dependent) backup paths to requests, in a way which makes it possible to survive a single shared risk link group (SRLG) failure at any given time. The routing strategy used is Available Shortest Path (ASP), where Dijkstra's algorithm is used to find shortest paths, first checking the resource availability on them. In the study a restricted reconfiguration is assumed, i.e., only a failing primary connection is reassigned to its backup path, while unaffected connections remain in place.

2.1   This Work

This study extends and enhances investigations of the recovery and routing strategy presented in [7]. The main focus in this paper is on performance comparisons of the proposed strategy to various combined on-line/off-line protection strategies. The proposed protection and routing strategy is related to the studies reviewed above in several ways. First of all, the chosen failure-dependent backup path protection mechanism with restricted reconfiguration corresponds to that presented in [6], except that in our case the proposed method is for off-line computation and for IP/MPLS network, while the one in [6] is on-line and is dedicated to WDM network. Also, the routing strategy we propose is different from ASP in [6] and takes load balancing on links into account. The off-line network design with failure-dependent routing also resembles the $L + 1$ approach in [5], except that in our case routing in all the failure states is designed simultaneously, and the reconfiguration is restricted.

Secondly, the primary and backup paths used in our protection and routing strategy are *failure-disjoint*. This means that the primary and backup paths for a given demand do not share the failing element in the given failure scenario. The non-failing resources on a primary path may be reused by the backup paths. Also, the unused backup resources can be used by other demands. In this way, our work relates to [1], except that in our case sharing of resources is also allowed between demands. Also, there is a similarity to QPR schemes [2,3,8]. However, differently from QPR, the proposed strategy does not restrict how the non-failing resources of the primary path are reused by its backup path. There are also similarities with the primary-backup multiplexing method in [4].

Also, the method we propose differs from all the reviewed studies in that we assume traffic requests with different priorities.

3   MPLS Network Model

The structure of an IP/MPLS network is modeled as a graph. Network nodes (label switching routers–LSRs) are interconnected by links. Links (indexed with subscript e) are assumed to be undirected and have limited capacities c_e, which are given constants. A demand (indexed with d) between a pair of nodes (ingress and egress) is a requirement for a certain amount of bandwidth, called demand volume and denoted by h_d. Each demand has a certain associated holding priority. For each demand, a list of *candidate paths* (routing list) is specified. The paths are label switched paths (LSPs) and are indexed with p. An LSP is identified with the set of its links. An LSP of demand with a higher priority level can preempt LSPs of demands with the lower priority levels. In this study flows are assumed to be unsplittable, i.e., the whole demand volume is carried on a selected single path.

Failures of network elements are modeled using link availability coefficients $\beta_{es} \in \{0, 1\}$. A *failure situation* (indexed by s) is a situation (state) when one or more links completely fail. For the failing links β_{es} is set to 0, whereas for the working links $\beta_{es} = 1$. The situation where all network resources are fully functional will be called *normal (nominal) situation*. Node failures can be modeled by setting $\beta_{es} = 0$ for all links e adjacent to the failing node and discarding the demands to/from the node.

4   Protection of HP Demands with Failure-Dependent Backup Paths

The idea of failure-dependent backup paths is as follows. For a given list of possible failure scenarios affecting the nominal path, backup paths depending on particular failures are preplanned. When, during the network operation, a particular failure occurs, specific backup paths matching the failure are activated and used to carry the flows until the nominal paths are repaired. The nominal and backup paths are failure-disjoint. And non-failing resources of a given nominal path can be reused by a backup path of a demand.

In general, failure-dependent backup paths (FDBPs) allow for a more effective way of using the network resources than using single (failure-independent) backup paths (SBPs) in all failure situations. This is because a space of feasible solutions for NDP with SBP protection is in general more constrained than the solution space for NDP with FDBP protection, and an optimal solution to the SBP problem is also a feasible solution to the FDBP problem. The efficiency of failure-dependent recovery mechanisms was illustrated in e.g., [5,6].

4.1 Mathematical Problem Formulation

The following mathematical problem is used to select primary and backup paths for HP demands for all predefined failure situations. The paths for each demand are chosen from predefined candidate path lists. A feasible solution to this problem guarantees 100% survivability for the HP demands.

Problem FD-LSP: an offline design problem for IP/MPLS network with single-path routing, failure-dependent backup paths (LSPs) and restricted reconfiguration of flows (unaffected flows are not rerouted).

indices

$d = 1, 2, ..., D$ demands between source–destination pairs of the IP/MPLS layer

$e = 1, 2, ..., E$ links of the IP/MPLS layer

$p = 1, 2, ..., P_d$ candidate paths (LSPs) for demand d

$s = 0, 1, 2, ..., S$ failure situations; $s = 0$ is the normal state

constants

δ_{edp} link-path incidence matrix

β_{es} binary availability coefficient of link e in situation s

τ_{dps} binary availability coefficient equal to 1 if path p of demand d is available in situation s, 0 otherwise; $\tau_{dps} \in \{0, 1\}$,
$\tau_{dps} = \min\{\beta_{es} : \delta_{edp} = 1\}$, where $\beta_{es} \in \{0, 1\}$

h_d demand volume for demand d

c_e capacity of link e

ξ_e cost of link e; assumed to be equal to 1, if not specified otherwise

variables

x_{dps} continuous flow allocated to path p of demand d in situation s

v_{dps} binary variable equal to 1 if all the flow of demand d is routed on path p in situation s, 0 otherwise

y_{es} load of link e in situation s

constraints

$$\sum_p v_{dps} = 1 , \qquad\qquad \forall d, s \qquad (1)$$

$$x_{dps} = v_{dps} h_d , \qquad\qquad \forall d, p, s \qquad (2)$$

$$x_{dps} \geq \tau_{dps} x_{dp0} , \qquad\qquad \forall d, p, s \qquad (3)$$

$$\sum_d \sum_p \delta_{edp} x_{dps} = y_{es} , \qquad\qquad \forall e, s \qquad (4)$$

$$y_{es} \leq \beta_{es} c_e , \qquad\qquad \forall e, s. \qquad (5)$$

Constraint (1) assures that only one single path p for each demand d is used in each situation s, while constraint (2) assures that the whole demand volume h_d is assigned to the chosen path p. For each demand d constraint (3) forces using the normal path p (used in normal situation $s = 0$) in all situations in which this path is not affected by a failure. Due to this, flows unaffected by failures are not rerouted. Path availability coefficients τ_{dps} are used in (3) to indicate if a given path p is affected by a failure of a link e, and are calculated from link availability coefficients β_{es} as shown above. Constraints (4) and (5) are essentially one constraint; it is split into two constraints for the clarity of presentation when referring to link loads given by the left hand side of constraint (4). Thus, the load of each link e in situation s is given by a sum of all flows traversing that link in the given situation. Then constraint (5) assures that the link loads do not exceed available link capacities in each situation expressed as $\beta_{es}c_e$. Note that in the problem formulation above variables x_{dps} are auxiliary and are included in the model only for clarity of presentation.

FD-LSP is a $\mathcal{NP}$-hard (mixed integer programming problem–MIP), since a simpler problem of single path allocation for only normal state is already $\mathcal{NP}$-hard [9]. General purpose MIP solvers such as CPLEX or XPRESS-MP can be used for resolving the problem. Observe that solving problem FD-LSP without an objective function results in paths assigned to demands that may not necessary be proper in some desirable sense.

5   Routing Strategy for HP Demands

By choosing an appropriate objective function for problem FD-LSP, different routing strategies can be implied. The strategy proposed in this paper combines load balancing on the links with shortest paths, as opposed to the widely used shortest path (minimum hop) routing alone. In this paper by shortest path (SP) routing we mean that an arbitrary chosen shortest path (if there are few) is used for demand flows.

In the proposed strategy the shortest path routing is used for backup LSPs of HP demands. This decision stems from the natural reasoning that an HP demand with the shorter backup path will preempt less low priority LSPs than the demand with the longer backup path. Therefore minimizing the length of the backup path(-s) possibly decreases the number of disrupted low priority LSPs.

For routing the nominal paths a criterion implying load balancing on the links is used. The idea behind the load balancing is to distribute traffic in a network in order to reduce the number of rejected future requests (blocking probability) because of the insufficient available capacity on some links. This is related to the principle of "minimum interference routing" [10]. What is also important in the resilience context is that the load balancing mechanism reduces the risk of several flows that traverse the same network segment between two given endpoints being disrupted by the same failure. That is due to spreading the flows on different paths, as opposed to a solution where all of them use the same shortest (sub-)path through that network segment.

The proposed routing strategy has not been considered before (to the best of our knowledge). As it was mentioned above, this strategy combines balancing of the load on the links implied by nominal paths and minimization of the length of the protection paths. First, the two criteria will be presented, and then, ways of combining them will be discussed.

5.1 Load Balancing on the Links

Various objective functions can be used for problem FD-LSP in order to improve load balancing on the links. Before presenting the load balancing criteria considered in this study, let us first define *residual bandwidth* (RB) on each link in the nominal situation as: $r_{e0} = c_e - y_{e0}$.

It is important what criterion is chosen for distribution of load on the links. There exist several possibilities, e.g., maximizing the total residual bandwidth on the links (expressed as $\sum_e r_{e0}$), max-min fair [11] allocation of the residual bandwidths to the links, etc. In this paper we suggest using the principle of proportional fairness (PF), see [12] and Section 8.3 in [9]. PF allocation is achieved by using the following objective:

$$\max \quad \sum_e \log r_{e0}. \tag{6}$$

Because of the particular properties of the logarithmic function, this objective will tend to distribute the residual capacity to links equally, at the same time avoiding allocation of very small residual bandwidth (this is the common case in the maximization of the total residual bandwidth), and not forcing maximization of the minimal allocation (the case of max-min fairness). Thus, PF is a reasonable trade-off between maximization of the total residual capacity and its max-min fair allocation.

Objective (6) is non-linear, which, combined with the non-convex solution space of FD-LSP, makes the problem very difficult to solve. In order to somewhat simplify the problem solution process, the logarithmic function in (6) can be linearized by using a piece-wise linear approximation $\Gamma(X)$ of the logarithmic function $\log(X)$ [13]: $\Gamma(X) = \min\{F_i(X) = a_i X + b_i : i = 1, 2, \ldots, I\}$, for some I to be specified. The linear approximation consists in introducing auxiliary variables f_e and a set I of constraints corresponding to the linear pieces of the approximation $\Gamma(X)$, which replace the logarithm of the residual bandwidth ($\log r_{e0}$):

$$f_e \leq a_i r_{e0} + b_i , \ \forall i, \ e. \tag{7}$$

Then the objective function (6) is replaced with:

$$\max \quad \sum_e f_e. \tag{8}$$

The approximation of the objective function introduces some error to the solution. However, the error can be made as small as desired by using sufficiently many linear pieces for the approximation and assuring that they are sufficiently well spread out in the desired region. Besides, using logarithm or its piece-wise approximation is generally not important in practice.

5.2   Shortest Path Routing of Backup Paths

Minimizing the length of backup paths is desirable in order to minimize a number of disrupted low priority LSPs. The following objective function, when used with problem FD-LSP, implies minimization of the length of backup paths:

$$\min \quad \sum_{s\neq 0}\sum_{e}\sum_{d}\sum_{p} \xi_e \delta_{edp} v_{dps}. \tag{9}$$

The objective function (9) implies minimization of the lengths of the paths for all demands in all failure situations ($s \neq 0$), when link costs $\xi_e \equiv 1$ (and thus can be interpreted as a hop-count). If the costs are different than 1, the objective function implies using the cheapest (in some desirable sense) paths instead.

Combining the two criteria. Sections 5.1 and 5.2 have dealt with two important aspects of the proposed routing strategy, namely–load balancing in the nominal network state and using shortest backup path(-s) in case of failures.

One way of combining the PF load balancing (8) for nominal state and minimization of lengths of backup paths (9) is by solving a problem FD-LSP in two stages. First the problem (8), (1)-(5) and (7) is solved for the nominal situation. Then flow allocation pattern for nominal situation is fixed, and problem (9),(1)-(5) is solved for all failure situations $s \neq 0$.

Another approach (it was used for the experiments reported in this paper) is to combine the two criteria in a single objective function as follows:

$$\min \quad -A\sum_{e} f_e + (1 - A)\sum_{s\neq 0}\sum_{e}\sum_{d}\sum_{p} \xi_e \delta_{edp} v_{dps}. \tag{10}$$

Coefficient A is used for weighting the two criteria in the objective function. In the numerical experiments reported in the next section A was set to 0.5. The final NDP is then defined as:

Problem FD-LSP-COMB: objective (10), **s.t.** (1)-(5),(7).

6   Numerical Experiments

Several numerical experiments have been conducted in order to test the effectiveness of the proposed protection and load balancing strategy. The networks used for experiments were: network N_{12} with 12 nodes, 22 links, 6 to 14 paths per demand and 66 demands; network N_{25} with 25 nodes, 56 links, 10 paths per demand and 300 demands. The experiments were composed of two stages. First, the optimization problem FD-LSP-COMB, defined in AMPL, was solved (also in the two-phase form described in Section 5.2) by CPLEX 9.1 solver for the considered networks. The solution time was below one minute for the N_{12} network and below two minutes for the N_{25} network on a PC with Pentium 4 HT 3GHz CPU and 2GB of RAM. All the single-link failure states were assumed in the experiments. The solution of the optimization problem has found, for every high priority (HP) demand, a primary path and backup paths for all failure situation. All the HP demand volumes were set to the same value and were equal to about 2% of a single link capacity. Link capacities were the same for all links. The demand volumes have been chosen such that in all situations the network is loaded by nominal flows of all the HP demands to about 25%.

In the second stage the routing information of the HP demands, produced by the optimization program, was sent to the Advanced CSPF Simulator [14], which, based on the input information, performed a setup of the HP LSPs. Then additional demand requests were generated, with priorities raging from 1 to 4 (hence, not with the high priority 0) and demand volume uniformly distributed between 1% and 5% of a link capacity. In the sequel we will refer to these as low priority (LP) demands. The probability of existence of a demand between a given source-destination pair was the same for all the pairs. There were 1500 such demands generated for network N_{12} and 2300 for the network N_{25}. Just nominal paths were found using CSPF and provisioned for these dynamically arriving LP LSPs and no backup paths provisioned. The number of the LP LSPs was chosen so that after placing them the network would be (nearly) saturated. Simulation section describes in detail different strategies used for routing these additional LSPs. When all the additional LSPs that could be placed were placed, the network failure states were simulated one by one. All the single link failure scenarios were considered. Our goal was to measure the number of surviving LSPs in the network in the case of each failure scenario. No new arrivals occurred in the failure states of the network. However, the failures required backup paths of HP LSPs to be activated, preempting in turn some of the LP LSPs. Then an attempt was made by the simulator to restore as many of the preempted LP LPSs as possible, by dynamically finding new paths for them. Only when all the preemption and possible restorations were over, the number of surviving LSPs in the network was recorded. The different strategies for restoration of the LP LSPs are presented in the section below.

In the experimental framework described above we have compared the proposed FDBP protection mechanism with off-line preplanned SBP protection [7], as well as with the dynamic online SBP recovery mechanism. For the online routing mechanism, we have considered two options for link metric.

6.1   Link Metrics in Simulations

Two different link metrics for the on-line routing of LP LSPs were considered in the simulator. They were used by the CSPF protocol for finding a path for the dynamically arriving LP LSP request, as well as when a restoration path for the LP LSPs had to be found.

The first metric is an interior gateway protocol (IGP) cost. It is simply some administrative cost ξ_e, assigned to links. In the experiments we have assumed $\xi_e = 1$ for all links, thus IGP merely representing a hop count.

The second metric is an inverse of the priority-based residual bandwidth R_{es}^j, which for an LSP request with priority j and link e in situation s is expressed as: $R_{es}^j = r_{es} + \mu_{es}^{H(j)} + \eta_{es}^{L(j)}$. Here, r_{es} is, as before, the free (unused) bandwidth on link e in situation s, $\mu_{es}^{H(j)}$ is a bandwidth reserved on link e for all backup paths of HP demands (in general, LSPs with priority levels higher than j). This bandwidth can be used by LP LSPs if the protection paths of HP LSPs are not activated in a given situation s. Similarly, $\eta_{es}^{L(j)}$ is a bandwidth reserved on link e in situation s for all primary LSPs of the demands with priority levels lower than j, and thus can be preempted by j (mind that lower priority number implies higher priority level). Finally, when finding a constrained shortest path, the quantity $1/R_{es}^{j}$ is used as the link metric.

6.2 Test Scenarios

Because of multiple dimensions of the input data and possible results, further in this paper we focus only on the case of the larger network N_{25}. The test scenarios are summarized in Table 6.2.

In Cases I and II, both nominal and backup paths are calculated off-line. Problem FD-LSP-COMB is used to find nominal and protection paths in Case I, whereas in Case II the problem FD-LSP-COMB is modified to imply single backup paths. In cases III and IV only the nominal paths were calculated off-line by using problem FD-LSP-COMB, and the (single) protection paths were found on-line and provisioned together with the nominal path. Cases III and IV differ by the link metric (see Table 6.2), which was used for finding the protection paths. In case III, protection paths were calculated using $1/R_{e0}^{j}$ as link metric, whereas in case IV the metric was the IGP cost. In cases V and VI both nominal and single protection paths were calculated on-line, where the link metric for finding primary paths was $1/R_{e0}^{j}$. The protection paths were calculated using $1/R_{e0}^{j}$ as a link metric in Case V and IGP cost in Case VI.

Thus, in Cases I and II the calculation of primary and backup paths for HP demands was performed completely off-line. Cases III and IV present a combined approach, where primary paths were calculated off-line and the protection paths were calculated on-line. And in Cases V and VI all the paths were calculated on-line. Nominal paths for the LP traffic were always calculated on-line using the quantities $1/R_{e0}^{j}$ as link metrics, and no backup paths for the LP traffic were provisioned.

Table 1. Test scenarios

Case	Placement strategy	
	HP nominal paths	HP backup paths
I	off-line	off-line, FDBP
II	off-line	off-line, SBP
III	off-line	on-line, $1/R_{e0}^{j}$
IV	off-line	on-line, IGP
V	on-line, $1/R_{e0}^{j}$	on-line, $1/R_{e0}^{j}$
VI	on-line, $1/R_{e0}^{j}$	on-line, IGP

Table 2. Comparison of the recovery mechanisms

Case	DNF	+F,%	CL	HCL,%
I	39.43	–	2.84	57.34
II	43.95	11.46	2.93	61.73
III	43.70	10.83	2.93	58.79
IV	42.57	7.96	2.89	57.96
V	42.59	8.01	2.91	57.39
VI	43.05	9.18	2.95	57.38

6.3   Results and Discussion

Table 2 presents results for the cases described in Section 6.2. The considered recovery mechanisms are compared in terms of the average number of failing LSPs per failure situation (column DNF in the Table 2). For each case it is calculated as a difference between number of LSPs placed in the nominal situation and the average number of LSPs surviving after the failures. Second column (+F) in the table shows increase in failing LSPs for Cases II-VI as related to case I (it is calculated from the values in the DNF column). The number of LSPs placed in the nominal situation was very close in all the cases and was on average equal to 1546.33. The table also shows the maximum cascading level (column CL) and the percentage of heavily congested links (column HCL) for each case (both numbers are averages over all the failure situations). Cascading level is related to preemption mechanism and LSP priorities. Each LSP can preempt any of the lower priority LSPs, which in turn can preempt yet lower priority LSPs and so on. Thus CL gives how many priority levels down does the preemption process propagate to. A link is regarded heavily congested if it is loaded not less than by 70% of its capacity.

As can be seen from Table 2, using the proposed protection mechanism with failure-dependent backup paths resulted in almost 11% less disrupted LSPs per failure situation (on average), as compared to the single backup path protection, where the paths are calculated off-line (Case II). Thus, when using the FDBP protection mechanism it is possible to retain more traffic in the network in the case of failures as compared to SBP protection, using the same network resources. This results in better network resource utilization. Actually, SBP with off-line calculated backup paths appeared to be the worst among all the considered options. The closest result to the FDBP case is achieved by using an off-line calculated nominal path and on-line calculated single disjoint backup path (see entry IV in Table 2), using IGP as the link metric. Since all the IGP costs are equal to 1, this implies shortest path routing for the backup paths. This result also confirmed that choosing to minimize the length of the failure-dependent backup paths in the problem FD-LSP-COMB is beneficial. The results for complete on-line routing cases V and VI were worse than for Case IV but better than for case III. The CL and HCL values were also best (lowest) for the FDBP case. The worst CL was for Case VI, and the worst HCL was again for Case II (SBP). The lowest HCL value for the FDBP case is due to the load balancing. The lower CL value can in general imply lower number of preempted LSPs.

7   Conclusion

The paper considers a combined problem of load balancing and traffic recovery in an IP/MPLS network carrying traffic with different priority classes. An efficient recovery mechanism (protection with failure-dependent backup paths) is studied and a routing strategy combining shortest path routing and load balancing is proposed. A combined numerical and simulation-based study has revealed that for the considered network examples using the proposed protection mechanism and routing strategy, the average number of discarded LSPs per failure situation is lower by at least 7% than for the other considered recovery options. Unexpectedly, the off-line single backup path protection was worst among all the considered cases, outperformed even by the complete on-line routing and recovery strategies. Also, the experiments confirmed that using the shortest (hop count) paths for protection of nominal paths results in a better performance.

References

1. Gruber, C.G., Koster, A.M., Orlowski, S., Wessäly, R., Zymolka, A.: A new model and computational study for demand-wise shared protection. Technical Report ZIB-Report 05-55, Konrad-Zuse-Zentrum für Informationstechnik Berlin, Berlin-Dahlem, Germany (2005)
2. Jain, V., Alagar, S., Baig, S.I., Venkatesan, S.: Optimal quasi-path restoration in telecom backbone networks. In: 13th International Conference on System Engineering, Las Vegas, USA, CS–175–CS–180 (1999)
3. Patel, M., Chandrasekaran, R., Venkatesan, S.: Improved quasi-path restoration scheme and spare capacity assignment in mesh networks. Technical Report UTDCS-14-03, University of Texas at Dallas, Richardson, TX, USA (2003)
4. Mohan, G., Murthy, C.S.R., Somani, A.K.: Efficient algorithms for routing dependable connections in WDM optical networks. IEEE/ACM Transactions on Networking 9(5), 553–566 (2001)
5. Fredric, M.T., Somani, A.K.: A single-fault recovery strategy for optical networks using subgraph routing. In: 7th IFIP Working Conference on Optical Network Design & Modelling (ONDM-2003), Budapest, Hungary (2003)
6. Ramasubramanian, S.: On failure dependent protection in optical grooming networks. In: International Conference on Dependable Systems and Networks (DSN'04), Florence, Italy, pp. 440–449 (2004)
7. Kubilinskas, E., Aslam, F., Dzida, M., Pióro, M.: Network protection by single and failure-dependent backup paths–modeling and efficiency comparison. In: 4th Polish-German Teletraffic Symposium (PGTS 2006), Wroclaw, Poland (2006)
8. Patel, M., Chandrasekaran, R., Venkatesan, S.: A comparative study of restoration schemes and spare capacity assignment in mesh networks. In: The 12th International Conference on Computer Communications and Networks (ICCCN 2003), pp. 399–404. Dallas, TX, USA (2003)
9. Pióro, M., Medhi, D.: Routing, Flow and Capacity Design in Communication and Computer Networks. Morgan Kaufmann, San Francisco (2004)
10. Kodialam, M.S., Lakshman, T.V.: Minimum interference routing with applications to MPLS traffic engineering. Proc. of INFOCOM 2, 884–893 (2000)
11. Pióro, M., Dzida, M., Kubilinskas, E., Nilsson, P., Ogryczak, W., Tomaszewski, A., Zagożdżon, M.: Applications of the max-min fairness principle in telecommunication network design. In: First Conference on Next Generation Internet Networks Traffic Engineering (NGI 2005), Rome, Italy (2005)
12. Kelly, F., Mauloo, A., Tan, D.: Rate control for communication networks: Shadow prices, proportional fairness and stability. Journal of the Operations Research Society 49, 2006–2017 (1997)
13. Kubilinskas, E., Pióro, M., Nilsson, P.: Design models for multi-layer next generation Internet core networks carrying elastic traffic. In: Proc. 4th International Workshop on the Design of Reliable Communication Networks (DRCN 2003), Banff, Canada (2003)
14. Advanced CSPF Simulator. (http://suraj.lums.edu.pk/~te/code.htm)

Performability Analysis of Multi-layer Restoration in a Satellite Network

K.N. Oikonomou[1], K.K. Ramakrishnan[2], R.D. Doverspike[1], A. Chiu[1],
M. Martinez-Heath[3], and R.K. Sinha[1]

[1] AT&T Labs Research, Middletown, NJ, U.S.A.
[2] AT&T Labs Research, Florham Park, NJ, U.S.A.
[3] AT&T Labs, Middletown, NJ, U.S.A.

Abstract. The ability of an IP backbone network to deliver robust and dependable communications relies on quickly restoring service after failures. Service-level agreements (SLAs) between a network service provider and customers typically include overall availability and performance objectives. To achieve the desired SLA, we have developed a methodology for the combined analysis of performance and reliability (performability) of networks across multiple layers by modeling the probabilistic failure state space in detail and analyzing different restoration alternatives. This methodology has been used to analyze large commercial IP-over-Optical layer networks. In this paper we extend our methodology to evaluate restoration approaches for an IP-based satellite backbone network. Because of the environment in which they operate (long delay links, frequent impairments), satellite networks pose an interesting challenge to typical restoration strategies. We describe the potential multi-layer restoration alternatives and compare their performability. Interestingly, while it is commonly thought that SONET ring restoration at the lower layer improves overall reliability, we find that it may not always improve performability in this environment.

1 Introduction

The ability of an IP backbone network to deliver robust and dependable communications relies on restoring service quickly after failures. Service-level agreements (SLAs) between a network service provider and customers typically include overall availability and performance objectives. Backbone networks are implemented in multiple layers and include complex interactions between switches or routers and protocols at each layer. Usually, requirements are specified for each layer independently, without considering that end-to-end availability and performance depend on how the overall, multi-layer system functions as a whole, and therefore the layers are not really separable. It is thus critical to understand how the layers interact, and to consider restoration from failures as a fundamental component of the network architecture.

In a layered telecommunications model, each layer can be thought of abstractly as a graph consisting of nodes (switching or routing points) and links

L. Mason, T. Drwiega, and J. Yan (Eds.): ITC 2007, LNCS 4516, pp. 77–91, 2007.

(fixed transmission signals between the nodes). The links of one layer are carried as connections at a lower layer. A given layer experiences congestion and loss from failures that originate at that layer or at lower layers. An important characteristic of layered networks is that failures that originate at a lower layer can be restored at a higher layer, but not vice-versa.

It is important to have a methodology that can model and evaluate the impact of failures on performance and reliability, and assess the benefits of various restoration mechanisms. For this purpose we have developed nperf, a network *performability* analyzer. Performability analysis (see, e.g. [4] and §3) examines both the reliability/availability and performance of a network, given the topology, restoration scheme, component failure probabilities, and traffic (demand). The analysis evaluates a set of performance measures under all possible failures, including multiple simultaneous failures.

nperf has been used extensively to study terrestrial metropolitan and long-distance IP-over-optical layer networks. Such networks typically consist of an IP layer over a SONET or SDH layer. The SONET/SDH layer routes over a Wavelength Division Multiplexed (WDM) layer, which in turn routes over a fiber layer, although sometimes the SONET/SDH layer routes directly over the fiber layer. We have explored the most efficient restoration architectures for such terrestrial, commercial multi-layered networks in [5], [3], [13], and [7]. A further development is the performability analysis in [10] and [11].

Satellite networks pose a challenge with respect to restoration and restoration strategies because of their long latencies. In addition, inter-satellite links are prone to frequent impairments because of the environment in which they operate. With newer technologies such as Free Space Optics yielding high-capacity links (see [2]), the capacity of the satellite backbone can be quite significant, and failures may affect substantial amounts of traffic. Thus, performability analysis of satellite networks is important for understanding the effects of the various possible restoration approaches. In this paper, we further extend our nperf methodology to analyze performability of a 3-layer satellite network.

In §2 we present the satellite network and discuss how to achieve a balance between feasibility and accuracy in modelling the restoration. In §3 we describe the 4-level performability formalism used by nperf and show how restoration at two layers can be accommodated within it. In §4 we present our results for three restoration alternatives for the satellite network. Some of the results are surprising. Finally, §5 gives our conclusions.

2   Satellite Network, Restoration Schemes, and Protocols

The network we study consists of five satellites in high stationary orbit $\approx$ 30,000 km above the earth, each connected to two ground stations. (This is an example network, and does not represent any specific architecture or traffic pattern of any commercial or government network.) Our network consists of 3 layers: an IP layer, a SONET layer, and a transport technology layer[1].

The top layer consists of IP routers and Packet-over-SONET (POS) links between them in the form of concatenated STS-nc signals. Below this is a SONET cross-connect layer, whose nodes are stand-alone SONET add/drop multiplexers (ADMs), or ADM interfaces integrated into digital cross-connect systems, connected by SONET OC-n links. The links of the IP-layer form connections (demand) that are routed over the SONET layer. The bottom layer consists of hard-configured technology (i.e., the nodes are not switches) to transport the SONET OC-n links; for example Free Space Optics (FSO) or RF technology (with appropriate interfacing and encapsulation of optical SONET signals). In addition, the bottom layer includes ground-to-space links to the satellites. Since the bottom layer is fixed, restoration (rerouting) takes place only in the top two layers, so most of our analysis addresses those layers. Consequently, the details of the technologies of the bottom layer and uplinks are not critical to our analysis.

We model and compare three different network restoration architectures. The first uses Bidirectional Line Switched Ring (BLSR) at the SONET layer, a typical architecture in SONET terrestrial carrier networks. The second relies on IP-layer rerouting alone based on OSPF (see [9]), typical of long-distance carrier IP networks. Finally, the third uses SONET BLSR restoration, supplemented by IP-layer restoration whenever needed.

2.1   Protocols

Both SONET BLSR and IP OSPF restoration rely on complex protocols to detect a failure and notify the network nodes to take appropriate action. Because performability analysis requires the evaluation of the performance measures over a very large number of network states (see §3.1), it is impractical to incorporate the details of these protocols into the analysis. For this reason we abstract many of these details, while ensuring that we model the *effect* of the restoration mechanism on the performance measures to acceptable accuracy. The most important parameters are the *durations* of the restorations, which we assume to be constants (upper bounds), denoted τ_{son} and τ_{ospf} for the SONET and IP layers respectively.

Because of the long distances between the satellites, τ_{son} in our topology can be substantially longer than the 50 ms, which is typical in terrestrial SONET networks. We also assume that the SONET restoration is revertive, so there is one interval of duration τ_{son} during restoration and one upon completion of the repair[2].

For IP-only restoration all nodes in our topology are within the same OSPF area, so the predominant time is for OSPF to detect and recover from the failure

---

[1] Both the actual network and its performability model consist of a number of levels. We use the term "layer" in connection with the network, and the term "level" in connection with the model.

[2] The 2nd interval may be eliminated with a technique known as "bridge and roll", but we do not consider this here.

(converge). Whereas the failure detection time is relatively immune to topology [8], the convergence time depends on several timers which have to be set conservatively because of the length of the links. The default value for the OSPF `Hello` timer is 10 seconds, which we assume is retained in this environment. The `RouterDeadInterval` (the number of `Hello` timer intervals that pass before an adjacency is declared broken) is 4 `Hello` intervals. Thus, overall, the OSPF convergence time τ_{ospf} can be several tens of seconds.

For the case of IP plus SONET restoration, we model the individual restorations as described above. However, standard BLSR implementations in terrestrial networks use only half the bandwidth under no-failure conditions, which is unacceptable in the satellite environment where bandwidth is limited. We have adapted the BLSR implementation to the space environment where we allow IP to use the full capacity of the ring under no-failure conditions. Half of each IP link's capacity is assigned to service slots (restorable) and the other half is assigned to restoration (pre-emptable) slots. Higher priority traffic (services that have associated guarantees) is routed over the service slots, with admission control to ensure that the total traffic does not exceed 50% of the link capacity. Although we model the above scenario appropriately, we do not model the distinct classes or the admission control function in `nperf`.

3   The Performability Model

We describe the multi-level model used by `nperf` to represent a wide variety of networks with restoration at a single layer, and how this model can accommodate a network with restoration at two layers. The `nperf` performability model has a *demand* level, a *graph* level, a *component* level, and a *reliability* level.

3.1   The 4-Level Model

The graph level represents the network's "transport" layer. The network routing and restoration algorithms operate at this level, so graph edges have associated *capacities* and (routing) *costs*. The edges are directed, and there may be more than one edge between a pair of nodes. The demand or traffic level specifies the amount (and possibly type) of traffic flowing between pairs of graph nodes via the edges.

At the component level, each component corresponds to an *independent failure mechanism* (ensuring this requires some care in modelling); failure of a component may affect a whole set of graph-level elements. A component may represent an actual network element, or may be of a more abstract nature, and may have an arbitrary number of failure modes. In general, the effect of a component entering one of its failure modes is to *change some of the attributes* of a set of graph nodes and edges.

Finally, the reliability level specifies the failure modes of the components by their mean time between failures (MTBF) and mean time to repair (MTTR). It is assumed that each component is a continuous-time Markov process with a working (good) state G, and $m \geq 1$ failure (bad) states $B_1, \ldots, B_m$; the only transitions allowed are between G and the B_i. nperf assumes that the components are independent, and that they operate in their steady state.

If we have n components with $m_1, \ldots, m_n$ modes respectively, they define a state space S of size $m_1 m_2 \cdots m_n$ (2^n if all components are binary). This is the space of all possible *network states*. Each state $s \in S$ has a (steady-state) probability $\Pr(s)$ found by multiplying together an appropriate set of component mode probabilities. If F is a performance measure (function) that maps each network state to a real number, nperf evaluates the *expectation* of F over S, i.e.

$$\bar{F} = \sum_{s \in S} F(s) \Pr(s). \tag{3.1}$$

Because finding $\bar{F}$ exactly is computationally hard ([4], [14]), nperf produces lower and upper *algebraic bounds* to its value; see §3.4. With our assumption that components fail independently, the underlying global Markov process is ergodic, so $\bar{F}$ can be interpreted as the *long-run time average* of F.

We emphasize that what is "visible" to the measure is the graph and demand level of the model. Even though at the component level events occur independently, *this is not so* at the (higher) graph and demand levels. E.g., a mode change of a component may affect a whole set of graph-level elements at the same time.

The measures we use are $F_{\mathrm{lnr}}(s)$, the amount of traffic that, after restoration, is lost because it has no route in state s, and $F_{\mathrm{lcg}}(s)$, the amount of traffic that is lost in s because of congestion on the links. Since a performance measure has to be evaluated on a large number of states, some approximations to computing its value on a given state are usually necessary to render the overall computation feasible. In particular, for the F_{lcg} measure we ignore the fact that TCP "throttles" a link when congestion is detected, and instead compute what would be more precisely called "loss in total network bandwidth" by solving a network flow problem in which the link capacities are set to a fraction of their nominal values; this fraction corresponds to a link utilization threshold beyond which we assume that routers begin to drop packets.

3.2 The Model with Restoration at a Single Layer

The demand and graph levels. Fig. 3.1 shows the graph and traffic levels of our model. The graph depicts the logical connectivity at the IP layer (see §2). Each node r_i is a satellite, and g_{iA}, g_{iB} are its ground stations. The edges between satellites have capacity 5Gbps, whereas the up/down links are 10Gbps, except those for r_1, which are 12Gbps. All demands are between pairs of ground stations. There are "local" demands that go between the ground stations of a given satellite, and "long" demands whose ground stations belong to different satellites. For the symmetric traffic pattern all local demands are 2Gbps, and all long demands are 1.6Gbps. For the asymmetric pattern the local demands are again 2Gbps, and the long demands are 1Gbps, except for the 4 demands to g_{1A} and the 4 to g_{1B}, which are 2Gbps.

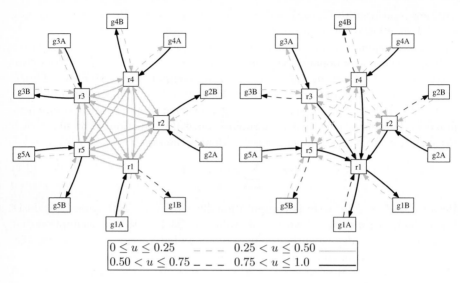

Fig. 3.1. The graph and traffic levels. Symmetric traffic (left), asymmetric traffic (right). Edges are colored according to utilization. The minimum, average, and maximum are $(0.167, 0.416, 0.840)$ and $(0.2, 0.412, 0.9)$, respectively.

The component and reliability levels. For each node r_i we include a router component, and, if the option with SONET-layer restoration is being modeled, a SONET cross-connect component and two links between it and the router. There is also one component per inter-satellite link, and one per satellite-to-ground link. Finally, there is an "interference" component for each inter-satellite link, which models impairments due to sunlight, sun spots, radiation, etc.

For the MTBFs and MTTRs of the components we use FIT and repair data from commercial fiber network equipment (some of it proprietary), and the satellite free-space optics interference estimates given in §4.1.

3.3   The Model'S Component Level for Two-Layer Restoration

To represent the architecture in which restoration occurs at both the IP and SONET network layers, one could consider adding a new SONET routing level to the performability model of §3.1. However, the BLSR protocol is so rigid that a useful approximate model can be constructed within the 4-level formalism of §3.1 by exploiting the modelling power of *multi-mode* components within the `nperf` tool. The basic idea is that, as pointed out in §2, the SONET layer's *demands* constitute the *links* (edges) of the IP layer. This means that the effect of the SONET ring can be captured by a multi-mode component that affects the attributes of the graph of the 4-level model in a certain way. To create this component we make the following two assumptions:

Full ring. As stated in §2.1, we assume that each link of the ring has the service slots allocated to high-priority traffic and the restoration slots allocated to lower-priority traffic. When a link (or node) fails, the segment of each service path that routes over the failed component is looped back at each node surrounding the failure onto the restoration slots of the unaffected links around the ring. This means that the lower-priority traffic is then lost.

Balanced allocation. In addition, we assume that *half* of each IP link capacity is routed over the service slots and *half* over the restoration slots (this can be accomplished by a SONET feature called Virtual Concatenation; see [1]). Thus a *restorable* failure of the ring (i.e., excluding multiple failures that divide the ring into isolated segments) will cause half of the capacity of each IP link to be lost: the half routed over the restoration channels will be pre-empted in order to restore the half routed over the service channels. An *unrestorable* failure will cause the IP links with no path to be lost in their entirety. In reality, depending on the exact failure, some of the links that route over the restoration channels may not be dropped; our assumption simplifies the multi-mode component significantly and gives a conservative (upper bound) approximation on performability.

Our model for two-layer restoration is then constructed as follows:

1. First we construct a 4-level model for the SONET ring by itself. The performance measure is a *vector* $F = (c_1, \ldots, c_m, \nu_1, \ldots, \nu_5)$, with one element for each demand $d_1, \ldots, d_m$ and one element for each node r_i (recall §3.2). The c_i elements describe the capacities of the final model's graph edges, and the ν_i the states of its r_i nodes. In state s of the SONET model:
   - $\nu_i = 1$ if node r_i is working, 0 if it is failed.
   - $c_i = 0$ if d_i has no path after restoration, $1/2$ the size (volume) of d_i otherwise.
2. We analyze this model to total probability $1 - \varepsilon_1$; say that this requires N_1 states to be generated, see Fig. 3.2, left.
3. Now construct the model shown in the r.h.s. of Fig. 3.2. Using the F values of the previous step, create a component with N_1 modes that represents the effect of the SONET layer on the graph layer of the combined model. (The number of modes can be reduced below N_1 by combining into one modes that have the same effect on the IP graph layer.)
4. Run the combined model to an accuracy $1 - \varepsilon_2$ to obtain an overall state space coverage of $(1 - \varepsilon_1)(1 - \varepsilon_2)$. See §3.4.

3.4 Performance Measure Details

As mentioned in §3.1, at each network state we compute the measures F_{lnr} and F_{lcg}. Two aspects of this computation require further explanation: bounds, and the inclusion of restoration time.

The size of the network state space is normally very large, so it can be explored only to a certain total probability $1 - \varepsilon$. This means that what we really get from (3.1) is a *lower bound* on $\bar{F}$. If $F(s)$ is normalized to lie in $[0, 1]$, a corresponding *upper bound* can be obtained simply by adding ε to the lower bound. Further,

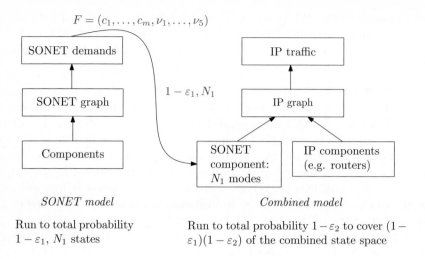

$F = (c_1, \ldots, c_m, \nu_1, \ldots, \nu_5)$

SONET demands

SONET graph

Components

IP traffic

IP graph

$1 - \varepsilon_1, N_1$

SONET component: N_1 modes

IP components (e.g. routers)

SONET model

Run to total probability $1 - \varepsilon_1$, N_1 states

Combined model

Run to total probability $1 - \varepsilon_2$ to cover $(1 - \varepsilon_1)(1 - \varepsilon_2)$ of the combined state space

Fig. 3.2. Representing the SONET restoration layer by a multi-mode component

the same thing applies to bounds on the points of the *distribution* of F: if the set of states E has been explored, we have for any $x \in [0, 1]$

$$\sum_{s \in E: F(s) = x} \Pr(s) \leq \Pr(F = x) \leq \sum_{s \in E: F(s) = x} \Pr(s) + \varepsilon. \tag{3.2}$$

It is explained in [10] that for a relatively reliable network, if the states are generated in order of decreasing probability, E need be only a very small portion of the state space to achieve sufficiently good accuracy. For example, the IP-over-SONET model has 25 binary components and 1 component with 55 modes, leading to 1.85×10^9 states, of which just the 309 most probable achieve $\varepsilon = 10^{-6}$; our IP-only model has 40 binary components and about 1.1×10^{12} states, but just the 466 most probable states suffice to achieve an $\varepsilon = 10^{-6}$. This illustrates the scalability of our approach. Finally, tighter bounds than (3.2) are also possible: see [11], [12].

To include the times $\tau_{\mathrm{ospf}}, \tau_{\mathrm{son}}$ of §2.1 in the value of F at state s, we regard traffic lost due to no path or congestion as a reward, accumulated at a constant rate (Gbps) during the deterministic interval τ, and at a different constant rate during the rest of the (random) stay in state s; $F(s)$ is then the expected accumulated reward during the stay in s.

4   Results

We present our results in the form of histograms for the probability distributions of the lost traffic measures F_{lnr} and F_{lcg}. We normalize the measures so that their values are in $[0, 1]$, and to make the histograms more intuitive, we label the y axis by "minutes per year", by multiplying the probability value by the number of minutes in a year (the long-term time average interpretation of probabilities, §3.1). Finally, recalling that our results are really lower and upper bounds on each point of the distribution, defined in (3.2), we present only the lower bound and report the ε.

4.1   SONET-Layer Restoration Only

We will abbreviate inter-satellite link interference (recall §3.2) to "islint". We consider low and high levels of islint: low islint occurs once a week on the average, while high islint occurs twice a day, and both types last an average of 2 minutes, which corresponds to the well-known effect of sunrise and sunset at low earth angles.

Fig. 4.1 shows the lower bound on the distribution of the F_{lnr} metric, computed with $\varepsilon_1 = 10^{-7}$. There is never any loss due to congestion, $F_{\mathrm{lcg}} \equiv 0$, because SONET routing takes link capacities into account, unlike OSPF. The main observation is the very significant loss of 50% of the traffic for about 7000 mins/yr, and of 70% of the traffic for about 80 mins/yr. [The sparseness of the histogram is due to (a) the discreteness of the state space, and (b) the fact that only *some* events at the bottom (component) level of the model affect the top (traffic) level. Only these events give rise to bars in the histogram, and different restoration mechanisms affect which bottom-level events "percolate" to the top level.]

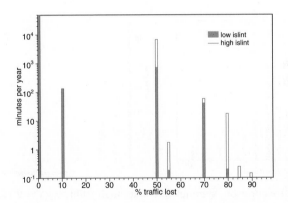

Fig. 4.1. SONET ring restoration, % traffic lost due to no path, low vs. high islint

4.2   IP Layer Restoration Only

We begin by examining the sensitivity of the performability measures as τ_{ospf} varies. The computations have been carried out with $\varepsilon_2 = 10^{-6}$.

We first consider symmetric traffic, and distinguish two cases: low islint in Fig. 4.2 and high islint in Fig. 4.3. We see that the traffic lost due to no path increases significantly when τ_{ospf} goes from 2 to 60 seconds. The increase is both in the amount of traffic affected (e.g., in the region of 30% to 40% lost) and in the amount of time a fraction of traffic does not find a path (e.g., about 9% of the traffic does not find a path for about 8500 minutes/year). We also see that the effect of τ_{ospf} on traffic lost due to no path is mostly independent of the frequency of islint. With IP-layer restoration, the traffic lost solely due to no path is significantly less than with SONET-only restoration. We then observe that traffic lost from congestion (the r.h.s.'s of Figs. 4.2 and 4.3) is not very sensitive to τ_{ospf}. Thus, with restoration that uses only the IP layer, faster OSPF convergence does not necessarily relieve congestion, either in magnitude or in time.

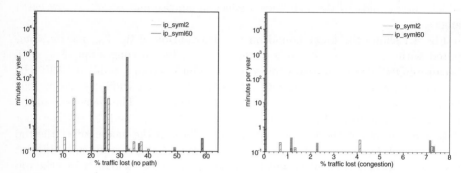

Fig. 4.2. IP-only restoration, symmetric traffic, low islint, $\tau_{ospf} = 2$ vs. $\tau_{ospf} = 60$

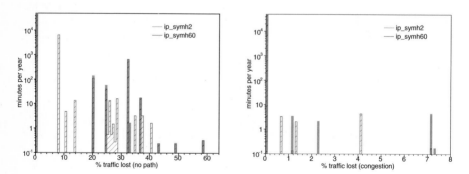

Fig. 4.3. IP-only restoration, symmetric traffic, high islint, $\tau_{ospf} = 2$ vs. $\tau_{ospf} = 60$

When we look at the results of the symmetric case shown in Figs. 4.2 and 4.3, the events that cause traffic to experience no path or congestion increase with the frequency of interference, but the relative influence of the parameter τ_{ospf} does not change, as would be expected. From both Figs. 4.2 and 4.3, traffic lost due to congestion is fairly small. To compare against SONET-only restoration, both kinds of traffic losses have to be considered. Even then, IP-based restoration affects less traffic than SONET-based restoration.

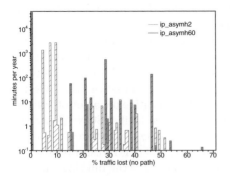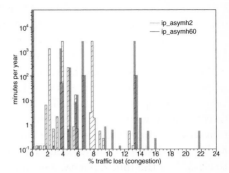

Fig. 4.4. IP-only restoration, asymmetric traffic, high islint, $\tau_{\mathrm{ospf}} = 2$ vs. $\tau_{\mathrm{ospf}} = 60$

Because of the above observations, we now focus only on the high islint cases and try to understand the influence of the traffic pattern (symmetric vs. asymmetric) as τ_{ospf} changes. It can be seen from Fig. 4.4 that the asymmetric pattern causes substantially more traffic to be affected than the symmetric one. [E.g., the bar at 10% loss due to no path is due to a set of failures with elements of the type islint, (islint, rslpair), (islint, isl), where "rslpair" is the failure of the router-to-SONET link inside a satellite, and "isl" is the failure of an inter-satellite link.] τ_{ospf} affects traffic with no path similarly, regardless of the traffic pattern. Also, significant amounts of traffic experience congestion for significant periods of time, in contrast to the symmetric case. [E.g., the bar at 8% loss due to congestion is due to failures of the type (islint, rn), (islint, rslpair), "rn" being a router failure.] Once again, reducing τ_{ospf} does not fundamentally improve the congestion situation.

4.3   IP vs. IP over SONET

We now investigate the IP over SONET restoration approach, where the SONET layer restoration happens on a faster timescale ($\tau_{\mathrm{son}} = 2\,\mathrm{sec}$), and, if unsuccessful, is complemented with the slower OSPF restoration at the IP layer. We also look at what happens if the OSPF restoration is assumed to be as fast as SONET: $\tau_{\mathrm{ospf}} = \tau_{\mathrm{son}}$. The computations have been carried out with $\varepsilon_2 = 10^{-6}$, leading to an overall bound of $\approx \varepsilon_1 + \varepsilon_2 = 1.1 \times 10^{-6}$, equivalent to 0.5 mins./yr.

As before, we begin with the symmetric traffic pattern. Fig. 4.5 shows that by adding SONET restoration we remove the events that have a long-lasting impact on traffic having no path. The traffic lost due to no path is also significantly reduced. We also see that adding SONET reduces the amount of, and time that, traffic experiences congestion. However, Fig. 4.6 shows that if OPSF is fast with $\tau_{\mathrm{ospf}} = \tau_{\mathrm{son}} = 2\,\mathrm{sec}$, then adding SONET restoration does not give us any benefit in the amount of traffic lost due to no path, nor a significant reduction in the amount of time that we see such an impact. Turning to congestion, Fig. 4.5 shows that with IP-only restoration, even when $\tau_{\mathrm{ospf}} = 2$, a very large fraction of the traffic experiences congestion. But from Figs. 4.6 and 4.7

IP restoration complemented by SONET restoration dramatically reduces, and effectively eliminates, network congestion.

When we look at *asymmetric* traffic, we see different and interesting results. With $\tau_{ospf} = 60$ and $\tau_{son} = 2$, Fig. 4.7 shows that there is a benefit for traffic without a path similar to the symmetric case. Adding SONET restoration to IP helps, although the benefit is somewhat less than in the symmetric case. What is very interesting to observe in Fig. 4.7 is the counter-intuitive behavior of loss due to congestion when we add SONET. With asymmetric traffic, SONET restoration has the undesirable effect of dramatically *increasing congestion*, potentially for long periods of time as well. This is because the SONET restoration effectively *halves* the link capacity. Recall from 2.1 that IP uses the full ring capacity under no-failure conditions, so this congestion loss will be experienced by the lower-priority preemptable traffic that is routed over the restoration slots. With asymmetric traffic, shown in Fig. 4.8, when τ_{ospf} is reduced to 2 seconds and is comparable to τ_{son}, the addition of SONET slightly improves the lost traffic due to no path. However, for congestion, adding SONET yields mixed results: congestion is reduced at the low end but it is exacerbated at the high end. With

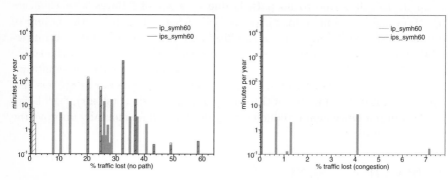

Fig. 4.5. IP (ip_) vs. IP over SONET (ips_), symmetric traffic, high islint, $\tau_{ospf} = 60, \tau_{son} = 2$

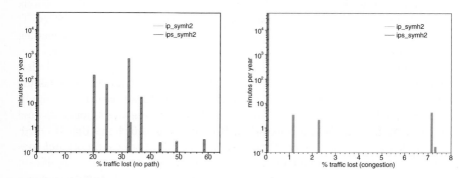

Fig. 4.6. IP vs. IP over SONET, symmetric traffic, high islint, $\tau_{ospf} = 2, \tau_{son} = 2$

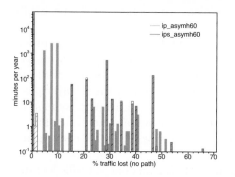
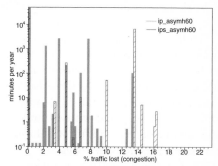

Fig. 4.7. IP vs. IP over SONET, asymmetric traffic, high islint, $\tau_{\mathrm{ospf}} = 60, \tau_{\mathrm{son}} = 2$

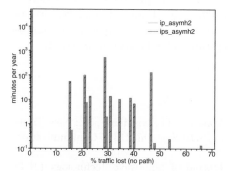
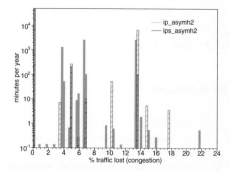

Fig. 4.8. IP vs. IP over SONET, asymmetric traffic, high islint, $\tau_{\mathrm{ospf}} = \tau_{\mathrm{son}} = 2$

IP and SONET layer restoration, traffic lost due to congestion increases because OSPF is not sensitive to capacity changes on a link. A more effective approach in satellite networks (where bandwidth is precious) could be to implement OSPF-TE which is adaptive to capacity changes resulting from lower layer restoration. We anticipate that it will help to reduce the impact on traffic both due to the lack of a path as well as congestion. In addition, we will incorporate support for distinct classes in `nperf` so that the capacity of the network can be fully utilized during non-failure conditions. Work on these aspects is in progress.

5   Conclusion

Although a backbone network normally comprises multiple layers, the SLA requirements relating to performance and reliability are typically specified at each layer independently. However, end-end performance and availability depend on how the multi-layered system functions as a whole. Satellite networks, because of their long latencies and environmental impairments, pose challenges for the design of restoration techniques. Failures can affect substantial amounts of traffic. In this paper we analyzed the performability of an IP/SONET satellite backbone network across multiple layers, given its topology, component failure probabilities, restoration alternatives and traffic pattern.

While it is commonly thought that adding SONET ring restoration improves availability, we found that this this may not always hold. Depending on the traffic pattern, SONET restoration can cause significant amounts of traffic to be lost for lack of a path. On the other hand, restoration only at the IP layer poses a different problem. While it is better for some traffic patterns (e.g., symmetric traffic), we showed that with asymmetric traffic, significant amounts of traffic are affected by congestion after a failure, and that reducing the OSPF convergence time does not offer any relief.

By complementing IP restoration with SONET, the amount of traffic lost due to no path is reduced over the individual alternatives of SONET-only or IP-only restoration. Congestion in the network can be limited, even in the stressful asymmetric traffic case, to lower priority traffic using protocols such as OSPF-TE, that are adaptive to capacity changes. Thus, the IP and SONET restoration approaches would properly complement each other and minimize the amount of traffic lost due to no-path, as well as congestion. This is work in progress.

References

[1] Cavendish, D., Murakami, K., Yun, S., Matsuda, O., Nishihara, M.: New Transport Services for Next-Generation SONET/SDH Systems. IEEE Communications Magazine, (May 2002)

[2] Chan, V.: Optical Satellite Networks. Journal of Lightwave Technology (JLT), Special Issue on Optical Networks, pp. 2811–2827 (2003)

[3] Chiu, A., Strand, J.: Joint IP/Optical Layer Restoration after A Router Failure, OFC2001, Anaheim, CA (March 2001)

[4] Colbourn, C.J.: Reliability Issues in Telecommunications Network Planning. In: Telecommunications Network Planning, Kluwer Academic Press, Boston (1999)

[5] Doverspike, R., Phillips, S., Westbrook, J.: Future Transport Network Architectures. IEEE Communications Magazine 37(8), 96–101 (August 1999)

[6] Hubenko Jr., V.P., Raines, R.A., Mills, R.F., Baldwin, R.O., Mullins, B.E., Grimaila, M.R.: Improving the Global Information Grid's Performance through Satellite Communications Layer Enhancements. IEEE Communications Magazine (November 2006)

[7] Li, G., Wang, D., Doverspike, R., Kalmanek, C., Yates, J.: Economic Analysis of IP/Optical Network Architectures. Optical Fiber Communications Conference (OFC 2004), Los Angeles, CA, (March 2004)

[8] Goyal, M., Ramakrishnan, K.K., Feng, W.-c.: Achieving Faster Failure Detection in OSPF Networks. In: Proceedings ICC 2003, Anchorage, Alaska, (May 2003)

[9] Moy, J.T.: OSPF: Anatomy of an Internet Routing Protocol. Pearson Education (1998)

[10] Oikonomou, K.N., Sinha, R.K., Doverspike, R.: Multi-Layer Network Performance and Reliability Analysis. (Submitted for publication)

[11] Oikonomou, K.N., Sinha, R.K.: A New State Generation Algorithm for Evaluating the Performability of Networks with Multimode Components. In: 8th INFORMS Telecommunications Conference, Dallas, Texas (March 2006)

[12] Oikonomou, K.N., Sinha, R.K.: Improved Bounds for Network Performability Evaluation Algorithms. In: 8th INFORMS Telecommunications Conference, Dallas, Texas (March 2006)
[13] Phillips, S., Reingold, N., Doverspike, R.: Network Studies in IP/Optical Layer Restoration. Optical Fiber Communications Conference (OFC 2002), Anaheim, CA (March 2002)
[14] Shier, D.R.: Network Reliability and Algebraic Structures. Oxford (1991)

Adaptive Congestion Control Under Dynamic Weather Condition for Wireless and Satellite Networks

Hongqing Zeng[1,2], Anand Srinivasan[1], Brian Cheng[1], and Changcheng Huang[2]

[1] EION Inc., 945 Wellington Street, Ottawa, ON, Canada
[2] Dept. of Systems and Computer Engineering, Carleton University, Ottawa, ON, Canada
{hzeng, anand, brian}@eion.com, {hzeng, huang}@sce.carleton.ca

Abstract. Broadband satellite-based IP networks have been considered as the technology to enable a strong and promising next-generation market. In satellite communication systems, the channel performance might be severely degraded due to the dynamic weather conditions such as the precipitation, and thereby lead to network congestion. Therefore, congestion control is critical in such networks to satisfy QoS based Service Level Agreement (SLA). Moreover, the inherent large bandwidth-delay product of satellite channels impedes the deployment of existing numerous congestion control schemes. In this paper, we propose a modified Random Early Detection (RED) based congestion avoidance/control mechanism that incorporates a fuzzy logical controller to tune the queue thresholds of RED according to the dynamic weather conditions. We will show using analysis and simulations that the newly developed congestion control method is effective and efficient for broadband satellite-based IP networks.

Keywords: Satellite network, congestion control, active queue management, fuzzy logic.

1 Introduction

Due to extensive geographic reach, satellite communication systems are attractive for providing anywhere, anytime pervasive Internet access, especially in rural areas where the cost of alternative Internet access is high. In the satellite communications family, a new concept, broadband satellite networks have been proposed [1]. In the last few years, the broadband satellite network has gained tremendous research interest [2]. The broadband satellite network is IP-based and provides a ubiquitous means of communications for multimedia and high-data rate Internet-based applications, such as audio and video applications.

Dynamic weather conditions, such as the precipitation-caused channel fading, may severely affect the performance of satellite communications and thereby leads to network congestions. The congestion control is even more important in satellite communications networks due to the inherent large channel latency, or high bandwidth-delay product [3]. Therefore, similar to territorial IP networks, the network congestion control is critical in broadband satellite-based IP networks, for committing service-level-agreement (SLA) with regard to quality-of-service (QoS).

L. Mason, T. Drwiega, and J. Yan (Eds.): ITC 2007, LNCS 4516, pp. 92–103, 2007.

Existing approaches for network congestion control in territorial IP networks reported in literatures cover a broad range of techniques, including window flow control [4], source quench [5], slow start [6], schedule-based control [7], binary feedback [8], rate-based control [9], etc. Recently, the active queue management (AQM) has been proposed to support the SLA QoS [10] and attracted much attention in the research community. AQM is a packet dropping/marking mechanism for router queue management. It targets to reduce the average queue length and thereby decrease the packet delay, while reduces the packet loss to ensure efficient network resource utilization. The most important and popular AQM mechanisms fall into two categories: Random Early Detection (RED) [11] and its variants, and Proportional Integral Derivative controllers (PID) [12]. PID mechanisms emerge from control theory and have better performance than RED in terms of the control of end-to-end delay. However, PID involves larger queue length oscillation, which leads to delay fluctuation (larger jitter). On the other hand, RED mechanisms are widely employed in today's IP routers, though they are often disabled by the users in the real world due to drawbacks such as the sensitivity to network configuration parameters and states, queue fluctuation, low throughout, etc [13]. Alternatively, the simple Drop-Tail scheme is applied as the *de facto* router queue management method.

Network congestion is caused by saturation of network resources, which is a dynamic resource allocation problem, rather than a static resource shortage problem. Such an optimal control problem of network resources is proven to be intractable even for the simplest cases [14]. The behavior of networks is usually highly non-linear, time varying, and chaotic (e.g., the observed self-similar traffic pattern [15]). Measurements on the state of networks are always incomplete, relatively poor and time-delayed for congestion control. Therefore, there is an inherent fuzziness in the network congestion control. In recent years, some fuzzy logic based mechanisms have been proposed for congestion control [16–20] in various territorial networks. In this paper, we propose an integrated RED-based congestion control mechanism using a fuzzy logic controller (FLC) with inference rules base. In this mechanism, the FLC tunes the RED queue thresholds to adapt to the change of weather conditions. Simulations show that it overcomes the drawbacks of original RED, such as the sensitivity to parameter configuration.

The remainder of this paper is organized as the following: Section II briefly introduces the background of RED and fuzzy logic, including membership functions and the inference rules base. Section III gives the architecture of our satellite-based IP network and the system design with the functional block diagram. In Section IV, we describe the simulation environment and scenarios for our system, in order to evaluate the newly developed congestion control mechanism: RED with fuzzy logic control (RED-FL). Finally, some conclusive remarks are listed in Section V.

2 Theory on RED and Fuzzy Logic

2.1 Random Early Detection (RED)

The key point of RED is to avoid network congestions by controlling the average queue length in a reasonable range. RED sets the minimum and maximum thresholds for queues and handle newly arrived packets according to the following rules:

1) if the queue length falls in the range from the minimum threshold (q_{min}) to the maximum threshold (q_{max}), then RED drops newly arrived packets with the probability that is calculated using exponential weighted moving average (EWMA) function;

2) if the queue length is larger than q_{max}, then RED drops all newly arrived packets (i.e., Drop Tail)

3) if the queue length is smaller than q_{min}, then no packet dropping;

RED was proven to be stable [21]. However its performance is sensitive to the dynamically parameter tuning, especially the tuning of q_{min} and q_{max}. The efforts for accurately tuning the RED parameters achieved limited success so far. In this paper, we applied a fuzzy inference system (FIS) to the traditional RED scheme for parameter tuning, based on the inherent fuzziness in network congestion control. Especially, such a RED scheme with fuzzy logic control (RED-FL) is applied to making our satellite-based IP network more adaptive to the dynamic weather condition change, e.g., channel rain-fade.

2.2 Fuzzy Logic

Fuzzy Logic was first developed by L. A. Zadeh in 1965 to represent various types of "approximate" knowledge, which cannot otherwise be represented by crisp methods. Fuzzy logic is an extension of crisp two-state logic and it provides a better platform to handle approximate knowledge. Fuzzy logic is based on fuzzy sets and a fuzzy set is represented by a *Membership Function (MF)*. A membership function is a curve that defines how each point in the input space is mapped to a membership value between 0 and 1. It provides the degree of membership within a set of any element that belongs to the universe of discourse (simply put the universal set). It maps the elements of the universal set onto numerical values within the interval [0, 1]. Specifically if X is the universe of discourse and the elements in X are denoted by x, then a fuzzy set A in X is defined as a set of ordered pairs.

$$A = \{x, \mu_A(x) \mid x \in X\} \tag{1}$$

$\mu_A(x)$ is a membership function of x in A. The membership function maps each element of X to a MF value between 0 and 1. The membership function could be either piecewise linear or quadratic.The AND, OR, and NOT operators of Boolean logic exist in fuzzy logic as well. The Boolean logic is usually defined as the minimum, maximum, and complement. OR, AND and NOT are the so-called Zadeh operators since they were first defined as such in Zadeh's seminal papers. So for any fuzzy variables x and y:

NOT x = (1 - truth (x))
x AND y = minimum(truth(x), truth(y))
x OR y = maximum(truth(x), truth(y))

Fuzzy logic usually uses IF/THEN rules that are statements used to formulate the conditional statements that comprise the fuzzy logic. A single IF-THEN rule assumes the format of "*if x is A then y is B*". Where *A* and *B* are *linguistic values* and they are defined by fuzzy sets on the ranges X and Y (where X and Y are the universes of discourse). The if-part of the rule "*x is A*" is called the *antecedent* or *premise* and the then part of the rule "*y is B*" is called the *consequence* or *conclusion*.

3   Satellite-Based IP Network Model and Implementation

3.1   The Architecture of the Satellite-Based IP Network

The architecture of the satellite communication system for the project is depicted in Fig 1. In the architecture, all ground terminals communicate only with the central hub gateway, and vice versa. All communications must go through the relay satellite and the hub gateway. The satellite works as a channel relay. Bandwidth allocation algorithms may be required in both the hub gateway and ground terminals to maintain appropriate SLA QoS requirements. The bandwidth allocation granularity in the hub gateway is terminal-based while in ground terminals are user-based. Packet-switch is the main form of data communication running on our project of Intelligent Satellite System for Broadband Services (ISSBS), e.g., voice over IP (VoIP), video stream, and data (TCP/UDP), etc.

Fig 2 describes the functional block diagram for the high-level system design for our satellite-based IP network. The motivation of the design is to achieve high performance communications and congestion adaptation to impacts of dynamically weather change, on system performance (SLA QoS), e.g., channel rain-fade. Functional blocks of intelligent software in the system are defined as the following,

Hub Correlator
  – Correlate relevant precipitation data with fuzzy inference engine to predict service degradation within specific timeframe
  – Initiate appropriate proactive or reactive required actions
Link Detector:
  – Assign traffic on forward and return link timeslots
  – Adapted to fade & congestion
QoS Arbitrator
  – Active queue management to discard packets/flows and notify impending network congestion
Traffic Shaper
  – Connection admission control
  – Packet conditioning and scheduling

3.2   Design of the Fuzzy Logic Controller (Fuzzy Inference System)

Fuzzy control uses the principles of fuzzy-logic based decision making to arrive at the control actions. A fuzzy controller is usually built of four units: Fuzzifier, Knowledge Base, Inference Engine and Defuzzifier, as shown in Fig 3.

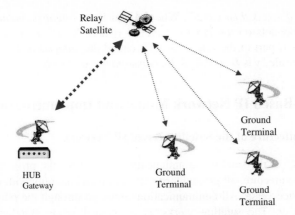

Fig. 1. The architecture of the satellite-based IP network

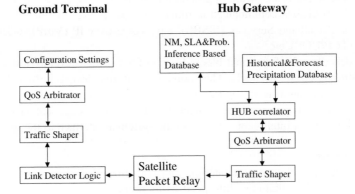

Fig. 2. Functional diagram of the satellite-based IP network

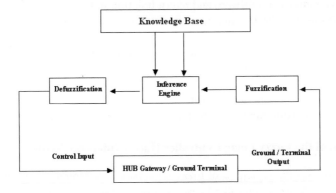

Fig. 3. Fuzzy Logic Controller showing the functional blocks

Fuzzifier - In the fuzzifier, crisp values are transformed into grades of membership functions for linguistic terms of fuzzy sets. The membership function is used to associate a grade to each linguistic term. In other words, the fuzzifier maps each of the crisp values of the inputs into its corresponding linguistic values, which may be viewed as the labels of fuzzy sets.

Knowledge Base (Fuzzy Rule Base) - The knowledge base contains domain-specific facts and heuristics that are useful to solve the domain related problem. They are represented in form of a set of linguistic control rules (IF-THEN rules) that characterize the goals and policies of the problem considered.

Fuzzy Inference Engine - This is the main "driver" of the Fuzzy Logic Controller and provides the decision-making logic to the system. It responds to the fuzzy input provided from the fuzzifier (also possibly previous inferences from the rule-base itself) and scrutinizes the knowledge base to identify one or more possible control actions or conclusions.

Defuzzifier - The control action of the Inference Engine encompasses a range of output values and is represented by membership functions. Defuzzification has to be done on this to resolve them into crisp non-fuzzy control signals. The most popular defuzzification strategy is the center of area method that yields a superior result.

In the design of our fuzzy logic controller for weather adaptation, three different scenarios (control problems) are dealt with:

1) Adjustment of the Bandwidth
2) Adjustment of the minimum queue length q_{min}
3) Adjustment of the maximum queue length q_{max}

The current weather conditions, allocated bandwidth and the queue length are the input parameters to the above control problems. The universe of discourse for the input and output parameters are assumed to be:

1) Bandwidth range (input in scenario #1) is selected to vary between 16 and 128 kbps per user. The packet size is fixed at 512 bytes.
2) The minimum queue length q_{min} (input in scenario #2) is varied between 4 and 16 packets.
3) The maximum queue length q_{max} (input in scenario #3) is varied between 8 and 20 packets.
4) Current weather condition (input in all cases) is based on the total precipitation rate.
5) The bandwidth adjustment (output in scenario #1) is done between 0 and 48 kbps per user.
6) The output parameters in scenarios #2 and #3 are adjusted to fall within the ranges already mentioned under 2) and 3).

The fuzzy sets associated with the input and output parameters are depicted in Fig. 4 and 5. The synthesis of the membership functions depend on the choice of thresholds of the weather condition parameters, e.g., the amount of precipitation. Such thresholds may be derived from historical data and empirical analysis. Also, the number of fuzzy sets for each weather parameter is flexible. In the design of our fuzzy logic controller for weather adaptation, we choose the triangular membership functions, which were proven to be extremely effective in territorial networks [19].

The fuzzy inference engine is built of numerous control rules that are fuzzy in nature and contain the linguistic variables associated with each input and output parameters.

Some of the sample rules for scenario #1 are listed below:

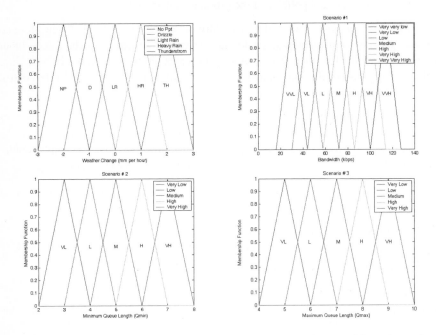

Fig. 4. Input parameters to the FIS (a) Change in current weather conditions (mm of rain drops/hour) (b) Scenario #1 where bandwidth ranges are shown (c) Scenario #2 where ranges of q_{min} are shown (d) Scenario #3 wherein ranges of q_{min} are shown

Scenario # 1 [Bandwidth Adjustment]

- *IF (the current allocated bandwidth is low) AND (weather becomes worse) THEN {heavily increase the bandwidth to commit the SLA QoS};*
- *IF (the current allocated bandwidth is medium) AND (weather becomes worse) THEN {slightly increase bandwidth to commit the SLA QoS};*
- *IF (the current allocated bandwidth is high) AND (weather becomes worse) THEN {do not change bandwidth};*
  ………..

Defuzzification converts the output fuzzy result into a crisp and a singular output. Center of area method, as mentioned in the earlier section is used as the defuzzification method.

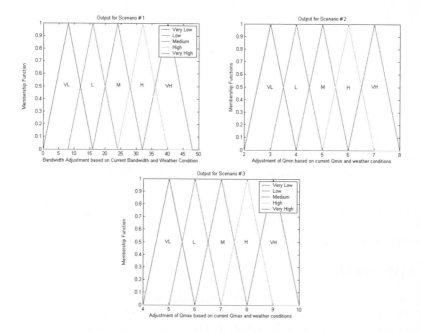

Fig. 5. Output parameters for the three control problems considered

4 Performance Evaluation

4.1 Simulation Setup

In order to evaluate the performance and robustness of our proposal, we map the architecture shown in Figure 2 into a simulation setup given in Figure 6, using the open source simulation tool NS-2.

Within the setup, two edge routers are deployed to simulate a ground terminal and the hub gateway, respectively. The intermediate node, relay satellite, is not really necessary for the setup but is included and reserved for the future use, e.g., IP routing and switching for multiple satellites. The link capacity between the edge router and the relay satellite is set to be 2 Mbps, which is typical for satellite links. The guaranteed minimum bandwidth for each session is 64 Kbps, and the bandwidth could burst up to the link capacity if all link capacity is not used.

In our satellite communication system, the traffic includes both TCP and UDP packets. In the research report, the traffic composition classified by protocols is as the following [23],

TCP: 83% of packets, 91% of bytes
UDP: 14% of packets, 5% of bytes

In our satellite-based IP network, due to the widely used video application, the percentage of UDP traffic is much higher than that in general Internet environment.

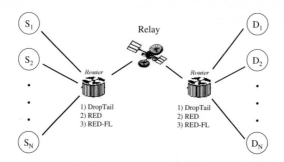

Fig. 6. Simulation setup for the satellite-based IP network

Therefore, we set half of the traffic be TCP and the other half UDP. The number of sessions is set to be 16/32/64 for various scenarios respectively. Note that the parameters of RED are set as recommended in [11].

4.2   Evaluation Metrics

With the system running, we monitor the instantaneous queue length for all scenarios. Also, the queue delays are measured and system throughputs are calculated for performance evaluation upon all scenarios. The dropping probabilities, which can be considered as the indicator of network congestion status, are also collected.

In addition to the inspection of RED-FL, we also implement the RED scheme as recommended in [11] and compare the performance of the two schemes in terms of the above performance metrics: instantaneous queue length, queue delay, dropping probability, and throughput.

4.3   Results and Analysis

A weather condition change pattern (precipitations in mm/time a.u.), where a.u. represents an arbitrary unit, is given in Fig. 7. Please note that the weather change is mapped into the change of TCP sessions. Since rain-fading effects cannot be reflected in the simulation directly, but an increasing channel fading can be considered equivalently as the increase of TCP session number. The corresponding simulation results of queue length, queue delay, packet dropping probabilities, as well system throughput, are collected from scenarios described above for both RED and RED-FL schemes. Figure 8 (a – d) compares the evaluation metrics for the two schemes. In the simulation, there are some TCP bursty sessions (to simulate the weather change) between the simulation time 0 to 20 second, and between 50 and 60 second. The bursty TCP sessions are obviously reflected in the instantaneous queue length, as well as 3 other metrics, in Figure 8. The packet size is fixed at 512 bytes.

The simulation results in Figure 8 show that all network congestion control methods we investigated in this paper have good performance in queue length, but the RED-FL improves the system throughput and packet dropping probabilities. Although it introduces somehow an increased queuing delay (Figure 8b), this queuing delay in RED-FL is more constant and stable.

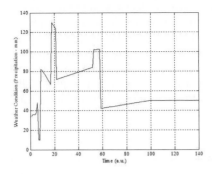

Fig. 7. Weather condition (Precipitation) change pattern

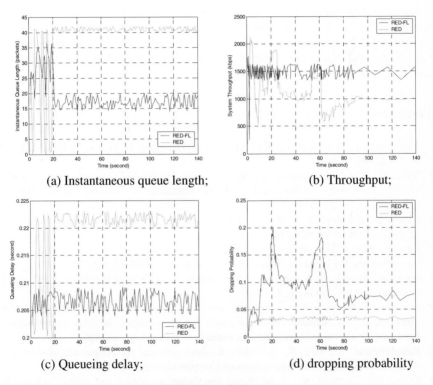

(a) Instantaneous queue length;

(b) Throughput;

(c) Queueing delay;

(d) dropping probability

Fig. 8. Simulation results

5 Conclusion

Broadband satellite-based IP networks have become more and more attractive for the construction of anywhere, anytime pervasive networking. However, due to the inherent impacts of dynamical weather change on system performance, e.g., channel rain-fading, intelligent network congestion control mechanisms are highly expected for such networks. In this paper, we introduced the architecture of broadband satellite-based IP network, together with the functional high-level design. A network congestion mechanism based on RED with the fuzzy logic controller is proposed for such satellite-based IP networks. Finally, the performance evaluation in terms of instantaneous queue length, queue delay, dropping probability, and throughput, is analyzed for the proposed RED-FL congestion control. The discussion implies that the newly proposed RED-FL is an effective congestion control method for satellite-based IP networks.

Acknowledgements

We wish to thank Ms. Banupriya Vallamsundar Prof. K. Ponnambalam of University of Waterloo and Mr. Kamal Harb of Carleton University for their participation in the discussion towards fuzzy control design.   The authors also wish to thank Mr. Bharat Rudra of OCE, Mr. Derek Best of Precarn, Dr. Kalai Kalaichelvan and Dr. Rama Munikoti of EION for supporting this research. The authors wish to acknowledge Mr. Andre Bigras and Mr. Abdul Lakhani of Telesat for verifying the applicability and usefulness in satellite systems and networks.

References

1. Jamalipour, A.: Broadband satellite networks — The global IT bridge. In: Proceedings of the IEEE, vol. 1 (2001)
2. Taleb, T., Kato, N., Nemoto, Y.: Recent trends in IP/NGEO satellite communication systems: transport, routing, and mobility management concerns, IEEE Wireless Communications, 5 (2005)
3. Katabi, D., Handley, M., Rohrs, C.: Congestion control for high bandwidth-delay product networks, SIGCOMM'02 (2002)
4. Lam, S., Reiser, M.: Congestion control of store and forward network by input buffer limits analysis, IEEE Transactions on Communications, COM-27, 1 (1979)
5. Prue, W., Postel, J.: Something a host could do with source quench, RFC- 1016, Networking Working Group, 7 (1987)
6. Jacobson, V.: Congestion avoidance and control. In: Proceedings of SIGCOMM'88 (1988)
7. Mukherji, U.: A schedule-based approach for flow-control in data communication Networks, Ph.D. Thesis, Massachusetts Institute of Technology, 2 (1986)
8. Ramakrishnan, K., Jain, R.: A binary feedback scheme for congestion avoidance in computer networks, ACM Transactions on Computer Systems, 2 (1990)
9. Comer, D., Yavatkar, R.: A rate-based congestion avoidance and control scheme for packet switched networks. In: Proceedings of 10th IEEE ICDCS (1990)
10. Braden, B., et al.: Recommendations on Queue Management and Congestion Avoidance in the Internet, RFC 2309 (1998)
11. Floyd, S., Fall, K.: Random early detection gateways for congestion avoidance, IEEE/ACM Transactions on Networking, 4 (1993)
12. Fan, Y., Ren, F., Lin, C.: Design a PID controller for active queue management. In: Proceedings of the 8th IEEE ISCC, 7 (2003)
13. Bitorika, A.: A comparative study of active queue management schemes. In: Procedddings of IEEE ICC 2004, 6 (2004)

14. Papadimitriou, C.H., Tsitsiklis, J.N.: The complexity of optimal queueing network control, Mathematics of Operations Research, 2 (1999)
15. Paxon, V., Floyd, S.: Wide area traffic: the failure of Poisson modeling, IEEE/ACM Transactions on Networking, 3 (1995)
16. Ghosh, S., Razouqi, Q., Schumacher, H.J., Celmins, A.: A survey of recent advances in fuzzy logic in telecommunications networks and new challenges, IEEE Transactions on Fuzzy Systems, 3 (1998)
17. Chrysostomou, C., Pitsillides, A., Rossides, L., Sekercioglu, A.: Fuzzy logic controlled RED: congestion control in TCP/IP differentiated services networks, Soft Computing, 2 (2003)
18. Hadjadj Aoul, Y., Nafaa, A., Negru, D., Mehaoua, A., FAFC,: fast adaptive fuzzy AQM controller for TCP/IP networks. In: Proc. of IEEE Globecom 2004, Dallas, TX, USA (2004)
19. Fan, Y., Ren, F., Lin, C.: Design an active queue management algorithm based on fuzzy logic decision. In: Proceedings of IEEE ICCT 2003, Beijing, PRC (2003)
20. Lim, H.H., Qiu, B.: Fuzzy logic traffic control in broadband communication networks. In: Proceedings of IEEE International Fuzzy Systems Conference 2001 (2001)
21. Ohsaki, H., Murata, M., Miyahara, H.: Steady state analysis of the RED gateway: stability, transient behavior, and parameter setting, IEICE Trans. on Comm., E85-B (2002)
22. Zadeh, L.A.: Fuzzy sets, Information and Control, 8 (1965)
23. Roberts, J.: IP traffic and QoS control, ENST, 12 (2002)

Distributed Path Computation Without Transient Loops: An Intermediate Variables Approach[*]

Saikat Ray[1], Roch Guérin[2], and Rute Sofia[3]

[1] Department of Electrical Engineering, University of Bridgeport
[2] Department of Electrical and Systems Engineering, University of Pennsylvania
[3] Siemens AG Corporate Technology, Information and Communications, Munich, Germany

Abstract. Paths with loops, even transient ones, pose significant stability problems in networks. As a result, much effort has been devoted over the past thirty years to designing distributed algorithms capable of avoiding loops. We present a new algorithm, *Distributed Path Computation with Intermediate Variables* (DIV), that guarantees that no loops, transient or steady-state, can ever form. DIV's novelty is in that it is *not* restricted to shortest paths, can easily handle arbitrary sequences of changes and updates, and provably outperforms earlier approaches in several key metrics. In addition, when used with distance-vector style path computation algorithms, DIV also prevents *counting-to-infinity*; hence further improving convergence. The paper introduces DIV and its key properties. Simulation quantifying its performance gains are also presented.

1 Introduction

Distributed path computation is a central problem in modern communication networks, and has therefore received much attention. Its importance together with the lack of a "generic" solution is what motivated this paper. In distributed path computations, end-to-end paths are formed by concatenating individual node decisions, where for each destination a node chooses one or more successors (next-hop) based only on local information and so as to optimize some global objective function. The use of inconsistent information across nodes can then have dire consequences, including possible formation of transient routing loops[1]. Loops can severely impact performance, especially in networks with no or limited loop mitigation mechanisms (such as Time-to-Live (TTL)), where a routing loop often triggers network-wide congestion. The importance of avoiding transient routing loops remains a key requirement for path computation in both existing and emerging network technologies, e.g., see [1] for recent discussions, and is present to different extents in both link-state and distance-vector algorithms.

Link-state algorithms (e.g., OSPF [2]) decouple information dissemination and path computation, so that routing loops, if any, are short-lived[2], but the algorithms overhead is high in terms of communication (broadcasting updates), storage (maintaining a

---

[*] Research supported in part by a gift to the University of Pennsylvania by the Siemens AG Corporate Technology, Information and Communications, Munich, Germany.

[1] By "routing" we mean creation of forwarding tables, irrespective of their "layer,"e.g., 2 or 3.

[2] Even such *micro-loops* may be undesirable; see [3].

L. Mason, T. Drwiega, and J. Yan (Eds.): ITC 2007, LNCS 4516, pp. 104–116, 2007.

full network map), and computation (changes anywhere in the network trigger computations at all nodes). By combining information dissemination and path computation, distance-vector algorithms (cf. RIP [4], EIGRP [5]) avoid several of these disadvantages, which make them potentially attractive, especially in situations of frequent local topology changes and/or when high control overhead is undesirable. However, they can suffer from frequent and long lasting routing loops and slower convergence (cf. the *counting-to-infinity* problem [6]). Thus, making usable distance-vector based solutions calls for overcoming these problems. Since the 70's, several works [7,8,9,10,11] have targeted this goal in the context of shortest path computations (cf. Section 2), but the problem remains fundamental and timely (e.g., see the efforts for introducing distributed shortest path algorithms in lieu of a distributed spanning tree algorithm [12]). Furthermore, as we briefly allude to in Section 3.4, extending this capability to other types of path computation is also becoming increasingly important.

In this paper, we introduce the D*istributed Path Computation with* I*ntermediate* V*ariables* (DIV) algorithm which enables generic, distributed, light-weight, loop-free path computation. DIV is *not* by itself a routing protocol; rather it can run on top of any routing algorithm to provide loop-freedom. DIV generalizes the *Loop Free Invariant* (LFI) based algorithms [10,11] and outperforms previous solutions including known LFI and *Diffusing Computation* based algorithms, such as the *Diffusing Update Algorithm* [9][3]:

1. *Applicability*: DIV is not tied to shortest path computations. It can be integrated with other distributed path computation algorithms, e.g., the link-reversal mechanisms of [13,14] or algorithms targeting path redundancy or distributed control as outlined in Section 3.4.
2. *Frequency of Synchronous Updates*: When applied to shortest path computations, DIV triggers synchronous updates less frequently as well as reduces the propagation of synchronous updates (cf. Theorem 4), where synchronous updates are updates that potentially must propagate to all upstream[4] nodes before the originator is in a position to update its path. They are time and resource consuming. Thus, the less frequent the synchronous updates, the better the algorithm.
3. *Maintaining a path*: A node can potentially switch to a new successor without forming a loop provably faster with DIV than with earlier algorithms (cf. Section 3.2). This is especially useful when the original path is lost due to a link failure.
4. *Convergence Time*: When a node receives multiple overlapping updates[5] from neighbors, DIV allows the node to process them in an arbitrary manner. Thus, in DIV, the node can respond only to the latest or the best update, converging potentially faster. (cf. Theorem 2).
5. *Robustness*: DIV can tolerate arbitrary control packet reordering and losses without sacrificing correctness (cf. Theorem 3).

---

[3] The authors gratefully acknowledge J.J. Garcia-Luna-Aceves for introducing them to the LFI-based algorithms.

[4] Upstream nodes of a node x for destination z are the nodes whose path to z includes x.

[5] Two updates are overlapping if the latter appears before the algorithm has converged in response to the first.

Finally, the rules and update mechanism of DIV and their correctness proofs are rather simple, which hopefully will facilitate correct and efficient implementations.

2   Previous Works

The Common Structure. The primary challenge in avoiding transient loops lies in handling inconsistencies in the information stored across nodes. Otherwise, simple approaches can guarantee loop-free operations at each step [13,15]. Most previous distance-vector type algorithms free from transient loops and convergence problems follow a common structure: Nodes exchange update-messages to notify their neighbors of any change in their own cost-to-destination (for any destination). If the cost-to-destination decreases at a node, the algorithms allow updating its neighbors in an arbitrary manner; these updates are called *local* (asynchronous) updates. However, following an increase in the cost-to-destination of a node, these algorithms require that the node potentially update all its upstream nodes before changing its current successor; these are *synchronous* updates.

The algorithm proposed in [7] follows the above broad structure and is one of the earliest work guaranteeing loop-free operations with inconsistent information. For handling multiple overlapping updates, it relies on unbounded sequence numbers that mark update epochs. An improvement to this algorithm is presented in [8], which handles multiple overlapping updates by maintaining *bit vectors* at each node.

Diffusing Update Algorithm (DUAL). DUAL, a part of CISCO's popular EIGRP protocol, is perhaps the best known algorithm. In DUAL, each node maintains, for each destination, a set of neighbors called the *feasible successor set*. The feasible successor set is computed using a *feasibility condition* involving *feasible distances* at a node. Several feasibility conditions are proposed in [9] that are all tightly coupled to the computation of a shortest path. For example, the *Source Node Condition* (SNC) defines the feasible successor set to be the set of all neighbors whose current cost-to-destination is *less* than the minimum cost-to-destination seen so far by the node. A node can choose any neighbor in the feasible successor set as the successor (next-hop) without having to notify any of its neighbors and without causing a routing loop regardless of how other nodes in the network choose their successors, as long as they also comply with this rule.

If the neighbor through which the cost-to-destination of the node is minimum is in the feasible successor set, then that neighbor is chosen as the successor. If the current feasible successor set is empty or does not include the best successor, the node initiates a synchronous update procedure, known as a *diffusing computation* (cf. [16]), by sending *queries* to all its neighbors and waiting for acknowledgment before changing its successor. Multiple overlapping updates—i.e., if a new link-cost change occurs when a node is waiting for replies to a previous query—are handled using a *finite state machine* to process these multiple updates sequentially.

Loop Free Invariance **(LFI) Algorithms.** A pair of invariances, based on the cost-to-destination of a node and its neighbors, called *Loop Free Invariances* (LFI) are introduced in [10] and it is shown that if nodes maintain these invariances, then no transient loops can form (cf. Section 3.2). Update mechanisms are required to maintain the LFI conditions: [10] introduces *Multiple-path Partial-topology Dissemination Algorithm* (MPDA) that uses a link-state type approach whereas [11] introduces *Multipath Distance Vector Algorithm* (MDVA) that uses a distance vector type approach. Similar to DUAL, MDVA uses a diffusing update approach to increase its cost-to-destination, thus it also handles multiple overlapping cost-changes sequentially. The primary contribution of LFI based algorithms such as MDVA or MPDA is a unified framework applicable to both link-state and distance-vector type approaches and multipath routing.

Comparative Merits of Previous Algorithms. DUAL supersedes the other algorithms in terms of performance. Specifically, the invariances of MPDA and MDVA are based directly on the cost of the shortest path. Thus, every increase in the cost of the shortest path triggers synchronous updates in MDVA or MPDA. In constrast, the feasibility conditions of DUAL are *indirectly* based on the cost of the shortest path. Consequently, an increase in the cost of the shortest path may not violate the feasibility condition of DUAL, and therefore may not trigger synchronized updates—an important advantage over MDVA or MPDA. Because of the importance of this metric, we consider DUAL the benchmark against which to compare DIV (cf. Section 4).

DIV combines advantages of both DUAL and LFI. DIV generalizes the LFI conditions, is not restricted to shortest path computations and, as LFI-based algorithms, allows for multipath routing. In addition, DIV allows for using a feasibility condition that is strictly more relaxed than that of DUAL, hence triggering synchronous updates less frequently than DUAL (and consequently, than MPDA or MDVA) as well as limiting the propagation of any triggered synchronous updates. The update mechanism of DIV is simple and substantially different from that of previous algorithms, and allows arbitrary control packet reordering/losses. Last but not least, unlike DUAL or LFI algorithms, DIV handles multiple overlapping cost-changes *simultaneously* without additional efforts resulting in simpler implementation and potentially faster convergence.

3   DIV

3.1   Overview

DIV lays down a set of rules on existing path computation algorithms to ensure their loop-free operation at each instant. This rule-set is not predicated on shortest path computation, so DIV can be used with other path computation algorithms as well.

For each destination, DIV assigns a *value* to each node in the network. To simplify our discussion and notation, we fix a particular destination and speak of *the* value of a node. The values could be arbitrary—hence the independence of DIV from any underlying path computation algorithm. However, usually the value of a node will be related to the underlying objective function that the algorithm attempts to optimize and the network topology. Some typical value assignments include: (i) in shortest path computations, the value of a node could be its cost-to-destination; (ii) as in DUAL, the value could be the minimum cost-to-destination seen by the node from time $t = 0$; (iii) as in TORA [14], the value could be the *height* of the node; etc.

As in previous algorithms, the basic idea of DIV is to allow a node to choose a neighbor as successor only if the value of that neighbor is less than its own value: this is called the *decreasing value property* of DIV, which ensures that routing loop can never form. The hard part is enforcing the decreasing value property when network topology changes. Node values must be updated in response to changes to enable efficient path selection. However, how does a node know the *current* value of its neighbors to maintain the decreasing value property? Clearly, nodes update each other about their own current value through update messages. Since update messages are asynchronous, information at various nodes may be inconsistent, which may lead to the formation of loops. This is where the non-triviality of DIV lies: it lays down specific update rules that guarantee that loops are never formed even if the information across nodes is inconsistent.

3.2   Description of DIV

There are four aspects to DIV: (i) the variables stored at the nodes, (ii) two ordering invariances that each node maintains, (iii) the rules for updating the variables, and (iv) two semantics for handling non-ideal message deliveries (such as control packet loss or reordering). A separate instance of DIV is run for each destination, and we focus on a particular destination.

The Intermediate Variables. Suppose that a node x is a neighbor of node y. These two nodes maintain intermediate variables to track the value of each other. There are three aspects of each of these variables: whose value is this? who believes in that value? and where is it stored? Accordingly, we define $V(x; y|x)$ to be the value of node x as known (believed) by node y stored in node x; similarly $V(y; x|x)$ denotes value of node y as known by node x stored in node x.

Thus, node x with n neighbors, $\{y_1, y_2, \ldots, y_n\}$, stores, for each destination:

1. its own value, $V(x; x|x)$;
2. the values of its neighbors as known to itself, $V(y_i; x|x)$ [$y_i \in \{y_1, y_2, \ldots, y_n\}$],
3. and the value of itself as known to its neighbors $V(x; y_i|x)$ [$y_i \in \{y_1, y_2, \ldots, y_n\}$].

That is, $2n + 1$ values for each destination. The variables $V(y_i; x|x)$ and $V(x; y_i|x)$ are called intermediate variables since they endeavor to reflect the values $V(y_i; y_i|y_i)$ and $V(x; x|x)$, respectively. In steady state, DIV ensures that $V(x; x|x) = V(x; y_i|x) = V(x; y_i|y_i)$.

The Invariances. DIV requires each node to maintain at all times the following two invariances based on its set of *locally stored variables*.

Invariance 1. *The value of a node is not allowed to be more than the value the node thinks is known to its neighbors. That is,*

$$V(x; x|x) \leq V(x; y_i|x) \text{ for each neighbor } y_i. \tag{1}$$

Invariance 2. *A node x can choose one of its neighbors y as a successor only if the value of y is less than the value of x as* known *by node x; i.e., if node y is the successor of node x, then*

$$V(x; x|x) > V(y; x|x). \tag{2}$$

Thus, due to Invariance 2, a node x can choose a successor only from its *feasible successor set* $\{y_i | V(x; x|x) > V(y_i; x|x)\}$. The two invariances reduces to the LFI conditions if the value of a node is chosen to be its current cost-to-destination.

Update Messages and Corresponding Rules. There are two operations that a node needs to perform in response to network changes: (i) decreasing its value and (ii) increasing its value. Both operations need notifying neighboring nodes about the new value of the node. DIV uses two corresponding update messages, Update::Dec and Update::Inc, and acknowledgment (ACK) messages in response to Update::Inc. ACKs in reponse to Update::Dec messages do not play any role in loop-avoidance; so we do not consider them. Both Update::Dec and Update::Inc contain the new value (the destination), and a sequence number[6]. The ACKs contain the sequence number and the value (and the destination) of the corresponding Update::Inc message. DIV lays down precise rules for exchanging and handling these messages which we now describe.

Decreasing Value. Decreasing value is the simpler operation among the two. The following rules are used to decrease the value of a node x to a new value V_0:

- Node x first simultaneously decreases the variables $V(x; x|x)$ and the values $V(x; y_i | x) \forall i = 1, 2, \ldots, n$, to V_0,
- Node x then sends an Update::Dec message to all its neighbors that contains the new value V_0.
- Each neighbor y_i of x that receives an Update::Dec message containing V_0 as the new value updates $V(x; y_i | y_i)$ to V_0.

Increasing Value. In the decrease operation a node first decreases its value and then notifies its neighbors; in the increase operation, a node first notifies its neighbors (and wait for their acknowledgments) and then increases its value. In particular, a node x uses the following rules to increase its value to V_1:

- Node x first sends an Update::Inc message to all its neighbors.
- Each neighbor y_i of x that receives an Update::Inc message sends an acknowledgment (ACK) when able to do so according to the rules explained in details below (Section 3.2). When y_i is ready to send the ACK, it first modifies $V(x; y_i | y_i)$, changes successor if necessary (since the feasible successor set may change), and then sends the ACK to x; the ACK contains the sequence number of the corresponding Update::Inc message and the new value of $V(x; y_i | y_i)$. Note that it is essential that node y_i changes successor, if necessary, *before* sending the ACK.
- When node x receives an ACK from its neighbor y_i, it modifies $V(x; y_i | x)$ to V_1. At any time, node x can choose any value $V(x; x|x) \leq V(x; y_i | x) \forall i = 1, 2, \ldots, n$.

Rules for Sending Acknowledgment. Suppose node y_i received an Update::Inc message from node x. Recall that node y_i must increase $V(x; y_i | y_i)$ before sending an ACK. However, increasing $V(x; y_i | y_i)$ may remove node x from the feasible successor set at node y_i. If node x is the only node in the feasible successor set of node y_i, node y_i may lose its path if $V(x; y_i | y_i)$ is increased without first increasing $V(y_i; y_i | y_i)$. Node y_i

---

[6] For simplicity, sequence numbers are assumed large enough so that rollover is not an issue.

then has two options: (i) first increase $V(y_i; y_i|y_i)$, increase $V(x; y_i|y_i)$, and then send the ACK to node x; or (ii) increase $V(x; y_i|y_i)$, send ACK to node x, and then increase $V(y_i; y_i|y_i)$. We call option (i) the *normal mode*, and option (ii) the *alternate mode*.

In the normal mode, node y_i keeps the old path while it awaits ACKs from its neighbors before increasing $V(y_i; y_i|y_i)$, since it keeps x in the feasible successor set until then. Thus the update request propagates to upstream nodes in the same manner as in DUAL and other previous works. However, note that DIV allows a node to respond with an ACK in response to Update::Inc messages sent its neighbors, if there is any, without the fear of any loop creation. This guarantees that DIV never enters any deadlock situations.

In the alternate mode, node y_i may have no successor for a period of time (until it is allowed to increase its value). At a first glance, it may seems unwise to use the alternate mode. However, note that if node x originated the value-increase request in the first place because the link to its successor was *down* (as opposed to only a finite cost change), then the old path does not exist and the normal mode has no advantage over the alternate mode in terms of maintaining *a* path. In fact, in the alternate mode, the downstream nodes get ACKs from their neighbors more quickly and thus can switch earlier to a new successor (which hopefully has a valid path) than in the normal mode.

Semantics for Handling Message Reordering. We maintain the following two semantics that account for non-zero delays between origination of a message at the sender and its reception at the receiver and possible reordering of messages and ACKs.

Semantic 1. *A node ignores an update message that comes out-of-order (i.e., after a message that was sent earlier).*

Semantic 2. *A node ignores outstanding ACKs after issuing an Update::Dec message.*

These semantics are enforced using the embedded sequence numbers in update messages (an ACK includes the sequence number of the Update::Inc that triggered it).

3.3   Properties of DIV

The two main properties of DIV are: (i) it prevents loops at every instant, and (ii) it prevents counting-to-infinity in the normal mode. Due to space constraint, we only prove Property (i) (see [17] for details). Note that even in the specific case of shortest path computations where the value of a node is set to its current cost-to-destination, these properties of DIV cannot be deduced from those for the LFI conditions since DIV operates without any assumption on control packet reordering, delay or losses.

Loop-free Operation at Every Instant. The following is the key proposition based on which our result follows.

Proposition 1. *For any two neighboring nodes x and y, we always have*

$$V(x; y|x) \leq V(x; y|y) \tag{3}$$

Proof. The proof is by contradiction. Suppose at time $t = 0$ condition (3) is satisfied and at time $t = t_4$ condition (3) is violated for the first time. I.e., at time $t = t_4$, we have

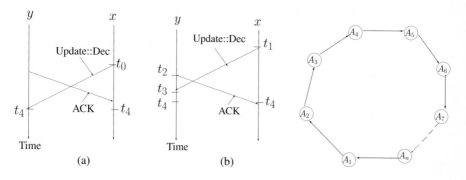

Fig. 1. Two cases of possible message exchanges between two neighboring nodes which would violate Eq. (3). Both cases are shown to be contradictory.

Fig. 2. A possible loop in the successor graph

$V(x; y|x) = V_1$ and $V(x; y|y) = V_0$ with $V_1 > V_0$. Thus, at time t_4 either $V(x; y|y)$ decreases or $V(x; y|x)$ increases. We consider these two cases separately.

Case (i): $V(x; y|y)$ decreases at time t_4 to V_0. Thus node y receives an Update::Dec message from node x at time t_4. As shown in Fig. 1(a), suppose that this message originated at node x at time t_0. Therefore, at time t_0, we have $V(x; y|x) = V_0$. But as per our assumption, $V(x; y|x) = V_1 > V_0$ at time t_4. Thus, node x must receive an ACK from node y that increases $V(x; y|x)$ during the period (t_0, t_4) (cf. Fig. 1(a)). Suppose t_2 denotes the time when node x sent the update message that triggered this ACK. We then have two cases:

- $t_2 < t_0 < t_4$: In this case, the Update::Inc message that triggered the ACK was outstanding at t_0; the time when node x sent an Update::Dec message. Thus node x would disregard this ACK due to Semantic 2, and therefore not increase $V(x; y|x)$.
- $t_0 < t_2 < t_4$: In this case, the Update::Inc message that triggered the ACK was sent by node x after the Update::Dec message, but node y received the Update::Inc message before the Update::Dec message; i.e., the Update::Dec message arrived node y out of order and thus node y would disregard the Update::Dec message due to Semantic 1, and therefore not decrease $V(x; y|y)$.

We therefore have a contradiction in both cases.

Case (ii): $V(x; y|x)$ increases at time t_4 to V_1. Thus node x receives an ACK from node y at time t_4. As shown in Fig. 1(b), suppose that this ACK originated at node y at time t_2. Thus, we have $V(x; y|y) = V_1$ at time t_2. But by assumption, $V(x; y|y) = V_0$ at time t_4. Thus, node y must receive an Update::Dec message during the period (t_2, t_4), say at time t_3. Suppose that node x originated this Update::Dec message at time t_1 (cf. Fig. 1(b)). Moreover, suppose node x originated at time t_0 the Update::Inc message that triggered the ACK it receives from node y at time t_4. Then, there are two possibilities:

- $t_0 < t_1 < t_4$: In this case, the Update::Inc message that triggered the ACK was outstanding at t_1; the time when node x sent an Update::Dec message. Thus node x would disregard this ACK due to Semantic 2, and not increase $V(x; y|x)$ to V_1 at time t_4.

– $t_1 < t_0 < t_4$: In this case, the Update::Inc message that triggered the ACK was sent by node x after the Update::Dec message, but node y received the Update::Inc message before the Update::Dec message; i.e., the Update::Dec message arrived node y out of order and thus node y would disregard the Update::Dec message due to Semantic 1, and not decrease $V(x; y|y)$ to V_0 at time t_3.

We therefore again have a contradiction in both cases.

Thus we have shown that both case (i) and case (ii) lead to contradictions. Hence, we conclude that it is not possible to violate Eq. (3).   □

Theorem 1. *The successor graph created following DIV's update algorithm is an acyclic graph at each instant.*

Proof. The proof is again by contradiction. Suppose at some instant of time there is a loop in the successor graph, as shown in Fig. 2. Since the number of nodes in this loop is finite, there is a node in this loop whose value is smaller than or equal to the value of its successor. Without any loss of generality, let A_n be this node and let A_1 be its successor. Thus,

$$V(A_1; A_1|A_1) \geq V(A_n; A_n|A_n). \tag{4}$$

But since node A_1 maintains the first invariance, we have

$$V(A_1; A_1|A_1) \leq V(A_1; A_n|A_1). \tag{5}$$

Also since node A_n maintains the second invariance, we have

$$V(A_n; A_n|A_n) > V(A_1; A_n|A_n). \tag{6}$$

But equations (4), (5) and (6) together imply that $V(A_1; A_n|A_1) > V(A_1; A_n|A_n)$, which contradicts Proposition 1.   □

Multiple Overlapping Updates and Control Packet Losses. Unlike earlier algorithms [7,8,9,10,11], DIV can handle multiple updates without additional efforts. A node can send multiple Update::Inc or Update::Dec messages in any order; a neighbor can hold on to sending an ACK for an arbitrary time—e.g., use a hold-down timer—and when replying with an ACK, it can choose to respond to only a subset of pending updates—even just one; none of these actions results in routing loops. This is because these policies of handling multiple overlapping updates still preserve the Semantics and the Invariances. Semantics are satisfied at each node using the sequence numbers of updates, and invariances depend only on the locally stored variables. Thus they are never violated. We summarize this important property in the following theorem, which establishes the tremendous flexibility DIV gives in choosing policies for replying with ACKs to optimize different criteria.

Theorem 2. *The correctness of DIV remains valid under arbitrary policies for handling multiple overlapping updates.*

DIV can also handle an arbitrary sequence of lost control packets without jeopardizing its correctness. If an Update::Dec message sent by node x to neighbor y is lost, then

$V(x; y|x)$ is lowered (by x), but not $V(x; y|y)$; i.e., we have $V(x; y|x) < V(x; y|y)$. But this still satisfies Proposition 1, hence does not affect DIV's correctness. If an Update::Inc message sent by node x to neighbor y is lost, then node x cannot increase its value, but the invariances remains valid. Finally, if an ACK is lost, then $V(x; y|y)$ is increased (by y), but not $V(x; y|x)$; i.e., we have $V(x; y|x) < V(x; y|y)$. Again, this satisfies Proposition 1 and DIV remains correct. When combined with the fact that Semantics 1 and 2 handle arbitrary reordering and delay of messages, this leads to the following important property of DIV:

Theorem 3. *The correctness of DIV remains valid under arbitrary sequence of loss, reordering or delay of messages.*

Frequency of Synchronous Updates: A Comparison with DUAL

Claim. Suppose x and y are neighbors. If SNC is true at x through y, then with DIV x can choose y as a successor.

Proof. We need to show that SNC is true at x through y implies $V(x; x|x) > V(y; y|y)$. From the definition of SNC (cf. Section 2), since SNC is satisfied, we have the minimum cost-to-destination of x, $V(x; x|x)$, is more than the *current* cost-to-destination of y. However, the current cost-to-destination of y is clearly as large as the minimum cost-to-destination of y, $V(y; y|y)$; i.e., $V(x; x|x) > V(y; y|y)$.   □

However, the other direction is clearly not true. Suppose $V(x; x|x) = 2$, $V(y; y|y) = 1$ and the current cost-to-destination of y is 3. Then SNC is not satisfied, but with DIV, x can still choose y as its successor. Since the condition of DIV is strictly more relaxed than SNC, and a synchronous update is issued only when the condition of DIV (or SNC for DUAL) is not satisfied, we have

Theorem 4. *DIV issues synchronous updates less frequently than DUAL under SNC.*

Note that this cannot be remedied simply by replacing SNC in DUAL with the conditions of DIV since without DIV's update mechanisms, these are not sufficient to guarantee loop-free operation.

Theorem 4 is stated in terms of SNC as it is the most common condition used in practice (e.g., it is used in EIGRP). However, the theorem remains true if SNC is replaced by other conditions of DUAL, such as CSC or DIC [9]; the proof is similar.

3.4   Other Applications of DIV

By decoupling loop-freedom from the path computation metric (e.g., shortest-path), DIV opens up new possibilities. Due to space limitations, we only mention two such applications; these are discussed further in [17].

Redundant-Path Routing. An assignment of values (along with the Invariances) induces an acyclic successor graph (i.e., a routing), and if each node other than the destination has at least one outgoing link, the destination is reachable from every node. By an appropriate choice of values, DIV can be used to maximize a measure of the multitude of paths to the destination. This can be appealing when bandwidth is cheap and reliability takes a higher priority.

Distributed Vehicular Formation. Forming rigid patterns, e.g., of unmanned aerial vehicles, using localized sensing is an important problem. It is known that stabilizing the pattern is easy if the underlying *formation graph* is acyclic [18]. DIV, especially its alternate mode, can be used to ensure acyclicity of the formation graph in a distributed fashion under dynamic environments.

4   Performance Evaluation

This section presents simulation results comparing the performances of DIV (with normal mode used with DBF to compute shortest paths) in terms of routing loops, convergence times and frequency of synchronous updates against DUAL (cf. Section 2). The performance of DBF without DIV is also presented as a reference. The simulations are performed on random graphs with fixed average degree of 5. The number of nodes are varied from 10 to 90 in increments of 10. For each graph-size, 100 random graphs are generated. Link costs are drawn from a bi-modal distribution: with probability 0.5 a link cost is uniformly distributed in [0,1]; and with probability 0.5 it is uniformly distributed in [0,100]. For each graph, 100 random link-cost changes are introduced, again drawn from the same bi-modal distribution. All three algorithms are run on the same graphs and sequence of changes. Processing time of each message is random: it is 2 s with probability 0.0001, 200 ms with probability 0.05, and 10 ms otherwise.

The table in Fig. 3 shows the average loop-retention time in seconds, T_{loop}—the time from when a routing loop is detected to when it eventually subsides—given that a loop is formed, as the size of the graphs varies. As expected, no loops were found with DUAL or DIV, so the table only shows results for DBF, which illustrate that without loop-prevention mechanisms, loops can be retained for a significant time.

Fig. 5 shows average convergence times—the time from a cost change to when no more updates are exchanged—for all three algorithms as the size of the graphs varies.

Nodes	10	20	30	40	50
$T_{\text{loop}}(s)$	2.282	2.456	2.344	2.702	2.108
Conf. (s)	0.259	0.365	0.259	0.391	0.276

Nodes	60	70	80	90	—
$T_{\text{loop}}(s)$	2.126	2.339	2.290	2.354	—
Conf. (s)	0.237	0.273	0.311	0.250	—

Fig. 3. Average loop-retention time, T_{loop}, in seconds

Nodes	10	20	30	40	50
Fraction	0.717	0.784	0.823	0.843	0.846

Nodes	60	70	80	90	—
Fraction	0.832	0.843	0.846	0.840	—

Fig. 4. Fraction of times DIV is satisfied given that SNC is not

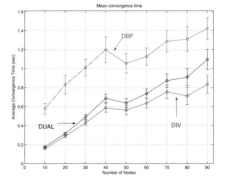

Fig. 5. Mean convergence time

The vertical bars show standard deviations. Both DIV and DUAL converge faster than DBF; however, DIV performs better, especially for larger graphs. This is because DIV's conditions are satisfied more easily, so that synchronous updates often complete earlier (recall that a node with a feasible neighbor replies immediately). This is supported by the table in Fig. 4, which shows the fraction of times the condition of DIV is satisfied given that SNC is not satisfied; this fraction exceeds 80% for larger graphs.

5   Conclusion

Distance-vector path computation algorithms are attractive candidates not only for shortest path computations, but also in several important areas involving distributed path computations due to their simplicity and scalability. Leveraging those benefits, however, calls for eliminating several classical drawbacks such as transient loops and slow convergence. The algorithm proposed in this paper, DIV, meets these goals, and which unlike earlier solutions is not limited to shortest path computations. In addition, even in the context of shortest path computations, DIV outperforms earlier approaches in several key performance metrics, while also providing greater operational flexibility, e.g., in handling lost or out-of-order messages. Given these many benefits and the continued and growing importance of distributed path computations, we believe that DIV can play an important role in improving and enabling efficient distributed path computations.

References

1. Francois, P., Filsfils, C., Evans, J., Bonaventure, O.: Achieving sub-second IGP convergence in large IP networks. ACM SIGCOMM Computer Communication Review (July 2005)
2. Moy, J.: OSPF version 2, Internet Engineering Task Force, RFC 2328 (April 1998) [Online]. Available: http://www.rfc-editor.org/rfc/rfc2328.txt
3. Francois, P., Bonaventure, O., Shand, M., Bryant, S., Previdi, S.: Loop-free Convergence using oFIB, INTERNET-DRAFT draft-ietf-rtgwg-ordered-fib-00 (December 2006)
4. Malkin, G.: RIP version 2, Internet Engineering Task Force, RFC 2453 (November 1998) [Online]. Available: http://www.rfc-editor.org/rfc/rfc2453.txt
5. Albrightson, R., Garcia-Luna-Aceves, J.J., Boyle, J.: EIGRP–A fast routing protocol based on distance vectors. In: Proceedings of Network/Interop, Las Vegas, NV (May 1994)
6. Bertsekas, D., Gallager, R.: Data Networks, 2nd edn. Prentice Hall, Englewood Cliffs (1991)
7. Merlin, P.M., Segall, A.: A failsafe distributed routing protocol. IEEE Transactions on Communications COM-27(9), 1280–1288 (1979)
8. Jaffe, J.M., Moss, F.M.: A responsive routing algorithm for computer networks. IEEE Transactions on Communications COM-30(7), 1758–1762 (1982)
9. Garcia-Lunes-Aceves, J.J.: Loop-free routing using diffusing computations. IEEE/ACM Transactions on Networking 1(1), 130–141 (1993)
10. Vutukury, S., Garcia-Luna-Aceves, J.J.: A simple approximation to minimum-delay routing. In: Proceedings of ACM SIGCOMM, Cambridge, MA (September 1999)
11. ——, MDVA: A distance-vector multipath routing protocol. In: Proceedings of IEEE INFOCOM, Anchorage, AK (April 2001)
12. Elmeleegy, K., Cox, A.L., Ng, T.S.E.: On count-to-infinity induced forwarding loops in Ethernet networks. In: Proceedings of IEEE INFOCOM, Barcelona, Spain (April 2006)

13. Gafni, E., Bertsekas, D.: Distributed algorithms for generating loop-free routes in networks with frequently changing topology. IEEE/ACM Transactions on Communications (January 1981)
14. Park, V.D., Corson, M.S.: A highly adaptive distributed routing algorithm for mobile wireless networks. In: Proceedings of IEEE INFOCOM (1997) [Online]. Available: citeseer.ifi.unizh.ch/park97highly.html
15. Gallager, R.G.: A minimum delay routing algorithm using distributed computation. IEEE Transactions on Communications (January 1977)
16. Dijkstra, E.W., Scholten, C.S.: Termination detection for diffusing computations. Information Processing Letters 11(1), 1–4 (1980)
17. Ray, S., Guérin, R., Rute, S.: Distributed path computation without transient loops: An intermediate variables approach, University of Pennsylvania, Tech. Rep, [Online]. Available (2006), http://www.seas.upenn.edu/~saikat/loopfree.pdf
18. Baillieul, J., Suri, A.: Information patterns and hedging Brockett's theorem in controlling vehicle formations. In: Conference on Decision and Control, Maui, Hawaii (December 2003)

A Robust Routing Plan to Optimize Throughput in Core Networks

Ali Tizghadam and Alberto Leon-Garcia

University of Toronto, Electrical and Computer Engineering Department
{ali.tizghadam, alberto.leongarcia}@utoronto.ca

Abstract. This paper presents an algorithm for finding a robust routing plan in core networks that takes into consideration network topology, available capacity, traffic demand, and quality of service (QoS) requirements. The algorithm addresses the difficult problem in routing and traffic engineering of optimal path selection. Our approach is inspired by the concept of "between-ness" from graph theory, from which we introduce quantitative metrics for link and path criticality. Paths are ranked according to path criticality and the algorithm tries to avoid placing flows on the most critical paths, maximizes throughput over the short term in the presence of QoS constraints, and attempts to increase the bandwidth of the critical paths for future use. The proposed approach shows promise relative to previous proposals in simulations on benchmark and experimental networks.

Keywords: Flow Assignment; Graph Theory; Quality of Service (QoS); Routing; Traffic Engineering.

1 Introduction

An abundance of work has already been done in the research community and industry to address the routing and flow assignment problem and the traffic engineering issues in core network systems, especially MPLS based networks [1], [2], [3] but still far from an ideal routing scheme. The main goal of the research reported in this paper is to examine the problem from a new standpoint, motivated by the definition of "between-ness" from graph theory [4], [5]. We introduce the notion of link and path criticality, and use them to identify the most critical paths, to build the routing plan on less critical paths in the short-term and optimize the throughput, as well as to plan the increase of bandwidth in the paths with high criticality index. We will show the extension of our algorithm to incorporate the quality of service (QoS) based routing. Our simulations on benchmark networks and realistic topologies discussed in the research literature show that the approach is promising.

The paper is organized as follows. Section 2 reviews the state of the art in routing plan and flow assignment problems. We describe the issues with existing methods and the associated research challenges. In section 3, we give the formal description of the problem, then in section 4, we propose our path-criticality based routing scheme.

L. Mason, T. Drwiega, and J. Yan (Eds.): ITC 2007, LNCS 4516, pp. 117–128, 2007.

Section 5 provides an extension of our approach to cover situations that have additive QoS constraints on the links of the network. Section 6 provides a "proof of concept" for our proposal. We assess the proposed method on benchmark networks as well as experimental scenarios and present encouraging results. Finally we conclude with a discussion of open issues and future work.

2 Previous Works

The most popular algorithm used in the research community for routing is the *shortest path routing algorithm* (SP*)*. While the shortest path algorithm enjoys the benefit of simplicity, it can suffer from major problems to the network due to the lack of any load balancing mechanism. Widest shortest path (WSP) [7], improves the performance of SP, but the possibility of having bottlenecks remains. Furthermore both SP and WSP do not impose any form of admission control to control the flow in the network.

Ref. [1] introduced the minimum interference routing algorithm (MIRA). In contrast to the prior methods, MIRA considers the effect of source-destination pairs on the routing plan. A number (max-flow) is assigned to every source-destination pair (S, D), indicating the maximum amount of traffic that can be sent from S to D through the network. MIRA operates based on the notion that running traffic on some of the links may decrease the max-flow of (S, D). This process is called "interference".

In brief MIRA tries to build the paths in such a way as to minimize interference. Introducing the notion of "interference" is significant, but some problems are still present with the algorithm introduced in [1]. MIRA concentrates on the effect of interference on just one source-destination pair, but there are situations where some links can cause bottlenecks on a cluster of node pairs. Ref. [3] investigates three benchmark networks: parking-lot (Fig. 1), concentrator and distributor, and shows that MIRA is unable to respond to the network flow requests appropriately and causes blocking for a large number of incoming flows in these networks. Finally MIRA is only designed to provide bandwidth guaranteed paths. It does not account for any other QoS constraint in the network.

Profile-based routing (PBR) is another proposal for routing of bandwidth guaranteed flows in MPLS networks [3]. PBR assumes that the source-destination pair and the traffic-profile between them are known. According to PBR, a traffic profile is the aggregate bandwidth demand for a specific traffic class between a source-destination pair. PBR has two phases. In the offline phase a multicommodity flow assignment problem is solved with the goal of routing as much commodity as possible. PBR also has some problems. Like MIRA, PBR deals only with bandwidth and does not consider QoS. Furthermore in [8] the authors introduce a network called "Rainbow Topology" and show that the performance of PBR in this network is much worse than MIRA and WSP. The main reason behind this is that PBR relies on the results of the offline phase which are not always correct.

In more recent works, the concentration is on oblivious routing to make the routing scheme independent of traffic but all of these approaches are far from an optimal solution due to over-provisioning [6] or because of considering a special case of core networks such as mesh and two-hop routes [15].

3   Problem Statement

Our goal is to achieve a robust routing plan. Robust routing plan in this work means a routing strategy that can cope with variations in traffic matrices as much as possible, as well as link/node failures and changes in community of interest (source-destination pairs). We start with the traditional formulation of the problem which is an LP formulation to optimize the selected metric or metrics. This formal presentation of the problem is adopted from [6]. Assume that the network is modeled as a directed graph $G(V, E), |V| = n, |E| = m$. In [13] the "Hose Model" for the network is proposed and used to manage VPN service. According to this model, one does not need to know the exact traffic matrix, but the maximum ingress/egress capacity of a node. This condition means that one can have any traffic matrix as long as the sum of its columns does not exceed γ_i^{in} (maximum ingress capacity for node i) and the sum of its rows is not more than γ_i^{out} (maximum egress capacity for node i) for any node in the network.

$$\sum_{j, j \neq i} \gamma_{ij} \leq \gamma_i^{in}, \quad \sum_{j, j \neq i} \gamma_{ji} \leq \gamma_i^{out} \tag{a}$$

We denote the set of all traffic matrices which are satisfying (a) with Γ. Now we are looking for maximum multiplier θ such that all traffic matrices in $\theta \times \Gamma$ can be routed. Assuming that the maximum utilization of all the links in the network is u, then maximizing the throughput is equal to minimizing u. To find link-based routing (path-based can be easily obtained then [12]), we run one unit of bandwidth into the network between a source-destination pair (i, j) to obtain fraction of traffic that traverses link l. We show this fraction by $x_{ij}(l)$. Now we can write the linear programming (LP) problem to give us these fractions:

$$\text{Minimize } u$$

Subject to:

$$\sum_{l \in T^{in}(k)} x_{ij}(l) \; - \sum_{l \in T^{out}(k)} x_{ij}(l) \; = \; \begin{cases} 1 & k = i \\ -1 & k = j \\ 0 & otherwise \end{cases} \quad i, j, k \in V \tag{1}$$

$$\sum_{i,j} \gamma_{ij} x_{ij}(l) \; \leq \; u \times c_l \quad \forall l \in E, \; [\gamma_{ij}] \in \Gamma \tag{2}$$

$$x_{ij}(l) \; \geq \; 0 \quad \forall l \in E, \; \forall i, j \in V \tag{3}$$

where $T^{in}(k)$, $T^{out}(k)$ are sets of incoming and outgoing links for node k respectively.

The set of constraints in (2) lead to an infinite number of constraints due to the hose model (traffic matrices could change as long as they meet requirements of (a)). Although it is possible to change this problem to an equivalent one with limited number of constraints using the dual of this LP problem [6], we do not choose this way as we want to provide a dynamic routing scheme that is robust to changing the traffic matrices as well as link failure and changing community of interest (source-destination pair).

4 Path Criticality Routing (PCR)

In this paper our goal is to find a robust routing plan for the core network that allows the network service provider to manage the assignment of flows to the paths primarily at the edge of the core network and obtain close to maximum throughput. To achieve the goal first we need to identify the important factors affecting the routing plan and flow assignment. One can summarize these factors as:

1. Network topology and connectivity.
2. Community of interest
3. Capacity of the links.
4. Traffic Matrix.

In order to have a robust routing plan we need to recognize the effect of link and node on network connectivity. Connectivity is a well studied subject in graph theory [4], [5], [10] allowing us to define some useful metrics to measure the sensitivity of the network to node or link failures. Capacity of a network is another key issue in flow assignment problem. Clearly the paths with more capacity are desired since the low capacity paths are prone to congestion. Hence an intelligent routing plan should avoid routing the flows onto the low capacity paths and should request for capacity increases for those paths if possible. Finally traffic demand directly affects the routing plan. The traffic demand profile may change from time to time (e.g. week-day traffic profile). Traffic changes might be predictable and periodical or chaotic. We need to find a routing scheme which is robust to the predicted traffic patterns and unpredicted ones to the extent possible.

We now introduce two metrics to estimate the effect of the aforementioned characteristics: link criticality index (LCI) and path criticality index (PCI) which are built based on the theory of graphs [4], [5]. We will subsequently propose our routing algorithm based on PCI.

A. Link Criticality Index
Freeman [4] introduced a useful measure in graph theory called "between-ness centrality." Suppose that we are measuring the centrality of node k. The between-ness centrality is defined as the share of times a node i needs a node k in order to reach a node j via the shortest path. We can modify the definition of between-ness centrality to introduce a useful measure for criticality of links in a network. Suppose

p_{sd} is the number of paths between source-destination pair (s,d) and p_{sld} is the number of paths between (s,d) containing the specific link l. Inspired by the definition of between-ness, one can quantify the effect of network topology by dividing p_{sld}/p_{sd} over all source-destination pairs. This gives an indication of how critical the link l is in the network topology. This topological aspect of the criticality of link l is then:

$$LCI_{top} = \sum_{s,d} p_{sld}/p_{sd} \tag{4}$$

The effect of link capacity and average demand for the source-destination (s,d) (provided by the traffic matrix) is accounted for in the residual bandwidth of the link. The residual bandwidth of link l is the capacity available after considering the flows already traversing the link, and is denoted by $c_r(l)$. Obviously the link criticality has an inverse relation with available bandwidth, and so we can account for residual bandwidth by multiplying equation (4) by $1/c_r(l)$.

The ability of a link to handle a given offered volume of data flow also has to be reflected in the link criticality. For example, suppose a new request offers flow γ_l to the link. Let the indicator function $I(x)$ and the modified link criticality be:

$$I(x) = 1 \ \ if \ x\rangle 0 \ \ otherwise \ \ 0 \tag{5}$$

$$LCI_{trc}(l) = 1/c_r(l) \ \times \ 1/I(c_r(l) - \gamma_l) \tag{6}$$

In addition: $$\frac{1}{c_r(l)} = \frac{1}{c_l - c_{used}} = \frac{1}{c_l \times (1 - u_l)} \tag{7}$$

In this formula c_l is the link capacity, c_{used} is the used bandwidth by link l, c_l is the link capacity, c_{used} is the used bandwidth by link l, and u_l is the utilization of link l.

Hence: $$LCI(l) = \sum_{s,d} \frac{p_{sld}}{p_{sd}} \times \frac{1}{c_l \times (1 - u_l)} \times \frac{1}{I(1 - u_l)} \tag{8}$$

The indicator function can be written as above because $c_l \geq 0$. In this section we assume that there is no QoS constraint other than bandwidth. We will consider the case of multi-constraint routing in section 5. One can clearly see the effect of topology and connectivity p_{sld}/p_{sd} as well as capacity and traffic matrix $c_r(l)$ in the definition of LCI. In addition, this formulation is inline with our goal which was to minimize the maximum link utilization or maximizing throughput of the network. Indeed according to (8) when the utility of the link l increases and u_l gets near to 1, the criticality of the link increases dramatically, forcing the routing plan to find an alternative path.

B. Path Criticality Index (PCI)

Now we are ready to move to the next step and define the path criticality index (PCI). For every path between a source-destination pair (s,d) consisting of the links $(l_1, l_2, ..., l_n)$, the path criticality index is defined as the average of the link criticality index of l_1 to l_n. In other words:

$$PCI(s,d) = \frac{s(P)}{|P|} = \frac{\sum_{i=1}^{n} LCI(l_n)}{n} \quad \text{Where } |P| = n \text{ is the order of the path} \qquad (9)$$

In general PCI is a function LCIs of the path. Finding the best form of this function is one of our ongoing research topics, but the average function (9) which is introduced here works well for all of the benchmark networks and experimental topologies that we have examined.

C. Path Criticality Routing Algorithm (PCR)

The basic idea of our routing algorithm is to accommodate new requests for connections along paths that have a low PCI. This requires that we find the link criticality indices. To do this, we need to obtain all possible paths for each source-destination pair. This is not feasible since the number of paths grows rapidly with the number of network nodes and links. Although the shortest path is not necessarily the path with the lowest PCI, one can expect that the path or paths with lowest PCI are among the *k-shortest paths* of the network. Hence we use the k-shortest path method proposed by Eppstein [11] with a modification to avoid loops.

Our algorithm begins with a predefined value of k (default is 1), but the value may be increased during the course of running the routing algorithm if the desired number of paths to route the traffic cannot not be found. We use thresholds tr_1 (the default value is infinity) and tr_2 (the default value is zero) for PCI. The first threshold defines the lower confidence boundary for the path criticality index. All the paths with path criticality index less than tr_1 are considered eligible to route traffic. On the other hand all the paths with the criticality index larger than tr_2 are considered too risky and may be identified to the (offline) core network management system for increased capacity assignment. The paths with criticality index in between the thresholds will share traffic based on their criticality index as long as they remain within the boundaries. We note that when a path accepts traffic, the residual capacity of its links will decrease for the duration of the traffic flow. This means that the criticality index of this path must be increased. In other words a constant monitoring (not real-time necessarily but in reasonable time slots) of the PCI for all the paths is necessary.

PCR:

Input: A network or more formally a graph $G(V, E)$, a set of capacities (residual capacities if we are not in the initial stage), a set of source-destination pairs (s, d) and traffic matrix Γ for these source-destination pairs.

Output: A set of routes (LSPs in case of MPLS) between all source-destination pairs meeting the demand requirements according to the traffic matrix.

Algorithm: (By default we are in idle state)

1. *Go to the initial state, Select k, tr_1 and tr_2 (PCR uses default values of $tr1$, $tr2$, and k not concerned about the thresholds).*

2. *Compute the k-shortest paths for all source-destination pairs to meet the demand requirements and measure their PCI.*

3. *If a path with $PCI \leq tr_1$ exists, choose that path to route the demand (in case there are more than one path meeting the threshold requirements then choose the one with lowest PCI).*

> *i.Adjust residual capacities*
> *ii.Adjust path criticality indexes accordingly*

   If there were still more than one path, choose one in round robin fashion.

4. *If there are paths with $PCI \geq tr_2$ send a message to the core management system requesting additional bandwidth for these paths.*

5. *If there is no path satisfying the condition of step 3 then increase k by one and go to step 2.*

6. *In case that no path with criticality index less than tr_1 can be found (this happens when k keeps increasing without any satisfactory path results) then use the path with minimum PCI so that $tr_1 \leq PCI \leq tr_2$. If still there were more than one path, choose one in round robin fashion.*

The most time-consuming part of the algorithm is the k-shortest path calculation. According to [11] the complexity of the proposed k-shortest path algorithm is $O(m + n \log n + k)$ where, m is the number of links and n is the number of the nodes. The algorithm is polynomial time and as a result PCR is also a polynomial time algorithm if we set a maximum value for k such as k_{max}. The complexity of the other parts of the algorithm (without k-shortest path) is $O(m^2)$. The problem in this case is that we might not get the desired path from the algorithm by k_{max} iteration. On the other hand if we do not place any upper bound for k then we can find a desired path (if one exists) but not necessarily in polynomial time. This comes from the fact that the problem we are trying to solve by nature is an NP-Hard one [12].

In MIRA the time complexity without considering the max-flow algorithm is also $O(m^2)$. But if we compare the time complexity of the Tarjan max-flow method [12] that is being used in MIRA ($O(n \times m \times \log(n))$ where n and m are node and link dimensions) with the Eppstein k-shortest path method [11] used in PCR algorithm

$(O(m+n \times \log n+k))$ we notice that the complexity of the k-shortest path algorithm is less than max-flow one.

5 Multiple QoS Constraints Case

We briefly discuss the situation in which the QoS constraints are included in the path routing (LSP routing in case of MPLS) problem. We can assume that the QoS constraints are additive without loss of generality [14].

    To describe the multi-constrained routing we consider a graph $G(V,E)$ and assume each link $l = (u,v)$ is characterized by link weight vector $\vec{W}(l) = [w_1(l), w_2(l), ...w_r(l)]^T$ where the component $w_i \succ 0$ is an additive QoS measure such as delay. There is also a vector of r constraints $\vec{R} = [R_1, R_2, ..., R_r]^T$ determining the upper bounds on QoS measures (multi-constrained problem or MCP [14]). For any path P we have:

$$w_i(P) \equiv \sum_j w_i(l_j) \quad \forall \quad 1 \le i \le r \text{ where } l_j \text{s are constituent links of } P.$$

    A path is feasible if it satisfies all the constraints. To quantify these constraints we need to use a path length or "norm" in its mathematical sense. In other words any function $L(.)$ satisfying:

- $L(\vec{p}) > 0$ for all non zero vectors and $L(\vec{p}) = 0$ *iff* $\vec{p} = 0$.

- For all $\vec{p}$ and $\vec{q}$ $L(\vec{p}+\vec{q}) \le L(\vec{p}) + L(\vec{q})$.

We are using Holder's q-vector norm [14]:

$$L_q(P) = (\sum_{i=1}^{r} [\frac{w_i(P)}{R_i}]^q)^{\frac{1}{q}} \quad \text{for } q \to \infty \text{ we have: } L_\infty = \max_{1 \le i \le r} (\frac{w_i(P)}{R_i}) \tag{10}$$

    Now we can use this norm function to quantify the QoS measure on paths and our formula for PCI of a path will be multiplied by length (10):

$$PCI_q(P) = \sum_i LCI(l_i)/n \times L_\infty(P) \tag{11}$$

6 Proof of Concept

The PCR algorithm has been implemented with C++ and tested for many network configurations. Among these we have chosen one benchmark topology (parking lot in fig. 1) as well as a realistic networks (fig.2., a part of US network map [1] as well as Alilene [16]) to show the effectiveness of PCR algorithm to address problem of finding the best routing plan for networks that have been found difficult to be handled by previous proposals.

A. Parking-Lot Topology

The parking-lot network topology, shown in Fig. 1, is an interesting example. If one unit of bandwidth is requested to be sent from S_0 to D_0, all the previous routing approaches such as SP, WSP and MIRA will choose the straight path and run the flow resulting in the blocking of demands of one unit coming from any other source such as S_i to the destination D_i [3]. A wiser decision is to block the first request from S_0 to D_0 so the network will be able to route the other n source-destination pair requests.

To investigate the behavior of our *PCR* algorithm we suppose n = 10 and different combinations of source-destination pairs are possible. Our experiment results show that the criticality of the path $S0 \rightarrow D0$ is much higher than the other combinations. In Fig. 1 the results of our experiment with n=10 is reflected and clearly shows thate $S0 \rightarrow D0$ is the most critical path.

In general case of the parking-lot topology with n nodes the same approach can be followed and again the proposed routing plan will choose the straight path $S0 \rightarrow D0$ as the most critical one.

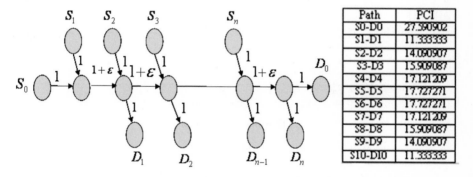

Path	PCI
S0-D0	27.590902
S1-D1	11.333333
S2-D2	14.090907
S3-D3	15.909087
S4-D4	17.121209
S5-D5	17.727271
S6-D6	17.727271
S7-D7	17.121209
S8-D8	15.909087
S9-D9	14.090907
S10-D10	11.333333

Fig. 1. Parking Lot Network (n=10)

B. Simulation results for KL-Topology and Abilene

We ran a set of simulations on the network of Ref. [1] that we refer to as the KL-topology (Fig. 2(a). We assume the bandwidth of the thin links is 1200 units and that the thick links have 4800 units of bandwidth. In order to compare the results with [1] we implemented exactly the same simulations.

In the first experiment the requests for bandwidth (which is our main QoS measure in these simulations) arrive with Poisson distribution and stay for ever (no departures). In our tests the bandwidth requests for paths are taken to be uniformly distributed between 1 to 3 units. In Fig. 2(b) we show the number of rejected calls for the KL-topology and we compare the performance to that of shortest path, widest shortest path and PCR (with initial value $k = 1$ and possible subsequent increments based on PCR). We measured the number of blocked requests for from $S1$ to $D1$. As one can see after about 1200 trials the SP algorithm starts to experience blocking while the PCI-based algorithm can adapt itself and still accept bandwidth requests

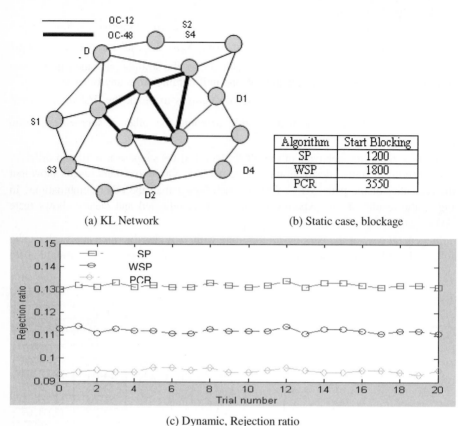

Algorithm	Start Blocking
SP	1200
WSP	1800
PCR	3550

(a) KL Network   (b) Static case, blockage

(c) Dynamic, Rejection ratio

Fig. 2.

without significant blockage. We observed that the PCR begins increasing the value k when the knee in the curve is reached.

In another experiment we examined the behavior of the algorithms in the presence of dynamic traffic. Fig. 2(c) shows the proportion of the path requests rejected in 20 experiments for the following scenario. Path requests arrive between each source-destination point according to a Poisson process with an average rate λ, and the holding times are exponentially distributed with mean $1/\mu$.

We assume $\lambda/\mu = 150$ in our experiments. In this scenario, we scale down the bandwidth of each link in the KL-network with the ratio of 10 to have bandwidths of 120 and 480 units for thin and thick links respectively. Next we generate about 1,000,000 requests and measure the rejection ratio for each one of the algorithms. The results are shown in Fig. 2(c) and again the results are very close to MIRA.

In last experiment we compare the response of PCR algorithm with the optimal solution which is obtained by solving the Linear Programming (LP) equations (1), (2), (3) using the method described in [6]. We conducted our algorithm on Abeline

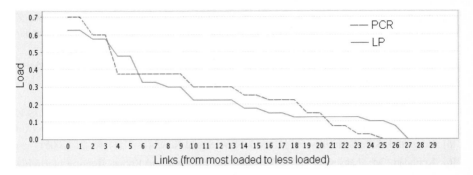

Fig. 3. Abilene network load distribution

network [16] (11 nodes, 28 links) using the traffic matrices obtained from [17] and observed the load distribution on different links of the Abilene network using LP and PCR methods. Figure 3 shows the result. The deviation from the optimal response using PCR was always less than 8%.

7   Concluding Remarks

In this paper we have proposed a new approach for path setup and routing of flows in core networks. The most important problem with the existing approaches is that each one of them solves a part of the overall problem but fails with other parts. We have tried to consider different aspects of the network (i.e. topology, capacity, and demand) and quantified these aspects using measures inspired by the mathematics of graphs. The essence of our work is based on determining a path criticality index for each path showing how critical that path is to the changes in the topology and traffic demand of a network. Our algorithm identifies the least critical paths for allocation of new traffic flow requests. The results from applying the proposed algorithm to networks that are difficult to handle by existing approaches are very encouraging.   These results confirm the validity of the notion of path criticality. The simulation results show that PCR matches the performance of MIRA in typical networks. We also showed that the complexity of PCR relative to MIRA shows improvement.

However there are many issues that remain to be investigated in the new approach. We need to investigate more on the effect of the threshold parameters. The PCI is a function of link criticality indexes and in our first algorithm we used "average function" to obtain PCI but more elaboration is necessary.   As another research challenge we need to look into the back up paths and the efficient algorithms to find them again with the goal of having less critical paths and back up paths.

References

[1] Kar, K., Kodialam, M., Lakshman, T.V.: Minimum Interference Routing of Bandwidth Guaranteed Tunnels with MPLS Traffic Engineering Applications. IEEE Journal on Selected Areas in Communications 18(12), 2566–2579 (2000)

[2] Awduche, D.O., Berger, L., Gan, D., Li, T., Swallow, G., Srinivasan, V.: Extensions to RSVP for LSP tunnels. Internet Draft draf-ietf-mpls-rsvp-lsp-tunnel-04.txt (September 1999)

[3] Suri, S., Waldvogel, M., Warkhede, P.R.: Profile-Based Routing: A New Framework for MPLS Traffic Engineering. In: Quality of Future Internet Services. LNCS, vol. 2156, Springer Verlag, Heidelberg (2001)

[4] Freeman, L.C.: Centrality in Networks: I. Conceptual Clarification. Social Networks 1, 215–239 (1978/79)

[5] Borgatti, L.S.P.: Centrality and Network Flow. Social Networks 27(1), 55–71 (2005)

[6] Kodialam, M., Lakshman, T.V., Sengupta, S.: Efficient and Robust Routing of Highly Variable Traffic, HotNets-III (November 2004)

[7] Guerin, R., Orda, A., Williams, D.: QoS routing mechanisms and OSPF extensions. In: Proceedings of 2nd Global Internet Miniconference (November 1997)

[8] Yilmaz, S., Matta, I.: On the Scalability-Performance Tradeoffs in MPLS and IP Routing. In: Proceedings of SPIE ITCOM (May (2002)

[9] Ott, T., Bogovic, T., Carpenter, T., Krishnan, K.R., Shallcross, D.: Algorithms for Flow Allocation for Multi Protocol Label Switching. MPLS International Conference (October 2000)

[10] Dekker, H., Colbert, B.D.: Network Robustness and Graph Topology. In: Proceedings of 27th Australasian Computer Science Conference, Vol. 26, pp. 359-368 (January 2004)

[11] Eppstein, D.: Finding the k Shortest Paths. Society for Industrial and Applied Mathematics (SIAM) Journal of Computing 28(2), 652–673 (1998)

[12] Ahuja, R.K., Magnanti, T.L., Orlin, J.B.: Network Flows: Theory, Algorithms, and Applications. Prentice Hall, Englewood Cliffs (1999)

[13] Duffield, N.G., Goyal, P., Greenberg, A.G., Mishra, P.P., Ramakrishnan, K.K., Van der Merwe, J.E.: A flexible model for resource management in virtual private network, ACM SIGCOMM 1999 (August 1999)

[14] Van Mieghem, P., Kuipers, F.A.: Concepts of Exact QoS Routing Algorithms. IEEE/ACM Transactions on Networking 12(5) (October 2004)

[15] Zhang-Shen, R., McKeown, N.: designing a Predictable Internet Backbone Networks. HotNet-III (November 2004)

[16] http://abilene.internet2.edu/

[17] http://totem.info.ucl.ac.be/

Stable and Robust Multipath Oblivious Routing for Traffic Engineering*

Yuxi Li, Baochun Bai, Janelle Harms, and Robert Holte

Department of Computing Science, University of Alberta

Abstract. Intra-domain traffic engineering is essential for the operation of an Internet Service Provider. Demand-oblivious routing [2] promises excellent performance guarantee with changing and uncertain traffic demands. However, it is difficult to implement it. We investigate an efficient and deployable implementation of oblivious routing. We study its performance by both numerical experiments and simulation. The performance study shows that the multipath implementation achieves a close approximation to oblivious routing [2], especially when approximate knowledge of traffic is available. The study also shows its robustness under varying traffic demands, link failures and an adversary attack. Its performance is excellent even with a 100% error in traffic estimation.

1 Introduction

Intra-domain traffic engineering is essential for the operation of an Internet Service Provider (ISP). It is desirable to design a routing protocol that can balance network utilization, mitigate the impact of failures and attacks, and thus provide good quality of service to network users, with economic provisioning of network resources. However, it is challenging to design such a routing protocol due to traffic changes and uncertainty. Network traffic is inherently changing and uncertain, due to factors such as the diurnal pattern, dynamic inter-domain routing, link failures, and attacks. Adaptive traffic resulting from overlay routing or multihoming further aggravates the problems.

There are three classes of solutions: link weight optimization [4, 13], traffic-adaptive approaches [3, 5, 10] and demand-oblivious routing [1, 2, 12]. The approach of link weight optimization guarantees performance only for a limited set of traffic demands. An adaptive approach is responsive to traffic changes, so that the issues of stability and convergence have to be addressed both in theory and in practice. Demand-oblivious routing is particularly promising; it promises excellent performance guarantee with changing and uncertain traffic demands. Its performance is particularly good with approximate knowledge of traffic demands, which is made available by the recent great progress in traffic

---

* Y. Li received honorary Izaak Walton Killam memorial scholarship and Informatics Circle of Research Excellence (iCore) graduate student scholarship. B. Bai received Province of Alberta Graduate Fellowship. This research was partially supported by the Natural Science and Engineering Research Council of Canada (NSERC).

L. Mason, T. Drwiega, and J. Yan (Eds.): ITC 2007, LNCS 4516, pp. 129–140, 2007.

estimation, e.g. [14]. In [12], the performance is optimized for expected scenarios and is guaranteed for unexpected scenarios.

However, it is difficult to implement oblivious routing in [2]. A straightforward implementation is for each node to forward incoming packets according to the routing fractions computed by [2]. However, without careful attention, such a distributed implementation may lead to loops. Furthermore, an oblivious routing may involve a large number of paths between each origin-destination (OD) pair, which requires a large number of labels in an MPLS deployment. It is thus desirable to route traffic on a small number of paths. However, since there are many paths between each OD pair, it may be difficult to select a small set of paths that gives good performance.

We investigate an efficient and deployable implementation of oblivious routing. We design MORE, Multipath Oblivious Routing for traffic Engineering, to obtain a close approximation to [2]. MORE achieves a very excellent performance guarantee when combined with approximate knowledge of traffic demands. However, it does not need frequent collection of network information like an adaptive approach. An oblivious routing guarantees the performance for much broader traffic variability. Oblivious routing optimizes a worst case performance metric. However, our empirical study will show that MORE achieves a performance close to the optimal. Its performance is excellent even with a 100% error in traffic estimation. In addition, as a quasi-static solution, MORE can be static on an hourly, multi-hourly or even daily basis. Thus, MORE is much less concerned with stability and convergence issues than an adaptive approach, which is responsive on a small time-scale, like seconds. MORE does not need changes to core routers, thus it can be efficiently implemented and gradually deployed.

We are the first to investigate a feasible implementation of demand-oblivious routing [2]. We design MORE, a multipath approximation to [2]. Through extensive numerical experiments and simulation, we show the excellent performance of MORE under varying traffic matrices, link failures and an adversary attack. Our work is complementary to [1, 2] and [12]. MORE is a promising option for traffic engineering, along with link weight optimization [4, 13] and adaptive approaches, like MATE [3], TeXCP [5] and [10].

2   Preliminaries

A traffic matrix (TM) specifies the amount of traffic between each OD pair over a certain time interval. An entry d_{ij} denotes the amount of traffic for OD pair $i \rightarrow j$. The capacity of edge e is denoted as $c(e)$.

Routing. A routing specifies how to route the traffic between each OD pair across a given network. OSPF and IS-IS, two popular Internet routing protocols, follow a destination-based evenly-split approach. The MPLS architecture allows for more flexible routing. Both OSPF/IS-IS and MPLS can take advantage of path diversity. OSPF/IS-IS distributes traffic evenly on multiple paths with equal cost. MPLS may support arbitrary routing fractions over multiple paths. Our work is applicable to MPLS, which is widely deployed by ISPs.

An *arc-routing* $f_{ij}(e)$ specifies the fraction of traffic demand d_{ij} on edge e [2]. An arc-routing is not readily implementable for either OSPF or MPLS.

Link Utilization. For a given arc-routing $\mathbf{f}$ and a given traffic demand $\mathbf{tm}$, the maximum link utilization (MLU) measures the goodness of the routing, i.e., the lower the maximum link utilization, the better the routing:

$$\mathrm{MLU}_{\mathrm{arc}}(\mathbf{tm}, \mathbf{f}) = \max_{e \in E} \sum_{i,j} d_{ij} f_{ij}(e)/c(e) \qquad (1)$$

Given a TM $\mathbf{tm}$, an *optimal arc-routing* minimizes the maximum link utilization:

$$\mathrm{OPTU}_{\mathrm{arc}}(\mathbf{tm}) = \min_f \max_{e \in E} \textstyle\sum_{i,j} d_{ij} f_{ij}(e)/c(e) \qquad (2)$$

Performance Ratio. The routing computed by (2) does not guarantee performance for other traffic matrices. Applegate and Cohen [2] developed LP models to compute an optimal routing that minimizes the oblivious ratio with a weak assumption on the traffic demand. We present the metric of performance ratio.

For a given routing $\mathbf{f}$ and a given traffic matrix $\mathbf{tm}$, the *performance ratio* is defined as the ratio of the maximum link utilization of the routing $\mathbf{f}$ on the traffic matrix $\mathbf{tm}$ to the maximum link utilization of the optimal routing for $\mathbf{tm}$. The performance ratio measures how far routing $\mathbf{f}$ is from the optimal routing for traffic matrix $\mathbf{tm}$.

$$\mathrm{PERF}(\mathbf{f}, \{\mathbf{tm}\}) = \frac{\mathrm{MLU}(\mathbf{tm}, \mathbf{f})}{\mathrm{OPTU}_{\mathrm{arc}}(\mathbf{tm})} \qquad (3)$$

This applies to both an arc- and a path-routing, thus we do not add a subscript to MLU. The performance ratio is usually greater than 1. It is equal to 1 only when the routing $\mathbf{f}$ is an optimal routing for $\mathbf{tm}$.

When we are considering a set of traffic matrices $\mathbf{TM}$, the performance ratio of a routing $\mathbf{f}$ is defined as

$$\mathrm{PERF}(\mathbf{f}, \mathbf{TM}) = \max_{\mathbf{tm} \in \mathbf{TM}} \mathrm{PERF}(\mathbf{f}, \{\mathbf{tm}\}) \qquad (4)$$

The performance ratio with respect to a set of traffic matrices is usually strictly greater than 1, since a single routing usually can not optimize link utilization over the set of traffic matrices.

When the set $\mathbf{TM}$ includes all possible traffic matrices, $\mathrm{PERF}(\mathbf{f}, \mathbf{TM})$ is referred to as the *oblivious performance ratio* of the routing $\mathbf{f}$. This is the worst performance ratio the routing $\mathbf{f}$ achieves with respect to all traffic matrices. An *optimal oblivious routing* is the routing that minimizes the oblivious performance ratio. Its oblivious ratio is the *optimal oblivious ratio* of the network.

3   Multipath Oblivious Routing for Traffic Engineering

As discussed in the Introduction, there are obstacles to the implementation of oblivious routing in [2], such as potential routing loops and a large number of

MPLS labels. We investigate a deployable oblivious routing, MORE, Multipath Oblivious Routing for traffic Engineering.

We use a quasi-static routing, so that the fractions of traffic on the multiple paths between an OD pair do not change over a large time period, in contrast to an adaptive routing. As well, MORE alleviates the reliance on global network information: it can achieve excellent performance with a large time-scale traffic estimation, but it does not need to collect the instantaneous link load. The oblivious ratio can be computed by the reformulation of the oblivious routing on K paths in LP (12), which gives the worst case performance guarantee.

3.1 Multipath Routing

Each OD pair $i \rightarrow j$ is configured with up to K_{ij} paths. For notational brevity, we use K paths for each OD pair. The set of paths for OD pair $i \rightarrow j$ is denoted as $P_{ij} = \{P_{ij}^1, ..., P_{ij}^K\}$. A multipath routing computes, for each OD pair $i \rightarrow j$, a routing fraction vector, defined as $< f_{ij}^1, ..., f_{ij}^K >, \sum_k f_{ij}^k = 1, f_{ij}^k \geq 0$ on the set of paths for OD pair $i \rightarrow j$. A *path-routing* f_{ij}^k specifies the fraction of traffic demand d_{ij} on path P_{ij}^k. A path-routing is readily implementable for MPLS. Given path-routing $\mathbf{f}$ and traffic demand $\mathbf{tm}$, the maximum link utilization is:

$$\text{MLU}_{\text{path}}(\mathbf{tm}, \mathbf{f}) = \max_{l \in E} \sum_{ij} d_{ij} \sum_k \delta_{ij}^k(l) f_{ij}^k / c(l) \tag{5}$$

Here $\delta_{ij}^k(l)$ is an indicator function, which is 1 if $l \in P_{ij}^k$, 0 otherwise. We use $l \in P_{ij}^k$ to denote edge l is on path P_{ij}^k. Given $\mathbf{tm}$, an *optimal path-routing* that minimizes the maximum link utilization is:

$$\text{OPTU}_{\text{path}}(\mathbf{tm}) = \min_{\mathbf{f}} \max_{l \in E} \sum_{ij} d_{ij} \sum_k \delta_{ij}^k(l) f_{ij}^k / c(l) \tag{6}$$

3.2 LP Formulation

We give LP models for multipath oblivious routing. We replace the arc formulation in Applegate and Cohen [2] with a path formulation to compute an optimal oblivious routing and its ratio. In an arc formulation, routing variables are on links and flow conservation constraints are at each node for each OD pair. In a path formulation, routing variables are on paths and flow conservation constraints are implicitly satisfied on each path. We start with the case in which there is approximate knowledge of traffic demand.

Similar to Applegate and Cohen [2], the optimal oblivious routing can be obtained by solving an LP with a polynomial number of variables, but infinitely many constraints. With the approximate knowledge that d_{ij} is in the range of $[a_{ij}, b_{ij}]$, we have the "master LP":

$$\begin{aligned}
&\min_{r,f,d} r \\
&\mathbf{f} \text{ is a path-routing} \\
&\forall \text{ edges } l, \forall \alpha > 0 : \forall \text{ TMs } \mathbf{tm} \text{ with OPTU}_{\text{arc}}(\mathbf{tm}) = \alpha, a_{ij} \leq d_{ij} \leq b_{ij} : \\
&\quad \sum_{ij} d_{ij} \sum_k \delta_{ij}^k(l) f_{ij}^k / c(l) \leq \alpha r
\end{aligned} \tag{7}$$

The oblivious ratio is invariant with the scaling of TMs or the scaling of the edge capacity. Thus, when computing the oblivious ratio, it is sufficient to consider TMs with $\text{OPTU}_{\text{arc}}(\mathbf{tm}) = 1$. Another benefit of using TMs with $\text{OPTU}_{\text{arc}}(\mathbf{tm}) = 1$ is that the objective of the LP, the oblivious ratio r, is equal to the maximum link utilization of the oblivious routing.

Since the oblivious ratio r is invariant with respect to the scaling of TMs, we can consider a scaled TM $\mathbf{tm}' = \lambda \cdot \mathbf{tm}$. With $\lambda = 1/\text{OPTU}_{\text{arc}}(\mathbf{tm})$, we have $\text{OPTU}_{\text{arc}}(\mathbf{tm}') = 1$. Under these conditions, the master LP (7) becomes:

$$
\begin{aligned}
&\min_{r,f,d} r \\
&\mathbf{f} \text{ is a path-routing} \\
&\forall \text{ edges } l : \forall \text{ TMs } \mathbf{tm} \text{ with } \text{OPTU}_{\text{arc}}(\mathbf{tm}) = 1, \lambda > 0, \lambda a_{ij} \leq d_{ij} \leq \lambda b_{ij} : \\
&\quad \sum_{ij} d_{ij} \sum_k \delta_{ij}^k(l) f_{ij}^k / c(l) \leq r
\end{aligned}
\tag{8}
$$

For the condition "$\forall$ TMs $\mathbf{tm}$ with $\text{OPTU}_{\text{arc}}(\mathbf{tm}) = 1$", we need the flow definition on edges. Flow $\mathbf{g}$ is defined as,

$$
\begin{cases}
\forall \text{ pairs } i \to j, k \neq i, j : \sum_{e \in out(k)} g_{ij}(e) - \sum_{e \in in(k)} g_{ij}(e) = 0 \\
\forall \text{ pairs } i \to j : \sum_{e \in out(j)} g_{ij}(e) - \sum_{e \in in(j)} g_{ij}(e) + d_{ij} = 0 \\
\forall \text{ pairs } i \to j, \forall \text{ edges } e : g_{ij}(e) \geq 0, d_{ij} \geq 0
\end{cases}
\tag{9}
$$

LP formulations can be simplified by collapsing flows g_{ij} on an edge e with the same origin by $g_i(e) = \sum_j g_{ij}(e)$.

Given a path-routing $\mathbf{f}$, the constraint of the master LP (8) can be checked by solving the following "slave LP" for each edge l to examine whether the objective is $\leq r$ or not. In (10), routing f_{ij}^k are constant and flow $g_{ij}(e)$, demand d_{ij} and λ are variables.

$$
\begin{aligned}
&\max_{g,d,\lambda} \sum_{ij} d_{ij} \sum_k \delta_{ij}^k(l) f_{ij}^k / c(l) \\
&\forall \text{ pairs } i \to j : \sum_{e \in out(j)} g_i(e) - \sum_{e \in in(j)} g_i(e) + d_{ij} \leq 0 && \Leftarrow w_l(i,j) \\
&\forall \text{ edges } e : \sum_i g_i(e) \leq c(e) && \Leftarrow \pi_l(e) \\
&\forall \text{ pairs } i \to j : d_{ij} - \lambda b_{ij} \leq 0 && \Leftarrow \kappa_l^+(i,j) \\
&\forall \text{ pairs } i \to j : -d_{ij} + \lambda a_{ij} \leq 0 && \Leftarrow \kappa_l^-(i,j) \\
&\forall \text{ pairs } i \to j : d_{ij} \geq 0, g_{ij}^k \geq 0, \lambda > 0
\end{aligned}
\tag{10}
$$

The flow conservation constraint is relaxed from equality to ≤ 0, which allows for OD pair $i \to j$ to deliver more flow than demanded, and does not affect the maximum link utilization of 1. The constraints of LP (10) guarantee the traffic can be routed with maximum link utilization of 1.

The dual of LP (10) is LP (11). To help make the derivation of the dual LP (11) clearer, we use leftarrow $\Leftarrow$ to indicate dual variables corresponding with primal constraints in LP (10). In dual LP (11), we indicate primal variables corresponding to dual constraints.

$$\min_{\pi, w, \kappa^+, \kappa^-} \sum_e c(e)\pi_l(e)$$

$\forall$ pairs $i \to j : w_l(i,j) + \kappa_l^+(i,j) - \kappa_l^-(i,j) \geq \sum_k \delta_{ij}^k(l) f_{ij}^k/c(l)$ $\Leftarrow d_{ij}$

$\forall$ nodes $i, \forall$ edges $(u,v) : \pi_l(u,v) + w_l(i,u) - w_l(i,v) \geq 0$ $\Leftarrow g_i(u,v)$

$\sum_{i,j}\{a_{ij}\kappa_l^-(i,j) - b_{ij}\kappa_l^+(i,j)\} \geq 0$ $\Leftarrow \lambda$

$\forall$ edges $e : \pi_l(e) \geq 0$

$\forall$ pairs $i \to j : w_l(i,j) \geq 0, \kappa_l^+(i,j) \geq 0, \kappa_l^-(i,j) \geq 0$

$\forall$ nodes $i : w_l(i,i) = 0, \kappa_l^+(i,i) = 0, \kappa_l^-(i,i) = 0$

$$(11)$$

According to the LP duality theory, the primal LP and its dual LP have the same optimal value if they exist. That is, LP (10) and LP (11) are equivalent. Because LP (11) is a minimization problem, we can use its objective in place of the objective of LP (10) in the "$\leq r$" constraints of LP (8). Replacing the constraint in the master LP (8) with LP (11), we obtain a single LP to compute the oblivious performance ratio using K paths.

$$\min_{r, f, \pi, w, \kappa^+, \kappa^-} r$$

$\mathbf{f}$ is a path-routing

$\forall$ edges $l :$

  $\sum_e c(e)\pi_l(e) \leq r$

  $\forall$ pairs $i \to j : w_l(i,j) + \kappa_l^+(i,j) - \kappa_l^-(i,j) \geq \sum_k \delta_{ij}^k(l) f_{ij}^k/c(l)$ (12)

  $\forall$ nodes $i, \forall$ edges $(u,v) : \pi_l(u,v) + w_l(i,u) - w_l(i,v) \geq 0$

  $\sum_{i,j}\{a_{ij}\kappa_l^-(i,j) - b_{ij}\kappa_l^+(i,j)\} \geq 0$

  $\forall$ edges $e : \pi_l(e) \geq 0$

  $\forall$ pairs $i \to j : w_l(i,j) \geq 0, \kappa_l^+(i,j) \geq 0, \kappa_l^-(i,j) \geq 0$

  $\forall$ nodes $i : w_l(i,i) = 0, \kappa_l^+(i,i) = 0, \kappa_l^-(i,i) = 0$

When there is no knowledge of the traffic demand, i.e., the range $[a_{ij}, b_{ij}]$ for d_{ij} becomes $[0, \infty)$, the LP to compute the oblivious routing is obtained by removing the variables $\kappa_l^+(i,j)$ and $\kappa_l^-(i,j)$.

3.3   MultiPath Selection

We discuss three approaches, sp$K$, mix$K$ and focus$K$, for multiple paths selection for each OD pair, to achieve a low oblivious ratio.

In sp$K$, we select K shortest paths w.r.t. hop count for each OD pair.

In mix$K$, we first find K shortest paths with respect to hop count, as in sp$K$. These shortest paths serve as base paths. Then, we sort the K paths in increasing order of their hop counts. After that, for each shortest path, we search for its edge-disjoint paths and record them, until K paths are found. Long paths are not preferred, so that we only search for disjoint paths that are not M hop longer than the base paths ($M = 3$). We use the name "mix$K$" to reflect that it is a mixture of shortest paths and their disjoint paths. We find K shortest paths first, in case none of them has an eligible disjoint path. In this case, the K shortest paths are chosen as the mix$K$ paths.

The method focus$K$ is based on our previous work [7], where we design a method to implicitly reduce the the number of paths and path lengths, with only negligible increase of the oblivious ratio. The basic idea is to put a penalty on using an edge far away from the shortest path for an OD pair. Thus, this method essentially focuses on short paths for each OD pair. We make an extension to [7] by considering range restrictions on traffic demands.

After computing the modified oblivious routing using the extended LP to [7], we extract K paths. In the performance study, we extract up to 20 shortest paths from the resultant oblivious routing with routing fractions ≥ 0.001.

4   Performance Study

We evaluate the performance of MORE by numerical experiments and simulation. We use the oblivious ratio of a routing and the maximum link utilization (MLU) a routing incurs as performance metrics. We solve LPs with CPLEX.[1]

Topology. ISP topologies and traffic demands are regarded as proprietary information. The Rocketfuel project [11] deployed new techniques to measure ISP topologies and made them publicly available. The OSPF weights on the links are also provided. The capacities of links are assigned according to the CISCO heuristics as in [2], referred to as InvCap, i.e., the link weight is inversely proportional to the link capacity. POP 12 is the tier-1 ISP topology in Nucci et al. [8], with the scaled link capacity provided in [8]. We also use random topologies generated by GT-ITM.[2]

Gravity TM. Similar to [2, 5], we use the Gravity model [14] to determine the estimated traffic matrices. The Gravity model is developed in [14] as a fast and accurate estimation of traffic matrices, in which, the traffic demand between an OD pair is proportional to the product of the traffic flowing into/out of the origin/the destination. We use a heuristic approach similar to that in [2], in which the volume of traffic flowing into/out of a POP is proportional to the combined capacity of links connecting with the POP. Then we extrapolate a complete Gravity TM.

Lognormal TM. We also use the log-normal model in Nucci et al. [8] to generate synthetic TMs. In the first step, we generate traffic entries using a log-normal distribution. Then these entries are associated with OD pairs according to a heuristic approach similar to that recommended in [8]. That is, OD pairs are ordered by the first metric of their fan-out capacities. The fan-out capacity of a node is the sum of the capacities of links incident with it. The fan-out capacity of an OD pair is the minimum of the fan-out capacities of the two nodes. Ties are broken by the second metric of connectivity, defined as the number of links incident to a node. Similarly, the minimum is taken for the two nodes.

---

[1] Mathematical programming solver. http://www.cplex.com
[2] http://www.cc.gatech.edu/projects/gtitm/

Similar to [2], in the experiments, when approximate knowledge is available, we consider a base TM, with the entry d_{ij} for OD pair $i \rightarrow j$, and an error margin $w > 1$, so that the traffic for $i \rightarrow j$ is in the range of $[d_{ij}/w, w * d_{ij}]$.

4.1   MultiPath Selection

First, we study the performance of the path selection methods, namely, sp$K$, mix$K$ and focus$K$. The benchmark is the method in Applegate and Cohen [2], which can achieve the lowest oblivious ratio for a given topology. Hereafter, we refer to the method in Applegate and Cohen [2] as **AC**. Recall that it is non-trivial to implement the routing computed by AC. Thus a close multipath approximation to AC is desirable.

In Figure 1, we show the performance of the various path selection methods, when approximate knowledge of the TM is available, with a Gravity base TM and $w = 2.0$. For AS 1755, all path selection methods have good performance when the error margin is small, with sp20 jumping up when error margin increases and mix20 maintaining the best performance. For AS3967 and AS6461, focus20 has overall good performance. For POP12 (see [6]), sp$K$ and mix$K$, for $K = 10, 20$, have similar results, with performance very close to AC.

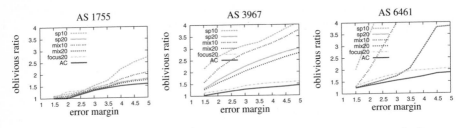

Fig. 1. Oblivious ratio vs. error margin for various path selection methods

Experiments also show the multipath selection methods outperform the link weight optimization [4] and InvCap on ISP and random topologies. See [6] for details.[3] In later studies, we use path selection methods as follows: mix20 for AS 1755, focus20 for AS 3967 and AS 6461, and mix10 for POP 12. When there is approximate knowledge of traffic demands, we use error margin $w = 2.0$, which can be interpreted as a tolerance of 100% error in traffic estimation.

4.2   Simulation

We analyze the performance of LP models for MORE in previous sections. In this section, we study the performance of MORE using packet-level simulation with NS2[4]. We implement the robust weighted hashing by Ross [9], so that traffic can be split into multiple paths according to the routing fraction of each path.

---

[3] See [6] also for numerical experiments on link failures and adversary attacks.
[4] http://www.isi.edu/nsnam/ns/

We use either the Gravity or the Lognormal model to generate synthetic TMs. Then, with the synthetic TMs, we generate Pareto traffic to obtain variability in the actual traffic. Note that although a TM may not change, traffic varies due to the Pareto distribution. For every 0.5 second, we average the link utilization and take the maximum to obtain the maximum link utilization (MLU).

Robust under varying TMs and routings. MORE is a quasi-static solution, it may have to change the routing when necessary. We attempt to study the robustness of MORE over changing TMs and routings by simulation. We generate 10 Lognormal TMs [8]. Each TM lasts 10 seconds. MORE computes an optimal multipath oblivious routing for a given TM with error margin $w = 2.0$. Thus there are potentially different routings for different TMs. AdaptiveK computes an optimal routing with K-shortest paths for each TM, with $K = 20$. We assume both MORE and adaptiveK know a new TM and reoptimize the routing for it instantaneously. AdaptiveK represents an adaptive scheme on K-shortest paths that can respond to traffic changes without any delay, i.e., it is an unachievable best case for adaptive schemes.

Results are shown in Figure 2.[5] We scale the TMs, so that optimal arc-routings of these TMs have the same MLU. The results show that MORE incurs similar MLUs over varying TMs and routings. We also observe that MORE achieves similar performance to adaptiveK. MORE also has similar performance for AS 6461 and POP 12, see [6].

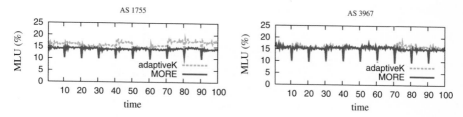

Fig. 2. Robustness of MORE over varying TMs and routings. For each 10 seconds, a random TM is generated, and MORE responds with an optimal multipath oblivious routing over the same set of paths. For adaptiveK, it computes, for each TM, an optimal routing on K-shortest paths ($K = 20$).

TeXCP vs. MORE. We compare MORE with TeXCP, an adaptive multipath routing approach [5]. TeXCP collects network load information and adjusts routing fractions on pre-selected multiple paths for each OD pair to balance the network load. TeXCP also uses MLU as the performance metric. For comparison with TeXCP, we set link capacity in a way similar to [5], i.e., links with high-degree nodes have large capacity and links with low-degree nodes have small capacity. We use the setting for TeXCP as suggested in [5]. Traffic is generated according to a Gravity TM. During time intervals $[25, 50]$ and $[75, 100]$,

---

[5] For Figure 2 and 5, there are downward spikes for both adaptiveK and MORE. These are due to the transition of stopping and starting TMs.

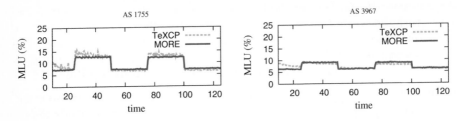

Fig. 3. TeXCP vs. MORE. During time interval [25,50] and [75,100], extra random traffic is generated.

the traffic volume is doubled for each OD pair. Figure 3 shows the comparison results. We show the results after 10 seconds, so that TeXCP may have passed the "warm-up" phase. We see both TeXCP and MORE respond to traffic increases. The results show that MORE has a comparable performance to TeXCP. When TeXCP is in the transition of adapting to its optimal routing, MORE may have better performance, e.g. in the time interval [25, 50] for AS 1755. However, TeXCP may adapt to a better routing than MORE, e.g., in the time interval [75, 100] for AS 3967. MORE, being oblivious to traffic changes, saves resources consumed by TeXCP for frequently collecting network information. MORE has similar performance for AS 6461 and POP 12, see [6]. With a longer time period (35 seconds) for the "warm-up", TeXCP has similar performance.

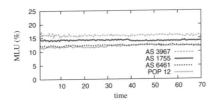

Fig. 4. Robustness of MORE over failures. At each 10's second, a random link failure occurs, and MORE uses the augmentation strategy for failure restoration over the same set of paths. The TM does not change.

Link failure. We study the robustness of MORE over link failures using simulation. We investigate two restoration strategies: *reoptimization* and *augmentation*. In reoptimization, we reoptimize multipath oblivious routing for the new topology after link failures occur. In augmentation, we reoptimize only for the affected OD pairs, which use the link(s) with failure. At each 10's second, a random link failure occurs with 20% link capacity reduction. After each link failure, the augmentation strategy for failure restoration is used to optimize the oblivious routing for the affected paths. The TM keeps unchanged, generated according to a Gravity TM. Figure 4 shows the results. We observe that the networks have rather stable performance, after several consecutive link failures. Reoptimization has similar performance.

Adversary attack. We introduce an attack which can exploit a routing **f**, by generating a TM for **f** to incur a high MLU. We will show that an oblivious routing is robust to such an attack. However, an adaptive routing may suffer much higher MLU. An adversary TM can be obtained using LP (10).

We compare MORE and adaptive$K$ under an adversary attack.[6] Adaptive$K$ computes an optimal routing on K-shortest paths ($K = 20$) for a given TM. An adversary attack can exploit an adaptive routing for the last TM, by generating a new TM. MORE does not change paths and routing fractions.

The simulation runs in iteration, each with 20 seconds. For the first 10 seconds, adaptive$K$ encounters an adversary attack; while for the second 10 seconds, it uses the optimal routing for the adversary in the last 10 seconds. We assume adaptive$K$ can know the exact TM, and deploys the new optimal routing instantaneously in the middle point of an iteration. The oblivious routing does not change over the whole run of the simulation.

The results are shown in Figure 5. When adaptive$K$ is under the adversary attack, it has much larger MLU than MORE. However, when adaptive$K$ operates in optimal, its performance is comparable to or slightly better than that of MORE. The results show that, MORE is robust under an adversary attack, and it has a performance close to adaptive$K$ when adaptive$K$ is not under attack. More has similar performance for AS 6461 and POP 12, see [6].

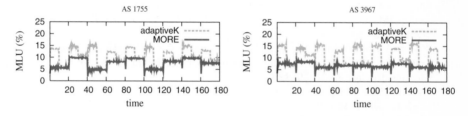

Fig. 5. Adaptive$K$ vs. MORE. During each iteration (20 seconds), for the first half, adaptive$K$ encounters an adversary attack; while for the second half, adaptive$K$ operates with an optimal routing. MORE does not change the routing over the whole run of the simulation.

5   Conclusions

We investigate a promising approach for stable and robust intra-domain traffic engineering in a changing and uncertain environment. We present MORE, a multipath implementation of demand-oblivious routing [2]. We evaluate the performance of MORE by both numerical experiments and simulation. The performance study shows that MORE can obtain a close multipath approximation to [2]. The results also show the excellent performance of MORE under varying traffic demands, link failures and an adversary attack. Its performance is

---

[6] Adaptive$K$ responds to traffic changes instantaneously, while TeXCP takes time for convergence, thus we do not compare MORE with TeXCP here.

excellent even with a 100% error in traffic estimation. See [6] for more discussions and experimental results.

References

[1] Applegate, D., Breslau, L., Cohen, E.: Coping with network failures: Routing strategies for optimal demand oblivious restoration. In: Proceedings of SIGMETRICS'04, pp. 270–281, New York (June 2004)

[2] Applegate, D., Cohen, E.: Making intra-domain routing robust to changing and uncertain traffic demands: understanding fundamental tradeoffs. In: Proceedings of SIGCOMM'03, pp. 313–324, Karlsruhe, Germany (August 2003)

[3] Elwalid, A., Jin, C., Low, S., Widjaja, I.: MATE: MPLS adaptive traffic engineering. In: Proceedings of INFOCOM'01, pp. 1300–1309, Anchorage (April 2001)

[4] Fortz, B., Thorup, M.: Optimizing OSPF/IS-IS weights in a changing world. IEEE Journal on Selected Areas in Communications 20(4), 756–767 (2002)

[5] Kandula, S., Katabi, D., Davie, B., Charny, A.: Walking the tightrope: Responsive yet stable traffic engineering. In: Proceedings of SIGCOMM'05, pp. 253–264, Philadelphia (August 2005)

[6] Li, Y., Bai, B., Harms, J., Holte, R.: Multipath oblivious routing for traffic engineering - stable and robust routing in changing and uncertain environments. Technical Report, TR06-11, Department of Computing Science, University of Alberta (May 2006)

[7] Li, Y., Harms, J., Holte, R.: A simple method for balancing network utilization and quality of routing. In: Proceedings of IEEE ICCCN'05, pp. 71–76, San Diego (October 2005)

[8] Nucci, A., Sridharan, A., Taft, N.: The problem of synthetically generating ip traffic matrices: initial recommendations. ACM SIGCOMM Computer Communication Review 35(3), 19–32 (2005)

[9] Ross, K.W.: Hash routing for collections of shared web caches. IEEE Network 11(7), 37–44 (1997)

[10] Shaikh, A., Rexford, J., Shin, K.G.: Load-sensitive routing of long-lived IP flows. In: Proceedings of SIGCOMM'99, pp. 215–226, Cambridge, MA, USA (August 1999)

[11] Spring, N., Mahajan, R., Wetherall, D.: Measuring ISP topologies with Rocketfuel. In: Proceedings of SIGCOMM'02, pp. 133–146, Pittsburgh (August 2002)

[12] Wang, H., Xie, H., Qiu, L., Yang, Y.R., Zhang, Y., Greenberg, A.: COPE: Traffic Engineering in dynamic networks. In SIGCOMM'06, Pisa, Italy (September 2006)

[13] Zhang, C., Ge, Z., Kurose, J., Liu, Y., Towsley, D.: Optimal routing with multiple traffic matrices: Tradeoff between average case and worst case performance. In: Proceedings of ICNP'05, Boston (November 2005)

[14] Zhang, Y., Roughan, M., Duffield, N., Greenberg, A.: Fast accurate computation of large-scale IP traffic matrices from link loads. In: Proceedings of SIGMETRICS'03, pp. 206–217, San Diego (June 2003)

Modeling BGP Table Fluctuations

Ashley Flavel, Matthew Roughan, Nigel Bean, and Olaf Maennel

School of Mathematical Sciences
University of Adelaide

Abstract. In this paper we develop a mathematical model to capture BGP table fluctuations. This provides the necessary foundations to study short- and long-term routing table growth. We reason that this growth is operationally critical for network administrators who need to gauge the amount of memory to install in routers as well as being a potential deciding factor in determining when the Internet community will run out of IPv4 address space.

We demonstrate that a simple model using a simple arrival process with heavy tailed service times is sufficient to reproduce BGP dynamics including the "spiky" characteristics of the original trace data. We derive our model using a classification technique that separates newly added or removed prefixes, short-term spikes and long-term stable prefixes. We develop a model of non-stable prefixes and show it has similar properties in their magnitude and duration to those observed in recorded BGP traces.

1 Introduction

The Border Gateway Protocol (BGP) [1] automatically discovers paths within the Internet, allowing end-hosts to communicate, whilst respecting policy requirements of Autonomous Systems (ASs). However, events such as link failures, newly added networks and policy changes can alter the path towards a particular destination in routers throughout the Internet, which in turn changes traffic flow and performance [2]. There exists a need for network operators to understand which events may lead to performance disruptions and traffic shifts [3, 4, 5] or whether a change in routing configuration may lead to an unforeseen interaction between policies [6, 7]. Despite this, the properties of routing updates [8, 9, 10] and the extent to which routers actually scale [11,12] are still poorly understood. Several questions operators are still struggling to answer [13] include:

- What is the maximum BGP table a router can handle?
- How much memory is needed to store the Forwarding Information Base (FIB) on the line-cards?
- What are the future hardware requirements for routers?

One of the reasons answers to these questions are missing, is that it is hard to create field conditions for realistic tests [14]. Efforts within the IETF, for example from the Benchmarking Methodology Working Group (BMWG) [15],

L. Mason, T. Drwiega, and J. Yan (Eds.): ITC 2007, LNCS 4516, pp. 141–153, 2007.
© Springer-Verlag Berlin Heidelberg 2007

try to overcome the discrepancy between the field and testing conditions by recommending metrics and test setups for test-beds. However, we argue that a good model of BGP events is necessary to understand BGP dynamics, which in turn will lead to the development of superior test tools as well as improved estimation of future Internet trends such as IPv4 address space usage.

In contrast to existing tools [16] we create a model characterizing BGP table fluctuations that is extendible to estimate the size of a BGP table at any particular point in time along with confidence interval estimates which are needed to understand the likelihood of extreme fluctuations. A typical example often cited in literature is the incident in April 1997 [17] where $AS7007$ accidentally announced almost all prefixes in the Internet (belonging to all other ASes) for approximately two hours. Although events of such a magnitude are rare, our data analysis shows that short-term fluctuations in the order of up to several thousand prefixes are not abnormal. It is unclear whether hardware limitations [12, 14], protocol interactions [18], BGP implementations [19] or other factors [20] are to blame for this behavior. Consequently, measuring magnitudes of previous events without understanding their underlying nature will not predict the likelihood of future events. A mathematical model, however, is capable of being trained using current behavior, to provide insight into possible future behavior, e.g. the likelihood of large routing events.

Large routing events can have a serious effect on routers and potentially even cause service interruptions. Nowadays, routers in the Internet have a specially designed data structure to store forwarding information. This Forwarding Information Base (FIB) is often stored in separate memory across the various line-cards to improve packet lookup times and thus forwarding performance (see for example the design of the Cisco GSR [21]). The FIB itself needs to be constructed from routing information comprising manual router configuration (including static routes), Interior Gateway Protocols (IGPs) and the BGP table or Routing Information Base (RIB). Typically the RIB contributes the largest proportion of the FIB table. It is also an "unknown factor", as a network administrator has limited control over what is learned from the outside. However, if the memory limit on the line-cards is reached, the router cannot perform its designed tasks and service outages occur.

In Section 4 we present a new classification technique to separate prefix behavior. By studying recorded RIBs, and applying our classification technique, we derive statistical properties of routing tables and in Section 5 we introduce a model capable of capturing the RIBs short-term dynamics. We concentrate our efforts on the previously un-examined short-term fluctuations, however, our model provides a mechanism to fully characterize all changes and predict the future components of the BGP table based on its current state.

2 Background and Related Work

Routing in the Internet is accomplished on a per-prefix basis. Routing protocols, such as OSPF and IS-IS, are used to find the shortest path internally within an AS, while BGP [1] is used to exchange reachability information between ASs. As

BGP is a policy-routing protocol, it gives operators some freedom to express their company requirements and policies. To accomplish this, BGP allows attachment of several attributes for each route, and is based on a path-vector protocol. Upon startup, a router establishes sessions with all configured neighbors and all appropriate table-entries are exchanged. Hence, each neighbor sends its RIB to adjacent routers, which in turn store each table in memory (sometimes referred to as the Adj-RIB-In). Next, the router may modify or filter attributes in the Adj-RIB-In before selecting a single "best path" used to create its own RIB. The RIB, together with static and IGP routes, are combined to form the Forwarding Information Base (FIB). The FIB is typically a high-speed lookup data structure which consists of prefix-next-hop pairs enabling forwarding of packets to the appropriate next-hop: Special memory is used for storage of the FIB directly on the line-cards of the routers. The selected best routes, if not subjected to out-bound filtering, are then propagated to other neighbors.

BGP's flexibility, coupled with the fact that network administrators (mis-)use BGP in numerous ways means it is often difficult to determine the underlying cause of routing behavior [9]. As BGP propagates changes to the best path, a single router may send multiple updates based on one triggering event [8]. Further, the propagation of one update may cause induced updates at other locations [22]. It is even possible for policy conflicts to occur that can potentially disrupt the entire Internet [7] which has led to a considerable body of research (see [10] for further details).

Pioneering work related to the size of the Internet RIB was undertaken by Fuller *et al.* [23] who measured the number of routes in the RIB on a monthly granularity over the period 1988-1992. Additionally, Huston [24] has used information obtained from the University of Oregon's RouteViews project [25] to display the long-term growth of numerous Autonomous Systems' RIBs since 1994 [26]. In contrast, our work examines the fluctuations within the Internet RIB on a fine timescale.

Several *prefix*-clustering techniques based on time correlations between updates have been previously described [27,28] while other methods which cluster *updates* based on their likelihood of being caused by a single event have also been considered [29,22]. Clustering updates can provide insight into underlying features, however, determining boundaries of clusters can be a difficult task. In contrast, we describe a simple classification technique in Section 4 which has clearly defined boundaries separating prefixes with different behaviors.

3   Data Sets

RouteViews[1] uses software [31] capable of conducting BGP sessions with routers throughout the Internet to collect BGP data. RouteViews archive all routing information and make their data publicly available to benefit the entire Internet community. Fig. 1 provides an overview of the data used in the ensuing analysis

---

[1] RIPE [30] collects similar data to that in RouteViews with a European focus. For this investigation we have only used RouteViews data.

Name	Verio-Trace	Verio-Trace-Prediction
Start Time (UTC)	1 June 2004 01:23:03	1 August 2004 01:48:06
Finish Time (UTC)	1 August 2004 01:48:06	1 October 2004 00:20:56
Start RIB size	137820 entries	140484 entries
Finish RIB size	140484 entries	129191 entries

Fig. 1. Detailed Data Sources Information: All traces obtained from RouteViews [25] for a Verio router (IP: 129.250.0.85, AS: 2914)

and unless otherwise stated, we use the **Verio-Trace** as an example dataset throughout this paper.

As we cannot obtain the actual RIB of any router in the Internet, we need to approximate a *potential* RIB from the available raw data. Each monitored router (or peering router) provides a snapshot of their perspective every 2 hours (hence the granularity of [24]). In addition to the snapshots, updates with timestamps to the nearest second are recorded. A snapshot, with updates applied in chronological order, provides a finer granularity than simply using 2 hourly snapshots. The *potential* RIB is *not* an exact representation of the RIB on a remote router, it is however a good approximation. Route monitors typically do not collect internal prefixes of remote peers and the Minimum Route Advertisement Interval (MRAI) [1] reduces the frequency of update messages recorded. We assume that a majority of routing table fluctuations occur externally and on time scales longer than 30 seconds (a typical setting for MRAI). Hence, we argue our approximation provides a good representation to an actual RIB stored on a remote router. Further, session resets between the route monitor and peering router can cause discrepancies between the constructed and recorded tables. We use intermediate table dumps to infer where missing withdrawals are likely to have occurred, and the impact of these session resets for the data analyzed in this paper is minimal.

4 Classification

Fig. 2 depicts the change in size of the RIB on a per second basis over a two-month interval. Huston [24] has previously shown that the RIB experiences a growth trend when monitored at hourly intervals. However, visual examinination of the time series for the number of entries in the RIB reveals several other key features such as "spikes" which are not evident at a coarser granularity. These spikes indicate short-term availability of prefixes and occur in highly varying magnitudes and durations. In Fig. 2 (b) a large upward spike in the early morning hours (UTC) is clearly visible. The short-term event consisted of approximately 2,000 prefixes which appeared in the routing table for less than 10 minutes.

Recall that our end-goal is to predict the size of the RIB at any given point in time in the future, as well as estimate intermediate table fluctuations. However, the observed time series are a complex combination of various components (trend, upward spikes and downward spikes). To achieve a clean simple model, we need to extract the various components so that we can identify their key

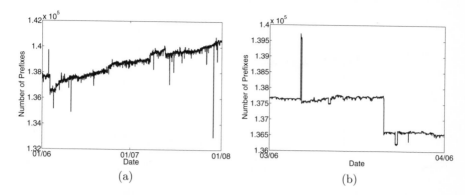

Fig. 2. Short-term RIB fluctuations from **Verio Trace** for (a) 1 June 2004 - 1 August 2004 and (b) a closer examination of 3 June 2004

characteristics and build models. Thus, in this section we develop a simple, straightforward classification technique of the raw BGP data that, at the same time, eases the construction of our model.

Given a number of observations at times $t_1, ..., t_n$, we first require our model to predict the number of table entries at some future time t_{n+i}. Then we are able to estimate the probability a router reaches a predefined memory limit between t_n and t_{n+i}. To achieve this, we need to know:

1. How many new prefixes are added?
2. What happens to existing prefixes?
3. What short-term changes prefixes exhibit?

Consequently, we need to understand the behavior of the RIB over the entire interval $[t_1, t_{n+i}]$. This behavior leads us to the following classifications:

Definition 1 (Stable Prefix). *Let RIB_t be the set of prefixes for which we have an explicit route at time t. If a prefix $p \in RIB_{t_1}$ and $p \in RIB_{t_2}$ then $p \in S_{stable}^{[t_1,t_2]}$ over time interval $[t_1, t_2]$.*

Within the RIB, there is a large proportion of prefixes which are permanently or almost permanently routable. Definition 1 is designed to separate the majority of prefixes which exhibit this behavior from the entire set of prefixes. A stable prefix is present within the RIB at the start and end of the time interval under consideration. Consequently, the number of stable prefixes can never exceed the initial (and final) number of prefixes. Stable prefixes can, however, leave the RIB during the time interval. As a result, the number of stable prefixes within the RIB captures the downward spikes within the time series (see Fig. 3 (a)). Not surprisingly, a large proportion of prefixes within the RIB are classified as stable.

Definition 2 (Transient Prefix). *If a prefix $p \in RIB_{t_1}$ and $p \notin RIB_{t_2}$ or $p \notin RIB_{t_1}$ and $p \in RIB_{t_2}$ then $p \in S_{transient}^{[t_1,t_2]}$ over time interval $[t_1, t_2]$.*

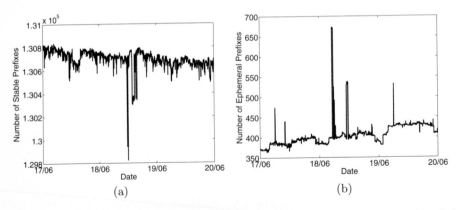

Fig. 3. (a) Stable prefixes and (b) Ephemeral prefixes from 17 - 20 June 2004 classified on interval 1 June - 1 August 2004 from **Verio-Trace**

New prefixes are announced and others permanently withdrawn when new networks are created and old networks aggregated or dismantled. The transient prefixes as defined in Definition 2 captures these newly announced and withdrawn prefixes. As previously shown [24], the RIB grows over time and thus the transient prefixes capture this long term growth (plot not shown as we focus on the short term process in this paper).

Definition 3 (Ephemeral Prefix). *If a prefix $p \notin RIB_{t_1}$ and $p \notin RIB_{t_2}$ and $p \in RIB_t$ for some $t \in (t_1, t_2)$ then $p \in S_{ephemeral}^{[t_1, t_2]}$ over time interval $[t_1, t_2]$.*

A majority of routing dynamics are caused by a minority of prefixes [32, 33, 34]. We aim to separate these prefixes with our definition of ephemeral prefixes. These prefixes as shown in Fig. 3 (b) tend to have short lifetimes although we show later that their lifetimes are best modeled as a heavy-tailed distribution. Consequently, ephemeral prefixes form the upward spikes within the RIB timeseries and may cause router line-cards to exceed their memory limit (before any expected long-term trend would exceed the bound).

All classifications described above are merely approximations to what operators may intuitively label as stable, transient or ephemeral. The times t_1 and t_2 are arbitrary times, hence the interval $[t_1, t_2]$ can be of any length. If we let the interval be infinite, every prefix would be classified as ephemeral. Conversely, if $t_1 = t_2$, every prefix would be classified as stable. In Section 5, we demonstrate that the approximate nature of the classifications, and the arbitrary time interval is not vital to our model's success.

5   Model

In this section we develop our model. We focus on the ephemeral prefixes, as they are most relevant in terms of short-term memory consumption. As described above, the classification of ephemeral prefixes is conceptually aimed at separating those prefixes which experience short-term presence in the RIB.

We define a *spike of prefixes* as a group of prefixes which arrive and depart at the same time. Visual detection of these spikes is somewhat trivial, however determining a simple and effective technique to automatically identify such spikes is not. The difficulty arises as updates relating to multiple events may overlap.

Definition 4 (A Spike of Prefixes). *The spike, $S_{a,w}$, is the set of prefixes which enter the table at time a and leave at time w where $a \leq w$. The spike duration is $r = w - a$ and the spike size is the number of prefixes in the set $S_{a,w}$.*

Updates are witnessed to occur in bursts which may last a number of seconds, after which the MRAI timer prevents any announcements from being sent, but updates caused by a single event may not be advertised or withdrawn at exactly the same time. As Cisco uses a jittered 30 seconds as their default MRAI, we use bins of 30s to capture the start and conclusion of spikes, sacrificing the ability to detect spikes at a finer granularity than 30s, however, large spikes which take several seconds to announce or withdraw are identified as a single set which is especially critical in identifying the probability of large spikes.

Our model assumes that the three components of spikes, (1) spike arrival times; (2) spike sizes; and (3) spike durations, are independent. Superimposing independent spikes defined by the three components above forms the basis for our model to predict the future statistical properties of the RIB.

5.1   Spike Arrival Times

Fig 4 (a) depicts the Complementary Cumulative Distribution Function (CCDF) for spike arrival times. It can be seen that the arrival times are uniformly distributed across the interval. Also, the mean inter-arrival time between spikes can be shown to be approximately 19.6 seconds. In this section, the uniform distribution of spike arrival times is satisfactory for our purposes. In Section 6, however, we require a spike arrival process. Given the difficulty in accurately determining the actual arrival time of each announcement, let alone spike, the precise form of the spike arrival process is impossible to identify. For simplicity, we have chosen to use the classical telecommunications arrival process, namely the Poisson process, to model the spike arrival process.

5.2   Spike Sizes

Fig. 4 (b) shows the CCDF for the spike sizes on log-log axes. We can see that the distribution of the size of spikes is heavy-tailed and consequently, we model the size of spikes as a Pareto distribution. We estimate parameters using logarithmic transformed data (dashed line). Note that, although we use the Pareto distribution, we do not claim this is the ideal distribution as there remains some disparity between the actual and fitted curves. Note a single outlier is responsible for the large discrepancy from the model in bottom right of the plot. We do, however, assert the distribution is heavy-tailed, and the Pareto distribution is a simple, parsimonious distribution with this feature.

5.3 Spike Durations

Recall from Definition 3, ephemeral prefixes are not present at the start and end of the time period. They may experience multiple lifetimes within the time period $[t_1, t_2]$ through multiple announcements and withdrawals. Consequently, some prefixes with multiple short lifetimes will provide more data points than other prefixes that have long lifetimes and contribute one or few data points. For our model we assume independence between events. The CCDF, plotted on log-log axes, for the lifetimes of ephemerals is shown in Fig. 4 (c).

The lifetimes of ephemeral prefixes are artificially limited by the size of the time period we use for classification. As a result, the CCDF representing the lifetimes of ephemeral prefixes is dependent on the arbitrary choice of time period $[t_1, t_2]$. Thus, we require a model to separate the censorship or truncation effect caused by the arbitrary choice of time interval and the underlying process defining the lifetimes of ephemeral prefixes.

We consider, without loss of generality, an arbitrary time interval $[0, T]$ and assume the start times are distributed according to the uniform distribution on

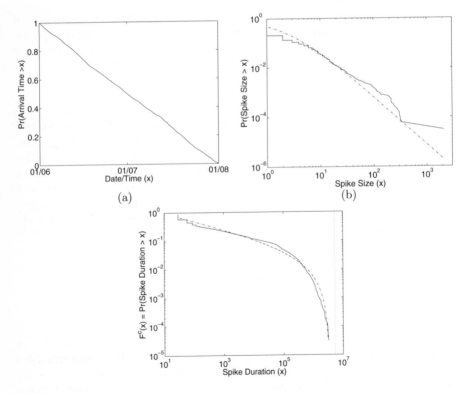

Fig. 4. Spike components from **Verio-Trace**. (a) Spike Arrival time CCDF (b) Empirical (solid) and Pareto Model (dashed) of Ephemeral Spike Size CCDF ($Mode = 2.26, Exponent = 1.93$). (c) Empirical (solid) and Truncated Pareto Model (dashed) of Ephemeral Spike Durations ($Mode = 8.24, Exponent = 0.28$).

$[0, T]$. Hence the probability an individual arrival occurs before time u obeys the probability distribution function

$$Pr\{t \leq u\} = u/T. \tag{1}$$

We also assume independent durations of spikes, r and let a, w denote the start and stop times, respectively. If we further assume the duration of spikes are Pareto distributed, then the probability density function is given by

$$f(x) = \frac{cb^c}{x^{c+1}}, \quad x \geq b, \tag{2}$$

where b is the mode and c is the exponent. However, a prefix will only be classified as ephemeral if it is absent at the end of the classification interval $[0, T]$, i.e., $w \leq T$. So the distribution we observe is the conditional distribution

$$F^c(x) = Pr\{r \leq x | w \leq T\} = \frac{Pr\{r \leq x \cap w \leq T\}}{Pr\{w \leq T\}}. \tag{3}$$

First, consider the numerator, so

$$Pr\{r \leq x \cap w \leq T\} = Pr\{r \leq x \cap a + r \leq T\}$$
$$= \int_b^x Pr\{r = u\} Pr\{a \leq T - u\} du. \tag{4}$$

Substituting (1) and (2) into (4) yields

$$Pr\{r \leq x \cap w \leq T\} = \int_b^x \left(\frac{cb^c}{u^{c+1}} \frac{T - u}{T} \right) du$$
$$= \frac{T - b}{T} - \frac{T - x}{T} \left(\frac{b}{x} \right)^c - \frac{b^c}{T} \left(\frac{x^{c+1} - b^{-c+1}}{1 - c} \right). \tag{5}$$

The normalization factor $Pr\{w \leq T\}$ is the particular case of (4) where $x = T$. Thus, if we set

$$Q_{b,c}(x) = \frac{T - b}{T} - \frac{T - x}{T} \left(\frac{b}{x} \right)^c - \frac{b^c}{T} \left(\frac{x^{c+1} - b^{-c+1}}{1 - c} \right) \tag{6}$$

then from (3)

$$F(x) = Pr\{r \leq x | w \leq T\} = \frac{Q_{b,c}(x)}{Q_{b,c}(T)}. \tag{7}$$

Using a nonlinear regression based on (7) on a log scale, we are able to estimate parameters for ephemeral prefixes to fit our model CCDF ($F^c(x) = 1 - F(x)$) to the empirical data. The example shown in Fig 4 (c) demonstrates that our model is successful in capturing the distribution for the lifetimes of ephemerals including the truncation caused by the classification. Furthermore, a Pareto distribution with parameters estimated using our *truncated* model of spike durations provides an intuitive description for the *non-truncated* duration short-lived prefixes spent in the table.

6   Results

Fig. 5 (a) shows the timeseries for the number of ephemeral prefixes in **Verio Trace**. Two model generated time series (Fig. 6) based on fitted parameters of **Verio Trace** found in Section 5, contain the same features as in the empirical data (Fig. 5 (a)). The empirical data contains a single large spike of magnitude greater than 2000 prefixes. The model generated time series in Fig. 6(a) also predicts (in one case) a large spike of magnitude greater than the limits of the plot. These large spikes are particularly important from a modeling perspective as they potentially cause memory capacity problems in router line cards. Our model provides the predictive ability to determine the probability an abnormally large spike will occur. Also, many short duration spikes and few spikes of several hundred prefixes that last from seconds to weeks are witnessed. Most clearly seen in Fig. 6(b) is performance of the model when capturing overlapping different duration spikes (i.e. the 15 day period in June). An obvious artifact caused by the classification technique is the 'bump' in the center of the timeseries. As seen in each realization (Fig. 6) our model is able to successfully account for this.

Recall our goal is to predict statistical characteristics of the future timeseries as shown in Fig. 5 (b). Other than a single spike lasting approximately 20 days towards the end of the timeseries, the Figs. 5 (a) and (b) are similar: - they have similar spike sizes, durations and truncation effect (the 'bump'). Note we are not aiming to predict exact locations of spikes, but rather their statistical characteristics. We believe *based only on parameters estimated from* **Verio-Trace**, our model is able to predict the statistical properties of future short-term fluctuations in the RIB.

The marginal distribution for our model is shown in Fig. 7. We used 500 model realizations to find a numerical approximation to the mean, maximum and minimum marginal distributions. The empirical data from **Verio-Trace** and **Verio-Trace-Predicion** are also plotted, both of which fall inside the ranges for

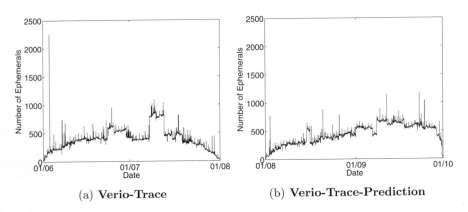

(a) **Verio-Trace**   (b) **Verio-Trace-Prediction**

Fig. 5. Empirical number of ephemeral prefixes: Each plot demonstrates how the number of ephemeral prefixes within the RIB changes over two 2 month periods

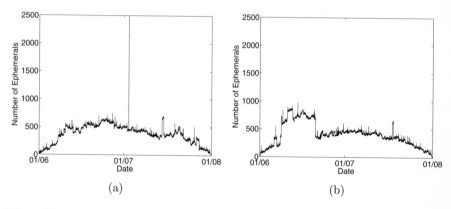

Fig. 6. Model generated number of ephemeral prefixes: Each plot demonstrates a different realization of the model for the number of ephemeral prefixes within the RIB over a 2 month period. The parameters for the model are obtained from **Verio-Trace**.

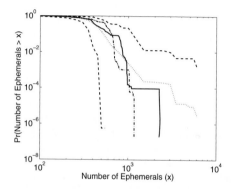

Fig. 7. Mean marginal distribution (dotted) for the number of ephemeral prefixes in RIB found from 500 realizations of our model. Also shown is the minimum and maximum marginal distributions (dashed) together with the marginal distribution for the empirical data from **Verio Trace** (solid) and **Verio-Trace-Prediction** (solid-dotted)

the marginal distribution. It is thus arguable that our model is able to reproduce the highly varying nature in the number of ephemeral prefixes.

More validation of the model is needed, but space limitations prevent us presenting more extensive results. These preliminary results indicate our model is capable of encapsulating the non-stable prefixes and extendable to the entire set of routable prefixes. We reiterate that although we use the Pareto distribution for lifetimes and sizes of spikes, we do not claim that this is the ideal distribution to use. However, we do assert that they exhibit heavy-tailed properties and the Pareto distribution is just one common distribution which has this property.

7   Conclusion and Future Work

In this paper we presented a classification technique to separate long-term growth trends from short-term state changes arising from newly added or removed BGP prefixes. We demonstrated the efficacy of our technique over a 2-month time interval using RouteViews BGP data. Our analysis confirmed the results of previous work such as [34] which supported the validity of our model. We further elicited the presence of heavy-tailed features in the number of short-term fluctuations in terms of size and duration.

The main contribution of this paper was to demonstrate that a simple arrival process with heavy-tailed service times is sufficient to capture BGP routing dynamics. To this end, we derived the parameters for our model from observable BGP data and successfully reproduced BGP table fluctuations including the "spiky" characteristics of the original traces.

In future work we will expand our model to estimate long-term table growth as well as the probability that short-term spikes do not exceed a fixed number of table entries. Also, we will investigate changing the classification interval start times and durations together with considering the evolution of prefixes as they enter the table. This issue is critically important in order to successfully predict when the Internet community will run out of IPv4 address space as well as the amount of memory needed for BGP tables on router line-cards.

References

1. Rekhter, Y., Li, T.: A Border Gateway Protocol 4, RFC 1771 (1995)
2. Li, J., Bush, R., Mao, Z.M., Griffin, T., Roughan, M., Stutzbach, D., Purpus, E.: Watching Data Streams Toward a Multi-Homed Sink Under Routing Changes Introduced by a BGP Beacon, PAM (2006)
3. Teixeira, R., Shaikh, A., Griffin, T.G., Rexford, J.: Dynamics of Hot-Potato Routing in IP Networks. In: Proc. ACM SIGMETRICS (2004)
4. Teixeira, R., Shaikh, A., Griffin, T.G., Voelker, G.M.: Network sensitivity to hot-potato disruptions. In: Proc. ACM SIGCOMM (2004)
5. Teixeira, R., Duffield, N.G., Rexford, J., Roughan, M.: Traffic matrix reloaded: Impact of routing changes. In: Proc (PAM) (2005)
6. Griffin, T.G., Shepherd, F.B., Wilfong, G.: Policy Disputes in Path Vector Protocols. In: Proc. ICNP (1999)
7. Griffin, T.G., Huston, G.: BGP Wedgies, RFC 4264 (2005)
8. Mao, Z.M., Bush, R., Griffin, T.G., Roughan, M.: BGP Beacons. In: Proc. ACM IMC (2003)
9. Griffin, T.G.: What is the Sound of One Route Flapping?, IPAM (2002)
10. (T. G. Griffin Interdomain routing links) http://www.cl.cam.ac.uk/users/tgg22/interdomain/
11. Chang, D.F., Govindan, R., Heidemann, J.: An empirical study of router response to large BGP routing table load. In: IMW'02: Proc. 2nd ACM SIGCOMM Workshop on Internet measurment (2002)
12. Agarwal, S., Chuah, C.N., Bhattacharyya, S., Diot, C.: Impact of BGP dynamics on router CPU utilization. In: Proc (PAM) (2004)
13. Jaeggli, J., NANOG 39 BOF: Pushing the FIB limits, perspectives on pressures confronting modern routers http://www.nanog.org/mtg-0702/jaeggli.html

14. Feldmann, A., Kong, H., Maennel, O., Tudor, A.: Measuring BGP Pass-Through Times. In: Proc (PAM) (2004)
15. (IETF Benchmarking Methodology Working Group (bmwg))
    http://www.ietf.org/html.charters/bmwg-charter.html
16. Maennel, O., Feldmann, A.: Realistic BGP traffic for test labs. In: Proc. ACM SIGCOMM (2002)
17. Misel, S.A.: (Wow, AS7007!)
    http://merit.edu/mail.archives/nanog/1997-04/msg00340.html
18. Griffin, T.G., Wilfong, G.: An analysis of BGP convergence properties. In: Proc. ACM SIGCOMM (1999)
19. Labovitz, C.: Scalability of the Internet backbone routing infrastructure. In: PhD Thesis, University of Michigan (1999)
20. Wetherall, D., Mahajan, R., Anderson, T.: Understanding BGP misconfigurations. In: Proc. ACM SIGCOMM (2002)
21. (Cisco 12000 Series Routers)
    http://www.cisco.com/en/US/products/hw/routers/ps167/
22. Feldmann, A., Maennel, O., Mao, M., Berger, A., Maggs, B.: Locating Internet Routing Instabilities. In: Proc. ACM SIGCOMM (2004)
23. Fuller, V., Li, T., Yu, J., Varadhan, K.: Supernetting: an Address Assignment and Aggregation Strategy, RFC 1338 (1992)
24. Huston, G.: Analyzing the Internet BGP Routing Table. The Internet Protocol Journal 4(1) (2001)
25. (University of Oregon RouteViews project) http://www.routeviews.org/
26. Huston, G.: http://bgp.potaroo.net
27. Zhang, J., Rexford, J., Feigenbaum, J.: Learning-Based Anomaly Detection in BGP Updates. In: Proc. of SIGCOMM Workshops (2005)
28. Andersen, D., Feamster, N., Balakrishnan, H.: Topology Inference from BGP Routing Dynamics. In: 2nd ACM SIGCOMM Internet Measurement Workshop, Boston, MA (2002)
29. Caesar, M., Subramanian, L., Katz, R.H.: Root cause analysis of Internet routing dynamics. Technical report, UCB/CSD-04-1302 (2003)
30. (RIPE's Routing Information Service) http://www.ripe.net/ris/
31. Ishiguro, K. (Zebra routing software) http://www.zebra.org/
32. Rexford, J., Wang, J., Xiao, Z., Zhang, Y.: BGP routing stability of popular destinations. In: Proc. ACM IMW (2002)
33. Wang, L., Zhao, X., Pei, D., Bush, R., Massey, D., Mankin, A., Wu, S.F., Zhang, L.: Observation and analysis of BGP behavior under stress. In: Proc. ACM IMW (2002)
34. Huston, G.: The BGP Instability Report (2006)
    http://bgpupdates.potaroo.net/instability/bgpupd.html

Performance Evaluation of a Reliable Content Mediation Platform in the Emerging Future Internet*

Simon Oechsner and Phuoc Tran-Gia

University of Würzburg
Institute of Computer Science
Germany
{oechsner,trangia}@informatik.uni-wuerzburg.de

Abstract. In todays Internet, a trend towards distributed content delivery systems can be observed. These systems still have to offer the same functionality as centralized architectures, e.g. content localization. While several advantages like load distribution and cost reduction could be gained by using a decentralized distribution platform, they come with a tradeoff in resource consumption. In this paper, we analyze and evaluate a decentralized Content Distribution Application (CDA) with respect to the time needed to locate specific content. We introduce a node model based on queueing theory and provide methods to compute important characteristics like the mean search time. In this model, we do not only consider the external offered load, but also the internal traffic created as a consequence of decentralization.

1 Introduction

Content application and content providing applications are significant emerging services in the evolution towards the next-generation Internet. Two trends can be observed: the decentralization of the application in content providing and the emerging field of scientific computing. The reasons for these trends are manifold. It is estimated that the volume of information doubles every year, i.e. one has to deal with a factor of one thousand in ten years. The complexity of the geographical structure of the content in such applications is increasing while the dynamic of the information provided is more stochastic. Further on, security issues have to be considered to provide integrity of the content, which adds another dimension to the operation of future content providing applications.

In several current and future content delivery applications, a trend of decentralization can be observed. Large content (genetic research data, customer-related data of mobile network providers, etc), which used to be stored in one single location with standby facilities, is to be divided structurally and stored in highly distributed architectures. The design aims of such structures are to i) balance the query load, ii) minimize the search delay and incurred network traffic and iii) provide higher reliability and resilience, among others. These structures should support application-layer routing in future content providing networks.

---

* This work was funded by Siemens AG, Berlin. The authors alone are responsible for the content of the paper.

L. Mason, T. Drwiega, and J. Yan (Eds.): ITC 2007, LNCS 4516, pp. 154–165, 2007.

The aim of this paper is to provide a performance discussion of content distribution platforms regarding search delay in a generic context. For this purpose we take into account a content providing and distribution platform (CDA: Content Distribution Application) with a Content C, which is segmented in a number of N Content Segments (CS) $C_1, C_2, \ldots, C_N$. These CS are hosted on Content Segment Nodes (CSN), where each node holds one or more CS (see Figure 1(a)). These nodes also serve as the interface to users of the CDA, accepting queries and locating content. The actual specific content, i.e., data documents, are part of the CS and in general assumed to be of much smaller size than a complete segment.

Therefore, the main functions of a CSN are: i) provide access to the content application platforms, ii) routing and lookup of requested Content Segments and iii) provide local data contained in the locally stored Content Segment to attached clients.

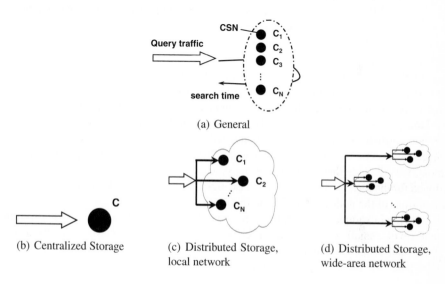

(a) General

(b) Centralized Storage

(c) Distributed Storage, local network

(d) Distributed Storage, wide-area network

Fig. 1. System architecture

Figures 1(b) to 1(d) show different realizations of such an architecture. One alternative is the storage of the complete content on one central server (or several redundant ones), shown in Figure 1(b). In contrast to this, the architectures shown in Figures 1(c) and 1(d) distribute the Content Segments to different nodes. The difference between those two solutions lies in the geographical distribution of the CSNs. In this paper, we concentrate on the analysis of the architecture shown in Figure 1(c), i.e., a system where all nodes are connected to the same local network. Details about the considered architecture will be given in Section 3.

One reason why we consider this variant is the assumption that the described CDA is to be employed in a performance-sensitive, i.e., corporate environment. This also means that the performance of the system is of great importance while the system is dimensioned for a pre-defined grade of service. This is a big difference to 'best-effort' applications, where the partaking nodes usually use only a fraction of their resources (especially processing power), and where virtually no hard timing constraints exist. These efficiency considerations make it interesting to conduct a performance evaluation. The remainder of the paper is organized as follows. In Section 4 we will show analytical results as well as a numerical example. We conclude this work in Section 5.

2   Related Work

In the past, several works have been published that consider efficient overlays for content storage and content distribution. Gupta et al. [1,2] evaluate a fully meshed overlay architecture, which is also the basis for the architecture used in this paper. They show that the additional overhead needed to keep the complete routing table in each node up-to-date can be handled even for large overlays.

In [3], a different scheme based on tokens for maintaining the global routing table in a one-hop overlay was presented and compared to the mechanism of Gupta et al.. The authors showed that less bandwidth was used in their architecture, resulting in a more efficient system.

Another architecture and its analysis of a one-hop DHT was presented in [4]. Again, it was shown that this kind of system is indeed feasible in terms of message overhead and response times to keep the complete routing tables needed at each peer up-to-date.

However, while [1,2,3,4] all analyze the bandwidth consumption at the nodes, they do not consider the processing load generated by the queries on the nodes themselves. Since we assume that the system considered here is positioned in a corporate environment, we conclude that the nodes of the system are not underutilized, but are dimensioned for the load offered by the database users. Therefore, we take a closer look at the conditions of the nodes themselves in this work.

In [5], an example of an application of the described CDA architecture is given in the form of a small-scale publish/subscribe system. It is also emphasized that the availability and usage of resources in a small-scale, highly utilized network is of critical importance for the performance of the system.

3   Architecture

The architecture considered in this paper consists of a number of CSNs that are physically fully meshed by a Local Area Network (LAN). When a CS is requested by a client, one of these lookup nodes is contacted. The choice of this initial node is arbitrary and for load distribution reasons assumed to be random. In reality, this could be accomplished by, e.g., a round robin load distributor.

Since we distributed the data to the CSNs, it is probable that the needed information is not stored on this node. Therefore, the node first queried has to identify the correct storage location of the data and request it from that node. Once this is done, either the first or the second node can forward it to the client that originated the query. However, in this work we assume the search process to be independent from the actual content delivery and therefore consider the search finished once the query reaches the node holding the requested content.

To ensure that the content segment is located quickly, we assume an addressing scheme for the nodes that is based on structured P2P algorithms, specifically Chord [6]. The basic principle is as follows: Each node gets an ID from a one-dimensional identifier space, typically of size $[0; 2^m - 1]$. The $<key, value>$ pairs of data are also placed in that space, with the hashed key as an identifier. Therefore, the hash function that transforms the original information of the data into the lookup key also has a codomain of $[0; 2^m - 1]$. The data is now stored on the node which has the next higher ID than the key of the data set in a clockwise direction. This rule also determines and defines the Content Segments. Content belongs to the same segment if it is hashed to the same interval between two nodes in the identifier space.

When a piece of content in a CS is searched for, the node with the next higher ID to the hash value of the content searched for is polled for that information. In a traditional structured P2P overlay, this would be accomplished by an internal routing operation that forwards the message closer to the target node with each hop. However, since the number of nodes used in this architecture is small for such an overlay (in the range of 10^2 nodes instead of the typically assumed 10^5 or 10^6), we can hold the addresses of all nodes in the 'routing table' of every node and therefore reach the target of the internal lookup in one hop. This makes the lookup efficient in terms of hops. Of course, if a client had knowledge about the addressing scheme it could query the according node directly, but we assume that the CDA is transparent to a client in the sense that the internal structure is not known outside of the system.

We assume that the nodes are distributed uniformly over the identifier space, so that the CS each node has to store has approximately the same size. Furthermore, we assume that each CS is as popular as the others, resulting in an equal number of queries for each segment. In case this assumption should be invalid, i.e., if there are CS queried very frequently in comparison to others, these CS could be copied to all CSNs in order to cope with the high demand. However, we do not consider this case here.

In this paper, we will not evaluate the described architecture during node failures. However, since it has to be considered that one or several nodes of the network can fail, some kind of redundancy mechanism should be implemented. This can happen either in the form of partial redundancy like known from P2P systems, where nodes replicate their data to neighboring peers, or via full redundancy (i.e., the whole ring system is copied), resulting in several networks.

4   Performance Evaluation and Comparison

4.1   Overall System Model

To evaluate the presented architecture, we employ the system model shown in Figure 2. The total initial load offered to the system is created by the clients connected to the CSNs. This query arrival process is assumed to be Poisson with rate λ_0. We further assume an equal distribution of this load on each node in the system, resulting from the client model described in the last section. Each node is modelled by means of a $M/GI/1$ waiting system with an arrival rate λ_{node} and a processing time B with mean $E[B]$. Thus, the normalized offered load per node is $\rho = \frac{\lambda_0 E[B]}{N}$, which is the utilization

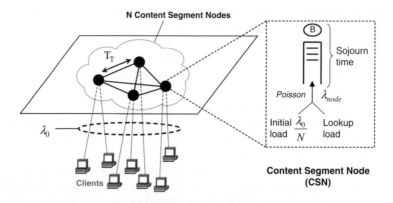

Fig. 2. System model

of one node caused by initial queries. The waiting time W is implicitly given by the node model and parameters. Query forwarding duration over the network is modeled by the transmission time distribution T_T.

The search procedure follows the phase diagram depicted in Figure 3. The first node is traversed in any case. With a probability $1 - p$, this node holds the queried data and the search is finished (the upper path of the diagram). With probability p however, the query has to be forwarded to the correct node, meaning one hop and an additional node traversal (corresponding to the lower path in the diagram). Since we do not consider content replication, the sum of all data is partitioned equally between the nodes. For a number N of nodes, therefore the probability to find the queried data in the first node is $\frac{1}{N}$ and the probability for an additional hop is $p = \frac{N-1}{N}$.

An important aspect of the architecture under consideration is that not each query is answered by the first node it encounters, but may spawn a subsequent internal query. Therefore, the query arrival rate at one node is not only the total external rate on the system λ_0 divided by the number of nodes N, but additionally the re-routed queries from all other nodes. Since each node forwards queries with a rate of $p \cdot \frac{\lambda_0}{N}$, it receives the same amount for symmetry reasons and under our assumption that an equal share

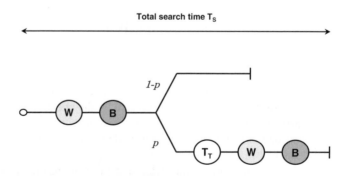

Fig. 3. Phase diagram for the search procedure

of the queries is forwarded to each node except the local node. We assume that the resulting total query flow arriving at one node is still Markovian with an adapted rate. We will later show simulation results that suggest that this assumption is valid.

Figure 4 shows the resulting traffic flows for one node. While this has no great impact on systems like P2P content storage or distribution platforms, where node load is assumed to be low, it may have a significant influence on systems which operate with a high utilization. Since we assume that our architecture falls in the latter category, we have to evaluate the nodes using the total arrival rate per node λ_{node} instead of $\frac{\lambda_0}{N}$.

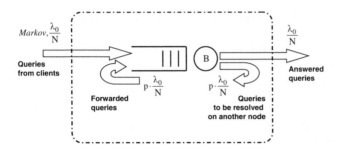

Fig. 4. Model for the query traffic at a node

The normalized offered initial load of one node is $\rho = \frac{\lambda_0 E[B]}{N}$. With the given system model, the total load on one node then is this inital load plus the additional re-routed requests:

$$\lambda_{node} = \frac{\lambda_0}{N}(1+p) = \frac{\lambda_0}{N}(1+\frac{N-1}{N}). \tag{1}$$

With a service time B, the final node utilization ρ^* can be derived as

$$\rho^* = \lambda_{node} \cdot E[B] = \rho \cdot (1+p). \tag{2}$$

4.2   Analytical Approaches

In this section, we start to describe how different parameters affect the system performance. We focus on the search time, i.e., the time needed to locate a specific content segment, in the architecture described in Section 3. An approximate analysis of the search time will be presented, followed by numerical examples.

First, we will analyze the distribution function T_S of the total search time. Using an independent assumption of waiting times at the first and second nodes we arrive at the Laplace-Transform of the search time T_S:

$$\Phi_S(s) = (1-p) \cdot \Phi_W(s)\Phi_B(s) + p \cdot (\Phi_W(s)^2 \Phi_B(s)^2 \Phi_T(s)). \tag{3}$$

As a CSN is modelled with a $M/GI/1$ system, we can use the Pollaczek-Khintchine formula [7]

$$\Phi_W(s) = \frac{s(1-\rho)}{s - \lambda_{node} + \lambda_{node}\Phi_B(s)} \tag{4}$$

and finally obtain

$$\Phi_S(s) = (1-p)\frac{s(1-\rho)\Phi_B(s)}{s - \lambda_{node} + \lambda_{node}\Phi_B(s)} + p\left(\frac{s(1-\rho)\Phi_B(s)}{s - \lambda_{node} + \lambda_{node}\Phi_B(s)}\right)^2 \Phi_T(s).$$
(5)

In general it is numerically difficult to quickly obtain values in the time domain in order to gain basic insights into the system behaviour. Thus, we will first compute directly the mean value and variance of the search time.

From the phase diagram shown in Figure 3, the mean search time can be given as

$$E[T_S] = E[W] + E[B] + p(E[T_T] + E[W] + E[B]).$$
(6)

Again, under the assumption of a $M/GI/1$ system, we can express the mean waiting time in the queue of a node according to Takács [8] as

$$E[W] = \frac{\lambda_{node}E[B^2]}{2(1-\rho)},$$

which leads us to

$$E[T_S] = \frac{2N-1}{N}\left(\frac{\lambda_{node}E[B^2]}{2(1-\rho)} + E[B]\right) + \frac{N-1}{N}E[T_T].$$
(7)

Note that we need the second moment of the service time distribution in (7) in order to compute the mean search time. The variance of the search time can also be derived from Figure 3 as

$$VAR[T_S] = VAR[W] + VAR[B] + p(VAR[T_T] + VAR[W] + VAR[B]),$$
(8)

where $VAR[W]$, $VAR[B]$ and $VAR[T_T]$ are the variances of the waiting time, service time and transmission time distribution, respectively. The second moment of the waiting time is

$$E[W^2] = 2E[W]^2 + \frac{\lambda_{node}E[B^3]}{3(1-\rho)},$$
(9)

yielding

$$VAR[W] = \frac{3\lambda_{node}^2E[B^2]^2 + 4(1-\rho)\lambda_{node}E[B^3]}{12(1-\rho)^2}.$$
(10)

Combining equations (8) and (10), we finally obtain the variance of the search time as

$$VAR[T_S] = (1+p)\left(\frac{3\lambda_{node}^2E[B^2]^2 + 4(1-\rho)\lambda_{node}E[B^3]}{12(1-\rho)^2} + E[B^2] - E[B]^2\right)$$
$$+ p(E[T_T^2] - E[T_T]^2).$$
(11)

This allows us to compute the variance of the search time from the first moments of the service time distribution and the transmission time distribution.

4.3   Parameter Studies

In this section, we will show some results derived from the formulas (7) and (11). First, we take a look at the effect of the variance of the service time of the nodes on the mean search time. Figure 5 shows the mean search times normalized by $E[B]$ for different utilizations and coefficients of variation c_B of the service time. The parameter c_B gives us the second moment of the service time and also allows us to derive the third moment as described e.g., in [9]. We chose coefficients of variation that describe different distributions, ranging from deterministic ($c_B = 0$) to strongly varying processes ($c_B = 2$). We considered a smaller ($N = 5$) as well as a larger system ($N = 20$).

For some parameter settings, simulations have been conducted which do not assume that the combined arrival process at the nodes is Markovian, just the arrival rate of the initial queries is modeled with an exponential distribution. These results are given with confidence intervals for a 95% confidence level over 5 different runs. They suggest that our assumption about the node arrival process are valid and that the approximation is very accurate.

It should be noted that due to Equation (2), the node utilization in this model can be up to twice the utilization by external queries alone. Therefore, we will only cover the range $[0, 0.5]$ for ρ in the following analysis.

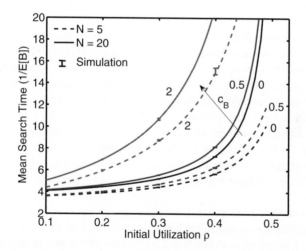

Fig. 5. Effect of the coefficient of variation of the service time on the mean search times

The results show an increase in the mean search time for service processes with a higher variance for both system sizes. This is due to the longer waiting times created by the service time distribution. It can also be observed that the mean search times are lower for smaller systems. The reason for this is twofold. First, a lower number of nodes also leads to a smaller probability for a second hop, shortening the search time. Second, and also as a consequence of a lower probability p, a smaller system incurs less additional internal load and therefore shorter waiting times.

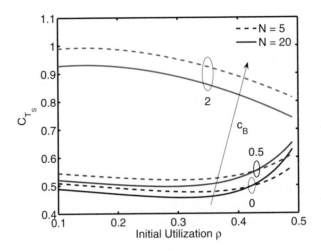

Fig. 6. Effect of the coefficient of variation of the service time on the variance of the search times

Figure 6 shows the coefficients of variation from the same experiments. A higher variance in the service time also leads to a higher variance in the total search time, as expected.

Another observation is that the coefficient of variation of the search time grows for higher utilizations and a low variance of the service time. This can be explained by the fact that ρ^* is close to 1 for these values of ρ and therefore the waiting times have a larger influence on the total system behaviour. The same can be observed for a value of $c_B = 2$, where the effect on the coefficient of variation is reversed.

To see how the system size influences the search time, we take a look at the mean search time for different numbers of nodes. Figure 7 again shows the medium search times depending on the utilization of the system and for systems with 10, 20 and 30 nodes. We additionally compare these values for coefficients of the service time $c_B \in \{0, 1\}$. Again, we have validated these results with simulations for chosen parameter settings.

We can again observe an increase in the mean search time for larger systems due to the higher probability for an additional hop in systems with more nodes, which in turn leads to a larger fraction of searches that have to visit two nodes to find their results. Additionally, the higher internal traffic again leads to longer waiting times. Also, the results show longer search times for a higher coefficient of variation of the service time, as was to be expected from the results discussed above.

Similarly, the coefficients of variation of the search time from the same experiments, shown in Figure 8, exhibit the same behaviour as discussed in Figure 6.

We can also observe a trend towards lower coefficients of variation for larger systems up to a certain point. A smaller system size leads to a higher variance, since a smaller amount of queries traverses two nodes. However, since we assume that the waiting and service time distributions of the two nodes are independent, the additional hop would lower the variance in the search time. This effect holds until the higher internal traffic

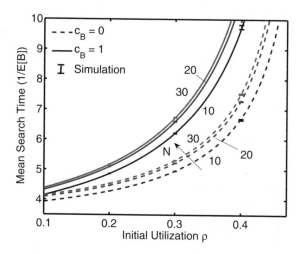

Fig. 7. Effect of the system size on the mean search times

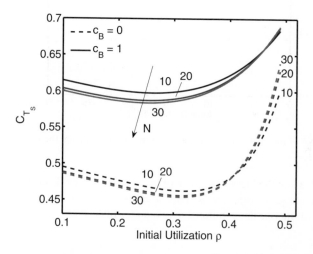

Fig. 8. Effect of the system size on the variance of the search times

in the larger system leads to longer waiting queues than in the smaller system, at least for small coefficients of variation of the service time.

4.4 Numerical Example

In this section, we present results taken from a numerical computation of the search times, using a time-discrete GI/GI/1 model as described in [10] with a Markovian arrival process with parameter λ_0 and a two-parameter representation of the service time. The reason why we use the GI/GI/1 computation algorithm instead of the analyzed

M/GI/1 model is its general applicability, allowing for studies of systems with different arrival processes. However, in this work we have adapted it to fit the system model described in Section 4.1. The service time is thought to be distributed following a negative binomial distribution with a mean of $E[B]$ and coefficient of variation $c_B = 1$, while a deterministic distribution is used for $c_B = 0$.

In Figure 9, we take a look at the impact of the coefficient of variaton of the service time on the total search times for systems with 5 and 20 CSNs, respectively. The complementary cumulative distribution function (CCDF) of the total search time is shown for coefficients of variation $c_B \in \{0, 1\}$. The system utilization is set to $\rho = 0.3$.

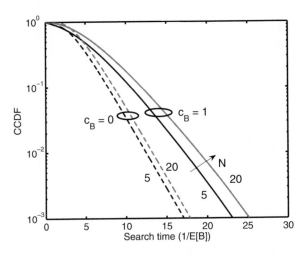

Fig. 9. Effect of the coefficient of variation of the service time on the search times

We can observe that the higher variance of the service process leads to longer search times, due to the longer waiting times. This effect is stronger for larger systems, since a larger number of queries have to traverse two nodes instead of one.

5   Conclusion and Outlook

In this work, we described and evaluated a distributed, fully meshed content storage system. We provided an analysis of the search times in such a system and showed some of the most important parameters that influence it. In this context, we concentrated on the query load that the nodes in such a system have to handle, in contrast to the network traffic considerations made in other papers.

Most importantly, we considered in our analytical model the additional load generated by internal routing, which may be unimportant in distributed architectures with underutilized nodes, e.g., large scale P2P platforms, but have an impact on the dimensioning of dedicated systems. We showed that this additional load can not be neglected and has to be considered when designing the system.

While our assumption of a fully meshed DHT-like structure limits the direct applicability of the results to these systems, we believe that our approach of characterizing the load experienced by a node through modeling the traffic flows can be also used to analyze more general architectures, such as current overlay networks.

In the future, it could therefore be of interest to adapt the analytical model presented here to other distributed architectures, e.g., relaxing the assumption that the queried content is located assuredly after the internal hop. Also, a more detailed model of the query processing could be implemented, where the process times are not independent of the fact whether a node actually has the content or has to forward the query to another node.

Acknowledgement

The authors would like to thank Robert Henjes for his valuable comments and the insights gained in the fruitful discussions.

References

1. Gupta, A., Liskov, B., Rodrigues, R.: One hop lookups for peer-to-peer overlays. In: Proceedungs of the Ninth Workshop on Hot Topics in Operating Systems, Lihue, Hawaii (May 2003)
2. Gupta, A., Liskov, B., Rodrigues, R.: Efficient routing for peer-to-peer overlays. In: Proceedings of the First Symposium on Networked Systems Design and Implementation, San Francisco, CA (March 2004)
3. Leong, B., Lik, J.: Achieving one-hop dht lookup and strong stabilization by passing tokens. 12th International Conference on Networks 2004 (ICON 2004), Singapore (November 2004)
4. Monnerat, L.R., Amorim, C.L.: D1ht: A distributed one hop hash table. Rhodes, Greece, IEEE International Parallel & Distributed Processing Symposium (April 2006)
5. Muthusamy, V., Jacobsen, H.-A.: Small-scale peer-to-peer publish/subscribe. San Diego, CA, USA, P2PKM'05: Peer-to-Peer Knowledge Management (July 2005)
6. Stoica, I., Morris, R., Liben-Nowell, D., Karger, D.R., Kaashoek, M.F., Dabek, F., Balakrishnan, H.: Chord: a scalable peer-to-peer lookup protocol for internet applications. IEEE/ACM Trans. Netw. 11(1), 17–32 (2003)
7. Takagi, H.: Queueing Analysis: A Foundation of Performance Evaluation, vol.1, North-Holland, Amsterdam, Netherlands (1991)
8. Takacs, L.: A single server queue with poisson input. Operat. Res. 10, 388–397 (1962)
9. Raith, T.: Leistungsuntersuchung von Multi-Bus-Verbindungsnetzwerken in lose gekoppelten Systemen. PhD thesis, Universitat Stuttgart (1986)
10. Tran-Gia, P.: Discrete-time analysis technique and application to usage parameter control modelling in ATM systems. In: Proceedings of the 8th Australian Teletraffic Research Seminar, Melbourne, Australia, vol. 12 (1993)

Fair Scheduling in Optical Burst Switching Networks*

Lin Liu and Yuanyuan Yang

Dept. of Electrical & Computer Engineering,
State University of New York, Stony Brook, NY 11794, USA
{linliu, yang}@ece.sunysb.edu

Abstract. Optical Burst Switching (OBS) is a promising switching technique that has received much attention in recent years. A fair scheduling algorithm can ensure fair bandwidth allocation among different users and isolate ill-behaved users from normal users in a network. However, fair scheduling in OBS networks is difficult to implement due to lack of inexpensive and large buffers in optical domain. In this paper, we propose a new scheme to provide fairness in OBS scheduling by considering the time-slotted version of OBS, the Slotted OBS or SOBS. We give a fair scheduling algorithm with $O(1)$ time complexity called the *Almost Strictly Proportional Fair Scheduling (ASPFS)*. Our simulation results demonstrate that by applying the algorithm to the control packets (CP) at the end of each time slot, fair scheduling can be achieved. The scheme significantly improves the fairness among different flows compared to other scheduling algorithms in OBS such as LAUC-VF and Pipeline with VST. We also give an approach to supporting Quality of Service (QoS) in OBS based on the proposed scheme.

Keywords: Optical networks, scheduling algorithms, optical burst switching (OBS), fair scheduling, time-slotted.

1 Introduction and Previous Work

All optical networking with Wavelength Division Multiplexing (WDM) is widely regarded as the most promising candidate as the backbone network for high speed communications because of its huge bandwidth [1]. So far three types of optical networks have been proposed based on WDM – Wavelength Routed (WR), Optical Packet Switching (OPS), and Optical Burst Switching (OBS). The first two types of networks are similar to traditional circuit-switching and packet-switching networks, respectively, while OBS is a combination of them. OBS provides the possibility for more dynamic wavelength configurations than WR, while keeping the implementation complexity at a relatively low level compared to OPS [2] [3] [6].

In OBS networks, traffic is accumulated and assembled into bursts at edge nodes. When a burst is ready for transmission, a control packet (CP) is sent out through a control channel. The control packet contains the offset time (the interval between sending the control burst and the corresponding data burst), the size of the data burst, etc., and is intended to reserve a wavelength for the coming data burst (DB) at each intermediate

---

* The research work was supported in part by NSF grant number CCR-0207999.

L. Mason, T. Drwiega, and J. Yan (Eds.): ITC 2007, LNCS 4516, pp. 166–178, 2007.

core node it traverses. After the offset, the DB will be sent. If successfully arriving at the destination, the DB will then be dissembled and sent to the upper layers.

In recent years, wavelength (or channel) scheduling in OBS networks has attracted intensive interests in optical networking community due to its significant impact on the system performance. The simplest scheduling algorithm is First Fit [3], where the data channels are searched in a fixed order, and the data burst will be sent to the first idle channel. The Latest Available Unscheduled Channel (LAUC) [3] or Horizon [4] algorithm keeps track of the horizon of each channel – the time after which the channel is always available, and schedules a burst on the channel with the largest horizon among all eligible channels. More complex algorithms include LAUC-VF [3] and Min-SV [6], which make use of void intervals and try to schedule a DB in a void interval that is large enough.

More recently, window-based channel reservation has been considered. It is noticed that some burst losses in OBS networks are due to insufficient information when scheduling is done. If the scheduling decision can be delayed for a certain time, during which more information is gathered, a better scheduling solution may be worked out. Algorithms falling into this category include virtual fix offset time [7] and pipeline buffering [8]. A scheduling scheme called Virtual Scheduling Technique (VST) was also proposed in [8], which schedules control packets in a FIFO manner while trying to give some priority to CPs whose data burst will arrive earlier than others.

However, none of the above scheduling algorithms have taken fairness into consideration, which is a very important aspect in broadband networks. A fair scheduling algorithm ensures fair allocation of network resources among different users. Without fairness, when the network is congested by some abnormal traffic flow, there will be significant performance degradation since the well-behaved flows are not protected. The MDGP algorithm proposed in [9] tries to add fairness at the core nodes, but it only deals with the unfairness caused by the difference in the length of paths that bursts need to traverse. It is more effective to provide fairness among flows. The classical definition of a flow is a stream of packets which traverse the same route from the source to the destination and require the same grade of service at each intermediate node in the path [15]. In this paper, a flow is similarly defined, except that in OBS networks the unit of a flow is a burst instead of a packet. A scheme has been proposed in [10], which provides fair bandwidth allocation by queueing and applying a fair packet queueing algorithm on CPs as if directly applying the algorithm to their corresponding data bursts. However, maintaining such a virtual data burst queue seems quite complex and the authors did not give much detail on it. In the algorithm, a CP can start transmission as soon as the last CP scheduled on the same channel completes the transmission, instead of the corresponding DB completes the transmission. Thus, the average queueing delay of CPs is almost independent of the output link bandwidth and tends to be short. As a result, the flows would often become unbacklogged, hence no fairness bounds can be guaranteed. On the other hand, if the virtual queue is maintained as an actual queue, i.e. no CP can be scheduled on the same channel before the corresponding DB of the last CP completes the transmission, the queueing delay will be long, which leads to another dilemma: either to increase the offset time – consequently, the total delay, or to risk the possibility that many bursts arrive before their CP get scheduled and have to be dropped.

In this paper we present a new scheme to provide fair scheduling in OBS networks. The scheme is based on a time-slotted version of OBS and a simple yet efficient fair scheduling algorithm. The rest of the paper is organized as follows. Section 2 presents the proposed architecture for fair scheduling in OBS networks. Section 3 gives the details of the new fair scheduling algorithm based on this architecture and the methods of QoS provisioning under this algorithm. Section 4 provides and discusses simulation results. Section 5 concludes this paper.

2   The Proposed Time-Slotted OBS Architecture

2.1   Time-Slotted OBS

The time-slotted version of OBS networks was first introduced in [11]. The signaling part of slotted OBS is identical to its unslotted counterpart. At an edge node, when a DB is ready for transmission, a CP is first sent through a control channel for the wavelength reservation purpose. The main difference is that in slotted OBS, all DBs are of the same length and can only be sent at the boundaries of each time slot. In general, the length of a time slot is set to be the transmission delay of a data burst.

It is worth mentioning that a variant of time-slotted OBS was proposed in [14], named Time Sliced Optical Burst Switching (TSOBS). In TSOBS, switching is done solely in the time domain rather than the wavelength domain, hence the need of wavelength converters is eliminated. However, to make fully use of the high flexibility of wavelength assignment in OBS, in this paper we assume that the switching is done in wavelength domain and full wavelength converters are available.

2.2   Why Slotted OBS

The main difficulty of implementing fair scheduling in OBS networks is lack of optical buffer in current optical networks. In all traditional fair scheduling algorithms, if a packet cannot be scheduled for transmission immediately, it is buffered for scheduling later. Unfortunately, in optical networks, including OBS, inexpensive, large optical buffers are not readily available. In OBS, only the control packets that have arrived and have not been scheduled are buffered. Consequently, at any instant only the information of a very limited number of data bursts is known. Thus, fair channel allocation becomes rather difficult since available information is much less than that in traditional networks, where packet queues are maintained and bandwidth allocation can be regulated by adjusting the service order among these queues.

To overcome this problem, we consider the time-slotted OBS architecture. The rationale behind this is as follows. As mentioned above, the foremost obstacle to providing fair scheduling in OBS networks is the lack of optical RAM, consequently, the queues of buffered packets in the core OBS network. The CPs can be buffered in electronic domain at each intermediate node during their transmission, however, buffering them to maintain the virtual data burst seems not practical either. On the other hand, in slotted OBS systems, if there are N wavelength channels on each output fiber link, when the network is fully-loaded or overloaded (the typical situation under which fair scheduling needs to be considered), for each time slot there will be roughly N CPs arriving at each output link waiting for scheduling. According to current optical technology, N is usually 64 to 128 or even larger [5]. By buffering such a large number of CPs during each time slot and scheduling them at the end of the slot, we will be able to provide better fairness. The basic idea is somewhat similar to that of the window-based scheduling algorithms mentioned above, in the sense that if more information is available when scheduling, better performance can be achieved. However, the difference is that, instead of delaying each scheduling decision for a small time interval and expecting more CPs to arrive during this interval, we try to gather as much information as possible in each time slot, then make all scheduling decisions together.

2.3 The SOBS Scheme

The proposed SOBS is a variant of slotted OBS, with some modifications to make it suitable for fairness scheduling. We describe the details of the SOBS scheme in this section. First, unlike in classical fair scheduling algorithms where a flow can only be either backlogged (some packets are buffered) or unbacklogged, in SOBS, there are three possible states for a flow: active, unbacklogged and pending. The difference is due to the fact that currently we cannot handle buffering in OBS as efficient as that in traditional packet switching networks. As a result, it is no longer reasonable to classify a flow into unbacklogged merely because this flow has no CPs buffered at this moment. Thus a new pending status is introduced to denote the status when we cannot determine whether a flow is backlogged or not.

In our scheme, at each core node two lists are maintained: the *ActiveList* containing all active flows and the *PendingList* containing all pending flows. A separate control packet queue is maintained for each flow passing through the core node. During a time slot, when a CP of flow f arrives, it is inserted into the corresponding CP queue. The status of f is set to "active," unless it already is. If f was unbacklogged, we also need to append f to the *PendingList*. Note that we cannot directly insert f to the *ActiveList*, since the scheduling of last slot's CPs may not be completed yet. At the end of the j_{th} time slot, which is also the end of the $(j-1)_{th}$ scheduling cycle, we first append the entire *ActiveList* to the *PendingList*. If the CP queue of any of the flows in *ActiveList*, for example f_1, is now empty, i.e. no CP of f_1 has arrived in the j_{th} time slot, then the status of f_1 is set to "pending." Next we need to check all flows in *PendingList*. If the status of a flow in *PendingList* is "active," we must move this flow to *ActiveList*. Note that since in each time slot at most N bursts can be transmitted, it is meaningless to maintain an *ActiveList* of length larger than N if each flow is guaranteed to send at least one burst during its turn of service, in other words, at most N flows can be served in each time slot. In this case, no more flow will be added to the *ActiveList* once the length of *ActiveList* reaches N. After that, the j_{th} scheduling cycle starts. All buffered CPs competing for the same output link are scheduled by a fair scheduling algorithm, and the switch matrix is configured correspondingly. Data bursts whose CP cannot be scheduled will have to be dropped. Note that during slot j when CPs arrived in slot $j-1$ are being scheduled, new CPs will be added to the end of the queues. Hence we also need to record the number of CPs of each flow arriving in each slot so that we can tell which time slot a CP belongs to.

Since the scheduling in OBS must be done at high speed, scheduling complexity is a major concern. Therefore, we believe the round-robin approach is a better choice for SOBS rather than time-stamped-based [17] or credit-based [18] approaches. Though the fairness bound provided by the former is looser, its $O(1)$ time complexity is more desirable for OBS schedulers.

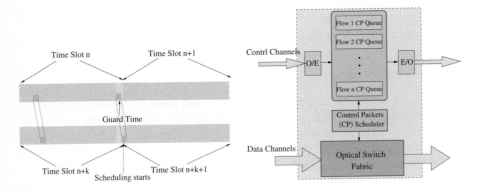

Fig. 1. The guard time to avoid unnecessary burst drops

Fig. 2. Core node architecture

Since the system works in a time-slotted manner, the timing becomes more important. If we start the scheduling of a time slot right at the end of this slot, then some guard time must be set during which no CP is allowed to be sent. Otherwise these CPs cannot arrive at the core node in time for scheduling. This can be illustrated in Fig. 1. Assume that the propagation delay is as long as k time slots (if it is not exactly a multiple of the time slot length, fiber delay lines can be appended). Due to the transmission delay, a CP sent in slot j may arrive in slot $j + k + 1$ and miss the scheduling opportunity. Since the scheduling cycle is about a time slot long and the guard time is about the transmission delay for a CP, the existence of the guard time will hardly affect the overall efficiency.

The architecture of a core node in SOBS is shown in Fig. 2. When a CP arrives, it first goes through an optical-electronic conversion before it is ready to be processed in electronic domain. After being scheduled, the CP will be converted back to optical domain and transmitted. The CP scheduler module works on the CP queues and sets the optical switch fabric accordingly.

3   Almost Strictly Proportional Fair Scheduling for Slotted OBS (ASPFS)

3.1 How ASPFS Works

In this section, we present a new round-robin based fair scheduling algorithm for slotted OBS, called the Almost Strictly Fair Scheduling (ASPFS). This algorithm is "strictly proportional fair" because that for each slot it allocates the service quantum to all the active and pending flows strictly proportional to their weights, or negotiated bandwidth. We say it is "almost" because a flow i can loan credits from the future as long as its credit is larger than some bottom line, denoted as B_i. Since each flow has different weight, the bottom line of each flow should also be different, preferably proportional to its weight, i.e., $B_i = -B \cdot w_i$, where B is a small positive constant and w_i is the weight of flow i.

The allocated service quantum for an active flow i of slot k, denoted as $Q_i^{(k)}$, is calculated at the end of this slot as follows:

$$Q_i^{(k)} = N \cdot w_i / W^{(k)},$$

where N is the number of data channels on an output link, and $W^{(k)}$ is the sum of the weights of all flows in $ActiveList$ at the end of slot k and after the update of $ActiveList$. $Q_i^{(k)}$ is then added to flow i's credit, denoted as $Credit_i$.

When flow i is at the head of $ActiveList$, if its credit is larger than B_i, the head of its CP queue will be scheduled – an idle channel will be reserved for the corresponding DB. After that, if the queue is still not empty and its credit is above the bottom line, flow i will be moved to the end of $ActiveList$ and wait for the next service round. Otherwise flow i will be removed from $ActiveList$ and appended to $PendingList$. When there is no more available data channel or unscheduled control packet, the scheduling of this time slot ends. All unscheduled control packets that belong to this slot will be dropped. All flows in $ActiveList$ will be checked. If a flow becomes unbacklogged, it will be removed from the list and its credit will be set to zero.

Now we return to the question of how to define "unbacklogged." It is very natural in traditional packet switched networks to define it as no packet buffered. However, in OBS the situation is different. The time a control packet can be buffered is limited, since it has to be scheduled before the DB arrives. Thus, an empty queue may be caused by packet dropping rather than no arrival, and the classical definition of "unbacklogged" becomes less reliable. Our solution is that, if the credit of a flow i exceeds some threshold T_i, it is then very likely that this flow has accumulated too much credits while having no bursts for transmission. Then it is reasonable to classify flow i as unbacklogged and reset its credit to 0. Similar to B_i, T_i should also be proportional to w_i so that the normalized threshold values of all flows are the same. We let $T_i = T \cdot w_i$, where T is a positive constant.

The ASPFS algorithm is described in Table 1. The main differences between ASPFS and the classical round-robin based fair scheduling algorithms such as Deficit Round-Robin (DRR) [15] include: (1) There are now three possible states instead of two; (2) In ASPFS, the credits of flows are increased once a slot, while in DRR it is once a service round; (3) In DRR, a flow can keep transmitting as long as it has enough credit; but in ASPFS, at most one burst can be transmitted for a flow during a service round. We make this change in order to reduce the jitter effect caused by the round-robin; (4) In ASPFS, some flows in $ActiveList$ may actually have an empty CP queue. This may occur at the end of a scheduling cycle if a flow in the $ActiveList$ has no control packets arrived in this time slot. As a result, before the next scheduling cycle starts, extra work is required to reconstruct a valid $ActiveList$ that contains only active flows, as shown in the constructing phase in Table 1. This is done by first appending the entire

Table 1. Almost Strictly Proportional Fair Scheduling for Slotted OBS (ASPFS)

$CPthisslot_i$: the number of CPs of flow i arriving in current slot
$CPtosche_i$: the number of CPs of flow i that need to be scheduled in current slot
$idlechannel$: the number of unscheduled channels
$Queue_i$: the CP queue of flow i
N: the number of wavelength channels per fiber
Appendlist(): append a flow index to the end of a list
Delist(): return and remove the head of a list
Del(): delete a flow index from the list
Enqueue() and Dequeue() are standard queue operations.

During each time slot:
  On arrival of a Control Packet CP of flow i
    $CPthisslot_i^{++}$;
    Enqueue($Queue_i, CP$);
    if $Status_i = Pending$ **then**
      $Status_i = Active$;
    else if $Status_i = Unbacklogged$ **then**
      $Status_i = Active$; Appendlist($PendingList, i$);

At the end of each time slot:
  $ActiveList$ Constructing Phase:
    for each flow i in $ActiveList$
      if $(CPthisslot_i == 0)$ **then**
        $Status_i = Pending$; Appendlist($Pending, i$); Del($ActiveList, i$);
    for each flow i in $PendingList$
      if $(Status_i == Active)$ **then**
        Appendlist($ActiveList, i$); Del($PendingList, i$);
        if (Length($ActiveList$) == N) **then**
          Set the status of the rest flows in $PendingList$ all to pending;
          break;
  Scheduling Phase:
    $idlechannel = N$;
    for each flow j in $ActiveList$
      $CPtosche_j = CPthisslot_j$;
      $CPthisslot_j = 0$;   //reset $CPthisslot$
      $Credit_j = Credit_j + Q_j$;
    while ($ActiveList$ is not empty & $idlechannel \geq 1$)
      $h = $ Delist($ActiveList$);
      if $(Credit_h \geq B_h$ & $CPtosche_h \geq 1)$ **then**
        $headCP = $ Dequeue($Queue_h$);
        Schedule $headCP$ to next available channel;
        $Credit_h^{--}$; $CPtosche_h^{--}$; $idlechannel^{--}$;
      if $(Credit_h \geq B_h$ & $CPtosche_h \geq 1)$ **then**
        Appendlist($ActiveList, h$);
      else Appendlist($PendingList, h$); $Status_h = Pending$;
    for each flow i in $ActiveList$
      if $(Credit_i > Threshold_i)$ **then**
        Del($ActiveList, i$); $Status_i = Unbacklogged$;
  Dropping Phase:
    All unscheduled CPs are discarded.

ActiveList to the *PendingList*, and marking the flows with no CP arrived in the current time slot as "pending." Then starting from the head of the *PendingList*, flows whose status is "active" are moved to the *ActiveList*. As explained earlier, if each flow is guaranteed to send at least one burst during its turn of service, once the length of *ActiveList* reaches N, the status of the rest of the flows in *PendingList* is all set to pending, and the construction of the new *ActiveList* is completed. In this case, since the positions of the active flows in *PendingList* are already known and we only have to select the first N of them, the amortized overhead increased due to the construction of *ActiveList* is only $O(1)$. As shown in [15], as long as $Q_{min}^{(k)} = \min_i\{Q_i^{(k)}\} \geq b$, where b is the burst length, it is guaranteed that at least one burst will be sent every time a CP queue is visited. This requirement can always be met by scaling all $Q_i^{(k)}$ if necessary. Consequently, the overall time complexity of the ASPFS algorithm is $O(1)$, the same as the DRR algorithm.

3.2   The Relative Fairness Bound of ASPFS

We now consider the fairness bound of ASPFS. Generally speaking, the fairness bound indicates how "unfair" an algorithm could be, thus a smaller bound usually means better fairness. There are two types of fairness bounds, the absolute fairness bound and the relative fairness bound. Denote the number of bursts of flow i scheduled during the interval (t_1, t_2) under some scheduling discipline D as $S_i^D(t_1, t_2)$, then the absolute fairness bound (AFB) is defined as the maximum value of $|S_i^D(t_1, t_2) - S_i^{GPS}(t_1, t_2)|$ for all flow i that are backlogged and over all possible intervals (t_1, t_2), where GPS (generalized processor sharing) is the ideal fair scheduling algorithm that can provide perfect fairness [17]. AFB is a preferred measure of fairness but difficult to obtain in most cases. On the other hand, the relative fairness bound (RFB) is defined as the maximum value of $|S_i(t_1, t_2)/w_i - S_j(t_1, t_2)/w_j|$, over all pairs of flows i, j that are backlogged and all possible intervals (t_1, t_2) [16]. In ASPFS, it is the maximum possible difference in service quantum between two flows whose states are either backlogged or pending. For most virtual-time based algorithms the RFB is 1, while for DRR it is 3. To calculate the RFB of ASPFS, suppose the interval (t_1, t_2) starts in slot m and ends in slot n, then we have

$$\left| \frac{S_i(t_1, t_2)}{w_i} - \frac{S_j(t_1, t_2)}{w_j} \right|$$

$$= \left| \frac{\sum_{k=m}^n Q_i^{(k)} - Credit_i}{w_i} - \frac{\sum_{k=m}^n Q_j^{(k)} - Credit_j}{w_j} \right|$$

$$= \left| \frac{\sum_{k=m}^n N w_i / W^{(k)} - Credit_i}{w_i} - \frac{\sum_{k=m}^n N w_j / W^{(k)} - Credit_j}{w_j} \right|$$

$$= \left| \frac{Credit_j}{w_j} - \frac{Credit_i}{w_i} \right| \leq B + T$$

Thus $RFB = B + T$. Generally the RFB of ASPFS is not so tight as that of the virtual-time based algorithms, but it is still a small constant which is independent of the length of (t_1, t_2). In other words, the difference between the service quantum received by any two active flows under ASPFS is bounded by $B+T$, no matter how long the time interval is on which the difference is measured. Meanwhile, compared to the $O(\log(n))$

complexity of virtual-time based algorithms, the $O(1)$ complexity of ASPFS is more desirable in a high speed environment such as OBS.

3.3   Implementing QoS Provisioning in ASPFS

Quality of Service (QoS) provisioning is indispensable for a WDM-based network if it is to support IP traffic, which integrates different types of services in nature. Conventionally, QoS requirements are in terms of end-to-end delay. However, in OBS networks the end-to-end delay is mainly determined by the assembling algorithm at edge nodes and the distance between the source and the destination rather than the scheduling algorithm at core nodes, since the data bursts are usually not buffered at any intermediate node. As a result, QoS in OBS networks generally refers to burst loss probabilities.

There exist several approaches for providing QoS in OBS networks, such as the offset time based approach [12] and strict priority approach [13]. The latter can be implemented in ASPFS without much difficulty – simply attach the priority of a flow to each of its control packet, and schedule those with high-priority before low-priority ones for each time slot. However, the potential problem of this approach is that, if a flow with high priority does not conform to its reserved bandwidth, all lower-priority flows will starve.

A more reasonable approach for ASPFS to provide QoS is to adjust the reserved bandwidth, or the weight of a flow according to the QoS requirement of this flow. If a flow requires low packet loss probability, we can assign it a bandwidth larger than its actual need. By doing so we guarantee that bursts of this flow have a better chance to be scheduled than those whose reserved bandwidth equals their actual rate. If the QoS requirement of a flow is looser, a small weight will suffice. We will examine how the packet loss probability changes with the weight of a flow through simulations in the next section.

4   Performance Evaluations

We have conducted simulations on an event-driven simulator to evaluate the performance of the proposed fair scheduling algorithm. For comparison purpose, we also implemented two representative OBS scheduling algorithms, Latest Available Unscheduled Channel with Void Filling (LAUC-VF) [3] and Pipeline buffering with Virtual Scheduling Technique (Pipeline with VST) [8]. The former is a well-known OBS scheduling algorithm with void filling, while the latter is a typical window-based scheduling scheme. The simulated topology is shown in Fig. 3. There are totally 11 edge nodes, all connected to the core node. Each link has N channels, among which one is the control channel and all others are data channels. We have conducted simulations for both $N = 64$ and $N = 128$. The bandwidth of a channel is $2.5Gb$. Edge node 0 to edge node 9 on the left side each generates a flow to edge node 10 on the right side. All of the flows pass through the core node. The guaranteed bandwidth of each flow is the same, i.e., $(N - 1) \times 2.5/10Gb/s$. The time slot is assumed to be the time to transmit a data burst plus the guard band. Flow 0 and flow 2 to 9 are sending Poisson traffic with the average rate equal to their guaranteed rate, but the ill-behaved flow 1 sends at three times of its guaranteed rate.

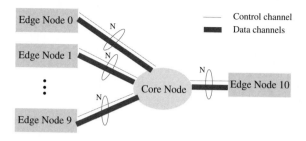

Fig. 3. Simulation topology

The simulation results are shown in Fig. 4 – Fig. 7. Since the core OBS network is bufferless, the criteria considered here are burst loss probability (BLP) and throughput. As can be seen, when the network is fully-loaded the performances of LAUC-VF and Pipeline with VST are very close to each other. Under both algorithms, the BLPs of all flows are very high for both $N = 64$ and $N = 128$, and the bandwidth grasped by flow 1 is much larger than that is preassigned to it, i.e., 10% of the total bandwidth. In other words, all flows suffer from the ill-behaved flow except itself, and the scheduler fails to provide fairness and isolation among flows. Since these two algorithm represent the void filling and window-based scheduling mechanisms, respectively, we can fairly predict that other OBS scheduling algorithms which do not take fairness into account would encounter the similar problem here. On the other hand, with ASPFS, the non-conforming flow 1 has a much larger burst loss ratio than other flows, and the throughput of all flows are almost equal, i.e., fairness is achieved. The reason why flow 1 still occupies a bandwidth slightly larger than the rest of the flows is due to the credit reset of some flows which actually received less service than they should but did not generate enough traffic before the credit threshold was reached. Most possibly, the bandwidth that should have been allocated to them was occupied by flow 1, since it sent much more bursts than other flows.

We also evaluated how the selection of the value of B and T affects the performance. On one hand, as shown in the simulation results, when B increases from 10 to 100, the bandwidth occupied by the ill-behaved flow also increases, and more bursts belonging to the normal flows are dropped. This observation conforms with our theoretical result that the fairness bound becomes looser when B is larger. Thus when $B = 0$ the smallest bound will be achieved, however, at the cost of low scheduling flexibility, since $B = 0$ means no credit loan is allowed. On the other hand, we can see that when T increases, the overall performance is almost unchanged. The reason is that the flows that will be affected by the change in T are those whose status keeps altering among unbacklogged, pending and active. And in this simulation, all flows are set to active in most of the time slots, which is the typical environment under which fair scheduling algorithms are evaluated [15] [18].

We also conducted simulations to test the QoS performance of ASPFS. Flows are randomly assigned a high or low priority. The reserved bandwidth for low-priority flows are equal to their actual rates, while for each high-priority flow the actual rate is smaller than the reserved bandwidth, with a ratio denoted as r. The BLPs of both high-priority

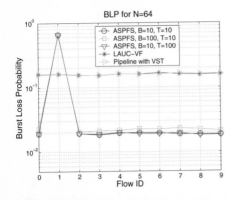

Fig. 4. Burst loss ratio of each flow under ASPFS and LAUC-VF when $N = 64$

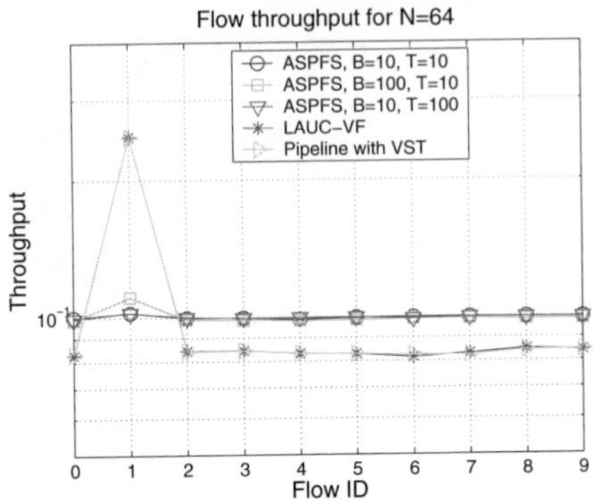

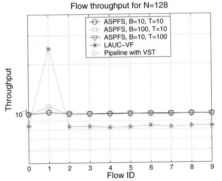

Fig. 5. Throughput of each flow under ASPFS and LAUC-VF when $N = 64$

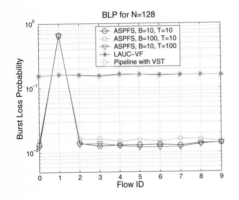

Fig. 6. Burst loss ratio of each flow under ASPFS and LAUC-VF when $N = 128$

Fig. 7. Throughput of each flow under ASPFS and LAUC-VF when $N = 128$

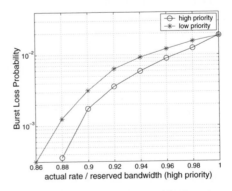

Fig. 8. Burst loss ratio of flows with different priorities

and low-priority flows under different r values are illustrated in Fig. 8. From the figure, we can see that not only the BLPs of high-priority flows are always smaller than that of low-priority ones when $r < 1$ (when $r = 1$, actually there is no service differentiation), but also the BLPs of high-priority flows change with the value of r. Hence ASPFS can satisfy the service requirement of a flow with regard to packet loss probability by adjusting the weight of the flow. The reason the BLPs of low-priority flows also become lower when r is smaller is that the total traffic load is lighter.

5 Conclusions

In this paper we have presented a new scheduling algorithm, ASPFS, for implementing fair scheduling in Optical Burst Switching (OBS) networks. The Slotted OBS (SOBS) architecture is adopted to overcome the difficulty of lack of large optical buffers in current optical networks. The newly proposed algorithm has shown a significant fairness improvement in bandwidth allocation compared to existing algorithms. We have also given an approach for providing QoS provisioning under ASPFS. Based on our theoretical analyses and simulation results, we believe ASPFS is a promising solution to provide fairness in OBS networks.

References

1. Choa, F.-S., Zhao, X., et al.: An optical packet switch based on WDM technologies. J. Lightwave Technology 23, 994–1014 (2005)
2. Qiao, C., Yoo, M.: Optical burst switching (OBS) - a new paradigm for an optical Internet. J. High Speed Networks 8(1), 69–84 (1999)
3. Xiong, Y., Vandenhoute, M., Cankaya, H.: Control architecture in optical burst-switched WDM networks. IEEE J. Sel. A. Comm. 18, 1838–1851 (2000)
4. Turner, J.: Terabit burst switching. J. High Speed Networks 8, 3–16 (1999)
5. Turner, J.: Terabit burst switching progress report (10/00-3/01) (2001)
6. Xu, J., Qiao, C., Li, J., Xu, G.: Efficient channel scheduling algorithms in optical burst switched networks. In: Proc. INFOCOM 2003, vol. 3, pp. 2268–2278 (March 2003)
7. Li, J.K., Qiao, C., Xu, J.H., Xu, D.H.: Maximizing throughput for optical burst switching networks. In: Proc. INFOCOM 2004, vol. 3, pp. 1853–1863 (2004)
8. Li, H.L., Neo, H.M., Ian, T.L.: Performance of the implementation of a pipeline buffering system in optical burst switching networks. In: Proc. GLOBECOM 2003, vol. 5, pp. 2503–2507 (2003)
9. Li, H., Liak, M.T.W., Li-Jin, I.T.: A distributed monitoring-based fairness algorithm in optical burst switching networks. In: Proc. ICC 2004, vol. 3, pp. 1564–1568 (June 2004)
10. Kaheel, A., Alnuweiri, H.: Quantitative QoS guarantees in labeled optical burst switching networks. In: Proc. GLOBECOM 2003, vol. 3, pp. 1747–1753 (2004)
11. Yao, S., Mukherjee, B., Dixit, S.: Advances in photonic packet switching: an overview, IEEE Commun. Mag., vol. 38(2) (February 2000)
12. Yoo, M., Qiao, C., Dixit, S.: Optical burst switching for service differentiation in the nex generation optical internet. IEEE Commun. Mag., pp. 98–104 (2001)
13. Vokkarane, V.M., Jue, J.: A strict priority scheme for quality of service provisioning in optical burst switching networks. In: Proc. ISCC 2003, pp. 16–21 (2003)

14. Ramamirtham, J., Turner, J.: Time-sliced Optical Burst Switching. In: Proc. INFOCOM 2003, pp. 2030–2038 (April 2003)
15. Shreedhar, M., Varghese, G.: Efficient fair queueing using deficit round-robin. IEEE/ACM Trans. Networking 4, 375–385 (1996)
16. Zhou, Y., Sethu, H.: On the relationship between absolute and relative fairness bounds. IEEE Commun. Letters 6(1), 37–39 (2002)
17. Parekh, A.K., Gallager, R.G.: A generalized processor sharing approach to flow control in integrated services networks: the single-node case. IEEE/ACM Trans. Networking 1(3), 344–357 (1993)
18. Bensaou, B., Tsang, D.H., Chan, K.T.: Credit-based fair queueing (CBFQ): a simply service-scheduling algorithm for packet-switched networks. IEEE/ACM Trans. Networking 9(5), 591–604 (2001)

Optical Packet Switches Enhanced with Electronic Buffering and Fixed Wavelength Conversion*

Zhenghao Zhang and Yuanyuan Yang

Dept. of Electrical and Computer Engineering,
State University of New York, Stony Brook, NY 11794, USA

Abstract. Optical networks with Wavelength Division Multiplexing (WDM), especially Optical Packet Switching (OPS) networks, have attracted much attention in recent years. However, to make OPS practical, it is crucial to reduce the packet loss ratio at the switching nodes. In this paper, we study a new type of optical switching scheme for OPS which combines optical switching with electronic buffering. In the new scheme, the arrived packets that do not cause contentions are switched to the output fibers directly; other packets are switched to *shared* receivers and converted to electronic signals and will be stored in the buffer until being sent out by *shared* transmitters. Since full range wavelength converters are difficult to implement with available technology, we focus on the more practical fixed wavelength conversion, with which one wavelength can be converted to another fixed wavelength. We will show with both analytical models and simulations that the packet loss ratio can be greatly reduced with few additional receivers and transmitters. Thus, we believe that the proposed hybrid switching scheme can greatly improve the practicability of OPS networks.

Index Terms: Wavelength-division-multiplexing (WDM), Slotted, Loss ratio, Optical Packet Switching, Buffering, Fixed wavelength conversion.

1 Introduction

Optical networks have the potential of supporting ultra fast future communications because of the huge bandwidth of optics. In recent years, Optical Packet Switching (OPS) has attracted much attention since it is expected to have better flexibility in utilizing the huge bandwidth of optics than other types of optical networks [9,3,11,10]. However, OPS in its current form is not yet practical and appealing enough to the service providers because typical OPS switching nodes suffer heavy packet losses due to the difficulty in resolving packet contentions. The common methods for contention resolution include wavelength conversion and all-optical buffering [3,11], where wavelength conversion is to convert a signal on one wavelength to another wavelength, and all-optical buffering is to use Fiber-Delay-Lines (FDL) to delay an incoming signal for a specific amount of time proportional to the length of the FDL. Wavelength conversion is very effective but it alone cannot reduce the packet loss to an acceptable level, thus buffering has to be used. However, FDLs are expensive and bulky and can provide only

---

* The research work was supported in part by the U.S. National Science Foundation under grant number CCR-0207999.

L. Mason, T. Drwiega, and J. Yan (Eds.): ITC 2007, LNCS 4516, pp. 179–191, 2007.

very limited buffering capacity. Thus the main challenge in designing an OPS switch is to find more practical methods to buffer the packets to resolve contention.

For this reason, we propose to use electronic buffers in OPS switches. In our switch, the arrived packets that do not cause contentions are switched to the output fibers *directly*; other packets are switched to receivers and converted to electronic signals and will be stored in an electronic buffer until being sent out by transmitters. It is important to note that not all packets need be converted to electronic signals; such conversion is needed only for packets that cause contentions. Thus, the advantage of this scheme is that far less high-speed receivers and transmitters are needed compared to switches that convert every incoming packet to electronic signals, since if the traffic is random, it is likely that the majority of the arrived packets can leave the switch directly without having to be converted to electronic signals.

In addition to electronic buffering, the switch will also incorporate wavelength conversion techniques, as wavelength conversion has been proved to be very effective in improving the performance of optical switches. With wavelength conversion, packets on more congested wavelengths can be spread smoothly to other wavelengths. Wavelength conversion can be *full range*, with which a wavelength can be converted to any other wavelengths. However, full range wavelength converters are difficult to implement at current time [7]. Wavelength conversion can also be *limited range*, with which a wavelength can be converted to several, not all, wavelengths [7,5,6]. However, even limited range wavelength converter is expensive and difficult to implement because it still needs an inside tunable laser, which is expensive since it must be tuned at very high speed, usually in the order of several tens of ns with high accuracy and high reliability. Thus in this paper we will focus on a more practical wavelength conversion scheme, i.e., *fixed wavelength conversion*, with which a wavelength can be converted to another fixed wavelength. Unlike the other two types of wavelength converter, fixed wavelength converters can be readily implemented with currently existing technologies such as Cross Phase Modulation in semiconductor optical amplifiers (SOA) [4,14]. The advantage of a fixed wavelength converter is that it does not need tunable laser and uses only a fixed laser, thus is much less expensive. In addition, it can be very accurate and very reliable since the inside fixed laser can be optimized for the specific wavelength conversion. We will show that with fixed wavelength conversion, the packet loss ratio can be greatly reduced with few additional receivers and transmitters.

2   Related Works

Optical Packet Switching has been studied extensively in recent years and many switch architectures have been proposed and analyzed. For example, [11,12] considered all-optical switches with output buffer implemented by FDLs and gave analytical models for finding the relations between the size of the buffer and packet loss ratio. However, the results in [11,12] show that to achieve an acceptable loss ratio, the load per wavelength channel has to be quite light if the number of FDLs is not too large. Switches with shared all-topical buffer have been proposed, for example, in [1,2], in which all output fibers share a common buffer space implemented by FDLs. However, in this type of switches, to achieve an acceptable packet loss ratio, the number of FDLs is still large and is often no less than the number of input/output fibers, which increases the size of the switching fabric. To avoid the difficulty of all-optical buffering, the recent OSMOSIS project [13] proposed an optical switch with OEO conversion for every channel. In the OSMOSIS switch, all arriving signals are converted to electronic forms and stored in electronic buffer, and then they will be converted back to optical form before entering the switching fabric. The advantage of such an approach is that it needs no optical buffer and does not increase the size of the switching fabric; however, the disadvantage is the expected high cost since it needs high speed receivers, high speed electronic memories and high speed tunable transmitters for every channel. The switch proposed in this paper also uses electronic buffer, however, less buffer, transmitters and receivers are needed because they are *shared* by all channels.

3   Architecture of the Hybrid Switch

The hybrid switch to be studied in this paper is shown in Fig. 1. It has N input/output fibers, and on each fiber there are k wavelengths. It operates in a time slotted manner and receives packets of one time slot long at the beginning of time slots. The composite signal coming from one input fiber will first be sent to a demultiplexer, in which signals on different wavelengths are separated from one another. The separated signal on one wavelength will then be sent to a *wavelength converter* to be converted to another wavelength if needed. The signal is then sent to a switching fabric. If there is no contention on the destination fiber, i.e., no two packets on the same wavelength want to go to the same destination fiber at the same time, the signal will be routed by the switching fabric to the destination fiber to allow the packet to leave the switch. Otherwise, the signal will be routed to a *receiver fiber*, which is shown in the right side of the figure under the output fibers. All signals sent to the receiver fiber will be first combined by the multiplexer and then be demultiplexed, and each of the demultiplexed signal will be sent to a receiver to be converted to electronic forms and stored in the electronic buffer. The packets stored in the buffer can be sent back to the switching fabric by the transmitters, which are *fast tunable lasers* that can be tuned to *any* wavelength. If there are U receiver fibers and T transmitters, the switching fabric should be able to route any of the Nk signals from the input fibers to any of the $N + U$ output fibers and receiver fibers. It should also be able to route any of the T signals from the transmitters to any of the N output fibers.

Note that so far we have assumed that all arrived packets should be sent to the output fibers. However, some arrived packets need not be sent to the output fibers because they have already reached their destination, which is the node where the switch is located. We call such packets the "to-local" packets. There are also some packets collected by this node from local area networks attached to it, which should be sent to the output fibers of the switch to reach remote nodes in the network. We call such packets the "from-local" packets. Accordingly, we call all other packets that are only passing by this node the "non-local" packets. To receive the to-local packets, the switch must be equipped with some receivers to which these packets can be routed to; similarly, to send the from-local packets, the switch must be equipped with some transmitters that can be used to send the packets to the output fibers. (The receivers and transmitters are also

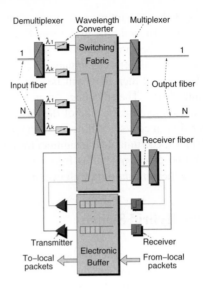

Fig. 1. A WDM switch enhanced with shared electronic buffer

referred to as the "droppers" and the "adders" in some optical networks, respectively.) In previous works on optical switches, the receivers and transmitters are only used for the to-local packets or the from-local packets. Thus, what we are proposing in this paper can be understood as *opening such resources to the non-local packets*, i.e., to allow the non-local packets to be received by the receivers and sent by the transmitters to help improving the performance of the switch.

The switch will try to route the non-local packets to the output fibers first, and will route packets stored in the buffer to the output fiber only if there are unused wavelengths on the output fibers. We call the non-local packets that cannot be sent to the output fibers directly the "leftover" packets. The switch will route the to-local packets to the receiver fibers first, and will route the leftover packets to the receiver fibers only if there are unused wavelengths on the receiver fibers. This is because on average, the to-local packets have traveled longer distance than the leftover packets before reaching this node, thus dropping to-local packets will waste more network resources than dropping leftover packets. The buffer is organized into N queues, one for each output fiber. The leftover packets and the from-local packets are stored in the same queue if they are for the same destination. The leftover packets and the from-local packets will be sent by the transmitters to the output fibers with equal probability.

4   WDM Switch with Fixed Wavelength Conversion

In this section we study the performance of WDM switches with fixed wavelength conversion. We wish to find the number of receivers and transmitters needed to reduce the packet loss ratio to an acceptable level.

First, we give a convenient method to represent fixed wavelength conversion.

4.1   Systematic Wavelength Conversion

If wavelength conversion is full range, any wavelength can be converted to any other wavelength. However, if wavelength conversion is fixed, one wavelength can only be converted to one other wavelength, and we should decide which wavelength it should be converted to. We only consider *balanced* wavelength conversion, in which one wavelength can be converted *to* and *from* only one other wavelength. Wavelength conversion can be illustrated by a bipartite graph, with left side vertices representing incoming wavelengths and right side vertices representing outgoing wavelengths. A left side vertex is connected to a right side vertex if they represent the same wavelength or if this incoming wavelength can be converted to this outgoing wavelength. For example, Fig. 2 (a) and (b) show two balanced wavelength conversions. Fig. 2(b) is called *systematic* wavelength conversion, in which λ_i is converted to λ_{i+1} for $1 \le i \le k-1$ and λ_k is converted to λ_1, while Fig. 2(a) shows a non-systematic wavelength conversion. Apparently, systematic wavelength conversion is easier to handle than non-systematic wavelength conversions because of its regularity. We show that all balanced wavelength conversions are equivalent to systematic wavelength conversion and thus only systematic wavelength conversion needs to be considered.

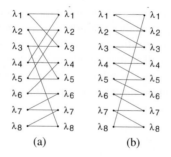

(a)   (b)

Fig. 2. (a) A balanced wavelength conversion for 8 wavelengths. (b) A systematic wavelength conversion.

Theorem 1. *After rearranging the indices of the wavelengths, any balanced wavelength conversion can be represented by systematic wavelength conversion or several non-overlapping "smaller" systematic wavelength conversions.*

Proof. The process of rearranging the indices of the wavelengths is described as follows. We use λ_i^{old} to denote the ith wavelength under the original indices and use λ_i to denote the ith wavelength under the new indices. Let $conv[\lambda_i]$ represent the wavelength that λ_i can be converted to. First, let λ_1 be λ_1^{old} and let $i = 1$. Then, *repeatedly*, let λ_{i+1} be $conv[\lambda_i]$ and increment i by one, until $conv[\lambda_i]$ has already been given a new index.

Note that the process must stop somewhere since every time it is repeated, a wavelength must be given a new index and there are a finite number of wavelengths. Suppose when the process ends, $i = I$. We claim that $conv[\lambda_I]$ must be λ_1. This is because at this time, only $\lambda_1, \lambda_2, \ldots, \lambda_I$ have been given new indices. Note that $\lambda_2, \lambda_3, \ldots, \lambda_I$

can be converted from a wavelength *other* than λ_I; indeed, λ_j can be converted from λ_{j-1} for $2 \leq j \leq I$. Since in balanced wavelength conversion, one wavelength can be converted from exactly one other wavelength, the wavelength λ_I can be converted to must be λ_1. Thus by definition λ_1, λ_2, ..., λ_I constitute a systematic wavelength conversion with I wavelengths.

If $I = k$, the balanced wavelength conversion can be rearranged into a systematic wavelength conversion with k wavelengths. Otherwise this reindexing process can be executed one more time for the rest of the wavelengths, and each time it is executed a systematic wavelength conversion with less than k wavelengths is found.   ∎

For example, consider the balanced wavelength conversion shown in Fig. 2(a). A systematic wavelength conversion with 6 wavelength can be found first, in which λ_1 to λ_6 correspond to λ_1^{old}, λ_3^{old}, λ_8^{old}, λ_7^{old}, λ_6^{old} and λ_4^{old}, respectively. After that, a systematic wavelength conversion with 2 wavelengths can be found, which includes λ_2^{old} and λ_5^{old}.

4.2   Assigning Packets to Output Fibers

When non-local packets arrived at the switch, we wish to send the maximum number of them directly to the output fibers, such that the packet loss and packet delay can be minimized. If the wavelength conversion is full range, this problem is trivial, since we need only to count the number of packets destined for each output fiber: if the number is no more than k, grant all packets; otherwise arbitrarily grant k packets. When the wavelength conversion is not full range this problem becomes more complicated because the wavelengths of the packets will matter. However, fortunately, with systematic wavelength conversion the scheduling algorithm is relatively simple and can be described as follows. Note that we are only considering non-local packets since local packets need not to be sent to the output fibers.

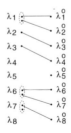

Fig. 3. A schedule found by the algorithm in Table 1 when there are $[2, 1, 1, 0, 0, 2, 2, 0]$ packets arrived on λ_1 to λ_8, respectively, when λ_8 was chosen as the pivot wavelength. The vertices on the left represent packets and vertices on the left represent output wavelength channels.

To grant maximum number of packets at the output fibers, packets destined to different output fibers can be scheduled independently. Thus, we will describe the scheduling algorithm for only one output fiber, as shown in Table 1. For clarity, the input wavelength is denoted as λ_i and the output wavelength is denoted as λ_i^o. The number of packets arrived for this output fiber on λ_i is denoted as X_i. For example, in Fig. 3

Table 1. Scheduling Algorithm at an Output Fiber

---

if $X_i > 0$ for all $1 \leq i \leq k$
       Send one packet on each wavelength; **exit**;
end if;
From wavelengths on which 0 packet arrives,
randomly choose one as the pivot wavelength, say, λ_k.
for $i := 1$ to $k - 1$ **do**
       if $X_i = 0$ **continue**;
       if λ_i^o has not been used
              Assign one packet on λ_i to λ_i^o.
              $X_i := X_i - 1$.
       end if
       if $X_i = 0$ **continue**;
       Assign one packet on λ_i to λ_{i+1}^o.
end for

---

where left side vertices represent the arrived packets and right side vertices represent output wavelength channels, and X_1 to X_8 are $[2, 1, 1, 0, 0, 2, 2, 0]$, respectively. First note that if $X_i > 0$ for all $1 \leq i \leq k$, the scheduling is trivial, since we can simply pick one packet on each wavelength and send it out on its own wavelength without wavelength conversion. Thus we need only to consider when there are some $X_i = 0$. If $X_i = 0$, λ_i can be chosen as the "pivot" wavelength, which means that the algorithm will scan the arrived packets according to their wavelengths in the order of $\lambda_{i+1}, \lambda_{i+2}$, $\ldots, \lambda_k, \lambda_1, \ldots \lambda_{i-1}$. Without loss of generality, suppose λ_k has been chosen as the pivot wavelength. The basic rule of assigning packets to the output wavelength channels is very simple: When scanning to λ_i, *try to use λ_i^o first; if λ_i^o has been used, use λ_{i+1}^o*. For example, in Fig. 3, since λ_1 is the first wavelength to be scanned, a packet on λ_1 is first assigned to λ_1^o and the remaining packet on λ_1 is assigned to λ_2^o. Then, when scanning to λ_2, since λ_2^o has already been assigned, the packet on λ_2 is assigned to λ_3^o, and so on. Since the wavelengths scanned first have more chances to send packets, to ensure fairness to each wavelength channel, the pivot wavelength is chosen randomly which can be realized in a round-robin manner. We can show that this algorithm finds a schedule that can grant maximum number of packets in $O(k)$ time. Due to the limited space, the proof is omitted here.

4.3   The Minimum Number of Transmitters

With the algorithm in Table 1, the maximum number of non-local packets are sent to the output fibers. The leftover packets should be sent to the electronic buffers and will be sent to the output fibers by the transmitters when there are free wavelength channels on the output fibers. Clearly, if there are not enough transmitters, packets will accumulate in the buffer which will eventually cause the buffer to overflow. We wish to find the minimum number of transmitters needed in the switch such that the buffer will not overflow.

Note that since the from-local packets are also sent by the transmitters, what we focus on here is the *additional* number of transmitters that are needed to serve the leftover packets. By queuing theory, the number of additional transmitters should be no less than the average number of packets that are leftover, denoted by $E(L)$. We will give a closed-form solution for $E(L)$ by taking advantage of the symmetry of the systematic wavelength conversion.

We denote the average arrival rate at an input channel, defined as the average number of packets arrived at a wavelength channel in a time slot, as $\rho = \rho_l + \rho_n$, where ρ_l and ρ_n are the arrival rates of to-local packets and non-local packets, respectively. We assume that at a time slot, the arrived packets have independent random destinations. Note that this assumption is likely to be true in large networks where traffics have a wide varieties of destinations. We also assume that the traffic arriving at different time slots are independent of each other.

We will derive $E(L)$ by focusing on one wavelength, say, λ_1, since all wavelengths are equivalent. Let the number of packets arrived on λ_1 for this output fiber and sent out directly to the output fiber be W, where, apparently, W can be 0, 1 or 2 since there are at most two output wavelength channels the arrived packets can be assigned to. Note that $E(L) = Nk\rho_n - NkE(W)$, thus we will only focus on finding $E(W)$.

First, let F be the event that there is at least one wavelength on which zero packet arrives. We have

$$E(W) = E(W|F)P(F) + E(W|\bar{F})[1 - P(F)], \tag{1}$$

where $\bar{F}$ is the event that F did not occur. Note that $E(W|\bar{F}) = 1$, since if on each wavelength there is at least one packet arrived, by the algorithm, one packet will be sent to the output fiber on each wavelength. Hence, Eq. (1) reduces to $E(W) = E(W|F)$ $P(F) + [1 - P(F)]$. To find $P(F)$, let the total number of non-local packets arrived on one wavelength be X. X is a Binomial random variable $B(N, \rho_n/N)$. Since the numbers of packets arrived on different wavelengths are independent of each other, we have

$$P(F) = 1 - [1 - P(X = 0)]^k. \tag{2}$$

Thus only one term, $E(W|F)$, needs to be determined.

To find $E(W|F)$, note that when F occurs, if λ_1 is chosen as the pivot wavelength, $W = 0$, since only the wavelength on which 0 packet arrives can be chosen as the pivot wavelength. Thus, let G be the event that given F occurs, λ_1 was not chosen as the pivot wavelength. We have $E(W|F) = E(W|G)$.

For the moment suppose we know $P(X = x|G)$ which is the probability that there are x packets on λ_1 given G occurs. $E(W|G)$ will be determined by $P(X = x|G)$ and whether the output wavelength λ_1^o has been assigned to packets on another wavelength (in this case, λ_k) or not when scanning to λ_1. Formally speaking, let $\lambda_{i'}$ be the ith scanned wavelength where $1 \leq i \leq k - 1$. For example, if the pivot wavelength is λ_2, $i' = 3$ for $i = 1$, $i' = 4$ for $i = 2$, etc. Let q_i be the probability that output wavelength $\lambda_{i'}^o$ has been assigned when scanning to $\lambda_{i'}$ (before assigning packets on $\lambda_{i'}$). Since the pivot wavelength is randomly chosen, the probability that a wavelength is the ith scanned wavelength is $1/(k - 1)$ for all i. Thus the average number of packets arrived on λ_1 that are sent out directly to the output fiber given G occurs is

$$E(W|G) = \frac{1}{k-1} \sum_{i=1}^{k-1} [q_i A + (1 - q_i) B],\tag{3}$$

where $A = P(X \geq 1|G)$ and $B = 2P(X \geq 2|G) + P(X \geq 1|G)$, since if λ_1^o has been assigned, $W = 1$ if $X \geq 1$; if λ_1^o has not been assigned, $W = 2$ if $X \geq 2$ and $W = 1$ if $X \geq 1$.

Apparently, $q_1 = 0$ since there is no packet arrived on the pivot wavelength. Note that if $\lambda_{i'}^o$ has been assigned at step i, the event that $\lambda_{(i+1)'}^o$ is assigned at step $i + 1$ occurs when there are more than one packets arrived on $\lambda_{i'}$, i.e., $P(X \geq 1|G)$. If $\lambda_{i'}^o$ has not been assigned at step i, the event that $\lambda_{(i+1)'}^o$ is assigned at step $i + 1$ occurs when there are more than two packets arrived on $\lambda_{i'}$, i.e., $P(X \geq 2|G)$. Thus, there is a recursive relation

$$q_{i+1} = q_i P(X \geq 1|G) + (1 - q_i) P(X \geq 2|G).$$

which can be reduced to $q_{i+1} = aq_i + b$, where $a = P(X = 1|G)$ and $b = P(X \geq 2|G)$. Given the boundary condition $q_1 = 0$, we have $q_2 = b$, $q_3 = ab + b$, $q_4 = a^2 b + ab + b$, ..., and thus $q_i = \frac{b}{1-a}\left(1 - a^{i-1}\right)$. Eq. (3) can now be simplified to a closed form

$$E(W|G) = (k - 1)B + \frac{b(A - B)}{1 - a}\left[(k - 1) - \frac{1 - a^{k-1}}{1 - a}\right].\tag{4}$$

The only remaining problem is thus to find $P(X = x|G)$. First let $P(X = x|F)$ be the probability that $X = x$ given event F occurs. By Bayesian's law, we have

$$P(X = x|F) = \frac{P(X = x, F)}{P(F)} = \frac{P(F|X = x)P(X = x)}{P(F)},$$

where, apparently,

$$P(F|X = x) = \begin{cases} 1 & x = 0 \\ 1 - [1 - P(X = 0)]^{k-1} & \text{otherwise} \end{cases}\tag{5}$$

$P(X = x|G)$, which is the probability that $X = x$ given event G occurs, can be found in a similar way:

$$P(X = x|G) = \frac{P(X = x, G)}{P(G)} = \frac{P(G|X = x)P(X = x)}{P(G)}.$$

Thus, to find $P(X = x|G)$, we need $P(G|X = x)$ and $P(G)$. Note that

$$P(G|X = x) = \begin{cases} \frac{z}{z+1} P(F|X = x) & x = 0 \\ 1 \cdot P(F|X = x) & \text{otherwise,} \end{cases}\tag{6}$$

where z is the number of wavelengths among λ_2 to λ_k on which zero packet arrives, given F occurs. We have

$$P(Z = z) = \binom{k-1}{z} P(X = 0|F)^z [1 - P(X = 0|F)]^{k-1-z},$$

and thus

$$P(G) = P(X = 0) \sum_{z=1}^{k-1} P(Z = z) \frac{z}{z+1} + P(X > 0) \left\{ 1 - [1 - P(X = 0)]^{k-1} \right\}$$

Combining the discussions above, the average number of leftover packets is

$$E(L) = Nk \left\{ \rho_n - E(W|G)P(G)P(F) - [1 - P(F)] \right\}, \tag{7}$$

where $P(F)$, $E(W|G)$ and $P(G)$ can be found in Eq. (2), Eq. (4) and Eq. (7), respectively.

The numerical results of the analytical model and the simulations are shown in Fig. 4, from which we can see that the analytical model is accurate.

4.4   Assigning Packets to Receiver Fibers

After assigning wavelength channels on the output fibers to the non-local packets, we should assign the wavelength channels on the receiver fibers to the to-local packets and the leftover packets. As mentioned earlier, we will first assign the to-local packets to the receivers since they have a higher priority, after that, the leftover packets will be assigned to the remaining receivers. The methods for assigning the to-local packets and non-local packets are very much the same, and thus in the following only the method for to-local packets is described. The algorithm is shown in Table 2 which is similar to the algorithm shown in Table 1.

We use V_j to denote the number of receivers on λ_j that have not been assigned to any packets. Initially, $V_j = U$ for all j. At first we will find the minimum number of to-local packets arrived on a wavelength, denoted by X_{min}, and without loss of generality, let λ_1 be this wavelength. Apparently, if $X_{min} \geq U$, no scheduling is needed. If $X_{min} < U$, it can be shown that there must be a schedule granting maximum number of packets in which the X_{min} packet on λ_1 are granted. There are $X_{min} + 1$ ways of assigning these packets: assign l of them to receivers on λ_1^o and the rest to receivers on λ_2^o, where $0 \leq l \leq X_{min}$. The algorithm will try all such possible assignments and count the number of packets that have to be dropped in each possible assignment, which are stored in D_l for $0 \leq l \leq X_{min}$. Then the one with minimum D_l will be chosen as the schedule. Note that since $X_{min} < U$, the algorithm needs to try at most U times, and to further reduce the scheduling time, it can be carried out in *parallel* by U decision making units and the time complexity can be reduced to $O(k)$.

4.5   The Minimum Number of Receivers

The number of receivers under fixed wavelength conversion is very difficult to obtain analytically because many random variables are dependent upon one another. Thus we have used simulation methods to find it. The numerical results and simulation results will be provided in the next section.

Table 2. Scheduling Algorithm at Redundant Fibers

Let λ_1 be the wavelength arrived X_{min} packets. Let $D_l = 0$ for all $0 \leq l \leq X_{min}$.
for $l := X_{min}$ **downto** 0 **do**
    Let $V_i = U$ for all $1 \leq i \leq k$. Assign l packets on λ_1 to λ_1^o.
    $V_1 := V_1 - l$.
    Assign $X_1 - l$ packets on λ_1 to λ_2^o.
    $V_2 := V_2 - X_1 + l$.
    for $i := 2$ to k **do**
        if $X_i = 0$ **continue**;
        $t_0 := \min(X_i, V_i)$.
        Assign t_0 packet on λ_i to λ_i^o.
        $V_i := V_i - t_0$. $X_i := X_i - t_0$.
        if $X_i = 0$ **continue**;
        $t_1 := \min(X_i, V_{i+1})$.
        Assign t_1 packet on λ_i to λ_{i+1}^o.
        $X_i := X_i - t_1$. $V_{i+1} := V_{i+1} - t_1$. $D_l := D_l + X_i$.
    end for
end for
Choose the lth schedule if D_l is minimum.

5 Numerical and Simulation Results

In this section we present our numerical results and simulation results shown in Fig. 4 and Fig. 5. The results are for a switch where $N = 8$, $k = 16$, and $\rho_n/\rho = 0.9$. We have conducted numerical studies and simulations on switches of other sizes and other traffic loads, and the results are similar. As a comparison, switches with full range wavelength conversion and no wavelength conversion are also shown.

Fig. 4 shows the average number of leftover packets in the switch, which is the minimum number of additional transmitters needed. It can be seen that unsurprisingly, the switch with full range wavelength conversion needs the fewest number of transmitters and the switch with no wavelength conversion needs the most number of transmitters. It can also be seen that by using fixed wavelength conversion, the number of transmitters can be reduced by a great amount comparing to no wavelength conversion. For example, when $\rho = 0.8$, only 9 transmitters are needed for the former while 25 are needed for the latter.

Fig. 5 shows the packet loss ratio of the to-local packets and the non-local packets. Again, full range wavelength conversion has the best performance and no wavelength conversion has the worst. In fact, the performance of no wavelength conversion is hardly acceptable, for example, it will need 96 receivers when the traffic load is 0.8 to make the packet loss ratio of both the to-local packets and the non-local packets less than 10^{-6} when $N = 8$, $k = 16$, that is, 6 receiver fibers are needed in the switch. However, to achieve the same performance, fixed wavelength conversion needs 64 receivers. Note that even with full range wavelength conversion at least 28 receivers are needed to make the packet loss ratio less than 10^{-6}. Considering the tremendous difficulties in making full range wavelength converters, we believe fixed wavelength conversion is a viable

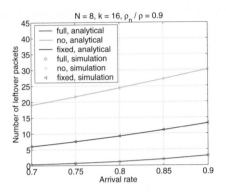

Fig. 4. The average number of leftover packets

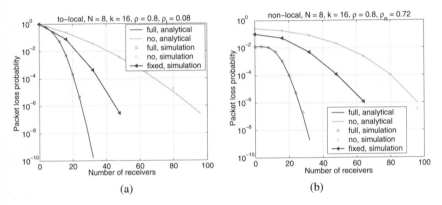

Fig. 5. (a) Packet loss probability of to-local packets. (b) Packet loss probability of non-local packets.

means for providing wavelength conversion abilities in optical switches and should be used.

We can also make a comparison between the proposed switch and a switch without electronic buffer. We note that the performance of the latter is very poor, since its packet loss probability (PLP) which can be observed when the number of receivers is 0 in Fig. 5(b) is higher than 10^{-2} even with full range wavelength conversion, while in the proposed switch the PLP can be reduced to lower than 10^{-6} with only 4 redundant fibers along with fixed wavelength conversion.

6   Conclusions

In this paper we have studied a new type of optical switch which combines electronic buffering with optical switching. In this switch not all optical packets need to be converted to electronic form and only those that cannot be sent to the output fibers are converted by shared receivers to be stored in the buffer. We focused on fixed wavelength conversion as it is the most practical type of wavelength conversion today. We showed that the performance of the switch is far better than switches without electronic buffer. We also showed with fixed wavelength conversion, the cost of the switch can be expected to be low since it needs a relatively small number of receivers and even a smaller number of transmitters. We therefore believe that this new switching scheme can greatly improve the practicability of OPS networks and should be used in future optical networks.

References

1. Bendeli, G. et al.: Performance assessment of a photonic ATM switch based on a wavelength controlled fiber loop buffer, OFC'96 Technical Digest, pp. 106–107, OFC (1996)
2. Hunter, D.K., et al.: WASPNET: a wavelength switched packet network. IEEE Communications Magazine 37(3), 120–129 (1999)
3. Xu, L., Perros, H.G., Rouskas, G.: Techniques for optical packet switching and optical burst switching. IEEE Communications Magazine, pp.136–14 (January 2001)
4. Ramaswami, R., Sivarajan, K.N.: Optical Networks: A Practical Perspective, 1st edn. Academic Press, San Diego (2001)
5. Zhang, Z., Yang, Y.: Optimal scheduling in WDM optical interconnects with arbitrary wavelength conversion capability. IEEE Trans. Parallel and Distributed Systems 15(11), 1012–1026 (2004)
6. Zhang, Z., Yang, Y.: Optimal scheduling in buffered WDM packet switching networks with arbitrary wavelength conversion capability. IEEE Trans. Computers 55(1), 71–82 (2006)
7. Ramaswami, R., Sasaki, G.: Multiwavelength optical networks with limited wavelength conversion. IEEE/ACM Trans. Networking 6, 744–754 (1998)
8. http://www.emcore.com/assets/fiber/ds00-306_EM.pdf
9. Mukherjee, B.: WDM optical communication networks:progress and challenges. JSAC 18(10), 1810–1824 (2000)
10. El-Bawab, T.S., Shin, J.-D.: Optical packet switching in core networks: between vision and reality. IEEE Communications Magazine 40(9), 60–65 (2002)
11. Danielsen, S.L., Joergensen, C., Mikkelsen, B., Stubkjaer, K.E.: Analysis of a WDM packet switch with improved performance under bursty traffic conditions due to tunable wavelength converters. J. Lightwave Technology 16(5), 729–735 (1998)
12. Danielsen, S.L., Mikkelsen, B., Joergensen, C., Durhuus, T., Stubkjaer, K.E.: WDM packet switch architectures and analysis of the influence of tunable wavelength converters on the performance. J. Lightwave Technology 15(2), 219–227 (1998)
13. Hemenway, R., Grzybowski, R.R., Minkenberg, C., Luijten, R.: Optical-packet-switched interconnect for supercomputer applications. J. Optical Networking 3(12), 900–913 (2004)
14. Ma, B., Nakano, Y., Tada, K.: Novel all-optical wavelength converter using coupled semiconductor optical amplifiers. Lasers and Electro-Optics, CLEO 98, pp. 477–478 (1998)

Impact of Complex Filters on the Message Throughput of the ActiveMQ JMS Server*

Robert Henjes, Michael Menth, and Valentin Himmler

University of Würzburg, Institute of Computer Science
Am Hubland, D-97074 Würzburg, Germany
Phone: (+49) 931-888 6644; Fax: (+49) 931-888 6632
{henjes,menth,himmler}@informatik.uni-wuerzburg.de

Abstract. In this paper we investigate the maximum message throughput of the ActiveMQ server in different application scenarios. We use this throughput as a performance criterion. It depends heavily on the installed filters and the message replication grade. In previous work, we have presented measurement results and an analytical model for simple filters. This work extends these studies towards more complex configuration options. It provides measurement results and analytical performance models for complex AND-, OR-, and IN-filters. The results are useful to understand the performance of JMS servers and help to dimension large distributed JMS-based systems.

1 Introduction

The Java Messaging Service (JMS) is a communication middleware for distributed software components. It is an elegant solution to make large software projects feasible and future-proof by a unified communication interface which is defined by the JMS API provided by Sun Microsystems [1]. A salient feature of JMS is that applications can communicate with each other without knowing their communication partners as long as they agree on a uniform message format. Information providers publish messages to the JMS server and information consumers subscribe to certain types of messages at the JMS server to receive a certain subset of these messages. This is known as the publish/subscribe principle.

In the non-durable and persistent mode, JMS servers efficiently deliver messages reliably to subscribers that are presently online. Therefore, they are suitable as backbone solution for large-scale realtime communication between loosely coupled software components. For example, some user devices may provide presence information to the JMS. Other users can subscribe to certain message types, e.g. the presence information of their friends' devices. For such a scenario, a high performance routing platform needs filter capabilities and a high capacity to be scalable to a large number of users. In particular, the throughput capacity of the JMS server should not suffer from a large number of clients or filters.

---

* This work was funded by Siemens AG, Munich. The authors alone are responsible for the content of the paper.

L. Mason, T. Drwiega, and J. Yan (Eds.): ITC 2007, LNCS 4516, pp. 192–203, 2007.

In previous work we have measured and modelled the message throughput of the ActiveMQ server depending on the number of installed simple filters n_{fltr} and the replication grade r of the messages. OR- and AND-filters are more complex as they may have different numbers of filter components. We also observed that the message throughput of the server decreases significantly with an increasing length of these complex filters. In this paper, we design suitable experiment series, perform a large number of measurements, and extend the previously found model to cover the server behavior in the presence of complex filters. The formula is still simple and can be used by engineers to predict the server performance for special use cases.

The paper is organized as follows. In Section 2 we present JMS basics that are important for our study and consider related work. In Section 3 we explain our test environment and measurement methodology. Section 4 develops the experiment design, shows measurement results, and Section 5 proposes a model for the processing time of a simple message depending on the server configuration and validates it by the obtained measurement data. Finally, we summarize our work in Section 6.

2   Background

In this section we describe the Java messaging service (JMS) and discuss related work.

2.1   The Java Messaging Service

Messaging facilitates the communication between remote software components. The Java Messaging Service (JMS) is one possible standard of this message exchange. So-called publishers connect to the JMS server and send messages to it. So-called subscribers connect to the JMS server and consume available messages or a subset thereof. So the JMS server acts as a relay node [2], which controls the message flow by various message filtering options. This architecture is depicted in Figure 1. Publishers and subscribers rely on the JMS API [1] and the JMS server decouples them by acting as a broker. As a consequence, publishers and subscribers do not need to know each other.

The JMS offers two different connection modes: a durable and a non-durable connection type. If a subscriber connects in the durable mode, the messages will be stored for delivery if this client disconnects. All stored messages will be delivered when the client connects next time to the JMS server. In the non-durable mode, messages are forwarded only to subscribers who are presently online. Persistence is another option for JMS. If the persistent option is set, each message has to be delivered reliably to all actively connected clients, which is ensured by confirming reception with acknowledgments. In the non-persistent mode the JMS server must deliver the message only with an at-most-once guarantee. This means that the message can be lost, but it must not be delivered twice according to [1]. In this study, we only consider the persistent but non-durable mode.

Information providers with similar themes may be grouped together by making them publish to a so-called common „topic"; only those subscribers having subscribed for that specific topic receive their messages. Thus, topics virtually separate the JMS server into several logical sub-servers. Topics provide only a very coarse and static method for

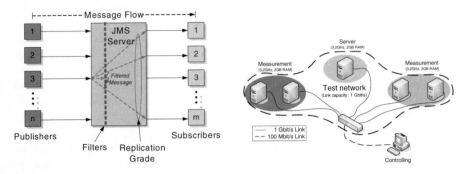

Fig. 1. The JMS server delivers messages from the publishers to all subscribers with matching filters

Fig. 2. Testbed environment

message selection due to the fact that publishers and subscribers have to know which topics they need to connect to. This results in a slight loose of the decoupling feature in the publish/subscribe context. In addition, topics need to be configured on the JMS server before they can be used actively. If no topics are explicitly introduced at the JMS server, exactly one default topic is present, to which all subscribers and publishers are connected.

Filters are another option for message selection. A subscriber may install a message filter on the JMS server. Only the messages matching the filter rules are forwarded to the respective subscriber instead of all messages. In contrast to topics, filters are installed dynamically during the operation of the server by each subscriber.

A JMS message consists of three parts: the fixed header, a user defined property header section, and the message payload itself [1]. So-called correlation IDs are ordinary strings that can be set in the fixed header of JMS messages as the only user definable option within this header section. Correlation ID filters try to match these IDs. Several application-specific properties may be set in the property section of the JMS message. Application property filters try to match these properties whereby wildcard filtering is possible, e.g., in the form of ranges like $[\#7; \#13]$, which means all IDs between $\#7$ and $\#13$ are matched including $\#7$ and $\#13$. Unlike correlation ID filters, a combination of different properties may be specified which leads to more complex filters with a finer granularity.

In this work we consider only application property filters, which search for so called StringProperties. Further investigations on this topic are published in [3]. We call a filter, which is searching only for one StringProperty (value), simple filter. If a filter contains logical operators, like "OR" or "AND" as concatenating elements of different components of this filter, we call it a complex filter. A complex filter searching for StringProperties is structured like the following example:

```
ID_1 = "0000" AND ID_2 = "0001" AND ... AND ID_x = "0000"
```

Corresponding to the structure of the filter the sent messages contain matching pairs of keys and values, which are set in the application property header part of a message.

2.2   Related Work

The JMS is a wide-spread and frequently used middleware technology. Therefore, its throughput performance is of general interest. Several papers address this aspect already but from a different point of view and in different depth.

The throughput performance of four different JMS servers is compared in [4]: FioranoMQ [5], SonicMQ [6], TibcoEMS [7], and WebsphereMQ [8]. The study focuses on several message modes, e.g., durable, persistent, etc., but it does not consider filtering, which is the main objective in our work. The authors of [9] conduct a benchmark comparison for the SunMQ [10] and IBM WebsphereMQ. They tested throughput performance in various message modes and, in particular, with different acknowledgement options for the persistent message mode. They also examined simple filters, but they did not conduct parametric studies, and no performance model was developed. The objective of our work is the development of such a performance model to forecast the maximum message throughput for given application scenarios. A proposal for designing a "Benchmark Suite for Distributed Publish/Subscribe Systems" is presented in [11] but without measurement results. The setup of our experiments is in line with these recommendations. General benchmark guidelines were suggested in [12] which apply both to JMS systems and databases. However, scalability issues are not considered, which is the intention of our work. A mathematical model for a general publish-subscribe scenario in the durable mode with focus on message diffusion without filters is presented in [13] but without validation by measurements. The same authors present in [14] an enhanced framework to analyze and simulate a publish/subscribe system. In this work also filters are modeled as a general function of time but not analyzed in detail. The validation of the analytical results is done by comparing them to a simulation. In contrast, our work presents a mathematical model for the throughput performance in the non-durable mode including several filter types and our model is validated by measurements on an existing implementation of a JMS server. Several other studies address implementation aspects of filters. A JMS server checks for each message whether some of its filters match. If some of the filters are identical or similar, intelligent optimizations may be applied to reduce the filter overhead [15].

The Apache working group provides the generic test tool JMeter for throughput tests of the ActiveMQ [16]. However, it has only limited functionality such that we rely on an own implementation to automate our experiments.

In previous work [3,17,18,19] we already examined the message throughput performance behavior of different JMS servers, e.g. the FioranoMQ, WebsphereMQ, SunMQ, and the ActiveMQ. These investigations cover the dependency of the server performance on the number of installed publishers, subscribers, and we provided for each of the servers an analytical model to predict the message processing time based on the message replication grade and the number of installed simple filters. The current work differentiates from these studies that an analytical model for joint impact of the message replication grade and complex AND- and OR-filters is developed. Since complex filters may have different length, the experiment design is more complex and a significantly larger amount of experiments is required.

3   Test Environment

Our objective is the assessment of the message throughput of the ActiveMQ JMS server with various filter configurations. For comparability and reproducibility reasons we describe our testbed, the server installations, and our measurement methodology in detail.

3.1   Testbed

Our test environment consists of five computers that are illustrated in Figure ??. Four of them are production machines and one is used for control purposes, e.g., controlling jobs like setting up test scenarios and monitoring measurement runs. The four production machines have a 1 Gbit/s network interface which is connected to one exclusive Gigabit switch. They are equipped with 3.2 GHz single "Intel PIV" CPUs and 2048 MB system memory. Their operating system is SuSe Linux 9.1 with kernel version 2.6.5-smp installed in standard configuration. The "smp"-option enables the support of the hyper-threading feature of the CPUs. Hyper-threading means that a single-core-CPU uses multiple program and register counters to virtually emulate a multi-processor system. In our case we have two virtual cores. To run the JMS environment we installed Java JRE 1.5.0 [20], also in default configuration. The control machine is connected over a 100 Mbit/s interface to the Gigabit switch. In our experiments one machine is used as a dedicated JMS server. Our test application is designed such that JMS subscribers or publishers can run as Java threads. Each thread has an exclusive connection to the JMS server component and represent a so-called JMS session. A management thread collects the measured values from each thread and appends these data to a log file in periodic intervals.

In our test environment the publishers run on one or two exclusive publisher machines, and the subscribers run on one or two exclusive subscriber machines depending on the experiment. If two publisher or subscriber machines are used, the publisher or subscriber threads are distributed equally between them.

3.2   Server Configuration

The ActiveMQ server version 4.0 stable [21] is an open source software provided by the Apache group. We installed it on one of the above described Linux machines in default configuration such that the hyper-threading feature of the Linux kernel is used and the internal flow control is activated. To ensure that the ActiveMQ JMS server has enough buffer memory to store received messages and filters we set explicitly the memory for the Java Runtime Environment to 1024 MB.

3.3   Measurement Method

Our objective is the measurement of the JMS server capacity and we use the overall message throughput of the JMS server machine as performance indicator. We keep the server in all our experiments as close as possible to 100% CPU load. We verify that no other resources on the server machine like system memory or network capacity are bottlenecks. The publisher and subscriber machines must not be bottlenecks. Therefore, their CPU load must be lower than 75%. To monitor these side conditions, we use the information provided in the Linux „/proc" path. We monitor the CPU utilization, I/O, memory, and network utilization for each measurement run. Without a running server, the CPU utilization of the JMS server machine does not exceed 2%, and a fully loaded server must have a CPU utilization of at least 95%.

Experiments are conducted as follows. The publishers run in a saturated mode, i.e., they send messages as fast as possible to the JMS server. However, the message through-put is slowed down by the flow control of the server such that we observe publisher-side message queueing. We count the overall number of sent messages at the publishers and the overall number of received messages by the subscribers to calculate the server's rate of received and dispatched messages. Our measurement runs take 10 minutes whereby we discard the first and last seconds, where the system is not in a stable condition. For verification purposes we repeat the measurements several times, but their results hardly differ such that confidence intervals are very narrow even for a few runs. Therefore, we omit them in the figures of the following sections. The following experiments use the non-durable and persistent messaging mode as described in the Section 2.

4   Impact of Filters on the Message Throughput

Our main objective is to characterize the impact of different filter types on the message throughput. We conduct suitable experiments, perform throughput measurements, pro-pose an analytical model to capture the performed behavior, and fit its parameters based on the measurement.

We focus on three different kind of filter types: simple filters, complex OR-filters, and complex AND-filters. For all experiments we use one dedicated ActiveMQ JMS server machine. We connect 20 publishers, distributed over two publisher machines, each of them carrying 10 publisher threads. Filters evaluate user defined message head-ers where we set searchable StringProperties as application properties. We use for the StringProperties a string representation of four digit numbers with potentially leading zeros. The following experiments are based on a common principle. The publishers send messages with a certain header value and n_{fltr}^{pos} subscribers filter for this value such that each message is replicated $r = n_{fltr}^{pos}$ times. The additional n_{fltr}^{neg} filters do not match, but they cause additional workload on the server. Thus, altogether $m = n_{fltr}^{pos} + n_{fltr}^{neg}$ subscribers are connected to the server and they are distributed over two subscriber ma-chines. Each of the m subscribers maintains one exclusive TCP connection to the JMS server.

4.1   Experiment Setup

In the following, we describe the experiments for the investigation of simple filters, complex OR-, and complex AND-filters.

Simple Filters. We already examined the impact of simple filters in [3] with the fol-lowing experimental setup. The publishers send only messages with ID #0. As depicted in Figure 3(a), we install n_{fltr}^{pos} matching filters searching for ID value #0. Additionally

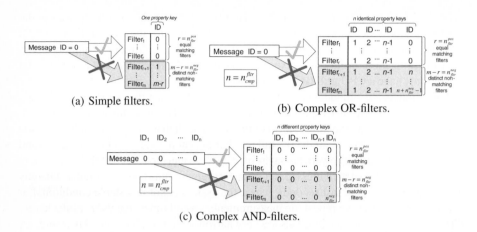

(a) Simple filters.

(b) Complex OR-filters.

(c) Complex AND-filters.

Fig. 3. Filter arrangements for the different experiments

we install n_{fltr}^{neg} different non-matching filters that search for values between #1 and #(n_{fltr}^{neg}).

Complex OR-Filters. We consider OR-filters with n_{cmp}^{fltr} components. As illustrated in Figure 3(b), we install n_{fltr}^{pos} equal complex OR-filters, searching for ID #0 set in the last component. As the matching filter component is in the last position, no early match can save processing power when the server evaluates the filter components from left to right. The publishers send messages with ID #0 to produce a message replication grade of $r = n_{fltr}^{pos}$. The last components of the n_{fltr}^{neg} non-matching filters take values from #(n_{cmp}^{fltr}) to #($n + n_{fltr}^{neg} - 1$) with $n = n_{cmp}^{fltr}$.

Complex AND-Filters. We consider AND-filters with n_{fltr}^{len} components. The publishers send messages with value #0 for each component ID_i. As illustrated Figure 3(c), n_{fltr}^{pos} subscribers install matching filters. The values set in the last component of the n_{fltr}^{neg} non-matching filters take values between #1 and #n_{fltr}^{neg}.

4.2   Results of the Measurement Experiments

We present the results for the experiments described in Section 4.1 for the parameters $n_{fltr}^{pos} \in \{1, 2, 5, 10, 20, 40\}$, $n_{fltr}^{neg} = \{1, 5, 10, 20, 40, 80, 160\}$, and $n_{cmp}^{fltr} = \{1, 2, 4, 8\}$ where applicable.

The solid lines plotted in Figures 4–6 show the measured message throughput of the ActiveMQ JMS server. The left figures present the received throughput and the right figures the overall throughput. We observe in all experimental studies a similar behavior. With an increasing number of filters the received and the overall throughput is reduced only slightly. An increasing message replication grade decreases the received message throughput, but it increases the overall message throughput. The figures for the overall throughput show a limitation of the overall throughput at approximately 50000

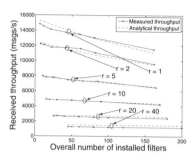 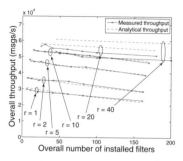

Fig. 4. Measured and analytical message throughput for simple filters depending on the message replication grade r

msgs/s. We take this observation into account when we fit the model parameters in the next section.

5 An Analytical Model for the Message Throughput in the Presence of Complex Filters

We use the measurement results of Section 4 as input for the analytical model of the message throughput. This model improves the understanding of the server performance of the ActiveMQ as well as the impact of different parameters like the number of filters, the filter type, and the replication grade.

Our model assumes three different parts of the processing time for a message. Each message requires a constant overhead t_{rcv}. The processing time t_{fltr} per installed filter depends on the overall number of installed filters $m = n_{fltr}^{pos} + n_{fltr}^{neg}$ and on their length n_{fltr}^{len}. Finally, the potential replication and transmission of a message takes t_{tx} time per outgoing message. Thus, the message processing time B can by calculated by

$$B = t_{rcv} + n_{cmp}^{fltr} \cdot m \cdot t_{fltr} + r \cdot t_{tx}. \qquad (1)$$

The empirical service time can be derived from the received message throughput of the measurement results in Section 4.2. The parameters t_{rcv}, t_{fltr}, and t_{tx} are fitted to the proposed model by a least square approximation. To that end we take only those curves into account that are not limited by the 50000 msgs/s margin. The parameters are derived separately for the simple, complex OR- and AND-filters. Table 1 summarizes

Table 1. Empirical values for the parameters of the model given in Equation 1

	t_{rcv}	t_{fltr}	t_{tx}
Simple filters	$4.88 \cdot 10^{-5}$ s	$1.62 \cdot 10^{-7}$ s	$1.55 \cdot 10^{-5}$ s
Complex OR-filters	$4.79 \cdot 10^{-5}$ s	$1.96 \cdot 10^{-7}$ s	$1.69 \cdot 10^{-5}$ s
Complex AND-filters	$5.19 \cdot 10^{-5}$ s	$1.86 \cdot 10^{-7}$ s	$1.71 \cdot 10^{-5}$ s

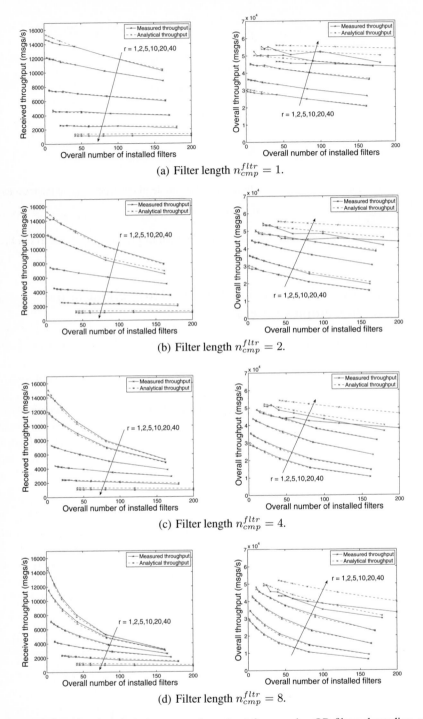

(a) Filter length $n_{cmp}^{fltr} = 1$.

(b) Filter length $n_{cmp}^{fltr} = 2$.

(c) Filter length $n_{cmp}^{fltr} = 4$.

(d) Filter length $n_{cmp}^{fltr} = 8$.

Fig. 5. Measured and analytical message throughput for complex OR-filters depending on the message replication grade r

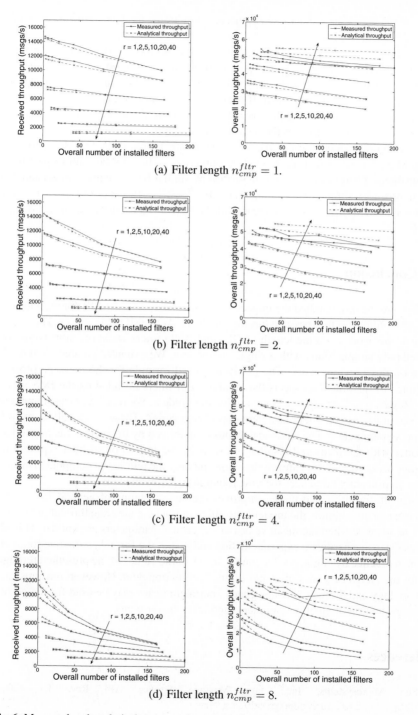

(a) Filter length $n_{cmp}^{fltr} = 1$.

(b) Filter length $n_{cmp}^{fltr} = 2$.

(c) Filter length $n_{cmp}^{fltr} = 4$.

(d) Filter length $n_{cmp}^{fltr} = 8$.

Fig. 6. Measured and analytical message throughput for complex AND-filters depending on the message replication grade r

their values. We observe that these empirical values of the model parameters are similar for all three experiment series.

Based on the model and the parameters we calculcate the analytical values for the received ($\frac{1}{B}$) and overall throughput ($\frac{r+1}{B}$). They are plotted as dashed lines in Figures 4–6. For small values of the replication grade $r = \{1, 2, 5, 10\}$ the analytical data approximate the measured data very well. If the replication grade is large, i.e. $r = \{20, 40\}$, the limit of 50000 msgs/s for the overall throughput of the server is reached and the analytical model overestimates the measured throughput.

A throughput comparison of the ActiveMQ with FioranoMQ, SunMQ, and WebsphereMQ [22] shows that the ActiveMQ outperforms them by far with respect to the simple filters. Their performance is described by similar but not equal models and their time to process a simple filter is about $1.46 \cdot 10^{-5}$ s, $2.11 \cdot 10^{-5}$ s, and $1.10 \cdot 10^{-5}$ s, respectively, which explains the superiority of the ActiveMQ for use cases with extensive filtering.

6 Conclusion

In this work we have studied the impact of simple filters, complex OR-filters, and complex AND-filters on the message throughput of the ActiveMQ JMS server. The special focus of this work is on the length of OR- and AND-filters while previous work considered only simple filters with a single component. We extended an analytical model based on this previous work. The newly proposed formula for the message processing time in Equation (1) respects the number m of filters installed on the JMS server, their lengths n_{fltr}^{cmp}, and the message replication grade r. We received measurement results based on appropriately designed experiment series. These results allow us to fit the model parameters. The analytical throughput derived by the model was in good accordance with the measured results. Surprisingly, the impact of all filter types on the message processing time is almost the same and only the number of components within a filter significantly influences the time required for its evaluation.

After all, the presented model improves the understanding of general JMS server performance. In addition, the model is useful to predict the message throughput for use cases in advance when the mean values of the critical parameters are known. This obsoletes extensive hardware experimentation and makes the formula attractive for application in practice by engineers. Of course, the absolute values of the presented throughput for the ActiveMQ are only valid in our test environment. However, our presented methodology and the specially designed experiment series may be used for the performance evaluation of other environments and other server types.

References

1. Sun Microsystems, Inc.: Java Message Service API Rev. 1.1 (2002) http://java.sun.com/products/jms/
2. Eugster, P.T., Felber, P.A., Guerraoui, R., Kermarrec, A.M.: The Many Faces of Publish/Subscribe. In: ACM Computing Surveys (2003)

3. Henjes, R., Schlosser, D., Menth, M., Himmler, V.: Throughput Performance of the Ac-tiveMQ JMS Server. In: ITG/GI Symposium Communication in Distributed Systems (KiVS), Bern, Switzerland (2007)

4. Krissoft Solutions: JMS Performance Comparison. Technical report (2004) http://www.fiorano.com/comp-analysis/jms_perf_comp.htm

5. Fiorano Software, Inc.: FioranoMQ^{TM}: Meeting the Needs of Technology and Business (2004) http://www.fiorano.com/whitepapers/whitepapers_fmq.pdf

6. Sonic Software, Inc.: Enterprise-Grade Messaging (2004) http://www.sonicsoftware.com/products/docs/sonicmq.pdf

7. Tibco Software, Inc.: TIBCO Enterprise Message Service (2004) http://www.tibco.com/resources/software/enterprise_backbone/message_service.pdf

8. IBM Corporation: IBM WebSphere MQ 6.0 (2005) http://www-306.ibm.com/software/integration/wmq/v60/

9. Crimson Consulting Group: High-Performance JMS Messaging. Technical report (2003) http://www.sun.com/software/products/message_queue/wp_JMSperformance.pdf

10. Sun Microsystems, Inc.: Sun ONE Message Queue, Reference Documentation (2005) http://developers.sun.com/prodtech/msgqueue/reference/docs/index.html

11. Carzaniga, A., Wolf, A.L.: A Benchmark Suite for Distributed Publish/Subscribe Systems. Technical report, Software Engineering Research Laboratory, Department of Computer Science, University of Colorado, Boulder, Colorado (2002)

12. Wolf, T.: Benchmark für EJB-Transaction und Message-Services. Master's thesis, Universität Oldenburg (2002)

13. Baldoni, R., Contenti, M., Piergiovanni, S.T., Virgillito, A.: Modelling Publish/Subscribe Communication Systems: Towards a Formal Approach. In: 8th International Workshop on Object-Oriented Real-Time Dependable Systems (WORDS 2003), pp. 304–311 (2003)

14. Baldoni, R., Beraldi, R., Piergiovanni, S.T., Virgillito, A.: On the Modelling of Publish/Subscribe Communication Systems. Concurrency and Computation: Practice and Experience 17, 1471–1495 (2005)

15. Mühl, G., Fiege, L., Buchmann, A.: Filter Similarities in Content-Based Publish/Subscribe Systems. Conference on Architecture of Computing Systems (ARCS) (2002)

16. Apache Incubator: ActiveMQ, JMeter Performance Test Tool (2006) http://www.activemq.org/jmeter-performance-tests.html

17. Henjes, R., Menth, M., Gehrsitz, S.: Throughput Performance of Java Messaging Services Using FioranoMQ. In: 13th GI/ITG Conference on Measuring, Modelling and Evaluation of Computer and Communication Systems (MMB), Erlangen, Germany (2006)

18. Henjes, R., Menth, M., Zepfel, C.: Throughput Performance of Java Messaging Services Using Sun Java System Message Queue. In: High Performance Computing & Simulation Conference (HPC&S), Bonn, Germany (2006)

19. Henjes, R., Menth, M., Zepfel, C.: Throughput Performance of Java Messaging Services Using WebsphereMQ. In: 5th International Workshop on Distributed Event-Based Sytems (DEBS) in conjuction with ICDCS 2006, Lisbon, Portugal (2006)

20. Sun Microsystems, Inc.: JRE 1.5.0 (2006) http://java.sun.com/

21. Apache: ActiveMQ, Reference Documentation (2006) http://www.activemq.org

22. Menth, M., Henjes, R., Gehrsitz, S., Zepfel, C.: Throughput Performance of Popular JMS Servers. In: ACM SIGMETRICS (short paper), Saint-Malo, France (2006)

Bio-inspired Analysis of Symbiotic Networks

Naoki Wakamiya and Masayuki Murata

Graduate School of Information Science and Technology, Osaka University
1-5 Yamadaoka, Suita, Osaka 565-0871, Japan
wakamiya@ist.osaka-u.ac.jp

Abstract. In the Internet, a variety of entities competes with each other. For example, coexisting overlay networks compete for physical network resources and they often disrupt each other. Service providers offer different services and compete for network resources and customers. If competitors could establish cooperative relationships, the collective performance can be improved and they can coexist peacefully and comfortably in a shared environment. In this paper, to understand the way that symbiosis emerges from direct and/or indirect interactions among coexisting entities in a shared environment, we adopt a mathematical model of symbiotic strains. Through thorough evaluations, we show that networks of different service rate and resource consumption can coexist by mutual interactions.

1   Introduction

In the Internet, a variety of entities competes and cooperates with each other to share limited resources including physical network resources such as link, router, and capacity, valuable information such as multimedia files and application data, and customers and users. A typical example of competitive and/or cooperative entities is coexisting overlay networks. With emerging needs for application-oriented network services, overlay networks have been widely deployed over physical IP networks. Since overlay networks share and compete for limited physical network resources, their autonomous behavior influence with each other. If they behave in a selfish manner to pursue their own benefit and to maximize their own utility, they disrupt each other and it leads to the degradation of the whole performance. Several papers [1,2,3] analyzed such harmful interactions and showed that selfish routing led to the instability of a system, degraded performance of competing overlay networks, and made traffic engineering meaningless. Speaking of competing service providers, a change in a charging system of one provider would influence the share of market among all providers. In some cases, an attempt to increase its revenue fails and an aggressive provider loses its customers, revenue, and credit.

When competitors cooperate with each other, we can expect that the collective performance is improved and they coexist peacefully and comfortably

L. Mason, T. Drwiega, and J. Yan (Eds.): ITC 2007, LNCS 4516, pp. 204–213, 2007.

in a shared environment. In the case of overlay network, cooperation leads to improvement in network QoS (Quality of Service) and resource utilization. In [4], they investigated a spectrum of cooperation among coexisting overlay networks. They proposed an architecture where overlay networks directly cooperated with each other in inter-overlay routing. In this architecture, a message emitted by a node in one overlay network is forwarded to another overlay network which provides a shorter path to the destination. They also briefly described perspectives on other kinds of cooperation, such as sharing measurement information, sharing control information, cooperative query forwarding, inter-overlay traffic engineering, and merging overlays. In [5], they considered a hierarchical overlay model, in which overlay networks were interconnected by an upper level network of representative peers, called super peers. They mentioned two types of composition of overlay networks, they were, absorption and gatewaying. By absorption, two overlay networks which accept each other are merged into one and represented by one super peer. On the other hand, if two overlay networks cannot agree to be merged for some reasons, e.g., incompatibility, they build a new upper level overlay which interconnects them by gatewaying. RON (Resilient Overlay Networks) [6] is another example of cooperation. In RON, nodes, called RON nodes, in different routing domains compose a mesh overlay network, which takes responsibility for inter-, and possibly intra-domain routing. Information originating in one routing domain will be sent to a destination node in another or the same routing domain through RON nodes in the overlay network, taking a better route than one in underlying IP networks. RON nodes communicate with each other to have a up-to-date view of the overlay network to offer better routing anytime. In addition, RON provides alternative paths in face of failures in underlying IP networks for the higher resilience.

As discussed in the above works, cooperation can bring benefits to each of competing entities and to the whole system. However, how does such peaceful cooperation emerge from direct and/or indirect interactions among entities in a shared environment? The analysis on coexistence and symbiosis of competitors in the environment has been investigated in the field of biology [7,8]. In the ecosystem, organisms in the same environment live together peacefully and comfortably through direct and/or indirect interactions with each other. In [7], they established a mathematical model of metabolic pathways of bacterial strains to elucidate mechanisms of coexistence of living organisms of closely related species. They revealed that the symbiosis emerged not only from interactions among competitors, but also from changes of their internal states. In this paper, we adopt their mathematical model to find an answer for the above question.

The rest of the paper is organized as follow. In Sect. 2, we introduce the mathematical model of coexistence of bacterial strains. Next in Sect. 3, we propose a model of symbiotic networks by adopting the biological model. Then we conduct thorough evaluations in Sect. 4. Finally, we summarize this paper and explain future directions in Sect. 5.

2   Mathematical Model of Symbiosis of Strains

In a cell, there is a network constituting of metabolic pathways, which describes chemical reactions in generating metabolites from other metabolites. Now assume that there are two types of bacterial strains A and B in a reactor. In [7], they focused on a metabolic network of two types of metabolites S_1 and S_2. The metabolites diffuse in and out of a cell depending on the difference in concentrations in the reactor and in the cell (Fig. 1).

Concentrations of metabolites S_1 and S_2 in a cell of type $i \in \{A, B\}$ are formulated as,

$$\frac{ds_1^{(i)}}{dt} = \frac{P}{V}(s_1^{(R)} - s_1^{(i)}) - (k_{1,2}^{(i)} + k_p)s_1^{(i)}, \tag{1}$$

$$\frac{ds_2^{(i)}}{dt} = \frac{P}{V}(s_2^{(R)} - s_2^{(i)}) + k_{1,2}^{(i)}s_1^{(i)} - k_p s_2^{(i)}, \tag{2}$$

where P stands for the permeation coefficient of cell membrane, V does for the average volume of a cell. $s_{\{1,2\}}^{(i)}$ and $s_{\{1,2\}}^{(R)}$ are concentrations of the metabolites in a cell of type i and in the reactor, respectively. k_p means the rate of consumption of the metabolites in a cell. $k_{1,2}^{(i)}$ is the rate of conversion of the metabolites in a cell of type i.

Next, concentrations of the metabolites in the reactor are formulated as,

$$\frac{ds_1^{(R)}}{dt} = D(s_1^{(0)} - s_1^{(R)}) + \sum_{i \in \{A,B\}} X^{(i)} P(s_1^{(i)} - s_1^{(R)}), \tag{3}$$

$$\frac{ds_2^{(R)}}{dt} = D(s_2^{(0)} - s_2^{(R)}) + \sum_{i \in \{A,B\}} X^{(i)} P(s_2^{(i)} - s_2^{(R)}), \tag{4}$$

where $X^{(i)}$ stands for the number of cells of type i per volume in the reactor. The fresh medium containing the metabolites of concentrations $s_{\{1,2\}}^{(0)}$ is added to the reactor at the constant rate while the culture is drained at the same rate. D is the resultant dilution rate.

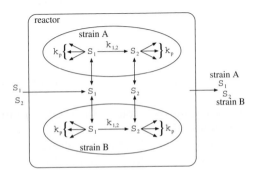

Fig. 1. Metabolic system of bacteria

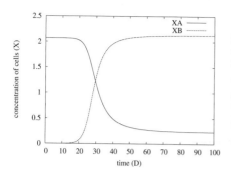

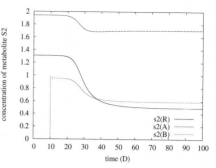

Fig. 2. Population of strains   **Fig. 3.** Concentration of metabolite S_2

The populations of cells are formulated as,

$$\frac{dX^{(i)}}{dt} = \mu^{(i)}X^{(i)} - DX^{(i)}, \tag{5}$$

where the growth rate $\mu^{(i)}$ is defined as,

$$\mu^{(i)} = \alpha s_1^{(i)} s_2^{(i)}. \tag{6}$$

Here, α is a constant.

In the model, cells interact with each other through permeation of the metabolites among cells and the culture in the reactor. An example of competition between a newly emerging mutant (strain B) and a wild type (strain A) is shown in Figs. 2 and 3 [7]. In this experiment, at first there is only strain A in the reactor. At time $10D$, strain B, which differs from strain A only in the rate of conversion, i.e., $k_{1,2}^B < k_{1,2}^A$, is introduced into the reactor. As shown in Fig. 2, they reach the stable condition where $X^{(A)} < X^{(B)}$. It means that the population of strain A, which consumes metabolite S_1 faster than strain B, becomes smaller than that of strain B. However, strain B does not completely dispel strain A. They live together. In Fig. 3, the concentration of metabolite S_2 in the reactor decreases as strain B appears. Although strain B produces metabolite S_2 from S_1 and its population is high, the rate of conversion is low and cannot compensate the decrease in the population of strain A. In this case, metabolite S_2 permeates cell membrane of both of strain A and B to the reactor.

In [7], they investigated conditions, i.e., parameter settings or environments, where two strains which differed in the conversion rate $k_{1,2}$ could coexist. The region is shown in Fig. 4 where each dot indicates parameter setting leading to the stable state. In addition, they found that one strain could live in an environment, where it could not exist alone, with a help of the coexisting strain. For example, when $k_{1,2}^{(B)}$ is smaller than 0.08, where there is a peak with high $k_{1,2}^{(A)}$, strain B cannot live alone for its low conversion rate. However, when strain A coexists,

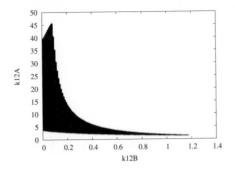

Fig. 4. Region of coexistence

strain B can live in the reactor with help of strain B. They also revealed that interactions among strains through changing metabolite concentrations of the reactor, that is, the shared environment or the external state, changed internal states, including metabolite concentrations, the conversion rate, and a network of metabolic pathways, of the strains. This led to the coexistence.

Here, we would show an example of intuitive interpretation of the mathematical bacterial model as a model of coexisting overlay networks. Assume that a cell corresponds to a node, a strain does to an overlay network, and the reactor does to the whole system including other overlay networks. Now, we regard metabolite S_1 as demand for information and S_2 as information. Therefore, the conversion rate $k_{1,2}$ corresponds to activity of information search and retrieval. More demand exists, more information a node searches for and retrieves. As time passes, new information appears and demand for new information increases (the fresh medium introduced to the reactor from outside and thus $s_1^{(R)}$ and $s_2^{(R)}$ increase). At the same time, information, demand, and nodes disappear from the system (the culture is drained from the reactor). If concentration $s_1^{(i)}$ of a cell of strain i is lower than that of the reactor, demands increases in cell i (penetration) being stimulated by the existence of attractive new information or the possibility of obtaining them. Depending on the aggressiveness $k_{1,2}$, a cell finds and retrieves information and $s_2^{(i)}$ increases to satisfy its demand. An overlay network of higher activity attracts more nodes and the size of the network becomes larger ($\frac{dX^{(i)}}{dt} = \alpha s_1^{(i)} s_2^{(i)} X^{(i)} - DX^{(i)}$). Information diffuses out from a node of an active overlay network to the system ($s_1^{(i)} > s_1^{(R)}$) and then it is found and retrieved by a node of another strain (penetration).

3   Analysis of Symbiotic Networks

Now, consider competition among networks which offer a service to users or customers. We can regard strains as competitors, i.e., networks, strain S_1 as a group of users or customers, strain S_2 as the shared resource. Then, temporal changes in the number of users $s_1^{(i)}$ in network i can be given by the following differential equation.

$$\frac{ds_1^{(i)}}{dt} = c(s_1^{(R)} - s_1^{(i)}) - (k_{1,2}^{(i)} + k_p)s_1^{(i)}, \tag{7}$$

where c is a constant. The first term in the right side of the equation means that unloaded network i with a small number of users s_1^i would accept new users $s_1^{(R)}$ ($s_1^i < s_1^R$). On the other hand, when network i has too many users and its load is high, users would not be satisfied with an offered service and leave the network ($s_1^i > s_1^{(R)}$). The second term explains the decrease in the number of users for being served at the rate of $k_{1,2}^{(i)}$ and the number of users dropping away at the rate of k_p. In this case, $k_{1,2}^{(i)}$ corresponds to the number of users served per unit of time, i.e., the service rate or service capacity of network i.

Next, temporal changes in the amount $s_2^{(i)}$ of shared resource that network i uses can be defined by the following differential equation.

$$\frac{ds_2^{(i)}}{dt} = c(s_2^{(R)} - s_2^{(i)}) + k_{1,2}^{(i)}s_1^{(A)} - k_p s_2^{(i)}. \tag{8}$$

The first term explains how the resource is shared among entities. When the system has plenty of residual resource, a network can occupy more shared resource ($s_2^{(i)} < s_2^{(R)}$). On the other hand, when the system is running out of resource, a rich network gives up some of its resource ($s_2^{(i)} > s_2^{(R)}$). The second term corresponds to the increase in the amount of the shared resource used by a network to answer user's requests ($k_{1,2}^{(i)}s_1^{(i)}$) and the decrease of the shared resource by consumption ($k_p s_2^{(i)}$).

Temporal changes in the population of users $s_1^{(R)}$ and the residual resource $s_2^{(R)}$ in the system are given by the following differential equations.

$$\frac{ds_1^{(R)}}{dt} = D\{(s_1^{(0)} - s_1^{(R)}) + \sum_i X^{(i)}(s_1^{(i)} - s_1^{(R)})\}, \tag{9}$$

and

$$\frac{ds_2^{(R)}}{dt} = D\{(s_2^{(0)} - s_2^{(R)}) + \sum_i X^{(i)}(s_2^{(i)} - s_2^{(R)})\}, \tag{10}$$

where D is a constant. The first term of Eq. (9) means join ($s_1^{(0)}$) and leave ($s_1^{(R)}$) of users for the system. The second term of Eq. (9) corresponds to join ($s_1^{(i)}$) and leave ($s_1^{(R)}$) of users for a network. On the other hand, the first term of Eq. (10) shows changes in the residual resource per unit time. The second term of Eq. (10) corresponds to exchange of the shared resource among the system and networks.

Temporal changes in the size $X^{(i)}$ of network i can be expressed by the following equation.

$$\frac{dX^{(i)}}{dt} = \mu^{(i)}X^{(i)} - DX^{(i)}. \tag{11}$$

Coefficient $\mu^{(i)}$ is further defined as a product of the number $s_1^{(i)}$ of users and the amount $s_2^{(i)}$ of occupied resource as,

$$\mu^{(i)} = \alpha s_1^{(i)} s_2^{(i)}, \tag{12}$$

where α is a constant. Therefore, the larger the number of users is and the more the occupied resource is, the larger a network grows.

4   Results and Discussions

In this section, we will see how two networks of different service capability compete or cooperate with each other in a variety of environments. We empirically set $c = 1.0$ and $\alpha = 1.0$. The rate $s_1^{(0)}$ that new users join in the system per unit time is set at 10.0 and the rate $s_2^{(0)}$ that resources, e.g., bandwidth, are additionally supplied to the system is set at 10.0. The number k_p of users dropping away from the system is set at 1.0.

When network A with service rate $k_{1,2}^{(A)}$ of 5.0 exists alone in the system, the size converges to about 10.1 from the arbitrary initial size as shown in Fig. 5. In comparison with the case of network B with smaller service rate $k_{1,2}^{(B)}$ of 1.0 shown in Fig. 6, network A is slightly smaller than network B at the stable condition. This is because that network A consumes more resources than network B and thus puts higher load on the system.

In Fig. 7, network B appears in the system where only network A exist. As shown in the figure, network A, which consumes more resources, becomes smaller than network B, but they coexist with each other. This result conforms to the case of a biological system shown in Fig. 2. In Fig. 8, changes in the amount of resources remaining in the system, that used by network A, and that used by network B are shown. It can be seen that network A refrains from consuming much resources as network B comes into the system.

To find conditions which make competing networks coexist with each other, we conducted thorough evaluations by changing $s_1^{(0)}$ and $s_2^{(0)}$ from 0.0 to 50.0,

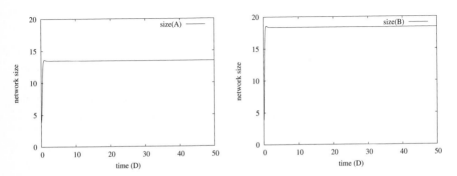

Fig. 5. Evolution of network A   **Fig. 6.** Evolution of network B

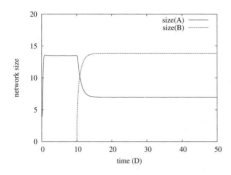

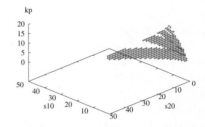

Fig. 7. Evolution of competing networks

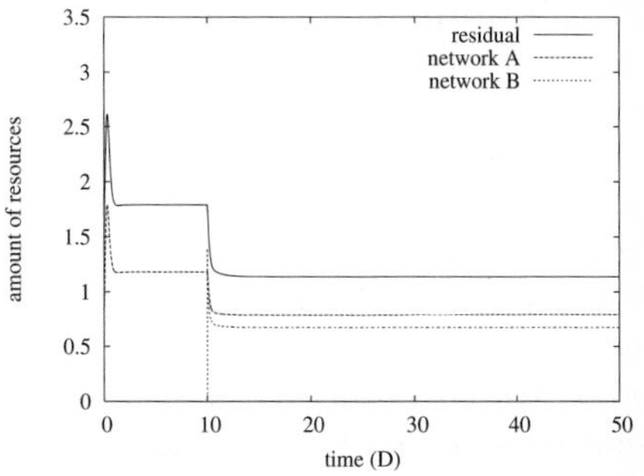

Fig. 8. Amount of remaining and occupied resources

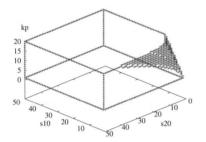

Fig. 9. Region of coexsistence of networks

Fig. 10. Region where a network can live with help of the other

$k_{1,2}^{(A)}$ and $k_{1,2}^{(B)}$ from 1.0 to 50.0, and k_p from 1.0 to 20.0. In Fig. 9, the region of $s_1^{(0)}$, $s_2^{(0)}$, and k_p for achieving symbiotic condition is illustrated. All points in a cube leads to symbiosis except for the corner whose cross-section surface is shown as the shade. When the number of users newly joining to the system per unit time and the amount of resources additionally introduced into the system per unit time are small, either or both of networks starve and collapse for insufficient nutrients, i.e., users and resources.

Fig. 10 shows the region of parameter settings, i.e., environments, where a network cannot live alone but it can live with help of a coexisting network. The implication of these results can be as follows. A network which offers a good service to customers but consumes much resource is short-lived. However, if customers have alternatives in choosing a network, some will join other networks which offer a lower-quality service but are modest in resource usage. Consequently, resource consumption of a big eater decreases as shown in Fig. 8. Then, all of networks can live together and share customers and resources peacefully. A similar phenomenon can be observed in the actual business market.

5   Conclusion

This paper is a part of our research work on "overlay network symbiosis", which we propose taking inspirations from the biological mathematical model [9,10]. In the model of overlay network symbiosis, we regard an overlay network as an organism, which evolves and expands as a new node joins, diminishes as a node leaves or halts, interacts with other overlay networks through direct and/or indirect communications, and changes its topology as a consequence of interactions. We also proposed mechanisms for P2P file-sharing networks to interact and cooperate with each other.

In this paper, we investigate symbiosis among competing networks by adopting the mathematical bacterial model. We showed how networks of different service rate coexist with each other. The model can be easily extended to multi-competitors. It can also be applied to other scenarios to investigate how competitors establish the cooperative relationship. For example, the model explains how a company should prepare services of different quality for different price to maximize revenues while sharing the market with other competitors. Another possible interpretation is to regard the conversion rate $k_{1,2}$ as the rate of flow. If information flows without disturbance in a part of a network, it activates other flows and other parts of the network. Consequently, the performance of the whole network in terms of, for example, total throughput, becomes higher by the cooperation. From a sociological point of view, metabolites are tasks and knowledge. A person obtains knowledge (S_2) by doing ($k_{1,2}$) a task (S_1). Knowledge is shared among people. Faster a person performs tasks, more he obtains knowledge. In addition, faster a person performs task, more he receives tasks. As future works, we will extend the model to treat more complex and concrete networks and investigate the way that symbiosis emerges.

Acknowledgment

This research was partly supported by "New Information Technologies for Building a Networked Symbiosis Environment" in the 21st Century Center of Excellence Program and the Grant-in-Aid for Scientific Research (A) 18200004 of the Ministry of Education, Culture, Sports, Science and Technology of Japan.

References

1. Seshadri, M., Katz, R.H.: Dynamics of simultaneous overlay network routing. Technical Report UCB//CSD-03-1291, Computer Science Division, University of California Berkeley (2003)
2. Qiu, L., Yang, Y.R., Zhang, Y., Shenker, S.: On selfish routing in Internet-like environments. In: Proceedings of ACM SIGCOMM, Karlsruhe, pp. 151–162 (2003)
3. Roughgarden, T.: Selfish Routing and the Price of Anarchy. The MIT Press, Cambridge (2005)
4. Kwon, M., Fahmy, S.: Synergy: An overlay internetworking architecture and its implementation. Technical report, Purdue University (2005)

5. Kersch, P., Szabo, R., Kis, Z.L.: Self organizing ambient control space – an ambient network architecture for dynamic network interconnection. In: Proceedings of the 1st ACM workshop on Dynamic Interconnection of Networks (DIN'05), pp. 17–21 (2005)

6. Andersen, D., Balakrishnan, H., Kaashoek, F., Morris, R.: Resilient overlay networks. In: Proceedings of 18th ACM Symposium on Operating Systems Principles (SOSP), Banff, pp. 131–145 (2001)

7. Yomo, T., Xu, W.Z., Urabe, I.: Mathematical model allowing the coexistence of closely related competitors at the initial stage of evolution. Researches on Population Ecology 38(2), 239–247 (1996)

8. Shimizu, H., Egawa, S., Wardani, A.K., Nagahisa, K., Shioya, S.: Microbial interaction in a symbiotic bioprocess of lactic acid bacterium and diary yeast. In: Proceedings of the Second International Workshop on Biologically Inspired Approaches to Advanced Information Technology (Bio-ADIT 2006), LNCS vol.3853, pp. 93–106 (2006)

9. Wakamiya, N., Murata, M.: Toward overlay network symbiosis. In: Proceedings of the Fifth International Conference on Peer-to-Peer Computing (P2P 2005), pp. 154–155, Konstanz (2005)

10. Wakamiya, N., Murata, M.: Overlay network symbiosis: Evolution and cooperation. In: to be presented at First International Conference on Bio-Inspired Models of Network, Information and Computing Systems (Bionetics 2006), Cavalese, Italy (2006)

Hierarchical Infrastructure-Based Overlay Network for Multicast Services

Josué Kuri and Ndiata Kalonji

France Télécom R&D
801 Gateway Blvd.
South San Francisco, CA 94080 USA
{josue.kuri,ndiata.kalonji}@orange-ftgroup.com

Abstract. This article proposes a hierarchical architecture for an infrastructure-based overlay network delivering multicast services. Such an overlay network is an alternative to IP multicast when the latter cannot be timely deployed for technical or business reasons. To make the overlay network scalable in terms of coverage and traffic volume, we propose a hierarchical architecture in which edge overlay routers are responsible for handing traffic to/from the end users, whereas core overlay routers handle transit traffic and perform the bulk of packet replication. We compare the hierarchical architecture to a flat architecture in which (edge) overlay routers are connected in a full mesh. We develop an asymptotic analysis to quantify the cost (in terms of additional required switching) and benefits of the hierarchical architecture over the flat architecture.

1 Introduction

The concept of overlay network has recently emerged among some service providers as a possible approach to alleviate the effects of slow deployment of new network services. This slow deployment typically results from the large capital investments required to upgrade a network and the inertia around the network *status quo* created by the business relationships between the users and the operator of the network.

With an overlay network, a service provider can deliver a new network service to an initial set of users in order to gauge the market for the service before committing to a complete network infrastructure upgrade. This incremental approach reduces the risks associated to network upgrades and makes the service provider more reactive.

Largely deployed overlay networks include not only popular Peer-to-Peer (P2P) systems such as Gnutella [1] and [2], but also content distribution networks like Akamai [3] and experimental networking platforms such as PlanetLab [4].

The nodes of an overlay network can be under the control of end users (like in P2P systems), of the service provider, or of both. When the nodes are exclusively in the service provider space, the system is referred to as a proxy-based or infrastructure-based overlay network. In the case of overlay networks offering a

L. Mason, T. Drwiega, and J. Yan (Eds.): ITC 2007, LNCS 4516, pp. 214–223, 2007.

multicast service, the functionality required to offer the service (group management, routing, packet replication, etc.) can be implemented in the user space, in the service provider space, or in both.

Overlay multicast systems implemented exclusively in the user space are an interesting alternative to IP multicast, since multicast capabilities do not need to be supported in the IP infrastructure for the service to be provided. The deployment can be incremental and cover a large geographical area in a short time. However, this approach excludes the service provider and often result in poor network performance since the multicast functionality is implemented in relatively unreliable nodes (commodity user equipment) and interconnection typically relies on best effort Internet connectivity.

In an infrastructure-based multicast overlay network, the nodes implementing the multicast functionality are reliable purpose-built equipment managed by a service provider, and the interconnection relies on SLA-backed IP connectivity supplied by one or more service providers. Thus, an infrastructure-based multicast overlay network provides not only deployment flexibility, but also higher reliability, better performance and an efficient use of network resources.

In this article we propose a hierarchical architecture for an infrastructure-based multicast overlay network running on top of an IP network. The goal of the architecture is to accommodate the growth of both the traffic volume and the number of users in a cost-effective manner by defining two types of overlay routers with different roles in the network: edge routers responsible of handing traffic to/from the end users, and core routers responsible of handling transit traffic and performing the bulk of packet replication.

In the next section we describe the proposed hierarchical architecture and explain the choice of the proposed connectivity pattern and location of routers. We also describe the flat architecture used as reference to assess the benefits of the proposed architecture. Section 3 describes previous work related to overlay multicast networks and highlights the contributions of the paper. Section 4 develops an asymptotic analysis to assess the scalability of the proposed architecture. In this section we focus first on the particular case in which each edge router is connected to exactly two core routers and the core routers are connected in full mesh among themselves. We then analyze a more general connectivity pattern. Section 5 summarizes the benefits of the hierarchical architecture highlighted by the analysis and concludes the paper.

2   Description of the Proposed Architecture

We define the architecture proposed in this article in terms of the different types of routers required, the topology prescribed for their interconnection, and the location of the routers with respect to the underlying IP infrastructure.

Figure 1(a) illustrates the proposed hierarchical architecture. Two types of overlay routers[1] are defined: edge and core routers. The former aggregate traffic

---

[1] An overlay router is a stand-alone piece of equipment, independent of the routers of the underlying IP network.

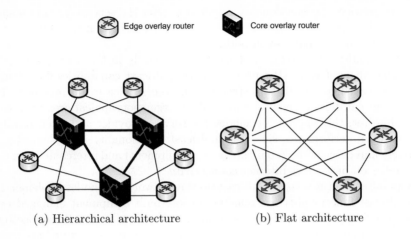

(a) Hierarchical architecture   (b) Flat architecture

Fig. 1. Two possible architectures for the overlay multicast network

coming from Customer Premises Equipment (CPE) or end-user proxies, feed this traffic into the network and, in the reverse direction, forward the traffic from the network to the CPE and end-user proxies. Edge routers have thus a "customer-facing" and a "network-facing" side. Core routers on the other hand are responsible for forwarding transit traffic between edge routers (neither CPE nor end-user proxies are attached to them). Each edge router is connected to a limited number of core routers (two in Figure 1(a)), and the core routers are highly connected among themselves (full mesh in the figure). Edge routers are colocated with edge routers of the underlying IP infrastructure, and core routers are colocated with core routers of the same IP infrastructure.

The connectivity pattern and the location of routers in the proposed architecture aim at shifting both the packet replication effort from the edge to the core routers, and the link stress (explained below) from the access links to the core links. The goal is to leverage the economies of scale in the core of the network. The cost of switching and transmission a unit of traffic is typically lower in the core than in the access [5] because of the inherent geographic dispersion of access networks, which results in higher CAPEX and OPEX.

Link stress refers to the redundant copies of the same information sent over a same IP link [6]. Link stress occurs when the same information is sent over two or more overlay links, and these links traverse the same IP link. The concept is illustrated in Figure 2: a 50 Kbps flow is sent from overlay router 1 to overlay routers 3 and 4 over two different overlay links. Both links share the (1, 3) IP link, which results on two copies of the same information being sent over the IP link.

The link stress tends to increase when the connectivity of the overlay network is much higher than the connectivity of the underlying IP network. This is the case, for example, of an overlay network with a full mesh topology instantiated over an IP network with a ring topology. Link stress is likely to occur in these situations because more overlay links cross a same IP link.

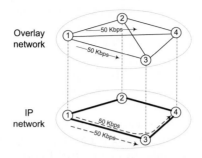

Fig. 2. Link stress: two copies of the same information are sent over the same IP link

In the proposed hierarchical architecture, link stress is more likely to occur in the core than in the access links because the overlay core routers are connected in full mesh. The shift of link stress to the core is in fact a desirable property of the proposed architecture: if link stress has to occur, it is better to have it in the core of the IP network, where switching and transmission are cheaper than in the access.

We use the flat architecture illustrated in Figure 1(b) as a reference to assess the benefits of the proposed hierarchical architecture. In the flat architecture, the full mesh provides direct connectivity between any pair of routers and hence eliminates transit traffic. However, the ingress overlay router receiving a multicast packet through its customer-facing interface is the only responsible for the replication of the packet. As the number of users and the traffic volume grow, adding a new router to the full mesh becomes more complicated and the burden of packet replication eventually outpaces the switching capacity of the routers. This type of architecture is used in overlay networks like OMNI [7], AMcast [8] and RON [9], and even in standardized architectures like VPLS [10].

3   Previous Work

As indicated in the introduction, an overlay network can be implemented in the user space, in the service provider space, or in both. Examples of overlay multicast networks implemented in the user space include: End System Multicast [11], RMX [12] and Overcast [13].

The Two-tier Overlay Multicast Architecture (TOMA) described in [14] is an example of an overlay multicast network implemented in the user and service provider spaces. In this architecture, a set of service nodes deployed by a service provider form a Multicast Service Overlay Network (MSON) [6]. Outside the MSON, groups of end users are organized into clusters and an application-level multicast tree rooted at a service node is used to deliver traffic to the end users in a cluster. Conceptually, the service nodes in the MSON correspond to the overlay edge routers of the architecture proposed in this paper. There are no overlay core routers in the MSON.

The IETF defines in [10] an architecture and standards for a Virtual Private LAN Service (VPLS) that extends an Ethernet LAN service over a metro or wide area network using a service provider's infrastructure. The architecture is based on a set of Provider Edge (PE) routers connected through a full mesh of pseudo-wires (PWs), as illustrated in Figure 1(b). Broadcast, Unknown-destination and Multicast (BUM) traffic received by a PE router is replicated over all the PWs connecting this router to the other PEs.

The Hierarchical-VPLS extension [10] addresses the scalability issues of the flat VPLS architecture by introducing a new type of equipment called MTU-s that aggregates and feeds traffic to one or two PE routers. This arrangement reduces the number of required PE routers and hence the complexity of their interconnection. Conceptually, the MTU-s and the PE routers in an H-VPLS network correspond to the overlay edge and core routers in the proposed architecture, respectively.

Unlike H-VPLS, the core routers in the proposed architecture are not constrained to be connected in full mesh. In addition to the generalization of the connectivity pattern, the contribution of this paper includes an asymptotic analysis to quantify the cost and benefits of the proposed architecture.

4   Asymptotic Analysis

For the sake of clarity we first analyze the particular case of a hierarchical architecture in which each edge router is connected to exactly two core routers, and the core routers are connected in full mesh. The analysis is extended later on to the general case.

We first compare the number of overlay links required in the hierarchical and the flat architectures. Since every link involves management overhead (commissioning, provisioning and monitoring), the smaller number of links in the hierarchical architecture results in lower OPEX.

Let N_e be the number of edge routers, N_c the number of core routers and $a = N_e/N_c$ the number of edge routers per core router. The number of (bidirectional) links required in the hierarchical architecture is:

$$\mu_h(N_e, N_c) = 2N_e + \frac{N_c(N_c - 1)}{2}. \tag{1}$$

The number of links required in the flat architecture is:

$$\mu_f(N_e) = \frac{N_e(N_e - 1)}{2}. \tag{2}$$

Finally, the reduction in links provided by the hierarchical architecture with respect to the flat architecture is:

$$\theta(N_e, N_c) = 1 - \frac{\mu_h(N_e, N_c)}{\mu_f(N_e)} \tag{3}$$

Asymptotically, the reduction in links for a constant value of a is:

$$\lim_{N_e \to \infty} \theta \left(N_e, \frac{N_e}{a} \right) = \frac{a^2 - 1}{a^2}. \tag{4}$$

The reduction in links approaches 100% as a increases. Figure 3 shows the reduction in links as a function of the number of edge routers for a constant $a = 4$. Although the actual reduction is smaller than the asymptotic value for a small N_e, it is worth to recall that the interest of the hierarchical network is to be able to support a very large number of edge routers. Therefore, we are interested in knowing the reduction for large N_e.

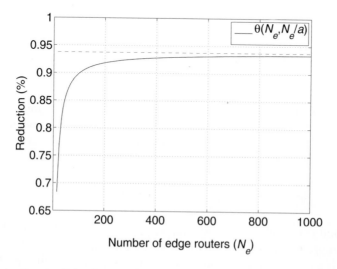

Fig. 3. Reduction in links due to the hierarchical architecture as a function of the number of edge routers for a constant $a = 4$

Since an edge router in the hierarchical architecture is connected to only two core routers, the traffic entering the network through this router is replicated at most twice, which is much less than the up to $N_e - 1$ replications required in the flat architecture. The bulk of the replication effort is thus shifted from the edge routers to the core routers, which are connected in full mesh. However, the hierarchical architecture introduces one or two intermediate core routers between any pair of edge routers, which means that, overall, more switching has to be performed in the network.

We now quantitatively relate the reduction in traffic switched in the edge routers to the overall additional switching induced by the hierarchical architecture to assess the cost/benefit trade-off of this architecture. For this, we first introduce the concepts of *edge router switched traffic* and *core router switched traffic*. The former refers to the volume of traffic (in bits per second) that must be handled by the network-facing side of a edge router. This includes the volume of traffic sent to and received from the network. In the example of Figure 4, the edge router switched traffic is 2.9 Mbps, which includes the 2 Mbps of outgoing traffic and the 900 Kbps of incoming traffic on the network-facing side. The incoming and outgoing traffic on the customer-facing side is not considered since this volume of traffic is the same for the flat and the hierarchical architectures.

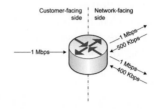

Fig. 4. The edge router switched traffic is the volume of traffic on the network-facing side of the edge router

The *core router switched traffic* refers to the volume of incoming and outgoing traffic in a core router. Assuming that each edge router receives one unit of traffic (*e.g.*, 1 Mbps) from its customer-facing side, and that this traffic must be received by the other $N_e - 1$ edge routers, the core router switched traffic is:

$$\varrho_h(N_e, N_c) = \frac{2N_e}{N_c}\left(1 + \frac{N_c - 2}{2}\right) + \frac{N_e(N_e - 1)}{N_c}, \tag{5}$$

and the total core router switched traffic is $\varphi_h(N_e, N_c) = N_c \varrho_h(N_e, N_c)$. Under these assumptions, the edge router switched traffic in the hierarchical architecture is $\rho_h(N_e) = 2 + N_e - 1$ and the total edge router switched traffic of a hierarchical network is $\phi_h(N_e) = N_e \rho_h(N_e)$.

Similarly, the edge router switched traffic in the flat architecture is $\rho_f(N_e) = 2(N_e - 1)$, and the total edge router switched traffic of a flat network is $\phi_f(N_e) = N_e \rho_f(N_e)$.

The percentage of reduction in traffic switched by the edge routers of the hierarchical architecture is:

$$\gamma(N_e) = \frac{\rho_f(N_e) - \rho_h(N_e)}{\rho_f(N_e)}, \tag{6}$$

and its asymptotic value is:

$$\lim_{N_e \to \infty} \gamma(N_e) = \frac{1}{2}. \tag{7}$$

On the other hand, the percentage of additional switching induced by the hierarchical architecture is:

$$\kappa(N_e, N_c) = \frac{\phi_h(N_e) + \varphi_h(N_e, N_c) - \phi_f(N_e)}{\phi_f(N_e)}, \tag{8}$$

and its asymptotic value for a constant a is:

$$\lim_{N_e \to \infty} \kappa \left(N_e, \frac{N_e}{a} \right) = \frac{1}{2a}. \tag{9}$$

Figure 5 shows the percentage of reduction in traffic switched by the edge routers and the percentage of additional switching induced by the hierarchical architecture for a constant $a = 4$. The 12.5% of additional switching in this case is reasonable considering that the traffic switched by every edge router will be reduced by about 50% and that there are $a = 4$ edge routers per core router.

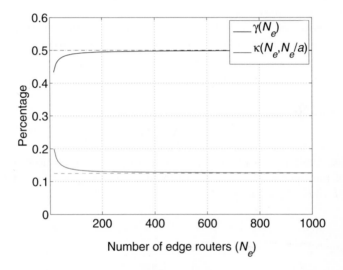

Fig. 5. Relationship between the reduction in traffic switched by the edge routers and the additional switching induced by the hierarchical architecture

4.1   General Case

We now analyze the reduction in links and the reduction in traffic switched by the edge routers for the general case in which each edge router is connected to M core routers and the core routers are not necessarily connected in full mesh among themselves.

Let $D = T/T_c$ be the core interconnection ratio, where T is the actual number of links between core routers and $T_c = N_c(N_c - 1)/2$ is the number of links when the core routers are connected in full mesh. The particular case analyzed before corresponds to the parameter values $M = 2$ and $D = 1$.

In the general case, the number of links required in the hierarchical architecture is:

$$\mu_h^*(N_e, N_c, M, D) = MN_e + \frac{N_c(N_c - 1)}{2} D \tag{10}$$

and the reduction in links provided by the hierarchical architecture is:

$$\theta^*(N_e, N_c, M, D) = 1 - \frac{\mu_h^*(N_e, N_c, M, D)}{\mu_f(N_e)}. \tag{11}$$

Asymptotically:

$$\lim_{N_e \to \infty} \theta^*\left(N_e, \frac{N_e}{a}, M, D\right) = \frac{a^2 - D}{a^2}. \tag{12}$$

Note that the reduction in links is independent of parameter M (the number of core routers an edge router is connected to) and that the reduction approaches 100% as a increases and D decreases.

The edge router switched traffic in the hierarchical architecture is $\rho_h^*(N_e, M) = M + N_e - 1$ and the reduction in traffic switched by edge routers is:

$$\gamma^*(N_e, M) = \frac{\rho_f(N_e) - \rho_h^*(N_e, M)}{\rho_f(N_e)}. \tag{13}$$

Asymptotically:

$$\lim_{N_e \to \infty} \gamma^*(N_e, M) = \frac{1}{2}. \tag{14}$$

Note that the reduction in traffic switched by the edge routers is also independent of parameter M. Figure 6 illustrates the independence of θ^* and γ^* vis-à-vis M when N_e is large.

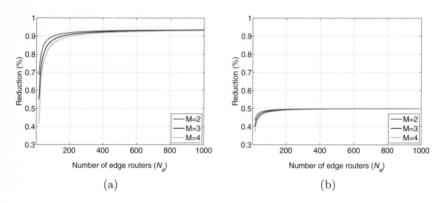

(a)   (b)

Fig. 6. Reduction of (a) links and (b) traffic switched by edge routers, for $D = 1$ and $M = 2$, $M = 3$ and $M = 4$

5   Conclusions

In this article we proposed a hierarchical architecture for an infrastructure-based overlay network delivering multicast services. The architecture aims at accommodating the growth of both the traffic volume and the number of users in a cost-effective manner.

The asymptotic analysis developed in the paper quantitatively shows that the percentage of additional switching induced by the hierarchical architecture is relatively small when compared to the significant reduction in the number of links and traffic switched by the edge routers. Moreover, not only the bulk of the replication effort is shifted from a large number of edge routers to a small number of core routers, but also the link stress is concentrated in the core of the underlying IP network, where transmission and switching costs are lower than in the access due to economies of scale.

Since the architecture of an H-VPLS network is similar to the one proposed in this paper, the asymptotic analysis can be adapted to assess the scalability of H-VPLS networks when Broadcast, Unknown-destination and Multicast (BUM) traffic is expected to be dominant. For a realistic assessment, however, unicast traffic needs to be considered in the model for this type of network.

References

1. The Gnutella website: http://www.gnutella.com Last accessed (November 2006)
2. The Kazaa website: http://www.kazaa.com Last accessed (November 2006)
3. Rahul, H., Kasbekar, M., Sitaraman, R., Berger, A.: Towards Realizing the Performance and Availability Benefits of a Global Overlay Network. Technical Report MIT-LCS-TR-1009, Massachusetts Institute of Technology, Computer Science and Artificial Intelligence Laboratory (2005)
4. The PlanetLab website: http://www.planet-lab.org Last accessed (November 2006)
5. Briscoe, B., Rudkin, S.: Commercial models for IP Quality of Service Interconnect. BT Technical Journal 23(2) (2005)
6. Lao, L., Cui, J.H., Gerla, M.: Multicast Service Overlay Design. In: Proceedings of SPECTS 05, Philadelphia, PA (2005)
7. Banerjee, S., Kommareddy, C., Kar, K., Bhattacharjee, B., Khuller, S.: Construction of an efficient overlay multicast infrastructure for real-time applications. In: Proceedings of IEEE Infocom, San Francisco, CA (2003)
8. Shi, S., Turner, J.: Routing in overlay multicast networks. In: Proceedings of IEEE Infocom, New York (2002)
9. Andersen, D., Balakrishnan, H., Kaashoek, M., Morris, R.: Resilient Overlay Network. In: Proceedings of ACM SOSP 2001, Banff, Canada (2001)
10. Lasserre, M., Kompella, V.: Virtual private lan services using ldp. Internet draft draft-ietf-l2vpn-vpls-ldp-09.txt (2006)
11. Cui, J.H., Rao, S.G., Zhang, H.: A Case for End System Multicast. In: Proceedings of ACM Sigmetrics, Santa Clara, CA (2000)
12. Chawathe, Y., McCanne, S., Brewer, E.: RMX: Reliable multicast for heterogeneous networks. In: Proceedings of IEEE Infocom, Tel-Aviv, Israel (2000)
13. Jannotti, J., Gifford, D., Johnson, K., Kaashoek, M.: Overcast: Reliable multicasting with an overlay network. In: Proceedings of USENIX, San Diego, CA (2000)
14. Lao, L., Cui, J.H., Gerla, M.: TOMA: A viable solution for large-scale multicast service support. In: Proceedings of IFIP Networking 2002, Waterloo, Canada (2005)

Network Capacity Allocation in Service Overlay Networks

Ngok Lam[1], Zbigniew Dziong[2], and Lorne G. Mason[1]

[1] Department of Electrical & Computer Engineering, McGill University, 3480 University Street, Montreal, Quebec, Canada H3A 2A7
[2] Department of Electrical Engineering, Ecole de technologie superieure,1100 Notre-Dame Street West, Montreal, Quebec, Canada H3C 1K3
ngok.lam@mcgill.ca, zdziong@ele.etsmtl.ca, lorne.mason@mcgill.ca

Abstract. We study the capacity allocation problem in service overlay networks (SON)s with state-dependent connection routing based on revenue maximization. We formulate the dimensioning problem as one in profit maximization and propose a novel model with several new features. In particular the proposed methodology employs an efficient approximation for state dependent routing that reduces the cardinality of the problem. Moreover, the new formulation also takes into account the concept of network shadow prices in the capacity allocation process to improve the efficacy of the solution scheme.

Keywords: Service Overlay networks, SON dimensioning, capacity allocations.

1 Introduction

The key components of Service Overlay Networks, (SON)s, are the SON gateways and the interconnecting logical links that lie on the top of one or more physical links. SON gateways can be treated as routers that relay service specific data and perform control functions. SON logical links provide connectivity to the SON network through existing physical links. The SON gateways are connected to adjacent gateways by the logical links. To provide SON service, the SON provider has to purchase bandwidth and QoS guarantees from the corresponding network infrastructure owners via Service Level Agreements (SLA). It is clear that the optimum amount of capacity to be purchased from the infrastructure owners, so as to maximize the net revenue, is an important issue to be faced by the SON providers.

The literature for the network dimensioning problem is usually related to circuit switched networks such as the telephone network. We shall introduce some of the works that are related to ours. In [21], Gavish and Neuman suggested a method based on Lagrangian relaxation to allocate network capacity and assign traffic in packet switching networks, but their model assumed that the traffic is routed through a single path. Medhi and Tipper did comparisons of four different approaches in [20] to a combinatorial optimization problem that describes a multi-hour network dimensioning problem for ATM networks, but their study was also based on the assumption that traffic is routed through a single path. Instead of maximizing the net

L. Mason, T. Drwiega, and J. Yan (Eds.): ITC 2007, LNCS 4516, pp. 224–235, 2007.

income generated from the network, both of the papers chose to compute the minimum capacity allocation costs for the networks. Duan et al [22] investigated the capacity allocation problem of the SON network in order to maximize the net income gained by the SON network. Their model was also confined to networks with single fixed routes for the traffic. Girard proposed in [6] an optimization framework for dimensioning circuit-switched networks employing a more flexible load sharing alternative routing scheme. This framework was applied in [5] for the dimensioning of telephone networks. The formulations in [5] and [6] were problem specific in that they dimension circuit-switched networks consisting of only one-link and two-link paths. Shi and Turner presented in [7] a heuristic approach to size SON multicast networks. Their main focus was on the routing algorithms that optimize the delays and the bandwidth usage on the multicast service nodes. The dimensioning uses a simple algorithm that equalizes the residual capacities across the multicast network.

In our study, we dimension the SON network based on revenue maximization. In this aspect we are not only considering the net income in the objective function, but we are also incorporating the notion of average network shadow price in the dimensioning process in order to reflect the sensitivity of net revenue to the dimensions of the links. We consider the SON network as a generic network and provide a framework for dimensioning based on the traffic rewards. The dimensioning problem is formulated as a constrained optimization problem for two distinct routing models. From the KKT conditions of the optimization formulation, we devise an iterative method that leads to near-optimal solutions. Compared with the previous studies reported in the literature, our models allow more flexible routing schemes whereas each path can be comprised of an arbitrary number of links. We also incorporate two sophisticated routing schemes to better approximate the state dependent routing scheme assumed in the SON environment. We present analytical optimization models, and include detailed discussions of the implementation issues, as well as numerical studies that verify the models' accuracy. A novelty of our study is that we provide an economic integration of the control layer and the dimensioning layer through the use of average shadow price concept.

This article is structured as follows: section 2 will be devoted to the description of routing algorithms used in later sections. The mathematical formulation is included in section 3 together with the details of the analytical models for the network dimensioning problem. We discuss implementation issues and present numerical results in section 4. The conclusion is given in section 5.

2 Routing Algorithms

As mentioned above, in our SON framework we apply the state dependent reward maximization routing strategy; such as the MDPD strategy [4], in order to achieve integrated economic framework. Nevertheless, to simplify analytical performance evaluation, in our dimensioning model we approximate MDPD routing strategy by a routing based on a load sharing concept. The pure load sharing routing strategy is inefficient as calls can be lost even when valid available paths are present which is not the case with MDPD approach. To overcome this issue we employ two relatively simple yet efficient load sharing routing strategies to provide conservative approximations to the MDPD strategy. The blocking performances of these two strategies provide upper bounds for the MDPD strategy. The dimensioning solutions based on them are therefore conservative.

The first routing strategy used here is known as the "combined load sharing and alternate routing" strategy [11]. We denote this strategy as routing strategy I and the corresponding optimization model as model I throughout the article. In this routing strategy, the potential paths for a traffic flow are ordered to form a set of routing sequences. Each of these routing sequences consists of all the potential paths for the traffic flow. The paths are arranged in different orders in different routing sequences. Every routing sequence bears a load sharing coefficient; the traffic flow is assigned to a routing sequence with probability proportional to the load sharing coefficient of that sequence. The traffic flow must attempt all the paths in its assigned sequence before declaring connection failure. A connection would fail if and only if all the paths in the assigned sequence are blocked. Figure 1 shows an instance of such a scheme for the traffic flow between the nodes S and D, f_{SD}. In that example, the first routing sequence carries a fraction $a_1/(a_1+a_2+a_3)$ of the total traffic between nodes S and D, and the flow f_{SD} must attempt paths in the order $P1$, $P2$, $P3$. The second sequence in the example carries $a_2/(a_1+a_2+a_3)$ of the traffic and the paths must be attempted in the order $P2$, $P3$, $P1$. The third sequence carries $a_3/(a_1+a_2+a_3)$ of the traffic flow f_{SD}, and the paths must be attempted in the order $P3$, $P1$, $P2$.

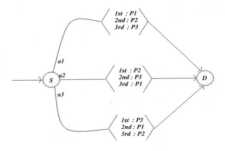

Fig. 1. Routing strategy I

In the second strategy considered for approximation of MDPD routing, each potential path for a particular traffic flow is assigned a routing coefficient. First, the traffic flow is assigned to a path with probability proportional to the routing coefficients. If this path is blocked, the scheme will attempt the remaining $n-1$ paths with probabilities proportional to the paths' original routing coefficients. If the new path chosen by the scheme also turns out to be blocked, the traffic will attempt the remaining $n-2$ paths with probabilities proportional to their original routing coefficients. This process continues until either the traffic flow is routed or until the scheme discovers that all n paths are blocked. Figure 2 depicts a case of such a routing scheme. In that scenario, path $P1$ is discovered blocked by a traffic flow assigned to it. The traffic therefore overflows to the remaining paths $P2$ and $P3$, with probabilities directly proportional to their routing coefficients a_2 and a_3. In the remainder of this paper the second strategy is referred to as routing strategy II and the corresponding optimization model as model II.

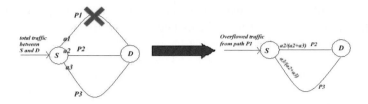

Fig. 2. Routing strategy *II*

3 The Optimization Models

We treat the SON network as a generic network and the SON gateways as generic networking nodes that ship data to generate revenues. In the following discussion we shall use "SON gateways" and "nodes" inter-changeably. The traffics considered here are homogenous traffics with the same bandwidth requirement following the exponential distribution for both their inter-arrival time and service time. We assume, at this stage, that the network topologies, the traffic intensities, the GoS requirements, and traffic revenues (or service prices) are all given parameters. We also assume that there exists at most one physical link between any pair of nodes in the underlying network, although it is possible that there exist more than one physical links. The GoS requirements are specified in the form of blocking probabilities. We formulate the problem as a dimensioning problem for the links in the network. For the sake of implementation, we leave most of the partial derivatives in the equations so as to enable numerical methods such as the finite difference method to be employed.

Without loss of generality, let's assume the paths in the sequence set q are indexed by the order they will be attempted in routing scheme *I*. It is easy to see that because we assumed every routing sequence for a particular flow f_{ij} contains all the corresponding end to end paths, the blocking probability for a particular traffic flow f_{ij} under strategy *I* can be written as:

$$\sum_q \alpha_q^{ij} (\prod_{k=2}^{|R_{ij}|} P(r_k^q \ blocked \mid r_i^q \ blocked, i = 1, 2, ...k - 1)) P(r_1^q \ blocked) \tag{1}$$

where α_q^{ij} is the probability of selecting the sequence q for traffic pair f_{ij}, r_k^q is a particular path indexed by k in the routing sequence q of the traffic pair f_{ij}, R_{ij} are all the end-to-end paths for the flow f_{ij}. To further simplify (*1*), we can assume independence of the paths r_k^q, $k=1,2,...|R_{ij}|$. As a consequence of this assumption, all the conditional probabilities of (*1*) are reduced to the unconditional probabilities, equation (*1*) is now given by:

$$\prod_{r \in R_{ij}} B_r \tag{2}$$

where B_r is the blocking probability for a particular path r. The approximation of (*2*) is exact when none of the paths r shares a common link. For the sake of simplicity we use (*2*) in the formulation of our model. A sidebar note is that the calculation of (*1*) can be equivalently viewed as finding the probability such that at least one of the cut sets with respect to R_{ij} has all its elements failed, in our case it is possible to use a recursive technique to tackle that without the need of finding the sets explicitly, we shall have a short discussion about this in section 3.

Now let's go to derive the overflow traffic generated by routing strategy I. We can see the amount of traffic overflowing to a particular link s, as a result of routing scheme I, is as follows:

$$\sum_{ij} \sum_{q \in Q_{ij}} \sum_{k \geq 1, r_k \in q} ((\lambda^{ij} \alpha_q^{ij} (\prod_{k > m, r_m \in q} B_{r_m})(1 - B_{r_k}))(\delta_{s \in r_k}) / (1 - B_s)) \tag{3}$$

where r_m are all the paths preceding r_k in the sequence q. We define r_0 to be a dummy path that does not consist of any physical links, and we artificially defined $B_{r_0} = 1$ for consistency. Note that we use the same independence assumption as that of (2) in formulating the blocking probabilities. In the expression (3), $\delta_A = 1$ if event A is true, and $\delta_A = 0$ if event A is false, q refers to a particular sequence of paths, and Q_{ij} is the set containing all the sequences of paths for the flow f_{ij}. A particular traffic flow f_{ij} will overflow to the link s if this link is used by a path r_k which is contained in one of the sequences, q, inside the set Q_{ij}, and all the paths r_m before the path r_k in the sequence q are blocked. This results in the expression (3) which will be useful in calculating the link blocking probabilities for our model.

We present the optimization model for routing strategy I first, and denote it as model I throughout this article. Optimization model II for the routing strategy II is similar although the expression for overflow traffic is more complicated. Let's first define the variables being used in model I in the table below:

Table 1. The set of expressions used

$C_s(N_s)$ = the cost function for having a capacity of N_s on link s.

w_r^{ij} = the revenue generated by traffic flow f_{ij}. (i.e. traffic from node i to node j) through path r.

B_r = the blocking probability of path r.

λ^{ij} = the offered traffic in terms of number of connections for the flow f_{ij}.

λ_r^{ij} = the carried traffic for flow f_{ij} on a path r, it is equal to $\lambda^{ij} \Sigma_q (\alpha_q^{ij} P(C_r^q))$, where α_q^{ij} is the load sharing coefficient corresponding to a sequence q, and $P(C_r^q)$ is the probability that the traffic is being admitted at route r of the sequence q, where $q \in Q_{ij}$.

$\overline{L}^{ij}$ = the upper bound for the end-to-end blocking probability of the flow f_{ij}.

α_q^{ij} = the probability of selecting sequence q for the flow f_{ij}.

$E(a_s, N_s)$= the Erlang-B equation for the link s, with offered traffic a_s and capacity N_s.

R_{ij} = the set containing all the possible end-to-end paths for the flow f_{ij}.

R^s = the average shadow price for the link s, this is a sensitivity measurement of the total revenue with respect to the link capacity of link s.

The optimization formulation is shown in equation (4), where x^{ij}, v^{ij}, u_q^{ij}, y_s, z_s are the KKT multipliers. The Lagrange equation for (4) is shown in equation (5). The first order KKT conditions of (5) are listed in equation (6). Equation (6.III) involves the term R^s, which is the sensitivity of the total revenue with respect to the link capacity of link s and is derived as a partial derivative of the revenue with respect to the link capacity. This term tends to be ignored in some of the literature, but we discovered that the addition of this term enables our methodology to yield better results, since it takes into account of the impact of the link capacities on the total revenue generated and reflects the knock-on effects of dimensioning link s over the total revenue. This term is also known as the average network shadow price [4] for the link s. Link shadow price is being used extensively in the routing literature as a control parameter to improve network resource utilization and therefore the incorporation of the average link shadow price in the dimensioning process forms an economic framework that integrates the dimensioning model with the control model of the SON network.

$$Min(\sum_s C_s(N_s) - \sum_{i,j,r} w_r^{ij} \lambda_r^{ij})$$

s.t.

$$\prod_{r \in R_{ij}} B_r \leq \overline{L}^{ij} \qquad ---(x^{ij})$$

$$\sum_{q \in Q_{ij}} \alpha_q^{ij} = 1 \qquad ---(v^{ij})$$

$$\alpha_q^{ij} \geq 0 \qquad ---(u_q^{ij})$$

$$E(a_s, N_s) = B_s \ ---(y_s)$$

$$N_s \geq 0 \qquad ---(z_s)$$

(4)

$$L = \sum_s C_s(N_s) - \sum_{i,j,q,r_k \in q} w_{r_l}^{ij} \lambda^{ij} \alpha_q^{ij} ((\prod_{k > m, r_m \in q} B_{r_m})(1 - B_{r_k})) + \sum_{ij} x^{ij} (\prod_{r \in R_{ij}} B_r - \overline{L}^{ij})$$

$$+ \sum_{ij} v^{ij} (\sum_{q \in Q_{ij}} \alpha_q^{ij} - 1) - \sum_{ij} \sum_{q \in Q_{ij}} u_q^{ij} \alpha_q^{ij} + \sum_s y_s (E(a_s, N_s) - B_s)$$

$$- \sum_s z_s N_s$$

(5)

Note that the KKT multiplier v^{ij} at left hand side of equation (6.I) is independent of the path taken, r. As a consequence, the expression at the right hand side of the equation should have the same value for all the paths carrying a non-zero traffic portion of the flow f_{ij}. This equation can be treated as the optimality equation for the optimal routing problem. This implies that all the paths with a positive share of the traffic f_{ij} should have the same marginal cost for the flows they carry. This is a well known fact for system optimality in the literature. Because of the complementary slackness conditions, we can further conclude from (4) that all paths with positive shares of traffic f_{ij} should have the multiplier u_q^{ij} being 0. Together with the multipliers y_s from the previous iteration (or from the initial values) we can solve (6.I) for α_q^{ij}. With the routing coefficients α_q^{ij} and the multipliers x^{ij} (from the previous iteration or from the initial values) we can calculate the multipliers y_s from (6.II). With the multipliers y_s we can solve the optimal dimension sub-problem in (6.III).

$$v^{ij} = u_q^{ij} + \sum_{r_k \in q} w_{r_k}^{ij} \lambda^{ij} (\prod_{k>m, r_m \in q} B_{r_m})(1 - B_{r_k}) - \frac{\partial (\sum_s y_s E(a_s, N_s))}{\partial \alpha_q^{ij}} \qquad (I)$$

$$y_s = \sum_{i,j,q,r_k \in q} w_{r_k}^{ij} \lambda^{ij} \alpha_q^{ij} (\frac{\partial (\prod_{k>m, r_m \in q} B_{r_m})(1 - B_{r_k})}{\partial B_s}) + \sum_{ij} x^{ij} (\frac{\partial (\prod_{r \in R_{ij}} B_r)}{\partial B_s}) + \sum_{s'} y_{s'} \frac{\partial E(a_{s'}, N_{s'})}{\partial B_s} \qquad (II)$$

$$C_s(N_s) = R^s - \frac{\partial (\sum_{ij} x^{ij} (\prod_{r \in R_{ij}} B_r))}{\partial N_s} - y_s \frac{\partial E(a_s, N_s)}{\partial N_s} + z_s \qquad (III)$$

$$(6)$$

Then with all the multipliers and the optimization variables, we can solve the dual of (5) for the multipliers x^{ij} which is: $\max_x (L(x))$, where $L(x)$ is the function $\min_{N, \alpha} L$ with x as the variable, $L(x)$ is continuous and concave for any primal, but if the primal problem has non-unique solutions. $L(x)$ is non-differentiable. To get around with this, we employ the sub-gradient method to maximize $L(x)$. With the new multipliers x^{ij}, we can go back to equation (6.I) and restart the whole process again until the solution converges. When the solution converges, that means equations (6.I), (6.II) and (6.III) will all be satisfied, which implies the first order KKT conditions of the optimization problem are satisfied and the solution of the dimensioning problem arrives at a stationary point. It is always possible to perform a second order optimality condition check to test for local optimality, although the computation of the Hessian matrix for the Lagrange equation (5) can be expensive as the size of the Hessian is of order $|S|^2$, where S is the set containing all the links of the network. The main solution given by this model is the dimension of the individual links. The iterative solution scheme employed here is similar to that of [6], but the formulation of [6] would be exceedingly complicated if it is restructured to suit the multi-link paths in the SON environment. Our formulation, on the other hand, can tackle paths consisting of an arbitrary number of links without making any modification. Moreover, we take into account the notion of average shadow price in dimensioning each of the individual links. This is something missing in other studies. Additionally, the more sophisticated routing schemes used here also improve the network resource utilization, which in turn helps to alleviate the problem of over-dimensioning in the final solutions.

Though routing strategy II appears to be more complicated in the sense that it attempts the paths to choose a route, it can be proven that the blocking performance expressions are the same as strategy I. Both strategies would declare connection failure for a particular traffic pair f_{ij} if and only if all the possible paths are blocked. In other words, we can use the same approximation (2) to represent the blocking performance for strategy II. The expression for overflow traffic of traffic f_{ij}, however turns out to be more complicated for strategy II. We assume that strategy II only overflows unblocked paths, which is the result of maintaining an up-to-date path status table. We can show that the expression for overflow traffic to a link s, is represented by (7). Note that in this equation, $|R_{ij}|$ is the cardinality of the set R_{ij}. Θ_k^{ij} is a set that contains some particular sets as elements - each of the elements is itself a set that contains k different paths for the traffic f_{ij}, we denote these elements by b_k^{ij}, and Θ_k^{ij} holds all the possible b_k^{ij}.

$$\sum_{ij}\sum_{k=1}^{|R_{ij}|-1}\sum_{r\in b_k^{ij},b_k^{ij}\in\Theta_k^{ij}}[[P(A_r)P(B_k\mid A_r)(\sum_{r'\in b_k^{ij}}\alpha_{r'}^{ij}.\lambda^{ij})(\frac{\alpha_r^{ij}}{\sum_{r'\in R_{ij},r'\in b_k^{ij}}\alpha_{r'}^{ij}})]\delta_{s\in r}]$$ (7)

A_r in the above equation denotes the event "path r is not blocked", and B_k denotes the event "only the k paths in b_k^{ij} are blocked".

$$v^{ij} = u_r^{ij} + w_r^{ij}\lambda^{ij}P(A_r) + w_r^{ij}\sum_{k=1}^{|R_{ij}|-1}\sum_{r\in b_k^{ij},b_k^{ij}\in\Theta_k^{ij}}P(A_r)P(B_k\mid A_r)\frac{\sum_{r'\neq r,r'\in b_k^{ij},r'\in R_{ij}}\alpha_{r'}^{ij}}{(\sum_{r'\in b_k^{ij},r'\in R_{ij}}\alpha_{r'}^{ij})^2}\sum_{r''\in b_k^{ij}}\lambda^{ij}\alpha_{r''}^{ij}$$

$$+ \sum_{i,j,\bar{r}\in R_{ij}}w_{\bar{r}}^{ij}\sum_{k=1}^{|R_{ij}|-1}\sum_{\bar{r}\in b_k^{ij},b_k^{ij}\in\Theta_k^{ij}}[P(A_{\bar{r}})P(B_k\mid A_{\bar{r}})\frac{\alpha_{\bar{r}}^{ij}\times\lambda^{ij}}{\sum_{r'\in b_k^{ij},r'\in R_{ij}}\alpha_{r'}^{ij}}]\delta_{r\in b_k^{ij}}$$

$$-\sum_s y_s\frac{\partial E(a_s,N_s)}{\partial\alpha_r^{ij}}$$ (I)

$$y_s = \sum_{i,j,r\in R_{ij}}w_r^{ij}\lambda^{ij}\alpha_r^{ij}\frac{\partial(P(A_r))}{\partial B_s}$$

$$+ \sum_{i,j,r\in R_{ij}}w_r^{ij}\sum_{k=1}^{|R_{ij}|-1}\sum_{r\in b_k^{ij},b_k^{ij}\in\Theta_k^{ij}}\frac{\partial(P(A_r)P(B_k\mid A_r))}{\partial B_s}(\frac{\alpha_r^{ij}}{\sum_{r'\in b_k^{ij},r'\in R_{ij}}\alpha_{r'}^{ij}})\sum_{r''\in b_k^{ij}}\lambda^{ij}\alpha_{r''}^{ij}$$

$$-\sum_{ij}x^{ij}\frac{\partial\prod_{r\in R_{ij}}B_r}{\partial B_s}-\sum_s y_s\frac{\partial E(a_s,N_s)}{\partial B_s}$$ (II)

$$C_s'(N_s) = R^s - \frac{\partial(\sum_{ij}x^{ij}(\prod_{r\in R_{ij}}B_r))}{\partial N_s} - y_s\frac{\partial E(a_s,N_s)}{\partial N_s} + z_s$$ (III)

(8)

The optimization model for routing strategy *II* is similar to that of strategy *I* although the equations are more complicated because of the overflow pattern. To save space, we shall only list the set of first order KKT conditions in (8). Again R^s is the sensitivity of the total revenue with respect to the link capacity of link s. The above KKT conditions can be solved by using an iterative approach as was done in model *I*.

4 Numerical Results and Discussions

We conducted a series of numerical studies with the mathematical models. We feel it is useful to give a brief discussion of some of implementation issues of both models. In our implementations we use the Frank-Wolfe method to compute the load sharing coefficients. As the Frank-Wolfe method may converge very slowly when it is close to the optimal solution; we artificially supply an upper bound for the number of iterations. This slightly decreases the accuracy of the solution, but in general the efficiency of Frank-Wolfe method is improved.

We use the Erlang B formula extensively in the models here. Direct implementation of the Erlang B formula suffers from two major problems. First, the magnitude of its components explode with the capacity and the offered traffic. Second, direct calculation of the Erlang formula would require a time complexity of $O(n^2)$ where n is the capacity of the link. Both problems can be circumvented by the method mentioned in [18], and the time complexity is reduced to $O(n)$ in our implementation. One further difficulty related to the original Erlang formula is that it is a discrete function in the capacity. A continuous version of Erlang B equation available in the literature[8] involves complicated components that make the computation inefficient. We take advantage of the fact that the Erlang B formula is a strictly decreasing function in the capacity, and use the linear interpolation method to approximate the continuous valued capacity.

For model II, a great deal of the difficulty lies in the computation of the expression $P(B_k|A_r)$. A recursive style algorithm can be used to calculate the overflow traffic elegantly. We employ a recursive DFS (Depth First Search) algorithm [19] to search through all combinations of path failures that can result in k path failures so as to avoid the complexities involved in calculating the sets Θ_k^j explicitly. The same code can also be exploited to find all the cut sets between node i and node j.

Preliminary numerical studies were conducted on two relatively small sample networks for both of the models. The first sample network is illustrated in figure 3. For this network, we assume there are two pairs of traffic, one is from node A to node B, with an average connection rate of 6 units and revenue of 7 units for each carried connection, and the other traffic is from node A to node C, with an average connection rate of 5 units and revenue of 8 units per connection. The GoS requirement is 0.1 for both traffic demands.

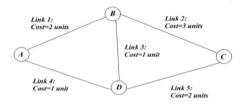

Fig. 3.

The following table summarizes results obtained by the two models, the results are rounded to integers:

Table 2. Results for the sample network in figure 3

Link index	1	2	3	4	5	Cost	Net reward
Model I	0	0	13	20	9	51	28
Model II	0	0	12	20	9	50	29

The GoS constraints are satisfied by both assignments. Model II generates slightly more net reward than model I. A plausible explanation to this phenomenon is that in general the routing strategy I can generate different load distribution on the considered path when compared with the routing strategy II. This fact, combined with significantly larger number of variables to optimize in case of strategy I, may lead to a more suboptimal solution in model I. The convergence graphs for both models are shown in figure 4 below, the y-axis corresponds to the net revenue, while the x-axis corresponds to the iteration number. As we can see, both of them converge in approximately 5 iterations, and each iteration takes around 6 seconds of time on a 1.4Ghz P4 machine for the both models.

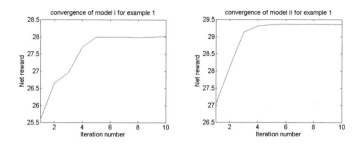

Fig. 4.

We also considered a larger problem as shown in figure 5. This problem has 5 pairs of traffic flows, the traffic details are listed in table 3 and the dimensioning results are depicted in table 4.

Table 3. Traffic matrix for the network in figure 5

	Average Rate	Revenue per connection	Possible routes (indexes of the links)	GoS
Traffic A-> B	25 units	18 units	1	0.1
			4->5->2	
Traffic B->A	15 units	12 units	1	0.1
Traffic A->C	18 units	25 units	1->2	0.1
			4->5	
			4->7->6	
Traffic B->C	30 units	17 units	2	0.1
			1->4->5	
Traffic E->D	12 units	18 units	7->8	0.1
			5->3	
			7->6->3	

Table 4. Dimensioning results for the network in figure 5

Link index	1	2	3	4	5	6	7	8	Cost	Net reward
Model I	74	51	9	0	0	24	30	13	563	1209
Model II	75	55	0	0	0	0	30	16	524	1227

It takes approximately 30 iterations for both models to converge in this larger network, with each iteration taking approximately 8 seconds. Model *II* again gives higher net revenue while satisfying all the GoS requirements. This is to be expected as according to our performance models, model *II* generates less overflow traffic. As a result, less resource is needed for model *II* to meet the GoS Constraints, which becomes more trivial in this example. Depending on the implementation, model *I* might further suffer from the problem of large cardinality in generating sequences, as the number of sequence grows as a factorial function of the possible paths. One may have to limit the number of sequences generated by model I in large examples and this could be another disadvantage of model I for large-size real-world networks,

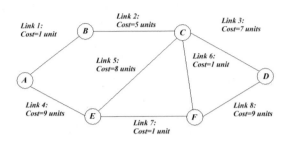

Fig. 5.

5   Conclusions

We studied the problem of SON dimensioning problem by employing an iterative process based on two different routing models. A major contribution of this study is that we provide an approach to dimension the SON network by considering the SON network as a generic network based on the traffic revenues. Moreover the concept of average link shadow price is also incorporated in the SON dimensioning models. We also provided numerical results to offer insights into the efficacy of the theoretical models. The numerical results are promising on the small sample networks tested, and convergence usually occurs within a few iterations. The current effort is to capture the key features of the solution scheme so as to improve computational performance. The verification of the solution quality through state dependent routing simulations and the convergence studies are both in progress. Overall the study reported here provides, under the new perspective of profit maximization, an economic integration of the control and dimensioning layers when allocating capacities in the SON network.

References

1. Dziong, Z.: ATM network resource management. McGraw-Hill, New York (1997)
2. Pioro, M., Medhi, D.: Routing flow and capacity design in communication and computer networks. Morgan Kaufmann Publishers, Washington (2004)
3. Kelly, F.P.: Routing in circuit-switched networks: Optimization, Shadow Prices and Decentralization. Advanced In Applied Probability 20(1), 112–144 (1988)
4. Dziong, Z., Mason, L.G.: Call admission and routing in multi-service loss networks. IEEE Transactions On Communications 42(2/3/4), 2011–2022 (1994)
5. Girard, A., Liau, B.: Dimensioning of adaptively routed Networks. IEEE Transactions On Networking 1(4), 460–468 (1993)
6. Girard, A.: Revenue optimization of telecommunication networks. IEEE Transactions On Communications 41(4), 583–591 (1993)
7. Shi, S., Turner, J.S.: Multicast routing and bandwidth dimensioning in overlay networks. IEEE Journal On Selected Areas In Communications 20(8), 1444–1455 (2002)

8. Farmer, R.F., Kaufman, I.: On the numerical evaluation of some basic traffic formulae. Networks 8(2), 153–186 (1978)
9. Bertsekas, D., Gallager, R.: Data Networks. Prentice Hall, Englewood Cliffs, NJ (1991)
10. Ross, S.M.: Introduction to Probability Models. Academic Press, San Diego (1993)
11. Girard, A.: Routing and dimensioning in circuit-switched networks. Addison-Wesley Publishing Company, New York (1990)
12. Medhi, D., Guptan, S.: Network dimensioning and performance of multi-service, multi-rate Loss networks with dynamic routing. IEEE Transactions On Networking 5(6), 944–957 (1997)
13. Liu, M., Baras, J.S.: Fixed Point approximation for multi-rate multi-hop Loss networks with state-dependent routing. IEEE Transactions On Networking 2(2), 361–374 (2004)
14. Girard, A., Sanso, B.: "Multicommodity flow models, failure propagation, and reliable loss network design", IEEE Transactions On Networking, Vol. IEEE Transactions On Networking 6(1), 82–93 (1998)
15. Chung, S.-P., Ross, K.W.: Reduced load approximations for multi-rate loss networks. IEEE Transactions On Communications 41(8), 1222–1231 (1993)
16. Shah, S.I.A., Girard, A.: Multi-Service network design: a decomposition approach. In: Proc. Globecom, vol. 98, pp. 3080–3085
17. Greenberg, A.G., Srikant, R.: Computational techniques for accurate performance evaluation of multi-rate, multi-hop communication networks. IEEE Transactions On Networking 5(2), 266–277 (1997)
18. Qiao, S., Qiao, L.: A robust and efficient algorithm for evaluating Erlang B formula. Technical Report CAS98-03, Department of Computing and Software, McMaster University (1998)
19. Cormen, T.H., Leiserson, C.E., Rivest, R.L., Stein, C.: Introduction to Algorithms, 2nd edn. MIT Press, Cambridge (2001)
20. Medhi, D., Tipper, D.: Some approaches to solving a multi-hour broadband network capacity design problem with single-path routing. Telecommunication Systems 13(2), 269–291 (2000)
21. Gavish, B., Neuman, I.: A system for routing and capacity assignment in computer communication networks. IEEE Transactions On Communications 37, 360–366 (1989)
22. Duan, Z., Zhang, Z.-L., Thomas, Y.: Service Overlay Networks: SLAs, QoS, and bandwidth provisioning. IEEE/ACM Transactions On Networking 11(6), 870–883 (2003)

Hybrid Topology for Multicast Support in Constrained WDM Networks*

Saurabh Bhandari, Baek-Young Choi, and Eun Kyo Park

Department of Computer Science & Electrical Engineering
University of Missouri, Kansas City, MO 64110, USA
{sbz6d,choiby,ekpark}@umkc.edu

Abstract. Supporting multicast at WDM layer is an important requirement for high bandwidth multicast applications emerging in IP over WDM optical networks. Light-splitting technique has been proposed to provide light-trees, i.e., multicast trees in an optical layer. Many recent studies have been focused to efficiently build and configure light-trees without existing light-paths for unicast traffic into consideration. In this paper we identify and explore the optimal design problem of multicast configuration for realistic and constrained WDM networks. In such a network, both unicast and multicast are supported, and WDM switches have limited light splitting capability. Using wavelength sharing among traffic demands of unicast and multicast, we build a hybrid virtual topology which exploits both light-trees and light-paths. By optimizing WDM resources in addition to resource sharing with unicast, we truly maximize the WDM layer capability and efficiently support multicast traffic demands. We validate the efficiency of our approach with extensive simulations.

Keywords: Multicast, IP over WDM network, Light-tree, Light-path.

1 Introduction

As the Internet traffic continues to grow exponentially and Wavelength Division Multiplexing (WDM) technology matures, WDM network with tera-bits per second bandwidth links becomes a dominant backbone for IP networks. Continuously emerging bandwidth-intensive applications however, presents the need of efficient network support. Particularly, it is increasingly important for an underlaying network to facilitate, in an efficient and scalable manner, high bandwidth multicast applications such as web cache updating, transfer of software upgrade, transfer of video, audio, and text data of a live lecture to a group of distributed participants, whiteboard and teleconferencing [6,13].

In-network replication or branching of multicast traffic may be done in either an optical WDM domain or an electronic IP domain. In IP over WDM networks, mere IP layer multicasting is not efficient enough without the support of WDM

* The authors acknowledge the financial support of the National Science Foundation (NSF Grant SGER-04-43257) for this research.

L. Mason, T. Drwiega, and J. Yan (Eds.): ITC 2007, LNCS 4516, pp. 236–247, 2007.

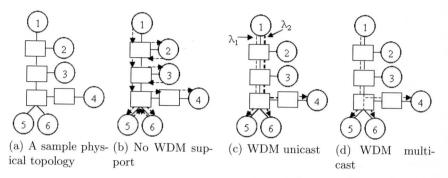

(a) A sample phys- (b) No WDM sup- (c) WDM unicast (d) WDM multi-
ical topology   port   cast

Fig. 1. IP multicast over WDM (Rectangles and circles represent WDM and the corresponding IP nodes respectively. Node 1 is a source and Nodes 4, 5 and 6 are multicast members.)

layer. Enabling multicasting at WDM layer has clear advantages: 1) with the available optical layer resources (i.e., availability of light splitters, wavelength converters and wavelengths) we can utilize a more efficient in-network replication via an optical layer multicast tree than an IP layer multicast tree created without understanding of underlaying physical network. With the inherent light splitting capability of optical switches, it is more efficient to do light splitting than copying IP datagrams in electronic domain. IP multicasting creates copies of data packets at intermediate routers from optical into electronic domain and then converts them into optical signal, called O/E/O conversion. On the intermediate non-member nodes, this process introduces delays and consumes IP resources unnecessarily. 2) performing multicast in optics is desirable and secure, as it provides consistent support of format and bit-rate transparencies across both unicast and multicast transmissions.

An illustration of different ways to realize IP multicast over WDM networks are shown Figure 1. The IP routers (circles) are co-located with the incident WDM switches (rectangles). With no WDM layer multicast support, IP multicast sessions can be realized by having IP router on a multicast tree make copies of a data packet in an electronic domain and transmit another copy to the downstream routers as in Figure 1(b). However, this requires O/E/O conversion of every data packet at intermediate routers on the tree incurs extra-latency and requires the data format to be known to the upper layer. Figure 1(c) depicts the case where IP multicast is supported via light-paths, i.e., WDM multiple unicasts. This avoids the delay of O/E/O conversion at intermediate nodes. However this scheme is not scalable with large number of multicast members and number of groups in the network. An ideal approach to supporting multicasting at the WDM layer is to create multicast trees in the optical layer directly. This can be achieved by light-tree [17], which uses optical light splitting at intermediate nodes as needed in order to replicate an optical signal to multiple downstream paths as in Figure 1(d). It minimizes the use of wavelengths and bandwidth as well as O/E/O conversion. The minimal use of physical resources enables us to support larger sessions as compared to other approaches.

In supporting large multicast groups and members, however, WDM layer multicast may not be possible as the number of multicast demands increases since WDM nodes have limited resources. Note that a light-tree is designed for a specific group and the wavelength is not desirable to be shared by different groups with different set of members. It is because the sharing of a light-tree for other multicast group will cause all the destinations on the tree receive data packet sent on the tree unnecessarily. To further enhance the issue of scalability, we propose to use unicast light-path along with light-tree for WDM layer multicast. It is from the observation that a high bandwidth capacity per wavelength is abundant enough to be shared by multiple traffic demands.

In our approach, we first attempt to create a light-tree for traffic demands of large multicast groups and members. However, if desirable or physical resources are in short, we allow the use of unicast light-paths in conjunction with (partial) light-trees. We find this hybrid (light-trees and light-paths) virtual topology design enables us to establish multicast trees when otherwise impossible. Thus more multicast traffic demands are supported under practical network condition of limited wavelength constraints. Furthermore, our solution maximally utilizes the available resources of existing light-paths whose traffic demand does not reach full wavelength capacity. The idea of using unicast light-paths gives enormous flexibility in terms of resource sharing as compared to a pure light-tree approach.

The concept of wavelength sharing has been proposed before in the context of unicast or multicast individually. To the best of our knowledge, this is the first work that proposes the sub-wavelength resource sharing among unicast and multicast traffic demands and provides a generic solution under practical constraints. By optimizing WDM layer multicast as well as resource sharing with unicast, we truly maximize the WDM layer capability.

The remainder of this paper is organized as follows. In Section 2 we provide the background of our study and summarize related works. In Section 3, we formally state the problem and discuss our approach. The evaluation and validation of our scheme is presented in Section 4. We conclude the paper in Section 5.

2   Background and Related Work

In Wavelength Division Multiplexing) (WDM) networks, each directional fiber optical link is partitioned into multiple data channels each of which operates on a separate wavelength, permitting high bandwidths. Routers or switches are connected via semi-permanent optical pipes called 'light-paths' that may extend over several physical channels via wavelength routing. At intermediate nodes, incoming channels belonging to in-transit light-paths are transparently coupled to outgoing channels through a passive wavelength router, thus avoids the unnecessary IP layer interruption with O/E/O conversion. Meanwhile, at a node terminating a light-path, the incoming signal from the channel is converted to the electronic domain so that packets can be extracted and processed and may be re-transmitted on an outgoing light-path(s) after electronic IP routing. The concept of light-tree can be extended using optical light splitters, in order to replicate an optical signal to multiple downstream paths. The light-paths and/or light-trees establish a virtual topology on top of a physical topology made of optical fibers and switches/routers.

A virtual topology configuration is constrained by a number of physical resource limitations: 1) the establishment of each light-path requires the reservation of WDM channel on the physical links along the paths and the number of available WDM channels are limited on a link. 2) the number of transmitter and receiver at each node limits the number of light-path initiating and terminating on the node. 3) the maximum length of a lightpath without signal regeneration may be limited by the signal attenuation along the light-path. Therefore, optimizing the use of WDM network resources is a crucial task in order to process traffic demand efficiently.

In recent years, many studies have been conducted in regards to the problem of designing virtual (or logical) topology for WDM networks. The problem of multicasting for IP over WDM networks can be decomposed into two subproblems, namely multicast-tree design, and routing and wavelength assignment (RWA) for the designed multicast-tree. To the problem of multicast-tree design, two classes of approaches have been taken namely optimization and heuristics.

The multicast-tree design problem has been modeled as a linear optimization problem often to minimize the O/E/O conversions, as it is the main bottleneck in utilizing true potential of optical networks. Other objectives such as minimizing the average hop count or average number of transceivers used in the network [17] and the total link weight of the light tree [18] have also been used. In [12], the problem of optimal virtual topology design for multicast traffic is studied using light-paths. The authors try to minimize the maximum traffic flowing on any light-path in the network while designing the logical topology for the multicast traffic. In the same work the authors have presented several heuristics for topology design such as Tabu search, simulated annealing. Linear optimization techniques have been also used for unicast single shortest virtual topology design ([1]), RWA ([20,4]), restoration and reconfiguration ([3,5,16]) problems in WDM networks.

Several heuristics have been proposed to design the multicast tree in WDM layer. Although a minimum Steiner tree [7] (which is obtained by solving MILP) is more desirable, finding one for an arbitrary network topology is an NP-complete problem [10], thus heuristics are often used to obtain a near-minimum cost multicast tree. Authors in [21] have presented four heuristic algorithms: namely Re-route-to-Source & Re-route-to-Any, Member-First, and Member-Only, for designing a multicast forest for a given multicast group. The minimum spanning tree (MST) heuristic or the shortest-path tree (SPT) heuristic [7] are also commonly used for designing the multicast tree. Authors in [9] have also presented two such algorithms. One is to use Breadth First Search (BFS) algorithm for a given multicast demand. The other is a dynamic and incremental tree construction. Given an existing multicast tree with a large number of members, new member nodes perform an operation of join/graft as similar in CBT [2] and DVMRP [14].

Once a multicast tree is designed, it can be implemented with either wavelength-routing ([11,17]) or Optical Burst/Label Switching (OBS/OLS) ([15,22]). In the former case, multicast data will be switched to one or more outgoing wavelengths according to the incoming wavelength that carries it. That is a wavelength needs to be reserved on each branch of a multicast tree. In IP over WDM multicast using label switching, multicast label switched paths are set up first. Afterwards, only the optical labels carried by the bursts need O/E conversions for electronic processing, whereas the burst payload always remains in the optical domain at intermediate nodes. The major disadvantage of the wavelength routing approach is that it may not utilize the bandwidth efficiently in case traffic demand is not up to wavelength capacity. It also has large setup latency and it is not efficient under bursty traffic conditions. Meanwhile, with OBS/OLS, a burst dropping (loss) probability may be potentially significant in a highly loaded OBS network which can lead to heavy overheads such as a large number of duplicate retransmissions in IP layer. In addition, it may take a longer time for an end host to detect and then recover from burst dropping (loss) [8].

We address the problem of multicast tree design using optimization in a unique manner. In particular, by bringing the concept of wavelength sharing among traffic demands of unicast and multicast, we build a hybrid virtual topology which exploits both light-trees and light-paths while minimizing WDM resources.

3   Problem Formulation

In this section, we formally state the problem of designing an optimal hybrid topology of light-tree and light-path for a given traffic request and existing network physical and virtual topologies. We first describe given input, parameters and variables. We then formulate our objective function and discuss constrains required for the design problem.

We assume the followings are given as input:

- A Physical topology $G^P = (V^P, E^P)$. It is a weighted bi-directional graph, where V^P is a set of network nodes and E^P is the set of links connecting the nodes. The link weight, w_{mn} is the cost of transferring from node m to n such as such as number of hops or equipment operating cost. The maximum number of wavelengths on each link is given as C^P.
- A Virtual (IP) topology $G^V = (V^V, E^V)$. It is a weighted bi-directional graph, where $V^V \subseteq V^P$. An edge in E^V represents a single hop light-path between the incident nodes for existing unicast traffic demand. The excess bandwidth unit that can be used for multicast traffic is limited by C^V.
- Multicast sessions or traffic demands $S^i(s^i, D^i)$, $i = 1, 2, \ldots, k$. s^i is the source node of multicast group i and D^i is the set of destination or multicast member nodes.

We further assume nodes have wavelength conversion and light-splitting capabilities and transceivers as needed.

Our goal is to build a topology to support the multicast sessions as many as possible given the resource constraints while minimizing the total link cost. In formulating our problem, we define a set of variables described as follows:

- M^i_{mn}: A Boolean variable which is equal to 1 if the link between nodes m and n in E^P is occupied by the multicast session i, and 0 otherwise.
- Y^i_{mn}: A Boolean variable which is equal to 1 if the light-path between nodes m and n in E^V is occupied by the multicast session i, and 0 otherwise.
- V^i_m: A Boolean variable which is equal to 1 if node m in V^P is on the multicast tree for session i, and 0 otherwise.
- $|S^i|$: The cardinality of the multicast session i.
- f^i_{mn}: An integer variable used for physical topology commodity constraint [19]. It represents the number of commodities flowing on the link mn for session i. Each destination node for a session needs one unit of commodity. Therefore, there are $|S^i| - 1$ units of commodity flowing out of source s^i for the session i.
- y^i_{mn}: An integer variable used for IP topology commodity constraint. It represents the number of commodities flowing on the virtual link mn for session i.

The objective function is to minimize the total cost combining physical and virtual link weights as follows:

$$\min \left\{ \sum_i \sum_{mn} w_{mn} M^i_{mn} + \sum_i \sum_{mn} \alpha_{mn} Y^i_{mn} \right\} \tag{1}$$

From the physical topology, the objective function attempts to select links of minimum weights, while in IP virtual topology it chooses paths with minimum number of hops.

We formulate a number of constraints to create light-tree and assign wavelengths as needed as follows.

- Constraints for light-tree generation:

$$\sum_m M^i_{mn} + \sum_m Y^i_{mn} = V^i_n, \forall i, \forall n \neq s^i \tag{2}$$

$$\sum_m M^i_{ms^i} + \sum_m Y^i_{ms^i} = 0, \forall i \tag{3}$$

$$V^i_j = 1, \forall i, \forall j \in S^i \tag{4}$$

$$\sum_n M^i_{mn} + \sum_n Y^i_{mn} \geq V^i_m, \forall i, \forall m \neq d^i, j \geq 1 \tag{5}$$

$$\sum_n M^i_{mn} + \sum_n Y^i_{mn} \leq (C^P + C^V) \cdot V^i_m, \forall i, \forall m \tag{6}$$

We use mn to denote a physical or virtual link between node m and n. Eq. (2) ensures that if a node is on the multicast tree then it has at least one incoming edge. The incoming edge can be in the form of an existing light-path in virtual topology or a physical link. Eq. (3) states that the source does not need the data from the multicast group as it is the source of data. Eq. (4) expresses the members of the corresponding multicast group. Eq. (5) ensures that if a node on the tree is not a destination node then it should have outgoing edges as it is not a terminal node of the tree. Eq. (6) is a constraint on the number of wavelengths which can be created over a link.

– Flow conservation constraints:

$$\sum_m f^i_{mn} + \sum_m y^i_{mn} = \sum_m f^i_{nm} + \sum_m y^i_{nm}, \; \forall i, \forall n \notin D^i \tag{7}$$

$$\sum_m f^i_{mn} + \sum_m y^i_{mn} = \sum_m f^i_{nm} + \sum_m y^i_{nm} + 1, \; \forall i, \forall n \in D^i \tag{8}$$

$$\sum_m f_{nm} + \sum_m y_{nm} = |S^i| - 1, \; \forall i, \forall n = s^i \tag{9}$$

$$\sum_m f^i_{mn} + \sum_m y^i_{mn} = 0, \; \forall i, \forall n = s^i \tag{10}$$

$$M^i_{mn} \leq f^i_{mn}, \; \forall i, \forall mn \tag{11}$$

$$f^i_{mn} \leq (|S^i| - 1) M^i_{mn}, \; \forall i, \forall mn \tag{12}$$

$$Y^i_{mn} \leq y^i_{mn}, \; \forall i, \forall mn \tag{13}$$

$$y^i_{mn} \leq (|S^i| - 1) Y^i_{mn}, \; \forall i, \forall mn \tag{14}$$

$$\sum_i M^i_{mn} \leq C^P, \; \forall mn \tag{15}$$

$$\sum_i Y^i_{mn} \leq C^V, \; \forall mn \tag{16}$$

The flow conservation constraints are to satisfy the traffic demand with a connected path and tree from a source to every multicast destination. Eq. (7) ensures that intermediate nodes have the same amount incoming and outgoing flow, thus not taking the flow to themselves. Eq. (8) secures the traffic to itself if a node belongs to the multicast group. Eq. (9) points out that the source sends one flow amount for each destination node in the multicast group. Eq. (10) states that a source node does not need any traffic amount for itself. Eq. (11) means that a physical link will be on the multicast tree if and only if some commodity is flowing through that link for the multicast session. Eq. (12) expresses that a commodity will only flow through a link if that link is a part of the multicast tree. Eq. (13) and Eq. (14) are similar to Eq. (11) and Eq. (12) respectively, in the context of virtual topology. Eq. (15) limits the number of multicast sessions which can be created using a unicast light-path between node m and n. Similarly Eq. (16) constrains the number of multicast sessions which can be created over link mn.

The above constraints of light-tree generation and flow-conservation enable us to use light-trees and light-paths within the resources available and to meet the given multicast demands. The composite objective function of physical and virtual topology resources given in Eq. (1) provides a generic abstraction for capturing a wide variety of resource and performance optimization such as wavelength, hop count and delays.

4 Evaluation

We have conducted extensive simulations to validate and investigate the feasibility and efficiency of the proposed technique with various topologies. We carried out simulations and evaluations using C and perl languages as well as CPLEX. For a concise illustration however, we discuss the results using the topology in Figure 2.

On the physical topology in Figure 2(a) there is one bidirectional fiber link between adjacent node pairs and each link carries limited number of wavelengths in both directions. We initially set the maximum with two wavelengths ($C^P = 2$). Figure 2(b) shows the existing virtual topology and its virtual links are mapped with physical ones on Figure 2(c).

We are given a group of five multicast sessions $S_i = \{s_i, D_i\}$: i.e., $S_1 = \{1, 2, 3, 5, 6\}$, $S_2 = \{1, 2, 3, 6\}$, $S_3 = \{1, 2, 4, 5, 6\}$, $S_4 = \{1, 2, 4\}$, and $S_5 =$

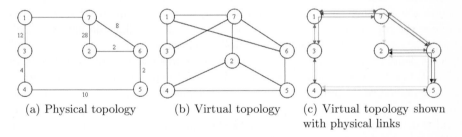

(a) Physical topology   (b) Virtual topology   (c) Virtual topology shown with physical links

Fig. 2. An input topology

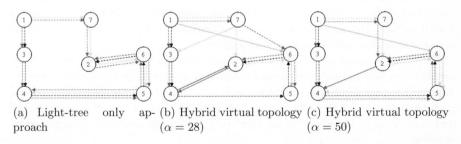

(a) Light-tree only approach   (b) Hybrid virtual topology ($\alpha = 28$)   (c) Hybrid virtual topology ($\alpha = 50$)

Fig. 3. Virtual topologies (solid and dotted lines represent physical and virtual links respectively)

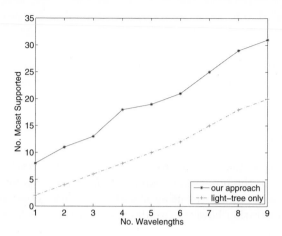

Fig. 4. The number of multicast groups supported: our hybrid vs. light-tree only optimization ($\alpha = 28$)

$\{1, 5, 6\}$. We used the value of $C^V = 2$, that is an existing unicast lightpath can be a branch only for maximum of two light-trees. We formulate the optimization problem as in Section 3 and solve it using CPLEX.

We first demonstrate that our proposed solution indeed can support more multicast traffic demands than pure light-tree approach. For a pure light-tree approach, we have used a similar problem formulation as in [18]. Figure 3 illustrates the topologies built for the multicast demands. Using the light-tree only approach we were able to satisfy only first 4 multicast session requests. The multicast demand S_5 had to be blocked, since there was no available wavelength on link 1-3 or 1-7 as shown in Figure 3(a). However, exploiting the existing virtual links enabled us to support all the multicast demands. Figure 3(b) and 3(c) show the results from our approach using the values of α, 28 and 50 respectively. The value of α controls the preference between light-path and light-tree. The value 28 is the weight sum of the given physical topology, thus it puts higher preference to light-tree in the solution, if the resource is available. By having uniform values (α) for α_{mn} of all links mn, our aim in the experiment is to reduce the hop count of light-path when used. However, note that our problem formulation shown in Eq. (1) is generic so that the optimization objective can be modified depending on the operational circumstances.

We further examine the efficiency of our approach extensively with varied number of wavelengths available on each link. In Figure 4, we compare our hybrid approach with a light-tree only approach in terms of the number of supported multicast groups. In all cases, our approach significantly outperforms, and with more resources of wavelengths the increased number of multicast groups can be satisfied.

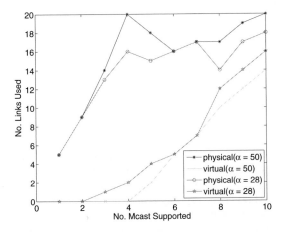

Fig. 5. The number of physical and virtual links used

Next, we investigate on the selection of physical and virtual links and the impact of the parameter α. Figure 5 depicts the number of physical and virtual links used in supporting multicast demands with different α values. In all cases, physical links are preferred thus mostly used. With a large number of multicast demands however, more available virtual links are used as the physical link resources are limited. The value of α controls the degree of favor in using existing light-paths and virtual links. The figure displays that the higher value of α leads to more penalty on use of virtual links, thus physical links are more preferred.

5   Conclusions

We proposed and validated a hybrid virtual topology design scheme for multicast support in constrained WDM networks. In establishing optimal multicast trees, we utilize excess bandwidth of light-paths for unicast as well as light-trees. This approach maximally utilizes the available resources of existing light-paths for unicast whose traffic demand does not reach full wavelength capacity. We show this hybrid virtual topology design enables us to establish multicast trees when otherwise impossible with a pure light-tree approach.

To the best of our knowledge, this is the first work that proposes the sub-wavelength resource sharing among unicast and multicast traffic demands and provides a generic solution under practical constraints.

As an extension of this work, we are currently working on schemes that satisfies multicast demands incrementally exploiting existing light-trees as well as light-paths. Particularly if multiple trees were to be shared, minimizing excess bandwidth usage due to redundant traffic becomes an issue. Other extensions include effective and simple heuristics to fast find solutions for large networks.

References

1. Agrawal, G., Medhi, D.: Single Shortest Path-based Logical Topologies for Grooming IP Traffic over Wavelength-Routed Networks. In: Proc. of 2nd IEEE/Create-Net International Workshop on Traffic Grooming (2005)
2. Ballardie, T., Francis, P., Crowcroft, J.: Core Based Trees (CBT): An Architecture for Scalable Inter-Domain Multicast Routing. In: ACM SIGCOMM, pp. 85–95 (October 1993)
3. Banerjee, D., Mukherjee, B.: Wavelength-Routed Optical Networks: Linear Formulation, Resource Budget Tradeoffs and a Reconfiguration Study. IEEE ACM Transactions on Networking 8(5), 598–607 (2000)
4. Cavendish, D., Sengupta, B.: Routing and wavelength assignment in WDM rings with heterogeneous wavelength conversion capabilities. In: IEEE Infocom (2002)
5. Gencata, A.E., Mukherjee, B.: Virtual-topology adaptation for WDM mesh networks under dynamic traffic. In: IEEE INFOCOM (June 2002)
6. Hastings, K., Nechita, N.: Challenges and opportunities of delivering IP-based residential television service. IEEE Communications Magazine 38(11), 86–92 (2000)
7. Hwang, F.K., Richards, D.S., Winter, P.: The Steiner Tree Problem. Elsevier, New York (1992)
8. Jeong, M., Qiao, C., Xiong, Y.: Reliable WDM Multicast in Optical Burst-Switched Networks. Opticomm, pp. 153–166 (October 2000)
9. Jeong, M., Xiong, Y., Cankaya, H.C., Vandenhoute, M., Qiao, C.: Efficient Multicast Schemes for Optical Burst-Switched WDM Networks. In: IEEE ICC, pp. 1289–1294 (2000)
10. Karp, R.: Reducibility among combinatorial problems. Complexity of Computer Computations (1972)
11. Malli, R., Zhang, X., Qiao, C.: Benefit of Multicasting in All-Optical Networks. In: SPIE Conf. All-Optical Networks, pp. 196–208 (1998)
12. Mellia, M., Nucci, A., Grosso, A., Leonardi, E., Marsan, M.A.: Optimal Design of Logical Topologies in Wavelength-Routed Optical Networks with Multicast Traffic. IEEE Globecomm 3, 1520–1525 (2001)
13. Pankaj, R.K.: Wavelength requirements for multicasting in all-optical networks. IEEE/ACM Transactions on Networking 7, 414–424 (1999)
14. Pusateri, T.: DVMRP version 3. draft-ietf-idmr-dvmrp-v3-07. IETF(August 1998)
15. Qiao, C.: Labeled optical burst switching for IP-over-WDM integration. IEEE Communications Magazine 38(9), 104–114 (2000)
16. Ramamurthy, B., Ramakrishnan, A.: Virtual topology reconfiguration of wavelength-routed optical WDM networks. In: Global Telecommunications Conference (GLOBECOM), vol.2 (2000)
17. Sahasrabuddhe, L., Mukherjee, B.: Light-trees: Optical multicasting for improved performance in wavelength-routed networks. IEEE Communications Magazine 37(2), 67–73 (1999)
18. Singhal, N.K., Mukherjee, B.: Protecting Multicast Sessions in WDM Optical Mesh Networks. Journal of Lightwave Technology, 21(4) (April 2003)
19. Vazirani, V.V.: Approximation Algorithms. Springer-Verlag, Berlin Heidelberg (2001)

20. Yang, D.-N., Liao, W.: Design of Light-Tree Based Logical Topologies for Multicast Streams in Wavelength Routed Optical Networks. In: IEEE INFOCOM (2003)
21. Zhang, X., Wei, J., Qao, C.: Constrained multicast routing in WDM networks with sparse light splitting. Journal of Lightwave Technology 18, 1917–1927 (2000)
22. Zhang, X., Wei, J., Qiao, C.: On Fundamental Issues in IP over WDM Multicast. In: IEEE International Conference on Computer Communications and Networks (October 1999)

Understanding IP Traffic Via Cluster Processes

Ian W.C. Lee and Abraham.O. Fapojuwo

Department of Electrical and Computer Engineering, University of Calgary
Calgary, AB, Canada T2N 1N4
iwclee@ucalgary.ca, fapojuwo@ucalgary.ca

Abstract. In this paper we investigate the characteristics of network traffic via the cluster point process framework. It is found that the exact distributional properties of the arrival process within a flow is not very relevant at large time scales or low frequencies. We also show that heavy-tailed flow duration does not automatically imply long-range dependence at the IP layer. Rather, the number of packets per flow has to be heavy-tailed with infinite variance to give rise to long-range dependent IP traffic. Even then, long-range dependence is not guaranteed if the interarrival times within a flow are much smaller than the interarrival times of flows. In this scenario, the resulting traffic behaves like a short-range dependent heavy-tailed process. We also found that long-range dependent interflow times do not contribute to the spectrum of IP traffic at low frequencies.

1   Introduction

Numerous time-series, stochastic process or point process models have been proposed to model computer network traffic [1][2]. Of these, the point process approach is the most physically appealing as it describes the instances of packet arrivals over a network link or node.

A ubiquitous characteristic of packet traffic is the presence of correlations that span multiple large time scales. This property, called long-range dependence is present in virtually all computer networks. Several point process models such as the so-called fractal point process models [1] and Cox process [3] with fractional Gaussian noise rate are able to capture this characteristic of computer network traffic. However, it is difficult to relate these models to networking properties. The physically motivated ON/OFF and $M/G/\infty$ models [4], on the other hand, cannot be applied to the Internet protocol (IP) layer [5]. Recently, [6] discovered that IP flows of large networks could be modeled as a Bartlett-Lewis cluster process (BLCP) [7][8]. Unlike the fractal point process and Cox process, the BLCP is modeled based on the structure and characteristics of IP flows.

A cluster point process is formulated in two steps, arrival of cluster centers and the distribution of points around the cluster center. This construction parallels the generation of IP flows where the cluster centers correspond to the arrival of IP flows and the distribution of points around the cluster center is the workload generated per flow. In the BLCP model, flows arrive in accordance with a Poisson

L. Mason, T. Drwiega, and J. Yan (Eds.): ITC 2007, LNCS 4516, pp. 248–259, 2007.

process. Each flow generates packets (workload) in the form of a finite renewal process.

The phenomenon of long-range dependence is intimately tied with heavy-tailed distributions with infinite variance. It has been proven many times, that if the ON duration or flow duration is heavy-tailed with tail exponent $\alpha \in (1,2)$, then the resulting traffic is long-range dependent with $H = 0.5(3 - \alpha)$ [4].

In this paper we investigate the properties of network traffic via the BLCP model. Specifically, we study the impact of the arrival process within a flow, the random number of packets per flow, and correlated flow arrival process on IP traffic. The results obtained in this paper will improve our understanding on the characteristics of IP traffic since we consider network centric features. In addition, the results here can be used to aid in parsimonious modeling of network traffic at the IP layer.

2   Preliminaries

2.1   Long-Range Dependence

Long-range dependence (LRD) is characterized by a slowly decaying autocorrelation at large lags or a power spectral density $S(w)$ that tends *rapidly* towards infinity as $w \to 0$ in a second-order stationary stochastic process $Z(k)$. This characteristic imply that there are non-negligible positive correlations in the order of magnitude of the length of the observation period. In the context of network traffic, LRD implies that small time scale (e.g. milliseconds) burstiness are not smoothed out at higher time scales (e.g. hours). The presence of LRD in packet traffic degrades the performance of network entities [9]. LRD is defined as [10]:

$$S(w) \sim c_w |w|^{-(2H-1)}, w \to 0, 0.5 < H < 1 . \tag{1}$$

where H is the Hurst parameter which controls the intensity of LRD, c_w is a constant with dimension of variance, and $f(x) \sim g(x), x \to a$ is taken to mean $\lim_{x \to a} f(x)/g(x) = 1$.

2.2   Heavy-Tailed Distributions

The hallmark of heavy-tailed distributions is the existence of a small number of extremely large values interspersed among the far more numerous smaller values. A random variable X is said to be heavy-tailed with infinite variance if its survivor function $\bar{F}(x)$ has a Pareto like tail:

$$\bar{F}(x) := \Pr(X > x) \sim cx^{-\alpha}, 1 < \alpha < 2, x \to \infty . \tag{2}$$

where α is the tail-exponent and c is a constant.

A broader class of distributions that subsumes (2) are the subexponential distributions. Subexponential distributions are distributions whose tail is heavier than the tail of an exponential distribution. The tails of subexponential distributions have the following closure properties [11]:

$$\lim_{x \to \infty} \bar{F}^{m*}(x) \sim m\bar{F}(x), m \geq 2 \ . \tag{3}$$

where $\bar{F}^{m*}(x)$ denotes the M-fold convolution of $\bar{F}(x)$.

The above definition of subexponential distributions is defined for continuous random variables. A summable nonnegative sequence $\{h_p\}$ is a discrete subexponential distribution if [12]:

$$\lim_{p \to \infty} h_{p+1}/h_p = 1, \ \lim_{p \to \infty} h_p^{\oplus 2}/h_p = 2 \sum_{i=0}^{\infty} h_i < \infty \ . \tag{4}$$

where $h_p^{\oplus 2} := \sum_{i=0}^{p} h_i h_{p-i}$. Then, if the distribution function F is subexponential, for every $c > 0$ the sequence $\{f_p\} \equiv \{\bar{F}(cp)\}$ is discrete subexponential [13, Lemma 3.4]. Thus the properties of subexponential distributions are passed down to its sampled version. Therefore, we are justified in using discrete samples of continuous heavy-tailed random variables such as the Pareto2 distribution to approximate heavy-tailed discrete variables later on. Analogous to (4), we shall denote a discrete distribution to be light-tailed if $\lim_{p \to \infty} \Pr(P > p + 1)/\Pr(P > p) < 1$.

2.3   Bartlett-Lewis Cluster Process

In the Bartlett-Lewis cluster process (BLCP) model, the arrivals of flows follow a Poisson process of rate λ_c while the arrival of packets within a flow follows a *finite* renewal process [7][8]. The packet arrival process within a flow is referred to as the subsidiary process of the flow arrival process. Denote $N_c(t)$, $N_S(t)$, and $N(t)$ as the arrival process of the cluster centers, subsidiary process, and overall process, respectively. Define also $F_Y(t)$ to be the cummulative distribution function (cdf) of the interarrival time Y and π_p the probability mass function (pmf) for the number of packets in a subsidiary process $\{P = 1, 2, ...\}$. The moment generating function of P is denoted as $\varphi_P(z)$. The BLCP can be formulated in terms of Dirac-delta function $\delta(t)$ as $N(t) = \sum_{j=1}^{N_c(t)} \sum_{i=1}^{P_j} \delta(t - (W_j + Y_i))$, where $W_j = X_1 + ... + X_j$ is the time for the jth arrival, X_j is the interarrival time of the jth cluster center, and P_j is the number of packets in the jth subsidiary process. The random variables P, Y, and X are independent of each other.

3   Analysis of the BLCP

3.1   Duration Per Flow

Several papers have shown that the flow duration of IP traffic is heavy-tailed or subexponential. Denote D as the random variable representing the duration per flow. By the law of total probability, $\Pr(D > t) = \sum_{p=1}^{\infty} \pi_p \Pr(D > t | P = p)$.

For a fixed p, D is just the summation of i.i.d. random variables, i.e., $\Pr(D > t | P = p) = \bar{F}_Y^{p*}(t)$. For subexponential Y, from (3),

$$\Pr(D > t) \sim \Pr(Y > t) \sum_{p=1}^{\infty} p\pi_p = \Pr(Y > t)E(P), t \to \infty \ . \tag{5}$$

Thus, D is subexponential (resp. heavy-tailed) if Y is subexponential (resp. heavy-tailed) provided that $E(P) < \infty$. Furthermore, if Y is heavy-tailed with index α, D will inherit the same tail exponent α of Y.

Another way to view D is to consider it as a random sum indexed by a random variable P, i.e., $D = \sum_{i=1}^{P} Y_i$. For a fixed $Y = y$, we have $D = Py$. Unconditioning and assuming $E(Y) < \infty$, we then have $D = E(Y)P$. Therefore, $\Pr(D > t) = \Pr(P > t/E(Y))$. If $\bar{F}_P(p) \sim cp^{-\alpha}, p \to \infty$,

$$\Pr(D > t) \sim cE(Y)^{\alpha} t^{-\alpha}, t \to \infty \ . \tag{6}$$

A question that one may ask is, is it possible for D to be heavy-tailed if neither P nor Y is heavy-tailed? The answer is the negative. This is easily seen by applying the theorem of Willmot and Lin [14]:

Theorem 1. *If there exists a positive $\phi < 1$ such that $\Pr(P > p+1) \leq \phi \Pr(P > p)$, $p = 1, 2, ...$, then the tail of the random sum D, $\Pr(D > t) \leq \frac{1-\pi_0}{\phi} e^{-\kappa_2 t}$, where $\kappa_2 > 0$ satisfies the Laplace-Stieltjes transform $\phi^{-1} = \int_0^{\infty} e^{\kappa_2 t} dF_Y(t) > 1$.*

In other words, the theorem above requires $\Pr(P > p + 1)/\Pr(P > p) < 1$ when $F_Y(t)$ is light tailed (since its Laplace-Stieltjes transform exists). Since we have also assumed that P is light-tailed, making use of the above theorem, $\Pr(D > t) \leq c(e^{-\kappa_2 t})$, for some constant c.

We have looked at 2 mechanisms for the generation of a heavy-tailed duration. The first if Y is heavy-tailed and the second if P is heavy-tailed. However, we will show that only the latter case results in long-range dependence.

3.2   The Subsidiary Process

Many properties of the BLCP can be obtained from the properties of the subsidiary process $N_S(t)$. The renewal function of the subsidiary process is denoted as $H_S(t) := E[N_S(t)]$. Its Laplace transform is denoted as $H_S^*(s)$ and is given as [8]:

$$H_S^*(s) = h^*(s) \left(1 - \varphi_P(f_Y^*(s))\right) \ . \tag{7}$$

where $h^*(s) = \frac{f_Y^*(s)}{1 - f_Y^*(s)}$ is the Laplace transform of the renewal density of an infinite renewal process [15]. The expected empirical spectrum of the subsidiary process observed for a time duration T, $E\left[\hat{S}(w, T)\right]$ is given as [16]:

$$E\left[\hat{S}(w, T)\right] = \frac{\mu_P}{T} + \frac{2}{T}\text{Re}\left\{\frac{f_Y^*(iw)\left[\mu_P\left(1 - f_Y^*(iw)\right) - 1 + \varphi_P(f_Y^*(iw))\right]}{\left(1 - f_Y^*(iw)\right)^2}\right\} \tag{8}$$

where $\mu_P = E(P)$. Equations (7) and (8) show the relationship between the random variables Y and P. In the subsequent sections we will see how they affect the renewal function and spectrum of the subsidiary process.

We will be using four different random variables to model P. If P follows a Pareto2 distribution with survivor function $\bar{F}(p) = \left(\frac{\lambda}{\lambda+p}\right)^{\alpha}$, then from [17, Theorem 8.1.6]

$$\varphi_P(z) \sim 1 + \mu_P \ln(z) - \lambda^{\alpha} \left(-\ln(z)\right)^{\alpha} \Gamma(1-\alpha), z \to 1 . \tag{9}$$

The Geometric $(\varphi_P(z) = z\rho/\left[1 - (1 - \rho)z\right])$, and Poisson distributions $(\varphi_P(z) = z \exp[\lambda(z - 1)])$ are also chosen to model P. These distributions represent different hazard or failure rates [18]. For example, the Geometric distribution has a constant hazard rate while the Poisson and Pareto2 distributions have an increasing and decreasing hazard rate, respectively. Of these distributions, the Pareto2 distribution is the most pertinent in networking traffic. In addition to those random distributions, we also consider a constant P $(\varphi_P(z) = z^P)$. It should be noted that for $P \to \infty$, the subsidiary process becomes an ordinary renewal process.

Effect of Coeficient of Variation. An important delineator of the behavior of a point process is the coefficient of variation (CV) defined as $CV = \text{std}(Y)/E(Y)$. A totally random process such as the Poisson process has $CV = 1$. A CV that is greater than one is indicative of clustering or bursty arrivals while a CV less than one indicates more regularity than a totally random process. Consequently, the larger (smaller) the CV, the more bursty (smooth) is the point process. The Gamma distribution, whose probability density function (pdf) is given by $f(y) = y^{\kappa-1}\beta^{\kappa}e^{-\kappa y}/\Gamma(\kappa)$ with parameter $\kappa = 1, 20, 0.05$, is used to produce $CV = 1$, $CV < 1$ and $CV > 1$ cases, respectively.

As shown in Fig 1, we see that $H_S^*(s)$ increases with decreasing s and peaks at $\lim_{s \to 0} H_S^*(s) = \mu_P$ from L'hospitals rule, regardless of the distribution of P or Y. We also see that the renewal function is larger (smaller) for high (resp. low) coefficient of variation. However, this difference is only noticable at small time scales since $t \propto 1/s$. In addition, the difference is smaller for smaller $E(Y)$. To understand the cause for this difference, we must note that for higher coefficient of variation with mean fixed, both the probabilities of large and small interarrival times increase. Therefore, we will observe more interarrivals in a fixed time period if Y has a larger coefficient of variation.

To see the effect of different coefficient of variation on the spectrum, we kept $E(Y)$ and μ_P to be the same for different P and Y. As shown in Fig. 2 the spectrum is strongly influenced by the distribution of P. The spectrum for large/small coefficient of variation can be larger or lower than the exponential case depending on the distribution of P, unlike the renewal function. For a Pareto2 distributed P, the spectrum with higher coefficient of variation has larger spectrum at higher frequencies as shown in Fig. 2a. At low frequencies, i.e., $w \to 0$, they exhibit the same long-range dependent behavior. It should be noted that the discrepancy at high frequencies decreases with decreasing $E(Y)$

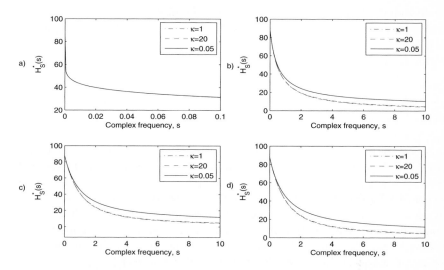

Fig. 1. Laplace transform of renewal function $H_S^*(s)$ for the subsidiary process with Gamma distributed interarrival times ($\kappa = [1, 20, 0.05]$ and $\beta = \kappa/0.02$) for different distributions of P with $\mu_P = 88$. Distribution of P a) Pareto2 ($\alpha = 1.1$) b) Geometric c) Poisson d) Constant $P = \mu_P$.

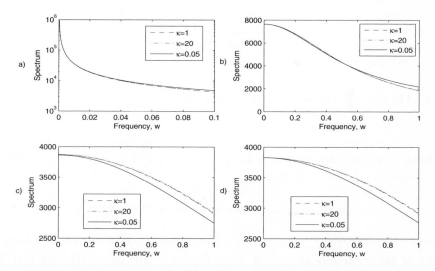

Fig. 2. Spectrum of the subsidiary process with Gamma distributed interarrival times ($\kappa = [1, 20, 0.05]$ and $\beta = \kappa/0.02$) for different distributions of P with $\mu_P = 88$. Distribution of P a) Pareto2 ($\alpha = 1.1$) b) Geometric c) Poisson d) Constant $P = \mu_P$.

or increasing μ_P. It is interesting to note that irrespective of the distribution of P, the peak of the spectrum is located at $w = 0$, indicating some form of long-term correlation.

Periodicity. By periodicity we mean that the packets within a flow arrive at a predetermined fixed time. Hence, periodicity corresponds to a coefficient of variation that is zero or very small. Therefore, the results from the previous section holds and will not be repeated again.

Effect of Wrong Distribution for Y. The empirical distribution of the inter-arrivals within a flow is very complicated [19]. In this section we seek to study the effect of modeling the empirical distribution with a parameteric distribution that just matches the first 2 moments of the empirical distribution. In other words, we study the effect of the interarrival distribution on the renewal function and spectrum. Assume that the interarrivals within a flow is a mixed exponential given by a pdf $f(y) = p_1\lambda_1 e^{-\lambda_1 y} + (1 - p_1)\lambda_2 e^{-\lambda_2 y}$. This distribution is chosen because its Laplace transform is available in close form and it exhibits clustering similar to actual observed empirical distribution [19]. We will attempt to model the mixed exponential with a Gamma distribution by matching the first two moments. Therefore, if Y is mixed exponential, then the parameters for the Gamma distribution is $\kappa = E^2(Y)/\text{Var}(Y)$ and $\beta = \kappa/E(Y)$.

The spectrum of the subsidiary process with mixed exponential and Gamma distribution is shown in Fig. 3. It is observed that the distributional properties of Y only significantly affect the spectrum at high frequencies. The difference at low frequencies, on the other hand, is negligible. The difference between the renewal function (not shown for brevity) for the mixed exponential and Gamma distributed Y is also small and only noticeable at small time scales.

In summary, we see from Sect. 3.2 that the distributional properties of Y is only important at small time scales. Even then, the effect on the renewal function and spectrum is small.

3.3   On the Spectrum of Cluster Processes

Spectral analysis of a stochastic process complements traditional time based methods of analysis. In this section we look at conditions for the BLCP to exhibit long-range dependence, a key characteristics of IP traffic. The Bartlett spectrum for a BLCP is given as [8]:

$$S_B(w) = 2\left\{\lambda_c(1 + \mu_P) + \gamma^*(iw) + \gamma^*(-iw)\right\} . \tag{10}$$

where $\gamma^*(s) = \frac{\lambda_c\mu_P f_Y^*(s)}{1-f_Y^*(s)} - \lambda_c \left(\frac{f_Y^*(s)}{1-f_Y^*(s)}\right)^2 [1 - \varphi_P(f_Y^*(s))]$ is the Laplace transform of the covariance density of $N(t)$. Equation (10) shows that the arrival rate of flows, λ_c only serves to determine the magnitude of $S_B(w)$. Note that the Bartlett spectrum does not exist in the pathological case $\mu_P = \infty$ but the empirical spectrum still shows a LRD like spectrum [16]. At high frequencies, $S_B(w) \to 2\lambda_c(1 + \mu_P), w \to \infty$ as $f_Y^*(iw) \to 0, w \to \infty$.

The $M/G/\infty$ model predicts LRD at the flow level if the flow duration D is heavy-tailed (which can be obtained if P is light-tailed and Y heavy-tailed or P heavy-tailed and Y light-tailed as shown in section 3.1). In the following we will see whether this result holds at the IP layer.

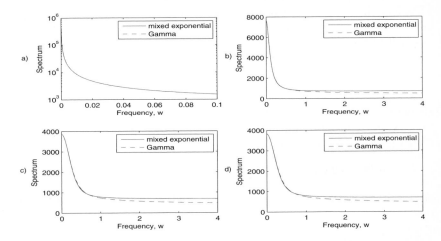

Fig. 3. Spectrum of the subsidiary process for mixed exponential ($p_1 = 0.1, \lambda_1 = 1, \lambda_2 = 1000$) and approximated Gamma distributed interarrival times for different distributions of P with $\mu_P = 88$. Distribution of P a) Pareto2 ($\alpha = 1.1$) b) Geometric c) Poisson d) Constant $P = \mu_P$.

Light-Tailed P. First off we consider a light-tailed P and heavy-tailed Y. It is known that for an infinite renewal process, a sufficient condition for the renewal process to be LRD would be for Y to be heavy-tailed with infinite variance [20]. For simplicity we consider P to be geometric distributed. Therefore, the spectrum reduces to [8, pg. 417]:

$$S_B(w) = 2\lambda_c \left[1 + \mu_P + \frac{f_Y^*(iw)}{(1-\rho)(1-\rho f_Y^*(iw))} + \frac{f_Y^*(-iw)}{(1-\rho)(1-\rho f_Y^*(-iw))} \right] \quad (11)$$

Applying the limit $w \to 0$,

$$\lim_{w \to 0} S_B(w) = 2\lambda_c \left[1 + \mu_P + \frac{2}{(1-\rho)^2} \right] < \infty \ . \quad (12)$$

Heavy-Tailed P. Assume for simplicity that Y is exponentially distributed with mean $1/\mu$ and P is Pareto distributed. From (9):

$$\varphi_P \left(\frac{\mu^2 - iw\mu}{\mu^2 + w^2} \right) \approx 1 + \mu_P \ln \left(\frac{\mu^2 - iw\mu}{\mu^2 + w^2} \right) - \lambda^\alpha \Gamma(1-\alpha) \left(-\ln \left(\frac{\mu^2 - iw\mu}{\mu^2 + w^2} \right) \right)^\alpha \ .$$

We have:

$$\left(-\ln \left(\frac{\mu^2 - iw\mu}{\mu^2 + w^2} \right) \right)^\alpha = A^\alpha \left(\cos(\alpha\theta_1) + i\sin(\alpha\theta_1) \right) \ .$$

where $A = \sqrt{\ln^2 \left(\frac{\mu}{\sqrt{\mu^2+w^2}} \right) + \left(\tan^{-1} \left(\frac{-w}{\mu} \right) \right)^2}$ and $\theta_1 = \tan^{-1} \left(\frac{-\tan^{-1}(-w/\mu)}{-\ln \left(\frac{\mu}{\sqrt{w^2+\mu^2}} \right)} \right)$.

Therefore the second and third terms in (10) become:

$$\lambda_c \left(\frac{\mu}{w}\right)^2 \left\{2 - \left[\varphi_P\left(\frac{\mu^2 - iw\mu}{\mu^2 + w^2}\right) + \varphi_P\left(\frac{\mu^2 + iw\mu}{\mu^2 + w^2}\right)\right]\right\} =$$

$$\lambda_c\mu^2 \left\{-\frac{1}{\mu w^2}\ln\left(\frac{\mu^2}{\mu^2+w^2}\right) + 2\lambda^\alpha\Gamma(1-\alpha)\frac{A^\alpha}{w^2}\cos(\alpha\theta_1)\right\} \quad . \quad (13)$$

Applying the limit $w \to 0$, the first term on the right hand side of (13) goes to $\lambda_c/\mu < \infty$ by L'hospital rule and applying the Maclaurin series expansion for $\tan^{-1}(x)$ leaves us with:

$$S_B(w) \sim 2\lambda_c\mu^{2-\alpha}\lambda^\alpha\Gamma(1-\alpha)\cos(\alpha\pi/2)w^{-(2-\alpha)}, w \to 0 \quad .$$

These results show that LRD at the IP layer is only achieved if P is heavy-tailed.

4   Simulation

In the previous section we derived some statistical properties of the BLCP. In this section, we verify some of the results presented previously and explore some properties of cluster processes which are not analytically tractable. In the following, we simulated 2^{14} flows.

4.1   From Long-Range Dependence to White Noise

Of particular importance in network traffic analysis and performance evaluation is the packet count time series $Z(k) = N((k + 1)\Delta_t) - N(k\Delta_t), k = 0, 1, 2, \ldots$ which counts the number of packets that fall in a time interval of length Δ_t. We saw previously that if P is heavy-tailed with infinite variance, then we would expect the spectrum of $Z(k)$ to be long-range dependent with $H = (3 - \alpha)/2$. Let P be Pareto2 distributed with mean $\mu_P = \lambda\alpha/(\alpha - 1)$, and X and Y exponentially distributed with mean $1/\lambda_c$ and $1/\mu = 1/(\lambda_c K)$, respectively. The variable $K > 1$ determines how small the interarrivals within a flow is compared to the interarrival times of the flow.

Fig. 4a shows the estimated Hurst parameter using the Abry-Veitch estimator [21] for different values of K. We see that the theoretical relation $H = (3-\alpha)/2$ holds only when K is comparable to μ_P. If $K >> \mu_P$, then the Hurst parameter is around 0.5, i.e., the spectrum is white at low frequencies. The reason for this result is that for $K >> \mu_P$, the duration of the flows are much smaller than the interarrival time of the flows. Consequently, at large time scales, the BLCP looks like (but is not) a Poisson process. Only when there is significant overlap between flows do we get LRD behavior. In Fig. 4b we see the survivor function of $Z(k)$ for different values of K. When K is large, the survivor function of $Z(k)$ exhibits a power-law behavior at intermediate ranges, i.e., the marginal distribution of $Z(k)$ inherits some of the properties of the random variable P. Moreover, the power-law behavior extends further to the tail when K increases. These results are reminiscent of the two limiting forms of network traffic, Gaussian and long-range dependent or heavy-tailed and white [22].

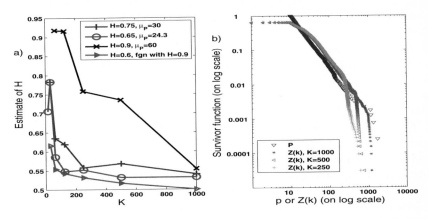

Fig. 4. a) Estimate of H for $Z(k)$ with $k \in (1, 16411)$, $\Delta_t = 2\,\mathrm{s}$, $\lambda = 10$, and $\alpha = 1.5, 1.7, 1.2$. b) Survivor function of P and $Z(k)$ for different values of K.

4.2   Long-Range Dependent Flow Arrivals

Analysis of collected WLAN data have shown that the arrivals of flows are not renewal but long-range dependent [23]. It is known that the spectrum of the counting process is proportional to the spectrum of its interarrival times at low frequencies [24, pg. 78]. Unfortunately, analysis is not possible in this case. We assume now that the interarrival times of flows follow a fractional Gaussian noise with Hurst parameter $H_A = 0.9$ (where the subscript A denotes Hurst parameter from arrival process) and a large enough mean such that all values are positive. The random variable P was also set to be Pareto2 distributed with $\alpha = 1.8$ so that the Hurst parameter caused by P, $H_P = (3 - \alpha)/2 = 0.6$. Consequently, we have two competing factors influencing the spectrum of $Z(k)$ with a predicted Hurst value of $H_A = 0.9$ or $H_P = 0.6$. Surprisingly, long-range dependent flow arrivals do not affect the spectrum of $Z(k)$ as shown in Fig. 4a. At large values of K, $H \approx 0.5$ while at low values of K, $H \approx 0.6$. Hence, LRD IP traffic stems from power-law distributed P only and not from LRD flow arrivals.

5   Discussion

We have seen that the duration of a flow D is heavy-tailed if either Y or P is heavy-tailed. We know from theory that the $M/G/\infty$ model which is a simplified version of the BLCP or rather a transport layer model predicts LRD if the duration is heavy-tailed. However, this result can be misleading if we incorrectly extrapolate to IP traffic. The origin of LRD at the IP layer rests on the number of packets per flow being heavy-tailed with infinite variance.

The distributional properties of the arrival process within a flow only affects low time scales of the renewal function and the high frequency part of the spectrum. In addition, only the low frequency components of the spectrum is pertinent for queueing performance [25]. Consequently, the main focus of modeling should be on the random variable P and less effort can be put on modeling Y. However, as the simulation results show, if Y is too small such that there is little overlap between flows, then LRD will not be manifested. Instead, the spectrum of IP traffic will be flat at low frequencies. At first glance, one might conclude that LRD will be eliminated with the use of high speed links, and hence queueing performance would improve. However, though burstiness resulting from LRD is eliminated, we gain burstiness in the form of heavy-tailed marginal distributions. The interplay between the rate of new flows and rate of packets per flow via K also serves to explain why we observe Gaussian like marginals in one network and non-Gaussian marginals in another. Another noteworthy observation is that analyzing the spectrum of the aggregate process $Z(k)$ alone does not tell us the value of α. In other words, $\alpha \neq 3 - 2\hat{H}$ and $H \neq (3 - \hat{\alpha})/2$, where $\hat{\alpha}$ and $\hat{H}$ stands for the estimated value of α and H, respectively.

We have also seen that the arrival process of flows is not influential in affecting the spectrum at low frequencies. Therefore, even if the arrival of TCP flows is long-range dependent, we can neglect this and assume a renewal process for practical modeling of TCP/IP traffic at the IP layer.

6   Conclusion

Cluster point processes facilitate the understanding of the interrelation between different aspects of network traffic. Through a detailed analysis of the different components of the cluster process such as the interflow arrival times X, intraflow interarrivals Y and the number of packets per flow P, we conclude that only P is of significant importance in modeling the spectrum of IP traffic. Consequently, more effort should be placed on modeling P while simpler models for Y and X are sufficient.

LRD is solely caused by P while the statistical properties of Y and X (which are strongly influenced by protocol and user behavior) do not affect the low frequency spectrum. This result accounts for the ubiquitously observed LRD phenomena in different network architectures.

References

1. Ryu, B.K., Lowen, S.B.: Point-process approaches to the modeling and analysis of self-similar traffic, IEEE INFOCOM (1996)
2. Lee, I.W.C., Fapojuwo, A.O.: Stochastic processes for computer network traffic. Computer Communications 29(1), 1–23 (2005)
3. Cox, D., Isham, V.: Point processes - Monographs on applied probability and statistics. Chapman and Hal, Sydney (1980)
4. Willinger, W., Paxson, V., Taqqu, M.: Self-similarity and heavy-tails: Structural modeling of network traffic. In: Adler, R.J., Feldman, R.E., Taqqu, M. (eds.) A practical guide to heavy-tails: Statistical techniques and applications, Birkhauser, pp. 27–54 (1998)
5. Stevens, W.: TCP/IP Illustrated: the protocols. Addison-Wesley, London (1994)

6. Hohn, N., Veitch, D., Abry, P.: Cluster processes: A natural language for network traffic. IEEE Transactions on Signal Processing, Special Issue on Signal Processing in Networking 51(8), 2229–2244 (2003)
7. Bartlett, M.: The spectral analysis of point processes. Journal of the Royal Statistical Society. B. 25, 264–296 (1963)
8. Lewis, P.: A branching Poisson process model for the analysis of computer failure patterns. Journal of the Royal Statistical Society: B 26(3), 398–456 (1964)
9. Park, K., Willinger, W.: Self-similar network traffic and performance evaluation. Wiley-Interscience, New York (2000)
10. Beran, J.: Statistics for long-memory processes. Chapman and Hall, Sydney (1994)
11. Goldie, C.M., Kluppelberg, C.: Subexponential distributions. In: Adler, R.J., Feldman, R.E., Taqqu, M. (eds.) A practical guide to heavy-tails: Statistical techniques and applications. Birkhauser, pp. 27–54 (1998)
12. Embrechts, P., Hawkes, J.: A limit theorem for the tails of discrete infinitely divisible laws with applications to fluctuation theory. Journal Australian Mathematical Society (Series A) 32, 412–422 (1982)
13. Baltrunas, A., Daley, D., Kluppelberg, C.: Tail behaviour of the busy period of GI/GI/1 queue with subexponential service times. Stoch. Proc. Appl. 111(2), 237–258 (2004)
14. Willmot, G., Lin, X.: Lundberg bounds on the tails of compound distributions. Journal of Applied Probability 31(3), 743–756 (1994)
15. Cox, D.: Renewal theory. Chapman and Hall, Sydney (1962)
16. Lee, I.W.C., Fapojuwo, A.O.: Network traffic under very heavy-tails. submitted for review to IEEE Trans. Sig. Proc. (February 2007)
17. Bingham, N.H., Goldie, C.M., Teugels, J.L.: Regular variation. In: Encyclopedia of mathematics and its applications, Cambridge University press, Cambridge (1987)
18. Barlow, R., Proschan, F.: Statistical theory of reliability and life testing. Holt, Rinehart and Winston (1975)
19. Lee, I.W.C., Fapojuwo, A.O.: Wireless TCP/IP traffic: Analysis and modeling. submitted for review to Computer networks (February 2007)
20. Daley, D.J.: The Hurst index of long-range dependent renewal processes. The. Annals of Probablity 27(4), 2035–2041 (1999)
21. Abry, P., Veitch, D.: Wavelet analysis of long-range dependent traffic. IEEE Transaction on Info. Theory 44(1), 2–15 (1998)
22. Mikosch, T., Resnick, S., Rootzen, H., Stegeman, A.: Is network traffic approximated by stable levy motion or fractional brownian motion? Ann. Appl. Probab. 12(1), 23–68 (2002)
23. Lee, I.W.C., Fapojuwo, A.O.: Modeling wireless TCP connection arrival process, IEEE GLOBECOM (2006)
24. Cox, D.R., Lewis, P.A.W.: The statistical analysis of series of events. Methuen (1966)
25. Li, S., Hwang, C.: Queue response to input correlation functions: Continuous spectral analysis. IEEE/ACM Transactions on Networking 1(6) (1993)

State-Space Modeling of Long-Range Dependent Teletraffic

Alexandre B. de Lima and José R. de A. Amazonas

Communications and Signals Laboratory, Department of Telecommunications and
Control Engineering, School of Engineering, University of São Paulo
{ablima,jra}@lcs.poli.usp.br

Abstract. This paper develops a new state-space model for long-range
dependent (LRD) teletraffic. A key advantage of the state-space
approach is that forecasts can be performed on-line via the Kalman
predictor. The new model is a finite-dimensional (i. e., truncated)
state-space representation of the FARIMA (fractional autoregressive integrated moving average) process. Furthermore, we investigate, via simulations, the multistep ahead forecasts obtained from the new model and
compare them with those achieved by fitting high-order autoregressive
(AR) models.

Keywords: Forecast, Kalman filter, long-range dependence, prediction,
self-similar, traffic.

1 Introduction

Measurements [1] [2] have shown that heterogeneous data traffic traces have
fractal properties such as self-similarity, burstiness (or impulsiveness) and *long
memory* or *long-range dependence* (LRD) (i. e., strong sample autocorrelations
over large lags). This fractal behavior spans several time scales, from milliseconds to minutes. The power spectral density (PSD) of a long memory process
tends to infinity at zero frequency ($1/f^\alpha$ shape near the origin, $0 < \alpha < 1$).
Due to historical reasons, we often express the degree of LRD and impulsiveness
of traffic is terms of the *Hurst coefficient* H [3], $0 < H < 1$ ($\alpha = 2H - 1$).
When $1/2 < H < 1$, traffic is (asymptotically second-order) self-similar and
LRD. The greater the H, the higher the degree of LRD and self-similarity. It
is worth noting that self-similarity and LRD are different concepts, as pointed
out by Petropulu and Yang [4]. However, when a process is second-order self-
similar, self-similarity implies LRD, and vice-versa, which is the case of data
traffic. Long memory can lead to higher packet loss rates than those predicted
by classical queueing theory [1] [5]. Hence, on-line prediction of aggregate LRD
teletraffic plays a key role in preventive congestion control schemes for broadband
networks such as measurement-based connection admission control (MBAC)
algorithms [6].

L. Mason, T. Drwiega, and J. Yan (Eds.): ITC 2007, LNCS 4516, pp. 260–271, 2007.
© Springer-Verlag Berlin Heidelberg 2007

Heterogeneous data traffic with LRD behavior is often modeled by means of the so called fractional autoregressive integrated moving average (FARIMA) process [7] [8] [9] [10] [11], which was independently proposed by Granger and Joyeux [12], and Hosking [13]. They showed that a LRD process can be modeled parametrically by extending the ARIMA (autoregressive integrated moving average) model of Box and Jenkins [14], which is an integrated process, to a fractionally integrated process. FARIMA has the fundamental ability to model the autocorrelation function of real network traffic at short and large lags (which can not be achieved by nonparametric models like FGN - fractional Gaussian noise [15] - that can only model correlations at large lags).

Recently, there has been significant research on state-space models and the Kalman filter [16] in the field of financial time series analysis (Tsay [17] provides a good survey of the topic and the authoritative book of Durbin and Koopman [18] gives a detailed treatment of the subject) inasmuch as the state-space approach is very flexible and capable of handling a wider range of problems than the classical Box-Jenkins ARIMA system [18]. By incorporating a state-space model with the Kalman filter, the state-space method allows for: a) an efficient model estimation, and b) the filtering, prediction, and smoothing of the state vector given past observations. When we consider the issue of on-line traffic prediction, a key advantage of the state-space approach is that forecasts can be performed recursively, using the Kalman predictor (being suitable for an on-line forecasting scheme). However, state-space modeling of an LRD model is not straightforward, as a long memory process has an infinite-dimensional state-space representation.

Bhansali and Kokoszka [19] distinguish two types of methods of time series prediction: a) Type-I methods, which obtain predictions without first estimating the parameter H, and b) Type-II methods, which explicitly seek to estimate H. In this study, we investigate the use of both methods and compare them on the basis of their impact on forecast accuracy. For the Type-I method, we postulate that heterogeneous teletraffic follows AR (autoregressive) models and obtain the forecasts via the Kalman predictor. For the Type-II method, we assume that data traffic follows FARIMA models and develop a finite-dimensional state-space representation for stationary LRD traffic traces, which we call T-FARIMA (truncated FARIMA). We then also obtain multistep ahead forecasts via the Kalman filter algorithm. It is worth noting that, to the best of our knowledge, we have identified only a few papers on the subject of Kalman prediction of aggregate data traffic, such as [20] [21] and [22]. Yet, none of them consider the use of LRD traffic models.

The rest of this article is organized as follows. Section 2 provides a brief overview of the theory of LRD time series. In Section 3, we develop a truncated state-space form for the FARIMA model (the T-FARIMA model) and present the Kalman predictor. Some experimental results and comparisons are exhibited in Section 4. Finally, section 5 summarizes the contributions of the paper and highlights the future work.

2 Background

Box and Jenkis [14] introduced the class of stationary ARMA(p, q) (autoregressive moving average) models[1]

$$x[n] - \mu = \sum_{j=1}^{p} \phi_j (x[n-j] - \mu) + w[n] - \sum_{j=1}^{q} \theta_j w[n-j], \qquad (1)$$

where $E\{x[n]\} = \mu$ is the mean of $x[n]$, and $w[n]$ is a wide-sense stationary white noise process with zero mean and power σ^2, i. e., $w[n] \sim (0, \sigma^2)$. In a more compact form, we have

$$\phi(B)x'[n] = \theta(B)w[n], \qquad (2)$$

where $x'[n] = x[n] - \mu$, B is the backward shift operator ($Bx[n] = x[n-1]$), $\phi(B)$ is the autoregressive operator of order p

$$\phi(B) = 1 - \phi_1 B - \phi_2 B^2 - \ldots - \phi_p B^p \qquad (3)$$

and $\theta(B)$ denotes the moving average operator of order q

$$\theta(B) = 1 - \theta_1 B - \theta_2 B^2 - \ldots - \theta_q B^q. \qquad (4)$$

In the rest of the paper, we will assume $\mu = 0$ without loss of generality.

The process $x[n]$ can be viewed as the output of a digital filter (ARMA filter) with system function

$$H(z) = \frac{1 - \theta_1 z^{-1} - \theta_2 z^{-2} - \ldots - \theta_q z^{-q}}{1 - \phi_1 z^{-1} - \phi_2 z^{-2} - \ldots - \phi_p z^{-p}}, \qquad (5)$$

where $H(z)$ denotes the z-transform of the impulse response $h[n]$ of the ARMA filter. An ARMA(p, q) process $x[n]$ is said to be *wide-sense stationary* if the poles of $H(z)$ in (5) lie inside the complex unit circle ($|z| = 1$), and it is *invertible* if the zeros of $H(z)$ in (5) lie inside the unit circle. The autocorrelation function[2] $\rho[h]$ of an ARMA(p, q) process shows exponentially decay to zero, i. e., at lag h converges rapidly to zero as $h \to \infty$ (short memory property) [14].

A random process $x[n]$ exhibits LRD (or strong dependence or long memory) if its autocorrelation function $\rho[h]$, $h = 0, \pm 1, \pm 2, \ldots$, is nonsummable, i.e., $\sum_h \rho[h] = \infty$ [23]. The autocorrelation $\rho[h]$ of a long memory process converges to zero as $h \to \infty$ at a much slower rate than $\rho[h]$ for an ARMA model.

---

[1] In this work, we use the simplified notation $x[n]$ to denote a discrete-time stochastic process $\{x[n]\}$.

[2] We assume that the *autocorrelation function* is given by $\rho[h] = \frac{\gamma[h]}{\gamma[0]}$, where $\gamma[h]$ corresponds to the autocovariance of $x[n]$ at lag h. Engineers usually define autocorrelation as $R[h] = E\{x[n]x[n+h]\}$.

Alternatively, LRD can be defined in the frequency domain [24]. $x[n]$ is said to be a $1/f^\alpha$ noise if there are numbers α and C_P satisfying $0 < \alpha < 1$ and $C_P > 0$ such that

$$\lim_{f \to 0} \frac{P^x(f)}{C_P |f|^{-\alpha}} = 1, \tag{6}$$

where $P^x(f)$ denotes the PSD of $x[n]$ and f is the normalized frequency, i. e., $-1/2 \le f \le 1/2$. Hence, long memory correponds to the blow up of the spectrum at the origin.

A stationary FARIMA(p, d, q) process $x[n]$, $-1/2 < d < 1/2$, satisfies a difference equation of the form

$$\nabla^d \phi(B) x[n] = \theta(B) w[n], \tag{7}$$

where $\nabla^d = (1 - B)^d =$

$$= \sum_{k=0}^{\infty} \binom{d}{k} (-B)^k = 1 - dB + \frac{1}{2!} d(d-1) B^2 - \frac{1}{3!} d(d-1)(d-2) B^3 + \dots$$

is the fractional difference operator. We thus define FARIMA(p, d, q) as the process

$$x[n] = \phi(B)^{-1} \theta(B) \nabla^{-d} w[n], \tag{8}$$

which is a generalized ARIMA(p, d, q) model by allowing for noninteger d. The FARIMA(p, d, q) model of (7) can be interpreted as the output of a cascade connection of two systems $h_1[n]$ and $h_2[n]$ with impulse response $h[n] = h_1[n] * h_2[n]$ ($*$ denotes convolution), where $h_1[n]$ is a *LRD filter* with system function

$$H_1(z) = 1 - dz^{-1} + \frac{1}{2!} d(d-1) z^{-2} - \frac{1}{3!} d(d-1)(d-2) z^{-3} + \dots , \tag{9}$$

and $h_2[n]$ has a system function which is the inverse of (5) (ARMA filter).

A FARIMA model has long memory if $0 < d < 0.5$. The parameter d models the autocorrelation function of FARIMA at large lags whereas p and q at small lags. H and d are related by

$$d = H - 1/2 . \tag{10}$$

3   State Space Modeling

3.1   T-FARIMA

The definition of the T-FARIMA model is motivated by the practical issue of predicting LRD series. Consider the FARIMA model of (7), which can be rewritten as

$$\frac{(1 - B)^d \phi(B)}{\theta(B)} x[n] = w[n]. \tag{11}$$

The lag polynomial on the left side of (11) can be expressed as an *infinite* order polynomial so that a FARIMA process can be expressed as an AR(∞) model.

In practice, forecasting from the AR(∞) representation usually *truncates* the AR(∞) model to an AR(L) model with a very large value of L (in Section 4, we empirically derive a value for L that gives good approximations of FARIMA models)

$$\frac{(1-B)^d \phi(B)}{\theta(B)} \approx 1 - \psi_1 B + \psi_2 B^2 - \ldots - \psi_L B^L. \qquad (12)$$

Hence, a truncated stationary representation AR(L) for a FARIMA process $\boldsymbol{x}[n]$ satisfies a difference equation of the form

$$\psi(B)\boldsymbol{x}[n] \approx \boldsymbol{w}[n], \qquad (13)$$

where $\psi(B) = 1 - \psi_1 B + \psi_2 B^2 - \ldots - \psi_L B^L$.

ARMA models may be represented in *state-space* form. The general *linear Gaussian state-space model* (LGSSM) can be written as the system of equations [17, p. 508][3]

$$\underset{m \times 1}{\boldsymbol{x}[n+1]} = \underset{m \times 1}{d[n]} + \underset{m \times m}{T_n} \cdot \underset{m \times 1}{\boldsymbol{x}[n]} + \underset{m \times r}{H_n} \cdot \underset{r \times 1}{\boldsymbol{\eta}[n]}, \qquad (14)$$

$$\underset{l \times 1}{\boldsymbol{y}[n]} = \underset{l \times 1}{c[n]} + \underset{l \times m}{Z_n} \cdot \underset{m \times 1}{\boldsymbol{x}[n]} + \underset{l \times 1}{\varepsilon[n]}, \qquad (15)$$

where $n = 1, 2, \ldots, N$ and

$$\boldsymbol{x}[1] \sim N(\mu_{1|0}, \Sigma_{1|0}), \text{(initial state)} \qquad (16)$$
$$\boldsymbol{\eta}[n] \sim iid\ N(0, I_r), \qquad (17)$$
$$\varepsilon[n] \sim iid\ N(0, I_l). \qquad (18)$$

and it is assumed that

$$E[\varepsilon[n]\boldsymbol{\eta}[n]^{\mathrm{T}}] = \boldsymbol{0}. \qquad (19)$$

The *state* (or *transition*) *equation* (14) describes the evolution of the state vector $\boldsymbol{x}[n]$ over time using a first order Markov structure. The *observation equation* (15) describes the vector of observations $\boldsymbol{y}[n]$ in terms of the state vector $\boldsymbol{x}[n]$ and a vector of Gaussian (zero mean) white noise $\varepsilon[n]$ (*measurement noise*) with covariance matrix I_l. It is assumed that the Gaussian (zero mean) white sequence $\boldsymbol{\eta}[n]$ (*innovations sequence*) with covariance matrix I_r in (14) is uncorrelated with $\varepsilon[n]$. The deterministic matrices T_n, H_n, Z_n are called *system*

---

[3] In order to avoid misunderstandings, we find it useful to adopt the following notational rules in the remainder of this article: a) **X** denotes a matrix with random quantities, X is a deterministic matrix, $\boldsymbol{x}$ can denote a random vector or a random variable (it will be clear from the context which one is), and x would be a deterministic vector or an observation, b) square brackets denote the time-dependency of vector and scalar quantities, whereas subscripts denote the time-dependency of matrices, and c) all vectors, unless stated otherwise, are columns vectors, as $\boldsymbol{y}[n] = [\boldsymbol{y}[n], \boldsymbol{y}[n-1], \ldots, \boldsymbol{y}[n-N+1]]^{\mathrm{T}}$, where T denotes transposition.

matrices and the vectors $d[n]$ and $c[n]$ contain fixed components and may be used to incorporate known effects or patterns into the model (otherwise they are equal to zero). For univariate models, $l = 1$ and, consequently, Z_n is a row vector.

There are many ways to transform a truncated AR model

$$y[n] = \phi_1 y[n-1] + \ldots + \phi_L y[n-L] + \xi[n] \tag{20}$$

into a state-space form [17]. Here, we choose to transform (20) following Harvey's approach [25]. So, (20) may be put in a state-space form, which we call the state-space T-FARIMA model, with state and observation equations

$$\boldsymbol{x}[n+1] = T\boldsymbol{x}[n] + H\xi[n], \quad \xi[n] \sim N(0, \sigma_\xi^2) \tag{21}$$

$$y[n] = Z\boldsymbol{x}[n], \tag{22}$$

where

$$T = \begin{pmatrix} \phi_1 & 1 & 0 & \cdots & 0 \\ \phi_2 & 0 & 1 & & 0 \\ \vdots & & & \ddots & \vdots \\ \phi_{L-1} & 0 & 0 & & 1 \\ \phi_L & 0 & 0 & \cdots & \end{pmatrix}, \quad H = \begin{pmatrix} 1 \\ 0 \\ \vdots \\ 0 \end{pmatrix}, \quad Z = (1 \quad 0 \quad \cdots \quad 0 \quad 0). \tag{23}$$

The state vector has the form

$$\boldsymbol{x}[n] = \begin{pmatrix} y[n] \\ \phi_2 y[n-1] + \ldots + \phi_L y[n-L+1] \\ \phi_3 y[n-1] + \ldots + \phi_L y[n-L+2] \\ \vdots \\ \phi_L y[n-1] \end{pmatrix}. \tag{24}$$

The model in Eqs. (21) and (22) has no measurement errors. It has an advantage that the AR coefficients are directly used in the system matrix T.

3.2 Multistep Prediction with the Kalman Filter

Linear estimation theory [26] states that the MMSE (minimum mean-square error) estimate $\hat{\boldsymbol{x}}[n]$ of the state $\boldsymbol{x}[n]$ in (21), given knowledge of observations $\mathcal{F}_n = [y[n], y[n-1], \ldots, y[1]]$, is the corresponding conditional mean $E\{\boldsymbol{x}[n]|\mathcal{F}_n\}$, which we denote by $\boldsymbol{x}_{n|n}$. The target of the Kalman filter algorithm for the state-space model defined by (21) and (22) is to estimate recursively the conditional mean $\boldsymbol{x}_{n+1|n}$ and covariance matrix $\Sigma_{n+1|n}$ of the state vector $\boldsymbol{x}[n+1]$ given the data $\mathcal{F}[n]$ and the model, that is

$$\boldsymbol{x}_{n+1|n} = E\{T\boldsymbol{x}[n] + H\xi[n]|\mathcal{F}_n\} = T\boldsymbol{x}_{n|n}, \tag{25}$$

$$\Sigma_{n+1|n} = \text{Var}\{T\boldsymbol{x}[n] + H\xi[n]|\mathcal{F}_n\} = T\Sigma_{n|n}T^\mathrm{T} + \sigma_\xi^2 HH^\mathrm{T}. \tag{26}$$

Let

- $y_{n|n-1}$ denote the conditional mean of $y[n]$ given $\mathcal{F}_{n-1}$,
- $\tilde{y}[n] = (y[n] - y_{n|n-1})$ be the 1-step ahead prediction error of $y[n]$ given $\mathcal{F}_{n-1}$,
- $\sigma^2[n] = \text{Var}\{\tilde{y}[n]|\mathcal{F}_{n-1}\}$ be the variance of the 1-step ahead prediction error,
- $C_n = \text{Cov}(\boldsymbol{x}[n], \tilde{y}[n]|\mathcal{F}_{n-1})$, and,
- $K[n] = TC_n/\sigma^2[n] = T\Sigma_{n|n-1}Z^{\text{T}}/\sigma^2[n]$ be the so called *Kalman gain*.

For the sake of brevity, we write below the equations of the Kalman filter algorithm directly (see [17], [18] and [25] for more details):

$$\begin{cases} \tilde{y}[n] & = y[n] - Z\boldsymbol{x}_{n|n-1}, \\ \sigma^2[n] & = Z\Sigma_{n|n-1}Z^{\text{T}}, \\ K[n] & = T\Sigma_{n|n-1}Z^{\text{T}}/\sigma^2[n], \\ L_n & = T - K[n]Z, \\ \boldsymbol{x}_{n+1|n} & = T\boldsymbol{x}_{n|n-1} + K[n]\sigma^2[n], \\ \Sigma_{n+1|n} & = T\Sigma_{n|n-1}L_n^{\text{T}} + \sigma_\xi^2 HH^{\text{T}}, \quad n = 1, \ldots, N. \end{cases} \quad (27)$$

As the state-space model given by (21) and (22) is stationary, one can prove that the matrices $\Sigma_{n|n-1}$ converge to a constante steady state matrix Σ_*. Once the steady state is achieved, $\sigma^2[n]$, $K[n]$, and $\Sigma_{n+1|n}$ are constants.

Assume that the forecast origin is n and we want to predict $y[n + h]$ for $h = 1, 2, \ldots, H$. The Kalman filter prediction equations (last two equations of (27)) produces 1-step ahead forecasts. In general, the forecasts $\hat{y}[h]$, $h = 1, \ldots, H$, can be obtained via the Kalman filter prediction equations by extending the data set $\mathcal{F}[n]$ with a set of missing values [18].

4   Experimental Results and Comparisons

We fitted FARIMA models, using the S+FinMetrics function FARIMA [27], to three simulated Gaussian FARIMA$(2, d, 0)$ traces ($\phi_1 = 0.3, \phi_2 = -0.4$, and $d = \{0.4, 0.3, 0.2\}$). The series have $N = 4096$ data points and were synthesized via the wavelet-based traffic generator developed by Mello, Lima, Lipas and Amazonas [28]. It is worth noting that the simulated series present LRD and SRD behavior. The T-FARIMA models are of order $L = 50$. Table 1 summarizes the estimated models.

We assess goodness-of-fit of the T-FARIMA and FARIMA models via spectral analysis. The fit of the estimated models to the periodograms (estimated Welch's averaged modified periodogram method [29]) of the simulated traces on log-log plots are shown in Fig.1. For $f > 10^{-3}$, we can observe a good fit of the T-FARIMA models (as well as of the FARIMA's), which endorses T-FARIMA as a good traffic model.

We produce Type-I and Type-II h-step ahead predictions $\hat{y}[n + h]$, $h = 1, 2, \ldots, 7$, with time origin $n = 4089$, for a simulated FARIMA$(0, 0.4, 0)$ trace

Table 1. Fitted FARIMA$(p, \hat{d}, 0)$ models for simulated series with $d = \{0.4, 0.3, 0.2\}$, $\phi_1 = 0.3$, $\phi_2 = -0.4$

	$\hat{d}$	$\hat{\phi_1}$	$\hat{\phi_2}$
FARIMA$(2, 0.4, 0)$	0.3859	0.3014	-0.3988
FARIMA$(2, 0.3, 0)$	0.2927	0.2663	-0.4007
FARIMA$(2, 0.2, 0)$	0.1714	0.3332	-0.4162

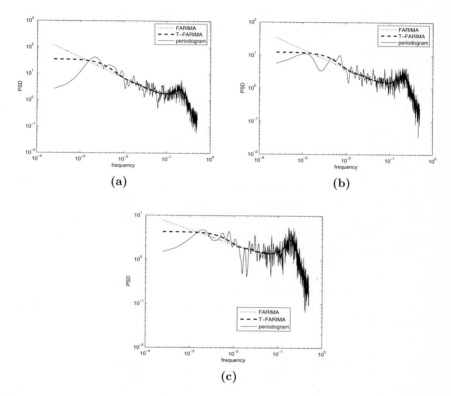

(a)   (b)

(c)

Fig. 1. In (a), (b), and (a), the thin solid curves are the periodograms for simulated realizations of FARIMA$(2, d, 0)$ models with $d = 0.4$, $d = 0.3$, and $d = 0.2$, respectively. Dotted and dashed lines are the PSDs for the estimated FARIMA (see Table 1) and T-FARIMA models, respectively.

via the Kalman filter algorithm, as explained in subsection 3.2. Table 2 summarizes the Type-I method forecasts from an AR(15) fitted model (best fitting process using the well known *Akaike information criterion* - AIC [30]) while Table 3 contains the Type-II multistep predictions obtained from an estimated T-FARIMA(50) model ($\hat{d} = 0.3684$) [4]. In Tables 2 and 3, $\hat{y}[n+h](-)$ and $\hat{y}[n+h](+)$

---

[4] The forecasts of Tables 2 and 3, which have several digits, were obtained with the S+FinMetrics software. However, that degree of precision is not mandatory.

Table 2. Forecasts from an AR(15) fitted model (best fitting process using AIC) for the simulated FARIMA$(0, 0.4, 0)$ model, and time origin at $n = 4089$

h	$\hat{y}[n+h](-)$	$\hat{y}[n+h]$	$\hat{y}[n+h](+)$	$y[n+h]$	$\tilde{y}[n+h]$
1	-0.20099872	0.1116940402	0.4056620	0.268	0.15631
2	-0.23251719	0.1112096416	0.4128297	0.17228	0.06107
3	-0.24856752	0.0850594357	0.4161012	0.1491	0.064041
4	-0.25964510	0.0894804283	0.4176313	0.00023861	-0.089242
5	-0.26726952	0.0718029120	0.4179018	-0.014208	-0.086011
6	-0.27351718	0.0707935404	0.4184437	0.24498	0.17419
7	-0.27802141	0.0606032489	0.4182850	0.33441	0.27381

$\text{EMSE}_{n=4089} = 0.021848$

Table 3. Forecasts from an estimated T-FARIMA(50) model ($\hat{d} = 0.3684$) for the simulated FARIMA$(0, 0.4, 0)$, and time origin at $n = 4089$

h	$\hat{y}[n+h](-)$	$\hat{y}[n+h]$	$\hat{y}[n+h](+)$	$y[n+h]$	$\tilde{y}[n+h]$
1	-0.20110861	0.102331663	0.4057719	0.268	$0, 16567$
2	-0.23322034	0.090156258	0.4135329	0.17228	$0, 082124$
3	-0.24853173	0.083766829	0.4160654	0.1491	$0, 065333$
4	-0.25874697	0.078993081	0.4167331	0.00023861	$-0, 078754$
5	-0.26623012	0.075316145	0.4168624	-0.014208	$-0, 089524$
6	-0.27196006	0.072463238	0.4168865	0.24498	$0, 17252$
7	-0.27657753	0.070131781	0.4168411	0.33441	$0, 26428$

$\text{EMSE}_{n=4089} = 0.021754$

denote, respectively, the lower and upper bounds of the 95% confidence bands. We compare the forecasts of the two models in terms of the *empirical mean-square error* (EMSE) at time n

$$\text{EMSE}_n = \left[\sum_{h=1}^{7} (y[n+h] - \hat{y}[n+h])^2 \right] / 7, \qquad (28)$$

which is a useful performance criterion should one wants to compare models on the basis of their impact on forecast accuracy [31]. As we can see in Tables 2 and 3, both models have similar forecast performance. However, T-FARIMA performs slightly better, since it has a lower EMSE_n. Though AR model provided competitive forecasts, we must stress the fact that we used a high-order AR(15) model, which is by no means a parsimonious representation (in practice, we use ARMA models with orders $\{p, q\} < 3$ [27]). This procedure is suitable only because N, the realization length, is large (4096) [19, p. 358]. Hence, T-FARIMA is the preferred forecasting model. Fig.2 illutrates actual values, h-step predictions and 95% confidence intervals for $y[n]$ from AR(15) and T-FARIMA(50).

Last but not least, we would like to point out that a main practical issue in working with the FARIMA models is that the order (p, q) is usually unknown, which was not the case in this study, as we have used simulated traces (so, the

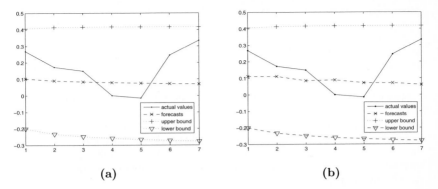

Fig. 2. (a) and (b) illutrate: actual values, h-step predictions and 95% confidence intervals for $y[n]$ from the fitted models of tables 3 and 2, respectively

orders p and q are known a priori). Beran, Bhansali, and Ocker [32] demonstrate that the standards order selection procedures based on the use of AIC, BIC (Schwarz Bayesian Information Criteria) and related order selection criteria still apply for the class of FARIMA$(p, d, 0)$ processes.

5   Conclusion

In this paper, we presented a new finite dimensional state-space traffic model called T-FARIMA(L), which is a truncated version of the FARIMA(p, d, q) model. We then derived the Kalman filter algorithm for T-FARIMA and investigated its forecast performance. Preliminary simulations indicate that T-FARIMA offers a slightly better forecast performance than high-order AR models (the latter models are suitable only when N, the series length, is large). Hence, we conclude that the T-FARIMA model with the Kalman predictor may be useful in on-line applications such as CAC or dynamic bandwidth allocation.

In future work, it will be investigated the main practical problem of model selection and forecasting (via the Kalman predictor) by means of a large scale simulation study. On-line prediction of impulsive data traffic with nonGaussian marginals is also a key topic that must be addressed. The use of SEMIFAR (semiparametric fractional autoregressive) models [33] seems to be promising in teletraffic modeling as this class of models includes deterministic trends, difference stationarity and stationarity with short- and long-range dependence.

References

1. Leland, W., Taqqu, M., Willinger, W., Wilson, D.: On the self-similar nature of Ethernet traffic (extended version). IEEE/ACM Transactions on Networking 2(1), 1–15 (1994)
2. Paxson, V., Floyd, S.: Wide-area traffic: The failure of Poisson modeling. IEEE/ACM Transactions on Networking 3(3), 226–244 (1995)

3. Hurst, H.E.: Long-term storage capacity of reservoirs. Trans. Am. Soc. Civil Engineers 116, 770–799 (1951)
4. Petropulu, A.P., Yang, X.: Data traffic modeling - a signal processing perspective. In: Barner, K.E., Arce, G.R. (eds.) Nonlinear Signal and Image Processing - Theory, Methods, and Applications, CRC Press, Boca Raton (2004)
5. Erramilli, A., Narayan, O., Willinger, W.: Experimental queueing analysis with long-range dependent traffic. IEEE/ACM Transactions on Networking 4, 209–223 (1996)
6. Qiu, J., Knightly, E.W.: Measurement-based admission control with aggregate traffic envelopes. IEEE/ACM Transactions on Networking 9(2) (2001)
7. Shu, Y., Jin, Z., Zhang, L., Wang, L., Yang, O.W.W.: Traffic prediction using FARIMA models. In: International Conference on Communications (icc'99) vol 2, pp. 891–895 (1999)
8. Shu, Y., Jin, Z., Zhang, L., Wang, L., Yang, O.W.W.: Prediction-based admission control using FARIMA models. In: International Conference on Communications 2000 (ICC'00) vol.2, pp. 1325–1329 (2000)
9. Ilow, J.: Forecasting network traffic using farima models with heavy tailed innovations. In: International Conference on Acoustic, Speech and Signal Processing 2000 (icassp'00). vol.6, pp. 3814–3817 (2000)
10. Corradi, M., Garroppo, R.G., Giordano, S., Pagano, M.: Analysis of FARIMA processes in the modeling of broadband traffic. In: International Conference on Communications 2000 (icc'01). vol.3, pp. 964–968 (2001)
11. Sadek, N., Khotanzad, A., Chen, T.: ATM dynamic bandwidth allocation using F-ARIMA prediction model. In: 12th International Conference on Computer Communications and Networks (iccn) 2003, pp. 359–363 (2003)
12. Granger, C.W.J., Joyeux, R.: An introduction to long-memory time series models and fractional differencing. Journal of Time. Series Analysis 1, 15–29 (1980)
13. Hosking, J.R.M.: Fractional differencing. Biometrika 68, 165–176 (1981)
14. Box, G.E.P., Jenkins, G.M., Reinsel, G.C.: Time Series Analysis: Forecasting and Control, 3rd edn. Prentice-Hall, Englewood Cliffs (1994)
15. Mandelbrot, B.B., Ness, J.V.: Fractional brownian motions, fractional noises and applications. SIAM Rev. 10, 422–437 (1968)
16. Kalman, R.E.: A new approach to linear filtering and prediction problems. Trans. ASME, J. Basic Eng. 82, 35–45 (1960)
17. Tsay, R.S.: Analysis of Financial Time Series, 2nd edn. John Wiley and Sons, Hoboken, New Jersey (2005)
18. Durbin, J., Koopman, S.J.: Time Series Analysis by State Space Models. Oxford University Press, Oxford (2001)
19. Bhansali, R.J., Kokoszka, P.S.: Prediction of long-memory time series. In: Doukan, P., Oppenheim, G., Taqqu, M.S. (eds.) Theory and Applications of Long Range Dependence. Birkhäuser, Boston, MA (2003)
20. Kolarov, A., Atai, A., Hui, J.: Application of Kalman filter in high-speed networks. In: Global Telecommunications Conference (globecom'94), vol.1, pp. 624–628 (1994)
21. Lim, A.O., Ab-Hamid, K.: Kalman predictor method for congestion avoidance in ATM networks. In: tencon 2000, pp. I346–I351(2000)
22. Zhinjun, F., Yuanhua, Z., Daowen, Z.: Kalman optimized model for MPEG-4 VBR sources. IEEE Transactions on Consumer Electronics 50(2), 688–690 (2004)
23. Beran, J.: Statistics for Long-Memory Processes. Chapman & Hall, Sydney (1994)
24. Percival, D.B., Walden, A.T.: Wavelet Methods for Time Series Analysis. Cambridge University Press, Cambridge (2000)

25. Harvey, A.C.: Time Series Models, 2nd edn. MIT Press, Cambridge (1993)
26. Stark, H., Woods, J.W.: Probability and Random Processes with Applications to Signal Processing, 3rd edn. Prentice Hall, Upper Saddle River, NY (2002)
27. Zivot, E., Wang, J.: Modeling Financial Time Series with S-PLUS. Springer, Heidelberg (2003)
28. de Mello, F.L., de Lima, A.B., Lipas, M., de Almeida Amazonas, J.R.: Generation of Gaussian self-similar series via wavelets for use in traffic simulations (IEEE Latin America Transactions) in press.
29. Percival, D.B., Walden, A.T.: Spectral Analysis for Physical Applications. Cambridge, New York (1993)
30. Akaike, H.: Information theory and an extension of the maximum likelihood principle. In: Petrov, B.N., Csaki, F. (eds.) 2nd International Symposium on Information Theory, Akademia Kiado, Budapest, pp. 267–281 (1973)
31. Morettin, P.A., Toloi, C.M.C.: Análise de Séries Temporais. Edgard Blucher ltda., São Paulo, SP (2004)
32. Beran, J., Bhansali, R.J., Ocker, D.: On unified model selection for stationary and nonstationary short- and long-memory autogressive processes. Biometrika 85, 921–934 (1998)
33. Beran, J.: Semifar models, a semiparametric framework for modelling trends, long-range dependence and nonstationarity. Technical report, University of Konstanz, Germany (1999)

Generating LRD Traffic Traces Using Bootstrapping

Shubhankar Chatterjee[1], Mike MacGregor[2], and Stephen Bates[1]

[1] Department of Electrical and Computer Engineering,
University of Alberta
[2] Department of Computing Science,
University of Alberta

Abstract. Long-range dependence (LRD) or second-order self-similarity has been found to be an ubiquitous feature of internet traffic. In addition, several traffic data sets have been shown to possess multifractal behavior. In this paper, we present an algorithm to generate traffic traces that match the LRD and multifractal properties of the parent trace. Our algorithm is based on the decorrelating properties of the discrete wavelet transform (DWT) and the power of stationary bootstrap algorithm.

To evaluate our algorithm we use multiple synthetic and real data sets and demonstrate its accuracy in providing a close match to the LRD, multifractal properties and queueing behavior of the parent data set.We compare our algorithm with the traditional fractional gaussian noise (FGN) model and the more recent multifractal wavelet model (MWM) and establish that it outperforms both these models in matching real data.

1 Introduction

Computer simulations are widely used by the networking community to evaluate the performance of new protocols and network infrastructure like network topology, buffer size in routers, link bandwidth, etc. One of the most important requirement for obtaining useful results from such simulations is to generate traffic that captures the characteristics of real traffic. One such important characteristic of network traffic is self-similarity or long-range dependence (LRD). Multiple studies of high precision and large volume data sets have shown this to be an ubiquitous feature in data traffic [1, 2, 3, 4]. Additional research has proven that LRD can lead to higher packet losses and greater queueing delay than what is predicted by short-range dependent (SRD) models like the Markov or Poisson models that have been traditionally used for traffic modeling [5]. Thus, it is necessary to account for LRD while generating synthetic traffic traces to be used for simulations.

The most commonly employed technique for generating traffic traces is the two step procedure of fitting a stochastic model to the observed traffic and then use it to generate traffic traces during simulation. However, the step of model fitting requires user intervention and can be complicated. For instance, for Fractionally

L. Mason, T. Drwiega, and J. Yan (Eds.): ITC 2007, LNCS 4516, pp. 272–283, 2007.

Integrated Autoregressive Moving Average (FARIMA) models with alpha stable innovations, there are a number of parameters that need to be estimated using complicated estimation procedures spanning multiple steps with continuous user involvement [6]. Using simple models like the Fractional Gaussian Noise (FGN) model [7], allow the long-range dependence to be matched, but not higher order properties or the marginal distribution.

Our proposed algorithm for generating aggregate traffic traces does not involve any model fitting, thereby avoiding the complexity of parameter estimation, and making it easy it use. The algorithm (Section 3) consists of employing the stationary bootstrap technique [8] after taking the Box-Cox transformation of the coefficients obtained by the discrete wavelet transformation (DWT) (Section 2.1) of the time domain data. We evaluate the performance of our algorithm by using synthetic FGN traces, as well as multiple real traces, and exhibit its capability in generating traces that are accurate in terms of the Hurst parameter, H (characterizing LRD), the multifractal spectrum and the queuing behavior (Section 4).

We believe this is the first attempt at using the bootstrapping technique in the wavelet domain for the generation of traffic traces, while preserving the Hurst parameter and the multifractal behavior. The comparison with related work is made in Section 5 and conclusions are presented in Section 6.

2   Background

In this section, we provide a brief overview of Wavelet Transforms and bootstrapping.

2.1   Discrete Wavelet Transform (DWT)

DWT is used to create a scale-space representation of a signal to enable the study of the different frequency components and their periods of occurrence in the signal by using translated and shifted version of a bandpass wavelet function $\psi(t)$ and a low-pass scaling function, $\phi(t)$.

On taking the inner products of the signal $x(t)$ with the shifted and translated versions of the wavelet and scaling functions, we get the wavelet and scaling coefficients respectively, as follows:

$$d_{j,k} = \langle x(t), \psi_{j,k}(t) \rangle$$
$$a_{j,k} = \langle x(t), \phi_{j,k}(t) \rangle \tag{1}$$

where $d_{j,k}$ are the detail or wavelet coefficients, and $a_{j,k}$ are the scaling or approximation coefficients. The signal $x(t)$ can then be represented by a collection of details at different resolutions and a low level approximation as

$$x(t) = \sum_k a_{J,k}\phi_{J,k}(t) + \sum_{j=1}^{k}\sum_k d_{j,k}\psi_{J,k}(t) , \tag{2}$$

where j indexes the scale of resolution; with higher j indicating higher resolution, and k indexes the spatial location of analysis. Eqn. (2) presents a multiresolution analysis of $x(t)$ up to scale J (or J is the number of levels of decomposition for $x(t)$). For our study we have used the Wavelet toolbox in Matlab to compute the wavelet coefficients.

DWT is useful in the study of network traffic data, which has been shown to possess long-range dependence in LAN [1], WAN [2], VBR [3], and other kinds of data traffic. This is due to the fact that DWT de-correlates the long memory data [9, 10] so that the correlation structure of the wavelet coefficients is not LRD, even though the original data $x(t)$ has LRD. Similarly, the wavelet coefficients for data possessing only SRD or a combination of LRD and SRD is also decorrelated, as shown in [11]. Thus it can be concluded that for all traffic traces, the wavelet coefficients will be decorrelated. In addition, it is also shown that the wavelet coefficients are wide-sense stationary [10]. These properties of decorrelation and stationarity of the wavelet coefficients will be used for developing our bootstrapping algorithm.

2.2   Bootstrapping

The idea of bootstrapping is to resample the original data with replacement to obtain a new series. This resampling procedure is repeated a number of times to obtain multiple data sets (which are known as the surrogate series) from the original data. The original bootstrap method was developed for i.i.d. data [12]. When the observations are not independent, the original bootstrap scheme fails to capture the dependency structure of the data [13]. To address this issue, a number of variants of the original bootstrapping technique have been proposed. Some examples are the residual bootstrap, sieve bootstrap, moving block bootstrap, stationary bootstrap, threshold bootstrap, etc. (see [13] and the references therein). All the above bootstrap techniques work for SRD data but not for LRD data.

To deal with LRD data sets, it has been proposed to transform the data into another domain, by using either the Fourier Transform [14] or Wavelet Transform [15]. In [16], the authors demonstrate that bootstrapping in the wavelet domain (also referred as wavestrapping) works better than in the Fourier domain when attempting to capture the characteristics of non-Gaussian long-memory processes. Aggregate data traffic also falls in this category of data, and hence we employ the Wavelet transform for our case.

As discussed in Section 2.1, the DWT decorrelates the LRD data. Our experiments reveal that for most traffic traces the wavelet coefficients within a given level have significant correlation up to lag 2. Hence, we cannot use the standard bootstrap mechanism. In addition, as stated earlier, the wavelet coefficients at each scale are stationary and hence it is necessary to use a bootstrapping scheme that will also produce a stationary series. To account for the stationarity of wavelet coefficients and their SRD, we select the stationary bootstrap algorithm [8] to be used on the wavelet coefficients $d_{j,k}$.

3   Proposed Algorithm

In this section, we describe the algorithm used to create the surrogate data sets from any given data set. As mentioned in the previous section, we apply the stationary bootstrap scheme independently to each wavelet decomposition scale. However, for data possessing long-tailed distribution, the wavelet coefficients will also possess a long-tailed distribution [17]. If the wavelet coefficients with the long-tailed distribution are sampled, then the long-tailed nature of the wavelet coefficients will be destroyed and the regenerated series will not possess the long-tailed nature of the parent series. In order to avoid this, we apply the Box-Cox transformation [18] to the wavelet coefficients at each level before resampling them.

The algorithm can be summarized as:

1. Compute the DWT of the data set to obtain the wavelet and scaling coefficients.
2. Apply the Box-Cox transformation to the wavelet coefficients at each level.
3. Re-sample the wavelet coefficients at each level by using the stationary bootstrap algorithm. The parameter p defining the geometric distribution in the stationary bootstrap algorithm is chosen separately for each scale as

$$p_j = 2^{((-log(T_j)/log(2))/3)} \ , \tag{3}$$

where p_j is value of p at scale j having T_j wavelet coefficients.
4. Take the inverse Box-Cox transformation of the resampled wavelet coefficients.
5. Take the inverse DWT of the resampled and transformed wavelet and scaling coefficients to obtain the surrogate data set.
6. Round off the series obtained in the previous step, and change the negative values to 0 to obtain the required data set. This transformation of the negative values to 0 might create an artifact in the surrogate data sets. However, after implementation, we found the percentage of data points having a value of zero ranges from 0.004% to 0.3% which is probably low enough to not cause any impact in simulations. This was observed when trying to model both FGN and real data sets, with one thousand surrogate series generated for each parent data set. Ideally, it will be nice to avoid this and we are currently investigating techniques to achieve this goal.
7. Steps 3, 4, 5 and 6 are repeated to obtain as many surrogate data sets as required.

There are two variables in our algorithm, for which we need to find appropriate values. The first is the number of vanishing moments of the mother wavelet and the second is the number of levels of decomposition for the DWT. We conducted a number of experiments and determined that the *db6* wavelet with six vanishing moments performs the best for our case. The number of levels of decomposition is determined by the length of the series and will have to be varied accordingly; with the criteria being that there are a sufficient number (about *30*) of wavelet coefficients at the highest level of decomposition.

4   Evaluation of the Algorithm

We use a combination of synthetic and real data sets for evaluating our algorithm. The synthetic traces are generated by using the FGN generator proposed by Paxson in [7]. Each data set is generated to be of length 16384, with the Hurst value, $H \in [0.52, 0.97]$. The real data sets are obtained from publicly available trace repositories and a trace recorded at the gateway to our university in 2001. The first two traces belong to the well known set of Bellcore traces analyzed in [1]. From this set, we use the pAug89 and pOct89 traces and generate data sets representing the number of bytes per 12 ms and 10 ms respectively. Next we have the 20030424-000000-0 trace obtained from CAIDA [19]. This is a 5-minute trace of packet headers on an OC48 backbone link. From this trace, we created a data set containing the number of packets per millisecond for the initial 107622 milliseconds. The fourth trace was collected on April 09, 2002 at 0300 hours by the Network Data Analysis Study Group at the University of North Carolina at Chapel Hill [20]. This trace contains the number of packets and bytes per millisecond for 732247 milliseconds. From this we created a data set of the number of packets per 100 milliseconds giving us 73225 data points. The final trace was collected in 2001 on the outgoing link connecting the University of Alberta's campus network to the Internet. The trace is a record of 100000 packets passing the gateway. From this trace, we formed a data set of length 71391 representing the number of packets per millisecond. The period of aggregation for each trace is randomly chosen. The details of the series formed from the traces are summarized in Table 1.

Table 1. Data Sets formed from actual traces

Series Name	Data Type	Aggregation Period (ms)	Length of Series
pAug89	Bytes	12	261902
pOct89	Bytes	10	175962
CAIDA	Packets	1	107622
UNC	Packets	100	73225
UofA	Packets	1	71391

4.1   Performance Tests

The criteria we use for evaluation of our algorithm are the Hurst parameter (H), the Linear Multiscale diagram (LMD) and the queue tail probability. It can be argued that either the Hurst parameter or the LMD should be used on any dataset, as the former is used for monofractal data and the later is applicable for multifractal data. However, that requires categorizing the dataset as either monofractal or multifractal. It is not our intention to discuss that in this paper, and hence we have used both these criteria for all the datasets that we have tested.

We compare our algorithm with the FGN model which has been traditionally used for comparison of all traffic models and a multifractal wavelet model that is one of the latest models proposed to capture the multifractal behavior for real traffic data having a non-gaussian distribution. The FGN data sets are generated by Paxson's algorithm [7] and the β-multifractal wavelet model (MWM) [21] is used as the multifractal model. For each of the three techniques, we generate one thousand data sets and for comparison we randomly pick one data set generated by each method and run the corresponding tests.

Hurst Values. For estimating the Hurst value, we have used the Wavelet based estimator [22] with three vanishing moments for the mother wavelet (It should be noted that the number of vanishing moments used for the estimator has no relation to the number of vanishing moments used in our algorithm). The wavelet estimator gives confidence intervals along with the estimates, however we report only the point estimates.

The Hurst values for the parent series and the means and variances of the Hurst values for the surrogate series are reported in Table 2. The values in the table indicate that the Hurst values of the surrogate series closely match those of the original data set; thereby demonstrating that our algorithm is able to capture the second order self-similarity of the parent data set (as measured by the Hurst parameter).

Table 2. Hurst values for multiple data sets

Parent Series		Surrogate Series	
		Hurst Value	
Name	Hurst Value	Mean	Variance
FGN-1	0.5214	0.5147	0.000093
FGN-2	0.7262	0.7181	0.000088
FGN-3	0.9281	0.9204	0.000095
pAug89	0.7958	0.7831	0.000052
pOct89	0.7739	0.7630	0.000039
CAIDA	0.5915	0.5869	0.000081
UNC	0.9339	0.9138	0.000563
UofA	0.6112	0.6273	0.000078

MultiScaling Behavior. As mentioned above, the Hurst parameter is a measure of the second order scaling properties of the data set. This is done by studying the variance $S_2(j)$ of the wavelet coefficients at octave j which has the behavior $S_2(j) \sim Cj^\alpha$. In addition to the scaling in the second order, the data set will very often also have scaling for all moments, which can be denoted as:

$$S_q(j) = E[|d(j)|^q] \sim C_q j^{\alpha_q} , \qquad (4)$$

with q signifying the scaling order.

If α_q is a linear function, the process is said to be monofractal. However, if α_q is not linear, then the process is said to exhibit multiscaling. It has been discovered [23] that data traffic in general and WAN traffic in particular possess multiscaling behavior, and a number of multifractal models have been proposed for capturing the behavior of such traffic [21, 24]. In order to accurately characterize our algorithm we study the higher order scaling behavior of the traces generated by our model and compare them to the original traces as well as to traces generated by Paxson's model and the β-MWM model.

The multiscaling behavior is studied by plotting a Linear Multiscale Diagram (LMD) representing α_q as a function of q. In order to study this behavior we have used the Multiscaling tool developed by Veitch et. al. [25].

In Figs. 1 and 2 we have plotted the LMD for the UofA and CAIDA data sets respectively. In both the figures, we have plotted the LMD for the parent data set, the data sets generated by our algorithm (referred to as Bootstrap series in the figures), the FGN model and the β-MWM model. In both the figures, it is seen that the LMD of the data set generated by our algorithm has a closer match to the LMD of the original trace, as compared to the β-MWM model and the FGN model. Similar results were obtained for the UNC, pAug89 and pOct89 data sets as well, but are not reported here due to lack of space.

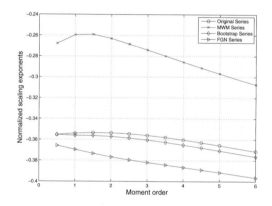

Fig. 1. Linear Multiscale diagram for UofA data set

This demonstrates that our algorithm is able to capture even the higher order scaling properties of the original data sets.

Queueing Behavior. The final test is a simple queueing experiment. The setup consists of a single server with infinite buffer size, servicing the incoming data at a rate of 1.1 times the mean arrival rate of the data set. We then compute the queue length at the end of each interval and from this series compute the queue tail probability, as the probability of the queue length exceeding a certain buffer size, i.e. we compute $P(Q > x)$, where Q is the queue length, and x is the buffer size. Once again, we use the parent data set, and a randomly picked data set each from the set generated by our algorithm (referred to as Bootstrap series) and the FGN and β-MWM model.

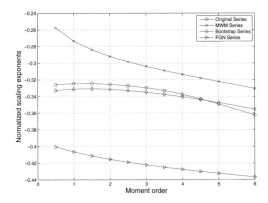

Fig. 2. Linear Multiscale diagram for CAIDA data sets

The queue tail probability for the UofA trace is plotted in Fig. 3. It is seen that the data set generated by our algorithm has almost similar behavior as the original data set, while the other two data sets show a very different behavior. Our algorithm performs better than the other two models even for the UNC data set as seen in Fig. 4.

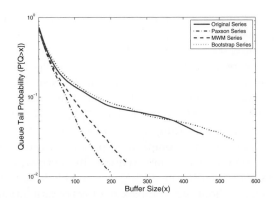

Fig. 3. Queue Tail Probability for UofA data set

For the CAIDA, pAug89 and pOct89 data sets, similar results are obtained for the queuing behavior of the surrogate data sets. Thus, it is shown that the surrogate series generated by our algorithm have similar queuing behavior as the parent data set for different data sets.

The results reported in this section conclusively demonstrate that our algorithm is able to retain the Hurst parameter of the parent data set and performs better than the FGN and β-MWM models, while considering the multifractal spectrum and the queue tail behavior. This performance is achieved without

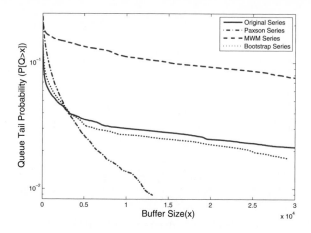

Fig. 4. Queue Tail Probability for UNC data set

using any complex parameter fitting procedure and is completed automated, thereby making it easy to use for simulations.

5   Previous Work

The most common technique for generating traffic traces is to use the FGN model [7, 26]. Like our proposed algorithm, the process of generating aggregate traffic traces can be automated when using the FGN model. However, the FGN model is not useful for most real traffic traces since they are usually non-Gaussian, whereas we have demonstrated that our algorithm gives satisfactory results for real traffic traces. Traces generated by using the alpha-stable models [27] or FARIMA models with alpha-stable distributions [6] have been shown to perform better than FGN models in capturing the characteristics of Internet data. However, using these models involves complicated parameter estimation techniques requiring significant time and user intervention. In contrast, as demonstrated in Section 4, traces generated by our technique are able to capture the characteristics of real traces, without any complex parameter estimation process. One model which does not involve very complex parameter estimation is the multifractal wavelet model (MWM) that has been proposed for efficient synthesis of non-Gaussian LRD traffic [21]. This method involves the use of a multiplicative cascade coupled with the Haar wavelet transform. This method ensures the generation of positive output, thereby making it appropriate for traffic modeling which always contains positive data. However, as demonstrated in Section 4 our algorithm significantly outperforms this model for real traffic traces in terms of the multifractal and queueing behavior.

The use of bootstrapping in the wavelet domain (referred to as wavestrapping), as used by us was first proposed in [15]. It was subsequently employed in [28] for testing the nonlinearity of data sets by generating surrogate data sets using three different resampling strategies. However, neither of these two papers on wavestrapping deal with LRD data. The use of bootstrapping for LRD data was addressed in [29]. In this paper, the authors propose the use of post-blackening moving block bootstrap to generate surrogate data sets for testing the effectiveness of some commonly used estimators of the Hurst parameter. However, in this paper, the bootstrapping technique is applied in the time domain, rather than in the wavelet domain as we have proposed here. As a result, the AR (auto regressive) model fitted to the data in [29] has a very high order, making it hard to implement. In addition, the block size selected also needs to be high enough to capture the dependency in the data, making the selection of the block length difficult.

The closest resemblance to our work is seen in [13], in which the authors use the residual bootstrap technique in the wavelet domain to create multiple surrogate series for LRD data. Their algorithm involves fitting a Markov model to capture the SRD in the wavelet domain, and then modeling the residuals by Efron's i.i.d bootstrap [12] to generate the bootstrap residuals. The bootstrap samples of the residuals are then combined with the Markov model to generate the bootstrap wavelet coefficients which is then used to produce bootstrap data sets in time domain. The authors demonstrate that their technique can be used to estimate the sample unit lag autocorrelation and standard deviation for Gaussian data sets. Another similar technique is discussed in [11], where the authors propose the use of an independent model for the wavelet coefficients and capturing the variance of these coefficients at each level of decomposition. They demonstrate that their technique is able to capture the autocorrelation function and the queue loss rate for heterogenous traffic possessing both long-range dependence (LRD) and short-range dependence (SRD). In our algorithm, we do not fit any model to the wavelet coefficients, but use the resampling technique to obtain the new wavelet coefficients. In addition, we are interested in capturing the Hurst parameter, multifractal scaling and the queuing behavior for Gaussian and non-Gaussian data sets as demonstrated in Section 4.

6   Conclusions

In this paper, we have combined the power of bootstrapping with the efficiency of wavelet transforms to develop an algorithm that can be used for the unsupervised generation of multiple traces by using a given data set. The algorithm is shown to produce data sets matching the original sequence in terms of its Hurst parameter and multifractal spectrum both for synthetic and real data sets. The queuing behavior of the traces generated by our algorithm is also studied, and is found to be close to that of the actual trace. The behavior of our algorithm is compared to that of the traditional FGN model and the more recent β-MWM model and is found to perform better than both of them. The traces generated by our algorithm can be used for simulations for network dimensioning and other purposes. That work is part of a different research and we leave that for future work.

References

[1] Leland, W., Taqqu, M., Willinger, W., Wilson, D.: On the Self-Similar Nature of Ethernet Traffic (Extended Version). IEEE/ACM Transactions on Networking 2, 1–15 (1994)

[2] Paxson, V., Floyd, S.: Wide-Area Traffic: The Failure of Poisson Modeling. IEEE/ACM Transactions on Networking 3, 226–244 (1995)

[3] Beran, J., Sherman, R., Taqqu, M., Willinger, W.: Long-Range Dependence in Variable-Bit-Rate Video Traffic. IEEE Transactions on Communications 43, 1566–1579 (1995)

[4] Crovella, M., Bestavros, A.: Self-similarity in world wide web traffic: Evidence and possible causes. IEEE/ACM Transactions on Networking 6, 835–846 (1997)

[5] Erramilli, A., Narayan, O., Willinger, W.: Experimental queueing analysis with long-range dependent packet traffic. IEEE/ACM Transactions on Networking 4, 209–223 (1996)

[6] Harmantzis, F., Hatzinakos, D.: Heavy network traffic modeling and simulation using stable farima processes. In: Proceedings of the 19th International Teletraffic Congress (2005)

[7] Paxson, V.: Fast, Approximate Synthesis of Fractional Gaussian Noise for Generating Self-Similar Network Traffic. Computer Communications Review 27, 5–18 (1997)

[8] Politis, D., Romano, J.: The stationary bootstrap. Journal of the American Statistical Association 89, 1303–1313 (1994)

[9] Tewfik, A.H., Kim, M.: Correlation structure of the discrete wavelet coefficients of fractional brownian motion. IEEE Transactions on Information Theory 38, 904–909 (1992)

[10] Flandrin, P.: Wavelet analysis and synthesis of fractional brownian motion. IEEE Transactions on Information Theory 38, 910–917 (1992)

[11] Ma, S., Ji, C.: Modeling heterogeneous network traffic in wavelet domain. IEEE/ACM Transactions on Networking 9(6), 634–649 (2001)

[12] Efron, B.: Bootstrap methods: Another look at the jackknife. Annals of Statistics 7, 1–26 (1979)

[13] Feng, H., Willemain, T.R., Shang, N.: Wavelet-based bootstrap for time series analysis. Communications in Statistics: Simulation and Computation 34(2), 393–413 (2005)

[14] Theiler, J., Eubank, S., Longtin, A., Galdrikian, B., Farmer, J.: Testing for nonlinearity in time series: The method of surrogate data. Physica D 58, 77–94 (1992)

[15] Percival, D.B., Sardy, S., Davison, A.C.: Wavestrapping time series: Adaptive wavelet-based bootstrapping. In: Fitzgerald, W.J., Smith, R.L., Walden, A.T., Young, P.C. (eds.) Nonliner and Nonstationary Signal Processing, pp. 442–470. Cambridge University Press, Cambridge, England (2001)

[16] Angelini, C., Cada, D., Katul, G., Vidakovic, B.: Resampling hierarchical processes in the wavelet domain: A case study using atmospheric turbulence. Physica D: Nonlinear Phenomena 207(1-2), 24–40 (2005)

[17] Stoev, S., Taqqu, M.S.: Asymptotic self-similarity and wavelet estimation for long-range dependent fractional autoregressive integrated moving average time series with stable innovations. Journal of Time. Series Analysis 26(2), 211–249 (2005)

[18] Box, G.E.P., Cox, D.R.: An analysis of transformations. Journal of Royal Statistical Society, Series B 26(2), 211–252 (1964)

[19] Cooperative   Association   for   Internet   Data   Analysis   (2006)
     http://www.caida.org/data/passive/index.xml
[20] University of North Carolina - Network Data Analysis Study Group (2006)
     http://www-dirt.cs.unc.edu/unc02_ts
[21] Riedi, R., Crouse, M., Riberiro, V.V., Baraniuk, R.: A multifractal wavelet model
     with application to network traffic. IEEE Transactions on Information Theory,
     vol. 45, pp. 992–1019 (1999)
[22] Veitch, D., Abry, P.: A wavelet based joint estimator of the parameters of long-
     range dependence. IEEE Transactions on Information Theory 45, 878–897 (1999)
[23] Riedi, R., Vehel, J.: Multifractal properties of tcp traffic: a numer-
     ical study. INRIA, Tech. Rep. RR-3129, [Online]. Available (1997)
     http://www.inria.fr/rrrt/rr-3129.html
[24] Nogueira, A., Salvador, P., Valadas, R.: Modeling network traffic with multifractal
     behavior. 10th International Conference on Telecommunications, vol.2, pp. 1071–
     1077 (2003)
[25] Veitch, D.: Matlab code for the estimation of MultiScaling Exponents (2004)
     http://www.cubinlab.ee.mu.oz.au/~darryl/MS_code.html
[26] Jeong, H.-D., McNickle, D., Pawlikowski, K.: Fast Self-Similar Teletraffic Gener-
     ation Based on FGN and Wavelets. IEEE International Conference on Networks,
     pp. 75–82 (1999)
[27] Karasaridis, A., Hatzinakos, D.: Network Heavy Traffic Modeling Using α-Stable
     Self-Similar Processes. IEEE Transactions On. Communications 49, 1203–1214
     (2001)
[28] Breakspear, M., Brammer, M., Robinson, P.A.: Construction of multivariate sur-
     rogate sets from nonlinear data using the wavelet transform. Physica D: Nonlinear
     Phenomena 182(1-2), 1–22 (2003)
[29] Grau-Carles, P.: Tests of long memory: A bootstrap approach. Computational
     Economics 25(1-2), 103–113 (2005)

On the Flexibility of M/G/∞ Processes for Modeling Traffic Correlations

M.E. Sousa-Vieira, A. Suárez-González, J.C. López-Ardao,
M. Fernández-Veiga, and C. López-García

Department of Telematics Engineering, University of Vigo (Spain)

Abstract. With the increasing popularity of multimedia applications, video data represents a large portion of the traffic in modern networks. Consequently, adequate models of video traffic, characterized by a high burstiness and a strong positive correlation, are very important for the performance evaluation of network architectures and protocols. This paper presents new models for traffic with persistent correlations based on the M/G/∞ process. We derive two new discrete distributions for the service time of the M/G/∞ queueing system, flexible enough to give rise to processes whose correlation structure is able to exhibit both Short-Range Dependence (SRD) and Long-Range Dependence (LRD). The corresponding simulation models are easy to initialize in steady state. Moreover, as the two distributions have subexponential decay, we can apply a highly efficient and flexible generator of synthetic traces of the resulting M/G/∞ processes.

1 Introduction

VBR video traffic represents an important part of the load in modern networks. So, adequate models for this type of traffic are needed for network design and performance evaluation.

It is well known that VBR video is characterized by a high burstiness and a strong positive correlation, induced by the presence in this traffic of self-similarity, along with a closely related property called Long-Range Dependence [4,8,1], that involves long-range correlations over multiple time scales. On the other hand, several traffic measurement results have convincingly shown the existence of persistent correlations in other kinds of traffic [12,17,2]. These findings have contributed to a very important revolution in the stochastic modeling of traffic, since the impact of the correlation on the performance metrics may be drastic [13,15,14,6] and the validity of traditional processes, like markovian or autoregressive, is in doubt because modeling long-term correlations through these processes requires many parameters, whose interpretation becomes difficult. Because of this, the use of new classes of stochastic processes for network modeling purposes, that can display forms of correlation as diverse as possible by making use of few parameters (parsimonious modeling) is essential.

We consider the class of M/G/∞ processes. In essence, the M/G/∞ process is a stationary version of the occupancy process of an M/G/∞ queueing model.

L. Mason, T. Drwiega, and J. Yan (Eds.): ITC 2007, LNCS 4516, pp. 284–294, 2007.

Its viability for modeling network traffic can be attributed to several factors. Firstly, many forms of time dependence can be obtained by means of varying the service time distribution. This makes this process a good candidate for modeling many types of correlated traffic, as VBR video. Secondly, queueing analytical studies are sometimes feasible [5,22,10,19,16], but when they are not, it has important advantages in simulation studies [11,18], such as the possibility of on-line generation. Furthermore, there exists a trivial method of producing exact sample paths of the process with complexity $\mathcal{O}(n)$: it suffices to simulate the M/G/∞ queue, sampling the occupancy of the system at integer instants.

In this work, we derive two new discrete distributions for the service time of the M/G/∞ queueing system that give rise to processes whose correlation structure can exhibit both SRD and LRD. Besides, their corresponding simulation models are easy to initialize in steady state. As the two distributions have subexponential decay, we can apply the method proposed in [20] to build a highly efficient generator of synthetic traces.

The remainder of the paper is organized as follows. We begin reviewing the main concepts related to SRD, LRD and self-similarity in Section 2. The M/G/∞ process and the discrete-time model used by the generator are described in Section 3. In Sections 4 and 5 we present the two discrete distributions for the service time of the M/G/∞ system that we propose. Finally, concluding remarks and guidelines for further work are given in Section 6.

2   Overview of SRD, LRD and Self-similarity

It is said that a process exhibits SRD when its autocorrelation function is summable.

Conversely, it is said that a process exhibits LRD [4] when its autocorrelation function is not summable, i.e., $\sum_{k=0}^{\infty} r[k] = \infty$, like in those processes whose autocorrelation function decays hyperbolically:

$$\exists \rho \in (0,1) \left| \lim_{k \to \infty} \frac{r[k]}{k^{-\rho}} = c_r \in (0, \infty) \right. . \tag{1}$$

Let $X = \{X_n; n = 1, 2, \dots\}$ be a stationary stochastic process with finite variance and let $X^{(m)}$ be the corresponding aggregated process, with aggregation level m, obtained by averaging the original sequence X over non-overlapping blocks of size m: $X^{(m)} = \{\overline{X}_i[m]; i = 1, 2, \dots\}$, where:

$$\overline{X}_i[m] = \frac{1}{m} \sum_{n=(i-1)m+1}^{im} X_n .$$

The covariance stationary process X is called exactly second-order self-similar, with self-similarity parameter H [9], if the aggregated process $X^{(m)}$ scaled by m^{1-H} has the same variance and autocorrelation as X for all m, that is, if the aggregated processes possess the same non-degenerate correlation structure as the original stochastic process.

The autocorrelation function of both X and $X^{(m)}$ is:

$$r[k] = r_{\mathsf{H}}[k] \overset{\triangle}{=} \frac{1}{2} \left[(k+1)^{2\mathsf{H}} - 2k^{2\mathsf{H}} + (k-1)^{2\mathsf{H}} \right] \quad \forall k \geq 0 \ , \tag{2}$$

where: $\lim_{k \to \infty} \frac{r_{\mathsf{H}}[k]}{k^{2\mathsf{H}-2}} = \mathsf{H}(2\mathsf{H}-1)$, that is, it decays hyperbolically as in (1), with $\rho = 2 - 2\mathsf{H}$, and so the process exhibits LRD if $\mathsf{H} \in (0.5, 1)$.

If (2) is satisfied asymptotically by the autocorrelation function of the aggregated process, $r^{(m)}[k]$, then the process is called asymptotically second-order self-similar:

$$\lim_{m \to \infty} r^{(m)}[k] = r_{\mathsf{H}}[k] \quad \forall k \geq 1 \ .$$

A covariance stationary process whose autocorrelation function decays hyperbolically as in (1) is asymptotically second-order self-similar [23].

3   M/G/∞ Process

The M/G/∞ process is a stationary version of the occupancy process of an M/G/∞ queueing model. In such queueing model, customers arrive according to a Poisson process with rate λ to a system of infinitely many servers, and their service times constitute a sequence of continuous i.i.d. random variables distributed as the random variable S with finite mean value $\mathsf{E}[S]$.

In [3] the authors show that the number of customers, or busy servers, in the system at any instant t, $\{X(t); t \in \Re^+\}$, has a Poisson marginal distribution with mean value $\lambda \mathsf{E}[S]$ and its autocorrelation function depends on the cumulative distribution function of the service time, $\mathsf{F}_S(x)$:

$$\mathsf{r}(t) = \frac{1}{\mathsf{E}[S]} \mathsf{E}\left[[S-t]^+ \right] = \frac{1}{\mathsf{E}[S]} \int_t^\infty [1 - \mathsf{F}_S(s)] \, \mathsf{d}s \quad \forall t \in \Re^+ \ , \tag{3}$$

where $[x]^+ = x$ if $x \geq 0$, and 0 otherwise.

Because:

$$\mathsf{f}_{\widehat{S}}(t) = \frac{1 - \mathsf{F}_S(t)}{\mathsf{E}[S]} \quad \forall t \in \Re^+ \ , \tag{4}$$

being $\mathsf{f}_{\widehat{S}}(t)$ the probability density function of the residual life of the service time, the relation between the autocorrelation function and the cumulative distribution function of $\widehat{S}$ is:

$$\mathsf{r}(t) = 1 - \mathsf{F}_{\widehat{S}}(t) \quad \forall t \in \Re^+ \ . \tag{5}$$

We are interested on the discrete-time version of $\{X(t); t \in \Re^+\}$, that is, $X \overset{\triangle}{=} \left\{ X_n \overset{\triangle}{=} X(n); n = 1, 2, \dots \right\}$. The most natural approach to generate it is to simulate the queue in discrete-time, since its simulation will be more efficient [21].

3.1   Discrete-Time Model

Let $A = \{A_n; n = 1, 2, \dots\}$ be a renewal stochastic process, where A_n is a Poisson random variable with mean value λ_d representing the number of arrivals at instant n; let $\{\{S_{dn,i}; i = 1, \dots, A_n\}; n = 1, 2, \dots\}$ be a renewal stochastic process where $S_{dn,i}$ is distributed as a positive-valued discrete random variable S_d with finite mean value $E[S_d]$, and corresponds to the service time of the i-th arrival at instant n. If the following conditions hold:

- the initial number of users X_0 is a Poisson random variable of mean value $\lambda_d E[S_d]$,
- their service times $\left\{ \widehat{S_d j}; j = 1, \dots, X_0 \right\}$ are mutually independent and have the same distribution as the residual life of S_d, $\widehat{S_d}$:

$$\Pr\left[\widehat{S_d} = k\right] = \frac{\Pr[S_d \geq k]}{E[S_d]} \quad,$$

then the stochastic process $X = \{X_n; n = 1, 2, \dots\}$ is strict-sense stationary and ergodic, and satisfies:

- it has Poisson marginal distribution and mean value:

$$\mu \overset{\Delta}{=} E[X] = \lambda_d E[S_d] \quad, \tag{6}$$

- its autocorrelation function is:

$$r[k] = \Pr\left[\widehat{S_d} > k\right] \quad \forall k \quad. \tag{7}$$

So, the autocorrelation structure of X is completely determined by the distribution of $\widehat{S_d}$ or, ultimately, by the distribution of S_d. Varying this distribution many forms of dependence can be obtained.

In particular, the M/G/∞ process belongs to the class of LRD processes if S_d has infinite variance, as it may happen in heavy-tailed distributions.

A discrete random variable S_d has a heavy-tailed distribution if it complies with:

$$\lim_{k \to \infty} \frac{\Pr[S_d > k]}{k^{-\rho}} = L[k] \quad 0 < \rho < \infty \quad,$$

with L a function that varies slowly, i.e.:

$$\lim_{k \to \infty} \frac{L[kx]}{L[k]} = 1 \quad \forall x > 0 \quad.$$

On the other hand, in [11] the authors show that an $\Re^+$-valued sequence $r[k]$ can be the autocorrelation function of the stationary M/G/∞ process, with integrable S_d, if and only if it is decreasing and integer-convex, with $r[0] = 1 > r[1]$ and $\lim_{k \to \infty} r[k] = 0$, in which case the distribution of S_d is given by:

$$\Pr[S_d = k] = \frac{r[k-1] - 2r[k] + r[k+1]}{1 - r[1]} \quad \forall k > 0 \quad. \tag{8}$$

Its mean value is:

$$E[S_d] = \frac{1}{1 - r[1]} \ .$$

And the cumulative distribution function is:

$$\Pr[S_d \leq k] = 1 - \frac{r[k] - r[k+1]}{1 - r[1]} \quad \forall k \geq 0 \ . \tag{9}$$

4   A New Discrete Distribution for the M/G/∞-Based Generation of LRD Processes

As we have seen in Section 3 the M/G/∞ process is LRD if the distribution of the service time has infinite variance. The Pareto is an example of distribution that for some values of its parameters has infinite variance.

The most general form of the Pareto random variable, P, has three parameters [7]: the shape parameter α, the scale parameter β and the location parameter γ. Its cumulative distribution function is:

$$F_S(x) = F_P(x) = \begin{cases} 0 & x \leq \gamma \\ 1 - \left(\dfrac{\beta}{x + \beta - \gamma}\right)^{\alpha} & x \geq \gamma \ , \end{cases}$$

with $\alpha > 0$, $\beta > 0$ and $\gamma \geq 0$.

For $1 < \alpha < 2$, S has finite mean value:

$$E[S] = E[P] = \gamma + \frac{\beta}{\alpha - 1} \ ,$$

and infinite variance.

Its probability density function is:

$$f_S(x) = f_P(x) = \begin{cases} 0 & x \leq \gamma \\ \dfrac{\alpha \beta^{\alpha}}{(x + \beta - \gamma)^{\alpha+1}} & x \geq \gamma \ . \end{cases}$$

As the simulation of the M/G/∞ queue is more efficient considering a discrete-time distribution for the service time, in this section we develop a discrete-time M/G/∞ model with an occupancy process statistically identical to that of the continuous-time M/P/∞ model, but we only consider the useful arrivals, that is, the arrivals that modify the state of the system. For this purpose, we will specify the two input processes to the discrete-time model:

- $A = \{A_n; n = 1, 2, \dots\}$ and
- $\{\{S_{dn,i}; i = 1, \dots, A_n\}; n = 1, 2, \dots\}$,

both as defined in Section 3.

We select as the arrival process, A, for the discrete-time model, that of the continuous-time model discarding all the arrivals that depart within the same time-slot, because they do not modify the system state. The most efficient form of generating the arrival times in a continuous-time simulation model is, given the number of arrivals in each time-slot $[n-1, n]$, to compute the instant of the i-th arrival as $n - U_i$, with U_i a uniform random variable in $[0,1]$. So, an arrival will be relevant with probability $\Pr[S > U]$ independent of other arrivals, and A will be a collection of Poisson random variables with mean value $\lambda_{\mathsf{d}} = \lambda \Pr[S > U]$, being λ the rate of the Poisson process of the continuous-time simulation model. Also, we select i.i.d. service times distributed as the number of partially completed slots that a relevant user remains in the system in the equivalent continuous-time model, without counting the slot containing the arrival.

In order to obtain a M/G/∞ process with the same mean value that of the M/P/∞ process, we do:

$$\lambda \mathsf{E}[S] = \lambda_{\mathsf{d}} \mathsf{E}[S_{\mathsf{d}}] \ ,$$

and we obtain:

$$\mathsf{E}[S_{\mathsf{d}}] = \frac{\mathsf{E}[S]}{\Pr[S > U]} \ ,$$

where $S_{\mathsf{d}} = \lceil S - U \rceil$ whenever $S > U$.

Fig. 1 illustrates the relationship between S and S_{d}.

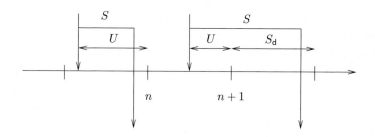

Fig. 1. Relationship between S and S_{d}

For $\gamma \geq 1$, $\Pr[S > U] = 1$, and for $\gamma \leq 1$ we have:

$$\Pr[S > U] = 1 - \Pr[S \leq U] = 1 - \int_0^\infty \Pr[S \leq x] \, \mathsf{dF}_U(x)$$

$$= 1 - \int_0^\infty \mathsf{F}_S(x) \, \mathsf{dF}_U(x) = 1 - \int_0^\infty \mathsf{F}_S(x) \, \mathsf{f}_U(x) \, \mathsf{d}x$$

$$= 1 - \int_0^1 \mathsf{F}_S(x) \, \mathsf{d}x = \gamma + \frac{\beta}{\alpha - 1} - \frac{\beta^\alpha}{(\alpha - 1)(1 + \beta - \gamma)^{\alpha - 1}} \ .$$

So:

$$E\left[S_{\mathsf{d}}\right] = \begin{cases} \dfrac{\gamma + \dfrac{\beta}{\alpha - 1}}{\gamma + \dfrac{\beta}{\alpha - 1} - \dfrac{\beta^{\alpha}}{(\alpha - 1)(1 + \beta - \gamma)^{\alpha - 1}}} & \gamma \leq 1 \\[4mm] \gamma + \dfrac{\beta}{\alpha - 1} & \gamma \geq 1 \ . \end{cases} \tag{10}$$

From (8) we have the probability mass function of S_{d}.

And from (9) we obtain its cumulative distribution function. For $\gamma \leq 1$:

$$\Pr\left[S_{\mathsf{d}} \leq k\right] = \begin{cases} 1 - \dfrac{\dfrac{\beta^{\alpha}(1 + \beta - \gamma)^{\alpha - 1}}{(k + \beta - \gamma)^{\alpha - 1}} - \dfrac{\beta^{\alpha}(1 + \beta - \gamma)^{\alpha - 1}}{(k + 1 + \beta - \gamma)^{\alpha - 1}}}{[\beta + \gamma(\alpha - 1)]\left(1 + \beta - \gamma\right)^{\alpha - 1} - \beta^{\alpha}} & \forall k > 0 \ , \end{cases}$$

and for $\gamma \geq 1$ we have:

$$\Pr\left[S_{\mathsf{d}} \leq k\right] = \begin{cases} 1 - \dfrac{\beta^{\alpha}}{\alpha - 1}\left[\dfrac{1}{(k + \beta - \gamma)^{\alpha - 1}} - \dfrac{1}{(k + 1 + \beta - \gamma)^{\alpha - 1}}\right] & k \leq \gamma \leq k + 1 \\[4mm] 1 - \gamma + k - \dfrac{\beta}{\alpha - 1} + \dfrac{\beta^{\alpha}}{(\alpha - 1)(k + 1 + \beta - \gamma)^{\alpha - 1}} & k \geq \gamma \ . \end{cases}$$

We will denote this random variable by P_{d}. As the distribution has subexponential decay, we can apply the method proposed in [20] in order to obtain a highly efficient generator of synthetic traces.

Since $\widehat{S_{\mathsf{d}}} = \left\lceil \widehat{S} \right\rceil$, we can obtain its cumulative distribution function from (4) and (5):

$$\Pr\left[\widehat{S_{\mathsf{d}}} \leq k\right] = \mathsf{F}_{\widehat{S}}\left(k\right) = \begin{cases} \dfrac{\alpha - 1}{\beta + \gamma(\alpha - 1)}k & \forall k \leq \gamma \\[4mm] 1 - \dfrac{\beta^{\alpha}}{[\beta + \gamma(\alpha - 1)]\left(k + \beta - \gamma\right)^{\alpha - 1}} & \forall k \geq \gamma \ . \end{cases} \tag{11}$$

And finally from (7) and (11) we have:

$$r[k] = r_{\alpha,\beta,\gamma}[k] \overset{\triangle}{=} 1 - \Pr\left[\widehat{S_{\mathsf{d}}} \leq k\right] = \begin{cases} 1 - \dfrac{\alpha - 1}{\beta + \gamma(\alpha - 1)}k & \forall k \leq \gamma \\[4mm] \dfrac{\beta^{\alpha}}{[\beta + \gamma(\alpha - 1)]\left(k + \beta - \gamma\right)^{\alpha - 1}} & \forall k \geq \gamma \ . \end{cases}$$

As an example, in Fig. 2 we plot the probability mass functions for the service time for different combinations of the parameters α, β and γ.

And in Fig. 3 we can see the autocorrelation matching of synthetic traces of the resulting $\mathrm{M/G/\infty}$ processes. We can observe a good fit with the analytical ones.

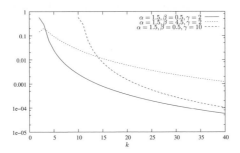

Fig. 2. Probability mass functions for the service time

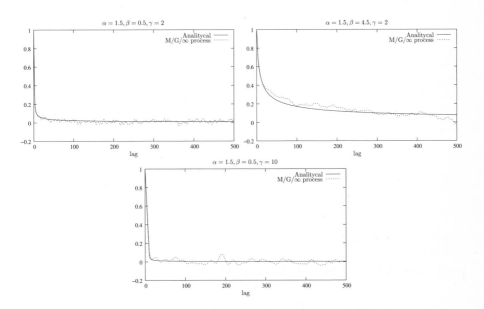

Fig. 3. Autocorrelation matching (M/P$_d$/∞ process)

4.1   Parameters of the M/P$_d$/∞ Model

In order to get the desired Hurst parameter of the resulting M/P$_d$/∞ process, form (1) we have:

$$\rho = 1 - \alpha \ ,$$

and so:

$$\mathsf{H} = \frac{3 - \alpha}{2} \ .$$

For $1 < \alpha < 2$, $\mathsf{H} \in (0.5, 1)$ and we have an LRD process.

And β and γ can be used to give rise to the desired correlation structure at first lags.

Finally, from the required mean value of the process, μ, and from (6) and (10), we obtain the value of λ_{d}.

5   A New Discrete Distribution for the M/G/∞-Based Generation of SRD Processes

In order to generate SRD processes, we consider the following autocorrelation function:

$$r[k] = \delta^{k^\tau} \quad 0 < \delta < 1,\, 0 < \tau < 1 \ .$$

Applying (8) we have the mass probability function for the service time of the M/G/∞ queueing system.

And from (9) we obtain the cumulative distribution function, that results:

$$\Pr\left[S_{\mathsf{d}} \leq k\right] = 1 - \frac{\delta^{k^\tau} - \delta^{(k+1)^\tau}}{1 - \delta} \quad \forall k > 0 \ .$$

We will denote this random variable by Z. Again, as the distribution has subexponential decay, we can apply the method proposed in [20] in order to obtain a highly efficient generator of synthetic traces.

The resulting cumulative distribution function for the residual life, that we need for initializing the process in steady state is:

$$\Pr\left[\widehat{S_{\mathsf{d}}} \leq k\right] = 1 - \delta^{k^\tau} \quad \forall k > 0 \ .$$

We have that:

$$\sum_{k=0}^{\infty} r[k] \sim (\tau^{-1})!\,[-\ln(\delta)]^{\frac{1}{\tau}} \ ,$$

so this method gives rise to SRD processes.

In Fig. 4 we plot the probability mass functions for the service time for different combinations of the parameters δ and τ.

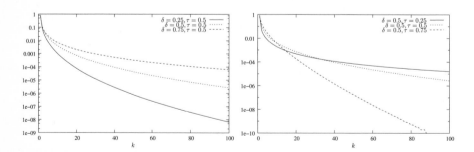

Fig. 4. Probability mass functions for the service time

Finally, in Fig. 5 the autocorrelation functions of synthetic traces of the resulting M/G/∞ processes are shown. Again, we can observe a good fit with the analytical ones.

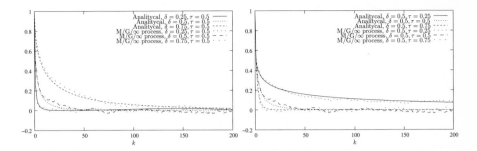

Fig. 5. Autocorrelation matching (M/Z/∞ process)

6 Conclusions and Further Work

In this paper, we have proposed new models for traffic with persistent correlations based on the M/G/∞ process. Our solutions enjoy several interesting features: the possibility of on-line generation, a highly efficient simulation model of the M/G/∞ queue and the possibility to capture both the short-term and the long-term correlation behavior of traffic in a parsimonious way. Now, we are working in the application of our models to capture the empirical correlation structure of real traffic, as VBR video traffic. Our results will be presented in a following work.

References

1. Beran, J., Shreman, R., Taqqu, M.S., Willinger, W.: Long-Range Dependence in Variable-Bit-Rate video traffic. IEEE Transactions on Communications 43(2/4), 1566–1579 (1995)
2. Crovella, M.E., Bestavros, A.: Self-similarity in World Wide Web traffic: Evidence and possible causes. IEEE/ACM Transactions on Networking 5(6), 835–846 (1997)
3. Cox, D.R., Isham, V.: Point Processes. Chapman and Hall, Sydney (1980)
4. Cox, D.R.: Long-Range Dependence: A Review. Statistics: An Appraisal. Iowa State University Press, pp. 55–74 (1984)
5. Duffield, N.: Queueing at large resources driven by long-tailed M/G/∞ processes. Queueing Systems 28(1/3), 245–266 (1987)
6. Erramilli, A., Narayan, O., Willinger, W.: Experimental queueing analysis with Long-Range Dependent packet traffic. IEEE/ACM Transactions on Networking 4(2), 209–223 (1996)
7. Fischer, M.J., Bevilacqua Masi, D.M., Gross, D., Shortle, J.F.: One-parameter Pareto, two-parameter Pareto, three-parameter Pareto: Is there a modeling difference? The Telecommunications Review, pp. 79–92 (2005)
8. Garrett, M.W., Willinger, W.: Analysis, modeling and generation of self-similar VBR video traffic. In: Proc. ACM SIGCOMM'94, London, UK, pp. 269–280 (1994)
9. Hurst, H.E.: Long-term storage capacity of reservoirs. Transactions of the American Society of Civil Engineers 116, 770–799 (1951)

10. Jelenković, P.R.: Subexponential loss rates in a GI/GI/1 queue with applications. Queueing Systems 33(2/4), 91–123 (1999)
11. Krunz, M., Makowski, A.: Modeling video traffic using M/G/∞ input processes: A compromise between markovian and LRD models. IEEE Journal on Selected Areas in Communications 16(5), 733–748 (1998)
12. Leland, W.E., Taqqu, M.S., Willinger, W., Wilson, D.V.: On the self-similar nature of Ethernet traffic (extended version). IEEE/ACM Transactions on Networking 2(1), 1–15 (1994)
13. Li, S.Q., Hwang, C.L.: Queue response to input correlation functions: Discrete spectral analysis. IEEE/ACM Transactions on Networking 1(5), 317–329 (1993)
14. Likhanov, N., Tsybakov, B., Georganas, N.D.: Analysis of an ATM buffer with self-similar (fractal) input traffic. In: Proc. IEEE INFOCOM'95, Boston, MA, EEUU, pp. 985–992 (1995)
15. Norros, I.: A storage model with self-similar input. Queueing Systems 16, 387–396 (1994)
16. Parulekar, M.: Buffer engineering for M/G/∞ input processes. Ph.D. Thesis, University of Maryland, College Park, MD, EEUU (2001)
17. Paxson, V., Floyd, S.: Wide-area traffic: The failure of Poisson modeling. IEEE/ACM Transactions on Networking 3(3), 226–244 (1995)
18. Poon, W., Lo, K.: A refined version of M/G/∞ processes for modeling VBR video traffic. Computer Communications 24(11), 1105–1114 (2001)
19. Resnick, S., Samorodnitsky, G.: Activity periods of an infinite server queue and performance of certain heavy-tailed fluid queues. Queueing Systems 33(1/3), 43–71 (1999)
20. Sousa, M.E., Suárez, A., Fernández, M., López, C., Rodríguez, R.F.: A highly efficient M/G/∞ generator of self-similar traces. In: Proc. 2006 Winter Simulation Conference, Monterey, CA, EEUU, pp. 2146–2153 (2006)
21. Suárez, A., López, J.C., López, C., Fernández, M., Rodríguez, R.F, Sousa, M.E.: A new heavy-tailed discrete distribution for LRD M/G/∞ sample generation. Performance Evaluation 47(2/3), 197–219 (2002)
22. Tsoukatos, K.P., Makowski, A.M.: Heavy traffic analysis for a multiplexer driven by M/G/∞ input processes. In: Proc. 15th International Teletraffic Congress, Washington, DC, EEUU, pp. 497–506 (1997)
23. Tsybakov, B., Georganas, N.D.: On self-similar traffic in ATM queues: Definitions, overflow probability bound and cell delay distribution. IEEE/ACM Transactions on Networking 5(3), 397–409 (1997)

Analytic Modeling of Ad Hoc 802.11 Networks with Hidden Nodes Applied to a Split Channel Solution for Performance Improvement

Andrea Baiocchi, Alfredo Todini, and Andrea Valletta

INFOCOM Department - University of Roma "Sapienza"
Via Eudossiana 18, 00184 Roma, Italy
{baiocchi,todini,valletta}@infocom.uniroma1.it

Abstract. Several simulation studies have shown that the performance of IEEE 802.11 DCF in an ad hoc scenario strongly depends on the coverage and interference radii. We state and solve an analytical model for an 802.11 DCF ad hoc network, with an interference radius larger than the coverage radius. The model is developed for the study of a split channel solution, where RTS/CTS signaling is conveyed via a separate, orthogonal channel with respect to data and ACK frames. By exploiting the model we can optimize the bandwidth split of the control and data channels. Further, we compare single channel, split channel and multichannel solutions, thus highlighting that the simple split channel achieves most of the performance advantage potentially offered by a multi-channel 802.11 DCF.

1   Introduction

Random access CSMA/CA of IEEE 802.11 DCF and spatial re-use of radio channels can bring about a large number of collisions when stations cannot reliably sense the channel state (idle/busy). A standard way to deal with hidden/exposed nodes is the RTS/CTS signaling. In general a given station can reach a number n_C of other stations out of the n stations making up the entire network and it can successfully exchange packets with them; the signals transmitted from the tagged station cause interference to $n_I \geq n_C$ stations, even though the most distant ones cannot successfully receive packets. A simplified propagation model that aims to capture the different ranges of successful reception and of interference is the so called double radii model (e.g. see [1]). When taking this propagation model into account in a dense ad hoc network, RTS/CTS alone appears to be quite ineffective in avoiding hidden/exposed node related collisions: a major performance drop comes from collisions between RTS/CTS frames and data frames. This observation motivated a multichannel MAC approach for CSMA based protocols, that has been shown to give performance benefits [2][3][4][5][6].

Multi-channel protocols can be exploited in WLANs in order to make use of additional bandwidth, or to decouple different transmissions, so as to reduce collisions in distributed access modes. A protocol for wireless LANs, in which

L. Mason, T. Drwiega, and J. Yan (Eds.): ITC 2007, LNCS 4516, pp. 295–308, 2007.

each node listens to a unique channel, is proposed in [7]. A different kind of protocol, called multichannel CSMA, is proposed in [2], and extended in [3]. The number of channels is fixed and much smaller than the number of nodes, and a terminal can transmit or receive on any channel. However, this protocol does not solve the hidden terminal problem. An RTS/CTS-like reservation mechanism is introduced in [4], where the IEEE 802.11 MAC layer is adapted to operate with multiple channels. One channel is used for control messages, the others for data. The sender includes in the RTS frames the list of free data channels, based on signal power measurements. The receiver then selects the channel on which it wants the communication to take place, and it sends this information on a CTS. The data and ack frames are then sent on the selected channel. Simulations show that this protocol outperforms the single channel 802.11 MAC, even in the case of fixed total bandwidth. Similar protocols are described in [5][6].

To gain insight into which scenarios can potentially benefit the most from multi-channel, a thorough simulation analysis of a reference ad hoc network topology has been carried out in [8]; multichannel based MAC protocols are shown to bring about no capacity gain if $n_C = n$; this is referred to as the "full visibility" case in the following. A significant performance improvement can instead be achieved if $n_C \ll n_I \ll n$.

A limitation of these works is that no satisfactory analytical model has been defined. Available analytical models of 802.11 DCF refer mostly to saturation traffic, symmetric stations (same packet length distribution, same channel error statistics, same MAC parameter values) and full visibility, so that carrier sensing guarantees no collision can take place once the RTS gets through. Few works have tried to extend those models to *non* full visibility scenarios, where RTS/CTS becomes ineffective and multichannel solutions can offer performance improvements. In [9], the IEEE 802.11 DCF is analyzed in a scenario with randomly scattered nodes, also taking the hidden node problem into account. In [10] and in [11] the authors extend Bianchi's model [12] with the results in [13]. These models seem to be valid for the IEEE 802.11 standard but they are not useful for the multi-channel protocol. In [14] and [15] authors investigate the performance of a split-channel and a multi-channel ALOHA MAC.

The present work aims to extend the analytic modeling of ad hoc 802.11 networks and to exploit the model to improve the performance at MAC layer. The analytic modeling of ad hoc 802.11 DCF is extended to the double radii propagation model, a simple way of modeling the general setting of the Defer Threshold (DT) and Carrier Threshold (CT) of the 802.11 stations [16]. On the performance improvement side, we show that the key to reap most of the performance gain potentially offered by multi-channel resides in a split channel solution, where RTS/CTS signaling is separated from data and ACK flows. This separation does not require any modification of MAC layer, it only requires PHY to sense both control and data channels at the same time; yet, a major increase in saturation throughput is obtained. Moreover, the packet delivery delay at low traffic load remains very close to that achieved by the single channel.

The rest of this paper is so organized. Section 2 describes the multichannel protocol analyzed in this work. In Section 3 the analytical model is defined, while model and simulations performance results are presented in Section 4. Section 5 draws the conclusions.

2   System Model and Multi-channel Protocol

We denote the overall capacity as C; this is divided into one control channel, with bandwidth C_{cch}, used to carry RTS and CTS frames, and M data channels each with a bandwidth $C_{dch} = (C - C_{cch})/M$ used to carry data packets and acknowledgments. Stations are assumed to be able to receive control packets and data packets, though they cannot receive multiple data packets on different data channels at the same time. Also, a station transmitting on a channel cannot receive anything on any other channel.

In the performance evaluation we adopt a reference network topology, where stations of the ad hoc network are placed on the vertices of an (infinite) hexagonal grid with inter-spacing d between any two neighboring stations . This is a worst case as to re-use of radio resource and interference, since the hexagonal grid is the tightest possible station packing over a plane with the constraint that the minimal distance between any two stations be d.

We define two coverage radii, R_D and R_I. Successful decoding of a MAC frame is assumed to have probability 1 for stations closer than R_D, 0 otherwise. Any transmission originating from a station closer than R_I to the receiving station **S** is assumed to cause enough interference to prevent correct decoding at **S**; interference coming from transmitters at a distance greater than R_I from the receiver is assumed to be negligible. In the hexagonal grid scenario, the two radii can be defined in terms of tiers around a tagged station; thus we denote the normalized radii as $r_D = R_D/d$ and $r_I = R_I/d$. The transmission link is assumed to be error free, thus the only cause of failed reception is a collision.

We refer to the multichannel 802.11 protocol defined in [8], where the protocol is described for the general case of M channels, $M \geq 1$. Here we focus on the special case $M = 1$, which makes it possible to simplify the model, while retaining most of the performance advantage of the general case (see Section 4).

In the split channel 802.11 DCF, a backlogged station **A** has to sense the state of *both* control and data channels. If idle, after completing the back-off countdown, **A** can send an RTS message to the intended recipient **B**. If **B** receives the RTS message without error (i.e. no collision on RTS in our model), and senses the data channel as idle, it sends a CTS message on the control channel. If this is successfully received, within a SIFS time **A** starts transmitting its data frame to **B**, then waits for the ACK from **B**, *on the data channel*. After the data channel has been idle for a DIFS time, the backoff countdown can start again, eventually resulting in a new RTS sent on the control channel. Stations need not listen to the control channel while they are either transmitting or receiving a frame on the data channel. So, implementing the multi-channel with $M = 1$ (split channel) only requires a single rx-tx hardware per station. Also, any station that can successfully decode the RTS message sent by a station A can be safely addressed if the control and data channels are checked idle, i.e. it will not be transmitting or receiving. This occurs thanks to the usual physical and virtual carrier sensing mechanisms *since* there is only a single data channel. Note that, apart from doing physical carrier sensing on two channels (control and data), no special modification is needed in the split channel MAC with respect to the standard 802.11 DCF.

3   Analytical Model

We introduce an analytical model for the *split channel* CSMA protocol. Our model is derived from the one presented in [9]. The two main novel contributions are the modeling of the interference area, which plays an important role in determining the performance of the multiple access protocols in an ad-hoc network, and the modeling of the split-channel scenario.

For the analytical model, it is assumed that each node always has packets to send (saturation traffic). In the following, we refer to a generic node $\mathbf{P}$ trying to send a packet to one of its neighbors, denoted as $\mathbf{Q}$. The set of stations in the transmission range of $\mathbf{P}$ is denoted as I_P. We assume $r_D = 1$ and $r_I = 2$, i.e. the reachable stations are only those in the first tier, the interfered ones are only those up to the second tier. While relieving most of the formal complexity of the model, this setting is a particularly interesting one. In the limit $r_D \to \infty$ the network model in saturation is the one in [12], where separating signaling and data channel for the same overall bandwidth cannot bring about any advantage (it actually worsens performance). The case $r_D = 1$ is just the opposite, as re-use can be maximized, but the hidden node problem is most relevant.

The state evolution of each station sampled at time epochs when the back off counter is decremented can be modeled by a bi-dimensional Markov chain $[s(t), b(t)]$ (see [12]), in which $s(t)$ and $b(t)$ are the back-off stage and the back-off counter at time t, respectively. For stage $s(t) = i$ we have a Contention Window size $W_i = \min\{CW_{min}2^i, CW_{max}\}$, $i = 0, 1, \ldots, m$, where m is the maximum number of retransmission allowed by the MAC layer for a data frame. The parameters CW_{min} and CW_{max} are set to 32 and 1024 respectively, according to the DSSS standard. If p denotes the collision probability conditional on a transmission attempt, the only non-null one-step transition probabilities in the Markov chain are:

$$
\begin{cases}
P\{i, j | i, j + 1\} = 1 & i \in [0, m], j \in [0, W_i - 1] \\
P\{i, j | i - 1, 0\} = p/W_i & i \in [1, m], j \in [1, W_i] \\
P\{0, j | i, 0\} = (1 - p)/W_0 & i \in [0, m - 1], j \in [1, W_0] \\
P\{0, j | m, 0\} = 1/W_0 & j \in [1, W_0]
\end{cases}
\tag{1}
$$

Let $b_{i,j} = \lim_{t \to \infty} P\{s(t) = i, b(t) = j\}, i \in [0, m], j \in [0, W_i]$ be the stationary state probability distribution of the chain. It can be easily found that:

$$
b_{i,j} = \frac{2p^i (W_i - \max\{0, j - 1\})/W_i}{\sum_{k=0}^{m} p^k (W_k + 3)} \qquad 0 \le i \le m, \quad 0 \le j \le W_i
\tag{2}
$$

We quantize the continuous time axis into back-off slots θ, taken as the time unit. Let τ be the steady state probability that a station starts a transmission attempt soon after a back-off slot time θ, conditional on the station not being transmitting already[1]. This is just the ratio of the time quantum divided by the the average duration of the overall back-off count down time (the *wait* time). The latter is $1 + Y\Theta$, where Y is the average number of back-off decrements before a new transmission attempt can start and Θ is the average time between two back-off counter decrements during the back-off count down. Y is derived from the average sojourn time in the *wait* state set of the Markov chain, yielding

$$Y = \frac{\sum_{i=0}^{m} \sum_{j=1}^{W_i} b_{i,j}}{\sum_{i=0}^{m} b_{i,1}} = \frac{1-T}{T} \qquad (3)$$

where T denotes the so called transmission probability and is given by

$$T = \sum_{i=0}^{m} b_{i,0} = \frac{2(1-p^{m+1})/(1-p)}{\sum_{k=0}^{m} p^k (W_k + 3)} \qquad (4)$$

So $\tau = 1/(1 + Y\Theta)$ i.e.

$$\tau = \frac{T}{T + (1-T)\Theta} \qquad (5)$$

The time Θ can be computed by observing that the probability Π_{ICh} that the tagged station senses the channel idle is the ratio between the back-off duration and the overall macro-slot duration Θ, i.e. $\Pi_{ICh} = \theta/\Theta = 1/\Theta$. In turn, we have[2] $\Pi_{ICh} \approx P_{ISt}^{N_I}$, where $N_I = 18$ is the number of stations within the interference area of the tagged station $\mathbf{P}$ and P_{ISt} is the continuous time probability of a station within $\mathbf{P}$'s interference area being idle during $\mathbf{P}$'s back off time. P_{ISt} can be found as the ratio between the average duration of the back-off count-down time (see (3)) and the average duration of the whole cycle encompassing a back-off count-down and a transmission attempt. Since the attempt is carried out by a station in the interference area of $\mathbf{P}$ between two successive back-off counter decrements of $\mathbf{P}$, most of the collision events due to stations hidden to the transmitting station are not possible. We can approximate the duration of such attempts with T_s, i.e. the duration of a successful transmission. Hence:

$$P_{ISt} \approx \frac{(1-T)\Theta/T + 1}{(1-T)\Theta/T + T_s} = \frac{1}{1 - \tau + \tau T_s} \qquad (6)$$

The values of τ and P_{ISt} can be found numerically from (4), (5) and $\Theta = 1/P_{ISt}^{N_I} = (1 - \tau + \tau T_s)^{N_I}$.

In the following, we assume that the conditional collision probability p is known (in Section 3.1 we show how to compute it in the scenario under study). Let $l_{SIFS}, l_{DIFS}, l_{RTS}, l_{CTS}, l_{DATA}, l_{ACK}$ denote the normalized duration of

---

[1] This is the probability that the station is observed in last time quantum, of size θ, of its *wait* time.

[2] The independence assumption leading to this equality holds for $\tau \ll 1$.

SIFS, DIFS, RTS, CTS, DATA and ACK respectively. We define the following time intervals:

1. $T_{cRTS} = l_{RTS} + l_{DIFS}{}^3$: the average duration of transmission resulting in a collision of the RTS frame, occurring with probability p_{RTS};
2. $T_{cCTS} = l_{RTS} + l_{SIFS} + l_{CTS} + l_{DIFS}$: the average duration of transmission resulting in a collision of the CTS frame, occurring with probability p_{CTS}, conditional on the RTS being received correctly;
3. $T_{cDATA} = l_{RTS} + l_{SIFS} + l_{CTS} + l_{SIFS} + l_{DATA} + l_{DIFS}$: the average duration of transmission resulting in a collision of the DATA frame, occurring with probability p_{DATA}, conditional on RTS/CTS exchange being successful;
4. $T_s = l_{RTS} + l_{SIFS} + l_{CTS} + l_{SIFS} + l_{DATA} + l_{SIFS} + l_{ACK} + l_{DIFS}$: the average duration of a successful transmission, occurring with probability $1 - p$;
5. $T_a = 1 + (1 - p)T_s + p_{RTS}T_{cRTS} + (1 - p_{RTS})p_{CTS}T_{cCTS} + (1 - p_{RTS})(1 - p_{CTS})p_{DATA}T_{cDATA}$: the average duration of a transmission activity, including the ensuing idle slot time.

In the above expressions, the duration of the RTS, CTS and data frame and ack transmissions are computed according to their respective standard format and to the bit rate assigned to control and data channel. It can be verified that the saturation throughput Λ is given by:

$$\Lambda = \frac{(1 - p)\mathrm{T}}{(1 - \mathrm{T})\Theta + \mathrm{T} \cdot T_a} \quad \text{(pkts/slot time)}. \tag{7}$$

3.1   Derivation of the Collision Probability

A key point for extension to non full visibility, specifically to the double radii propagation model, is the correct identification of the collision events, and the computation of the corresponding probabilities, which depend also on the access mode. A transmission attempt from node **P** to **Q** will be successful if:

1. the RTS from **P** does not collide with another RTS;
2. the RTS from **P** does not collide with a CTS;
3. the CTS from **Q** does not collide with a RTS;
4. the CTS from **Q** does not collide with another CTS;
5. the data frame from **P** does not collide with another data frame;
6. the data frame from **P** does not collide with an ACK;
7. the ACK from **Q** does not collide with a data frame;
8. the ACK from **Q** does not collide with another ACK frame.

All possible causes of collisions in the split-channel RTS/CTS access mode, as described in Section 2, are taken into account in the list above. The success probability of the above listed events are denoted as P_{sx} with $x = 1, \ldots, 8$. They are computed below by exploiting the station independence assumption, thus yielding the values of p_{RTS}, p_{CTS}, p_{DATA} and hence of p.

---

[3] In this and the following two expressions, DIFS should be replaced by appropriate EIFS or ack timeout intervals; we leave DIFS for simplicity and to keep consistent with the NS-2 based implementation of the 802.11 DCF employed for our simulations.

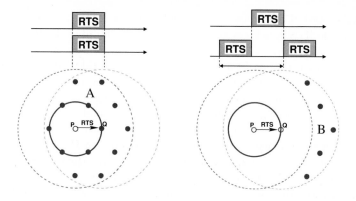

Fig. 1. Collision of a RTS with another RTS

As shown in Figure 1, a collision between RTS frames takes place if one of the $N_B = 5$ nodes in the area $B = I_Q \setminus I_P$, sends an RTS during the vulnerability interval of the RTS sent by node **P**; the collision also takes place if one of the $N_A = 13$ nodes in area $A = (I_P \cap I_Q) \cup \{Q\}$ sends a RTS starting in the same slot as the RTS sent by **P**. Then, the term P_{s1} is given by

$$P_{s1} = (1 - \tau)^{N_A + N_B 2l_{RTS}} \tag{8}$$

Figure 2 shows the collision of an RTS frame with a CTS. The sender of this CTS can only be one of the $N_B = 5$ nodes in the area $B = I_Q \setminus I_P$. We consider two cases.

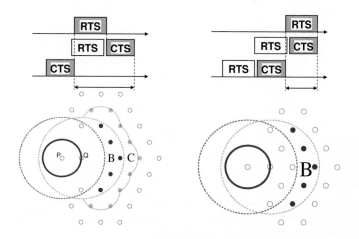

Fig. 2. Collision of a RTS with a CTS

1. The colliding CTS acknowledges an RTS sent by one of the $N_C = 9$ nodes in the area C (drawing on the left). Since these nodes are hidden to both **P** and **Q**, they can transmit while **P** sends the RTS. A collision with the CTS

will take place if one of the nodes in $\mathbf{C}$ sends an RTS to one of the nodes in B, and this RTS is acknowledged by a CTS overlapping the RTS sent by $\mathbf{P}$; the probability $P_{CB} = 13/54$ that an RTS generated by one of the nodes in C is addressed to one of the nodes in B has to be taken into account.

2. The CTS acknowledges an RTS sent by one of the nodes in B to another node in the same area. Since the nodes in B are in the interference range of $\mathbf{Q}$ the RTS cannot have been sent while $\mathbf{P}$ was sending an RTS, otherwise there would have been a collision (already taken into account in event # 1). A collision takes place if one of the $N_B = 5$ nodes in $\mathbf{B}$ sends a RTS to one of the other $N_B - 1$ nodes in $\mathbf{B}$ and the corresponding CTS overlaps the RTS sent by $\mathbf{P}$. As in the first case, the probability $P_{BB} = 7/30$ that an RTS generated by a node in B is addressed to another node in B has to be taken into account.

Thus we can write the following expression:

$$P_{s2} \approx \left[1 - P_{CB} + P_{CB}(1-\tau)^{N_C(l_{RTS}+l_{CTS})}\right]$$
$$\cdot \left[1 - P_{BB} + P_{BB}(1-\tau)^{N_B(l_{CTS}+l_{SIFS})}\right] \quad (9)$$

The quantity $1 - P_{s1}P_{s2}$ gives the probability of a collision disrupting the reception of the RTS sent by $\mathbf{P}$ to $\mathbf{Q}$. There may be no CTS back to $\mathbf{P}$ also when the RTS is correctly received on the control channel but the data channel is not sensed idle by the receiving node $\mathbf{Q}$. Let $\mathbf{X} \in B$ send an RTS at time t_X to $\mathbf{Y} \in C$; for no collision with the RTS sent at time t_0 by $\mathbf{P}$ to $\mathbf{Q}$ to occur, it must be $t_X + l_{RTS} < t_0$. For the data channel to be sensed busy by $\mathbf{Q}$ when it should send its CTS, it must be $t_X + l_{RTS} + l_{SIFS} + l_{CTS} + l_{SIFS} + l_{DATA} > t_0 + l_{RTS} + l_{SIFS}$. Thus the vulnerability interval for no answer is $\max\{0, l_{DATA} + l_{CTS} + l_{SIFS} - l_{RTS}\}$. Since there are $N_B = 5$ nodes in region B and the probability of a node in B to transmit to one in C is $P_{BC} = 7/10$, we can write the probability of no answer, p_{na}, on a correctly received RTS as

$$p_{na} = P_{BC}\left[1 - (1-\tau)^{N_B \max\{0, l_{DATA}+l_{CTS}+l_{SIFS}-l_{RTS}\}}\right] \quad (10)$$

A missed answer on RTS is dealt with by $\mathbf{P}$ as a collision. Hence, the probability of a collision on RTS can be evaluated as $p_{RTS} = 1 - P_{s1}P_{s2} + p_{na}$.

Figure 3 shows the collision of the CTS frame from $\mathbf{Q}$ to $\mathbf{P}$ with a RTS. Only the $N_D = 5$ nodes in D can send an RTS that disrupts the CTS and its transmission can only begin after at least a *DIFS* from the end of the RTS from $\mathbf{P}$ to $\mathbf{Q}$ (nodes in D sense the channel busy while $\mathbf{P}$ transmits). Thus, this term of the success probability can be written as:

$$P_{s3} = (1-\tau)^{N_D(l_{CTS}+l_{SIFS}-l_{DIFS})} \quad (11)$$

For a collision between two CTS frames to take place, the corresponding RTS frames must also have collided. Thus we can assert that collisions between CTS frames cannot happen and so $P_{s4} = 1$. To sum up, we have $p_{CTS} = 1 - P_{s3}$.

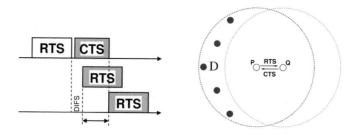

Fig. 3. Collision of a CTS with a RTS

As shown in Figure 4 a collision between two data frames can only take place if both the sender and the destination node of the second transmission do not correctly receive the RTS frame from **P** and the CTS frame from **Q**. Thus the data frame from **P** to **Q** can only collide with data frames transmitted by one of the $N_B = 5$ nodes in the area B. The probability $P_{BC} = 7/10$ that the destination is out of the interference range of node **Q** must be taken into account. This term of the overall success probability can be written as:

$$P_{s5} \approx P_{BC} + (1 - P_{BC}) \left[1 - (1 - \tau)^{N_B \cdot \max\{0, l_{DATA} - l_{DIFS} - l_{SIFS} - l_{RTS} - l_{CTS}\}} \right]$$
$$(12)$$

Collisions can also occur between data frames sent by **P** and ack frames sent by a node in the area B. Let **Y** $\in C$ be a node starting a transmission at time t_Y to a node **X** $\in B$. The ack eventually sent by **X** in case of success will collide with the data frame the tagged node **P** is sending to **Q** iff $t_Y < t_0 - l_{RTS} - i_{SIFS} - l_{CTS}$, where t_0 is the initial transmission time of **P**. This condition guarantees also that the CTS sent by **X** does not disrupt the reception of the RTS in **Q** On the other hand, it must also be $t_Y > t_0 - l_{DATA} - l_{SIFS} - l_{ACK}$. So, the vulnerability interval of the data frame sent by **P** is $\max\{0, l_{DATA} + l_{ACK} - l_{RTS} - l_{CTS}\}$. By considering that the probability of a node **Y** $\in C$ selecting a node **X** $\in B$ is $P_{CB} = 13/54$ and the number of nodes in C is $N_C = 9$, we get

$$P_{s6} = 1 - P_{CB} + P_{CB}(1 - \tau)^{N_C \max\{0, l_{DATA} + l_{ACK} - l_{RTS} - l_{CTS}\}} \qquad (13)$$

The probability of collision for a data frame is therefore $p_{DATA} = 1 - P_{s5}P_{s6}$.

The probability of collisions involving ACK frames is negligible, since ACKs are small compared to data frames; moreover, for an ACK to be transmitted, a successful RTS/CTS/DATA exchange must already have taken place, this makes a collision on ACK sent by **P** far less likely than one involving RTS, CTS or DATA frames. We can thus assume that $P_{s7} \approx P_{s8} \approx 1$.

This detailed analysis, though cumbersome, leads to quite accurate results (see the next Section) and ties the failure event probabilities to protocol parameters. Quantitative analysis shows that the major causes of failure of transmission attempts are RTS collisions and collision on data frames.

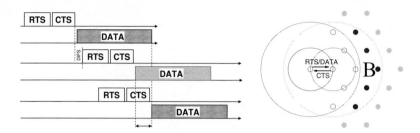

Fig. 4. Collision between two DATA packets

4 Performance Evaluation

4.1 Simulation Scenario

The results shown in the following sections have been obtained with a multichannel MAC module developed for the NS-2.26 network simulator [17]; this module has been derived from the 802.11 MAC as implemented in NS-2. A simplified physical layer, in which an interfering signal disrupts an ongoing reception with probability 1 if the interference power level is higher than a fixed threshold, and with probability 0 otherwise, has been adopted.

The placement of mobile nodes on a hexagonal grid and the choice of the transmission and interference ranges are all performed as in Section 3. In order to avoid edge effects, each side of the grid is seen as adjacent to its opposite side, thus obtaining a virtually infinite grid. All the simulations have been carried out in a scenario with 10×10 station grid. Data packets arrive according to independent Poisson processes with the same average rate λ at each station, and are then stored in a finite buffer of size B packets, waiting for service. Packet length is assumed fixed and equal to L (default value is $L = 1500$ bytes); packets are fit into the MAC data frame payload. The destination address of each packet is drawn at random among the six first tier neighbors of each station, independently per packet.

We choose $C = 2$ Mbps, $M = 1$ (split channel) or $M = 4$ (multi-channel). The value of C_{cch}/C is optimized in Section 4.3. To have a fair comparison, we set the transmission rate of all frames (including preamble and PLCP header) at the relevant channel capacity, also for the standard single channel case (see [8]). This choice allows us to find whether there are inherent performance gains due to the adoption of the split/multichannel protocol.

4.2 Model Validation

Figure 5 shows the collision probabilities of RTS and data frames evaluated with the analytical model and through simulations for different data frame payload sizes L. The left plot in Figure 5 shows the component of p_{RTS} due to collisions; when also considering the component p_{na} the overall probability of failure of RTS stays almost constant (comprised between 0.54 and 0.62) with a maximum relative error of 10% between model and simulation values. The left plot in Figure 6 shows the saturation throughput per node as a function of L. The analytical model appears to be quite accurate: the relative error between the analytically predicted throughput values and the simulation values is less than 10%.

Collision probabilities appear to be quite large. The overall collision probability (not plotted for space reasons) hovers around 0.75, with the biggest contributions coming from RTS and DATA collisions. The analysis in Section 3.1 highlights that such a high collision probability is essentially tied to the double radii propagation model, i.e. to the fact that some nodes are within the interference area of a station S but not within reach of S; this makes virtual carrier sensing ineffective.

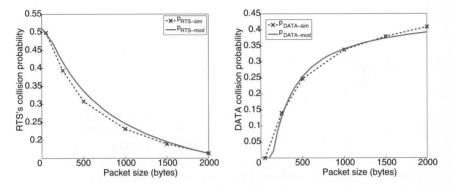

Fig. 5. Collision probability of RTS and DATA frames evaluated with the analytical model and through simulations

4.3   Determining the Share of Capacity for the Control Channel

Figure 6 shows the throughput per node for the split channel protocol, normalized to the total available bandwidth, computed with our model for different values of the packet size and of the bandwidth fraction assigned to the control channel. Too little bandwidth to the control channel makes RTS/CTS signaling the performance bottleneck, so that the data channel capacity is underutilized. On the other hand, increasing the control channel slows down data frames transmission, thus resulting again in a throughput penalty.

As expected, efficiency grows with the packet size: the overhead due to control information and to the collisions involving RTS or CTS frames remains constant for a single packet, while the amount of carried user data grows. For the same reason, the optimal control channel bandwidth is smaller for longer packets. Allocating 20% of the overall capacity C to the control channel appears to be a near-optimal choice for a wide range of packet sizes, so we set $C_{cch} = 0.2C = 400$ kbps in the following Subsections.

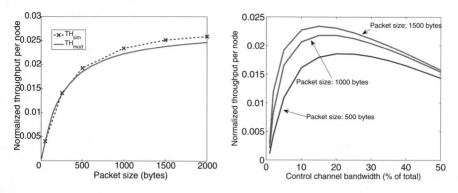

Fig. 6. Saturation throughput per node: comparison of analytical model and simulations (left plot); optimization of the control channel capacity C_{cch}

4.4 Comparison Among Single, Split and Multiple Channel Solutions

We compare simulation results obtained in the previously described scenario with three different protocols: IEEE 802.11 RTS/CTS (single channel), the proposed split channel protocol, and the multi-channel protocol described in [8], with $M = 4$ data channels. Figure 7 shows the achieved throughput (averaged over the nodes and normalized to C), and the mean packet delay (normalized to the value obtained for the single-channel protocol as the traffic load tends to zero[4]), obtained by varying the offered traffic load.

In terms of achieved throughput, both the split channel and the multi-channel protocols outperform the single-channel solution; moreover, most of the performance improvement takes place when moving from the single channel to the split channel solution. This indicates that the key factor is the reduction in data packet collisions brought about by separating the control and data channels. The analysis of simulation traces shows that collisions between RTS or CTS and data frames are relatively frequent in the single channel case.

The multi-channel protocol is able to achieve the highest throughput, even though it improves marginally over the split channel. This improvement comes at a price: at low traffic load the multi-channel solution introduces several times the delay imposed by the other two solutions (see the right plot in Figure 7). This happens because of the much smaller bandwidth available for sending a given data frame (at low loads, there is almost no queueing delay). As the traffic load increases, the packet delay saturates (since the station buffer has limited size) and all cases end up with quite similar average delays.

For high loads, significant metrics are packet loss probabilities. Figure 8 shows the probabilities of packet loss due to buffer overflow (left plot) and to repeated collisions[5] (right plot). Again we can see that the standard single channel pro-

---

[4] The normalizing delay is just $(W_0 + 1)/2 + T_s$ slot times.

[5] We assumed a maximum retransmission limit of 7 for the RTS frames and of 4 for the data frames.

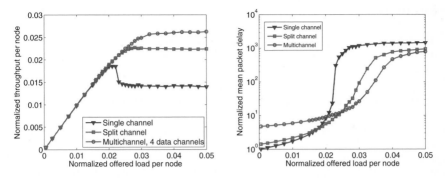

Fig. 7. Throughput per node and mean delay per packet vs. offered load per node

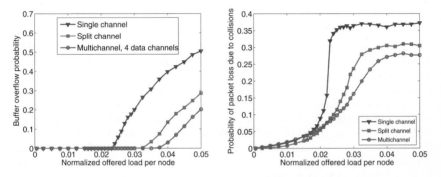

Fig. 8. Buffer overflow and packet loss probability due to collisions vs. offered load

tocol reaches saturation for a lower traffic load than the cases with separate control and data channels, while the simple split channel attains most of the performance advantage of a general multi-channel solution.

5   Conclusions

The contribution of this work is twofold. Firstly, we have defined a simple and accurate analytic model of the 802.11 DCF, shedding light on the issue of the interference area and on the protocol parameters that most affect saturation throughput performance and collision probability. Secondly, we have shown through analysis and simulations that capacity can be significantly improved with respect to the standard single channel solution by separating the control RTS/CTS channel from the data channel. The split channel achieves most of the performance gains brought about by multi-channel protocols, thus appearing as a practical means to alleviate collisions due to limited carrier sensing capabilities in ad hoc 802.11 wireless networks.

References

1. Ju, H.J., Rubin, I., Kuan, Y.C.: An adaptive RTS/CTS control mechanism for IEEE 802.11 MAC protocol. In: Proc. VTC 2003 Spring, vol. 2, pp. 1469–1473 (2003)
2. Nasipuri, A., Zhuang, J., Das, S.R.: A multichannel CSMA MAC protocol for multihop wireless networks. In: Proc. of IEEE WCNC'99, vol. 3, pp. 1402–1406 (1999)
3. Nasipuri, A., Das, S.R.: Multichannel CSMA with signal power-based channel selection for multihop wireless networks. In: Proc. of IEEE VTS-Fall VTC 2000, vol. 1, pp. 24–28 (2000)
4. Wu, S.L., Lin, C.Y., Tseng, Y.C., Sheu, J.P.: A new multi-channel MAC protocol with on-demand channel assignment for multi-hop mobile ad hoc networks. In: Proc. of I-SPAN 2000, pp. 232–237 (2000)
5. Jain, N., Das, S.R., Nasipuri, A.: A multichannel CSMA MAC protocol with receiver-based channel selection for multihop wireless networks. In: Proc. of IC3N 2001, pp. 432–439 (2001)
6. Li, J., Haas, Z.J., Sheng, M., Chen, Y.: Performance evaluation of modified IEEE 802.11 MAC for multi-channel multi-hop ad hoc network. In: Proc. of 17th IC-AINA 2003, pp. 312–317 (2000)
7. Kandukuri, S., Boyd, S.: A channel access protocol for multihop wireless networks with multiple channels. In: Proc. IEEE ICC '98, vol. 3, pp. 1617–1621 (1998)
8. Baiocchi, A., Todini, A., Valletta, A.: Why a multichannel protocol can boost ieee 802.11 performance. In: Proc. ACM MSWiM 2004, pp. 143–148 (2004)
9. He, J., Pung, H.: Performance modelling and evaluation of IEEE 802.11 distributed coordination function in multihop wireless networks. Computer Communications 29(9), 1300–1308 (2006)
10. Rahman, S.: Throughput analysis of IEEE 802.11 distributed coordination function in presence of hidden stations. Technical report, Stanford University project IEEE 802.11 DCF (2003)
11. Vassis, D., Kormentzas, G.: Throughput analysis for ieee 802.11 ad hoc networks under the hidden terminal problem. In: Proc. of IEEE CCNC 2006, vol. 2, pp. 1273–1276 (2006)
12. Bianchi, G.: Performance analysis of the IEEE 802.11 distributed coordination function. IEEE Journal Select. Areas Commun. 18(3), 535–547 (2000)
13. Kleinrock, L., Tobagi, A.: Packet switching in radio channels, Part II - the hidden terminal problem in carrier sense multiple access and the busy tone solution. IEEE Trans. Commun. 23(12), 1417–1433 (1975)
14. Deng, J., Han, Y.S., Haas, J.: Analyzing split channel medium access control schemes. IEEE Trans. Wireless Commun. 5(5), 967–971 (2006)
15. Han, Y., Deng, J., Haas, Z.J.: Analyzing multi-channel medium access control schemes with ALOHA reservation. IEEE Trans. Wireless Commun. 5(8), 2143–2152 (2006)
16. ANSI/IEEE Std. 802.11: Wireless LAN Medium Access Control (MAC) and Physical Layer (PHY) specification (1999)
17. McCanne, S., Floyd, S.: ns-2 2.26 (2003) http://www.isi.edu/nsnam/ns/

Providing QoS in Ad Hoc Networks with Distributed Resource Reservation

Ali Hamidian and Ulf Körner

Department of Communication Systems
Lund University, Sweden
Box 118, 221 00 Lund
{ali.hamidian, ulf.korner}@telecom.lth.se

Abstract. As the use of WLANs based on IEEE 802.11 increases, the need for QoS becomes more obvious. The new IEEE 802.11e standard aims at providing QoS, but its contention-based medium access mechanism, EDCA, provides only service differentiation, i.e. soft QoS. In order to provide hard QoS, earlier we have proposed an extension called *EDCA with resource reservation* (EDCA/RR), which enhances EDCA by offering also hard QoS through resource reservation. In this paper, we extend EDCA/RR to cope with the hidden terminal problem, outline a solution for multi-hop scenarios, and compare the proposed scheme with EDCA.

Keywords: QoS, IEEE 802.11e, ad hoc network, distributed resource reservation.

1 Introduction

Nowadays many of us use portable devices that support *wireless local area networks* (WLANs) based on the IEEE 802.11 standard [1] more or less frequently in universities, homes, cafés, train stations and airports. However, we usually do not reflect over the fact whether they support *quality of service* (QoS) - at least not until we start using an application that has strict QoS requirements. One important step towards solving the QoS issue is the IEEE 802.11e standard [2]. This standard enhances the *medium access control* (MAC) sublayer of 802.11 by specifying the *hybrid coordination function* (HCF) and its two medium access mechanisms: *enhanced distributed channel access* (EDCA) and *HCF controlled channel access* (HCCA). The contention-based EDCA provides QoS by delivering traffic based on differentiating user priorities while the contention-free HCCA provides QoS by allowing for reservation of transmission time.

Although HCCA is an important enhancement that aims at providing QoS guarantees in WLANs, it is EDCA that has received most attention so far. Thus, it is possible that EDCA's destiny will be similar to the one of its predecessor: the *distributed coordination function* (DCF). In other words, EDCA might be implemented by the majority of the vendors, whereas HCCA might be somewhat neglected just as the *point coordinator function* (PCF) - despite the fact that HCCA is a great improvement compared to its predecessor. In addition, EDCA is a distributed channel access method and can be used in ad hoc networks while HCCA is centralized and thus only usable in infrastructure networks. Hence, the focus of this paper lies on EDCA.

L. Mason, T. Drwiega, and J. Yan (Eds.): ITC 2007, LNCS 4516, pp. 309–320, 2007.

The motivation of our work is based on the fact that EDCA provides only service differentiation so it cannot guarantee any QoS. Our proposed scheme, called *EDCA with resource reservation* (EDCA/RR), extends EDCA by allowing for resource reservation with the aim of providing QoS guarantees. When talking about QoS guarantees, we must keep in mind that since a wireless medium is much more unpredictable and error-prone than a wired medium, QoS cannot be guaranteed as in a wired system, especially not in unlicensed spectra. However, it is possible to provide techniques that increase the probability that certain traffic classes get adequate QoS and that can provide QoS guarantees in controlled environments.

The remainder of this paper is organized as follows: Section 2 discusses some related work and Sect. 3 gives an overview of the 802.11e standard with focus on EDCA. In Sect. 4 we present EDCA/RR and also a solution for extending the scheme to handle multi-hopping and releasing unused reservations. The simulation results are presented and discussed in Sect. 5. Finally, Sect. 6 concludes the paper and gives some directions for future work.

2   Related Work

There has been a lot of research on providing QoS in ad hoc networks. However, some of these suggest proprietary protocols - based on time division multiple access, multiple channels, etc. It is our belief that any realistic proposal must be based on the widely spread de facto 802.11 standard(s). Among the work that is based on the existing standards, many are focused on infrastructure-based networks. Regarding ad hoc networks, most proposed solutions provide only service differentiation and not QoS guarantees. Below we shortly mention two such works but also two that aim at providing QoS guarantees through resource reservation.

To improve the performance of EDCA, Romdhani et al. [3] adjust the value of the *contention window* (CW) taking into account both application requirements and network conditions. Iera et al. [4] follow another, but similar, approach by dynamically adapting the priority class instead of the CW value, also taking into account both application requirements and network conditions. The improvement of EDCA is still based on service differentiation so it is not possible to provide QoS guarantees.

Hiertz et al. present the *Distributed Reservation Request Protocol* (DRRP) [5], which is a decentralized MAC scheme based on 802.11. The scheme is similar to EDCA/RR, allowing stations to reserve access to the medium. Whenever a station (A) needs to reserve medium access for communication with another station (B), it sends a data frame containing reservation request information. Upon reception of such a reservation request, B sends an *acknowledgment* (ACK) frame that also contains information about the reservation request. The ACK frame is overheard by the neighbors of B and thus, the stations hidden to A are also informed about its reservation request. The reservation request includes information regarding the duration and repetition interval of the next transmission. All neighbors receive the reservation information by overhearing the transmissions between A and B. However, since the neighbors do not acknowledge the overheard frames and these can be lost, the reservation request is transmitted periodically so the information about its periodicity is also included in the reservation request.

The main disadvantage with DRRP is that it does not have any admission control and does not aim to provide QoS guarantees. Moreover, DRRP introduces a new frame structure, which is not compatible with the existing standards.

The *Distributed end-to-end Allocation of time slots for REal-time traffic* (DARE) [6], by Carlson et al., is another decentralized MAC scheme based on 802.11. DARE allows stations to reserve periodic time slots. In particular, DARE extends the RTS/CTS reservation concept of 802.11 to a multi-hop end-to-end perspective. To reserve resources for a real-time flow over several hops, the routing protocol at the source must first find a route to the destination. The route is assumed to be symmetric. Once such a route is established, the source sends a *request-to-reserve* (RTR) frame, which includes the requested duration and periodicity of a time slot as well as the address of the destination node. When an intermediate node receives the RTR frame, it checks whether the request is conflicting with already existing reservations. If the intermediate node can make the requested reservation, it processes the RTR frame and forwards it; otherwise the request is rejected. Once the destination receives the RTR frame, it responds with a *clear-to-reserve* (CTR) frame. When the source receives the CTR frame, it can start transmitting real-time traffic at the next reserved interval. DARE is also able to repair and release unused reservations. One of the main disadvantages with DARE is the very complex and inefficient method for multiple reservations. A requested reservation may conflict with existing ones so, for example, if a requested receive slot cannot be admitted, an *Update-Transmit-Request* (UTR) frame must be sent back to the node that proposed the receive slot in order to re-schedule (shift in time) the slot. The UTR frame suggests at least another receive slot but the suggested slot might still be inappropriate. So nodes might have to send messages back and forth trying to find a suitable reservation slot, and this can happen at every hop! The authors mention that slot shifting becomes necessary more frequently as the number of reservations increases. Thus, new reservations can only be admitted if they can squeeze in between existing ones. Furthermore, although multi-hopping is one of the advantages of DARE, there is no mechanism for the routing protocol to consider the QoS requirements of the requested reservation during the route discovery process. Thus, the routing protocol might find a route that cannot support the requested reservation. Another disadvantage is that there is no RTS/CTS, ACK or retransmission for real-time frames (could have been optional), which can result in lost real-time frames.

In [7] we proposed a mechanism that supports QoS guarantees in a WLAN operating in ad hoc network configuration, i.e. in a single-hop ad hoc network. Although single-hop ad hoc networks might be seen as limited, we must remember that the main application area for EDCA is a WLAN and not a multi-hop ad hoc network. However, in this paper we extend our earlier proposal to, among others, cope with the hidden terminal problem and we also outline a solution for multi-hop scenarios. The scheme is based on the existing standards 802.11 and 802.11e and consequently, it can be integrated into existing systems without much difficulty. To give an example of the application area for single-hop ad hoc networks where our scheme can be used, we can mention network gaming where players can use their laptops to play demanding network games with each other at no cost anywhere they want; i.e. without needing to worry about (neither wired nor wireless) Internet connections.

3   An Overview of IEEE 802.11e

Since the 802.11 standard did not address the QoS issues sufficiently, the 802.11 working group wrote the new standard 802.11e. This new standard introduces the concepts
of *transmission opportunity* (TXOP) and *traffic specification* (TSPEC) in order to solve
a few problems in 802.11 related to the provisioning of QoS.

To solve the problem with unknown transmission times in 802.11, the concept of
TXOPs was introduced. A TXOP is a time interval defined by a starting time and a
maximum duration. During a TXOP, a station may send several frames as long as the
duration of the transmissions does not extend beyond the maximum duration.

In the 802.11 standard, there is no way for a station to send its QoS requirements
to an *access point* (AP). To solve this problem, the 802.11e standard allows stations
to use TSPECs to describe the QoS characteristics of a traffic stream. This is done by
specifying a set of parameters such as frame size, service interval, service start time,
data rate and delay bound. Most of the above-mentioned parameters are typically set
according to the requirements from the application while some are generated locally
within the MAC. The service interval specifies the time interval between the start of
two consecutive TXOPs. The service start time specifies the time when the service
period starts, i.e. when the station expects to be ready to send frames.

3.1   Enhanced Distributed Channel Access (EDCA)

The distributed and contention-based medium access mechanism of 802.11e, EDCA, is
an enhanced variant of DCF. The main problem with DCF, regarding QoS provisioning, is that it cannot provide any service differentiation since all stations have the same
priority. In addition, DCF uses one single transmit queue. To overcome this problem,
in EDCA each station has four *access categories* (ACs) and four transmit queues that
contend for TXOPs independently of the other ACs. Thus, each AC behaves like an enhanced and independent DCF contending for medium access. Before entering the MAC
sublayer, frames are assigned a *user priority* (UP) and based on these UPs each frame
is mapped to an AC. Besides using the UPs, the frames can be mapped to ACs based
on frame types. The management type frames, for example, shall be sent from AC_VO
(without being restricted by any admission control though). The four ACs can be used
for different kind of traffic: AC_BK for background traffic, AC_BE for best effort traffic, AC_VI for video traffic and AC_VO for voice traffic. Differentiated medium access
is realized by varying the contention parameters[1] for each AC:

- CWmin[AC] and CWmax[AC]: the minimum and maximum value of the CW used
  for calculation of the backoff time. These values are variable and no longer fixed
  per PHY as with DCF. By assigning low values to CWmin[AC] and CWmax[AC],
  an AC is given a higher priority.
- *arbitration interframe space number* (AIFSN[AC]): the number of time slots after a
  SIFS duration that a station has to defer before either invoking a backoff or starting
  a transmission. AIFSN[AC] affects the *arbitration interframe space* (AIFS[AC]),

---

[1] In our simulations, EDCA uses the default values for these parameters according to the 802.11e
standard.

which specifies the duration (in time instead of number of time slots) a station must defer before backoff or transmission: AIFS[AC] = SIFS + AIFSN[AC] × slot_time. Thus, by assigning a low value to AIFSN[AC], an AC is given a high priority.

- $TXOP_{limit}[AC]$: the maximum duration of a TXOP. A value higher than zero means that an AC may transmit multiple frames (if all belong to the same AC since a TXOP is given to an AC and not to a station) as long as the duration of the transmissions does not extend beyond the $TXOP_{limit}[AC]$. Thus, by assigning a high value to the $TXOP_{limit}[AC]$, an AC is given a high priority.

4 EDCA with Resource Reservation (EDCA/RR)

In a previous work we have enhanced EDCA to provide QoS guarantees by reserving TXOPs for traffic streams with strict QoS requirements [7]. In this section we start by giving an introduction to EDCA/RR in order to facilitate the reading and understanding of the rest of this paper. Then we describe our solution to a problem in EDCA/RR related to hidden stations. Next, we present a conceivable solution to extend EDCA/RR such that it can be used in multi-hop ad hoc networks. Finally, we identify some problems that might occur due to mobile stations leaving and entering a network.

The EDCA/RR scheme works like EDCA as long as there is no station that needs to reserve TXOPs for its high-priority traffic stream. Once a station (sender) wishes to reserve TXOPs to be able to send traffic with strict QoS requirements, it requests admission for its traffic stream. The request is not sent to any central station such as an AP, but is handled internally within the sender by an admission control algorithm. It should be mentioned that our scheme is not dependent on any specific admission control or scheduling algorithm; thus, it is possible to use any proposed enhancement (such as those presented in [8], [9], [10] and [11]) to the reference design algorithms provided in the 802.11e standard. However, in our EDCA/RR implementation, the reference admission control and scheduling algorithms have been used because they are specified in the standard, making them widely known and giving them certain acceptance.

In case the traffic stream is rejected, the sender can try to lower its QoS demands and retry. On the other hand, if the traffic stream is admitted, the sender schedules its traffic by setting the *service interval* (SI) and the *service start time* (SST) parameters. Details about the calculation of these parameters can be found in [7]. Next, the sender broadcasts an *add traffic stream* (ADDTS) request containing a TSPEC element with information such as mean data rate, nominal frame size, SST and SI. All stations that receive the ADDTS request store the information of the sender's SST and SI, and schedule the new traffic stream exactly as the sender. This ensures that no station starts a transmission that cannot be finished before a reserved TXOP starts and thus collision-free access to the medium is offered to the streams with reserved TXOPs. In order to make sure that all stations have similar schedules, all neighbors have to unicast an ADDTS response back to the sender to acknowledge a received ADDTS request.

Every time the sender receives an ADDTS response from a neighbor, it stores the address of the neighbor. After receiving a response from all neighbors, the sender waits until the SST specified in the TSPEC element and initiates a transmission. If the time instant when all responses are received occurs later than the advertised SST, the transmission is delayed until the next TXOP. Once the TXOP is finished, the station waits until the next TXOP, which occurs after an SI.

When a transmission failure occurs during a TXOP, the station does not start a back-off procedure. Instead, it retransmits the failed frame after SIFS if there is enough time left in the TXOP to complete the transmission.

4.1   The Hidden Station Problem in EDCA/RR

In the original version of EDCA/RR, the hidden station problem was handled through the exchange of *request to send* (RTS) and *clear to send* (CTS) frames, i.e. the same way as in 802.11/e. However, this method is not sufficient since in EDCA/RR, stations hidden to a station that has reserved TXOPs can cause other problems than the well-known hidden station problem (causing collisions). Contending stations that have received a TSPEC from the reserving station do not start a transmission unless it finishes before a TXOP starts. But unfortunately, stations hidden from the reserving station do not receive any TSPEC so they do not know when a TXOP starts. Therefore, they might start a transmission that extends across a TXOP.

To illustrate the problem, suppose there are three stations in a row (see Fig. 1): A, B and C, where A and B as well as B and C are within each others transmission range but A and C cannot hear each other. Assume further that A wants to send traffic with QoS requirements to B so it has broadcasted an ADDTS request and B has replied with an ADDTS response. However, C (that is hidden from A) is unaware of A's TXOP reservation since it has not received A's ADDTS request so there is a chance that C starts transmitting just before a TXOP reserved by A is about to start. In that case a collision would occur during A's reserved TXOP meaning that A would no longer have collision-free access to the medium. In order to prevent C from transmitting just before a reserved TXOP is about to start, it must become aware of A's TXOP reservation. In other words, the reservation schedule of any sender must be known by all stations within two hops from the sender.

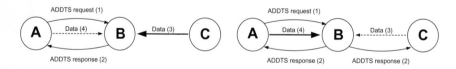

Fig. 1. C is hidden from A and can start transmitting just before A's TXOP starts

Fig. 2. C is informed about the TXOP reservation of A and defers during A's TXOP

There are different ways of achieving this goal, i.e. to spread the TSPEC to stations outside of the reserving station's transmission range. One approach can be to rebroadcast the ADDTS request sent by the reserving station during the TXOP reservation. Hence, in our example B would rebroadcast the ADDTS request of A to let also C receive the request frame. However, there are many problems related to this approach. First, should C send an ADDTS response to B just like B has to send an ADDTS response to A? We must remember that there might be many stations at the same distance from A as B and C respectively (i.e. one and two hops away from A respectively). This means that if C has to respond to B then every other station two hops away from A should also respond to B because those are also hidden stations. Moreover, this procedure would continue until all stations one hop away from A rebroadcast the ADDTS request from A, and all stations two hops away from A send back an ADDTS response. Obviously, this would lead to a lot of overhead and a significant increase in the reservation delay. On the other hand, if C does not have to send a response to B, then B cannot be sure whether the rebroadcasted ADDTS request was received by C or not.

Another approach to spread the TSPEC is to let the ADDTS responses contain the TSPEC and let all stations overhear these frames (see Fig. 2). This way, the TSPEC is known to all stations within two hops from the sender with no additional signaling and with limited increase of overhead. Thus, when B sends an ADDTS response back to A, C will hear this frame and save the information included in the TSPEC, i.e. the SST and SI of A. This approach is much less complex and results in less overhead than the previous approach. However, again B cannot be sure whether the ADDTS response was received correctly by C or not. Therefore, we let reserving stations transmit special RTS/CTS frames extended to contain a TSPEC (RTS_TSPEC and CTS_TSPEC), in the beginning of a TXOP. This way, a station with an out-of-date reservation schedule has the chance to update its schedule. Although one might think that this increases the overhead, we must remember that the RTS_TSPEC and CTS_TSPEC frames are sent only at the beginning of a TXOP and not for every single data frame.

4.2   Further Enhancements

The first goal of this paper was to enhance the EDCA/RR scheme operating in single-hop ad hoc networks. In particular, we wanted to solve the problems that could occur due to hidden stations. We have achieved this goal and presented our solution above. Another goal was to present an extension to the scheme such that it can be used to provide QoS guarantees in a multi-hop ad hoc network.

QoS Provisioning in Multi-hop Ad Hoc Networks.   To extend EDCA/RR for multi-hop ad hoc networks, we need an ad hoc routing protocol that can find a route between the communicating stations. In this paper we assume that the routing protocol is reactive, like for example *Ad hoc On-Demand Distance Vector* (AODV) [12] and *Dynamic MANET On-demand* (DYMO) [13]. Using one of these protocols, during the route discovery process the source broadcasts a *route request* (RREQ) throughout the network to find the destination. When the destination receives the RREQ, it responds with a *route reply* (RREP) unicast toward the source.

To illustrate the idea of our solution, let us assume there are three stations in a row: A, B and C, where A and B as well as B and C are within each other's transmission range but A and C cannot hear each other. Assume further that A wants to send high-priority traffic (with QoS requirements) to C.

To reserve resources along a multi-hop route, the QoS requirements of A's traffic stream must be known by the routing protocol so that it can use the requirements during the route discovery process. However, the routing protocol shall not start its route discovery process before the traffic stream has been admitted by the local admission control mechanism at A's MAC sublayer. In EDCA/RR, the resource reservation process (including admission control) starts when the first packet of a traffic stream with QoS requirements reaches the MAC sublayer, i.e. after it has been handled by the routing protocol at the network layer. To prevent the routing protocol searching for a route using its usual metrics (and thus not considering the QoS requirements of the traffic stream), it must be modified to co-operate with the protocol at the MAC sublayer (EDCA/RR in our case). Therefore, once the first high-priority packet of a traffic stream in station A reaches its network layer, the TSPEC information provided by the application (e.g. mean data rate and nominal frame size) is copied from the packet into the RREQ that is generated. These fields describe some characteristics of the application that are needed during the calculation of SI and TXOP duration at the MAC sublayer. In addition, the fields contain information about how the route discovery procedure should proceed in case there are not enough resources to reserve. This is necessary because although some applications need a certain minimum level of QoS for functioning, others can function despite that the QoS level is not sufficient. Therefore, if a traffic stream cannot be admitted, an application may prefer a normal route (not taking into account the QoS requirements of the application) rather than aborting.

When the RREQ extended with TSPEC information is sent down to the MAC sublayer, the admission control algorithm in EDCA/RR checks whether the requested traffic stream can be admitted. If the traffic stream is rejected, depending on the requirements of the application, there are two possible cases: either the application is notified to abort or the station can search for a normal route to the destination. On the other hand, if the traffic is admitted, the route discovery process is started to search for a route that can handle the QoS requirements. In that case, A broadcasts an extended RREQ and when B receives the message, its MAC sublayer uses the TSPEC information in the message to check whether the traffic can be admitted or not. In case the traffic is rejected, again depending on the QoS requirements of A's application, either it is notified to abort or B searches for a normal route to the destination. On the other hand, if the traffic is admitted, the extended RREQ is rebroadcasted and received by C. Station C processes the extended RREQ mainly as in B. However, if the traffic is admitted, the MAC sublayer schedules the traffic stream of A and notifies the routing protocol to send an extended RREP, containing information about the successful reservation, back to the source (i.e. station A).

When B receives the extended RREP, its MAC sublayer uses the TSPEC information in the message to schedule the traffic stream, which now has been admitted by all stations from the source to the destination. Then the RREP is forwarded to A, which processes the message just as in B. Finally, the resource reservation is finished and the traffic stream can start transmitting during its reserved TXOPs. Thus, to summarize, the RREQs check whether the requested resources are available while the RREPs confirm the TXOP reservations and do the actual scheduling of the traffic.

Leaving and Entering the Network. An important issue that needs special attention is mobility. If a station with reserved TXOPs leaves the network, the other stations must become aware of that because otherwise they will defer from transmitting although they should not and network capacity will be wasted. A conceivable solution to this problem is to let the stations in the network listen for frames in the beginning of each TXOP in order to determine whether the reserved TXOPs are still in use. (The station with the reserved TXOP may transmit dummy packets when it does not have any data to send.) If several consecutive TXOPs are determined to be unused, the receiver can ask the sender if it still has something to send. If the sender does not respond despite several attempts, the receiver and other stations can assume that the sender has left the network and delete the reservation by deallocating the TXOPs. Furthermore, possible TXOPs after the terminated TXOP, shall be moved back within the SI in order to avoid time gaps between TXOPs. Moving the streams is done pretty easily thanks to the distributed characteristic of EDCA/RR. There is no need for any signaling; each station performs the rescheduling itself in a distributed manner.

On the other hand, if a station enters a network with existing reserved TXOPs, it cannot start transmitting because such a transmission may collide with the transmission in a reserved TXOP. Instead, the station must update its schedule before it is allowed to transmit any frame. This can be done by setting a schedule update bit in beacons or other frames exchanged during the initialization process.

5   Evaluation

In order to evaluate the performance and effectiveness of EDCA/RR, we implemented our scheme in the widely-used network simulator ns-2 and compared it with EDCA. Since the 802.11 implementation in ns-2 is rather simple, we used another more advanced and detailed 802.11 implementation (developed by Mike Moreton), which also implements 802.11a/b/g and some features of 802.11e. Based on this detailed implementation, we implemented EDCA/RR. Various scenarios have been studied though just one is presented in this paper.

5.1   Simulation Setup

The studied scenario consists of a variable number of stations where each traffic stream is sent from a unique source to a unique destination. There is always one high-priority source while the number of low-priority sources is varied from zero to five. All low-priority sources are hidden to the high-priority source.

The stations use 802.11b DSSS (with short preamble) in the physical layer and 802.11e EDCA or EDCA/RR in the MAC sublayer. The routing protocol AODV is used at the network layer. At higher layers, the low-priority streams, which are sent from AC_BE, generate TCP segments with a size of 1000 bytes according to an FTP application that always has data to transmit. This is because we want to stress the compared MAC schemes regarding the QoS provisioning to a high-priority stream, by increasing the proportion of low-priority traffic load in the network. On the other hand, the high-priority stream generates UDP packets according to a *constant bit rate* (CBR) traffic generator. The high-priority stream is sent from AC_VO and models voice traffic encoded using the G.711 codec. The codec generates voice packets at a rate of 64 kbit/s, with packet size of 160 bytes and packet interarrival time of 20 ms. The protocols at each layer add some overhead to the voice packet so in total we have: G.711 (160 B) + RTP (12 B) + UDP (8 B) + IP (20 B) + 802.11 MAC (28 B) + 802.11b PHY (short preamble: 96 μs).

In this paper, we have studied the stationary behaviour of EDCA and EDCA/RR. More specifically, we studied the impact of an increasing number of low-priority streams on one high-priority stream. We calculated the average end-to-end delay[2], jitter and squared coefficient of variance of the end-to-end delay ($C^2[d]$) for the high-priority stream when the number of low-priority streams was varied between zero and five. For each of the six data points we ran 300 simulations each during 300 simulated seconds. Since we were interested in studying the behaviour of the network in steady state, the first 30 seconds of the simulations were ignored. Then, we calculated the average of the 300 averaged values for each data point. Because of the extensive simulations, we could calculate the 99% confidence interval of the average end-to-end delay.

5.2   Simulation Results

In Tables 1 and 2 we can see the average end-to-end delay and its corresponding 99% confidence interval for the high-priority stream. Table 1 shows the results for the case when the medium is error-free while Table 2 shows the corresponding results for the case when the packet error rate is 5 %. Both tables show that, as the number of low-priority streams increases, the average end-to-end delay for the high-priority stream increases rapidly for EDCA. This is a typical behaviour for contention-based medium access schemes and it is this kind of behaviour that we would like to avoid. Another typical, but more advantageous, behaviour for random-access schemes is that they have very low medium access delays when the network load is light. This is also shown in the tables. On the other hand, EDCA/RR provides an almost constant average end-to-end delay to the high-priority stream. The result shows that EDCA/RR succeeds offering periodic medium access to the high-priority stream no matter how large the network load is. Furthermore, it shows that EDCA/RR can handle hidden stations.

Table 1. 99% confidence interval of the average end-to-end delay of the high-priority stream - 0 % packet error

Table 2. 99% confidence interval of the average end-to-end delay of the high-priority stream - 5 % packet error

nbr of LP-streams	delay (ms) EDCA	delay (ms) EDCA/RR	confidence interval (ms) EDCA	confidence interval (ms) EDCA/RR
0	0.69	12.33	(0.69,0.69)	(12.13,12.53)
1	6.21	12.22	(6.20,6.22)	(12.02,12.42)
2	11.17	12.27	(11.14,11.19)	(12.08,12.47)
3	13.93	12.22	(13.90,13.96)	(12.01,12.42)
4	17.12	12.38	(17.08,17.16)	(12.19,12.57)
5	20.51	12.25	(20.46,20.56)	(12.06,12.45)

nbr of LP-streams	delay (ms) EDCA	delay (ms) EDCA/RR	confidence interval (ms) EDCA	confidence interval (ms) EDCA/RR
0	0.99	12.55	(0.99,0.99)	(12.37,12.73)
1	4.68	12.44	(4.68,4.69)	(12.27,12.61)
2	5.25	12.54	(5.24,5.25)	(12.35,12.73)
3	5.59	12.34	(5.58,5.60)	(12.16,12.52)
4	5.92	12.64	(5.91,5.93)	(12.45,12.82)
5	6.28	12.53	(6.27,6.29)	(12.34,12.72)

Comparing the two tables reveals that, under EDCA/RR, when the packet error increases to 5 % the end-to-end delay increases marginally. This is a result of the fact that retransmissions most often take place within the same TXOP as the discarded packet.

---

[2] The end-to-end delay is calculated as the time when a frame is received by the destination's application layer minus the time when the frame was generated at the application layer of the source.

However, with further increasing packet error probabilities, not even EDCA/RR may handle the QoS requirements. Under EDCA, the situation is a bit different. The tables show that the average end-to-end delay decreases with increasing packet error probabilities. This might seem surprising at first but the explanation is that, when the packet error rate is 5 % for EDCA, the low-priority frames are affected much more negatively than the high-priority frames. This is because the CW, which is doubled after each lost frame, rapidly becomes much larger for the low-priority streams than for the high-priority stream[3] resulting in longer medium access delays. Moreover, since the number of transmitted, and thus lost, low-priority frames are larger than the corresponding number of high-priority frames, the low-priority connections are broken more often. The consequence is that the number of simultaneously active low-priority streams is almost always lower than the actual number. Thus, to the high-priority stream the network is less loaded than it would be if the channel would be error-free. This also explains why EDCA shows a lower end-to-end delay than EDCA/RR. Still, one might argue that since there are no error-free channels in the real world, why use EDCA/RR when EDCA results in a lower average end-to-end delay. First, we could have set up much tougher QoS requirements, say 5 ms end-to-end delay and EDCA/RR would have met them. Second, we measure delays for those packets that really reach the receiver, though after a number of retransmissions. Under EDCA the number of packets that really reach the receiver, is much lower than for EDCA/RR.

Table 3 shows the end-to-end jitter and the corresponding squared coefficient of variance of the end-to-end delay ($C^2[d]$) for the high-priority stream when the channel is error-free. As expected, we see that when the high-priority stream is the only active stream in the network, the jitter and the $C^2[d]$ are very low. The explanation for this is the same as why the average end-to-end delay is very low in this situation; i.e. the frames get instant access to the medium. However, the general view is that while EDCA/RR provides very low variations those for EDCA increase significantly with the number of low-priority streams. The behaviour of EDCA is not acceptable for multimedia applications with QoS requirements on low jitter.

Table 3. Jitter and $C^2[d]$ for the high-priority stream - 0 % packet error

nbr of LP-	jitter ($10^{-6}s^2$)		$C^2[d]$	
streams	EDCA	EDCA/RR	EDCA	EDCA/RR
0	0.02	48	0.05	0.32
1	40	48	1.04	0.32
2	180	48	1.45	0.32
3	276	48	1.42	0.32
4	406	49	1.38	0.32
5	577	49	1.37	0.32

6   Conclusion

In this paper, we have extended the previously proposed EDCA/RR scheme, which allows multimedia applications to reserve medium time. EDCA/RR has been enhanced to

---

[3] The default CWmax value for AC_VO and AC_BE is equal to 15 and 1023 respectively.

prevent hidden stations causing collisions during reserved TXOPs. The main idea was to spread the information about the TXOP reservation such that also hidden stations become aware of the reservation and thus, defer during the reserved TXOPs. EDCA/RR was designed for single-hop ad hoc networks. Although such a scheme can be useful, our aim was to extend it to be useful also in multi-hop ad hoc networks since these networks are expected to offer new communication possibilities. Thus, we have presented an extension to the scheme such that it can be used together with an ad hoc routing protocol to find multi-hop QoS-enabled routes between the communicating stations. Thus, a station sending traffic with QoS requirements will be able to reserve TXOPs for deterministic medium access along a multi-hop route to the destination. The aim in future work will be to incorporate this extension into the existing EDCA/RR implementation.

References

1. ANSI/IEEE: Std 802.11 Part11: Wireless LAN Medium Access Control (MAC) and Physical Layer (PHY) Specifications (1999)
2. IEEE: Std 802.11e Part11: Wireless Medium Access Control (MAC) and Physical Layer (PHY) specifications - Amendment 8: Medium Access Control (MAC) Quality of Service (QoS) Enhancements (2005)
3. Romdhani, L., Ni, Q., Turletti, T.: Adaptive EDCF: Enhanced Service Differentiation for 802.11 Wireless Ad-Hoc Networks. In: Proceedings of IEEE Wireless Communication and Networking Conference (2003)
4. Iera, A., Molinaro, A., Ruggeri, G., Tripodi, D.: Improving QoS and Throughput in Single- and Multihop WLANs through Dynamic Traffic Prioritization. IEEE Network 19(4) (July/August 2005)
5. Hiertz, G.R., Habetha, J., May, P., Weib, E., Bagul, R., Mangold, S.: A Decentralized Reservation Scheme for IEEE 802.11 Ad Hoc Networks. In: Proceedings of the 14th International Symposium on Personal, Indoor and Mobile Radio Communications (2003)
6. Carlson, E., Prehofer, C., Bettstetter, C., Wolisz, A.: A Distributed End-to-End Reservation Protocol for IEEE 802.11-based Wireless Mesh Networks. Journal on Selected Areas in Communications (JSAC), Special Issue on Multi-Hop Wireless Mesh Networks (2006)
7. Hamidian, A., Körner, U.: An Enhancement to the IEEE 802.11e EDCA Providing QoS Guarantees. Telecommunication Systems 31(2-3) (March 2006)
8. Skyrianoglou, D., Passas, N., Salkintzis, A.: Traffic Scheduling in IEEE 802.11e Networks Based on Actual Requirements. Mobile Location Workshop Athens (2004)
9. Grilo, A., Macedo, M., Nunes, M.: A Scheduling Algorithm for QoS Support in IEEE 802.11e Networks. IEEE Wireless Communications Magazine (June 2003)
10. Fan, W.F., Gao, D., Tsang, D.H.K., Bensaou, B.: Admission Control for Variable Bit Rate Traffic in IEEE 802.11e WLANs. IEEE LANMAN (April 2004)
11. Ansel, P., Ni, Q., Turletti, T.: FHCF: A Fair Scheduling Scheme for 802.11e WLAN. Technical Report 4883, INRIA Sophia Antipolis (July 2003)
12. Perkins, C., Belding-Royer, E.M., Das, S.: Ad hoc On-Demand Distance Vector (AODV) Routing. IETF. RFC 3561
13. Ogier, R., Lewis, M., Templin, F.: Dynamic MANET On-demand (DYMO) Routing. IETF. Internet draft

Fluid-Flow Modeling of a Relay Node in an IEEE 802.11 Wireless Ad-Hoc Network

Frank Roijers[1,2,3], Hans van den Berg[1,4], and Michel Mandjes[2,3]

[1] TNO Information and Communication Technology, The Netherlands
[2] Korteweg-de Vries Institute, University of Amsterdam, The Netherlands
[3] Centre for Mathematics and Computer Science, The Netherlands
[4] Department of Design and Analysis of Communication Systems,
University of Twente, The Netherlands

Abstract. Wireless ad-hoc networks are based on shared medium technology where the nodes arrange access to the medium in a distributed way independent of their current traffic demand. This has the inherent drawback that a node that serves as a relay node for transmissions of multiple neighboring nodes is prone to become a performance "bottleneck". In the present paper such a bottleneck node is modeled via an idealized fluid-flow queueing model in which the complex packet-level behavior (MAC) is represented by a small set of parameters. We extensively validate the model by ad-hoc network simulations that include all the details of the widely used IEEE 802.11 MAC-protocol. Further we show that the overall flow transfer time of a multi-hop flow, which consists of the sum of the delays at the individual nodes, improves by granting a larger share of the medium capacity to the bottleneck node. Such alternative resource sharing strategies can be enforced in real systems by deploying the recently standardized IEEE 802.11E MAC-protocol. We propose a mapping between the parameter settings of IEEE 802.11E and the fluid-flow model, and validate the fluid-flow model and the parameter mapping with detailed system simulations.

1 Introduction

Developments in wireless communication technology open up the possibility of deploying wireless ad-hoc networks; these networks can be rolled out instantly without any fixed infrastructure or pre-advanced configuration. Currently, IEEE 802.11 wireless LAN [8] is the most popular technology used for wireless ad-hoc networks. Up to now mainly IEEE 802.11B was used for ad-hoc networking, but in 2005 IEEE 802.11E was standardized allowing for service differentiation.

Ad-hoc networks have two important characteristics: i) stations that cannot directly communicate with each other use other stations as relay nodes; ii) stations contend for access to the wireless medium in a distributed fashion. A consequence of the first characteristic is that certain nodes, in particular nodes that have a central location, are likely to become relay nodes having considerably higher traffic loads than other nodes. The second characteristic entails

L. Mason, T. Drwiega, and J. Yan (Eds.): ITC 2007, LNCS 4516, pp. 321–334, 2007.

that there is a lack of coordination between the nodes which may result in non-optimal sharing of the medium capacity. Therefore, a relay node can easily become a performance bottleneck. For example, when a relay node obtains the same share of the medium capacity as each of its active neighboring nodes, the input rate of traffic into the relay node regularly exceeds the output rate when more than one neighboring node sends traffic via the relay node. This results in the accumulation of backlogged traffic and consequently in increasing delays.

The vast majority of ad-hoc network performance studies available in the literature is based on simulation, see e.g. [6,7]. These studies usually capture great detail of the ad-hoc network protocols, but have the intrinsic drawback that they do not provide deeper understanding of the impact of the parameters on the realized performance. Moreover, simulation runtime may become prohibitively large, hampering, e.g., sensitivity analysis or parameter optimization. Analytical performance models usually capture less detail in order to retain tractability, but do provide insight into the behavior of the system.

In [1] we introduced a performance model for 2-hop flows (i.e., file transfer) relayed by a bottleneck node in an IEEE 802.11B ad-hoc network. The modeling approach is based on the principle of *separation of time scales*. The entire packet-level behavior is captured in the *net* medium capacity, which can be obtained by the well-known model of Bianchi [2]. Subsequently the net medium capacity is used as service capacity in a flow-level model that captures the "user dynamics", i.e., the initiation and completion of file transfer from neighboring source nodes to destinations via the bottleneck node. Another aspect of the modeling approach in [1] is that the flow-level model is based on so-called fluid flows, i.e., flows are modeled as if they *continuously* send traffic instead of sending individual *packets*. This modeling approach has been successfully used before for the single-hop case, see [5,10]; the model in [1] could be regarded as an extension to the situation of an additional hop that requires a share of the medium capacity.

The time-scale separation considerably reduces the complexity of the model in [1]. In particular, insightful, explicit formulas are obtained for the mean values of, e.g., the delay per hop and the overall transfer time of a 2-hop flow. These expressions can easily be evaluated in order to generate numerical results. However, [1] did not investigate whether the model really accurately describes the behavior of an IEEE 802.11 ad-hoc network. Furthermore, [1] focuses on the situation where the relay node obtains the *same* share of the medium capacity as its neighboring nodes. An extension to improve the overall flow transfer time is the option of granting a *larger* share of the medium capacity to the relay node than to each of the neighboring nodes.

Contribution. The present paper elaborates on the fluid-flow model, for short the *fluid model*, proposed in [1]. The present paper fills both above-mentioned gaps.

The first contribution is a validation of the fluid model for "equal resource sharing" of the medium capacity, i.e., the case where the bottleneck node can only obtain the same share of the medium capacity as each of its neighboring nodes. This model is validated for network nodes that operate according to the IEEE 802.11B MAC-protocol. The analytical results of [1], as summarized in Section 3.2, are numerically evaluated and compared to results obtained by a wireless ad-hoc network simulator with a detailed implementation of the IEEE 802.11 protocols.

The second contribution is the extension of the fluid model to an ad-hoc network scenario with so-called "unequal resource sharing" between the bottleneck and source nodes, i.e., the bottleneck node may obtain a larger share of the medium capacity. Importantly, we demonstrate that unequal resource sharing can considerably improve the transfer time of a flow. The fluid model with unequal resource sharing is validated by the earlier-mentioned IEEE 802.11E MAC-protocol, which can be used to implement the unequal resource-sharing policy.

Organization. This paper is organized as follows. In Section 2 we introduce the ad-hoc network scenario and the fluid model of a bottleneck node considered in this paper. Section 3 validates the fluid model for an equal resource sharing IEEE 802.11B bottleneck node; we obtain the service capacity used in the fluid model, present an excerpt of the analytical results and validate the fluid model by comparing it with ad-hoc network simulations. In Section 4 it is shown that unequal resource sharing between source and bottleneck nodes improves the overall flow transfer time. Section 5 presents the implementation of unequal resource sharing by an IEEE 802.11E bottleneck node; first we obtain the input parameters of the fluid model and second we validate this model. Finally, Section 6 concludes this paper and discusses some directions for further research.

2   Ad-Hoc Network Scenario and Fluid-Model Description

Section 2.1 introduces the ad-hoc network scenario considered in this paper. In Section 2.2 we present our fluid model that corresponds to this scenario. Section 2.3 briefly introduces the concept of the IEEE 802.11E EDCA-protocol.

2.1   The Ad-Hoc Network Scenario

As mentioned in the introduction, one of the goals of the present paper is to investigate whether the fluid-flow model, which we introduced in [1], accurately describes the behavior of a bottleneck node, both under equal and unequal resource-sharing policies. For that purpose we focus on a simple, special case of a wireless ad-hoc network: a two-hop ad-hoc network consisting of a number of source nodes that initiate flow transfers at random time instants, and a single relay node that forwards the traffic generated by the sources to the next-hop destination nodes, cf. Figure 1. The source nodes and destination nodes are all within the transmission range of the bottleneck node; the source and destination nodes are within each others sensing range, hence there are no hidden nodes.

The equal and unequal resource sharing policies correspond to data transmissions that are controlled by respectively the IEEE 802.11B DCF ([8]) and the IEEE 802.11E EDCA ([9], also see Section 2.3) mechanisms.

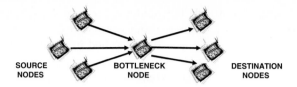

SOURCE   BOTTLENECK   DESTINATION
NODES   NODE   NODES

Fig. 1. Two-hop network with a single (bottleneck) node used as relay node by many sources

2.2   Fluid-Model Description

The ad-hoc network scenario described above is now modeled as a fluid-flow queueing system. We assume a large number of source nodes which become active and initiate flow transfers to destinations via the bottleneck node according to a Poisson process with rate λ ('flow arrival rate'). The bottleneck node relays all traffic of the source nodes in a first-come-first-served discipline. Active source nodes and the bottleneck node share the system capacity, which is conditional on the number of active source nodes n and is denoted by c_n. Once a source node has completed a flow transmission, the source node becomes inactive (although the last part of the flow may still be at the buffer of the bottleneck node waiting for service). Flow sizes (in terms of the amount of traffic/fluid) are i.i.d. random variables (denoted by F) with finite mean f and second moment f_2. A source node has at most one flow transfer in progress.

Resource sharing between bottleneck and source nodes. First we consider the case of equal resource sharing. If n source nodes have a flow transfer in progress, any source node transmits its traffic (fluid) into the buffer of the bottleneck node at rate $c_n/(n+1)$, while a rate $c_n/(n+1)$ is used by the bottleneck node to 'serve' the buffer (i.e., to forward the traffic stored in its buffer to the next node). The amount of work backlogged in the buffer is denoted by W_{buffer}. In case $W_{\text{buffer}} > 0$ and $n = 0$ the bottleneck receives the entire capacity c_0.

In case of unequal resource sharing, the maximum ratio between the share of the bottleneck node and a source node is denoted by $m_n \in \mathbb{R}$, and the bottleneck node *may* obtain capacity $m_n c_n/(n+m_n)$. The bottleneck node will only obtain the maximum share if it can actually use it, viz. the input rate exceeds the output rate ($n \geq m_n$) or if $W_{\text{buffer}} > 0$. Otherwise the input and output rates are coupled, resulting in capacity share of $c_n/2$ for the bottleneck node. The capacity share obtained by the bottleneck node is summarized as follows:

$$c_n \times \begin{cases} m_n/(n+m_n), & \{W_{\text{buffer}} > 0\} \vee \{n \geq m_n\}, \\ 1/2, & \{W_{\text{buffer}} = 0\} \wedge \{0 < n < m_n\}, \\ 1, & \{n = 0\}. \end{cases}$$

The source nodes always equally share the remaining capacity. Notice that IEEE 802.11B is a special case of IEEE 802.11E where $m_n = 1$.

Performance criteria. Our main performance measures of interest are the steady-state buffer workload W_{buffer} at the bottleneck node and the overall flow transfer time D_{overall}, i.e., the time required to completely transfer a flow from source to destination. The overall flow transfer time is the sum of two other performance measures: (i) the time (D_{source}) a source requires to completely transfer a particular flow to the bottleneck node, and (ii) the delay of the *last* particle of fluid of the flow at the bottleneck node (D^*_{buffer}).

2.3   IEEE 802.11E Enhanced Distributed Channel Access

IEEE 802.11E specifies the Enhanced Distributed Channel Access (EDCA) as the distributed contention mechanism that can provide service differentiation. Whereas an 802.11B station has only one queue for all traffic, an 802.11E station (QSTA) has multiple queues, so-called Access Categories (ACs), and traffic is mapped into one of the ACs according to its service requirements.

Each AC contends for a Transmission OPportunity (TXOP) using the CSMA/CA mechanism (see e.g. [8,10]) using its own set of EDCA parameters values. These EDCA parameters are CW$_{\text{min}}$, CW$_{\text{max}}$, AIFS and the TXOP$_{\text{limit}}$. The parameters CW$_{\text{min}}$ and CW$_{\text{max}}$ have the same functionality as under the DCF. The parameter AIFS (Arbitration InterFrame Space) differentiates the time that each AC has to wait before it is allowed to decrement its backoff counter after the medium has become idle. Under the DCF each station has to wait for a DIFS period while the duration of an AIFS is a SIFS period extended by a discrete number of time slots AIFSN, so AIFS=SIFS+AIFSN×timeslot (where AIFSN ≥ 2 for QSTAs). The TXOP$_{\text{limit}}$ (TXOP-limit) is the maximum duration of time that an AC may send after it has won the contention, so it may send multiple packets as long as the last packet is completely transmitted before the TXOP$_{\text{limit}}$ has passed. For simplicity we adapt the notation that TXOP$_{\text{limit}}$ is denoted as the maximum number of packets that may be transmitted per TXOP instead of the maximum duration.

Obviously, multiple ACs can obtain the TXOP at the same moment (i.e., the backoff counters of multiple ACs reach zero at the same moment), which is called a *virtual collision*. Each QSTA has an internal scheduler that handles a virtual collision. The AC with the highest priority is given the TXOP and may actually initiate a transmission. The ACs of lower priority are treated as if they experienced a collision, so they have to double their contention window CW and start a new contention for the medium.

3   Validation of the Fluid Model for IEEE 802.11B

This section validates the fluid model with equal resource sharing as an accurate description of the behavior of the source and bottleneck nodes in an IEEE 802.11B ad-hoc network scenario. First, we describe how to obtain accurate parameter values for the fluid model and we present a summary of the analytical results that were presented in [1]. Next, we introduce the traffic model and the ad-hoc network simulator that are used for the validation. Finally, we validate the fluid model by comparing the analytical results with simulation results of the ad-hoc network simulator.

3.1   Mapping of IEEE 802.11B Parameters

The fluid model of Section 2.2 for equal resource sharing only requires that the capacity c_n are determined; equal resource sharing implies that m_n equals 1 for all n.

The capacity c_n can be obtained by the model of Bianchi [2]. Recall that if at least one source node is active, then also the bottleneck node is active. As we assume that an active node is continuously contending for a TXOP, the ad-hoc network scenario satisfies the framework of Bianchi's model and c_n corresponds to Bianchi's saturated throughput for $n + 1$ nodes. Note that we are interested in the saturated throughput at flow level (i.e., excluding all overhead), although the overheads of all OSI-layers should be taken into account in the calculations.

As an example we consider an IEEE 802.11B bottleneck node with a *gross* bit rate of 11 Mbit/s with RTS/CTS-access. The *net* bit rate is 5.0 Mbit/s for all n, cf. the curve 802.11B in the left graph of Figure 5.

3.2   Analysis of the Fluid Model

In [1] insightful, explicit formulas for the mean values of the performance measures are presented. In the analysis it is assumed that c_n is constant for all n (cf. Section 3.1), for simplicity denoted by c, which allows us to define the load of the system by $\rho = \lambda f / c$.

The overall flow transfer time D_{overall} of a flow is the sum of its flow transfer time D_{source} and the buffer delay of its last particle D^*_{buffer}. Hence

$$D_{\text{overall}} = D_{\text{source}} + D^*_{\text{buffer}}. \tag{1}$$

Notice that D_{source} and D^*_{buffer} are *not* statistically independent.

The mean flow transfer time $\mathbb{E}D_{\text{source}}$ is easily obtained by considering the system as a generalized processor sharing queueing model (cf. Cohen [4]) for which the stationary distribution, here denoted by π_n, is known; Little's law on the mean number of active source nodes yields

$$\mathbb{E}D_{\text{source}} = \frac{\mathbb{E}N}{\lambda} = 2\,\frac{f/c}{1 - \rho}.$$

which is *insensitive* to the flow-size distribution apart from its mean.

The buffer delay D^*_{buffer} is derived from the buffer workload W^*_{buffer} seen by the last particle, which is the sum of the workload W_{buffer} upon flow arrival and the buffer increase ΔW_{buffer} during D_{source}. The amount of work in the buffer at the bottleneck node is the difference between the total amount of work in the system W_{total} (both at the sources and the buffer) and the work remaining at the source W_{sources}, hence

$$\mathbb{E}W_{\text{buffer}} = \mathbb{E}W_{\text{total}} - \mathbb{E}W_{\text{sources}} = \frac{2\rho^2 f_2}{f c}\,\frac{1}{(1 - 2\rho)(1 - \rho)}.$$

The expected workload increase during a flow transfer D_{source} is given by

$$\mathbb{E}\Delta W_{\text{buffer}} = \mathbb{E}D_{\text{source}} - 2f/\text{C} = \frac{2f\rho/\text{C}}{1-\rho}.$$

Therefore,

$$\mathbb{E}W^*_{\text{buffer}} = \mathbb{E}W_{\text{buffer}} + \mathbb{E}\Delta W_{\text{buffer}} = \frac{2\rho^2 f_2/f\text{C}}{(1-2\rho)(1-\rho)} + \frac{2f\rho/\text{C}}{1-\rho}.$$

Observe that the buffer delay of the last particle D^*_{buffer} is the time required to serve the amount of work W^*_{buffer} that is present at the buffer upon arrival of the last particle. As the capacity sharing between source nodes and bottleneck node is purely processor sharing, we approximate the buffer delay of the last particle by

$$\mathbb{E}D^*_{\text{buffer}} \approx \sum_{n=0}^{\infty} \pi_n \mathbb{E}X_n(\mathbb{E}W^*_{\text{buffer}}), \qquad (2)$$

where $\mathbb{E}X_n(\tau)$, the so-called *response time* for jobs in an M/M/1-PS queue presented by Coffman, Muntz, and Trotter (see [3]), is given by

$$\mathbb{E}X_n(\tau) = \tau + \frac{\rho\tau}{1-\rho} + (n(1-\rho) - \rho)\,(f/\text{C})\frac{1 - \exp(-(1-\rho)\tau\text{C}/f)}{(1-\rho)^2}. \qquad (3)$$

For further details about approximation (2) we refer to [1].

3.3   Validation Scenario

This section displays the traffic model and the ad-hoc network simulator that are used for the numerical validation in both Sections 3.4 and 5.2.

The traffic model considered in the ad-hoc network scenario is as follows. We examine the transfer of flows (files) with a mean size of 10 packets of 1500 Bytes each. The following flow-size distributions are considered: Deterministic, Exponential and Hyper-Exponential (with balanced means (e.g., see p. 359 of [12]) and a Coefficient of Variation (CoV) of 4 and 16). New flows are initiated according to a Poisson process and the flow initiation rate is varied between 1 and 20.

For the fluid model the traffic parameters specified above correspond to mean flow size $f = 0.12$ Mbit. The *net* capacity of the fluid model has a constant value of 5 Mbit/s; the variation of the initiation rate results in values for the load ρ between 0.024 and 0.48.

We do not consider mean flow sizes other than 10 packets, i.e., $f = 0.12$, as [1] showed that the performance measures are (almost) linear in the mean flow size. Results for other mean flow sizes can be directly obtained from the results of the specified traffic model.

The ad-hoc network simulator, which is used for the numerical validation of the ad-hoc network scenario, is an own-built simulation tool in programming-language Delphi. All the details of CSMA/CA contention of the EDCA are included in the simulator, e.g., the back-off mechanism, physical and virtual carrier sensing, and collision handling. The PHY-layer includes propagation- and

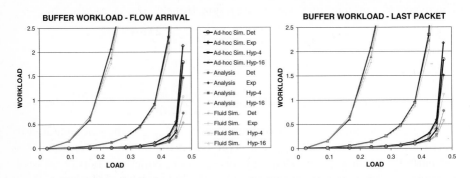

Fig. 2. Mean workload. Left: at flow arrival ($\mathbb{E}W_{\text{buffer}}$). Right: last packet ($\mathbb{E}W_{\text{buffer}}^*$).

fading-models and a clear channel assessment (CCA) procedure that results in limited ranges for successfully transmitting and receiving packets and sensing transmissions of other nodes. The PHY-parameters are set according to the IEEE 802.11B standard: RTS-, CTS- and ACK-frames are transmitted at 2 Mbit/s and the data-frames are transmitted at 11 Mbit/s.

3.4   Numerical Results

This section numerically validates the fluid model as an accurate description of the ad-hoc network scenario of Section 2.2. The validation consists of a comparison of i) detailed simulations of ad-hoc network scenario of Section 2.1, ii) simulation of the fluid-flow model of Section 2.2, and iii) the analytical results of Section 3.2.

The graphs of Figure 2 present the mean buffer workload $\mathbb{E}W_{\text{buffer}}$ at the bottleneck node for an arbitrary packet (left) and last packet of a flow $\mathbb{E}W_{\text{buffer}}^*$ (right). The graphs present three curves: ad-hoc network scenario simulations, fluid-model simulations, and fluid-model analysis. In both graphs it can be seen that the three curves more or less coincide. Only for loads close to the saturation load, the results are less accurate due to the imprecision of the estimated capacity C. Overall the curves indicates that the fluid model accurately describes the ad-hoc network scenario and that the analytical results of Section 3.2 are also very good. Further, it can be observed that the buffer occupancy seen by the last particle is only slightly higher than the buffer occupancy upon flow arrival; the relatively short flow transfer time and low number of active source nodes result in a minor increase of the buffer during the flow transfer time.

Figure 3 presents the results for the mean buffer delay $\mathbb{E}D_{\text{buffer}}^*$ of the last packet (left) and the mean overall flow transfer time $\mathbb{E}D_{\text{overall}}$ (right). Note that the analytically obtained buffer delay of the last particle in the left graph is based on an approximation (cf. (2)). The fluid model captures the behavior of an IEEE 802.11B bottleneck very well as the model reflects both the impact of the load and flow-size distribution, except for high loads the results are less accurate.

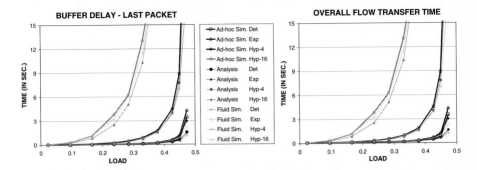

Fig. 3. Left: Mean buffer delay of the last packet/particle ($\mathbb{E}D^*_{\text{buffer}}$). Right: Mean overall flow transfer time ($\mathbb{E}D_{\text{overall}}$).

By comparing the graphs it is seen that the mean overall transfer time is almost completely determined by the buffer delay at the bottleneck node.

4   Benefits of Unequal Resource Sharing

The previous section studied the model where the system capacity is equally shared amongst the active source nodes and the bottleneck node (i.e., $m_n = 1$), this section studies the effects of assigning a different share of the system capacity to the bottleneck node (i.e., $m_n \neq 1$). The objective is to reduce the *overall* flow transfer time D_{overall}, which is the sum of the delays D_{source} and D^*_{buffer} (cf. (1)), by optimizing over m_n. Obviously the optimization is a trade-off: by granting a larger share of the capacity to the bottleneck node D^*_{buffer} reduces while D_{source} increases. We investigate the impact of the resource sharing ratio m_n.

Mathematical analysis of the fluid model with unequal resource sharing, i.e., $m_n > 1$, is significantly harder than for the model where $m_n = 1$. This is essentially due to the fact that in case $m_n > 1$ the resource sharing between source nodes and bottleneck node is not completely determined by the number of active source nodes (as is the case when $m_n = 1$), but also depends on the buffer workload W_{buffer} at the bottleneck node (cf. Section 2.2). Therefore, numerical results of the fluid model for $m_n > 1$ are obtained by simulations of the fluid model.

Figure 4 illustrates the impact of the resource sharing ratio m_n. The results are obtained by simulations of the fluid model and m_n is independently of n. The left graph presents the mean overall flow transfer time $\mathbb{E}D_{\text{overall}}$ for different m_n and loads, the right graph illustrates the trade-off between D_{source} and D^*_{buffer} for a given load (here chosen 0.43). When m_n increases, it becomes less probable that $W_{\text{buffer}} > 0$, and the bottleneck node will mostly obtain a share of $c/2$. Hence, there is hardly any queueing at the bottleneck node. From the right graph we conclude that resource sharing ratio $m_n = \infty$, i.e., always granting a share of $c/2$ to the bottleneck node, is optimal for the overall flow transfer time.

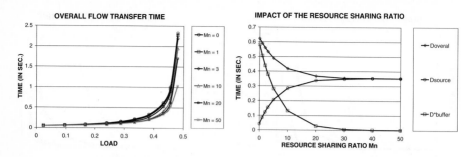

Fig. 4. Impact of resource sharing on $\mathbb{E}D_{\text{overall}}$. Left: $\mathbb{E}D_{\text{overall}}$ for different values of m_n. Right: trade-off between $\mathbb{E}D_{\text{source}}$ and $\mathbb{E}D^*_{\text{buffer}}$.

A special case is the resource sharing ratio m_n is *never* exceeded by number of active source nodes n. This occurs e.g., for very large m_n or when call admission control is applied to the maximum number of active source nodes n_{max} and $n_{\text{max}} \leq m_n$. The source nodes behave as a PROCESSOR SHARING model with service capacity $\text{C}/2$, flow arrival rate λ and mean flow size f; hence, the mean flow transfer delay is given by (independent of the flow-size distribution) $2f/(\text{C}(1 - 2\rho))$.

5   Validation of the Fluid Model for IEEE 802.11E

This section presents a validation of the fluid model for the case of unequal resource sharing. Unequal resource sharing can be achieved by an appropriate setting of the *differentiating parameters* in an IEEE 802.11E ad-hoc network. First, in Section 5.1 the IEEE 802.11E parameters are mapped onto the fluid model parameters, in particular, the capacity C_n and resource sharing ratio m_n. In Section 5.2 we validate our modeling approach by detailed ad-hoc network simulations.

5.1   Mapping of IEEE 802.11E Parameters

IEEE 802.11E provides four "differentiating parameters" (cf. Section 2.3), namely CW_{min}, CW_{max}, AIFS, and $\text{TXOP}_{\text{limit}}$. Unfortunately, the mapping of the IEEE 802.11E parameters onto the fluid-model parameters C_n and m_n is not self-evident, see e.g. [11].

In case the bottleneck node is *saturated*, i.e., $\{n \geq m_n\} \vee \{W_{\text{buffer}} > 0\}$, the fluid model parameters C_n and m_n can be estimated from an extension of the model of Bianchi [2] to two classes with different settings for the differentiating parameters, see e.g. [13]. In particular, the resource sharing in case of n active source nodes can be obtained from the model of [13] with n nodes in one class and a single node (representing our bottleneck node) in the other class. The

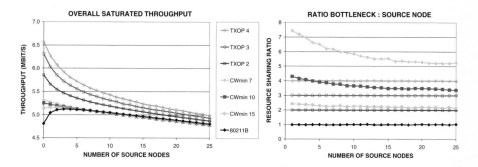

Fig. 5. Varying number of source nodes and a bottleneck node with a varying parameter setting. Left: Overall saturated throughput. Right: Resource sharing ratio.

parameters c_n and m_n for $\{n \geq m_n\} \vee \{W_{\text{buffer}} > 0\}$ are estimated by respectively the aggregate throughput and the ratio of the per node throughput in the two classes.

In case of a *non-saturated* bottleneck node, i.e., $\{n < m_n\} \wedge \{W_{\text{buffer}} = 0\}$, the above-mentioned approach would overestimate both the non-saturated capacity c'_n and resource sharing ratio m'_n due to the assumption that all nodes are saturated. Observe, the non-saturated resource sharing ratio m'_n equals n as this resource sharing ratio couples the input rate into the bottleneck node to the output rate (cf. Section 2.2). Next, the non-saturated capacity c'_n is estimated as follows: we consider the same differentiating parameter and its value is set such that it provides for the desired resource sharing ratio m'_n in the model of [13], the corresponding capacity c_n is used as an estimation of c'_n. For example, when we use differentiating parameter $\text{TXOP}_{\text{limit}} = 3$ and $\{n = 2\} \wedge \{W_{\text{buffer}} = 0\}$, then the resource sharing ratio m'_2 equals 2; therefore c'_2 is estimated by c_2 which is the saturated capacity for $\text{TXOP}_{\text{limit}} = 2$.

Figure 5 shows the saturated throughput (left graph) and the resource sharing ratio (right graph) as a function of the number of source nodes. First, we vary the value of CW_{min} at the bottleneck node; all other parameters of the bottleneck node and all parameters of the source nodes are set according to the IEEE 802.11B standard. Then we do a similar experiment in which we vary the $\text{TXOP}_{\text{limit}}$ of the bottleneck node, while all other parameters are set according to IEEE 802.11B. The left graph illustrates that higher overall throughputs are obtained by an IEEE 802.11E bottleneck node, especially for parameter $\text{TXOP}_{\text{limit}}$. In the right graph the throughput ratios for parameter $\text{TXOP}_{\text{limit}}$ are trivial, the ratios for parameter CW_{min} are examples of non-trivial resource sharing as, intuitively, the throughput ratio for CW_{min} is the inverse of the CW_{min} parameter-setting ratio. For example, when CW_{min} of the bottleneck node is set to 7, then the expected ratio is $31/7 \approx 4$, but the realized ratio is larger than 6 for a small number of active source nodes.

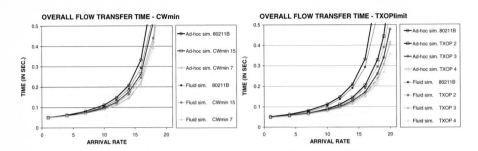

Fig. 6. System simulations and fluid-model simulations of the overall flow transfer time. Bottleneck node with varying parameter setting. Left: CW$_{min}$. Right: TXOP$_{limit}$.

5.2   Numerical Results

In the present section the fluid model for unequal resource sharing is numerically validated by ad-hoc network simulations. The validation scenario is as in Section 3.3 and the experiments coincide with those of the previous section: one of the parameters CW$_{min}$ or TXOP$_{limit}$ of the bottleneck node is varied, all other parameters of the bottleneck node and all parameters of the source nodes are set according to the IEEE 802.11B standard.

For fluid-flow simulations, of which results are presented in Figure 6, we use c_n and m_n estimated by the ad-hoc network simulator. The reason is that the fluid model is very sensitive for the used capacity if the offered load is close to the available capacity, cf. Section 3.4. Bianchi's model is proven to be accurate and the differences between results of the model and the ad-hoc network scenario simulations are small (just a few percent), but this approach ensures that deviations between the fluid model and the ad-hoc network scenario are solely due to fluid-modeling assumptions.

Figure 6 displays a comparison of ad-hoc network scenario simulations and fluid-model simulations. The fluid model simulations slightly underestimate the ad-hoc network scenario simulation results, but the behavior of the differentiating parameters is captured fairly well. The small deviations can be the result of modeling assumptions, e.g., in the fluid-model we assume that c_n and m_n are instantly valid after the number of active source nodes has changed. By slightly modifying the parameter values, i.e., a minimal reduction of c_n, the results coincide. We conclude that the fluid model of Section 2.2 accurately describes the behavior of unequal resource sharing in an IEEE 802.11E ad-hoc network.

6   Concluding Remarks and Directions for Further Research

In this paper we have shown that the fluid-flow model is an accurate description of multi-hop flows relayed by a performance bottleneck node in a wireless ad-hoc network. We have indicated how to map the parameter settings of both

IEEE 802.11B and IEEE 802.11E bottleneck nodes onto the fluid model, and the validation proves that the fluid modeling is accurate for both types of bottleneck nodes. It is also shown that the overall flow transfer time decreases by granting a larger share of the resource capacity to the bottleneck node, which can be obtained by an IEEE 802.11E bottleneck node.

Topics for further research include:

- Mathematical analysis of the fluid model with unequal resource sharing, i.e., $m_n > 1$. This model is significantly harder to analyze than the model with $m_n = 1$, where the resource sharing is entirely determined by the number of active sources; in case $m_n > 1$ also the buffer content of the bottleneck node has to be taken into account.
- Alternative service disciplines at the bottleneck node. In the above analysis it is assumed that the packet scheduling at the bottleneck node is FIRST COME FIRST SERVE. Alternative service disciplines, e.g. round robin, may yield considerably smaller mean overall flow transfer times.
- Investigation of the influence of higher-layer protocols, such as TCP, on the flow transfer time.
- A proper implementation of the unequal resource sharing policy $m_n = \infty$. Currently, obstacles for implementation are the lack of global knowledge of which nodes currently are bottlenecks and the absence of practical parameter settings to provide the desired resource sharing. A possible implementation is that each node is assigned an infinite TXOP$_{\text{limit}}$ for all *relay packets*, i.e., a node sends all packets that it has to relay for other nodes in a single TXOP.

Acknowledgements

This work has been carried out partly in the SENTER-NOVEM funded project EASY WIRELESS and the Dutch BSIK/BRICKS project.

References

1. van den Berg, J.L., Mandjes, M.R.H., Roijers, F.: Performance modeling of a Bottleneck Node in an IEEE 802.11 Ad-hoc Network. In: Proceedings of AdHoc-Now, LNCS 4104, pp. 321–336 (2006) Available as cwi-report at http://ftp.cwi.nl/CWIreports/PNA/PNA-E0607.pdf
2. Bianchi, G.: Performance analysis of the IEEE 802.11 Distributed Coordination function. ieee Journal on Selected Areas in Communications 18, 535–547 (2000)
3. Coffman jr., E.G., Muntz, R.R., Trotter, H.: Waiting time distributions for Processor-Sharing systems. Journal of the ACM 17, 123–130 (1970)
4. Cohen, J.W.: The multiple phase service network with Generalized Processor Sharing. Acta informatica 12, 245–284 (1979)
5. Foh, C.H., Zukerman, M.: Performance analysis of the IEEE 802.11 mac protocol. In: Proceedings of European Wireless '02, Florence, Italy (2002)

6. Fu, Z., Zerfos, P., Luo, H., Lu, S., Zhang, L., Gerla, M.: The impact of multihop wireless channel on tcp throughput and loss. In: Proceedings of infocom '03, San Francisco, USA (2003)
7. He, J., Pung, H.K.: Fairness of medium access control for multi-hop ad hoc networks. Computer Networks 48, 867–890 (2005)
8. IEEE p802.11b/d7.0, Supplement: higher speed physical layer extension in the 2.4 ghz band (1999)
9. IEEE p802.11e-2005, Amendment 8: Medium Access Control (Mac) Quality of Service Enhancements (November 2005)
10. Litjens, R., Roijers, F., van den Berg, J.L., Boucherie, R.J., Fleuren, M.J.: Analysis of flow transfer times in IEEE 802.11 Wireless LANs. Annals of Telecommunications 59, 1407–1432 (2004)
11. Roijers, F., van den Berg, J.L., Fan, X., Fleuren, M.J.: A Performance Study on Service Integration in IEEE 802.11E Wireless LANs. Computer Communications 29, 2621–2633 (2006)
12. Tijms, H.C.: Stochastic models: an algorithmic approach. Wiley & Sons, Chichester (1994)
13. Zhao, J., Guo, Z., Zhang, Q., Zhu, W.: Performance Study of mac for Service Differentiation in IEEE 802.11. In: Proceeding of IEEE GLOBECOM, vol. 33, pp. 778–782 (2002)

An Upper Bound on Multi-hop Wireless Network Performance

Tom Coenen[1], Maurits de Graaf[1,2], and Richard J. Boucherie[1]

[1]Department of Applied Mathematics,
Stochastic Operations Research Group,
University of Twente, The Netherlands
[2]Thales Division Land & Joint Systems,
Huizen, The Netherlands

Abstract. Given a placement of wireless nodes in space and a traffic demand between pairs of nodes, can these traffic demands be supported by the resulting network? A key issue for answering this question is wireless interference between neighbouring nodes including self interference along multi-hop paths. This paper presents a generic model for sustainable network load in a multi-hop wireless network under interference constraints, and recasts this model into a multicommodity flow problem with interference constraints. Using Farkas' Lemma, we obtain a necessary and sufficient condition for feasibility of this multicommodity flow problem, leading to a tight upper bound on network throughput. Our results are illustrated by examples.

Keywords: Wireless multi-hop network, interference, multicommodity flow problem, cut condition.

1 Introduction

Interference is an important aspect of wireless networks that seriously affects the capacity of the network. This is especially so in a wireless multi-hop network, where a transmission on one link interferes with transmissions on links in the vicinity. On a multi-hop path self-interference may result in substantial degradation of end-to-end network performance. In this respect, due to interaction among hops, multi-hop wireless networks considerably differ from wired networks thus calling for new modelling and analysis techniques that take into account interference constraints.

In the absence of interference, but including capacity constraints on the transmission rate of nodes, feasibility of a set of traffic demands between pairs of nodes can be determined by considering the flow allocation in the network as a multicommodity flow problem. The network is modelled as a graph, with the vertices representing the nodes where traffic is originated, terminated or forwarded. There is an edge between vertices if the corresponding nodes are within each others transmission range. Each edge has a capacity, representing the throughput that is possible over that edge. The multicommodity flow problem then addresses the

L. Mason, T. Drwiega, and J. Yan (Eds.): ITC 2007, LNCS 4516, pp. 335–347, 2007.

question whether there exists a set of paths and real numbers (fractions) so that: (1) for each traffic demand, there is a set of paths from the traffic source to the destination; (2) fractions of the traffic demand can be allocated to each path so that for each source-destination pair the traffic demand is realized and (3) the capacity constraints are taken into account.

This paper considers a generic wireless network configuration and traffic load specified via parameters such as nodes, transmission and interference ranges, as well as the traffic matrix indicating the demands between source nodes and sink nodes. We make no assumptions about the homogeneity of nodes with regard to transmission range or interference range, nor the capacity of the links. This is in contrast to previous work [3] that has focussed on asymptotic bounds under assumptions such as node homogeneity and random communication patterns. In [6] a conflict graph is used to address the problem of finding a feasible flow allocation to realize demands between pairs of nodes. While the conflict graph provides a more comprehensive modelling of the scheduling problem, it is also more complicated to deal with. A detailed discussion on the relation of our work with [6] is presented in Section 4, showing that we obtain a tight upper bound for an example provided in [6]. In [7] the question of a routing algorithm to find paths satisfying the traffic demands in a distributed setting is addressed. For an LP relaxation of the interference problem, [9] presents necessary conditions for link flow feasibility. This yields an upper bound similar to that of [6]. In addition [9] introduces an edge colouring problem in which each colour at an edge represents a time slot for transmission. This problem is solved using a FPTAS, yielding a lower bound for the link flow allocation. In [10] this work is extended to multi-radio and multi-channel networks. Our work does not solve any LP's, but provides a good characterization for the feasibility of the fractional multiflow problem with interference constraints which gives a fast way of finding upper bounds for a slightly different interference constraint setting.

In this paper we introduce a new approach to model interference in a carrier sensing multi-hop wireless network. To this end, we transform the sustainable load problem into a multicommodity flow problem that we extend with interference constraints. The main theorem of this paper states a condition for the feasibility of the multicommodity flow problem with interference constraints, given the demands between source and destination nodes. Using this theorem, we compute the maximal throughput between a single source and a single destination. We consider the following elements to be the key contribution of our work:

- The use of polyhedral combinatorics (Farkas' Lemma) to obtain a structural expression for feasibility of the multicommodity flow problem with interference constraints, in terms of a 'generalized cut condition' analogous to the 'max-flow min-cut' theorem of Ford and Fulkerson [1].
- The generality of our framework which incorporates the following realistic effects: the transmission range is not necessarily equal to the interference range, the network may consist of mixed wired and wireless connections and

wireless links have different capacities depending on distance, obstacles or transmission power.

The remainder of this paper is organized as follows. In Section 2 we introduce interference constraints to model the wireless parts of the network while taking the impact of the capacity of the links into account. Our main theorem is stated in Section 3, followed by examples and applications in Section 4. Section 5 concludes the paper.

2   Ad Hoc Interference Model

Ad-hoc networks use transmissions over a wireless channel to communicate between users. However, if multiple transmissions take place at the same time over the same wireless channel, transmissions may collide and the data will be lost. This interference limits the throughput of ad-hoc networks. Adopting the model of [5], we will model the interference constraints. To do so, we define the transmission range and the interference range of a node. When nodes in a wireless network want to communicate, they need to be close enough to receive each others signals. The *transmission range* of a node is the maximum distance from that node to where its received signal strength is insufficient for maintaining communications. Even though a signal may be too weak to be received correctly outside the transmission range, the signal can still cause interference preventing nodes from receiving other signals correctly. The *interference range* is the maximum distance from a node to where it prevents other nodes to maintain communications. Note that in general and in our model the transmission and interference range are not equal. In the following we adopt a graph representation (see e.g. [4],[6]) in which these ranges will be represented by arcs. Let V denote the set of nodes, A the set of arcs and let $\delta^+(u)$ denote the arcs leaving node u.

In a carrier sensing network, nodes within each others interference range will avoid transmitting at the same time. We will model this as follows: Let $R(v)$ denote the set of nodes within the interference range of node v (which includes v itself), that is: if one of the nodes in $R(v)\backslash\{v\}$ is transmitting, then v cannot receive other transmissions, nor can v transmit. Let $\rho(u)$ denote the fraction of time that a node u is transmitting, then

$$\sum_{u\in R(v)} \rho(u) \leq f(v)$$

where $f(v)$ denotes the interference capacity of the node. The interference capacity denotes the amount of interference a node can handle and still transmit data itself. For a wired network one could set $f(v) = \infty$, whereas $f(v) = 1$ ensures that no two nodes within each others interference range transmit at the same time. Taking $f(v) < 1$ can model a loss due to for instance non-ideal CSMA/CA effects.

Consider the set $I(v)$ of all arcs leaving v, entering v and leaving the nodes that are in v's interference range:

$$I(v) = \{a|a \in \delta^+(u), u \in R(v)\}. \tag{1}$$

It follows that for all the arcs that are within $I(v)$ the interference capacity $f(v)$ may not be exceeded. In particular, if $f(v) = 1$, simultaneous transmissions cannot take place over arcs $a_1, a_2 \in I(v)$. For later use, we also introduce here a dual notion of $I(v)$, viz. $J(a)$, the set of vertices that experience interference by a transmission over arc a:

$$J(a) = \{v \in V | a \in I(v)\}. \tag{2}$$

Consider node $v \in V$. The interference arc set $I(v)$ is denoted using bold arcs in Figure 1a and the set of nodes $J(a)$ affected by arc a by the grey nodes in Figure 1b.

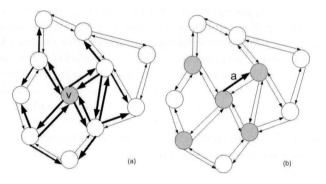

(a)   (b)

Fig. 1. $I(v)$ and $J(a)$

To each arc a a capacity $b(a) > 0$ is assigned. In actual networks, due to e.g. unequal distances among nodes or external disturbances such as noise, the link capacities may be different.

Consider a set of source and destination pairs $(r_1, s_1), ..., (r_k, s_k)$. When the net amount of flow between a source node r_i and a destination node s_i is d_i, we say that the value of the (r_i, s_i) flow is d_i. For an allocation, let $x_i(a)$ denote the amount of traffic for source destination pair (r_i, s_i) over link a. We will call this the flow of commodity i over arc a. To take the link capacities into account, we use the following interference constraints:

$$\sum_{i=1}^{k} \sum_{a \in I(v)} \frac{x_i(a)}{b(a)} \leq f(v) \quad \forall v \in V, \tag{3}$$

Note that the interference constraints indicate whether a node can receive correctly and that we assume that collisions are fatal, which is a worst case scenario as in practice caption may be possible. However, as we impose condition (3) on all nodes, for a transmission over link (u, v) to be successful, we have that (3) must hold for both u and v, so both sender and receiver must be free of interference. This closely resembles the behaviour of IEEE 802.11 under RTS-CTS, where both sender and receiver must be free of interference, see e.g. [11]. For a successful communication, the sender must be able to hear the link layer acknowledgement transmitted by the receiver.

3   The Extended Multicommodity Flow Problem

The multicommodity flow problem (MCFP) describes the problem of finding an allocation of flows over links such that all flows are transferred from their source to their destination, without exceeding the capacity of the links. The *multicommodity flow problem with interference constraints* is as follows.

Given a graph $G(V, A)$, with link capacities $b : A \to \mathbb{R}^+$, interference capacities $f : V \to \mathbb{R}^+$ and source and destination pairs $(r_1, s_1), ..., (r_k, s_k)$ with demands $d_1, ..., d_k \in \mathbb{R}^+$, find for each $i = 1, ..., k$ an (r_i, s_i) flow $x_i \in \mathbb{R}_+^{|A|}$ of value d_i, where $x_i(a)$ is the amount of traffic of commodity i sent via arc a, and so that for each arc $a \in A$ and vertex $v \in V$ the capacity and interference constraints are met. Let $\delta^+(U) = \{a = (u, v) \in A | u \in U, v \notin U\}$ and $\delta^-(U) = \{a = (u, v) \in A | u \notin U, v \in U\}$ so that $\delta^+(v)$ and $\delta^-(v)$ denote the arcs leaving and entering node v respectively. Our multicommodity flow problem with interference constraints has the following feasibility constraints:

$$\sum_{i=1}^{k} x_i(a) \leq b(a), \quad \forall a \in A \tag{4}$$

$$\sum_{a \in \delta^+(v)} x_i(a) = \sum_{a \in \delta^-(v)} x_i(a), \quad \forall v \in V, v \neq r_i, s_i \tag{5}$$

$$\sum_{a \in \delta^+(r_i)} x_i(a) - \sum_{a \in \delta^-(r_i)} x_i(a) = d_i, \quad \forall i \tag{6}$$

$$\sum_{a \in \delta^+(s_i)} x_i(a) - \sum_{a \in \delta^-(s_i)} x_i(a) = -d_i, \quad \forall i \tag{7}$$

$$\sum_{i=1}^{k} \sum_{a \in I(v)} \frac{x_i(a)}{b(a)} \leq f(v), \quad \forall v \in V \tag{8}$$

Equation (4) shows the capacity constraints on the arcs, equation (5) assures flow conservation, i.e. for each node the flow in must equal the flow out, and (6) and (7) define that the demands leaving the source and entering the destination. Note that (7) is redundant as it follows from (5) and (6), but is included here for completeness. Equation (8) is our interference constraint. Equations (4)-(7) define the multicommodity flow problem in its standard form, that is included in our formulation by setting the interference capacities of all nodes to infinity, that is $f(v) = \infty$ for all $v \in V$.

We now formulate a generalized cut condition for the multicommodity flow problem with interference constraints, so including (8). To this end, define length functions $l : A \to \mathbb{R}^+$ on all arcs and interference functions $w : V \to \mathbb{R}^+$ on all nodes.

For a given length function $l : A \to \mathbb{R}^+$ and interference function $w : V \to \mathbb{R}^+$, let $\mathbf{dist}_{l,w}(r_i, s_i)$ denote the distance function that incorporates both the length and interference, where the distance of a path is built up of the distance $\mathbf{q}_{l,w}(a)$ of the arcs in the path as follows

$$\mathbf{q}_{l,w}(a) = l(a) + \sum_{t \in J(a)} \frac{w(t)}{b(a)} \tag{9}$$

$$\mathbf{q}_{l,w}(P) = \sum_{a \in P} \mathbf{q}_{l,w}(a) \tag{10}$$

$$\mathbf{dist}_{l,w}(r,s) = \min_{P \in P_{r,s}} \mathbf{q}_{l,w}(P). \tag{11}$$

with $J(a)$ as in (2) and $P_{r,s}$ the set of all paths from r to s.

Theorem 1. *The multicommodity flow problem with interference constraints is feasible, if and only if for all length functions $l : A \to \mathbb{R}^+$ and node interference functions $w : V \to \mathbb{R}^+$ it holds that*

$$\sum_{i=1}^{k} d_i \mathbf{dist}_{l,w}(r_i, s_i) \leq \sum_{a \in A} l(a)b(a) + \sum_{v \in V} w(v)f(v). \tag{12}$$

The proof of Theorem 1 is given in the appendix. The interference function $w(v)$ is a dual variable that can be interpreted as the price paid for a unit of the interference capacity of a node v, which gives a weighted cut of the nodes by chosing the values of $w(v)$ to be zero or not (likewise the length function $l(a)$ can be interpreted as the price paid for a unit of capacity of an arc a, which also gives a weighted cut of the arcs). The distance of an arc is then the price paid for the capacity of that arc, together with the price paid for the interference capacity used by that arc for a flow unit. Note that our theorem is an extension of the cut condition for the multicommodity flow problem (without interference). This can be seen as follows. Setting all interference capacities to a very large value the condition of Theorem 1 reduces to

$$\sum_{i=1}^{k} d_i \mathbf{dist}_l(r_i, s_i) \leq \sum_{a \in A} l(a)b(a) \tag{13}$$

as the inequality only makes sense for $w = 0$. The cut condition by setting a length of 1 to all arcs in a cut and zero otherwise for the multicommodity flow problem states that

$$\sum_{a \in \delta^+(U)} b(a) \geq \sum_{\substack{r_i \in U \\ s_i \notin U}} d_i \ .$$

This cut condition is necessary for the existence of a solution, but not sufficient. The max-flow min-cut theorem states that if $k = 1$, for every network, there exists a flow (max-flow) for which the amount is equal to the total capacity of the smallest cut in the network (min-cut), see Ford and Fulkerson [1].

A direct consequence of Theorem 1 is that when there is only one commodity, a bound on the throughput d of the network can be determined by

$$d = \frac{\sum_{a \in A} l(a)b(a) + \sum_{v \in V} w(v)f(v)}{\mathbf{dist}_{l,w}(r,s)}.$$

When there are multiple commodities with demands $d_1, ..., d_k$, Theorem 1 can determine the maximal value $0 \leq \lambda \leq 1$ such that for all commodities a throughput of λd_i can be achieved using

$$\lambda = \frac{\sum_{a \in A} l(a)b(a) + \sum_{v \in V} w(v)f(v)}{\sum_{i=1}^{k} d_i \mathbf{dist}_{l,w}(r_i, s_i)}.$$

4 Examples

We can consider the network as depicted in Figure 2 as nodes in a network, connected by links using 802.11b with a maximum transmission rate of 11 Mbit/s, but with link 3 having a bad connection, due to distance or a disturbance, only reaching the minimal transmission rate of 5.5 Mbit/s.

We want to transmit data from node 1 to node 5. If we solve for the best solution without considering interference, it is clear that we can send at a speed of 5.5 Mbit/s, as this is limited by the slowest link. The interference constraints for the network with identical link capacities would imply that all links can be used one third of the time, leading to an overall throughput of 1.83 Mbit/s when considering the constraints separately.

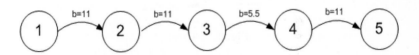

Fig. 2. Series of nodes with capacity constraints

For the example of Figure 2 we have the interference constraints

$$\frac{x(1)}{11} + \frac{x(2)}{11} + \frac{x(3)}{5.5} \leq 1$$
$$\frac{x(2)}{11} + \frac{x(3)}{5.5} + \frac{x(4)}{11} \leq 1.$$

From a direct solution of (4)-(8) it follows that $x(a) = 2.75$ is a feasible solution, higher than the earlier claimed 1.83 Mbit/s. Using Theorem 1, we find that

$$d_1\left(l(1) + l(2) + l(3) + l(4) + \frac{2w(1) + 4w(2) + 4w(3) + 3w(4) + w(5)}{11}\right) \leq$$
$$11l(1) + 11l(2) + 5.5l(3) + 11l(4) + w(1) + w(2) + w(3) + w(4) + w(5)$$

which gives by taking the cut $w(3) = 1$ and all other values (including $l(i)$) equal to zero

$$\frac{4}{11}d_1 \leq 1.$$

So $x(a) = 2.75$ is also the optimal solution. The value now found for $x(a)$ can be interpreted as the fraction of time a link is in use multiplied by the transmission rate of the link. This shows that links one, two and four get $\frac{1}{4}^{th}$ of the time, whereas link three gets $\frac{1}{2}$ of the time. So when considering four slots, link one, two and four each get one slot (where link one and four use the same slot!) and link three the other two. This way we have an accurate representation of the network incorporating both the capacity and interference constraints, together with the flow conservation laws.

We now consider the more sophisticated network used in [6] as depicted in Figure 3, where all arcs have capacity 1.

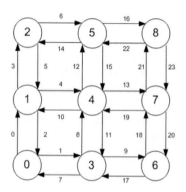

Fig. 3. 3x3 grid

The upper bound on the throughput from node 0 to node 8 for this network obtained in [6] is 0.667, opposed to the optimal 0.5, even though their algorithm has discovered all possible cliques in the conflict graph. Using Theorem 1 and taking the cut $w_1 = w_3 = 1$ or $w_4 = 1$ gives the lowest upper bound that can be achieved, resulting in $d_1 \leq 0.5$, which is tight. The reason we obtain a different result than in [6] is that we use constraints for all nodes, so that considering for example node 0 we have that $x(3) + x(9) \leq 1$, as both signals reach node 0. In the approach of Jain et al., arcs 3 and 9 are not connected in the conflict graph, as a simultaneous transmission over both arcs is possible.

There is an interesting relation between the approach presented in this paper and the results of [6]. Jain et al. use a *conflict graph* to determine lower and upper bounds for the throughput of the network. The conflict graph C has vertices corresponding to the arcs in the transmission graph, where there is an edge between two arcs if and only if the arcs are not allowed to transmit simultaneously. In our approach, C has as vertex set A, where there is an edge between a_i and a_j (for some $1 \leq i, j \leq |A|$) if and only if $\exists v \in V$ s.t. $a_i, a_j \in I(v)$. Note that $I(v)$ defines a clique in C for each v in V. In fact, our interference model adopted here resembles the *protocol model* of [6], but it is 'stronger' in the sense that for the same network, we have more interference constraints (edges in the conflict graph) than [6].

In [6] it is shown that a vector $x_i : A \to \mathbb{R}^+$ (corresponding to a flow i), can be scheduled without interference conflicts if and only if x_i lies in the stable set polytope of C. (The stable set polytope is the convex hull of the incidence vectors of the stable sets in the graph). It is well-known that the stable set polytope is contained in the *fractional* stable set polytope. (The fractional stable set polytope is defined by all constraints indicating that the total flow in a maximal clique in the conflict graph is at most 1.)

In this paper, instead of first defining the conflict graph C and then discovering its cliques, we directly formulate inequalities corresponding to the cliques $I(v)$ for all $v \in V$. The polytope defined by these inequalities will therefore contain the fractional stable set polytope. As a result, the upper bound obtained here cannot always be achieved using a flow allocation without interference conflicts.

5   Conclusion

In this paper we consider multi-hop wireless networks, for which we have stated a theorem giving a necessary and sufficient condition for the existence of a solution for the multicommodity flow problem with interference constraints, given a required throughput between nodes of the network. By extending the multicommodity flow problem with interference constraints, while still considering the capacity of the links, we have constructed an LP formulation for assigning flows to links to achieve the required throughput between certain nodes in the network. Using Farkas' Lemma, we have developed the corresponding general cut condition which gives an upper bound on the throughput that can be achieved in the network, given the source-destination pairs. The upper bounds are found by defining a cut, where the best cut will give a tight bound on the throughput, analogous to the max-flow min-cut theorem for the multicommodity flow problem. Several examples illustrate the use of the theorem in finding the optimal throughput for some simpler networks.

The applicability of this work for wireless multi-hop networks can for example be seen when designing a network that must be able to support a certain throughput for each user. Another possible application is for admission control, as our necessary and sufficient condition determines when adding a commodity will no longer result in a feasible solution using the same cutvector. If for any cutvector the condition does not hold, there exists no feasible solution for adding the commodity. If the cutvector however does not show infeasibility, this is no guarantee for feasibility as a different cutvector may give a tighter result for the situation with the added commodity. Due to the general nature of the theorem, any type of network with heterogeneous users and protocols can be modelled.

Acknowledgement. This research is partly supported by the Dutch Ministry of Economic Affairs under the Innovation Oriented Research Program (IOP-GenCom, QoS for Personal Networks at Home) and the ITEA Easy Wireless Project (grant: IS043014). This research is also supported by the NWO Casimir program 2005 (grant: 018.001.037).

References

[1] Ford, L.R., Fulkerson, D.R.: Maximal flow through a network. Canadian Journal of Mathematics 8, 399–404 (1956)

[2] Ford, L.R., Fulkerson, D.R.: Flows in Networks. Princeton University Press, Princeton, NJ (1962)

[3] Gupta, K., Kumar, P.R.: The capacity of wireless networks. IEEE Transactions on Information Theory 46(2), 388–404 (2000)

[4] Gupta, R., Musacchio, J., Walrand, J.: Sufficient rate constraints for QoS flows in ad-hoc networks, UCB/ERL Technical Memorandum M04/42 (2004)

[5] de Haan, R., Boucherie, R.J., van Ommeren, J-K.: Modelling multi-path signal interference in ad-hoc networks, Technical Memorandum, Dep. of Applied Mathematics, University of Twente (2006)

[6] Jain, K., et al.: Impact of interference on multihop wireless network performance. In: Proc. of ACM Mobicom 2003, pp. 66–80 (2003)

[7] Jia, Z., et al.: Bandwidth guaranteed routing for ad-hoc networks with interference constraints. In: Proc. of ISCC (2005)

[8] Johansson, T., Carr-Motyckova, L.: Reducing interference in ad-hoc networks through topology control. In: Proc. of DIALM-POMC '05, pp. 17–23 (2005)

[9] Kodialam, M., Nandagopal, T.: Characterizing achievable rates in multihop wireless networks: The joint routing and scheduling problem. In: Proc. of ACM Mobicom 2003, pp. 42–54 (2003)

[10] Kodialam, M., Nandagopal, T.: Characterizing the capacity region in multi-radio multi-channel wireless mesh networks. In: Proc. of ACM Mobicom (2005)

[11] Litjens, R., et al.: Analysis of flow transfer times in IEEE 802.11 wireless LANs. Annales de télécommunications 59, 1407–1432 (2004)

[12] von Rickenbach, P., et al.: A robust interference model for wireless ad-hoc networks. In: Proc. of IPDPS'05, vol. 13, p. 239.1 (2005)

[13] Schrijver, A.: Combinatorial Optimization: Polyhedra and efficiency. Springer, Berlin Heidelberg New York (2003)

Appendix

Proof of Theorem 1

Given a directed graph $G(V, A)$, arc capacities $b : A \to \mathbb{Q}^+$, node interference constraints $f : V \to \mathbb{Q}^+$, disjoint pairs $(r_1, s_1)...(r_k, s_k)$ and demands $d_1...d_k \in \mathbb{Q}^+$. The multicommodity flow problem with interference constraints can be written as an LP problem as follows:

Find $x = (x_1, ..., x_k)$ where $x_i : A \to \mathbb{Q}^+$ denotes the values of flow $r_i \to s_i$ assigned to an arc a s.t.

$$Ax \leq b$$
$$Cx = \hat{d}$$
$$Ex \leq f$$

for $A = m \times mk$, $C = nk \times mk$, $E = n \times mk$ where $m = |A|$ and $n = |V|$, with $b = m \times 1$ denoting the capacity constraints, $\hat{d} = (\hat{d}_1, ..., \hat{d}_k) = nk \times 1$ denoting the flow constraints and $f = n \times 1$ denoting the interference constraints with:

$$A = [I_m, I_m, ..., I_m]$$
$$C = \begin{bmatrix} M & 0 & 0 \\ 0 & ... & 0 \\ 0 & 0 & M \end{bmatrix}$$

where M is an $n \times m$ matrix defined by

$$M_{v,a} = \begin{cases} 1 & \text{when } a \text{ leaves } v \\ -1 & \text{when } a \text{ enters } v \\ 0 & \text{otherwise} \end{cases}$$

$$\hat{d}_i(v) = \begin{cases} d_i & \text{when } v \text{ is } r_i \\ -d_i & \text{when } v \text{ is } s_i \\ 0 & \text{otherwise} \end{cases}$$

$$E = [S, S, ..., S]$$

with S an $n \times m$ matrix defined by

$$S_{v,a} = \begin{cases} \frac{1}{b(a)} & \text{when } a \in I(v) \\ 0 & \text{otherwise} \end{cases}.$$

According to Farkas' Lemma there exists a solution $x \geq 0$ satisfying (4)-(8) iff for all vectors $y, w \geq 0$ and z, with $y \in \mathbb{R}^m$, $w = (w_1, ..., w_n) \in \mathbb{R}^n$ and $z = (z_1, ..., z_k) \in \mathbb{R}^{kn}$ where $z_i : V \to \mathbb{R}^n$:

$$yA + zC + wE \geq 0 \Rightarrow yb + z\hat{d} + wf \geq 0 \tag{14}$$

From the definitions of A, C, E we find

$$yA = [y, y, ..., y]$$
$$zC = [z_1 M, z_2 M, ..., z_k M]$$
$$wE = [wS, wS, ..., wS],$$

where $z_i M$ is given by

$$z_i M = \sum_{v \in V} z_i(v) M_{v,a} = [z_i(u_1) - z_i(v_1), ..., z_i(u_m) - z_i(v_m)]$$

with (u_j, v_j) denoting the starting and ending node of an arc a_j, and where the element of wS corresponding to arc a is given by

$$(wS)_a = \sum_{v \in V} w_v S_{v,a} = \sum_{\substack{v \in V \\ a \in I(v)}} \frac{w_v}{b(a)}.$$

As a consequence, (14) reads for all $y \geq 0, w \geq 0$ and $z_i \in \mathbb{R}^n$

$$z_i(v) - z_i(u) \leq y(a) + \sum_{t \in J(a)} \frac{w(t)}{b(a)}, \quad \forall i = 1...k, \forall a = (u,v) \in A \Rightarrow \quad (15)$$

$$\sum_{i=1}^{k} d_i(z_i(s_i) - z_i(r_i)) \leq \sum_{a \in A} y(a)b(a) + \sum_{v \in V} w(v)f(v),$$

We now show that there exists a feasible solution $x \geq 0$ if and only if for all length functions $l : a \to \mathbb{Q}^+$ and node interference functions $w : V \to \mathbb{R}^+$ it holds that

$$\sum_{i=1}^{k} d_i \mathbf{dist}_{l,w}(r_i, s_i) \leq \sum_{a \in A} l(a)b(a) + \sum_{v \in V} w(v)f(v). \quad (16)$$

Suppose there is a feasible solution, then (15) holds, now choose $l(a) = y(a)$ as the length function, and (for all i) $z_i(s) - z_i(r)$ as the distance between r and s, yielding

$$\sum_{i=1}^{k} d_i \mathbf{dist}_{l,w}(r_i, s_i) = \sum_{i=1}^{k} d_i(z_i(s_i) - z_i(r_i))$$

$$\leq \sum_{a \in A} l(a)b(a) + \sum_{v \in V} w(v)f(v)$$

Next suppose that (16) holds, we will now show that also (15) holds. Let the minimizing path use the arcs $(a_1...a_p)$, then

$$\sum_{a \in A} l(a)b(a) + \sum_{v \in V} w(v)f(v) \geq \sum_{i=1}^{k} d_i \mathbf{dist}_{l,w}(r_i, s_i) \quad (17)$$

$$= \sum_{i=1}^{k} d_i \sum_{j=1}^{p} \mathbf{q}_{l,w}(a_j)$$

$$= \sum_{i=1}^{k} d_i \left(\sum_{j=1}^{p} l(a_j) + \sum_{t \in J(a_j)} \frac{w(t)}{b(a_j)} \right)$$

Let $z_i : V \to \mathbb{R}$ be so that $z_i(v) - z_i(u) \leq y(a) + \sum_{t \in J(a)} \frac{w(t)}{b(a)}$, which in combination with $l(a) = y(a)$ gives that

$$\sum_{i=1}^{k} d_i \left(\sum_{j=1}^{p} l(a_j) + \sum_{t \in J(a_j)} \frac{w(t)}{b(a_j)} \right) \geq \sum_{i=1}^{k} d_i \sum_{j=1}^{p} z_i(v_j) - z_i(u_j)$$

where $a_j = (u_j, v_j)$ and $u_1 = r_i$, $v_j = u_{j+1}$ and $v_p = s_i$ so that the right hand side of the expression simplifies to

$$\sum_{i=1}^{k} d_i \sum_{j=1}^{p} z_i(v_j) - z_i(u_j) = \sum_{i=1}^{k} d_i(z_i(s_i) - z_i(r_i))$$

which taken together with (17) gives

$$\sum_{a \in A} l(a)b(a) + \sum_{v \in V} w(v)f(v) \geq \sum_{i=1}^{k} d_i(z_i(s_i) - z_i(r_i))$$

completing the proof.

On the Use of Accounting Data for QoS-Aware IP Network Planning

Alan Davy, Dmitri Botvich, and Brendan Jennings

Telecommunications Software & Systems Group,
Waterford Institute of Technology, Cork Road, Waterford, Ireland
{adavy, dbotvich, bjennings}@tssg.org

Abstract. We present an economically efficient framework for provision of essential input for QoS-aware IP network planning. Firstly, we define a process for reuse of network accounting data for construction of a QoS-aware network demand matrix. Secondly, we define a process for estimation of QoS-related effective bandwidth coefficients from packet traces collected per traffic classe. Taken together, these processes provide the necessary input required to plan a network in accordance with QoS constraints. We present results of a sensitivity analysis of the demand estimation process, and of an economic analysis of the relative merit of deployment of our approach in comparison to a traditional direct measurement-based approach. We conclude that although there is a degree of inaccuracy in our network demand estimation process this inaccuracy is within acceptable bounds, and that this is offset by the potential for significant cost reductions for the ISP.

Keywords: Network Planning, Accounting, Effective Bandwidth, QoS.

1 Introduction

Network planning typically involves the use of dedicated network monitoring hardware to gather and collate large amounts of network traffic data, which is then analysed to identify an optimal network configuration design reflecting estimated demand and specified Quality of Service (QoS) requirements. Use of dedicated hardware means that this approach is relatively expensive, incurring costs in hardware procurement and maintenance, in addition to significant training/operational costs. In previous work [1] we presented an initial QoS-aware network planning approach based on using pre-existing network accounting data as an alternative source of input. In this paper we augment that work by providing a detailed sensitivity evaluation of our approach. We propose and evaluate a light-weight process for estimation of QoS aware effective bandwidth coefficients, which can be used to construct a QoS-aware demand matrix, providing the necessary input for QoS-aware network planning. In addition, we offer an economic analysis comparing the cost of deploying a traditional direct measurement based system to our approach.

The paper is organised as follows: §2 discusses related work in the areas of demand matrix and effective bandwidth estimation. §3 describes our proposed network planning architecture. §4 presents our process for estimation of QoS aware

L. Mason, T. Drwiega, and J. Yan (Eds.): ITC 2007, LNCS 4516, pp. 348–360, 2007.
© Springer-Verlag Berlin Heidelberg 2007

demand matrices from accounting records, and §5 details our packet trace analysis process for calculating approximate QoS-related effective bandwidth coefficients. We present an experimental evaluation of our network demand estimation process, describing a number of sensitivity scenarios and results in §6. §7 provides an economic analysis of our method in comparison to traditional network planning architectures. Finally, §8 summarises our findings and outlines areas for future work.

2 Related Work

The network planning process generally requires three sources of input [2]: (1) attributes associated with the current traffic demands on the network which collectively specify their behavioural characteristics; (2) attributes associated with resource constraints on the network topology; and (3) a constraint based routing definition framework which plans routing of traffic subject to (1) and (2). Here we focus on how to estimate, with an acceptable degree of inaccuracy, attributes relating to the the current traffic demands on the network.

A traditional network planning process depends on direct network monitoring systems to collect large amounts of traffic data from the network – data that is then analysed to ascertain behavioural characteristics inherent within that traffic [3]. In contrast, we propose reuse of network accounting records for demand estimation. More specifically, this data will be processed and used to construct a network demand matrix, which, as shown in [4, 5] is an effective method of representing network demand. Both [4] and [5] propose methods of calculating the demand matrix from network flow records. We take this approach, extending it for QoS-awareness by generating a per traffic class network demand matrix.

The second element of our approach is the estimation of effective bandwidth coefficients of traffic carried over the network, which, when used in conjunction with per traffic class demand matrices facilitate QoS-aware network planning. Effective bandwidth is defined as the minimum amount of bandwidth required by a traffic stream to maintain specified QoS related targets [6]. There are many methods of theoretically estimating this value for different traffic models (for example [7, 8]); we adopt an empirical process based on packet trace analysis that is similar to those outlined in [9, 10].

3 Proposed Network Planning Architecture

The architecture illustrated in Fig. 1 extends the traditional usage based network accounting architecture to facilitate network demand and effective bandwidth estimation processes.

We reuse network accounting records to estimate a demand matrix, populated by the mean per traffic class demand between all edge node pairs. To support this estimation process we require configuration of additional mediation business logic – essentially rules outlining, how the information we require is to be extracted from raw metering records. For effective bandwidth coefficient estimation a selective traffic

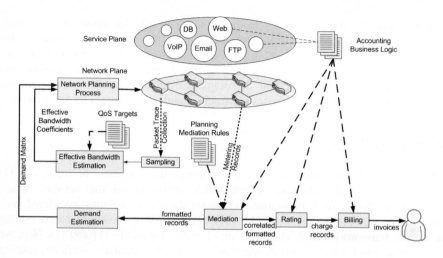

Fig. 1. Network Accounting / QoS-aware Planning Architecture

trace collection component is employed; this component collects data for a relatively small number of representative traffic traces over relatively short time frames which is then analysed offline to estimate effective bandwidth values.

4   QoS-Aware Demand Matrix Estimation

To estimate the demand matrix we use as input mediated accounting records generated by the Mediation component of the Accounting infrastructure (see Fig. 1.). We assume that each record received by the demand estimation component contains all required information relating to a single traffic flow in the network, specifically: source node, destination node, start time, end time, traffic class, and volume of traffic transferred. Hence, we use the term "flow records" to refer to these mediated records. We assume that all traffic transferred across the network is accounted for by such records, and that they are generated in real-time, i.e. once a flow terminates the associated accounting record is immediately transferred to the Mediation component (this assumption is valid for the majority of IP-based networks).

The Demand Estimation component constructs a demand matrix for a given measurement interval. Based on received flow records it estimates the total traffic, per traffic class, per source, destination pair for that interval. Of course some flows will have either started before and/or ended after the interval in question, so the flow records must be analysed to estimate the amount of traffic that occurred within the interval. For a given flow record there are four possibilities: (1) the flow *started before* the interval and *ended during* the interval; (2) the flow *started during* the interval and *ended after* the interval; (3) the flow *started during* and *ended during* the interval; and (4) the flow *started before* and *ended after* the interval. Note that for (2) and (4) we assume that the demand estimation takes place a sufficient time after the end of the interval that all flow records relating to flows that were active during the interval will have been received by the Demand Estimation component.

To calculate the traffic volume a flow contributes during the measurement interval we make the simplifying assumption that for all flows the packet arrival distribution during the flow duration is uniform with a constant inter-arrival time. Therefore the contributed traffic volume can be calculated as the total traffic volume associated with the flow multiplied by the fraction of the total flow duration that fell within the measurement interval. Of course, for real applications the packet inter-arrival distribution is rarely uniform, so this assumption will inevitably lead to inaccuracy. The simulation study described in §6 below quantifies this inaccuracy for a number of representative scenarios, in order to ascertain whether it is within acceptable bounds for the purposes of network planning.

Once traffic volumes associated with all flows active within the measurement interval are estimated, all volumes are summed on a per traffic class, per source/destination pair basis. These values are then used to populate an interim demand matrix for the network for the given measurement interval. Finally, the 95th percentile of the range of means for the measurement interval is taken, and these values are used to populate the final demand matrix.

5   Estimation of Effective Bandwidth Coefficients

Effective bandwidth is defined as the minimum amount of bandwidth required by a traffic stream to maintain specified QoS related targets [6]. Effective bandwidth can be defined for different types of QoS targets including delay, loss or both delay and loss targets together. Our method can be applied to all types of QoS targets but in this paper for simplicity we are interested in satisfaction of delay targets only. Delay targets specify both the nominal maximum delay experienced on the network and the proportion of traffic that is allowed exceed this maximum delay. A typical example of a delay target is (50ms, 0.001) which means that only 0.1% of traffic is allowed to be delayed by more than 50ms. Differing delay targets will lead to different effective bandwidths, with more stringent delay targets resulting in higher coefficients.

Our effective bandwidth estimation algorithm is defined as follows. Let $delay_{max}$ be the nominal maximum delay and let p_{delay} be the percentage of traffic which can exhibit delay greater than $delay_{max}$. Specifically we define effective bandwidth R_{eff} of a traffic source for delay QoS target $(delay_{max}, p_{delay})$ as the minimal rate R such that if we simulate a FIFO queue with unlimited buffer and processing rate R, the percentage of traffic which will exhibit delay greater than $delay_{max}$ will be less than p_{delay}. To estimate the effective bandwidth of a particular traffic source on the network, we take a recorded packet trace of that source. We observe that if we simulate a FIFO queue (initially assumed to be empty) with the same inputted traffic trace for different queue rates $R_1 > R_2$ and estimate the percentages p_1 and p_2 of traffic delayed more than $delay_{max}$ for different rates respectively, then $p_1 \leq p_2$. This means that the percentage of traffic, p, delayed more than $delay_{max}$ is a monotonically decreasing function of processing rate R. Using this observation it is straightforward to design, for example, a simple bisection algorithm for a recorded packet trace to find the minimal value of a queue rate such that the percentage of traffic delayed more than $delay_{max}$ is less than p_{delay}.

The effective bandwidth coefficients are estimated as follows. We assume that the QoS delay target ($delay_{max}$, p_{delay}) is fixed per traffic class. We take a large number of recorded traffic traces of more or less the same duration T_{max}. The choice of T_{max} is important. If T_{max} is too large or too small, the estimated "worst case" ratio of effective bandwidth to the mean rate may be inaccurate. Typically, T_{max} is chosen between 1 and 10 minutes. So suppose we have N traffic traces. For the ith traffic trace we estimate both effective bandwidth $R_{eff,i}$ and its mean rate $mean_i$ and calculate the ratio k_i of effective bandwidth to the mean rate as follows:

$$k_i = \frac{R_{eff,i}}{mean_i}$$

We note that the effective bandwidth is always larger or equal to the mean rate. Further considering a set of N ratios $\{k_1,..., k_N\}$ we first exclude any k_i with too small mean rate $mean_i$ using some appropriate threshold value and then we calculate "worst case" ratio of effective bandwidth to the mean rate K_{95} as the 95th percentile of left ratios k_i. K_{95} values are the effective bandwidth coefficient and are subsequently multiplied by the associated estimated demand values to populate the QoS-aware demand matrix.

6   Simulation Study

We use OPNET [11] to simulate the network topology depicted in Fig. 2 below, in which an ISP offers services to a number of customer groups over a DiffServ-enabled IP network. Each node is connected by 10Mbps Ethernet links. Customers access five service types: Web browsing, Email, Database access, Video-on-Demand (VoD) and Voice-over-IP (VoIP). We use the standard OPNET application models [12] to model the characteristics of traffic generated by users accessing these services. These traffic

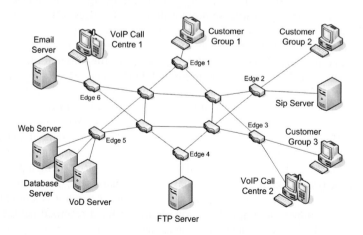

Fig. 2. Simulated ISP network topology

models are parameterised to model typical user behaviour in the work place. Web and Email are assigned to the Diffserv Best Effort (BE) class, Database to Assured Forwarding (AF), and VoD/VoIP to Expedited Forwarding (EF). All simulation runs were for a simulated duration of 2.5 hours, with data collected from time t = 1 hour to t = 2 hours. Three traffic patterns were set up for the collection of results, varying the proportion of QoS traffic per traffic pattern. This ensures our sensitivity tests are less biased to any particular proportional mix of QoS traffic.

6.1 Impact of Deterministic Sampling on Demand Estimation Accuracy

Due to the large number of traffic flows in typical IP network environments accounting systems employed by ISPs often employ deterministic sampling [13, 14] as a means of reducing the metering record collection overhead. Mediation components then use the sampled records to estimate the total traffic volume between source/destination nodes. To investigate the impact of such an approach on demand estimation we implement deterministic sampling in our simulation model.

We first look at the case of how a single sampling interval affects estimation of network demand from accounting data. We use a fixed deterministic sampling interval of 300 for collecting accounting data. A total of 12 simulations are executed varying the simulation seed value of each run to vary the results slightly. We calculate the edge to edge relative error demand from accounting records and directly measured records for each edge node pair per traffic class. Fig. 3a shows the distribution of global error calculated from these set of simulation runs. The mean of this distribution is 3.84, with a standard deviation of 2.7, suggesting global error in some cases can be as large as 10%. For the purpose of network planning we believe this is an acceptable level of error, as future network demands are in any case likely to differ from current demand levels.

Using the same simulation seed we now vary the sampling interval and calculate the relative error in estimation of network demand across the set of simulation runs. Fig. 3b illustrates the effect of varying the sampling interval on estimation of network demand on a per traffic class basis for particular edge node pairs in the network. We see that stream based traffic such as VoIP and VoD withstands coarser grained

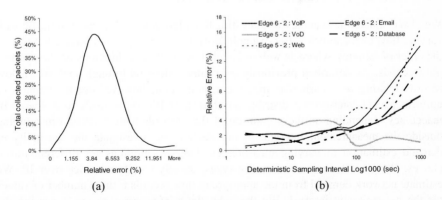

Fig. 3. Deterministic sampling of metering records; (a) Global relative error distribution, sampling 1 in every 300 records; (b) Relative error for different traffic types

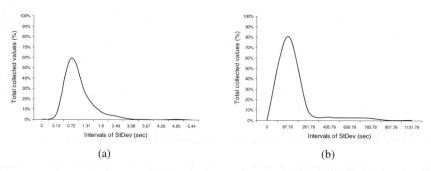

Fig. 4. Flow duration distribution of (a) email traffic and (b) VoIP traffic

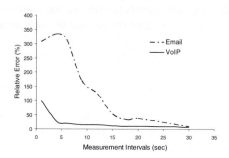

Fig. 5. Sensitivity to varying Flow Measurement Intervals

sampling as in general these traffic types have longer flow durations (as illustrated in Fig. 4.). Estimation of network demand for traffic such as web and email requires a smaller sampling interval, as a higher portion of their flows may not be detected with coarser grained sampling.

6.2   Sensitivity to Varying Measurement Interval

We also evaluate the accuracy of our estimation algorithm with respect to traffic with varying flow characteristics. For example VoIP flows will typically have durations on the order of minutes, where as web or email flows will have durations on the order of milliseconds. As described previously we estimate demand through analysis of flow records relating to a particular measurement interval, however our assumption of uniform packet inter-arrival distributions for all traffic types leads to a degree of inaccuracy, depending on the proportion of flow records with start/end times falling outside the measurement interval. In this section we investigate the sensitivity of demand estimation to varying measurement interval durations for different traffic classes. Here, we focus on two traffic types, namely email and voice over IP. We estimate network demand from the appropriate flow records using a number of values for the measurement interval. We then plot the relative error so see if varying the measurement interval affects estimation of network demand.

Table 1. Effective Bandwidth Coefficients from analysis of 1000 packet traces per traffic class

Traffic Class	QoS Target	Effective Bandwidth Coefficient	Standard Deviation
Expedited Forwarding	(10ms, 0.0001)	2.21	0.34
Assured Forwarding	(50ms, 0.001)	4.55	0.46
Best Effort	(100ms, 0.005)	3.87	0.62

Fig. 4 shows the distribution of flow durations for the relevant traffic class; both distributions show a heavy tail, thus a significant number of flow durations will be many times the mean. Fig. 5 shows that for small measurement intervals, Email yields a very large degree of relative error; VoIP on the other hand maintains a low level of relative error even at small sample intervals. This demonstrates that the measurement interval must be selected in cognisance of the prevailing flow duration distribution of network traffic.

6.3 Estimation of Effective Bandwidth Coefficients

To demonstrate our effective bandwidth estimation approach we take 1000 packet traces per traffic class from an ingress node on the simulated network. We calculate the effective bandwidth coefficient for each packet trace collected and find the 95th percentile of these range of values per traffic class. Within our simulation we specify a QoS Delay target for each traffic class, as specified in Table 1 above. If a higher percentage of traffic breaches this limit the QoS target is not being met within the network. The results in Table 1 illustrate certain characteristics of the different traffic classes. As EF traffic is generally streaming traffic, such as VoIP or video on demand (VoD), fluctuation in network demand is usually low over the course of a service session. We can see by the estimated effective bandwidth coefficient that EF traffic in our simulation does not fluctuate far from its mean network demand. Thus using this coefficient to reserve 2.21 times the mean EF traffic demand from edge to edge should ensure all EF traffic will maintain its specified QoS target.

On the other hand, both AF and BE traffic show signs of greater fluctuation in network demand as their estimated effective bandwidth coefficients are higher. We assigned database traffic as AF within this simulation; Database transactions commonly vary in size and duration, leading to fluctuations in network demand. We assigned Web and Email traffic as BE; both of these services also commonly show a high degree of network demand fluctuation. The main reason we require to reserve more bandwidth for these types of traffic, with respect to their mean network demand, is because delay within the network is caused by unexpected peaks in network

Table 2. Current operational cost per network

	Small ISP	Medium ISP	Large ISP
Network Nodes	5 edge, 3 core	20 edge 5 core	45 edge, 15 core
Support Costs	€20,000	€60,000	€200,000
DB Licence Fee	€20,000	€60,000	€200,000
Software Licence Fees	€20,000	€60,000	€200,000

demand, thus flooding queues with traffic and incurring delay. If we reserve bandwidth with this in mind, by using the effective bandwidth coefficients, we can reduce delay with the network per traffic class to the limit specified within the QoS delay target specified.

7 Economic Analysis

One of the central contentions of this paper is that the proposed network accounting-based planning approach will be significantly more cost effective then a traditional direct measurement-based approach. To support this contention we now present the results of a high level comparative economic analysis of the two approaches for representative small, medium and large sized ISPs. Specifically we compare the costs of extending an existing network accounting system to implement our approach to QoS-aware network planning with that of a traditional direct measurement-based system operating independently of the network accounting system.

7.1 Baseline Cost Assumptions

For the currently deployed network accounting system the ISP is required to pay a number of fees, most significantly database system licences and accounting system software licences. The ISP also has to pay for customer support fees for each of these software systems; in general these fees are directly proportional to the licence fees. Furthermore, ISPs often incur hardware related costs, for example hardware storage space rental, hardware-specific support costs, and costs associated with the ISP's replication policy. For simplicity, we assume here that the ISPs do not pay rental on hardware space, and that they do not implement a replication policy.

Given the above assumptions Table 2 outlines indicative costs for small/medium/large ISPs relating to their network accounting system. Of course, license fees are typically kept confidential, so the values we choose are based on anecdote. However these costs do indicate that costs grow as does the size of the ISP, which we believe is universally true. Note that our cost model is relatively simplistic, for example we disregard costs such as loss in revenue based on down time, data migration, operating system licensing.

Key to our proposed approach is a significant upgrade of the network accounting system. For any significant system upgrade all software, hardware and support costs are likely to increase; we assume that these costs cumulatively increase by 20%. In addition to the upgrade, specialised contractors must be hired for tasks such as installation, staff training, and on-site support. We assume a contractor charges a flat rate of €1,500 per day, and that he/she can work at a rate of one installation per day. However, as the network size increases, so too does the complexity of the installation; thus, we assume that for our medium and large sized networks, contractor fees raise to €1,750 and €2,000 per day, respectively. Finally, to support new functionality without degrading system performance, new hardware will typically be purchased and deployed; we assume that purchase and deployment of a single hardware platform is €5,000.

Table 3. Cost of accounting system upgrade

	Small	Medium	Large
Support costs	€4,000	€12,000	€40,000
DB licence fees	€0	€0	€0
Software licence fees	€4,000	€12,000	€40,000
Server costs	1 x €5,000	2 x €5,000	4 x €5,000
Network monitor costs	1 x €3,000	2 x €3,000	4 x €3,000
Contractor fees	2 x €1,500	4 x €1,750	8 x €2,000
Total Cost	€19,000	€47,000	€128,000

Table 4. Cost of network monitoring system deployment

	Small	Medium	Large
Support Costs	€20,000	€60,000	€200,000
DB licence fees	€4,000	€12,000	€40,000
Software licence fees	€20,000	€60,000	€200,000
Server costs	1 x €5,000	2 x €5,000	4 x €5,000
Network monitor costs	5 x €3,000	20 x €3,000	45 x €3,000
Contractor fees	6 x €1,500	22 x €1,750	49 x €2,000
Total Costs	€73,000	€240,500	€693,000

As well as upgrading the accounting system for network demand estimation our approach also requires use of a limited number of network monitoring devices for collection of traffic traces used in the effective bandwidth estimation process. For a small ISP we assume only 1 such device is required, for a medium sized ISP we assume 2, and for the large ISP we assume 4. Currently such devices cost approximately €3,000 each. In contrast for the measurement-based network planning approach network monitoring devices must be deployed permanently at all edge routers in the network, resulting in a significant cost overhead for larger ISPs.

7.2 Comparative Cost Analysis

Based on the cost assumptions outlined above we can now evaluate the cost of the network accounting and direct measurement-based approaches. We focus on how these deployments affect cost of customer support fees, software licence fees, specialised contractor fees, and hardware fees.

Firstly, we addresses the costs incurred in implementing our network accounting based approach. Additional servers will be required to host the new upgrades to the accounting system. As accounting system records are already stored within a deployed database system, there is no need to upgrade the database system. Licence fees and support costs will be increased by 20% as these are upgrade to an existing system. Depending on network size between 1 and 4 network monitoring devices must be purchased, and specialised contractors will be required to configure and install them. Table 3 shows the cost of an upgrade for the three ISP types.

Secondly we address the costs incurred in implementing a traditional direct measurement-based approach. The network monitoring system will include the installation of a larger number of network monitoring devices, each monitoring traffic at a single edge node. The deployment will require an extension to the existing

Table 5. Comparison of deployment cost

	Small	Medium	Large
QoS Network Planning based on Accounting Data	€19,000	€47,000	€128,000
QoS Network Planning based on Direct Measurement	€73,000	€240,500	€693,000
Difference	€54,000	€193,500	€565,000
(%Savings)	(74%)	(80%)	(82%)

database server, as a larger amount of new data will be collect and stored for subsequent analysis; hence database licence fees will increase by 20%. Additional servers will be required to host the network monitoring services and applications, which will themselves, incur new licence and support fees. Finally, installation of the new system and hardware will require specialised contractors. Table 4 outlines the cost on the ISP to this approach.

Based on our outlined set of assumptions, Table 5 shows a clear difference in the cost of deploying both approaches for network of different sizes, with the network accounting approach incurring significantly less costs, particularly as network size increases. This is a result of the greater level of reuse of existing systems in the network-accounting approach and the requirement for installation of significantly more hardware in the direct measurement-based approach.

8   Conclusions and Future Work

As ISPs grow it becomes increasingly important for them to plan their networks in a manner that minimises the likelihood of customer service level agreements being violated as the number of customers and the volume of generated traffic increases. In this paper we presented a framework for estimation of a demand matrix and QoS coefficients to be used as input into a QoS-aware network planning process. The approach is based on extension of pre-existing network accounting systems and, as such, is intended as a more economically beneficial solution then traditional direct measurement-based approaches.

Results of a sensitivity analysis of our approach presented in the paper revealed that a degree of accuracy in the estimation of network demand is lost, due both to sampling approaches adopted by accounting systems, and to the fact that accounting systems provide traffic information on a per flow, not on a per transmitted packet basis. We showed that for a representative scenario the degree of inaccuracy is in the range of ±10%, which we believe is sufficient for network planning purposes. However, the degree of inaccuracy is dependent on the type of traffic transferred across the network, so the system must be appropriately parameterised based on the prevailing traffic mix. Furthermore, a high level economic analysis indicates that our approach will be considerably less costly than an approach based on direct measurement, and that the cost disparity increases as the network grows in size. The main reasons for this disparity are reduced expenditure on network monitoring hardware purchase/maintenance and savings due to reuse of existing network accounting systems.

Future work includes an analysis of the accuracy of our estimation of effective bandwidth of a particular traffic type over a period of time to that of directly measured effective bandwidth approaches. An economic analysis comparing both approaches will also be prepared. We also intend to evaluate planning based on input from our approach to that of direct measurements approaches of estimating network demand and effective bandwidth.

Acknowledgement

This work has received support from Science Foundation Ireland via the Autonomic Management of Communications Networks and Services programme (grant no. 04/IN3/I4040C) and the 2005 Research Frontiers project Accounting for Dynamically Composed Services (grant no. CMS006).

References

1. Davy, A., Botvich, D.D., Jennings, B.: An Efficient Process for Estimation of Network Demand for QoS- Aware IP Network Planning. In: Parr, G., Malone, D., ÓFoghlú, M. (eds.) IPOM 2006. LNCS, vol. 4268, Springer, Heidelberg (2006) ISBN 978-3-540-47701-3
2. Awduche, D., Malcolm, J., Agogbua, J., O'Dell, M., McManus, J.: Requirements for Traffic Engineering over MPLS. Informational RFC, last accessed 27/02/2007, http://www.ietf.org/rfc/rfc2702.txt
3. Awduche, D., Chiu, A., Elwalid, A., Widjaja, I., Xiao, X.: Overview and Principles of Internet Traffic Engineering. Informational RFC 3272, last accessed 04/10/2006, http://www.ietf.org/rfc/rfc3272.txt
4. Feldman, A., Greenberg, A., Lund, C., Reingold, N., Rexford, J., True, F.: Deriving traffic demands for operational IP networks: methodology and experience. IEEE/ACM Transactions on Networking (ToN) 9(3), 265–280 (2001)
5. Papagiannaki, K., Tatf, N., Lakhina, A.: A Distributed Approach to Measure Traffic Matrices. In: Proceedings of the 4th ACM SIGCOMM Conference on Internet Measurement, pp. 161–174. ACM Press, New York, USA (2004) ISBN 1-58113-821-0
6. Kelly, F.: Notes on Effective Bandwidth. In: Kelly, F.P., Zachary, S., Ziedins, I.B. (eds.) in Stochastic Networks: Theory and Application. Royal Statistical Society Lecture Notes Series, vol. 4, pp. 141–168. Oxford University Press, Oxford (1996) ISBN 0-19-852399-8
7. Kesidis, G., Walrand, J., Chang, C.S.: Effective Bandwidths for Multiclass Markov Fluids and Other ATM Sources. IEEE Transactions on Networking 1(4), 1063–6692 (1993) ISSN 1063-6692
8. Spitler, S.L., Lee, D.C.: Integrating effective-bandwidth-based QoS routing and best effort routing. IEEE INFOCOM 2003, The. Conference on Computer Communications 2, 1446–1455 (2003) ISBN 0-7803-7752-4, ISSN 0743-166X
9. Botvich, D.D., Duffield, N.: Large deviations, the shape of the loss curve, and economies of scale in large multiplexers. In: Queueing Systems, vol, pp. 1573–9443. Springer, Heidelberg (1995) ISSN 1573-9443
10. Baras, J.S., Liu, N.X.: Measurement and simulation based effective bandwidth estimation. IEEE Global Telecommunications Conference GLOBECOM'04 4, 2108–2112 (2004) ISBN 0-7803-8794-5

11. OPNET Technologies, Inc., OPNET Modeler TM Product Overview, last accessed 27/02/2007, http://www.opnet.com/products/modeler/home.html
12. OPNET Technologies, Inc., OPNET ModelerTM, Discrete Event Simulation Model Library, last updated 27/02/2007, http://www.opnet.com/products/library/des_model_library.html
13. Claffy, K.C., Polyzos, G.C., Braun, H.-W.: Application of Sampling Methodologies to Network Traffic Characterization. In: Proceedings of ACM SIGCOMM'93, vol. 23(4), pp. 194–203. ACM Press, New York, USA (1993) ISBN 0-89791-619-0
14. Duffield, N.A: Framework for Packet Selection and Reporting. IETF Internet Draft, last accessed 27/02/2007, http://www.ietf.org/internet-drafts/draft-ietf-psamp-framework-10.txt

Quantification of Quality of Experience
for Edge-Based Applications

Tobias Hoßfeld[1], Phuoc Tran-Gia[1], and Markus Fiedler[2]

[1] University of Würzburg, Institute of Computer Science, Department of Distributed Systems,
Würzburg, Germany
{hossfeld,trangia}@informatik.uni-wuerzburg.de
[2] Blekinge Institute of Technology, Department of Telecommunication Systems,
Karlskrona, Sweden
markus.fiedler@bth.se

Abstract. In future Internet, multi-network services correspond to a new paradigm that intelligence in network control is gradually moved to the edge of the network. As a consequence, the application itself can influence or determine the amount of consumed bandwidth. Thus the user behaviour may change dramatically. This impacts the Quality of Service (QoS) and the Quality of Experience (QoE), a subjective measure from the user perspective of the overall value of the provided service or application. A selfish user or application tries to maximize its own QoE rather than to optimize the network QoS, in contrast to a legacy altruistic user.

In this paper we present the IQX hypothesis which assumes an exponential functional relationship between QoE and QoS. This contribution is a first step towards the quantification of the QoE for edge-based applications, where an example of VoIP is taken into account. Starting from a measurement of the Skype application, we show the basic properties of selfish and altruistic user behaviour in accordance to edge-based intelligence. The QoE is quantified in terms of MOS in dependence of the packet loss of the end-to-end connection, whereby Skype's iLBC voice codec is used exemplarily. It is shown that the IQX hypothesis is verified in this application scenario. Furthermore, selfish user behaviour with replicated sending of voice datagrams is investigated with respect to the obtained QoE of a single user. In addition, the impact of this user behaviour on congestion in the network is outlined by means of simulations.

1 Introduction

In future telecommunication systems, we observe an increasing diversity of access networks and the fixed to mobile convergence (FMC) between wireline and wireless networks. This implies an increasingly heterogeneous networking environment for networked applications and services. The separation of transport services and applications or services leads to *multi-network services*, i.e., a future service has to work transparently to the underlying network infrastructure. For such multi-network services, the Internet Protocol is the smallest common denominator. Still, roaming users expect theses services to work in a satisfactory way regardless of the current access technology such as WLAN, UMTS, WiMAX, etc. Thus, a true multi-network service must be able

L. Mason, T. Drwiega, and J. Yan (Eds.): ITC 2007, LNCS 4516, pp. 361–373, 2007.

to adapt itself to its "surroundings" to a much stronger degree than what is supported by the TCP/IP protocol suite.

Streaming multimedia applications for example face the problem that their predominant transport protocol UDP does not take any feedback from the network into account. Consequently, any quality control and adaptation has to be applied by the application itself at the edge of the network. Prominent examples of *edge-based applications* applying edge-to-edge control are peer-to-peer (P2P) applications such as eDonkey or BitTorrent, Skype VoIP, YouTube, etc. The network providers have to cope with the fact that these edge-based applications dynamically determine the amount of consumed bandwidth. In particular, applications such as Skype do their own network quality measurements and react to quality changes in order to keep their users satisfied. The edge-based intelligence is established via traffic control on application layer. Traffic engineering in future Internet has to consider this new paradigm.

The shift of the control intelligence to the edge is accompanied with the fact that the observed user behaviour changes. A user can appear altruistic or selfish. Selfish user behaviour means that the user or the application tries to maximize the user-perceived *Quality of Experience QoE* rather than to optimize the network *Quality of Service QoS*. Very often the selfish behaviour is implemented in the software downloaded by the user without his explicit notice. In contrast, altruistic users, whose behaviour is instructed by network provider traffic control protocols (like TCP) help to maximize the overall system performance in a fair manner. In the case of file-sharing platforms, an altruistic user is willing to upload data to other users, while a selfish user only wants to download without contributing to the network. For voice over IP (VoIP), altruistic users would reduce the consumed bandwidth in the case of facing congestion, while selfish users would continuously try to achieve a high goodput and QoE, no matter of consequences for other users.

User satisfaction with application and service performance in communication networks has attracted increased attention during the recent years. The notion of QoE was introduced in several white papers [1,2,3,4], mostly in the context of multimedia delivery like IPTV. Besides of objective end-to-end QoS parameters, QoE focuses on subjective valuations of service delivery by the end users. It addresses *(a)* service reliability comprising service availability, accessibility, access time and continuity, and *(b)* service comfort comprising session quality, ease of use and level of support [2]. The necessity of introducing QoE can be explained on the example of VoIP. A voice user is not interested in knowing performance measures like packet loss or received throughput, but mainly in the experienced speech quality and timeliness of the connection.

There is however a lack of quantitative descriptions or exact definitions of QoE. One particular difficulty consists in matching subjective quality perception to objective, measurable QoS parameters. Subjective quality is amongst others expressed through *Mean Opinion Scores* (MOS) [5]. Links between MOS and QoS parameters exist predominately for packetised voice such as VoIP. Numerous studies have performed measurements to quantify the effect of individual impairments on the speech quality to a single MOS value for different codecs, for example G.729 [6], GSM-FR [7], iLBC used by Skype [8], or a comparison of some codecs [9]. Additionally, the E-model [10] and extensions [11] exist that assess the combined effects of different influence factors on the voice quality. In [12], the logarithmic function is selected as generic function for mapping the QoE from a single parameter because of its mathematical characteristics.

This work, in contrast, motivates a fundamental relationship between the QoE and quality impairment factors such as packet loss and related jitter. An exponential solution is derived for the Interdependency of **Q**oE and **Q**oS hypothesis, referred to as IQX . This contribution is a first step towards the quantification of the QoE for edge-based applications, where an example of VoIP is taken into account.

The rest of this paper is organized as follows. Section 2 introduces multi-network services and the emerging of edge-based intelligence. Starting from a measurement of the Skype application, we show the basic properties of selfish and altruistic user behaviour due to edge-based intelligence in Section 3. This is realized among others by an adaptive bandwidth control triggered by QoE. Section 4 starts with the quantification of the QoE of a VoIP application. We discuss the IQX hypothesis and the exponential functional relationship between QoE and QoS. It is exemplarily verified in Section 4.1 in terms of MOS depending on the packet loss of the end-to-end connection, whereby the iLBC codec as used by Skype is taken. We assume that the selfish users of the VoIP application utilize replication of voice datagrams to maximize their QoE, while the altruistic users change to a codec with a lower quality to consume less bandwidth. As a result, the benefit of the replication is investigated from a single user's point of view in Section 4.2. The impact of this selfish user behaviour on the network congestion is briefly illustrated in Section 4.3. Finally, Section 5 summarizes this paper.

2 Edge-Based Intelligence and Quality of Experience

From traffic engineering viewpoint, the shift of intelligence to the edge is accompanied by a number of changes:

- Change of user behaviour and traffic profile: edge-based services (like Skype) perform QoS measurements itself and adapt the traffic process according to the perceived QoS (packet blocking probability or jitter). The traffic change of those applications could be quite selfish, i.e. it tries to maximize its own QoE no matter of the network overload condition.
- Change from Multi-service Networks to Multi-Networks Services: An edge-based application could use many networks with different technologies in parallel, raising the question which network has to maintain which portion of the agreed QoS. From this perspective, the QoE will be the major criterion for the subscriber of a service.
- Higher Dynamic of Network Topology: an edge-based application is often controlled by an overlay network, which can change rapidly in size and structure as new nodes can leave or join the overlay network in an distributed manner.

Multi-network services will be often customer originated services. Together with the edge-based intelligence, the change of bandwidth demand and consumption is observed which only depends on the user behaviour and the used software of that service. The bandwidth demand is no longer under control of the network provider. A good example for this paradigm change is illustrated by the huge amount of traffic for P2P file-sharing [13] compared to web traffic.

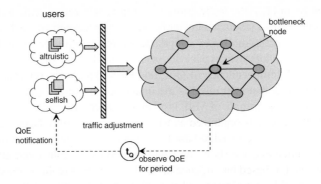

Fig. 1. Quality assessment mechanisms for QoE

However, the multi-network service has to maintain a certain QoE for each user. As a consequence, the edge-based application is responsible

(a) to evaluate the QoE at the end user's site and
(b) to react properly on the performance degradation, i.e., that the application adapts itself to the current network situation to maintain the QoE.

Figure 1 illustrates the QoE control scheme of such a multi-network service. Users are connected to each other via the corresponding access technologies. The QoE is assessed during a period t_Q of time. Accordingly, the altruistic users and the self-ish users react on feedback obtained from measurements. In this paper we observe the Skype VoIP service in more detail as an example for a service with edge-based intelligence. This example shows the change in user behaviour and bandwidth demand and discusses the QoE adaptation scheme, i.e. the way Skype reacts to keep the QoE.

3   Measurement of Skype VoIP

Skype is a proprietary VoIP application which is based on P2P technology. It offers rapid access to a large base of users, seamless service operation across different types of networks (wireline and wireless) with an acceptable voice quality [8], and a distributed and cost-efficient operation of a new service. The voice quality of the Skype service is achieved by using appropriate voice codecs, such as iSAC and iLBC [14], and by adapting the sender traffic rate according to the current packet loss and jitter of the end-to-end connection. The latter one is referred to as *QoE adaptation* in the following.

This QoE adaptation can be illustrated by a measurement study presented [15]. The general measurement setup is the following: Skype user A sends audio data to Skype user B. We used an English spoken text without noise of length 51 seconds, a sample rate of 8 kHz, encoded with 16 bits per sample which is a standard audio file for evaluating VoIP and available at [16]. The wav-file is played in a loop with a pause of 9 seconds in between using the Winamp audio player on machine A. The output of Winamp is used as input for Skype (instead of a microphone). On sender A and receiver B, Windows XP is the OS, Skype 2.0.0.81 (February, 2006) is installed and a packet trace is captured with TCPDump on each machine. In order to emulate various network conditions on the link between machine A and machine B, we use the Nistnet software [17]. Nistnet is installed on a separate machine with three network interfaces and operates as gateway for A and B and to the Internet, cf. Figure 2. With this measurement setup, both Skype user A and B have access to the Internet (which is required for using this service), while packet loss is only emulated on the direct connection from A to B. Here, Skype encodes audio with the iSAC codec due to the used hardware. If the power of the machines is below 600 MHz, Skype will use the iLBC codec.

Figure 3 shows the reaction of the Skype software on packet loss. Every 30 ms, a packet is sent from user A to user B (with a measured standard deviation of 6.65 ms). The measured packet loss ratio on the right y-axis denotes how many packet got lost, whereby we used the average for a window size of 6 s. On the left y-axis, the average size of the voice packets on application layer is plotted in bit. Again, we used a window size of 6 s corresponding to 200 voice packets. First the Skype call is established between user A and B and we start with no packet loss. The size of a packet varies between 90 bit and 190 bit with a measured average of 150 bit. It has to be noted that the oscillations of the packet size derive from the measurement setup. During the pause interval, Skype sends still packets, but only with a size of 50 Bit.

After 5 minutes the packet loss probability is increased about 5% every two minutes, until the packet loss probability reaches 30%. The time interval of two minutes was chosen to ensure that Skype reacts to changes. We have found out that Skype needs about one minute to change e.g. a voice codec. As we can see in Figure 3, Skype reacts on the experienced QoE degradation in terms of packet loss by increasing the packet size, whereas still every 30 ms a packet is sent. The size mainly ranges between 240 bit and 320 bit with an average of 280 bit. In contrast to before, the packet size is nearly doubled. This means that Skype sends now redundant information within every voice packets while experiencing packet loss in order to maintain the QoE. However, as a certain threshold is exceeded (here: about 20% packet loss), the packet size is decreased again and with 125 bit on average smaller than in the beginning. This indicates a change in the used voice codec. As soon as the packet loss probability is decreased again and falls below a certain threshold, the sender rate is again adapted by changing the packet size. In [15], we have also shown that Skype even does rerouting on application layer if the packet loss or the round trip time on the direct end-to-end connection is too high.

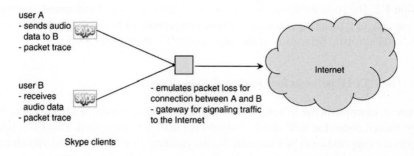

Fig. 2. Measurement setup for a Skype call

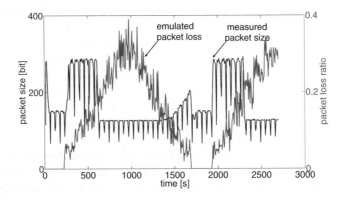

Fig. 3. Measurement of Skype's QoE adaptation on changes in the end-to-end link

This measurement points out that edge-based applications try in fact to keep the QoE above an acceptable threshold. In the case of Skype, this is done by adapting the amount of consumed bandwidth. If the receiver's application detects packet loss, it instructs the sender to increase the bandwidth. For a VoIP call, this is easily possible, since the connection is full duplex and the connection from user B to user A is used to send the feedback information. Here, a change of the bandwidth consumption and the user behaviour is observed. A user – or to be more precise, the application – behaves selfish to get the maximum QoE, irrespective of the network overload condition. This observation was the starting point for this study aiming at the estimation of the QoE.

4   Quantitative Observation of QoE

In this section we focus on a fundamental relationship between the QoE and quality impairment factors, like packet loss or jitter. As an analytical solution of the relationship between QoE and loss, we formulate the IQX hypothesis (exponential interdependency of QoE and QoS) in Section 4.1. A first verification of this hypothesis is done using real measurement of the iLBC codec. Regarding the single user's point of view, the benefit of replicating voice datagrams is analytically derived with respect to the QoE in Section 4.2. The costs for this achievement are a higher amount of consumed bandwidth and the risk of worsening potential network congestion. In Section 4.3, the impact of selfish and altruistic behaviour on the network itself is discussed.

4.1   The IQX Hypothesis for Quality of Experience

We use as example in the following the *Internet low bitrate codec iLBC* [18], which is a free speech codec for VoIP and is designed for narrow band speech. Two basic frame lengths are supported: (a) 304 bit each 20 ms, yielding 15.2 kbps, and (b) 400 bit each 30 ms, yielding 13.3 kbps, respectively. The latter is used in Skype when the CPU of the used machines is below 600 MHz [8].

We performed a measurement series in which the iLBC codec (b) is explicitly used. However, with a probability p_{loss} a packet gets lost on its way from user A to user B. We vary the packet loss probability from 0% to 90% in steps of 0.9%. The audio data as described in Section 2 is used as input speech file. At the receiver side, the audio stream is piped into an audio wav-file. Each experiment is repeated ten times, i.e. 1010 measurements were conducted.

In order to express the QoE of the VoIP call, the *Mean Opinion Score MOS* [5] is used. Therefore, the audio file sent is compared with the received wav-file using the Perceptual Evaluation of Speech Quality (PESQ) method described in ITU-T P.862 [19]. The resulting PESQ value can be mapped into a subjective MOS value according to ITU-T Recommendation ITU-T P.862.1 [20]. Figure 4 shows the obtained MOS values in dependence of the packet loss probability p_{loss} for the conducted experiments. The MOS can take the following values: (1) bad; (2) poor; (3) fair; (4) good; (5) excellent. Obviously, the higher the packet loss probability, the lower the MOS value is.

In general, the QoE is a function of n influence factors $I_j, 1 \leq j \leq n$:

$$QoE = \Phi(I_1, I_2, \cdots, I_n) . \tag{1}$$

However, in this contribution we focus on one factor indicating the QoS, the packet loss probability p_{loss}, in order to motivate the fundamental relationship between the QoE and an impairment factor corresponding to the QoS. Hence, the idea is to derive the functional relationship $QoE = f(p_{loss})$. In general, the subjective sensibility of the QoE is the more sensitive, the higher this experienced quality is. If the QoE is very high, a small disruption will decrease strongly the QoE, also stated in [12]. On the other hand, if the QoE is already low, a further disturbance is not perceived significantly. This relationship can be motivated when we compare with a restaurant quality of experience. If we dined in a five-star restaurant, a single spot on the clean white table cloth strongly disturbs the atmosphere. The same incident appears much less severe in a beer tavern.

On this background, we assume that the change of QoE depends on the current level of QoE – the expectation level – given the same amount of change of the QoS value. Mathematically, this relationship can be expressed in the following way. The

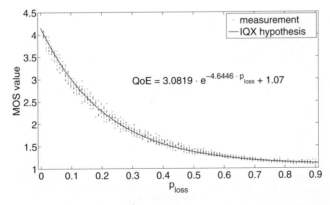

Fig. 4. Exponential estimation of QoE in dependence of packet loss probability p_{loss}

performance degradation of the QoE due to packet loss is $\frac{\partial QoE}{\partial p_{loss}}$. Assuming a linear dependence on the QoE level, we arrive at the following differential equation:

$$\frac{\partial QoE}{\partial p_{loss}} = -\tilde{\beta} \cdot (QoE - \gamma) . \tag{2}$$

The solution for this equation is easily found as an exponential function, which expresses the basic relation of the IQX hypothesis:

$$QoE = \alpha \cdot e^{-\beta \cdot p_{loss}} + \gamma . \tag{3}$$

For $p_{loss} \to 1$, the QoE in terms of MOS approaches its minimum of 1 from above. From the measured data, we obtain the following fit for iLBC voice codec (400 bits each 30 ms), following the IQX hypothesis:

$$QoE = 3.0829 \cdot e^{-4.6446 \cdot p_{loss}} + 1.07 . \tag{4}$$

It has to be noted that the packet loss is only one impairment factor indicating the QoS. For a general quantification of the QoE, additional factors like jitter have to be considered according to Eq. (1), which will be part of future work. Nevertheless, Eq. (4) will be used in the following section to derive analytically the impact of replication of voice datagrams on the QoE.

4.2   Impact of Replication of Voice Datagrams on QoE

Based on the experiences with Skype, we propose as one possibility the replication of voice datagrams to overcome a QoE degradation due to packet loss. Again, we consider the iLBC voice codec, as introduced in Section 4.1. This means that every $\Delta t = 30$ ms, a voice datagram of size $s_{voice} = 400$ bits is sent. A *replication degree* R means that the voice datagram is additionally sent in the following $R - 1$ packets. As a consequence, a packet contains now R voice datagrams with a total packet size of $s_{packet} = s_{header} + R \cdot s_{voice}$. The variable s_{header} denotes the overhead for each packet caused by TCP and IP headers (20 Byte + 20 Byte) and on link layer (e.g. 14 Byte for Ethernet). Hence, the required bandwidth is a linear function in R: $C_{req} = \frac{s_{header} + R \cdot s_{voice}}{\Delta t}$. The gain of this bandwidth consumption is the reduction of

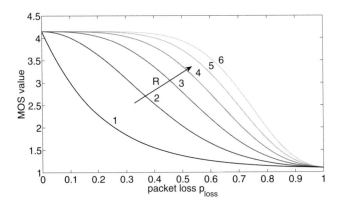

Fig. 5. QoE in dependence on the replication degree (w/o jitter)

the effective voice datagram loss probability $1 - p_{voice}$. For a given packet loss probability p_{loss} and a replication degree R , a voice datagram only gets lost if all R consecutive packets containing this voice datagram get lost. Thus, it holds

$$p_{voice} = 1 - p_{loss}^{R} \, . \tag{5}$$

The effect of the voice datagram replication can be seen in Figure 5 for a replication degree of $R = 1, \cdots, 6$. On the x-axis the packet loss probability p_{loss} is denoted. The QoE on the y-axis is computed according to Eq. (4) whereby the voice datagram probability in Eq. (5) is used. For $p_{loss} = 0.2$, the QoE is only 2.29 for $R = 1$. A replication degree of $R = 2$ and $R = 3$ leads to a QoE of 3.63 and 4.04 , respectively. This means the QoE could be improved from a poor quality to a good quality. A further increase of the replication degree only yields to a small gain as compared to the growth of the required bandwidth C_{req} .

Besides the increased bandwidth consumption, the replication also causes some jitter, as the voice datagrams are not received every $\Delta t = 30$ ms, but maybe in one of the $R - 1$ following packets. Next, we compute the probability $\tilde{y}(i)$ that a voice datagram is successfully transmitted in the i -th try, used to quantify the jitter.

$$\tilde{y}(i) = p_{loss}^{i-1} \cdot (1 - p_{loss}) \tag{6}$$

The probability that a voice packet is received follows as

$$p_{voice} = \sum_{i=1}^{R} \tilde{y}(i) = (1 - p_{loss}) + p_{loss}(1 - p_{loss}) + \cdots + p_{loss}^{R-1}(1 - p_{loss}), \tag{7}$$

which agrees with Eq. (5). The number Y of trials which is required to successfully transmit a voice datagram is a conditional random variable. It follows a shifted geometric distribution and is defined for $1 \le i \le R$:

$$Y \sim \frac{\mathrm{GEOM}_1(p_{loss})}{p_{voice}} \quad \text{with} \quad y(i) = \frac{\tilde{y}(i)}{p_{voice}} = \frac{p_{loss}^{i-1} \cdot (1 - p_{loss})}{1 - p_{loss}^{R}} \, . \tag{8}$$

We define the jitter σ to be the standard deviation $\sqrt{\mathrm{Var}[t_{rcvd}]}$ of the interarrival time of received packets, normalized by the average time Δt between any two sent

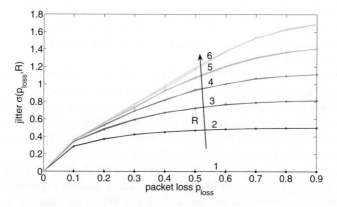

Fig. 6. Increase of jitter due to replication of voice datagrams

packets, $\sigma = \sqrt{\mathrm{Var}[t_{rcvd}]}/\Delta t$. For the sake of simplicity, we assume a deterministic inter packet sent time Δt and a deterministic delay $t_{s \to r}$ from the sender to the receiver. Then, the jitter can – after some algebraic transformations – be expressed as

$$\sigma = \frac{\sqrt{\mathrm{E}[t_{rcvd}^2] - \mathrm{E}[t_{rcvd}]^2}}{\Delta t} = \frac{\sqrt{\mathrm{E}[(Y\Delta t)^2] - \mathrm{E}[Y\Delta t]^2}}{\Delta t} = \sqrt{\mathrm{E}[Y^2] - \mathrm{E}[Y]^2}$$

$$= \sqrt{\frac{p_{loss}}{(p_{loss} - 1)^2} - \frac{p_{loss}{}^R \cdot R^2}{(p_{loss}{}^R - 1)^2}} . \tag{9}$$

Figure 6 shows the jitter σ for a replication degrees $1 \leq R \leq 6$ in dependence of the packet loss probability p_{loss} . Eq. (9) is an exact formula, which we also validated by implementing a simulation. The solid lines correspond to the analytical calculation of the jitter, while the solid lines with the dots as marker show the simulation results. Both curves agree and the confidence intervals are too small to be visible.

The cost of the voice datagram replication – beside the increased bandwidth consumption – is an increased jitter. But jitter also impacts the QoE and is of course one impairment factor in Eq. (1). As a result, a maximal degree R_{max} of replication exists and a further increase does not improve the QoE anymore. ITU-T G.114 recommends a latency of the end-to-end delay of 150 ms, referred to as toll quality, and a maximum tolerable latency of 400 ms. According to the end-to-end delay $t_{s \to r}$ and the inter packet sent time $\Delta t = 30$ ms, the following inequation has to hold

$$R \cdot \Delta t + t_{s \to r} < t_{max} \tag{10}$$

for a maximum allowed latency t_{max} . For example with $t_{max} = 200$ ms and $t_{s \to r} = 10$ ms, the maximum replication degree is limited by $R_{max} \leq 6$.

4.3   Network's Perspective for Edge-Based QoE Management

From the single user's point of view, the replication of voice data overcomes the degradation of packet loss and enables to keep a certain QoE. The cost for this achievement is a higher amount of consumed bandwidth. However, if the packet loss is caused by congestion in the network, this additionally required bandwidth worsens the network situation. We consider selfish and altruistic users which react on the perceived QoE. A single user measures the QoE during a period t_Q , the so called *QoE assessment period*. After each period t_Q , the user reacts on the obtained QoE value and adjusts the amount of consumed bandwidth, as illustrated in Figure 1. If the QoE is too low over some time, the user drops the call.

On one hand, the pure selfish user only looks on its own QoE which it tries to maximize by adjusting the throughput. This can be achieved *a)* by increasing the packet size by the replication degree R or *b)* by increasing the frequency of sending packets to $\frac{R}{\Delta t}$. On the other hand the altruistic user tries to minimize congestion in the network, i.e. the packet loss probability, in order to get a good QoE. Therefore, she uses a low-quality voice codec if packet loss, i.e. congestion, is detected.

In Figure 7, the consumed bandwidth over time of all altruistic and selfish users is considered in a congested system in which a bottleneck node of 110 kbps has to carry the traffic from six selfish and five altruistic users. While the altruistic users reduce

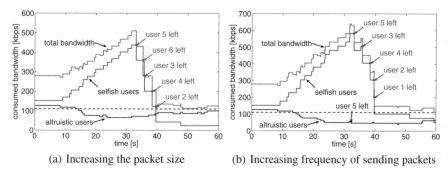

(a) Increasing the packet size   (b) Increasing frequency of sending packets

Fig. 7. Voice datagram replication to achieve a bit rate of sent audio information

their packet size, the selfish users increase the throughput. As a consequence, packets get dropped, the QoE decreases, and the users give up after some time.

In practice, however, we do not observe or at least expect that the selfish users will lose. First of all, an edge-based application would react more sensitive than discussed in this section. An important point is how the QoE is monitored and what are the optimal thresholds to react. In addition, there is different traffic traveling through the bottleneck. TCP traffic, e.g., will be pushed away by UDP traffic. In that case, the entire system behaviour will be changed. These aspects will be considered in future work.

5   Conclusions

Multi-network services with edge-based intelligence, like P2P file-sharing or the Skype VoIP service, impose a new control paradigm on future Internet: They adapt the amount of consumed bandwidth to reach different goals. A selfish behaviour tries to keep the Quality of Experience (QoE) of a single user above a certain threshold. Skype, for instance, repeats voice samples in view of end-to-end-perceived loss, which increases the consumed bandwidth. Altruistic behaviour, on the other side, would reduce the bandwidth consumption in such a case in order to release the pressure on the network and thus to optimize the overall network performance.

In order to study such behaviour, we first focus on the quantification of the QoE for edge-based applications as a function of network Quality of Service (QoS), where an example of VoIP is taken into account. The QoE is quantified in terms of MOS in dependence of the packet loss of the end-to-end connection, whereby the iLBC voice codec is used exemplarily. The IQX hypothesis (interdependency of QoE and QoS) is proposed and verified for packet loss as a QoS indicator. IQX assumes an exponential functional relationship between QoE and QoS: $QoE = \alpha \cdot e^{-\beta \cdot P_{loss}} + \gamma$.

The impact of the bandwidth adaptation on the QoE of a single user is then quantified. We consider a selfish user which replicates voice datagrams to overcome packet loss. The gain of this increased bandwidth consumption is the reduction of the effective voice datagram loss probability. The cost of the replication – beside the increased bandwidth consumption – is an increased jitter. The jitter also impacts the QoE. As a result, a maximal degree of replication can be derived up to which an increase of the QoE can be achieved. However, if the packet loss is caused by congestion in the network, this additionally required bandwidth worsens the network situation. Thus, we illustrated the impact of selfish and altruistic behaviour on the network itself by means of simulations. Summarizing, the emergence of edge-based applications and the resulting user behaviour open a new scientific field with a lot of challenges to be solved.

Acknowledgements

The authors would like to thank Andreas Binzenhöfer for his support and help in performing and evaluating the measurements; Dirk Staehle for his fruitful comments what to investigate (and what not to); Michael Menth for his help in modeling the system; Kurt Tutschku for the discussions on QoE; Prof. Helmke for his insights into control theory; and last, but not least, Michael Duelli and Melanie Brotzeller for performing the measurements.

References

1. Nortel Networks Coporation: Nortel and Microsoft Innovative Communications Alliance to transform business communications. White paper (2006) http://www.innovativecommunicationsalliance.com/collateral/
2. Nokia: Quality of experience (QoE) of mobile services: Can it be measured and improved? White paper (2006) http://www.nokia.com/NOKIA_COM_1/Operators/Downloads/Nokia_Services/Q oE_whitepaper_2006.pdf
3. Spirent: Delivering optimal quality of experience (QoE) for IPTV success. White paper (2006) www.spirentcom.com/documents/4002.pdf
4. Sandvine: Using network intelligence to provide carrier-grade VoIP. White paper (2006) www.sandvine.com/general/getfile.asp?FILEID=31
5. ITU-T Recommendation: ITU-T Rec. P.800.1: Mean Opinion Score (MOS) Terminology (2003)
6. Perkins, M.E., Evans, K., Pascal, D., Thorpe, L.A.: Characterizing the subjective performance of the ITU-T 8 kb/s speech coding algorithm-ITU-T G.729. IEEE Communications Magazine, vol. 35(9) (1997)
7. Kwitt, R., Fichtel, T., Pfeiffenberger, T.: Measuring perceptual VoIP speech quality over UMTS. In: 4th International Workshop on Internet Performance, Simulation, Monitoring and Measurement (IPS-MoMe 2006), Salzburg, Austria (2006)
8. Hoßfeld, T., Binzenhöfer, A., Fiedler, M., Tutschku, K.: Measurement and analysis of Skype VoIP traffic in 3G UMTS systems. In: 4th International Workshop on Internet Performance, Simulation, Monitoring and Measurement (IPS-MoMe 2006), Salzburg, Austria (2006)
9. Markopoulou, A.P., Tobagi, F.A., Karam, M.J.: Assessing the quality of voice communications over internet backbones. IEEE/ACM Trans. Netw., vol. 11(5) (2003)
10. ITU-T Recommendation: ITU-T Rec. G107: The E-model, a computational model for use in transmission planning (1998)
11. Ding, L., Goubran, R.A.: Speech quality prediction in VoIP using the extended E-model. In: GLOBECOM 2003 - IEEE Global Telecommunications Conference, San Francisco, USA (2003)
12. Richards, A., Rogers, G., Antoniades, M., Witana, V.: Mapping User Level QoS from a Single Parameter. In: Proceedings of MMNS '98, Versailles, France (1998)

13. Azzouna, N., Guillemin, F.: Experimental analysis of the impact of peer-to-peer application on traffic in commercial IP networks. European Transactions on Telecommunications, vol. 15(6) (2004)
14. Baset, S.A., Schulzrinne, H.: An analysis of the Skype peer-to-peer internet telephony protocol. In: Proceedings of the INFOCOM '06, Barcelona, Spain (2006)
15. Hoßfeld, T., et al: VoIP applications across mobile networks. In: 17th ITC Specialist Seminar, Melbourne, Australia (2006)
16. Signalogic: Speech codec wav samples (2005) http://www.signalogic.com/index.pl?page=codec_samples
17. National Institute of Standards and Technology: NIST Net - A Linux-based Network Emulation Tool (2006) http://www-x.antd.nist.gov/nistnet/
18. Andersen, S., Duric, A., Astrom, H., Hagen, R., Kleijn, W., Linden, J.: Internet Low Bit Rate Codec (iLBC). RFC 3951 (Experimental) (2004)
19. ITU-T Recommendation P.862: Perceptual evaluation of speech quality (PESQ), an objective method for end-to-end speech quality assessment of narrowband telephone networks and speech codecs (2001)
20. ITU-T Recommendation: ITU-T Rec. P.862.1: Mapping function for transforming P.862 raw result scores to MOS-LQO (2003)

A Comparative Study of Forward Error Correction and Frame Accumulation for VoIP over Congested Networks

Steffen Præstholm[1], Hans-Peter Schwefel[2], and Søren Vang Andersen[2]

[1] Motorola A/S, Mobile Devices, Denmark
[2] Aalborg University, Dept. of Electronic Systems, Denmark

Abstract. We compare Forward Error Correction (FEC) and frame AC-Cumulation (ACC) to see which of the two schemes most effectively reduce frame loss rate for an aggregate of VoIP flows, sharing a network bottleneck. We model this bottleneck by a M/M/1/K queue and we analytically show that given certain assumptions, FEC is the best choice for low initial load at the bottleneck. Then, as the initial load increases, a crossing point is reached after which applying ACC is the better choice. We study this crossing point through numerical examples. Furthermore, we present numerical examples indicating that ACC is better than FEC in bandwidth limited network scenarios, while performance is more equal for packet processing limited scenarios, with FEC being the slightly better choice. Finally, we introduce more general queue models, e.g. the MMPP/M/1/K queue, to model traffic scenarios like the aggregate of ON/OFF VoIP traffic.

1 Introduction

In packet based voice communication, like Voice over IP, schemes like Forward Error Correction (FEC) and Forward Error Protection (FEP) attempt to avoid frame losses by sending redundant information piggybacked to other packets in a speech flow. In the case of FEC, the scheme ensures exact reconstruction of each frame if enough information reaches the receiver, while for FEP the redundant information is often in the form of a low bit encoding with low quality. One possible drawback to FEC/FEP is the introduction of algorithmic delay (may decrease conversational quality) as the receiver have adjust the play-out scheduling, recognizing that it may have to wait for the last of the transmitted redundant information. Such FEC/FEP schemes have received a lot of attention [1,2,3,4,5], and one of the central questions is: when should FEC/FEP be applied? That is, even though FEC/FEP reduces the chance of losing a frame on account of the redundant information, the redundant information also adds to the load of the network and hereby increases the chance of losing packets due to increased congestion. The work in [3] showed that if packet losses are assumed to occur at a bottleneck point modeled by a queue model, then FEC/FEP may improve speech quality if VoIP traffic only accounts for a small part of the aggregate traffic. However, their results also showed that if only VoIP traffic flows

L. Mason, T. Drwiega, and J. Yan (Eds.): ITC 2007, LNCS 4516, pp. 374–385, 2007.

share the bottleneck, then speech quality may improve if the piggybacked redundant information only increases the total packet size slightly and still generates high quality speech. This result was obtained for a system where all the VoIP flows apply the same level of protection and thus share the bottleneck fairly.

In contrast to FEC/FEP, ACCumulation (ACC) of frames [6] attempt to avoid frame losses by reducing the load of the system. That is, frames are accumulated in fewer, but larger, packets, and hereby, there is less header information to be transmitted. This may be quite efficient, given the large header overhead encountered in VoIP. Similar to FEC, ACC also introduces algorithmic delay as the transmitter has to wait for the last frame before a packet can be transmitted, and in order to share the bottleneck fairly, all flows have to apply the same level of ACC.

In this work, we choose a FEC scheme and compare it to ACC to study which of the two schemes gives the best performance, given a bottleneck shared fairly by an aggregate of VoIP flows. That is, we assume a VoIP dedicated queue, e.g. in a differentiated service setup or in a voice exclusive network. We choose the FEC scheme in which encoded voice frames are repeated ($N - 1$ times) and piggybacked to each of the immediately following packets (frame m is sent with packet m to packet $m + (N - 1)$). Exact copies give us the best possible quality and, given the overhead of e.g. a RTP/UDP/IP header, we only increase packet sizes slightly, adding a redundant frame. When we compare FEC and ACC we, unless otherwise stated, assume that they are on the same level (both FEC and ACC packets contain N frames). This way we ensure that they do not differ in algorithmic delay.

Similar to previous work, we assume that packet losses occur at a single point of congestion, "the bottleneck". This is the router having the smallest available bandwidth, either in terms of Bytes per second (Bps) or in terms of packets per second (pps), where the former typically is the result of insufficient link capacity and the later typically is the result of insufficient router processing resources. For each of these two bottleneck scenarios we further distinguish between two buffer scenarios. Thus, the buffer size is used to control the tradeoff between packet loss rate and mean waiting time in the queue, and it may be given in either a fixed number of bytes or a fixed number of packets. In section 2, we model this bottleneck using a M/M/1/K queue model, assuming an aggregate of continuously transmitting VoIP sources. Based on this model, we show, in section 3, that for a given level of FEC/ACC, FEC is always the best choice, if the queue buffer is limited in size to one packet. If the buffer has room for more than one packet, FEC is the best choice for scenarios with light traffic load. Then, as the initial load increases, a crossing point is reached after which ACC becomes the better choice. In section 4, we study the behavior of the crossing point through numerical examples, and we compare FEC and ACC from a more practical perspective, considering the difference in performance and the significance with respect to VoIP quality. Most interestingly, it seems that ACC is the better choice for Bps scenarios while the two schemes are more equal in the pps scenario with FEC as the slightly better choice. In section 5, we introduce the Markov Modulated Poisson Process (MMPP) to model more complex traffic scenarios, like an aggregate of VoIP sources in discontinuous transmission mode, and illustrate in an example how this may affect the crossing point. Finally we summarize in section 6.

2   Modeling Approach

In this section we present our choice of queue model and models for the FEC and ACC schemes in both the Bps and pps scenarios, as well as their effect on either the packet or byte limited buffer.

2.1   The M/M/1/K Model

The bottleneck is modeled by an M/M/1/K model. That is, VoIP packets have to cross parts of the network with varying delay before arriving at the bottleneck. It is common to model this delay variation by exponentially distributed inter-arrival times constituting a Poisson arrival process, and the aggregation of such processes is again a Poisson process with arrival rate λ. Furthermore, the service time is also assumed to be exponentially distributed with service rate μ. This models the variation in service time as the router has to perform other services such as servicing other traffic, or exchanging routing table information. Finally, it is a single server with a finite buffer capacity, K. In section 5 we discuss the use of more general models such as the MMPP/D/1/K model.

The M/M/1/K queue and its properties are well known, see e.g. [7]. The packet loss rate is given as

$$P_L = \left[1 + \sum_{k=1}^{K} \rho^k \right]^{-1} \cdot \rho^K \,, \tag{1}$$

where K is the buffer size in number of packets and $\rho = \lambda/\mu$ is the load factor.

2.2   FEC and ACC Models

The packet loss rate depends on the on the load factor ρ and the buffer size K. Hence, these are the parameters we consider, when we apply FEC and ACC.

Load factor. For the Bps limited scenario the choice of scheme affects both the mean service rate, which is inverse proportional to the packet size, and the mean arrival rate. Both schemes increase the packet size, but ACC lowers the arrival rate, where the FEC arrival rate is unchanged. Hence, arrival rate, service time, and load factor change as follows:

$$
\begin{aligned}
\lambda_{ACC} &= \lambda/N \,, & \mu_{ACC} &= \mu/(H + Nf_s) \,, & \rho_{ACC} &= \rho_s(H/N + f_s) \text{ and} \\
\lambda_{FEC} &= \lambda \,, & \mu_{FEC} &= \mu/(H + Nf_s) \,, & \rho_{FEC} &= \rho_s(H/N + f_s)N \,,
\end{aligned} \tag{2}
$$

where $\rho_s = \lambda/\mu$ is the initial load (before applying FEC/ACC), H is the fraction occupied by header information in a packet containing one frame, N is the number of frames per packet, and f_s is the fraction occupied by frame information in a packet containing one frame. For the pps case, service time does not change and arrival rate, service rate, and load factor are therefore given by

$$\begin{aligned} \lambda_{ACC} &= \lambda/N \,, \quad \mu_{ACC} = \mu \,, \quad \rho_{ACC} = \rho_s/N \quad \text{and} \\ \lambda_{FEC} &= \lambda \,, \qquad \mu_{FEC} = \mu \,, \quad \rho_{FEC} = \rho_s \,. \end{aligned} \tag{3}$$

Later in this paper we also make use of the following useful relations. If we compare Equation (2) and Equation (3), then we recognize that the pps scenario can be seen as a special case of the Bps scenario in which f_s is set to zero (only header) irrespective of the actual frame size. Furthermore, in the case where f_s is equal to one (no header), ρ_{ACC} is equal to ρ_s in Equation (2). We also recognize that we can express ρ_{FEC} in terms of ρ_{ACC} as follows

$$\rho_{FEC} = N \cdot \rho_{ACC} \,. \tag{4}$$

Having modeled the effect of FEC and ACC on packet loss rate, frame loss rate is computed after post processing at the receiver, according to the protection schemes applied by either ACC or FEC, respectively. ACC does not apply any protection and frame loss rate equals packet loss rate, $P_{L,ACC} = P_L$. FEC sends repeated versions of each frame. Thus, frame loss probability is the probability that all redundant frames are lost. We assume that each of the redundant frames are lost with independent loss probability P_L. This is a best case for FEC [3], where independent losses generally result in a lower loss mean burst length than otherwise observed at the bottleneck. Hence the frame loss rate is given by

$$P_{L,FEC} = (P_L)^N \,. \tag{5}$$

Buffer. Until now we have assumed that the buffer is limited by the number of packets it may contain. This actually corresponds to having a buffer with potential unlimited storage capacity in bytes as the packet size grows with increasing N. The other case is that the buffer is limited in storage capacity in bytes. In this case the resulting buffer size, in number of packets, decreases as packets contain more frames. That is, K as a function of N is given as

$$K(N) = \lfloor K_s/(H + N \cdot f_s) \rfloor \,, \tag{6}$$

where K_s is the initial buffer size measured in packets, with one frame per packet. We observe that for the byte limited buffer we cannot just set f_s to zero to get the pps scenario as a special case of the Bps scenario. The buffer depends on f_s and should be adjusted before f_s is set to zero.

3 Analysis of Frame-Loss Rates

In this section we analyze the problem of choosing between FEC and ACC, for a given level of FEC/ACC, N, and a given buffer size, K. In this setup we show that FEC is better than ACC for initial light traffic load scenarios while ACC outperforms FEC as the initial load, ρ_s, increases, for $K > 1$. There is only one point where the two schemes perform equally and we call this the crossing point, $\rho_{s,C}$.

3.1   The Crossing Point

To show that there is exactly one crossing point, we first consider the difference in frame loss rate between FEC and ACC, given as

$$\Delta = \frac{(\rho_{FEC})^{K \cdot N}}{(1 + \sum\limits_{k=1}^{K} (\rho_{FEC})^k)^N} - \frac{(\rho_{ACC})^K}{1 + \sum\limits_{k=1}^{K} (\rho_{ACC})^k} \,, \tag{7}$$

from which we see that the best scheme is chosen according to the following rules:

$$\text{FEC if } \Delta < 0 \,, \ \text{ACC if } \Delta > 0 \,, \tag{8}$$

In Equation (7) the denominators are both positive and we can multiply Δ by the denominators without changing the rules. Thus

$$\Delta' = \left[(N \cdot \rho_{ACC})^{K \cdot N} \cdot (1 + \sum_{k=1}^{K} \rho_{ACC}{}^k) \right] - \\ \underbrace{\left[\left(1 + \sum_{k=1}^{K} (N \cdot \rho_{ACC})^k \right)^N \cdot (\rho_{ACC})^K \right]}_{P3} \,, \tag{9}$$

where ρ_{FEC} has been expressed in terms of ρ_{ACC} according to Equation (4). The first term is a polynomial, P1, of order $(NK + K)$ with coefficients

$$c_{i,P1}(N, K) = \begin{cases} N^{NK} & NK \leq i \leq NK + K \\ 0 & 0 \leq i < NK \end{cases} \,, \tag{10}$$

The second term is also a polynomial, P2, of order $(NK + K)$. However, in order to determine the coefficients, we first need to expand the polynomial P3 in terms of equal powers of $(N \cdot \rho_{ACC})$. From studies on extended Pascal triangles [8], it is well known that these coefficients are given by

$$c_{i,P3}(N, K) = \sum_j (-1)^j \binom{N}{j} \binom{N - 1 + i - (K + 1) \cdot j}{N - 1} \,, \tag{11}$$

These coefficients are all positive and they have symmetric properties similar to those for binomial coefficients [8]. It easily follows that the P2 coefficients are given by

$$c_{i,P2}(N, K) = \begin{cases} c_{i-K,m} \cdot N^{i-K} & K \leq i \leq NK + K \\ 0 & 0 \leq i < K \end{cases} \,. \tag{12}$$

Given the two polynomials, We add terms of equal power and see that the coefficients for the K+1 highest powers of the polynomial Δ' are given by

$$c_i(N, K) = N^{NK} - \binom{N - 1 + (NK + K - i)}{N - 1} \cdot N^{i-K} \,,$$

$$\text{for } NK \leq i \leq NK + K \,, \tag{13}$$

where we have used the symmetric property of the P3 coefficients. This is a sequence of coefficients

$$c_i(N, K) = \begin{cases} 0 & i = NK + K \\ 0 & i = NK + K - 1 \\ \frac{N^{NK}}{2} + \frac{N^{NK-1}}{2} & i = NK + K - 2 \\ \vdots & \vdots \\ N^{NK} - \binom{K+N-1}{N-1} \cdot N^{NK-K} & i = NK \end{cases} \,, \tag{14}$$

where the coefficients for $NK \leq i \leq NK + K - 2$ are positive and increasing for i decreasing. This can be proven by induction as follows. For $NK + 1 \leq i$, c_{i-1} is given as

$$c_{i-1} = c_i + \binom{N - 1 + (NK + K - i)}{N - 1} N^{i-K} \left(1 - \frac{N + (NK + K - i)}{(NK + K - i + 1)N}\right) \,, \tag{15}$$

which is clearly increasing. The remaining coefficients of the Δ' polynomial are negative (only P2 has coefficients different from zero). Δ' is therefore a polynomial where the coefficients have the following signs

$$c_i(N, K) = \begin{cases} + \text{ for } (NK) \leq i \leq (NK + K - 2) \\ - \text{ for } K \leq i < (NK) \\ 0 \text{ for } i < K \end{cases} \,. \tag{16}$$

We are now in a position to draw the following conclusions: First, we observe that for $K = 1$, FEC is always the best choice as the polynomial has no positive coefficients. Secondly, for $K > 1$, we have exactly one positive root, according to Descartes' rule of signs, and it is clear from inspection that for low ρ_{ACC}, Δ' is negative and hence FEC is the best scheme. As the load increases, we reach the point $(\rho_{ACC,C})$ where the schemes have equal performance, and after this point, ACC gives better performance. Given $(\rho_{ACC,C})$, the crossing point expressed with respect to the load factor before applying ACC, $\rho_{s,C}$, is given as

$$\rho_{s,C} = \frac{\rho_{ACC,C}}{H/N + f_s} \,, \tag{17}$$

according to Equation (2). This is a positive linear mapping and our conclusions for $(\rho_{ACC,C})$ still holds for $(\rho_{s,C})$. Furthermore, we observe that decreasing the frame size will move the crossing point towards higher initial loads and vice versa. Here, The pps scenario is represented through $f_s = 0$ (assuming that the buffer is properly adjusted).

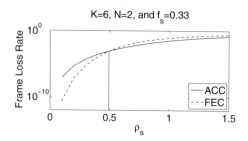

Fig. 1. Crossing point example

Finally, we observe that finding the crossing point corresponds to the problem of finding roots in an order NK polynomial. This problem has no general closed form solution. We solve for the root numerically.

4   Numerical Examples

In this section we first study the impact of N and K on the crossing point, an example of which is illustrated in Figure 1. In this example, we observe that we have a crossing point at approximately $\rho_s = 0.5$, for $K = 6$, $N = 2$ and $f_s = 0.33$. We further study the problem of choosing between FEC and ACC, considering the magnitude of the performance difference and recognizing that a small difference might not have a significant impact on speech quality.

4.1   The Crossing Point

We study the behavior of the crossing point for each of the two buffer cases separately. For each case we study $\rho_{s,C}(N, K)$ in three different scenarios. First we have the Bps scenario with $f_s = 1$ (No header), where $\rho_{ACC} = \rho_s$ and we get the crossing point as given by Equation (7). Secondly we consider the pps limited network scenario and finally, we consider the Bps scenario with $f_s = 0.33$, corresponding to AMR 12.2 frames sent in packets with a RTP/UDP/IPv6 header.

Packet limited buffer. Figure 2 contains plots of $\rho_{s,C}(N, K)$ for the three different scenarios. For $f_s = 1$, we observe that the crossing point moves monotonically towards zero as N and K increases. In the pps scenario (also representing $f_s = 0$), the crossing point again moves towards zero for larger K, but as N increases the crossing point moves towards infinity. In the last scenario, $f_s = 0.33$, we observe that the crossing point generally moves towards zero as K and N increases, but it does increase a little for low N. This increase is an effect of the relationship between ρ_{ACC} and ρ_s as given by Equation (17).

Byte limited buffer. Figure 3 contains the plots for the byte limited buffer case. We note that in the pps scenario, $f_s = 0.33$ is used to adjust the buffer

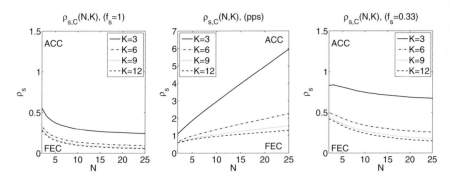

Fig. 2. Packet limited buffer: $\rho_{s,C}$ as a function of N, K, and for $f_s = \{1, 0, 0.33\}$

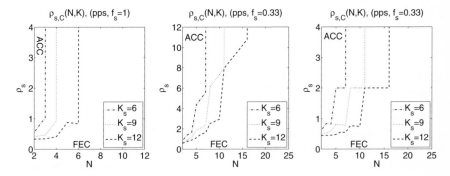

Fig. 3. Byte limited buffer: $\rho_{s,C}$ as a function of N, K_s, and for $f_s = \{1, (0.33, pps), 0.33\}$

size $K(N)$. That is, buffer size is affected by packet size while service time is not. From the plots, we observe how increasing K_s still moves the crossing point towards zero, but for this kind of buffer, increasing N favors FEC. This is because $K(N)$ decreases as N increases. We note that for large enough N, packets are too large to be contained in the buffer.

4.2 Performance Difference

We study the FEC and ACC performance difference to determine how much we gain from choosing one scheme over the other. We first consider the difference (Δ) as given by Equation (7), where the difference is computed for the same level of FEC and ACC. Secondly, we assume that an algorithmic delay within a given range of N does not have an impact on the VoIP quality. Hence we are free to choose the best N for each scheme: The schemes does not necessarily perform optimally for the same N, as illustrated in Figure 4.

In both setups we assume that $f_s = 0.33$ and $N \leq 6$. This corresponds to the scenario for an AMR 12.2 coder, allowing a maximum algorithmic delay of 120

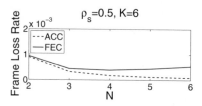

Fig. 4. ACC and FEC may perform optimally for different values of N

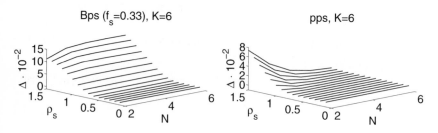

Fig. 5. Packet limited buffer: Δ as a function of ρ_s and N, for $K = 6$. The left plot represents the Bps scenario. The right plot represents the pps scenario.

ms. For such a coder, it is our experience that small changes in the frame loss rate do not change speech quality significantly, and we therefore assume that the two schemes are equal in performance if the difference in loss rate is below 0.005.

Same level, N, of FEC/ACC. Figure 5 shows how ACC always has equal or better performance compared to FEC for both network limitation scenarios, given a packet limited buffer ($K = 6$). This is also true for $K = \{9, 12\}$ and true for $K = 3$ in most cases, though FEC will be as much as 0.01 better in a few cases for $0.4 \leq \rho_s \leq 0.6$. The results for the byte limited buffer is presented in Figure 6. Here, FEC improves performance significantly compared to ACC for large N, while ACC is better for combinations of lower N and high ρ_s in both scenarios. For $K_s = 3$, FEC is always the best choice, while for for $K_s = \{9, 12\}$, ACC performs equally or better.

Best Choice of FEC/ACC. In the following we study the difference (Δ_b) between the best choice of FEC and the best choice of ACC, with respect to N. Additionally, we also study the difference (Δ_{NS}) between the $N = 1$ case, No Scheme (NS) applied, and the overall best choice.

The left plot in Figure 7 shows how ACC always has equal or better performance given a packet limited buffer in the Bps scenario, while the two schemes perform equally in the pps scenario. This is also true for $K = \{3, 6, 12\}$, though for $K = 3$, FEC improves performance by 0.006 to 0.007 in a few cases in both scenarios. Compared to NS, the best choice gives no significant gain for $\rho_s < 0.5$, as the frame loss rate is already low, but for $\rho_s \geq 0.5$ the gain increases from 0.01

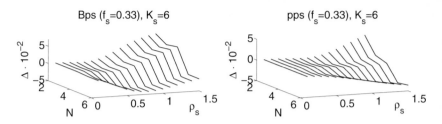

Fig. 6. Byte limited buffer: Δ as a function of ρ_s and N, for $K_s = 6$. The left plot represents the Bps scenario. The right plot represents the pps scenario.

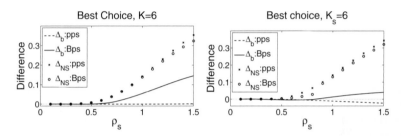

Fig. 7. Left plot: Δ_b and Δ_{NS} as functions of ρ_s, given a packet limited buffer in both network scenarios. Right plot: Δ_b and Δ_{NS} as functions of ρ_s, given a byte limited buffer in both network scenarios.

up to as much 0.35 as ρ_s increases. For increasing K, the point where the gain becomes significant moves towards 1, while it moves towards zero for decreasing K.

The results for a byte limited buffer is presented in the right plot in Figure 7. We observe that in the Bps scenario ACC is the best choice, having equal, very close to equal, or better performance. We make the same observation for $K_s = \{9, 12\}$, but for $K_s = 3$, FEC will always be the better choice. In the pps scenario the schemes perform equally for $\rho_s \leq 0.7$, while FEC improves performance with improvements from 0.005 to 0.02 as ρ_s increases. For $K_s = 3$ FEC is an even better choice compared to ACC, while for $K_s = \{9, 12\}$ ACC is the better choice in the Bps scenario and performance is equal in the pps scenario. In a comparison to NS we see that there is also much to gain for higher ρ_s, given a byte limited buffer.

5   Outlook: More General Models

The M/M/1/K queue model might not be the best choice in all VoIP scenarios. In particular, e.g., when VoIP sources transmit in discontinuous transmission mode. Discontinuous transmission makes each source transmit packets in a ON/OFF like pattern not well modeled by the Poisson arrival process. This kind of traffic is better modeled by e.g. the double stochastic Markov Modulated Poisson

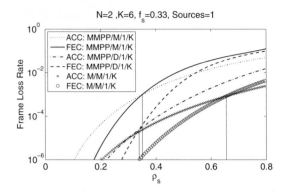

Fig. 8. Crossing points for different bottleneck models

Process (MMPP) [9]. In particular, we use a single source exponential ON/OFF model [10]. For this model, we have estimated the mean ON time, $\overline{ON}$, and mean OFF time, $\overline{OFF}$, based on packet patterns generated on conversational speech [11] using the AMR coder in discontinuous transmission mode. In order to illustrate the impact of the service times distribution we also consider the MMPP/D/1/K model [12] with deterministic service time.

To model the effect of ACC we use the same approach as in Equation (2), but instead of adjusting the total arrival rate, we adjust the source transmission rate during ON times. For FEC we recognize that extra, but smaller packets are introduce at the end of each ON burst. Hence the effect of FEC is modeled by re-estimating $\overline{ON}$, $\overline{OFF}$, and the mean packet size, used to determine the service time as N increases.

In Figure 8 we give an example of the crossing point for each of our three models, given the Bps and packet limited buffer scenario. From the figure, we observe how, for this example, the crossing point moves towards lower initial load as we change model from the M/M/1/K queue to the MMPP/M/1/K queue and then to the MMPP/D/1/K queue. Hence, we see that the choice of model and, more importantly, the choice of transmission mode will have an impact on the crossing point, which in this example is in favor of ACC if we change from continuous to discontinuous transmission mode.

6   Summary and Outlook

In this work we have compared Forward Error Correction (FEC), through repeated transmission of voice frames piggybacked to the immediately following packets, and ACCumulation (ACC) of frames in larger, but fewer packets. The two schemes are compared with respect to frame loss rate in a network bottleneck scenario, modeled by a M/M/1/K queue model. Given this model, we have analytically shown that, for the same level (number of frames per packet) of FEC and ACC and $K > 1$, there is a single crossing point between the light traffic scenarios, for which applied FEC results in the lowest frame loss rate, and the more congested scenarios, for which applied ACC is the best choice. We have further compared the two schemes numerically and results indicate that ACC generally is the better choice in bandwidth limited network scenarios, while the two schemes are more equal, with FEC being the slightly better choice, in packet processing limited network scenarios.

Additional interesting analysis can be obtained from comparing FEC and ACC within an optimization framework where both frame-loss rate and delay are considered jointly, in line with the work in [5]. The performance of the two schemes may further be analyzed with respect to differences in loss pattern characteristics and the impact of such differences on VoIP quality. Furthermore, the work started in section 5 can be extended to a more detailed study of the impact of bursty traffic and variations of the service times distribution. In this respect, utilizing more complex MMPP models, scenarios of heterogeneous traffic can easily be included.

References

1. Podolsky, M., Romer, C., McCanne, S.: Simulation of fec-based error control for packet audio on the internet. In: Proc. IEEE INFOCOM, vol. 2, pp. 505–515 (1998)
2. bolot, J.-C., Fosse-Parisis, S., Towsley, D.: Adaptive fec-based error control for internet telephony. In: Proc. IEEE INFOCOM, vol. 3, pp. 1453–1460 (1999)
3. Altman, E., Barakaat, C., Ramos, V.M.: Queueing analysis of simple fec schemes for ip telephony. In: Proc. IEEE INFOCOM, vol. 2, pp. 796–804 (2001)
4. Altman, E., Barakaat, C., Ramos, V.M.: On the utility of fec mechanisms for audio applications. LNCS, vol. 2156, pp. 45–56. Springer, Heidelberg (2001)
5. Boutremans, C., Le Boudec, J.-Y.: Adaptive joint playout buffer and fec adjustement for internet telephony. In: Proc. IEEE INFOCOM, vol. 1, pp. 652–662 (2003)
6. Hoene, C., Karl, H., Wolisz, A.: A perceptual quality model for adaptive voip applications. In: Proc. SPECTS (2004)
7. Cassandras, C.G., Lafortune, S.: Introduction to Discrete Event Systems, 1st edn. Kluwer Academic Publishers, Boston (2004)
8. Bollinger, R.C.: Extended pascal triangles. Mathematics Magazine 66, 87–94 (1993)
9. Fischer, W., Meier-Hellstern, K.: The markov-modulated poisson process (MMPP) cookbook. Performance Evaluation 18, 149–171 (1993)
10. Krieger, U., Naoumov, V., Wagner, D.: Analysis of a finite fifo buffer in an advanced packet-switched network. IEICE Trans. Commun. E81-B, 937–947 (1998)
11. Graff, D., Walker, K., Millier, D.: Switchboard Cellular Part 1 Transcribed Audio. Linguistic Data Consortium (2001)
12. Blondia, C.: The N/G/1 finite capacity queue. Commun. Statist.-Stochastic Models 5, 273–294 (1989)

Performance Optimization of Single-Cell Voice over WiFi Communications Using Quantitative Cross-Layering Analysis

Fabrizio Granelli[1], Dzmitry Kliazovich[1], Jie Hui[2], and Michael Devetsikiotis[3]

[1] DIT – University of Trento, Via Sommarive 14, I-38050 Trento, Italy
{granelli,klezovic}@dit.unitn.it
[2] Intel – Communication Technology Lab, Portland, Oregon, USA
Jie.Hui@intel.com
[3] ECE – North Carolina State University, Raleigh, NC 27695-7911, USA
mdevets@ncsu.edu

Abstract. Cross-layer design has been proposed to optimize the performance of networks by exploiting the inter-relation among parameters and procedures at different levels of the protocol stack. This may be particularly beneficial in wireless scenarios, and for quality-of-service support. This paper proposes a quantitative study of cross-layer performance optimization for Voice over WiFi communications, which enables design engineers to analyze and quantify inter-layer dependencies and to identify the optimal operating point of the system, by using cost-benefit principles. Furthermore, insight gained on the problem enables the proposal of design principles for a Call Admission Control scheme able to enhance the overall system performance by limiting the number of users in the system and signalling to the active terminals of the proper parameter settings to optimize overall performance.

Keywords: Cross-Layer design, metamodeling, VoIP over WiFi.

1 Introduction

The layering principle has been long identified as a way to increase the interoperability and to improve the design of telecommunication protocols, where each layer offers services to adjacent upper layers and requires functionalities from adjacent lower ones. Standardization of such protocol stacks in the past enabled fast development of interoperable systems, but at the same time limited the performance of the overall architecture, due to the lack of coordination among layers. This issue is particularly relevant for wireless networks, where the very physical nature of the transmission medium introduces several performance limitations (including time-varying behavior, limited bandwidth, severe interference and propagation environments) and thus, severely limits the performance of protocols (e.g. TCP/IP) designed for wired networks.

To overcome these limitations, a modification of the layering paradigm has been proposed, namely, *cross-layer design*, or "cross-layering." The core idea is to

L. Mason, T. Drwiega, and J. Yan (Eds.): ITC 2007, LNCS 4516, pp. 386–397, 2007.

maintain the functionalities associated to the original layers but to allow coordination, interaction and joint optimization of protocols crossing different layers.

Several cross-layering approaches have been proposed in the literature so far [1, 2, 3, 4]. Nevertheless, little formal characterization of the cross-layer interaction among different levels of the protocol stack is available yet, with the exception of [5], where the impact of different layers is studied in order to optimize service delivery in mobile ad-hoc networks, and [6], where the authors introduced a meta-modeling approach to study cross-layer scheduling in wireless local area networks.

A clear need is emerging for identifying approaches able to analyze and provide *quantitative guidelines* for the design of cross-layer solutions, and, even more important, to decide whether cross-layering represents an effective solution or not. In [7], we initiated a quantitative approach for calculating the sensitivity of system performance with respect to parameters across different layers for a simple Voice over WiFi system.

Voice over WiFi (VoWiFi) communications represent a challenging scenario, as, even in the simplest case of a single IEEE 802.11 cell, performance optimization requires the consideration of several parameters at different levels of the protocol stack. Indeed, codec parameters as well as link layer and physical parameters (and several others) clearly have impact on the overall quality of communication as it is perceived by the end user.

As limited quality-of-service strategies are employed on the wireless link, there is a need for a proper *Call Admission Control* (CAC) strategy [8-16], in order to provide ways to limit the number of users in the system and, more generally, to provide possible on-line adjustments to terminal parameters.

In view of the above, this paper describes the use of a formal framework to (1) identify and formalize the interactions crossing the layers of the standardized protocol stack; (2) systematically study cross-layer effects in terms of quantitative models; (3) support the design of cross-layering techniques for optimizing network performance; (4) define design principles of a CAC strategy for the system to work at, or near to, its optimal operating point. The presented approach, based on techniques well-established in operations research, allow engineers to identify correlations among different parameters and to estimate the potential advantages (if any) deriving from cross-layer interactions.

The structure of the paper is the following: Section 2 summarizes our formalization of parameters and measurements at different layers of the protocol stack in layering and cross-layering schemes. Section 3 describes a specific VoWiFi single-cell scenario, cross-layer signaling implementation and the process of system modeling. Furthermore, the section discusses a formal *cost-benefit analysis* to optimize the performance of the system from the service provider and the wireless terminal perspectives. Section 4 derives design principles of a candidate CAC mechanism to optimize the overall performance and, finally, Section 5 draws conclusions and outlines future work on the topic.

2 Cross-Layer Design and Cost-Benefit Analysis

In a previous paper [7] we outlined a formal framework for cross-layer design by identifying system parameter vectors as the merging of two sub-arrays representing, respectively, internal and external parameter subsets: $\overline{p}^N = \left[\overline{p}_i^N \mid \overline{p}_e^N \right]$, $\overline{m}^N = \left[\overline{m}_i^N \mid \overline{m}_e^N \right]$. Cross-layer design allows a large degree of flexibility, by enabling a higher level of interaction among the entities at any layer of the protocol stack. Layer N is enabled to control, depending on the specifics, a subset of all the parameters at any level $\overline{p}^{TOT} = \left[\overline{p}^1 \mid \overline{p}^2 \mid ... \mid \overline{p}^7 \right]$ and can acquire measurements as a subset of $\overline{m}^{TOT} = \left[\overline{m}^1 \mid \overline{m}^2 \mid ... \mid \overline{m}^7 \right]$ where we assume an OSI-like seven layer stack.

Cross-layer design derives from the observation that the performance of a network or other system depends on several mechanisms situated at different levels of the protocol stack interacting in a complex fashion. *Quantifying* the effect of these interactions is very important in order to be able to systematically relate such interactions to system outcomes and be able to quantify the decision to take such interactions into account – using a cost-benefit analysis, so that the benefits outweigh the cost of additional complexity and "layer violation" [17, 7].

In [7] we advocated the use of formal system modeling to express cross-layer interactions and their effect on system performance, based on sensitivity analysis. The system response with respect to the k-th performance metric is modeled as a function jointly of all parameters across the layers, $f_k()$. The sensitivity of the system response and the interactions among factors, within and across layers, can then be captured naturally as the partial derivatives $\dfrac{\partial f_k}{\partial p_i^j}$ and $\dfrac{\partial^2 f_k}{\partial p_i^j \partial p_l^m}$ where p_i^j is the i-th parameter at layer j. Subsequently, one can then strictly or nearly optimize the performance e_i with respect to a subset of p^{TOT} under general constraints by using any available method, such as steepest ascent, stochastic approximation, ridge analysis, and stationary points [18, 19].

The function $f_k()$ across the layers can be analytically calculated or empirically estimated. Since closed form mathematical expressions are often unattainable for real systems, in [7] we outlined a mathematical modeling procedure based on *metamodeling*. In this paper, we continue and extend our work on metamodeling of wireless systems, by (meta)modeling the performance of a multi-user VoWiFi system with several parameters, and across several layers.

Our "raw" performance metrics, e_i, are further incorporated into a utility or "benefit" function $U(e)$ that expresses how valuable the (net) system performance is to the system owner or user. In general, the exact functional form of the utility and resulting objective function are less important than their curvature (often convex, to denote a certain "saturation") and their ability to preserve a relative ordering of the engineering alternatives, to enable ultimate design decisions.

Results achieved during the system optimization phase are then employed to define guidelines for system design. By employing the proposed framework, it is possible to select 1) the sensitivity of the system utility with respect to individual parameters; 2) the optimal operating point of the system (direct consequence of the optimization process); 3) the proper cross-layer interactions to enable (based on sensitivity of the system); 4) and the proper signaling architecture to employ (allowing to identify the set of parameters and measurements to use).

3 A VoIP over WiFi Scenario

In this section we illustrate the application of the proposed modeling approach in a VoIP over WiFi setting. The model is built in a four-dimensional domain defined by a set of parameters considered crucial for the overall system performance, namely, physical bandwidth, link error rate, maximum number of link layer retransmissions, and VoIP frame generation interval. The chosen set of parameters is spread over several layers of the protocol stack, making it difficult to predict the optimal operation point using ad hoc or intuitive methods.

3.1 System Model

Network Model

The network model is shown in Fig. 1. The network is an infrastructure WLAN with one Access Point (AP) serving N client nodes. Each client node initiates a bidirectional VoIP call with the AP. As the result, there are N uplink and N downlink calls carried in the network simultaneously. For each call, we use the ITU G.711 64kbps codec [20] where frames are sent for transmission at regular time intervals.

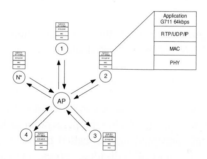

Fig. 1. Simulation scenario

The frames produced by the voice coder are then encapsulated by RTP/UDP/IP layers of the protocol stack adding an overhead of 40 bytes. In the MAC layer, IEEE 802.11 DCF basic access mode with no RTS/CTS exchange is used.

Inputs

The controllable or design variables of interest are the following:

Physical data rate (D) [1] is the data rate available for transmission at the physical layer. In order to comply with IEEE 802.11b, physical data rate values are taken equal to 1 Mbps, 2 Mbps, 5.5 Mbps and 11 Mbps.

---

[1] We acknowledge the fact that data rate D is determined by link adaptation algorithm on the basis of signal-noise ratio detected by the receiver. In the paper, we assume data rate can be controlled independently. The consideration of interaction of D with SNR will be exploited in future work with an appropriate link adaptation algorithm and accurate interference modeling at physical layer.

Packet Error Rate (PER). Wireless systems are usually characterized by a high error rate corrupting data transmitted at the physical layer. In fact, typical PERs for WLAN links are in the range between 10^{-3} and 10^{-1}. However, in order to evaluate system performance also for low error rate channels, we decided to vary PER between a lower value of 10^{-9} and 10^{-1}.

Maximum number of retransmissions (R). The task of link layer ARQ is to compensate high error rates on wireless channels. The crucial parameter for ARQ scheme performance is the maximum number of retransmission attempts performed before the link layer gives up and drops the frame. Each retransmission consumes the same physical resources as the original frame transmission, thus reducing the overall capacity of the cell. On the other hand, retransmissions increase packet delivery delay. In our network model, R is varied from 0 to 5, where 0 corresponds to the case when no retransmissions are performed at the link layer.

Voice packet interval (I) defines the time interval between frames generated by the voice codec. Voice packets are then encapsulated using RTP over UDP/IP protocols. Voice frames produced by the codec are relatively small (usually smaller than 100 bytes). As a result, a portion of network capacity is wasted on protocol overhead (40 bytes per packet). The parameter I is varied from 10 to 90 ms in our scenario.

Outputs

The output response of interest, $e=N^*$, is the maximum number of VoIP calls that can be supported by the WLAN cell with a satisfactory quality, which is defined by the following constraints.

Constraints

Several factors affecting VoIP performance can be mainly divided into *human* factors and *network* factors. Human factors define the perception of the voice quality by the end-user. The most widely accepted metric, called the Mean Opinion Score (MOS) [21], provides arithmetic mean of all individual scores, and can range from 1 (worst) to 5 (best).

The factors affecting the MOS ranking are related to network dynamics and include end-to-end propagation delay and frame loss [21, 22]. The delay includes the encoder's processing and packetization delay, queuing delay, channel access and propagation delay. For this reason, in order to ensure an acceptable VoIP quality, we limit the delay parameter to 100 ms measured between unpacketized voice data signal at codecs located at the sender and the receiver nodes. The second factor, frame loss rate, affects the VoIP quality due to non-ideal channel conditions. The chosen ITU G.711 64kbps codec [20] shows acceptable MOS rating (MOS=3) for frame loss rate up to 5% [23].

Cross layer Model of VoIP

Following from the above, we assume a quantitative model for the VoIP capacity as $N^* = f(D,PER,R,I)$ which we proceed to estimate via response surface (meta)modeling, since a closed form analytical model across the layers is clearly intractable.

3.2 Implementation and Cross-Layer Signaling

The network model is implemented in the ns2 network simulator (version 2.29) [24]. The simulation parameters are summarized in Table 1. The ITU G.711 64kbps codec [20] is emulated using Constant Bit Rate (CBR) generator source, producing blocks of data in regular intervals specified by the voice interval I input parameter. In addition to the voice codec, the Cross-Layer Control (CLC) module is added at the application layer of the protocol stack (see Fig.2). CLC is able to read the measured values of D and PER at the physical and link layers as well as internally I at the application layer. Moreover, it can set R, I, or D to the desired value.

Table 1. Simulation Parameters

Parameter Name	Value
Slot	20 μs
SIFS	10 μs
DIFS	50 μs
PLCP preamble + header	192 μs
Data Rate	1, 2, 5.5, or 11 Mb/s
Basic Data Rate	1 Mb/s
Propagation Model	two-ray ground
RTS/CTS	OFF

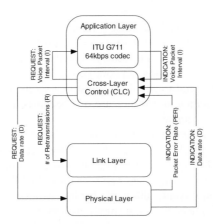

Fig. 2. Cross-Layer Control (CLC) module and cross-layer interactions

3.3 Model Definition

For each combination of input parameters, that is, D, PER, R, and I, we run a series of simulations with the number of VoIP flows incrementally set from 1 to 25. Then we find the maximum number of VoIP flows N^* accepted by the system as the output for which the quality of the voice signal remains above a satisfactory level (as defined in Section 3.1, with end-to-end delay less than 100 ms and frame error rate less than 5%), by checking every voice frame. Output capacity varies based on the controllable input variables change. Table 2 shows the values of input parameters used in the experiment.

In order to fit the simulation results with a model, we used the JMP [25] tool and a second order polynomial RSM model, with interactions which terms and corresponding coefficients are presented in the following equation (note that the interaction between I and R is not significant, therefore it is excluded from the model). Results show that R-square of the fitted model is equal to 0.81.

$$N^* = -5.1027 + 1.5575 * D + 292.8806 * I + 1.3677 * R - 157.3738 * PER$$
$$+5.9569 * D * I + 0.1980 * D * R - 5.1210 * D * PER - 891.6851 * I * PER \qquad (1)$$
$$+3.7706 * R * PER - 0.1186 * D^2 - 2710.813 * I^2 - 0.2935 * R^2 + 1644.7405 * PER^2$$

Table 2. Experiment Design Parameters

	Parameter Name	Abbreviation	Levels	Values
Inputs	Physical data rate	D	4	1, 2, 5.5, 11
	Packet Error Rate	PER	9	10^{-9}, 10^{-8}, 10^{-7}, 10^{-6}, 10^{-5}, 10^{-4}, 10^{-3}, 10^{-2}, 10^{-1}
	# of retransmissions	R	6	0, 1, 2, 3, 4, 5
	Voice packet interval	I	9	10, 20, 30, 40, 50, 60, 70, 80, 90
Constraints	Voice E2E delay	-	-	< 100 ms
	Frame error rate	-	-	< 5%

Fig. 3 illustrates the obtained metamodel N^* function in all four dimensions of D, I, R, and PER. The maximum of N^* with respect to I is located between 0.05 and 0.07 seconds at it is evident in Fig. 3a. Obviously, with the increase of I, client nodes generate fewer packets, thus increasing network capacity. However, the voice packet interval is included into the end-to-end delay constraint set to 100 ms. Consequently, after a certain threshold, an additional increase of I becomes unfavorable, leading to an overall network capacity decrease.

A similar observation can be made for the maximum number of retransmissions configured at the link layer. With a higher R, the system can sustain a higher error rate at the wireless link. However, each retransmission consumes bandwidth resources from the shared channel. For high data rate scenarios (11 Mb/s), retransmissions take just a small fraction entire bandwidth while for low data rate scenarios (1 or 2 Mb/s) the portion of bandwidth used for retransmissions becomes considerable (see Fig. 3b). As a result, the N^* is maximized at R equal to 3 for low data rates.

Fig. 3c illustrates that N^* is not sensitive with respect to low PERs (10^{-9} – 10^{-4}). However, when PER is high (> 10^{-4}) – which is often the case in WLAN networks – the system capacity dramatically decreases. The absolute maximum of N^* corresponds to $D=11$ Mb/s, $I=0.07$ s, $R=5$, $PER=10^{-9}$, and is equal to 20 bidirectional VoIP calls. The reader should note that this maximum corresponds to approximately 36% utilization of D provided at the physical layer. The remaining 64% is wasted on physical and link layer overhead which becomes especially relevant for small packets (like in VoIP). A detailed study of small packet performance under IEEE 802.11 WLAN as well as an optimization employing packet concatenation techniques is presented in [27].

Based on the above model, we proceed to quantify the sensitivity of the response $e=N*$ on the four cross-layer variables D, I, R, and PER by calculating the derivatives:

$$dN*/dD = -0.2372D + 5.9569I + 0.198R - 5.1209PER + 1.5575 \tag{2}$$

$$dN*/dI = 5.9569D - 5421.626I - 891.6851PER + 292.8815 \tag{3}$$

$$dN*/dR = 0.198D - 0.587R + 3.77PER + 1.3677 \tag{4}$$

$$dN*/dPER = -5.12D - 891.6851I + 3.77R + 3289.48PER - 157.3774 \tag{5}$$

The knowledge of the behavior of the first-order derivatives of $N*$ allows the estimation of the impact of each of the parameters. The absolute maximum values of the derivatives are presented in Table 3. In our case, the voice packet interval I and packet error rate PER at the physical layer have a higher impact on the maximum number of calls $N*$ that can be supported by the system.

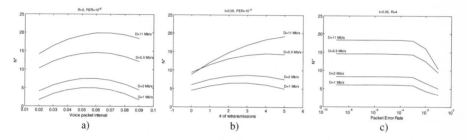

Fig. 3. Metamodel of the system VoIP call capacity ($N*$)

Table 3. Absolute maximum values for $N*$ derivatives

Derivative	Maximum	D [Mb/s]	I [s]	R	PER
max $\lvert dN*/dD \rvert$	2.84	1	0.09	5	0
max $\lvert dN*/dI \rvert$	278.27	1	0.09	≥ 0	0.1
max $\lvert dN*/dR \rvert$	3.92	11	≥ 0.02	0	0.1
max $\lvert dN*/dPER \rvert$	293.95	11	0.09	0	0

3.4 System Optimization

Once the metamodel is established, it is possible to exploit the information it contains to build a utility or similar function and, thus, enable a cost-benefit analysis of the problem. In the case under examination, it is possible to identify two optimization scenarios, focused on the service provider and the wireless terminal, respectively, which are described in the following sub-sections.

Service Provider Perspective

From the point of view of the service provider, the main concern is associated with maximization of the profit obtained from the operating network. The profit is directly proportional to the number of calls that can be supported by the system simultaneously deducing network setup and operating costs.

Here, we define a utility function for the VoWiFi system as

$$U(D,I,R,PER) = N*\cdot P_{call} - \frac{D_{wasted}}{D} \cdot N*\cdot P_{call} - P_{power} \cdot (D_{norm} + PER_{norm})^2 =$$

$$= N*\cdot P_{call} \cdot (1 - \frac{D_{wasted}}{D}) - P_{power} \cdot (D_{norm} + PER_{norm})^2 \qquad (6)$$

where P_{call} is the price charged for (or the marginal income from) a single call, P_{power} is the marginal cost of a unit of transmitted power, and $D_{wasted} = D \cdot PER \cdot R / R_{max}$ is the bandwidth wasted for re-transmissions, in packets/sec. The $N*\cdot P_{call} \cdot D_{wasted} / D$ term accounts for the portion of bandwidth used for re-transmissions instead of voice traffic that the owner could have charged for. The last term is similar to the one used in [7] and captures the quadratic relationship of D and PER with respect to the radiated power.

Fig. 4 presents the behavior of U. The P_{call}/P_{power} ratio is chosen to be equal to 100 in our example. This corresponds to the policy of service provider to charge one dollar per VoIP call while the price paid for a power resource unit is just one cent.

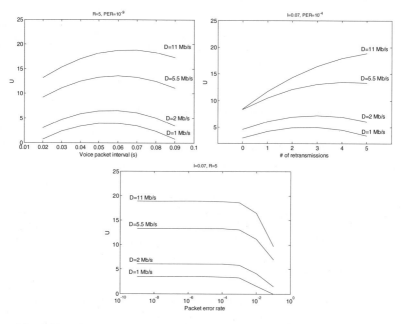

Fig. 4. VoIP network utility function U from the service provider point of view

The obtained utility function U is similar to the metamodel $N*$ describing VoIP system capacity. The max $U(D,I,R,PER)$ corresponds to the max of $N*$ at $D=11$ Mb/s, $I=0.07$ s, $R=5$, $PER=10^{-9}$ and is equal to 18.89 dollars. The difference with the maximum number of VoIP calls (max U) of one dollar is associated with the network operating costs.

Mobile Terminal Perspective

From the point of view of the mobile terminal, the main constraint can be identified in enabling long battery lifetime while providing acceptable voice performance. Since the latter is already included in the metamodel of the system, we can concentrate on the former to identify the cost related to the considered scenario. More specifically, the main parameters impacting on power consumption are the following:

- Transmission data rate D: in IEEE 802.11b, data rate selection at the physical level is based on the strength (or SNR) of the received signal. We use the same assumptions as in [26]: Power setting of Data Rate of 1Mbps = 30 mW; Power setting of Data Rate of 2Mbps = 35 mW; Power setting of Data Rate of 5.5 Mbps = 50 mW; Power setting of Data Rate of 11 Mbps = 100 mW. As a consequence, in order to support a given physical data rate D, the power should be adjusted to the appropriate value.
- Maximum number of retransmissions R: clearly, as the number of allowed retransmissions increases, more power is used for delivery of a single packet.

It is then possible to define a utility function to be maximized including such observations, where the relative weight of benefits against costs can be identify by varying the parameters α and β:

$$U = \alpha \cdot N^* - \beta \cdot [R + f(D)] \tag{7}$$

where the function $f(D) = 10^{1.4291 + 0.0515D}$ captures the above power constraint on receiver SNR in order to enable physical data rate D.

4   Design Principles for a VoWiFi Call Admission Control

The proposed framework represents a novel approach to cross-layer design and to the authors' knowledge it has not yet been addressed in the literature. Therefore, the purpose of this section is only to sketch a possible application scenario, beyond static performance optimization. In fact, on the basis of the analysis presented in the previous section, two empirical considerations are confirmed: (i) limitation on number of active nodes, and thus, an admission mechanism (CAC), is required in order to provide satisfactory performance to VoIP communications; (ii) the performance of the overall system significantly depends on several parameters, which can be recognized (and quantified) at different layers of the protocol stack, thus suggesting and justifying the use of a cross-layering scheme.

This motivates the introduction of a centralized call admission control to monitor the status of the overall VoIP system, which can exploit the metamodel information to provide the proper cross-layer parameter settings to perform run-time optimization of the system. Such CAC should be supported by the knowledge of the utility function (see Section 3) and could be implemented at the AP as the "central" point of the cell where all traffic converges. A example scenario could be the following:

- a new VoIP call is activated by a terminal;
- the CAC process on AP checks whether the cell already reached the maximum number of calls which can be supported (information inferred from the metamodel);

- if yes, the call is rejected;
- if no (i.e., there is room for an additional call), the CAC computes the optimal configuration of the cell to provide the best performance (using the information derived from the selected utility function);
- the optimal configuration is sent to all VoIP terminals (with WiFi beacons).

Such an architecture underlines two relevant aspects of the considered framework: 1) information captured by the metamodel and utility function is useful for supporting configuration / decision making processes in case of complex scenarios, such as the one considered in this paper; 2) cross-layering can be implemented in a *distributed* fashion (AP as reconfiguration manager for all the nodes).

5   Conclusions

This paper proposes a quantitative study of the problem of cross-layer performance optimization applied to a Voice over WiFi scenario, which enables to analyze and quantify inter-layer dependencies and to identify the optimal operating point of the system using cost-benefit principles. Achieved simulation results confirm some empirical considerations already available in the literature. The insight gained on the problem is then used to propose design principles for a Call Admission Control scheme able to enhance the overall system performance by limiting the number of users in the system and signalling to the active terminals the proper parameter settings to optimize overall performance.

Future work will deal with the actual implementation of the defined CAC scheme in order to provide performance analysis and validation of the proposed cross-layer optimization framework, as well as with the definition of proper signalling methods to support distributed cross-layering solutions.

References

1. Toumpis, S., Goldsmith, A.J.: Capacity Regions for Wireless Ad Hoc Networks. IEEE Trans. on Wireless Communications 4, 746–748 (2003)
2. Pollin, S., Bougard, B., Lenoir, G.: Cross-Layer Exploration of Link Adaptation in Wireless LANs with TCP Traffic. IEEE Benelux Ch. on Comms and V.T (2003)
3. Chen, L., Low, S.H., Doyle, J.C.: Joint Congestion Control and Media Access Control Design for Ad Hoc Wireless Networks. In: Proc. INFOCOM 2005 (March 2005)
4. Lin, X., Shroff, N.B.: The Impact of Imperfect Scheduling on Cross Layer Rate Control in Wireless Networks. In: Proc. INFOCOM 2005 (March 2005)
5. Vadde, K.K., Syrotiuk, V.R.: Factor Interaction on Service Delivery in Mobile Ad Hoc Networks. IEEE JSAC 22(7), 1335–1346 (2004)
6. Hui, J., Devetsikiotis, M.: Metamodeling of Wi-Fi Performance. In: Proc. IEEE ICC 2006, Istanbul, Turkey (June 11-15, 2006)
7. Granelli, F., Devetsikiotis, M.: Designing Cross-Layering Solutions for Wireless Networks: a General Framework and Its Application to a Voice-over-WiFi Scenario. In: Proceedings of CAMAD'06, Trento (June 8-9, 2006)

8. Medepalli, K., Gopalakrishnan, P., Famolari, D., Kodama, T.: Voice Capacity of IEEE 802.11b, 802.11a and 802.11g Wireless LANs. Globecom'04 (Nov.29-Dec.3, 2004)
9. Anjum, F., Elaoud, M., Famolari, D., Ghosh, A., Vaidyanathan, R., Dutta, A., Agrawal, R.: Voice Performance in WLAN Networks- An Experimental Study. Globecom 03 (2003)
10. Medepalli, K., Gopalakrishnan, P., Famolari, D., Kodama, T.: Voice Capacity of IEEE 802.11b and 802.11a Wireless LANs in the Presence of Channel Erros and Different User Data Rates. In: Proc. of VTC Fall 2004 (2004)
11. Clifford, P., Duffy, K., Leith, D., Matone, D.: On Improving Voice Capacity in 802.11 Infrastructure Networks. In: Proc. of IEEE Wireless-Com 2005 (2005)
12. Wang, W., Liew, S.C., Li, V.O.K.: Solutions to Performance Problems in VoIP Over a 802.11 Wireless LAN. IEEE Transactions on Vehicular Technology (2005)
13. Pong, D., Moors, T.: Call Admission Control for IEEE 802.11 Contention Access Mechanism. Globecom'03 (2003)
14. Hole, D.P., Tobagi, F.A.: Capacity of an IEEE 802.11b Wireless LAN supporting VoIP. In: International Conference on Communications (ICC'04) (2004)
15. Coupechoux, M., Kumar, V., Brignol, L.: Voice over IEEE 802.11b Capacity, in ITC Specialist Seminar on Performance Evaluation of Wireless and Mobile Systems (2004)
16. Ergen, M., Varaiya, P.: Throughput Analysis and Admission Control for IEEE 802.11a. MONET Special Issue on WLAN optimization at the MAC and Network Levels (2004)
17. Kawada, V., Kumar, P.R.: A Cautionary Perspective on Cross-Layer Design. IEEE Wireless Communications 12(1), 3–11 (2005)
18. Box, G.E.P., Draper, N.R.: Empirical Model Building and Response Surfaces (1987)
19. Kleijnen, J.P.C.: Experimental design for sensitivity analysis, optimization, and validation of simulation models (1998)
20. Ohrtman, F.: Voice over 802.11. Artech House (2004)
21. ITU-T Recommendation P.800, Methods for Subjective Determination of Speech Quality, International Telecommunication Union, Geneva, 2003.
22. Schulzrinne, H., Casner, S., Frederick, R., Jacobson, V.: RTP: A Transport Protocol for Real-Time Applications. RFC 1889 (1996)
23. Ding, L., Goubran, R.A.: Speech quality prediction in VoIP using the extended E-model. Globecom'03 (December 2003)
24. NS-2 simulator tool home page (2000) http://www.isi.edu/nsnam/ns/
25. JMP desktop statistical software from SAS. http://www.jmp.com
26. Obaidat, M.S., Green, D.G.: SNR-WPA: An Adaptive Protocol for Mobile 802.11 Wireless LANs. ICC 2004, Paris (2004)
27. Kliazovich, D., Granelli, F.: Packet Concatenation at the IP Level for Performance Enhancement in Wireless Local Area Networks, Wireless Networks (2007)

A Cross-Layer Approach for Evaluating the Impact of Single NEXT Interferer in DMT Based ADSL Systems[*]

Indradip Ghosh, Kalyan Basu, and Sumantra R. Kundu

Center for Research in Wireless Mobility and Networking (CReWMaN)
University of Texas at Arlington
Arlington, TX, USA
{ighosh, basu, kundu}@cse.uta.edu

Abstract. In this paper we aim to develop an analytical model of the cross-layer characteristics of a Near End Cross Talk (NEXT) limited ADSL channel and determine the effective channel capacity of such a system. The ADSL users often encounter impact of other users in the same cable system due to NEXT interference that reduces the transmission capacity of the ADSL link. At the same time, due to application behavior, the source traffic of the user introduces a level of uncertainty. This paper proposes a model for capturing the cross-layer dynamics of these two uncertainties of an ADSL system. The model provides a closed form analytical method for determining the effective capacity of the ADSL system limited by NEXT interference.

1 Introduction

In telecommunication (telco) networks, the ADSL line uses a dedicated copper pair to connect customers to central office (CO). In the telco outside plant, the copper pair is enclosed in the underground cable system that houses copper lines from other active ADSL users in the same cable network. Thus, the ADSL line of an user is subject to interference from other active ADSL lines in the same cable bundle. There are two major types of interferences (see Figure 1): *Near End Cross Talk* (NEXT), which is caused by interference of signals of the same frequency in between the receiving path of a subscriber loop and a transmitting path of another subscriber loop of ADSL transceivers at the same end of the binder cable, and *Far End Cross Talk* (FEXT), caused by interference of signals of the same frequency in between the receiving path of the subscriber loop and a transmitting path of another subscriber loop at the opposite ends of the ADSL link. However, in our analysis we have considered NEXT as the only source of interference since it is the most dominant source of noise in ADSL system.

Due to interference, the ADSL channel capacity is dependent on the dynamics of NEXT interference. This variation of channel capacity has a causal impact on

---

[*] This work is supported by NSF ITR grant IIS-0326505.

L. Mason, T. Drwiega, and J. Yan (Eds.): ITC 2007, LNCS 4516, pp. 398–409, 2007.

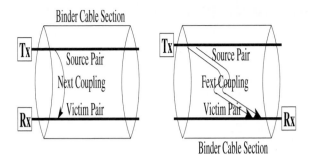

Fig. 1. A diagrammatic representation of NEXT and FEXT

the system throughput and on the queue size of the system buffer that houses the packets waiting for transmission. Therefore, intuitively we can perceive that the evolution of the buffer content has a tightly coupled relationship with the channel state and the source rate of the user application.

In this paper, we model the uncertainty in user traffic generation and the capacity of the ADSL system. Applying established approaches in fluid-flow modeling, it is possible to determine the QoS level available to the user application with varying ADSL channel capacity and at different levels of user traffic rates. Additionally we have considered *Discrete Multitone Modulation* (DMT) scheme of ADSL. The G.DMT divides the designated frequency spectrum into a number of sub-bands called *bins*, which can be duplex as well as simplex depending on individual bit loading capacity of the bin. There are a total of 255 bins distributed over the entire spectrum which carry variable number of bits per sample (see Figure 2). The number of bits that are loaded in a bin at any instant depends on the channel condition of the bin. The frequency spectrum of the copper media is allocated as follows for the ADSL system: (i) 0-4 kHz, voice over POTS, (ii) 4-25 kHz, unused guard band, (iii) 25-160 kHz, 32 duplex bins, and (iv) 200 kHz-1.1 MHz, 224 simplex bins (downstream only).

The rest of the paper is organized as follows: In Section 2 we formulate the ADSL channel capacity based on the dynamics of NEXT interference levels. In Sections 3 and 4 we formulate the uncertainties outlined before as a *Markov Modulated Rate Process* (MMRP) process. In Sections 4.1 and 4.2 we solve the fluid-flow equations. Numerical results are presented in Section 5 with conclusions in Section 6.

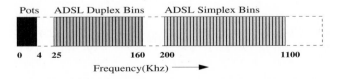

Fig. 2. G.DMT ADSL Frequency Spectrum Utilization

2   Evaluation of NEXT Limited Channel Capacity

We are specifically interested in the bin loading capacities of each bin in order to estimate the total capacity of the ADSL frequency spectrum. In our analysis, we have adopted the NEXT interference model by Kalet et. al. [1], where the channel capacity of NEXT limited ADSL channels is determined using a Gaussian model. It is also shown that the capacity of the channel is independent of the transmitted power spectral density. The dominant crosstalk transfer function $|H_x(f)|^2$, of NEXT noise due to N disturbers is given by [1] [2]:

$$|H_x(f)|^2 = \beta f^{3/2} \times N^{0.6} \tag{1}$$

where β is a constant that is characteristic of the copper cable and N is the number of proximal pair of disturbers in the cable for a single victim pair at any moment. Similarly, the attenuation transfer function $|H_0(f)|^2$, is given by [2]:

$$|H_0(f)|^2 = e^{-\alpha\sqrt{f}} \tag{2}$$

where f is the frequency in kHz and $\alpha = k\left(\frac{l}{l_0}\right)$, k being a constant of physical channel, l length of channel in ft, and l_0 the maximum reference length of cable (18,000 ft in our case) at which it is possible to support minimum channel speeds for given Quality of Service (QoS).

Let C_{Next} be the capacity of a NEXT only channel with no Gaussian interference. Using Shannon's model [3] for a conductor, we have:

$$C_{Next} = \int_{f \in A} df \log_2 \left(1 + \frac{|H_0(f)|^2 P_s(f)}{|H_x(f)|^2 P_s(f)}\right) bits/sec$$

$$= \int_{f \in A} df \log_2 \left(1 + \frac{|H_0(f)|^2}{|H_x(f)|^2}\right) bits/sec \tag{3}$$

where A is the frequency range in which $P_s(f) \neq 0$. If the frequency band of a bin ranges from f_1 to f_2 ($f_1 < f_2$), then the capacity of the individual bin using Equations (1), (2), and (3) is given by:

$$C_{Next} = \int_{f_1}^{f_2} df \log_2 \left(1 + \frac{e^{-\alpha\sqrt{f}}}{\beta f^{3/2} \times N^{0.6}}\right) bits/sec$$

$$\Rightarrow C_{Next} = \int_{f_1}^{f_2} df \ln \left(1 + \frac{e^{-\alpha\sqrt{f}}}{\beta f^{3/2} \times N^{0.6}}\right) nats/sec \tag{4}$$

3   NEXT Interference Model

The interference in a NEXT dominated system depends on:

1. The number of interference sources present in the cable system. For example, it has been observed that for N number interfering signals, the interference power is proportional to $N^{0.6}$ in cables of US origin and $N^{0.7}$ and $N^{0.8}$ in cables of European origin. In general, there are 25 to 100 pairs of copper wire housed in each cable binder out of which only 7 to 9 proximal pairs are potential sources of interference [2].
2. The number of ADSL sources in a binder group and their respective duration of ON-OFF times. The cumulative distribution of the ON-OFF sources govern the probability distribution of the number of active sources at any instant of time.

From the individual bin capacity given by Equation (4), the capacity of the entire frequency spectrum is obtained by algebraically summing the capacities for all the bins forming the ADSL spectrum for a given number of NEXT interferers. However, to reduce the computational complexity of our analysis, we have selected the following bins as *representative bins* for the entire channel. This is shown in Table-1 with Simplex Bins for up-link and the Duplex Bins for down-link capacities of the channel.

Table 1. Representative bins for calculating channel capacities

Duplex Bins Bin Nos.	Frequency Range (KHz)	Simplex Bins Bin Nos.	Frequency Range (KHz)
6	25.000-29.312	39	163.000-167.312
22	89.687-94.000	147	624.437-628.750
38	158.687-163.000	255	1090.187-1094.500

3.1 NEXT Uncertainty Model

We model the probability distribution of the Single NEXT Interference using the Engset distribution [4]. In this case the activation of the NEXT source (i.e., presence of ADSL traffic in the respective copper wire generating NEXT interference) is the *birth process* and deactivation of the NEXT source (i.e., absence of ADSL traffic in the respective copper wire generating no NEXT interference) is the *death process*. We consider each state of the Markov chain being identified by the number of active interferers i. Let N be the maximal number of interferers in the system, λ be the birth rate and μ be the death rate of each interferer. In Figure 3, a single active ADSL line is subjected to $(N-1)$ random interference sources in a binder group. The interfering sources are active for certain time after which they becomes inactive. When a source is active, it contributes towards the NEXT interference of the ADSL line under consideration. This physical process is modeled by the Engset Birth-Death model as illustrated in Figure 4. Thus:

$$\lambda_i = \lambda * (N - i) \tag{5}$$

$$\mu_i = i * \mu \tag{6}$$

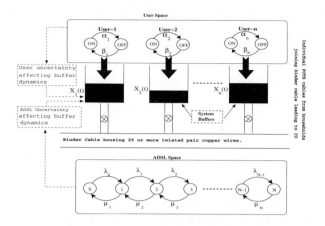

Fig. 3. NEXT System Reference Model

The probability for each state P_i, is given by:

$$P_i = P_0 \times \prod_{i=0}^{N-1} \left[\frac{\lambda * (N-i)}{(i+1) * \mu} \right]$$

$$\Rightarrow P_i = P_0 \times \left(\frac{\lambda}{\mu} \right)^i \binom{N}{i} \tag{7}$$

Applying $\sum_{i=0}^{N} P_i = 1$, and eliminating P_0 we have:

$$P_i = \frac{\left(\frac{\lambda}{\mu} \right)^i \binom{N}{i}}{\sum_{i=0}^{N} \left(\frac{\lambda}{\mu} \right)^i \binom{N}{i}} \tag{8}$$

The estimated probability P_i gives the probability of the ADSL channel affected by i NEXT interferers. We present the model in the next section.

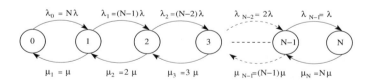

Fig. 4. Engset Markov Chain comprising N interferers

3.2   Single NEXT Interference Capacity Model

This is the simplest case where only two proximal active pairs of cables are assumed to exist and influence each other through NEXT interference. This system can be modeled by the simple birth-death Finite State Markov Chain as illustrated in Figure 5. The *inter-arrival time* (IAT) $1/\lambda$, depends on ON-OFF probability of the source generating user traffic and the *service rate* $1/\mu$, is dependent on the effective channel capacity limited by the presence of an active interferer. A tabulation of the rate dynamics of the system defined in Figure 5 for the 4 ON-OFF models is listed in Table 2. The 4 ON-OFF probabilities

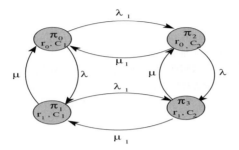

Fig. 5. Cross-layer state diagram for single NEXT interferer

Table 2. Parameters of the system shown in Figure 5

$P_{on/off}$	λ sec^{-1}	IAT	μ (sec^{-1})
50-50	250	0.0004	200
53-47	265	0.0037	200
56-44	280	0.0035	200
60-40	300	0.0044	200

($P_{on/off}$ in Table 2) have been chosen considering the fact that steady state behavior of the system generated should be stable and sustainable depending on the eigenvalues of the system matrix.

4 Cross-Layer View

In this section we form the cross-layer model of the ADSL channel. It consists of the NEXT limited capacity of the copper wire in the physical layer and the buffer capacity in the MAC layer of the system where the bits from applications are getting buffered. Figure 3 illustrates the cross-layer view of the system considered in our model. We are interested in finding the *effective bandwidth* (EB) of the ADSL system given a specific buffer size and fixed overflow probability of the system buffer. The model can be used to determine the cross-layer parameters of the ADSL system as a two dimensional MMP. Using this model along with the fluid flow approach of [5] [6] [7], we are able to estimate the EB of the ADSL channel and the overflow probability of the associated system buffer. The source traffic is a traffic stream consisting of an ON-OFF source flowing into a finite capacity system buffer in the NIC card. We assume a first-in-first-out (FIFO) system buffer with large capacity where all packets have equal priority so that the buffer content distribution represented by $X(t)$ is always a continuous random variable irrespective of the discrete nature of arriving and departing flows. The fluid flow model evaluates the overflow probability and effective bandwidth of the system.

4.1   Buffer Content Distribution

Let $X(t)$ be the buffer content at time instant t, $\mathbf{D}$ the fluid outflow matrix or the rate matrix, $\mathbf{M}$ the state transition matrix, and $\mathbf{\Pi}(x)$ the steady state probability vector of the MMP. We are interested in finding the solution to the condition $\mathbf{\Pi}_{1i}(x) = Prob[X(t) \leq x]$ which means a single source is in 'ON' state and the underlying ADSL channel is in state i. By [5], the solution is governed by the differential equation :

$$\mathbf{D\Pi}'(x) = \mathbf{M\Pi}(x) \tag{9}$$

By spectral decomposition, the general solution to Equation 9, in an infinite buffer system, is given by:

$$\mathbf{\Pi}(x) = \sum_{Re(z_i)<0} a_i \mathbf{\Phi}_i e^{z_i(x)}, z_i \geq 0 \tag{10}$$

where a_n are the unknown coefficients, z_n and $\mathbf{\Phi}_n$ are the left eigenvalues and eigenvectors of the matrix product $\mathbf{MD}^{-1}$. The initial condition to Equation 9 is obtained by balancing the steady state Markov chain $\mathbf{M\Pi}(0) = \mathbf{0}$; where $\mathbf{\Pi}(0)$ is the eigen vector corresponding to eigen value 0. Thus:

$$\mathbf{\Pi}(x) = \mathbf{\Pi}(0) + \sum_{Re(z_i)<0} a_i \mathbf{\Phi}_i e^{z_i(x)}, z_i \geq 0 \tag{11}$$

4.2   Overflow Probability and Effective Bandwidth

Equation (9) in the previous section provides the infinite buffer content distribution under steady state conditions. However, for the evaluation of overflow probability and EB, we need to determine the critical boundary conditions of the system governed by Equation (9). Thus, for a system with large buffer sizes and with small loss probability, packet loss rate (PLR),$\mathbf{G}(x)$ is determined from the tail distribution of the buffer system, $Prob\{(X > x)\}$:

$$\mathbf{G}(x) = 1 - \mathbf{1} * \mathbf{\Pi}(x) \tag{12}$$

With the increase in the number of states of the MMP, determination of the eigen values, eigen vectors and the coefficients of Equation (9) become computationally intensive. Hence, the value of $\mathbf{G}(x)$ is generally approximated from the dominant eigen value of the system. Consequently, $\mathbf{G}(x)$ is closely approximated as:

$$\mathbf{G} = Prob(\mathbf{X}) > x) \approx \mathbf{L}e^{-zx} \tag{13}$$

The EB is calculated for a given buffer overflow probability ϵ and a given buffer capacity b. The value of L may be approximated by 1. We are interested in $P(X > b) = \epsilon$, which is approximately given by:

$$G(b) = \mathbf{L}e^{-zx} \le \epsilon \Rightarrow z \ge \log(\mathbf{L}/\epsilon)/b \tag{14}$$

5   Numerical Results

In this section, we present numerical solutions for evaluating the dynamics of the ADSL system. We have calculated the bit loading capacities of each bin (4.3125 KHz each) of the ADSL channel. Using Equation (4) it is possible to gauge the variation of the loading capacities across the entire channel. Figure 6 shows the distribution of the bin capacities across the entire spectrum. From the bit loading capacities depicted in Figure 6, we have selected the NEXT limited capacities of 6 bins (3 duplex and 3 simplex) that are maximally spaced over the ADSL spectrum for ease of computation, and calculated an aggregate channel capacity for each type of bin. A specimen calculation considering 2 NEXT interferers is shown in Table 3.

From Table 3 we calculate the total link capacity of the channel in $bits/sec$ using the following formula:

Total up-link capacity is given by:

$$= Up - linkAvg. \times 32bins \times 8bits/symbol$$

$$= 64.219 \times 32 \times 8 = 16440.064bits/sec.$$

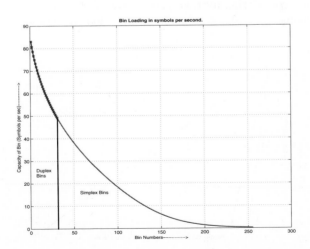

Fig. 6. Bit loading capacity variation of the entire ADSL spectrum as per Equation(4)

Table 3. A specimen calculation considering 2 NEXT interferers

Bin No	Frequency Range in KHz	Capacity for 2-NEXT Interferers in Symbols Sec-1	Average Capacity in Symbols Sec-1
	Duplex Bins		
6	25.000-29.312	83.0314	
22	89.687-94.000	61.1481	64.219
38	158.687-163.000	48.4779	
	Simplex Bins		
39	163.000-167.312	47.8256	
147	624.437-628.750	8.1010	18.749
255	1090.187-1094.500	0.3192	

The average down-link capacity is given by,

$$(Duplex Bins Avg. + Simplex Bins Avg.)/2$$

$$(64.219 + 18.749)/2 = 41.484 Symbols/Sec$$

Hence the down-link capacity may be given by,

$$Total Down - link capacity = Down - link Avg. \times 249 \times 8 bits/symbol$$

$$= 41.484 \times 249 \times 8 = 82038.528 bits/sec.$$

It is worth observing that the average bit loading capacities of a bin does not vary significantly in the presence of 3 or 4 NEXT interferers. Hence, we have assumed an average channel speed of 0.8 Mbps on the downstream channel which accounts for more than 85 percent of the traffic on the ADSL link. With this underlying assumption, it is possible to find out the effective service rate of the ADSL channel, considering average packet size of 500 bytes each. The service rate of the channel is $= (0.8 \times 10^6)/(500 \times 8) = 200 \ packets/sec.$

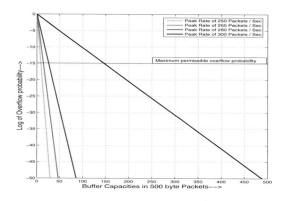

Fig. 7. Overflow probability plot of single NEXT model

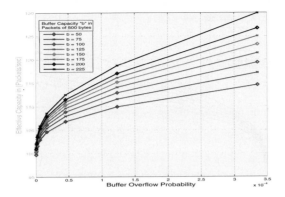

Fig. 8. Variation of Effective Capacity with respect to Buffer Overflow Probability for PSR of 265 Packets/sec for fixed Buffer Capacities in case of 2 interferers

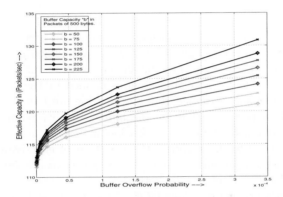

Fig. 9. Variation of Effective Capacity with respect to Buffer Overflow Probability for PSR of 280 Packets/sec for fixed Buffer Capacities in case of 2 interferers

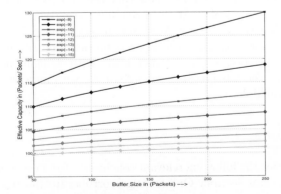

Fig. 10. Variation of Effective Capacity with respect to Buffer Capacity for PSR of 265 Packets/sec for fixed Buffer Overflow Probabilities in case of 2 interferers

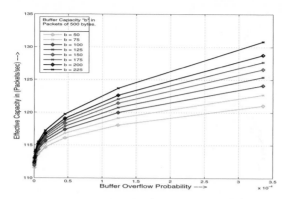

Fig. 11. Variation of Effective Capacity with respect to Buffer Capacity for PSR of 280 Packets/sec for fixed Buffer Overflow Probabilities in case of 2 interferers

In Figure 7, we plot overflow probability characteristics of the system in order to find out the minimum buffer provisioning required for a given probability of overflow for fixed peak source rates of 250, 265, 280, and 300 *packets/sec*. Here we have plotted the tail function of the buffer content distribution which signifies the overflow probability for a given system as defined by the eigen values and eigen vectors. It is possible to infer from graph of Figiure 7 that with increasing peak source rate of packet generation the buffer capacity required for such a system to maintain a given QOS for packet loss increases.

In Figures 8 and 9 we have shown the variation of effective capacity with varying buffer overflow probability and buffer capacity at different peak source rates of 265 and 280 per second. In both these graphs we observe that, lower the probability of overflow for a given QOS specification, the higher is the buffer provisioning required for the system under consideration. The graphs in Figures 8 and 9 throw light upon the behavior of the EB with changing overflow probability. The nature of these graphs show that the EB *increases monotonically* with increasing overflow probability for fixed buffer size. We also infer from these two plots that effective capacity of the system increases as we increase the buffer size of the system for constant overflow probabilities. The graphs in Figures 10 and 11 show the variation of effective bandwidth with changing buffer size. Here as well we see that the EB *increases monotonically* with increasing buffersize at fixed overflow probabilities.

6   Conclusions

We are able to draw the following relevant conclusions from the numerical results given in the previous section. Our findings corroborate the results stated by Elwalid et. al. in [8] where they claim that the EB is a monotonic increasing and convex function of all the state dependent rates of the source. This can be inferred by comparing the respective plots with the same buffer size with two different source rates in Figures 8 and 9. From the same two graphs we also observe that increase in effective capacity increases the overflow probability of the system for fixed buffer size, which at first might seem a little counter intuitive. The results from the last two sets of graphs in Figures 10 and 11 supports the intuition that increasing the buffercapacity increases the effective capacity of the system.

References

1. Kalet, I., Shamai, S.: On the capacity of a twisted-wire pair: Gaussian model. IEEE Transactions on Communications 38(3), 379–383 (1990)
2. Galli, S., Kerpez, K.J.: Methods of Summing Crosstalk from Mixed Sources Part I: Theoretical Analysis. IEEE Transactions on Communications vol.50(3) (March 2002)
3. Shannon, C.E.: A mathematical theory of communication. Univ. of Illinois Press, Urbana, IL (1949)
4. Kleinrock, L.: Queuing Systems Volume I: Theory. John Whiley and Sons Publications, New York (1974)
5. Anick, D., Mitra, D., Sondhi, M.M.: Stochastic theory of data handling system with multiple sources. Bell Syst. Tech. J. 61, 1871–1894 (1982)
6. Kim, J.G., Krunz, M.M.: Member. IEEE Bandwidth Allocation in Wireless Networks with Guaranteed Packet-Loss. Performance 8(3), 337–349 (2000)
7. Kundu, S.R., Basu, K., Das, S.K.: Finite State Markov Model for Effective Bandwidth Calculation in Wireless. In: Third International Symposium on Modeling and Optimization in Mobile, Ad Hoc, and Wireless Networks, pp. 351-357 (2005)
8. Elwalid, A.I., Mitra, D.: Effective Bandwidth of general Markovian traffic sources and admission control of high speed networks. IEEE/ACM Transactions on Networking 1, 329–343 (1993)

A Statistical Bandwidth Sharing Perspective on Buffer Sizing

J. Augé and J. Roberts

France Telecom
DRD/CORE/CPN
38, rue du Général Leclerc
92794 Issy-Moulineaux, France
{jordan.auge, james.roberts}@orange-ftgroup.com

Abstract. The issue of buffer sizing is rightly receiving increasing attention with the realization that the bandwidth delay product rule-of-thumb is becoming unsustainable as link capacity continues to grow. In the present paper we examine this issue from the light of our understanding of traffic characteristics and the performance of statistical bandwidth sharing. We demonstrate through simple analytical models coupled with the results of ns2 simulations that, while a buffer equivalent to the bandwidth delay product is certainly unnecessary, the recently advocated reduction to a few dozen packets is too drastic. The required buffer size depends significantly on the peak exogenous rate of multiplexed flows.

1 Introduction

The rule-of-thumb whereby router buffers should be sized to store a full bandwidth delay product (BDP) of data has recently come under considerable scrutiny [1,2,3,4,5,6,7]. It has been pointed out that to realize such large buffers for future 40Gb/s links is a significant design challenge for electronic routers and remains completely impractical for future optical routers. Moreover, the original reasoning behind the rule-of-thumb [8] is no longer valid for the present Internet backbone, both with respect to the size of links and their traffic and to the relative costs of memory and bandwidth.

Realized performance, in terms of packet loss and delay and flow throughput, for a given buffer size clearly depends on the assumed characteristics of link traffic. The previously cited papers differ significantly in their assumptions and consequently arrive at some conflicting conclusions. Our aim in the present paper is to identify the essential components of a typical mix of flows and to evaluate the buffer size performance trade-off under realistic traffic assumptions.

Internet traffic is composed of a dynamic superposition of finite size flows. An important characteristic of the flows sharing a given link is their exogenous "peak rate". This is the rate the flow would achieve if the link were of unlimited capacity. Some flows with a high peak rate will be bottlenecked by the considered link and will share bandwidth using end-to-end congestion control. However, the vast majority of flows are not bottlenecked because their peak rate, determined

L. Mason, T. Drwiega, and J. Yan (Eds.): ITC 2007, LNCS 4516, pp. 410–421, 2007.

for instance by a low speed access line, is much smaller than the current fair share. The number of bottlenecked flows is not an exogenous traffic characteristic but results from the dynamic statistical bandwidth sharing process and can be characterized as a function of the overall link load.

In the paper we review some simple models of statistical bandwidth sharing and identify two main link operating regimes relevant for buffer sizing. These are a "transparent" regime, where no flows are bottlenecked, and an "elastic" regime where a relatively small number of bottlenecked flows share bandwidth with a background traffic produced by many non-bottlenecked flows. While small buffers are sufficient in the transparent regime, it appears necessary to scale buffer size with link capacity in the elastic regime. We begin by reviewing the existing literature on the buffer sizing issue.

2   Related Work on Buffer Sizing

Appenzeller and co-authors were first to argue that the bandwidth delay product rule-of-thumb was both unsustainable, given the anticipated increase in network link capacity, and unnecessary [1]. Assuming flow congestion window (cwnd) evolutions are desynchronized they argued the required buffer should be proportional to $\sqrt{N}$ where N is the number of flows. It was noted, however, by Raina and Wischik [2] and Raina et al. [4] that the occurrence of flow synchronization depends on the buffer size and that, for a very large number of flows, the buffer size proposed in [1] is still too big. They suggest the buffer capacity needs to be as small as a few tens of packets. Instability was not observed in [1] because the authors only performed simulations for a few hundred flows whereas the phenomenon occurs for some thousands. In Section 3 we argue that it is not reasonable to suppose so many flows are actually *bottlenecked* at the link and therefore question the validity of this argument in favour of very small buffers.

Dhamdhere et al. [6] recognize the importance of distinguishing bottlenecked flows and non-bottlenecked flows. They suggest it is necessary to achieve a low packet loss rate while realizing high link utilization. Consequently they advocate a relatively large buffer that is proportional to the number of bottlenecked flows. The authors advance their analysis in [7], notably introducing open and closed loop dynamic flow level traffic models that correspond to our notion of statistical bandwidth sharing. However, the model in [6] is claimed to be valid when the bottlenecked flows constitute more than 80% of link load and performance is evaluated in [7] for a very high link load when some 200 bottlenecked flows are in progress. We again question the relevance of these traffic assumptions in evaluating buffer requirements.

The model proposed by Enachescu et al. [5,9] evaluates buffer requirements when none of the flows is bottlenecked. This is a valid assumption for many network links and corresponds to what we term the transparent regime (see Section 3). It is suggested in [9] that buffer size should be proportional to the log of the maximum TCP congestion window size. A necessary assumption is that packets are paced at the average rate determined by the window size rather than emitted as bursts. We note that the analysis in [2,4] is based on a fluid model and also therefore makes an implicit assumption that packet arrivals are not bursty. We believe buffer requirements must be evaluated for a mix of non-bottlenecked flows (where packets are "paced" to the flow peak rate) and bottlenecked flows that typically emit packets in bursts.

The present paper builds on our preliminary work [10]. We seek to evaluate buffer size based on our understanding of the statistical nature of traffic and accounting for the burstiness of TCP packet emissions.

3   Statistical Bandwidth Sharing

Consider a link of capacity C shared by a set of flows. The size and make up of this set of flows varies in time as flows of various types and characteristics arrive and depart. To understand what constitutes a typical traffic mix (e.g., for evaluating required buffer size), it is necessary to evaluate the performance of appropriate statistical bandwidth sharing models.

3.1   Processor Sharing Models

The processor sharing (PS) model for statistical bandwidth sharing provides insight into the way TCP flow-level performance depends on traffic characteristics [11,12]. In the PS model, flows are assumed to arrive according to a Poisson session model (a large population of users independently generate sessions, each session consisting of a succession of flows and think times) and to share link bandwidth perfectly fairly with any other concurrent flow. It can then be demonstrated that performance measures like expected flow throughput are largely insensitive to detailed traffic characteristics like the distribution of flow size or the number of flows in a user session. The essential characteristics are the mean load ρ, equal to flow arrival rate × mean flow size / link rate, and the flow "peak rate", the maximum rate a flow can attain independently of the considered link. Insensitivity is only approximate when flows have different peak rates or share the link unfairly due to different round trip times but the broad characteristics deduced from ideal symmetric models remain valid.

If flows can all individually attain the link rate, the number of flows in progress in the ideal fluid PS model has a geometric distribution of mean $\rho/(1-\rho)$. Despite the simplicity of the model, this is a good indication that the number of flows in contention at any instant would typically be quite small (e.g., less than 20 with probability .99 at 80% load). Of course, we know that the number of flows in progress on most network links is, on the contrary, very large (tens of thousands on a Gb/s link, say). This is because the vast majority of flows are peak rate limited and cannot realize a fair bandwidth share.

3.2   Throughput Performance

To illustrate statistical bandwidth sharing performance, we consider the following measure of flow throughput: γ = mean flow size / mean flow duration. For fair sharing, the parameter γ can also be interpreted as the expected instantaneous throughput of a flow in progress at an arbitrary instant [13]. We consider generalized PS models where flows fairly share an overall service rate that depends on the number in progress. This service rate depends, for example, on the link buffer size. Let the service rate when i flows are in progress be $\phi(i)C$ and write $\Phi(i) = \prod_1^i \phi(i)$. We have,

$$\gamma = \frac{\sum \rho^i / \Phi(i)}{\sum i \rho^{i-1} / \Phi(i)}. \tag{1}$$

3.3 Bandwidth Sharing Regimes

Figure 1 plots γ as a function of ρ when flows have no peak rate constraint ($\phi(i) = 1$) and when flows all have the same peak rate $p = 0.1C$ ($\phi(i) = \min(ip/C, 1)$). These simple cases illustrate two important points: i) the number of bottlenecked flows is very large only when link load is close to 1 (this number is proportional to $1/\gamma$), ii) when flows are peak rate limited, the link is transparent to throughput performance up to high loads (close to $(1 - p/C)$). More generally, in discussing buffer sizing, it is useful to distinguish three bandwidth sharing regimes:

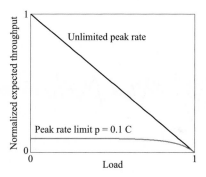

Fig. 1. Per flow throughput γ for PS model as fraction of link capacity C and as a function of load ρ

- an overload regime ($\rho > 1$) where realized flow throughput tends to zero as the number of competing flows increases,
- a transparent regime where the sum of peak rates of all flows remains less than link capacity (with high probability),
- an intermediate "elastic" regime where the majority of flows are peak rate limited but share the link with a small number of other flows capable of using all the residual capacity.

Buffer sizing is clearly inadequate for dealing with congestion in the overload regime. Alternative mechanisms to deal with this (i.e., traffic engineering and admission control) should also avoid situations of near overload when the number of bottlenecked flows grows rapidly. In the next sections we discuss buffer requirements for transparent and elastic regimes, respectively.

4 Buffer Sizing for the Transparent Regime

The transparent regime is characterized by the fact that the sum of flow peak rates is, with high probability, less than link capacity. The buffer must be sized to avoid significant loss due to coincident packet arrivals from independent flows. We assume the rate of flows is well defined at the time scale of packet emissions as, for instance, when it is limited by an upstream access line.

4.1 Locally Poisson Arrivals

Figure 2 depicts the overall rate of a superposition of peak rate limited TCP Reno flows using the simulation set-up specified in Figure 3. Without bottlenecked traffic, the flows arrive as a Poisson process and have a size drawn from an exponential distribution of 100 packet mean. The figures plot the average rate in successive 100ms slots. The rate is shown for two flow peak rates, $p = 200\text{Kb/s}$ and $p = 50\text{Kb/s}$, and that measured for a Poisson process.

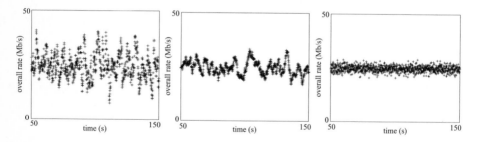

Fig. 2. Packet arrival process: rate of packets arriving in successive 100ms slots for flows of peak rate $p = 200\text{Kb/s}$, $p = 50\text{Kb/s}$ and $p = 0$

Visibly, a Poisson packet arrival process is not a good approximation unless the peak rate is very small relative to the link rate. However, in a small time interval (e.g., in each 100ms slot), the packet arrival process, as a superposition of a large number of periodic processes, is approximately Poisson. The figure depicts a realization of this modulated Poisson process. We denote its intensity by Λ_t.

Assuming Poisson arrivals allows buffer occupancy to be approximated locally by that of the M/G/1 queue. To simplify, one can further assume exponential packet sizes and approximate packet loss probability for a buffer of size B by $(\Lambda_t/C)^B$.

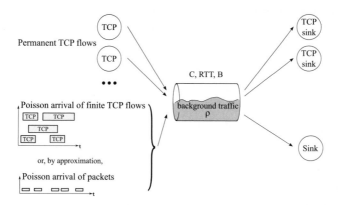

Fig. 3. Simulation set-up: unless otherwise stated, simulations use the following base set of parameters: $C = 50$ Mb/s, $B = 20$ packets, $RTT = 100$ ms, $\rho_b = 0.5C$, FIFO scheduling, 1000 bytes packets

4.2   Required Buffer Size

A possible approach for buffer sizing is to compute an average loss rate by conditioning on the distribution $F(\lambda)$ of Λ_t and requiring $\int (\lambda/C)^B dF(\lambda) < \epsilon$. This is reasonable when the rate variations are rapid so that ϵ is a good measure of the performance of any given flow. It turns out that, for a peak rate less than $.1C$ and $\epsilon > .001$, the required buffer size is the same as would be required for the Poisson packet arrival process. In other words, the M/M/1 formula $\rho^B < \epsilon$ is a useful sizing guideline. For example, a buffer of 20 packets limits admissible load to $\rho = .79$ for $\epsilon = .01$ or $\rho = .7$ for $\epsilon = .001$.

5   Buffer Sizing for the Elastic Regime

In general, there is no means to guarantee flow peak rates are limited and it is therefore important to understand the impact of buffer size on performance in the elastic regime (i.e., when one or several bottlenecked flows combine with background load to momentarily saturate the link for periods that are long compared to the time scale of packet emissions).

5.1   Unlimited Rate Bottlenecked Flows

To simplify analysis and discussion, we suppose a clear dichotomy between flows with unlimited peak rate and a background traffic composed by flows having a low peak rate. Furthermore, we assimilate the background traffic to a Poisson packet arrival process producing load $\rho_b C$. This simplification greatly facilitates the simulation experiments and reproduces the broad behavioural characteristics of more realistic background traffic.

To evaluate throughput performance we proceed as follows. For given link capacity, buffer size and background load, we successively simulate a number of permanent bottlenecked TCP flows (like in Fig. 3). For each number i (between 1 and 100), we evaluate the overall realized throughput $\phi(i)$ expressed as a fraction of residual capacity $C(1-\rho_b)$. We then derive the expected flow throughput γ by formula (1). This corresponds to a quasi-stationary analysis allowing us to ignore phenomena like loss of throughput in slow start and momentary unfairness.

Figures 4, 5, 6 depict the values of $\phi(i)$ and γ as a function of link load ρ for a range of configurations. Note that γ is only defined for loads greater than the background load ρ_b and its value at that load is determined by $\phi(1)$.

The results show that there is a significant loss of throughput with small buffers (Fig. 4) and that this loss is accentuated as link capacity increases (Fig. 5). The higher the background load, the more difficult it is for the TCP flows to fully use the residual capacity (Fig. 6).

To understand the loss in throughput it is necessary to explain the behaviour with just one bottlenecked flow. This determines $\phi(1)$ and consequently the form

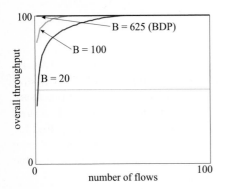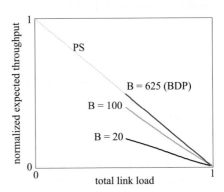

Fig. 4. Overall throughput $\phi(i)$ with i flows as fraction of residual bandwidth for a various number of flows, and expected flow throughput γ as function of load ρ, for various buffer sizes

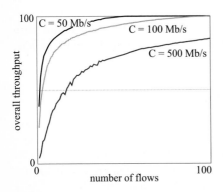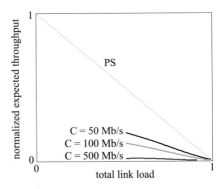

Fig. 5. Overall throughput $\phi(i)$ with i flows as fraction of residual bandwidth for a various number of flows, and expected flow throughput γ as function of load ρ, for various capacities

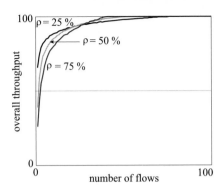
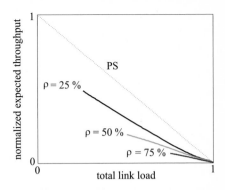

Fig. 6. Overall throughput $\phi(i)$ with i flows as fraction of residual bandwidth for a various number of flows, and expected flow throughput γ as function of load ρ, for various background loads

of γ which is approximately linear, decreasing from the maximum for $\rho = \rho_b$ to 0 for $\rho = 1$.

While the TCP window is small compared to the residual bandwidth delay product $C(1 - \rho_b) \times$ RTT, packets are emitted in bursts starting at instants separated by a Round Trip Time (RTT). By TCP self-clocking, the sum of the burst rate and the rate of the background traffic is very close to the link capacity. Buffer occupancy therefore tends to increase under this heavy load while the burst is in progress and then to empty when only background packets arrive. In the absence of loss, TCP increases cwnd by 1 packet per RTT, prolonging the period of overload. At some point background and bottlenecked flow packet arrivals combine to saturate the buffer and a packet is lost.

If the residual bandwidth delay product is sufficiently large and the buffer is small, the process describing the value of cwnd when the packet loss occurs depends only on ρ_b. This determines the average window and therefore the flow throughput in this regime. For a larger buffer size, cwnd is able to increase further and eventually attain the value that completely fills the residual bandwidth.

Figure 7 plots the product $\phi(1)C(1 - \rho_b)$RTT as a function of buffer size. For small buffers this is equal to the expected value of cwnd and depends only on ρ_b. As the buffer size increases, the residual bandwidth is used entirely and the function flattens to a horizontal line.

The form of $\phi(1)$ as a function of B suggests the buffer should be sized to at least avoid the initial high degradation in throughput. It is not necessary, however, to attain 100% efficiency and a buffer considerably smaller than the residual bandwidth delay product would be sufficient.

A possible approach would be to set B to the value where the common curve for small buffers and given background load ρ_b intersects with the horizontal line representing the residual bandwidth delay product.

Inspection of the form of the common curve for given background load suggests an approximate dependency in B^2. In other words, the required buffer size

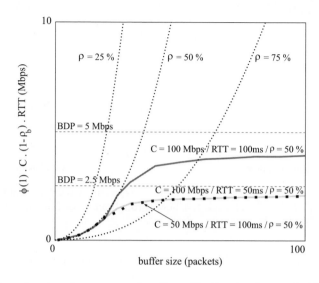

Fig. 7. Throughput performance as function of buffer size for one bottlenecked flow

according to the above approach would be roughly proportional to the square root of the residual bandwidth. This clearly requires further investigation, notably by more realistically modelling background traffic, but is an indication of the likely dependence of buffer size on link capacity for this elastic regime.

5.2   Peak Rate Limited Bottlenecked Traffic

The assumption of unlimited peak rate bottlenecked flows is not necessarily reasonable since even the rate of high speed access lines is generally only a fraction of link rate. To illustrate the impact of a limited peak rate, we assume flows of peak rate p share a link with Poisson background traffic. Figure 8 plots throughput $\phi(i)$, as a function of the number of bottlenecked flows i, and γ/C, as a function of link load, for a number of configurations.

Results show that $\phi(i)$ increases linearly while the total rate of bottlenecked flows is somewhat less than the residual capacity as each flow realizes its peak rate and the link operates in the transparent regime. When overall load attains a level where the bottlenecked flows begin to lose packets, however, the inefficiency of small buffers is again apparent. For example, for $p = 2$ Mb/s in Figure 8, the throughput $\phi(i)$ dips when there are more than 11 flows and only increases to nearly 100% for a much larger number.

The efficiency loss with small buffers in this case is less significant for throughput performance, however, as illustrated by the behaviour of γ as a function of link load. The loss in throughput is only visible at high loads where it accentuates the degradation occurring in the ideal PS model (seen here with B=625).

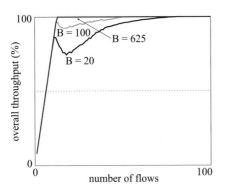
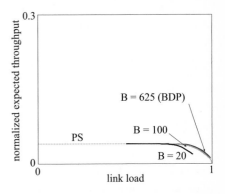

Fig. 8. Overall throughput $\phi(i)$ with i flows as fraction of residual bandwidth and expected flow throughput γ as function of load ρ, for peak rate limited bottlenecked flows (2Mb/s)

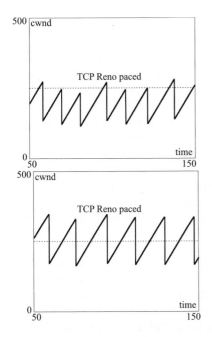
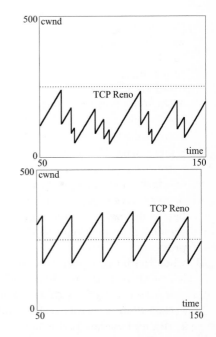

Fig. 9. Evolution of cwnd for one paced Reno flow (left) and one Reno flow (right), with $B = 20$ packets (top) and $B = 100$ packets (bottom), for $C = 50$ Mbps, $\rho_b = 0.5$

5.3 Paced TCP

In [9] it is proposed that flows that are not peak rate limited by an access line should use pacing. This would indeed attenuate the loss of throughput for

small buffers arising when a window is emitted as a burst at the start of each
RTT interval. Figures 9 illustrate the evolution of cwnd for a single bottlenecked
flow with[1] and without pacing with $B = 20$ and $B = 100$. The results confirm that pacing significantly improves the performance of the small buffer since
throughput is roughly 50% higher. The difference is negligible for the larger
buffer $B = 100$.

6   Conclusions

The relation between buffer size and realized performance clearly depends on
the assumed traffic characteristics. The most significant characteristic is the mix
of exogenous flow peak rates, the rates flows would attain if the considered link
were of unlimited capacity. The link load (flow arrival rate × mean flow size
/ link capacity) then determines which, if any, high peak rate flows are bottlenecked, the remainder constituting a background load. We distinguish three
main statistical bandwidth sharing regimes:

1. when all peak rates are a relatively small and load is not too close to 1,
   the sum of flow rates remains less than link capacity with high probability;
   we refer to this as the transparent regime; a simple $M/M/1$ queue model
   can be used to evaluate the relationship between buffer size and packet loss
   probability; a small buffer is then adequate; for example, a 20 packet buffer
   overflows with probability 0.01 at a load close to 80%;
2. when some flows can individually saturate the residual link bandwidth not
   used by the background load due to low peak rate flows, bandwidth sharing
   is controlled by end-to-end congestion control; we refer to this as the elastic
   regime; with the current practice of sending packets as soon as they are
   authorized by the receipt of an acknowledgement, a small buffer tends to
   overflow too early to allow full development of cwnd and utilization can
   be very low; required buffer size in this regime increases with the residual
   bandwidth delay product; preliminary empirical evidence suggests buffer size
   should be proportional to the square root of the residual bandwidth delay
   product;
3. when the highest peak rate flows must combine (i.e., several flows in parallel) to saturate the residual bandwidth, we have a more general intermediate
   transparent/elastic regime; when the peak rate of the (potentially) bottlenecked flows is a relatively small fraction of the residual bandwidth (e.g.,
   1/10), and overall load is not too close to 1, the link is rarely saturated and
   a small buffer sized for the transparent regime is adequate.

Since large buffers and fair queuing appear to be impractical propositions for
future optical routers, it appears important to ensure that these always operate
in the transparent regime. It may be sufficient to rely on the continuing large
disparity between flow peak rates and backbone link capacity, as suggested in [9].

---

[1] We used the paced TCP ns2 code made available by D.X. Wei : *A TCP pacing implementation for NS2*, available at http://www.cs.caltech.edu/~weixl/
technical/ns2pacing/index.html, with the *traditional pacing* option.

References

1. Appenzeller, G., Keslassy, I., McKeown, N.: Sizing router buffers. In: Proceeding of ACM SIGCOMM '04, Portland, Oregon (September 2004)
2. Raina, G., Wischik, D.: Buffer sizes for large multiplexers: TCP queueing theory and instability analysis. NGI'05, Rome (April 2005)
3. Wischik, D., McKeown, N.: Part I: buffer sizes for core router. ACM SIGCOMM Computer Communication Review 35(3) (July 2005)
4. Raina, G., Towsley, D., Wischik, D.: Part II: control theory for buffer sizing. ACM SIGCOMM Computer Communication Review 35(3) (July 2005)
5. Enachescu, M., Ganjali, Y., Goel, A., McKeown, N., Roughgarden, T.: Part III: routers with very small buffers. ACM SIGCOMM Computer Communication Review 35(3) (July 2005)
6. Dhamdhere, A., Dovrolis, C., Jiang, H.: Buffer sizing for congested internet links. In: Proceedings of IEEE INFOCOM, Miami FL (March 2005)
7. Dhamdhere, A., Dovrolis, C.: Open issues in router buffer sizing. ACM SIGCOMM Computer Communications Review (editorial section) (January 2006)
8. Villamizar, C., Song, C.: High performance TCP in ANSNET. Computer Communications Review 24(5), 45–60 (1994)
9. Enachescu, M., Ganjali, Y., Goel, A., McKeown, N., Roughgarden, T.: Routers with very small buffers. In: Proceedings of the IEEE INFOCOM'06, Barcelona, Spain (April 2006)
10. Augé, J., Roberts, J.: Buffer sizing for elastic traffic. NGI'06, València (April 2006)
11. Fredj, S.B., Bonald, T., Proutière, A., Régnié, G., Roberts, J.: Statistical bandwidth sharing: a study of congestion at flow level. SIGCOMM 2001, San Diego, CA, USA (August 2001)
12. Bonald, T., Proutière, A.: Insensitivity in processor-sharing networks. In: Proceedings of Performance (2002)
13. Bonald, T.: Throughput performance in networks with linear capacity constraints. In: Proceedings of CISS 2006 (2006)

Approximating Flow Throughput in Complex Data Networks

Juha Leino, Aleksi Penttinen, and Jorma Virtamo

Networking Laboratory, TKK Helsinki University of Technology,
P.O. Box 3000, FI-02015 TKK, Finland
`firstname.lastname@tkk.fi`

Abstract. Flow level analysis of data networks has recently taken a major step towards tractability with the introduction of a resource sharing scheme called balanced fairness. We consider the balanced fairness concept in analyzing per-flow throughput in complex networks with a large number of flow classes. The two existing practical approaches in the setting, namely performance bounds and asymptotic analysis, require that the capacity set of the network is given explicitly as a set of (linear) constraints. We extend the asymptotic analysis method by providing explicit expressions for the second order throughput derivative in the light traffic regime. We show how asymptotic analysis can be applied in multipath routing and wireless networks, where the linear constraints cannot be readily worked out in explicit form. Finally, we introduce a numerical throughput analysis scheme based on Monte Carlo method.

1 Introduction

Traffic in data networks consists primarily of elastic flows, i.e., file transfers using TCP protocol. These flows can adapt their transmission rate to the available resources and, indeed, the bandwidth shares of the flows are frequently adjusted as new flows arrive or others leave the network as the corresponding file transfers are completed. Analyzing the flow level performance of such a dynamic system is generally difficult, but has been recently brought within the realm of tractability by the introduction of the balanced fairness (BF) resource sharing scheme by Bonald and Proutière [1]. Balanced fairness allows deriving the equilibrium distribution of the dynamic flow system by a simple recursion and even leads to explicit solutions in certain cases, e.g., in tree networks [2].

We study a performance metric called flow throughput, which is defined as the ratio of mean flow size and mean flow duration. We focus on complex networks with a large number of flow classes, i.e., different types of flows, where the throughput analysis under BF resource sharing generally resorts to two approximate methods; throughput bounds [3,4] and asymptotic analysis [5]. The throughput bounds (both upper and lower) are available if the network capacity set, i.e., the set of the available flow rates, can be expressed as an intersection of linear half-spaces. The asymptotic analysis concentrates on characterizing the exact throughput behavior at very low and very high loads, respectively. This

L. Mason, T. Drwiega, and J. Yan (Eds.): ITC 2007, LNCS 4516, pp. 422–433, 2007.

information allows "educated" guesses on the throughput behavior between the extremes by an appropriate interpolation. Also the heavy load asymptotic analysis in practice requires that the (linear) capacity set constraints are known.

The contributions of this work to the throughput analysis of complex data networks are: I) extension of asymptotic analysis by providing second derivative formulas in the light traffic regime, II) extension of asymptotic analysis to two important network scenarios, namely multipath routing [6] and wireless networks [7], where the capacity set is given in an implicit form, III) introduction of a Monte Carlo based throughput approximation for low and medium loads.

The paper is organized as follows. Section 2 describes the flow level modeling and asymptotic analysis. Section 3 and Sect. 4 discuss multipath routing and wireless network scenarios, respectively. Section 5 outlines the Monte Carlo method for throughput computation before concluding in Sect. 6.

2 System Description

We study communication networks in a dynamic setting where flows (file transfers) are initiated randomly and depart upon completion. The flows are assumed elastic, i.e., the size of the flow is fixed and the transmission utilizes all the capacity allocated to it. The transmission rate of each flow depends on the network capacity, which is shared among contending flows according to the balanced fairness principle [1]. The rate received by a flow evolves during the flow lifetime stochastically as other flows enter the system or depart from it.

We are interested in the average flow throughput, defined as the ratio of the mean file size and mean flow duration for any pre-defined flow class. Our goal is to characterize the throughput behavior as traffic loads of different classes are increased from zero to the capacity limit of the network in given proportions.

2.1 Network Model

Consider a communication network used by N flow classes. A flow class represents a set of similar flows in terms of network resource usage. In a wired network, for instance, a flow class is characterized by a set of links representing a path in the network. Let x_i be the number of class-i flows in progress and denote the network state by $\mathbf{x} = (x_1, \ldots, x_N)^{\mathrm{T}}$.

Class-i flows arrive stochastically and have finite, random sizes. We denote by ρ_i the traffic intensity (in bit/s) of class-i flows, defined as the product of the flow arrival rate and the mean flow size. We use the notation $\boldsymbol{\rho} = (\rho_1, \ldots, \rho_N)^{\mathrm{T}}$.

In each state $\mathbf{x}$ the network resources are shared by the contending flows. Let $\boldsymbol{\phi} = (\phi_1(\mathbf{x}), \ldots, \phi_N(\mathbf{x}))^{\mathrm{T}}$ be the vector of capacities allocated to each flow class. The allocation $\phi_i(\mathbf{x})$ is assumed to be equally shared between the x_i flows in each class i. The network resources are defined by the capacity set $\mathcal{C}$, which is a collection of capacity allocations $\boldsymbol{\phi}$ that can be supported by the network. For example, a wired network has the capacity set $\mathcal{C} = \{\boldsymbol{\phi} : \mathbf{A}\boldsymbol{\phi} \leq \mathbf{c}\}$, where c_j is the capacity of link j, and $\mathbf{A}$ is the link-flow incidence matrix with $A_{ji} = 1$ if flow i uses link j and 0 otherwise.

2.2   Throughput in Balanced Fairness

The throughput experienced by flows depend on how the available capacity is shared between the flows. In this work we assume that the balanced fairness (BF) resource sharing scheme is utilized. BF is the most efficient resource allocation for which the Markov process on $\mathbf{x}$ is reversible [1] and thus makes the system considerably simpler to analyze than other sharing schemes such as max-min fairness of proportional fairness. Furthermore, with BF the throughput is insensitive to detailed traffic characteristics such as flow size distribution. In fact, the whole steady state distribution and, therefore, all performance metrics derivable from it are insensitive provided that the session arrival process is Poissonian [1]. Each session may consist of flows and idle times whose distributions can be arbitrary and even correlated within a session. Despite its somewhat non-intuitive definition, BF provides a reasonable approximation of flow throughput in systems where other fair resource sharing schemes are applied [8].

The steady state distribution of the system under BF is

$$\pi(\mathbf{x}) = \frac{1}{G(\boldsymbol{\rho})}\Phi(\mathbf{x}) \prod_i \rho_i^{x_i}, \tag{1}$$

where $\Phi(\mathbf{x})$ is the balance function defined by $\Phi(\mathbf{0})=1$, $\Phi(\mathbf{x})=0$, $\forall\, \mathbf{x} \notin \mathbb{Z}_+^N$ and

$$\Phi(\mathbf{x}) = \min\{\alpha : \frac{\tilde{\boldsymbol{\Phi}}(\mathbf{x})}{\alpha} \in \mathcal{C}\}, \tag{2}$$

where $\tilde{\boldsymbol{\Phi}}(\mathbf{x}) = (\Phi(\mathbf{x} - \mathbf{e}_1), \ldots, \Phi(\mathbf{x} - \mathbf{e}_N))^{\mathrm{T}}$ with $\mathbf{e}_i$ denoting a vector in which the element i is 1 and the others are all zero. $G(\boldsymbol{\rho})$ is the normalization constant,

$$G(\boldsymbol{\rho}) = \sum_{\mathbf{x}} \Phi(\mathbf{x})\rho_1^{x_1} \ldots \rho_N^{x_N}. \tag{3}$$

The key performance metric is the flow throughput, defined as the ratio of the mean flow size to the mean flow duration. By Little's result, the flow throughput γ_i of class-i flows is given by

$$\gamma_i = \frac{\rho_i}{\mathrm{E}[x_i]} = \frac{\rho_i}{\sum_{\mathbf{x}} x_i \pi(\mathbf{x})} = \frac{\rho_i}{\frac{\rho_i}{G(\boldsymbol{\rho})}\frac{\partial}{\partial\rho_i}G(\boldsymbol{\rho})} = \frac{G(\boldsymbol{\rho})}{\frac{\partial}{\partial\rho_i}G(\boldsymbol{\rho})}. \tag{4}$$

For some simple systems recursion (2) can be solved in closed form. Generally, however, one has to resort to numerical recursion. The benefit of BF is that the solution can indeed be obtained recursively state-by-state using (1) and (2), which is much easier than solving the global balance equations essentially entailing a matrix inversion. Thus BF makes it possible to study larger systems than is otherwise feasible. For very large systems, however, even the recursive solution ultimately becomes infeasible calling for approximate methods.

2.3   Throughput Asymptotics

In [5] the authors proposed the following method to approximate the throughput of a given flow class. By computing the *asymptotic* throughput values and its derivatives with respect to the load at the low load and heavy load cases, one can quickly sketch (e.g., using interpolation) the throughput behavior in the system when load is increased from zero to the capacity limit along a given load line.

Let $\mathbf{p} = (p_1, \ldots, p_N)^{\mathrm{T}}$, with $\sum p_i = 1$, be a given traffic profile. Load line is the set of all $\boldsymbol{\rho} \in \mathcal{C}$ such that the proportions of the loads in different classes are given by $\mathbf{p}$. Let $\hat{\boldsymbol{\rho}}$ be the end point of the load line on the boundary of $\mathcal{C}$, i.e. the maximum amount of traffic the system can sustain. Then the load line can be parameterized as $r\hat{\boldsymbol{\rho}}$, where $r \in [0, 1]$. We are interested in characterizing the class-i throughput $\gamma_i(r)$ along a given load line, where with slight abuse of notation we write $\gamma_i(r)$ instead of $\gamma_i(r\hat{\boldsymbol{\rho}})$.

Denote $G_i(\boldsymbol{\rho}) = \partial/\partial\rho_i G(\boldsymbol{\rho})$, $G(r) = G(r\hat{\boldsymbol{\rho}})$, and $G_i(r) = G_i(r\hat{\boldsymbol{\rho}})$. At low loads one can easily derive from (3) and (4) the following values for the throughput and its derivatives with respect to r at load $r = 0$:

$$
\begin{cases}
\gamma_i(0) = \dfrac{G(0)}{G_i(0)}, \\[2ex]
\gamma_i'(0) = \dfrac{G_i(0)G'(0) - G(0)G_i'(0)}{G_i(0)^2}, \\[2ex]
\gamma_i''(0) = \dfrac{2G_i'(0)\left(G(0)G_i'(0) - G'(0)G_i(0)\right) + G_i(0)\left(G_i(0)G''(0) - G(0)G_i''(0)\right)}{G_i(0)^3},
\end{cases}
\tag{5}
$$

where $G(0) = 1$ and the rest of the terms are

$$
G'(0) = \sum_{j=1}^{N} \Phi(\mathbf{e}_j)\hat{\rho}_j, \qquad\qquad G_i(0) = \Phi(\mathbf{e}_i),
$$

$$
G''(0) = 2\sum_{j=1}^{N}\sum_{k=j}^{N} \Phi(\mathbf{e}_j + \mathbf{e}_k)\hat{\rho}_j\hat{\rho}_k, \qquad G_i'(0) = \sum_{j=1}^{N} \Phi(\mathbf{e}_i + \mathbf{e}_j)\hat{\rho}_j + \Phi(2\mathbf{e}_i)\hat{\rho}_i,
$$

$$
G_i''(0) = 2\left(\sum_{j=1}^{N}\sum_{k=j}^{N} \Phi(\mathbf{e}_i + \mathbf{e}_j + \mathbf{e}_k)\hat{\rho}_j\hat{\rho}_k + \sum_{j=1}^{N} \Phi(2\mathbf{e}_i + \mathbf{e}_j)\hat{\rho}_i\hat{\rho}_j + \Phi(3\mathbf{e}_i)\hat{\rho}_i^2\right).
\tag{6}
$$

In the heavy load regime we are interested in the derivative $\gamma_i'(1)$ for those classes i such that $\gamma_i(1) = 0$. We assume a polytope capacity set, i.e., $\mathcal{C} = \{\boldsymbol{\phi} : \mathbf{B}\boldsymbol{\phi} \leq \mathbf{e}\}$, with $\mathbf{e} = (1, \ldots, 1)^{\mathrm{T}}$ and $\mathbf{B}$ a matrix representing the constraints. Let $\mathcal{L}$ represent the set of saturated constraints $l \in \{1, \ldots, L\}$ at $r = 1$. It is conjectured (proven if $|\mathcal{L}| = 1$, conjectured if $|\mathcal{L}| > 1$) in [5] that the heavy load derivative is given by

$$
\gamma_i'(1) = -\frac{1}{\sum_{l \in \mathcal{L}} b_{li}},
\tag{7}
$$

where the b_{li} are the corresponding elements in $\mathbf{B}$.

Using the values and the derivatives at the extreme traffic setting one can efficiently sketch the throughput behavior, see [5]. In the following, we apply this scheme in networks with multipath routing and in wireless networks. In both cases the explicit determination of the linear constraints is generally infeasible and the analysis has to be carried out with the implicit capacity set definition.

3   Multipath Routing

First, we study multipath routing problem in fixed networks, where each flow can use resources on multiple routes. This way, the resources of the network are used more efficiently and the performance is better than when only one route is used. The problem is formulated using two different approaches. First, we assume that each traffic class uses a set of predefined routes. Second, we assume that the traffic classes can utilize all the possible routes in the network. While the first approach has more practical value, for example in modelling of load balancing, the second approach gives an theoretically interesting upper limit of the performance. In both cases, the balanced fairness recursion (2) needed in the asymptotic approximation can be formulated as a network flow problem [6]. In addition, the capacity limit $\hat{\rho}$ and the set of saturated constraints needed in the heavy load derivative (7) can be derived using the same problem formulation.

3.1   Balanced Fairness Recursion

First, we formulate the balanced fairness recursion corresponding to (2) in a network with predefined routes. As before, $\mathbf{c}$ is a vector containing the link capacities. The set of links is denoted $\mathcal{J}$. $\boldsymbol{\phi}$ now denotes the flow matrix, where element $\phi_{i,r}$ is the amount of class-i traffic on route r. Each route r consists of a set of links $r \subset \mathcal{J}$. $\mathbf{R}$ is the routing matrix, where element $R_{r,j}$ is 1 if route r uses link j and 0 otherwise. Starting from $\varPhi(\mathbf{0}) = 1$, the balanced fairness recursion can be solved as an LP problem:

$$
\begin{aligned}
\varPhi(\mathbf{x})^{-1} = \quad &\max_{\boldsymbol{\phi}} \; \alpha, \\
\text{s.t.} \quad &\boldsymbol{\phi}\,\mathbf{e} = \alpha\,\tilde{\boldsymbol{\varPhi}}(\mathbf{x}), \\
&\mathbf{e}^{\mathrm{T}}\boldsymbol{\phi}\mathbf{R} \le \mathbf{c}^{\mathrm{T}}, \\
&\boldsymbol{\phi} \ge \mathbf{0},
\end{aligned}
\tag{8}
$$

with the convention $\varPhi(\mathbf{x}) = 0$ for all $\mathbf{x} \notin \mathbb{Z}_+^N$. The first constraint corresponds to equation (2) and ensures that the capacity allocation is BF. The second constraint is the link capacity constraint.

If the traffic classes can use all possible routes in the network, the previous approach is not feasible as the number of possible routes explodes with the size of the network. However, the BF recursion step problem corresponds to the well-known multicommodity flow problem and enumeration of all the paths can be avoided by formulating the problem in the LP setting. Define the link-node incidence matrix $\mathbf{S}$ such that element $S_{j,n}$ has the value -1 if link j originates from node n, 1 if the link ends in node n, and 0 otherwise. $\mathbf{D}$ is the divergence matrix, i.e. the difference between the incoming and outgoing traffic in each node. Element $D_{i,n}$ is ± 1 if n is the source (destination) node of class i and 0 otherwise. Matrix ϕ now contains the link traffics. Element $\phi_{i,j}$ is the class-i traffic along link j. The problem formulation reads

$$
\begin{aligned}
\Phi(\mathbf{x})^{-1} &= \max_{\phi} \alpha, \\
\text{s.t.} \quad \phi\mathbf{S} &= \alpha \, \mathrm{diag}(\tilde{\boldsymbol{\Phi}}(\mathbf{x})) \, \mathbf{D}, \\
\mathbf{e}^{\mathrm{T}} \phi &\leq \mathbf{c}^{\mathrm{T}}, \\
\phi &\geq \mathbf{0},
\end{aligned}
\tag{9}
$$

where $\mathrm{diag}(\mathbf{a})$ denotes the matrix with the elements of the vector $\mathbf{a}$ in the diagonal and 0 elsewhere. The first constraint is the divergence constraint expressing the balance of incoming and outgoing traffic of each class at each node, and the second is the link capacity constraint.

3.2 Throughput Asymptotics

Light traffic approximation is straightforward. Balance function values needed in equations (6) for calculating the derivatives (5) are solved using equations (8) or (9). Given a traffic profile $\mathbf{p}$ defining the proportions of the loads in different classes, the capacity limit of the network can be determined by solving the optimal α for the LP problems (8) and (9) with $\tilde{\boldsymbol{\Phi}}(\mathbf{x})$ being replaced by $\mathbf{p}$. The end point $\hat{\boldsymbol{\rho}}$ of the load line is then given by $\hat{\boldsymbol{\rho}} = \alpha\mathbf{p}$.

Next, we consider the heavy traffic approximation. In order to use the heavy load derivative (7), the constraints limiting the total traffic of the traffic classes need to be determined. Active constraints can be generated from the solutions of the LP problems (8) and (9) at the capacity limit. In both cases, the solution identifies a set of saturated links $\mathcal{J}_s$. The constraints limiting the flow classes can be generated by enumerating the subsets of $\mathcal{J}_s$. If the constraint corresponding to set $J \subset \mathcal{J}_s$ is active, subsets of J are omitted.

First, we consider the problem with predefined routes. Optimal solution of the LP problem (8) is denoted ϕ^*. The capacity of a subset $J \subset \mathcal{J}_s$, i.e. the sum of the link capacities, constraints flow class i if all the routes available for class i pass through at least one link in J. If every class-i route uses exactly n links in J, each unit of class-i traffic actually consumes n units of capacity, hence the capacity of J limits the term $n\phi_i$, where ϕ_i is the total rate allocated for class i. A more complex situation arises, if class i uses routes utilizing different number of links in J. In this case, the capacity of J constraints the term $n_{\max}\phi_i$, where $n_{\max}$ is the maximum number of common links among J and a route available for class i. However, only a part of class-i traffic uses all $n_{\max}$ links in J, hence the capacity in the constraint is increased with sum $\sum_r (n_{\max} - n_{J,r})\phi^*_{i,r}$, where $n_{J,r}$ is the number of set J links on route r.

When a network with arbitrary routing is considered, the same procedure can be used. Traffic classes not limited by link capacities are omitted. Interpreting class-i link flows as link capacities, a shortest path algorithm can be used

 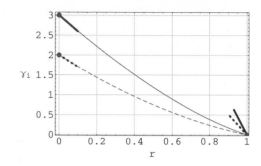

Fig. 1. Multipath network (left) and throughput asymptotics of the class denoted by the bigger nodes (right). Dotted lines illustrate routing with predefined routes, solid lines arbitrary routing.

iteratively to divide the total traffic on individual routes. Given the routes and the traffic amounts, the method used with predefined routes can be applied.

3.3   Example

Consider the fixed multipath network illustrated in Fig. 1. The network has 8 nodes and 19 links with unit capacities and an equal amount of traffic between all the 28 node pairs. Both multipath approaches are studied. For predefined routes, we choose two shortest link-disjoint routes between every node pair. The throughput between the extremes is sketched using a cubic polynomial fitted to the end points and to the first two derivatives at $r = 0$. Inclusion of the heavy load derivative would overestimate the throughput at high loads if a simple polynomial interpolation is used. At high loads the throughput is affected also by constraints that are close to saturation (but not saturated) and the effect of the nearly saturated constraints vanishes only in the asymptotical limit [5]. Existence of nearly saturated constraints is very likely in networks with many constraints.

4   Wireless Networks

Capacity set of a wireless network is generally difficult to determine due to various physical constraints and the effects of interference in the network. In the following, we model the wireless MAC layer functionality by STDMA [9]. In STDMA the capacity set is indeed an intersection of linear half-spaces, but these constraints are given only in an implicit form.

In STDMA, sets of links with specified transmission parameters are scheduled on a fast time scale to produce constant (virtual) link capacities on the flow time scale. Each such set, called a transmission mode, thus defines, which links are active and which transmission power is used on each active link. For each transmission mode there is a capacity vector which defines the capacity of each link when that particular mode is active. Assume that these vectors form the columns of rate matrix $\mathbf{R}$. When the modes are scheduled on a fast time scale, the resulting link capacities available for flows are defined by the convex combinations of the columns of $\mathbf{R}$. The capacity set in the space spanned by link capacities (referred to as the link space) is thus given by $\mathcal{C}_{\text{link}} = \{\mathbf{c} = \mathbf{Rt} : \mathbf{e}^T\mathbf{t} \leq \mathbf{1}, \mathbf{t} \geq \mathbf{0}\}$. See [7] for more extensive description of the model.

In principle, from $\mathbf{R}$ one can define the capacity set also as a collection of linear inequality constraints, but generally this approach is computationally infeasible. However, for any fixed link capacity proportions (i.e., a direction in the link space) one may compute the corresponding boundary point, which defines the maximum available link capacities with the predefined proportions. The boundary point to the direction of $\mathbf{b}$ is given by $\mathbf{c_b} = \mathbf{b}/\varUpsilon(\mathbf{b})$, where the function $\varUpsilon(\mathbf{b})$ is defined as the solution to the LP problem

$$\varUpsilon(\mathbf{b}) = \min_{\mathbf{q}} \mathbf{e}^T\mathbf{q},$$
$$\text{s.t. } \mathbf{Rq} \geq \mathbf{b},$$
$$\mathbf{q} \geq \mathbf{0}.$$

4.1 Throughput Asymptotics

The throughput behavior of wireless networks at low loads can be straightforwardly characterized using the formulae (5). Recall the definition of $\mathbf{A}$ as the link-flow incidence matrix, i.e. $A_{ji} = 1$ indicates the flow i uses link j, otherwise $A_{ji} = 0$. Given the traffic pattern $\mathbf{p}$, the boundary point in the flow space and the corresponding balance function (with, as usual, $\varPhi(\mathbf{0}) = 1$ and $\varPhi(\mathbf{x}) = 0$ for all $\mathbf{x} \notin \mathbb{Z}_+^N$) [7] are

$$\hat{\rho} = \frac{\mathbf{p}}{\varUpsilon(\mathbf{Ap})}, \qquad \varPhi(\mathbf{x}) = \varUpsilon(\mathbf{A}\tilde{\varPhi}(\mathbf{x})). \qquad (10)$$

In order to apply the heavy traffic derivative formula (7), we need to identify the saturated constraints at $r = 1$. Consider first the constraints in the link space. At the boundary of the capacity set in the link space any saturated constraint is a hyperplane the equation of which is directly available via duality. To the link space direction $\mathbf{Ap}$ the vector $\mathbf{u}$ defining the constraining hyperplane $\mathbf{u}^T\mathbf{c_{Ap}} = 1$ is given by the dual of the scheduling problem, i.e., it is the vector that solves

$$\max_{\mathbf{u}} \mathbf{u}^T\mathbf{Ap},$$
$$\text{s.t. } \mathbf{R}^T\mathbf{u} \leq \mathbf{e},$$
$$\mathbf{u} \geq \mathbf{0}.$$

In the case that the solution is not unique, i.e. several constraints are saturated simultaneously, one enumerates the spanning vectors of the solution space. This can be done by solving the dual problem by the simplex method, storing the optimal basis and carrying out, e.g., a depth-first search using simplex iterations over all other solutions that do not change the value of the objective function.

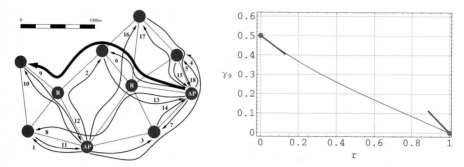

Fig. 2. Left: Wireless mesh network and the flow classes. Right: Throughput asymptotics of the class denoted by the thick line (right).

After the link space constraints have been determined, they need to be translated to the flow space. Let $\mathbf{u}_l$ be the lth extremal solution of the dual problem. Now, $b_{li} = \mathbf{a}_i^{\mathrm{T}} \mathbf{u}_l$ in (7), where $\mathbf{a}_i$ stands for the ith column of $\mathbf{A}$. Thus, the heavy load derivative for classes i such that $\gamma_i(1) = 0$ is given by

$$\gamma_i'(1) = -\frac{1}{\sum_j \mathbf{a}_i^{\mathrm{T}} \mathbf{u}_j}. \tag{11}$$

4.2   Example

Consider a wireless mesh network shown in Figure 2 (left), with two access points (AP), two relays (R) and 18 traffic classes. As the interference model, we assume that no node can participate in more than one transmission at a time. In other words, a feasible transmission mode is a matching on the network graph. We assume that each link has unit capacity when active. The traffic pattern is $p_i' = i$ (without normalization) for each class i. We study the throughput of class 9 (marked with thick line in the left figure) which has the load $\hat{\rho}_9 = 3/40$ at the capacity limit. Figure 2 (right) shows the asymptotic throughput behavior of class 9. The throughput between the extremes is sketched using a cubic polynomial fitted to the end points and to the first two derivatives at $r = 0$ neglecting the derivative at $r = 1$, similarly to the previous example.

5   Throughput Evaluation in Low to Medium Load Range

The asymptotic analysis is an efficient way to approximate flow-level performance in many large systems that are otherwise intractable. However, the heavy traffic information is not always available. The size and complexity of the system may prevent solving the capacity limit and even in simple systems the heavy traffic behavior is not known if the studied class does not saturate at $r = 1$.

In this section we propose an alternative method to approximate the throughput directly in low and medium load range. At small and moderate loads the number of flows in the network is often small. Consider, for example, a single unit capacity link. The mean number of active flows is $\rho(1-\rho)^{-1}$ and at the load $\rho = 0.5$ there is, on average, only one flow active in the system. Correspondingly, the performance of the network at this load range is dominated by the states with only a few active flows. This is the starting point for our computational scheme. The idea is to write the throughput expression in a suitable form and then approximate the sums contained in the expression by Monte Carlo method. The approach is based on the assumption that the capacity set can be constructed for any given set of flows *if* the number of flows is small.

5.1   Sparse Matrix Notation for Balanced Fairness

In systems with large number of flow classes it is useful to consider balanced fairness in *sparse matrix notation*. Let ξ be the set of indices to active flows, i.e., $\mathbf{x} = \sum_{i \in \xi} \mathbf{e}_i$. Note that the same index may appear in ξ more than once. With this notation the recursion (2) can be written:

$$\Phi(\xi) = \min\{\alpha : \frac{(\Phi(\xi \setminus \{\hat{\xi}_1\}), \ldots, \Phi(\xi \setminus \{\hat{\xi}_L\}))^{\mathrm{T}}}{\alpha} \in \mathcal{C}^{\hat{\xi}}\},$$

where $\hat{\xi} = \bigcup \xi$, i.e. the set of different flow classes in ξ, $L = |\hat{\xi}|$, and $\mathcal{C}^{\hat{\xi}}$ is the capacity set defined for flow classes $\hat{\xi}$ only. In other words, to compute the value of the balance function, we remove active flows one by one until we reach $\Phi(\emptyset)$, which is 1, by convention. The case with continuous index set is discussed in [5]. In this notation we may write the normalization constant in terms of increasing number of flows in the system. Accordingly, (4) becomes

$$\gamma_i(r) = \frac{1 + r \sum_{j=1}^{N} \Phi(j)\hat{\rho}_j + r^2 \sum_{j=1}^{N}\sum_{k \geq j}^{N} \Phi(j,k)\hat{\rho}_j\hat{\rho}_k + \ldots}{\Phi(i) + r \sum_{j=1}^{N} c_i(i,j)\Phi(i,j)\hat{\rho}_j + r^2 \sum_{j=1}^{N}\sum_{k \geq j}^{N} c_i(i,j,k)\Phi(i,j,k)\hat{\rho}_j\hat{\rho}_k + \ldots},$$

(12)

where the function $c_i(\cdot)$ gives the number of indices equalling i in its argument list. We use the notation $\Phi(i,j) = \Phi(\{i,j\}) = \Phi(\mathbf{e}_i + \mathbf{e}_j)$ for brevity.

Each n-fold summation in the expression (12) corresponds to going through all states where there are n active flows in total. Obviously, only the sums corresponding to a few active flows can be evaluated numerically. For others we use the Monte Carlo method: we draw n flows randomly, compute the corresponding term and repeat the procedure sufficiently many times to get an average which is then multiplied with the number of terms in the sum, $(N+n-1)!/N!/(n-1)!$.

The sparse matrix notation should be viewed as an alternative implementation of the recursion (2), which is especially suitable for evaluating values of balance function for a small set of flows when the total number of flow classes is large. In such a case most of the flow classes are empty and we can neglect all the resources used only by the empty classes in constructing $\mathcal{C}^{\hat{\xi}}$. This significantly reduces the computational burden and and memory consumption of the recursion compared to applying (2). The notation can often be applied even in cases where the whole capacity set $\mathcal{C}$ cannot be handled.

5.2 Example

Consider the fixed network shown in Fig. 3 (left), with 20 nodes and 40 links with unit capacity. The flows are characterized by the source-destination pairs with shortest path routing, totalling in 380 flow classes. Traffic pattern is uniform, $p_i = 1/380$, for all i. We study the throughput of the route shown in the figure. The boundary point of the capacity set is $\hat{\rho}_i = 1/112$, for all i. Figure 3 (right) shows the approximations of the route throughput. The curves (top-down) correspond to the throughput when the terms up to $\{1, 3, 5, 7, 9, 11\}$-fold sums are taken into account, respectively. Two first sums are computed by enumeration and the subsequent sums with Monte Carlo method using 10^5 samples. The figure shows also the asymptotic derivatives and the interpolated throughput using a cubic polynomial fitted to the end points and to the first two derivatives at $r = 0$. The interpolated curve fits well with the numerical results. Again, it can be seen that a simple polynomial curve would be misleading if the heavy load derivative was used in the fitting.

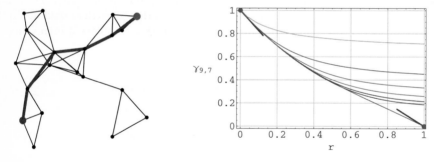

Fig. 3. Left: Network topology and the studied route (directed from top to down). Right: Throughput of the route by the Monte Carlo method with up to 11-fold sums.

6   Conclusions

Flow level performance evaluation of data networks is numerically demanding. While balanced fairness concept has made throughput analysis more tractable, approximative methods are still needed to the analysis of complex systems.

In this paper, we have extended the approximative asymptotic analysis in two ways. First, we provided explicit formulation of the second derivative in the light traffic regime. Second, we applied the analysis to two important systems, multipath routing and wireless networks, whose (polytope) capacity set is only known in implicit form instead of the easier constraint form. The light traffic approximation is very informative in the sense that, with small increments in the load, the throughput generally differs only a little from the tangent line at zero. In contrast, the heavy traffic derivative may not be the best overall descriptor for the heavy load behavior. As already discussed in [5], the throughput curve may change very quickly at the heavy load end of the curve if only one constraint is saturated but several others are close to saturation. One may expect that this case is common in large complex systems as studied in this work. The simple polynomial fitting applied in this paper should neglect the heavy load derivative to preserve the good match at low and medium loads. Inclusion of the heavy load derivative in the interpolation scheme would require deeper insight in the analytical form of the throughput curves. We identify this as an interesting topic for future work.

In addition, we presented a computational scheme based on Monte Carlo method applicable to systems with low or medium loads. Instead of recursively going through the state space in order, the state space is sampled randomly and the average throughput is calculated. Even in large systems, the approach converges relatively quickly at low and medium loads.

Acknowledgments

This work has been supported in part by the Academy of Finland (Grant No. 210275). Juha Leino has been partly supported by Nokia Foundation.

References

1. Bonald, T., Proutière, A.: Insensitive bandwidth sharing in data networks. Queuing Systems 44, 69–100 (2003)
2. Bonald, T., Virtamo, J.: Calculating the flow level performance of balanced fairness in tree networks. Performance Evaluation 58, 1–14 (2004)
3. Bonald, T., Proutière, A.: On performance bounds for balanced fairness. Performance Evaluation 55, 25–50 (2004)
4. Bonald, T.: Throughput performance in networks with linear constraints. In: Proc. 40th Annual Conference on Information Sciences and Systems (CISS) (2006)
5. Bonald, T., Penttinen, A., Virtamo, J.: On light and heavy traffic approximations of balanced fairness. In: Proc. of SIGMETRICS/Performance, pp. 109–120 (2006)
6. Leino, J., Virtamo, J.: Insensitive load balancing in data networks. Computer Networks 50, 1059–1068 (2006)
7. Penttinen, A., Virtamo, J., Jäntti, R.: Performance analysis in multi-hop radio networks with balanced fair resource sharing. Telecommunications Systems 31, 315–336 (2006)
8. Bonald, T., Massoulié, L., Proutière, A., Virtamo, J.: A queueing analysis of max-min fairness, proportional fairness and balanced fairness. Queueing Systems 53, 65–84 (2006)
9. Nelson, R., Kleinrock, L.: Spatial TDMA: A collision-free multihop channel access protocol. IEEE Transactions on Communications 33, 934–944 (1985)

Improving RED by a Neuron Controller

Jinsheng Sun and Moshe Zukerman

The ARC Special Research Centre for Ultra-Broadband Information Networks,
Department of Electrical and Electronic Engineering, The University of Melbourne,
Victoria, 3010, Australia
{j.sun,m.zukerman}@ee.unimelb.edu.au

Abstract. In this paper, we propose a novel active queue management (AQM) algorithm called Neuron Control RED (NC-RED) that overcomes the drawbacks of the original RED. NC-RED uses a neuron controller to adaptively adjust the maximum drop probability to stabilize the average queue length around the target queue length. We demonstrate by simulations that NC-RED maintains stable operation independent of traffic loading, round trip propagation delay, and bottleneck capacity. We also demonstrate that NC-RED is robust to non-responsive UDP traffic and HTTP traffic, and it is effective for networks with multiple bottlenecks. Comparison with other well-known AQM algorithms like PI, REM and ARED demonstrates the superiority of NC-RED in achieving faster convergence to queue length target and smaller queue length jitter.

1 Introduction

Active Queue Management (AQM) algorithms enable end hosts to adapt their transmission rates to network traffic conditions by providing them with congestion information. Recognizing the advantages of such feedback, the Internet Engineering Task Force (IETF) has recommended the use of AQM for network congestion control to reduce loss rate, to support low-delay interactive services, and to avoid TCP lock-out behavior [2]. Although the benefit of AQM feedback seems obvious, its widespread use in the Internet has not materialized because of the difficulty of configuring AQM algorithms in a dynamic networking environment to achieve a stable operation. As a result, the traditional DropTail mechanism is still being widely used.

A well-known AQM algorithm is the Random Early Detection (RED) [6]. It has been demonstrated in many studies that RED outperforms DropTail. RED is able to prevent global synchronization which is a drawback of DropTail. It achieves lower packet loss than DropTail and it also reduces another drawback of DropTail which discriminates against bursty sources [6]. However it is often difficult to parameterize RED to optimize its operation in a dynamic network environment under various congestion scenarios.

In the past few years, many RED variants and enhancements [4,5,9,11,14] or other AQM algorithms [1,7,10,12,13,15,16] have been proposed to overcome the weaknesses of the basis RED algorithms. In [4], a self-configuring RED mechanism was proposed, which varies the maximum drop probability max_p using

L. Mason, T. Drwiega, and J. Yan (Eds.): ITC 2007, LNCS 4516, pp. 434–445, 2007.

two different constant factors based on observed average queue length. Floyd *et al.* [5] further improved upon this proposal by using an Additive Increase Multiplicative Decrease (AIMD) to adjust max_p. In [11], a PD controller was used to adjust max_p to improve the performance of RED.

In this paper, we introduce a new adaptive RED algorithm which we call Neuron Control RED (NC-RED), where we use a neuron controller to adaptively adjust max_p. Neuron control has been widely applied to control non-linear, time-varying systems, so it is suitable for TCP congestion control, which is in fact a non-linear and time-varying system. We demonstrate by simulation results that NC-RED performs well independently of traffic loading, round trip propagation delay, and bottleneck capacity. We also demonstrate that NC-RED is robust to non-responsive UDP traffic and HTTP traffic, and maintains good performance in the presence of multiple bottlenecks. By comparing with other AQM algorithms we demonstrated the superiority of NC-RED over well-known algorithms like PI, REM and ARED in achieving faster convergence to queue length target, and smaller queue length jitter.

The remainder of the paper is organized as follows. We describe the proposed NC-RED algorithm in Section 2. Sections 3 and 4 present performance evaluation of NC-RED for the cases of single and multiple bottlenecks, respectively, and in Section 5 we compare NC-RED with other AQM algorithms. Finally, we present our conclusions in Section 6.

2   Neuron Controller

The neuron controller can be described by the following equation,

$$max_p(k) = max_p(k-1) + \Delta p(k) \tag{1}$$

where $max_p(k)$ is the maximum dropping probability of RED, $\Delta p(k)$ is the increment of maximum dropping probability given by a neuron

$$\Delta p(k) = K \sum_{i=1}^{3} w_i(k) x_i(k) \tag{2}$$

where $K > 0$ is the neuron proportional coefficient; $x_i(k)(i = 1, 2, 3)$ denote the neuron inputs, and $w_i(k)$ is the connection weight of $x_i(k)$ determined by the learning rule.

Let

$$e(k) = avq(k) - Q_T \tag{3}$$

denote the queue length error, where $avq(k)$ is the average queue length of RED, and Q_T is the target queue length. The inputs of the neuron are $x_1(k) = e(k) - e(k-1)$, $x_2(k) = e(k)$, $x_3(k) = e(k) - 2e(k-1) + e(k-2)$. According to Hebb [3], the learning rule of a neuron is formulated by

$$w_i(k+1) = w_i(k) + d_i y_i(k) \tag{4}$$

where $d_i > 0$ is the learning rate, and $y_i(k)$ is the learning strategy. The associative learning strategy given in [3] is as follows:

$$y_i(k) = e(k)\Delta p(k)x_i(k) \tag{5}$$

where $e(k)$ is used as teacher's signal. This implies that a neuron, which uses integrated Hebbian Learning and Supervised Learning, makes actions and reflections to the unknown outsides with associative search. It means that the neuron self-organizes the surrounding information under supervision of the teacher's signal $e(k)$. It also implies a critic on the neuron actions.

The adaptive neuron controller is based on the following six parameters.

1. Sampling time interval T; an appropriate value is $T = 0.1s$.
2. Target queue length Q_T; this parameter decides the steady-state queue length value, which will affect the utilization and the average queueing delay. A high target will improve link utilization, but will increase the queueing delay. The target queue length Q_T should be selected according the Quality of Service (QoS) requirements.
3. The neuron proportional coefficient K; a suggested value is $K = 0.01$.
4. The learning rate d_1; a suggested value is $d_1 = 0.0000001$.
5. The learning rate d_2; a suggested value is $d_2 = 0.0000001$.
6. The learning rate d_3; a suggested value is $d_3 = 0.0000001$.

The above values of K and $d_i(i = 1, 2, 3)$ have been chosen based on trial and error by a few simulations. Nevertheless, Zhang $et\ al.$ [18] showed that a neuron control system is very robust and adaptable, so the choice of values of K, $d_i(i = 1, 2, 3)$ is not that critical. The initial values of $w_i(i = 1, 2, 3)$ do not affect the performance significantly; we use: $w_i = 0.001, (i = 1, 2, 3)$.

3   Single Bottleneck

We study the performance of NC-RED via simulations. In our simulation testing we focus on stabilizing queue length at a target value Q_T as a key performance measure. If we can control the queue to stay close to a desirable target, we can achieve high throughput, predictable delay and low jitter. The low jitter enables meeting Quality of Service (QoS) requirements for real time services especially when the queue length target is achieved independently of traffic conditions [17]. Depending on Q_T, it also has the benefit of low buffer capacity requirement.

A number of simulations were performed to validate the performance of NC-RED using the network simulation tool $ns2$ [8]. The single bottleneck network topology used in the simulation is shown in Figure 1. The only bottleneck link is the Common Link between the two routers. The other links are assumed to have sufficient capacity to carry their traffic. Router B uses NC-RED and Router C uses DropTail. The sources use TCP/Reno.

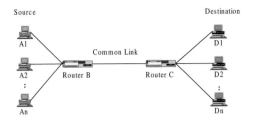

Fig. 1. The single bottleneck network topology

3.1   Constant Number of TCP Connections

In this simulation, we test whether NC-RED can stabilize the queue length at a chosen value for different loads and link capacities. In the following simulations, unless mentioned otherwise, the following parameters are used: the packet size is 1000 bytes, the common link capacity is 45 Mb/s, the round trip propagation delay is 100 ms, the buffer size is 1125 packets. The TCP connections always have data to send as long as their congestion windows permit. The receiver's advertised window size is set sufficiently large so that the TCP connections are not constrained at the destination. The ack-every-packet strategy is used at the TCP receivers. The parameters of RED are set as follows: $min_{th} = 15$, $max_{th} = 785$, $w_q = 0.002$, and target queue length: $Q_T = 400$. The parameters of the adaptive neuron controller are set as previous section given.

Figure 2 shows the instantaneous queue lengths of NC-RED for 100 and 2000 TCP connections. All sources start data transmission at time 0. We can see that NC-RED is effective at stabilizing and keeping the queue length around the target Q_T. In order to adequately show this ability of NC-RED, we also set Q_T at 200 and 600 for 500 TCP connections. The results are depicted in Figure 3. We can see from these figures that NC-RED is indeed successful in controlling the queue length at any arbitrary chosen target. Thus, we demonstrate that NC-RED is able to achieve stable queue length independent of loading, thus overcoming the parameter setting drawback of RED.

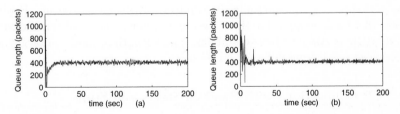

Fig. 2. Queue length variations for different numbers of greedy TCP connections: (a) 100 (b) 2000

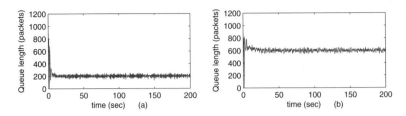

Fig. 3. Queue length variations for the following queue length targets: (a) $Q_T = 200$ (b) $Q_T = 600$

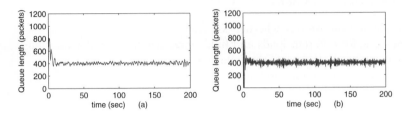

Fig. 4. Queue length variations for the following link capacities: (a)15 Mb/s (b)115 Mb/s

In order to test the performance of NC-RED for different link capacities, we vary the capacity from 45 Mb/s to 15 Mb/s and 115 Mb/s while the other parameters remain the same. The simulation results for 500 TCP connections are given in Figure 4. We can see that again stability is achieved.

3.2   TCP Connections Having Different Propagation Times

Here we test the performance of NC-RED with TCP connections having different round trip propagation times (RTPT). Two simulations have been performed. In the simulations there are 500 TCP connections with different RTPTs. In the first, The RTPTs are uniformly distributed between 50 and 500 ms, and in the second they are exponentially distributed with mean of 100 ms. Figure 5 presents the queue lengths for these two simulations. The results demonstrate that NC-RED is still effective to stabilize queue length around the target with TCP connections having different RTPTs.

3.3   TCP Connections with Random Starts and Stops

We first present results of two simulation runs where we dynamically vary the number of active TCP connections. The number of TCP connections is varied from 200 to 2000 in the first and from 2000 to 200 in the second. In each of the runs, a group of 200 connections are started (or stopped) at the same time at each 10 seconds interval. The instantaneous queue lengths are plotted in Figure 6. We

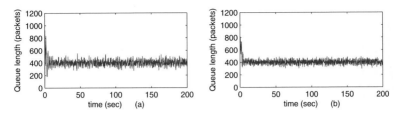

Fig. 5. Queue length variations in the following scenarios: (a) RTPTs are uniformly distributed between 50 and 500 ms (b) RTPTs are exponential distributed with mean of 100 ms

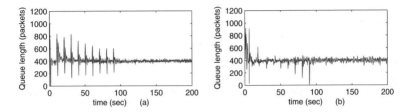

Fig. 6. The simulation results for two cases: (a) Number of TCP connections varies from 200 to 2000 (b) Number of TCP connections varies from 2000 to 200

can clearly see that NC-RED is able to stabilize the queue length around the control target even when the number of connections is dynamically varied over time.

We will now examine the performance of NC-RED when we have a traffic scenario involving random start and stop times, thus simulating staggered connection setup and termination. We performed two simulations. In the first, the initial number of connections is set to 200 and, in addition, 1800 connections have their start-time uniformly distributed over a period of 100 seconds. In the second simulation, the initial number of connections is set to 2000 out of which 1800 connections have their stop-time uniformly distributed over a period 100 seconds. The instantaneous queue lengths are plotted in Figure 7. We can clearly see that NC-RED is able to stabilize the queue length around the control target.

3.4   Long Delay Network

Recalling that simulations in [10] have shown that AQMs, such as PI, RED and REM, are unstable when RTPT is 400 ms, we test here the NC-RED for a long delay network. Two simulation tests have been performed. In both simulations, there are 1000 TCP connections. In the first, the RTPTs are 400 ms, and in the second they are uniformly distributed between 50 and 750 ms. Figure 8 presents the queue lengths for these two simulations. The results demonstrate that NC-RED is still effective in stabilizing the queue length around the target for TCP connections with large RTPTs.

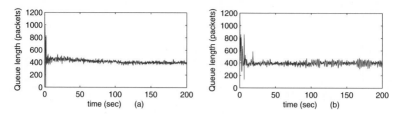

Fig. 7. Queue length variations under varying number of TCP connections (a) Number of TCP connections varies from 200 to 2000 (b) Number of TCP connections varies from 2000 to 200

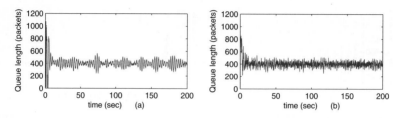

Fig. 8. The simulation results for long delay networks (a) RTPTs = 400 ms (b) RTPTs range between 50 and 750 ms

3.5   Mixing TCP, UDP and HTTP Connections

In this simulation experiment, we investigate the ability of NC-RED to cope with the disturbances caused by HTTP as well as UDP connections. We have performed two simulation runs. In the first, the 500 TCP connections are mixed with 100 UDP flows and 400 HTTP sessions, in the second they are mixed with 800 UDP flows and 400 HTTP sessions. The RTPTs of TCP connections are 100 ms. The number of pages per HTTP session is 250. The round trip propagation delay of the HTTP connections is uniformly distribute between 50 and 300 ms. The propagation delay of all the UDP flows are uniformly distributed between 30 and 250 ms. Each of the UDP sources follows an exponential ON/OFF traffic model, the idle and the burst times have mean of 0.5 second and 1 second, respectively. The packet size is set at 500 bytes, and the sending rate during on-time is 64 kb/s. Figure 9 provides the queue length, which demonstrate that NC-RED is robust.

4   Multiple Bottlenecks

Here we extend the simple single bottleneck topology to a case of multiple bottlenecks. We consider the network topology presented in Figure 10. There are two bottlenecks in this topology. One is between Router B and Router C, and the other is between Router D and Router E. The link capacity of the two bottlenecks is 45 Mb/s and the capacity of other links is 100 Mb/s. There are three

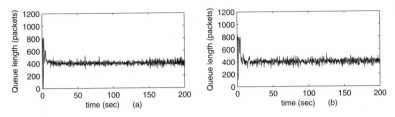

Fig. 9. Queue length variation in the following scenarios: (a)500 TCP connections mixed with 100 UDP flows and HTTP (b)500 TCP connections mixed with 800 UDP flows and HTTP

traffic group. The first group has N TCP connections traversing all links, the second group has N_1 TCP connections traversing the link between Router B and Router C, and the third group has N_2 TCP connections traversing the link between Router D and Router E. The RTPTs of the first group are uniformly distributed between 50 ms and 500 ms, and for the second and third groups, they are uniformly distributed between 50 ms and 300 ms and 50 ms and 400 ms, respectively. Two simulation tests have been performed. In the first, $N = 500$, $N_1 = 200$, and $N_2 = 200$, and in the second, $N = 500$, $N_1 = 800$, and $N_2 = 200$. Figures 11 and 12 present the queue lengths for these two simulations. The results demonstrate that NC-RED is effective in stabilizing the queue length around the target for TCP connections in multiple bottlenecks network.

5   Comparison with Other AQMs

In this section, we perform two simulations to compare the performance of NC-RED with ARED [5], PI controller [7], and REM [1]. The network topology used in the simulation is the same as in Figure 1. The same parameters as in Section 3 are used: packet size is 1000 bytes, common link capacity is 45 Mb/s, the buffer size is 1125 packets. The target queue length is set at 400 packets for all AQM algorithms. For ARED, we set the parameters: $min_{th} = 15$, $max_{th} = 785$ and $w_q = 0.002$, and other parameters are set the same as in [5]: $\alpha = 0.01$, $\beta = 0.9$, $intervaltime = 0.5s$. For PI controller, we use the default parameters in $ns2$: $a = 0.00001822$, $b = 0.00001816$ and the sampling frequency $w = 170$. For REM, the default parameters of [1] are used: $\phi = 1.001$, $\gamma = 0.001$. For NC-RED, the parameters are the same as in Section 3.

In the first simulation, there are 1000 TCP connections with RTPT of 100 ms. Figure 13 presents the queue lengths for all four AQMs. We can see that NC-RED reacts and converges to the target queue length of 400 faster than all three other AQMs. In order to evaluate the performance in steady-state, we calculate, the mean and the standard deviation of the queue length for the last 150 seconds. The results are presented in Table 1. We observe that NC-RED queue length has the mean of 400.0, which is the closest to the target of 400, and achieved the lowest standard deviation of all other AQMs.

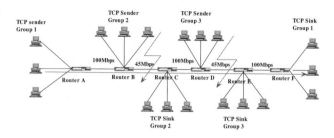

Fig. 10. The multiple bottleneck network topology

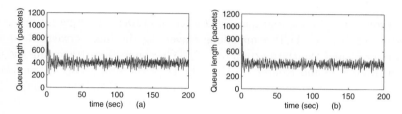

Fig. 11. Queue length variations in multiple bottleneck network for $N = 500$, $N_1 = 200$, and $N_2 = 200$: (a) Router B (b) Router D

In the second simulation test, we compare the performance of the AQMs under dynamic traffic loading and the disturbances caused by HTTP as well as UDP connections. The initial number of TCP connections is set to 200, and 1800 additional connections have their start-time uniformly distributed over a period of 100 seconds. The RTPTs of these TCP connections are uniformly distributed between 50 and 500 ms. The bursty HTTP traffic involves 400 sessions (connections), and the number of pages per session is 250. The round trip propagation delay of each of the HTTP connection is uniformly distributed between 50 and 300 ms. There are 800 UDP flows with propagation delay uniformly distributed between 30 to 300 ms. Each of the UDP sources follows an exponential ON/OFF traffic model, the idle and the burst times have mean of 0.5 second and 1 second, respectively. The packet size is set at 500 bytes, and the sending

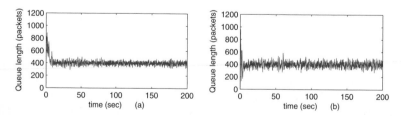

Fig. 12. Queue length variations in multiple bottleneck network for $N = 500$, $N_1 = 800$, and $N_2 = 200$: (a) Router B (b) Router D

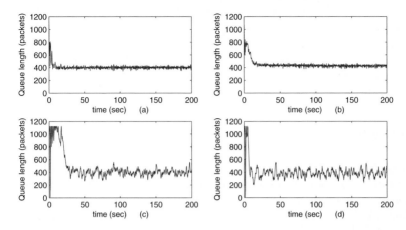

Fig. 13. Comparison of queue length variations for the following AQM algorithms: (a) NC-RED, (b) ARED, (c) PI and (d) REM

Table 1. Mean and standard deviation of the queue length for various AQM algorithms

	NC-RED	ARED	REM	PI
Mean	400.0	426.4	399.0	386.2
Standard Deviation	15.5	15.6	46.3	51.5

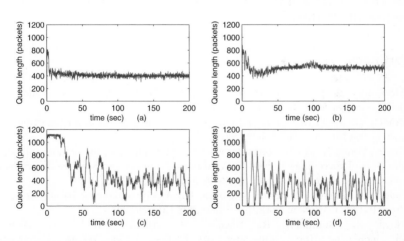

Fig. 14. Comparison of queue length variations for the following AQM algorithms: (a) NC-RED, (b) ARED, (c) PI and (d) REM

rate during on-time is 64 kb/s. Figure 14 presents the queue lengths for all four AQMs. We can see again that NC-RED reacts and converges to the target queue length faster than all three other AQMs. We also can see that REM and PI are nearly unstable under this scenario. The means of the queue lengths for the last 150 seconds are 400.9, 523.0, 401.3 and 279.6 for NC-RED, ARED, REM and PI, respectively, and the standard deviations are 23.1, 26.9, 148.7 and 166.3, respectively. We observe that NC-RED queue length had the mean of 400.9, which is the closest to the target of 400, and achieved the lowest standard deviation of all other AQMs.

6   Conclusions

We have developed an improved adaptive RED algorithm called NC-RED based on a neuron controller to improve the performance of RED. We have demonstrated by simulations that NC-RED is able to maintain the queue length around a given target under different traffic loading, different RTPTs, and different bottleneck link capacities. Further simulation testing involving non-TCP traffic types and a multiple bottleneck topology have further confirmed the robustness of NC-RED. Comparison with other AQM algorithms has demonstrated the superiority of NC-RED in achieving faster convergence to target queue length, and then maintaining the queue length closest to the target.

Acknowledgments

This work was jointly supported by grants from the Australian Research Council (Grant DP0559131), and the Natural Science Foundation of Jiangsu Province, China (No. BK2004132).

References

1. Athuraliya, S., Low, S.H., Li, V.H., Yin, Q.: REM: Active queue management. IEEE Network Mag. 15(3), 48–53 (2001)
2. Braden, B., et al.: Recommendations on queue management and congestion avoidance in the Internet. IETF RFC2309 (1998)
3. Du, Y., Wang, N.: A PID controller with neuron tuning parameters for multi-model plants. In: Proceedings of 2004 International Conference on Machine Learning and Cybernetics, Shanghai, vol. 6, pp. 3408–3411 (August 2004)
4. Feng, W., Kandlur, D., Saha, D., Shin, K.: A self-configuring RED gateway. In: Proc. INFOCOM, New York, pp. 1320–1328 (March 1999)
5. Floyd, S., Gummadi, R., Shenker, S.: Adaptive RED: An algorithm for increasing the robustness of RED's active queue management, http://www.icir.org/floyd/red.html
6. Floyd, S., Jacobson, V.: Random early detection gateways for congestion avoidance. IEEE/ACM Trans. Networking 1(4), 397–413 (1993)
7. Hollot, C.V., Misra, V., Towsley, D., Gong, W.: On designing improved controllers for AQM routers supporting TCP flows. Proc. IEEE INFOCOM 2001 3, 1726–1734 (2001)
8. The NS simulator and the documentation, http://www.isi.edu/nsnam/ns/

9. Ott, T.J., Lakshman, T.V., Wong, L.: SRED: Stabilized RED. In: Proc. IEEE INFOCOM, New York, pp. 1346–1355 (March 1999)
10. Ren, F., Lin, C., Wei, B.: A robust active queue management algorithm in large delay networks. Computer Communications 28, 485–493 (2005)
11. Sun, J., Ko, K.T., Chen, G., Chan, S., Zukerman, M.: PD-RED: to Improve the Performance of RED. IEEE Communications Letters 7(8), 406–408 (2003)
12. Sun, J., Chen, G., Ko, K.T., Chan, S., Zukerman, M.: PD-Controller: A new active queue management scheme. In: proc. IEEE Globecom, San Francisco, vol. 6, pp. 3103–3107 (December 2003)
13. Sun, J., Chan, S., Ko, K.T., Chen, G., Zukerman, M.: Neuron PID: A robust AQM scheme. In: Proceedings of ATNAC 2006, Melbourne, pp. 259-262, (Dec 2006)
14. Sun, J., Zukerman, M., Palaniswami, M.: Stabilizing RED using a fuzzy controller. In: Proc. ICC 2007, Glasgow (June 2007)
15. Sun, J., Zukerman, M.: An adaptive neuron AQM for a stable Internet. In: proc. IFIP Networking 2007, Atlanta (May 2007)
16. Sun, J., Zukerman, M.: RaQ: A robust active queue management scheme based on rate and queue length, to appear in Computer Communications (2007)
17. Wydrowski, B., Zukerman, M.: QoS in best-effort networks. IEEE Communications Magazine 40(12), 44–49 (2002)
18. Zhang, J., Wang, S., Wang, N.: A new intelligent coordination control system for a unit power plant. In: Proceedings of the 3rd World Congress on Intelligent Control and Automation, Hefei, pp. 313-317 (June 2000)

BFGSDP: Bloom Filter Guided Service Discovery Protocol for MANETs*

Zhenguo Gao[1], Xiang Li[2], Ling Wang[3], Jing Zhao[2], Yunlong Zhao[2], and Hongyu Shi[1]

[1] College of Automation, Harbin Engineering University, Harbin, 150001, China
gag@ftcl.hit.edu.cn
http://www.angelfire.com/planet/gaohit/gagCV.htm
[2] College of Computer Science and Technology, Harbin Engineering University, Harbin, 150001, China
[3] College of Computer Science and Technology, Harbin Institute of Technology, Harbin, 150001, China

Abstract. The ability to discovery services is the major prerequisite for effective usability of MANETs. Broadcasting and caching service advertisements is an essential component in service discovery protocols for MANETs. To fully utilize the advantage of the cached service advertisements, the cached service description can be advertised with the local service description in a vicinity, which however will increase the size of service advertisement packets. A better solution is to using some aggregating method to reduce the packet size. In this paper, utilizing Bloom Filter as the aggregating method, we propose Bloom Filter Guided Service Discovery Protocol (BFGSDP) for MANETs. Two salient characteristics of BFGSDP are Bloom Filter Guiding scheme (BFG) and Broadcast Simulated Unicast (BSU). BFG scheme decreases the risk of flood storm problem by guiding request packets with bloom filters to those nodes with potential matched services. BSU scheme benefits from the broadcast nature of wireless transmissions by replacing multiple unicast request packets with one request packet transmitted in broadcast mode with all unicast receivers enclosed. Extensive simulations show that BFGSDP is a more effective, efficient, and prompt service discovery protocol for MANETs.

Keywords: Bloom Filter, Service Discovery Protocol, MANET.

1 Introduction

Mobile Ad-Hoc Networks (MANETs) [1] are temporary infrastructure-less multi-hop wireless networks that consist of many autonomous wireless mobile nodes. Effective

* This work is supported by National Postdoctoral Scientific Research Program (No.060234), Postdoctoral Scientific Research Program of HeiLongJiang Province (No.060234), Fundamental Research Foundation of Harbin Engineering University (No.HEUFT06009), Foundation of Experimentation Technology Reformation of Harbin Engineering University (No.SJY06018).

L. Mason, T. Drwiega, and J. Yan (Eds.): ITC 2007, LNCS 4516, pp. 446–457, 2007.

exploitation of networked services provided in MANETs requires discovering the services that best match the application's requirements and invoking such services for the application. Service discovery protocol (SDP) which aims to advertise and discover the services available in the current environment plays an important role in service-oriented MANETs.

Service discovery is originally studied in the context of wired networks, such as IETF's Service Location Protocol (SLP) [2], Sun's Jini [3], etc.

With the development of MANETs, a number of service discovery protocols are proposed for MANETs. Some of them [4]-[5] are adapted from service discovery protocols of wired networks. Some others are designed specially for MANETs[6]-[9].

In general, the SDPs for MANETs work as follows. The service request packet is flooded from the source to the neighbors. When receiving a service request packet, if either its local services or the cached services match the service request, a node will respond with a service reply packet to the corresponding source, otherwise, each node will flood the service request packet to its neighbors.

A general technique to reduce the overhead of the flooding-based service discovery is caching service advertisements. Cached information is valid source for service matching. Different caching policies are studied in the literature, e.g., decentralized caching policy [6]-[9] where each node caches the service advertisements it hears in the vicinity or centralized caching policy where a central directory is elected to cache the service advertisements heard in a vicinity on behalf of all nodes in that vicinity [10].

Caching the service advertisements in vicinity can reduce the service discovery delay. However, if both the local services and detailed cached services are advertised in the vicinity, it will increase the length of service advertisement packet. A better solution is to using some aggregating method to reduce the packet size.

In Group-based Service Discovery (GSD) protocols [6], service group information is selected as the aggregation of the detailed service description. Utilizing the group information, service request packets are forwarded towards candidate nodes with potentially matched services. However, it implies that GSD can only support service type-based searches, not allowing sophisticated searches in terms of other various service attributes. Besides, when multiple candidate nodes are found, one request packet will be sent in unicast mode to each candidate node, individually, which degrades its efficiency badly.

Bloom Filter [11] can be a better replacement of the group-based aggregation method in GSD[6]. Considering GSD's problems, and borrowing the idea of intelligent forwarding of service request packets, we propose BFGSDP (Bloom Filter Guided Service Discovery Protocol) for MANETs. In BFGSDP, service request packets are guided by bloom filters of the descriptions of services in vicinity, which overcomes GSD's first problem. Further more, broadcast Simulated Unicast (BSU) is proposed to solve the second problem. BSU replaces several successive unicast request packets sent by one node with a broadcasted request packet that encloses the receivers of those unicast request packets. Mathematical analysis and extensive simulations shows the superiority of BFGSDP.

The rest of the paper is organized as follows. Section 2 gives an overview to related works. Section 3 introduces the bloom filter and the adaptation in BFGSDP. Section 4 explains the salient schemes of BFGSDP. Section 5 illustrates the basic operations of BFGSDP. Section 6 evaluates the performance of BFGSDP and some other typical service discovery protocols through extensive simulations. Finally, section 7 concludes the paper.

2 Related Works

Many research efforts have been focused on SDPs for MANETs. In current various approaches, to facilitate service discovery operation, nodes are organized into different structures. According to these structures constructed, application layer approaches are classified into three classes: one-layer approaches, two-layer approaches, and multi-layer approaches.

The dynamic nature of MANETs makes mutli-layer structure hard to maintain. Hence, one-layer approaches and two-layer approaches are more preferable. Compared to two-layer approaches, one-layer approaches are more suitable for highly dynamic MANETs. A more detailed survey can be found in [12].

Recently, a few works adopts bloom filter method in information discovery related applications. In [7], bloom filter is used for context discovery. More recently, in [8], bloom filter is used to guide service request packets forwarding operation in the proposed SDP for MAENTs. The authors proposes three methods to send request packets: parallel querying, sequential querying, and hybrid querying. In parallel querying method, a request packet is sent to multiple candidate nodes with potential matched services simultaneously; in sequential querying method, the request packet is sent to the best/first candidate nodes; the third method is the tradeoff between the first two methods. Parallel querying is intrinsically robust than sequential querying, but it is inefficient: in most cases, multiple candidate nodes with potential matched sevices will be found, which result in multiple unicast packets. This drawback is similar to that of GSD.

3 Bloom Filter in BFGSDP

In our BFGSDP, bloom filter is used to guide the forwarding of service request packets. Bloom filter is independent of particular service description methods.

3.1 Overview of Bloom Filter

Bloom filter is a method for representing a set $A=\{a_i|i=1, ..., n\}$ of n elements (also called keys) to support membership queries. It was invented by in 1970 [11].

The idea of bloom filter is to allocate a vector v of L bits, initially all set to 0, and then choose k independent hash functions h_i ($i=1, ..., k$), each with range $\{1, ..., L\}$. For each element $a \in A$, the bits at positions $h_i(a)$ ($i=1, ..., k$) in v are set to 1. A particular bit might be set to 1 multiple times. An example of bloom filter with 4 hash functions is shown in Fig. 1.

Given a query for b we check the bits at positions $h_i(b)$ ($i=1, ..., k$). If any of them is 0, then certainly b is not in the set A. Otherwise we conjecture that b is in the set

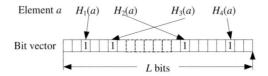

Fig. 1. An example of bloom filter with 4 hash functions

although there is a certain probability that we are wrong, which is called a "false positive". The parameters k and m should be chosen such that the probability of a false positive is acceptable.

The key property of bloom filters is that they provide summarization of a set of data, but collapse this data into a fixed-size table, trading off an increased probability of false positives for index size. The salient feature of bloom filters is the tradeoff between m and the probability of a false positive.

After inserting n keys into a table of size L, the probability that a particular bit is still 0 is

$$\left(1-\frac{1}{L}\right)^{k \cdot n}$$

Hence, the probability of a false positive in this case is

$$\left(1-\left(1-\frac{1}{L}\right)^{k \cdot n}\right)^k \approx \left(1-e^{k \cdot n / L}\right)^k$$

3.2 Bloom Filter in BFGSDP

To locate services based on subsets of attributes, just computing hashes over the whole service descriptions and queries is not sufficient for correct operation. Hence, the procedure to use bloom filter in BFGSDP is as follows:

Step1: extracting some basic attribute tags from service descriptions or query descriptions.
Step2: hashing these tags alone with a list of hash functions, each with range $\{1, ..., L\}$.
Step3: inserting hash results into a bloom filter.

4 Salient Schemes in BFGSDP

The two salient schemes in BFGSDP are Bloom filter Guiding scheme (BFG) and Broadcast Simulated Unicast (BSU).

4.1 Bloom Filter Guiding scheme (BFG)

In BFG, all service descriptions of the services that a server has seen in its neighbors are bloom filtered. The fix-sized bit vector is enclosed into service broadcast packets and broadcast to other nodes. Nodes that receive the packet will cache the content in the packet, including the bloom filter. Then, when forwarding service request packets, cached bloom filters are matched with the bloom filter of query descriptions, and then the request packet is forwarded to those nodes that have seen potential matched services. This scheme is called as Bloom filter Guiding scheme (BFG).

BFG scheme decreases the risk of flood storm problem by guiding request packets to those nodes with potential matched services. Furthermore, instead of enclosing the mass of whole service descriptions, only a fix-sized bit vector is inserted into broadcast packets. Hence, the size of service advertisement packets is extremely reduced.

4.2 Broadcast Simulated Unicast Scheme (BSU)

When forwarding service request packets, it will be very often that multiple nodes that have seen potential services are found. In such cases, instead of unicast one request packet to each of them, respectively, only one packet that encloses all these nodes is sent in broadcast mode. This scheme is called as Broadcast Simulated Unicast scheme (BSU). BSU eliminates much unicast transmissions by making use of the broadcast nature of wireless transmissions.

To implement BSU scheme, a compound field called *receiver-list* is inserted into service request packets. The *receiver-list* field contains a subfield *receiver-number* which indicates the number of valid receivers in a list of *node-id* subfields following the *receiver-number* subfield.

5   Basic Operations of BFGSDP

There are three basic operations in BFGSDP: 1) service advertisement packet spreading; 2) service request packet forwarding; and 3) service reply packet routing. They will be described one by one in following sections.

To facilitate the description of BFGSDP, some notations and definitions are defined in the first subsection.

5.1   Preliminaries

5.1.1   Notations
The following notations are used in the following discussion.

d: the maximum number of hops that advertisement packets can travel.

u: the current node.

req: the description of the requested service.

$N_x(u)$: the set of nodes that are at most x-hop away from node u, i.e., node u's x-hop neighbor set (excluding none u itself).

$S(u)$:   the set of servers that have corresponding valid entries in node u's Service Information Cache (SIC), which caches service advertisement packets. Each server in $N_d(u)$ has a corresponding entry in node u's SIC.

e(***u***, ***s***):   the entry in node *u*'s SIC that represents the service advertisement packet of server *s*.

E(***u***):   the set of entries in node *u*'s SIC.

r(***u***, ***s***):   the node from which node *u* receives the latest service advertisement packet of node *s*. That is, the first node on the path from node *u* to node *s*.

$BF_N(s)$:   the bit vector of entry *e*(*u*, *s*). This is just the bloom filter in a service broadcast packet sent by server *s*.

$BF_S(desc)$:   The bloom filter generated from the service description of service "desc" flowing the procedure described in section3.2.

S(***u***, ***req***): $S(u, req)=\{s \mid s \in S(w), BF_N(s)\&BF_S(req)>0\}$. "&" is the "and" bitwise logic operator.

5.1.2 Definitions
To facilitate the description of BFGSDP, we defined the basic definitions here.

Definition 1: Candidate Node. Nodes in $S(u, req)$ are all candidate nodes of node *u*.

Definition 2: Relay Node. $r(u, s)$ is the relay node of candidate node *s*. In other words, the relay node of a candidate node is the next-hop node on the path from the current node *u* to its candidate node *s*.

5.2 Service Advertisement Packet Spreading

In BFGSDP, each server generates service advertisement packets periodically. These packets can be forwarded further for a limited hop (denoted as *d*). Each node maintains a cache called Service Information Cache (SIC), which stores service advertisement packets temporally. A service advertisement packet contains not only the description of the services provided by the server, but also the bloom filter of all services in the server's cache, i.e., services provided by nodes in the server's *d*-hop neighbor set.

When generating new service advertisement packet, a server should update the packet's *BF* field basing on its SIC as follows:

$$BF_N(u)=\{ \&BF_S(e.local\text{-}service) \mid e \in E(u)\}.$$

Here "&" is the "and" bitwise logic operator, and *e.local-service* represents the description of the services provided by node corresponding to entry *e*.

Fig. 2 shows an example of service advertisement packet spreading operation in BFGSDP with *d*=1. Symbols in Fig. 2 are illustrated in Table 1.

In Fig. 2, after several operation cycles, each node has constructed its SIC. When new operation cycle comes, each node constructs new service advertisement packet basing on its SIC. Service advertisement packets of all servers are shown in the figure. Before sending a packet, the number of hops that the packet can travel is decreased by 1. Hence, although the hop limit of the service advertisement packets is 1, the remaining hop of these packets is 0.

Table 1. Tokens used in representing MANETs

Symbol	Representation	Example
circle	Indicates mobile node	
string in a circle	Indicates the identity of the node and the services it provides	The circle with string {B, b_1} in Fig. 2 indicates that: 1)the node is B; 2)node B provides a service "b_1"; which belongs to service group "b".
White table near a node	Indicates the Service Information Cache (SIC) of the node (not all fields are shown)	The first entry of A's SIC, {B, B, b_1, BF(a_1, b_2)}, in Fig. 2 indicates: 1) the server corresponds to the entry is node B; 2) node B provides service "b_1"; 3) combined bloom filter results of the descriptions of services provided by node B's 1-hop neighbors is BF(a_1, b_2).
double-headed arrow	indicates that two nodes on both ends are neighbors	In Fig. 2, nodes A and B are mutual neighbors, while nodes A and C are not neighboring.
Arcs around a node	indicate packet transmission	In Fig. 2, node B sends out a packet, while node A not.
grey table over arcs	represents the content of the packet being transmitted (not all fields are shown)	In Fig. 2, grey tables represent the content of service advertisement packet. The packet {B, b_1, BF(a_1, b_2), 0} sent by node B indicates that: 1) the sender is B; 2)node B provides a service "b_1"; 3) combined bloom filter results of the descriptions of services provided by node B's 1-hop neighbors is BF(a_1, b_2); 4) the packet can still travel 0 hops.

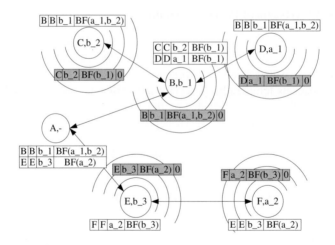

Fig. 2. Example of service advertisement packet spreading in BFGSDP

5.3 Service Request Packet Forwarding

When receiving an unduplicated service request packet, a node that does not know any matched service will forward the packet if either of the following two conditions is matched:

- The *receiver-number* subfield of the service request packet's *receiver-list* compound field is 0.
- The *receiver-number* subfield of the service request packet's *receiver-list* compound field is greater than 0 and the current node is in the *receiver-list* field.

The current node performs the following 4 steps in sequence to forward the service request packet:

Step 1. Determine candidate node set $S_{BFGSDP}(u, req)$.
$S_{BFGSDP}(u,req)=S(u, req)$.

Step 2. Determine Relay Node Set $R_{BFGSDP}(u, req)$.
$R_{BFGSDP}(u, req)=\{r(u, s) \mid s \square S_{BFGSDP}(u,req)\}$.

Step 3. Enclose the List of Relay Nodes.
Enclose nodes in $R_{BFGSDP}(u, req)$ into the service request packet's *receiver-list* compound field and set the *receiver-number* subfield of the *receiver-list* compound field to the number of nodes in $R_{BFGSDP}(u, req)$.

Step 4. Send the request packet in broadcast mode.

An example of service request packet forwarding operation in BFGSDP is shown in Fig. 3. In this figure, most symbols have the same meaning as in Fig. 2 except for grey tables. In Fig. 3, grey table over arcs represents the content of the service request packet being transmitted (not all fields are shown).

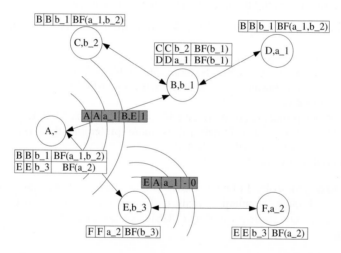

Fig. 3. Example of service request packet forwarding in BFGSDP

Suppose that node A needs a service "a_1", and there is no match in its SIC, then it has to select some nodes based on its SIC. Node A selects nodes B and E as candidate nodes through bloom filter match operation. They are just their corresponding relay nodes. Hence, two unicast request packets are sent to them, respectively. For example, the request packet $\{A, A, a\_1, \{B, E\}, 1\}$ sent by A indicates that: 1) the direct sender of the packet is node A; 2) the node that generates the service request is node A; 3) the requested service is "a_1"; 4) valid receivers of the packet are nodes B and E; 5) the request packet can still travel 1 hop.

When receiving the service request packet from node A, node B sends back a service reply packet since that there is a match in its SIC, whereas node E has to forward the service request packet further because of no match. Node E finds no candidate nodes, hence node E broadcasts a packet $\{E, A, a\_1, -, 0\}$. The receiver of the packet forwarded by node E is marked as "-", which indicates that all neighboring nodes are valid receivers. Valid receivers of a service request packet are all responsible for forwarding the packet further.

5.4 Service Reply Packet Routing

If the node that receives a new request packet finds any matched services, it should send a service reply packet to the direct sender of the service request packet in unicast mode. The service reply packet will be forwarded to the source of the service discovery request along the reverse path.

6 Simulation Research

6.1 Select Comparative Service Discovery Protocols

We select several typical service discovery protocols for MANETs and make comparative studies among them through extensive simulations.

- As in many papers, FLOOD is selected as the benchmark for our simulation studies.
- One-layer protocols are more suitable for highly dynamic MANETs. Among one-layer protocols, GSD is more preferable for its interesting semantic routing policy. Furthermore, basic operations of BFGSDP are some what same to that of GSD.
- Two-layer protocols are more suitable for more stable MANETs. Analysis in ref [12] showed that DSDP [9] is more preferable than other two-layer protocols. Hence, DSDP is selected.
- BFGSDP is also implemented.

Hence, four protocols, FLOOD, GSD, DSDP, BFGSDP are implemented and compared in through simulation studies. To facilitate the simulation process, BFGSDP is simulated using a variation of GSD that assumes a bloom filter with false positive ratio of 10%.

6.2 Performance Metrics

Four performance metrics are considered in our simulations.

- **Number of Service Request Packets Per Session:** It measures the number of service request packets sent in one simulation. It reflects the efficiency the policy of forwarding service request packets.
- **Ratio of Succeeded Requests:** It is the number of service discovery sessions in which the client has received at least one successful reply packet. It reflects the effectiveness (service discoverability) of service discovery protocols.
- **Response Time (m/s):** It is the interval between the arrival of the first reply packet and the generation of the corresponding request packet. This metric is averaged over all succeeded service discovery sessions. It measures the promptness of service discovery protocols. It also reflects the average distance between clients and the corresponding first repliers.

- **Ratio of Succeeded-SDP-Number to Total-packet-number (Suc2Packet):** This metric is the ratio of Succeeded-SDP-number to the sum of service request packets and service reply packets. It reflects the efficiency of service discovery protocols.

6.3 Simulation Models

Simulation studies are performed using Glomosim [13]. The distributed coordination function (DCF) of IEEE 802.11 is used as the underlying MAC protocol. RWM (Random Waypoint Model) is used as the mobility model.

In RWM mobility model, nodes move towards their destinations with a randomly selected speed $V \in [V_{min}, V_{max}]$. When reaching its destination, a node keeps static for a random period $T_P \in [T_{min}, T_{max}]$. When the period expires, the node randomly selects a new destination and a new speed, then moves to the new destination with the new speed. The process repeats permanently. In our simulations, $T_{min}=T_{max}=0$, $V_{min}=V_{max}=V$.

6.4 Simulation Settings

Node speed and radio range are two factors that have great effect on the performance of service discovery protocols. In this section, the effects of these two factors are inspected through extensive simulations.

Some basic parameters used in all the following simulations are set as shown in Table 2. Simulation scenarios are created with 100 nodes randomly distributed in the scenario area. At the beginning of each simulation, some nodes are randomly selected out to act as servers. These selected servers provide randomly selected services. During each simulation, 100 SDP sessions are started at randomly selected time by randomly selected nodes. In all the following figures showing simulation results, error bars report 95% confidence.

Table 2. Basic parameters

Parameters	Value	Parameters	Value
Scenario area	1000m×1000m	Number of service discovery requests	100
Node number	100	Number of service group	2
Simulation time	1000s	Number of service info in each group	5
Wireless bandwidth	1 Mbps	Maximum hop of request packets	3
Valid time of SIC item	21s	Maximum hop of advertisement packets	1
		Service advertisement interval	20s

6.5 Simulation Results

In this section, we inspect the effect of node speed on the selected protocols through simulations. To do so, we run 4 simulation sets that use the 4 selected service discovery protocols, respectively. In these simulations, 1) radio range is set to 150m, 2) the number of servers is fixed to 50. Each set includes 5 subsets of simulations, where $V=V_{MIN}=V_{MAX}$ and V is set to 0m/s, 5m/s, 10m/s, 15m/s, and 20m/s, respectively. Each subset consists of 50 similar simulations. Simulation results are averaged over 50 simulations. The results are shown in Fig. 4.

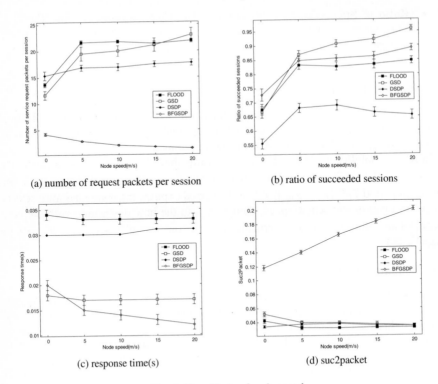

Fig. 4. The effects of node speed

Fig. 4 (a) shows the effect of node speed on the number of request packets per session. The number of request packets per session in BFGSDP is the lowest among the 4 protocols. All other three protocols have almost the same value in this metric, and DSDP is a little superior to other two protocols. Number of request packets per session in BFGSDP is about 7% to 23% of that in DSDP. Fig. 4 (b) shows the effect of node speed on ratio of succeeded sessions. The result shows that 1) DSDP has the lowest service discoverability; 2) two group-based protocols are superior to others, which is because that each server spread its services to neighbors. Fig. 4 (c) shows the effect of node speed on response time. GSD and BFGSDP are more prompt than FLOOD and DSDP. BFGSDP becomes more prompt as node speed increase, whereas GSD does not. This is because that 1) node's movement enlarges the spreading range of service advertisement packets, and 2) packet transmission in BFGSDP is more effective because of its fewer packet transmissions. Fig. 4 (d) shows that BFGSDP outperforms all other protocols distinctly. Suc2Packet of BFGSDP is about 4 to 6 times of other protocols. The superiority of BFGSDP becomes more significant as node speed increases.

7 Conclusions

In this paper, we propose BFGSDP (Bloom Filter Guided Service Discovery Protocol) for MANETs. Two salient characteristics of BFGSDP are BFG (Bloom

Filter Guiding) and BSU (Broadcast Simulated Unicast). BFG scheme decreases the risk of flood storm problem by guiding request packets with bloom filters to those nodes with potential matched services. BSU scheme benefits from the broadcast nature of wireless transmissions by replacing multiple unicast request packets with one request packet transmitted in broadcast mode with all unicast receivers enclosed. Extensive simulations show that 1) BFGSDP has the lowest request packet overhead; 2) the efficiency of BFGSDP can be several times of other tested service discovery protocols; and 3) response time of BFGSDP is much shorter than other tested protocols. In conclusion, BFGSDP is an effective, efficient, and prompt service discovery protocol for MANETs.

References

1. IETF, Mobile ad-hoc network (MANET) working group: Mobile ad-hoc networks (MANET), http://www.ietf.org/html.charters/manet-charter.html
2. Guttman, E., Perkins, C., Veizades, J., Day, M.: Service location protocol, version 2, IETF RFC 2608 (1999), http://www.faqs.org/rfcs/rfc2608.html
3. Sun Microsystems: Jini architecture specification jini-spec.pdf (1999), http://www.javasoft.com/products/jini/specs/
4. Motegi, S., Yoshihara, K., Horiuchi, H.: Service discovery for wireless ad hoc networks. In: Proc. 5th Int'l Symp. Wireless Personal Multimedia Communications (WPMC'02), pp. 232–236 (2002)
5. Engelstad, P.E., Zheng, Y.: Evaluation of service discovery architectures for mobile ad hoc networks. In: Proc. 2nd annual conference on Wireless On-demand Networks and Services (WONS'05), St. Moritz, Switzerland, pp. 2–15 (2005)
6. Chakraborty, D., Joshi, A., Yesha, Y., Finin, T.: GSD: a novel group-based service discovery protocol for MANETs. In: Proc. 4th IEEE Conf. Mobile and Wireless Communications Networks (MWCN'02), pp. 140–144 (2002)
7. Liu, F., Heijenk, G.: Context discovery using attenuated bloom filters in ad-hoc networks. In: Proc. 4th International Conference on Wired/Wireless Internet Communications (WWIC'06), pp. 13–25 (2006)
8. Goering, P., Heijenk, G.J.: Service discovery using bloom filters. In: Proc. 12th annual conference of the Advanced School for Computing and Imaging, Lommel, Belgium, pp. 219–227 (2006)
9. Kozat, U.C., Tassiulas, L.: Service discovery in mobile ad hoc networks: an overall perspective on architectural choices and network layer support issues. Ad. Hoc. Networks 2, 23–44 (2003)
10. Sailhan, F., Issarny, V.: Scalable service discovery for MANET. In: Proc. 3rd International Conference on Pervasive Computing and Communications (PerCom'05), pp. 235–244 (2005)
11. Bloom, B.: Space/time tradeoffs in hash coding with allowable errors. Communications of the ACM, pp. 422–426 (1970)
12. Gao, Z.G., Yang, Y.T., Zhao, J., Cui, J.W., Li, X.: Service discovery protocols for MANETs: a survey. In: Cao, J., Stojmenovic, I., Jia, X., Das, S.K. (eds.) MSN 2006. LNCS, vol. 4325, Springer, Heidelberg (2006)
13. Glomosim, Wireless Adaptive Mobility Lab.: Glomosim: a scalable simulation environment for wireless and wired network system, http://pcl.cs.ucla.edu/projects/domains/glomosim.html

Opportunistic Medium Access Control in MIMO Wireless Mesh Networks[*]

Miao Zhao, Ming Ma, and Yuanyuan Yang

Dept. of Electrical and Computer Engineering,
Stony Brook University, Stony Brook, NY, 11790
{mzhao,mingma,yang}@ece.sunysb.edu

Abstract. As a newly emerging technology with some inherent advantages over other wireless networks, wireless mesh networks (WMNs) have received much attention recently. In this paper, we present a cross-layer protocol design for WMNs, which combines the MIMO technique in the physical layer and the opportunistic medium access in the MAC layer into an integrated entity and jointly considers their interactions. In particular, we propose a protocol named opportunistic medium access control in MIMO WMNs (OMAC-MWMN). In an infrastructure/backbone WMN, mesh routers are grouped into clusters, in each of which a cluster head has multiple pending links with its neighbors. This traffic-driven clustering provides a great opportunity for a cluster head to locally coordinate the multiuser medium access. In each iteration of the scheduling, a cluster head opportunistically chooses some compatible neighbors among multiple candidates by utilizing the benefits of multiuser diversity. Then the cluster head simultaneously communicates with multiple selected compatible neighbors through multiuser spatial multiplexing. We formalize the problem of finding a scheduler for a cluster head to select compatible neighbors as finding a compatible pair with the maximum SINR (signal-to-interference and noise ratio) product. Our simulation results show that the proposed protocol can significantly improve system capacity with minimum extra overhead.

Keywords: Multiple-input multiple-output (MIMO), Wireless mesh networks (WMNs), Opportunistic medium access, Multiuser diversity, Transmit beamforming, Receive beamforming.

1 Introduction

Wireless mesh networks (WMNs), regarded as a promising solution for the next-generation wireless communication systems, have received a considerable amount of attention recently. WMNs are dynamically self-organized and self-configured [1]. This flexible feature in networking establishment and maintenance makes WMNs inherit the advantages of traditional ad hoc networks. Beyond the flexibility, minimum mobility of mesh routers brings the WMN many advantages of

---

[*] Research supported by NSF grant ECS-0427345 and ARO grant W911NF-04-1-0439.

L. Mason, T. Drwiega, and J. Yan (Eds.): ITC 2007, LNCS 4516, pp. 458–470, 2007.

its own, such as reliable connectivity, easy network maintenance, service coverage guarantee, low radio power consumption, etc. However, though WMNs have made a remarkable progress since their inception, some techniques are still at their infant stage and the performance is still far from satisfactory. Thus, further studies are much needed.

One of the critical factors affecting the performance of WMNs is radio techniques [1]. Multiple-input multiple-output (MIMO) techniques provide an effective way to boost up channel capacity significantly, along with more reliable communications. The MIMO system, which exploits the spatial diversity obtained by separated antennas, can be used to achieve either diversity gain to combat channel fading or capacity gain to achieve higher throughput. The operation mode for the capacity gain is called spatial division multiplexing (SDM). SDM creates extra dimensions in the spatial domain, which simultaneously carry independent information in multiple data streams. SDM is also applicable to multiuser MIMO systems, namely, multiuser SDM. It enables a node to concurrently communicate with multiple users in the same neighborhood [2]. As a multiuser system, WMNs are suitable to implement multiuser SDM, which can further increase the capacity and flexibility of WMNs. However, a good scheme for suppressing co-channel interference should be adopted in order to efficiently harvest the spatial multiplexing gain.

Another effective way to improve system capacity of WMNs is to exploit multiuser diversity. Since not all users are likely to experience deep fading at the same time in a multiuser system, the total throughput of the entire multiuser system is resilient to different users. Thus, diversity occurs not only across the antennas within each user, but also across different users. This type of diversity is referred to as multiuser diversity [3]. It lessens the effect of channel variation by exploiting the fact that different users have different instantaneous channel gains to the shared medium [4]. Opportunistic medium access [4] utilizes the physical layer information from multiple users to optimize the medium access control, in which the users with poor channel condition yield the channel access opportunity to the users with favorable channel quality so that the overall network performance can be improved.

The above observations encourage us to consider a cross-layer approach to improving the performance of WMNs. The key motivation of our work is to jointly implement multiuser MIMO in the physical layer and opportunistic medium access control in the MAC layer. In particular, we propose a protocol named opportunistic medium access control in MIMO WMNs (OMAC-MWMN). We consider homogeneous infrastructure/backbone mesh networks. In this architecture, multiple mesh routers, each of which is equipped with two antennas, form an infrastructure for clients. In the physical layer of the system, with specific transmit and receive beamforming, a mesh router can concurrently communicate with two neighbors. However, current MAC protocols were designed to allow only a single node pair to communicate at a time. In order to overcome the difficulties of applying MIMO in WMNs and also utilize the multiuser diversity gain to opportunistically schedule the medium access, in this paper we design a new MAC layer protocol for such systems. Our MAC solutions can be summarized as follows. In the WMN system, each mesh router in the system typically has multiple links[1]. Time is divided into beacon intervals with a fixed length. Within a short contention period T_{cp} at the beginning of each beacon interval, each mesh router with a certain number of output links contends to form a cluster and functions as a cluster head to coordinate multiuser communications locally [6]. For each data transmission during the beacon interval, by processing data according to the channel state which can be considered as transmit beamforming, the cluster head can send distinct packets to two distinct cluster members simultaneously. We call such two cluster members a pair of compatible members and each one is a compatible peer of the other. With specific receive beamforming, each compatible member can intelligently separate and decode the information for its own. To leverage the effect of multiuser diversity, for each data transmission, the cluster head will select multiple cluster members as candidate receivers. Since the total transmitting power is limited, not all pairs among the candidate members are compatible. The cluster head always transmits data to the pair of compatible members with the maximum SINR product among the candidate members. To the best of our knowledge, there is still a little such cross-layer design of WMNs in the literature. The main contribution of this paper is three folds. First, we formalize the problem of finding a scheduler for a cluster head to select receivers among multiple candidate members as finding a compatible pair with the maximum SINR product. Second, we modify IEEE 802.11 standard to efficiently cooperate with the advances of MIMO techniques in the physical layer. We also design new MAC solutions to allow the cluster head to simultaneously communicate with multiple cluster members in its neighborhood. Third, the proposed traffic-driven clustering approach helps the cluster head opportunistically optimize the medium access control by utilizing multiuser diversity such that better performance can be achieved.

2   System Model of MIMO WMNs

Consider a homogeneous infrastructure/backbone mesh network where each mesh router is equipped with two antennas. Since the spatial degrees of freedom are equal to $\min(n, K)$ where n is the number of antennas of a mesh router and K is the total number of mesh routers in its neighborhood, a mesh router with the complete knowledge of channel state information (CSI) can simultaneously communicate with two neighbors. Here, we take a cluster in the WMN shown in Fig. 1 as an example to explain how to implement MIMO. The baseband model of narrowband intra cluster head to two member transmission can be described as follows [2]. Without loss of generality, we assume that the two members are cluster member 1 and cluster member 2.

$$\mathbf{y}_i[m] = \mathbf{H}_i\mathbf{x}[m] + \mathbf{n}_i[m], \qquad i = 1, 2. \qquad (1)$$

---

[1] A link corresponds to a node pair with pending data traffic for transmission.

where $\mathbf{x}[m]$ is a 2×1 vector representing the transmitted vector at time m, $\mathbf{y}_i[m]$ is the received vector of size 2×1 at receiver i, $\mathbf{H}_i$ is a 2×2 channel matrix, with each entry of $\mathbf{H}_i$ representing the complex channel coefficient from an antenna of the cluster head to an antenna of cluster member i, and $\mathbf{n}_i[m]$ is an i.i.d. $\mathcal{CN}(0, N_0)$ noise vector.

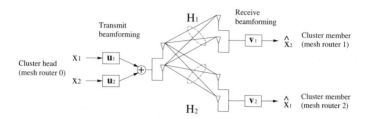

Fig. 1. The architecture of a cluster-based MIMO WMN with linear transmitting and receiving strategies

Suppose $x_i[m]$ denotes the transmitted data destined to cluster member i and P_i is the power allocation for cluster member i. $x_i[m]$ is assumed to be independent and normalized (i.e., $E|x_i[m]|^2 = 1$). The scalar data $x_i[m]$ is multiplied by an 2×1 transmit beamforming vector $\mathbf{u}_i$ before being transmitted over the channel. Then, the overall 2×1 transmitting signal vector can be written as

$$\mathbf{x}[m] = \mathbf{u}_1 x_1[m] + \mathbf{u}_2 x_2[m] \tag{2}$$

The estimation of the data stream $\hat{x}_i[m]$ at receiver i is obtained via receive beamforming as a weighted sum rate of the received signal at each receive antenna and can be expressed as follows.

$$\hat{x}_i[m] = \mathbf{v}_i^* \mathbf{y}_i[m] = \sum_{j=1}^{2} \mathbf{v}_i^* \mathbf{H}_i \mathbf{u}_j x_j[m] + \mathbf{v}_i^* \mathbf{n}_i[m], \quad i = 1, 2. \tag{3}$$

where $\mathbf{v}_i^*$ is a 2×1 row vector representing the receive beamforming vector of receiver i.

Thus, the SINR for receiver i is given by

$$\text{SINR}_i = \frac{E(|x_i|^2)\|\mathbf{v}_i^* \mathbf{H}_i \mathbf{u}_i\|^2}{E[\|\mathbf{v}_i^*(\sum_{j \neq i} \mathbf{H}_i \mathbf{u}_j x_j + \mathbf{n}_i)\|^2]} = \frac{\|\mathbf{v}_i^* \mathbf{H}_i \mathbf{u}_i\|^2}{\sum_{j \neq i} \|\mathbf{v}_i^* \mathbf{H}_i \mathbf{u}_j\|^2 + N_0\|\mathbf{v}_i^*\|^2}, \quad i, j = 1, 2. \tag{4}$$

In order to maximize the received signal strength of each cluster member, the transmit and receive beamforming vectors should be properly chosen. Our objective is to optimize the system capacity of MIMO WMNs. Hence, we follow the idea in [7] to use the product of SINR as the performance metric so that we can find the transmit and receive beamforming vectors at the cluster head and the members to ensure multiuser SDM. The problem formulation can be written as

$$(\mathbf{u}_i, \mathbf{v}_i)_{\text{opt}} = \arg \max_{(\mathbf{u}_i, \mathbf{v}_i)} \left(\prod_{i=1}^{2} SINR_i \right)$$
$$s.t. \quad \|\mathbf{v}_i\| = 1, \|\mathbf{u}_i\| = \sqrt{P_T} \quad \text{for } i = 1, 2. \tag{5}$$

Because optimal beamforming is not the main focus of our work, we simply follow the derivations in [7] and briefly state the results of the setting of the beamforming vectors. In order to optimize (5), the receive beamforming vector can first be set since it does not affect the SINR of other cluster member signals. The optimal receive beamforming can be found as

$$\mathbf{v}_i = \alpha \mathbf{\Phi}_i^{-1} \mathbf{H}_i \mathbf{u}_i \quad \text{where } \mathbf{\Phi}_i = \sum_{j \neq i} \mathbf{H}_i \mathbf{u}_j \mathbf{u}_j^* \mathbf{H}_i^* + N_0 \mathbf{I}, \quad \text{for } i, j = 1, 2. \tag{6}$$

where α is a real constant. Then, the SINR of receiver i can be rewritten as

$$\text{SINR}_i = \frac{\mathbf{u}_i^* \mathbf{H}_i^* \mathbf{H}_i \mathbf{u}_i}{\sum_{j \neq i} \mathbf{H}_i \mathbf{u}_j \mathbf{u}_j^* \mathbf{H}_i^* + N_0 \mathbf{I}} = \mathbf{u}_i^* \mathbf{H}_i^* \mathbf{\Phi}_i^{-1} \mathbf{H}_i \mathbf{u}_i, \quad \text{for } i, j = 1, 2. \tag{7}$$

Since $\mathbf{\Phi}_i^{-1}$ can be rewritten as $\mathbf{\Phi}_i^{-1} = \frac{1}{N_0} (\mathbf{I} - \frac{\mathbf{H}_i \mathbf{u}_j \mathbf{u}_j^* \mathbf{H}_i^*}{\mathbf{u}_j^* \mathbf{H}_i^* \mathbf{H}_i \mathbf{u}_j + N_0})(i \neq j)$, the SINR of the two cluster members can be expressed as

$$\text{SINR}_i = \frac{1}{N_0} \left[\mathbf{u}_i^* \mathbf{H}_i^* \mathbf{H}_i \mathbf{u}_i - \frac{\|\mathbf{u}_i^* \mathbf{H}_i^* \mathbf{H}_i \mathbf{u}_j\|^2}{\mathbf{u}_j^* \mathbf{H}_i^* \mathbf{H}_i \mathbf{u}_j + N_0} \right], i, j = 1, 2 \text{ and } i \neq j. \tag{8}$$

It is shown in [7] that

$$\zeta = \text{SINR}_1 \cdot \text{SINR}_2 \geq \left[\frac{\mathbf{u}_1^* \mathbf{H}_1^* \mathbf{H}_1 \mathbf{u}_1}{\mathbf{u}_2^* (\mathbf{H}_1^* \mathbf{H}_1 + \frac{N_0}{P_T} \mathbf{I}) \mathbf{u}_2} \right] \cdot \left[\frac{\mathbf{u}_2^* \mathbf{H}_2^* \mathbf{H}_2 \mathbf{u}_2}{\mathbf{u}_1^* (\mathbf{H}_2^* \mathbf{H}_2 + \frac{N_0}{P_T} \mathbf{I}) \mathbf{u}_1} \right] \tag{9}$$

By maximizing (9) which is the lower bound of the product of SINR, the optimum transmit beamforming vector can be set as

$$\mathbf{u}_i = d_i \mathbf{W}_i \mathbf{e}_i, \quad i = 1, 2. \tag{10}$$

where d_i is a real constant to ensure the power constraint, $\mathbf{W}_i$ is called whitening matrices which satisfies $\mathbf{W}_i^* [\mathbf{H}_j^* \mathbf{H}_j + \frac{N_0}{P_T} \mathbf{I}] \mathbf{W}_i = \mathbf{I}$ $(i \neq j \in (1, 2))$, and $\mathbf{e}_i$ is the eigenvector which corresponds to the largest eigenvalue of $\mathbf{W}_i^* \mathbf{H}_i^* \mathbf{H}_i \mathbf{W}_i$.

After choosing the beamforming vectors based on the above principles, any two cluster members among a total of K members which satisfy the following criteria can be called a pair of compatible members, with each one being the compatible peer of the other, denoted as $i \bowtie j$.

$$\begin{cases} \text{SINR}_i = \frac{1}{N_0} \left[\mathbf{u}_i^* \mathbf{H}_i^* \mathbf{H}_i \mathbf{u}_i - \frac{\|\mathbf{u}_i^* \mathbf{H}_i^* \mathbf{H}_i \mathbf{u}_j\|^2}{\mathbf{u}_j^* \mathbf{H}_i^* \mathbf{H}_i \mathbf{u}_j + N_0} \right] \geq \gamma_i, \quad i, j \in K \\ \text{SINR}_j = \frac{1}{N_0} \left[\mathbf{u}_j^* \mathbf{H}_j^* \mathbf{H}_j \mathbf{u}_j - \frac{\|\mathbf{u}_j^* \mathbf{H}_j^* \mathbf{H}_j \mathbf{u}_i\|^2}{\mathbf{u}_i^* \mathbf{H}_j^* \mathbf{H}_j \mathbf{u}_i + N_0} \right] \geq \gamma_j, \quad i, j \in K \end{cases} \tag{11}$$

The above criteria demonstrate that a cluster head can simultaneously forward packets to the two compatible members with $\mathbf{u}_i$ and $\mathbf{u}_j$, and the SINR of each link can satisfy the requirement of the QoS constraint for each member. In this paper, we assume $\gamma_i = \gamma_j = \gamma_0$, where γ_0 is the SINR threshold for the base rate of the system.

3   Principles of OMAC-MWMN

In the previous section, we have discussed how the cluster head makes concurrent communication with two compatible cluster members by using the advanced MIMO techniques (i.e., multiuser SDM) in the physical layer. In this section, we present our MAC protocol OMAC-MWMN in details, which mainly aims to solve the following problems: (1) How to group the mesh routers into clusters; (2) How to support multiple links simultaneously through utilizing MIMO technique in the physical layer; (3) How to utilize multiuser diversity to opportunistically schedule the medium access within a cluster.

3.1   Overview of the OMAC-MWMN Protocol

The proposed MAC timing structure is as shown in Fig. 2, where time is divided into identical beacon intervals. Each mesh router periodically sends out beacons to synchronize time in a distributed manner. The time for synchronization is not depicted in the figure. Synchronization in a multi-hop network is generally a challenging task. The simplest method can be described as follows [8,9]. Each mesh router includes a timestamp of its local timer when transmitting a beacon signal. Once it receives a beacon from another router, it cancels its beacon and adjusts its timer according to the timestamp included in the beacon. More sophisticated solutions can be used to achieve more accurate synchronization. Here, we simply assume that synchronization can be achieved through beacons. There are two phases in each beacon-interval after the synchronization beacon. Phase I is called contention access period, which performs cluster formation among the mesh routers. Phase II is called data transmission period which is used for data exchange.

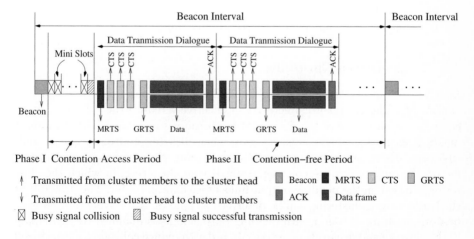

Fig. 2. The timing structure of proposed OMAC-MWMN

Phase I consists of several mini-slots [10]. Mesh routers with pending output links are qualified to contend to be a cluster head during Phase I. Each qualified mesh router intends to declare itself to be a cluster head by transmitting a busy signal during a selected mini-slot. Once a mesh router successfully transmits its busy signal, other qualified mesh routers in its neighborhood will refrain from the contention and return to the normal state. In this way, the length of phase I becomes adjustable. If each mesh router randomly selects a mini-slot to transmit its busy signal, collisions may occur when more than one mesh routers intend to advertise the pending links. Intuitively, this situation would become even worse with the increase of the number of participating mesh routers. In order to circumvent the problem, a scheduling scheme is needed. In our study, we adopt a distributed scheduler namely prioritized backoff algorithm to coordinate the contention for the cluster head, which will be discussed in detail in the next subsection. Once a mesh router successfully becomes a cluster head, it will take this responsibility for the rest time of the current beacon interval. The cluster head will locally coordinate the medium access within the cluster. Since the cluster will be reformed during each beacon interval, the cluster head rotates from one router to another. Therefore, our scheme will not severely affect the fairness in the system in the long run.

Phase II is used for data transmission. During this phase, the cluster head in each cluster forwards the backlogged packets to the selected compatible pair. To take the advantage of link dynamics across different users, opportunistic medium access control which utilizes the benefit of multiuser diversity is used in each data transmission dialogue. For each iteration of data transmission dialogue, the cluster head will first select some cluster members as receiver candidates. After querying whether these members are available or not, the cluster head will finally determine two compatible members based on a certain scheduling policy. The details of the scheduling policy will be discussed in the next subsection. With the help of multiuser SDM, the cluster head can concurrently transmit data to the two compatible members as discussed in Section 2.

3.2   Traffic-Driven Clustering

In order to improve the capacity of the WMN, mesh routers are grouped into clusters [1]. The cluster architecture can effectively reduce the collisions, especially in the cases where the density of mesh routers increases. Furthermore, it makes it easier to achieve capacity gain when utilizing multiuser spatial multiplexing. There are many approaches to forming clusters based on different factors. In our studies, we consider traffic-driven clustering, in which the clustering is based on the pending output links of each mesh router. The principle of the traffic-driven clustering can be described as follows. After synchronized with beacons, each mesh router with pending output links is qualified to participate in the cluster head declaration process. In order to reduce collisions, a prioritized backoff algorithm is employed. Each router needs first calculate its own priority P_i, which can be expressed as follows

$$P_i = \alpha \cdot (N - n_i) \tag{12}$$

where n_i is the number of pending links held by router i, and N and α are constant system coefficients. In order to ensure the value of the priority is always positive, N is set to the maximum number of neighbors in the system. α is a constant number to scale the quantity level of the priority.

In order to reduce the collisions as much as possible, it is desirable for mesh routers to employ different backoff timers. That is, each mesh router should refrain from promptly transmitting its busy signals before its own backoff timer expires [4]. The backoff timer $T_{backoff}^{(i)}$ of router i can be expressed as

$$T_{backoff}^{(i)} = \lceil P_i + rand(\sigma) \rceil \cdot T_{mini\_slot} \tag{13}$$

where T_{mini_slot} is the default mini-slot size, and σ is the seed to generate a random number. In order to circumvent the problem of more than one neighboring mesh routers having the same priority, it is necessary to add a small additional random priority to the original priority of each mesh router to distinguish one from another. Since we bind the priority of each router to its pending links, the router with more pending links will backoff less time. Consequently, it has more chance to capture the channel and successfully become the cluster head among its neighbors. The mesh routers in the neighborhood other than the cluster head will become its cluster members. Consider the worst case that all the routers in the same neighborhood have the same number of pending links. In order to effectively differentiate the routers, σ should be proportional to the total number of the routers in the neighborhood (i.e., $\sigma \propto N$). Otherwise, there will be severe collisions. If no mesh router successfully declares itself as the cluster head within a waiting duration (denoted as T_{out}), another cycle of the cluster head selection will take place.

Once a cluster is formed, the cluster head will create a cluster ID and advertise it to its cluster members. In addition, each cluster member is also assigned a local member ID as an intra-cluster identification. It is possible that a mesh router can be on the output links of two cluster heads. In order to keep our work simple, we assume a mesh router is only associated with one cluster.

3.3 Intra-cluster Opportunistic Medium Access Control

As mentioned earlier, the cluster architecture makes it possible for the cluster head to locally coordinate multiuser scheduling among its pending links. In this subsection, we describe the intra-cluster opportunistic medium access control in detail. As will be seen, this scheme not only can intelligently utilize multiuser diversity but also can effectively inter-operate with the multiuser spatial multiplexing in the physical layer.

The intra-cluster handshake of our proposed OMC-MWMNs is depicted in Fig. 3. The cluster head initiates the medium access for each iteration of data transmission dialogue during the current interval. In order to probe multiple output pending links simultaneously, the cluster head first sends out a MRTS (multicast request-to-send) [4] control packet. The main difference between MRTS and

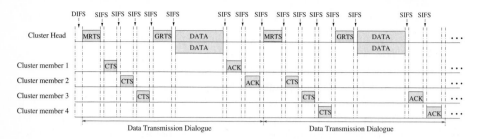

Fig. 3. The intra-cluster handshake of the proposed OMAC-MWMN protocol

regular RTS is that MRTS has multiple 6-byte RA (receiver address) fields. Multicast RTS for channel probing parallelizes the multiple serial unicast RTS/CTS control packet exchanges so that the overhead and time can be significantly reduced. In order to limit the additional overhead, only a subset of pending links will be considered as candidate receivers each time. A larger size of the subset means more diversity, but also means higher additional overhead on control packet exchange and computing complexity. Therefore, a proper size should be adopted. For each MRTS probing, among a total of K output links, the cluster head will select k ($k \leq K$) cluster members as the candidate receivers complying with the basic round robin scheduling. The addresses of k candidates are indicated in the RA list of MRTS. Once received MRTS from the cluster head, all available candidate cluster members will in turn reply a CTS (clear-to-send) control packet. The rank of an RA in the RA list indicates the order in which the candidate cluster members should respond CTS. Since the cluster head can concurrently transmit data to at most two compatible cluster members each time, the cluster head needs to select two members as the final receivers for the following data exchange among the multiple candidates. This selection is based on the scheduling policy of the maximum SINR product, i.e., the cluster head will preferentially serve the compatible pair with the maximum SINR product among multiple candidate cluster members. The maximum SINR product is achieved by searching over the pairs of compatible members (i, j) with transmit beamforming vectors $\mathbf{u}_i$, $\mathbf{u}_j$ as given below

$$(i, j)_{\mathrm{fnl}} = \arg\max_{(i,j)}(\zeta) = \arg\max_{(i,j)}(\mathrm{SINR}_i \cdot \mathrm{SINR}_j)$$

$$= \arg\max_{(i,j)}\left\{\frac{1}{N_0^2}\left[\mathbf{u}_i^*\mathbf{H}_i^*\mathbf{H}_i\mathbf{u}_i - \frac{\|\mathbf{u}_i^*\mathbf{H}_i^*\mathbf{H}_i\mathbf{u}_j\|^2}{\mathbf{u}_j^*\mathbf{H}_i^*\mathbf{H}_i\mathbf{u}_j + N_0}\right] \cdot \left[\mathbf{u}_j^*\mathbf{H}_j^*\mathbf{H}_j\mathbf{u}_j - \frac{\|\mathbf{u}_j^*\mathbf{H}_j^*\mathbf{H}_j\mathbf{u}_i\|^2}{\mathbf{u}_i^*\mathbf{H}_j^*\mathbf{H}_j\mathbf{u}_i + N_0}\right]\right\}$$

$$s.t. \quad P_i = P_j = P_T, (i, j) \in k, i \bowtie j$$

$$(14)$$

In this way, the cluster head can finally determine the compatible pair for data exchange. It will then advertise a GRTS (group RTS) control packet. GRTS is also an enhanced RTS introduced in [5], which contains two 6-byte RA fields to indicate the selected two compatible members. Finally, the cluster head can simultaneously communicate with the two compatible members with the help of multiuser spatial multiplexing. If in the rare cases that there is only one pending output link for the cluster head, or no compatible pair exists among the candidate cluster members, the cluster head will send to the only receiver in the former case, or can choose to send to the receiver with the most favorable channel condition among the multiple candidates in the latter case.

In the multi-hop scenario, the inter-cluster collision is also an important issue. The interference introduced by adjacent clusters may mostly depend on which sets of routers are actually transmitting. In our study in this paper, we simply use the traditional carrier sensing to keep the inter-cluster collisions low. Furthermore, the hidden-terminal and the exposed-terminal problems are inherently existing. Many previous studies [11,12,13] have spent much efforts on these problems, among which busy-tone based schemes possess great predominance over others. Therefore, we also employ dual busy tone to notify other routers of the transmissions currently ongoing to effectively ensure successful transmissions in the same neighborhood [13].

4   Simulation Results

We have conducted extensive simulations to evaluate the proposed protocol. In this section, we discuss the simulation results of OMAC-MWMN and compare it with the IEEE 802.11 standard.

4.1   A Single Cluster

We first consider the simple case where there is only one cluster in the system, which implies that all the mesh routers are within each other's radio range. We assume that a total of K mesh routers are randomly distributed over a $D \times D$ square area and all the routers always hold backlogged packets destined for all the neighbors. The packet size follows the exponential distribution with a mean L. If not specified otherwise, the average packet size L is set to 1000 bytes. A RTS packet is 40 bytes, CTS and ACK packets are 39 bytes, and the MAC header of a data packet is 47 bytes [14]. The beacon interval is fixed to $50ms$, $T_{mini\_slot} = 10\mu s$, $\alpha = 1$, $\sigma = K$, and $T_{out} = 2 * K * T_{mini\_slot}$. The candidate receiver list size (k) in each data transmission dialogue is set to 3 unless stated otherwise. The base rate of the system is $2Mbps$. Three sets of simulation experiments have been carried out, in each of which one of the following parameters varies: the total number of mesh routers K, the size of the candidate list k, and the average packet size L.

Fig. 4 (a) depicts the throughput performance of OMAC-MWMN and IEEE 802.11 with the number of mesh routers varying from 2 to 20, where D is set to $200m$. It is clear that IEEE 802.11 achieves the highest capacity with the 2-node scenario and decreases as the number of users increases since collision becomes more severe. Also as shown in the figure, the throughput of OMAC-MWMN first increases as K increases and then keeps relatively stable with a slight decrease when K further increases. This observation is reasonable. This is because that, at the initial stage, as K increases the cluster head has more chance to select a

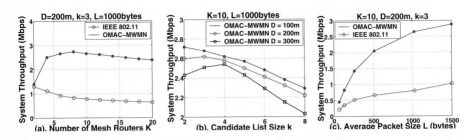

Fig. 4. Simulation results for single cluster case. (a) Throughput vs. K. (b) Throughput vs. k. (c) Throughput vs. L.

compatible pair with better channel conditions to communicate with in each cycle of data transmission dialogue. Therefore, the system throughput is improved, which mainly attributes to the benefits of multiuser diversity. However, as K further increases beyond the value that can fully extract the multiuser diversity gain, the system throughput will no longer be improved. Instead, the throughput has a slight decrease since there are more collisions during the contention period for the cluster head declaration. It is noticed that the performance of OMAC-MWMN always outperforms IEEE 802.11 since we take the advantage of simultaneous data transmission with the help of spatial multiplexing in our protocol. And also OMAC-MWMN is less sensitive to the increase of K due to the intra-cluster collision resolution in the protocol.

Fig. 4 (b) shows the system throughput varying with the candidate list size k, where D is equal to $100m$, $200m$ and $300m$ to represent different cases in WMNs. The total number of mesh routers K is set to 10. With the results of this set of experiments, we attempt to experimentally answer the question: What is the optimal number of candidate receivers a cluster head should query simultaneously in a transmission dialogue? From the results plotted in Fig. 4 (b), we can draw some observations. First, the optimal value of k increases with the distance D. Second, the throughput improvement becomes less obvious when k is larger than 3, and the throughput decreases in all the cases when k is larger than 4. Larger k means more diversity, but the overhead in the control packets and especially the computing complexity introduced to the cluster head will sharply increase to overshadow the multiuser diversity gain.

Fig. 4 (c) shows the impact of the average packet size L on the throughput performance of OMAC-MWMN and IEEE 802.11, where K is set to 10. More throughput can be achieved in all the cases we investigated as L increases since the proportion of effective time on data transmission is enhanced. OMAC-MWMN always performs better than IEEE 802.11 due to the opportunistic scheduling which utilizes the multiuser diversity gain and the simultaneous data transmission with the help of multiuser spatial multiplexing. It is shown that OMAC-MWMN improves the system throughput by up to 160% with respect to IEEE 802.11.

4.2   Multiple Clusters

In this subsection, we discuss the case of multiple clusters. A square lattice network is considered, in which all the mesh routers are static without mobility as shown in Fig. 5 (a). Both vertical and horizontal flows are present in the network, all traffic originates at the top and left edges of the network and is forwarded downward or rightward to the opposite edges, and the intermediate mesh routers do not generate any traffic [14]. We assume that each router is 200 meters from its adjacent neighbors in the east, west, north and south directions. The transmission range and the carrier sensing range of each mesh router are set to $250m$ and $550m$, respectively. Fig. 5 (b) shows the average per flow throughput obtained when the number of mesh routers (i.e., the lattice size) varies from 16 to 144. We can see that the throughput of two mechanisms drops with the increase of the lattice size. These inefficiencies are due to the interference inherently existed in the multi-hop scenario. However, OMAC-MWMN still improves the average per flow throughput by at least 40% compared to IEEE 802.11. Fig. 5 (c) depicts the average per flow throughput for a variety of the average packet size. The observations in the single-cluster case are still applicable here, that is, a larger average packet size will achieve higher channel utilization.

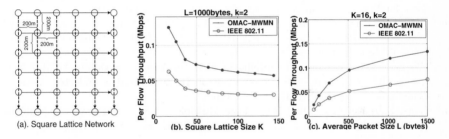

Fig. 5.   Simulation results for multiple cluster case. (a) Simulation Topology. (b) Throughput vs. K. (c) Throughput vs. L.

5   Conclusions

In this paper, we have proposed a cross-layer design on WMNs, which jointly combines the MIMO technique (i.e., multiuser spatial multiplexing) in the physical layer and opportunistic medium access control in the MAC layer. In particular, we have proposed a protocol named opportunistic medium access control in MIMO WMNs (OMAC-MWMN). In our protocol, mesh routers are grouped into clusters based on their pending output links. In each cycle of data transmission, the cluster head has the opportunity to select multiple output links as its candidate receivers and determines two compatible cluster members with the maximum SINR product among the multiple candidates to communicate with. By effectively exploiting the multiuser diversity and simultaneous data transmission, the channel utilization is greatly improved.

References

1. Akyildiz, I.F., Wang, X.: A Survey on Wireless Mesh Networks. IEEE Communications Magazine 43, 23–30 (2005)
2. Tse, D., Viswanath, P.: Fundamentals of Wireless Communication. Cambridge University Press, Cambridge (2005)
3. Yu, W.: Spatial Multiplex in Downlink Multiuser Multiple-Antenna Wireless Environments. IEEE GLOBECOM 4, 1887–1891 (2003)
4. Wang, J., Zhai, H., Fang, Y., Yuang, M.: Opportunistic Media Access Control and Rate Adaptation for Wireless Ad Hoc Networks. IEEE ICC 1, 154–158 (2004)
5. Ji, Z., Yang, Y., Zhou, J., Takai, M., Bagrodia, R.: Exploiting Medium Access Diversity in Rate Adaptive Wireless LANs. ACM Mobicom, pp. 345–359 (2004)
6. Wang, J., Zhai, H., Fang, Y., Shea, J.M., Wu, D.: OMAR: Utilizing Multiuser Diversity in Wireless Ad Hoc Networks. IEEE Trans. Mobile Computing
7. Wong, K.K., Murch, R.D., Letaief, K.B.: Performance Enhancement of Multiuser MIMO Wireless Communication Systems. IEEE Trans. Communications 50(12), 1960–1970 (2002)
8. Tseng, Y.C., Hsu, C.S., Hsieh, T.Y.: Power-Saving Protocols for IEEE 802.11-Based Multi-Hop Ad Hoc Networks. IEEE INFOCOM (2002)
9. So, J., Vaidya, N.H.: A Multi-channel MAC Protocol for Ad Hoc Wireless Networks. Technical Report (2003)
10. Wang, J., Fang, Y., Wu, D.: A Power-Saving Multi-radio Multi-channel MAC Protocol for Wireless Local Area Networks. IEEE INFOCOM (2006)
11. Wang, D., Tureli, U.: Joint MIMO-OFDM and MAC Design for Broadband and Multihop Ad Hoc Networks. EURASIP Journal on Wireless Communications and Networking (2006)
12. Wu, S., Tseng, Y., Sheu, J.: Intelligent Medium Access for Mobile Ad Hoc Networks with Busy Tones and Power Control. IEEE Journal on Selected Areas in Communications 18(9), 1647–1657 (2000)
13. Haas, Z.J., Deng, J.: Dual Busy Tone Multiple Access (DBTMA) - A Multiple Access Control Scheme for Ad Hoc Networks. IEEE Trans. Communications 50(6), 975–985 (2002)
14. Li, J., Blake, C., De Couto, S.J.D., Lee, H., Morris, R.: Capacity of Ad Hoc Wireless Networks. ACM Mobicom (2001)

A Delay Based Multipath Optimal Route Analysis for Multi-hop CSMA/CA Wireless Mesh Networks

Jiazhen Zhou and Kenneth Mitchell

Dept. of Computer Science Electrical Engineering
University of Missouri - Kansas City, Kansas City, MO, 64110
{jzmpc,mitchellke}@umkc.edu

Abstract. In this paper we present a method for determining optimal routes along selected paths in a wireless mesh network based on an interference aware delay analysis. We develop an analytic model that enables us to obtain closed form expressions for delay in terms of multipath routing variables. A flow deviation algorithm is used to derive the optimal flow over a given set of routes. The model takes into account the effects of neighbor interference and hidden terminals, and tools are provided to make it feasible for the performance analysis and optimization of large-scale networks. Numerical results are presented for different network topologies and compared with simulation studies.

1 Introduction

Wireless mesh networks are multi-hop access networks used to extend the coverage range of current wireless networks [1]. They are composed of mesh routers and mesh clients, and generally require a gateway to access backhaul links. Access to the medium is either centrally controlled by the base station or distributed, typically using some form of CSMA/CA protocol.

As pointed out by Tobagi [2], the exact throughput-delay analysis for multi-hop networks requires a large state space. For large topologies, an exact analysis is almost impossible, leading us to consider an approximate analysis. Similar work by Leiner [3] and Silvester and Lee [4], assume that frames are independently generated at each node for transmission to a neighbor, whereby the amount of traffic generated is a function of the topology, routing, and end-to-end traffic requirements. They also develop a model of the neighborhood around a node and characterize it with a number of parameters representing average behavior. Parameters for all nodes are then found through an iterative process. However, in Leiner's work, single-hop models are used for the neighborhood around each node, which means that all of the interfering nodes of a certain node interfere with each other. This makes the model relatively simple, but generally not applicable for most multi-hop networks.

In Boorstyn et al. [5], large networks are decomposed into smaller groups and Markov chains are constructed for those nodes that can transmit simultaneously.

L. Mason, T. Drwiega, and J. Yan (Eds.): ITC 2007, LNCS 4516, pp. 471–482, 2007.

The throughput of the network is then studied, based on the assumption of Poisson arrivals. The algorithm is iterative, and due to the need to compute all independent sets in the network, the complexity of the algorithm is prohibitive. Wang and Kar [6] present work based on the same architecture as Boorstyn, while paying special attention to the optimal min/max fairness and throughput with the RTS/CTS mechanism.

Other than the node-group based method proposed by Boorstyn, single node or flow based methods have also been used in recent research and are generally scalable to large networks. Carvalho and Garcia-Luna-Aceves [7] present a model that takes into consideration the effects of physical layer parameters, MAC protocol, and connectivity. They mainly focus on the throughput of nodes for the saturated case. Garetto, Salonidis, and Knightly [8] address fairness and starvation issues by using a single node view of the network that identifies dominating and starving flows and accurately predicts per-flow throughput in a large-scale network.

We can see that most recent studies mainly deal with throughput and fairness issues. In our work, the model we build is not only suitable for throughput analysis, but also for delay analysis. Based on our closed form solution for delay, multi-path route optimization becomes possible. The analytical model we introduce is based on a single node analysis. Interfering nodes and hidden terminals are taken into account when computing the probability that a node successfully transmits frames.

The rest of this paper is organized as follows: In section 2 we describe the basic model and exploit the neighbor relationships to derive solutions using iterative algorithms. In section 3, the closed form representation of delay at each node is derived and a corresponding optimization model is introduced. Examples using our method for the analysis and optimization of wireless mesh networks are shown in section 4, and section 5 concludes this paper.

2   Basic Model

Similar to work presented in [6] and [8], our model is based on a generic carrier sense multiple access protocol with collision avoidance (CSMA/CA). We generalize on the work of Kleinrock and Tobagi [9,10] and Boorstyn et al. [5] to include a finite number of nodes, multiple hops, and interference caused by routing. Nodes having frames to transmit can access the network if the medium is idle. If the medium is detected as being busy, a node will reattempt to access the medium after a specified time interval. We assume that there is some mechanism (such as RTS/CTS in the 802.11 standard) that allows the node to determine if the medium is available or if it must wait and reattempt access to the channel. We use a nodal decomposition method that relies on an iterative process to determine the probability that a transmission attempt is successful.

We assume that messages at each node i are generated according to a Poisson distribution with mean rate λ_i. All message transmission times are exponentially distributed with mean $1/\mu$. Likewise, the channel capacity is taken to be

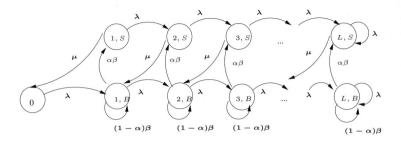

Fig. 1. Markov chain diagram of a single node

μ. We assume an ideal collision avoidance mechanism that can always detect if the medium is busy or free at the end of a transmission attempt waiting period. All waiting periods between transmission attempts (backoff periods) are exponentially distributed with mean $1/\beta$, resulting in a geometrically distributed number of backoff attempts (see Cali et al. [11]). An infinite number of backoff periods are possible. Each node backs off after a successful transmission to ensure fairness. The probability that node i finds the medium free and is able to successfully transmit a message is denoted as α_i. If node A interferes with node B, then node B also interferes with node A (symmetrical transmission range.) All successfully transmitted frames are received error free.

In multi-hop networks, some nodes directly interfere with each other and some indirectly interfere (hidden terminal problem [10].) Those nodes that directly interfere or are hidden terminals to each other cannot send messages at the same time. We refer to these nodes as "neighbors" in this paper and introduce a "neighbor matrix", N, in section 2.2, to derive these relationships.

Fig. 1 depicts the queueing model for a single node. For each state (l, S) or (l, B), S means the node is sending (transmitting), B means it's backing off, l represents the number of messages waiting in the queue, and L is the queue length. State 0 means there is no frame at this node, so the node is in idle state. This is an M/G/1/L model from which the steady state, busy probability, blocking probability, etc. can be easily derived [12]. Strictly speaking, for internal nodes in the network that relay messages, the arrivals from different sources can be correlated with each other, so the total arrival stream will not be Poisson. However, the assumptions we make allow us use the M/G/1/L model, which produces results that are extremely close to simulation.

2.1 Calculating Successful Transmission Probabilities

We have defined α_i as the probability that node i successfully accesses the medium during a transmission attempt, so α_i is a *statistical view* of the medium being idle when node i has a frame to send. Now consider the state of the medium in the region around node i. There are three possible states for node i: 1) being "idle", with probability $P_I[i]$, 2) being in "sending" state, with probability $P_S[i]$, and 3) being in "backoff" state, with probability $P_B[i]$. When a node is transmitting frames, we denote this node as being in its "sending" state. Let ρ_i be the queuing system utilization of node i, which means this node is either in its "backoff" or "sending" state, so $\rho_i = P_S[i] + P_B[i]$. Only when it is in "backoff", will node i sense the medium (attempt to transmit). The corresponding probability is $\rho_i - P_S[i]$. To make sure node i's attempt is successful, no neighbor of node i can be sending at that moment, so the probability of a successful attempt is $\rho_i - P_S[i] - \rho_i \cup_{k \in \omega_i} P_S[k]$, where ω_i represents all nodes that are neighbors of node i, and $\cup_{k \in \omega_i} P_S[k]$ represents the "sending" probability of neighbors as viewed by node i.

The parameter α_i can be interpreted as the probability that node i transmits successfully given that it *attempts* to do so.

$$\alpha_i = \frac{\rho_i - P_S[i] - \rho_i \cup_{k \in \omega_i} P_S[k]}{\rho_i - P_S[i]} = \frac{1 - P_S[i]/\rho_i - \cup_{k \in \omega_i} P_S[k]}{1 - P_S[i]/\rho_i}. \tag{1}$$

The value of α_i is determined by the "sending" probability of node i itself and its neighbors $k \in \omega_i$. Likewise, each neighbor k will have node i as its neighbor, and its successful transmission probabilities will depend on node i. Therefore, we need use an iterative method to find the value of α_i.

In order to compute $\cup_{k \in \omega_i} P_S[k]$ (the medium busy probabilities as seen by node i), we need to solve several problems first: Which nodes will prevent node i from sending? Will all the sending times of neighboring nodes k ($k \in \omega_i$) be mutually exclusive? If not, how should we decide the possible nodes that can transmit simultaneously? (We call them *simultaneously transmitting nodes*.) How do we calculate the corresponding simultaneous transmitting probability? In the following sections we will introduce ways to solve the above problems.

2.2   Neighbor Matrix

As mentioned by Jain et al. [13], an interference matrix, $\boldsymbol{F}$, can be easily configured based on the interference relationship between nodes. However, deriving hidden terminal relationships is not provided in their paper. Here we provide a way to identify the hidden terminal relationship based on the known routing information and the interference relationship. The hidden terminal relationship and the direct interference relationship will be combined into a "neighbor matrix", $\boldsymbol{N}$.

We define a binary routing matrix $\boldsymbol{R}$ to represent the routing relationship. Denote $\boldsymbol{R}_{ij} = 1$ if node i sends messages to j, otherwise $\boldsymbol{R}_{ij} = 0$. In the interference matrix $\boldsymbol{F}$, if node i and j are interfering with each other, we denote $\boldsymbol{F}_{ij} = 1$ and $\boldsymbol{F}_{ji} = 1$, otherwise $\boldsymbol{F}_{ij} = 0$ and $\boldsymbol{F}_{ji}=0$. The algorithm to derive the neighbor matrix is shown below, (note that all Multiply and Add operations are Boolean algebra operations.)

Algorithm 1: Neighbor matrix

Step 1: *Generate the hidden terminal relationship:* Multiply $\boldsymbol{R}$ by $\boldsymbol{F}$ to get a new matrix $\boldsymbol{H} = \boldsymbol{RF}$. The hidden terminal information is already embedded in

H, since if node i and j are hidden terminals to each other, there must exist one or more nodes k such that $R_{ik} = 1$ (node i wishes to talk to node k) and $F_{kj} = 1$ (node k and node j interferes with each other), so $H_{ij} = \sum_k R_{ik}F_{kj} = 1$;

Step 2: *Combine the hidden terminal relationship with the direct interference relationship:* let $Y = X + F$;

Step 3: *Remove the self-neighbor relationship:* Change all of the diagonal elements of Y to 0 (a node is not considered a neighbor to itself). The resulting matrix is the neighbor matrix N. This matrix incorporates both the interference relationship and the hidden terminal relationship. Note that $N_{ij} = 1$ means that node i and j are "neighbors" to each other; $N_{ij} = 0$ means that they are not "neighbors", allowing them to transmit simultaneously. Since the interference relationships and the hidden terminal relationships are both symmetrical, N is a symmetrical matrix. We can now define ω_i in equation (1) as the set of nodes represented by 1's in the ith row of the neighbor matrix N.

2.3 Simultaneously Transmitting Nodes

There may be nodes that are neighbors to node i that are neither hidden terminals nor directly interfering nodes with each other. Thus, the probability that two or more nodes can send messages simultaneously (they do not interfere with each other, but do interfere with node i) is very important information for calculating the medium busy probability around node i, which we defined as $\cup_{k \in \omega_i} P_S[k]$.

When there are 2 nodes that can transmit simultaneously, we call the set of those nodes *simultaneously transmitting pairs* denoted as STS_2; when there are 3 or more nodes that can transmit simultaneously, we call the set of those nodes *simultaneously transmitting groups* and denote them as STS_3, STS_4.... We define "group degree" as the number of nodes that can transmit simultaneously.

Algorithm 2: Simultaneously transmitting pairs/groups

Step 1: Take the complementary set of the neighbor matrix N to identify the simultaneously transmitting pairs. Since this matrix describes the relationship between any *two* nodes, we denote it as S_2, so $S_2 = \overline{N}$.

Step 2: For all node pairs (i, j) such that $S_{2,ij} = 1$ (to avoid the duplication, we just consider the upper diagonal part of S_2), list all possible nodes k (different from i, j) such that $S_{2,ik} = 1$ and $S_{2,kj} = 1$, and put all valid node groups (i, j, k) into set STS_3.

Step 3: For each node group (i, j, k) in STS_3, find all possible nodes m such that $S_{2,im} = 1$, $S_{2,jm} = 1$ and $S_{2,km} = 1$, and put all valid node groups (i, j, k, m) into STS_4. The above process continues until we reach n such that no n nodes can transmit simultaneously.

2.4 Simultaneous Transmitting Probabilities

The busy probability of the medium around each node i in the multi-hop environment can be calculated as

$$\cup_{k \in \omega_i} P_S[k] = \sum_{k \in \omega_i} P_S[k] - \sum_{(k_1, k_2) \in STS_2} P_S[k_1 k_2] + \sum_{(k_1, k_2, k_3) \in STS_3} P_S[k_1 k_2 k_3] - \dots (2)$$

where $k_1, k_2, k_3... \in \omega_i$. Now we need to calculate the overlapped sending probabilities $P_S[k_1 k_2], P_S[k_1 k_2 k_3] \ldots$.

For two nodes that are not neighbors to each other, if they also don't have shared neighbors, we assume that they can independently transmit; if they have shared neighbors, they are independent only during the period when no messages are being transmitted to or from the shared neighbors. In the latter case, these nodes can be viewed as "conditionally independent".

The neighbors of node k_1 will be $\omega_{k_1} = \{q : N_{k_1 q} = 1\}$, and the neighbors of node k_2 is $\omega_{k_2} = \{q : N_{k_2 q} = 1\}$. Denote $\omega_{k_1 k_2} = \omega_{k_1} \cup \omega_{k_2}$. When both node k_1 and k_2 are sending, none of the nodes in $\omega_{k_1 k_2}$ can be sending.

$$P_S[k_1, k_2] = P_S[k_1, k_2, \overline{\omega_{k_1 k_2}}] = P_S[k_1, k_2 | \overline{\omega_{k_1 k_2}}] P_S[\overline{\omega_{k_1 k_2}}]. \tag{3}$$

Since nodes k_1, k_2 are independent conditioned on the probability that none of the nodes in $\omega_{k_1 k_2}$ are sending, we have

$$P_S[k_1, k_2 | \overline{\omega_{k_1 k_2}}] = P_S[k_1 | \overline{\omega_{k_1 k_2}}] P_S[k_2 | \overline{\omega_{k_1 k_2}}] = \frac{P_S[k_1, \overline{\omega_{k_1 k_2}}]}{P_S[\overline{\omega_{k_1 k_2}}]} \frac{P_S[k_2, \overline{\omega_{k_1 k_2}}]}{P_S[\overline{\omega_{k_1 k_2}}]}. \tag{4}$$

$P_S[\overline{\omega_{k_1 k_2}}]$ represents the probability that no neighbor of node k_1, k_2 is sending, which can be written as $1 - P_S[\omega_{k_1 k_2}]$ instead. $P_S[k_1, \overline{\omega_{k_1 k_2}}]$ represents the probability that node k_1 is sending while all the neighbors of node k_1, k_2 are not sending. As we know, neighbors of node k_1 must not be sending when node k_1 is sending, if we denote $\omega_{k_2 \overline{k_1}}$ as the nodes that are neighbors of node k_2 but not of node k_1, we have $P_S[k_1, \overline{\omega_{k_1 k_2}}] = P_S[k_1, \overline{\omega_{k_2 \overline{k_1}}}] = P_S[k_1] - P_S[k_1, \omega_{k_2 \overline{k_1}}]$. Similarly we can get $P_S[k_2, \overline{\omega_{k_1 k_2}}] = P_S[k_2] - P_S[k_2, \omega_{k_1 \overline{k_2}}]$.

After combining equation (3), (4), and the computation for $P_S[k_1, \overline{\omega_{k_1 k_2}}]$, $P_S[k_1, \overline{\omega_{k_1 k_2}}]$, and $P_S[\overline{\omega_{k_1 k_2}}]$, the resulting expression is

$$P_S[k_1, k_2] = \frac{(P_S[k_1] - P_S[k_1, \omega_{k_2 \overline{k_1}}])(P_S[k_2] - P_S[k_2, \omega_{k_1 \overline{k_2}}])}{1 - P_S[\omega_{k_1 k_2}]}. \tag{5}$$

The calculation of $P_S[\omega_{k_1 k_2}], P_S[k_1, \omega_{k_2 \overline{k_1}}], P_S[k_2, \omega_{k_1 \overline{k_2}}]$ can be done similarly by using equation (2). We can get the exact solution by solving the system of equations, or by using iterative methods. After we get $P_S[k_1, k_2]$, $P_S[k_1, k_2, k_3]$ etc. can be computed similarly.

3   Path Delays and Optimization

If we make the assumption that the queue length is infinite and thus no loss occurs, we can get closed form solutions for α_i, which makes system optimization possible.

The service time distribution at each node consists of both the transmission time and the queueing delay (waiting time when frames ahead are transmitting or the node is in "backoff" state). It has a matrix exponential distribution representation

$$F(t) = 1 - \boldsymbol{p} \exp(-\boldsymbol{B}t) \boldsymbol{e}', \tag{6}$$

where $\boldsymbol{p}$ is the starting vector for the process, $\boldsymbol{B}$ is the progress rate operator for the process, and $\boldsymbol{e'}$ is a summing operator consisting of all 1's [12]. The moments of the matrix exponential distribution are

$$E[X^n] = n!\boldsymbol{p}\boldsymbol{B}^{-n}\boldsymbol{e'}. \tag{7}$$

Based on the Markov chain of Fig. 1, the matrix exponential representation of the service distribution at each node i is

$$\boldsymbol{p} = \begin{bmatrix} 1 & 0 \end{bmatrix}, \quad \boldsymbol{B} = \begin{bmatrix} \beta\,\alpha_i & -\beta\,\alpha_i \\ 0 & \mu \end{bmatrix}. \tag{8}$$

Using equation (7), the mean and the second moment of the of the service distribution at node i are

$$E[S_i] = \frac{1}{\mu} + \frac{1}{\alpha_i\,\beta} = \frac{\mu + \alpha_i\,\beta}{\alpha_i\,\beta\,\mu}, \qquad E[S_i^2] = 2\frac{\mu^2 + \alpha_i\,\beta\,\mu + \alpha_i{}^2\beta^2}{\alpha_i{}^2\beta^2\mu^2}. \tag{9}$$

When the queue is infinite, $\rho_i = \lambda_i\frac{\mu+\alpha_i\,\beta}{\alpha_i\,\beta\,\mu}$, where λ_i is the mean arrival rate to node i. Also, since there is no loss, the "sending" probability will be λ_i/μ (percentage of total channel capacity that node i is using). Substituting ρ_i and $P_S[i]$ into equation (1) and solving for α_i, we get

$$\alpha_i = \frac{\mu(1 - \cup_{k\in\omega_i} P_S[k])}{\mu + \beta \cup_{k\in\omega_i} P_S[k]} = \frac{1 - \cup_{k\in\omega_i} P_S[k]}{1 + \beta/\mu \cup_{k\in\omega_i} P_S[k]}. \tag{10}$$

Since $P_S[k] = \lambda_k/\mu$, α_i can now be represented in terms of λ_k – the arrival rate of each node.

Using the P-K formula for M/G/1 queues, the mean waiting time in the queue at each node is $E[W] = \frac{\lambda E[S^2]}{2(1-\lambda E[S])}$. By substituting the expressions for the mean and second moment of the service times and noting that the mean total time spent at node i is $E[T_i] = E[W_i] + E[S_i]$, we get

$$E[T_i] = \frac{\mu + \alpha_i\,\beta - \lambda_i}{\alpha_i\,\beta\,\mu - \lambda_i\,\mu - \lambda_i\,\alpha_i\,\beta}. \tag{11}$$

To express the delay as an optimization problem, we use the following notation:

$\mathcal{K}$ Set of all origin-destination nodes that have traffic.

$\mathcal{I}$ Set of communicating nodes in the network.

Λ_k Average arrival rate for origin destination pair k.

Λ Total arrival rate to the network, $\Lambda = \sum_{k\in\mathcal{K}} \Lambda_k$.

$\mathcal{P}_k$ Set of possible paths for o-d pair k.

λ_{kj} Amount of flow on path j for pair k.

α_i Transmission success probability at node i, which is expressed as a function of the path flow variables λ_{kj} using equation (10).

δ_{kj}^i Node path indicator: 1 if path j for pair k passes through node i.

F_i Total flow through node i, $F_i = \sum_{k\in\mathcal{K}} \sum_{j\in\mathcal{P}_k} \delta_{kj}^i \lambda_{kj}$.

The optimization problem for minimizing the mean delay a frame experiences in the network is

$$\min_{\lambda_{kj}, F_i} \frac{1}{\Lambda} \sum_{i \in \mathcal{I}} F_i \frac{F_i - \mu - \alpha_i \beta}{-\alpha_i \beta \mu + F_i \mu + F_i \alpha_i \beta}, \tag{12}$$

such that

$$\sum_{j \in \mathcal{P}_k} \lambda_{kj} = \Lambda_k, \quad k \in \mathcal{K}, \tag{13}$$

$$\sum_{k \in \mathcal{K}} \sum_{j \in \mathcal{P}_k} \delta_{kj}^i \lambda_{kj} - F_i = 0, \quad i \in \mathcal{I}, \tag{14}$$

$$\lambda_{kj} \geq 0, \ F_i \geq 0. \tag{15}$$

The objective function is rational, with polynomials in both the numerator and denominator. The constraints are linear, so we use a flow deviation algorithm using the projection method [14] to solve this problem. Convergence is very fast for the examples we present under the assumption that the network is stable and the starting point is feasible. Routing strategies based on optimal path delays and QoS requirements for different traffic classes can also be constructed, but here we only present the case for mean delay.

4   Numerical Results

4.1   Simulation Model

We use CSIM simulation tools to construct the simulation model. If a node has a frame to transmit, it will first wait one backoff period which is exponentially distributed with mean $1/\beta$. Upon completion of the backoff period, the node initiates an RTS to see if the medium is available. We use the same assumptions in the simulation as in the analytic model, namely, that RTS/CTS communication is instantaneous and that there are no errors. If the channel is not available, the node will go into backoff, otherwise the frame is transmitted with a mean time of $1/\mu$. Frames are forwarded based on the route indicated in the frame header.

4.2   Evaluation of Wireless Mesh Networks

In this subsection we show the effectiveness of our model by comparing analytical results to simulation. In the scenarios we show, we assume the maximum transmission rate is 10 Mbps and the average frame size is 1250 bytes ($10,000$ bits), resulting in a mean transmission rate of $\mu = 1000$ frames per second (fps).

For the multi-hop mesh topology depicted in Fig. 2, we show a simple example where both node 1 and 2 have frames to send through the gateway to the wired network and all other nodes have no frames to send. We use solid lines with arrows to specify the routing relationship between nodes and dashed lines between nodes that directly interfere with each other. No lines between nodes

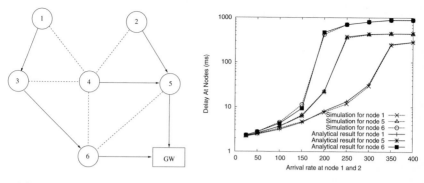

Fig. 2. Six node mesh network   **Fig. 3.** Comparison of the delay

means that they will not interfere directly with each other. For convenience, we assume node 1 and node 2 have the same arrival rate. We denote the gateway, GW, as node 7. Also, since the gateway does not send messages upstream on the same channel, we can omit the direct relationships in the neighbor matrix. Using the algorithm described in section 2.2, we get

$$
R = \begin{bmatrix} 0 & 0 & 1 & 0 & 0 & 0 & 0 \\ 0 & 0 & 0 & 0 & 1 & 0 & 0 \\ 0 & 0 & 0 & 0 & 0 & 1 & 0 \\ 0 & 0 & 0 & 0 & 1 & 0 & 0 \\ 0 & 0 & 0 & 0 & 0 & 0 & 1 \\ 0 & 0 & 0 & 0 & 0 & 0 & 1 \\ 0 & 0 & 0 & 0 & 0 & 0 & 0 \end{bmatrix}, \quad
F = \begin{bmatrix} 0 & 0 & 1 & 1 & 0 & 0 & 0 \\ 0 & 0 & 0 & 1 & 1 & 0 & 0 \\ 1 & 0 & 0 & 1 & 0 & 1 & 0 \\ 1 & 1 & 1 & 0 & 1 & 1 & 0 \\ 0 & 1 & 0 & 1 & 0 & 1 & 1 \\ 0 & 0 & 1 & 1 & 1 & 0 & 1 \\ 0 & 0 & 0 & 0 & 1 & 1 & 0 \end{bmatrix}, \quad \text{and } N = \begin{bmatrix} 0 & 0 & 1 & 1 & 0 & 1 & 0 \\ 0 & 0 & 0 & 1 & 1 & 1 & 1 \\ 1 & 0 & 0 & 1 & 1 & 1 & 1 \\ 1 & 1 & 1 & 0 & 1 & 1 & 1 \\ 0 & 1 & 0 & 1 & 0 & 1 & 1 \\ 1 & 1 & 1 & 1 & 1 & 0 & 1 \\ 0 & 0 & 0 & 0 & 1 & 1 & 0 \end{bmatrix}.
$$

The comparison of analytical results and simulation results are shown in Figs. 3 - 5. Since nodes 1 and 2 have the same offered load and interference, they exhibit the same behavior. Therefore, we just show the performance for node 1. For the same reason, we only show performance measures for node 5 and not node 3.

In Figs. 3, 4, and 5, we show the delay, mean backoff times, and blocking probabilities at various load for nodes having a finite capacity of size $L = 100$. We can see that for low load and heavy load, the analytical results match perfectly with the simulation results, while for moderate load, there is a slight difference. In total, the analytical results are very good at catching the abrupt increase in delay as the load increases.

A more complicated scenario is shown in Fig. 6. There is one gateway, nodes 1-5 are actively sending messages, and the other nodes are acting as mesh routers. The buffer size at each node is 100.

The comparison of simulation and analytical results for the delay at nodes 1, 6, 8, and 9 are shown in Fig. 7 and blocking probabilities are shown in Fig. 8. We can see that node 6 and node 8 are the bottlenecks in this network. Note that the results are accurate over a wide variety of offered loads.

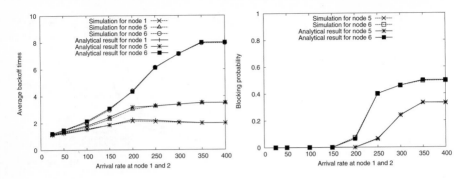

Fig. 4. Average backoff times at each node   **Fig. 5.** Blocking at each node

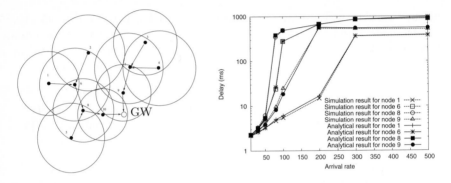

Fig. 6. Ten node multi-hop network   **Fig. 7.** Delay at nodes 1, 6, 8, and 9

4.3   Optimization Results

In a wireless mesh network, nodes can also communicate in ad-hoc mode, which means that they can transmit frames to peer nodes through other intermediate nodes. Referring to the topology in Fig. 2, we assume that there are two communicating pairs. Node 2 is sending to node 6 and node 3 is sending to node 5. The paths available to node 2 are path $2, 4, 6$ and $2, 5, 6$ and the paths available to node 3 are $3, 4, 5$ and $3, 6, 5$. The mean arrival rate at node 2 and 3 are denoted as λ_2 and λ_3 respectively. The path flow variables are denoted as λ_{246}, λ_{256}, λ_{345}, and λ_{365}. The constraints are $\lambda_{256} = \lambda_2 - \lambda_{246}$ and $\lambda_{365} = \lambda_3 - \lambda_{345}$. The load at node 4 is $F_4 = \lambda_{345} + \lambda_{246}$, the load at node 5 is $F_5 = \lambda_{256}$, and at node 6 the load is $F_6 = \lambda_{365}$. Solving the optimization problem (equations (12), (13), (14), and (15)), we get the optimal routing for different values of λ_2 and λ_3. We can see that the system delay function is convex, as shown in Fig. 9. We set $\lambda_2 = 200$ fps and $\lambda_3 = 230$ fps.

To show the effect of one node's traffic on the choice of routes, we set the traffic at node 2 fixed at 200 fps, and then watch the changes of routing and

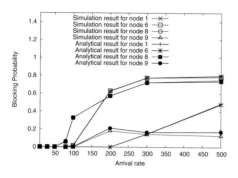

Fig. 8. Blocking at nodes 1, 6, 8, and 9

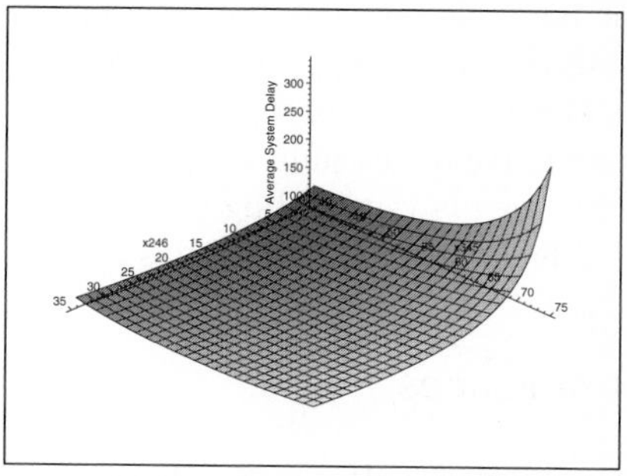

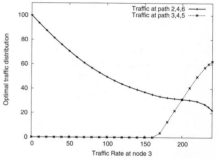

Fig. 9. Convex cost function

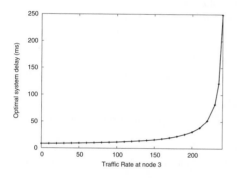

Fig. 10. Optimal system delay vs. load

Fig. 11. Traffic distribution vs. load

traffic distribution according to different traffic from node 3. In Fig. 10, we show the optimal system delay for different loads at node 3, and Fig. 11 shows the resulting traffic sent through path 2, 4, 6 and 3, 4, 5. We can see that when the traffic originating from node 3 is low, all segments will take path 3, 6, 5 because node 4 is heavily interfered and will result in high delay. But traffic from node 2 will take advantage of route 2, 4, 6 since the traffic from node 2 is already high. As the load at node 3 increases, part of traffic from node 3 will take path 3, 4, 5 and the volume will keep increasing. At the same time, traffic at path 2, 4, 6 will decrease due to the stronger interference at node 4.

5 Conclusion

In this paper, the neighbor concept is extended to incorporate both directly interfering nodes and hidden terminals of each node based on the topology and routing in the network. Based on the relationships of "neighbors", we use a node based analysis where an iterative process is used to find the probability of a successful transmission at each node. To facilitate neighbor identification, identifying algorithms are provided. For the infinite buffer case, we derive a means to identify the optimal multipath flow that minimizes the mean delay in the network. The comparison of simulation and analytical results show that our analytical method is accurate under both saturated and unsaturated cases.

The evaluation of wireless mesh networks shows that the system performance is sensitive to the number of interfering neighbors and route selection. For future work, we plan on adding cost functions to the multipath optimization to insure QoS fairness for different classes of traffic at each node.

References

1. Akyildiz, I., Wang, X., Wang, W.: Wireless mesh networks: a survey. Journal of Computer Networks 47, 455–487 (2005)
2. Tobagi, F.A.: Modeling and performance analysis of multihop packet radio networks. In: Proceedings of the IEEE, vol.75(1), pp. 135–155 (January 1987)
3. Leiner, B.M.: A simple model for computation of packet radio network communication performance. IEEE Transactions on Communications 28(12), 2020–2023 (1980)
4. Silvester, A., Lee, I.: Performance modeling of buffered csma-an iterative approach. In: Proceedings of the IEEE Conference on Global Communications (GLOBECOM), Miami, FL (1982)
5. Boorstyn, R., Kershenbaum, A., Maglaris, B., Sahin, V.: Throughput analysis in multihop CSMA packet radio networks. IEEE Transactions on Communications COM-35(3), 267–274 (1987)
6. Wang, X., Kar, K.: Throughput modelling and fairness issues in CSMA/CA based ad-hoc networks. In: Proceedings of the Conference on Computer Communications (IEEE Infocom), Miami, FL (2005)
7. Carvalho, M., Garcia-Luna-Aceves, J.J.: A scalable model for channel access protocols in multihop ad hoc networks. In: MOBICOM, Philadelphia, PA (2004)
8. Garetto, T.S.M., Knightly, E.: Modeling per-flow throughput and capturing starvation in CSMA multi-hop wireless networks. In: Proceedings of the Conference on Computer Communications (IEEE Infocom), Barcelona, Spain (2006)
9. Kleinrock, L., Tobagi, F.: Packet switching in radio channels: Part I-carrier sense multiple-access modes and their throughput-delay characteristics. IEEE Transactions on Communications 23(12), 1400–1416 (1975)
10. Tobagi, F., Kleinrock, L.: Packet switching in radio channels: Part II- the hidden terminal problem in carrier sense multiple-access and the busy-tone solution. IEEE Transactions on Communications 23(12), 1417–1433 (1975)
11. Cali, F., Conti, M., Gregori, E.: IEEE 802.11 wireless LAN: Capacity analysis and protocol enhancement. In: Proceedings of the Conference on Computer Communications (IEEE Infocom), San Francisco (March 1998)
12. Lipsky, L.: Queueing Theory - A Linear Algebraic Approach. Macmillan Publishing, NYC (1992)
13. Jain, K., Padhye, J., Padmanabhan, V., Qiu, L.: Impact of interference on multihop wireless network performance, in Mobicom'03, pp. 66–80 (2003)
14. Bertsekas, D., Gallager, R.: Data Networks. Prentice-Hall, Englewood Cliffs (1992)

Load Balancing by Joint Optimization of Routing and Scheduling in Wireless Mesh Networks

Riikka Susitaival

TKK, Helsinki University of Technology
P.O. Box 3000, FIN-02015 TKK, Finland
riikka.susitaival@tkk.fi

Abstract. In this paper we study load balancing in wireless mesh networks when the MAC layer of the network is modelled by STDMA. We formulate the linear problem for joint optimization of traffic allocation and transmission schedule. Both unconstrained path set, allowing arbitrary routing, and predefined paths are considered. In our numerical examples roughly third of the load of the most congested link can be reduced by load balancing. This reduction in the load decreases the delays of the network as well as increases the reliability of the system if link conditions change suddenly.

1 Introduction

Wireless Mesh Networks (WMNs) are dynamic networks that are constructed from two types of nodes, mesh routers and mesh clients [1]. By supporting multi-hop routing the mesh routers form a mesh backbone which enables cost-effective communication between the clients in the network as well as an access to other networks such as Internet. Originally, multi-hop wireless networks were considered primarily for military use but now there is also a growing interest in commercial exploitation of them.

Familiar from fixed networks, the term *load balancing* refers to optimization of usage of network resources by moving traffic from congested links to less loaded parts of the network based on knowledge of network state. By this approach QoS experienced by the users, such as transmission delay, is improved. Many load balancing algorithms proposed for IP networks, especially along development of new tunnelling technique MPLS (see [2], [3]). However, these algorithms are not directly usable in WMNs, since the interference restricts the simultaneous use of the links. However, by the scheduling the interference can be alleviated, and moreover, the resources of the congested links can be added. Combining scheduling and routing is obviously beneficial in optimization of WMNs.

In this paper we study load balancing problem in wireless mesh networks. In the problem our aim is to find such routing and MAC layer scheduling that the maximum link utilization of the network is minimized. We formulate an LP-problem both for free routing in which data can be split to arbitrarily many

L. Mason, T. Drwiega, and J. Yan (Eds.): ITC 2007, LNCS 4516, pp. 483–494, 2007.

routes as well as for constrained routing in which the paths used by data flows are predefined, but the traffic allocation to the paths can be optimized. We compare results of joint optimization of MAC layer and link layer parameters (also known as cross-layer optimization) to equal splitting of traffic onto the shortest paths and to the more conventional scenario where optimization is done separately at two layers. As a result we find that cross-layer approach is required to achieve significant improvements in maximum link utilization.

We assume that the MAC layer of the wireless network is be modelled by Spatial TDMA [4]. In STDMA the transmission resources are divided into time slots and the links that are spatially sufficiently separated, can transmit in the same slot. The set of links transmitting in same slot is called as the transmission mode. In addition, we assume that the interference affects only the set of active links, but the nominal capacity of the links remain fixed.

Optimization of resource usage in wireless mesh networks has gained lot of interest recently. Major part of them focuses on throughput maximization ([5], [6], [7]), where as some papers have studied delay minimization [8] or dimensioning [9]. Our work is most closely related to study of Wu et al. in [6], but has some differences: our view is on load balancing as a routing problem with given demands between source and destination pairs, where as paper [6] concentrates on multicast sessions. In addition, the formulation of the multipath routing problem with a predefined set of routes is missing from the above studies.

The paper is organized as follows: In Section 2 we introduce the wireless network model and, in particular, the interference model that we are assuming. The problem formulations for load balancing with arbitrary routing as well as with the predefined path set are given in Section 3. In section 4 we give numerical examples of load balancing. Finally, section 5 makes a short conclusion of the paper.

2   Modelling the Wireless Network

Let us consider a wireless mesh network, which consists of N nodes equipped with a single radio. The distance between nodes i and j is denoted by d_{ij}. The communication range of node i is denoted by R_i and the interference range by R_i'. Depending on the interference model, there might exists a link between two nodes i and j in the network. Also the capacity of link l, denoted by c_l, depends on the interference model. We assume that the link exists from node i node j if the receiver of node j is inside the communication range of the transmitter of node i. Let vector $\mathbf{b} = \{b_1, b_2, ..., b_L\}^T$ denote nominal bandwidths of the links without interference and let L denote the number of these links.

Adapting the interference model from [5], the transmission from a transmitter to a receiver is successful if there is not any other node transmitting in the interference range of the receiver simultaneously. That is to say, for successful transmission from node i to j the following constraints should be true:

1. $d_{ij} \leq R_i$.
2. No node k with $d_{kj} \leq R_k'$ is transmitting.

A set of the links that can transmit simultaneously according to the above constraints is called a *transmission mode*. The mode is said to be *maximal*, if we cannot insert any link to the mode without violating the aforementioned rules for successful transmission. Let M denote the number of transmission modes and let $\mathbf{S} \in \mathbb{R}^{L \times M}$ denote the transmission mode matrix, in which element $S_{l,m}$ denotes the capacity of link l in mode m. We assume that the capacities of the active links remain fixed over the transmission modes.

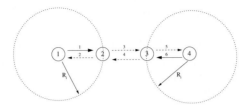

Fig. 1. Transmission mode in a network with 4 nodes and 6 links

As an example of transmission modes, consider a chain of four nodes with six one-directional links. The distances from each node to the neighboring nodes as well as communication and interference ranges of the nodes are assumed to be 1. For instance, if link 1 is transmitting the only link that can transmit without interference is link 6. The transmission mode of these links is depicted in Figure 1, in which a solid arrow refers to an active link and dashed arrow to non-active link. Altogether, there are four feasible and maximal transmission modes. The corresponding transmission mode matrix is:

$$\mathbf{S} = \mathbf{b}^T \begin{pmatrix} 1 & 0 & 0 & 0 \\ 0 & 1 & 0 & 0 \\ 0 & 0 & 1 & 0 \\ 0 & 0 & 0 & 1 \\ 0 & 1 & 0 & 0 \\ 1 & 0 & 0 & 0 \end{pmatrix}.$$

In general, finding all feasible transmission modes is known to be a hard problem. For example, in a network with 22 links, the feasible transmission modes have to be selected from over 4 million different link combinations. However, the maximal transmission modes can effectively be generated by a simple algorithm which adds non-interfering links to a tree in a depth-first search manner. The proposed algorithm is described in Appendix.

In wireless networks with STDMA type MAC layer, different transmission modes are scheduled in a TDMA manner. The transmission mode matrix can be associated with a schedule, which defines the proportion of time that each mode uses for transmission. Let $\mathbf{t} = \{t_1, t_2, ..., t_M\}^{\mathrm{T}}$ denote the schedule, which element t_m is the transmission time of mode m. The actual capacities of links, denoted by vector $\mathbf{c} = \{c_1, c_2, ..., c_L\}$, can be determined by weighting the capacity of a link in one transmission mode by the length of the mode's time slot in the schedule and taking the sum over all M modes:

$$\mathbf{c} = \mathbf{St}.$$

When scheduling is executed at sufficiently fast time scale, the flows using the links in effect experience a constant link capacity.

3   Load Balancing Problem

In this section we formulate the load balancing problem in wireless mesh networks with STDMA scheduling. The problem is well-known from the wired networks with fixed capacity (see [2], [3]). In the wireless context the new thing is that, in addition to routing, the resources of the network can be shifted from congested areas to other areas by scheduling. We consider two different formulations of the load balancing problem: In the one formulation we can split traffic to arbitrary many routes, whereas in the second one the path set is predefined.

The network is loaded by traffic flows travelling from origin nodes to destination nodes. Let s_k refer to the origin node and t_k referring to the destination node of OD-pair k. In addition, let K denote the number of OD-pairs and vector $\mathbf{d} = \{d_1, d_2, ..., d_K\}^{\mathrm{T}}$ the mean sending rates of the OD-pairs.

3.1   Paths Unconstrained

Let us consider a wireless network described in Section 2. In addition to earlier notation, let $\mathbf{A} \in \mathbb{R}^{L \times N}$ denote the link-node incidence matrix for which $A_{l,n} = -1$ if link l directs to node n, $A_{l,n} = 1$ if link l leaves from node n, and $A_{l,n} = 0$ otherwise; and let $\mathbf{R} \in \mathbb{R}^{K \times N}$ denote the demand matrix for which $R_{k,s_k} = d_k$, $R_{k,t_k} = -d_k$, and $R_{k,n} = 0$ for all other nodes n.

Let $\mathbf{X} \in \mathbb{R}^{K \times L}$ be the traffic allocation matrix, with element $X_{k,l}$ corresponding to the traffic of OD-pair k allocated to link l. The allocation matrix is unknown yet, but will be determined as a solution of the load balancing problem. Given the traffic allocation matrix $\mathbf{X}$, the induced link load vector $\mathbf{y} = \{y_1, y_2, ..., y_L\}^{\mathrm{T}}$ is

$$\mathbf{y} = \mathbf{X}^{\mathrm{T}} \mathbf{e}_K,$$

where vector $\mathbf{e}_K$ is K-dimensional column vector with all elements equalling 1 (same notation is used for other unit vectors also). In general, the mapping between the traffic matrix $\mathbf{X}$ and the link load vector $\mathbf{y}$ is not one-to-one, i.e. while $\mathbf{X}$ determines $\mathbf{y}$ uniquely, the opposite is not true.

Let variable α denote the maximum utilization of the links

$$\alpha = \max_l \frac{y_l}{c_l}.$$

The primal objective in the unconstrained load balancing problem is to minimize the maximum link utilization α by choosing an optimal traffic allocation $\mathbf{X}$ and transmission schedule $\mathbf{t}$. For obtaining linear constraints, we introduce also a new variable $\mathbf{q} = \{q_1, ..., q_M\}$ which elements satisfying $q_m = \alpha t_m$. The linear load balancing problem for unconstrained path set is as follows:

$$\text{Minimize } \alpha \text{ over } \mathbf{X} \text{ and } \mathbf{q}$$
$$\text{subject to the constraints}$$
$$\mathbf{X} \geq 0, \mathbf{q} \geq 0,$$
$$\alpha = \mathbf{e}_M{}^{\mathrm{T}}\mathbf{q}, \tag{1}$$
$$\mathbf{X}^{\mathrm{T}}\mathbf{e}_K \leq \mathbf{Sq},$$
$$\mathbf{XA} = \mathbf{R}.$$

In the problem above we have five constraints. The first two constraints state that all free variables should be positive. The third one, when divided by α, states that in a schedule the sum of the time proportions should equal to 1, fourth one is the link capacity constraint and the fifth one is the so-called *conservation of flow constraint*, which states that the traffic of each OD-pair incoming to a node has to equal the outgoing traffic from that node.

By problem formulation (1) we can find an unique solution for α, but the load allocation is not necessarily optimal in some sense. For example, if there is a bottleneck in the network, the rest of the network remains unbalanced. Ott et al. present in [3] that the routing $\mathbf{y}$ is *non-dominated* if there does not exist another routing with link-loads $\mathbf{y}'$ with property

$$y'_l \leq y_l \text{ for all } l \text{ and } y'_l < y_l \text{ for at least one } l. \tag{2}$$

For this reason, after finding the minimum of the maximal link utilization α^* by optimizing problem (1), we optimize the traffic allocation and scheduling again in the network, in which the link capacities are reduced by the factor α^*. The optimization problem in the second phase, where our objective is to minimize the overall usage of resources, is as follows:

$$\text{Minimize } \mathbf{e}_K{}^{\mathrm{T}}\mathbf{Xe}_L \text{ over } \mathbf{X} \text{ and } \mathbf{t}$$
$$\text{subject to the constraints}$$
$$\mathbf{X} \geq 0, \mathbf{t} \geq 0,$$
$$\mathbf{e}_M{}^{\mathrm{T}}\mathbf{t} = 1, \tag{3}$$
$$\mathbf{X}^{\mathrm{T}}\mathbf{e}_K \leq \alpha^*\mathbf{St},$$
$$\mathbf{XA} = \mathbf{R}.$$

As a solution of the second optimization problem we obtain the link loads of each OD-pair as well as the optimal schedule. For each OD-pair k, from the resulting link load vector $\mathbf{X}_k^*$, the used paths and the traffic volumes on the paths can be solved by the following procedure: First we construct a reduced link topology, which includes the links that has traffic in the solution $\mathbf{X}_k^*$, and then calculate all paths between origin node s_k and destination node t_k. As the link load vector $\mathbf{X}_k^*$ is optimal, the paths between the source and destination node are automatically loop-free. Let $\mathbf{Q}_k$ denote the path-link incidence matrix, which element $Q_{k,p,l}$ is 1, if path p uses link l in the new reduced topology and 0 otherwise. Let $\mathbf{X'}_k$ denote the traffic allocation vector to the links in the reduced topology. This new traffic allocation vector can directly be constructed from the original optimal solution $\mathbf{X}_k^*$. Finally, traffic allocation to each path in matrix $\mathbf{Q}_k$, denoted by vector $\mathbf{z}_k$, can be solved from the matrix equation:

$$\mathbf{Q}_k{}^T\mathbf{z}_k = \mathbf{X'}_k. \tag{4}$$

The system described by matrix equation (4) can be over- or underdetermined, because the number of paths often differs from the number of links. However, for us it is sufficient to find one path allocation that induces the link loads of the optimal solution.

3.2 Predefined Paths

Now we assume that the paths available for each OD-pair are defined explicitly. This type of formulation is needed in situations where the number of paths is constrained to some value or some paths are known to be preferable in advance.

In the problem each OD-pair k has a predefined set $\mathcal{P}_k$ of the paths available. Let $\mathcal{P} = \cup_{k \in \mathcal{K}} \mathcal{P}_k$ denote the set of all available paths and P the number of all paths. We also introduce a path-link incidence matrix $\mathbf{P} \in \mathbb{R}^{P \times L}$ for all paths, with element $P_{p,l}$ equalling 1 if path p uses link l and 0 otherwise; and OD-pair-path incidence matrix $\mathbf{O} \in \mathbb{R}^{K \times P}$ with element $O_{k,p}$ indicating whether OD-pair k uses path p or not.

As in the previous subsection, let $\mathbf{X} \in \mathbb{R}^{K \times P}$ denote the allocation matrix of traffic of each OD-pair onto paths. To ensure that all traffic of each OD-pair is allocated, the traffic allocation matrix must satisfy:

$$(\mathbf{X} * \mathbf{O})\mathbf{e}_P = \mathbf{d},$$

where operator $*$ means element-wise multiplication of two matrices. Given $\mathbf{X}$, the link load vector $\mathbf{y}$ is

$$\mathbf{y} = \mathbf{e}_K^T \mathbf{X} \mathbf{P}.$$

The objective in the load balancing problem with predefined path set is to minimize the link utilization coefficient α by choosing an optimal traffic allocation $\mathbf{X}^*$ and a scaled transmission schedule $\mathbf{q}^*$. The first phase of the problem, the minimization of the maximum link utilization is formulated as follows:

$$
\begin{aligned}
&\text{Minimize } \alpha \text{ over } \mathbf{X} \text{ and } \mathbf{q}\\
&\text{subject to the constraints}\\
&\mathbf{X} \geq 0, \mathbf{q} \geq 0\\
&\alpha = \mathbf{e}_M^T \mathbf{q},\\
&\mathbf{e}_K^T \mathbf{X} \mathbf{P} \leq \mathbf{S} \mathbf{q},\\
&(\mathbf{X} * \mathbf{O})\mathbf{e}_P = \mathbf{d}.
\end{aligned}
\tag{5}
$$

Having optimal α^* from the first phase, the second phase, minimization of capacity usage, is as follows:

$$\text{Minimize } \mathbf{e}_K^T \mathbf{XPe}_L \text{ over } \mathbf{X} \text{ and } \mathbf{t}$$

subject to the constraints

$$\mathbf{X} \geq 0, \mathbf{t} \geq 0$$

$$\mathbf{e}_M^T \mathbf{t} = 1,$$

$$\mathbf{e}_K^T \mathbf{XP} \leq \alpha^* \mathbf{St},$$

$$(\mathbf{X} * \mathbf{O})\mathbf{e}_P = \mathbf{d}.$$

$$(6)$$

The pathwise solution is not necessarily unique if the paths share common links. In addition, note that if path set $\mathcal{P}$ contains all possible paths between the origin and destination nodes, the path-constrained load balancing problem is similar to problems of the previous subsection.

4   Results

In this section we present numerical results for the load balancing problems presented in the previous section. To get an idea about good routing, we study first load balancing in a very small network with 4 nodes. After that more realistic network examples are studied. Since the gains achieved by cross-layer approach in resource allocation depends heavily on interaction between the topology and demands, we present results for both backbone and access type networks as well as for many OD-pair combinations. The feasible transmission modes are calculated beforehand once and for all. The optimal traffic allocation and the schedule can be then solved by an LP-solver with relatively small computational efforts.

Small Network Example. Let us consider a 2x2 grid network with $N = 4$ nodes and $L = 8$ links. The distances between the neighbor nodes as well as transmission ranges are assumed to be 1. The network is loaded by two OD-flows; one from node 1 to node 4 and another one from node 4 to 3. The mean volumes of the flows are 1 unit, whereas the nominal capacities of the links are 3 units. There are 4 transmission modes in the network. The modes are depicted in Figure 2.

We consider three different optimization policies; shortest path routing with MAC layer optimization, separate optimization of two layers, routing and scheduling, and joint optimization of them. Using single shortest paths, traffic of the first OD pair is routed through nodes 1-2-4 and the second through nodes 4-3 (see the left hand side of Figure 3). This routing induces link load vector $\mathbf{y} = \{1, 0, 0, 1, 0, 0, 0, 1\}^T$. To allocate required resources to these two paths, transmission modes 1, 2 and 3 have to be used. It is easy to see that the optimal schedule is $\mathbf{t} = \{1/3, 1/3, 1/3, 0\}^T$ producing actual capacities $\mathbf{c} = \{1, 1, 1, 1, 0, 1, 0, 1\}^T$. The maximum load is then 1. The problem in the traffic allocation produced by

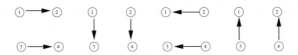

Fig. 2. Transmission modes in a 2x2 grid network

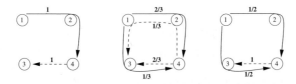

Fig. 3. Traffic allocations in a 2x2 grid network. Left: shortest path routing, middle: optimization at two layers, right: joint optimization.

shortest path routing is that in each schedule only one of two links is in use and thus the overall capacity usage is not optimal.

In the second approach we first balance the load without knowledge of the interference and then optimize the schedule. To obtain a balanced network, we move traffic from primary paths to secondary paths (not used by shortest path routing) until the maximum load does not decrease anymore (see the middle of Figure 3). Resulting traffic allocation is $\mathbf{y} = \{2/3, 2/3, 1/3, 2/3, 0, 1/3, 1/3, 2/3\}^T$. The schedule that minimizes the maximum load is $\mathbf{t} = \{2/7, 2/7, 2/7, 1/7\}^T$ and the corresponding maximum load is $7/9 \approx 0.78$.

Finally, we find that due to interference it is not optimal to spread traffic of OD-pair 2 to route 4-2-1-3 since there are bottlenecks in two different schedules. Instead, in optimal routing and scheduling, traffic of OD pair 1 is routed evenly over paths 1-2-4 and 1-3-4, but OD pair 2 uses only the shortest path 4-3 (see the right hand side of Figure 3). This produces quite uneven load distribution $\mathbf{y} = \{1/2, 1/2, 0, 1/2, 0, 1/2, 0, 1/2, 1\}^T$ but using the schedule $\mathbf{t} = \{1/4, 1/4, 1/2, 0\}^T$ load can be balanced. The resulting maximum load is $3/4 = 0.75$.

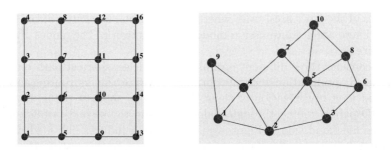

Fig. 4. Network topologies. Left: 4x4 grid topology, right: mesh topology.

4x4 Grid Topology. Next we study load balancing in more realistic-size networks (see the topologies in Figure 4). In the 4x4 grid topology there are 16 nodes, 48 links and 2934 maximal transmission modes. Generation of these modes by the DFS algorithm presented in Appendix takes 7 minutes (the algorithm implemented by Mathematica).

We study two different traffic scenarios. In the first scenario we consider that all nodes are identical and they form a wireless backbone network. To load the network, we generate random OD pairs, the source and destination nodes of which are randomly selected and the mean volumes are uniformly distributed between 0 and 1. In the second scenario we consider an access network, where all traffic is sent to one common access point located at the edge of the grid. We generate again random OD pairs, but now only source nodes and the mean volumes are random, whereas the destination node is the same for all OD pairs.

First we compare the optimal resource allocation with unconstrained routing to the scenario where traffic is optimally allocated to predefined paths with different lengths. Then the optimal resource allocation is compared to shortest path routing (referred as SP), optimization of the two layers separately (referred as 2-layer) and to so-called Equal Cost Multipath (ECMP) policy, in which traffic is distributed evenly to all shortest paths.

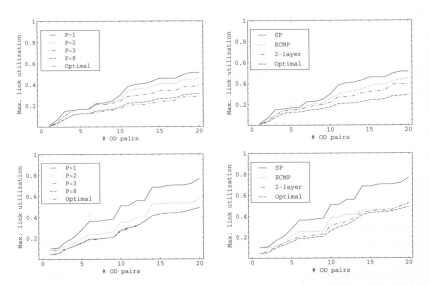

Fig. 5. The maximum link utilization as a function of the number of OD pairs. Top: backbone grid, bottom: access grid. Left: Path set constrained. Right: SP, ECMP, 2-layer optimization and optimum.

On the left hand side of Figure 5 we depict the maximum link utilization as a function of the number of random OD pairs in the backbone (top) and the access grid network (bottom), when the number of predefined paths varies from 1 to 8. Since finding N optimal paths is an integer optimization problem, the paths are just selected randomly from shortest to longest. From the figures we can see that adding one or two alternative paths improves the performance of the system. However, to obtain results very close to the optimum, we need considerably many paths. The optimal routing is compared to the single shortest path routing, ECMP and 2-layer optimization on the right hand side of the Figure 5. The maximal, minimal and mean differences between optimal resource allocation and other policies are also shown in Table 1. We can see that in some cases ECMP policy, for example, provides the same maximum link utilization as the optimal resource allocation, but on average the results of ECMP policy are 28% worse.

In general, as the backbone and access network cases are compared, the link utilizations are higher in the latter case, since there is a bottleneck around the access point. However, as seen from Table 1, by allocating the resources optimally the maximum link utilization can still be decreased 43 % on average as compared to shortest path routing.

Table 1. The difference between the optimal resource allocation and other policies

	SP	ECMP	2-layer
Backbone grid			
Min	25 %	0 %	14 %
Max	53 %	37 %	28 %
Mean	42 %	28 %	22 %
Access grid			
Min	35 %	5 %	3 %
Max	58 %	58 %	17 %
Mean	43 %	19 %	9 %
Mesh network			
Min	15 %	14 %	17 %
Max	43 %	42 %	42 %
Mean	32 %	32 %	31 %

Mesh Topology. The last studied network is a mesh network (also used in [9]). In the network, there are two access points (nodes 2 and 6) as well as two relay nodes (nodes 4 and 5). The number of links is 34 and the number transmission modes 168, and the nominal capacities of the links are 100 units. The set of source and destination pairs is fixed and contains 18 OD pairs, but the mean volume of each OD pair is random. In Figure 6, the maximum link utilization is depicted as a function of the mean traffic volume of the OD pairs. The performance of load balancing as a function of the number of predefined paths is close to the grid network. However, as seen from the right side of Figure 6, in the mesh network the joint optimization of routing and scheduling is the only approach that can

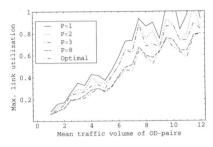 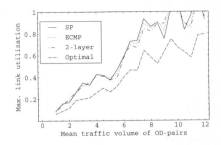

Fig. 6. The maximum link utilization as a function of the mean traffic volume in the mesh network. Left: Path set constrained. Right: SP, ECMP, 2-layer optimization and optimal.

fully exploit the resources of the network. The mean difference between optimal resource allocation and other methods is approximately 32 % (see the last three rows in Table 1).

5 Conclusion

In this paper we have studied load balancing in wireless mesh networks by formulating the linear optimization problem for the both unconstrained and predefined path sets and finding the optimal values for traffic allocation as well as the schedule. The main difference of load balancing in WMNs with STDMA as compared to traditional wired networks is that actual link capacities observed by the users are not fixed. In WMNs, in addition to taking advantage of spatial load balancing by moving traffic from loaded links to other routes, congestion can temporally be alleviated by allocating longer time shares for loaded links. Optimization of these two dimensions has to be done jointly to achieve the best results. In our numerical examples, when the joint optimization was other approaches, the average decrease in the maximum load was roughly 30%.

In this paper we considered only a simple boolean interference model. However, the LP formulations for balancing load can be used for other models just by modifying capacity matrix **S**.

Acknowledgements

I like to thank Prof. Jorma Virtamo and Dr. Aleksi Penttinen from TKK for their helpful advices and comments.

References

1. Akyildiz, I., Wang, X.: A Survey on Wireless Mesh Networks, IEEE Radio Communications (September 2005)
2. Wang, Y., Wang, Z., Zhang, L.: Internet Traffic Engineering without Full Mesh Overlaying. In: Proceeding of IEEE INFOCOM (2001)

3. Ott, T., Bogovic, T., Carpenter, T., Krishnan, K.R., Shallcross, D.: Algorithms for flow allocation for Multi Protocol Label Switching, TM-26027 (2001)
4. Nelson, R., Kleinrock, L.: Spatial-TDMA: A collision-free multihop channel access control. IEEE Transactions on Communications 33, 934–944 (1985)
5. Jain, K., Padhye, J., Padmanabhan, V., Qiu, L.: Impact of Interference on Multi-hop Wireless Network Performance. Wireless Networks 11, 471–487 (2005)
6. Wu, Y., Chou, P., Zhang, Q., Jain, K., Zhu, W., Kung, S.: Network Planning in Wireless Ad Hoc Networks: A Cross-Layer Approach. IEEE Journal on Selected Areas in Communication 23(1) (January 2005)
7. Johansson, M., Xiao, L.: Cross-layer Optimization of Wireless Networks Using Nonlinear Column Generation. IEEE Transactions on Wireless Communications 5(2) (February 2006)
8. Chen, C., Wu, W., Li, Z.: Multipath Routing Modeling in Ad Hoc Networks, IEEE ICC (2005)
9. Lassila, P., Penttinen, A., Virtamo, J.: Dimensioning of Wireless Mesh Networks with Flow-Level QoS Requirements, ACM PE-WASUN'06 (2006)

Appendix: Depth-First Search Algorithm

The algorithm searches all maximal transmission modes in a DFS manner starting from link 1 as the root and ending to link L. The modes are returned as a list M, in which element M_m consists of the links in mode m. In addition to earlier notations, let s_l and t_l denote the starting and destination node of link l, respectively.

Algorithm 1. DFS

$M \leftarrow \emptyset$ {set of feasible and maximal transmission modes}
$a \leftarrow \{0\}$ {A candidate for transmission mode}
while $a \neq \{L\}$ **do**
    $ind \leftarrow a(last) + 1$ {index ind set to value of the last element of a plus one}
    $a \leftarrow a \setminus \{a(last)\}$ {the last element of a dropped}
    for $i = ind$ to $i = L$ **do**
        if $(d_{s_l,t_i} > R'_l) \wedge (d_{s_i,t_l} > R'_i)$, $\forall l \in a$ **then**
            $a \leftarrow a \cup \{i\}$ {if no interference, link i added to transmission mode a}
        end if
    end for
    if $a \nsubseteq M(j)$, $\forall j$ **then**
        $M \leftarrow M \cup \{a\}$ {if not already a subset of some mode, candidate a added to M}
    end if
end while

Network Calculus Delay Bounds in Queueing Networks with Exact Solutions

Florin Ciucu

Department of Computer Science, University of Virginia, U.S.A.
florin@cs.virginia.edu

Abstract. The purpose of this paper is to shed light on the accuracy of probabilistic delay bounds obtained with network calculus. In particular, by comparing calculus bounds with exact results in a series of M/M/1 queues with cross traffic, we show that reasonably accurate bounds are achieved when the percentage of cross traffic is low. We use recent results in network calculus and, in addition, propose novel bounds based on Doob's maximal inequality for supermartingales. In the case of single M/M/1 and M/D/1 queues, our results improve existing bounds by $\Omega\left(\frac{\log(1-\rho)^{-1}}{1-\rho}\right)$ when the utilization factor ρ converges to one.[1]

1 Introduction

Stochastic network calculus is an extension of deterministic network calculus [1,2,3] for analyzing network performance in terms of probabilistic backlog and delay bounds. Compared to its deterministic counterpart, the advantage of stochastic network calculus is that it can account for statistical multiplexing [4,5,6,7]. In addition, the calculus can be applied to a wide class of traffic models including deterministically regulated, Markov modulated or fractional Brownian motion [8,9,7]. The 'pay-bursts-only-once' property [3] observed in deterministic network calculus holds in a probabilistic setting as well [10,11,12].

One of the main concerns in analyzing networks with performance bounds is whether the bounds are accurate enough to be applied to practical problems. As far as network calculus bounds are concerned, there are several approaches to estimate the bounds' accuracy. For example, the authors of [4,8,13] use simulation results as benchmarks for calculus bounds. The admissible region of connections requiring some performance guarantees is compared with two corresponding regions: the region obtained from simulations [5,6], and the maximal possible region based on an average rate admission control [6,7]. Asymptotic properties of end-to-end bounds are established in both networks where arrivals and service at the nodes are either statistically independent [11], or subject to correlations [10].

---

[1] Adopting Landau notation for two sequences f_n and g_n, we say that $f_n \in \mathcal{O}(g_n)$ and $f_n \in \Omega(g_n)$ if the fractions f_n/g_n and g_n/f_n, respectively, are bounded. Also, $f_n \in \Theta(g_n)$ if both $f_n \in \mathcal{O}(g_n)$ and $f_n \in \Omega(g_n)$.

L. Mason, T. Drwiega, and J. Yan (Eds.): ITC 2007, LNCS 4516, pp. 495–506, 2007.

In this paper we take a different approach to estimate the accuracy of network calculus bounds. We apply the calculus to the derivation of end-to-end delay bounds in a network of M/M/1 queues in series, with cross traffic at each queue. We then compare the obtained bounds with exact results that are readily available in M/M/1 queueing networks [14]. This comparative study leads to accurate estimations of the bounds' behavior, yet it dispenses with computationally expensive simulations. Moreover, the presented network calculus methodology to analyze M/M/1 queueing networks can be extended to more general queueing networks where exact results are usually not available.

Applying the calculus in scenarios specific to queueing network theory contributes to an understanding of some of the complementary features between the two analytical tools. For instance, queueing networks analysis applies to a small class of scheduling algorithms (of which we only consider FIFO), whereas network calculus applies to a broader scheduling class. We derive calculus bounds for static priority (SP) scheduling assuming higher priority for cross traffic. When the percentage of cross traffic is low, we show that the obtained bounds are reasonably accurate; however, when the cross traffic dominates the traffic across the network, then the bounds may degrade significantly.

Another complementary aspect between queueing networks and network calculus is that the former requires statistical independent arrivals, whereas the latter considers both independent and correlated arrivals. By accounting for the independence of arrivals in network calculus, we show that much smaller bounds can be achieved than those holding for correlated arrivals. This indicates that the independence of arrivals may play a significant role in network calculus for practical purposes. We mention that queueing networks and network calculus have been related before in [15] where the effects of traffic shaping on queueing networks analysis are considered.

In our derivations we use recent results in network calculus, and also propose novel bounds for the special class of Lévy processes. For the first time in the context of network calculus, where service is expressed with service curves, we invoke Doob's maximal inequality for supermartingales to estimate sample path bounds. Estimating sample path bounds is a difficult problem in network calculus [7], and existing solutions generally rely on approximations using extreme value theory [13,6], or the derivation of bounding sums with Boole's inequality [8,2]. By using Doob's inequality we can recover exact results in the M/M/1 queue; the bounds obtained in the M/D/1 queue numerically match the corresponding exact results. Moreover, our bounds improve those obtained with Boole's inequality by $\Omega\left(\frac{\log(1-\rho)^{-1}}{1-\rho}\right)$ when the utilization factor ρ converges to one.

We structure the rest of the paper as follows. In Section 2 we derive performance bounds in a network calculus with effective bandwidth and a formulation of a statistical service curve that generalizes several existing definitions. In Section 3 we improve these bounds by exploiting the special properties of Lévy processes. In Section 4 we apply the network calculus bounds from Sections 2 and 3 to queueing networks with exact solutions, and show numerical comparisons. Finally, in Section 5, we present brief conclusions.

2   Performance Bounds

We consider a discrete time domain with discretization step $\tau_0 = 1$. The cumulative arrivals and departures at a node are modelled with nondecreasing processes $A(t)$ and $D(t)$, where $A(0) = D(0) = 0$. We denote for convenience $A(s,t) = A(t) - A(s)$. The corresponding delay process at the node is denoted by $W(t) = \inf \{d : A(t - d) \leq D(t)\}$.

We assume that the moment generating function of the arrivals is bounded for all $s \leq t$ and some $\theta > 0$ by

$$E\left[e^{\theta A(s,t)}\right] \leq e^{\theta \rho_a(\theta)(t-s)} . \tag{1}$$

The quantity $\rho_a(\theta)$ is called *effective bandwidth* [16] and varies between the average and peak rate of the arrivals. Effective bandwidths can be obtained for a wide class of arrivals [16], or traffic descriptions with *effective envelopes* [7].

In network calculus, the service at a node is usually expressed with *service curves* that are functions specifying lower bounds on the amount of service received. We now introduce a service curve formulation that generalizes several existing formulations. This service curve is particularly useful in network scenarios where services at the nodes are either statistically independent or correlated; moreover, the service curve can account for the benefits of (partial) statistical independence. Let us first define the *convolution* operator for two doubly indexed processes f and g as $f * g(u,t) = \inf_{u \leq s \leq t} \{f(u,s) + g(s,t)\}$; also, we denote $f * g(t) := f * g(0,t)$.

We say that a nonnegative, doubly indexed random process $\mathcal{S}(s,t)$ is a *statistical service curve* if for all $t, \sigma \geq 0$

$$Pr\left(D(t) < A * [\mathcal{S} - \sigma]_+ (t)\right) \leq \varepsilon(\sigma) , \tag{2}$$

where we denoted $[x]_+ = \sup\{x, 0\}$. The nonnegative, nonincreasing function $\varepsilon(\sigma)$ is referred to as the *error* function. When $\varepsilon = 0$ then Eq. (2) recovers a service curve from $[2,11]^2$; if further $\mathcal{S}(s,t)$ is non-random and stationary (depending on $t - s$, and invariant of s or t alone), then $\mathcal{S}$ is a deterministic service curve [2]. If $\mathcal{S}(s,t)$ is non-random and stationary then Eq. (2) recovers a definition from [10]. In the most general form, $\mathcal{S}(s,t)$ is random and $\varepsilon \geq 0$.

Similar to the condition on the arrivals from Eq. (1), we assume that the Laplace transform of service curves is bounded [11] for some $\theta > 0$ by

$$E\left[e^{-\theta \mathcal{S}(s,t)}\right] \leq M(\theta)e^{-\theta \rho_s(\theta)(t-s)} . \tag{3}$$

The next theorem provides delay bounds at a node where the service is given with statistical service curves. The presented bounds generalize the bounds obtained in [11] for the special case when $\varepsilon = 0$.

---

[2] We say that $\varepsilon = 0$ whenever $\varepsilon(\sigma) = 0$ for all σ.

Theorem 1. (DELAY BOUNDS) *Consider a network node offering a statistical service curve $\mathcal{S}(s,t)$ with error function $\varepsilon(\sigma)$ to an arrival process $A(t)$. Assume that $A(t)$ and $\mathcal{S}(s,t)$ are statistically independent, and are bounded according to Eqs. (1), (3) with parameters $\rho_a(\theta)$, $M(\theta)$, $\rho_s(\theta)$, for some $\theta > 0$. If $\rho(\theta) = \rho_s(\theta) - \rho_a(\theta) > 0$, then a delay bound is given for all discrete $t, d \geq 0$ by*

$$Pr\Big(W(t) > d\Big) \leq \inf_{\sigma} \left\{ M(\theta) \frac{e^{-\theta \rho_s(\theta) d}}{\theta \rho(\theta)} e^{\theta \sigma} + \varepsilon(\sigma) \right\} . \tag{4}$$

Proof. In the first part of the proof we separate the estimation of the delay bound into a service curve bound and a sample path bound. The latter is estimated in the second part of the proof.

Fix σ and some discrete times t, d. Assume that for a particular sample path the following inequality holds

$$D(t) \geq A * [\mathcal{S} - \sigma]_+ (t) , \tag{5}$$

such that we can write

$$W(t) > d \implies A(t - d) > D(t) \implies A(t - d) > A * [\mathcal{S} - \sigma]_+ (t) .$$

It follows that

$$Pr\left(W(t) > d\right) \leq Pr\Big(A(t - d) > A * [\mathcal{S} - \sigma]_+ (t)\Big) + Pr\Big(\text{Eq. (5) fails}\Big)$$

$$\leq Pr\left(\sup_{0 \leq s < t-d} \{A(s, t - d) - \mathcal{S}(s,t)\} > -\sigma\right) + \varepsilon(\sigma) . \tag{6}$$

We remark that the points $s = t - d, \dots, t$ do not contribute to the supremum in Eq. (6) (due to the positivity constraint). Next, to estimate the sample path bound in Eq. (6), we apply Boole's inequality and the Chernoff bound

$$Pr\left(\sup_{0 \leq s < t-d} \{A(s, t - d) - \mathcal{S}(s,t)\} > -\sigma\right) \leq \sum_{s=0}^{t-d-1} E\left[e^{\theta(A(s,t-d) - \mathcal{S}(s,t))}\right] e^{\theta \sigma}$$

$$\leq M(\theta) \sum_{s=0}^{t-d-1} e^{\theta \rho_a(\theta)(t-d-s)} e^{-\theta \rho_s(\theta)(t-s)} e^{\theta \sigma} \leq M(\theta) \frac{e^{-\theta \rho_s(\theta) d}}{\theta \rho(\theta)} e^{\theta \sigma} . \tag{7}$$

In Eq. (7) we first used the independence of A and $\mathcal{S}$. Then we substituted the bounds from Eqs. (1) and (3), and finally applied the inequality $\sum_{s \geq 1} e^{-as} \leq 1/a$, for $a > 0$. The proof is completed by minimizing over σ. □

Consider now a flow along a network path with H nodes. Assume that the service given to the flow at each node is expressed by a statistical service curve $\mathcal{S}^h(s,t)$ with error function $\varepsilon^h(\sigma)$. Then, the service given to the flow by the network as a whole can be expressed using a *statistical network service curve*, such that end-to-end performance bounds can be derived using single node bounds. If $\varepsilon^h = 0$, then the network service curve is given by $\mathcal{S}^{net} = \mathcal{S}^1 * \dots * \mathcal{S}^H$ [11]. A similar expression can be constructed in the case when $\varepsilon^h \geq 0$ [10].

For the rest of the section we show how to construct *leftover* service curves for a tagged flow at a node, in terms of the capacity left unused by the remaining flows. Consider a workconserving network node operating at a constant rate R. We denote by $A(t)$ a tagged flow (or aggregate of flows) at the node, and by $A_c(t)$ the aggregate of the remaining flows; $A_c(t)$ is referred to as cross traffic. We assume SP scheduling with $A(t)$ getting lower priority.

If $A(t)$ and $A_c(t)$ are statistically independent, then a leftover service curve is given by

$$S(s,t) = R(t-s) - A_c(s,t) , \qquad (8)$$

with error function $\varepsilon = 0$ [11]. Assume now that $A(t)$ and $A_c(t)$ are not necessarily independent, and that $A_c(t)$ is bounded according to Eq. (1) with parameter $\rho_c(\theta) < R$, for some $\theta > 0$. Then for any choice of $\delta > 0$, a leftover service curve is given by

$$S(t) = (R - \rho_c(\theta) - \delta)\, t \quad \text{with} \quad \varepsilon(\sigma) = \frac{1}{\theta\delta}e^{-\theta\sigma} . \qquad (9)$$

The proof of Eq. (9) proceeds similarly as the proof of Theorem 3 in [10] and is omitted here (the main difference is that here we use effective bandwidth to describe arrivals, whereas [10] uses *statistical envelopes*).

Although leftover service curves give a worst case view on the per-flow service, they have the advantage of leading to simple, closed-form expressions for the performance bounds of interest. Tighter per-flow service curves can be derived for GPS or EDF schedulers, but their notation increases and the differences with SP service curves at a single-node are small [7].

3   Performance Bounds for Lévy Processes

In this section we assume that the arrivals and service curves are Lévy processes. Using the special properties of Lévy processes, i.e., independent and stationary increments, we show that we can improve the performance bounds obtained in the previous section. We discretize Lévy processes (that are defined in a continuous time domain) with discretization step $\tau_0 = 1$.

Theorem 2. (DELAY BOUNDS FOR LÉVY PROCESSES) *Consider the hypothesis from Theorem 1. In addition, assume that $A(t)$ and $S(s,t)$ are Lévy processes and that the following condition holds*

$$M(\theta)e^{-\theta\rho(\theta)} \leq 1 . \qquad (10)$$

Then, a statistical delay bound is given for all discrete $t, d \geq 0$ by

$$Pr\Big(W(t) > d\Big) \leq \inf_{\sigma} \Big\{ M(\theta)e^{-\theta\rho(\theta)}e^{-\theta\rho_s(\theta)d}e^{\theta\sigma} + \varepsilon(\sigma) \Big\} . \qquad (11)$$

The delay bounds obtained in Theorem 2 are smaller than those obtained in Theorem 1; the reason is that $e^{-\theta\rho(\theta)} < (\theta\rho(\theta))^{-1}$, for all $\theta > 0$. Note that the difference becomes large when $\theta\rho(\theta) \to 0$ (i.e. at very high utilizations).

The proof's main idea is to estimate sample path bounds using Doob's maximal inequality for supermartingales. This technique is applied in a classic note by Kingman [17] to the derivation of exponential backlog bounds in GI/GI/1 queues. Since Kingman's note, several works use related supermartingales techniques to derive exponential bounds (e.g. in queueing systems with Markovian arrivals [18,19], or in stochastic linear systems under the (max,+) algebra [20]). Here we integrate the technique with supermartingales in network calculus, where service is expressed with service curves. Using the properties of service curves, supermartingales can then be directly applied to analyze many scheduling algorithms and multi-node networks.

Proof. We adopt the first part of the proof of Theorem 1. The rest of the proof estimates the sample path bound from Eq. (6) by first constructing a supermartingale, and then invoking Doob's maximal inequality.

Fix t, d, σ. For positive s with $s \leq t - d$ we construct the process

$$T(s) = e^{\theta(A(t-d-s,t-d)-\mathcal{S}(t-d-s,t))},$$

with the associated σ-algebras $\mathcal{F}_s$ generated by $A(t - d - s, t - d)$ and $\mathcal{S}(t - d - s, t)$. We can write

$$E[T(s+1) \| \mathcal{F}_s] = E\left[T(s)e^{\theta(A(t-d-s-1,t-d-s)-\mathcal{S}(t-d-s-1,t-d-s))} \| \mathcal{F}_s\right]$$

$$= T(s)E\left[e^{\theta(A(1)-\mathcal{S}(1))}\right] \leq T(s)M(\theta)e^{-\theta\rho(\theta)} \leq T(s) . \quad (12)$$

In Eq. (12) we first used the fact that A and $\mathcal{S}$ are independent Lévy processes, then we substituted the bounds from Eqs. (1) and (3), and finally we used the condition from Eq. (10).

From Eq. (12) we obtain that $T(1), T(2), \ldots, T(t-d)$ form a supermartingale. We can now estimate the sample path bound from Eq. (6) as follows

$$Pr\left(\sup_{0 \leq s < t-d}\{A(s,t-d) - \mathcal{S}(s,t)\} > -\sigma\right) \leq Pr\left(\sup_s T(s) > e^{-\theta\sigma}\right)$$

$$\leq E[T(1)]e^{\theta\sigma} \leq M(\theta)e^{-\theta\rho(\theta)}e^{-\theta\rho_s(\theta)d}e^{\theta\sigma} \quad (13)$$

In Eq. (13) we first invoked Doob's inequality (see [17]) for the supermartingale $T(s)$, and the rest follows as in Eq. (12). The proof is completed by minimizing over σ. $\qquad\square$

Finally we show how to exploit the properties of Lévy processes to the construction of leftover service curves. Consider the scenario from the end of Section 2, with a node serving a tagged flow $A(t)$ and some cross traffic $A_c(t)$. If, in addition, the cross traffic $A_c(t)$ is a Lévy process, then a leftover service curve for the tagged flow $A(t)$ is now given by

$$\mathcal{S}(t) = (R - \rho_c(\theta))t \quad \text{with} \quad \varepsilon(\sigma) = e^{-\theta\sigma} . \quad (14)$$

The proof for Eq. (14) can be constructed by invoking Doob's maximal inequality, similarly as in the proof of Theorem 2. Note that the service curve in Eq. (14) is tighter than the one given in Eq. (9); the difference becomes significant when the rate of $A_c(t)$ approaches the rate R.

4   Applications to Queueing Networks with Exact Solutions

In this section we apply network calculus to the derivation of delay bounds in queueing networks. For single M/M/1 and M/D/1 queues we show that by using the special properties of Lévy processes, the derived bounds match the exact results. In the multi-node case we investigate the bounds' behavior depending on factors such as the traffic mix in the network and the statistical independence of arrivals.

We assume that exogenous flows at a node (queue) consist of packets arriving according to a Poisson process $N(t)$ with rate λ. Since a Poisson process is given in a continuous time domain, we discretize time as in Sections 2 and 3 with step $\tau_0 = 1$. Each node serves packets at rate μ and each flow is locally FIFO. For stability, we assume that the utilization factor $\rho = \lambda/\mu$ is less than one. To fit a queueing model with network calculus, we construct the arrival process $A(t) = \sum_{i=1}^{N(t)} X_i$, where X_i represents the service time of the i'th packet [16].

In the single node-case is sufficient to model the service with a deterministic service curve $\mathcal{S}(t) = t$ that induces a fluid view of the service (infinitesimal service unit). However, in the multi-node case the output from a node h may be the input at the next node $h + 1$. Consequently, we introduce packetizers [21] to enforce that, for each packet, the starting processing time at node $h + 1$ can be no sooner than the completion time at node h. Packetizers can be ignored at the last node [21], hence no packetizer is needed in the single-node case. A packetizer can be described with the service curve $\mathcal{S}(s, t) = \left[t - s - 1 - X^f(t) \right]_+$, where $X^f(t)$ denotes the time already spent in service by the packet currently in service at time t. The substraction of 1 in the expression of $\mathcal{S}(s, t)$ is a consequence of discretizing continuous time processes.

Along with the derivation of bounds, we provide numerical comparisons with exact results for the following setting. Each node has a service rate $R = 100\,Mbps$ and the average size of packets is $400\,Bytes$. We optimize the delay bounds over the parameter τ_0. We show the delays on a milliseconds time scale, and with violation probability $\varepsilon = 10^{-6}$. The numerical comparisons reflect the sensitivity of the bounds to factors such as different network loads or number of nodes.

4.1   Single M/M/1 and M/D/1 Queues

Here we apply network calculus to the analysis of two of the most common queueing models, namely the M/M/1 and M/D/1 queues.

In the M/M/1 queue the service times X_i are exponentially distributed ($X_i \sim exp(\mu)$). The distribution of the steady state delay $W = \lim_{t \to \infty} W(t)$ is given by [14]

$$P\left(W > d\right) = e^{-\mu(1-\rho)d} . \tag{15}$$

Next we derive two network calculus delay bounds for the M/M/1 queue. First, the conditions from Eqs. (1) and (3) yield $\rho_a(\theta) = \frac{\lambda}{\mu - \theta}$, $M(\theta) = 1$, and $\rho_s(\theta) = 1$ for all $0 < \theta < \mu$. One delay bound is obtained by plugging in these values into Eq. (4) from Theorem 1 (recall that $\varepsilon(\sigma) = 0$). Moreover, since $A(t)$ is a Lévy process, a second delay bound can be obtained with Eq. (11) from Theorem 2. Remarkably, by choosing $\theta = \mu - \lambda$, the latter delay bound recovers the exact result from Eq. (15). We note that the same bound is obtained by Kingman in [17], but for the waiting time in the *queue*.

For some fixed violation probability ε, let us solve for the ε-quantiles in Eqs. (4) and (11) yielding d_1 and d_2, respectively. Then we have $d_1 - d_2 \geq \frac{1}{\theta} \log \frac{1}{\theta(1-\rho_a(\theta))}$, implying that $d_1 - d_2 \in \Omega\left(\frac{\log(1-\rho)^{-1}}{1-\rho}\right)$ as $\rho \to 1$.

In the M/D/1 queue the service times X_i are constant. The distribution of the steady state delay W is given by [22]

$$P(W > d) = 1 - (1 - \rho)e^{\lambda d} \sum_{k=0}^{T} \frac{(k\rho - \lambda d)^k}{k!} e^{-(k-1)\rho} , \qquad (16)$$

where $T = \lfloor d\mu \rfloor$ denotes the largest integer less than or equal to $d\mu$. This formula poses numerical complications when ρ is close to unity, due to the appearance of large alternating, very nearly cancelling terms (note that the factor $k\rho - \lambda d$ is negative). We evaluate Eq. (16) using a numerical algorithm from [22].

Next we derive delay bounds for the M/D/1 queue with network calculus. The conditions from Eqs. (1) and (3) give $\rho_a(\theta) = \frac{\lambda}{\theta}\left(e^{\frac{\theta}{\mu}} - 1\right)$, $M(\theta) = 1$, and $\rho_s(\theta) = 1$ for all $\theta > 0$ satisfying $\rho_s(\theta) - \rho_a(\theta) > 0$. One delay bound is obtained by plugging in these values into Eq. (4) from Theorem 1. Since $A(t)$ is a Lévy process, a second delay bound is obtained with Eq. (11) from Theorem 2. As shown above, the latter delay bound improves the former by $\Omega\left(\frac{\log(1-\rho)^{-1}}{1-\rho}\right)$.

Figures 1.(a) and (b) show that the bounds obtained with Theorem 2 improve the bounds obtained with Theorem 1 at very high utilizations, as a consequence of accounting for the special properties of Lévy processes. For small to high utilizations, the bounds closely match and are not depicted. This indicates that

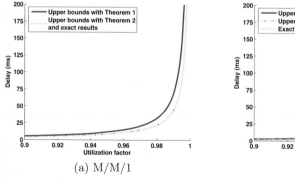
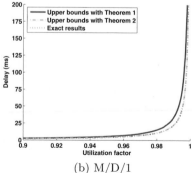

(a) M/M/1   (b) M/D/1

Fig. 1. Delay bounds at a node as a function of the utilization factor (node's service rate $R = 100$ *Mbps*, average packet size 400 B, violation probability $\varepsilon = 10^{-6}$)

the use of Boole's inequality in estimating sample path bounds can lead to conservative bounds, but only at very high utilizations. Figure 1.(b) also shows that, at all utilizations, the M/D/1 delay bounds obtained with Theorem 2 exactly match the exact results from Eq. (16).

4.2 M/M/1 Queues in Series

Now we analyze a network with H nodes arranged in series. A Poisson *through* flow $A(t)$ with rate λ traverses the entire network; moreover, a Poisson *cross* flow $A_h(t)$ with rate λ_c transits each node h, and exits the network thereafter. Each packet has independent and exponentially distributed service times at each traversed node [14]; also, the flows and the service times of packets are assumed independent. The utilization factor is now $\rho = (\lambda + \lambda_c)/\mu$.

This network is an M/M/1 queueing network where exact results are available. In particular, considering FIFO scheduling, the steady-state end-to-end delay W^{net} of the through flow has a Gamma distribution $\Gamma(\mu(1-\rho), H)$ [14]:

$$P(W^{net} > d) = \left(\sum_{k=0}^{H-1} \frac{(\mu(1-\rho)d)^k}{k!} \right) e^{-\mu(1-\rho)d} . \tag{17}$$

Next we derive two end-to-end delay bounds for SP scheduling ($A(t)$ gets lower priority) with network calculus. The first one uses the independence of $A(t)$ and $A_h(t)$, and is constructed using techniques from [11]. The second bound is obtained using techniques from [10], that apply for both independent or correlated arrivals. Observe first that the condition from Eq. (1) gives $\rho_a(\theta) = \frac{\lambda}{\mu-\theta}$.

Using the statistical independence of arrivals: Using Eq. (8), a leftover service curve for the through flow at node h is given by $T^h(s,t) = [t-s-A_h(s,t)]_+$. Convolving $T^h(s,t)$ with the service curve corresponding to the packetizer at each node, we obtain that the service at node h is described with the service curve $\mathcal{S}^h(s,t) = \left[t - s - A_h(s,t) - 1 - X_h^f(t)\right]_+$. Therefore, the service given by the network to the through flow can be expressed with the network service curve $\mathcal{S}^{net}(s,t) = \mathcal{S}^1 * \mathcal{S}^2 * \ldots * \mathcal{S}^H(s,t)$ [11]. Using $E\left[e^{\theta X_h^f(t)}\right] = \frac{\mu}{\mu-\theta}$ and denoting $K = \frac{e^\theta \mu}{\mu-\theta}$, the Laplace transform of $\mathcal{S}^{net}(s,t)$ gives

$$E\left[e^{-\theta \mathcal{S}^{net}(s,t)}\right] \leq \sum_{s=x_0 \leq x_1 \leq \cdots \leq x_H = t} E\left[e^{-\theta\left(t-s-\sum A_h(x_h,x_{h+1})-H-\sum X_h^f(x_{h+1})\right)}\right]$$

$$\leq \binom{t-s+H-1}{H-1} K^H e^{-\theta\left(1-\frac{\lambda_c}{\mu-\theta}\right)(t-s)} . \tag{18}$$

The binomial coefficient is the number of combinations with repetitions. Matching the last equation with Eq. (3) yields $M(\theta) = \binom{t-s+H-1}{H-1} K^H$ and $\rho_s(\theta) = 1 - \frac{\lambda_c}{\mu-\theta}$. Since $M(\theta)$ depends on $t-s$, Theorems 1 and 2 do not apply. However, we can use the proof of Theorem 1 and plug $M(\theta)$ into Eq. (7). Using

$\sum_s \binom{s+H-1}{H-1} a^s = \left(\frac{1}{1-a}\right)^H$ for $0 < a < 1$ [11], $\left(1 + \frac{1}{x}\right)^x \le e$ for $x > 0$, and optimizing $\tau_0 = \frac{1}{\theta \rho(\theta)} \log(1 + \rho(\theta))$ where $\rho(\theta) = \rho_s(\theta) - \rho_a(\theta) > 0$, we obtain

$$Pr\left(W^{net}(t) > d\right) \le \left(e \frac{\mu}{\mu - \theta} \frac{1 + \rho(\theta)}{\rho(\theta)}\right)^H e^{-\theta \rho_s(\theta) d} . \tag{19}$$

Lastly, the parameter θ is optimized numerically.

Without the statistical independence of arrivals: Now we derive delay bounds that hold for both independent and correlated arrivals. Using the Lévy properties of $A_h(t)$, we first get the leftover service curve $T^h(s,t) = \left(1 - \frac{\lambda_c}{\mu - \theta_c}\right)(t - s)$ with error function $\varepsilon^h(\sigma) = e^{-\theta_c \sigma}$ for some $\theta_c > 0$ (see Eq. (14)). Taking into account the packetizers, the service at node h is given by the service curve $S^h(s,t) = \left[\left(1 - \frac{\lambda_c}{\mu - \theta_c}\right)(t - s) - 1 - X_h^f(t)\right]$ with error function $\varepsilon^h(\sigma)$. Then, the network service curve [10] is given by $S^{net}(s,t) = \left[\rho_s(\theta_c)(t - s) - H - \sum Y_h\right]_+$ with error function $\varepsilon^{net}(\sigma) = H\left(\frac{1}{\theta_c \delta}\right)^{\frac{H-1}{H}} e^{-\frac{\theta_c}{H}\sigma}$, where $\delta > 0$, $\rho_s(\theta_c) = 1 - \frac{\lambda_c}{\mu - \theta_c} - (H-1)\delta$ and $Y_h \sim exp(\mu)$. Proceeding as before and optimizing $\tau_0 = \frac{1}{\theta \rho(\theta, \theta_c)} \log(1 + \rho(\theta, \theta_c))$ where $\rho(\theta, \theta_c) = \rho_s(\theta_c) - \rho_a(\theta) > 0$, we obtain

$$Pr\left(W^{net}(t) > d\right) \le \frac{\alpha}{\theta_c}\left(\frac{1}{\delta}\right)^{\frac{H\theta}{\alpha}}\left(e \frac{\mu}{\mu - \theta} \frac{1 + \rho(\theta, \theta_c)}{\rho(\theta, \theta_c)}\right)^{\frac{H\theta_c}{\alpha}} e^{-\frac{\theta \theta_c}{\alpha} \rho_s(\theta_c) d} , \tag{20}$$

where $\alpha = H\theta + \theta_c$. The parameter δ can be optimized as in [10]. Lastly, the parameters θ and θ_c are optimized numerically.

Figure 2.(a) illustrates the delay bounds from Eqs. (19), (20) for fixed $\rho = 75\%$, through traffic percentages of 50% and 90%, and different number of nodes H. The bounds approach the exact results from Eq. (17) when the percentage of cross traffic is low (less than 10%), and when accounting for the independence of arrivals (Eq. (19)). Increasing the cross traffic mix leads to more conservative bounds, due to the higher priority given to cross traffic. The decay of the bounds is more visible when dispensing with the independence of arrivals (Eq. (20)). This indicates that the leftover service curves holding for adversarial arrivals give much smaller service than those holding for independent arrivals.

In Figure 2.(b) we illustrate the delay bounds for 10 nodes, 90% through traffic, and variable utilization factor ρ. This figure shows that, at all utilizations, the independence of arrivals leads to much smaller bounds than those holding for adversarial arrivals. Therefore, the independence of arrivals appears to play a significant role in network calculus for practical purposes. From an asymptotic point of view, the delay bounds from Eq. (19) grow as $\Theta(H)$, whereas the delay bounds from Eq. (20) grow as $\Theta(H \log H)$; the extra logarithmic factor stems from dispensing with the independence of arrivals [23].

Finally, we remark that the network calculus bounds derived in this section can be extended to more general queueing networks where exact results are usually not available. Such an extension reduces to the derivation of bounds on the

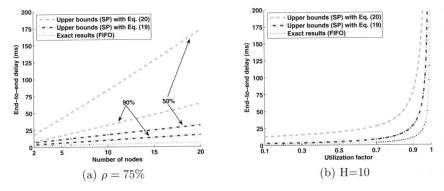

Fig. 2. End-to-end delay bounds in a M/M/1 network as a function of (a) number of nodes and (b) utilization factor; the through traffic percentages are 50%, 90% in (a), and 90% in (b); nodes' service rate $R = 100\ Mbps$, average packet size 400 B, $\varepsilon = 10^{-6}$

moment generating functions of $A(t) = \sum_{i=1}^{N(t)} X_i$ and X_i. Moreover, one can adapt the presented calculus to analyze queueing networks with constant service times of packets at each traversed node. A solution consists in describing packetizers with non-random service curves whose convolution can be analytically expanded without independence requirements, as needed for Eq. (19) (see [23]).

5   Conclusions

We have explored the accuracy of stochastic network calculus bounds by comparing them with exact results available in product-form networks. The single-node analysis showed that the bounds are tight at most utilizations and, by using the independent increments property of arrivals we could recover exact M/M/1 results and numerically match M/D/1 results. The multi-node analysis showed that for some scenarios (low percentage of cross traffic and accounting for independence of arrivals), the obtained bounds are reasonably accurate. Nevertheless, there exist complementary scenarios where the calculus may yield conservative bounds, due to the worst-case representation of service with leftover service curves and dispensing with statistical independence.

Acknowledgements

The research in this paper is supported in part by the National Science Foundation under grant CNS-0435061. The author gratefully acknowledges valuable comments from Almut Burchard and Jörg Liebeherr. Also, the author thanks the anonymous reviewer for pointing out a technical error in an earlier manuscript.

References

1. Cruz, R.: A calculus for network delay, parts I and II. IEEE Transactions on Information Theory 37, 114–141 (1991)
2. Chang, C.S.: Performance Guarantees in Communication Networks. Springer Verlag, Heidelberg (2000)
3. Le Boudec, J.-Y., Thiran, P.: Network Calculus. LNCS, vol. 2050. Springer, Heidelberg (2001)
4. Kurose, J.: On computing per-session performance bounds in high-speed multi-hop computer networks. In: ACM Sigmetrics, pp.128–139 (1992)
5. Knightly, E.W.: Second moment resource allocation in multi-service networks. In: ACM Sigmetrics, pp. 181–191 (1997)
6. Boorstyn, R., Burchard, A., Liebeherr, J., Oottamakorn, C.: Statistical service assurances for traffic scheduling algorithms. IEEE Journal on Selected Areas in Comm. Special Issue on Internet QoS 18, 2651–2664 (2000)
7. Li, C., Burchard, A., Liebeherr, J.: A network calculus with effective bandwidth. Tech. Rep. CS-2003-20, University of Virginia (November 2003)
8. Yaron, O., Sidi, M.: Performance and stability of communication networks via robust exponential bounds. IEEE/ACM Trans. on Net. 1, 372–385 (1993)
9. Starobinski, D., Sidi, M.: Stochastically bounded burstiness for communication networks. IEEE Transactions of Information Theory 46, 206–212 (2000)
10. Ciucu, F., Burchard, A., Liebeherr, J.: Scaling properties of statistical end-to-end bounds in the network calculus. IEEE Transactions on Information Theory 52, 2300–2312 (2006)
11. Fidler, M.: An end-to-end probabilistic network calculus with moment generating functions. In: IEEE 14th Int. Workshop on Quality of Service, pp. 261–270 (2006)
12. Jiang, Y.: A basic stochastic network calculus. In: ACM Sigcomm, pp. 123–134 (2006)
13. Knightly, E.: Enforceable quality of service guarantees for bursty traffic streams. In: IEEE Infocom, pp. 635–642 (1998)
14. Kleinrock, L.: Queueing Systems. John Wiley and Sons, Chichester (1975)
15. Pandit, K., Schmitt, J., Steinmetz, R.: Network calculus meets queueing theory - a simulation based approach to bounded queues. In: IEEE 12th International Workshop on Quality of Service, pp. 114–118 (2004)
16. Kelly, F.: Notes on effective bandwidths. In: Kelly, F.P., Zachary, S., Ziedins, I.B. (eds.) Stochastic Networks: Theory and Applications Royal Statistical Society. Lecture Notes Series, vol. 4, pp. 141–168. Oxford University Press, New York (1996)
17. Kingman, J.F.C.: A martingale inequality in the theory of queues. In Cambridge Philos. Soc. 59, 359–361 (1964)
18. Buffet, E., Duffield, N.G.: Exponential upper bounds via martingales for multiplexers with Markovian arrivals. J. of App. Prob. 31, 1049–1060 (1994)
19. Liu, Z., Nain, P., Towsley, D.: Exponential bounds with applications to call admission. J. of the ACM 44, 366–394 (1997)
20. Chang, C.S.: On the exponentiality of stochastic linear systems under the max-plus algebra. IEEE Trans. on Automatic Control 41, 1182–1188 (1996)
21. Boudec, J.-Y.L.: Some properties of variable length packet shapers. IEEE/ACM Transactions on Networking 10, 329–337 (2002)
22. Iversen, V., Staalhagen, L.: Waiting time distribution in $M/D/1$ queueing systems. Electronics Letters 35(25), 2184–2185 (1999)
23. Burchard, A., Liebeherr, J., Ciucu, F.: On Θ (H log H) scaling of network delays. In: Proceedings of IEEE Infocom (2007)

Simple Approximations of Delay Distributions and Their Application to Network Modeling

Romas Armolavicius

Nortel Networks
romas@nortel.com

Abstract. This paper presents a simple approximation to the delay distributions commonly used in traffic engineering and network planning. A key advantage of the approximation is its ability to estimate end-to-end delay without requiring convolutions or Laplace transforms. Two example applications to network planning are given: first, directly to the estimation of percentiles of end-to-end delay from component subnetwork measurements, and second, to network element characterization for computing delays in signaling or other systems that exhibit complex multiple message exchanges. A consideration of the accuracy of the method is included for both examples.

Keywords: Delay distribution, network planning, Normal-Power approximation.

1 Introduction

Network planning often requires the estimation of delays in situations where many network elements are involved. In packet networks the accumulated effects of traversing multiple queues operating under different loading and processor scheduling disciplines contribute significantly to such delays. This effect is compounded in signaling networks where many messages are exchanged among devices that carry a variety of traffic types. It is desirable to have available simple models of network elements which enable the ready prediction of delay under a variety of operating conditions, and in particular, under different traffic mix and loading assumptions. Ideally, these models should have a small number of parameters that can be used to "tune" the models to represent a particular modeling situation while having sufficient accuracy to provide useful results. Finally, because of the importance of "end-to-end" delays, or in signaling networks, delays from an initial "stimulus" message until a corresponding "response" message, the models should provide a simple method for combining delays to produce these end-to-end quantities. As a random variable, the end-to-end delay $T = T_1 + T_2 + \cdots + T_n$ requires the joint distribution of the component delays T_i for its computation. It is customary out of convenience or necessity to assume independence of the T_i's, in which case the joint distribution is theoretically straightforward to construct, only requiring multiple convolutions. If the distributions of the T_i's do not have a form that can be exploited, such convolutions

L. Mason, T. Drwiega, and J. Yan (Eds.): ITC 2007, LNCS 4516, pp. 507–518, 2007.

will typically be done in the Laplace transform domain, requiring numerical inversion to recover T. Although such an approach is feasible, it is advantageous to have a computationally simpler approach to estimate the distribution T of end-to-end delay. This paper describes an approximation method that is almost as simple as possible while meeting all of the the preceding requirements. The paper includes a discussion of the method's accuracy for several delay distributions that are typical of those that arise in practice. It also includes two examples that apply the method to practical planning situations.

2 A Simple Approximation of Delay Distributions

2.1 Description of the Method

Let T be a real-valued random variable with finite mean $\mu = E[T]$ variance $\sigma^2 = E[T - \mu]^2$, and skewness $\gamma = E[T - \mu]^3/\sigma^3$. If we define the standardized variable $X = (T - \mu)/\sigma$ then we have approximately [1]:

$$X \approx Z + \frac{\gamma}{6}(Z^2 - 1) \tag{1}$$

where Z is a standard normal (mean 0, variance 1) random variable. The approximation is interpreted in the sense that the distributions of both sides are similar. More specifically, for $t > \mu$ we have

$$\Pr(T < t) \approx \Pr(Q(t) < 0) \tag{2}$$

where $Q(t) = Z + \frac{\gamma}{6}(Z^2 - 1) - \frac{t - \mu}{\sigma}$. If $\gamma > 0$ then $Q(t) < 0$ is equivalent to the requirement that $r_1 < Z < r_2$ where r_1, r_1 are the roots of the quadratic $Q(t)$. We have

$$r_1, r_2 = \pm \frac{1}{\gamma}\sqrt{9 + 6\gamma\left(\frac{t - \mu}{\sigma}\right) + \gamma^2} - \frac{3}{\gamma} \tag{3}$$

From this we get $\Pr(T < t) \approx \Pr(r_1 < Z < r_2) = \Phi(r_2) - \Phi(r_1)$ where Φ is the standard normal distribution function $\Phi(x) = \frac{1}{\sqrt{2\pi}} \int_{-\infty}^{x} e^{-x^2/2} dx$. If $\gamma > 0$ (the most common and important case that arises in practice) $\Phi(r_1)$ will be negligible for the purposes of our approximation and we can write

$$\Pr(T < t) \approx \Phi\left(\frac{1}{\gamma}\sqrt{9 + 6\gamma\left(\frac{t - \mu}{\sigma}\right) + \gamma^2} - \frac{3}{\gamma}\right) \tag{4}$$

We define the q-th quantile of a random variable T to be the value t satisfying $\Pr(T < t) = q$. From (4) we can deduce that

$$\frac{1}{\gamma}\sqrt{9 + 6\gamma\left(\frac{t-\mu}{\sigma}\right) + \gamma^2} - \frac{3}{\gamma} = \xi \tag{5}$$

where $\xi = \Phi(q)$. An estimate of the q-th quantile is then:

$$t = \mu + \sigma\left[\xi + \frac{\gamma}{6}(\xi^2 - 1)\right] \tag{6}$$

Now let T_i, $1 \le i \le n$ be random variables that are mutually independent (but not necessarily identically distributed) and which have finite means μ_i, variances σ_i^2 and skewnesses γ_i. By the independence of the T_i, $T = T_1 + T_2 + \cdots + T_n$ has the approximate distribution specified by (3) with mean $\mu = \sum_{i=1}^{n} \mu_i$, variance $\sigma^2 = \sum_{i=1}^{n} \sigma_i^2$, and skewness $\gamma = E[D - \mu]^3 / \sigma^3$ where $E[D - \mu]^3 = \sum_{i=1}^{n} E[D_i - \mu_i]^3$.

The estimated q-th quantile of end-to-end delay is then given by (6).

2.2 Applications to Network Planning

Define the q-th quantile of a random variable T to be the value t satisfying $\Pr(T < t) = q$. Here we are required to estimate the q-th quantile of end-to-end delay T experienced by packets that sequentially traverse n component devices or networks N_i, $1 \le i \le n$ [2]. Assume that for each N_i, delay measurements $d_k, 1 \le k \le m_i$ are available where m_i is sufficiently large that reliable sample means μ_i, variances σ_i^2, and skewnesses γ_i can be computed using the usual formulas. The q-th quantiles t_i satisfying $\Pr(T_i < t_i) = q$ can be computed by sorting the delays d_k and re-indexing them so that $1 \le k \le m_i$ and $d_k \le d_j$ if and only if $k \le j$ and then setting $t_i = d_k$ where k is the largest integer with $k/m_i \le q$. Two main approaches are now possible:

1. **Method 1:** Compute the end-to-end mean μ, variance σ^2, and skewness γ from the corresponding component values as in the previous section, or
2. **Method 2:** Use the q-th quantile t_i to estimate the skewness γ_i from (3) by setting $\Pr(D_i < t_i) = q = \Phi(\xi)$ and setting

$$\frac{1}{\gamma_i}\sqrt{9 + 6\gamma_i\left(\frac{t_i - \mu_i}{\sigma_i}\right) + \gamma_i^2} - \frac{3}{\gamma_i} = \xi \tag{7}$$

Solving for the quantile t_i produces

$$t_i = \mu_i + \sigma_i \left[\xi + \frac{\gamma_i}{6} \left(\xi^2 - 1 \right) \right] \qquad (8)$$

Compute the end-to-end mean μ, variance σ^2, and skewness γ from the corresponding component values as above.

The only difference between these approaches is how the skewness γ_i is estimated.

Note 1: If we multiply (5) by σ_i^2 and (7) by σ^2 and add we can eliminate ξ and deduce the following formula for the direct composition of q-th quantiles.

$$\frac{\sigma^2}{\delta} \cdot \left\{ \sqrt{1 + 2\delta \cdot \left(\frac{t - \mu}{\sigma_i} \right) + \delta^2} - 1 \right\} = \sum_{i=1}^{n} \frac{\sigma_i^2}{\delta_i} \cdot \left\{ \sqrt{1 + 2\delta_i \cdot \left(\frac{t_i - \mu_i}{\sigma_i} \right) + \delta_i^2} - 1 \right\} \qquad (9)$$

Here $\delta = \gamma/3$ and $\delta_i = \gamma_i/3$.

Note 2: Define the *jitter J* for a random variable T as $J = T_1 - T_2$ where the T_i are independent and have the same distribution as T: then $E[J] = E[T_1] - E[T_2] = 0$, $E[J^2] = E[T_1^2 + T_2^2 - 2T_1T_2] = 2E[T^2] - 2E[T]^2 = 2 \cdot E[(T - E[T])^2] = 2 \cdot \sigma^2$, and $E[J^3] = E[T^2] - 3E[T^2]E[T] + 3E[T]E[T^2] - E[T^2] = 0$. Therefore the component and end-to-end distributions are normal and from (9) the q-th quantiles add directly as

$$t = \sum_{i=1}^{n} \frac{\sigma_i}{\sigma} \cdot t_i \quad \text{or equivalently} \quad (t - \mu)^2 = \sum_{i=1}^{n} (t_i - \mu_i)^2 . \qquad (10)$$

2.3 Accuracy of the Method

This section examines three numerical examples where packets traverse a number of identical "component" devices and the end-to-end delays are computed using Methods 1 and 2 presented earlier. The underlying component delay distributions are M/M/1, M/D/1, and the stochastic fluid model presented in [3]. The quantiles corresponding to probabilities ranging from 0.75 to 0.99 were computed for each of 5 and 10 components end-to-end and the resulting relative errors are presented in Table 1 where the relative error is defined

$$Re\,lative\_error = (Estimate - Exact) / Exact \qquad (11)$$

The exact delays for each component were generated from implementations of the exact models while the exact end-to-end delays were computed by numerically inverting the Laplace transform of the end-to-end delay. The moments and quantiles used for the approximations were computed from the exact models or from their transforms (in a real setting measured delay values would be used but the approach taken here eliminates measurement error from the relative errors).

Table 1. Relative errors for selected end-to-end quantile approximations for M/M/1, M/D/1, and fluid flow models. In all cases the component capacity is 1250.83 packets per sec. with a mean packet arrival rate of 1137.5 packets per sec. The occupancy of the devices is 0.9094. The fluid model has 65 on-off sources with an exponential on period of 350 msec. and an exponential off period of 650 msec. and a peak rate of 50 packets per sec. when on.

		5 Components			10 Components		
	Probability	Method 1	Method 2	Average	Method 1	Method 2	Average
M/M/1	0.75	-0.0087	0.0083	-0.0002	-0.0058	0.0031	-0.0013
	0.8	-0.0106	0.0095	-0.0006	-0.0071	0.0036	-0.0018
	0.85	-0.0122	0.0104	-0.0009	-0.0084	0.0040	-0.0022
	0.9	-0.0134	0.0108	-0.0013	-0.0093	0.0043	-0.0025
	0.95	-0.0136	0.0102	-0.0017	-0.0097	0.0041	-0.0028
	0.96	-0.0134	0.0097	-0.0018	-0.0096	0.0040	-0.0028
	0.97	-0.0130	0.0091	-0.0020	-0.0094	0.0037	-0.0029
	0.98	-0.0123	0.0080	-0.0021	-0.0090	0.0033	-0.0028
	0.99	-0.0107	0.0059	-0.0024	-0.0079	0.0024	-0.0028
M/D/1	0.75	-0.0083	0.0080	-0.0002	-0.0055	0.0030	-0.0013
	0.8	-0.0102	0.0091	-0.0005	-0.0068	0.0035	-0.0017
	0.85	-0.0118	0.0100	-0.0009	-0.0080	0.0039	-0.0021
	0.9	-0.0129	0.0104	-0.0013	-0.0090	0.0041	-0.0024
	0.95	-0.0132	0.0099	-0.0017	-0.0094	0.0040	-0.0027
	0.96	-0.0130	0.0095	-0.0018	-0.0093	0.0038	-0.0027
	0.97	-0.0127	0.0089	-0.0019	-0.0091	0.0036	-0.0028
	0.98	-0.0120	0.0078	-0.0021	-0.0087	0.0032	-0.0028
	0.99	-0.0104	0.0057	-0.0024	-0.0076	0.0023	-0.0027
Fluid	0.75	-0.0573	0.0325	-0.0124	-0.0361	0.0117	-0.0122
	0.8	-0.0573	0.0337	-0.0118	-0.0374	0.0127	-0.0123
	0.85	-0.0546	0.0340	-0.0103	-0.0372	0.0135	-0.0119
	0.9	-0.0497	0.0332	-0.0082	-0.0355	0.0139	-0.0108
	0.95	-0.0418	0.0307	-0.0055	-0.0318	0.0137	-0.0091
	0.96	-0.0396	0.0298	-0.0049	-0.0307	0.0135	-0.0086
	0.97	-0.0371	0.0285	-0.0043	-0.0292	0.0132	-0.0080
	0.98	-0.0339	0.0268	-0.0036	-0.0273	0.0126	-0.0074
	0.99	-0.0295	0.0239	-0.0028	-0.0245	0.0116	-0.0065

Considering their simplicity, both methods are remarkably accurate for M/M/1 and M/D/1, and only slightly less so for the fluid model case. In these examples, Method 1 consistently underestimates while Method 2 overestimates (so the average of the two estimates produces a better estimate): it is not known how general this effect is.

3 Application to Product Planning

3.1 Network Element Characterization

Here we are again required to estimate the end-to-end delay $T = T_1 + T_2 + \cdots + T_n$ experienced by packets that sequentially traverse component devices D_i, $1 \le i \le n$ which may be links or nodes such as packet gateways, routers, or application servers. In a product planning situation the behavior of such devices will be characterized by a model which employs a number of parameters to represent characteristics such as traffic loading that the planner may wish to modify over the course of analysis. The approach taken here is to try to model each device D_i as an M/G/1 queue and then apply the approximations presented earlier to allow calibration of the model to real data and then subsequent computation of the quantiles $\Pr(T_i < t_i) = q$ for a specific probability q. The end-to-end quantiles $\Pr(T < t) = q$ can then be computed using the component models by the methods presented in Section 2. Therefore the only question remaining to consider in what follows is how the component device approximations should be determined.

In the previous section, the underlying component distributions were not known, but measurements were available that allowed estimates to be made of the three parameters required for approximation. In this section, the problem posed is one of product planning so it is likely that few real measurements are available (particularly in the early part of the product development cycle), but that the mean service time is known. For example, measurements of earlier versions of an underlying network element may produce a base service time which can extrapolated to a new service time using anticipated processor speed improvements. As before, the goal is to produce an approximation of the distribution $P_T(t) = \Pr(T \le t)$ of total system time T (= waiting + service) for the component device D (component subscripts have been dropped as we only focus on a single device from now on) being modeled. based on using the first few moments of T as described earlier. In general, the form of the distribution of T will not be known explicitly. The approach taken here is relate the moments of T in to those of the underlying service time B. Capacity improvements can then be reflected by changing the modeled characteristics of B, while performance under different loading assumptions can be studied by manipulating the packet arrival rates which as it turns out will appear explicitly in the model.

Given this overall approach, there is still a lot of freedom in how the approximations could be constructed depending mainly on what performance data is initially available to "calibrate" the models. The next few sections explore this for different assumptions about what is known about the network elements. The goal is to provide examples of how to adapt the method to different data sets and the coverage of alternatives is not exhaustive. Again, for all cases presented, since the computation of end-to-end delays has already been described, we focus on characterizing a single network element.

3.2 Known Service Time Distribution

Suppose for the moment that the service time distribution $P_B(t) = \Pr(B \le t)$ and its

Laplace transform [4] $B^*(s) = \int_0^\infty e^{-st} dP_B(t)$ are known. In this case, we can

calculate $E[B^k]$, $1 \le k \le 4$ by differentiating and evaluating the transform B^* at 0 as follows

$$E[B^k] = \int_0^\infty t^k \, dP_B(t) = (-1)^k \frac{d^k}{ds^k} B^*(0) \tag{12}$$

The central moments $\mu_k(B) = E[(B - E[B])^k]$, $k \geq 2$ (for notational convenience we define $\mu_1(B) = E[B]$) can then be found from the expansions

$$\mu_2(B) = E[B^2] - E[B]^2$$

$$\mu_3(B) = E[B^3] - 3E[B^2]E[B] + 2E[B]^3 \tag{13}$$

$$\mu_4(B) = E[B^4] - 4E[B^3]E[B] + 6E[B^2]E[B]^2 - 3E[B]^4$$

Recall that our goal is to model the device as an M/G/1 queue. The Pollaczek-Khinchin formula [4] for an M/G/1 queue relates the Laplace transform T^* of the system time to the transform B^* of the service time distribution as follows:

$$T^*(s) = B^*(s) \frac{s(1 - \lambda h)}{s - \lambda + \lambda B^*(s)} \tag{14}$$

Just as for the service time, the k-th non-central moment $E[T^k]$ of T can be obtained by differentiating T^* and evaluating the result at 0. Denoting

$$\phi_k = \frac{\lambda}{k} \frac{E[B^k]}{1 - \rho} \quad \text{where} \quad \rho = \lambda E[B] \tag{15}$$

a tedious calculation produces the following expressions for the central moments of T

$$\mu_1(T) = \mu_1(B) + \phi_2$$

$$\mu_2(T) = \mu_2(B) + \phi_3 + \phi_2^2 \tag{16}$$

$$\mu_3(T) = \mu_3(B) + \phi_2(3\phi_3 + 2\phi_2^2) + \phi_4$$

We can now use our approximation to compute the p-th quantile of delay as

$$t_p = \mu + \sigma \left\{ \xi_p - \frac{\gamma}{6}\left(1 - \xi_p^2\right) \right\} \tag{17}$$

where $\mu = \mu_1(T)$, $\sigma^2 = \mu_2(T)$, $\gamma = \mu_3(T)/\sigma^3$, and $\Phi(\xi_p) = p$.

The formulas of this section are the basis for all of the more involved methods described in what follows. This approach can be used directly even in those cases where little is known about the performance of the device being characterized. If only the mean service time $E[B]$ is known (from separate modeling, measurement, or extrapolation of existing data) it may be reasonable from a knowledge of the traffic and mode of device operation to assume, for example, that the service time distribution B is exponential so that $E[B^k] = k! E[B]^k$, or that B is deterministic and $E[B^k] = E[B]^k$. Plugging these into (13) and evaluating (16) allows the delay to be computed using (17). An alternative approach that produces a more accurate characterization even for the same input data is given in the next section.

Note 3: The discussion so far has concentrated on system (waiting + service) time. A pure waiting time analysis using the methods under discussion is possible by recalculating from (14) with the impact of service time removed, that is

$$T^*(s) = \frac{s(1 - \lambda h)}{s - \lambda + \lambda B^*(s)} \tag{18}$$

In this case the formulas (15) simplify to the following

$$\mu_1(T) = \phi_2$$

$$\mu_2(T) = \phi_3 + \phi_2^2 \tag{19}$$

$$\mu_3(T) = \phi_2(3\phi_3 + 2\phi_2^2) + \phi_4$$

All other calculations are unchanged.

3.3 Unknown Service Time Distribution

In general, the distribution of B may not be known so the computations above cannot be used directly. Instead, we may have some limited delay measurements or estimates based on modeling or extrapolation. For this example we assume that we know:

- The mean service time $h = E[B]$
- The mean system time $E[T]$ corresponding to the submitted load $\rho = \lambda h$.
- Two quantiles $\Pr(T < t_1) = p_1$ and $\Pr(T < t_2) = p_2$ for the load $\rho = \lambda h$.

The load ρ applied during performance testing is called the *reference* load. In fact, data may be available for a range of loads $\rho_1, \rho_2, \ldots, \rho_m$, but the above is the minimum required and is the only case considered here. For this example, we will require the delay for the device operating at traffic loads $\hat{\rho}$ different (typically larger) than the load ρ used for model "calibration". We may also require a new mean service time $\hat{h} \neq h$ as a result of, say, a different mean packet size. Theoretically this implies the appearance of a new service time distribution $\hat{B}$, but in practice we want to limit the available modeling choices to those that have the same "structural characteristics" as B. In fact, we may want the mean service time and load to be the only degrees of freedom that can be changed by the analyst.

Recall that the mean delay $\mu_T = E[T]$ expected (or measured) at the reference load ρ is known as are two quantiles. From the quantile information and (7), (8) we get

$$t_i = \mu_T + \sigma_T \left\{ \xi_i - \frac{\gamma_T}{6} \left(1 - \xi_i^2 \right) \right\} \qquad i = 1,2 \tag{20}$$

where the parameters σ_T and γ_T are unknown. Solve for σ_T and γ_T producing

$$\sigma_T = \frac{(t_2 - \mu_T)(1 - \xi_1^2) - (t_1 - \mu_T)(1 - \xi_2^2)}{\xi_2(1 - \xi_1^2) - \xi_1(1 - \xi_2^2)}$$

$$\frac{\gamma_T}{6} = \frac{(t_2 - \mu_T)\xi_1 - (t_1 - \mu_T)\xi_2}{(t_2 - \mu_T)(1 - \xi_1^2) - (t_1 - \mu_T)(1 - \xi_2^2)} \tag{21}$$

where $\sigma_T^2 = \mu_2(T)$, $\gamma_T = \mu_3(T)/\sigma_T^3$, and $\Phi(\xi_i) = p_i$, $i = 1,2$. Use this information to find the non-central moments of B recursively using

$$E[B^2] = \frac{2}{\lambda} \{1 - \rho\} \cdot \{\mu_1(T) - \mu_1(B)\}$$

$$E[B^3] = \frac{3}{\lambda} \{1 - \rho\} \cdot \left\{ \mu_2(T) - \mu_2(B) - \phi_2^2 \right\} \tag{22}$$

$$E[B^4] = \frac{4}{\lambda} \{1 - \rho\} \cdot \left\{ \mu_3(T) - \mu_3(B) - \phi_2(3\phi_3 + 2\phi_2^2) \right\}$$

These are just (16) written in an alternative form. Here we have also used the facts expressed in formulas (13) and (15) to support the computations and compute in the following order: $E[B^2]$, ϕ_2, $\mu_2(B)$, $E[B^3]$, ϕ_3, $\mu_3(B)$, $E[B^4]$, ϕ_4, $\mu_4(B)$. Now recall that these values correspond to the mean service time $E[B] = h$ and load $\rho = \lambda h$. In practice, the service time could depend on the load through packet size, for example, link service (transmission) time increases with packet size and the service time through a router could be a non-linear function of packet size (limited by processing capacity for small packets due to the need for per-packet header inspection and by switching fabric capacity for large packets).

We want to be able to adjust the models, once calibrated as above, for different service times $E[\hat{B}] = \hat{h}$, packet arrival rates $\hat{\lambda}$, and therefore different loads $\hat{\rho} = \hat{\lambda} \cdot \hat{h}$. To this end, using (13) we define the ratios

$$C_2 = \mu_2(B)/h^2 = E[B^2]/h^2 - 1$$

$$C_3 = \mu_3(B)/h^3 = E[B^3]/h^3 - 3C_2 - 1 \tag{23}$$

$$C_4 = \mu_4(B)/h^4 = E[B^4]/h^4 - 4C_3 - 6C_2 + 1$$

Using these definitions it is possible to write

$$E[B^k] = Q_k \cdot h^k \tag{24}$$

$$Q_2 = 1 + C_2$$

$$Q_3 = 1 + 3C_2 + C_3 \tag{25}$$

$$Q_4 = 1 + 6C_2 + 4C_3 + C_4$$

We now take it as a *modeling assumption* that the ratios Q_k, $k = 2,3,4$ associated with the service time are unchanged as we vary the service time $h = E[B]$ or underlying load $\rho = \lambda h$. We then model a service time distribution $\hat{B}$ that differs from B only in that it has a different service time $\hat{h} = E[\hat{B}]$ through the formulas

$$E[\hat{B}^k] = Q_k \cdot \hat{h}^k \tag{26}$$

We can compute the central moments $\mu_k(\hat{B}) = E[(\hat{B} - \hat{h})^k]$, $1 \le k \le 4$ in the usual way using (16) and compute the new system time model parameters as.

$$\mu_1(\hat{T}) = \mu_1(\hat{B}) + \hat{\phi}_2$$

$$\mu_2(\hat{T}) = \mu_2(\hat{B}) + \hat{\phi}_3 + \hat{\phi}_2^2 \tag{27}$$

$$\mu_3(\hat{T}) = \mu_3(\hat{B}) + \hat{\phi}_2(3\hat{\phi}_3 + 2\hat{\phi}_2^2) + \hat{\phi}_4$$

Where $\hat{\phi}_k = \dfrac{\hat{\lambda}}{k} \dfrac{E[\hat{B}^k]}{1 - \hat{\rho}}$ reflecting the impact of a new packet arrival rate $\hat{\lambda}$.

Finally, we can compute the p-th quantile of delay $P(\hat{T} < \hat{t}_p) = p$ as

$$\hat{t}_p = \hat{\mu} + \hat{\sigma} \left\{ \xi_p - \frac{\hat{\gamma}}{6} \left(1 - \xi_p^2 \right) \right\} \tag{28}$$

where $\hat{\mu} = \mu_1(\hat{T})$, $\hat{\sigma}^2 = \mu_2(\hat{T})$, $\hat{\gamma} = \mu_3(\hat{T}) / \hat{\sigma}^3$, and $\Phi(\xi_p) = p$.

3.4 A Simple Example

The underlying delays in this example will be generated by M/M/1 (exponential service time) and M/D/1 (deterministic service time) models with an average packet size of 500 bytes and the reference load of $\rho = 0.5$ erlangs. We also assume that:

- The mean service time is $E[B] = 0.04$ ms.
- $E[T] = 0.08$ ms. for exponential service, 0.06 ms. for constant service,
- $\Pr(T < 0.368) = 0.99$ and $\Pr(T < 0.553) = 0.999$ for exponential service, $\Pr(T < 0.214) = 0.99$ and $\Pr(T < 0.287) = 0.999$ for constant service.

The packet arrival rate is $\lambda = \rho / E[B] = 12.5$ packets per ms. We want the quantiles of delay for a mean packet size of 1500 bytes and an operating load of $\hat{\rho} = 0.9$ erlangs. The exact and estimated quantiles and the associated relative errors for the reference "calibration" load ρ and for the target operating load $\hat{\rho}$ are shown in the Table 2 (exponential service time) and Table 3 (deterministic service time).

Table 2. Relative errors for deterministic service time case

| Probability | Reference $\rho = 0.5$ | | | Operating $\hat{\rho} = 0.9$ | | |
| | Quantiles | | | Quantiles | | |
	Exact	Estimate	Error	Exact	Estimate	Error
0.990	0.737	0.737	0.0000	3.649	3.684	-0.0096
0.991	0.754	0.754	0.0002	3.725	3.768	-0.0115
0.992	0.773	0.773	0.0005	3.810	3.863	-0.0136
0.993	0.794	0.794	0.0007	3.906	3.969	-0.0159
0.994	0.819	0.819	0.0009	4.017	4.093	-0.0185
0.995	0.849	0.848	0.0010	4.147	4.239	-0.0216
0.996	0.884	0.883	0.0011	4.306	4.417	-0.0252
0.997	0.930	0.929	0.0011	4.509	4.647	-0.0297
0.998	0.995	0.994	0.0008	4.794	4.972	-0.0357
0.999	1.105	1.105	0.0000	5.276	5.526	-0.0453

Table 3. Relative errors for constant service time case

| Probability | Reference $\rho = 0.5$ | | | Operating $\hat{\rho} = 0.9$ | | |
| | Quantiles | | | Quantiles | | |
	Exact	Estimate	Error	Exact	Estimate	Error
0.990	0.213	0.213	0.0000	2.072	1.912	0.0838
0.991	0.217	0.217	0.0009	2.114	1.953	0.0827
0.992	0.221	0.221	0.0017	2.161	1.998	0.0815
0.993	0.225	0.225	0.0024	2.214	2.050	0.0800
0.994	0.230	0.230	0.0030	2.274	2.109	0.0783
0.995	0.236	0.236	0.0033	2.346	2.180	0.0762
0.996	0.243	0.243	0.0036	2.433	2.266	0.0736
0.997	0.253	0.252	0.0036	2.544	2.377	0.0703
0.998	0.265	0.265	0.0027	2.699	2.533	0.0654
0.999	0.573	0.573	0.0000	2.962	2.801	0.0573

From the tables it can be seen that, at least for these examples, the relative error of the approximations is negligible at the reference load but increases in the operating region. For planning purposes, if performance data for the operating region are not available, these estimates may be sufficiently accurate. The estimates can be improved by recalibrating the models for new, higher load data as it is acquired.

4 Conclusions

This paper presented a simple approximation that can be used to estimate quantiles of the distributions of delay that arise in many network and product planning problems. The approximation has the advantage of providing a convolution and transform-free method of generating end-to-end delays. It also lends itself well to a variety of ways of calibration to available data, two of which were presented in this paper, both based on using performance data for a particular reference load ρ. An extension to the case where performance data are available for a set of loads $\rho_1, \rho_2, ..., \rho_m$ is beyond the scope of this paper and may be published elsewhere.

References

1. Ramsay, C.M.: A Note on the Normal Power Approximation. ASTIN Bulletin, Vol. 21(1) (April 1991)
2. ITU-T Recommendation Y.1541, Network performance objectives for IP-based services (February 2006)
3. Anick, D., Mitra, D., Sondhi, M.M.: Bell System Technical Journal 61(8), 1871–1894, (October 1982)
4. Kleinrock, L.: Queueing Systems, vol.1: Theory, John Wiley & Sons (1975)

Modeling and Predicting End-to-End Response Times in Multi-tier Internet Applications

Sandjai Bhulai, Swaminathan Sivasubramanian, Rob van der Mei,
and Maarten van Steen

Vrije Universiteit Amsterdam
Faculty of Sciences
De Boelelaan 1081a
1081 HV Amsterdam
The Netherlands
{sbhulai,swami,mei,steen}@few.vu.nl

Abstract. Many Internet applications employ multi-tier software architectures. The performance of such multi-tier Internet applications is typically measured by the end-to-end response times. Most of the earlier works in modeling the response times of such systems have limited their study to modeling the mean. However, since the user-perceived performance is highly influenced by the variability in response times, the variance of the response times is important as well.

We first develop a simple model for the end-to-end response times for multi-tiered Internet applications. We validate the model by real data from two large-scale applications that are widely deployed on the Internet. Second, we derive exact and approximate expressions for the mean and the variance, respectively, of the end-to-end response times. Extensive numerical validation shows that the approximations match very well with simulations. These observations make the results presented highly useful for capacity planning and performance prediction of large-scale multi-tiered Internet applications.

Keywords: end-to-end response times, Internet services, multi-tier architectures, variance approximation.

1 Introduction

The past few years have witnessed a tremendous growth in Internet services and applications. Modern Web sites such as amazon.com, yahoo.com, and ebay.com do not simply deliver static pages but generate content on the fly using multi-tiered applications, so that the pages can be customized for each user. For example, a single request to amazon.com's home page is served by hundreds of internal applications [1]. These enterprises build their software systems out of many Web applications, usually known as services. Typically, such a service performs certain business logic that generates queries to its associated database and requests to other services. They are usually exposed through well-defined client interfaces accessible over the network. Examples of services include online

L. Mason, T. Drwiega, and J. Yan (Eds.): ITC 2007, LNCS 4516, pp. 519–532, 2007.

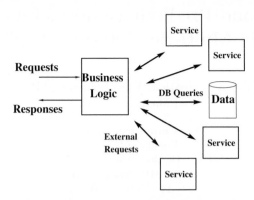

Fig. 1. Application Model of an Internet Service

entertainment services, order processing services, PC banking, online shopping. The application model of one typical service is shown in Figure 1.

For online services, providing a good client experience is one of the primary concerns. Human computer interaction studies have shown that frequent users prefer response times of less than a second for most tasks, and that human productivity improves more than linearly as computer system response times fall in the sub-second range [2]. Hence, for such online services, providing low response times to their clients is a crucial business requirement.

A key factor for providing good client response times is the ability to predict and control the performance in terms of the end-to-end response times. For the user-perceived quality, both the mean and the variance of the response times are key metrics. Therefore, we first focus on deriving a simple analytical model for such multi-tier Internet services. Second, with the use of such a model we aim to estimate the end-to-end response time of a service and also its variance. Such a model is important for several reasons. First, it allows service administrators to provision their system with enough resources at the right tiers so that the service response times are within the expected limits. In addition, it allows them to employ appropriate request policing so that excess requests can be rejected during overload scenarios. Finally, it allows service designers to predict the performance of their applications for different load conditions.

In the past few years, several groups have looked at modeling Internet applications. Most of them focus on modeling single-tier applications such as Web servers [3,4,5,6] and databases [7]. However, very few have studied models for multi-tier applications, like those given in Figure 1, which are more commonplace in the Web. Applying simple single-tier models to multi-tiered systems leads to poor solutions as different tiers have different characteristics. Therefore, there is a need to model multi-tier systems directly such that performance analysis is tractable.

Closest to our approach is recent work by Urgaonkar et al. [8]. They use mean-value analysis based techniques for estimating the mean response time of applications modeled as a network of queues. However, they cannot provide any bounds on variances. In practice, most e-commerce organizations measure client experience based on percentiles and not on means [1]. Hence, obtaining such bounds, even if approximate, is highly beneficial.

Our contribution in this paper is that we model a multi-tiered Internet service within a queueing-theoretical framework. Our model consists of $N+1$ nodes with a single entry node which receives the requests and sends requests to the other N nodes in a deterministic order. The request arrival distribution is assumed to be a Poisson process, which is usually true for arrivals from a large number of independent sources and shown to be realistic for Internet systems [9]. The entry node can be equated to the business logic (shown in Figure 1) that receives the service requests, and the service nodes correspond to databases/other services that are queried upon for serving the request. Secondly, using such a model, we derive expressions for the mean end-to-end response time and an approximation to its variance. We show the effectiveness of our variance approximations in two steps. In the first step, we compare the analytical results with simulations. In the second step, we validate our complete model using real Web services for different numbers of tiers. In our experiments, we find that our model is very effective in predicting the mean end-to-end response time and its variances (with an error margin less than 10%).

The rest of the paper is organized as follows. Section 2 presents the related work and Section 3 presents the system model. Section 4 presents the estimations for the mean response time and its variances. Section 5 validates the approximations we derived for variances using numerical simulations. Section 6 presents the validation experiments, where we validate the model with two Web service applications and Section 7 concludes the paper.

2   Related Work

2.1   Modeling Internet Systems

Various groups in the past have studied the problem of modeling Internet systems. Typical models include modeling Web servers, database servers, and application servers [3,4,7,5]. For example, in [3], the authors use a queueing model for predicting the performance of Web servers by explicitly modeling the CPU, memory, and disk bandwidth in addition to using the distribution of file popularity. Bennani and Menasce [10] present an algorithm for allocating resources to a single-tiered application by using simple analytical models. Villela et al. [9] use an M/G/1/PS queueing model for business logic tiers and to provision the capacity of application servers. A G/G/1 based queueing model for modeling replicated Web servers is proposed in [11], which is to perform admission control during overloads. In contrast to these queueing-based approaches, a feedback control based model was proposed in [6]. In this work, the authors demonstrate that by determining the right handles for sensors and actuators, the Web servers can provide stable performance and be resilient to unpredictable traffic. A novel approach to modeling and predicting the performance of a database is proposed in [7]. In this work, the authors employ machine learning and use an K-nearest neighbor algorithm to predict the performance of database servers during different workloads. However, the algorithm requires substantial input during the training period to perform effective prediction.

All the aforementioned research efforts have been applied only to single-tiered applications (Web servers, databases, or batch applications) and do not study complex multi-tiered applications which is the focus of this paper. Some recent works have focused on modeling multi-tier systems. In [5], the authors model a multi-tiered application as a single queue to predict the performance of a 3-tiered Web site. As mentioned, Urgaonkar et al. [8] model multi-tier applications as a network of queues and assume the request flows between queues to be independent. This assumption enables them to assume a product-form network so that they can apply a mean value analysis (MVA) to obtain the mean response time to process a request in the system. Although this approach can be very effective, MVA approaches can be limiting in nature as they do not allow us to get variances which are also of crucial importance in large scale enterprises [1].

2.2   Performance Analysis

There are few works that study the performance of Internet systems in the context of multi-tier applications. Although the response time was made explicit for the first time in [12], a lot of research on response times has already been done in other systems. Results for the business logic, modeled as a processor sharing (PS) node, are given in [13], where the Laplace-Stieltjes Transform (LST) is obtained for the $M/M/1/PS$ node. In [14] an integral representation for this distribution is derived, and in [15] the distribution is derived for the more general $M/G/1/PS$ system (in case a representation of the service times is given). The individual services, behind the business logic, are usually modeled as first-come-first-served (FCFS) queueing systems for which results are given in [16].

The first important results for calculating response times in a queueing network are given in [17], in which product-form networks are introduced. A multi-tier system modeled as a queueing network is of product-form when the following three conditions are met. First, the arrival process is a Poisson process and the arrival rate is independent of the number of requests in the network. Second, the duration of the services (behind the business logic) should be exponentially distributed, and is not allowed to depend on the number of requests present at that service. Finally, the sequence in which the services are visited is not allowed to depend on the state of the system except for the state of the node at which the request resides. Multi-tier systems that satisfy these properties fall within the class of so-called Jackson networks and have nice properties.

In [18], the authors give an overview of results on response times in queueing networks. In particular, they give expressions for the LST of the joint probability distribution for nodes that are traversed by requests in a product-form network according to a pre-defined path. Response times in a two-node network with feedback (such as at the business logic) were studied in [19]. The authors propose some solid approximations for the response times with iterative requests. They show that the approximations perform very well for the first moment of the response times. In [20], a single PS node is studied with several multi-server FCFS nodes. The authors derive exact results for the mean response time as well as estimates for the variance. The performance analysis in this paper is an extension of their work.

3   System Model

In this section we develop a model for multi-tier Internet services in the context of a queueing-theoretical framework. For this purpose, consider a queueing network with $N + 1$ nodes. Requests, that are initiated by an end-user, arrive according to a Poisson process with rate λ to a dedicated entry node. This node can be identified with the business logic depicted in Figure 1. The other N nodes in the queueing network represent the existing basic services delivered by the service provider.

A request traverses the network by first visiting the entry node. This node serves requests in a processor-sharing fashion with service durations that are drawn from a general probability distribution having a mean service time of β_{ps} time units. After service completion, the request is routed to each of the N service nodes in sequence. At service node i, requests receive service with a duration that is exponentially distributed with a mean of $\beta_{fcfs,i}$ time units. Requests are served on a first-come first-served (FCFS) basis by one of c_i servers dedicated to node i. When no idle servers are available upon arrival, the request joins an infinite buffer in front of node i, and waits for its turn to receive service. After having received service at node i, the results are sent back for processing to the entry node (which takes place with the same parameters as upon first entry). Thus, every request visits the entry node $N + 1$ times, and finally leaves the system after having visited all N service nodes.

The mean response time of the service is modeled as the sojourn time of the request in the system. Let $S_{ps}^{(k)}$ and $S_{fcfs,i}$ be the sojourn times of the k-th visit to the entry node and the sojourn time of the visit to service node i, respectively. Then the expected sojourn time $\mathbb{E}S$ of an arbitrary request arriving to the system is given by

$$\mathbb{E}S = \mathbb{E}\left[\sum_{k=1}^{N+1} S_{ps}^{(k)} + \sum_{i=1}^{N} S_{fcfs,i} \right].$$

Note that the system is modeled such that it satisfies the conditions of a product-form network. First, the arrivals occur according to a Poisson process with a rate that is not state dependent. This is not unrealistic in practice, since arrivals from a large number of independent sources do satisfy the properties of a Poisson process. Second, the service times follow an exponential distribution which do not depend on the number of requests at that node. Even though this modeling assumption may seem unrealistic, in our validation results we will show that it describes the performance of real systems very well. Finally, since the sequence in which the service nodes are visited is fixed, and thus does not depend on the state of the system, the network is of product-form. Also note that when c_i is large (i.e., hardly any queueing occurs at the basic services) the basic services resemble $\cdot/G/\infty$ queues for which insensitivity with respect to the service times is known to hold. In cases where queueing is significant, we loose the product form which results in more complex models (see [19]).

4   The Mean Response Time and Its Variance

In the previous section we have seen that the queueing network is of product-form. Consequently, when L_{ps} and $L_{\mathrm{fcfs,i}}$ denote the stationary number of requests at the entry node and service node i for $i = 1, \ldots, N$, respectively, we have

$$\mathbb{P}(L_{\mathrm{ps}} = l_0; L_{\mathrm{fcfs},1} = l_1, \ldots, L_{\mathrm{fcfs,N}} = l_N) = \mathbb{P}(L_{\mathrm{ps}} = l_0) \prod_{i=1}^{N} \mathbb{P}(L_{\mathrm{fcfs},i} = l_i),$$

with $l_i = 0, 1, \ldots$ for $i = 0, \ldots, N$. From this expression, the expected sojourn time at the entry node and the service nodes can be determined. First, define the load on the entry node by $\rho_{\mathrm{ps}} = (N+1)\lambda\beta_{\mathrm{ps}}$ and the load on service node i by $\rho_{\mathrm{fcfs},i} = [\lambda \cdot \beta_{\mathrm{fcfs},i}]/c_i$. Then, by Little's Law, the sojourn time at the entry node for the k-th visit is given by

$$\mathbb{E}S_{\mathrm{ps}}^{(k)} = \frac{\beta_{\mathrm{ps}}}{1 - \rho_{\mathrm{ps}}},$$

for $k = 1, \ldots, N + 1$. The expected sojourn time at service node i is given by $\beta_{\mathrm{fcfs},i}$ if upon arrival the request finds a server idle. However, when no idle server is available, the request also has to wait for the requests in front of him to be served. The probability π_i that this occurs, can be calculated by modeling the service node i as a birth-death process with birth rate λ and death rate $\min\{c_i, l_i\}/\beta_i$ when l_i requests are present at the node and in the queue. From the equilibrium equations $\lambda\mathbb{P}(L_{\mathrm{fcfs},i} = l_i - 1) = \min\{c_i, l_i\}\mathbb{P}(L_{\mathrm{fcfs},i} = l_i)/\beta_i$, the probability of delay is given by

$$\pi_i = \frac{(c_i \cdot \rho_{\mathrm{fcfs},i})^{c_i}}{c_i!} \left[(1 - \rho_{\mathrm{fcfs},i}) \sum_{l=0}^{c_i-1} \frac{(c_i \cdot \rho_{\mathrm{fcfs},i})^l}{l!} + \frac{(c_i \cdot \rho_{\mathrm{fcfs},i})^{c_i}}{c_i!} \right]^{-1}.$$

Given that a request has to wait, the expected waiting time is equal to the expected waiting time in an FCFS queueing system with one server and service time β_i/c_i. This expression is given by $\beta_{\mathrm{fcfs},i}/(1 - \rho_{\mathrm{fcfs},i})c_i$. Finally, combining all the expressions for the expected sojourn time at each node in the network, we derive that the expected response time is given by

$$\mathbb{E}S = \mathbb{E}\left[\sum_{k=1}^{N+1} S_{\mathrm{ps}}^{(k)} + \sum_{i=1}^{N} S_{\mathrm{fcfs},i} \right] = (N+1)\mathbb{E}S_{\mathrm{ps}}^{(1)} + \sum_{i=1}^{N} \mathbb{E}S_{\mathrm{fcfs},i}.$$

$$= \frac{(N+1)\beta_{\mathrm{ps}}}{1 - \rho_{\mathrm{ps}}} + \sum_{i=1}^{N} \left[\frac{\beta_{\mathrm{fcfs},i}}{(1 - \rho_{\mathrm{fcfs},i}) \cdot c_i} \pi_i + \beta_{\mathrm{fcfs},i} \right].$$

Let us now focus our attention to the variance of the sojourn time. It is notoriously hard to obtain exact results for the variance. Therefore, we approximate the total sojourn time of a request at the entry node by the sum of $N + 1$ independent identically distributed sojourn times. Moreover, we approximate the variance by imposing the assumption that the sojourn times at the entry node and the sojourn times at the service nodes are uncorrelated. In that case, we have

$$\mathbb{V}\text{ar } S = \mathbb{V}\text{ar} \left[\sum_{k=1}^{N+1} S_{\text{ps}}^{(k)} + \sum_{i=1}^{N} S_{\text{fcfs},i} \right] = \mathbb{V}\text{ar} \left[\sum_{k=1}^{N+1} S_{\text{ps}}^{(k)} \right] + \mathbb{V}\text{ar} \left[\sum_{i=1}^{N} S_{\text{fcfs},i} \right].$$

To approximate the variance of the sojourn times at the entry node, we use the linear interpolation of Van den Berg and Boxma in [21] to obtain the second moment of the sojourn time of an M/G/1/PS node. We adapt the expression by considering the $N + 1$ visits together as one visit with a service time that is a convolution of $N + 1$ service times. Therefore, we have that the second moment of the sojourn time is given by

$$\mathbb{E}S_{\text{ps}}^2 \approx (N + 1)c_{\text{ps}}^2 \left[1 + \frac{2 + \rho_{\text{ps}}}{2 - \rho_{\text{ps}}} \right] \left[\frac{\beta_{\text{ps}}}{1 - \rho_{\text{ps}}} \right]^2 +$$
$$\left((N + 1)^2 - (N + 1)c_{\text{ps}}^2 \right) \left[\frac{2\beta_{\text{ps}}^2}{(1 - \rho_{\text{ps}})^2} - \frac{2\beta_{\text{ps}}^2}{\rho_{\text{ps}}^2(1 - \rho_{\text{ps}})}(e^{\rho_{\text{ps}}} - 1 - \rho_{\text{ps}}) \right],$$

where c_{ps}^2 is the squared coefficient of variation of the service time distribution at the entry node, which should be derived from real data. The variance at the entry node is therefore given by

$$\mathbb{V}\text{ar} \left[\sum_{k=1}^{N+1} S_{\text{ps}}^{(k)} \right] = \mathbb{E}S_{\text{ps}}^2 - \left((N + 1)\mathbb{E}S_{\text{ps}}^{(1)} \right)^2.$$

Let $W_{\text{fcfs},i}$ and $c_{\text{fcfs},i}^2$ denote the waiting time and the coefficient of variation of the service distribution at service node i, respectively. Then, the variance of the total sojourn times at the service nodes can be expressed as follows

$$\mathbb{V}\text{ar} \left[\sum_{i=1}^{N} S_{\text{fcfs},i} \right] = \sum_{i=1}^{N} \mathbb{V}\text{ar } S_{\text{fcfs},i} + 2 \sum_{i=1}^{N} \sum_{j=i+1}^{N} \mathbb{C}\text{ov} \left[S_{\text{fcfs},i}, S_{\text{fcfs},j} \right]$$
$$\approx \sum_{i=1}^{N} \left(\mathbb{E}W_{\text{fcfs},i}^2 - [\mathbb{E}W_{\text{fcfs},i}]^2 + \beta_{\text{fcfs},i}^2 \right)$$
$$= \sum_{i=1}^{N} \left[\pi_i \frac{2\beta_{\text{fcfs},i}^2}{c_{\text{fcfs},i}^2(1 - \rho_{\text{fcfs},i})^2} - \pi_i^2 \frac{\beta_{\text{fcfs},i}^2}{c_{\text{fcfs},i}^2(1 - \rho_{\text{fcfs},i})^2} + \beta_{\text{fcfs},i}^2 \right]$$
$$= \sum_{i=1}^{N} \left[\frac{\pi_i(2 - \pi_i)\beta_{\text{fcfs},i}^2}{c_{\text{fcfs},i}^2(1 - \rho_{\text{fcfs},i})^2} + \beta_{\text{fcfs},i}^2 \right].$$

Finally, by combining all the expressions for the variances of the sojourn time at each node in the network, we derive that the variance of the response time is given by

$$\text{Var } S \approx (N+1)c_{\text{ps}}^2 \left[1 + \frac{2 + \rho_{\text{ps}}}{2 - \rho_{\text{ps}}}\right]\left[\frac{\beta_{\text{ps}}}{1 - \rho_{\text{ps}}}\right]^2 +$$

$$\left((N+1)^2 - (N+1)c_{\text{ps}}^2\right)\left[\frac{2\beta_{\text{ps}}^2}{(1 - \rho_{\text{ps}})^2} - \frac{2\beta_{\text{ps}}^2}{\rho_{\text{ps}}^2(1 - \rho_{\text{ps}})}(e^{\rho_{\text{ps}}} - 1 - \rho_{\text{ps}})\right] -$$

$$\left[\frac{(N+1)\beta_{\text{ps}}}{1 - \rho_{\text{ps}}}\right]^2 + \sum_{i=1}^{N}\left[\beta_{\text{fcfs},i}^2 + \frac{\pi_i(2 - \pi_i)\beta_{\text{fcfs},i}^2}{c_{\text{fcfs},i}^2(1 - \rho_{\text{fcfs},i})^2}\right].$$

5 Numerical Experiments

In this section we assess the quality of the expressions of the mean response time and the variance that was derived in the previous section. First, we perform some numerical experiments to test validity of the expressions against a simulated system. In this case, the mean response time does not need to be validated, because the results are exact due to [17]. Therefore, we can restrict our attention to validating the variance only. Then, we validate the expressions by using real data from Web services. Note that real systems are notoriously complex, while the model we use is simple. Therefore, the validation results should be judged from that perspective.

5.1 Accuracy of the Variance Approximation

We have performed numerous numerical experiments and checked the accuracy of the variance approximation for many parameter combinations. This was achieved by varying the arrival rate, the service time distributions, the asymmetry in the loads of the nodes, and the number of servers at the service nodes. We calculated the relative error by $\Delta\text{Var } \% = 100\% \cdot \left(\text{Var } S - \text{Var}_s S\right)/\text{Var}_s S$, where $\text{Var}_s S$ is the variance based on the simulations.

We have considered many test cases. We started with a queueing network with exponential service times at the entry node and two service nodes. In the cases where the service nodes were equally loaded and asymmetrically loaded, we observed that the relative error was smaller than 3% and 6%, respectively. We also validated our approximation for a network with five single-server service nodes. The results demonstrate that the approximation is still accurate even for very highly loaded systems. Based on these results, we expect that the approximation will be accurate for an arbitrary number of service nodes. The reason is that cross-correlations between different nodes in the network disappear as the number of nodes increases. Since the cross-correlation terms have not been included in the approximation (because of our initial assumptions), we expect the approximation to have good performance in those cases as well.

Table 1. Response time variances of a queueing network with general service times at the entry node and two asymmetrically loaded single-server service nodes

β_{ps}	c_{ps}^2	β_{fcfs}		$\mathbb{V}\mathrm{ar}_s\,S$	$\mathbb{V}\mathrm{ar}\,S$	$\Delta\mathbb{V}\mathrm{ar}\,\%$
0.1	0	0.1	0.9	82.51	81.05	-1.33
0.3	0	0.8	0.5	85.89	86.81	1.06
0.1	0	0.5	0.3	1.25	1.22	-1.76
0.3	0	0.9	0.1	147.49	150.82	2.25
0.1	4	0.1	0.9	80.83	81.33	0.62
0.3	4	0.8	0.5	274.00	278.46	1.63
0.1	4	0.5	0.3	1.54	1.50	-2.68
0.3	4	0.9	0.1	331.30	342.47	3.37
0.1	16	0.1	0.9	81.55	82.16	0.75
0.3	16	0.8	0.5	831.15	853.41	2.68
0.1	16	0.5	0.3	2.37	2.33	-1.86
0.3	16	0.9	0.1	871.49	917.42	5.27

Table 2. Response time variances of a queueing network with general service times at the entry node and two asymmetrically loaded multi-server service nodes

β_{ps}	c_{ps}^2	β_{fcfs}		$\mathbb{V}\mathrm{ar}_s\,S$	$\mathbb{V}\mathrm{ar}\,S$	$\Delta\mathbb{V}\mathrm{ar}\,\%$
0.1	0	0.2	2.7	80.82	85.66	5.99
0.3	0	1.6	1.5	88.82	89.70	0.99
0.1	0	1.0	0.9	2.43	2.43	-0.02
0.3	0	1.8	0.3	149.31	152.38	2.05
0.1	4	0.2	2.7	88.50	85.94	-2.90
0.3	4	1.6	1.5	272.00	281.35	3.44
0.1	4	1.0	0.9	2.71	2.71	-0.23
0.3	4	1.8	0.3	330.14	344.03	4.21
0.1	16	0.2	2.7	89.60	86.77	-3.16
0.3	16	1.6	1.5	820.26	856.30	4.39
0.1	16	1.0	0.9	3.57	3.57	-0.79
0.3	16	1.8	0.3	920.45	918.98	-0.16

Since the approximation for different configurations with single-server service nodes turned out to be good, we turned our attention to multi-server service nodes. We carried out the previous experiments with symmetric and asymmetric loads on the multi-server service nodes while keeping the service times at the entry nodes exponential. Both cases yielded relative errors smaller than 6%. Finally, we changed the service distribution at the entry node. Table 1 shows the results for a variety of parameters, where the coefficient of variation for the service times at the entry nodes is varied between 0 (deterministic), 4 and 16 (Gamma distribution). These results are extended in Table 2 with multi-server service nodes. If we look at the results, we see that the approximation is accurate in all cases. To conclude, the approximation covers a wide range of

different configurations and is therefore reliable enough to obtain the variance
of the response time.

6   Model Validation

In the previous section, we showed using simulation results that the approximation we derived for the variance was effective. In this section, we validate
our model with two Web applications: a Web service application and a popular
bulletin board Web application benchmark. We present our experimental setup
followed by the validation results.

6.1   Experimental Setup

For validating our model, we experimented with two types of applications: a
promotional Web service and the RUBBoS benchmark. The choice of these two
applications was motivated by their differences in their behavior. We hosted the
business logic of these applications in the Apache Tomcat/Axis platform and
used MySQL 3.23 for database servers. We ran our experiments on Pentium *III*
machines with a 900 Mhz CPU and 2 GB memory running a Linux 2.4 kernel.
These servers belonged to the same cluster and network latency between the
clusters was less than a millisecond. We implemented a simple client request
generator that generated Web service requests as a Poisson arrival process. In
our experiments, we varied the arrival rate and measured the mean end-to-end
response and its variance and compared them with the values predicted by the
model.

In our model, an application (or a service) is characterized by the parameters
β_{ps} and β_{fcfs}. Therefore, to accurately estimate the mean and the variance of the
response times for a given application, we first need to obtain these values. Note
that all measurements of execution times should be realized during low loads
to avoid measuring the queueing latency in addition to the service times. This
method has been successfully employed in similar problems [8].

6.2   Experiment 1: A Promotional Service

The first type of service we experimented with is a *promotional service* modeled
after the *"Recommended for you"* service in amazon.com. This service recommends products based on the user's previous activity. The service maintains two
database tables: (i) the item table contains details about each product (title,
author, price, and stock); (ii) the customer table contains information regarding
customers and the list of previous items bought by each customer. The business
logic of this service executes two steps to generate a response. First, it queries
the customer table to find the list of product identifiers to which a customer
has shown prior interest. Second, it queries the item table to generate the list of
items related to these products and their information. The database tables are
stored in different database servers running at different machines. This makes

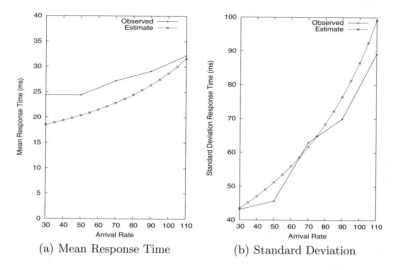

(a) Mean Response Time   (b) Standard Deviation

Fig. 2. Comparison between the observed and the predicted values for the mean and the standard deviation of the response times for the promotional service

the application a 3-node system with $N = 2$. The business logic is exposed as a Web service and receives the requests in XML format (over HTTP) and returns responses also in XML. In our experiments, we populated the database server with $500,000$ product records and $200,000$ customer records. The appropriate indices were created to enhance the query performance. We restricted ourselves to read-only workloads in our experiments[1].

The results of our experiments are given in Figure 2. As seen in the figure, the model does a reasonable job in predicting the mean response times. The margin of error is less than 5% in most of the cases. In Figure 2(b), we can see that the model does a commendable job in getting reasonable approximations to the standard deviation. This is especially remarkable considering the fact that the service times of the database do not strictly adhere to any specific statistical distribution. This observation also explains the deviation of the predicated mean from the observed mean at high loads. At higher loads, considerable queueing occurs in our model so that the results become more sensitive to the service distribution. Moreover, in practice, the service nodes are multi-threading systems that are best modeled by processor sharing queues.

6.3 Experiment 2: RUBBoS – A Bulletin Board Web Application

In our next set of experiments, we experimented with the RUBBoS benchmark, a bulletin board application that models `slashdot.org`. The RUBBoS application models a popular news Web site. RUBBoS's database consists of five tables,

---

[1] The read and write query characteristics of databases vary a lot and is highly temporal in nature. By far, there are few models that have modeled database systems with complex workloads. Hence, in this study we restrict ourselves to read-only workloads.

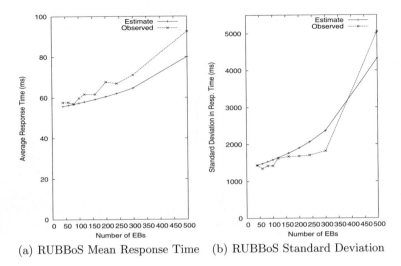

(a) RUBBoS Mean Response Time   (b) RUBBoS Standard Deviation

Fig. 3. Comparison between the observed and the predicted values for the mean and the standard deviation of the response times for RUBBoS benchmark

storing information regarding users, stories, comments, submissions, and moderator activities. Each request to the Web site is handled by the bulletin board application which in turn issues one or more requests to the underlying database to generate the HTML responses. In our experiments, we filled the database with information on 500, 000 users and 200, 000 comments on the stories. Our experiments were performed with the browse-only workload mix. For our experiments, we chose the open source implementation of these benchmarks[2]. In this implementation, the application is written in Java and runs on the Apache Tomcat servlet engine.

The client workload for the benchmark is generated by Emulated Browsers (EBs). The run-time behavior of an EB models a single active client session. Starting from the home page of the site, each EB uses a Customer Behavior Graph Model (a Markov chain with various interactions with the Web pages in a Web site as nodes and transition probabilities as edges) to navigate among Web pages, performing a sequence of Web interactions [22]. The behavior model also incorporates a think time parameter that controls the amount of time an EB waits between receiving a response and issuing the next request, thereby modeling a human user more accurately. The mean think time between subsequent requests of a client session is set to 7 seconds.

In our experiments, we varied the number of EBs and measured the end-to-end response time for the requests. We compared the measurements against the estimated response time from our model and the results are given in Figure 3. As in the previous experiment, the results in Figure 3 show that the model reasonably predicts both the mean and the variance. The margin of error is less

---

[2] http://jmob.objectweb.org/rubbos.html

than 10% in most of the cases. Given that the RuBBoS application does not conform to all of the assumptions of our model, this is a commendable result.

7   Conclusion

In this paper, we presented a simple analytical model for multi-tiered Internet applications based on a queueing-theoretical framework. Based on this model, we can estimate the mean response time of a service and also provide a reasonable approximation to its variance. We verified the effectiveness of our approximations using numerical simulations and validated the complete model with real data using two large-scale Web service applications that are widely deployed on the Internet. Our validation experiments suggest that the model provides high accuracy in estimating not just the mean response time but also its standard deviation. We believe this work serves as a good starting point in modeling complex multi-tiered Internet applications.

There are many interesting avenues for further research on the modeling side. We plan to derive accurate approximations when deterministic routing of requests to the service nodes is replaced by a mixture of Markovian and deterministic routing. This will enlarge the scope of applicability of our model considerably as we can take into effect various caching tiers that are typically used as performance optimization tiers. Moreover, we also plan to enhance the model by modeling the service nodes as processor sharing nodes, which we believe will model servers more accurately. In addition to these extensions, we plan to investigate web admission control models that will allow us to determine the best strategy to handle overloads.

References

1. Vogels, W.: Learning from the Amazon technology platform. ACM Queue 4(4) (2006)
2. Shneiderman, B.: Response time and display rate in human performance with computers. ACM Comput. Surv. 16(3), 265–285 (1984)
3. Menasce, D.A.: Web server software architectures. IEEE Internet Computing 7(6), 78–81 (2003)
4. Doyle, R., Chase, J., Asad, O., Jin, W., Vahdat, A.: Web server software architectures. In: Proc. of USENIX Symp. on Internet Technologies and Systems (2003)
5. Kamra, A., Misra, V., Nahum, E.: Yaksha: A controller for managing the performance of 3-tiered websites. In: Proceedings of the 12th IWQoS (2004)
6. Abdelzaher, T.F., Shin, K.G., Bhatti, N.: Performance guarantees for web server end-systems: A control-theoretical approach. IEEE Trans. Parallel Distrib. Syst. 13(1), 80–96 (2002)
7. Chen, J., Soundararajan, G., Amza, C.: Autonomic provisioning of backend databases in dynamic content web servers. In: Proceedings of the 3rd IEEE International Conference on Autonomic Computing (ICAC 2006) (2006)
8. Urgaonkar, B., Pacifici, G., Shenoy, P., Spreitzer, M., Tantawi, A.: An analytical model for multi-tier internet services and its applications. In: Proc. of the ACM SIGMETRICS conference, pp. 291–302 (2005)

9. Villela, D., Pradhan, P., Rubenstein, D.: Provisioning servers in the application tier for e-commerce systems. In: Proceedings of the Twelfth IEEE International Workshop on Quality of Service (IWQoS 2004), Montreal, Canada (2004)

10. Bennani, M.N., Menasce, D.A.: Resource allocation for autonomic data centers using analytic performance models. In: ICAC '05: Proc. of the Second Int. Conf. on Automatic Computing, pp. 229–240, Washington, DC, USA (2005)

11. Urgaonkar, B., Shenoy, P.: Cataclysm: policing extreme overloads in internet applications. In: WWW '05: Proceedings of the 14th international conference on World Wide Web, pp. 740–749. ACM Press, New York, USA (2005)

12. van der Mei, R., Meeuwissen, H.: Modelling end-to-end quality-of-service for transaction-based services in a multi-domain environment. In: Proceedings IEEE International Conference on Web Services ICWS, Chicago, USA (2006)

13. Coffman, E., Muntz, R., Trotter, H.: Waiting time distributions for processor-sharing systems. Journal of the ACM 17(1), 123–130 (1970)

14. Morrison, J.: Response-time distribution for a processor-sharing system. SIAM Journal on Applied Mathematics 45(1), 152–167 (1985)

15. Ott, T.: The sojourn time distribution in the $M/G/1$ queue with processor sharing. Journal of Applied Probability 21, 360–378 (1984)

16. Cooper, R.: Introduction to Queueing Theory. North Holland (1981)

17. Jackson, J.: Networks of waiting lines. Operations Research 5, 518–521 (1957)

18. Boxma, O., Daduna, H.: Sojourn times in queueing networks. Stochastic Analysis of Computer and Communication Systems, pp. 401–450 (1990)

19. Boxma, O., van der Mei, R., Resing, J., van Wingerden, K.: Sojourn time approximations in a two-node queueing network. In: Proc. of ITC 19, pp. 112–1133 (2005)

20. van der Mei, R., Gijsen, B., Engelberts, P., van den Berg, J., van Wingerden, K.: Response times in queueing networks with feedback. Performance Evaluation 64 (2006)

21. van den Berg, J., Boxma, O.: The $M/G/1$ queue with processor sharing and its relation to a feedback queue. Queueing Syst. Theory Appl. 9(4), 365–402 (1991)

22. Smith,   W.:   (TPC-W: Benchmarking   an   e-commerce   solution), http://www.tpc.org/tpcw/tpcw_ex.asp

Delay Bounds in Tree Networks with DiffServ Architecture

Jinoo Joung[1], Jongtae Song[2], and Soonseok Lee[2]

[1] Sangmyung University, Seoul, Korea
jjoung@smu.ac.kr
[2] BcN Research Division, ETRI, Daejon, Korea

Abstract. We investigate the end-to-end delay bounds in large scale networks with Differentiated services (DiffServ) architecture. It has been generally believed that networks with DiffServ architectures can guarantee the end-to-end delay for packets of the highest priority class, only in lightly utilized cases. We focus on tree networks with DiffServ architecture and obtain a closed formula for delay bounds for such networks. We further show that, in tree networks with DiffServ architecture, the delay bounds for highest priority packets exist regardless of the level of network utilization. These bounds are quadratically proportional to the maximum hop counts in heavily utilized networks; and are linearly proportional to the maximum hop counts in lightly utilized networks. We argue that based on these delay bounds DiffServ architecture is able to support real time applications even for a large tree network. Considering that tree networks, especially the Ethernet networks, are being adopted more than ever for access networks and for provider networks as well, this conclusion is quite encouraging for real-time applications. Throughout the paper we use Latency-Rate ($\mathcal{LR}$) server model, with which it has been proved that First In First Out (FIFO) and Strict Priority schedulers are $\mathcal{LR}$ servers to each *flows* in certain conditions.

1 Introduction

QoS characteristics of the network with Integrated services (IntServ) [1,2] architecture have been well studied and understood by numerous researches in the past decade. Providing the allocated bandwidths, or service rates, or simply rates of an output link to multiple sharing *flows* plays a key role in this approach. A flow is usually defined to be a set of packets that have the same 5-tuples (Destination and Source addresses, Port numbers, and Protocol number). A myriad of scheduling algorithms have been proposed. The Packetized Generalized Processor Sharing (PGPS) [3] and Deficit Round Robin (DRR) [4], and many other rate-providing servers are proved to be a Latency-Rate server [5], or simply $\mathcal{LR}$ server. All the work-conserving servers that guarantee rates to each flow can be modeled as $\mathcal{LR}$ servers. The behavior of an $\mathcal{LR}$ server is determined by two parameters, the latency and the allocated rate. The latency of an $\mathcal{LR}$ server may be considered as the worst-case delay seen by the first packet of the busy period

L. Mason, T. Drwiega, and J. Yan (Eds.): ITC 2007, LNCS 4516, pp. 533–543, 2007.
© Springer-Verlag Berlin Heidelberg 2007

of a flow. It was shown that the maximum end-to-end delay experienced by a packet in a network of $\mathcal{LR}$ servers can be calculated from only the latencies of the individual servers on the path of the flow, and the traffic parameters of the flow that generated the packet. More specifically for a leaky-bucket constrained flow,

$$D_i \leq \frac{\sigma_i - L_i}{\rho_i} + \sum_{j=1}^{N} \Theta_i^{S_j}, \tag{1}$$

where D_i is the delay of flow i within a network, σ_i and ρ_i are the well known leaky bucket parameters, the maximum burst size and the average rate, respectively, L_i is the maximum packet length and $\Theta_i^{S_j}$ is the latency of flow i at the server S_j.

We consider networks with DiffServ [6] architecture, especially the QoS characteristics of the highest priority class traffic with the strict priority scheduling scheme. We focus on the queueing and scheduling behaviors of the flows or aggregated flows, and investigate the delay characteristics of them. In networks with DiffServ architecture, packets in a same class are enqueued to a single queue and scheduled in a FIFO manner within the class. The higher priority class may be served with a strict priority over the lower classes, or a certain amount of bandwidth may be assigned to each class. It is proved that FIFO schedulers and strict priority schedulers used in this network is also an $\mathcal{LR}$ server to the individual flows within a class, under conditions that every flow conforms to leaky-bucket model and the aggregated rate is less than or equal to the link capacity [7]. We will apply this result with the $\mathcal{LR}$ server model to DiffServ networks for further analysis.

The current belief on such networks is that only with the low enough network utilization, one can guarantee delay bounds. Otherwise the bounds *explode* to infinity with just sufficient number of hops or the size of the network [8]. Based on this argument, there have been a trend of aborting DiffServ for delay sensitive real-time applications such as Voice over IP (VoIP) [9,10,11,12,13,14,15]. One notable research direction is to aggregate flows selectively [9,10] so that effective compromises are achieved in the area between the extreme points with the completely flow-state aware IntServ and the unaware DiffServ. A series of implementation practices in this regard are being proposed, including flow aggregation using Multi-Protocol Label Switching (MPLS) framework [11,12]. Another direction is aligned with the argument that achieving absolute performance guarantee is hard with conventional DiffServ so that only a *relative* differentiation is meaningful [13]. Some even go further by arguing that in a core network, traditional approach of requesting, reserving, and allocating a certain rate to a flow or the flow aggregates (e.g. a class in DiffServ) is too burdensome and inefficient so that the flows should not explicitly request for a service, but rather they should be implicitly detected by the network and treated accordingly [14,15]. The exploding nature of the delay bounds that has stimulated such diverse research activities, however, is the conclusion with general network topology. If we can somehow avoid the burst size accumulation due to the loops formed in a network, which is suspected as the major reason for the explosion, the delay bounds may be still useful with the simple DiffServ architecture. We investigate this possibility throughout the paper. We focus on the tree networks where the loops are avoided. The tree networks can be found in many types of Local Area Network (LAN), especially Ethernet network. Ethernet networks are gaining momentum to be extensively used more and more in access networks as well as provider networks [16]. In a large scale, these networks together form a tree topology, by means of spanning tree algorithms or manual configurations [17,18].

In the next section, a brief review on delay bounds with DiffServ architecture is given. In the third section, we focus on tree networks, and obtain the closed formula for delay bounds. We discuss our result in section 4. The conclusion is given in the final section.

2   Previous Works on Delay Bounds in Networks with DiffServ Architecture

Under a condition that there is no flow that violates the leaky bucket constraint specified, we can guarantee the delay upper bound for highest priority flows, even with relatively simple scheduling strategies. For example if there is only a single class of flows, and the flows meet the preset leaky bucket constraints so that sum of the average rates does not exceed the link capacity, then a simple First-in-first-out (FIFO) server can guarantee the delay bounds. If in the network there are best-effort traffic that does not specify its rate, the minimum protection against such best-effort traffic is necessary. In this regard the Strict Priority (SP) scheduling can guarantee delay bounds for flows of the real-time traffic. A Strict Priority (SP) server is a server that maintains at least two queues. The queue with highest priority, which is for the flows with real-time constraints, transmits packets whenever it has one, right after the completion of the current packet that is being served.

In this environment, the sum of all flows that want to pass through a server is compared with the link capacity, and if it's less than the link capacity then delay upper bounds will be prescribed, as it can be calculated with the method explained in this section. We consider a network of packet level servers.

Theorem 1. *A FIFO server or an SP server, under conditions that all the input flows are leaky bucket constrained and the sum of average rates is less than the link capacity, is an $\mathcal{LR}$ server for individual flows with latency given as the following:*

$$\Theta_i^S = \frac{\sigma^S - \sigma_i^S}{r^S} + \Theta^S, \tag{2}$$

where σ^S is the sum of all the σ_i^S within the server $\mathcal{S}$, r^S is the link capacity of $\mathcal{S}$, and

$$\Theta^S = \begin{cases} L/r^S & when \ \mathcal{S} \ is \ FIFO \\ 2L/r^S & when \ \mathcal{S} \ is \ SP. \end{cases}$$

Proof. See the proof in section 3 of [7].   □

Corollary 1. *The output traffic of flow i from a FIFO server or an SP server S conforms to the leaky bucket model with parameters $(\sigma_i^S + \rho_i\Theta_i^S, \rho_i)$, where σ_i^S is the maximum burst size of flow i into the server $\mathcal{S}$.*

The end-to-end delay of a network with DiffServ architecture with FIFO servers and/or Strict Priority servers can be obtained by the following sets of equations:

$$D_i \leq \frac{\sigma_i - L_i}{\rho_i} + \sum_{n=1}^{N} \Theta_i^{In},$$

$$\Theta_i^{In} = \frac{\sigma^{In} - \sigma_i^{In}}{r^{In}} + \Theta^{In},$$

$$\sigma_i^{In+1} = \sigma_i^{In} + \rho_i\Theta_i^{In}, \text{ for } n \geq 1, \tag{3}$$

where I_n is the nth server of the network for i, N is the number of servers that i traverses in the network; and $\sigma_i^{I1} = \sigma_i$.

3   Delay Bounds in Tree Networks with DiffServ Architecture

While (3) gives a tight bound through iterative computation of the latencies and the maximum burst sizes of each server, we still have to make assumptions about the burst sizes of other flows. With a reasonable restriction on the network topology, the other flows' burst sizes can be inferred, and the delay bound for a whole network can be obtained. In this section we consider delay bounds in tree networks, which are defined to be acyclic connected graphs. Tree networks appear in a broad range of networks. For example in Ethernet, both in LAN and in wide area networks, the logical tree-based network topology is achieved by running Spanning Tree Protocol (STP) or static configuration (e.g. configuring Virtual LANs) [17,18].

Let us define *hop* by the server and the accompanying link through which packets are queued and serviced and then transmitted. First we start by observing an important property of tree networks. Let i be the flow under observation.

Lemma 1. *Consider a tree network with the flow under observation, i. Assume that the flow i traverses the path of the maximum possible hop counts. Then at any server in the i's path, other flows that are confronted by i have traversed less number of hops than i just have.*

Proof. Let us denote the i's path by $(I_1, I_2, \ldots, I_H)$, where I_n is the nth server in the i's path and H is the maximum number of hops possible in the given tree network. Similarly let us denote another flow j's path by $(J_1, J_2, \ldots, J_m)$, where $m \leq H$. Let $I_k = J_l$, for some k and l where $1 \leq k \leq H$ and $1 \leq l \leq m$, that is the flow i and j confront each other at a server in the path. A path is defined to be a sequence of nodes, with no repeated nodes, in which each adjacent node pair is linked. We will show that $k \geq l$ for any case, by contradiction. Assume $k < l$.

Case 1: $\{J_1, J_2, \ldots, J_{l-1}\}$ is disjoint with $\{I_{k+1}, \ldots, I_H\}$. This is to say that there is no server that takes part of both the remaining path of i and the traveled path of j. Then the path $(J_1, J_2, \ldots, J_{l-1}, I_k, I_{k+1}, \ldots, I_H)$ exists that has more hops than H. This contradicts to the assumption that H is the maximum possible hop counts.

Case 2: There is at least one server in the remaining path of i, $(I_{k+1}, \ldots, I_H)$, that is also a server in the traveled path of j, $(J_1, \ldots, J_{l-1})$. Let us call this server J_p. Then $(J_p, J_{p+1}, \ldots, J_{l-1}, I_k, I_{k+1}, \ldots, J_p)$ forms a cycle, which contradicts the assumption that the network is a tree.

In both cases the statement $k < l$ contradicts the assumption. Therefore $k \geq l$ for any case and the lemma follows. $\qquad\square$

Lemma 1 lets us infer about the maximum burst sizes of the confronted flows in the path. Therefore the end-to-end delay bound can be obtained from a few network parameters.

Now consider a server I_n in the path of i. Let the set of flows $\mathcal{F}_{In}$, including i, are competing for service in I_n. For any flow j, $j \in \mathcal{F}_{In}$, which has traveled $(n-m-1)$ hops until reaching I_n, imagine a corresponding flow j' with m more hops from the starting node of the flow j. Moreover j' has entered the network with the same parameter with j. Further imagine that for additional hops for each flows, not the existing nodes but new nodes are attached to the starting nodes of each flows, so that the numbers of flows in the upstream nodes of I_n are intact. See figure 1. Now we have constructed an imaginary network, in which at I_n, the flows in $\mathcal{F}_{In}$ all have traveled exactly $(n-1)$ hops until reaching I_n. We claim the following.

Lemma 2. *The maximum burst size of any flow at the entrance of I_n in the constructed network is always greater than or equal to that of the original network. That is,*

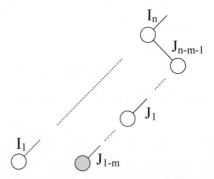

Fig. 1. Construction of an imaginary network. The shaded nodes are added ones.

$$\sigma_j^{In} \le \sigma_{j'}^{In}, \tag{4}$$

for all $j \in \mathcal{F}_{In}$.

Proof. It is enough to show that $\sigma_j^{J1} \le \sigma_{j'}^{J1}$ for any $j, j \in \mathcal{F}_{In}$. Since σ_j^{Jn} is a nondecreasing function of n,

$$\sigma_{j'}^{J1} \ge \sigma_{j'}^{J(1-m)} = \sigma_j = \sigma_j^{J1}. \tag{5}$$

The lemma follows.   □

We argue the main result of the paper as the following.

Theorem 2. *The end-to-end delay of a tree network with DiffServ architecture with FIFO servers is bounded by*

$$D_i \le \frac{\sigma_i - L_i}{\rho_i} + \tau \frac{(1+\alpha)^H - 1}{\alpha}, \tag{6}$$

where τ and α are defined as

$$\sum_{j \in \mathcal{F}_S} \sigma_j \le \tau r^S, \quad \sum_{j \in \mathcal{F}_S} \rho_j \le \alpha r^S, \tag{7}$$

for any server S in the network, in which there is a set of flows, $\mathcal{F}_S$.

The parameters τ and α is similarly defined in [8]. We will call τ the *burst allowance level* measured in time for their transmission, and α the *network utilization*. Note that $0 < \alpha < 1$ in our network configurations.

Proof. First, we will show that the following inequalities hold for any I_n, $1 \le n \le H$.

$$\Theta_i^{In} \le \tau(1+\alpha)^{n-1},$$
$$\sigma_i^{In} \le \sigma_i + \tau\rho_i \frac{(1+\alpha)^{n-1} - 1}{\alpha}. \tag{8}$$

Let us first assume that (8) is true for I_n. We will show that it holds for I_{n+1} as well, and then it holds for I_1, therefore it holds for any I_n.

If (8) is true for I_n, from (3) we get

$$\sigma_i^{In+1} = \sigma_i^{In} + \rho_i\Theta_i^{In}$$
$$\le \sigma_i + \tau\rho_i((1+\alpha)^n - 1)/\alpha. \tag{9}$$

From (9), and from lemma 2, the sum of the maximum burst sizes of all the incoming priority flows in I_{n+1},

$$\sigma^{In+1} = \sum_{j \in \mathcal{F}_{In}} \sigma_j^{In+1} \le \sum_{j' \in \mathcal{F}_{In}} \sigma_{j'}^{In+1}$$
$$\le \sum_{j'} \{\sigma_j + \tau\rho_j((1+\alpha)^n - 1)/\alpha\}, \tag{10}$$

since any j, $j \in \mathcal{F}_{In}$, has traveled $(n-1)$ hops, therefore $\sigma_j^{In+1} = \sigma_j + \tau\rho_j((1+\alpha)^n - 1)/\alpha$ by the assumption at the beginning of the proof. We obtain

$$\sigma^{In+1} \leq \tau r^{In+1}(1+\alpha)^n \tag{11}$$

from (7). Equation (3) yields

$$\Theta_i^{In+1} = \frac{\sigma^{In+1} - \sigma_i^{In+1}}{r^{In+1}} + \Theta^{In+1}. \tag{12}$$

Note that Θ^{In+1} is L/r^{In+1} for a FIFO server. The maximum burst size of a flow, by definition, is always greater than or equal to the maximum packet length, that is $\sigma_i^{In+1} \geq L$. Therefore, and from (11),

$$\Theta_i^{In+1} \leq \frac{\sigma^{In+1}}{r^{In+1}} \leq \tau(1+\alpha)^n. \tag{13}$$

With (9) and (13), we have shown that (8) holds for I_{n+1}. Now we'll consider the case for I_1. For I_1, $\sigma^{I1} = \sum_{j \in \mathcal{F}_{I1}} \sigma_j \leq \tau r^{I1}$, and

$$\Theta_i^{I1} \leq \tau - \frac{\sigma_i}{r^{I1}} + \frac{L}{r^{I1}} \leq \tau, \tag{14}$$

which shows that (8) holds for I_1 as well. From (3),

$$D_i \leq \frac{\sigma_i - L_i}{\rho_i} + \sum_{n=1}^{H} \tau(1+\alpha)^{n-1}$$

$$\leq \frac{\sigma_i - L_i}{\rho_i} + \tau \sum_{n=0}^{H-1} (1+\alpha)^n$$

$$\leq \frac{\sigma_i - L_i}{\rho_i} + \tau \frac{(1+\alpha)^H - 1}{\alpha}, \tag{15}$$

for $H \geq 1$. $\qquad\square$

Similar conclusion can be claimed for the network with SP servers.

Theorem 3. *The end-to-end delay of a tree network with DiffServ architecture with SP servers is bounded by*

$$D_i \leq \frac{\sigma_i - L_i}{\rho_i} + \tau' \frac{(1+\alpha)^H - 1}{\alpha}, \tag{16}$$

where τ' and α are defined as

$$L + \sum_{j \in \mathcal{F}_S} \sigma_j \leq \tau' r^S, \quad \sum_{j \in \mathcal{F}_S} \rho_j \leq \alpha r^S, \tag{17}$$

for any server S in the network, in which there is a set of flows, $\mathcal{F}_S$.

Proof. The proof of this theorem is exactly the same with that of the previous theorem except that Θ^{In+1} in (12) is $2L/r^{In+1}$ for an SP server, therefore (13) is still valid with the modified τ'. We omit the detail. $\qquad\square$

4   Discussion

We first examine the extreme cases; in which $\alpha \to 0$ or $\alpha \to 1$. When $\alpha \to 0$, the delay bound becomes $(\sigma_i/\rho_i + H\tau')$. For the worst case delay we can say that it is $\left(\max_j(\sigma_j/\rho_j) + H\tau'\right)$. When $\alpha \to 1$, the delay bounds becomes $\left(\max_j(\sigma_j/\rho_j) + (2^H - 1)\tau'\right)$. The delay bounds increases linearly as hop count increases, when the utilization is low. When the utilization is high, however, the delay bounds increases quadratically with hop counts.

The delay bounds in a general topology network with DiffServ architecture have been obtained in the literatures [8,19]. In [8], it was concluded that unless the link utilizations are kept under a certain level, the end-to-end delay explodes to infinity. The delay bound obtained in [8] is, only under condition that $\alpha < 1/(H-1)$,

$$D \leq \frac{H}{1-(H-1)\alpha}\tau', \tag{18}$$

for a case with infinite incoming links' capacity, which is also the case considered in this paper. Equation (18) becomes, as $\alpha \to 0$, $D \leq \tau' H$. As it was already noted in [19], however, (18) has apparently not taken the non-preemptive nature of SP servers into consideration, therefore was later corrected in [19] to, under the same condition,

$$D \leq \frac{H}{1-(H-1)\alpha}\left(\tau' + \max_j(L/\sum_j \rho_j)\right). \tag{19}$$

Equation (19) becomes, as $\alpha \to 0$, $D \leq \left(\tau' + \max_j(L/\sum_j \rho_j)\right)H$. Table 1 summarizes the delay bounds obtained from this paper and from two previous works, with the suggested network parameters both in [8,19]. The reason that

Table 1. The bounds from this paper and from the related works, in milliseconds: $H=10$, $\sigma_j=100$ bytes for all flows, $\rho_i=32$kbps for all flows, $L=1500$ bytes, $r^S = 149.760$Mbps for all S

α	0.04	0.08
Bound by (16)	12.97	30.13
Bound in [8]	16.88	74.29
Bound in [19]	48.18	110.06

the bounds obtained in this paper is less than the one from [19] is twofold. The first and obvious reason is that our network is strictly confined in a tree topology. This restriction inhibits the burst accumulation in loop, so that the bound still exists even in high network utilization. The second reason is that, even in the low network utilization cases, the bounds in previous works are simply the summation of nodal delays. Our bound, on the other hand, is from considering the whole network as a single virtual node, so that the correlation among consecutive nodes are taken into account. This characteristic is the primary virtue of the analysis based on the $\mathcal{LR}$ server [5]. In the following table, we examine the bounds with varying H, the maximum number of hops, and with varying α, the network utilization. As table 2 suggests, even in a tree network, delay bounds

Table 2. The bounds obtained from (16), in seconds, with varying H and α: $\sigma_j = L$ = 1500 bytes for all flows, ρ_i=32kbps for all flows, r^S = 149.760Mbps for all S

α	0.1	0.3	0.5	0.7
$H = 6$	0.290	1.436	3.898	8.679
$H = 8$	0.430	2.686	9.240	25.792
$H = 10$	0.599	4.798	21.258	75.245
$H = 12$	0.804	8.368	48.301	218.175

with moderate to high network utilization seems quite unacceptable. Since in numerous standardization bodies it is suggested that the end-to-end delay bound for voice to be less than 400ms [20], only with α much less than 0.1 can only meet those standards, if network parameters in table 2 are to be used.

Consider now where the maximum burst size, one of primary contributor to delay bound increment, and the maximum packet length can be controlled to be much less than the ordinary IP packet length. This assumption on the maximum packet length, therefore on the maximum burst size, is not extravagant since if we are to transmit the MPEG-2 Transport Streams (TS) data whose lengths are fixed at 188 bytes [21], with 12 bytes RTP fixed header [22], 4 bytes RTP video-specific header [23], 8 bytes UDP header, 20 bytes IP header and finally 26 bytes Ethernet header and trailer including preamble, then the maximum packet length in this case becomes 258 bytes. Considering the extended headers fields and Ethernet inter-frame gap, the maximum packet length will be about 300 bytes. In such a network of limited packet length, the 400ms requirement

Table 3. The bounds obtained from (16), in seconds, with varying H and α: $\sigma_j = L$ = **300** bytes for all flows, ρ_i=32kbps for all flows, r^S = 149.760Mbps for all S

α	0.05	0.1	0.2	0.3
$H = 6$	0.0256	0.0580	0.149	0.287
$H = 8$	0.0360	0.0859	0.248	0.537
$H = 10$	0.0474	0.120	0.390	0.960
$H = 12$	0.0600	0.161	0.594	1.674
$H = 14$	0.0738	0.210	0.889	2.880
$H = 16$	0.0891	0.270	1.313	4.919

can be reasonably met as table 3 suggests. For example, if we manage to restrict the maximum packet length to be 300 bytes and send one packet at a time to the network (that is to restrict the maximum burst size at the entrance of the network), then at the 10% network utilization even a very large network with 16 hop counts can successfully support the voice applications. Considering that the applications requesting Premium Service in DiffServ are usually a small portion of the total traffic, table 3 suggests DiffServ may indeed be useful in some cases.

5   Conclusion

In this paper we have investigated the end-to-end delay bounds of FIFO servers and Strict Priority servers in DiffServ architecture, especially in tree networks. Based on the observation that the FIFO and SP servers are $\mathcal{LR}$ servers for each *micro flows* under the condition that all the flows in the high priority class conform to the arranged leaky bucket parameters and the sum of allocated rates is less than the link capacity for all the links, and on the established analysis technique with $\mathcal{LR}$ servers, we suggested the iterative computation method to calculate the tight end-to-end delay bounds with FIFO or SP servers, i.e. in Expedited Forwarding DiffServ architecture. This iterative computation sequence, however, requires an assumption on the burst size increase on other flows. Concurrent iteration on every flow in the network may be possible. Instead, however, we focused on tree networks, a special type but widely used networks, and derived a closed formula for the delay bound.

Contrary to the traditional belief that without a very small network utilization delay bound goes to infinity, we have shown that in tree networks delay bounds always exist. We have shown also that this bound is linearly proportional to the number of hop counts when the utilization is small; and is quadratically proportional to the number of hop counts when the utilization is large. It may be argued, however, that the existing bound is not acceptable for moderate to large sized networks with moderate to large network utilization, even in tree networks. On the other hand, with a manipulation to the network configuration such as maximum packet length restriction, we have shown that DiffServ architecture can support the real-time application even in large networks with moderate network utilization.

References

1. Braden, R., Clark, D., Shenker, S.: Integrated Services in the Internet Architecture: an Overview. IETF RFC 1633 (1994)
2. White, P.P.: RSVP and Integrated services in the Internet: A tutorial. IEEE Communications Mag. 35, 100–106 (1997)
3. Parekh, A.K., Gallager, R.G.: A Generalized Processor Sharing Approach to Flow Control in Integrated Services Networks: The Single-Node Case. IEEE/ACM Trans. Networking 1(3) (June 1993)
4. Shreedhar, M., Varghese, G.: Efficient fair queueing using deficit round-robin. IEEE/ACM Trans. Networking 4(3), 375–385 (1996)

5. Stiliadis, D., Varma, A.: Latency-Rate servers: A general model for analysis of traffic scheduling algorithms. IEEE/ACM Trans. Networking 6(5) (October 1998)
6. Blake, S., Black, D., Carlson, M., Davies, E., Wang, Z., Weiss, W.: An architecture for Differentiated Services. IETF RFC 2475 (1998)
7. Joung, J., Choe, B.-S., Jeong, H., Ryu, H.: Effect of Flow Aggregation on the Maximum End-to-End Delay. In: Gerndt, M., Kranzlmüller, D. (eds.) HPCC 2006. LNCS, vol. 4208, pp. 426–435. Springer, Heidelberg (2006)
8. Chanry, A., LeBoudec, J.-Y.: Delay bounds in a network with aggregate scheduling. In: Proc. First International Workshop of Quality of Future Internet Services (QOFIS2000) (2000)
9. Cobb, J.A.: Preserving quality of service guarantees in spite of flow aggregation. IEEE/ACM Trans. on Networking 10(1), 43–53 (2002)
10. Sun, W., Shin, K.G.: End-to-End Delay Bounds for Traffic Aggregates Under Guaranteed-Rate Scheduling Algorithms. IEEE/ACM Trans. on Networking 13(5) (October 2005)
11. Le Faucheur, F., et al.: Multi-Protocol Label Switching (MPLS) Support of Differentiated Services. IETF RFC 3270 (May 2002)
12. Rong, B., et al.: Modeling and Simulation of Traffic Aggregation Based SIP over MPLS Network Architecture. In: Proc. IEEE 38th Annual Simulation Symposium (ANSS05) (2005)
13. Dovrolis, C., Stiliadis, D., Ramanathan, P.: Proportional Differentiated Services: Delay Differentiation and Packet Scheduling. IEEE/ACM Trans. on Networking 10(1) (Febuary 2002)
14. Kortebi, A., Oueslati, S., Roberts, J.: Cross-protect: implicit service differentiation and admission control. In: Proc. IEEE HPSR (April 2004)
15. S. Oueslati, J. Roberts, A new direction for quality of service: Flow-aware networking. In: Proc. Conference on Next Generation Internet Networks (NGI) (April 2005)
16. Layered Network Architecture and Implementation for Ethernet Services, White Paper, Fujitsu Network Communications, Inc. (2004)
17. 802.1ad Virtual Bridged Local Area Networks – Amendment 4: Provider Bridges. IEEE Higher Layer LAN Protocols Working Group (IEEE 802.1), Ammedment to IEEE 802.1Q (May 2006)
18. Metro Ethernet Service – A Technical Overview, White Paper, R. Santitoro, Metro Ethernet Forum (2003), [Online]. Available: http://www.metroethernetforum.org/metro-ethernet-services.pdf
19. Jiang, Y.: Delay Bounds for a Network of Guaranteed Rate Servers with FIFO. Computer Networks, Elsevier Science 40(6), 683–694 (2002)
20. Network performance objectives for IP-based services, International Telecommunication Union – Telecommunication Standadization Sector (ITU-T), Recommendation Y.1541 (Febuary 2006)
21. Information Technology – Generic Coding of Moving Pictures and Associated Audio Information Part 1: Systems. ISO/IEC International Standard IS 13818 (November 1994)
22. Schulzrinne, H., Casner, S., Frederick, R., Jacobson, V.: RTP: A Transport Protocol for Real-Time Applications. IETF RFC 3550 (July 2003)
23. Hoffman, D., Fernando, G., Goyal, V., Civanlar, M.: RTP Payload Format for MPEG1/MPEG2 Video. IETF RFC 2250 (January 1998)

Polling Models with Two-Stage Gated Service: Fairness Versus Efficiency[*]

R.D. van der Mei[1,2] and J.A.C. Resing[3]

[1] CWI, Advanced Communication Networks
P.O. Box 94079, 1090 GB Amsterdam, The Netherlands
[2] Vrije Universiteit, Faculty of Sciences
De Boelelaan 1081a, 1081 HV Amsterdam, The Netherlands
[3] Eindhoven University of Technology, Mathematics and Computer Science
P.O. Box 513, 5600 MB Eindhoven, The Netherlands

Abstract. We consider an asymmetric cyclic polling system with general service-time and switch-over time distributions with so-called two-stage gated service at each queue, an interleaving scheme that aims to enforce fairness among the different customer classes. For this model, we (1) obtain a pseudo-conservation law, (2) describe how the mean delay at each of the queues can be obtained recursively via the so-called Descendant Set Approach, and (3) present a closed-form expression for the expected delay at each of the queues when the load tends to unity (under proper heavy-traffic scalings), which is the main result of this paper. The results are strikingly simple and provide new insights into the behavior of two-stage polling systems, including several insensitivity properties of the asymptotic expected delay with respect to the system parameters. Moreover, the results provide insight in the delay-performance of two-stage gated polling compared to the classical one-stage gated service policies. The results show that the two-stage gated service policy indeed leads to better fairness compared to one-stage gated service, at the expense of a decrease in efficiency. Finally, the results also suggest simple and fast approximations for the expected delay in stable polling systems. Numerical experiments demonstrate that the approximations are highly accurate for moderately and heavily loaded systems.

1 Introduction

This paper is motivated by dynamic bandwidth allocation schemes in an Ethernet Passive Optical Network (EPON), where packets from different Optical Network Units (ONUs) share channel capacity in the upstream direction. An EPON is a point-to-multipoint network in the downstream direction and a multi-point to point network in the upstream direction. The Optical Line Terminal (OLT) resides in the local office, connecting the access network to the Internet. The OLT allocates the bandwidth to the Optical Network Units (ONUs) located at the

---

[*] A preliminary version of this paper has also been presented at the Second Korea-Netherlands Conference on Queueing Theory and its Applications to Telecommunication Systems (Amsterdam, October 24-27, 2006).

L. Mason, T. Drwiega, and J. Yan (Eds.): ITC 2007, LNCS 4516, pp. 544–555, 2007.
© Springer-Verlag Berlin Heidelberg 2007

customer premises, providing interfaces between the OLT and end-user network to send voice, video and data traffic. In an EPON the process of transmitting data downstream from the OLT to the ONUs is broadcast in variable-length packets according to the 802.3 protocol [6]. However, in the upstream direction the ONUs share capacity, and various polling-based bandwidth allocation schemes can be implemented. Simple time-division multiplexing access (TDMA) schemes based on fixed time-slot assignment suffer from the lack of statistical multiplexing, making inefficient use of the available bandwidth, which raises the need for dynamic bandwidth allocation (DBA) schemes. A dynamic scheme that reduces the time-slot size when there is no data to transmit would allow excess bandwidth to be used by other ONUs. However, the main obstacle of implementing such a scheme is the fact the OLT does not know in advance how much data each ONU has to transmit. To overcome this problem, Kramer et al. [7,8] propose an OLT-based interleaved polling scheme similar to hub-polling to support dynamic bandwidth allocation. To avoid monopolization of bandwidth usage of ONUs with high data volumes they propose an interleaved DBA scheme with a maximum transmission window size limit.

Motivated by this, in this paper we analyze the effectiveness of this interleaved DBA scheme in a queueing-theoretical context. To this end, we quantify fairness and efficiency measures related to the expected delay figures at each of the queues, providing new fundamental insight in the trade-off between fairness and efficiency by implementing one-stage and two-stage gated service policies. The two-stage gated service policy was introduced in Park et al. [10] where the authors study a symmetric version of the model described in this paper.

A polling system is a multi-queue single-server system in which the server visits the queues in cyclic order to process requests pending at the queues. Polling models occur naturally in the modeling of systems in which service capacity (e.g., CPU, bandwidth) is shared by different types of users, each type having specific traffic characteristics and performance requirements. Polling models find many applications in the areas of computer-communication networks, production, manufacturing and maintenance [9]. Since the late 1960s polling models have received much attention in the literature [12,13]. There are several good reasons for considering heavy-traffic asymptotics, which have recently started to gain momentum in the literature, initiated by the pioneering work of Coffman et al. [2,3] in the mid 90s. Exact analysis of the delay in polling models is only possible in some cases, and even in those cases numerical techniques are usually required to obtain the expected delay at each of the queues. However, the use of numerical techniques for the analysis of polling models has several drawbacks. First, numerical techniques do not reveal explicitly how the system performance depends on the system parameters and can therefore contribute to the understanding of the system behavior only to a limited extent. Exact closed-form expressions provide much more insight into the dependence of the performance measures on the system parameters. Second, the efficiency of each of the numerical algorithms degrades significantly for heavily loaded, highly asymmetric systems with a large number of queues, while the proper operation of the system is particularly critical when the system is heavily loaded. These observations raise the importance of an exact asymptotic analysis of the delay in polling models in heavy traffic.

We consider an asymmetric cyclic polling model with generally distributed service times and switch-over times. Each queue receives so-called two-stage gated service, which works as follows: Newly incoming customers are first queued at the stage-1 buffer. When the server arrives at a queue, it closes a gate behind the customers residing in the stage-1 buffer, then serves all customers waiting in the stage-2 buffer on a FCFS basis, and moves all customers before the gate at the stage-1 buffer to the stage-2 buffer before moving to the next queue. We focus on the expected delay incurred at each of the queues. First, we derive a so-called pseudo-conservation law for the model under consideration, giving a closed-form expression for a specific weighted sum of the mean waiting times. Then, we describe how the individual expected delays at each of the queues can be calculated recursively by means of the so-called Descendant Set Approach (DSA), proposed for the case of standard one-stage gated and exhaustive service in [5]. The two-stage gated model introduces several interesting complications that do not occur for the standard gated/exhaustive case. Denoting by $X_i^{(k)}$ the number of customers at Q_i in stage k at an arbitrary polling instant at Q_i ($k = 1, 2$), the mean delay at Q_i does not only depend on the first two marginal moments of $X_i^{(2)}$, but also depends on the *cross*-moments $E\left[X_i^{(1)}X_i^{(2)}\right]$ (see (7)). To this end, we need to consider the numbers of customers at a queue in both stage 1 and stage 2 at polling instants at that queue, which leads to a two-dimensional analysis.

The main result of this paper is the presentation of a closed-form expression for $(1 - \rho)E[W_i]$, referred to as the scaled expected delay at Q_i, when the load tends to 1. The expression is strikingly simple and shows explicitly how the expected delay depends on the system parameters, thereby explicitly quantifying the trade-off between the increase in fairness and decrease of efficiency introduced by implementing two-stage gated service policies. In particular, the results provide new fundamental insight with respect to mean waiting times for one-stage versus two-stage gated service policies. Furthermore, the results reveal a variety of asymptotic insensitivity properties, which provide new insights into the behavior of polling system under heavy load. The validity of these properties is illustrated by numerical examples. In addition, the expressions obtained suggest simple and fast approximations for the mean delay at each of the queues in stable polling systems. The accuracy of the approximations is evaluated by numerical experiments. The results show that the approximations are highly accurate when the system load is significant.

The remainder of this paper is organized as follows. In section 2 the model is described. In section 3 we present the pseudo-conservation law for the model under consideration. In section 4 we describe how the expected delay figures can be obtained by the use of the DSA. In section 5 we present closed-form expressions for the scaled expected delay at each of the queues, and discuss several asymptotic properties. Finally, in section 6 we propose and test simple approximations for the moments of the waiting times in heavy traffic, and address the practicality of the asymptotic results.

2   Model Description

Consider a system consisting of $N \geq 2$ stations $Q_1, \ldots, Q_N$, each consisting of a stage-1 buffer and a stage-2 buffer. A single server visits and serves the queues in cyclic order. Type-i customers arrive at Q_i according to a Poisson arrival process with rate λ_i, and enter the stage-1 buffer. The total arrival rate is denoted by $\Lambda = \sum_{i=1}^{N} \lambda_i$. The service time of a type-i customer is a random variable B_i, with Laplace-Stieltjes Transform (LST) $B_i^*(\cdot)$ and with finite k-th moment $b_i^{(k)}$, $k = 1, 2$. The k-th moment of the service time of an arbitrary customer is denoted by $b^{(k)} = \sum_{i=1}^{N} \lambda_i b_i^{(k)} / \Lambda$, $k = 1, 2$. The load offered to Q_i is $\rho_i = \lambda_i b_i^{(1)}$, and the total offered load is equal to $\rho = \sum_{i=1}^{N} \rho_i$. Define a polling instant at Q_i as a time epoch at which the server visits Q_i. Each queue is served according to the two-stage gated service policy, which works as follows. When the server arrives at a queue, it closes the gate behind the customers residing in the stage-1 buffer. Then, all customers waiting in the stage-2 buffer are served on a First-Come-First-Served (FCFS) basis. Subsequently, all customers before the gate at the stage-1 buffer are instantaneously forwarded to the stage-2 buffer, and the server proceeds to the next queue. Upon departure from Q_i the server immediately proceeds to Q_{i+1}, incurring a switch-over time R_i, with LST $R_i^*(\cdot)$ and finite k-th moment $r_i^{(k)}$, $k = 1, 2$. Denote by $r := \sum_{i=1}^{N} r_i^{(1)} > 0$ the expected total switch-over time per cycle of the server along the queues. All interarrival times, service times and switch-over times are assumed to be mutually independent and independent of the state of the system. Necessary and sufficient condition for the stability of the system is $\rho < 1$ (cf. [4]).

Let W_i be the delay incurred by an arbitrary customer at Q_i, defined as the time between the arrival of a customer at a station and the moment at which it starts to receive service. Our main interest is in the behavior of $E[W_i]$. It will be shown that, for $i = 1, \ldots, N$,

$$E[W_i] = \frac{\omega_i}{1 - \rho} + o((1 - \rho)^{-1}), \quad \rho \uparrow 1. \tag{1}$$

where the limit is taken such that the arrival rates are increased, while keeping both the ratios between the arrival rates and the service-time distributions fixed. The main result of the paper is a closed-form expression for ω_i, in a general parameter setting (see section 5). Throughout, the following notation is used. For each variable x that is a function of ρ, we denote its values *evaluated at $\rho = 1$* by $\hat{x}$.

3   Pseudo-conservation Law

In this section we present a pseudo-conservation law (PCL) for the model described above. On the basis of the principle of work decomposition, Boxma and Groenendijk [1] show the following result: For $\rho < 1$,

$$\sum_{i=1}^{N} \rho_i E[W_i] = \rho \frac{\rho}{1-\rho} \frac{b^{(2)}}{2b^{(1)}} + \rho \frac{r^{(2)}}{2r} + \frac{r}{2(1-\rho)} \left[\rho^2 - \sum_{i=1}^{N} \rho_i^2 \right] + \sum_{i=1}^{N} E[M_i], \quad (2)$$

where M_i stands for the amount of work at Q_i at an arbitrary moment at which the server departs from Q_i. Then

$$M_i = M_i^{(1)} + M_i^{(2)}, \quad (3)$$

where $M_i^{(k)}$ is the amount of work at stage k at a server departure epoch from Q_i, $k = 1, 2$. Simple balancing arguments lead to the following expression for $E[M_i]$: For $i = 1, \ldots, N$,

$$E[M_i] = E\left[M_i^{(1)}\right] + E\left[M_i^{(2)}\right] = \rho_i^2 \frac{r}{1-\rho} + \rho_i \frac{r}{1-\rho}. \quad (4)$$

4   The Descendant Set Approach

For $i = 1$, define the two-dimensional random variable $\underline{X}_i := \left(X_i^{(1)}, X_i^{(2)} \right)$, where $X_i^{(k)}$ is the number of stage-k customers at Q_i at an arbitrary polling instant at Q_i when the system is in steady state ($k = 1, 2$), and denote the corresponding Probability Generating Function (PGF) by

$$X_i^*(z_1, z_2) := E\left[z_1^{X_i^{(1)}} z_2^{X_i^{(2)}} \right]. \quad (5)$$

Denoting the Laplace-Stieltjes Transform (LST) of the waiting-time distribution at Q_i by $W_i^*(\cdot)$, the waiting-time distribution at Q_i is related to the distribution of $\underline{X}_i$ through the following expressions (cf. [10]): For $Re\ s \geq 0$, $i = 1, \ldots, N$, $\rho < 1$,

$$W_i^*(s) = \frac{X_i^*(1 - s/\lambda_i, B_i^*(s)) - X_i^*(1 - s/\lambda_i, 1 - s/\lambda_i)}{E\left[X_i^{(2)}\right] (B_i^*(s) - 1 + s/\lambda_i)}. \quad (6)$$

Then it is easy to verify that $E[W_i]$ can be expressed in terms of the first two (cross-)moments of $\underline{X}_i$ as follows: for $i = 1, \ldots, N$, $\rho < 1$,

$$E[W_i] = \frac{1}{\lambda_i E\left[X_i^{(2)}\right]} \left(\frac{1 + \rho_i}{2} E\left[X_i^{(2)}(X_i^{(2)} - 1)\right] + E\left[X_i^{(1)} X_i^{(2)}\right] \right). \quad (7)$$

Note that the first part of (7) is similar to the formula for the mean delay in the model with one-stage gated service policy. For this model we have (see Takagi [11])

$$E[W_i^{(one-stage)}] = (1 + \rho_i) \frac{E[X_i(X_i - 1)]}{2\lambda_i E[X_i]}, \quad (8)$$

with X_i the steady-state number of customers at Q_i at an arbitrary polling instant at Q_i.

To derive $E[W_i]$, it is sufficient to obtain the first two factorial moments of $X_i^{(2)}$, and the cross-moments $E\left[X_i^{(1)}X_i^{(2)}\right]$. To this end, note first that straight-forward balancing arguments indicate that the first moments $E[X_i^{(k)}]$ $(k = 1, 2)$ can be expressed in following closed form:

$$E[X_i^{(1)}] = E[X_i^{(2)}] = \frac{\lambda_i r}{1 - \rho}. \tag{9}$$

However, in general the second-order moments can not be obtained explicitly. In the literature, there are several (numerical) techniques to obtain the moments of the delay. In this section we focus on the Descendant Set Approach (DSA), an iterative technique based on the concept of so-called descendant sets [5]. The use of the DSA for the present model is discussed below.

The customers in a polling system can be classified as originators and non-originators. An originator is a customer that arrives at the system during a switch-over period. A non-originator is a customer that arrives at the system during the service of another customer. For a customer C, define the children set to be the set of customers arriving during the service of C; the descendant set of C is recursively defined to consist of C, its children and the descendants of its children. The DSA is focused on the determination of the moments of the delay at a fixed queue, say Q_1. To this end, the DSA concentrates on the determination of the distribution of the two-dimensional stochastic vector $\underline{X}_1(P^*) := \left(X_1^{(1)}(P^*), X_1^{(2)}(P^*)\right)$, where $X_1^{(k)}(P^*)$ is defined as the number of stage-k customers at Q_1 present at an arbitrary fixed polling instant P^* at Q_1 $(k = 1, 2)$. P^* is referred to as the reference point at Q_1. The main ideas are the observations that (1) each of the customers present at Q_1 at the reference point P^* (either at stage-1 or stage-2) belongs to the descendant set of exactly one originator, and (2) the evolutions of the descendant sets of different originators are stochastically independent. Therefore, the DSA concentrates on an arbitrary tagged customer which arrived at Q_i in the past and on calculating the number of type-1 descendants it has at both stages at P^*. Summing up these numbers over all past originators yields $\underline{X}_1(P^*)$, and hence $\underline{X}_1$, because P^* is chosen arbitrarily.

The DSA considers the Markov process embedded at the polling instants of the system. To this end, we number the successive polling instants as follows. Let $P_{N,0}$ be the last polling instant at Q_N prior to P^*, and for $i = N-1, \ldots, 1$, let $P_{i,0}$ be recursively defined as the last polling instant at Q_i prior to $P_{i+1,0}$. In addition, for $c = 1, 2, \ldots$, we define $P_{i,c}$ to be the last polling instant at Q_i prior to $P_{i,c-1}$, $i = 1, \ldots, N$. The DSA is oriented towards the determination of the contribution to $\underline{X}_1(P^*)$ of an arbitrary customer present at Q_i at $P_{i,c}$. To this end, define an (i, c)-customer to be a customer present at Q_i at $P_{i,c}$. Moreover, for a tagged (i, c)-customer $T_{i,c}$ at stage 1, we define $\underline{A}_{i,c} := \left(A_{i,c}^{(1)}, A_{i,c}^{(2)}\right)$, where $A_{i,c}^{(k)}$ is the number of type-1 descendants it has at stage k at P^*, $k = 1, 2$. In this way, the two-dimensional random variable $\underline{A}_{i,c}$ can be viewed as the contribution of $T_{i,c}$ to $\underline{X}_1(P^*)$. Denote the joint PGF of $\underline{A}_{i,c}$ by

$$A_{i,c}^*(z_1, z_2) := E\left[z_1^{A_{i,c}^{(1)}} z_2^{A_{i,c}^{(2)}}\right].\tag{10}$$

To express the distribution of $\underline{X}_1$ in terms of the distributions of the DS variables $\underline{A}_{i,c}$, denote by $R_{i,c}$ the switch-over period from Q_i to Q_{i+1} immediately after the service period at Q_i starting at $P_{i,c}$. Moreover, denote $\underline{S}_{i,c} := \left(S_{i,c}^{(1)}, S_{i,c}^{(2)}\right)$, where $S_{i,c}^{(k)}$ is the total contribution to $X_1^{(k)}$ of all customers that arrive at the system during $R_{i,c}$ (note that, by definition, these customers are original customers), and denote the joint PGF of $\underline{S}_{i,c}$ by

$$S_{i,c}^*(z_1, z_2) := E\left[z_1^{S_{i,c}^{(1)}} z_2^{S_{i,c}^{(2)}}\right].\tag{11}$$

In this way, $\underline{S}_{i,c} = \left(S_{i,c}^{(1)}, S_{i,c}^{(2)}\right)$ can be seen as the joint contribution of $R_{i,c}$ to $\underline{X}_1$. It is readily verified that we can write

$$\underline{X}_1 = \left(X_1^{(1)}, X_1^{(2)}\right) = \sum_{i=1}^N \sum_{c=0}^\infty \left(S_{i,c}^{(1)}, S_{i,c}^{(2)}\right) = \sum_{i=1}^N \sum_{c=0}^\infty \underline{S}_{i,c}.\tag{12}$$

Note that $S_{i,c}^{(1)}$ and $S_{i',c'}^{(2)}$ are dependent if $(i, c) = (i', c')$ but independent otherwise. Hence we can write, for $|z_1|, |z_2| \le 1$,

$$X_1^*(z_1, z_2) = \prod_{i=1}^N \prod_{c=0}^\infty S_{i,c}^*(z_1, z_2).\tag{13}$$

Because $\underline{S}_{i,c}$ is the total joint contribution to $\underline{X}_1$ of all (original) customers that arrive during $R_{i,c}$, the joint distribution of $\underline{S}_{i,c}$ can be expressed in terms of the distributions of the DS-variables $\underline{A}_{i,c}$ as follows: For $i = 1, \ldots, N$, $c = 0, 1, \ldots$, and $|z_1|, |z_2| \le 1$,

$$S_{i,c}^*(z_1, z_2) = R_i^*\left(\sum_{j=i+1}^N [\lambda_j - \lambda_j A_{j,c}^*(z_1, z_2)] + \sum_{j=1}^i [\lambda_j - \lambda_j A_{j,c-1}^*(z_1, z_2)]\right).\tag{14}$$

To define a recursion for the evolution of the descendant set, note that a customer at stage-1 present at Q_1 at the polling instant at Q_1 during cycle c is served during the *next* cycle, which lead to the following relation: For $i = 1, \ldots, N$, $c = 0, 1, \ldots$, and $|z_1|, |z_2| \le 1$,

$$A_{i,c}^*(z_1, z_2) = B_i^*\left(\sum_{j=i+1}^N [\lambda_j - \lambda_j A_{j,c-1}^*(z_1, z_2)] + \sum_{j=1}^i [\lambda_j - \lambda_j A_{j,c-2}^*(z_1, z_2)]\right),\tag{15}$$

supplemented with the basis for the recursion

$$A_{i,-1}^*(z_1, z_2) = z_1\, I_{\{i=1\}}, \quad \text{and} \quad A_{i,-2}^*(z_1, z_2) = z_2\, I_{\{i=1\}},\tag{16}$$

where I_E is the indicator function of the event E. In this way, relations (13)-(16) give a complete, characterization of the distribution of $\underline{X}_1$. Similarly, recursive relations to calculate the (cross-)moments of $\underline{X}_1$ can be readily obtained from those equations.

5   Results

In this section we will present heavy-traffic results that can be proven by exploring the use of the DSA. For compactness of the presentation, the details of the proofs are omitted.

Theorem 1. *For $i = 1, \ldots, N$,*

$$\lim_{\rho \uparrow 1} (1-\rho)^2 E\left[X_i^{(2)}\left(X_i^{(2)} - 1\right)\right] = \lim_{\rho \uparrow 1} (1-\rho)^2 E\left[X_i^{(1)} X_i^{(2)}\right] = \hat{\lambda}_i^2 \left[r^2 + \frac{r}{\delta}\frac{b^{(2)}}{b^{(1)}}\right],$$

(17)

where

$$\delta := \sum_{i=1}^{N} \hat{\rho}_i (3 + \hat{\rho}_i).$$

(18)

Proof. The result can be obtained along the lines similar to the derivation of the results for the one-stage polling models in [14,15]. □

Theorem 2 (Main result). *For $i = 1, \ldots, N$,*

$$\omega_i = \frac{(3 + \hat{\rho}_i)}{\sum_{j=1}^{N} \hat{\rho}_j (3 + \hat{\rho}_j)} \frac{b^{(2)}}{2b^{(1)}} + \frac{r(3 + \hat{\rho}_i)}{2}.$$

(19)

Proof. The result follows directly by combining (7), (9) and Theorem 1. □

Theorem 2 reveals a variety of properties on the dependence of the limit of the scaled mean waiting times with respect to the system parameters.

Corollary 1 (Insensitivity). *For $i = 1, \ldots, N$,*

(1) ω_i *is independent of the visit order,*
(2) ω_i *depends on the switch-over time distributions only through r, i.e., the total expected switch-over time per cycle,*
(3) ω_i *depends on the second moments of the service-time distributions only through $b^{(2)}$, i.e., the second moment of the service time of an arbitrary customer.*

Corollary 1 is known to be not generally valid for stable systems (i.e., for $\rho < 1$), where the visit order, the second moments of the switch-over times and the individual second moments of the service-time distributions *do* have an impact on the mean waiting times. Hence, Corollary 1 shows that the influence of these parameters on the mean waiting times vanishes when the load tends to unity, and as such can be viewed as lower-order effects in heavy traffic.

Let us now discuss the trade-off between efficiency and fairness for the one-stage and two-stage gated service policies, using the exact asymptotic results presented in Theorem 2. To this end, denote by $\omega_i^{(one-stage)}$ and $\omega_i^{(two-stage)}$ the heavy-traffic residues of the mean waiting times for the case of one-stage and two-stage gated service at all queues, respectively, defined in (1). For the case of one-stage gated service at all queues, the following results holds (cf. [14]): For $i = 1, \ldots, N$,

$$\omega_i^{(one-stage)} = \frac{(1 + \hat{\rho}_i)}{\sum_{j=1}^N \hat{\rho}_j (1 + \hat{\rho}_j)} \frac{b^{(2)}}{2b^{(1)}} + \frac{r(1 + \hat{\rho}_i)}{2}. \tag{20}$$

Also, for the one-stage gated polling model the pseudo-conservatiom law is given by equation (2), supplemented with

$$E[M_i] = \rho_i^2 \frac{r}{1 - \rho}. \tag{21}$$

Denote by V the amount of waiting work in the system. Then using Little's Law and straightforward arguments it is readily verified that, for $\rho < 1$,

$$E[V] = \sum_{i=1}^N \rho_i E[W_i]. \tag{22}$$

Throughout, we will use $E[V]$ to be the measure of efficiency: for a given set of parameters, a combination of service policies (at each of the queues) is said to be more efficient than another combination of policies if the resulting value of $E[V]$ is smaller. Denoting by $V^{(one-stage)}$ and $V^{(two-stage)}$ the amount of work in the one-stage and two-stage gated model, respectively, the following result follows directly from (2)-(4), (21) and (22).

Corollary 2 (Efficiency). *Two-stage gated service is less efficient than one-stage gated service in the sense that*

$$E[V^{(one-stage)}] < E[V^{(two-stage)}]. \tag{23}$$

Definition of unfairness:
For a given polling model, the unfairness is defined as follows:

$$\mathcal{F} := \max_{i,j=1,\ldots,N} \left| \frac{E[W_i]}{E[W_j]} - 1 \right|. \tag{24}$$

Note that according to this definition, the higher $\mathcal{F}$, the *less* fair is the service policy. Note also that the symmetric systems are optimally fair, in the sense that $\mathcal{F} = 0$. The following result follows directly from Theorem 2 and (20).

Corollary 3 (Fairness). *Two-stage gated service is asymptotically more fair than one-stage gated service in the following sense: For $i, j = 1, \ldots, N$,*

$$\left| \frac{\omega_i^{(two-stage)}}{\omega_j^{(two-stage)}} - 1 \right| = \left| \frac{3 + \hat{\rho}_i}{3 + \hat{\rho}_j} - 1 \right| < \left| \frac{1 + \hat{\rho}_i}{1 + \hat{\rho}_j} - 1 \right| = \left| \frac{\omega_i^{(one-stage)}}{\omega_j^{(one-stage)}} - 1 \right|. \quad (25)$$

Corollaries 2 and 3 give asymptotic results on the relative fairness and efficiency between one-stage and two-stage gated service. To assess whether similar results also hold for stable systems (i.e., whith $\rho < 1$) we have performed extensive numerical validation. The results are outlined below. Consider the asymmetric model with the following system parameters: $N = 2$; the service times at both queues are deterministic with mean $b_1^{(1)} = 0.8$ and $b_2^{(1)} = 0.2$; the switch-over times are deterministic with $r_1^{(1)} = r_2^{(1)} = 1$, and the arrival rates at both queues are equal. For this model, we have calculated the expected waiting times at both queues and the unfairness measure (24), for different values of the load, both for one-stage and two-stage gated service at all queues. Table 1 below shows the results.

Table 1. Mean waiting times and fairness

	one-stage gated			two-stage gated		
ρ	$E[W_1]$	$E[W_2]$	$\mathcal{F}$	$E[W_1]$	$E[W_2]$	$\mathcal{F}$
0.50	3.159	2.465	0.28	7.158	6.468	0.11
0.60	4.241	3.186	0.33	9.239	8.196	0.13
0.70	6.045	4.386	0.38	12.705	11.080	0.15
0.80	9.653	6.788	0.42	19.633	16.868	0.16
0.90	20.475	13.998	0.46	40.400	34.299	0.18
0.95	42.119	28.424	0.48	81.919	69.225	0.18
0.98	107.048	71.708	0.49	206.456	174.075	0.19
0.99	215.262	143.851	0.50	413.997	348.837	0.19

Note that for the model analyzed in Table 1 we have $\hat{\rho}_1 = 4/5$, $\hat{\rho}_2 = 1/5$, so that it is readily seen that for the one-stage gated model $\mathcal{F}$ tends to $1/2$ as ρ goes to 1, whereas for two-stage gated $\mathcal{F}$ tends to $3/16 = 0.1875$ in the limiting case. The results shown in Table 1 show that the two-stage gated service is indeed more "fair" than the one-stage gated service for all values of the load, which suggests that Corollary 3 is also applicable to stable systems. We suspect that this type of results may be proven rigorously; however, a more detailed analysis of the relative fairness is beyond the scope of this paper.

6   Approximation

Equation (1) and Theorem 2 suggest the following approximation for $E[W_i]$ in stable polling systems: For $i = 1, \ldots, N$, $\rho < 1$,

$$E[W_i^{(app)}] := \frac{\omega_i}{1 - \rho}, \quad (26)$$

where ω_i is given by Theorem 2. To assess the accuracy of the approximation in (26), in terms of "How high should the load be for the approximation to be accurate?", we have performed numerical experiments to test the accuracy of the approximations for different values of the load of the system. The relative error of the approximation of $E[W_i]$ is defined as follows: For $i = 1, \ldots, N$,

$$\Delta\% := \text{abs} \left(\frac{E[W_i^{(app)}] - E[W_i]}{E[W_i]} \right) \times 100\%. \tag{27}$$

For the model considered in Table 1 above, Table 2 shows the exact (obtained via the DSA discussed in section 3) and approximated values (obtained via (26)) of $E[W_1]$ for different values of the load, both for the model in which all queues receive one-stage gated service, and for the model in which all queues receive two-stage gated service. The results in Table 2 demonstrate that the

Table 2. Exact and approximated values for $E[W_1]$

	one-stage gated			two-stage gated		
ρ	$E[W_1^{(app)}]$	$E[W_1]$	$\Delta\%$	$E[W_1^{(app)}]$	$E[W_1]$	$\Delta\%$
0.50	4.329	3.159	37.04	8.302	7.158	15.98
0.60	5.411	4.241	27.59	10.378	9.239	12.33
0.70	7.214	6.045	19.34	13.837	12.705	8.91
0.80	10.821	9.653	12.10	20.755	19.633	5.71
0.90	21.643	20.475	5.70	41.510	40.400	2.75
0.95	43.286	42.119	2.77	83.020	83.020	1.34
0.98	108.215	107.048	1.09	207.550	206.456	0.53
0.99	216.429	215.262	0.54	415.100	413.997	0.27

relative error of the approximations indeed tends to zero as the load tends to 1, as expected on the basis of Theorem 2. Moreover, the results show that the approximation converges to the limit rather quickly when $\rho \uparrow 1$. Roughly, the results are accurate when the load is 80% or more, which demonstrates the applicability of the asymptotic results for practical heavy-traffic scenarios.

Acknowledgment. The authors wish to thank Onno Boxma for interesting discussions on this topic.

References

1. Boxma, O.J., Groenendijk, W.P.: Pseudo-conservation laws in cyclic-service systems. J. Appl. Prob. 24, 949–964 (1987)
2. Coffman, E.G., Puhalskii, A.A., Reiman, M.I.: Polling systems with zero switchover times: a heavy-traffic principle. Ann. Appl. Prob. 5, 681–719 (1995)
3. Coffman, E.G., Puhalskii, A.A., Reiman, M.I.: Polling systems in heavy-traffic: a Bessel process limit. Math. Oper. Res. 23, 257–304 (1998)

4. Fricker, C., Jaïbi, M.R.: Monotonicity and stability of periodic polling models. Queueing Systems 15, 211–238 (1994)
5. Konheim, A.G., Levy, H., Srinivasan, M.M.: Descendant set: an efficient approach for the analysis of polling systems. IEEE Trans. Commun. 42, 1245–1253 (1994)
6. Kramer, G., Muckerjee, B., Pesavento, G.: Ethernet PON: design and analysis of an optical access network. Phot. Net. Commun. 3, 307–319 (2001)
7. Kramer, G., Muckerjee, B., Pesavento, G.: Interleaved polling with adaptive cycle time (IPACT): a dynamic bandwidth allocation scheme in an optical access network. Phot. Net. Commun. 4, 89–107 (2002)
8. Kramer, G., Muckerjee, B., Pesavento, G.: Supporting differentiated classes of services in Ethernet passive optical networks. J. Opt. Netw. 1, 280–290 (2002)
9. Levy, H., Sidi, M.: Polling models: applications, modeling and optimization. IEEE Trans. Commun. 38, 1750–1760 (1991)
10. Park, C.G., Han, D.H., Kim, B., Jung, H.-S.: Queueing analysis of symmetric polling algorithm for DBA schemes in an EPON. In: Choi, B.D., (ed.) Proceedings 1st Korea-Netherlands Joint Conference on Queueing Theory and its Applications to Telecommunuication Systems, Seoul, Korea, pp. 147–154 (2005)
11. Takagi, H.: Analysis of Polling Systems. MIT Press, Cambridge, MA (1986)
12. Takagi, H.: Queueing analysis of polling models: an update. In: Takagi, H., (ed.) Stochastic Analysis of Computer and Communication Systems, North-Holland, Amsterdam, pp. 267–318 (1990)
13. Takagi, H.: Queueing analysis of polling models: progress in 1990-1994. In: Dshalalow, J.H. (ed.) Frontiers in Queueing: Models, Methods and Problems, pp. 119–146. CRC Press, Boca Raton, FL (1997)
14. van der Mei, R.D., Levy, H.: Expected delay in polling systems in heavy traffic. Adv. Appl. Prob. 30, 586–602 (1998)
15. van der Mei, R.D.: Polling systems in heavy traffic: higher moments of the delay. Queueing Systems 31, 265–294 (1999)
16. van der Mei, R.D.: Delay in polling systems with large switch-over times. J. Appl. Prob. 36, 232–243 (1999)

On a Unifying Theory on Polling Models in Heavy Traffic

R.D. van der Mei

CWI, Probability and Stochastic Networks, Amsterdam, The Netherlands
Vrije Universiteit, Faculty of Sciences, Amsterdam, The Netherlands
mei@few.vu.nl

Abstract. For a broad class of polling models the evolution of the system at specific embedded polling instants is known to constitute multi-type branching process (MTBP) with immigration. In this paper it is shown that for this class of polling models the vector $\underline{X}$ that describes the state of the system at these polling instants satisfies the following heavy-traffic behavior, under mild assumptions:

$$(1 - \rho)\underline{X} \to_d \underline{\gamma}\, \Gamma(\alpha, \mu) \quad (\rho \uparrow 1), \tag{1}$$

where $\underline{\gamma}$ is a known vector, $\Gamma(\alpha, \mu)$ has a gamma-distribution with known parameters α and μ, and where ρ is the load of the system. This general and powerful result is shown to lead to exact - and in many cases even closed-form - expressions for the Laplace-Stieltjes Transform (LST) of the complete asymptotic queue-length and waiting-time distributions for a broad class of branching-type polling models that includes many well-studied polling models policies as special cases. The results generalize and unify many known results on the waiting times in polling systems in heavy traffic, and moreover, lead to new exact results for classical polling models that have not been observed before. As an illustration of the usefulness of the results, we derive new closed-form expressions for the LST of the waiting-time distributions for models with a cyclic globally-gated polling regime. As a by-product, our results lead to a number of asymptotic insensitivity properties, providing new fundamental insights in the behavior of polling models.

Keywords: polling models, queueing theory, heavy traffic, unfication.

1 Introduction

Polling systems are multi-queue systems in which a single server visits the queues in some order to serve the customers waiting at the queues, typically incurring some amount of switch-over time to proceed from one queue to the next. Polling models find a wide variety of applications in which processing power (e.g., CPU, bandwidth, manpower) is shared among different types of users. Typical application areas of polling models are computer-communication systems, logistics, flexible manufacturing systems, production systems and maintenance systems;

L. Mason, T. Drwiega, and J. Yan (Eds.): ITC 2007, LNCS 4516, pp. 556–567, 2007.

the reader is referred to [24,13] for extensive overviews of the applicability of polling models. Over the past few decades the performance analysis of polling models has received much attention in the literature. We refer to the classical surveys by [22,23], and to a recent survey paper by Vishnevskii and Semenova [34] for overviews of the available results on polling models. One of the most remarkable results is that there appears to be a striking difference in complexity between polling models. Resing [18] observed that for a large class of polling models, including for example cyclic polling models with Poisson arrivals and exhaustive and gated service at all queues, the evolution of the system at successive polling instants at a fixed queue can be described as a multi-type branching process (MTBP) with immigration. Models that satisfy this MTBP-structure allow for an exact analysis, whereas models that violate the MTBP-structure are often more intricate.

In this paper we study the heavy-traffic behavior for the class of polling models that have an MTBP-structure, in a general parameter setting. Initiated by the pioneering work of Coffman et al. [6,7], the analysis of the heavy-traffic behavior of polling models has gained a lot of interest over the past decade. This has led to the derivation of asymptotic expressions for key performance metrics, such as the moments and distributions of the waiting times and the queue lengths, for a variety of model variants, including for example models with mixtures of exhaustive and gated service policies with cyclic server routing [25], periodic server routing [31,32], simultaneous batch arrivals [28], continuous polling [11], amongst others. In this context, a remarkable observation is that in the heavy-traffic behavior of polling models a central role is played by the gamma-distribution, which occurs in the analysis of these different model variants as the limiting distribution of the (scaled) cycle times and the marginal queue-lengths at polling instants. This observation has motivated us to develop a unifying theory on the heavy-traffic behavior of polling models that includes all these model instances as special cases, where everything falls into place. We believe that the results presented in this paper are a significant step towards such a general unifying theory.

The motivation for studying heavy-traffic asymptotics in polling models is twofold. First, a particularly attractive feature of heavy-traffic asymptotics (i.e., when the load tends to 1) for MTBP-type models is that in many cases they lead to strikingly simple expressions for queue-length and waiting-time distributions, especially when compared to their counterparts for arbitrary values of the load, which usually leads to very cumbersome expressions, even for the first few moments (e.g., [12]). The remarkable simplicity of the heavy-traffic asymptotics provides fundamental insight in the impact of the system parameters on the performance of the system, and in many cases attractive insensitivity properties have been observed. A second motivation for considering heavy-traffic asymptotics is that the computation time needed to calculate the relevant performance metrics usually become prohibitively long when the system is close to saturation, both for branching-type [5] and non-branching-type polling models [3], which raises the need for simple and fast approximations. To this end, heavy-traffic asymptotics form an excellent basis for developing such approximations, and in fact, have been found to be remarkably accurate in several cases, even for moderate load [25,27,32].

To develop a unifying theory on the heavy-traffic behavior of branching-type polling models, it is interesting to observe that the theory of MTBPs, which was largely developed in the early 1970s, is well-matured and powerful [17,9]. Nonetheless, the theory of MTBPs has received remarkably little attention in the literature on polling models. In fact, throughout this paper we will show that the following result on MTBPs can be used as the basis for the development of a unifying theory on branching-type polling models under heavy-traffic assumptions: the joint probability distribution of the M-dimensional branching process $\{\underline{Z}_n, n = 0, 1, \ldots\}$ (with immigration in each state) converges in distribution to $\underline{v}\Gamma(\alpha, \mu)$ in the sense that Quine [17]:

$$\lim_{n \to \infty} \frac{1}{\pi_n(\xi)} \underline{Z}_n \to_d \underline{v}\Gamma(\alpha, \mu) \quad (\xi \uparrow 1), \tag{2}$$

where ξ is the maximum eigenvalue of the so-called mean matrix, $\pi_n(\xi)$ is a scaling function, $\underline{v}$ is a known M-dimensional vector and $\Gamma(\alpha, \mu)$ is a gamma-distributed random variable with known shape and scale parameters α and μ, respectively. We emphasize that (2) is valid for general MTBPs under very mild moment conditions (see Section 2 for details). In this paper, we show that this result (2) can be transformed into equation (1), providing an asymptotic analysis for a very general class of MTBP-type polling models. Subsequently, we show that equation (1) leads to exact asymptotic expressions for the scaled time-average queue-length and waiting-time distributions under heavy-traffic assumptions; for specific model instances, basically all we have to do is calculate the parameters $\underline{v}$, α and μ, and the derivative of ξ as a function of ρ at $\rho = 1$, which is usually straightforward. In this way, we propose a new and powerful approach to derive heavy-traffic asymptotics for polling model that have MTBP-structure. To demonstrate the usefulness of the results we use the approach developed in this paper to derive new and yet unknown closed-form expressions for the complete asymptotic waiting-time distributions for a number of classical polling models. To this end, we derive closed-form expressions for the asymptotic waiting-time distributions for cyclic polling models with the Globally-Gated (GG) service policy, and for models with general branching-type service policies. As a by-product, the results also lead to asymptotic insensitivity properties providing new fundamental insights in the behavior of polling models. Moreover, the results lead to simple approximatons for the waiting-time distributions in stable polling systems.

The remainder of this paper is organized as follows. In Section 2 we give a brief introduction on MTBPs and formulate the limiting result by Quine [17] (see Theorem 1) that will be used throughout. In Section 3 we translate this result to the context of polling models, and give an approach for how to obtain heavy-traffic asymptotics for MTBP-type polling models. To illustrate the usefulness of the approach, we derive closed-form expressions for the LST of the scaled asymptotic waiting-time distributions for cyclic models with GG service. The implications of these results are discussed extensively.

2 Multitype Branching Processes with Immigration

We consider a general M-dimensional multi-type branching process $\mathbf{Z} = \{\underline{Z}_n, n = 0, 1, \ldots\}$, where $\underline{Z}_n = (Z_n^{(1)}, \ldots, Z_n^{(M)})$ is an M-dimensional vector denoting the state of the process in the n-th generation, and where $Z_n^{(i)}$ is the number of type-i particles in the n-th generation, for $i = 1, \ldots, M$, $n = 0, 1, \ldots$. The process $\mathbf{Z}$ is completely characterized by (1) its one-step offspring function and (2) its immigration function, which are assumed mutually independent and to be stochastically the same for each generation. The one-step offspring function is denoted by $f(\underline{z}) = (f^{(1)}(\underline{z}), \ldots, f^{(M)}(\underline{z}))$, with $\underline{z} = (z_1, \ldots, z_M)$, and where for $|z_k| \leq 1$ $(k = 1, \ldots, M), i = 1, \ldots, M$,

$$f^{(i)}(\underline{z}) = \sum_{j_1, \ldots, j_M \geq 0} p^{(i)}(j_1, \ldots, j_M) z_1^{j_1} \cdots z_M^{j_M}, \tag{3}$$

where $p^{(i)}(j_1, \ldots, j_M)$ is the probability that a type-i particle produces j_k particles of type k $(k = 1, \ldots, M)$. The immigration function is denoted as follows: For $|z_k| \leq 1$ $(k = 1, \ldots, M)$,

$$g(\underline{z}) = \sum_{j_1, \ldots, j_M \geq 0} q(j_1, \ldots, j_M) z_1^{j_1} \cdots z_M^{j_M}, \tag{4}$$

where $q(j_1, \ldots, j_M)$ is the probability that a group of immigrant consists of j_k particles of type k $(k = 1, \ldots, M)$. Denote

$$\underline{g} := (g_1, \ldots, g_M), \quad \text{where } g_i := \frac{\partial g(\underline{z})}{\partial z_i}\Big|_{\underline{z}=\underline{1}}, \tag{5}$$

and where $\underline{1}$ is the M-vector where each component is equal to 1. A key role in the analysis will be played by the first and second-order derivatives of $f(\underline{z})$. The first-order derivatives are denoted by the mean matrix

$$\mathbf{M} = (m_{i,j}), \quad \text{with } m_{i,j} := \frac{\partial f^{(i)}(\underline{z})}{\partial z_j}\Big|_{\underline{z}=\underline{1}} \quad (i, j = 1, \ldots, M). \tag{6}$$

Thus, adopting the standard notion of "children", for a given type-i particle in the n-th generation, $m_{i,j}$ is the mean number of type-j children it has in the $(n+1)$-st generation. Similarly, for a type-i particle, the second-order derivatives are denoted by the matrix

$$\mathbf{K}^{(i)} = \left(k_{j,k}^{(i)}\right), \quad \text{with } k_{j,k}^{(i)} := \frac{\partial^2 f^{(i)}(\underline{z})}{\partial z_j \partial z_k}\Big|_{\underline{z}=\underline{1}}, \quad i, j, k = 1, \ldots, M. \tag{7}$$

Denote by $\underline{v} = (v_1, \ldots, v_M)$ and $\underline{w} = (w_1, \ldots, w_M)$ the left and right eigenvectors corresponding to the largest real-valued, positive eigenvalue ξ of $\mathbf{M}$, commonly referred to as the maximum eigenvalue [2], normalized such that

$$\underline{v}^\top \underline{1} = \underline{v}^\top \underline{w} = 1. \tag{8}$$

The following conditions are necessary and sufficient conditions for the ergodicity of the process $\mathbf{Z}$ (cf. [18]): $\xi < 1$ and

$$\sum_{j_1+\cdots+j_M>0} q(j_1,\ldots,j_M)\log(j_1+\cdots+j_M) < \infty. \tag{9}$$

Throughout the following definitions are convenient. For any variable x that depends on ξ we use the hat-notation $\hat{x}$ to indicate that x is evaluated at $\xi = 1$. Moreover, for $\xi \geq 0$ let

$$\pi_0(\xi) := 0, \quad \text{and} \quad \pi_n(\xi) := \sum_{r=1}^{n} \xi^{r-2}, \quad n = 1, 2, \ldots. \tag{10}$$

A non-negative continuous random variable $\Gamma(\alpha, \mu)$ is said to have a gamma-distribution with shape parameter $\alpha > 0$ and scale parameter $\mu > 0$ if it has the probability density function

$$f_\Gamma(x) = \frac{1}{\Gamma(\alpha)} x^{\alpha-1} e^{-\mu x} \quad (x > 0) \quad \text{with} \quad \Gamma(\alpha) := \int_{t=0}^{\infty} t^{\alpha-1} e^{-t} dt, \tag{11}$$

and Laplace-Stieltjes Transform (LST)

$$\Gamma^*(s) = \left(\frac{\mu}{\mu + s}\right)^\alpha \quad (Re(s) > 0). \tag{12}$$

Note that in the definition of the gamma-distribution μ is a scaling parameter, and that $\Gamma(\alpha, \mu)$ has the same distribution as $\mu^{-1}\Gamma(\alpha, 1)$. Using these definitions, the following result was shown in [17]:

Theorem 1
Assume that all derivatives of $f(\underline{z})$ through order two exist at $\underline{z} = \underline{1}$ and that $0 < g_i < \infty$ $(i = 1, \ldots, M)$. Then

$$\lim_{n\to\infty} \frac{1}{\pi_n(\xi)} \begin{pmatrix} Z_n^{(1)} \\ \vdots \\ Z_n^{(M)} \end{pmatrix} \to_d A \begin{pmatrix} \hat{v}_1 \\ \vdots \\ \hat{v}_M \end{pmatrix} \Gamma(\alpha, 1) \quad (\xi \uparrow 1) \tag{13}$$

where $\underline{\hat{v}} = (\hat{v}_1, \ldots, \hat{v}_M)$ is the normalized the left eigenvector of $\hat{\mathbf{M}}$, and where $\Gamma(\alpha, 1)$ is a gamma-distributed random variable with scale parameter 1 and shape parameter

$$\alpha := \frac{1}{A}\underline{\hat{g}}^\top \underline{\hat{w}} = \frac{1}{A}\sum_{i=1}^{M} \hat{g}_i \hat{w}_i, \quad \text{with} \quad A := \sum_{i=1}^{M} \hat{v}_i \left(\underline{\hat{w}}^\top \hat{\mathbf{K}}^{(i)} \underline{\hat{w}}\right) > 0. \tag{14}$$

3   Heavy-Traffic Asymptotics for Polling Models

In this section we show how Theorem 1 can be transformed to derive new closed-form expressions for the LST of the queue-length and waiting-time distributions for a broad class of polling models, under heavy-traffic scalings. As an illustration, we consider a cyclic polling model with globally-gated (GG) service, in a general parameter setting. For this model, we show how Theorem 1 can be used to obtain derive the LST of the asymptotic waiting-time distribution at each of the queues. In Section 3.1 we discribe the GG-model, in Section 3.2 we derive HT-asymptotics for the waiting-time distributions at each of the queues and in Section 3.3 we extensively discuss the implications of the results.

3.1   Model

Consider an asymmetric cyclic polling model that consists of $N \geq 2$ queues, $Q_1, \ldots, Q_N$, and a single server that visits the queues in cyclic order. Customers arrive at Q_i accoring to a Poisson process with rate λ_i, and are referred to as type-i customers. The total arrival rate is $\Lambda := \sum_{i=1}^{N} \lambda_i$. The service time of a type-i customer is a random variable B_i, with LST $B_i^*(\cdot)$ and k-th moment $b_i^{(k)}$, which is assumed to be finite for $k = 1, 2$. The k-th moment of the service time of an arbitrary customer is $b^{(k)} := \sum_{i=1}^{N} \lambda_i b_i^{(k)} / \Lambda$ $(k = 1, 2)$. The total load of the system is $\rho := \sum_{i=1}^{N} \rho_i$. We define a polling instant at Q_i to be the moment at which the server arrives at Q_i, and a departure epoch at Q_i a moment at which the server depart from Q_i. The visit time at Q_i is defined as the time elapsed between a polling instant and its successive departure epoch at Q_i. Moreover, as i-cycle is the time between two successive polling instants at Q_i. The GG service discipline works as follows (cf. [4]). At the beginning of a 1-cycle, marked by a polling instant at Q_1 (see above), all customers present at $Q_1, \ldots, Q_N$ are marked. During the coming 1-cycle (i.e., the visit of queues $Q_1, \ldots, Q_N$), the server serves all (and only) the marked customers. Customers that meanwhile arrive at the queues will have to wait until being marked at the next cycle-beginning, and will be served during the next 1-cycle. Since at each cycle the server serves all the work that arrived during the previous cycle, the stability condition is $\rho < 1$, which is both necessary and sufficient (cf. [8,4]). Throughout this paper, this model will be referred to as the GG-model. Upon departing from Q_i the server immediately proceeds to Q_{i+1}, incurring a switch-over time R_i with LST $R_i^*(\cdot)$ and first two moments $r_i^{(k)}$ $(k = 1, 2)$, which are assumed to be finite. Denote by $r > 0$ and $r^{(2)} > 0$ be the first two moments of the switch-over time per 1-cycle of the server along the queues. The interarrival times, service times and switch-over times are assumed to be mutually independent and independent of the state of the system.

Throughout, we focus on the behavior of the model when the load ρ tends to 1. For ease of the discussion we assume that as ρ changes the total arrival rate changes while the service-time distributions and ratios between the arrival rates are kept fixed; note that in this way, the limit for $\rho \uparrow 1$, which will be used frequently throughout this paper, is uniquely defined. Similar to the hat-notation for the MTBPs defined in Section 2, for each variable x that is a function of ρ we use the hat-notation $\hat{x}$ to indicate its value *at* $\rho = 1$.

For GG-model model the joint queue-length vector at successive moments when the server arrives at a fixed queue (say Q_k) consitutes an MTBPs with immigration. To this end, the following notation is useful. Let $X_{i,n}^{(k)}$ be the number of type-i customers in the system at the n-th polling instant at Q_k, for $i, k = 1, \ldots, N$ and $n = 0, 1, \ldots$, and let $\underline{X}_n^{(k)} = (X_{1,n}^{(k)}, \ldots, X_{N,n}^{(k)})$ be the joint queue-length vector at the n-th pollling instant at Q_k. Moreover, $\mathbf{X}^{(k)} = \{\underline{X}_n^{(k)}, n = 0, 1, \ldots\}$ is the MTBP describing the evolution of the state of the system at successive polling instants at Q_k. For $\rho < 1$, we have $\underline{X}_n^{(k)} \rightarrow_d X^{(k)}$ for $n \rightarrow \infty$, where $X^{(k)}$ denotes the steady state joint queue-length vector at an arbitrary polling instant at Q_k.

3.2   Analysis

To analyze the HT-behavior of the GG-model, we proceed along a number of steps. First, we establish the relation between the GG-model and the general MTBP-model described in Section 2. Then, we use Theorem 1 to obtain HT-limits for the joint queue-length vector at polling instants at a fixed queue (Theorem 2). Finally, this result is transformed into an expression for the asymptotic scaled waiting-time distribution at an arbitrary queue (Theorem 3). For compactness of the presentation, the proofs of the various results are omitted.

To start, we consider the MTBP $\mathbf{X}^{(1)} := \{\underline{X}_n^{(1)}, n = 0, 1, \ldots\}$ describing the evolution of the joint queue-length vector at successive polling instants of the server at Q_1. Then the process $\mathbf{X}^{(1)}$ is characterized by the offspring generating functions, for $i = 1, \ldots, N$,

$$f^{(i)}(z_1, \ldots, z_N) = B_i^* \left(\sum_{j=1}^N \lambda_j (1 - z_j) \right) \tag{15}$$

and the immigration function

$$g(z_1, \ldots, z_N) = \prod_{i=1}^N R_i^* \left(\sum_{j=1}^N \lambda_j (1 - z_j) \right). \tag{16}$$

Note that it follows directly from (16) that, for $j = 1, \ldots, N$,

$$g_j = \sum_{i=1}^N r_i \lambda_j = r \lambda_j. \tag{17}$$

To derive the limiting distribution of the joint queue-length vector at polling instants at Q_1, we need to specify the following parameters: (a) the mean matrix $\mathbf{M}$ and its corresponding left and right eigenvectors $\hat{\underline{v}}$ and $\hat{\underline{w}}$ at $\rho = 1$ (normalized according to (8)), and (b) the parameters A and $\hat{g}$. These parameters are obtained in the following two lemmas.

Lemma 1
For the GG-model, the mean matrix $\hat{\mathbf{M}}$ is given by the following expression:

$$\hat{\mathbf{M}} = \begin{pmatrix} b_1^{(1)}\hat{\lambda}_1 & b_1^{(1)}\hat{\lambda}_2 & \cdots & b_1^{(1)}\hat{\lambda}_N \\ b_2^{(1)}\hat{\lambda}_1 & \cdots & \cdots & b_2^{(1)}\hat{\lambda}_N \\ \vdots & \vdots & \vdots & \vdots \\ b_N^{(1)}\hat{\lambda}_1 & \cdots & \cdots & b_N^{(1)}\hat{\lambda}_N \end{pmatrix}. \tag{18}$$

Moreover, the right and left eigenvectors of $\hat{\mathbf{M}}$ are

$$\underline{\hat{w}} = |\underline{b}|^{-1}\begin{pmatrix} b_1^{(1)} \\ b_2^{(1)} \\ \vdots \\ b_N^{(1)} \end{pmatrix}, \quad and \quad \underline{\hat{v}} = |\underline{b}|\begin{pmatrix} \hat{\lambda}_1 \\ \hat{\lambda}_2 \\ \vdots \\ \hat{\lambda}_N \end{pmatrix}, \quad respectively, \tag{19}$$

with

$$\underline{b} := (b_1^{(1)}, \ldots, b_N^{(1)})^\top, \quad and \quad |\underline{b}| := \sum_{i=1}^N b_i^{(1)}. \tag{20}$$

Lemma 2
For the GG-model, we have

$$\hat{\underline{g}}^\top \underline{\hat{w}} = |\underline{b}|^{-1}r, \quad and \quad A = |\underline{b}|^{-1}\frac{b^{(2)}}{b^{(1)}}. \tag{21}$$

Let us consider the heavy-traffic behavior of the maximum eigenvalue ξ of $\mathbf{M}$. Note that in general, ξ is a non-negative real-valued function of ρ (cf. [2]), say

$$\xi = \xi(\rho), \tag{22}$$

for $\rho \geq 0$. Then the following result describes the behavior of $\xi(\cdot)$ in the neighbourhood of $\rho = 1$.

Lemma 3
For the GG-model, the maximum eigenvalue $\xi = \xi(\rho)$ has the following properties:

(1) $\xi < 1$ *if and only if* $0 \leq \rho < 1$, $\xi = 1$ *if and only if* $\rho = 1$, *and* $\xi > 1$ *if and only if* $\rho > 1$;
(2) $\xi = \xi(\rho)$ *is a continuous function of* ρ;
(3) $\lim_{\rho\uparrow 1} \xi(\rho) = f(1) = 1$;
(4) *the derivative of* $\xi(\cdot)$ *at* $\rho = 1$ *is given by*

$$\xi'(1) := \lim_{\rho\uparrow 1} \frac{1 - \xi(\rho)}{1 - \rho} = 1. \tag{23}$$

The following result transform Theorem 1 into an HT-result for the GG-model under consideration.

Theorem 2
For the GG-model, the steady-state joint queue-length distribution at polling instants at Q_k $(k = 1, \ldots, N)$ satisfies the following limiting behavior:

$$(1-\rho)\begin{pmatrix} X_1^{(k)} \\ \vdots \\ X_N^{(k)} \end{pmatrix} \to_d \frac{b^{(2)}}{b^{(1)}} \left[(\hat\rho_1 + \cdots + \hat\rho_{k-1}) \begin{pmatrix} \hat\lambda_1 \\ \vdots \\ \hat\lambda_{k-1} \\ \hat\lambda_k \\ \vdots \\ \hat\lambda_N \end{pmatrix} + \begin{pmatrix} 0 \\ \vdots \\ 0 \\ \hat\lambda_k \\ \vdots \\ \hat\lambda_N \end{pmatrix} \right] \Gamma(\alpha, 1) \quad (\rho \uparrow 1),$$

$$\tag{24}$$

where

$$\alpha = r \frac{b^{(1)}}{b^{(2)}}. \tag{25}$$

We are now ready to present the main result for the GG-model.

Theorem 3
For the GG-model, the waiting-time distribution satisfies the following limiting behavior: For $i = 1, \ldots, N$,

$$(1 - \rho)W_i \to_d \tilde{W}_i \quad (\rho \uparrow 1) \tag{26}$$

where the LST of $\tilde{W}_i$ is given by, for $Re(s) > 0$,

$$\tilde{W}_i^*(s) = \frac{1}{(1-\hat\rho_i)rs}\left\{ \left(\frac{\mu}{\mu + s(\hat\rho_1 + \cdots + \hat\rho_i)}\right)^\alpha - \left(\frac{\mu}{\mu + s(1 + \hat\rho_1 + \cdots + \hat\rho_{i-1})}\right)^\alpha \right\},$$

where

$$\alpha = r\frac{b^{(1)}}{b^{(2)}}, \ and \ \mu = \frac{b^{(1)}}{b^{(2)}}. \tag{27}$$

3.3   Discussion

Theorem 3 leads to a number of interesting implications that will be discussed below.

Corrollary 1 (Insensitivity properties)
For $i = 1, \ldots, N$, the asymptotic waiting-time distribution $\tilde{W}_i$,

(1) *is independent of the visit order (assuming the order is cyclic),*
(2) *depends on the variability of the service-time distributions only through $b^{(2)}$,*
(3) *depends on the switch-over time distributions only through r.*

Note that similar insensitivity properties are generally not valid for stable systems (i.e., $\rho < 1$), in which case the waiting-time distributions *do* depend on the visit order, the complete service-time distributions and each of the individual switch-over time distributions. Apparently, these dependencies are of lower order, and hence their effect on the waiting-time distributions becomes negligible, in heavy traffic.

We end this session with a number of remarks.

Remark 1 (Model extensions): The results presented for the GG-model described in Section 3.1 mainly serve as an illustration, and can be readily extended to a broader set of models. The requirements for the derivation of heavy-traffic limits similar to Theorems 2 and 3 are that (a) the evolution of the system at specific moments can be described as a multi-dimensional branching process with immigration, and (b) that the system is work conserving. In addition to the models addressed above, this class of models includes as special cases for example models with gated/exhaustive service and non-cyclic periodic server routing [31], models with (simultaneous) batch arrivals [28,14], continuous polling models [11], models with customer routing [20], globally-gated models with elevator-type routing [1], models with local priorities [19], amongst many other model variants. Basically, all that needs to be done for each of these model variants is to determine the parameters α, $\hat{\underline{u}}$ and the derivative of $\xi = \xi(\rho)$ at $\rho = 1$, which is usually straightforward.

Remark 2 (Assumptions on the finiteness of moments): Theorems 2 and 3 are valid under the assumption that the second moments of the service times and the first moments of the switch-over times are finite; these assumptions are an immediate consequence of the assumptions on the finiteness of the mean immigration function g and the second-order derivatives of the offspring function $K_{j,k}^{(i)}$, defined in (5) and (7), respectively. It is interesting to observe that the results obtained in by Van der Mei [25] via the use of the Descendant Set Approach (DSA) assumes the finiteness of *all* moments of the service times and switch-over times; these assumptions were required, since the DSA-based proofs in [25] are based on a *bottom-up* approach in the sense that the limiting results for the waiting-time distributions are obtained from the asymptotic expressions for the moments of the waiting times obtained in [27,26]. Note that in this way the DSA-based approach differs fundamentally from the *top-down* approach taken in the present paper, where the asymptotic expressions for the moments can be obtained from the expressions for the asymptotic waiting-time distributions in Theorem 3.

Remark 3 (Approximations): The results presented in Theorem 3 suggest the following simple approximations for the waiting-time distributions for stable systems: For $\rho < 1$, $i = 1, \ldots, N$,

$$\Pr\{W_i < x\} \approx \Pr\{\tilde{W}_i < x(1 - \rho)\}, \tag{28}$$

and similarly for the moments: for $\rho < 1$, $i = 1, \ldots, N$, $k = 1, 2 \ldots$,

$$E[W_i^k] \approx \frac{E[\tilde{W}_i^k]}{(1 - \rho)^k}, \tag{29}$$

where closed-form expressions for $E[\tilde{W}_i^k]$ can be directly obtained from Theorem 3 by k-fold differentiation. Extensive validation of these appoximations fall beyond the scope of this paper. We refer to [25,29,30] for extensive discussions about the accuracy of these approximations for the special case of exhaustive and gated service.

References

1. Altman, E., Khamisy, A., Yechiali, U.: On elevator polling with globally gated regime. Queueing Systems 11, 85–90 (1992)
2. Athreya, K.B., Ney, P.E.: Branching Processes. Springer, Berlin (1972)
3. Blanc, J.P.C.: Performance evaluation of polling systems by means of the power-series algorithm. Ann. Oper. Res. 35, 155–186 (1992)
4. Boxma, O.J., Levy, H., Yechiali, U.: Cyclic reservation schemes for efficient operation of multiple-queue single-server systems. Ann. Oper. Res. 35, 187–208 (1992)
5. Choudhury, G., Whitt, W.: Computing transient and steady state distributions in polling models by numerical transform inversion. Perf. Eval. 25, 267–292 (1996)
6. Coffman, E.G., Puhalskii, A.A., Reiman, M.I.: Polling systems with zero switchover times: a heavy-traffic principle. Ann. Appl. Prob. 5, 681–719 (1995)
7. Coffman, E.G., Puhalskii, A.A., Reiman, M.I.: Polling systems in heavy-traffic: a Bessel process limit. Math. Oper. Res. 23, 257–304 (1998)
8. Fricker, C., Jaïbi, M.R.: Monotonicity and stability of periodic polling models. Queueing Systems 15, 211–238 (1994)
9. Joffe, A., Spitzer, F.: On multitype branching processes with $\rho \leq 1$. Math. Anal. Appl. 19, 409–430 (1967)
10. Konheim, A.G., Levy, H., Srinivasan, M.M.: Descendant set: an efficient approach for the analysis of polling systems. IEEE Trans. Commun. 42, 1245–1253 (1994)
11. Kroese, D.P.: Heavy traffic analysis for continuous polling models. J. Appl. Prob. 34, 720–732 (1997)
12. Kudoh, S., Takagi, H., Hashida, O.: Second moments of the waiting time in symmetric polling systems. J. Oper. Res. Soc. of Japan 43, 306–316 (2000)
13. Levy, H., Sidi, M.: Polling models: applications, modeling and optimization. IEEE Trans. Commun. 38, 1750–1760 (1991)
14. Levy, H., Sidi, M.: Polling systems with simultaneous arrivals. IEEE Trans. Commun. 39, 823–827 (1991)
15. Morris, R.J.T., Wang, Y.T.: Some results for multi-queue systems with multiple cyclic servers. In: Bux, W., Rudin, H. (eds.) Performance of Computer Communication Systems, North-Holland, Amsterdam, pp. 245–258 (1984)
16. Park, C.G., Han, D.H., Kim, B., Jun, H.-S.: Queueing analysis of symmetric polling algorithm for DBA scheme in an EPON. In: Choi, B.D. (ed.) Proc. Korea-Netherlands joint conference on Queueing Theory and its Applications to Telecommunication Systems, Seoul, pp. 147–154 (June 22-25, 2005)
17. Quine, M.P.: The multitype Galton-Watson process with ρ near 1. Adv. Appl. Prob. 4, 429–452 (1972)

18. Resing, J.A.C.: Polling systems and multitype branching processes. Queueing Systems 13, 409–426 (1993)
19. Shimogawa, S., Takahashi, Y.: A note on the conservation law for a multi-queue with local priority. Queueing Systems 11, 145–151 (1992)
20. Sidi, M., Levy, H.: Customer routing in polling systems. In: King, P.J.B., Mitrani, I., Pooley, R.B., (eds.) Proc. Performance '90, North-Holland, Amsterdam, pp. 319–331(1990)
21. Takagi, H.: Analysis of Polling Systems. MIT Press, Cambridge, MA (1986)
22. Takagi, H.: Queueing analysis of polling models: an update. In: Takagi, H., (ed.) Stochastic Analysis of Computer and Communication Systems, North-Holland, Amsterdam, pp. 267–318 (1990)
23. Takagi, H.: Queueing analysis of polling models: progress in 1990-1994. In: Dshalalow, J.H. (ed.) Frontiers in Queueing: Models and Applications in Science and Technology, pp. 119–146. CRC Press, Boca Raton, FL (1997)
24. Takagi, H.: Application of polling models to computer networks. Comp. Netw. ISDN Syst. 22, 193–211 (1991)
25. Van der Mei, R.D.: Distribution of the delay in polling systems in heavy traffic. Perf. Eval. 31, 163–182 (1999)
26. Van der Mei, R.D.: Polling systems in heavy traffic: higher moments of the delay. Queueing Systems 31, 265–294 (1999)
27. Van der Mei, R.D.: Polling systems with switch-over times under heavy load: moments of the delay. Queueing Systems 36, 381–404 (2000)
28. Van der Mei, R.D.: Waiting-time distributions in polling systems with simultaneous batch arrivals. Ann. Oper. Res. 113, 157–173 (2002)
29. Van der Mei, R.D., Levy, H.: Expected delay analysis in polling systems in heavy traffic. J. Appl. Prob. 30, 586–602 (1998)
30. Van der Mei, R.D., Levy, H.: Polling systems in heavy traffic: exhaustiveness of service policies. Queueing Systems 27, 227–250 (1997)
31. Olsen, T.L., Van der Mei, R.D.: Periodic polling systems in heavy-traffic: distribution of the delay. J. Appl. Prob. 40, 305–326 (2003)
32. Olsen, T.L., Van der Mei, R.D.: Periodic polling systems in heavy-traffic: renewal arrivals. Oper. Res. Lett. 33, 17–25 (2005)
33. Vatutin, V.A., Dyakonova, E.E.: Multitype branching processes ans some queueing systems. J. of Math. Sciences 111, 3901–3909 (2002)
34. Vishnevskii, V.M., Semenova, O.V.: Mathematical methods to study the polling systems. Automation and Remote Control 67, 173–220 (2006)

Performance Analysis of a Fluid Queue with Random Service Rate in Discrete Time

Onno J. Boxma[1], Vinod Sharma[2], and D.K. Prasad[2]

[1] Dept of Mathematics and Computer Science and EURANDOM
Eindhoven University of Technology
P.O. Box 513, 5600 MB Eindhoven
The Netherlands
boxma@win.tue.nl
[2] Dept Elect. Comm. Engg., Indian Istitute of Science,
Bangalore, 560012, India
{vinod, dkp}@ece.iisc.ernet.in

Abstract. We consider a fluid queue in discrete time with random service rate. Such a queue has been used in several recent studies on wireless networks where the packets can be arbitrarily fragmented. We provide conditions on finiteness of moments of stationary delay, its Laplace-Stieltjes transform and various approximations under heavy traffic. Results are extended to the case where the wireless link can transmit in only a few slots during a frame.

Keywords: Fluid queue, discrete-time queue, wireless link, moments of delay, heavy-traffic approximations.

1 Introduction

We consider a discrete-time queue served by a wireless link. The input to the queue is a stochastic fluid. The link rate of the wireless link changes randomly with time. We will assume that the link rate stays constant during a slot. Such a queue has been used to model wireless systems in several previous studies ([9], [18], [19], [22]). Although in practical wireless systems the input arrives at a wireless link as packets, due to varying link rate, the packets need to be fragmented (for efficient utilization of the link) or several packets may be transmitted within a slot, as the case may be. Thus, the packets loose their identity (from the point of view of service at the queue) and it may be convenient to consider all the contents in a queue as a fluid. As is usually done, we will ignore the overheads due to packet fragmentation.

Discrete-time queues, where the identity of packets is retained (i.e., the packets are not fragmented) have been studied extensively ([5], [8], [27]). The discrete-time queue that we study has two differences from the discrete queues studied in the above literature. The packets in our queues can be fragmented and thus packet boundaries have no relevance, and the number of bits that can be served during a slot is variable (due to wireless link). The discrete-time fluid queue we

L. Mason, T. Drwiega, and J. Yan (Eds.): ITC 2007, LNCS 4516, pp. 568–581, 2007.

study has many similarities to the queues usually studied in literature, like the GI/GI/1 queue in continuous time ([2], [7]) and fluid queues in continuous time ([1], [4], [13]), but there are also significant differences. Fortunately, the similarities between our model and continuous and discrete queues studied previously are so strong that we can borrow significantly from previous studies. Interestingly the similarities are stronger with a continuous time queue than with a discrete queue.

Although, as pointed out above, the queue we study has been considered in wireless literature, it has not been studied extensively. We provide conditions for stability, finiteness of moments of queue length and stationary delay, various approximations under heavy traffic and the exact Laplace-Stieltjes transform (LST) of the delay for our queue.

We also consider a generalization of the queue which is useful for practical wireless systems. Often, a wireless link is not available at all slots, i.e., the time axis is divided into frames made of several slots and only in some of the slots the link can transmit data ([23]). For example, this happens in uplink and downlink of cellular systems using Time Division Multiple Access (TDMA) and in multihop wireless networks ([12], [17]) due to interference in neighbouring links. The performance of such a link will certainly be different from the queue considered so far. Such a system in wireline context (e.g., when the service rate is fixed) has been studied in [5], [24], [25] (see also the references there-in). We will extend most of our results to this queue.

We are currently working on feed-forward networks of the queues studied in this paper.

The paper is organized as follows. Section 2 presents the model and the preliminaries. Section 3 studies stationary delay. It provides conditions for the finiteness of moments of stationary delay, its LST and various approximations under heavy traffic. Section 4 considers the discrete queue embedded in a frame and extends most of the results provided in Section 3. Section 5 provides simulation results to show the closeness of approximations.

Due to lack of space, most of the proofs in this paper are omitted; they are available in ([6], [21]).

2   The Model and Preliminaries

We consider a discrete-time queue with infinite buffer. At time k, the queue length is q_k, the new arrivals are X_k and the link (service) rate is r_k during the time slot $(k, k+1)$. We will denote by X and r r.v.s (random variables) with the distribution of X_1 and r_1 respectively. Then

$$q_{k+1} = (q_k + X_k - r_k)^+ \tag{1}$$

where $(x)^+$ denotes $\max(0, x)$. We will assume $\{X_k, k \geq 0\}$ and $\{r_k, k \geq 0\}$ are iid (independent, identically distributed) and independent of each other. However for the stability results in the paper we will only assume that $\{X_k\}$ and $\{r_k\}$ are stationary, ergodic sequences.

Sometimes one can assume X_k and r_k to be nonnegative integers (bits). But the granularity of X_k and r_k can often be fine enough such that these can be taken nonnegative real valued. In the following we will do that.

Equation (1) is the well studied Lindley equation ([2], [7]) and $\{q_k\}$ in (1) corresponds to the waiting time process in a G/G/1 queue. If $\mathbb{E}[X] < \mathbb{E}[r] < \infty$ (this assumption will be made in the rest of the paper) and $\{X_k, r_k\}$ is strictly stationary and ergodic, there is a unique stationary distribution of q_k. Let q be a r.v. with this stationary distribution. If the queue starts at $k = 0$ with any initial distribution, q_k converges to q in total variation.

From now on we will make the above mentioned independence assumptions on $\{X_k\}$ and $\{r_k\}$. Then ([11], [26], [28]), $\mathbb{E}[q^{\alpha-1}] < \infty$ if and only if $\mathbb{E}[X^\alpha] < \infty$ for $\alpha \geq 1$ and if X has finite moment generating function (mgf) in a neighbourhood of 0 then so does q.

The epochs when $q_k = 0$, are the regeneration epochs for the process $\{q_k, k \geq 0\}$. Let τ be a regeneration length (it corresponds to the number of customers served in a busy period in a GI/GI/1 queue). Then ([10], [26], [28]) $\mathbb{E}[\tau^\alpha] < \infty$ for $\alpha \geq 1$, if and only if $\mathbb{E}[X^\alpha] < \infty$. Also, τ has a finite mgf in a neighbourhood of 0 if and only if X has. This provides rates for the convergence of the distribution of q_k to that of q. Combined with results on $\mathbb{E}[q^\alpha] < \infty$, one can obtain various functional limit theorems ([26]).

For a queueing system the stationary delay distribution is a key performance measure. We use the results provided above to study the delay distribution in the next section.

3   Delay Distribution

In this section we study the delay distribution for the system described in Section 2. We will study delay for the FCFS (First Come First Served) discipline.

For the fluid X_k arriving at time k, the first time its contents are served (i.e., the delay of the first bit) is

$$D_k = \inf\{n : r_k + r_{k+1} + \ldots + r_{k+n-1} > q_k\}. \tag{2}$$

The last bit of X_k will wait for

$$\bar{D}_k = \inf\{n : r_k + r_{k+1} + \ldots + r_{k+n-1} \geq q_k + X_k\}. \tag{3}$$

Since the bits served in a slot may belong to packets corresponding to different flows in a wireless system, to ensure QoS (Quality of Service) it is important to study both $\{D_k\}$ and $\{\bar{D}_k\}$ and not just the delay of some average (typical) bit in $\{X_k\}$. Observe that τ is also a regeneration length for $\{D_k\}$ and $\{\bar{D}_k\}$. Thus, if $\mathbb{E}[X_k] < \mathbb{E}[r_k]$ then $\mathbb{E}[\tau] < \infty$ and τ will also be aperiodic. Hence $\{D_k\}$ and $\{\bar{D}_k\}$ have unique stationary distributions. We denote by D and $\bar{D}$ r.v.s with the stationary distributions of $\{D_k\}$ and $\{\bar{D}_k\}$ respectively. We can define D and $\bar{D}$ from (2) and (3) by replacing q_k and $q_k + X_k$ by q and $q + X$ (because of iid $\{X_k\}$, the stationary distribution of queue length seen by arriving bits is the same as that of q), where q, X and $\{r_k\}$ can be taken independent of each other. For convenience we write

$$D = \inf\{n : r_1 + r_2 + ... + r_n > q\}, \tag{4}$$
$$D(t) = \inf\{n : r_1 + r_2 + ... + r_n > t\} \tag{5}$$

for any $t \geq 0$. To avoid trivialities we assume $P[r > 0] > 0$ and $P[X > 0] > 0$.

From (4), if $\mathbb{E}[q] < \infty$ (then from Proposition 1 in Section 3.1 below, $\mathbb{E}[D] < \infty$ and $\mathbb{E}[\bar{D}] < \infty$),

$$\mathbb{E}[D]\mathbb{E}[r] > \mathbb{E}[q] \quad \text{and} \quad \mathbb{E}[\bar{D}]\mathbb{E}[r] \geq \mathbb{E}[q] + \mathbb{E}[X]. \tag{6}$$

We remark that $\mathbb{E}[D]$ and $\mathbb{E}[\bar{D}]$ do not satisfy Little's law (although mean delay experienced by a *typical* bit does). Thus the bounds provided in (6) are useful if $\mathbb{E}[q]$ is known. One can also obtain bounds on higher moments. For example,

$$\mathbb{E}[D^2] \geq \frac{\mathbb{E}[q^2]}{B_2\mathbb{E}[r^2]} \tag{7}$$

where B_2 is a known constant. Thus if we know moments of q or have bounds/ approximations for them, we will obtain bounds on moments of D and $\bar{D}$. We will study the accuracy of the bounds in (6) in Section 3.3.

If we assume r_k to be exponentially distributed, then $\mathbb{E}[q]$ equals the mean delay in the M/GI/1 queue and hence is $\mathbb{E}[X^2]/(2\mathbb{E}[r](1-\rho))$, where $\rho = \mathbb{E}[X]/\mathbb{E}[r]$. For this case we will provide an exact expression for $\mathbb{E}[D]$ in Section 3.2. Exponentially distributed r is of particular importance in wireless channels because a Rayleigh distributed channel at low SNR could lead to an exponential r. Thus we will pay special attention to this case throughout the paper.

For the GI/GI/1 queue several approximations for the mean waiting time are available. For example from [14] we get

$$\mathbb{E}[q] \approx \frac{\rho g \mathbb{E}[X](C_X^2 + C_r^2)}{2(1 - \rho)} \tag{8}$$

where

$$\rho = \mathbb{E}[X]/\mathbb{E}[r], \quad C_X^2 = \frac{\text{var}(X)}{(\mathbb{E}[X])^2},$$
$$g = \exp\left[-2\frac{1-\rho}{3\rho}\frac{(1-C_r^2)^2}{C_r^2 + C_X^2}\right], \quad if \quad C_r^2 < 1$$
$$= \exp\left[-(1-\rho)\frac{C_r^2 - 1}{C_r^2 + 4C_X^2}\right], \quad if \quad C_r^2 \geq 1$$

and C_r^2 is defined similarly as C_X^2. This approximation can be used with (6) to obtain approximations/bounds for $\mathbb{E}[D]$. For exponential r this approximation reduces to the exact formula provided above. Under heavy traffic it is close to the exponential approximation provided at the end of Section 3.3. If we add $\mathbb{E}[X]$

on the right side of (8), we get approximations for $\mathbb{E}[\bar{D}]$. In Section 5 we will provide some simulation results to check the accuracy of these approximations.

The LST of q is also available if X_k is of phase type. We obtain from this the LST of D and $\bar{D}$ in Section 3.2.

In the rest of the section we study D and $\bar{D}$ in more detail. Section 3.1 provides conditions for finiteness of moments and mgf. Section 3.2 provides the LST. Section 3.3 shows that the bounds in (6) can be tight in heavy traffic and in fact the heavy traffic analysis provides a correction term for these bounds.

3.1   Finiteness of Moments

In this section we provide conditions for finiteness of moments of D and $\bar{D}$. The proofs are available in ([6], [21]).

Proposition 1. If $\mathbb{E}[\tau^{\alpha+1}] < \infty$ for some $\alpha \geq 1$ then $\mathbb{E}[D^\alpha] < \infty$ and $\mathbb{E}[\bar{D}^\alpha] < \infty$. Also, if $\mathbb{E}[\exp(\gamma\tau^\alpha)] < \infty$ for some $\gamma > 0$ and $\alpha > 0$ then $\mathbb{E}[\exp(\gamma'D^\alpha)] < \infty$ and $\mathbb{E}[\exp(\gamma'\bar{D}^\alpha)] < \infty$ for $\gamma' < \gamma$.

Thus we obtain that if $\mathbb{E}[X^{\alpha+1}] < \infty$ for some $\alpha > 0$ then $\mathbb{E}[D^\alpha] < \infty$ and $\mathbb{E}[\bar{D}^\alpha] < \infty$. Also, D and $\bar{D}$ have finite mgf in a neighborhood of 0 if X has. The next proposition provides a partial converse.

Proposition 2. (i) If $\mathbb{E}[D^\alpha] < \infty$ for some $\alpha \geq 1$ then $\mathbb{E}[q^\alpha] < \infty$.
(ii) If $\mathbb{E}[\exp(\gamma D)] < \infty$ for some $\gamma > 0$ then $\mathbb{E}[\exp(\gamma'q)] < \infty$ for some $\gamma' > 0$.

From Propositions 1 and 2, using the previously known results for q mentioned in Section 2, we obtain that $\mathbb{E}[D^\alpha] < \infty$ ($\mathbb{E}[\bar{D}^\alpha] < \infty$) if and only if $\mathbb{E}[X^{\alpha+1}] < \infty$ for some $\alpha \geq 1$. Also, $D(\bar{D})$ has an mgf in a neighbourhood of 0 if and only if q has.

3.2   LST of the Delay Distribution

Let us first consider the case of $\exp(\lambda)$ distributed r_k. Then, according to (4), $D - 1 =$ number of Poisson(λ) arrivals in $[0, q]$. Hence

$$\mathbb{E}[z^{D-1}] = \mathbb{E}[e^{-\lambda(1-z)q}]. \tag{9}$$

Using the well-known Pollaczek-Khintchine formula for the LST of the waiting time distribution in the $M/G/1$ queue, it then follows, with $\beta(\cdot)$ denoting the LST of the distribution of X:

$$\mathbb{E}[z^D] = \frac{(1 - \lambda\mathbb{E}[X])(1 - z)z}{\beta(\lambda(1 - z)) - z}.$$

Next we consider the case when X_k has a rational LST $\beta(s) = \beta_1(s)/\beta_2(s)$ where $\beta_2(s)$ is a polynomial of degree m and $\alpha(s)$ is the LST of the distribution of r. From [7], Section III 5.10,

$$\mathbb{E}[e^{-sq}] = \frac{\beta_2(s)}{\beta_2(0)} \prod_{i=1}^{m} \frac{\xi_i}{\xi_i - s}, \quad \Re(s) \geq 0,$$

where ξ_i, $i = 1, ..., m$ are the zeros of $1 - \beta(s)\alpha(-s)$ in the left half plane. If the ξ_i are different, then we rewrite

$$\mathbb{E}[e^{-sq}] = C_0 + \sum_{i=1}^{m} \frac{C_i \xi_i}{\xi_i - s}$$

and then

$$P[q = 0] = C_0 = 1 - \sum_{i=1}^{m} C_i, \quad P[q > t] = \sum_{i=1}^{m} C_i \, e^{\xi_i t}, \; t > 0 .$$

Next we consider the distributions of D and $\bar{D}$. We have

$$P[D > n] = P[q \geq r_1 + ... + r_n] = \sum_{i=1}^{m} C_i \, \alpha^n(-\xi_i). \tag{10}$$

Hence the distribution of D is a mixture of m geometric distributions with parameters $\alpha(-\xi_1), ..., \alpha(-\xi_m)$. In particular,

$$\mathbb{E}[D] = \sum_{n=0}^{\infty} P[D > n] = \sum_{i=1}^{m} C_i \, \frac{1}{1 - \alpha(-\xi_i)} .$$

If r is exponential then we can get a more explicit expression for $\mathbb{E}[D]$. Indeed, from (10)

$$\mathbb{E}[D] = 1 + \sum_{n=1}^{\infty} P[q \geq r_1 + ... + r_n] \tag{11}$$

$$= 1 + \int_0^{\infty} \sum_{n=1}^{\infty} P[r_1 + ... + r_n \leq t] \, dP_q(t) = 1 + \frac{\mathbb{E}[q]}{\mathbb{E}[r]} \tag{12}$$

where P_q is the distribution of q and the last equality follows from the fact that $\sum_{n=1}^{\infty} P[r_1 + ... + r_n \leq t] = t/\mathbb{E}[r]$. Of course, (12) could also be obtained by taking the derivative of $\mathbb{E}[Z^D]$ at $z = 1$ in (9).

One can similarly obtain the LST of $\bar{D}$ and $\mathbb{E}[\bar{D}]$ by replacing q with $q + X$, q and X being independent of each other.

3.3   Heavy Traffic Approximations

In this section we show that in heavy traffic the bound in (6) will indeed be tight. We also obtain upper bounds on higher moments of D which are valid in heavy traffic. In fact we first obtain bounds which are valid under congestion for any traffic intensity. Hence our bounds are valid whenever there is congestion in the queue, thus covering all the cases where the delay is of real concern.

Proposition 3.   For $p \geq 1$,

$$\lim_{t \to \infty} \frac{\mathbb{E}[D^p | q > t]}{\mathbb{E}[q^p | q > t]} \leq \frac{1}{(\mathbb{E}[r])^p}.$$

From (4) we obtain $\mathbb{E}[D \mid q > t]\mathbb{E}[r] \geq \mathbb{E}[q \mid q > t]$. Then using Proposition 3

$$\lim_{t \to \infty} \frac{\mathbb{E}[D \mid q > t]}{\mathbb{E}[q \mid q > t]} = 1/\mathbb{E}[r]. \tag{13}$$

In the following we see that under heavy traffic the conditioning on $\{q > t\}$ in (13) can be removed.

Consider a sequence of queues where the distribution of sequence $\{r_k\}$ is fixed but the nth queue is fed an iid input sequence $\{X_k^{(n)}, k \geq 0\}$ such that $X_k^{(n)} \leq_{st} X_k^{(n+1)}$ and $\rho^{(n)} = \mathbb{E}[X_1^{(n)}]/\mathbb{E}[r_1] \nearrow 1$ as $n \to \infty$. In the following $q^{(n)}$ and $D^{(n)}$ denote the stationary queue length and delay in the nth queue and then $n \to \infty$ provides us a heavy traffic result.

Proposition 4. Under the above assumptions, for $p \geq 1$,

$$\lim_{n \to \infty} \frac{\mathbb{E}[(D^{(n)})^p]}{\mathbb{E}[(q^{(n)})^p]} \leq \frac{1}{(\mathbb{E}[r])^p}. \tag{14}$$

From (4) and Proposition 4 we obtain $\lim_{n \to \infty} \mathbb{E}[D^{(n)}]/\mathbb{E}[q^{(n)}] = 1/\mathbb{E}[r]$ and hence $(\mathbb{E}[D^{(n)}]\mathbb{E}[r] - \mathbb{E}[q^{(n)}])/\mathbb{E}[q^{(n)}] \to 0$.

The above results show that under heavy traffic the relative error in approximating $\sum_{k=1}^{D} r_k$ with q goes to zero in mean. However, the error itself does not go to zero. Now we provide some information on it. Let $R(t) = \sum_{k=1}^{D(t)} r_k - t$. If $\mathbb{E}[r^2] < \infty$, then from Gut [10], Chapter 3, Section 10,

$$\lim_{t \to \infty} \mathbb{E}[R(t)] = \frac{\mathbb{E}[r^2]}{2\mathbb{E}[r]} \tag{15}$$

if r is nonarithmetic. If r has an arithmetic distribution on a lattice with span d then $\lim_{k \to \infty} \mathbb{E}[R(kd)] = \mathbb{E}[r^2]/2\mathbb{E}[r] + d/2$. In the following we provide the results for only the nonarithmetic case. For the arithmetic, just add $d/2$. Using the above techniques, one can easily show from (15) that

$$\lim_{T \to \infty} \mathbb{E}[R(q) \mid q > T] = \frac{\mathbb{E}[r^2]}{2\mathbb{E}[r]}. \tag{16}$$

The next proposition removes the conditioning in heavy traffic.

Proposition 5. If $\mathbb{E}[r^2] < \infty$,

$$\lim_{n \to \infty} \mathbb{E}[R^{(n)}] = \frac{\mathbb{E}[r^2]}{2\mathbb{E}[r]}.$$

Thus $(\mathbb{E}[q] + \mathbb{E}[r^2]/2\mathbb{E}[r])/\mathbb{E}[r])$ provides a better approximation of $\mathbb{E}[D]$ under heavy traffic. For exponential r, this reduces to the exact formula (11).

One can similarly obtain results for higher moments of $R^{(n)}$. We also know that $R(t) \xrightarrow{w} Y$ where $P[Y \leq x] = \int_0^x (1 - F_r(s))ds/\mathbb{E}[r]$ and $\xrightarrow{w}$ denotes weak convergence. This then will provide us with $R^{(n)} \xrightarrow{w} Y$.

We can also exploit the standard heavy traffic approximations on the GI/GI/1 queue. For example, ([2], Chapter X) we know that if $\mathbb{E}[r] - \mathbb{E}[X^{(n)}] \to \nu$ and $Var(X^{(n)}) + Var(r) \to \sigma^2$ then

$$2\nu q^{(n)}/\sigma^2 \xrightarrow{w} \exp(1). \tag{17}$$

Since $t \to D(t)$ is a continuous function, this implies that $D(2\nu q^{(n)}/\sigma^2) \xrightarrow{w} D(Y)$ where Y is exponentially distributed with mean 1. In particular, then from (10) in heavy traffic (but with $\rho < 1$),

$$P[D > m] = \alpha^m(2\nu/\sigma^2) \text{ and } \mathbb{E}[D] = 1/(1 - \alpha(2\nu/\sigma^2)) \tag{18}$$

where $\alpha(s)$ is the LST of r.

Comparing these results with those in Section 3.2, one observes that these results are simpler and hold under general assumptions on the distribution of X but of course provide good approximations only under heavy traffic (interestingly we will see in Section 5 via simulations that although the approximation for $\mathbb{E}[q]$ is not good under light traffic, for $\mathbb{E}[D]$ it is).

Finally consider $\mathbb{E}[D^2]$. From (7) and (14), under heavy traffic

$$\frac{1}{(\mathbb{E}[r])^2} \geq \frac{\mathbb{E}[D^2]}{\mathbb{E}[q^2]} \geq \frac{1}{B_2\mathbb{E}[r^2]}$$

whereas the above exponential approximation provides $\mathbb{E}[q^2] = \sigma^4/2\nu^2$.

4 Discrete Queue Embedded in a Frame

In a practical wireless system it will often happen that a wireless link gets the opportunity to transmit data from its queue only in some of the slots and often these slots appear periodically. For example, this can happen if several queues share a common wireless link in TDMA fashion (e.g., GSM cellular system, the subscriber stations in a WiMAX uplink ([20]) and also in a multihop wireless network ([12])). These slots can also appear randomly: if the wireless channel is bad in a slot it may be decided not to transmit in that slot in order to save transmit power. In this section we extend the results obtained so far to this setup when a queue is served at periodic intervals.

We assume a frame is made up of T slots. The queue under consideration gets to serve in the first $L \leq T$ slots of each frame. In the last $T - L$ slots the fluid can arrive at the queue but it will need to wait till the next frame for transmission. Let q_k be the queue length at the beginning of the kth frame. Let $X_{k,i}$ be the new arrivals to the queue in the beginning of the ith slot of the kth frame and let $r_{k,i}$ be the link rate in that slot, $i = 1, ..., T$. Then

$$\begin{aligned}
q_k &= q_{k,1}, \\
q_{k,i+1} &= (q_{k,i} + X_{k,i} - r_{k,i})^+, \quad i = 1, ..., L, \\
q_{k+1} &= (q_{k,L+1} + X_{k,L+1} + ... + X_{k,T}).
\end{aligned} \tag{19}$$

We denote $Y_k = X_{k,L+1} + ... + X_{k,T}$. We assume $X_{k,i}$ and $Y_{k,i}$ to be iid for each k and i (for the stability result we need to assume them to be only stationary, ergodic, sequences). We will also denote

$$X_k = (X_{k,1}, X_{k,2}, ..., X_{k,T}), \quad r_k = (r_{k,1}, r_{k,2}, ..., r_{k,T}).$$

Then $q_{k+1} = f(q_k, (X_k, r_k))$ where f can be found from (19). One can easily see from (19) that f is nondecreasing and continuous in q. Thus, from Loynes [16], if $\{(X_k, r_k)\}$ forms a stationary, ergodic sequence then q_k has a stationary distribution (q_k may be infinite with positive probability). Let q be a r.v. with this stationary distribution. Since sequence $\{q_k\}$ can be lower bounded by sequence

$$\bar{q}_{k+1} = \left(\bar{q}_k + \sum_{i=1}^{T} X_{k,i} - \sum_{i=1}^{L} r_{k,i} \right)^+, \quad \bar{q}_0 = q_0 \tag{20}$$

and $\bar{q}_k$ converges a.s. to ∞ if $T\mathbb{E}[X_{k,1}] > L\mathbb{E}[r_{k,1}]$, q_k also converges to ∞ a.s. under these conditions.

Using Loynes's construction ([16]) we can also show that when $T\mathbb{E}[X_{k,1}] < L\mathbb{E}[r_{k,1}]$, then $P[q < \infty] = 1$, the stationary distribution is unique and starting from any initial distribution, q_k converges in total variation to it. See details in [6], [21]. From now on in this section we will assume $T\mathbb{E}[X_{k,1}] < L\mathbb{E}[r_{k,1}]$.

From now onwards we impose the independence assumptions on $\{(X_k, r_k)\}$. We can rewrite (19) slotwise as in (1) with $r_{k,i} = 0$ for $i = L+1, ..., T$. Now however the rate sequence is no longer iid but periodic and the results obtained in the previous section cannot be directly used. However, this can be taken as a regenerative sequence with regeneration epochs the frame boundaries. Then we are within the framework of [26]. Thus we obtain $\mathbb{E}[\tau^\alpha] < \infty$ (τ has a finite mgf in a neighborhood of 0), if $\mathbb{E}[X_1^\alpha] < \infty$ (X has a finite mgf in a neighborhood of 0) for $\alpha \geq 1$ where τ is a regeneration length for this system, the regeneration epochs being the frame boundaries where $q_k = 0$. We also obtain $\mathbb{E}[q^\alpha] < \infty$ whenever $\mathbb{E}[X_1^{\alpha+1}] < \infty$ for $\alpha \geq 1$ where q is the stationary queue length at frame boundaries. Finiteness of the mgf of q in a neighborhood of 0 is also implied by that of X_1.

Various functional limit theorems for the process $\{q_k\}$ and rates of convergence to the stationary distribution are also obtained ([26]). Although [26] provides these results for the process observed at slot boundaries, these results at frame boundaries can be obtained easily in the same way.

The stationary distribution of the queue length process is different at the ith slot than at the jth slot, $j \neq i$ within a frame but one can easily relate these stationary distributions. Also finiteness of moments of their stationary distributions holds under the same conditions. For example, for $1 < i \leq L$ (denoting by q_i a r.v. with the stationary distribution of $q_{k,i}$)

$$q_i = ((q + X_1 - r_1)^+ + ... + X_{i-1} - r_{i-1})^+ \leq q + \sum_{j=1}^{i-1} X_j$$

and hence $\mathbb{E}[q_i^\alpha] < \infty$ if $\mathbb{E}[q^\alpha] < \infty$ and $\mathbb{E}[X^\alpha] < \infty$. Similarly we obtain the finiteness of the exponential moments.

Let D_i and $\bar{D}_i$ be r.v.s obtained from (2) and (3) by replacing q_k by q_i. Thus the Propositions 1-5 and the asymptotics on the tail distributions hold as for the system in Section 2. Let $\hat{D}_i$ and $\hat{\bar{D}}_i$ be the delay of the first bit of $X_{k,i}$ under stationarity. Let $D, \bar{D}, \hat{D}, \hat{\bar{D}}$ denote these quantities when $i = 1$. Then,

$$\left(\frac{D}{L} - 1\right)T \le \hat{D} = \left[\frac{D}{L}\right]T + \left(D - \left[\frac{D}{L}\right]L\right) \le D\,\frac{T}{L} \tag{21}$$

where $[x]$ denotes the largest integer $< x$. Similarly one can relate $\hat{D}_i$ and D_i. Thus we obtain the finiteness of moments of $\hat{D}_i$ and $\hat{\bar{D}}_i$ from that of D_i and $\bar{D}_i$, for which we obtain these results as in Proposition 1 from that of q_i. Also the distribution of $\hat{D}$ is of regular variation of index $-\alpha$ if and only if that of D is. Similarly $\hat{D}$ has a mgf in a neighbourhood of 0 if and only if D has. If we know the distribution of D then we can use the equality in (21) to obtain the distribution of $\hat{D}$. If we have only moments and/or bounds or approximations on moments of D then the inequalities in (21) can be used to obtain the corresponding bounds/approximations for $\hat{D}$.

Under heavy traffic, one expects that the queue will not be empty most of the time. Then, one can approximate (19) by removing $()^+$ on the RHS of (19). Now observe this queue only at frame boundaries. Consider the frame as a slot and then use the results of Section 3 with $X_k = X_{k,1} + ... + X_{k,T}$ and $r_k = r_{k,1} + ... + r_{k.L}$.

Finally we obtain the LST of q and D when r is exponentially distributed. As commented before, this could correspond to Rayleigh fading channels and hence is of practical concern. The corresponding results for q_i, D_i and $\bar{D}_i$ can then be easily obtained by relating them to q, D and $\bar{D}$.

Let $\beta(s)$ be the LST of X. Then

$$\mathbb{E}[\,e^{-sq_{k,2}} \mid q_{k,1} = y] = \mathbb{E}[e^{-s(q_{k,1}+X_{k,1}-r_{k,1})^+} \mid q_{k,1} = y]$$

$$= \mathbb{E}[e^{-s(y+X-r)}] - \mathbb{E}[e^{-s(y+X-r)}\,1_{\{y+X-r\le 0\}}] + P[y+X-r \le 0]$$

$$= e^{-sy}\,\beta(s)\,\frac{\lambda}{\lambda - s} - \frac{s}{\lambda - s}\,\beta(\lambda)\,e^{-\lambda y}.$$

Thus

$$\mathbb{E}[e^{-sq_{k,2}}] = \beta(s)\,\frac{\lambda}{\lambda - s}\,\mathbb{E}[e^{-sq_{k,1}}] - \beta(\lambda)\,\frac{s}{\lambda - s}\,\mathbb{E}[e^{-\lambda q_{k,1}}].$$

We can iterate this equation to obtain

$$\mathbb{E}[e^{-sq_{k,L+1}}] = h^L(s)\,\mathbb{E}[e^{-sq_{k,1}}] - g(s)\sum_{j=0}^{L-1} h^j(s)\,\mathbb{E}[e^{-\lambda q_{k,L-j}}]$$

where $h(s) = \beta(s)\lambda/(\lambda - s)$, $g(s) = \beta(\lambda)s/(\lambda - s)$. Furthermore, we obtain $\mathbb{E}[e^{-sq_{k,T+1}}] = \beta^{T-L}(s)\mathbb{E}[e^{-sq_{k,L+1}}]$. In steady state $\mathbb{E}[e^{-sq_{k,T+1}}] = \mathbb{E}[e^{-sq_{k,1}}] \stackrel{\triangle}{=} \mathbb{E}[e^{-sq}]$. Thus we obtain

$$\mathbb{E}[e^{-sq}] = -\left[1 - \beta^T(s)\left(\frac{\lambda}{\lambda - s}\right)^L\right]^{-1}$$

$$\left\{\beta^{T-L}(s)\beta(\lambda)\frac{s}{\lambda - s}\sum_{j=0}^{L-1}\left(\beta(s)\frac{\lambda}{\lambda - s}\right)^j y_{L-j}(\lambda)\right\}. \tag{22}$$

The right hand side contains L unknowns y_{L-j}, $j = 0, ..., L-1$. From Rouché's theorem ([7], p 652), we know that under stability, $1 - \beta^T(s)\ (\lambda/(\lambda - s))^L$ has exactly L zeros $s_0, ..., s_{L-1}$ in the right half plane. The term in curly brackets in the right side of (22) should also be zero for $s = s_0, ..., s_{L-1}$. This yields L equations that can be used to obtain $y_1(\lambda), ..., y_L(\lambda)$. There are several algorithms available to compute the zeros $s_0, ..., s_{L-1}$ (see [15]). It follows, after a lengthy calculation, that

$$\mathbb{E}[q] = (T - L)\mathbb{E}[X] + \frac{1}{L - T\lambda\mathbb{E}[X]}\left[\beta(\lambda)\sum_{j=0}^{L-1} y_{L-j}(\lambda)(j\mathbb{E}[X] + \frac{L-j-1}{\lambda})\right.$$

$$+\ \lambda T(T-1)\frac{\mathbb{E}[X]^2}{2} + \lambda T\frac{\mathbb{E}[X^2]}{2} - \frac{L(L-1)}{2\lambda}\right]. \tag{23}$$

It should be noted that (23) reduces to the familiar expression for the mean waiting time in the $M/G/1$ queue when $T = L = 1$.

The mgf of D follows from the LST of q in (22), using (9):

$$\mathbb{E}[z^D] = \frac{\beta^{T-L}(\lambda(1-z))\beta(\lambda)(1-z)\sum_{j=0}^{L-1}\beta(\lambda(1-z))^j z^{L-j}y_{L-j}(\lambda)}{\beta^T(\lambda(1-z)) - z^L}.$$

If we are able to invert this mgf to obtain the distribution of D (there are various techniques to perform such a numerical inversion) then the mgf of $\hat{D}$ follows from (21):

$$\mathbb{E}[z^{\hat{D}}] = \sum_{j=0}^{\infty}\sum_{k=0}^{L-1} z^{jT+k}P(D = jL + k).$$

In fact one can show that

$$\mathbb{E}[z^{L\hat{D}}] = \frac{1}{L}\sum_{k=0}^{L-1}\mathbb{E}[z^{TD}a^k]\frac{1 - (a^{-k}z^{L-T})^L}{1 - a^{-k}z^{L-T}}.$$

where $a = exp(\frac{2\pi i}{L})$, $i = \sqrt{-1}$. Thus one can avoid inverting the mgf of D. One can then obtain

$$\mathbb{E}[\hat{D}] = \frac{1}{L}\frac{d}{dz}\mathbb{E}[z^{L\hat{D}}]\big|z = 1. \tag{24}$$

5   Simulation Results

In this section we provide some simulation results to verify the accuracy of approximations provided in Sections 3 and 4.

We first consider the single queue studied in Sections 2 and 3. For X we have taken a few discrete distributions: Poisson (examples 9, 10) and finite valued while for r we have taken Rayleigh (examples 9, 10), exponential (examples 7, 8) and a few discrete finite valued distributions. The ρ has been taken as 0.3, 0.5, 0.7, 0.9, 0.95 and 0.98. Each simulation was done for 20 million slots, long enough to have negligible estimation error. The simulated values of $\mathbb{E}[q]$, $\mathbb{E}[D]$, $\mathbb{E}[q]$ from formula (8) and its heavy traffic approximation $= (\text{var}(X) + \text{var}(r))/(2(\mathbb{E}[r] - \mathbb{E}[X]))$ and $\mathbb{E}[D]$ via the lower bound $\mathbb{E}[q]/\mathbb{E}[r]$ and with the heavy traffic correction $\mathbb{E}[q]/\mathbb{E}[r] + \mathbb{E}[r^2]/(2(\mathbb{E}[r])^2)$ (with $d/(2\mathbb{E}[r])$ added if r has an arithmetic distribution) and as $1/(1 - \alpha(2\nu/\sigma^2))$ are provided in Table 1. As commented before, for r exponential, (8) and $\mathbb{E}[D]$ with heavy traffic correction are in fact exact formulae for all traffic intensities. This is also seen from Table 1. For other distributions, the theoretical formulae for $\mathbb{E}[q]$ and $\mathbb{E}[D]$ are very close to the simulated values for $\rho \geq 0.90$. For lower values of ρ the approximations are not good as expected (but the approximation (18) for $\mathbb{E}[D]$ is still working quite well). We also observe that $\mathbb{E}[q]$ from (8) is quite close to simulations even for small ρ and is always more accurate than the heavy traffic approximation.

Next we provide an example for the queue in Section 4. We consider exponential service and arrival rates. We fix the service rate at $\mathbb{E}[r] = 7.5$ and take different values for $\mathbb{E}[X]$. Also, we take $T = 10$ and $L = 4$. We obtained $\mathbb{E}[q]$ and $\mathbb{E}[\hat{D}]$ via simulations and compared with the theory. The theoretical results are obtained via (23) and (24). We also obtained $\mathbb{E}[q]$ via the heavy traffic approximation mentioned in Section 3.

The results of the simulations and theory are presented in Table 2. The simulated values $\mathbb{E}[\hat{D}]$ are closer to the upper bounds than to the lower bounds for $\hat{D}$ when we used (23) and hence we present only the upper bounds. We

Table 1. Simulation results for single queue of Sections 2 and 3

examples	ρ	$\mathbb{E}X$	$\mathbb{E}r$	$\text{var}(X)$	$\text{var}(r)$	$\mathbb{E}q$			$\mathbb{E}D$			
						simulated value	From (8)	From heavy traffic appr. (17)	simulated value	$\mathbb{E}q/\mathbb{E}r$	with heavy traffic correction	from (18)
1	0.30	0.36	12.00	17.09	21.00	0.61	0.52	2.27	1.12	0.04	0.82	1.12
2	0.50	5.00	10.00	20.10	33.75	2.65	2.21	5.38	1.39	0.22	1.14	1.44
3	0.70	7.13	10.18	37.36	23.39	6.45	6.69	9.95	1.47	0.66	1.32	1.68
4	0.90	7.13	7.92	37.36	32.33	40.23	39.58	44.01	6.01	5.00	6.51	6.35
5	0.95	7.13	7.50	37.36	18.75	70.39	71.33	74.81	10.40	9.51	10.84	10.66
6	0.98	7.35	7.50	12.23	18.75	100.08	99.70	103.26	14.13	13.29	14.63	14.45
7	0.95	7.13	7.50	37.36	56.25	118.00	117.50	124.81	16.74	15.67	16.67	17.64
8	0.98	7.35	7.50	12.23	66.25	220.99	220.83	228.26	30.46	29.44	30.44	31.43
9	0.95	7.13	7.50	7.13	15.41	27.10	26.81	30.04	4.30	3.58	4.21	4.66
10	0.98	7.35	7.50	7.35	15.41	73.48	72.54	75.85	10.46	9.67	10.31	10.76

Table 2. Example for queue embedded in a frame

ρ	$\mathbb{E}[X]$	$\mathbb{E}[r]$	Simu-lated $\mathbb{E}[q]$	$\mathbb{E}[q]$ from (23)	$\mathbb{E}[q]$ via HT	$\mathbb{E}\hat{D}$ Simu-lated	via (24)	via (23) $\hat{D}_{\text{upper}}$	via HT $\hat{D}_{\text{upper}}$
0.80	2.40	7.50	28.45	28.44	23.55	9.00	9.60	10.35	10.35
0.90	2.70	7.50	55.06	55.34	49.65	17.48	18.63	18.45	19.05
0.95	2.85	7.50	107.40	108.15	102.17	34.74	36.27	36.05	36.56
0.98	2.94	7.50	262.17	265.83	259.53	86.21	88.85	88.61	89.01

further observe that the heavy traffic approximations are also providing good approximations.

Acknowledgement. The work of the second author was done while he was visiting EURANDOM. The authors gratefully acknowledge the numerical support provided by Dr. J. S. H van Leeuwaarden(EURANDOM) and his remark leading to a part of Section 4.

References

1. Anick, D., Mitra, D., Sondhi, M.M.: Stochastic theory of a data handling system with multiple sources. Bell Syst. Tech. J. 61, 1871–1894 (1982)
2. Asmussen, S.: Applied Probability and Queues, 2nd edn. Springer, Heidelberg (2003)
3. Bingham, N.H., Goldie, C.M., Teugels, J.L.: Regular Variation. Cambridge Univ. Press, Cambridge (1987)
4. Boxma, O.J.: Fluid queues and regular variation. Performance Evaluation 27/28, 699–712 (1996)
5. Bruneel, H., Kim, B.G.: Discrete-Time Models for Communication Systems including ATM. Kluwer Academic Publishers, Boston (1993)
6. Boxma, O.J., Sharma, V., Prasad, D.K.: Scheduling and performance analysis of queue with random service rate in discrete time, EURANDOM Technical Report 2006-019
7. Cohen, J.W.: The Single Server Queue, 2nd edn. North-Holland Publishing Company, Amsterdam (1982)
8. Daduna, H.: Queueing Networks with Discrete-Time Scale. LNCS. Springer, Heidelberg (2001)
9. Eryilamz, A., Srikant, R., Perkins, J.R.: Stable Scheduling policies for fading wireless channels. IEEE/ACM Trans. Networking 13, 411–424 (2005)
10. Gut, A.: Stopped Random Walk. Springer, N. Y. (1991)
11. Kiefer, J., Wolfowitz, J.: On the characterization of the general queueing process with applications to random walks. Annals Math. Stat. 27, 147–161 (1956)
12. Kodialam, M., Nandogopal, T.: Characterizing the capacity region in multi-radio, multi-channel wireless mesh networks. In: Proc. ACM Conf. Mobicom (2005)
13. Kosten, L.: Stochastic theory of data handling systems with groups of multiple sources. In: Bux, W., Rudin, H. (eds.) Performance of Computer Communication Systems, pp. 321–331. North Holland, Amsterdam (1984)
14. Krämer, W., Langenbach-Beltz, M.: Approximate formulae for the delay in the queueing system GI/G/1, Eighth International Telemetric Congress, Melbourne, pp. 235-1-8 (1976)

15. Van Leeuwaarden, J.S.H.: Queueing Models for Cable Access Networks, Ph.D Thesis, Technische Universiteit Eindhoven, The Netherlands (2005)
16. Loynes, R.M.: The Stability of a queue with nonindependent interarrival and service time distributions. In: Proc. Cambridge Phil. Society, Vol. 58, 479–520 (1962)
17. Mergen, G., Tong, L.: Stability and Capacity of regular wireless networks. IEEE Trans. Inf. Theory 51, 1938–1953 (2005)
18. Neely, M.J., Modiano, E., Li, C.-P.: Fairness and optimal stochastic control for heterogeneous networks. In: Proc. Conf. INFOCOM (2005)
19. Neely, M.J., Modiano, E., Rhors, C.E.: Dynamic power allocation and routing for time varying wireless networks. IEEE INFOCOM 1, 745–755 (2003)
20. Ohrtman, F.: WiMAX Handbook: Building 802.16 Wireless Networks. McGraw-Hill, Professional (2005)
21. Prasad, D.K.: Scheduling and performance analysis of queues in wireless systems. M.E. Thesis, Dept. of Electrical Communication Engg., Indian Institute of Science, Bangalore, July (2006)
22. Radunović, B., LeBoudec, J.-Y.: Rate performance objectives of multihop wireless networks. IEEE Trans. Mobile Computing 3, 334–349 (2004)
23. Roberts, J.W.: Performance Evaluation and Design of Multiserver networks. COST Project 224, Final report (1991)
24. Rom, R., Sidi, M.: Message delay distribution in generalized time division multiple access. Prob. Eng. Inform. Sci. 4, 187–202 (1990)
25. Rubin, I., Zhang, Z.: Message delay analysis for TDMA schemes using contiguous slot assignments. IEEE Trans. Communications 40, 730–737 (1992)
26. Sharma, V.: Some limit theorems for regenerative queues. Queueing Systems 30, 341–363 (1998)
27. Sharma, V., Gangadhar, N.D.: Asymptotics for transient and stationary probabilities for finite and infinite buffer discrete-time queues. Queueing Systems 26, 1–22 (1997)
28. Sigman, K.: A primer on heavy tailed distributions. Queueing Systems 33, 261–275 (1999)

Performance of a Partially Shared Priority Buffer with Correlated Arrivals

Dieter Fiems, Joris Walraevens, and Herwig Bruneel

SMACS Research Group, Department TELIN (IR07), Ghent University
Sint-Pietersnieuwstraat 41, B-9000 Gent, Belgium
{df,jw,hb}@telin.UGent.be

Abstract. In this paper, we analyse a finite sized discrete time priority buffer with two types of packet arrivals, referred to as class 1 and class 2 packets. Packets of both classes arrive according to a discrete batch Markovian arrival process, taking into account the correlated nature of arrival processes in heterogeneous telecommunication networks. Packets of class 1 are assumed to have both space priority and transmission priority over packets of class 2. In particular, the partial buffer sharing acceptance policy is adopted as space priority mechanism. Using a matrix analytic approach, the buffer content is analysed and through some numerical examples, the impact of the priority scheduling, the threshold and the correlation in the arrival process is demonstrated.

Keywords: queueing theory, priority, partial buffer sharing, matrix analytic methods.

1 Introduction

In recent years, much research has been devoted to the transmission of multimedia streams over packet based (IP) networks. Different types of traffic require different Quality of Service (QoS) standards. E.g. for real time (interactive) applications (telephony, multimedia, gaming, ...), it is important that the mean delay and delay jitter are minimal, while this is of minor importance for data applications (file transfers, email, ...).

Further, different packets within the same traffic stream may have different QoS constraints as well. This is in particular the case when one makes use of scalable coding techniques (see a.o. Radha et Al. [1]). A scalable video encoder produces both base layer and enhancement layer packets. Only base layer packets are required to decode and playback the video, although at a poor quality. Combined with the enhancement packets, the video can be played back at full quality. Therefore, *loss* of base layer packets in the network should be avoided. It is also absolutely necessary for the receiver to receive the base packets *in time* to be able to maintain an uninterrupted flow of video packets to the end user. As little loss and delay as possible is not strictly required for the enhancement packets, although receiving most of those (in time) obviously increases the quality of the received video stream.

L. Mason, T. Drwiega, and J. Yan (Eds.): ITC 2007, LNCS 4516, pp. 582–593, 2007.

In order to guarantee acceptable delay boundaries to real time traffic, several scheduling schemes — for switches, routers, etc — have been proposed and analysed, each one with its own specific algorithmic and computational complexity. The most drastic scheduling scheme is head of the line (HOL) priority. With this scheduling discipline, as long as high priority packets are present in the system, they are transmitted. Low priority packets can thus only be transmitted when the buffer is void of high priority traffic. Discrete time HOL priority queues with deterministic transmission times equal to one slot and infinite buffer size have been studied in [2,3,4,5,6]. In [2] the steady state system content and delay in the case of a multiserver queue are studied. Mehmet Ali and Song [3] analyse the system content in a multiplexer with two state on-off sources. The steady state system content and the delay for Markov modulated high priority interarrival times and geometrically distributed low priority interarrival times are presented in [4]. Walraevens et al. [5] study the steady state system content and packet delay in the special case of an output queueing switch with Bernoulli arrivals. Finally in [6], a HOL priority queue with train arrivals is analysed.

Although an infinite sized buffer is a realistic model for a large buffer, it may not be for a small one. Buffers in access networks for instance are typically much smaller than the large buffers in core networks. Buffers of e.g. 140 packets or less are quite normal. Therefore, we here analyse a finite sized discrete time priority buffer using matrix analytic methods [7]. The arrival process is assumed to be a two class discrete batch Markovian arrival process (2-DBMAP), which can take into account the bursty nature within packet streams as well as the correlation between different packet streams (and more specifically between high and low priority arrivals).

A finite sized priority queue with DBMAPs was already analysed in [8]. In this paper, each priority class has its own dedicated queue, i.e., the system consists of two (finite sized) queues, one for the high priority and one for the low priority traffic. In the current paper, we assume that a single queue is shared by the packets of both classes. We furthermore adopt the following acceptance policy: when the total buffer content reaches a certain threshold T, low priority packets are no longer accepted. So when the system content is lower than T, packets of both classes are regularly accepted, while only class 1 packets can enter the buffer when the buffer content is larger than or equal to T. Obviously, when the buffer is completely full, packets of both classes are lost upon arrival. This acceptance policy is generally referred to as partial buffer sharing or PBS and is quite easily implementable in practice (see e.g. [9]).

The described acceptance policy is prompted by the already mentioned scalable video coding: base (high priority) packets have to be received as soon as possible and with as little loss as possible. Therefore these packets are given priority in transmission as well as in acceptance over enhancement (low priority) packets. In that way, uninterrupted video can be provided to the end user, even if the network is heavily congested. Note that for other applications, the acceptance priority could be switched: when the delay sensitive traffic is loss insensitive and the delay insensitive traffic is loss sensitive, one class may have transmission priority while the other one has acceptance priority. The analysis of this paper can be modified to be useful in the latter case.

The remainder of this contribution is organised as follows. In the next section, the queueing model is described in more detail. Section 3 then concerns the analysis of the queueing system under consideration. We illustrate our approach by means of some numerical examples in Section 4 and draw conclusions in Section 5.

2   Queueing Model

We consider a discrete time single server priority queueing system. Time is divided into fixed length intervals or slots and arrivals and departures are synchronised with respect to slot boundaries. There are two classes of packets, say class 1 and 2, whereby class 1 packets have transmission priority over class 2 packets. Transmission times of packets of both classes are assumed to be fixed and equal to the slot length.

The buffer capacity of the queueing system under investigation is finite. The buffer can store up to N packets simultaneously — including the packet being transmitted — but only accommodates class 2 packets if there are no more than T packets present upon arrival. Here the packets that are present "upon arrival" of some packet include the packets that arrived at the same slot boundary but that entered the buffer before this packet. The parameter T is referred to as the threshold. Such a buffer acceptance policy is referred to as partial buffer sharing.

Class 1 and class 2 arrivals are modelled by means of a 2 class discrete time batch Markovian arrival process (2-DBMAP), see also Zhao et Al. [10]. Such an arrival process is completely characterised by the $Q \times Q$ matrices $A(n, m)$ governing the transitions of the underlying discrete time Markov chain from slot to slot when there are n class 1 arrivals and m class 2 arrivals at the slot boundary $(n, m = 0, 1, 2, \ldots)$. Here Q denotes the size of the underlying state space of the 2-DBMAP. Thus

$$A(n, m) = \left[\Pr[A_{1,k} = n, A_{2,k} = m, S_{k+1} = j | S_k = i] \right]_{i,j=1..Q}, \qquad (1)$$

with $A_{1,k}$ and $A_{2,k}$ the number of class 1 and class 2 packet arrivals respectively at the kth slot boundary and S_{k+1} the state of the system during the following slot.

For further use, let ρ_i denote the mean number of packets of class i that arrive at a slot boundary $(i = 1, 2)$. We have,

$$\rho_1 = \sum_{n,m=0}^{\infty} n\psi A(n, m)e, \quad \rho_2 = \sum_{n,m=0}^{\infty} m\psi A(n, m)e. \qquad (2)$$

Here e is a column vector of ones and ψ is the steady state probability row vector of the underlying Markov chain, i.e., it is the unique non-negative solution of

$$\psi = \psi \sum_{n,m=0}^{\infty} A(n, m), \quad \psi e = 1. \qquad (3)$$

We further let $\rho = \rho_1 + \rho_2$ denote the total arrival load.

Due to the possible simultaneity of arrivals of both classes and departures at slot boundaries, one needs to specify the order in which these arrivals and departures are processed at a boundary. We here assume the following order: (1) departures (2) arrivals of class 1, (3) arrivals of class 2. In the remainder observation of the queue "at slot boundaries" means after possible departures but before the arrivals.

3   Queueing Analysis

We first relate the number of class 1 and class 2 in the buffer at consecutive slot boundaries. These relations are then used to obtain a set of balance equations which can be solved numerically. Finally, expressions for various performance measures are obtained in terms of the solution of the balance equations.

3.1   System Equations

Consider a random slot boundary, say slot boundary k and let U_k and V_k denote the total queue content and the class 2 queue content — i.e., the total number of packets and the number of class 2 packets in the queue — at this slot boundary. The buffer has a finite capacity and a threshold for class 2 packets and cannot accommodate all arrivals. Therefore let $\tilde{A}_{1,k}$ and $\tilde{A}_{2,k}$ denote the number of class 1 and class 2 packet arrivals at the kth slot boundary that the buffer can accommodate. Since the buffer accommodates arriving class 1 packets until there are N packets in the buffer, we find,

$$\tilde{A}_{1,k} = \min(A_{1,k}, N - U_k)\,. \tag{4}$$

Similarly, the buffer accommodates arriving class 2 packets until there are T packets in the buffer. One therefore retrieves,

$$\tilde{A}_{2,k} = \min(A_{2,k}, (T - U_k - A_{1,k})^+)\,. \tag{5}$$

Here $(\cdot)^+$ is standard shorthand notation for $\max(\cdot, 0)$. Notice that we here used the assumption that class 1 packets arrive before class 2 arrivals as indicated before.

We may now relate the total buffer content and the class 2 buffer content at two consecutive slot boundaries as follows. A packet departs at the $(k+1)$st boundary when there are packets present during the preceding slot. We have,

$$U_{k+1} = (U_k + \tilde{A}_{1,k} + \tilde{A}_{2,k} - 1)^+\,. \tag{6}$$

If $U_k - V_k + A_{1,k} > 0$, a class 1 packet departs at the $(k+1)$st slot boundary and therefore no class 2 packets can leave. We have,

$$V_{k+1} = V_k + \tilde{A}_{2,k}\,. \tag{7}$$

If $U_k - V_k + A_{1,k} = 0$, there are no class 1 packets present and therefore a class 2 packet departs, if any. As such, we find,

$$V_{k+1} = (V_k + \tilde{A}_{2,k} - 1)^+\,. \tag{8}$$

3.2   Balance Equations

Let $\tilde{A}_u(n,m)$ denote the matrix governing the transitions of the underlying Markov chain at a slot boundary when there are n class 1 arrivals and m class 2 arrivals that the buffer can accommodate, given that there are u packets in the buffer at that slot boundary. That is,

$$\tilde{A}_u(n,m) = \left[\Pr[\tilde{A}_{1,k} = n, \tilde{A}_{2,k} = m, S_k = j | S_{k-1} = i, U_k = u]\right]_{i,j=1..Q} \quad (9)$$

From (4) and (5) and by conditioning on the number of class 1 and class 2 arrivals at a slot boundary, we find,

$$\tilde{A}_u(n,m) = \sum_{i,j=0}^{\infty} A(i,j) 1\{n = \min(i, N-u)\} 1\{m = \min(j, (T-u-i)^+)\}, \quad (10)$$

for $u = 0, 1, \ldots, N-1$, $n = 0, 1, \ldots, N$ and $m = 0, 1, \ldots, T$. Here $1\{\}$ denotes the indicator function which evaluates to 1 if its argument is true and to 0 if this is not the case. Notice that for $n < N - u$ and $m < T - u - n$, we have $\tilde{A}_u(n,m) = A(n,m)$. For $n = N-u$ ($m = (T-u-n)^+$ respectively), the buffer is full (exceeds the threshold respectively). Also, $\tilde{A}_u(n,m)$ equals the zero matrix for all (n,m) outside the set $\{(i,j)|0 \le i \le N - u \wedge 0 \le j \le (T - u - i)^+\}$.

Clearly, the triple (U_k, V_k, S_k) describes the state of the queueing system at the kth slot boundary (in the Markovian sense). Therefore, let $\pi_k(n,m)$ denote the row vector whose ith entry is the probability to have $n - m$ class 1 and m class 2 packets in the queue at the kth slot boundary while the arrival process is in state i, i.e.,

$$\pi_k(m,n) = \left[\Pr\left[V_k = m, U_k = n, S_k = i\right]\right]_{i=1,\ldots,Q}, \quad (11)$$

for $n = 0, \ldots, N-1$ and for $m = 0, \ldots, \min(T,n)$.

In view of (6), (7) and (8), these vectors satisfy the following set of equations,

$$\pi_{k+1}(m,n) = \sum_{j=0}^{\min(n+1,N-1)} \sum_{i=0}^{\min(j,T)} \pi_k(i,j) \tilde{A}_j(n - m + i + 1 - j, m - i)$$

$$+ 1\{n = m\} \sum_{j=0}^{\min(n+1,N-1)} \pi_k(j,j) \tilde{A}_j(0, n - j + 1)$$

$$+ 1\{n = m = 0\} \pi_k(0,0) \tilde{A}_0(0,0), \quad (12)$$

for $n = 0, \ldots, N-1$ and for $m = 0, \ldots, \min(T,n)$.

We now group the vectors $\pi_k(m,n)$ for every $n = 0, 1, \ldots, N-1$. I.e., we define the following row vectors,

$$\pi_k(n) = [\pi_k(0,n), \pi_k(1,n), \ldots, \pi_k(\min(n,T),n)], \quad (13)$$

for $n = 0, \ldots, N - 1$. The set of equations (12) then has the following block matrix representation,

$$\pi_{k+1}(n) = \sum_{j=0}^{\min(n+1,N-1)} \pi_k(j)C(j,n), \qquad (14)$$

with,

$$C(0,0) = \tilde{A}_0(0,0) + \tilde{A}_0(1,0) + \tilde{A}_0(0,1), \qquad (15)$$

and where $C(j,n)$ denotes a $(\min(T,j)+1) \times (\min(T,n)+1)$ block matrix for all $n = 1, \ldots, N-1$ and $j = 0, \ldots, \min(n+1, N-1)$ (with blocks of size $Q \times Q$). The (block) elements of the matrix $C(j,n)$ are given by,

$$c_{i+1,m+1}(j,n) = \tilde{A}_j(n+1-j-m+i, m-i) + 1\{m = n\}1\{i = j\}\tilde{A}_j(0, n-j+1), \qquad (16)$$

for $i = 0, \ldots, \min(T,j)$ and $m = 0, \ldots, \min(T,n)$.

Under mild assumptions, the Markov chain under consideration has only one ergodic class (see [11]). There then exists a unique stationary distribution (a non-negative normalised vector), say $\pi = [\pi(0), \pi(1), \ldots, \pi(N-1)]$ — with $\pi(n) = [\pi(0,n), \pi(1,n), \ldots, \pi(\min(n,T),n)]$ — satisfying the balance equations,

$$\pi(n) = \sum_{j=0}^{\min(n+1,N-1)} \pi(j)C(j,n), \qquad (17)$$

for $n = 0, \ldots, N - 1$.

Equation (17) now shows that the transition matrix of the priority queueing system under consideration has an upper Hessenberg block structure with varying block sizes. Such a system of equations is efficiently solved by means of a linear level reduction algorithm (see e.g. Blondia and Casals [12]).

3.3   Performance Measures

Given the vectors $\pi(n)$, we may now obtain various performance measures.

The supported arrival load of a class is defined as the average number of packets of that class that arrive at a slot boundary and that can be accommodated by the buffer. The total supported load is defined as the average number of packets that arrive at a slot boundary and can be accommodated by the buffer. In view of (10), the supported class 1 load $\tilde{\rho}_1$, the supported class 2 load $\tilde{\rho}_2$ and the total supported load $\tilde{\rho}$ are given by,

$$\tilde{\rho}_1 = \sum_{i=0}^{N-1} \sum_{n=0}^{N-i} \sum_{m=0}^{T-n-i} n\tau(i)\tilde{A}_i(n,m)e,$$

$$\tilde{\rho}_2 = \sum_{i=0}^{N-1} \sum_{n=0}^{N-i} \sum_{m=0}^{T-n-i} m\tau(i)\tilde{A}_i(n,m)e,$$

$$\tilde{\rho} = \tilde{\rho}_1 + \tilde{\rho}_2. \qquad (18)$$

Here e is a column vector of ones and $\tau(n) = \sum_{m=0}^{\min(n,T)} \pi(m,n)$ denotes a row vector of size Q whose ith element denotes the probability that there are n packets in the queue and that the arrival process is in state i.

Alternatively, the supported arrival load can also be retrieved by observing the departure process. In particular, the total supported load equals the probability that there is a departure at a random slot boundary. Since there is a departure whenever there are packets in the queue, we have,

$$\tilde{\rho} = 1 - \pi(0)A(0,0)e\,. \tag{19}$$

Similarly, the class 1 supported load equals the probability that there is a class 1 departure at a random slot boundary. Since a class 1 packet leaves the queue if there are class 1 packets present in the buffer, we find,

$$\tilde{\rho}_1 = 1 - \sum_{m=0}^{T} \sum_{n=0}^{T-m} \pi(m,m)\tilde{A}_0(0,n)e\,. \tag{20}$$

Finally, the class 2 supported load equals $\tilde{\rho}_2 = \tilde{\rho} - \tilde{\rho}_1$.

The packet loss ratio of a class is defined as the fraction of packets of a class that the buffer cannot accommodate. In view of the definition of the supported load and the packet loss ratio, one easily obtains the following expressions for the packet loss ratio of class 1 packets (plr_1), of class 2 packets (plr_2) and of all packets (plr),

$$\mathrm{plr}_1 = 1 - \frac{\tilde{\rho}_1}{\rho_1}\,, \quad \mathrm{plr}_2 = 1 - \frac{\tilde{\rho}_2}{\rho_2}\,, \quad \mathrm{plr} = 1 - \frac{\tilde{\rho}}{\rho}\,. \tag{21}$$

Moments of the class 1 and class 2 queue content at random slot boundaries are easily written in terms of the steady state vectors $\pi(i,n)$. In particular, the mean class i queue content $\overline{u}_i$ at random slot boundaries is given by,

$$\overline{u}_1 = \sum_{n=0}^{N-1} \sum_{m=0}^{\min(n,T)} (n-m)\pi(m,n)e\,, \quad \overline{u}_2 = \sum_{n=0}^{N-1} \sum_{m=0}^{\min(n,T)} m\pi(m,n)e\,. \tag{22}$$

The mean total queue content at random slot boundaries is denoted by $\overline{u} = \overline{u}_1 + \overline{u}_2$.

Let $\theta(m,n)$ denote the probability to find $n-m$ class 1 and m class 2 packets in the queue during slots, that is, the queue is observed after all departures and arrivals occurred at the preceding slot boundary. Since all packets that are present in the queue during a random slot, either were already present at the preceding slot boundary or arrived at this slot boundary, we have,

$$\theta(m,n) = \sum_{j=0}^{n} \sum_{i=0}^{m} \pi(i,j)\tilde{A}_j(n-m-j+i, m-i)e\,, \tag{23}$$

for $n = 0,\ldots, N$ and for $m = 0,\ldots,\min(n,T)$. From (23), we now easily retrieve expressions for the moments of the queue content at random points in time.

In particular, the mean class i queue content at random points in time $\overline{v}_i$ is given by,

$$\overline{v}_1 = \sum_{n=0}^{N} \sum_{m=0}^{\min(n,T)} (n-m)\theta(m,n)e, \quad \overline{v}_2 = \sum_{n=0}^{N} \sum_{m=0}^{\min(n,T)} m\theta(m,n)e. \quad (24)$$

Since there are $\tilde{\rho}_i$ class i arrivals at a slot boundary on average, we also have,

$$\overline{v}_i = \overline{u}_i + \tilde{\rho}_i. \quad (25)$$

That is, one does not need to obtain $\theta(m,n)$ to retrieve these mean values.

Let a packet's waiting time be defined as the number of slots this packet spends in the queueing system. By means of Little's result, we find that the mean class i ($i = 1, 2$) waiting time is given by,

$$\overline{w}_i = \frac{1}{\tilde{\rho}_i}\overline{v}_i = \frac{1}{\tilde{\rho}_i}\overline{u}_i + 1. \quad (26)$$

Notice that Little's result does not relate the mean waiting time to the mean queue content at random slot boundaries in this case (see Fiems and Bruneel [13]). This is due to the chosen order of arrival, observation and departure epochs in our queueing model.

4   Numerical Examples

To illustrate our approach, we now consider some numerical examples. In particular, the packets that arrive in the buffer are generated by M on/off sources. When a source is on at a slot boundary, it generates a class 1 packet with probability p and a class 2 packet with probability q. A source does not generate packets when it is off at a slot boundary. Further, given that a source is on (off) at a slot boundary, it is still on (off) at the following slot boundary with probability α (β). That is, the consecutive on-periods (off-periods) constitute a series of geometrically distributed random variables with mean $1/(1-\alpha)$ ($1/(1-\beta)$). The arrival process at the buffer is thus completely characterised by the quintuple (M, p, q, α, β). For further use, we also introduce the following parameters,

$$\sigma = \frac{1-\beta}{2-\alpha-\beta}, \quad K = \frac{1}{2-\alpha-\beta}. \quad (27)$$

The parameter σ denotes the fraction of time a source is on and K is a measure for the absolute lengths of the on- and off-periods. The parameter K takes values between $\max(\sigma, 1-\sigma)$ and ∞. For $K < 1$, $K = 1$ and $K > 1$ there is negative correlation, no correlation and positive correlation respectively in the arrival process. The quintuple (M, p, q, σ, K) also completely characterises the arrival process.

Consider now a buffer with capacity $N = 100$ and the arrival process as described above.

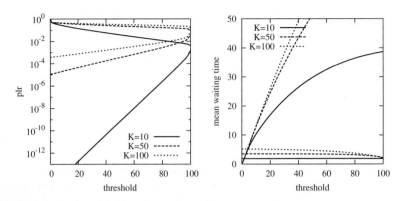

Fig. 1. Packet loss ratio (left) and mean waiting time (right) of class 1 and class 2 packets vs. the threshold T for various values of K. For each K, lower and upper curves correspond to class 1 and class 2 respectively.

In Fig. 1 the packet loss ratio (left) and the mean waiting time (right) of the class 1 and class 2 are depicted versus the threshold T for different values of K. For each K, the lower and upper curves correspond to the packet loss ratio (mean waiting time) of the class 1 packets and class 2 packets respectively. For all curves, packets are generated by $M = 4$ sources that are on during a fraction $\sigma = 0.25$ of the time. When a source is on it produces a class 1 packet with probability $p = 0.4$ and a class 2 packet with probability $q = 0.4$. The load of the system is thus 80%, half of which is class 1 traffic.

A first important observation is that service differentiation is clearly provided, in terms of the packet loss as well as in terms of the delay. This is a conclusion that can also be taken from all the following plots: the priority scheduling provides the delay differentiation while the threshold for class 2 packets takes care of the packet loss differentiation.

Further it can be seen from the plots in Fig. 1 that an increase of the threshold yields a decrease of the packet loss ratio of class 2 at the cost of an increase of the packet loss ratio of class 1. Indeed, for larger thresholds, less class 2 packets are dropped while the buffer is not yet full. As such, the total buffer content increases which in turn implies that the probability that the buffer cannot accommodate a class 1 packet arrival increases. Notice that when the threshold equals the buffer size, the packet loss ratio of class 1 and class 2 traffic are not equal. This comes from the fact that class 1 packets enter the queue *before* class 2 packets at a slot boundary. Further, an increase of the threshold leads to a decrease of the mean class 1 waiting time and an increase of the mean class 2 waiting time. When the threshold increases, class 2 packets are also admitted when there are more packets in the buffer. The waiting times of these packets are obviously longer. The decrease of the mean waiting time of the class 1 packets is explained by the fact that more class 1 packets are dropped which would have had a long waiting time when the threshold is set to a larger value. From Fig. 1 it can be seen that choosing the value of the threshold is a compromise between the packet

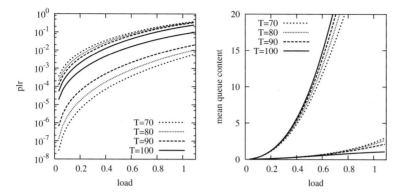

Fig. 2. Class 1 and class 2 packet loss ratio (left) and mean queue content (right) vs. the arrival load for various values of the threshold T. For each T, lower and upper curves correspond to class 1 and class 2 respectively.

loss ratio and the mean waiting time of the (high and low priority) packets. A higher threshold means that more high priority packets are dropped, but the ones getting through have a smaller delay, while a reverse effect is noticeable for the low priority packets. Finally note that the performance of the buffer system deteriorates for increasing arrival correlation. Both the packet loss ratio and mean waiting time of both class 1 and class 2 increases for larger K.

In Fig. 2 the class 1 and class 2 packet loss ratio (left) and mean queue content (right) are depicted versus the arrival load ρ for various values of the threshold T. For each T the lower and upper curve correspond to the class 1 and 2 packet loss ratio (or mean queue content) respectively. Packets are again generated by $M = 4$ sources. During on-periods a source generates a packet at 80% of the slot boundaries and half of these packets are class 1 packets ($p = q = 0.4$). Further, the burstiness factor is fixed and equal to $K = 50$ and the source is on during a fraction $\sigma = \rho/(M(p+q))$ of the slots.

As expected, both the packet loss ratio and the mean queue content increase when the load increases. Further, the class 2 packet loss ratio decreases for increasing values of the threshold as opposed to the class 1 packet loss ratio which increases. As was already noted in Fig. 1, this is explained by the fact that more packets of class 2 are allowed in the buffer such that more class 1 packets find the buffer fully occupied. This also explains that the average class 1 queue content decreases and the average class 2 queue content increases for increasing values of the threshold T.

In Fig. 3 the packet loss ratio (left) and mean waiting time (right) of class 1 and class 2 traffic is depicted versus the fraction $\gamma = \rho_1/\rho$ of the total arrival load that belongs to class 1. Different values of the total arrival load ρ are assumed as depicted. For each ρ the upper curve corresponds to the class 2 packet loss ratio (mean waiting time), the middle curve corresponds to the packet loss ratio (mean waiting time) of all traffic and the lower curve corresponds to the class 1 packet loss ratio (mean waiting time). As before, the arrival traffic is generated

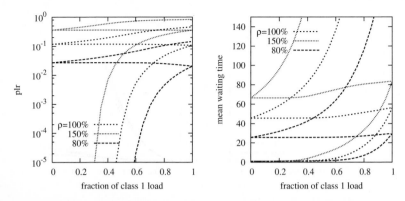

Fig. 3. Packet loss ratio (left) and mean waiting time (right) of class 1 and class 2 packets vs. the fraction of class 1 packets for different values of the total load ρ. For each ρ, lower, middle and upper curves correspond to class 1, to all traffic and to class 2 respectively.

by $M = 4$ sources. The threshold is fixed to $T = 80$ and the burstiness factor of the arrival process to $K = 100$. Each source is on during half of the slots ($\sigma = 0.5$).

For both class 1 and class 2, the packet loss ratio and mean waiting time increase for increasing γ. For class 1 packets, increasing γ means that more packets share preferential treatment and thus overall class 1 performance is worse. For class 2 packets, increasing γ means that more class 1 packets block acceptance or transmission of class 2 packets. This is a typical observation in the behaviour of priority queues. Further, the overall packet loss ratio decreases whereas the overall waiting time increases. It is immediately clear that the total packet loss ratio decreases when γ increases since less packets are affected by the threshold T. Also, the mean waiting time increases since more packets can enter the buffer when the buffer size exceeds the threshold when γ increases.

5   Conclusions

We considered a priority queueing system with partial buffer sharing and batch Markovian arrivals. By means of matrix analytic techniques we obtained various performance measures such as the packet loss ratio and mean delay of both the high and low priority class. We then illustrated our approach by means of some numerical examples. We observed that the buffer system under consideration clearly provides service differentiation, both in terms of the packet loss ratio and in terms of the mean delay.

Acknowledgements

This work has been partly carried out in the framework of the CHAMP project sponsored by the Flemish Institute for the Promotion of Scientific and Technological Research in the Industry (IWT). The second author is a Postdoctoral Fellow with the Fund for Scientific Research, Flanders (F.W.O.-Vlaanderen), Belgium.

References

1. Radha, H., Chen, Y., Parthasarathy, K., Cohen, R.: Scalable internet video using MPEG-4. Signal Processing: Image communication 15(1-2), 95–126 (1999)
2. Laevens, K., Bruneel, H.: Discrete-time multiserver queues with priorities. Performance Evaluation 33(4), 249–275 (1998)
3. Mehmet Ali, M., Song, X.: A performance analysis of a discrete-time priority queueing system with correlated arrivals. Performance Evaluation 57(3), 307–339 (2004)
4. Takine, T., Sengupta, B., Hasegawa, T.: An analysis of a discrete-time queue for broadband ISDN with priorities among traffic classes. IEEE Transactions on Communications 42(2-4), 1837–1845 (1994)
5. Walraevens, J., Steyaert, B., Bruneel, H.: Performance analysis of a single-server ATM queue with a priority scheduling. Computers & Operations Research 30(12), 1807–1829 (2003)
6. Walraevens, J., Wittevrongel, S., Bruneel, H.: Performance analysis of a priority buffer with train arrivals. In: Proceedings of the, International Symposium on Performance Evaluation of Computer and Telecommunication Systems (SPECTS 2006), Simulation Series 38(3), pp. 586–594. Calgary (July 2006)
7. Latouche, G., Ramaswami, V.: Introduction to matrix analytic methods in stochastic modeling. Series on statistics and applied probability. ASA-SIAM (1999)
8. Van Velthoven, J., Van Houdt, B., Blondia, C.: The impact of buffer finiteness on the loss rate in a priority queueing system. In: Horváth, A., Telek, M. (eds.) EPEW 2006. LNCS, vol. 4054, pp. 211–225. Springer, Heidelberg (2006)
9. Kröner, H., Hébuterne, G., Boyer, P., Gravey, A.: Priority management in ATM switching nodes. IEEE Journal on Selected Areas in Communications 9(3), 418–427 (1991)
10. Zhao, J., Li, B., Cao, X., Ahmad, I.: A matrix-analytic solution for the DBMAP/PH/1 priority queue. Queueing Systems 53(3), 127–145 (2006)
11. Spaey, K.: Superposition of Markovian traffic sources and frame aware buffer acceptance. PhD thesis, Universitaire Instelling Antwerpen (2002)
12. Blondia, C., Casals, O.: Statistical multiplexing of VBR sources: A matrix-analytic approach. Performance Evaluation 16, 5–20 (1992)
13. Fiems, D., Bruneel, H.: A note on the discretization of Little's result. Operations Research Letters 30(1), 17–18 (2002)

Unfairness in the e-Mule File Sharing System

Sanja Petrovic, Patrick Brown, and Jean-Laurent Costeux

France Telecom R&D,
Sophia-Antipolis, France
{sanja.petrovic, patrick.brown, jeanlaurent.costeux}@orange-ft.com

Abstract. Measurement studies have shown that the probabilities of
file possession and the arrival rates of file requests have an important
impact on the file downloading times in p2p file sharing systems resulting in unfair performances. A model has been proposed to capture this
phenomenon. This model is used to derive the qualitative impact of system parameters on performances. However to obtain results for realistic
p2p networks one is confronted with the large number of files in these
systems. To show the influence of the number of files shared in the system we present results obtained by simulations in the case of small to
medium size systems.

1   Introduction

Sixty percent of the Internet traffic belongs to peer-to-peer (p2p) file sharing
applications, according to our recent measurements made in France Telecom.
Moreover, this huge percentage has a growing tendency, reenforcing the motivation of better understanding these applications. Our motivation in this study
is to predict and explain the performances obtained by users downloading files
with these applications.

A file sharing application is composed of a network which defines a common
set of protocols to find and exchange files between peers. It is also composed of
client programs which are different programs which may be used by peers on
their computers to access the network. Among the p2p networks, eDonkey [3]
has the biggest number of users according to the statistics in [2]. Furthermore,
in Europe, this network is by far the most popular p2p network according to [1].
Its most popular client is eMule [4]. In addition the source of eMule is accessible
which allows to understand how it operates.

The eDonkey network is a file sharing network used predominantly to share
large files of tens or hundreds of megabytes, such as music, images, films, games,
and software programs. The network is decentralized, files are not stored on a
central server but are exchanged directly between users. eDonkey servers only
act as communication hubs for the users and allow users to locate files within
the network. Using the eDonkey client programs (with e-Mule among them),
peers connect to the network to share files, i.e. to offer files and to request files.
An important goal in peer-to-peer networks is that all clients provide resources,
including bandwidth, storage space, and computing power so that the system
may scale with the number of users.

L. Mason, T. Drwiega, and J. Yan (Eds.): ITC 2007, LNCS 4516, pp. 594–605, 2007.

A critical parameter with p2p file sharing systems is the upload bandwidth resource. Each user gets requests from the other users who want to download any file among the files he is sharing. Thus his upload bandwidth will be shared over time by the users requesting the files he owns. Having in mind that the majority of the eDonkey users utilize ADSL connections, which possess much bigger download bandwidth than upload bandwidth, the bottleneck is thus usually the upload bandwidth. We show that although users share their upload bandwidth fairly[1] among requesting peers the resulting performances are unfair.

In [11], we proposed a fluid model for eMule-like systems in order to study the effect of file popularity and arrival request rate on the file downloading time and the fairness issues which these performance raise (as we observed in [14]). However the complexity of the model increases quickly with the number of files in the system which may typically reach several million files in real systems. We present some numerical results of our fluid model on small test systems to characterize the qualitative influence of the system parameters on file download times. We then present simulations of small to medium size systems. We deduce some conclusions on the influence of the number of files shared on system performance.

This paper is organized as follows. In Section 2 we recall the fluid model used to predict the influence of the file popularity and the file arrival request rate on the downloading times. In Section 3 we present measurements of file possession probabilities, i.e. popularity, and of file arrival request rates. In Section 4 these measurements are used to define realistic input parameters to our model and to simulations to obtain numerical examples of file download performance.

2 Fluid Model for eMule-Like File Sharing System

2.1 Main Notations

Let us consider n files $\{f_i\}_{(i=1..n)}$, with the same size. We consider here an eMule-like system with n groups of peers, $\{g_i\}_{(i=1..n)}$, where the group g_i is composed of peers wanting to download file f_i.

The probability that peers in group i ($i = 1..n$) possess file j ($j \neq i$) is denoted by p_j. This probability describes the availability of replicas of a file or, in other words, a storage popularity of a file.

Peer arrivals to each group are modeled as a Poisson process with rates equal to $\{\lambda_i\}_{(i=1..n)}$. These arrival rates could also be seen as a request file popularity and it could differ from a storage file popularity. When a peer completes its download, it leaves the system. We consider that a peer shares the file it is currently downloading and the files that it already possesses if any. In our model, it is considered that λ_i are constant in time, which is coherent with our measurement. We observed that the file request popularity tends to increase suddenly and then to decrease gradually (see also [9]). This decrease is very slow compared

---

[1] eMule implements a mechanism favoring less popular files which will not be discussed here and which does not completely succeed in giving uniform performances.

to the observation time interval in our modeling, so that we can approximate it with a constant, i.e., the arrival rate changes sufficiently slowly so that the system is always near steady state.

We assume the same upload bandwidth for each peer, denoted by μ, meaning that the average uploading time of a file of unit size is equal to $1/\mu$.

The number of downloaders of file f_i, i.e., the population size of group g_i in the system at time t, is denoted by $x_i(t)$.

2.2 Assumptions

Let us present the assumptions and properties of the model regarding the connectivity among peers, the downloading policy and the uploading policy on each peer in the model.

- Connectivity: All peers are mutually connected. A request of any peer for the file is directed to all peers sharing the file.
- Downloading policy: Each peer downloads exactly one file at the time. This assumption is justified by the fact that most peers in the real eMule system are ADSL users thus often the bottleneck is in the upload bandwidth and not in download bandwidth. We do not consider peers that abort downloading.
- Uploading policy: Each peer can share the files that it already possesses and the file which is currently being downloaded. Shared files can be uploaded to other peers who request them. The existence of a bottleneck in the upload bandwidth will cause the concurrence and hence the influence between uploaded files. The uploading policy is modeled with the processor sharing policy: the upload bandwidth is equally shared among all peers that request shared files. The motivation for this kind of modeling lies in observing the real eDonkey system.

2.3 Model for Two Groups

We start with the case of two groups ($n = 2$), in order to better explain the construction of our model. So, each group of users is modeled with a fluid, in such a way that the whole system is modeled with two fluids that mutually interact. Each user gets requests from the other users who want to download any file among the files he is sharing. The upload file policy is modeled as processor sharing with a share proportional to the number of requests for each shared file.

Let us consider the evolution of x_1 (see Figure 1). For the first group, we have the following evolution

$$\frac{dx_1}{dt} = \underbrace{\lambda_1}_{1} - \underbrace{(1 - p_2)x_1\mu}_{2} - \underbrace{\frac{x_1}{x_1 + x_2}p_2 x_1\mu}_{3} - \underbrace{\frac{x_1}{x_1 + x_2}p_1 x_2\mu}_{4},$$

where each term can be interpreted in the following manner:

- Term 1 represents the arrival rate of requests.
- Term 2 represents contribution of users in the first group who do not posses f_2 and who thus dedicates all their upload resources for uploading f_1.
- Term 3 represents contribution of users in the first group who possesses f_2. Hence these users have $x_1 + x_2$ requests pending so they proportionally dedicate their upload resources for uploading f_1.
- Term 4 represents contribution of users in the second group who posses f_1. Hence these users have $x_1 + x_2$ requests pending so they proportionally dedicate their upload resources for uploading f_1.

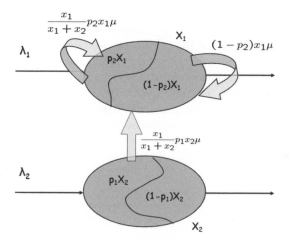

Fig. 1. Description of the different downloading events which occur in our emule-like model

We proceed in the same way with group g_2 and after some simple calculations we obtain the system in the following form:

$$\begin{cases} \dfrac{dx_1}{dt} = \lambda_1 - \mu x_1 + \mu(p_2 - p_1)\dfrac{x_1 x_2}{x_1 + x_2} \\ \dfrac{dx_2}{dt} = \lambda_2 - \mu x_2 + \mu(p_1 - p_2)\dfrac{x_1 x_2}{x_1 + x_2}. \end{cases} \qquad (1)$$

We studied this system in [11] where we proved its convergence to the unique stable point. In the following we present some conclusions.

Applying Little law's formula to the population of each group, we obtain:

$$\overline{x_1} = \lambda_1 T_1, \quad \overline{x_2} = \lambda_2 T_2, \qquad (2)$$

where T_1 and T_2 are the average downloading times respectively for groups g_1 and g_2. Also, the following conservation law ties T_1 and T_2:

$$T = \frac{\lambda_1}{\lambda}T_1 + \frac{\lambda_2}{\lambda}T_2 = \frac{1}{\mu}, \qquad (3)$$

i.e., the average downloading time scales perfectly with the request arrival rate λ_i, and depends only on the upload bandwidth μ.

We note also that equation (3) expresses a conservation law between T_1 and T_2: any improvement in T_1 with a value of T_1 smaller than $\frac{1}{\mu}$ results in a value of T_2 higher than $\frac{1}{\mu}$. A change in the system parameters cannot improve or deteriorate the average system performance but only increase or decrease the fairness in performance.

We distinguished a request popularity (i.e., the arrival rates of downloading requests) from a storage popularity (i.e., a probability for a peer to have a file) and investigated the influence of these popularity parameters on the system performance (i.e., average downloading times for different files). First, the average times depend only on α (with $\alpha = \frac{\lambda_1}{\lambda_2}$) and p (with $p = p_1 - p_2$). Second, it shows that the higher value of the parameter p_i determines which group has better performance but the amplitude of this difference in performance depends additionally on the ratio of the arrival rates of requests for each group (α) (Figure 2 and 3). Finally, for $p_1=p_2$, x_1 converges to $\frac{\lambda_1}{\mu}$ and x_2 to $\frac{\lambda_2}{\mu}$. Thus $T_1=T_2=\frac{1}{\mu}$. Interestingly, this is also the performance which would be obtained if both groups were disjoint and only shared their common file as in BitTorrent [15].

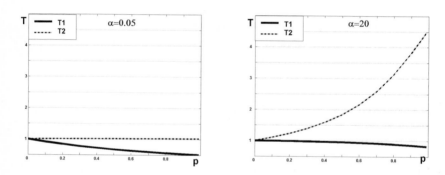

Fig. 2. Downloading times for groups 1 and 2, as a function of the difference in storage popularity (p_1-p_2). Upload capacity is normalized, $\mu=1$. The ratio of group arrival rates are kept constant at a small value ($x \ll 1$) on the left and at a large value ($x \gg 1$) on the right.

2.4   General Case

Indeed, we can follow the same reasoning with n groups and propose a dynamical system with n variables. Also, similarly to (3) the T_i are tied by the following conservation law:

$$T = \sum_{i=1}^{n} \frac{\lambda_i}{\sum_j \lambda_j} T_i. \qquad (4)$$

As a consequence the average system performance may not be improved (or worsened). In this context the criterium of interest is the fair redistribution

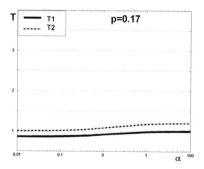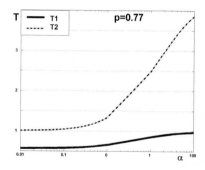

Fig. 3. Downloading times of groups 1 and 2 as a function of the ratio of groups arrival rates $(\frac{\lambda_1}{\lambda_2})$. Upload capacity is normalized, $\mu=1$. The difference in storage capacity $(p_1 - p_2)$ is kept small on the left graph and large on the right.

of system resources brought by the users, i.e., the fairness in performance dedicated to different groups. However, studying such a system for large n becomes impossible since the evolution of each variable is defined by a sum of $2^{n-2}(n-1)+2$ terms (see [11]). We claim that the previous analysis of case $n = 2$ gives an intuition of what happens with more users, and helps us to predict the behavior of the system. However, the system with n files presents some differences: the order of the performance among the groups (T_i) can no longer be precisely predicted by observing the order among p_i. In the case with n files, the influences on one group by the other groups can be with a different direction: one group may speed up group i's download, while an other group may slow it down. The bigger influence between the two will determine the direction of the evolution of the performance of the considered group. However, we show that for large values of n and for sufficiently distinct file storage popularity, p_i and p_j, the order of popularity (say $p_i > p_j$) dictates the order of performances $(T_i < T_j)$.

This analysis helped us to conclude that the uploading policy on peers, which treats all requests equally, results in an unfair sharing of the upload resources in the system. We observe also that the improvement in downloading time for very popular files is limited while the deterioration in downloading time for unpopular files can be unlimited. It can lead to the situation where even with the existence of a requested file in the network it becomes in practice impossible to download it.

3   Measurements

3.1   Settings and Methodology

The probe, which is used in our measurements, collects ADSL packets from a switch between a BAS (Broadband Access Server) and the IP backbone. All TCP packets without any sampling or loss are captured by using the adapted version of tcpdump. We denote as *local* peers or users the ADSL hosts connected to the observed BAS, and as *non-local* or *distant* peers the remainder of the hosts.

The identification of the eDonkey protocol is done through packet inspection. All eDonkey control packets are wrapped in a special header started with the eigth-bit character *0xe3*. Only several eDonkey message types are used for our measurements: LOGIN message (*0x01*), HELLO message (*0x01*), FILE_REQUEST message (*0x58*) and FILE_STATUS (*0x50*).

The distinction among the p2p users was based on eDonkey user ID, i.e., USER_HASH. USER_HASH is a unique 16 byte number which is used in eDonkey system in order to identified a user across all its sessions. Contrarily, a user identification based on IP address can be wrong because of existence of NATs (Network Address Translator) and dynamic IP addresses.

Note that data is made anonymous and there is no stable storage of connections after the analysis. We analyzed the traffic over a period of 3.5 days in September 2006 with 4861 local eDonkey users and 4041825 non-local eDonkey users. During the observed period, these users requested for 51553 different files. The files were identified using their MD4 hash, which is globally unique ID computed by hashing the file's content.

3.2   File Request Popularity Versus File Storage Popularity

In the previous studies (see [5], [9]), up to our knowledge, file popularity was measured either by the number of replicas in the system, or by the number of requests for the file, with the hypothesis that the two measures lead to the similar popularity characterization. Contrarily, we distinguish here file request popularity from file storage popularity, and we give their correlation.

The measurements based on the collection of 19944 files and shown in Figure 4. Note that considering the file request popularity only the first request per user is taken into account. Also, in Figure 4 values of file request popularity are normalized by the value for the most requested file which is 5025 times for 3.5 days.

The measurements reveal that the increase of file request popularity is followed by the increase of a range of the corresponding possible values of the file storage popularity. Additionally, files with a small request popularity may have only small storage popularity, these correspond to old files or globally non popular new files. Contrarily, files with the high file request popularity may have from small (for recently introduced, new and very popular files) to high storage popularity.

Further, in Figure 5 we show the distribution of file storage popularity obtained from the previous measurements. This distribution is in accordance with the previous studies in literature (see [9]). After a moderately flat region the popularity curb is very close to a Pareto distribution, revealing that in the network there are many unpopular files, i.e. files with few replicas.

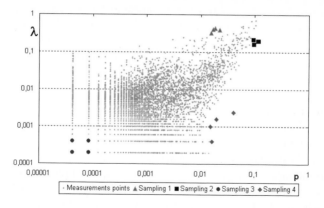

Fig. 4. File request popularity versus file storage popularity. Points labeled *sampling* were obtained with Equation (5).

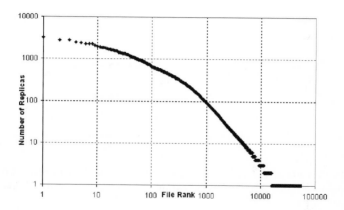

Fig. 5. File storage popularity

4 Numerical Analysis

Our intention is to better understand what the model predicts and the consequences of this mechanism for real values of p_i and λ_i (Figure 4). However, we must turn to numerical analysis due to the complexity of the analytical analysis and the high dimension model in such a case.

Here we present results of six different experiments done by numerically solving of the system of equations with $n = 20$ in order to discuss which values of system parameters have dominant influence over system performance.

In the first experiment a *basic sampling* is performed, i.e., for parameters p_i we choose a sequence with a distribution as observed in the measurements

(Figure 5). Then, for each chosen value of parameter p_i, we choose the parameter λ_i by approximating the measurement with the polynomial of 3 degree defined by:

$$\lambda = 20.29p^3 + 14.56p^2 + 2.56p + 0.0009, \qquad (5)$$

Note that the sampling points are shown in Figure 4.

In order to examine the influence of the different category of files, samplings for the four next experiments are done in the following way. First, we start with the 16 points obtained as in the *basic sampling*. Then, for each experiment we add 4 points (Figure 6) which we call *additional points*.

In *Sample 1* these *additional points* are taken with high values of both storage and request popularity while in *Sample 4* similar values of parameters p_i are used but with much smaller values of parameters λ_i. In *Sample 2 additional points* are chosen with very high values of p_i and values of λ_i as in *Sample 1*. Finally, in *Sample 3, additional points* represent files with small values of both parameters.

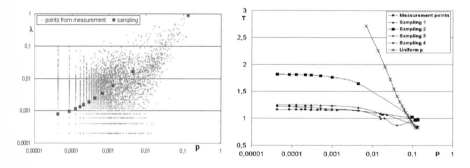

Fig. 6. Left: Position of the *additional points* for each sample with respect to the measurement points. Right: Average downloading time for each sample. Upload capacity is normalized, $\mu=1$.

Finally for the last experiment parameters p_i are chosen with uniform distribution instead of the observed distribution in Figure 5.

Results of the simulations of each of the previous experiments are presented in Figure 6 where we have normalized the upload capacity, $\mu = 1$. Note then that $T = 1$ corresponds to the overall average download time as seen in Equation (3).

The results reveal that the system does not produce fair performance, meaning that the performance for files with different popularity are not the same. More precisely, only a small number of files improve their performance while a lot of less popular files have performance below average.

Further, comparing the first five experiments, it can be noticed that *additional points* with larger p_i increase the unfairness in the system. Variations in parameters p_i and λ_i produce the same qualitative effect on performance (see also Section 2) but the influence of parameters p_i is much sharper. This can be seen by comparing *Sample 1* and *Sample 4* (i.e. similar p_i, different λ_i), and *Sample 1* and *Sample 2* (similar λ_i, different p_i).

Now, let us compare the first five experiments with the last experiment with uniformly distributed p_i parameters. We observe that the increase in concentration of more popular files results in much more unfair redistribution of upload resources in the system.

We remark that because of the Pareto distribution of file storage popularity, but also because of the upload bandwidth resource sharing defined in the fluid model, most files will have bad performance (i.e, $T_i/T > 1$), and only the small number of popular files will benefit from that. More popular files will take an unfair share of upload resources from less popular groups. The size of this share depends on the difference in parameters p_i but also on the ratio of λ_i parameters. As a consequence, it is interesting to note that each newly introduced file will have worse performance $(T_i/T > 1)$ because its storage probability p_i is very small. Also, among the newly introduced files the one with the bigger arrival rate will have better performance (see [11]).

As it is previously said the points with higher parameter p_i have a big influence on the redistribution of system resources. In order to see what happens when a less popular file is introduced in the system, we analyzed a system with the first n points with the largest value of p_i (n=19 and 20, Figure 7).

It can be noticed that introducing a single *test point* with small p_i, the performance of the first 19 points do not change significantly while the performance of less popular file shows an important deterioration.

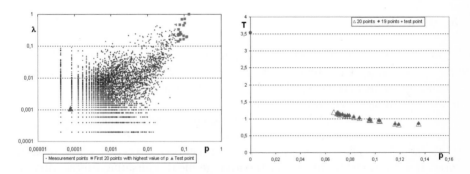

Fig. 7. Left: File popularity and arrival rates used to represent the n most popular files and the test point in the graph on the right. Right: File download times obtained when considering population composed of the n most popular files for n=19 and 20 points. For n=19 a test point is added for less popular file.

When turning to larger n, the numerical analysis of the fluid model reaches its limitation as a consequence of the large number of terms that then appear. Thus we run discrete event simulations in order to perform a qualitative analysis for larger systems. Results for systems with sizes of n=50, 100 and 200, are presented in Figure 8. In each simulation, the considered set of n points are constructed taking the first n points with the highest values of p_i from the measurement result (Figure 5). As the n points with largest p_i are chosen at each step n,

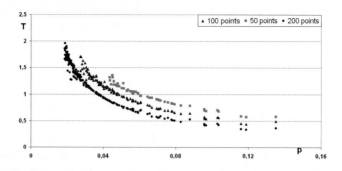

Fig. 8. Simulation results for the system size equals to n=50, 100 and 200

the simulation provides a worst case performance of all possible n-sized systems based on measurement data. We note that the performance do not stabilize when going from 50 to 100 files exchanged, and from 100 to 200 files. Our fluid model intrinsically assumes a large number of peers. In addition these results show that the very large number of files exchanged has a major impact on the system behavior. Interestingly, the results also reveal that the amplification of an average number of files shared per users leads to more unfair sharing of the upload capacities in the system among files with different popularity.

5   Conclusion

Measurement studies have shown that the probabilities of file possession and the arrival rates of file requests have an important impact on the file downloading times in p2p file sharing systems. This results in unfair performances attributed to users which contribute identically to the p2p network by sharing their resources. This contributes to the unequal performances observed on these networks when downloading different files.

To model the phenomenon a fluid model has been proposed which takes as input these parameters to predict the average file downloading times. After recalling explicit results for a system where two files are shared, we present numerical results for a system of n files. We show that large differences in file storage popularity (i.e. number of file replicas) result in pronounced unfairness in downloading times. When file request rates are correlated to file storage popularity, as can be seen in the measurements presented, the system unfairness is further amplified.

References

1. Cachelogic corporation. peer-to-peer in (2005) `http://www.cachelogic.com`
2. Slyck: News site with regular articles and statistics on the state of major p2p networks, `http://www.slyck.com`

3. User's guide with descriptions of the features of edonkey, application (2000) http://www.edonkey2000.com
4. User's guide with descriptions of the features of emule application, http://www.emule-project.com
5. Andreolini, M., Lancellotti, R., Yu, P.S.: Analysis of peer-to-peer systems: Workload characterization and effects on traffic cacheability. In: MASCOTS '04: Proceedings of the The IEEE Computer Society's 12th Annual International Symposium on Modeling, Analysis, and Simulation of Computer and Telecommunications Systems (MASCOTS'04), pp. 95–104. IEEE Computer Society, Washington, DC, USA (2004)
6. Clevenot, F., Nain, P., Ross, K.: Multiclass p2p networks: Static resource allocation for service differentiation and bandwidth diversity. In: Proc. PERFORMANCE 2005, pp. 32–49, Juan-les-Pins, France (October 2005)
7. Gaeta, R., Gribaudo, M., Manini, D., Sereno, M.: Fluid stochastic petri nets for computing transfer time distributions in peer-to-peer file sharing applications. Electr. Notes Theor. Comput. Sci. 128(4), 79–99 (2005)
8. Guo, L., Chen, S., Xiao, Z., Tan, E., Ding, X., Zhang, X.: Measurements, analysis, and modeling of bittorrent-like systems. In: Proceedings of Internet Measurement Conference 2005 (IMC 2005), pp. 35–48. Berkeley, California, USA (October 2005)
9. Handurukande, S.B., Kermarrec, A.M., Massoulie, L., Le Fessant, F., Patarin, S.: Peer sharing behaviour in the edonkey network, and implications for the design of server-less file sharing systems. Technical Report 5506, Inria (February 2005)
10. Kulbak, Y., Bickson, D.: The eMule protocol specification. Technical report, School of Computer Science and Engineering, Hebrew University of Jerusalem (January 2005)
11. Petrovic, S., Brown, P.: Fluid model for emule file sharing system. In: ECUMN, Toulouse, France (February 2007)
12. Piccolo, F., Neglia, G., Bianchi, G.: The effect of heterogeneous link capacities in bittorrent-like file sharing systems. In: Int. Workshop on Hot Topics in Peer-to-Peer Systems (HOT-P2P 2004), pp. 40–47, Volendam, Nederlands (October (2004)
13. Plissonneau, L., Costeux, J.L., Brown, P.: Analysis of peer-to-peer traffic on adsl. In: Passive and Active Network Measurement, pp. 69–82. Boston, USA (March/April 2005)
14. Plissonneau, L., Costeux, J.L., Brown, P.: Detailed analysis of edonkey transfers on adsl. In: 2nd EuroNGI Conference on Next Generation Internet Design and Engineering, Valencia, Spain (2006)
15. Qiu, D., Srikant, R.: Modeling and performance analysis of bittorrent-like peer-to-peer networks. In: SIGCOMM, pp. 367–378 (2004)

On Uncoordinated File Distribution with Non-altruistic Downloaders

Ilkka Norros[1], Balakrishna Prabhu[2], and Hannu Reittu[1]

[1] VTT Technical Research Centre of Finland, P.O. Box 1000, 02044 VTT, Finland
`firstname.lastname@vtt.fi`
[2] CWI, Postbus 94079, 1090 GB Amsterdam, The Netherlands
`bjprabhu@gmail.com`

Abstract. We consider a BitTorrent-like file sharing system, where the peers interested in downloading a large file join an overlay network. The seed node possessing the file stays in the system, whereas all other peers are non-altruistic in the sense that they leave the system as soon as they have downloaded the whole file. We consider a flash crowd scenario, where the peers join the overlay simultaneously. We show that the chunk selection algorithm is critical, propose an analytic approach to the process, and find that the encounters can be restricted to neighbours in a Chord overlay without losing much in performance.

1 Introduction

BitTorrent [1] revolutionized the technique of distributing a large file to a large number of recipients. The file is chopped into small chunks that the recipients can immediately upload further. A salient feature of such networks is the scalability of the service capacity with offered load [2], resulting in efficient utilization of network resources. Thus, even a seed node with a relatively small available upload bandwidth is able to distribute a file to a large number of nodes in a reasonable amount of time. In the original BitTorrent, a "tracker" keeps certain centralized control over the chunk transfer process. This paper studies file distribution systems with fully distributed and symmetric architecture.

The Finnish research project PAN-NET (2004-2005) designed an experimental prototype of a BitTorrent-counterpart with completely distributed architecture. This system is based on random encounters, i.e., the peers contact each other randomly and transfer one (or several) chunk(s) if available. The PAN-NET prototype uses Chord [3] as the underlying structure where random contacts can be realised.

Massoulie and Vojnovic [4] studied a similar system as an abstract model and obtained remarkable results by considering a limiting differential system in a linear scaling of the initial state of the system. In particular, they could conclude that a random encounter based file sharing system can be essentially insensitive to load balancing strategies like the principle to transfer always the rarest available chunk first, which is implemented in BitTorrent. This is interesting,

L. Mason, T. Drwiega, and J. Yan (Eds.): ITC 2007, LNCS 4516, pp. 606–617, 2007.

because several simulation studies have shown that the *rarest first* policy outperforms other policies such as *random* selection [5], [6], and hence could be one of the main reasons for the good performance of BitTorrent. The above results are not contradictory, because the insensitivity conclusion of [4] presupposes that the system has already reached a sufficiently balanced state. In particular, they show that a sufficient condition for the preservation of a balanced chunk distribution is that the nodes join the overlay with a uniformly distributed first chunk in possession.

With this background, we focused in a recent paper [7] on the so called *flash crowd* scenario, where all nodes except the seed join the file distribution overlay simultaneously and empty-handed. This is an extreme case of the so-called flash-crowd scenario in which a large number of peers try to simultaneously acquire a copy of a popular software. It is also a model for distributing a new file in an existing network, say, within a company. We assumed that the nodes are *non-altruistic* in the sense that they stay in the system only as long as they have not downloaded all the chunks. This is an interesting scenario both as a pessimistic extreme and as being complementary to steady-state-oriented scenarios.

The main findings of [7] were: (i) The *random* chunk selection policy easily results in one chunk (usually exactly one!) becoming rare in the system. (ii) The replication of chunks resembles the growth of the number of balls in Pólya's urn model. This analogy helps understanding how the early history of the random copying process has a long-lasting influence on the chunk distribution. (iii) We provided an extremely simple distributed load balancing algorithm, the "Deterministic last K chunks policy", which outperforms the *random* selection policy without using any state information.

The present paper continues the work started with [7] mainly in two respects. First, we draw into consideration the principles of the optimal chunk distribution algorithms (thanking Michela Meo and Marco Mellia for raising this question), despite the fact that they require complete coordination and are as such unusable in a distributed control scenario. Using the optimal solution as a heuristic, we experimented by replacing unrestricted random encounters by chunk downloads from Chord topology neighbours only and found encouraging results. Second, we provide new analytical insight into the random process of the system's big bang -like evolution.

Related Work

Performance of file sharing systems has been studied both by simulations and through mathematical models. The large number of parameters and complicated algorithms make accurate mathematical models intractable. Hence, articles such as [4], [8] and [9] have modelled the system in limiting regimes. Fundamental performance limits [10,11] in this connection are discussed in Section 2. The models in [8] and [9] mainly deal with systems in which the entire file is considered as a single chunk, and the steady state performance measures (e.g., time to download the file) are then studied through fluid models. Simulation studies such as [5] and [6] allow to study the effect of various parameters (e.g., the chunk selection policy, the peer selection policy, and heterogenous upload and download bandwidths) on the performance of the system. Our observation regarding the existence of rare chunks in random chunk selection policy is consistent with that in [5].

Organisation of the Paper
We start in Section 2 by analysing an optimal solution to the file distribution problem and identify four interesting features of it. In Section 3, we present the models of uncoordinated file distribution strategies that will be studied in the sequel. Section 4 proposes a novel analytic scheme to describe the evolution of the "chunk universe" in the "big bang" of the flash crowd scenario. In Section 5, we present and discuss the results of several simulation setups. Finally, we draw the conclusions and propose items for further work in Section 6.

2   About the Optimal File Distribution Speed

Although our focus is on completely distributed algorithms, it is interesting to note how fast a coordinated procedure can distribute C chunks of same size to N peers with identical and finite upload capacities. The answer is $C + \log_2 N$, with single chunk copying time as the time unit. In the case $N = 2^L$, this has been known for long [11], but a proof for general N was given only quite recently by Mundinger *et al.* [10].

Remark 2.1. *We assume that chunks must be downloaded completely before they can be copied further to other peers. If the chunks were infinitely small, the distribution time would be independent of N.*

Although the algorithm for general N is quite complicated, it is based on some simple but important observations, formulated as lemmas in [10]. The most important one is that

(i) one should always transmit chunks at full speed, one at the time.

Interestingly, the classical BitTorrent uses typically four simultaneous downloads. Maybe this number is motivated by the heterogeneity of the peers' transmission speeds, but we don't know rigorous arguments for this choice. For small chunk size, other delays like queueing delays in the network start to affect along with the transfer delay. Then it may become favorable to split bandwidth to reduce these delays by parallel transmission.

When $N = 2^L$ with integer L, optimal file distribution algorithms need only the hypercube topology for chunk tranfer. Here is one of the variants. Let the node IDs be $(\iota_1, \ldots, \iota_L)$, $\iota_i \in \{0, 1\}$. Call two nodes ℓ-pairs if their IDs differ only in the ℓth coordinate. Now, the "copying machine" runs as follows:

Algorithm 2.2 *([11], see also [12].) Assume $C > L$.*
for $\ell = 1, \ldots, L, 1, \ldots, L, 1 \ldots$ *(cyclically)* **do**
  if *node $(0, \ldots, 0)$ has less than C chunks*
    then *node $(0, \ldots, 0)$ downloads next chunk from seed*

else *the ℓ-pair of node $(0,\ldots,0)$ downloads chunk C from seed
every other node downloads its ℓ-pair's highest-numbered chunk*

For any C and L with $C \geq L + 1$, this "machine" distributes all chunks to all nodes in $C + L$ cycles. We note additional interesting features of this algorithm:

(ii) each node needs only $\log_2 N$ neighbours with whom to exchange chunks
(iii) most of the transfers are bi-directional, chunks moving along the hypercube edges in both directions
(iv) no nodes (except the seed) need to be altruistic. (For this we required $C > L$ — see what happens if $C = 1$ and the peers are non-altruistic!)

The reason for the sufficiency of the hypercube topology is that the number of copies of each chunk can at most be doubled in each period. It is easy to see that copying cyclically to hypercube neighbours, once to each, distributes a single chunk to all nodes. With appropriate organization, this process can run for $\log_2 N$ chunks in parallel and even so that all nodes except one finish simultaneously.

It is clear that only property (i) is a useful rule in most designs: the faster the chunks are transferred, the faster proceeds the copying process. A stochastic variant of property (ii) will be studied in Section 5 by simulations. Although this Chord-based solution does not reach optimal performance, the results are encouraging. The usefulness of property (iii) is questionable in an uncoordinated setup, but we have not experimented on it yet. Finally, property (iv) is obviously not true in general, on the contrary — the presence of altruistic peers is certainly advantageous in fully distributed schemes like ours. On the other hand, it is useful to know that if the copying process is sufficiently well organized, altruism is not necessary for fast file distribution.

3   Fully Distributed File Sharing Strategies in Overlays

The basic scenario of this paper is the following. One peer, called the seed, possesses a file, chopped into chunks, and stays in the system as long as it wants to distribute the file. The seed maintains also an overlay network, being initially the network's only node. Each peer interested in downloading the file joins the overlay and obtains the possibility to contact randomly selected peers of the overlay. (An algorithm for finding almost uniformly random peers was given in [7].) The peers then initiate independently such "random encounters" and, if the counterpart possesses chunks which the initiator does not have, a chunk is downloaded by the initiator.

This scenario leaves many algorithms and their parameters to be specified:

(i)   the overlay structure and its maintenance mechanism
(ii) the peer selection mechanism for encounters
(iii) the rule for selecting the chunk to be downloaded
(iv) the number of chunks tranferred in one encounter
(v)   chunks can also be transferred in both directions within the same encounter

(vi) transfer speed

(vii) number of parallel down- and uploads involving one peer

(viii) distribution of the time a peer stays in the system after having downloaded the whole file.

We make comments on each item and explain the choices made in this paper.

About (i): The overlay structure used in our experimental prototype is Chord [3], which is a popular example overlay systems based on a Distributed Hash Table. In such networks, the nodes take random or hashed identifiers from a hash value space $I = \{0, 1, \ldots, 2^M\}$, and the routing and resource discovery mechanisms are based on these identifiers. In the case of Chord, the hash value space is interpreted as a ring (in mathematical modelling it could be represented as the unit circle S^1). For $x \in I$, let $\texttt{succ}(x)$ denote the node whose identifier is closest to x in clockwise direction. A node with identifier i maintains connections to nodes with identifiers $\texttt{succ}((i+2^k) \bmod 2^M)$, $k = 0, 1, \ldots, M-1$. These nodes are called the *fingers* of node i. With N nodes, a typical number of separate fingers of a node is $\log_2(N)$. The finger connections are the edges of the Chord graph.

About (ii): In our original setup, random peers were encountered uniformly over the whole population of $N + 1$ peers. A novelty in this paper, motivated by the considerations of Section 2, is the restriction to encountering only the fingers of each node, and using cyclic instead of random selection among them.

About (iii): A good rule would be to choose the chunk whose copies are rarest in the whole population. This information is not available in a distributed algorithm. The strategies we compare in this paper are random choice and the "deterministic last chunk" policy, introduced in [7] (see Section 5.2).

About (iv), (v), (vi), (vii): We did not study variations of these parameter. On the contrary, we adopted in both analysis and simulations the abstract model of [4], where nodes make encounters according to independent Poisson processes with unit rate, and the download time is set to zero.

About (viii): Throughout this paper, we assume that the peers are non-altruistic, i.e., the last parameter is concentrated at zero. This is a natural pessimistic assumption. (We have studied the effect of altruism elsewhere and found it significant in the last phase of the process.)

4   Towards an Asymptotical Analytical Model

We wish to extend our previous results [7] on Pólya's urn models as simple models for the starting phase of the random chunk copying process. We take the case of two chunks, 0 and 1. In total there is a fixed number of $n+2$ nodes in the system. At the start, one random node has chunk 1 and another distinct random node has 0. The remaining n nodes do not have any chunks, they are the empty nodes. At each time step, a uniformly random ordered pair of nodes is selected and called a random encounter. The first node downloads a chunk that it does not have, if the second has it but does not have both chunks. The last condition means unaltruistic behavior. As soon as a node has both chunks it becomes uncooperative and does not share its file. For large n, this copying process is likely to start as a Pólya's urn: first there are two balls (0 and 1) in the urn, then, at each time step, one ball is selected uniformly randomly from the urn and the selected ball is returned to the urn along with its copy. At any moment of time the distribution of all possible outcomes is uniform. For a large time it stabilizes to some value that is still uniformly distributed. However, this model is a pure growth process. We would rather have a process that has a steady state, meaning that, eventually, no encounters can change the state of the system. This means, for instance, that one chunk has become extinct outside the nodes that have both chunks and all empty nodes have also disappeared. Another option is that both chunks have disappeared in a similar way, in which case some empty nodes could be present also.

Our idea is to assume n large and use first Pólya's model to cope with the starting phase of the process. Then, for considerable populations of chunks, a continuous model in the form of differential equations is probably plausible. Finally, we must fuse these approaches to describe the distribution of the final states as $n \to \infty$. So far, we have made only the first steps rigorously, but we have some conjectures how to complete this scenario. First, it can be shown that the probability $p_0(t)$ that after t steps from the start (we count only steps that result in the change of the state) there have been no encounters between two nodes possessing 0 and 1, respectively (in other words, at each step an empty node has encountered a node having 0 or 1):

$$p_0(t) = \frac{1}{2(n+1)} \frac{n!}{(n-t)!} \times \sum_{a \in \{0,1,\cdots,2^t-1\}} \{f_t(a)!(t - f_t(a))!$$

$$\times \prod_{s=2}^{t} \frac{1}{(s+1)(n-s+1) + 2(f_{s-1}(a)+1)(s - f_{s-1}(a))}\},$$

where a runs through all integer values from 0 to $2^t - 1$, and the functions f_s, $s = 1, 2, \cdots, t$ of a are defined as follows: write $a = \sum_{k=1}^{t} a_k 2^{t-1}$ with $a_k \in \{0,1\}$, then $f_s(a) = \sum_{k=1}^{s} a_k$. Using this presentation we can show:

Proposition 4.1. *Assume that* $t, n \to \infty$ *in such a way that* $t/n^{\frac{1}{3}} \to 0$. *Then* $p_0(t) \to 1$.

This means that, asymptotically for large n, the system behaves at the beginning for long time like a Pólya urn, and only after a "macroscopic" time the encounters between non-empty chunk holders become likely. Another limiting case is obtained when the number of chunks is already very large. Then it is reasonable to shift to continuous modelling and continuous time t and assume densities of the chunk holders: $n_0 = |(\text{nodes with only chunk 0})| / (n+2)$ and similarly n_1 and $n_\emptyset$ for chunk 1 holders and the empty nodes, correspondingly. They would evolve according to an autonomous system of differential equations:

$$n'_0(t) = n_0(t)(n_\emptyset(t) - n_1(t)) \tag{1}$$
$$n'_1(t) = n_1(t)(n_\emptyset(t) - n_0(t))$$
$$n'_\emptyset(t) = -n_\emptyset(t)(n_0(t) + n_1(t)),$$

where $'$ denotes the time derivative.

Now, initial conditions are needed. Returning to the discrete system, at the moment when the number of chunks reaches the value n^α with $0 < \alpha < \frac{1}{3}$, the distribution of the number of chunk holders is uniform. The precise value of α is not principal. Denote $c = n^\alpha/n$. At this instance, we heuristically fuse the two approximations: take the initial conditions as $n_0(0) = \beta(c)c$, $n_1(0) = (1 - \beta(c))c$, and $n_\emptyset(0) = 1 - c$, where $\beta(c)$ is a random variable taking values in the interval $[0, 1]$. We conjecture that, as $n \to \infty$, $\beta(c)$ converges to a uniformly distributed random variable $\beta(+0) \sim U[0, 1]$, as it should reflect the Pólya-like phase of evolution. The steady state is reached at time $t \approx \log n$, due to the exponential growth at the start. Numerical simulations indicate that the system of equations above really have such a limit for the steady state, see Figure 1 below. On the other hand, this limiting scheme seems to be a quite realistic approximation already for such a small number of nodes as 100. This approach could probably be extended to more realistic scenarios with more chunks and the seed. This would result in a different distribution for $\beta(+0)$.

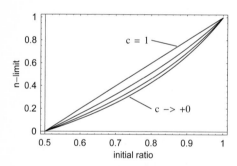
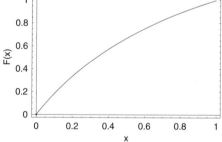

Fig. 1. Steady state solution for the larger component of the chunk holders' density as a function of initial conditions on the interval $(0, c]$. As $c \to +0$, this function seems to converge. The values were taken as $1, 0.5, 10^{-3}, 10^{-6}, 10^{-20}$ with seemingly monotonous convergence. The curves of two smallest values merge visually.

Fig. 2. Distribution function $F(x)$ for the steady state value of the successful (possessing both chunks) node density x. Solid line corresponds to the solution of the differential equations with $c = 10^{-20}$ and the points for a simulated system with $n = 98$ with experiments repeated 1000 times.

5   Simulation Results

In our simulations, we always assume that N nodes arrive simultaneously to the system at time zero. The file to be broadcast is available as a set of C chunks at the seed node. In order to download these chunks, each node initiates encounters with other nodes in the system, and downloads missing chunks. Having downloaded the whole file, the node leaves the system. Each node initiates encounters as a Poisson process with rate one, and the download time of a chunk is zero. The main metric of interest in this model is the average time to broadcast the file to N nodes.

We shall simulate the above system for two different topologies. In the *fully-connected* topology, a node can contact any other node present in the system. The second topology is the Chord topology which was described in Section 3.

5.1   The Rare Chunk Phenomenon

We shall first assume that the network graph is fully connected, i.e., a node can initiate an encounter with any other node present in the system.

In Figure 3, we plot the expected number of uploads of a chunk by the seed node in ascending order of the number of uploads for $C = 100$.

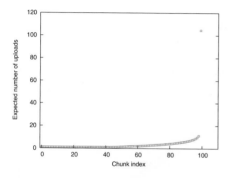

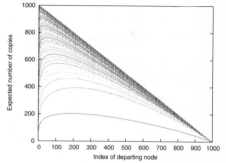

Fig. 3. Expected number uploads of chunks by the seed node. *Random chunk policy.* $N = 1000$. $C = 100$.

Fig. 4. Expected number of copies of the chunks as seen by departing nodes. *Random chunk* policy. $N = 1000$. $C = 100$.

From this plot, we can observe that there exists one chunk which the seed node has to upload a disproportionate number of times compared to the other chunks. This suggests that there exists an imbalance in the number of replicas of the chunks in the system. This phenomenon becomes critical towards the end of the system lifetime when the nodes would need to download this rare chunk from the seed node. In Figure 4, we plot the expected (ordered) number of replicas of chunks in the system as seen by the nth departing node. We observe that the number of replicas of the chunks which are distributed first tend to grow faster than those which enter the system later. Moreover, we see that, after a while, the very last chunk becomes more and more rare, whereas the others tend to good balance with each other. A similar observation was made in [5].

5.2   Deterministic Last K Chunks Policy

We can summarize the observations from the previous sections as follows.

(i) The imbalance in the number of replicas of chunks in the system can be traced to the first few moments which determine which chunks replicate faster than the others. It would be desirable to have a more balanced number of replicas of chunks at the start of the system as this would lead to a more predicatable number of replicas of chunks in the system at later instants.

(ii) It would be very desirable to have high diversity in the last missing chunk of each node, say, to be uniformly random and independent of that of all other nodes. This would make the seed node less critical to the downloading process.

A decentralized way to ensure a better balance in the number of replicas of chunks during the first few moments and to ensure a high diversity in last missing chunk was proposed in [7]:

> **Deterministic last K chunks policy.** *Every node selects $K < C$ chunk indices at random. These chunks are downloaded only after the remaining $C - K$ chunks have been downloaded.*

The order in which these K chunks will be downloaded is arbitrary and is not decided beforehand.

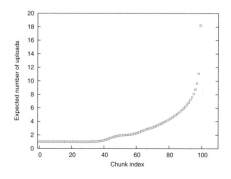

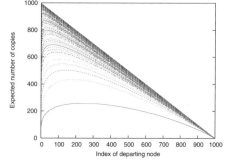

Fig. 5. Expected number uploads of chunks by the seed node. *Deterministic last chunk* policy. $N = 1000$. $C = 100$. $K = 1$.

Fig. 6. Expected number of copies of the chunks as seen by departing nodes. *Deterministic last chunk* policy. $N = 1000$. $C = 100$. $K = 1$.

For $K = 1$, we compare the performance measures of this policy with those of the random selection policy. The initial number of nodes in the system is the same as before (i.e., $N = 1000$). Figure 5 and 6 show the number of expected (ordered) number of uploads by the seed node, and the expected number of copies of each chunk at departure instants, respectively, for $C = 100$. On comparing these with

Figures 3 and 4 we see that the maximum uploads by the seed node has reduced compared with the random selection policy. The better performance of this policy can be ascribed to two reasons. Firstly, by rejecting some chunks that may be on offer in the initial stages of downloading, the policy delays the departure of the first few nodes from the system. Thus, there are more uploaders present in the system for longer duration, thereby reducing the load on the seed. Secondly, in the *random* chunk policy, the chunks which entered the system initially would spread quickly to all nodes. However in the *deterministic last K chunks* policy a proportion of nodes would reject these chunks thereby reducing the imbalance between the rapidly multiplying chunks and rare chunks. However, one cannot expect to improve the performance by increasing K a lot. As K increases, the number of unsuccessful encounters, i.e., encounters in which chunks are rejected, would also increase, and this would increase the broadcast time again.

5.3   Expected Broadcast Time

We now compare the time to broadcast all the C chunks to the N nodes using the *random* and *deterministic last K* policies. We are interested in the scaling of the broadcast time with the number of nodes. In order to do so we simulate the system for $N = 2^j, j = 1, ..., 9$. For the Chord topology, we shall always assume that the number of addresses available, 2^M, is twice the number of initial nodes in the system. In these simulations we considered an augmented Chord topology where each node was given two additional, randomly distributed fingers. The finger tables are updated instantaneously as soon as a node leaves the system. Each node initiates encounters by selecting one finger either at random or, as another variant, cyclically. Since the Chord topology is a random graph (it is random because the $N + 1$ different addresses are chosen randomly), the results for the Chord graph are obtained by averaging over different topologies.

We recall that in this model each node is initiating an encounter at rate one. Hence, the expected time to broadcast would give the average number of encounters initiated by the last departing node. In Figures 7 and 8, we plot the expected time to broadcast the file in both the Chord and fully-connected topologies and for the *random* and *deterministic last* chunk policy. A surprising and counter-intuitive observation from these figures is that the expected time to broadcast on Chord topology is smaller than on the fully connected one for the *random* chunk selection policy. The result is surprising because one would expect restricted connectivity to increase the expected time to broadcast. However, the restricted connectivity slows down the spread of the intially released chunks as they are unable to spread to all the network quickly. This reduces the imbalance in the chunk distribution, and also retains the nodes for a longer duration in the system, thereby reducing the expected broadcast time. We note that the larger the number of nodes in the system, the faster the system evolves and the greater is the number of uploaders in the system. The effect of the Chord topology is quite similar to the *deterministic last K chunks* policy in that there would be some set of nodes where the initially released chunks would not reach quickly enough, thereby reducing the chunk imbalance. The trade-off between slowing

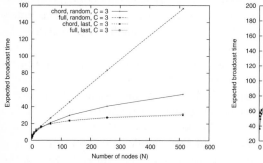
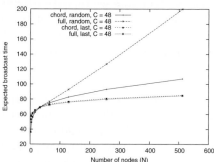

Fig. 7. Expected broadcast time for different values of N. $K = 1$. $C = 3$.

Fig. 8. Expected broadcast time for different values of N. $K = 1$. $C = 48$.

down the chunk replication (increasing the broadcast time) and retaining the nodes for a longer duration is apparent when we compare the expected broadcast time for small and large values of N. For smaller values of N, the *deterministic last K* chunk policy performs worse than the *random* chunk policy because the number of "unsuccessful encounters" may outweigh the benefits of retaining the nodes for a longer duration.

In Figures 9 and 10, we compare the performance of cyclical finger selection in the Chord topology to that of random finger selection. We observe that cyclical finger selection yields slightly better performance than random finger selection.

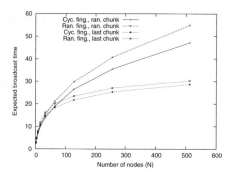
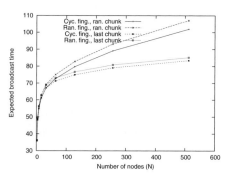

Fig. 9. Comparing cyclical and random finger selection. Expected broadcast time for different values of N. $K = 1$. $C = 3$.

Fig. 10. Comparing cyclical and random finger selection. Expected broadcast time for different values of N. $K = 1$. $C = 48$.

6 Conclusions

We have shown that even in a flash crowd scenario, chunk-wise distribution of large files can be based on random encounters within an overlay. However, attention must be paid to the avoidance of the rare chunk phenomenon by applying some other chunk selection algorithm than random choice. The extremely simple deterministic last chunk policy was shown to improve the situation already.

To analyse the flash crowd scenario mathematically, we sketched an approach where the early evolution of the system is modelled by Pólya's urn model, and after the stabilization of the chunk distribution the evolution is described by a system of differential equations.

As a novel idea, suggested by the hypercube-based optimal solution to the chunk distribution problem, we experimented by replacing random encounters by encounters restricted to the neighbour nodes in the Chord topology. The results were promising and prompt for further study of this kind of systems.

Acknowledgement. Part of this work was done in the EU FP6 projects Euro-NGI/REDLARF and project 034413 NET-ReFOUND.

References

1. Cohen, B.: BitTorrent specification (2006) http://www.bittorrent.org
2. Yang, X., de Veciana, G.: Service Capacity in Peer-to-Peer Networks. In: Proc. IEEE INFOCOM, Hong Kong (2004)
3. Stoica, I., Morris, R., Liben-Nowell, D., Karger, D.R., Kaashoek, M.F., Dabek, F., Balakrishnan, H.: Chord: A Scalable Peer-to-peer Lookup Protocol for Internet Applications. IEEE/ACM Transactions on Networking 11(1), 17–32 (2003)
4. Massoulie, L., Vojnovic, M.: Coupon Replication Systems. In: Proc. ACM SIG-METRICS, Banff, Canada (2005)
5. Felber, P., Biersack, E.: Cooperative Content Distribution: Scalability through Self-Organization. In: Babaoğlu, Ö., Jelasity, M., Montresor, A., Fetzer, C., Leonardi, S., van Moorsel, A.P.A., van Steen, M. (eds.) Self-star Properties in Complex Information Systems. LNCS, vol. 3460, Springer, Heidelberg (2005)
6. Bharambe, A.R., Herley, C., Padmanabhan, V.N.: Analyzing and Improving a BitTorrent Network's Performance Mechanisms. In: Proc. IEEE INFOCOM, Barcelona (2006)
7. Norros, I., Prabhu, B., Reittu, H.: Flash crowd in a file sharing system based on random encounters. In: Inter-Perf, Pisa, Italy (2006) http://www.inter-perf.org
8. Qiu, D., Srikant, R.: Modeling and Performance Analysis of BitTorrent-Like Peer-to-Peer Networks. In: Proc. ACM Sigcomm, Portland, OR (2004)
9. Clévenot-Perronnin, F., Nain, P., Ross, K.W.: Multiclass P2P Networks: Static Resource Allocation for Service Differentiation and Bandwidth Diversity. Performance Evaluation 62(1-4), 32–49 (2005)
10. Mundinger, J., Weber, R., Weiss, G.: Analysis of peer-to-peer file dissemination. To appear in Performance Evaluation Review, Special Issue on MAMA 2006 (2006)
11. Kwon, C.H., Chwa, K.Y.: Multiple message broadcasting in communication networks. Networks 26, 253–261 (1995)
12. Ho, C.: Optimal broadcasting on SIMD hypercubes without indirect addressing capability. J. Parallel Distrib. Comput. 13, 246–255 (1991)

TCPeer: Rate Control in P2P over IP Networks

Kolja Eger and Ulrich Killat

Institute of Communication Networks
Hamburg University of Technology (TUHH)
21071 Hamburg, Germany
{eger, killat}@tu-harburg.de

Abstract. The prevalent mechanism to avoid congestion in IP networks is the control of the sending rate with TCP. Dynamic routing strategies at the IP layer are not deployed because of problems like route oscillations and out-of-order packet deliveries.

With the adoption of P2P technology, routing is done also in these overlay networks. With multi-source download protocols peers upload and download to/from other peers in parallel.

Based on congestion pricing for IP networks this paper proposes a rate control algorithm for P2P over IP networks. A peer adopts the functionality of TCP and extends the congestion window mechanism with information from the overlay network. Thus, a sending peer is able to shift traffic from a congested route to an uncongested one. This change in the rate allocation will be balanced by other peers in the overlay. Hence, the receiving peers experience no degradation of their total download rate.

Keywords: Congestion pricing, Cross-layer optimisation, Rate control, P2P networks.

1 Introduction

Peer-to-peer (P2P) networks gained in popularity in recent years. In particular file-sharing applications are used in these overlay networks. The advantages are obvious. While in the client-server-architecture the total load must be carried by the server(s), it is distributed among the users in a P2P network. Namely, each peer acts as a client and a server at the same time.

A very interesting approach is the multi-source download (or swarming principle), which normally is applied for large files like movies or music. The file of interest is fragmented into blocks. When a peer completes the download of a single block, it offers it to other peers that so far have not downloaded this block. Thus, a file request of a user results in several requests at different peers for different parts of the file and consequently parts of the file are downloaded from multiple peers in parallel. Most of the research in this area studies the efficiency and fairness of these P2P protocols at the overlay network. For example, [1] and [2] study the popular BitTorrent application [3] analytically and by simulation, respectively. Thereby, the order in which blocks are downloaded at different peers or to which peers data is uploaded is discussed.

L. Mason, T. Drwiega, and J. Yan (Eds.): ITC 2007, LNCS 4516, pp. 618–629, 2007.
© Springer-Verlag Berlin Heidelberg 2007

In this paper we study the interaction between the P2P overlay and (in the case of the Internet) the underlying IP network. With multi-source download protocols peers upload and download to/from multiple peers in parallel. Thus, by controlling the sending rates between the peers we may achieve two desirable properties: Load balancing at the peers, which upload data, and the avoidance of congested links in the IP network. Considering load balancing, P2P technology has found already its way to content distribution networks [4]. But most of the work assumes that a file request is routed to a single server (in a sophisticated way). Multi-source download protocols adapt faster to changes at the servers or in the network because data is requested at multiple peers in parallel. Hence, the total bandwidth of all serving peers can be allocated in a fair manner to the client peers. Furthermore, also congestion can be avoided. When a connection between two peers becomes overloaded, the uploading peer can favour another peer on an uncongested route. Thus, it uses its upload bandwidth more efficiently. On the other hand the downloading peer could ask for a higher download rate at other providers of the file. Thus, it could receive the same download rate as in an uncongested network.

In this paper we model the allocation of resources for P2P over IP networks as a cross-layer optimisation problem. Our approach extends congestion pricing for IP networks proposed in [5,6,7]. In IP networks the TCP sender controls the sending rate. The corresponding TCP flow consumes bandwidth at each hop along its path to the sink. But the routers charge a price for the usage of bandwidth, which is summed up to a path price for each TCP flow and controls the future sending rate of the flow. Besides single-path routing also multi-path routing is studied with congestion pricing [8,9]. In these papers traffic is sent between a specific source-destination pair over multiple paths. To support this [9] proposes the usage of an overlay network.

In our approach we take advantage of the overlay formed by a P2P application. Furthermore, with multi-source downloads routing decisions in the overlay are done with respect to blocks of the file rather than specific source-destination pairs. We denote our approach as *TCPeer*, because a peer adopts the functionality of TCP and extends it with information from the overlay network. The service requesting peers in the overlay make price offers depending on their download rates. The upload bandwidths of the serving peers are allocated based on the price offers of the remote peers minus the path price charged by the links along the routes of these flows. Flows with high price offers and low path prices receive higher rates than others.

This work builds on previous work by the authors [10,11,12], in which only the overlay network is taken into account. To the best of our knowledge this paper is the first which studies the allocation of resources of P2P over IP networks in a single model. The paper is structured as follows. In Section 2 we model the resource allocation problem of P2P over IP networks as optimisation problem. A distributed implementation is discussed in Section 3. First simulation results are presented and discussed in Section 4. Section 5 concludes the paper.

2   Network Model

We model the allocation of resources in the network as a cross-layer optimisation problem. In general, we say the network consists of three different entities. These are the servers[1] or service providers, the clients or service customers and the links or service carriers. We associate a client with a user, which requests a file for download. To differentiate between the different entities in a mathematical model we introduce the set of servers $\mathcal{S}$, the set of clients $\mathcal{C}$, and the set of links $\mathcal{L}$. The capacity of the servers and the links is finite and is denoted by C. In a P2P application that supports a multi-source download a client uses several flows to different servers to download a file. We denote the set of servers that can be used by a client c as $\mathcal{S}(c)$. The other way around $\mathcal{C}(s)$ is the set of clients which can be served by server s. The client/server architecture can be interpreted as a special case of a P2P application where the set $\mathcal{S}(c)$ consists of one single server. Furthermore, we define a flow as a rate allocation from a server $s \in \mathcal{S}$ to a client $c \in \mathcal{C}$ over a route, which is a non-empty subset of all links, denoted as $\mathcal{L}(s, c)$. To identify a flow that uses the link l we introduce δ_{sc}^{l}, where $\delta_{sc}^{l} = 1$ if $l \in \mathcal{L}(s, c)$ and $\delta_{sc}^{l} = 0$ otherwise.

Suppose the utility of a client c is defined by a concave, strictly increasing utility function U, which depends on the total download rate y_c. Thereby, y_c is the sum of the rates x_{sc} from all servers which are known to c and have parts of the file in which c is interested in. Similar to the rate control algorithm for IP networks [5] we model the resource allocation for an IP network with a P2P overlay as a global optimisation problem

SYSTEM :

$$\text{maximise} \sum_{c \in \mathcal{C}} U_c(y_c) \tag{1}$$

$$\text{subject to} \sum_{s \in \mathcal{S}(c)} x_{sc} = y_c, \quad \forall c \in \mathcal{C} \tag{2}$$

$$\sum_{c \in \mathcal{C}(s)} x_{sc} \leq C_s, \quad \forall s \in \mathcal{S} \tag{3}$$

$$\sum_{s \in \mathcal{S}} \sum_{c \in \mathcal{C}(s)} \delta_{sc}^{l} x_{sc} \leq C_l, \quad \forall l \in \mathcal{L} \tag{4}$$

$$\text{over} \quad x_{sc} \geq 0 \tag{5}$$

Hereby, maximising the aggregated utility of the service rate y_c over all service customers is the objective of the whole system. The problem is constrained by the capacity at the servers and the links. Although servers and links face the same constraint, we differentiate between both in the model, because of their different functionalities: A server can freely allocate its bandwidth over its

---

[1] We use the terms server and client also in the context of P2P networks. Thus, we mean by server a peer which offers a service and by client a peer which requests a service. A peer can be a server and a client at the same time.

clients, whereas a link only forwards or discards packets. With a concave utility function this optimisation problem has a unique optimum with respect to y_c. The rates x_{sc} are not necessarily unique at the optimum, because different rate allocations may sum up to the same total download rate y_c. Thus, many possible rate allocations may exist with respect to x_{sc}.

The Lagrangian of (1-5) is

$$L\left(x, y; \lambda, \mu, \nu; m, n\right) = \sum_{c \in \mathcal{C}} \left(U_c(y_c) - \lambda_c y_c\right) + \sum_{s \in \mathcal{S}} \sum_{c \in \mathcal{C}(s)} x_{sc} \left(\lambda_c - \mu_s - \sum_{l \in \mathcal{L}(s,c)} \nu_l\right)$$
$$+ \sum_{s \in \mathcal{S}} \mu_s \left(C_s - m_s\right) + \sum_{l \in \mathcal{L}} \nu_l \left(C_l - n_l\right), \tag{6}$$

where λ_c, μ_s and ν_l are Lagrange multipliers and $m_s \geq 0$ and $n_l \geq 0$ are slack variables. Like in [6] Lagrange multipliers can be interpreted as prices for the usage of resources. Hence, we say λ_c is the price per unit offered by the customer c and μ_s and ν_l are prices per unit charged by the server s and the link l, respectively.

By looking at the Lagrangian in (6) we see that the global optimisation problem in (1-5) is separable into sub-problems for the clients, the servers and the links. Furthermore, maximising the total of a sum is equivalent to maximising each summand. Hence, we can decompose the Lagrangian into the sub-problems

$$\text{CLIENT } c: \qquad \text{maximise } U_c(y_c) - \lambda_c y_c \tag{7}$$
$$\text{over} \quad y_c \geq 0 \tag{8}$$

$$\text{SERVER } s: \qquad \text{maximise } \sum_{c \in \mathcal{C}(s)} \left(\lambda_c - \sum_{l \in \mathcal{L}(s,c)} \nu_l\right) x_{sc} \tag{9}$$

$$\text{subject to } \sum_{c \in \mathcal{C}(s)} x_{sc} \leq C_s \tag{10}$$

$$\text{over} \quad x_{sc} \geq 0 \tag{11}$$

$$\text{LINK } l: \qquad \text{maximise } \nu_l \left(C_l - n_l\right) \tag{12}$$

$$\text{over} \quad \nu_l \geq 0 \tag{13}$$

Thereby, for each sub-problem only locally available information is needed. Only a server needs to be informed about the price offers λ_c of all connected clients minus the prices charged by the links along the used route.

The economical interpretation of the sub-problems is as follows. A client is selfish and tries to maximise its own utility, which depends on the rate y_c. However, the client has to pay a price for using bandwidth. Since λ_c is a price per unit bandwidth, the product $\lambda_c y_c$ reflects the total price that client c pays. On the other hand a server gets paid for the allocation of bandwidth. It is interested in maximising its total revenue. The total revenue is the sum of the revenue for each client, which is computed by price times quantity. The price the server earns for a unit bandwidth is the price paid by the client minus the cost charged by the links which are needed for forwarding the traffic to the client. Thus, also the server behaves selfishly since it is interested in its own revenue only. Therefore, it will allocate bandwidth preferentially to flows, where clients pay a high price and costs for using the routes to the clients are low. Also the links maximise their revenue. Since the slack variable n_l can be interpreted as the spare capacity on the link, subtracting n_l from the capacity C_l reflects the total rate of forwarded traffic of this link. By multiplying it with the price per unit charged by the link we obtain the revenue of that link.

As already well studied for IP networks [5] we set the utility function for all clients to

$$U_c(y_c) = w_c \ln(y_c), \tag{14}$$

where w_c is the willingness-to-pay. This utility function ensures a weighted proportional fair resource allocation. Furthermore, in the context of P2P networks the willingness-to-pay can be interpreted as the contribution of a peer to the network. Thus, we could set w_c to the upload bandwidth, which peer c allocates to the P2P application (see [12] for details).

Using (14) the optimum (y^*, λ^*) of (1-5) can be computed by differentiating (6) with respect to y_c and x_{sc}. Hence,

$$y_c^* = \sum_{s \in \mathcal{S}(c)} x_{sc} = \frac{w_c}{\lambda_c^*} \quad \text{if} \quad \mathcal{S}(c) \neq \{\} \tag{15}$$

$$\lambda_c^* = \mu_s^* + \sum_{l \in \mathcal{L}(s,c)} \nu_l^* \quad \text{if} \quad x_{sc} > 0 \tag{16}$$

$$\leq \mu_s^* + \sum_{l \in \mathcal{L}(s,c)} \nu_l^* \quad \text{if} \quad x_{sc} = 0 \tag{17}$$

Furthermore, we deduce from $\partial L/\partial m_s = -\mu_s$ and $\partial L/\partial n_l = -\nu_l$ that the price at a server or at a link is only greater zero, if the corresponding slack variable is zero. Hence, a price is charged only when the resources are fully used and competition for this resource is present.

In case the bottlenecks of the network are the uplink connections of the servers and all link prices are zero, a detailed derivation of the equilibrium can be found in [10].

3   Implementation

An implementation in a decentralised architecture can only use locally available information. Thus, the problems CLIENT, SERVER and LINK from Section 2 provide a good starting point for the development of a distributed algorithm. CLIENT is an optimisation problem with respect to y_c, but in a real implementation a client has no direct influence on the total download rate. It can only vary its price offer λ_c. Similar, a link has only influence over its load by varying the charged price. This price depends on the level of congestion. We assume that a server receives the difference of the price offered by a client minus the price charged by the links along the used route and adapts its sending rate to the client. Thus, we propose the following distributed algorithm

CLIENT c :

$$\lambda_c(t+1) = \frac{w_c}{\max\left(\eta, y_c(t)\right)} = \frac{w_c}{\max\left(\eta, \sum_{s \in \mathcal{S}(c)} x_{sc}(t)\right)} \tag{18}$$

SERVER s :

$$x_{sc}(t+1) = \max\left(\epsilon, x_{sc}(t) + \gamma x_{sc}(t)\left(p_{cs}(t) - \frac{\sum_{d \in \mathcal{C}(s)} x_{sd}(t)p_{ds}(t)}{C_s}\right)\right) \tag{19}$$

LINK l :

$$\nu_l(t+1) = \max\left(0, \nu_l(t) + \kappa\left(\sum_{s \in \mathcal{S}} \sum_{c \in \mathcal{C}(s)} \delta_{sc}^l x_{sc}(t) - C_l\right)\right) \tag{20}$$

where

$$p_{cs}(t) = \lambda_c(t) - \sum_{l \in \mathcal{L}(s,c)} \nu_l(t) \tag{21}$$

is the price from client c received by server s and the parameters ϵ and η are small positive constants. Hence, the service rate x_{sc} does not fall below ϵ. This models a kind of probing, which is necessary to receive information about the price p_{cs}. Similar, η ensures a bounded price offer, if the total download rate is zero.

(19) can be interpreted as follows. Since p_{cs} is the price per unit bandwidth, which server s receives from client c, the product $x_{sc}p_{cs}$ is the total price which c pays to s. Thus, the sum in the numerator of (19) represents the total revenue of server s. Multiplying the total revenue by x_{sc}/C_s specifies the average revenue server s generates by allocating the bandwidth x_{sc}. If the total price of client c is higher than the average price (i.e. the left term in the inner bracket of (19) is greater than the right one), then server s increases the rate to client c. Otherwise, server s decreases the rate.

The distributed algorithm represents a market for bandwidth where prices control the rate allocations. Since the capacity of a server or a link is a non-storable commodity the market is a spot market, where the current supply and demand affect the future price.

A prove that the stable point of the distributed algorithm corresponds to the optimum of (1-5) is presented in [10] for the case when all link prices are zero (i.e. the bottlenecks in the network are the uplink capacities of the servers). Similarly, it can be proven for the extended model in this paper.

The link prices in (20) are greater zero when the input rate exceeds the capacity of a link and thus the link is overloaded. (20) corresponds to the link price rule PC1 in [13]. Thus, although we extend the congestion pricing model to P2P networks, we do not change the router implementation of previous work for IP networks. Furthermore, we show in the following that the rate adaptation of a server in (19) reduces to a differential equation very similar to the primal algorithm (Equation 5) in [6] for the case of a single client-server pair. A client-server pair is a server handling a single client, which is served only by this server. Hence, $\mathcal{C}(s) = \{c\} \wedge \mathcal{S}(c) = \{s\}$ holds for a server s and a client c and (19) reduces to

$$\frac{d}{dt}x_{sc}(t) = \gamma\left(\left(w_c - x_{sc}(t)\sum_{l\in\mathcal{L}(s,c)}\nu_l(t)\right)\cdot\left(1 - \frac{x_{sc}(t)}{C_s}\right)\right) \qquad (22)$$

When the uplink of the server is not the bottleneck the term $(1 - x_{sc}/C_s)$ is just a scaling of the step size γ and (22) corresponds to Equation 5 in [6].

Packet-level implementation. The distributed algorithm discussed above considers the problem at flow-level. Now, we transform the algorithm to packet-level. Like in any other TCP implementation the sending rate is specified by the congestion window $cwnd$. $cwnd$ is the number of packets being sent per round-trip time (RTT). Hence, the sending rate can be computed with $x = cwnd/RTT$. The congestion window is adapted on each acknowledgement. These are received with rate $cwnd/RTT$ by assuming that the RTT is constant for small time intervals. Therefore, updating the congestion window with

$$\Delta cwnd_{sc}(t) = \gamma RTT_{sc}(t)\left(p_{cs}(t) - \frac{\sum_{d\in\mathcal{C}(s)}\frac{cwnd_{sd}(t)}{RTT_{sd}(t)}p_{ds}(t)}{C_s}\right) \qquad (23)$$

at packet-level corresponds to (19) at flow-level. The link rule in (20) computes the price based on the current utilisation only and does not take the queue size into account. To penalise large queues (20) can be extended to

$$\nu_l(t+T_l) = \max\left(0, \nu_l(t) + \kappa\left(\alpha_l\left(b_l(t) - b_0\right) + \sum_{s\in\mathcal{S}}\sum_{c\in\mathcal{C}(s)}\delta_{sc}^l x_{sc}(t) - C_l\right)\right), \quad (24)$$

where b_0 is the target and $b_l(t)$ the current queue size. $\alpha_l \geq 0$ weights the influence of the queue size compared to the current utilisation. Furthermore, the time between two updates is T_l. (Similarly, we assume that (18) is updated every T_c in a packet-level implementation.) (24) corresponds to PC3 in [13].

In this work we assume explicit prices of the links and clients are communicated to the server (e.g. in the header of the packets). Although this does not conform to the IP and TCP standards, it demonstrates the functionality of the proposed approach in general. For IP networks a large body of research (e.g. [13,14]) study marking strategies where a single bit (see explicit congestion notification (ECN) [15]) is used to convey the pricing information. We assume this approach is applicable also for our model, but this is part of future research (e.g. ECN is used to transport the price information of the links and the P2P protocol is used to communicate the price offers from the clients).

4   Evaluation

To demonstrate the functionality of the distributed algorithm we study two simple examples in ns-2. The network of interest is depicted in Figure 1. The first example consists of the two servers $s1$ and $s2$ and the two clients $c1$ and $c2$. The bottlenecks in the network are the uplink capacities of the servers and $C_{s1} = C_{s2} = 10\,$Mbps. The capacity of the links in the core (i.e. in Figure 1 the links connecting routers denoted by r) and of the downlinks to the clients is much larger, e.g. $C = 100\,$Mbps. The delay of all links is $10\,$ms.

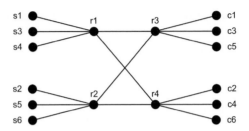

Fig. 1. Network

Assume both clients can download from both servers in parallel, if they are online. This represents a small P2P overlay. At $t = 0\,$s only server $s1$ and client $c1$ are online. $c2$ and $s2$ are switched on at $t = 20\,$s and $t = 40\,$s, respectively. At $t = 60\,$s client $c1$ goes offline, followed by server $s1$ at $t = 80\,$s.

The total download rate of the two clients and the price information received by the servers (see (21)) are depicted in Figure 2. For the first 20 seconds only $c1$ and $s1$ are active and the download rate of $c1$ converges to the capacity of $s1$ (which is around 874 packets per second when using a packet size of $1500\,$B). With the link rule in (20) the queue size at the uplink of $s1$ converges to 12 packets. Also the price p_{c1s1} converges to a steady-state value of around 1.15, which can be computed by dividing the willingness-to-pay of $c1$ with the server capacity in packets [10].

When $c2$ gets active at $t = 20\,$s the server $s1$ receives higher payments because the total willingness-to-pay in the network doubles. The download rates of both clients converge to a fair share of the bottleneck capacity in the following. Hereby, oscillations occur since the estimated download rate (and consequently its price offer) of the new client $c2$ is inaccurate at the beginning. As depicted in Figure 2(b) the price received by the server can fall below zero because the server allocates to much bandwidth and the uplink bandwidth is exceeded such that the link price of the server's uplink increases.

When a new server gets online at $t = 40\,$s and one client leaves at $t = 60\,$s the spare capacity is used efficiently by the online client(s). Thus, this example shows that the proposed algorithm can be used for load balancing because it realises a fair allocation of the servers' capacities. For the case, where the uplink

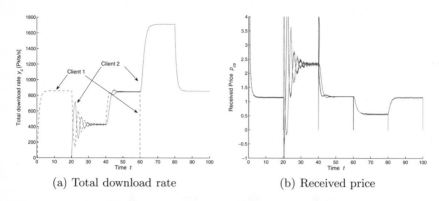

(a) Total download rate   (b) Received price

Fig. 2. Example 1 ($w_c = 1000 \ \forall c \in \mathcal{C}$, $\gamma = 0.1RTT$, PC1: $\kappa = 0.1$, $T_l = T_c = 0.1\,\mathrm{s}$)

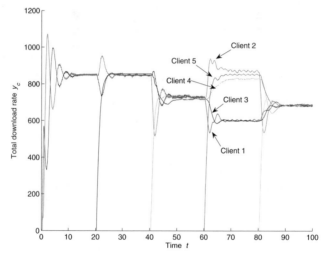

Fig. 3. Example 2: Total download rate ($w_c = 1000$, $\gamma = 0.1RTT$, PC3: $\kappa = 0.1$, $\alpha = 0.1$, $b_0 = 5$, $T_l = T_c = 0.1\,\mathrm{s}$)

capacities are the only bottlenecks in the network, further simulation results at flow-level can be found in previous work of the authors [11,12].

In the following we investigate the influence of congestion in the core of the network. Therefore, we reduce the capacity for the links $r1 - r3$, $r1 - r4$, $r2 - r3$ and $r2 - r4$ in Figure 1 to $C_{\mathrm{core}} = 12\,\mathrm{Mbps}$. To avoid large queuing delays we compute link prices with (24) in this example. We start with the two servers and the two clients from the previous example, but new client-server pairs get active during the simulation. Assume $s3 - c3$, $s4 - c4$, $s5 - c5$ and $s6 - c6$ start at 20 s, 40 s, 60 s and 80 s, respectively. Furthermore, only $c1$ and $c2$ are served in parallel by $s1$ and $s2$. The other clients are served by a single server only (indicated by the same number).

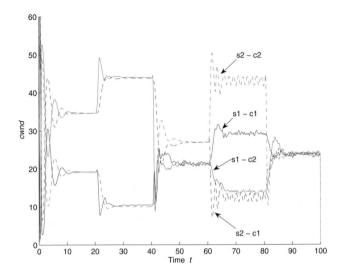

Fig. 4. Example 2: Congestion window

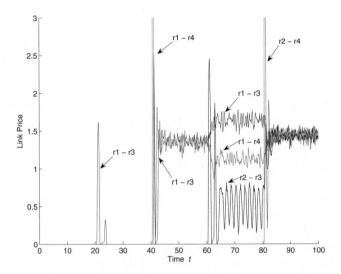

Fig. 5. Example 2: Link price

The total download rate for each client is depicted in Figure 3. At the beginning the clients $c1$ and $c2$ share the available bandwidth equally between each other as in the previous example. Looking at the congestion window of the four connections in Figure 4 we see that not all connections are used equally. It suffices that the total download rate of the two clients is the same.

When $s3$ and $c3$ start at $t = 20\,$s the link $r1 - r3$ gets congested. This can be seen by an increasing link price in Figure 5. Thus, the congestion window between $s1 - c1$ as well as the total download rate of client 1 decrease. However, this increases the price offer of client 1 and therefore the congestion window between $s2 - c1$. The servers $s1$ and $s2$ shift traffic from one connection to the other one (see Fig. 4) and the total download rates of the three active clients converge to a fair and efficient allocation (see Fig. 3). The link price of $r1 - r3$ falls to zero again, which indicates that it is not a bottleneck in the network.

When the next client-server pair starts at $t = 40\,$s prices for the links $r1 - r3$ and $r1 - r4$ increase and stagnate. The servers $s1$ and $s2$ restart to shift the traffic but cannot avoid congestion in the core network. However, by changing the rate they still ensure a fair rate allocation since all four clients receive roughly the same total download rate.

At $t = 60\,$s the connection $s5 - c5$ starts. Thereby, the link $r2 - r3$ becomes overloaded and its link price increases. Now, the network is in a congested state, where a trade-off between fairness and efficiency has to be done. Only connection $s2 - c2$ uses the uncongested core link $r2 - r4$. Thus, $c2$ has the largest download rate. Since most of the server capacity is allocated to the connection $s2 - c2$, client 5 and client 4 benefit from small link prices at $r2 - r3$ and $r1 - r4$. The link price on $r1 - r3$ is the highest one. Hence, client 1 and client 3 get smaller download rates as compared to the other clients.

Finally, with $s6 - c6$ starting at $t = 80\,$s all core links get congested and the servers $s1$ and $s2$ control the rate to $c1$ and $c2$ such that bandwidth is allocated in a fair manner. The price oscillates around 1.43, which is the equilibrium price.

This example demonstrates that the proposed algorithm is able to shift traffic from congested to uncongested routes. Thus, resources of the network are used efficiently while preserving fairness between clients.

5   Conclusion and Future Work

This paper models the allocation of resources in an IP network when it is used by a P2P overlay. It extends the well-known congestion pricing approach for IP networks to multi-source download P2P protocols. Furthermore, a distributed algorithm is proposed, where sending peers control the rate depending on the price offers by the remote peers and the path prices charged by the routers in the IP network. Since sending rates depend on the price offers of the remote peers the algorithm ensures fairness between peers. Furthermore, indicated by the path price a peer is aware of the level of congestion in the underlying IP network. Thus, it can shift upload bandwidth from congested to uncongested flows. Because of the change in price offers at the receiving peers other senders will change their sending rates accordingly. The rate allocation of all peers converges to an efficient and fair allocation of bandwidth in the network. Thereby, P2P traffic avoids congested links when uncongested routes between some peers exist.

We validate the functionality of the new approach analytically and by simulations for a small network. Thereby, explicit price information is assumed.

Part of future work is to show the functionality of the algorithm for limited price information using ECN to be compliant with the IP protocol. Since the deployment of ECN is slow, another possible direction is to study the interaction between P2P and predominantly used TCP versions. These can be modelled by different utility functions as described in [16].

References

1. Qiu, D., Srikant, R.: Modeling and performance analysis of BitTorrent-like peer-to-peer networks. Computer Communication Review 34(4), 367–378 (2004)
2. Bharambe, A., Herley, C., Padmanabhan, V.: Analyzing and improving BitTorrent performance. Technical Report MSR-TR-2005-03, Microsoft Research (2005)
3. Cohen, B.: Incentives build robustness in BitTorrent. In: Proc. 1st Workshop on Economics of Peer-to-Peer Systems, Berkeley (2003)
4. Turrini, E., Panzieri, F.: Using P2P techniques for content distribution internet-working: A research proposal. In: Proc. IEEE P2P 2002 (2002)
5. Kelly, F.: Charging and rate control for elastic traffic. In: European Transactions on Telecommunications. vol. 8, 33–37 (1997)
6. Kelly, F., Maulloo, A., Tan, D.: Rate control in communication networks: shadow prices, proportional fairness and stability. In: Journal of the Operational Research Society. vol.49, pp. 237–252 (1998)
7. Low, S., Lapsley, D.: Optimization flow control, I: basic algorithm and convergence. IEEE/ACM Transactions on Networking 7(6), 861–874 (1999)
8. Wang, W., Palaniswami, M., Low, S.: Optimal flow control and routing in multi-path networks. Perform. Eval. 52(2-3), 119–132 (2003)
9. Han, H., Shakkottai, S., Hollot, C.: Overlay TCP for multi-path routing and congestion control (2003)
10. Eger, K., Killat, U.: Resource pricing in peer-to-peer networks. IEEE Communications Letters 11(1), 82–84 (2007)
11. Eger, K., Killat, U.: Fair resource allocation in peer-to-peer networks. In: Proc. SPECTS'06, Calgary, Canada, pp. 39–45 (2006)
12. Eger, K., Killat, U.: Bandwidth trading in unstructured P2P content distribution networks. In: Proc. IEEE P2P 2006, pp. 39–46. Cambridge, UK (2006)
13. Athuraliya, S., Low, S.: Optimization flow control, II: Implementation. Technical report, Melbourne University (2000)
14. Zimmermann, S., Killat, U.: Resource marking and fair rate allocation. In: Proc. ICC 2002, vol.2, pp. 1310–1314. New York (2002)
15. Ramakrishnan, K., Floyd, S., Black, D.: The addition of explicit congestion notification (ECN) to IP. RFC 3168 (2001)
16. Low, S., Srikant, R.: A mathematical framework for designing a low-loss, low-delay internet. Networks and Spatial Economics 4, 75–101 (2004)

Estimating Churn in Structured P2P Networks

Andreas Binzenhöfer[1] and Kenji Leibnitz[2]

[1] University of Würzburg, Institute of Computer Science
Chair of Distributed Systems, Würzburg, Germany
binzenhoefer@informatik.uni-wuerzburg.de
[2] Graduate School of Information Science and Technology
Osaka University, 1-5 Yamadaoka, Suita, Osaka 565-0871, Japan
leibnitz@ist.osaka-u.ac.jp

Abstract. In structured peer-to-peer (P2P) networks participating peers can join or leave the system at arbitrary times, a process which is known as churn. Many recent studies revealed that churn is one of the main problems faced by any Distributed Hash Table (DHT). In this paper we discuss different possibilities of how to estimate the current churn rate in the system. In particular, we show how to obtain a robust estimate which is independent of the implementation details of the DHT. We also investigate the trade-offs between accuracy, overhead, and responsiveness to changes.

1 Introduction

With the recent development of new peer-to-peer (P2P) architectures, P2P has evolved from simple file-sharing networks to efficient alternatives to the classic client-server architecture. This is accomplished by each peer participating in a logical overlay structure, simultaneously acting as client and as server. The peer is responsible for maintaining its share of information and providing it to the other peers requesting this data.

Additionally, P2P networks have no static network topology and each participating peer may join or leave the overlay at any time. This process is referred to as *churn* [1]. However, this freedom of having a highly dynamic network structure comes at a cost. The higher the churn rate is, the more difficult it becomes for the network to maintain its consistency [2]. Too high churn can cause routing failures, loss of stored resources or the entire overlay structure, or inconsistent views of the peers on the overlay.

Thus, it is essential that the overlay network structure is maintained even in the presence of high churn. Especially in structured P2P architectures, such as Chord [3], where all peers are arranged in a ring structure, the integrity of the neighborhood relationship among the peers must be kept at all times. As a consequence, these networks require more maintenance traffic when the churn rate is high. However, P2P networks operate without a centralized control unit and each peer has only a limited view of the entire network, usually not being aware of the current churn rate in the network. Thus, a peer should be able to estimate the churn rate from the limited information that is available and autonomously react to high churn situations by increasing the maintenance traffic.

In this paper, we propose a fully distributed algorithm for peers to estimate the churn rate by exchanging measurement observations among neighbors. The overlay network

L. Mason, T. Drwiega, and J. Yan (Eds.): ITC 2007, LNCS 4516, pp. 630–641, 2007.

itself is used as a memory for the estimate while each online peer contributes to updated measurements of the estimator. The advantage of this method is that it operates passively, i.e., there are no additional entities required to monitor online and offline periods of the peers and no further overhead is necessary. While we mainly consider Chord-based DHT networks, our method is not restricted to any type of structured P2P overlay since it operates independently of the underlying DHT protocol. Wherever necessary, we will point out the corresponding differences to other types of structured P2P networks, e.g. Kademlia [4] or Pastry [5].

The paper is organized as follows. In Section 2, we discuss some existing models for estimating the churn rate in P2P networks. This is followed by Section 3 where we give a detailed description of our proposed estimation scheme. Section 4 will show that our algorithm is capable of retrieving accurate estimates and we will study the impact of the parameters, e.g. the number of monitored neighbors or the stabilization interval, on the performance of our approach. Finally, we conclude the paper in Section 5 and elaborate on possible extensions.

2   Discussion of Different Churn Models

The impact of joining peers is usually the less problematic aspect of churn, since it mainly results in temporary failures like routing inconsistencies or resources which might be temporarily located at a wrong position in the overlay. The process of peers leaving the system, however, can result in irreparable damage like loss of the overlay structure or loss of data stored in the overlay. In general, node departures can be divided into *friendly leaves* and *node failures*. Friendly leaves enable a peer to notify its overlay neighbors to restructure the topology accordingly. Node failures, on the other hand, seriously damage the structure of the overlay by causing stale neighbor pointers or data loss. In this paper we therefore concentrate on node failures.

There are two predominant ways to model churn. The first assumes churn per network by specifying a global join and leave rate [1]. This is also very similar to the half-life of a system as defined in [6]. Usually the global join process is modeled by a Poisson process with rate λ. One of the main problems of this model is that the number of nodes joining the system within a given time interval is independent of the current size of the system. However, while a join rate of 50 peers per second is quite significant for small networks, it might have no noticeable influence in very large networks.

Another way to model churn is to specify a distribution for the time a peer spends in the system (online time) or outside the system (offline time). This way the churn rate can be considered per node and thus generates a churn behavior, which is comparable in networks of different size. As in [7] we turn our main attention to scenarios where the join and failure rate are both described per node. To be able to model the offline time of a peer, we assume a global number of n peers, each of which can either be online or offline. Joins are then modeled by introducing a random variable T_{off} describing the duration of the offline period of a peer. Accordingly, leaves are modeled by a random variable T_{on} describing the online time of a peer. Usually, T_{on} and T_{off} are exponentially distributed with mean $E[T_{on}]$ and $E[T_{off}]$, respectively. However, this may not hold in realistic scenarios where distributions tend to become more skewed [8]. Therefore, in Section 3 we design our estimator independent of the distribution of T_{on} and T_{off}.

The actual user behavior in a real system heavily depends on the kind of service which is offered. For example, Gummadi et al. [9] showed that P2P users behave essentially different from web users. Additionally, Bhagwan et al. [10] argue that availability is not well-modeled by a single-parameter distribution, but instead is at least a combination of two time-varying distributions. This is supported by the observation that failure rates vary significantly with both daily and weekly patterns and that the failure rate in open systems is more than an order of magnitude higher than in a corporate environment [11]. Finally, to be able to compare the performance of different selection strategies for overlay neighbors, Godfrey et al. [8] present a definition of churn which reflects the global number of changes within a time interval Δt. While the definition is very useful in simulations which permit a global view on the system, it cannot be used by an estimator which can only rely on local information.

3   Estimating the Churn Rate

In general, an estimator for the churn in the system must in some way capture the fluctuations in the overlay structure and then deduce an estimate for the churn rate from these observations. In structured P2P networks, each peer has periodic contact to a specific number of overlay neighbors. Those overlay neighbors are called *successors* in Chord, *k-bucket entries* in Kademlia, or *leafs* in Pastry. The basic principle of the estimator described here is to monitor the changes in this neighbor list and use them to derive the current churn rate.

3.1   Obtaining Observations

We model the behavior of a peer using two random variables T_{on} and T_{off} which describe the duration of an online session and an offline session. This model assumes that offline peers will rejoin the overlay network at a later point in time. While this is a very reasonable assumption for closed groups like corporate networks or distributed telephone directories (Skype), other applications like content distribution (BitTorrent) might have no recurring customers. For the latter case, an estimator for the global join rate λ is presented in [12] based on the average age of peers in the neighbor list. The main problem is that such estimators require an additional estimate of the current system size [13].

Each online peer p stores pointers to c well defined overlay neighbors (or contacts) which are specified by the individual DHT protocols. To maintain this structure of the overlay, peer p periodically contacts a special subset of its neighbors every t_{stab} seconds and runs an appropriate *stabilization* algorithm. This corresponds, e.g., to *bucket refreshes* in Kademlia or the stabilization with the direct successor in Chord. At each of these stabilization instants the peer synchronizes its neighbor list with those of its contacts. Our estimator monitors the changes in this neighbor list and collects different realizations of the random variables T_{on} and T_{off}. Thereby, $obs(i)$ is the ith observation made by the peer and $time(i)$ is the time when the observation was made. The observation history is stored in a list which contains up to k_{max} entries. Furthermore, a peer

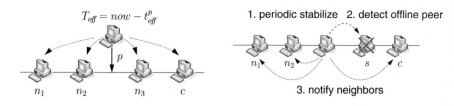

Fig. 1. Peer p rejoins the network and sends its offline duration to its c neighbors

Fig. 2. Peer p only monitors its direct neighbor s but distributes its observations

stores the time stamps t_{on}^p and t_{off}^p which correspond to the time peer p itself joined or departed from the overlay, respectively.

The shorter a peer stays offline on average the higher is the join rate. To obtain realizations of T_{off}, a peer stores the time t_{off}^p when it last went offline. The next time it goes online it calculates the duration of its offline session as $now - t_{off}^p$ and sends this value to its c overlay neighbors n_i (cf. Fig. 1). Note that the information can be piggybacked on other protocol messages to avoid unnecessary overhead. In Chord, a joining peer contacts its successors and possibly its fingers, in Pastry its leaf set or neighborhood set, and in Kademlia it refreshes its closest bucket. These messages can be used to disseminate the observed offline time to the overlay neighbors.

To obtain realizations of T_{on} we proceed as follows. In a DHT system, a peer p periodically contacts at least one neighbor s to stabilize the overlay structure (cf. Step 1 in Fig. 2). In Chord this would be the direct successor in a clockwise direction, in Kademlia the closest peer according to the XOR-metric. If, during one of its stabilization calls, p notices that s has become offline (cf. Step 2 in Fig. 2), it calculates the duration of the online session of peer s as $now - t_{on}^s$, where t_{on}^s is the time when peer s went online. Peer p then distributes this observation to all its overlay neighbors as shown in Step 3 in Fig. 2. If the DHT applies some kind of *peer down alert mechanism* [1, 11], the information could also be piggybacked on the corresponding notify messages.

An obvious problem of this approach is that peer p does not always naturally know t_{on}^s, the time when peer s went online. This is, e.g., true if p went online after s or if s became the successor of p due to churn. For this reason each peer s memorizes the time t_{on}^s when it went online and sends this information to its new predecessor whenever it stabilizes with a new peer. To cope with the problem of asynchronous clocks it sends its current online duration $now - t_{on}^s$. This way the error is in the order of magnitude of a network transmission and thus negligible in comparison to the online time of a peer.

When a peer joins the network, it first needs to obtain some observations before it can make a meaningful estimate of the churn rate. Therefore, we use the overlay network as a memory of already obtained observations. If a new peer joins the overlay it downloads the current list of observations from its direct successor. This way the observations persist in the overlay and a new peer can already start with a useful estimate which reflects the current churn rate in the network. An alternative is to invest more overhead by periodically contacting a number of peers instead of just one. Mahajan et al. [14] present an algorithm which relies on the fact that a peer continuously observes c overlay

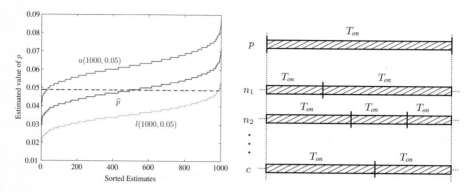

Fig. 3. Sorted estimates ($k = 10^3$, $\alpha = 95$)   **Fig. 4.** Observations in lifetime of peer p

neighbors. Such a peer should on average observe one failure every $\Delta t = \frac{1}{c} \cdot E[T_{on}]$. Thus, if a peer observes k failures in Δt the mean online time of a peer can be estimated as:

$$\widehat{E}[T_{on}] = \frac{c \cdot \Delta t}{k} = \frac{c \cdot (time(k) - time(1))}{k}.$$

In addition to the periodic contact to c neighbors, the algorithm also has to struggle with the correctness of the neighbor pointers and the problem of obtaining enough observations during the lifetime of the peer.

3.2 Derivation of the Churn Rate

In this section we will use the following notation: For a random variable X, we denote $x(t)$ as the pdf, $X(t)$ as the cdf, and $E[X]$ as the mean. Estimated values will be marked using a hat. Once a peer has obtained a list of observations $obs(i), i = 1, \ldots, k$ of the random variables T_{on} and T_{off}, it can rely on robust estimates like the empirical mean and the empirical standard deviation.

The larger we set k, i.e. the more observations a peer maintains in its history, the more accurate the estimate is going to be. However, if k is chosen too large, it will take longer for the estimator to react to changes in the current churn rate. In this context, the limits of the corresponding confidence interval can be used to autonomously derive an optimal value of k. If the calculated confidence interval is larger than a predefined threshold, a peer can increase k accordingly.

While the mean of T_{on} and T_{off} give a first idea about the churn in the system, the main purpose of the estimator is to self-tune the parameters of the DHT or to calculate the probability of certain events. This usually requires knowledge of the entire distribution or at least of some important quantiles. For example, to calculate the probability that an overlay neighbor will no longer be reachable at the next stabilization instant, we need to know the probability p that this contact will stay online for less than t_{stab} seconds. An unbiased point estimator for this probability is given by:

$$\widehat{p} = \widehat{P}\left(T_{on} < t_{stab}\right) = \frac{1}{k}\left|\left\{T_{on}^i : T_{on}^i < t_{stab} \quad \text{for } i = 1, 2, ..., k\right\}\right|, \qquad (1)$$

where $|\cdot|$ indicates the cardinality of a set. The $100(1 - \alpha)$ confidence interval for $\widehat{p}$ can be calculated using the following bounds:

$$u(k, \alpha) = \widehat{p} + z_{1-\frac{\alpha}{2}} \cdot \sqrt{\frac{\widehat{p}(1 - \widehat{p})}{k}} \qquad l(k, \alpha) = \widehat{p} - z_{1-\frac{\alpha}{2}} \cdot \sqrt{\frac{\widehat{p}(1 - \widehat{p})}{k}} \qquad (2)$$

where $z_{1-\frac{\alpha}{2}}$ is the $1 - \frac{\alpha}{2}$ critical point for a standard normal random variable. In case over- or underestimating has serious consequences for the applied application, the limits of the confidence interval can be used as estimates themselves.

We simulated an overlay with $t_{stab} = 30\,\text{s}$ where the online time of a peer was exponentially distributed with mean $E[T_{on}] = 600\,\text{s}$. Under these conditions, the probability p that a specific peer goes offline before the next stabilization instant is 4.88%. Fig. 3 shows the sorted estimates of p and the corresponding upper and lower bounds from 1000 peers. The upper bound $u(k, \alpha)$ tends to overestimate and the lower bound $l(k, \alpha)$ tends to underestimate. Note, that due to the denominator in Eqn. (1) the estimate is discretized into steps of $\frac{1}{k}$.

In some cases an application requires knowledge of the entire distribution function of the online time. If the type of distribution is known a priori, the peer can use the corresponding *Maximum Likelihood Estimator* (MLE) to estimate the parameters of the distribution. However, there is always the danger of assuming an incorrect distribution which would lead to correspondingly distorted results. A possibility to reduce this risk is to perform a hypothesis test [15] to verify that the type of distribution is actually the assumed one and only use an MLE if the test delivers a positive result. In general, however, the actual type of distribution is not known or a superposition of multiple distributions. In this case, a peer has to rely on an estimate of the quantiles [16] of the online distribution.

To show the importance of using the overlay network as a memory for already made observations, we regard the random variable X which describes the number of observations a peer makes during its lifetime provided that it continuously observes c overlay neighbors. This concept is visualized in Fig. 4 where we assume that a neighbor n_i which went offline is immediately replaced by another peer. The random variable X corresponds to the number of leave events in the figure and can be computed as

$$P(X = i) = \int_0^\infty t_{on}(t) \cdot P(X = i | T_{on} = t)\, dt. \qquad (3)$$

In the case of exponentially distributed online times, this can be written as

$$P(X = i) = \int_0^\infty \lambda e^{-\lambda t} \cdot \frac{(c\lambda t)^i}{i!} \cdot e^{-c\lambda t}\, dt = \frac{c^i}{(c + 1)^{i+1}}, \qquad (4)$$

since the number of departures in a fixed interval of length t is Poisson distributed with parameter $c \cdot \lambda$.

To compare this theoretical approximation to practical values, we simulated an overlay network with $T_{on} = 300\,\text{s}$, $t_{stab} = 30\,\text{s}$, and $c = 40$, where the online/offline time of a peer is exponentially distributed. The maximum size of the history was set to $k_{max} = 100$. Fig. 5 shows the probability density function of X for both the analysis

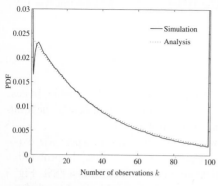

Fig. 5. Expected observations for $c = 40$ **Fig. 6.** Response to churn changes

and the simulation. It can be seen that the analysis matches the simulation very well except for the two peaks at the left and the right of the figure. The peak at 100 clearly results from the maximum size of the history. That is, all probabilities for $P(X > 100)$ are added to $P(X = 100)$. The peak at 0 arises from the fact that while the analysis immediately takes offline peers into account, the first stabilization instant in the simulation occurs 30 s after the peer joined the network. Thus, all peers which stay online for less than 30 s, can never make an observation. In conclusion, both the analytical and the simulation results show that a peer does not make enough observations during its lifetime in order to derive a meaningful estimate and a good estimator should therefore utilize the overlay network as a memory for already made observations.

The more observations a peer makes per time unit, the faster it can react to changes in the global churn rate. This can be measured by looking at T_{obs}^{leave}, the time between two observed leave events, or T_{obs}^{join}, the time between two observed join events. If a peer shares its observations with c overlay neighbors, the next observation is made as soon as one of these $c + 1$ peers goes offline. Thus, the distribution of T_{obs}^{leave} can be calculated as the minimum of $c + 1$ forward recurrence times of T_{on}. Due to the memoryless property, the forward recurrence time of an exponentially distributed online time T_{on} is also exponentially distributed with the same parameters. In this case the distribution of T_{obs}^{leave} can be calculated as:

$$P\left(T_{obs}^{leave} < t\right) = 1 - P\left(T_{on} \geq t\right)^{c+1} = 1 - e^{-(c+1)\lambda t}. \tag{5}$$

If the distribution is not known, we can still easily compute the mean of T_{obs}^{join} as $E\left[T_{obs}^{join}\right] = \frac{E[T_{off}]}{c+1}$. The calculation is a little more complicated for T_{obs}^{leave} since the time when a peer actually observes that another peer is offline differs from the actual time the node left the overlay. Assuming that overlay neighbors are updated every t_{stab} seconds, the average error is $\epsilon_{on} = \frac{t_{stab}}{2}$ which leads to

$$E\left[T_{obs}^{leave}\right] = \frac{E[T_{on}] + \epsilon_{on}}{c + 1} \tag{6}$$

The above considerations can be used to approximate the expected time it takes the estimator to respond to a global change of the churn rate. When the mean online time of the peers changes from $E_{old}[T_{on}]$ to $E_{new}[T_{on}]$, we approximate the expected response time $E[R]$ by the time needed to collect k_{max} new observations.

$$E[R] = E_{old}[T_{on}] + \frac{k_{max}}{c+1} \cdot (E_{new}[T_{on}] + \epsilon_{on}) \tag{7}$$

Fig. 6 compares the analytical response time to that obtained from a simulation run. In the simulation we again used exponentially distributed online/offline times, set $k_{max} = 100$, $c = 10$, $t_{stab} = 30$ s, and changed $E[T_{on}]$ from 10 min to 5 min to 15 min and back to 10 min after 8.33 h, 16.66 h, and 25 h of simulation time, respectively. The simulated curve shows the mean of the estimated $E[T_{on}]$ values of all peers, which were online at the corresponding time. The error bars represent the interquartile range. It can be seen that the estimator is able to capture the changes in the churn rate and that the time it takes to adjust to the new value complies with the analysis. Note that, due to the stabilization period of 30 s, the estimated values lie $\epsilon_{on} = 15$ s above the actual value.

4   Numerical Results

In this section we will evaluate the proposed estimator using simulations. Unless stated otherwise, we will always consider that the online and offline times of the users are exponentially distributed with mean $E[T_{on}]$ and $E[T_{off}]$, respectively. The default stabilization interval is $t_{stab} = 30$ s and the size of neighbor list is $c = 20$. We will further assume that there are 40000 initial peers with $E[T_{on}] = E[T_{off}]$, resulting in an average of 20000 online peers at a time. Although our estimator yields results for both online and offline time, we will concentrate on estimating the online time T_{on}, since this is usually a more important parameter for the system performance and T_{off} can be calculated in an analogous way.

4.1   Proof of Concept

The main purpose of this section is to show that the theoretic concept of the proposed estimator as described in Section 3 does work equally well in practice. We focus on Chord since it is the currently most studied DHT network architecture. Additionally, we will provide analytical calculations verified by simplified simulations, focusing on properties which are important to our estimator. That is, we mainly disregard all mechanisms dealing with document management or replication. To model the stabilization algorithm, a peer synchronizes its neighbor list every $t_{stab} = 30$ s with its direct successor. When a peer notices that another peer is offline, it notifies the peers in its neighbor list, piggybacking the observed online time in these messages. We consider a symmetric neighbor list, i.e. the number of peers in the successor list is the same as that of the predecessor list. This improves the stability of the Chord overlay and provides a better comparability of the result to symmetric overlays like Kademlia.

In practice too high or too low estimates might have critical consequences in terms of performance or even functionality. In such a case it should be avoided that the estimator underestimates or overestimates the actual churn rate. This can be achieved by using the upper or lower bound of a specified confidence level instead of the estimated value

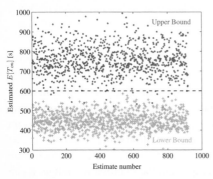

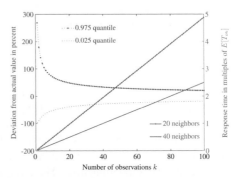

Fig. 7. Upper and lower confidence levels   **Fig. 8.** Accuracy vs. responsiveness

itself. Fig. 7 shows the upper and lower bounds of the 99% confidence interval for the mean. As expected, the upper bound overestimates the actual value, while the lower bound underestimates it. The frequency at which the upper bound underestimates or the lower bound overestimates the actual value can be influenced by the confidence level. The higher the confidence level is chosen, the smaller is the probability for this to happen at the cost of more inaccurate values.

4.2 Accuracy and Responsiveness

We now take a closer look at the trade-off between accuracy and responsiveness in dependence of the size of the history. To express accuracy, we consider how much the 97.5% and 2.5% quantiles of the estimated values based on k observations differ from the actual value in percent. This is plotted as the dotted blue curves in Fig. 8 using the left y-axis. It can be recognized that increasing the history size results in more accurate estimates which decreases exponentially over k.

An increased accuracy, however, comes at the drawback of reducing the responsiveness of the estimator. *Responsiveness* is defined as the time it takes to collect k fresh results when there is a change in the global churn rate. It is expressed in multiples of $E[T_{on}]$ and approximated by Eqn. (7). Responsiveness increases linearly with k (cf. green solid curves of Fig. 8 with right y-axis) and its slope is determined by the number of overlay neighbors. The more neighbors there are, the more results are obtained per time unit and the faster the estimator reacts to the change. The study shows that depending on the application requirements, a trade-off can be made between higher accuracy and faster responsiveness by changing the number of considered observations.

In order to provide a more comprehensive study of the responsiveness of the estimator and to validate our analytical approximation in Eqn. (7), we perform simulation runs with different churn rates and measure the time between two successive observations. Obviously, the smaller this inter-observation time is, the faster the reaction to changes of the churn rate, see Fig. 9. For different churn rates of $E[T_{on}] = 300\,s, 600\,s$, and $900\,s$, the inter-observation time is shown over the number of overlay contacts. The dashed lines are the results obtained by the approximation, cf. Eqn. (6). It can be seen that the inter-observation time decreases exponentially and that the analytical curves

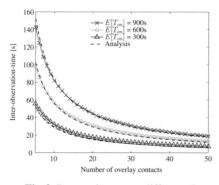

Fig. 9. Responsiveness to different churn

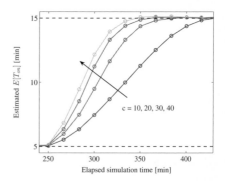

Fig. 10. Reaction to global churn changes

match well with those obtained by simulations. A greater number than 20 neighbors is not justified due to the small improvement in responsiveness and the higher overhead in maintaining those neighbors. Smaller values of $E[T_{on}]$ result in smaller values of the inter-observation time, but the number of overlay contacts has an even greater influence on the inter-observation time. Note, that the responsiveness also depends on the quality of the stabilization algorithm. If a simple algorithm is used, the neighbor lists might be inaccurate, which in turn results in a loss of updates and a higher inter-observation time.

To show how the inter-observation time translates into the actual response time and how the estimator behaves during these reaction phases, we simulated a network where the mean online time of all peers was globally changed from the initial value of 5min to 15min after a simulation time of 250 min. In Fig. 10 each data point shows the average of the estimated $E[T_{on}]$ values of all online peers at the same time instant. Again the more neighbors there are, the faster the estimator approaches the new churn rate. However, increasing the number of neighbors beyond $c = 20$ does not justify its additional overhead. Thus, using 20 overlay neighbors, as e.g. suggested in Kademlia, is a reasonable choice.

4.3   Practicability and Implementation Aspects

In practice, it is desirable that all peers obtain equal estimates in order to derive similar input parameters for the maintenance algorithms of the P2P network. However, those algorithms are performed between direct neighbors of the DHT. Since these direct overlay neighbors also exchange their measured observations, their churn estimates derived from this data are expected to be highly correlated. To quantify the degree of this correlation, we took a global snapshot during the simulation and had a closer look at the estimates of 5000 consecutive peers on the Chord ring. We then investigated the correlation between these peers by applying methods from time series analysis. Fig. 11 depicts the autocorrelation over the number of neighbors and shows that there is a high correlation among neighboring peers. The curves for the different numbers c of overlay neighbors among which the measurement values are exchanged show that the correlation extends to at least c neighbors in both directions of the ring.

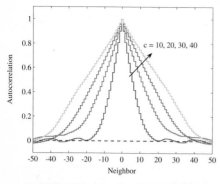

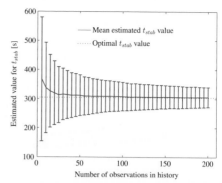

Fig. 11. Correlation among neighbor peers   **Fig. 12.** Choice of stabilization interval

A possible application of the proposed estimator is self-tuning the stabilization of the overlay structure. In practice, the stabilization interval, i.e. the frequency at which overlay neighbors are contacted to update the neighbor lists, is a fixed value. This results in unnecessary overhead when there is no churn in the network, but when there is a high churn rate, the stabilization overhead may not be sufficient to maintain the stability of the overlay. For self-adaptive selection of t_{stab}, a peer should therefore estimate the current churn rate to derive the probability that the overlay structure becomes instable, i.e. that all neighbors will be offline before the next stabilization call. For example, given a mean online time of $E[T_{on}] = 600$ s, a peer needs to stabilize at least every 300 s in order to maintain the overlay stability with a probability of 99.99%. In Fig. 12, the mean and standard deviation of the t_{stab} derived from estimation is shown over the size of the observation history. It can be seen that the standard deviation decreases exponentially and that a history size of 100 again results in a good value for practical purposes.

5   Conclusion

Structured P2P networks apply different maintenance mechanisms to guarantee the stability of the overlay network and the redundancy of stored documents. Ideally, the parameters of these mechanisms should be adapted to the current churn rate. The more churn there is in the system, the more overhead is needed to keep the system stable. As a first step toward a self-organizing overlay network, we introduced a method which enables a peer to estimate the current churn rate in the system and can be used to autonomically adapt the overhead.

The estimator is based on the changes a peer observes in its list of overlay neighbors. The more observations a peer makes, the better is the quality of its estimate. Therefore, a peer shares observed events with its direct overlay neighbors by piggybacking the corresponding information in regular protocol messages. Both analytical and simulation results show that the estimator is able to capture the current churn rate. The accuracy, the required overhead, and the responsiveness to changes can be adjusted by the number of observations considered in the estimation process and by the number of overlay neighbors which share the results. We investigated the corresponding trade-offs and deduced values which are suitable for practical purposes. For applications which are sensitive to an overestimation or underestimation of the actual value, we showed how to use the upper and lower bounds of a confidence interval as estimates themselves.

In future work, we intend to use the estimator to enable a peer to autonomously adapt the number of overlay neighbors and the number of replicas to the current churn rate. This way, the functionality of the overlay network will still be guaranteed in times of high churn while the maintenance overhead will be reduced in times of no churn.

Acknowledgments

The authors would like to thank Dirk Staehle and Simon Oechsner for their many ideas, the input, and the insightful discussions during the course of this work.

References

1. Rhea, S., Geels, D., Roscoe, T., Kubiatowicz, J.: Handling Churn in a DHT. In: USENIX Annual Technical Conference, Boston, MA (2004)
2. Binzenhöfer, A., Staehle, D., Henjes, R.: On the Stability of Chord-based P2P Systems. In: GLOBECOM 2005, vol.5, St. Louis, MO, USA (2005)
3. Stoica, I., Morris, R., Karger, D., Kaashoek, M.F., Balakrishnan, H.: Chord: A Scalable Peer-to-peer Lookup Service for Internet Applications. In: Proc. of ACM SIGCOMM'01, San Diego, CA (2001)
4. Maymounkov, P., Mazieres, D.: Kademlia: A peer-to-peer information system based on the xor metric. In: Proc. of IPTPS'02, Cambridge, MA (2002)
5. Rowstron, A., Druschel, P.: Pastry: Scalable, distributed object location and routing for large-scale peer-to-peer systems. In: Proc. of Middleware'01, Heidelberg, Germany (2001)
6. Liben-Nowell, D., Balakrishnan, H., Karger, D.: Analysis of the Evolution of Peer-to-Peer Systems. In: Proc. of ACM PODC, Monterey, CA (2002)
7. Krishnamurthy, S., El-Ansary, S., Aurell, E., Haridi, S.: A statistical theory of chord under churn. In: Proc. of IPTPS'05, Ithaca, NY (2005)
8. Godfrey, P.B., Shenker, S., Stoica, I.: Minimizing churn in distributed systems. In: Proc. of ACM SIGCOMM, Pisa, Italy (2006)
9. Gummadi, K.P., Dunn, R.J., Saroiu, S., Gribble, S.D., Levy, H.M., Zahorjan, J.: Measurement, modeling, and analysis of a peer-to-peer file-sharing workload. In: Proc. of ACM SOSP'03, Bolton Landing, NY (2003)
10. Bhagwan, R., Savage, S., Voelker, G.: Understanding availability. In: Proc. of IPTPS'03, Berkeley, CA (2003)
11. Castro, M., Costa, M., Rowstron, A.: Performance and dependability of structured peer-to-peer overlays. In: Proc. of DSN'04, Washington, DC (2004)
12. Ghinita, G., Teo, Y.: An adaptive stabilization framework for distributed hash tables. In: Proc. of IEEE IPDPS, Rhodes Island, Greece (2006)
13. Binzenhöfer, A., Staehle, D., Henjes, R.: On the Fly Estimation of the Peer Population in a Chord-based P2P System. In: 19th International Teletraffic Congress (ITC19), Beijing, China (2005)
14. Mahajan, R., Castro, M., Rowstron, A.: Controlling the cost of reliability in peer-to-peer overlays. In: Proc. of IPTPS'03, Berkeley, CA (2003)
15. Stephens, M.A.: Edf statistics for goodness of fit and some comparisons. In: Journal of the American Statistical Association, vol. 69, pp. 730–739 (1974)
16. Chen, E.J., Kelton, W.D.: Quantile and histogram estimation. In: Proc. of 33nd Winter Simulation Conference, Washington, DC (2001)

Performance Analysis of Peer-to-Peer Storage Systems

Sara Alouf, Abdulhalim Dandoush, and Philippe Nain

INRIA – B.P. 93 – 06902 Sophia Antipolis – France
{salouf,adandous,philippe.nain}@sophia.inria.fr

Abstract. This paper evaluates the performance of two schemes for recovering lost data in a peer-to-peer (P2P) storage systems. The first scheme is centralized and relies on a server that recovers multiple losses at once, whereas the second one is distributed. By representing the state of each scheme by an absorbing Markov chain, we are able to compute their performance in terms of the delivered data lifetime and data availability. Numerical computations are provided to better illustrate the impact of each system parameter on the performance. Depending on the context considered, we provide guidelines on how to tune the system parameters in order to provide a desired data lifetime.

Keywords: Peer-to-Peer systems, performance evaluation, absorbing Markov chain, mean-field approximation.

1 Introduction

Traditional storage solutions rely on robust dedicated servers and magnetic tapes on which data are stored. These equipments are reliable, but expensive. The growth of storage volume, bandwidth, and computational resources has fundamentally changed the way applications are constructed, and has inspired a new class of storage systems that use distributed peer-to-peer (P2P) infrastructures. Some of the recent efforts for building highly available storage system based on the P2P paradigm are Intermemory [6], Freenet [3], OceanStore [13], CFS [4], PAST [16], Farsite [5] and Total Recall [1]. Although inexpensive compared to traditional systems, these storage systems pose many problems of reliability, confidentiality, availability, routing, etc.

In a P2P network, peers are free to leave and join the system at any time. As a result of the intermittent availability of the peers, ensuring high availability of the stored data is an interesting and challenging problem. To ensure data reliability, redundant data is inserted in the system. Redundancy can be achieved either by replication or by using erasure codes. For the same amount of redundancy, erasure codes provide higher availability of data than replication [18].

However, using redundancy mechanisms without repairing lost data is not efficient, as the level of redundancy decreases when peers leave the system. Consequently, P2P storage systems need to compensate the loss of data by continuously storing additional redundant data onto new hosts. Systems may rely on a centralized instance that reconstructs fragments when necessary; these systems will be referred to as *centralized-recovery systems*. Alternatively, secure agents running on new peers can reconstruct by themselves the data to be stored on the peers disks. Such systems will be referred to as

L. Mason, T. Drwiega, and J. Yan (Eds.): ITC 2007, LNCS 4516, pp. 642–653, 2007.

distributed-recovery systems. A centralized server can recover at once multiple losses of the same document. This is not possible in the distributed case where each new peer thanks to its secure agent recovers only one loss per document.

Regardless of the recovery mechanism used, two repair policies can be adopted. In the *eager* policy, when the system detects that one host has left the system, it immediately repairs the diminished redundancy by inserting a new peer hosting the recovered data. Using this policy, data only becomes unavailable when hosts fail more quickly than they can be detected and repaired. This policy is simple but makes no distinction between permanent departures that require repair, and transient disconnections that do not. An alternative is to defer the repair and to use additional redundancy to mask and to tolerate host departures for an extended period. This approach is called *lazy* repair because the explicit goal is to delay repair work for as long as possible.

In this paper, we aim at developing mathematical models to characterize fundamental performance metrics (lifetime and availability – see next paragraph) of P2P storage systems using erasure codes. We are interested in evaluating the centralized- and distributed-recovery mechanisms discussed earlier, when either eager or lazy repair policy is enforced. We will focus our study on the quality of service delivered to each block of data. We aim at addressing fundamental design issues such as: *how to tune the system parameters so as to maximize data lifetime while keeping a low storage overhead?*

The *lifetime* of data in the P2P system is a random variable; we will investigate its distribution function. *Data availability* metrics refer to the amount of redundant fragments. We will consider two such metrics: the expected number of available redundant fragments, and the fraction of time during which the number of available redundant fragment exceeds a given threshold. For each implementation (centralized/distributed) we will derive these metrics in closed-form through a Markovian analysis.

In the following, Sect. 2 briefly reviews related work and Sect. 3 introduces the notation and assumptions used throughout the paper. Sections 4 and 5 are dedicated to the modeling of the centralized- and distributed-recovery mechanism. In Sect. 6, we provide numerical results showing the performance of the centralized and decentralized schemes, under the eager or the lazy policy. We conclude the paper in Sect. 7.

2   Related Work

There is an abundant literature on the architecture and file system of distributed storage systems (see [6,13,4,16,5,1]; non-exhaustive list) but only a few studies have developed analytical models of distributed storage systems to understand the trade-offs between the availability of the files and the redundancy involved in storing the data.

In [18], Weatherspoon and Kubiatowicz characterize the availability and durability gains provided by an erasure-resilient system. They quantitatively compare replication-based and erasure-coded systems. They show that erasure codes use an order of magnitude less bandwidth and storage than replication for systems with similar durability. Utard and Vernois perform another comparison between the full replication mechanism and erasure codes through a simple stochastic model for node behavior [17]. They observe that simple replication schemes may be more efficient than erasure codes in presence of low peers availability. In [10], Lin, Chiu and Lee focus on erasure codes analysis under different scenarios, according to two key parameters: the peer availability level and the storage overhead. Blake and Rodrigues argue in [2] that the cost of dynamic membership makes the cooperative storage infeasible in transiently available peer-to-peer environments. In other words, when redundancy, data scale, and dynamics are all high, the needed cross-system bandwidth is unreasonable when clients desire to download files during a reasonable time. Last, Ramabhadran and Pasquale develop in [14] a Markov chain analysis of a storage system using full replication for data reliability, and a distributed recovery scheme. They derive an expression for the lifetime of the replicated state and study the impact of bandwidth and storage limits on the system.

3   System Description and Notation

We consider a distributed storage system in which peers randomly join and leave the system. Upon a peer disconnection, all data stored on this peer is no longer available to the users of the storage system and is considered to be lost. In order to improve data availability it is therefore crucial to add redundancy to the system.

In this paper, we consider a single block of data D, divided into s equally sized fragments to which, using erasure codes (e.g. [15]), r redundant fragments are added. These $s+r$ fragments are stored over $s+r$ different peers. Data D is said to be *available* if any s fragments out of the $s + r$ fragments are available and *lost* otherwise. We assume that at least s fragments are available at time $t = 0$. Note that when $s = 1$ the r redundant fragments will simply be replicas of the unique fragment of the block; replication is therefore a special case of erasure codes.

Over time, a peer can be either *connected* to or *disconnected* from the storage system. At reconnection, a peer may still or may not store one fragment. Data stored on a connected peer is available at once and can be used to reconstruct a block of data. We refer to as *on-time* (resp. *off-time*) a time-interval during which a peer is always connected (resp. disconnected). Typically, the number of connected peers at any time in a storage system is much larger than the number of fragments associated with a given data D. Therefore, we assume that there are always at least r connected peers – hereafter referred to as *new* peers – which are ready to store fragments of D. For security issues, a peer may store at most one fragment.

We assume that the successive durations of on-times (resp. off-times) of a peer form a sequence of independent and identically distributed (iid) random variables (rvs), with an exponential distribution with rate $\alpha_1 > 0$ (resp. $\alpha_2 > 0$). We further assume that peers behave independently of each other, which implies that on-time and off-time sequences associated with any set of peers are statistically independent. We denote by p the probability that a peer that reconnects still stores one fragment and that this fragment is different from all other fragments available in the system.

As discussed in Sect. 1 we will investigate the performance of two different repair policies: the *eager* and the *lazy* repair policies. In the eager policy a fragment of D is reconstructed as soon as one fragment has become unavailable due to a peer disconnection. In the lazy policy, the repair is triggered only when the number of unavailable fragments reaches a given threshold $k \geq 1$. Note that $k \leq r$ since D is lost if more than r fragments are not available in the storage system at a given time. Both repair policies can be represented by a threshold parameter $k \in \{1, 2, \ldots, r\}$, where $k = 1$ in the eager policy, and where k can take any value in the set $\{2, \ldots, r\}$ in the lazy policy.

Let us now describe the fragment recovery mechanism. As mentioned in Sect. 1, we will consider two implementations of the eager and lazy recovery mechanisms, a *centralized* and a *(partially) distributed* implementation.

Assume that $k \leq r$ fragments are no longer available due to peer disconnections, triggering the recovery mechanism. In the centralized implementation, a central authority will: (i) download s fragments from the peers which are connected, (ii) reconstruct at once the k unavailable fragments, and (iii) transmit each of them to a new peer for storage. We will assume that the total time required to perform these tasks is exponentially distributed with rate $\beta_c(k) > 0$ and that successive recoveries are statistically independent.

In the distributed implementation, a secure agent on *one* new peer is notified of the identity of the k unavailable fragments. Upon notification, it downloads s fragments of D from the peers which are connected, reconstructs *one* out of the k unavailable fragments and stores it on its disk; the s downloaded fragments are then discarded so as to meet the security constraint that only one fragment of a block of data is held by a peer. We will assume that the total time required to perform the download, reconstruct and store a new fragment follows an exponential distribution with rate $\beta_d > 0$; we assume that each recovery is independent of prior recoveries.

The exponential distributions have mainly been made for the sake of mathematical tractability. We however believe that these are reasonable assumptions due to the unpredictable nature of the node dynamics and of the variability of network delays.

We conclude this section by a word on the notation: a subscript/superscript "c" (resp. "d") will indicate that we are considering the centralized (resp. distributed) scheme.

4   Centralized Repair Systems

In this section, we address the performance analysis of the centralized implementation of the P2P storage system, as described in Sect. 3. We will focus on a single block of data and we will only pay attention to peers storing fragments of this block.

Let $X_c(t)$ be a $\{a, 0, 1, \ldots, r\}$-valued rv, where $X_c(t) = i \in \mathcal{T} := \{0, 1, \ldots, r\}$ indicates that $s + i$ fragments are available at time t, and $X_c(t) = a$ indicates that less than s fragments are available at time t. We assume that $X_c(0) \in \mathcal{T}$ so as to reflect the assumption that at least s fragments are available at $t = 0$. Thanks to the assumptions made in Sect. 3, it is easily seen that $\boldsymbol{X}_c := \{X_c(t), t \geq 0\}$ is an absorbing homogeneous Continuous-Time Markov Chain (CTMC) with transient states $0, 1, \ldots, r$ and with a single absorbing state a representing the situation when the block of data is lost. Non-zero transition rates of $\{X_c(t), t \geq 0\}$ are shown in Fig. 1.

4.1   Data Lifetime

This section is devoted to the analysis of the data lifetime. Let $T_c(i) := \inf\{t \geq 0 : X_c(t) = a\}$ be the time until absorption in state a starting from $X_c(0) = i$, or equivalently the time at which the block of data is lost. In the following, $T_c(i)$ will be referred

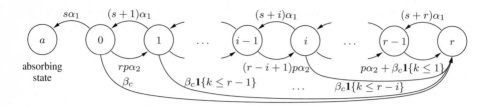

Fig. 1. Transition rates of the absorbing Markov chain $\{X_c(t), t \geq 0\}$

to as the *conditional block lifetime*. We are interested in $P(T_c(i) < x)$, the probability distribution of the block lifetime given that $X_c(0) = i$ for $i \in \mathcal{T}$, and the expected time spent by the absorbing Markov chain in transient state j, given that $X_c(0) = i$.

Let $\boldsymbol{Q}_c = [q_c(i, j)]_{0 \leq i,j \leq r}$ be a matrix, where for any $i, j \in \mathcal{T}$, $i \neq j$, $q_c(i, j)$ gives the transition rate of the Markov chain $\boldsymbol{X}_c$ from transient state i to transient state j, and $-q_c(i, i)$ is the total transition rate out of state i. Non-zero entries of $\boldsymbol{Q}_c$ are

$$
\begin{aligned}
q_c(i, i-1) &= c_i \ , & i &= 1, 2, \ldots, r \ , \\
q_c(i, i+1) &= d_i + \mathbf{1}\{i = r-1\}u_{r-1} \ , & i &= 0, 1, \ldots, r-1 \ , \\
q_c(i, r) &= u_i \ , & i &= 0, 1, \ldots, \min\{r-k, r-2\} \ , \\
q_c(i, i) &= -(c_i + d_i + u_i) \ , & i &= 0, 1, \ldots, r \ ,
\end{aligned}
\tag{1}
$$

where $c_i := (s + i)\alpha_1$, $d_i := (r - i)p\alpha_2$ and $u_i := \beta_c(r - i)\mathbf{1}\{i \leq r - k\}$ for $i \in \mathcal{T}$. Note that $\boldsymbol{Q}_c$ is not an infinitesimal generator since entries in its first row do not sum up to 0. From the theory of absorbing Markov chains we know that (e.g. [11, Lemma 2.2])

$$
P(T_c(i) < x) = 1 - \boldsymbol{e}_i \cdot \exp(x\boldsymbol{Q}_c) \cdot \mathbf{1} \ , \quad x > 0 \ , i \in \mathcal{T} \ ,
\tag{2}
$$

where $\boldsymbol{e}_i$ and $\mathbf{1}$ are vectors of dimension $r + 1$; all entries of $\boldsymbol{e}_i$ are null except the i-th entry that is equal to 1, and all entries of $\mathbf{1}$ are equal to 1. In particular [11, p. 46]

$$
\mathrm{E}[T_c(i)] = -\boldsymbol{e}_i \cdot \boldsymbol{Q}_c^{-1} \cdot \mathbf{1} \ , \quad i \in \mathcal{T} \ ,
\tag{3}
$$

where the existence of $\boldsymbol{Q}_c^{-1}$ is a consequence of the fact that all states in $\mathcal{T}$ are transient [11, p. 45]. Let $T_c(i, j) = \int_0^{T_c(i)} \mathbf{1}\{X_c(t) = j\}dt$ be the total time spent by the CTMC in transient state j given that $X_c(0) = i$. It can also be shown that [7]

$$
\mathrm{E}[T_c(i, j)] = -\boldsymbol{e}_i \cdot \boldsymbol{Q}_c^{-1} \cdot \boldsymbol{e}_j \ , \quad i, j \in \mathcal{T} \ .
\tag{4}
$$

Even when $\beta_c(0) = \cdots = \beta_c(r)$ an explicit calculation of either $P(T_c(i) < x)$, $\mathrm{E}[T_c(i)]$ or $\mathrm{E}[T_c(i, j)]$ is intractable, for any k in $\{1, 2, \ldots, r\}$. Numerical results for $\mathrm{E}[T_c(r)]$ and $P(T_c(r) > 10$ years$)$ are reported in Sect. 6 when $\beta_c(0) = \cdots = \beta_c(r)$.

4.2 Data Availability

In this section we introduce different metrics to quantify the availability of the block of data. The fraction of time spent by the absorbing Markov chain $\{X_c(t), t \geq 0\}$ in state

j with $X_c(0) = i$ is $\mathrm{E}[(1/T_c(i)) \int_0^{T_c(i)} \mathbf{1}\{X_c(t) = j\}dt]$. However, since it is difficult to find a closed-form expression for this quantity, we will instead approximate it by the ratio $\mathrm{E}[T_c(i,j)]/\mathrm{E}[T_c(i)]$. With this in mind, we introduce

$$M_{c,1}(i) := \sum_{j=0}^{r} j \frac{\mathrm{E}[T_c(i,j)]}{\mathrm{E}[T_c(i)]} \quad , \quad M_{c,2}(i) := \sum_{j=m}^{r} \frac{\mathrm{E}[T_c(i,j)]}{\mathrm{E}[T_c(i)]} \quad , \quad i \in \mathcal{T} . \quad (5)$$

The first availability metric can be interpreted as the expected number of available redundant fragments during the block lifetime, given that $X_c(0) = i \in \mathcal{T}$. The second metric can be interpreted as the fraction of time when there are at least m redundant fragments during the block lifetime, given that $X_c(0) = i \in \mathcal{T}$. Both quantities $M_{c,1}(i)$ and $M_{c,2}(i)$ can be (numerically) computed from (3) and (4). Numerical results are reported in Sect. 6 for $i = r$ and $m = r - k$ in (5).

Since it is difficult to come up with an explicit expression for either metric $M_{c,1}(i)$ or $M_{c,2}(i)$, we make the assumption that parameters k and r have been selected so that the time before absorption is "large". This can be formalized, for instance, by requesting that $P(T_c(r) > q) > 1 - \epsilon$, where parameters q and ϵ are set according to the particular storage application(s). Instances are given in Sect. 6. In this setting, one may ignore the absorbing state a and represent the state of the storage system by a new irreducible and aperiodic – and therefore ergodic – Markov chain $\tilde{X}_c := \{\tilde{X}_c(t), t \geq 0\}$ on the state-space $\mathcal{T}$. Let $\tilde{Q}_c = [\tilde{q}_c(i,j)]_{0 \leq i,j \leq r}$ be its infinitesimal generator. Matrices $\tilde{Q}_c$ and Q_c, whose non-zero entries are given in (1), are identical except for $\tilde{q}_c(0,0) = -(u_0 + d_0)$. Until the end of this section we assume that $\beta_c(i) = \beta_c$ for $i \in \mathcal{T}$.

Let $\pi_c(i)$ be the stationary probability that $\tilde{X}_c$ is in state i. Our objective is to compute $\mathrm{E}[\tilde{X}_c] = \sum_{i=0}^{r} i\pi_c(i)$, the (stationary) expected number of available redundant fragments. To this end, let us introduce $f_c(z) = \sum_{i=0}^{r} z^i \pi_c(i)$, the generating function of the stationary probabilities $\pi_c = (\pi_c(0), \pi_c(1), \ldots, \pi_c(r))$. Starting from the Kolmogorov balance equations $\pi_c \cdot \tilde{Q}_c = 0$, $\pi_c \cdot \mathbf{1} = 1$, standard algebra yields

$$(\alpha_1 + p\alpha_2 z) \frac{df_c(z)}{dz} = rp\alpha_2 f_c(z) - s\alpha_1 \frac{f_c(z) - \pi_c(0)}{z} + \beta_c \frac{f_c(z) - z^r}{1-z} - \beta_c \sum_{i=r-k+1}^{r} \frac{z^i - z^r}{1-z} \pi_c(i).$$

Letting $z = 1$ and using the identities $f_c(1) = 1$ and $df_c(z)/dz|_{z=1} = \mathrm{E}[\tilde{X}_c]$, we find

$$\mathrm{E}[\tilde{X}_c] = \frac{r(p\alpha_2 + \beta_c) - s\alpha_1(1 - \pi_c(0)) - \beta_c \sum_{i=0}^{k-1} i\pi_c(r-i)}{\alpha_1 + p\alpha_2 + \beta_c} . \quad (6)$$

Unfortunately, it is not possible to find an explicit expression for $\mathrm{E}[\tilde{X}_c]$ since this quantity depends on the probabilities $\pi_c(0), \pi_c(r - (k-1)), \pi_c(r - (k-2)), \ldots, \pi_c(r)$, which cannot be computed in explicit form. If $k = 1$ then

$$\mathrm{E}[\tilde{X}_c] = \frac{r(p\alpha_2 + \beta_c) - s\alpha_1(1 - \pi_c(0))}{\alpha_1 + p\alpha_2 + \beta_c} , \quad (7)$$

which still depends on the unknown probability $\pi_c(0)$.

Below, we use a mean field approximation to develop an approximation formula for $\mathrm{E}[\tilde{X}_c]$ for $k = 1$, in the case where the maximum number of redundant fragments r is large. Until the end of this section we assume that $k = 1$. Using [9, Thm. 3.1] we know that, when r is large, the expected number of available redundant fragments at time t, $E[\tilde{X}_c(t)]$, is solution of the following first-order differential (ODE) equation

$$\dot{y}(t) = -(\alpha_1 + p\alpha_2 + \beta_c)y(t) - s\alpha_1 + r(p\alpha_2 + \beta_c) \ .$$

The equilibrium point of the above ODE is reached when time goes to infinity, which suggests to approximate $E[\tilde{X}_c]$, when r is large, by

$$E[\tilde{X}_c] \approx y(\infty) = \frac{r(p\alpha_2 + \beta_c) - s\alpha_1}{\alpha_1 + p\alpha_2 + \beta_c} \ . \tag{8}$$

Observe that this simply amounts to neglect of the probability $\pi_c(0)$ in (7) for large r.

5   Distributed Repair Systems

In this section, we address the performance analysis of the distributed implementation of the P2P storage system, as described in Sect. 3. Recall that in the distributed setting, as soon as k fragments become unreachable, secure agents running on k new peers simultaneously initiate the recovery of one fragment each.

5.1   Data Lifetime

Since the analysis is very similar to the analysis in Sect. 4 we will only sketch it. Alike in the centralized implementation, the state of the system can be represented by an absorbing Markov chain $X_d := \{X_d(t), t \geq 0\}$, taking values in the set $\{a\} \cup \mathcal{T}$ (recall that $\mathcal{T} = \{0, 1, \ldots, r\}$). State a is the absorbing state indicating that the block of data is lost (less than s fragments available), and state $i \in \mathcal{T}$ gives the number of available redundant fragments. The non-zero transition rates of this absorbing Markov chain are displayed in Fig. 2. Non-zero entries of the matrix $Q_d = [q_d(i, j)]_{0 \leq i,j \leq r}$ associated with the absorbing Markov chain X_d are given by

$$\begin{aligned}
q_d(i, i-1) &= c_i \ , & i &= 1, 2, \ldots, r \ , \\
q_d(i, i+1) &= d_i + w_i \ , & i &= 0, 1, \ldots, r-1 \ , \\
q_d(i, i) &= -(c_i + d_i + w_i) \ , & i &= 0, 1, \ldots, r \ ,
\end{aligned}$$

with $w_i := \beta_d \mathbf{1}\{i \leq r - k\}$ for $i = 0, 1, \ldots, r$, where c_i and d_i are defined in Sect. 4. Note that letting $s = 1$, $p = 0$ and $k = 1$ yields the model of [14]. Introduce $T_d(i) := \inf\{t \geq 0 : X_d(t) = a\}$ the time until absorption in state a given that $X_d(0) = i$, and let $T_d(i, j)$ be the total time spent in transient state j given that $X_d(0) = i$. The distribution $P(T_d(i) < x)$, $E[T_d(i)]$ and $E[T_d(i, j)]$ are given by (2), (3) and (4), respectively, after replacing Q_c by Q_d. Alike for Q_c it is not tractable to explicitly invert Q_d. Numerical results for $E[T_d(r)]$ and $P(T_d(r) > 1$ year$)$ are reported in Sect. 6.

5.2   Data Availability

As motivated in Sect. 4.2 the metrics

$$M_{d,1}(i) := \sum_{j=0}^{r} j \frac{E[T_d(i, j)]}{E[T_d(i)]} \ , \quad M_{d,2}(i) := \sum_{j=m}^{r} \frac{E[T_d(i, j)]}{E[T_d(i)]} \ , \tag{9}$$

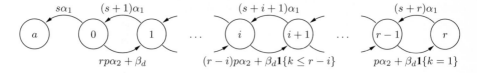

Fig. 2. Transition rates of the absorbing Markov chain $\{X_d(t), t \geq 0\}$

can be used to quantify the data availability in distributed-recovery P2P storage systems. Numerical results are given in Sect. 6. Similar to what was done in Sect. 4.2, let us assume that parameters r and k have been tuned so that the time before absorption is "long". If so, then as an approximation one can consider that absorbing state a can no longer be reached. The Markov chain X_d becomes an irreducible, aperiodic Markov chain on the set $\mathcal{T}$, denoted $\tilde{X}_d$. More precisely, it becomes a birth and death process (see Fig. 2). Let $\pi_d(i)$ be the stationary probability that $\tilde{X}_d$ is in state i, then (e.g. [8])

$$\pi_d(i) = \left[1 + \sum_{i=1}^{r} \prod_{j=0}^{i-1} \frac{d_j + w_j}{c_{j+1}} \right]^{-1} \cdot \prod_{j=0}^{i-1} \frac{d_j + w_j}{c_{j+1}} , \quad i \in \mathcal{T} . \tag{10}$$

From (10) we can derive the expected number of available redundant fragments through the formula $\mathrm{E}[\tilde{X}_d] = \sum_{i=0}^{r} i\pi_d(i)$. Numerical results for $\mathrm{E}[\tilde{X}_d]$, or more precisely, for its deviation from $M_{d,1}(r)$ are reported in Sect. 6.

6   Numerical Results

In this section we provide numerical results using the Markovian analysis presented earlier. Our objectives are to characterize the performance metrics defined in the paper against the system parameters and to illustrate how our models can be used to engineer the storage systems.

Throughout the numerical computations, we consider storage systems for which the dynamics have either one or two timescales, and whose recovery implementation is either centralized or distributed. Dynamics with two timescales arise in a *company context* in which disconnections are chiefly caused by failures or maintenance conditions. This yields slow peer dynamics and significant data losses at disconnected peers. However, the recovery process is particularly fast. Storage systems deployed over a wide area network, hereafter referred to as the *Internet context*, suffer from both fast peer dynamics and a slow recovery process. However, it is highly likely that peers will still have the stored data at reconnection.

The initial number of fragments is set to $s = 8$, deriving from the fact that fragment and block sizes in P2P systems are often set to $64KB$ and $512KB$ respectively (block sizes of $256KB$ and $1MB$ are also found). The recovery rate in the centralized scheme is made constant. The amount of redundancy r will be varied from 1 to 30 and for each value of r, we vary the threshold k from 1 to r. In the company context we set $1/\alpha_1 = 5$ days, $1/\alpha_2 = 2$ days, $p = 0.4$, $1/\beta_c = 11$ minutes and $1/\beta_d = 10$ minutes.

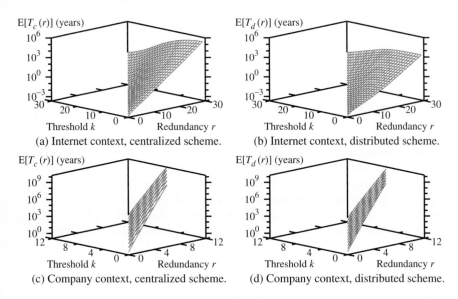

Fig. 3. Expected lifetime $E[T_c(r)]$ and $E[T_d(r)]$ (expressed in years) versus r and k

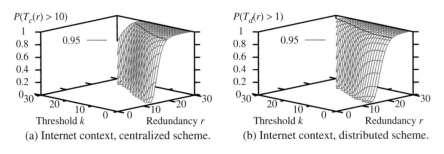

Fig. 4. (a) $P(T_c(r) > 10$ years) and (b) $P(T_d(r) > 1$ year) versus r and k

In the Internet context we set $1/\alpha_1 = 5$ hours, $1/\alpha_2 = 3$ hours, $p = 0.8$, $1/\beta_c = 34$ minutes and $1/\beta_d = 30$ minutes. This setting of the parameters is mainly for illustrative purposes. Recall that the recovery process accounts for the time needed to store the reconstructed data on the local (resp. remote) disk in the distributed (resp. centralized) scheme. Because of the network latency, we will always have $\beta_c < \beta_d$.

The Conditional Block Lifetime. We have computed the expectation and the *complementary* cumulative distribution function (CCDF) of $T_c(r)$ and $T_d(r)$ using (3) and (2) respectively. The results are reported in Figs. 3 and 4 respectively. The discussion on Fig. 4 comes later on.

We see from Fig. 3 that $E[T_c(r)]$ and $E[T_d(r)]$ increase roughly exponentially with r and are decreasing functions of k. When the system dynamics has two timescales like in a company context, the expected lifetime decreases exponentially with k whichever the

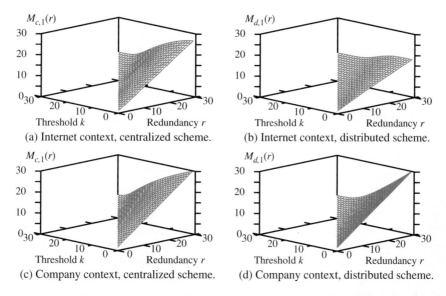

Fig. 5. Availability metrics $M_{c,1}(r)$ and $M_{d,1}(r)$ versus r and k

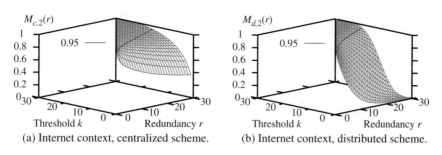

Fig. 6. Availability metrics (a) $M_{c,2}(r)$ and (b) $M_{d,2}(r)$ versus r and k with $m = r - k$

recovery mechanism considered. Observe in this case how large the block lifetime can become for certain values of r and k. Observe also that the centralized scheme achieves higher block lifetime than the distributed scheme unless $k = 1$ and $r = 1$ (resp. $r \le 6$) in the Internet (resp. company) context.

The Availability Metrics. We have computed the availability metrics $M_{c,1}(r)$, $M_{d,1}(r)$ and $M_{c,2}(r)$ and $M_{d,2}(r)$ with $m = r - k$ using (5) and (9). The results are reported in Figs. 5 and 6 respectively. The discussion on Fig. 6 comes later on.

We see from Fig. 5 that alike for the lifetime, metrics $M_{c,1}(r)$ and $M_{d,1}(r)$ increase exponentially with r and decrease as k increases. The shape of the decrease depends on which recovery scheme is used within which context. We again find that the centralized scheme achieves higher availability than the distributed scheme unless $k = 1$ and $r = 1$ (resp. $r \le 26$) in the Internet (resp. company) context.

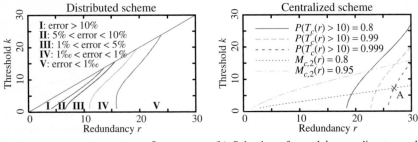

(a) Relative error $|M_{d,1} - \mathrm{E}[\tilde{X}_d]/M_{d,1}$.   (b) Selection of r and k according to predefined requirements.

Fig. 7. Numerical results for the Internet context

Regarding $M_{c,2}(r)$ and $M_{d,2}(r)$, we have found them to be larger than 0.997 for any of the considered values of r and k in the company context. This result is expected because of the two timescales present in the system. Recall that in this case the recovery process is two-order of magnitude faster than the peer dynamics. The results corresponding to the Internet context can be seen in Fig. 6.

Last, we have computed the expected number of available redundant fragments $\mathrm{E}[\tilde{X}_c]$ and $\mathrm{E}[\tilde{X}_d]$. The results are almost identical to the ones seen in Fig. 5. The deviation between $\mathrm{E}[\tilde{X}_d]$ and $M_{d,1}(r)$ in the Internet context is the largest among the four cases. Figure 7(a) delimits the regions where the deviation is within certain value ranges. For instance, in region V the deviation is smaller than 1‰. If the storage system is operating with values of r and k from this region, then it will be attractive to evaluate the data availability using $\mathrm{E}[\tilde{X}_d]$ instead of $M_{d,1}(r)$.

Engineering the system. Using our theoretical framework it is easy to tune the system parameters for fulfilling predefined requirements. As an illustration, Fig. 7(b) displays three contour lines of the CCDF of the lifetime $T_c(r)$ at point $q = 10$ years (see Fig. 4(a)) and two contour lines of the availability metric $M_{c,2}(r)$ with $m = r - k$ (see Fig. 6(a)). Consider point A which corresponds to $r = 27$ and $k = 7$. Selecting this point as the operating point of the storage system will ensure that $P(T_c(r) > 10) = 0.999$ and $M_{c,2}(r) = 0.8$. In other words, when $r = 27$ and $k = 7$, only 1‰ of the stored blocks would be lost after 10 years and for 80% of a block lifetime there will be $20 (= r - k)$ or more redundant fragments from the block available in the system.

One may be interested in only guaranteeing large data lifetime. Values of r and k are then set according to the desired contour line of the CCDF of data lifetime. Smaller threshold values enable smaller amounts of redundant data at the cost of higher bandwidth utilization. The trade-off here is between efficient storage use (small r) and efficient bandwidth use (large k).

7   Conclusion

We have proposed simple Markovian analytical models for evaluating the performance of two approaches for recovering lost data in distributed storage systems. One approach relies on a centralized server to recover the data; in the other approach new peers perform this task in a distributed way. We have analyzed the lifetime and the availability of data achieved by both centralized- and distributed-repair systems through Markovian analysis and fluid approximations. Numerical computations have been undertaken to support the performance analysis. Using our theoretical framework it is easy to tune the system parameters for fulfilling predefined requirements. Concerning future work, current efforts focus on modeling storage systems where peer lifetimes are either Weibull or hyperexponentially distributed (see [12]).

References

1. Bhagwan, R., Tati, K., Cheng, Y., Savage, S., Voelker, G.M.: Total Recall: System support for automated availability management. In: Proc. of ACM/USENIX NSDI '04, pp. 337–350. San Francisco, California, (March 2004)
2. Blake, C., Rodrigues, R.: High availability, scalable storage, dynamic peer networks: Pick two. In: Proc. of HotOS-IX, Lihue, Hawaii (May 2003)
3. Clarke, I., Sandberg, O., Wiley, B., Hong, T.W.: Freenet: A distributed anonymous information storage and retrieval system. In: Federrath, H. (ed.) Designing Privacy Enhancing Technologies. LNCS, vol. 2009, pp. 46–66. Springer, Heidelberg (2001)
4. Dabek, F., Kaashoek, M.F., Karger, D., Morris, R., Stoica, I.: Wide-area cooperative storage with CFS. In: Proc. of ACM SOSP '01, pp. 202–215. Banff, Canada (October 2001)
5. Farsite: Federated, available, and reliable storage for an incompletely trusted environment (2006) http://research.microsoft.com/Farsite/
6. Goldberg, A.V., Yianilos, P.N.: Towards an archival Intermemory. In: Proc. of ADL '98, pp. 147–156. Santa Barbara, California (April 1998)
7. Grinstead, C., Laurie Snell, J.: Introduction to Probability. American Math. Soc. (1997)
8. Kleinrock, L.: Queueing Systems, vol. 1. J. Wiley, New York (1975)
9. Kurtz, T.G.: Solutions of ordinary differential equations as limits of pure jump markov processes. Journal of Applied Probability 7(1), 49–58 (1970)
10. Lin, W.K., Chiu, D.M., Lee, Y.B.: Erasure code replication revisited. In: Proc. of IEEE P2P '04, Zurich, pp. 90–97. Switzerland, (August 2004)
11. Neuts, M.F.: Matrix Geometric Solutions in Stochastic Models. An Algorithmic Approach. John Hopkins University Press, Baltimore (1981)
12. Nurmi, D., Brevik, J., Wolski, R.: Modeling machine availability in enterprise and wide-area distributed computing environments. Technical Report CS2003-28, University of California Santa Barbara (2003)
13. The OceanStore project: Providing global-scale persistent data (2005), http://oceanstore.cs.berkeley.edu/
14. Ramabhadran, S., Pasquale, J.: Analysis of long-running replicated systems. In: Proc. of IEEE INFOCOM '06, Barcelona, Spain (April 2006)
15. Reed, I.S., Solomon, G.: Polynomial codes over certain finite fields. Journal of SIAM 8(2), 300–304 (June 1960)
16. Rowstron, A., Druschel, P.: Storage management and caching in PAST, a large-scale, persistent peer-to-peer storage utility. In: Proc. of ACM SOSP '01, pp. 188–201. Banff, Canada (October 2001)
17. Utard, G., Vernois, A.: Data durability in peer to peer storage systems. In: Proc. of IEEE GP2PC '04, Chicago, Illinois (April 2004)
18. Weatherspoon, H., Kubiatowicz, J.: Erasure coding vs. replication: A quantitative comparison. In: Proc. of IPTPS '02, Cambridge, Massachusetts (March 2002)

Traffic Matrix Estimation Based on Markovian Arrival Process of Order Two (MAP-2)

Suyong Eum[1], Richard J. Harris[2], and Irena Atov[2]

[1] RMIT University
Melbourne GPO 2476V Victoria 3001, Australia
suyong@catt.rmit.edu.au
[2] Massey University
Palmerston North Private Bag 11 222, New Zealand
R.Harris@massey.ac.nz, i.atov@ieee.org

Abstract. Traffic matrices constitute essential inputs in a wide variety of network planning and management functions as they provide the traffic volumes that flow between the node pairs in a network. In operational IP networks, it is desirable that traffic matrix (TM) estimation relies on information that is directly obtainable from SNMP system measurements i.e., link counts data. Existing approaches for TM estimation based on link counts have been shown to have limited accuracy and cannot be generally applied to practical IP networks. In this paper, we propose a new method for TM estimation which makes more accurate assumptions about the traffic characteristics of the flows between node pairs and, specifically, a Markovian Arrival Process of order two (MAP-2) has been applied for this purpose. The presented evaluation study shows the ability of the method to accurately capture the correlation and burstiness statistics of real IP flows and, therefore, can be successfully applied in IP network management functions.

1 Introduction

Traffic matrix (TM) provides an overall picture of traffic information exchanged between node pairs in a network and many different network functions, such as network capacity planning, congestion control, and inter & intra routing configuration, critically depend on this information. It is possible to measure TM directly using a network measurement tool such as NetFlow [1]. However, the deployment of such measurement system in a large-scale network along with its associated procedures for collection of a complete set of traffic data is too costly and renders the direct measurement approach impractical. Instead, research interests have focused on "Internet Tomography" where by using inference methods one can estimate the size of the traffic flows in the TM based on easily obtainable SNMP (Simple Network Management Protocol) system measurement link counts data.

If a TM is estimated systematically, every given fixed time period (e.g., as considered in [7], in order to capture more accurately the dynamics of the flows

L. Mason, T. Drwiega, and J. Yan (Eds.): ITC 2007, LNCS 4516, pp. 654–665, 2007.

between the OD pairs) the elements of the resulting TMs will form counting processes. Existing statistical approaches for TM estimation have been founded on the modelling assumption for these counting processes either being Poisson [2] or Gaussian [3], driven by the attractive theoretical properties of these traffic models. These approaches, however, when applied to real IP networks have been shown to have limited accuracy as they fail to capture the bursty and correlated nature of IP flows [7]. Thus, increasingly, there is a need for new methods that make more accurate assumptions about the traffic characteristics of the flows between node pairs in operational networks.

In [5] a moment matching model was developed for populating a Markovian Arrival Process of order two (MAP-2) using traffic count statistics obtainable from SNMP system measurements. Through an extensive evaluation study, using both synthetic correlated traffic and real traffic data obtained from backbone networks, we showed that this model captures well the impact that real IP traffic has on a queueing performance across various traffic load scenarios and, therefore, can be successfully applied in the context of TM estimation. MAPs have also been successfully applied in IP network planning functions and it has been shown that these traffic descriptors provide a very good balance between modelling accuracy (i.e., the capability of capturing burstiness and correlation characteristics of real IP traffic) and efficiency (as provided by the powerful matrix-analytic computational procedures for queueing systems) [4].

In light of these developments, in this paper we propose a new method for TM estimation in IP networks that provides an improvement on accuracy over existing methodologies by basing the assumptions about the traffic characteristics of real IP flows on the MAP-2 model. More specifically, the proposed method solves the TM estimation problem by dividing it into two main subproblems. First, from a multiple set of link counts obtainable from a consecutive set of SNMP measurements and a given routing matrix, we obtain a series of TMs by applying a deterministic approach. In the second step, the counting processes formed from the series of estimated TMs are matched to the MAP-2 traffic model by applying the method we proposed in [5].

The rest of this paper is organized as follows. In Section 2, the problem of TM estimation is illustrated. This is followed by a description of symbols and definitions that we use throughout the paper in Section 3. Section 4 provides a detailed description of the proposed TM estimation method. In Section 5, we introduce the experimental setup used in our evaluation study and present the validation results of the method. Finally, we conclude the paper in Section 6.

2   Traffic Matrix Estimation Problem

TM estimation problem can be described by the vector equation: (1).

$$Y = AX \tag{1}$$

where Y is the vector of measured link loads, A is the routing matrix, and X is the vector of traffic demands. In an IP network, the routing matrix can be derived from the OSPF or IS-IS link weights by computing the shortest paths between all node pairs and the link volumes are available from the SNMP data. The traffic demand vector X need to be estimated from the given Y and A, however, it turns out that there may be an infinite number of feasible solutions for X because the system of linear equations (1) is highly under-constrained.

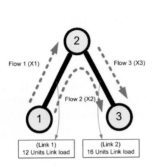

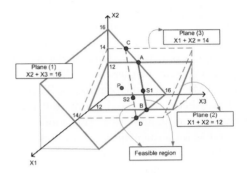

Fig. 1. Three node network example

Fig. 2. Solution space with time varying

Figure 1 illustrates this problem by means of a simple three node and two link network which carries three flows. These flows need to be estimated from the load measurements of the two links, which in this example are 12 and 16 units of load, respectively (measured at a time point t_1). For this network, the measured load on link 1 is equal to the sum of flows 1 and 2 and, similarly, the measured load on link 2 equals the sum of the flows 2 and 3. Figure 2 illustrates the two constraints defined by the equations $X_1 + X_2 = 12$ (Plane(2)) and $X_2 + X_3 = 16$ (Plane(1)). This is an under-constrained problem because the number of unknown variables (the three flows X_1, X_2 and X_3) is higher than the number of constraints. Therefore, the problem defines a *solution plane (hyperplane)* rather than having an unique point for the solution. In Fig. 2, the line AB represents the solutions which satisfy both constraints. Whatever technique is used for the TM estimation, a solution from the method should lie on the line AB to satisfy the inter-link measurement constraints.

Now, let us assume that a second set of link load measurements is obtained during the next time interval at point t_2. Further, if we suppose that the measured load on link 1 increased from 12 to 14, this will consequently move the hyperplane AB up to CD. The solutions at time point t_1 and t_2 are denoted by $S1$ and $S2$, respectively in Fig. 2, and more solutions ($S1, S2, S3, \ldots$) can be further obtained by measuring link loads at subsequential time intervals. In this case, the series of solutions need to be expressed through a mathematical model rather than a deterministic solution. The idea of the statistical approach for TM estimation is to assume that the series of the solutions are generated from an underlying mathematical model. The problem with this approach is thus to find a model that can describe the real solutions (statistical data) accurately.

3   Symbols and Definitions

Assuming a network of n nodes and r links, the TM of the network is a square matrix of dimension $n \times n$, with diagonal elements equal to zero. The number of origin-destination (OD) pairs, denoted by c, is obtained as $c = n \times (n-1)$. For the elements of the TM (i.e., traffic flows between the OD pairs) we use both matrix and vector representations interchangeably. In the first case, the flows are denoted by a lowercase $x_{ij}^{k}(i \neq j)$ where i and j represent the origin and destination for the flow. In the vector representation case, the flows are denoted by an uppercase X_{i}^{k} where i shows the OD pair index i=1, 2, ..., c. The symbol k in both cases indicates the index of different TMs obtained in measurement period k, $k = 1, 2, \ldots, K$. For instance, both x_{ij}^{k} and X_{i}^{k} represent an element of a TM obtained in the k-th measurement period which is turn is denoted by x^{k} and X^{k}, respectively. The set of link loads obtained in the k-th measurement period is represented by Y^{k}. In this paper, the terms, "fan-out

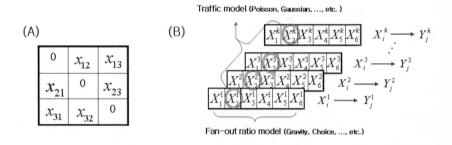

Fig. 3. The definition of the Traffic Matrix

model" and "traffic model" are used. The fan-out model is defined as a model to estimate the ratio of the size of each element in a single TM, and the traffic model is used for modelling of the components of the TMs obtained in a series, at various time points, forming a stochastic counting process.

Figure 3 shows a TM of a 3 node network with $6 = (3 \times 2)$ OD pairs. The OD pair flows are shown in matrix and vector representation in (A) and (B), respectively. The fan-out models, such as the gravity model [6] and the choice model [7], show the relation among elements $(X_{1}^{1}, X_{2}^{1}, X_{3}^{1}, \ldots, X_{6}^{1})$ in a single TM, whereas the traffic models, such as Poisson [2] and Gaussian [3], focus on modelling the series of elements $(X_{2}^{1}, X_{2}^{2}, X_{2}^{3}, \ldots, X_{2}^{k})$ illustrated by the circles along the continuous TMs in Fig. 3.

4   Proposed Method

TM estimation solutions have largely been obtained based on deterministic techniques and statistical inference methods. The former approach regards the link counts as hard constraints (constants), while the latter approach regards these as random variables and further makes assumptions about their distributions e.g., the traffic flow between each OD pair varies every point in time according to a Poisson, or Gaussian distribution. However, as discussed in Section 1, these traffic modelling assumptions are generally not valid for Internet traffic [7]. Moreover, they significantly impact on the accuracy of the TM estimation techniques [8]. Therefore, rather than estimating TM under assumption that

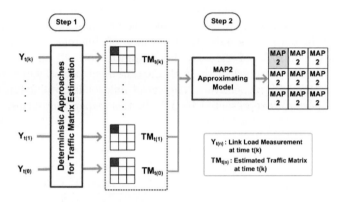

Fig. 4. The overview of the proposed method

each OD pair flow follows a certain distribution, (as per the statistical inference methods), to achieve higher accuracy but at the same time without increasing the complexity of the solution, we propose a *hybrid* model. This model first uses a deterministic approach, to find a series of TMs, continuously, at every fixed time period and, subsequently, fits the obtained statistical data for each OD pair flow into a traffic model that is shown can capture well the correlated and burstiness characteristics of IP flows. More specifically, our proposed TM method is illustrated in Fig. 4. In step one, from a multiple set of link counts obtainable from a consecutive set of k SNMP measurements and a given routing matrix, we obtain a series of TMs by applying a deterministic approach. The components of these matrices represent estimates of the OD pair flows at specific measurement points and they form stochastic counting processes, respectively. For example, Fig.4 shows a counting process arising from the first element of each TM (i.e., $TM_{t(0)}, TM_{t(1)}, \ldots, TM_{t(k)}$). In our second step, the counting processes formed from this series of estimated TMs are fitted into a MAP-2 model by applying the procedures developed in [5].

4.1   Step 1. Deterministic Approaches

Since the linear vector equation (1) is an under-constrained system, as explained in Section 2, the solutions generate a hyperplane.

The basic idea of the deterministic approach is how to choose one solution on the hyperplane with a given prior solution. A prior solution can be obtained

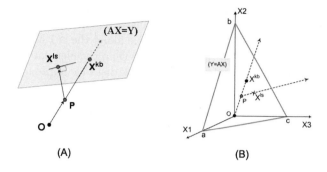

Fig. 5. Geometric Analysis

from a fan-out model (defined in Section 3) or, alternatively, an old TM can be used. Since the prior solution contains certain characteristics of the real solution, the deterministic approach uses these characteristics to identify one point on the hyperplane.

Figure 5 illustrates the basic idea of the deterministic approaches. The linear equation $AX = Y$ forms a hyperplane and this is shown in (A). When a prior solution P is given, one can find the solution of the least square, which in Fig. 5 (A) is represented by X^{ls}. The least square solution [6] is obtained by finding a point, which is the closest point from the prior solution P on the hyperplane i.e., by drawing a perpendicular line from the prior solution to the hyperplane. The distance between P and X^{ls} is defined as Euclidean distance and the least square solution can be found by minimising this distance from the prior P to a point on the hyperplane. The least square solution has a very interesting property that any point on the hyperplane is closer to X^{ls} than P. This means that the solution of the least square approach will always be closer (less absolute error) to the real solution than the prior solution. Therefore, if a prior solution is chosen close to the hyperplane, the least square approach will be the preferred method to use because it will guarantee smaller absolute error.

Represented by X^{kb} in Fig. 5 (A), on the other hand, is the solution obtained by minimising the Kullback distance [9] [10] (rather than Euclidean distance). The Kullback distance determines the probabilistic closeness between two probability distributions and, if the TM components are normalised, one can also calculate the probabilistic closeness between two TMs. The solution of this approach keeps the ratio of the size of each element in the prior TM P. As illustrated in Fig. 5 (A), if the extended line $\overline{OP}$ goes through the hyperplane, one would get the X^{kb} solution i.e., the minimum Kullback distance from the prior solution P. If a TM from few years ago is a prior TM, each element of the prior TM is likely to increase proportionally in size and, therefore, the Kullback distance approach is likely to be more suitable than the Euclidean approach in this case.

Figure 5 (B) illustrates how both the Euclidean and the Kullback distance approaches use the prior solution P through an example of three unknown flows. Assuming that the plane surrounded by points (a-b-c) is the hyperplane $(Y=AX)$, then this hyperplane indicates that an element along $X2$ axis of the real solution is likely to be larger than the one along the $X1$ and $X3$ axis. If an element of $X2$ of the prior solution P is larger than the elements of $X1$ and $X3$, as is the case in our example, then the Kullback distance approach solution X^{kb} has the largest element along the $X2$ axis also.

Suppose that there are two TMs P and X. The former is a prior TM, and the latter is a TM on the hyperplane. Both matrices have n rows and n columns and elements of both TMs are represented by lowercase p_{ij} and x_{ij} respectively. The sums of the total elements in each TM P and X are denoted by T_X and T_P i.e., $\sum_{i=1}^{n} \sum_{j=1}^{n} x_{ij}$ and $\sum_{i=1}^{n} \sum_{j=1}^{n} p_{ij}$.

Table 1. Two deterministic approaches for TM estimation

	Objective Functions	Constraint
Euclidean Distance	$(Min) \sum_{i=1}^{n} \sum_{j=1}^{n} (x_{ij} - p_{ij})^2$	$Ax = Y$
Kullback Distance	$(Min) \sum_{i=1}^{n} \sum_{j=1}^{n} \frac{x_{ij}}{T_X} \{ \log \frac{x_{ij}}{p_{ij}} - \log \frac{T_X}{T_P} \}$	

Table 1 shows the problem formulation for the two deterministic approaches with their objective functions and constraints. Detailed discussion, implementation and analysis of the computation efficiency of these approaches can be found in [6] [10]. While both approaches can be used in our TM estimation framework, we chose to implement the least square approach combined with a prior solution obtained from the gravity model for computation efficiency reasons. Also, in order to increase the accuracy of the solution, we include additional constraints, the row and column total sums $\sum_{j=1}^{n} x_{ij}$ and $\sum_{i=1}^{n} x_{ij}$, in the formulation of the problem. A row sum of a TM represents the amount of traffic entering the network through a node and a column sum represents the amount of traffic leaving the network through a node. The rows and columns sums of a TM can also be easily obtained from the edge links using SNMP measurement data. In addition, to ensure non-negativity of TM elements, the Iterative Proportional Fitting (IPF) method (suggested in [3]) was applied.

4.2   Step 2. MAP-2 for the Correlated Counting Process

In step 2 of our proposed TM framework, we apply the model developed in [5] to model the resulting counting processes for the OD pair flows, which are formed from the series of TMs estimated in Step 1. The matching method developed in [5] takes six moments of the counting process and translates them into an equivalent MAP-2 process by solving a minimisation problem. This model was specifically built for use in the context of TM estimation framework discussed in this paper and, consequently, it takes at input traffic count statistics which can be readily obtained from SNMP system measurement data.

Here we only summarise the six characteristics that the MAP-2 model uses at input and for more detailed elaboration on the solution of the parameter estimation problem we refer the reader to [5]. For a counting process $N(t)$, representing the number of counts in $(0, t]$, the following six statistical characteristics are used to match the series of traffic counts to the parameters of MAP-2, viz 1) Mean arrival rate, 2) Index of dispersion for counts, 3) Limiting index of dispersion for counts, 4) Squared coefficient of variation of the counts, 5) Covariance of the counts in intervals $(0, t]$ and $(t, 2t]$, and 6) Third moment of the counts.

5   Results and Discussion

A simulation program using C^{++} was built to simulate a network and to generate traffic distributions for the network. One network topology, comprising of 4 nodes and 7 links, was created as shown in Fig. 6.

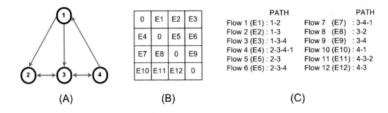

	E1	E2	E3
E4	0	E5	E6
E7	E8	0	E9
E10	E11	E12	0

PATH
Flow 1 (E1) : 1-2
Flow 2 (E2) : 1-3
Flow 3 (E3) : 1-3-4
Flow 4 (E4) : 2-3-4-1
Flow 5 (E5) : 2-3
Flow 6 (E6) : 2-3-4

PATH
Flow 7 (E7) : 3-4-1
Flow 8 (E8) : 3-2
Flow 9 (E9) : 3-4
Flow 10 (E10) : 4-1
Flow 11 (E11) : 4-3-2
Flow 12 (E12) : 4-3

(A)   (B)   (C)

Fig. 6. (A) Network Topology (B) TM for the network (C) Path for each flow

Ideally, a real TM is required for assessing the accuracy of TM estimation methods, however, such information is generally not available to the public. Since real TM wasn't available, we resorted to using a synthetic TM in the following analysis. This was generated based on our observations from the extensive analysis in [5] that the MAP-2 model captures very well the properties of real IP flows. In [5], we assessed the accuracy of the MAP-2 matching model by using real traffic data for a flow exchanged between a pair of POPs in two Tier-1 backbone networks. Since our test network has 12 OD pairs, for the synthetic TM, 12 different sets of MAP-2 processes were generated. Once the traffic flows from the synthetic TM were offered to the network, the link count data was obtained every fixed period of time and step 1 of our TM estimation method was applied to generate one estimated TM in every measurement period. These were used as input for step 2 of our method which produced the final estimated TM (Fig. 4). In evaluating our method, our approach was to investigate whether the estimated TM is capable of capturing and conveying the impact that the real TM has on a queueing performance of a network, as discussed next.

5.1   Queueing Behavior of the Real and the Estimated Traffic Matrices

In order to evaluate how well the estimated TM matches the queueing behavior of the real TM, the performance of a MAP/PH/1 system was analysed. The mean and the variance of the queue waiting times are measured by feeding the real traffic streams and the estimated traffic streams into the queueing system, which are generated from the real TM and the estimated TM, respectively. Figure 7 plots the mean of the waiting time as a function of the link utilization, ranging from 0.05 to 0.9, for the 12 elements (OD pairs) of the real and the estimated TMs. It can be seen that the results are very accurate over the entire range of link utilisation for elements $E1$, $E7$, $E8$, $E9$, $E10$, $E11$, $E12$, while $E2$, $E3$, $E4$, $E5$, $E6$ show some discrepancies. The larger discrepancies are found in $E3$, $E4$, and $E6$. This is because these flows ($E3$, $E4$, $E6$) traverse longer paths, and the paths include links which are shared by larger number of flows. For instance, $E4$, representing the flow from node 2 to node 1, traverses the longest path

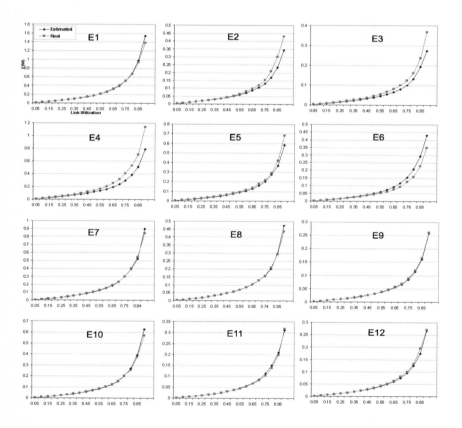

Fig. 7. The mean waiting time of MAP/PH/1 queueing system with the real and the estimated TMs as inputs as a function of link utilisation

(2-3-4-1) and the links (2-3, 3-4) along the path are shared by many other flows, which suggests that the flow $E4$ becomes more difficult to estimate from the deterministic approach. Similar observation, that flows associated with longer paths, as well as paths comprised of links that are heavily shared with other flows, are more difficult to estimate than others, was reported in [7]. Table 2 shows the numerical results of the Fig.7. The average errors of the mean and the variance are obtained by averaging the means and the variances of the waiting times at different link utilisations. As observed previously from Fig. 7, $E4$ has the largest error - 25.54% and 52.06% for the mean and the variance of the waiting time, respectively. The overall mean waiting time of the real TM is estimated by the proposed method with less than 10% error.

Table 2. The average error (%) of the mean and the variance of waiting times for each element of the TM

	E1	E2	E3	E4	E5	E6	E7	E8	E9	E10	E11	E12	Total AVE
Err(E)	3.26	15.78	20.16	25.54	10.50	23.45	0.15	2.10	2.47	3.07	2.67	4.70	9.49 (%)
Err(V)	16.62	45.53	41.27	52.06	28.02	56.60	6.36	7.41	0.96	43.72	3.87	10.71	26.09 (%)

5.2 Does the Proposed Method Need to Be Iterated ?

An iteration scheme was introduced in [11] when a traffic model is defined with its probability density function (pdf). This scheme consists of two main steps. The first step obtains a series of TMs using Markov Chain Monte Carlo (MCMC) algorithm [12] from a given prior distribution for the OD pair flows and link load measurements. To obtain a prior distribution for the OD pair flows, required by the MCMC algorithm, a gravity model was used in the first step of the iteration in [11]. The second step fits the successive values of each OD pair flow into a traffic model. The method iterates over these two steps until convergence is reached.

Iteration procedure, as suggested above, can be incorporated in our proposed method if the pdf of MAP-2 is given. MAP-2 model is completely characterised by six parameters representing its state transition rates $(\lambda_1, \lambda_2, \lambda_3, \lambda_4, \mu_1, \mu_2)$, and the steady state probability vector π is defined by [5]:

$$\pi(1,1) = \frac{\lambda_3 + \mu_2}{\lambda_3 + \mu_2 + \lambda_2 + \mu_1}, \qquad \pi(1,2) = \frac{\lambda_2 + \mu_1}{\lambda_3 + \mu_2 + \lambda_2 + \mu_1} \qquad (2)$$

The pdf of MAP-2 can be derived from the pdf of its special case (Poisson) and it is given by:

$$\mathbf{P}(6 parameters, x) = \frac{e^{-(\pi_1(\lambda_1 + \lambda_2) + \pi_2(\lambda_3 + \lambda_4))}(\pi_1(\lambda_1 + \lambda_2) + \pi_2(\lambda_3 + \lambda_4))^x}{x!} \qquad (3)$$

One can verify the validity of equation (3) by cross-checking whether the pdf satisfies the following conditions.

1. $\sum_{x=0}^{\infty} \mathbf{P}(\text{6parameters}, x) = 1$
2. $\sum_{x=0}^{\infty} x\mathbf{P}(\text{6parameters}, x) = \mathbf{E}[x]$
3. $\mathbf{P}(\text{6parameters}, x)$ has a maximum probability when x is $\mathbf{E}[x]$

Figure 8 plots the pdf of MAP-2 (equation (3)) defined by ($\lambda_1=3$, $\lambda_2=4$, $\lambda_3=5$, $\lambda_4=6$, $\mu_1=1$, $\mu_2=3$) parameters. As observed, the mean value for this process computed from the pdf matches exactly the theoretical mean (8.53684), thus, satisfying the above stated conditions.

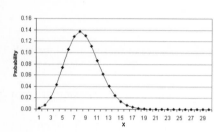
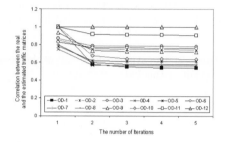

Fig. 8. PDF of MAP-2 with $\lambda_1=3$, $\lambda_2=4$, $\lambda_3=5$, $\lambda_4=6$, $\mu_1=1$, $\mu_2=3$

Fig. 9. Correlation coefficients of the real and estimated TMs

With the given pdf of MAP-2, we have iterated our proposed method five times as per the iterative procedure described in [11], and the results are plotted in Fig. 9. Interestingly, the correlation coefficients for the OD-pair flows are reduced after the first iteration, which is contrary to the results reported on in [11]. This is because we use a deterministic approach to obtain the prior distribution in the first iteration step rather than using the gravity model as suggested in [11]. In other words, the MCMC algorithm which is used to obtain a prior distribution, seems to be less accurate than the deterministic approach used in our method. The results in Fig. 9 show that, in our case, the iteration scheme may not always improve on the accuracy of our method and, given that the iteration procedure increases significantly the computational costs of the method, we conclude that our proposed method should be run in one iteration step as described in Section 4.

6   Conclusions

A new method for TM estimation has been proposed which uses a MAP-2 model to achieve more accurate representation of the correlated and bursty characteristics of IP flows. The method uses a deterministic approach to estimate a series of TMs from a given set of link load measurements (obtained at various points in time), and then approximates the components of the TMs by a MAP-2 model by applying the matching method introduced in [5]. The evaluation study showed that the estimated TM from the proposed method matches well the impact that the real TM has on a queueing performance of a network and, therefore, can be efficiently used in the dimensioning and planning functions of operational IP networks.

In addition, we showed how the proposed method can be extended to include an iterative procedure, as suggested in [11]. However, it was shown that the iteration scheme can not always guarantee improvement of the accuracy of the method since our deterministic approach was able to estimate the TMs in the first step more accurately than the MCMC algorithm. The addition of the iteration procedure significantly increases the computational cost of the method and, therefore, for efficiency reasons we suggested the use of the method without the iterative scheme.

References

1. Cisco Netflow Documentation, [Online]. Available:
   http://www.cisco.com/warp/public/732/Tech/netflow
2. Vardi, Y.: Network Tomography: Estimating Source-Destination Traffic Intensities from Link Data, J. Amer Stat Assoc. pp. 365–377 (1996)
3. Cao, J., Davis, D., Vander Wiel, S., Yu, B.: Time-varying network tomography: router link data, J. Amer Stat Assoc. pp.1063–1075 (2000)
4. Atov, I.: Design of IP Networks with End-to-End Performance Guarantees, PhD Thesis, RMIT University, Melbourne, Australia (2003)
5. Eum, S., Harris, R.J., Atov, I.: A Matching Model for a MAP-2 using Moments of the Counting Process. International Network Optimization Conference (INOC 2007), Spa, Belgium (Accepted for publication) (April 2007)
6. Zhang, Y., Roughan, M., Duffield, N., Greenberg, A.: Fast accurate computation of large-scale IP traffic matrices from link loads. In: ACM SIGMETRICS (June 2003)
7. Medina, A., Taft, N., Salamatian, N., Bhattacharyya, S., Diot, C.: Traffic Matrix Estimation: Existing Techniques and New Directions, ACM SIGCOMM (August 2002)
8. Eum, S., Murphy, J., Harris, R.J.: A Failure Analysis of the Tomogravity and EM Methods, TENCON, Melbourne, Australia (2005)
9. Zhang, Y., Roughan, M., Lund, C., Donoho, D.: An Information-Theoretic Approach to Traffic Matridx Estimation, SIGCOMM (August 2003)
10. Eum, S., Harris, R., Kist, A.: Generalized Kruithof Approach for Traffic Matrix Estimation, ICON, Singapore (2006)
11. Vaton, S., Bedo, J., Gravey, A.: Chapter 8 - Advanced Methods for the Estimation of the Origin Destination Traffic Matrix, Next Generation Internet, p.196 (2004)
12. Tebaldi, C., West, M.: Bayesian Inference on network traffic using link count data (with discussion). J. Amer Stat Assoc. pp.557–576 (June 1998)

Survey on Traffic of Metro Area Network with Measurement On-Line

Gaogang Xie[1,2,4], Guangxing Zhang[1,3], Jianhua Yang[1], Yinghua Min[1], Valerie Issarny[2], and Alberto Conte[4]

[1] Institute of Computing Technology, Chinese Academy of Sciences,
Beijing, 100080, P.R. China
[2] INRIA-Rocquencourt, Domaine de Voluceau, 78153 Le Chesnay, France
[3] Computer and Communication School, Hunan University, ChangSha, 410082, P.R. China
[4] Alcatel Research & Innovation Center, Marcoussis, France
{xie,guangxing,jhyang,min}@ict.ac.cn,
Valerie.issarny@inria.fr, alberto.conte@alcatel.fr

Abstract. Network traffic measurements can provide essential data for network research and operation. While Internet traffic has been heavily studied for several years, there are new characteristics of traffic having not been understood well brought by new applications for example P2P. It is difficult to get these traffic metrics due to the difficulty to measurement traffic on line for high speed link and to identify new applications using dynamic ports. In this paper, we present a broad overview of Internet traffic of an operated OC-48 export link of a metro area network from a carrier with the method of measurement on-line. The traffic behaves a daily characteristic well and the traffic data of whole day from data link layer to application layer is presented. We find the characteristics of traffic have changed greatly from previous measurements. Also, we explain the reasons bringing out these changes. Our goal is to provide the first hand of traffic data that is helpful for people to understand the change of traffic with new applications.

Keywords: Internet Traffic, Metro Area Network, Measurement.

1 Introduction

Traffic metrics not only provide insight into the design, operation and usage patterns of operated network [1, 2, 3, 4], but also are used in network research and device development [5, 6]. Since Ramon Caceres captured the first published measurements of a size's wide area Internet traffic in 1989 [7], great effort and progress have been made in the research area of traffic measurement [8-13]. For instance, the characterization of self-similar nature of traffic is revealed from traffic measurement [4]. New applications, for example P2P-based VoIP, have not only changed characteristics of traffic and behavior of network, but also proposed more different requirements on QoS provisioning ability of network infrastructure. To understand the behavior of network and applications, firstly we should investigate the traffic characteristic of different applications. But, it is far away enough to just measure the

L. Mason, T. Drwiega, and J. Yan (Eds.): ITC 2007, LNCS 4516, pp. 666–677, 2007.

traffic rate at data link layer in order to reveal how the characterizations of traffic to form. More metrics such as traffic proportion at application layer, concurrent flows and packets dynamic with different priorities, are necessary for network planning, operating and research to meet QoS requirement of every application.

Although Pang and others [9] look at enterprise traffic , P2P traffic is not included in their networks under test. In fact, P2P traffic dominates the traffic in other enterprise networks [14]. In addition, enterprise network is often affected by a few hosts. Some statistic characteristics is then not so obvious. Metro area network (MAN) as PoPs of Internet connecting millions of users into backbone plays key role in Internet. It affects the performance of application directly. Most complicated traffic engineering policies and operating rules are also deployed in MAN. Thus, it is important to characterize traffic of MANs.

On the other hand, the difficulty in characterizing MAN traffic comes from high speed of links [15] and new application identifications [16]. It is very difficult with the most powerful hardware to map every packet into the list of millions flows using the methodology with the time complexity $O(l)$ in line speed to measure the flow-level metrics where l is the count of concurrent flows, for up to OC-192/OC-768 backbone link [15]. On the other hand, the determinate ports in the headers of packets are used to identify applications associated with a particular port and thus of a particular application before. It is well known that such a process is likely to lead to inaccurate estimates of the amount of traffic carried by different applications given that specific protocol [14, 16]. For example, some P2P applications tends to use dynamic protocol ports for not being blocked by firewall or bandwidth management device, even tunneled through HTTP with 80 protocol port [16]. These are likely reasons why MAN traffic has not been studied extensively.

This paper explores the traffic characteristics from 5 months measurement experimentations on an export link of a carrier's operating MAN. Thanks to the accurate MPI traffic identification technology in our measurement infrastructure NetTurbo [14, 16], we get a large amount of data to characterize traffic from packet-level to flow-level, from data link layer traffic to application layer traffic. To the best of our knowledge, this is the first paper giving the overview of the characteristics of MAN traffic based on long term on line measurement. Since the traffics behave obviously daily characteristic, only the traffics in one day is analyzed. The main goal of our research is to provide a broad overview of traffic characteristic of MAN.

- The traffic is analyzed according to inbound, outbound and bidirectional.
- The traffic breakdown is made from data link layer to application layer with our accurate traffic identification method MPI.
- The traffic metrics of packet level, byte level and flow level are listed.
- P2P and VoIP traffic rates are presented.
- Traffic rate and traffic amount are presented.
- The traffic of OC-48 link in one whole day is analyzed instead of several minutes trace.

The rest of this paper is organized as follows. Section 2 describes our measurement experiment and the traffic dataset. In Section 3, the makeup of traffic from data link layer to application layer is discussed. We also introduce the flows and IP pairs in this section. In Section 4, the relative work is presented. Finally, we conclude and discuss our future work in Section 5.

2 Measurement Experiment and Dataset

A traffic measurement system named NetTurbo supporting traffic monitoring from data link layer to application layer, which is developed by ourselves and is composed several measurement probes named NetPro and a GUI based data collection and analysis software named NetCom, is deployed in a carrier's metro area network as fig. 1. The measurement probe has been installed in the OC-48 POS export link through an optical splitter and monitored the bidirectional traffics since May 18 2006. All traffic metrics in this paper are produced in NetPro, pushed to NetCom every 10 seconds through management network port, and stored into the database server for history analysis. Our traffic identification method named MPI [16] is used in NetPro to identify new appearing applications such as VoIP and P2P in order to understand the makeup of traffic in application level. The link under tested is one of two export links of a metro area network of a main carrier in China connected to the national backbone between two Cisco's GSR 12416 routers. The population is about 3.8 million living in the region, and among them 1.4 million in the city and the others in the countryside. All of accessing the resource of Internet outside the city is transmitted through the two export links.

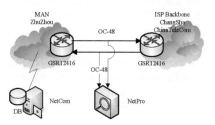

Fig. 1. Measure Experiment

The measurement experiment is performed in two phases. The first phase is from May 18 to September 20 and the second phase is from October 8 to November 1. The main difference in the two data sets is more application protocols identified in the second phases with an improvement over our MPI. So, only the data set of the second phase is analyzed in this paper.

3 Traffic Breakdown at Different Layers

3.1 Traffic Overview

We select traffic in a week randomly and the traffic data present obviously daily characteristic as fig. 2.

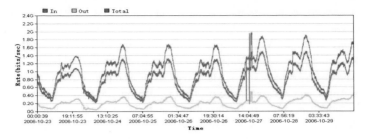

Fig. 2. Traffic Rate in a Week

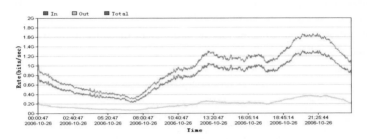

Fig. 3. Traffic Rate on Oct. 26 2006

The maximum traffic rate is 1507Mbps for inbound direction, 499.1Mbps for outbound direction and 1987Mbps in total of bidirection, those of the minimum traffic rates are 181.6Mbps, 41.8Mbps and 260.1Mbps and those of average traffic rates are 808.4Mbps, 205.6Mbps and 1014Mbps in the week. The peak traffic rate appears from 9:00 pm 10:00 pm usually and the bottom appears from 6:30am to 7:30 am every day. The traffic burst in Fig.2 in October 27 is caused by the mechanism of traffic load balance between two export links. Since there is the characteristic of day, the traffic from on October 26 is selected for advanced analysis as fig. 3. The overview of traffic rate of the traffic is shown in table 1.

Table 1. Summary of Traffic Rate on October 26

	BPS			pps		
	Max	**Min**	**Avg**	**Max**	**Min**	**Avg**
Inbound	1311	230.7	760.9	297520	49267	167111
Outbound	374	68.27	192.1	127765	21558	68441
Bidirection	1671	304.3	953	424043	72211	235552

About 19.17% of packets, i.e. about 1.76% of bytes, are with 64 byte in a packet, whereas about 16.32% of packets, i.e. 42.90% in bytes are with from 1024 to 1518 bytes in a packet. The characteristic of tri-modal for packet size distribution in our measurement is not as obvious as previous research [19, 20]. Thompson et al. [19] developed a high performance monitoring system with OC-3 interface capturing

traffic and performed measurement experiments on the OC-3 trunks within internet MCI's backbone and also within the NSF-sponsored vBNS. They reported on measurements from two OC-3 trunks in the presence of up to 240000 flows. The Packet size distribution is tri-modal in their result: 40-44 bytes(20%), <=552 or 576 bytes(65%) and 1500 bytes(15%). McReary [20] also reported the similar result from the NASA Ames Internet Exchange over 10 months in 2000. The average packet size for every application category is shown as Table 2.c. New applications such as P2P and VoIP change the packet size distribution. It is easily to be understood for HTTP the average packet size 959.87 bytes for inbound link and 155.71 bytes for outbound due to more big packets for data in inbound link and 64 byte packets for acknowledgement in outbound link. It is also well to be understood the packet sizes are similar in inbound link and outbound link for P2P applications. But, it is unknown why the packet size in inbound link is larger than that in outbound link for instant messenger and VoIP.

Table 2. Distribution of Packet Size

(A) IN BYTE AND PACKETS(%)

Pkt_size	In Bytes (%)			In Packets (%)		
	Inbound	Outbound	Bidirection	Inbound	Outbound	Bidirection
0~64	1.76	6.21	2.66	19.17	42.54	25.96
64~128	3.25	5.01	3.61	23.10	22.03	22.79
128~256	5.59	6.31	5.73	15.83	11.14	14.47
256~512	10.11	10.07	10.10	8.90	5.38	7.88
512~1024	36.39	39.52	37.02	16.68	11.17	15.08
1024~1518	42.90	32.88	40.88	16.32	7.74	13.83
>1518	0.00	0.00	0.00	0.00	0.00	0.00

(B) PACKET SIZE FOR EVERY APPLICATION CATEGORY (BYTES)

Apps	Inbound	Outbound	Bidirection
Protocols in RFCs	959.87	155.71	756.43
P2P File Sharing	779.42	616.76	724.37
Instant Messenger	359.89	131.83	285.07
Games	220.47	203.54	217.30
Streaming Media	791.42	309.70	643.95
VoIP	241.02	193.35	232.60
Uncategorized	322.24	275.55	307.84

In network layer, almost all the traffic is IPv4 and the summation of other protocols traffic such as PPP, GSMP and others is less than 0.001%. As to transport layer, the traffic of TCP is still dominating over UDP both in inbound and outbound, and both in packets and in bytes as Table 3. There are 95% of bytes and 90% packets for TCP and 5% bytes, 10% packets for UDP in Thompson's experiment in 1997 [19]. McReary [20] in 2000 and Fraleigh [22] in 2003 also got the similar results with Thompson. The reason why UDP traffic increases is more applications such VoIP and P2P are undertaken by UDP. The UDP traffic becomes no more ignorable. Additionally, Comparing Table (3.a) and (3.b), it is easy to know that the average packet size of UDP is smaller than that of TCP. It is reasonable that some UDP applications for example VoIP tend to pack data with small packets.

Table 3. Traffic in Transport Layer

	In bytes (%)			in packets (%)		
	Inbound	**Outbound**	**Bidirection**	**Inbound**	**Outbound**	**Bidirection**
TCP	74.09	69.81	73.24	59.17	66.45	61.24
UDP	25.83	30.06	26.67	40.44	32.95	38.31
ICMP	0.06	0.13	0.07	0.36	0.60	0.43
Others	0.02	0.00	0.02	0.03	0.00	0.02

3.2 Traffic in Application Layer

We classify all applications into 7 application categories according to protocol definitions and traffic characteristics as table 4 [16].

Table 4. Application categories

Application	Example Applications
Protocols in RFCs	HTTP, FTP, DNS, Telnet, SNMP, DHCP, SMTP, POP
P2P File Sharing	Bittorrent, eDonkey, Gnutella, KAD, Fasttrack, Freenet, Poco, Xunlei
Instant Messenger	ICQ, MSN Messenger, Napster, QQ, Skype, Yahoo Messenger, Google Talk
Games	Balltefield 1942, Doom, Quake, Need for Speed, Unreal, Xbox Live, Counter-Strike
Streaming Media	RTSP, PNM, MMS, QuickTime
VoIP	H.323, SIP, MEGACO, Google Talk, QQ
Uncategorized	VPN, Lotus Notes, Radius, pcAnywhere, Oracle, CodeRed, Nimda, Worm

There are about 3.7 TB traffic for P2P file sharing applications, among them 2.6 TB for inbound and 1.1 TB for outbound. While for traditional applications such as HTTP and FTP, there are about 1.8 TB traffic in total and 1.7 TB for inbound and 0.1 TB for outbound. The traffic inbound and outbound are symmetrical for P2P file sharing due to hosts are playing as the roles of both severs and clients in P2P applications. But, there are few servers providing traditional services inside the MAN, so the traffic amount of outbound is much less than that of inbound for traditional applications. The traffic ratio of P2P application is not so much traffic amount for P2P applications as that reported by others in our measurement [14]. This phenomenon is unlikely caused by our traffic identification method MPI since the uncategorized traffics are impossible belonging to P2P applications from their characteristics.

The packet size of the inbound traffic of traditional application is larger than that of outbound since the packets of inbound mostly are requested data while acknowledgements packets for outbound. Then, it is well understood that the traditional applications holds 19.56% in packets and 33.42% in bytes in inbound, but 16.70% in packets and 7.40% for outbound. The packet size is also very symmetrical for P2P file sharing applications in both directions. The average packet size of uncategorized traffics is small and these traffics hold 32.59% in packets but only 19.99% in bytes.

Table 5. Traffic in Applications Categories

	In Bytes (%)			In Packets (%)		
	Inbound	**Outbound**	**Bidirectional**	**Inbound**	**Outbound**	**Bidirectional**
Protocols in RFCs	33.42	7.40	28.25	19.56	16.70	18.75
P2P File Sharing	32.59	53.22	36.69	23.48	30.34	25.43
Instant Messenger	0.30	0.22	0.29	0.48	0.59	0.51
Games	3.13	2.73	3.05	7.97	4.71	7.04
Streaming Media	7.46	5.21	7.01	5.29	5.91	5.47
VoIP	5.03	3.53	4.73	11.71	6.41	10.21
Uncategorized	18.07	27.71	19.99	31.50	35.34	32.59

The traffic rates of all application categories are shown as Fig. 4. The time characteristic of traffic rate of P2P applications is some different with others. In most time the P2P file sharing traffic is the largest part of all in bps, but the traffic rate of uncategorized applications is the largest one in *pps* (pakcets per second) since the packet size of the uncategorized traffic is smaller. The characteristic of period of every application category is very obvious and similar. But, the peak of traffic rate of P2P delays back to others since more P2P files are downloaded during sleeping time.

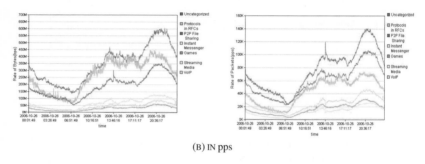

(B) IN pps

Fig. 4. Traffic Rate of Application Categories

A more detailed traffic amounts list of all applications is listed in Table 6. HTTP as an application, its traffic amount still is the largest both in byte and packet.

According to Table 6, there are about 9.39% packets for VoIP. The distribution of calling times is shown as Fig. 5. The x axis 1 to 14 denote less than 1, 1-2, 2-3, 3-4, 4-5, 5-6, 6-7, 7-8, 8-9, 9-10, 10-20, 20-30, 30-60 and more than 60 minutes. There are about 5% callings more than 10 minutes and about 69.93% callings in 1 minute. One of the reasons of many sessions of VoIP less than 1 minute is the failure to set up the connection.

3.3 IP Pair and Flow

IP pair and flow are also checked in this paper to describe the behaviors of hosts and sessions.

The concurrent IP pairs are shown in Fig. 6. The maximum count of concurrent IP pairs is 196638, the minimum 35779 and average 10767. The time characteristic of concurrent IP pairs is similar to traffic rate as in Fig. 3.

Table 6. Traffic in Applications

Apps	Inbound	Outbound	Bidirectional	Inbound	Outbound	Bidirectional
Http	30.38	6.33	25.61	16.78	14.27	16.07
FTP Data	0.69	0.0	0.55	0.42	0.01	0.30
SMTP	0.0	0.02	0.01	0.01	0.01	0.01
POP3	0.03	0.0	0.02	0.01	0.01	0.01
DNS	0.04	0.04	0.04	0.16	0.25	0.18
HTTPS	0.62	0.65	0.63	0.86	1.06	0.92
Edonkey	2.42	4.64	2.86	1.88	2.49	2.05
BitTorrent	4.69	9.88	5.72	4.27	6.05	4.78
Poco	1.01	1.68	1.14	1.15	1.05	1.12
lunlei	4.66	9.69	5.66	3.83	5.23	4.23
P2P_Others	19.78	27.32	21.28	12.33	15.51	13.23
MSN	0.02	0.06	0.03	0.02	0.03	0.02
QQ	0.28	0.15	0.26	0.46	0.56	0.49
Games	3.13	2.73	3.05	7.97	4.71	7.04
PPLive	1.44	1.84	1.52	1.80	1.89	1.83
PPstream	1.80	2.71	1.98	1.60	2.16	1.76
Rtsp	2.61	0.15	2.12	1.08	1.17%	1.10
SM_Other	0.69	0.38	0.63	0.36	0.26	0.33
Sip	0.88	0.11	0.73	0.88	0.19	0.68
h323	0.09	0.17	0.10	0.09	0.13	0.10
VoIP_Others	4.03	3.23	3.87	10.71	6.06	9.39
Total	79.29	71.78	77.81	66.67	63.10	65.64

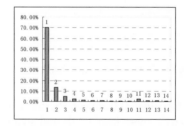

Fig. 5. Time Distribution of VoIP

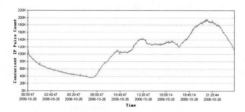

Fig. 6. Concurrent IP Pairs

The characteristic of concurrent flows as Fig. 7 is close to that of concurrent IP pairs as Fig.6. The maximum, minimum and average count of concurrent flows is 293591, 81663 and 163235. So, comparing with the count of concurrent IP pairs, there are only 1.5 concurrent flows for every IP pair averagely.

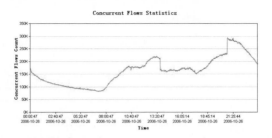

Fig. 7. Concurrent Flows

The rate of new added and closed flows is in direct proportion with the concurrent IP pairs shown in Fig. 8. The rates of new added and closed flows are almost same. The maximum, minimum and average rate of new added flows is 5967, 971, 3352 and that of closed is 6123, 968 and 3356 respectively.

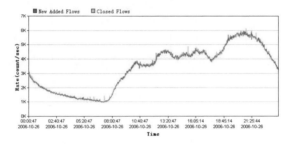

Fig. 8. The Rate of New Added and Closed flows

The 73.25% flows are less than 1kB, about 19.56% from 1 KB to 10 KB, 5.58% from 20 KB to 160 KB, 0.61% more than 1 MB and about 0.08% flows more than 5 MB as Fig. 9.a. There are 80% flows are less than 10 packets and 0.59% flows are more than 1000 packets. This result is also some different with previous research of Brownlee et al. [21] which shows for Web streams 87% under 1kB, 8% between 1 and 10 kB and 4.8% between 10 and 100 kB. As to Non-web streams there is 89% under 1kB, 7% between 1 and 10 kB and 1.5% between 10 and 100 kB. Comparing with previous result, the flows with lager workload increase with the possible reason the web pages tend to get larger with pictures and the flows of P2P and VoIP increase.

(A) BYTES   (B) LASTING TIME

Fig. 9. Distribution of Flows

The sessions of 84.35% flows are less than 10 seconds, 8.43% lasting 10 seconds to 30 seconds, 3.14% lasting 30 seconds to 60 seconds, 3.7% lasting 1 minute to 10 minutes and only 0.39% flows lasting more than 10 minutes.

4 Related Work

Traffic metrics are essential for network research, planning and operating. Due to its fundamental nature, the research of traffic measurement and modeling has maintained continuous interest.

There are lots of works on the methodologies of traffic identification recent years since there are more and more applications using dynamic protocol ports bringing great challenge to it [13, 14, 16]. To address the inaccurate classification of network traffic just by a simple inspection of protocol port number used by flows, Andrew W. Moore and Konstantina Papagiannaki devise a contect-based classification scheme utilizing the full packet payload [13]. But, this methodology is not automated and may require human intervention. Consequently, the methodology is not suitable for real-time measurement system. So, it is difficult to get long term traffic metrics with detailed breakdown with this method. In this paper we present these metrics with on line measurement thanks to the traffic identification ability of MPI.

There are also lots of works on traffic models and characteristics for WAN or LAN, for packet-level and flow-level [5, 6, 10, 19], or applications. These works analyzed one of features and characterize with the theory of stochastic processes. Ruoming [9] presents a broad overview of internal enterprise traffic recorded at a medium sized site based on packet traces which describes some characteristic of traditional applications such as HTTP, Email, DNS and other enterprise-only applications. But, the traffic composing is highly different between different enterprises. That decreases the significance for measurement on enterprise traffic.

More related works are presented in paper [19, 20, 21, 22]. The traffic characteristics have been changed greatly with new applications appearing as we point out in previous sections. The newest traffic measurement from Fraleigh is published in 2003 [22]. In their experiment, a packet captured and analysis system is developed with 10 terabyte storage area network for packets storage and with a computing cluster for analyzing the trace. The results include traffic workload, TCP flow round-trip time, out of sequence packet rates and packet delay. The traffic characteristics for example traffic breakdown in different protocol layers look extraordinary different with our result, even for traffic rate which is more smooth than that in their result. It goes without saying for these changes due to applications and access networks also change greatly. So, the traffic should be measured and analyzed with more powerful online measurement system and more accurate traffic identification methods. It is impossible to measure traffic based on packet trace for high speed network today as previous works [22, 23].

5 Conclusions

Metro area network (MAN) as PoPs of Internet connecting millions of users into backbone plays key role in Internet. Network traffic measurements can provide essential data for network research and operation. Under the characteristic of MAN traffic is very important for performance traffic engineering to improve performance in MAN. The major contribution of this paper is to provide a broad view of MAN network traffic from packet, host and flow. We present a detailed breakdown of traffic from data link layer to application layer based on measurement the export link of MAN on line. To the best of our knowledge, this is the first work introducing these so broad metrics with on line and long term measurement which can provide first hand metrics about traffic of MAN. We find the characteristics of traffic have changed greatly from previous measurements. Also, we explain the reasons for these changes.

Obviously, our investigation is only the first step on the research of traffic characteristic. We are divided ascend, descend, peak, and bottom phase according to the traffic rate and discussing the traffic characteristic during different phases. More profound traffic models on flows and applications will also be researched in future work.

Acknowledgment

This research is funded by National Natural Science Foundation of China with grant no. 60403031 and 90604015 and the Foundation of French-Chinese for Sciences and their Applications (FFCSA 2005-2006). This paper is also supported by France Telecom R&D with grant no. 46135216 and China Mobile. Thanks for others efforts on development the measurement system, especially thanks for Liang SHUAI, Zhenbao ZHOU, Shuangguang WEN and Da PENG.

References

1. Ibrahim Hamdy, A., Nossier Bahnasy, M., Darwish Mohamed, G.: Billing system for Internet Service Provider (ISP), IEEE MELECON 2002: Mediterranean electrotechnical conference (2002)
2. Mortier, R.: Internet Traffic Engineering, Technical Report of University of Cambridge, UCAM_CL_TR_532 (2002)
3. Mai, J., Chuah, C.-N., Sridharan, A., Ye, T., Zang, H.: Is Sampled Data Sufficient for Anomaly Detection? In: IMC 2006. Rio de Janeiro, Brazil (October 2006)
4. Duffield, N.G., Lund, C.: Predicting Resource Usage and Estimation Accuracy in an IP Flow Measurement Collection Infrastructure. ACM SIGCOMM Internet Measurement Conference 2003, Miami Beach, Fl, (October 27-29, 2003)
5. Leland, W., Taqqu, M., Willinger, W., Wilson, D.: On the self-similar nature of Ethernet traffic. In: Proceedings of SIGCOMM '93, pp. 183–193 (September 1993)
6. Shakkottai, S., Brownlee, N., Claffy, K.C.: Study of Burstiness in TCP Flows, PAM (2005)
7. Ramon Caceres, Measurement of wide area internet traffic. Technical report of UC Berkeley (1989) http://www.kiskeya.net/ramon/work/pubs/ucb89.pdf
8. Leland, W.E., Wilson, D.V.: High time-resolution measurement and analysis of LAN traffic: Implications for LAN interconnection. In: Proceeedings of IEEE Infocomm '91, pp. 1360–1366, Bal Harbour, FL (1991)
9. Pang, R., Allman, M., Bennett, M., Lee, J., Paxson, V., Tierney, B.: A First Look at Modern Enterprise Traffic. In: Proceeding of Internet Measurement Conference 2005 (2005)

10. Mah, B.: An empirical model of HTTP network traffic. In: Proceedings of INFOCOM 97 (April 1997)
11. Fraleigh, C., Moon, S., Lyles, B., Cotton, C., Khan, M., Moll, D., Rockell, R., Seely, T., Diot, C.: Packet-Level Traffic Measurements from the Sprint IP Backbone. In: IEEE Network (2003)
12. Jaiswal, S., Iannaccone, G., Diot, C., Kurose, J., Towsley, D.: Measurement and Classification of Out-of-Sequence Packets in a Tier-1 IP Backbone. In: IEEE Infocom. San Francisco (March 2003)
13. Moore, A., Papagiannaki, K.: Toward the Accurate Identification of Network Applications. In: Proceedings of PAM (2005)
14. Karagiannis, T., Broido, A., Brownlee, N., Claffy, Kc., Faloutsos, M.: Is P2P Dying or Just Hiding. In: Proceeding of IEEE Globecom, Dallas (2004)
15. Xie, G., Yang, J., Issarny, V., Conte, A.: An Accurate and Efficient 3-Phases Measurement Method for IP Traffic Flow on High Speed Link (to Submitted)
16. Xie, G., Zhang, G., Yang, J., Issarny, V., Conte, A.: Accurate Online Traffic Classification with Multi-phases Identification Methodology (to Submitted)
17. Duffield, N.G., Lund, C., Thorup, M.: Learn more, sample less: control of volume and variance in network measurement. IEEE Transactions in Information Theory 51(5), 1756–1775 (2005)
18. Barakat, C., Thiran, P., Iannaccone, G., Diot, C., Owezarski, P.: Modeling Internet Backbone Traffic at the Flow Level. In: IEEE Transactions on Signal Processing (Special Issue on Networking) (August 2003)
19. Thompson, K., Miller, G.J., Wilder, R.: Wide-area Internet traffic patterns and characteristics. IEEE Network, 11(6) (November 1997)
20. McCreary, S., Claffy, K.: Trends in wide area IP traffic patterns - A view from Ames Internet Exchange, ITC Specialist Seminar /AIX0005/ (2000) http://www.caida.org/publications/papers/
21. Brownlee, N., Claffy, K.C.: Internet stream size distributions. In: Proc. ACM SIGCOMM, pp.282–283 (2002)
22. Fraleigh, C., Moon, S., Lyles, B., Cotton, C., Khan, M., Moll, D., Rockell, R., Seely, T., Diot, C.: Packet-level traffic measurements from the Sprint IP backbone. IEEE Network 17(6), 6–16 (2003)
23. Azzouna, N.B., Guillemin, F.: Impact of peer-to-peer applications on wide area network traffic: an experimental approach. GLOBECOM '04. IEEE. 29 Nov.-3 Dec. 2004 3, 1544–1548 (2004)

Deterministic Versus Probabilistic Packet Sampling in the Internet

Yousra Chabchoub[1], Christine Fricker[1], Fabrice Guillemin[2],
and Philippe Robert[1]

[1] INRIA, RAP project, Domaine de Voluceau, 78153 Le Chesnay, France
{Yousra.Chabchoub,Christine.Fricker,Philippe.Robert}@inria.fr
http://www-rocq.inria.fr/~robert
[2] France Telecom, Division R&D, 2 Avenue Pierre Marzin, 22300 Lannion, France
Fabrice.Guillemin@orange-ft.com
http://perso.rd.francetelecom.fr/guillemin

Abstract. Under the assumption that packets are sufficiently inter-leaved and the sampling rate is small, we show in this paper that those characteristics of flows like the number of packets, volume, etc. obtained through deterministic 1-out-of-k packet sampling is equivalent to random packet sampling with rate $p = 1/k$. In particular, under mild assumptions, the tail distribution of the total number of packets in a given flow can be estimated from the distribution of the number of sampled packets. Explicit theoretical bounds are then derived by using technical tools relying on bounds of Poisson approximation (Le Cam's Inequality) and refinements of the central limit theorem (Berry-Essen bounds). Experimental results from an ADSL traffic trace show good agreement with the theoretical results established in this paper.

1 Introduction

Packet sampling is an efficient method of reducing the amount of data to re-trieve and to analyze in order to study the characteristics of IP traffic (cf. the drafts of IPFIX [11] and PSAMP [12] working groups at the IETF). The sim-plest approach to packet sampling is certainly the so-called 1-out-of-k sampling technique, which consists of capturing and analyzing one packet every other k packets. This method will be referred to in the following as deterministic sam-pling, which has been implemented, for instance, in CISCO routers (NetFlow facility [6]) and is widely used in today's operational networks, even if it suffers from several shortcomings identified in [8]. In particular, recovering original flow statistics from sampled data is a difficult task (see [7] for instance). Different solutions have been introduced to overcome these limitations (e.g., the "sample and hold" technique by Estan and Varghese [9], adaptive sampling [5,8], etc.). It is worth noting that recovering original flow statistics is utmost important for network operators for estimating the bit rate of flows since it reflects more or less the quality of service perceived by end users and gives an indication on the bit rate really needed by applications.

L. Mason, T. Drwiega, and J. Yan (Eds.): ITC 2007, LNCS 4516, pp. 678–689, 2007.

Because deterministic sampling may introduce some synchronization and then some bias in sampled data, which bias is not easy to determine because it depends upon the realization of flows (i.e., the relative position of packets between each other), several studies and IETF drafts [15] recommend probabilistic sampling. In its most basic version, random sampling consists of picking up a packet, independently from other packets, with a given probability p. The major advantage is that random sampling provides isolation between flows: the selection of a packet does not depend upon the relative position of flows between each other.

In this paper, it is shown that if packets are sufficiently interleaved (which is definitely the case on a transmission link of a backbone network), then 1-out-of-k deterministic sampling is equivalent to random sampling with $p = 1/k$. More precisely, an explicit estimation of the distance (for the total variation norm) between the distributions of the numbers of packets in a flow sampled with the two sampling techniques is obtained.

On the basis of this result, bounds on the difference between the distributions of the original flow size and of the sampled flow size rescaled by the sampling factor are established. While the estimation of the size of a flow by using the number of sampled packets scaled by the sampling factor is natural and frequently used in the literature, it is not always accurate and can lead to severe overestimation errors. A bound to estimate the accuracy of this estimation is therefore important in practice. Provided that the flow size is sufficiently heavy tailed, it can be shown that the original size of a flow can indeed be estimated from the number of sampled packets.

The different theoretical results obtained in this paper are illustrated on a traffic trace from the France Telecom backbone network carrying ADSL traffic. For this purpose, we introduce a flow decomposition technique based on an ad-hoc mouse/elephant dichotomy. The theoretical results are applied to elephants and experimental data show good agreement with theoretical results. It is worth noting that mice appear as background noise in sampled data and their flow size distribution is of less interest, since their volume represents only a small fraction of global traffic.

The paper is organized as follows: In Section 2, we describe the traffic analysis methodology. The comparison between deterministic sampling and random sampling is discussed in Section 3 and results on random sampling are then established. These results are compared in Section 4 against experimental results. Concluding remarks are presented in Section 5.

2   Traffic Analysis Methodology

Let us consider a high speed transmission link carrying Internet traffic and let us divide time into slots of length T. The constant T may range from a few seconds to several tens of minutes (say, from one to two hours).

In this paper, we are interested in the characteristics of TCP traffic since it still represents today 95% of the total amount of traffic in IP networks, even though the proportion of UDP traffic is growing with the development of streaming applications (VoIP, video, peer-to-peer streaming, etc.). To analyze TCP traffic, we adopt a flow based approach, a flow being defined as the set of those packets with the same source and destination IP addresses together with the same source and destination port numbers (and of course the same protocol type, in this case TCP). In the literature on Internet traffic characterization, it is well known that all flows are not all equivalent: there are flows with many packets, possibly transmitted in bursts, and small flows comprising only a few packets. Many small flows are composed of single SYN segments corresponding to unsuccessful TCP connection establishments attempts.

To simplify the notation, small flows will be referred to in the following as mice and long flows as elephants. This notation corresponds more or less to the elephant/mouse dichotomy introduced by Paxson and Floyd [14], even if clear definitions for mice and elephants do not exist (see the discussion in [13]). To be more specific, we shall use the following definitions.

Definition 1 (Mouse/Elephant). *A mouse is a flow with less than b packets in a time window of length T. An elephant is a flow with at least b packets in a time window of length T.*

We do not claim that the above definitions should be *the* definitions for mice and elephants; they are introduced for convenience to split the flow population into two distinct sets. In particular, they depend upon the length T of the measurement window and the threshold b. In previous studies (see [1] for instance), a threshold $b = 20$ packets yields a neat delineation between mice and elephants when dealing with ADSL traffic even for large observation windows.

To illustrate the above definitions, we consider a traffic trace from the France Telecom IP backbone network carrying ADSL traffic. This traffic trace has been captured on a Gigabit Ethernet link in October 2003 between 9:00 pm and 11:00 pm (this time period corresponding to the peak activity by ADSL customers); the link load was equal to 43.5%. The complementary cumulative distribution function (ccdf) of the number N_{mice} of packets in mice is displayed in Figures 1(a) and 1(b) for $T = 5$ seconds and $T = 3200$ seconds, respectively. We see that for $T = 5$ seconds, the distribution of the random variable N_{mice} can reasonably be approximated by a geometric distribution (i.e., $\mathbb{P}(N_{\mathrm{mice}} > n) \approx r_1^n$). By using a standard Maximum Likelihood Expectation (MLE) procedure, we find $r_1 = 0.75$. For $T = 3200$ seconds, only the tail of the distribution can be approximated by a geometric distribution; experimental results give $\mathbb{P}(N_{\mathrm{mice}} > n) \approx c_2 r_2^n$ for large n, with $c_2 = .1$ and $r_2 = .6$.

The distribution of the number N_{eleph} of packets in elephants is displayed in Figure 1(c) and 1(d) for $T = 5$ and $T = 3200$ seconds, respectively. Now, we see that elephants clearly exhibit a behavior, which is significantly different from that of mice. The random variable N_{eleph} has a slowly decreasing distribution, which can reasonably be approximated by a Pareto distribution, at least for moderate values of N_{eleph} for $T = 5$ seconds.

We specifically have $\mathbb{P}(N_{\mathrm{eleph}} > n) \approx (b/n)^a$ for $n \geq b = 20$. For $T = 5$ seconds, we find by means of a standard MLE procedure $a = 1.95$. When $T = 3200$ seconds, the distribution of N_{eleph} is more complicated and can be

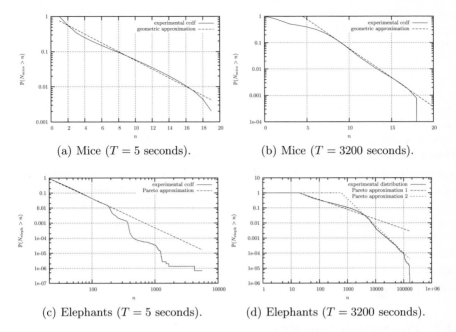

Fig. 1. Ccdf of the number of packets in mice and elephants for $T=5$ seconds and $T = 3200$ seconds ($b = 20$ packets)

approximated by two Pareto distributions, namely $\mathbb{P}(N_{\text{eleph}} > n) \approx (20/n)^{a_2}$ for $20 \leq n \leq 2000$ with $a_2 = .55$, and $\mathbb{P}(N_{\text{eleph}} > n) \approx (600/n)^{a'_2}$ for $n \geq 2000$ with $a'_2 = 1.8$.

It is worth noting from above experimental data that taking a rather small time window for computing the statistics of a flow gives a much more robust statistical description in the sense that the Pareto and the geometric distributions appear as invariant laws in traffic. Additional works has to be done to recover complete information on flows since flows may span over several time windows.

In this paper, we are interested in comparing the random variables describing the number of packets in a sampled flow, when deterministic or random sampling is performed.

3   Properties of Random and Deterministic Sampling

3.1   Deterministic Sampling

In the case of deterministic sampling, one packet is selected every other $1/p$ (integer) packets, where p is the sampling rate. If packets of flows are back to back, then there is little chance of seeing flows more than once if their number of packets is not significantly larger than the sampling coefficient $1/p$. Fortunately, on a high speed backbone link, the number of simultaneous flows is very large and packets of the different competing flows are highly interleaved, i.e., there is some randomness in the relative position of packets and a given pattern does not reproduce for ever. Hence, consecutive packets of a given flow are separated by a random number of packets of other flows. This introduces some randomness in the selection of packets of a given flow.

More precisely, assuming that flows are permanent in a time window of length T, deterministic sampling consists of drawing $\lfloor pM(T) \rfloor$ packets out of the total number $M(T)$ of packets in the time window. If packets are sufficiently interleaved, a sampled packet belongs to a given flow f with probability $N_f/M(T)$ if flow f has originally N_f packets. Under this assumption, the number of sampled packets from flow f is $n_f = B_1^f + B_2^f + \cdots + B_{pM(T)}^f$, where the quantities B_j^f are independent Bernoulli random variables equal to one if and only if the jth sampled packet is from flow f. Note that if f and g are distinct flows, then the variables (B_j^f) and (B_j^g) are *not* independent.

The assumption of permanent flows is reasonable, when the observation window length T is small. When T is large, however, flows may be bursty and alternate between on and off periods. This phenomenon has been observed in particular when analyzing elephants in ADSL traffic [1].

3.2   Probabilistic Sampling

It is assumed in this section that random sampling is performed: each packet of a given flow f with N_f packets is taken with a probability p and the number of packets in the sampled flow is exactly given by $\tilde{n}_f = \widetilde{B}_1^f + \widetilde{B}_2^f + \cdots + \widetilde{B}_{N_f}^f$, where the random variables $(\widetilde{B}_i^f)$ are Bernoulli with mean p. The key property of this sampling mode is that it provides isolation between flows. Mathematically speaking, it amounts to the fact that the Bernoulli variables $(\widetilde{B}_i^f)$ and $(\widetilde{B}_i^g)$ are independent for distinct flows f and g.

The comparison between the two sampling methods is done through the estimation of the *total variation distance* between the distributions of n_f and $\tilde{n}_f$,

$$\|\mathbb{P}(n_f \in \cdot) - \mathbb{P}(\tilde{n}_f \in \cdot)\|_{tv} \overset{\text{def.}}{=} \sup_{A \subset \mathbb{N}} |\mathbb{P}(n_f \in A) - \mathbb{P}(\tilde{n}_f \in A)|.$$

Proposition 1 (Probabilistic vs. Deterministic Sampling). *Under the above assumptions, for a flow f with N_f packets with $\mathbb{E}(N_f^2) < +\infty$, we have*

$$\|\mathbb{P}(n_f \in \cdot) - \mathbb{P}(\tilde{n}_f \in \cdot)\|_{tv} \leq p\frac{\mathbb{E}(N_f^2)}{M(T)} + p^2\mathbb{E}(N_f). \tag{1}$$

Moreover, as $M(T)$ goes to infinity, the number of sampled packets n_f converges in distribution to $\mathbb{Q}$ defined by

$$\mathbb{Q}(k) = \frac{p^k}{k!}\mathbb{E}\left(N_f^k e^{-pN_f}\right).$$

Proof. The proof relies on Le Cam's inequality conditionally on the value of N_f, see Chapter 1 of Barbour [3]. If $\mathrm{Pois}(\lambda)$ denotes the Poisson distribution with parameter λ, then

$$\|\mathbb{P}(n_f \in \cdot \mid N_f) - \mathrm{Pois}(pN_f)\|_{tv} \leq pN_f^2/M(T). \tag{2}$$

By integrating this relation, we obtain $\|\mathbb{P}(n_f \in \cdot) - \mathbb{Q}\|_{tv} \leq p\mathbb{E}(N_f^2)/M(T)$. For the random variable $\tilde{n}_f$, similar arguments yield $\|\mathbb{P}(\tilde{n}_f \in \cdot) - \mathbb{Q}\|_{tv} \leq p^2\mathbb{E}(N_f)$. Relation (1) is hence proved. The convergence in distribution is a direct consequence of Inequality (2).

Equation (1) implies that when the sampling parameter p is small, the distribution of the number of sampled packets of a given flow is close to the analogue quantity obtained by probabilistic sampling.

Considering that if we deal with an elephant, the number of packets of the flow is quite large, the law of large numbers would suggest the following approximation $\tilde{B}_1^f + \tilde{B}_2^f + \cdots + \tilde{B}_{N_f}^f \overset{\text{dist.}}{\sim} pN_f$, so that the total number of packets of a flow can be recovered from the number of sampled packets. In spite of the fact that this approximation is quite appealing and natural, it turns out that it has to be handled with care. Indeed, if N_f is geometrically distributed, then it is easily checked that the above approximation is not valid. The fact that N_f is very likely heavy-tailed helps to establish such an approximation. This is the subject of the rest of the section. The following result is a first step in this direction.

Proposition 2. *If $h_k(x) = x^2/4p^2 \left(\sqrt{1 + 4kp/x^2} - 1\right)^2$ for $x \in \mathbb{R}$ and $k > 0$, and the random variables B_i are Bernoulli with mean p, then*

$$\left| \mathbb{P}\left(\sum_{i=1}^{N_f} B_i \geq k\right) - \mathbb{P}\left(N_f \geq h_k\left(\sqrt{p(1-p)}\mathcal{G}\right) \vee k\right) \right| \leq c\mathbb{E}\left(\frac{1}{\sqrt{N_f}} \mathbb{1}_{\{N_f \geq k\}}\right),$$

where $\mathcal{G}$ is a standard Gaussian random variable, for real numbers $a \vee b = \max(a, b)$, and $c = 3(p^2 + (1-p)^2)/\sqrt{p(1-p)}$.

Proof. Let $\sigma^2 = \mathrm{Var}(B) = p(1-p)$, $S_n = B_1 + \cdots + B_n$, $\bar{S}_n = S_n/n$ and $\hat{S}_n = \sqrt{n}(\bar{S}_n - np)/\sigma$. By Berry-Essen's theorem [10], for each $n \in \mathbb{N}$ and $k > 0$,

$$\left| \mathbb{P}\left(\hat{S}_n \geq \frac{k - pn}{\sigma\sqrt{n}}\right) - \mathbb{P}\left(\mathcal{G} \geq \frac{k - pn}{\sigma\sqrt{n}}\right) \right| \leq \frac{c}{\sqrt{n}}$$

where $c = 3E((p - B)^3)/\sigma^3 = 3(p^2 + (1-p)^2)/\sqrt{p(1-p)}$. Thus, multiplying by $\mathbb{1}_{\{n \geq k\}}$, using the independence of S_n and N_f and Fubini's theorem, noticing that $\mathbb{P}(\hat{S}_{N_f} \geq (k - pN_f)/\sqrt{N_f}) = \mathbb{P}(S_{N_f} \geq k)$ and that, if $S_{N_f} \geq k$ then $N_f \geq k$, we obtain

$$\left| \mathbb{P}\left(\hat{S}_{N_f} \geq \frac{k - pN_f}{\sigma\sqrt{N_f}}\right) - \mathbb{P}\left(\mathcal{G} \geq \frac{k - pN_f}{\sigma\sqrt{N_f}}, N_f \geq k\right) \right| \leq c\mathbb{E}\left(\frac{1}{\sqrt{N_f}} \mathbb{1}_{\{N_f \geq k\}}\right).$$

Now, we prove that

$$\mathbb{P}\left(\mathcal{G} \geq \frac{k - pN_f}{\sigma\sqrt{N_f}}, N_f \geq k\right) = \mathbb{P}(pN_f + \sqrt{N_f}\sigma\mathcal{G} \geq k, N_f \geq k)$$

$$= \mathbb{P}(N_f \geq f_k(\sigma\mathcal{G}) \vee k).$$

Indeed, denoting $z = \sqrt{y}$, the equation $pz^2 + zx - k = 0$ has two roots in $\mathbb{R}$, equal to $z_1 = (-x - \sqrt{x^2 + 4pk})/2p < 0$ and $z_2 = (-x + \sqrt{x^2 + 4pk})/2p > 0$. Thus, for every $x \in \mathbb{R}$, $pz^2 + zx - k \geq 0$ and $z \geq 0$ is equivalent to $z \geq z_2$, i.e., $y \geq h_k(x)$. The result then readily follows.

From the above result, under mild assumptions on the distribution of N_f, the tail distribution of $B_1 + B_2 + \cdots + B_{N_f}$ is related to the tail distribution of N_f. In particular, if N_f has a Pareto distribution, we have the following result.

Corollary 1. *If the random variable N_f has a Pareto distribution, i.e. for some $b > 0$ and $a > 1$, $\mathbb{P}(N_f \geq k) = (b/k)^a$, and if the random variables B_i are Bernoulli with mean p, then*

$$\lim_{k \to +\infty} \frac{\mathbb{P}\left(B_1 + B_2 + \cdots + B_{N_f} \geq k\right)}{\mathbb{P}(N_f \geq k/p)} = 1.$$

Proof. We have

$$\mathbb{P}(N_f \geq h_k(\sqrt{p(1-p)}\mathcal{G}) \vee k) = \mathbb{E}((b/(h_k(\sqrt{p(1-p)}\mathcal{G}) \vee k))^a) \sim (bp/k)^a,$$

since $h_k(x) = k/p(1 + O(\frac{1}{\sqrt{k}}))$ for large k.

The above asymptotic results have been established for a random variable N_f, which has a Pareto distribution. But it is straightforwardly checked that similar results hold, when only the tail of N_f is Pareto as in the traffic trace described in Section 2. With regard to the comparison between the original flow size distribution and the rescaled sampled size distribution, let us mention that Berry-Essen bound based on the normal approximation is accurate only around the mean value. To obtain a tighter bound on the tail of the distribution, it is possible to establish the following result (see [4] for details).

Theorem 1. *For $\alpha \in (1/2, 1)$, there exist positive constants C_0 and C_1 such that for any $p \in (0,1)$ and $k \geq 1/p$,*

$$\left|\frac{\mathbb{P}\left(\sum_{i=1}^{N_f} B_i \geq k\right)}{\mathbb{P}(N_f \geq k/p)} - 1\right| \leq \sup_{-C_1 \leq u \leq C_1} 3\left|\frac{\mathbb{P}\left(N_f \geq \frac{k}{p} + u\left(\frac{k}{p}\right)^\alpha\right)}{\mathbb{P}(N_f \geq k/p)} - 1\right|$$

$$+ \frac{C_0}{\mathbb{P}(N_f \geq k/p)}\exp\left(-\frac{p}{4(1-p)}k^{2\alpha-1}\right).$$

From the above result, we see that the quantity $\mathbb{P}\left(\sum_{i=1}^{N_f} B_i \geq k\right)$ related to the probability that a sampled flow contains at least k packets is exponentially close for sufficiently large k to $\mathbb{P}(N_f \geq k/p)$. In Section 4, the above theoretical results are used to interpret the experimental results when performing deterministic and random sampling on the France Telecom ADSL traffic traces.

3.3 Refinements

To prove Proposition 1, it has been assumed that flows are permanent. This assumption is reasonable, when the observation window length T is small. When T is large, however, flows may be bursty and alternate between on and off periods. To take into account this phenomenon, convergence to Poisson distributions as in Proposition 1 can be proved, when flows have different transmission rates.

More precisely, let us assume that there are L classes of flows. For a class $\ell \in \{1, \ldots, L\}$, let $r_\ell(x)$ be the transmission rate of a flow of class ℓ at time x. The quantity $C_\ell = \frac{1}{T}\int_0^T r_\ell(u)\, du$ is the average transmission rate of a flow on $[0, T]$. Flows are assumed to arrive uniformly in $[0, T]$. Consequently, for each flow f in class ℓ, the number of packets transmitted up to time $t \in [0, T]$ is $N_f(t) = \int_0^t r_\ell((u - \tau_f) \mod T)\, du$, where the random variables τ_f are independent and uniformly distributed in $[0, T]$. It follows that the different processes $N_f(t)$ for flows f in class ℓ have the same distribution.

For $\ell \in \{1, \ldots, L\}$, let K_ℓ be the number of flows of class ℓ in $[0, T]$ and $K = \sum_\ell K_\ell$. The total number of transmitted packets up to time u is denoted by $M(u) = \sum_{i=1}^K N_i(u)$. Let $K = \sum_\ell K_\ell$ and $pM(T)$ be the number of sampling times between 0 and T, p denoting the sampling rate. When K becomes large, assume that for every ℓ, K_ℓ/K tends to a constant α_ℓ. By the law of large numbers, $M(u)/K$ converges almost surely to $C = \sum_{\ell=1}^L \alpha_\ell C_\ell$ for all $u \in [0, T]$. The numbers of packets n_f in the sampled flows f of class ℓ have the same distribution. We have the following result, whose proof is given in Appendix A.

Proposition 3. *If $pM(T)/K \to x$, the distribution of the number $n(\ell)$ of packets in a sampled flow of class ℓ converge to a Poisson distribution with parameter xC_k/C.*

The above proposition shows that the distribution of the number of sampled packets of a flow in class ℓ depends only upon the ratio of the average rate of class ℓ to the total average rate in the observation window. This indicates that we could have considered the flows permanent at the average rate in the observation window.

4   Experimental Results

In this section, we consider the traffic trace from the France Telecom backbone network described in Section 2 and we fix the length of the observation window equal to $T = 3200$ seconds and the sampling rate $p = 1/100$. The complementary cumulative distribution functions (ccdf) of the number of packets in original mice and elephants are displayed in Figure 1(b) and 1(d), respectively. In the original trace, there were 252,854 elephants and in the sampled trace, we found 132,352 and 132,185 of the original elephants with deterministic and random sampling, respectively.

From the above experimental results, we see that the probabilities of seeing elephants after sampling in the different cases are very close one to each other, about 0.523.

If N_f has a Pareto distribution of the form $\mathbb{P}(N_f > k) = (b/k)^a \mathbb{1}_{\{k \geq b\}}$, the probability of seeing an elephant by random sampling is

$$\nu \overset{\text{def}}{=} \mathbb{P}\left(\sum_{i=1}^{N_f} B_i > 0\right)$$

$$= 1 - (1-p)^b + p\sum_{k=b}^{\infty}(1-p)^k \mathbb{P}(N_f > k) \sim bp + (bp)^a \Gamma(1-a, bp),$$

when p is small. With $a = a_2 = 0.55$, $b = 20$, and $p = 1/100$, we find that the probability of seeing an elephant is approximately equal to 55%, which is very close to the experimental value. Hence, estimating the exponent a of the Pareto distribution allows us to estimate the probability of seeing an elephant. This quantity is critical for the estimation of the parameters of flows. For instance, for estimating the original duration of flows, a method is presented in [2], but the estimation of ν is critical because it relies on the tails of some probability distributions. The method based on the estimation of the exponent of the Pareto distribution is more reliable.

The major difficulty for exploiting the sampled trace comes from the fact that we do not know if a sampled flow is really an elephant or not. If we had adopted the convention that a sampled flow corresponds to an elephant as soon as it is composed of at least two packets, we would have found 143,517 and 144,000 elephants with deterministic and random sampling, respectively. We see that this convention leads to slightly overestimating the number of elephants.

Figure 2 represents the ccdf of the number of packets in elephants after probabilistic and deterministic sampling, along with the rescaled original distribution $\mathbb{P}(N > k/p)/\nu$. We can observe that the three curves coincide, which is in agreement with the results obtained in Section 3. By using Proposition 2 and Theorem 1 and assuming that random and deterministic sampling are sufficiently close one to each other, we can recover the distribution of the original elephants from the distribution of sampled elephants with known bounds.

For the volume V (expressed in bytes) of elephants, we can first compute the mean number of bytes in packets. For instance, for the traffic trace considered in this paper, the mean number of bytes in packets of elephants is equal to $\bar{V} = 1000$. Then, we can verify that multiplying the number of packets in elephants by the mean number $\bar{V}$ of bytes in packets give a fair estimate of the volume of elephants, as illustrated in Figure 3(a). From the results established for the number of packets in elephants and under the assumption that

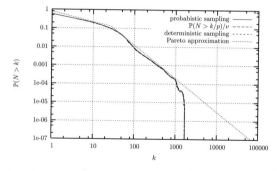

Fig. 2. Number of packets in elephants after sampling and comparison with the rescaled original size $\mathbb{P}(N > k/p)/\nu$ along with the Pareto approximation

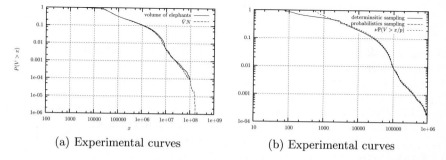

(a) Experimental curves   (b) Experimental curves

Fig. 3. Volume (in bytes) of elephants after deterministic and probabilistic sampling and comparison with the rescaled original volume $\mathbb{P}(V > x/p)/\nu$

random sampling is sufficiently close to deterministic sampling, we can estimate the volume of original elephants with known bounds; Figure 3(b) shows that the rescaled distribution $\mathbb{P}(V > x/p)/\nu$ is close to the distribution of the volume of sampled elephants.

5   Conclusion

We have shown in this paper that as far as the volume and the number of packets in elephants are concerned, random and deterministic sampling are very close one to each other, when the sampling rate becomes small. Several results for the number of packets contained in randomly sampled flows have been established. In particular, bounds between the distribution of the number of packets in a randomly sampled elephant and the rescaled original distribution have been established. Experimental results obtained by using a traffic trace from the France Telecom IP backbone network show good agreement with theoretical results.

References

1. Ben Azzouna, N., Clérot, F., Fricker, C., Guillemin, F.: A flow-based approach to modeling ADSL traffic on an IP backbone link. Annals of Telecommunications 59(11-12), 1260–1299 (2004)
2. Ben Azzouna, N., Guillemin, F., Poisson, S., Robert, P., Fricker, C., Antunes, N.: Inverting sampled ADSL traffic. In: Proc. ICC 2005, Seoul, Korea (May 2005)
3. Barbour, A.D., Holst, L., Janson, S.: Poisson approximation. The Clarendon Press Oxford University Press, New York, Oxford Science Publications (1992)
4. Chabchoub, Y., Fricker, C., Guillemin, F., Robert, P.: Bounds for packet sampling in the Internet. In Preparation.
5. Choi, B.Y., Park, J., Zhang, Z.L.: Adaptive packet sampling for accurate and scalable flow measurement. In: Proc. Globecom'04, Dallas, TX (December 2004)
6. CISCO, http://www.cisco.com/warp/public/netflow/index.html
7. Duffield, N., Lund, C., Thorup, M.: Properties and prediction of flow statistics. In: ACM SIGCOMM Internet Measurement Workshop, pp. 6–8 (November 2002)
8. Estan, C., Keys, K., Moore, D., Varghese, G.: Building a better NetFlow. In: Proc. ACM Sigcomm'04, Portland, Oregon, USA (August 30 – September 3, 2004)
9. Estan, C., Varghese, G.: New directions in traffic measurement and accounting. In: Proc. Sigcomm'02, Pittsburgh, Pennsylvania, USA (August 19-23, 2002)
10. Feller, W.: An introduction to probability theory and its applications. John Wiley and Sons, New York (1996)
11. IETF, IPFIX Working Group. IP flow information export. For information, see the url, http://www.ietf.org/html.charters/ipfix-charter.html
12. IETF, PSAMP Working Group. Packet sampling working group. See the url, https://ops.ietf.org/lists/psamp
13. Papagiannaki, K., Taft, N., Bhattachayya, S., Thiran, P., Salamatian, K., Diot, C.: On the feasibility of identifying elephants in Internet backbone traffic. Technical Report TR01-ATL-110918, Sprint Labs, Sprint ATL (November 2001)
14. Paxson, V., Floyd, S.: Wide area traffic: The failure of the Poisson assumption. IEEE/ACM Trans. on Networking, pp. 226–244 (1995)
15. Zseby, T., Molina, M., Duffield, N., Niccolini, S., Raspall, F.: Sampling and filtering techniques for IP packet selection (January 2006)

A   Appendix: Proof of Proposition 3

Let $(t_j)_{1 \leq j \leq pM(T)}$ be the sequence of the $pM(T)$ sampling times in $[0, T]$. We have for any flow in class ℓ, say, flow i

$$\mathbb{P}(n_i = 0) = \mathbb{E}\left(\prod_{j=1}^{pM(T)} \left(1 - \frac{N_i(t_j)}{M(t_j)}\right)\right) = \mathbb{E}\left(e^{\sum_{j=1}^{pM(T)} \log(1 - N_i(t_j)/M(t_j))}\right),$$

where n_i is the number of packets in the sampled flow i. First, note that

$$\sum_{j=1}^{pM(T)} \log\left(1 - \frac{N_i(t_j)}{M(t_j)}\right) = -\sum_{j=1}^{pM(T)} \frac{N_i(t_j)}{M(t_j)} + O\left(\frac{1}{K}\right).$$

Second, if f is a twice continuously differentiable function in $[0, T]$, we have

$$\sum_{j=1}^{pM(T)} f(t_j) = \frac{pM(T)}{T} \int_0^T f(u)\, du + \frac{f(T) - f(0)}{2} + O\left(\frac{1}{pM(T)}\right),$$

since the points (t_j) are distributed more or less uniformly in $[0, T]$. Hence, we have

$$\sum_{j=1}^{pM(T)} \log\left(1 - \frac{N_i(t_j)}{M(t_j)}\right) = -\frac{pM(T)}{T} \int_0^T \frac{N_i(u)}{M(u)}\, du + \frac{1}{2}\frac{N_i(T)}{M(T)} + O(\frac{1}{K}).$$

The first term in the right-hand side of the above equation is equal to

$$-\frac{x}{T}\int_0^T \frac{r_\ell(u - \tau_i)}{M(u)/K}\, du \rightarrow -\frac{x}{CT}\int_0^T r_\ell(u - \tau_i)\, du = -x\frac{C_\ell}{C} \quad \text{a.s.}$$

when K tends to $+\infty$. It follows that, when K tends to $+\infty$,

$$\mathbb{P}(n_i = 0) \rightarrow \exp\left(\frac{-xC_\ell}{C}\right). \tag{3}$$

For $k \geq 1$,

$$\mathbb{P}(n_i = k) = \mathbb{E}\left(\sum_{i_1 < \ldots < i_k} \prod_{m=1}^k \frac{N_i(t_{i_m})}{M(t_{i_m})} \prod_{j \notin \{i_1 < \ldots < i_k\}} \left(1 - \frac{N_i(t_j)}{M(t_j)}\right)\right)$$

$$= \mathbb{E}\left(\prod_{j=1}^{pM(T)}\left(1 - \frac{N_i(t_j)}{M(t_j)}\right) \Sigma_k(g_i(t_1), \ldots, g_i(t_{pM(T)}))\right),$$

where $g_i(u) = N_i(u)/(M(u) - N_i(u))$ and $\Sigma_k = \sum_{i_1 < \ldots < i_k} \prod_{j=1}^k X_{ij}$ is the symmetric homogeneous polynomial of degree k. Denoting $S_i = \sum_{j=1}^{pM(T)} X_j^i$ for $i > 1$, Newton's formula

$$(-1)^k k \Sigma_k + \sum_{p=0}^{k-1}(-1)^p \Sigma_p S_{k-p} = 0 \quad (1 \leq k \leq pM(T))$$

establishes that Σ_k can be expressed as a function of $S_1, \ldots, S_k$. It is clear that $S_q(g_i(t_1), \ldots, g_i(t_{pM(T)}))$ is the Riemann sum with $pM(T)$ terms associated to g_i^q on $[0, T]$. Using Newton's formula, it can be proved that when K tends to $+\infty$,

$$\Sigma_k(g_i(t_1), \ldots, g_i(t_{pM(T)})) \sim \frac{\left(\sum_{j=1}^{pM(T)} g_i(t_j)\right)^k}{k!}.$$

Taking into account approximation (3), we obtain that, for flow i of class ℓ,

$$\mathbb{P}(n_i = k) \rightarrow e^{-x\frac{C_\ell}{C}}\frac{1}{k!}\mathbb{E}\left(\left(\frac{x}{CT}\int_0^T r_l(u - \tau_i)\, du\right)^k\right) = \frac{e^{-x\frac{C_\ell}{C}}}{k!}\left(\frac{xC_\ell}{C}\right)^k$$

and Proposition 3 follows.

Simple and Accurate Forecasting of the Market for Cellular Mobile Services

Åke Arvidsson[1], Anders Hederstierna[2], and Stefan Hellmer[2]

[1] Ericsson AB, Box 1505, SE-125 25 Älvsjö, Sweden
Ake.Arvidsson@ericsson.com
[2] School of Management, Blekinge Institute of Technology,
SE-372 27 Ronneby, Sweden
{Anders.Hederstierna,Stefan.Hellmer}@bth.se

Abstract. We consider the problems of explaining and forecasting the penetration and the traffic in cellular mobile networks. To this end, we create two regression models, *viz.* one to predict the penetration from service charges and network effects and another one to predict the traffic from service charges and diffusion and adoption effects. The results of the models can also be combined to compute the likely evolutions of essential characteristics such as Minutes of Use (MoU), Average Revenue per User (ARPU) and total revenue. Applying the models to 26 markets throughout the world we show that they perform very well. Noting the significant qualitative differences between these markets, we conclude that the model has some universality in that the results are comparable for all of them.

1 Introduction

This paper deals with econometric methods for forecasting the number of subscribers and the voice traffic in cellular mobile networks.

Forecasting is an essential part of planning and maintaining telecommunications networks. For this reason forecasting has been a concern mainly for network operators. Today, however, forecasting is a concern for a wider community including equipment manufacturers, service providers and government regulators. Equipment manufacturers need to support their customers and plan their development projects; service providers need to plan and maintain their systems to ensure adequate quality and stay competitive; and regulators need to make independent estimates of the demand to secure the availability of essential services and to adequately judge complaints from operators, providers or end users. At the same time, forecasting is becoming increasingly complex as deregulation and competition not only encourages faster development and deployment of new services and technologies but also tend to reduce their life times.

Our primary aim is to consider a market for cellular mobile services and *forecast* penetration and traffic with a *long-term* perspective. As a secondary aim, we also wish to *explain* the evolution of these quantities over time and

L. Mason, T. Drwiega, and J. Yan (Eds.): ITC 2007, LNCS 4516, pp. 690–706, 2007.

to *predict* other, related quantities of interest. In this context, a "market" is a country or an operator and "long-term" refers to a couple of years.

In principle, we are interested in all kinds of services, such as voice, video, messaging and data hence our models do not include service specific characteristics. As for numerical examples, however, voice (and possibly SMS) is (are) the only service(s) for which sufficient statistics are available.

In this context it is worth recalling that voice still is the most important service in just about all cellular networks. To see this, consider the possibly most advanced markets in the world, *viz.* South Korea (represented SK Telecom [1]) and Japan (represented NTT DoCoMo [2]).[1] In terms of *revenues* [1] reports a ratio of three to one between voice ARPU and data ARPU (KRW 29,516 to KRW 10,689). We also note that about 40% of the data ARPU is related to SMS. The same figure in [2] is about four to one (JPY 5,030 to JPY 1,880). In terms of *volumes* [1] reports a per-user monthly voice volume of 20 Megabytes (MoU 197 and 13.4 kbps). The same figure in [2] is 15 Megabytes (MoU 149 and 13.4 kbps). We also note that these numbers double if inbound calls are included and double again if IP overhead is added. We are not aware of any official figures for data volumes, but it is clear that the volumes from services like SMS are extremely small. Finally, in terms of *growth* [1] reports voice volumes up by 5.6% (MoU up from 194 to 197 and users up from 18.789 million to 19.530 million). The same figure in [2] is 3.4% (MoU down from 151 to 149 and users up from 48.825 million to 51.144 million). We are not aware of any official figures for data growth rates.

Considering these numbers and the fact that other markets are far behind, we may safely conclude that voice is important, both in terms of revenues and volumes, and that voice is growing. Having said that, we firmly believe that other services will dominate in the future and we are looking forward to obtaining data which allows us to apply our models to these services.

The remainder of the paper is organised as follows: We begin in Sec. 2 with brief overviews of established forecasting methods and of the relevant literature. In Sec. 3 we formulate our problem and present our initial model. The parameters are discussed further in Sec. 4 and Sec. 5 after which the final models are obtained in Sec. 6. The results are applied to real data in Sec. 7 where we explain and forecast penetration, traffic and other quantities of interest. Finally we sum up our results and discuss further work in Sec. 8.

2   Preliminaries

2.1   Methods

The number of forecasting methods is possibly as large as (or larger than) the number of persons involved in forecasting. Using the taxonomy of financial markets, there are *analytical* methods and *technical* methods. In simple terms, analytical models (like regression models) are based on mathematical models of

---

[1] These operators were chosen with no other intention but to provide real numbers from the *leading* operators on the *leading* markets.

the *relationship between observed and explanatory data* whereas technical methods (like curve fitting, ARIMA or Kalman) are based on mathematical models of the *observed data itself*. See, *e.g.*, [3,4,5] for a general introduction. In addition there are models for *proposed* future services. For more details see, *e.g.*, [6,7].

We consider existing services with a known history. Analytical models of these services add an understanding of the factors behind the observations and this tends to make them more robust. Technical models, on the other hand, are simple to apply and some of them do not require long series of historic observations. The two methods thus both have their *pros* and *cons* hence they complement each other rather than compete with each other.

2.2   Literature

Because of our long-term perspective and our ambition to explain the evolution, we settle for an analytical method. Like most such methods we rely on product form models solved by ordinary least squares linear regression. The literature in this area is surprisingly sparse.

The problem of explaining the evolution of the penetration can be referred to as a diffusion problem. The classical work in this area is the Bass model [8] where a saturation level and two diffusion parameters called innovation and imitation are estimated from observed data. A thorough review of diffusion models and their applications, with particular emphasis on telecommunications, is found in [9]. In addition, the Bass model was used by [10] but with an externally estimated saturation level. We have tried the pure model with unsatisfactory results.

As for economic approaches, a thorough survey of economic studies modelling cross-country mobile telephone diffusion is found in [11]. This paper also examines the significance of 18 different socio-economical factors in 180 different countries. The study concludes that the most significant factor is the network effect. An explanatory model with almost the same variables as in our model is proposed by [12]. The main differences are that [12] treats income and cost as two variables whereas we use cost relative to income as one variable and that the per-minute costs in [12] are based on official model calls whereas we use actual *de facto* expenses.

The problem of explaining the evolution of the traffic by economic models is even less well represented in the literature. The works in [4,5] consider international, fixed telephony. These specific models, which include variables like "number of automatic exchanges" and "bilateral trade", are too old and too specific for our purposes. The only other work we have come across is our own [13] which we now replace by a model which is more straight forward yet more accurate.

We conclude the literature survey by mentioning the two special issues of "Telektronikk" devoted to forecasting telecommunications, one from 1994 [14] and another one from 2004 [15].

2.3   Contribution

The major contribution of the present work is that (i) we consider *both* penetration and traffic, (ii) we consider a *combination* of (relative) service charges network effects, and diffusion and adoption effects and that (iii) our models are applied to *several* markets, (iv) our models yield very *accurate* results yet (v) our models only need *one* external variable.

3   Basic Model

Let X denote the quantity of interest, *i.e.* penetration or traffic. Based on our earlier work [13] and inspired by the examples in [4, pp. 11–13] and [5, pp. 19–20], we selected a basic model of X with both economic and non-economic variables, *viz.* the actual cost C of the service and the penetration P on the market

$$X = g(C) \cdot h(P) \tag{1}$$

where g and h are functions and X, C and P are indexed by time (*e.g.*, year). It is noted that, although it may be desirable to include more variables, we are limited by the fact that the number of parameters must be significantly lower than the number of historic observations available in order for a model to be meaningful. We are also limited by the kind of data that is publicly available.

4   Modelling Cost

The decision to subscribe to and to use mobile services depends on the cost for using the mobile services, including subscription fees, usage tariffs and the cost of the terminal. We can assume that mobile services are like most other goods in that they have a downward sloping demand curve, implying that penetration and traffic increase when prices go down.

The question is how to measure the cost for mobile services. In most markets, there are several operators each of which offers a large number of different subscription fees and tariff structures. Moreover, many agreements for customers in the business segment are not even disclosed to the public. These circumstances make it very difficult in practice to measure average costs on mobile markets.

However, we have used the fact that the cost for the users must be equal to the operators' revenues — they are just two sides of the same coin. Revenues are often measured as (monthly) ARPU and usage is often measured as (monthly) MoU. A simple and approximate measure of the actual cost per minute may thus be obtained as ARPU divided by MoU.

Since ARPU often includes both subscription fees and usage tariffs (and sometimes also terminal costs), our measure includes the combined effect from changes in the total cost for subscription and usage (and sometimes also equipment).

Cost is, however, also a relative measure in the sense that people have limited financial resources and need to make decisions on how much of the available resources to spend on mobile services. Spending on mobile services thus depends not only on cost as such, but also on the size of the consumption budget and on individual preferences between consumption alternatives.

The consumption budget may change over time and the preferences may differ depending on the size of this budget. We have used the gross national product per capita (GDP), as the measure of the overall spending power in the market. GDP has advantages over many other measurements like, *e.g.*, consumer spending index, as it is commonly defined and publicly available in most countries. An approximate measure of the actual cost per minute relative to the spending budget may thus be obtained as (ARPU/MoU)/GDP.

The suggested metric thus "aggregates" price and GDP into one variable whereas they typically are treated as two separate variables. The advantage with this approach is that one variable less means one more degree of freedom, and this will increase the precision of the computed parameters. In addition, the suggested metric will by itself be adjusted for inflation and it is independent of the currency used. The approach may also be viewed as a compromise between the standard approach of including the GDP and the finding in [11] that GDP is relatively insignificant. Similar conflicts are also mentioned in [12].

A possibly more important aspect is, however, the fact that we consider *actual costs* as reported by operators whereas other works tend to rely on *official statistics* such as (some) peak-time charge per minute (*e.g.*, [11]) or (some) subscription fee (*e.g.*, [12]). This is a significant difference because our metric will account for a whole range of important factors such as form of payment (prepaid, *e.g*, typically implies additional per-minute charges but no subscription fees), charging scheme (different forms of "flat rate" are, *e.g*, available on many markets) and user segment (different discounts may be offered to, *e.g*, large corporations).

The relationship between cost C and demand D can be modelled in different ways. The classical model with a constant elasticity reads

$$D = K \cdot C^a \tag{2}$$

where K is a scaling constant and a corresponds to the elasticity of demand with respect to cost. With $a < 0$ the model may be interpreted such that the demand will increase by $a\%$ if the cost decreases by 1% and *vice versa*. The two unknown parameters K and a can be estimated from historic observations of D and C.

The model in Eq. (2) is a special case of willingness to pay (WTP) models. (To avoid possible confusion we point out that the *concept* of WTP is used in *hypothetical* situations whereas we will use *models* of WTP to characterise *actual* outcomes.) This class of models also includes models based on distributions

$$D = K \cdot \bar{F}(C) \tag{3}$$

where K again is a scaling constant and $\bar{F}$ is the survivor function of a probability distribution. Typical WTP distributions include, *e.g.*, the log normal distribution and the negative exponential distribution,

$$D = K \cdot e^{aC} \tag{4}$$

where $-1/a$ is the average price a user is willing to pay.

We adopt the simpler model Eq. (2) for traffic because it is a continuous entity where a marginal increase may have a marginal value which is hard to model by distributions. On the other hand, we adopt the more advanced model Eq. (4) for penetration because a subscription is a discrete entity and the result of elaborated decisions based on cost, budget and personal preferences.

5   Modelling Penetration

The decision to subscribe to and to use mobile services also depends on the penetration. Like most works in this area, we can assume that there is a network effect (which means that the higher the penetration, the more attractive is the network) and diffusion and adoption effects (which mean that the most keen users will be the first to join and *vice versa*).

A simple approach to characterise the network effect is attributed to Metcalf [16] who suggested that the value of a network is proportional to the square of the number of users connected to it. This statement can be generalised to a model of constant elasticities similar to the one in Eq. (2),

$$D = K \cdot P^b \tag{5}$$

where, as before, K is a scaling constant and b is the elasticity of demand with respect to penetration. The model may be interpreted such that the demand will increase by $b\%$ if the penetration increases by 1% and *vice versa*. The two unknown parameters K and b can again be estimated from historic observations of D and P.

A similar approach may be used to model diffusion and adoption effects. A more advanced version, however, is to apply such a model to a weighted penetration where the penetration is recomputed to reflect not just "total users" but rather "total user interest". To this end, we suggest to characterise the interest of a user by the cost of the service at the time the user joins. Such a model may then be written as

$$D = K \cdot \left(\sum_y C_y \Delta P_y \right)^b = K \cdot W^b \tag{6}$$

where K again is a scaling constant, C_y is the cost year y, ΔP_y is the change in penetration during year y and b as before is the elasticity of demand with respect to weighted penetration.

We prefer the simpler model Eq. (5) for penetration because increasing penetration is not only a result of increasing value (the network effect) but also related to increasing awareness (the imitation effect). On the other hand, we prefer the more advanced model Eq. (6) for traffic because it contains more information and we can assume that less keen subscribers always will be less active and *vice versa*.

6   Final Models

The suggested models of penetration P (average penetration over a year) and traffic T (average number of traffic minutes per month) are thus obtained by inserting Eq. (4) and (5) and Eq. (2) and (6) respectively into Eq. (1)

$$P_y = K \cdot e^{aC_y} \cdot P_{y-1}^b \tag{7}$$
$$T_y = K \cdot C_y^a \cdot W_y^b \tag{8}$$

where subscript y refers to year.

It is noted that Eq. (7) can be used recursively to forecast the penetration P_{y+1} for different assumptions regarding costs C_{y+1} after which Eq. (8) can be used to forecast the traffic T_{y+1} for the same assumptions regarding costs C_{y+1}.

The results of these forecasts can be used to calculate the corresponding evolutions of MoU U_{y+1}, adjusted voice ARPU A'_{y+1} and adjusted total voice revenues R'_{y+1} as follows

$$U_y = T_y/P_y \tag{9}$$
$$A'_y = C_y \cdot U_y \tag{10}$$
$$R'_y = P_y \cdot A'_y \tag{11}$$

Note that voice ARPU A' and total voice revenues R' are not given in nominal currency but adjusted with respect to inflation and GDP per capita.

The unknown parameters K, a and b can be solved for by rewriting Eq. (7) and Eq. (8) as multiple linear regression models

$$\ln(P_y) = \ln(K) + a(C_y) + b\ln(P_{y-1}) \tag{12}$$
$$\ln T_y = \ln(K) + a\ln(C_y) + b\ln\left(\sum_y C_y \Delta P_y\right) \tag{13}$$

from which the unknowns can be computed by the ordinary least squares technique.

We require that $K > 0$, $a < 0$ and $b > 0$. The first requirement is obvious and the other requirements are motivated by the facts that interest is discouraged by higher costs and encouraged by increasing penetration.

If required, we repeatedly reject the oldest remaining observation until these requirements are met. The reason for this is that "invalid" models most likely are attributed to "invalid" data. There are many reasons for why data can be invalid.

- Some simple reason are, *e.g.*, that operators tend to change the way parameters like ARPU and penetration are reported. Examples of such changes include the reporting of late payment fees and the policies towards passive subscribers. Sometimes these changes may pass unnoticed or be impossible to correct.
- Some more complex reasons are, *e.g.*, that operators may have expanded the coverage or governments may have changed the regulations. Such changes may be very significant but lie outside our model.

– Finally, customers may change behaviour when prices and penetration change
by several orders of magnitude. (Note, however, that this does not happen in
case of, *e.g.*, electricity.)

The idea behind rejecting old observations is thus to remove observations regarding old reporting, old coverage, old regulations *etc.* until the model makes
sense. This will not guarantee but at least improve the chances that all "external"
conditions are stable in the remaining time period.

7   Examples

We will now apply the models to $M = 26$ different markets. It is emphasised
that the markets and the numbers are provided by the authors only as examples
and that they do not, in any way, represent official business outlooks.

The starting point is tables of penetration $P_{m,y}$, ARPU $A_{m,y}$, MoU $U_{m,y}$ and
GDP $G_{m,y}$ for all markets $m = 1, \ldots, 26$ and over ten years $y = 1995, \ldots, 2004$
(except for some markets where the tables span shorter times).

This data is used to compute additional tables of cost

$$C_{m,y} = A_{m,y}/U_{m,y}/G_{m,y}, \tag{14}$$

weighted penetration

$$W_{m,y} = C_{m,1995}P_{m,1995} + \sum_{v=1996}^{y} C_{m,v}\left(P_{m,v} - P_{m,v-1}\right) \tag{15}$$

and traffic

$$T_{m,y} = M_{m,y}P_{m,y} \tag{16}$$

for these years. For the penetration model, the costs for each market are finally
normalised by the highest cost on that market to enable comparisons between
different markets.

7.1   Market Models

The results of fitting the coefficients as suggested by Eq. (12) and Eq. (13) to
all M markets are given in Tab. 1 and Tab. 2 respectively. The tables give the
adjusted R^2 values, the fitted coefficients a and b with t-ratios, and the number
of available and rejected rejected data points for each market respectively.

Tab. 1 shows that the R^2 values are very close to unit which suggests a high
degree of explanation. Moreover, most t-ratios are high which means that the
uncertainty of the coefficients tends to be small and that the coefficients typically
are significantly distinct from zero. Finally it is noted that one observation was
rejected fro Colombia, Korea and South Africa whereas three observations were
rejected for Japan and Poland.

Tab. 2 also shows very high R^2 values which again suggests a high degree of
explanation for all markets. Moreover, most t-ratios are even higher which means

Table 1. Results from fitting the penetration model

Market	R^2	K	$-a$	t-rat.	b	t-rat.	Val.	Rej.
Canada	1.00	5.55	0.98	6.29	0.46	5.67	8	0
United States	1.00	3.11	0.25	2.20	0.75	11.3	9	0
Argentina	0.99	5.97	2.21	3.65	0.44	4.14	9	0
Brazil	1.00	2.36	0.60	1.90	0.84	23.8	9	0
Colombia	0.96	2.62	0.58	0.87	0.75	6.15	9	1
Chile	1.00	5.27	2.19	10.8	0.49	11.8	9	0
Mexico	1.00	5.14	2.12	9.05	0.57	13.8	9	0
China	1.00	2.81	0.56	1.58	0.80	11.9	9	0
India	0.98	5.50	2.24	3.01	0.50	4.00	9	0
Japan	1.00	2.65	0.02	0.22	0.77	51.1	9	3
Korea	0.98	8.16	0.61	0.57	0.24	0.49	7	1
Malaysia	0.99	4.78	1.32	3.60	0.55	4.90	9	0
Thailand	0.99	7.11	2.64	5.59	0.43	3.56	7	0
France	1.00	6.26	1.36	3.18	0.45	4.70	8	0
Germany	1.00	9.74	3.42	6.22	0.30	3.26	8	0
Italy	1.00	7.06	1.68	4.06	0.44	5.34	8	0
Netherlands	0.99	6.08	1.58	4.08	0.42	4.46	8	0
Spain	0.99	5.31	1.30	2.57	0.56	5.39	8	0
Sweden	1.00	3.38	0.54	0.67	0.66	1.96	9	0
United Kingdom	1.00	5.77	1.49	8.61	0.53	11.9	8	0
Hungary	0.99	3.80	1.33	2.58	0.66	6.71	9	0
Israel	0.99	2.21	0.46	1.61	0.81	20.6	8	0
Poland	1.00	3.18	0.48	1.26	0.74	11.1	9	3
Russia	1.00	1.00	0.17	1.12	0.98	45.6	9	0
South Africa	0.99	1.82	0.10	0.09	0.84	6.31	7	1
Turkey	0.98	2.49	0.04	0.07	0.78	6.74	7	0

that the coefficients tend to be even less uncertain and even more distinct from zero. It may appear odd that all markets contain one more observation than before, but this is because no lagged values are required. Finally it is noted that one observation was rejected for South Africa.

The above results suggest that the models work very well in *explaining* the observed evolutions. When it comes to *forecasting*, however, the true test of the models is to perform ex-post forecasting. To this end, the last two observations were removed from the data sets (one observation was removed for data sets of less than eight observations) after which the models were refitted. The new coefficients and the actual input values C and P were then used to compute forecasts.

The results for all markets are given in Tab. 3. The tables give the average absolute errors in percent.

It is seen that the differences between the actual and forecast values typically are on the order of 10% or less. The clear exceptions to this are India (penetration) and Argentina, Colombia and Mexico (traffic). We believe that the larger errors for these markets, at least to some extent, is caused by invalid data

Table 2. Results from fitting the traffic model

Market	R^2	K	$-a$	t-rat.	b	t-rat.	Val.	Rej.
Canada	1.00	5.68	1.59	10.5	0.16	0.85	9	0
United States	1.00	5.26	1.35	25.5	0.67	6.42	10	0
Argentina	0.97	6.81	0.65	1.29	0.80	4.17	10	0
Brazil	0.99	8.78	0.31	1.51	0.74	23.7	10	0
Colombia	0.96	6.87	0.93	3.85	0.61	9.17	10	0
Chile	1.00	5.34	1.25	4.96	0.63	4.76	10	0
Mexico	0.99	6.12	1.30	1.67	0.61	1.65	10	0
China	1.00	9.97	0.92	9.24	0.51	5.79	10	0
India	1.00	5.04	1.17	17.1	0.94	18.0	10	0
Japan	0.99	7.59	1.02	7.60	0.78	4.85	10	0
Korea	1.00	6.36	0.89	14.3	0.87	14.7	8	0
Malaysia	1.00	6.34	0.62	3.64	0.97	8.52	10	0
Thailand	1.00	7.39	0.57	4.45	0.84	18.2	8	0
France	1.00	5.40	1.23	8.58	0.70	6.54	9	0
Germany	0.99	7.71	0.84	1.12	0.57	2.51	9	0
Italy	1.00	6.28	1.03	2.90	0.70	4.62	9	0
Netherlands	0.99	5.67	0.99	2.87	0.76	3.44	9	0
Spain	1.00	5.54	1.04	10.6	0.80	21.6	9	0
Sweden	1.00	6.75	0.37	19.0	1.17	77.3	10	0
United Kingdom	1.00	5.15	1.27	6.70	0.70	5.96	9	0
Hungary	1.00	8.41	0.48	4.92	0.65	19.3	10	0
Israel	0.98	9.00	0.55	2.06	0.75	17.2	9	0
Poland	1.00	7.67	0.23	3.54	0.83	40.7	10	0
Russia	0.99	7.28	0.73	8.03	0.73	32.0	10	0
South Africa	0.98	8.37	0.80	0.96	0.55	3.62	8	1
Turkey	0.97	7.24	0.73	1.70	0.69	4.91	8	0

Table 3. Results from ex-post forecasting

Market	Penetration		Traffic		Market	Penetration		Traffic	
	Error	Val.	Error	Val.		Error	Val.	Error	Val.
Canada	1.17	2	4.04	2	France	1.80	2	7.49	2
United States	1.63	2	7.45	2	Germany	3.86	2	4.54	2
Argentina	10.9	2	22.8	2	Italy	2.47	2	6.71	2
Brazil	8.45	2	1.17	2	Netherlands	7.44	2	9.18	2
Colombia	10.4	1	39.8	2	Spain	7.90	2	7.04	2
Chile	2.86	2	14.0	2	Sweden	0.03	2	1.92	2
Mexico	9.55	2	20.8	2	United Kingdom	1.02	2	1.07	2
China	11.3	2	8.24	2	Hungary	14.3	2	5.22	2
India	37.6	2	7.40	2	Israel	13.2	2	2.31	2
Japan	0.53	1	3.12	2	Poland	4.64	1	3.42	2
Korea	2.21	1	1.05	2	Russia	14.7	2	16.6	2
Malaysia	11.3	2	5.44	2	South Africa	6.70	1	3.31	1
Thailand	6.41	1	10.0	2	Turkey	0.08	1	5.98	1

due to, *e.g.* changed reporting, changed coverage or changed regulations as discussed above.

7.2   Market Forecasts

The models of Eq. (12) and Eq. (13) were fitted as above and used to produce forecasts as suggested by these equations and by Eq. (9)–(11). Forecasts were made for three very different cost evolution scenarios.

- A basic one where the annual price drop amounts to the average annual price drop in the observed data sets. This scenario is indicated by solid lines and intended to represent a continuation of the present evolution.
- A fast one where the annual price drop amounts to twice the average annual price drop in the observed data sets. This scenario is indicated by dashed lines and means noticeable cost reductions and is intended to model increasing competition on price.
- A slow one where the annual price drop amounts to the average annual increase in GDP in the observed data sets. This scenario is indicated by dotted lines and means virtually no nominal cost reductions and is intended to represent decreased competition on price.

We emphasise that a more realistic set of scenarios would consider much smaller differences in the cost evolution.

The results for three markets, *viz.* United States, China and Germany are given below. All numbers are given in percent relative to 2004. The markets were chosen because they are big and because they are different. Again it is emphasised that the markets and the numbers are provided by the authors only as examples and that they do not, in any way, represent official business outlooks.

The forecasts for penetration are given in Fig. 1.

- For the United States all three scenarios result in modest growth. This indicates that that the network effect is weak at the present price level and only aggressive price cuts can drive up penetration.
- For China all three scenarios result in strong growth. This indicates that the network effect is very strong at the present price level hence no price cuts are necessary to drive up penetration.
- For Germany the three scenarios result in very different growths. This indicates that the network effect is weak at the present price level but that even modest price cuts can drive up penetration.

To understand these differences, we consider the penetration saturation limit P_∞. The penetration saturation limit P_∞ with respect to year v is the penetration at which the network effect with respect to the price of year v is completely exhausted.

Formally we consider the costs of year v and take the limit as $y \to \infty$ in Eq. (7),

$$\ln P_\infty = K + aC_v + b\ln P_\infty \tag{17}$$

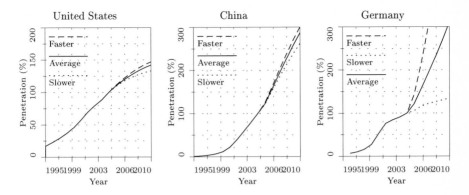

Fig. 1. Forecasting penetration for the three markets

which gives

$$\ln P_\infty = \frac{K + aC_v}{1 - b}. \tag{18}$$

The ratios between the limits for $v = 2004$ and the corresponding, actual values amount to 1.42, 4.03 and 1.31 for the three markets respectively. Although these values may be very uncertain, we may conclude that the network effects are stronger in China than in the United States and in Germany. The growth in China is thus "self-driven" to a larger extent than in the United States and in Germany.

Next consider the price sensitivity defined as

$$-\frac{\frac{\partial P_y}{\partial C_y}}{P_y} = -a \tag{19}$$

The numerical values for the three markets amount to 0.254, 0.558 and 2.42 respectively. Again these values may be very uncertain but we can conclude that the market in Germany is more price sensitive than in the United States and in China. The growth in Germany thus depends more on price cuts than in the United States and in China.

Summing up, for the United States small price cuts are sufficient, for China no price cuts are necessary and for Germany large price cuts are necessary. These findings may not be surprising considering that prices, as defined in Eq. (14), have fallen from one to 0.14 in the United States, to 0.07 in China but only to 0.50 in Germany. Another factor is, of course, the penetration relative to the population which is about 2/3 in the United States, 1/4 in China and 15/16 in Germany. (Note, however, that ratios above unit already are seen in, *e.g.*, Sweden.)

The forecasts traffic are given in Fig. 2.

- For the United States price cuts will have a strong impact with an explosive growth in the fast scenario and almost no growth in the slow scenario.

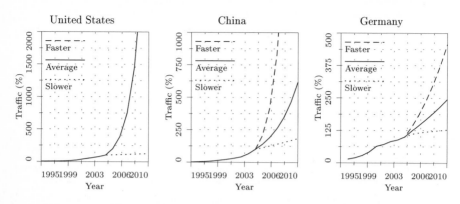

Fig. 2. Forecasting traffic for the three markets

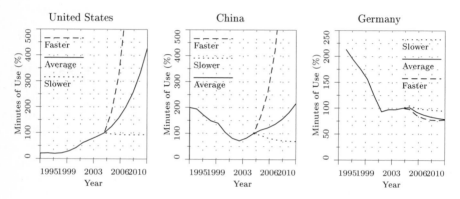

Fig. 3. Corresponding evolution of the minutes of use for the three markets

- For China price cuts will have a less pronounced impact.
- For Germany price cuts will have an even smaller impact.

To understand these differences, we consider the price elasticities a in Tab. 2 which amount to 1.35, 0.916 and 0.842 for the United States, China and Germany respectively. It is immediately noted that the degree of sensitivity is directly coupled to the price elasticity. The differences in traffic increase between China and Germany are also affected by the differences in penetration increase. In China, new subscribers are attracted by network effects and will be quite active. In Germany, however, new subscribers are attracted by cost and will be less active.

The corresponding evolution of the minutes of use are given in Fig. 3.

- For the United States an explosive growth is noted in the fast scenario and a slight negative growth in the slow scenario.
- For China the growth in the fast scenario is less pronounced and the growth in the slow scenario is even more negative.

– For Germany the growth is negative in all cases, and more negative the more
  aggressive the price cuts.

To understand these findings, we note that price cuts will have two effects.
Firstly, lower prices will increase the traffic from existing subscribers. Secondly,
lower prices will attract new subscribers. The key point is that an average "new"
subscriber will generate less traffic than an average "old" subscriber. The first
phenomenon will thus increase the minutes of use whereas the second phenomenon will decrease the minutes of use. The net result of a price cut therefore
depends on the relative strengths between the two phenomena.

For the United States we noted that penetration is relatively stable no matter
the prices. This means there is a strong, price dependent positive contribution
from many old subscribers and a weak negative contribution from a few new
subscribers.

For China we saw that penetration grows very fast no matter the prices. This
means that there is a moderate, price dependent positive contribution from old
subscribers and an almost independent, negative contribution from new subscribers.

For Germany we found that penetration grows only in response to price
changes. This means that there is a moderate, price dependent positive contribution from old subscribers and a heavily dependent, negative contribution
from new subscribers.

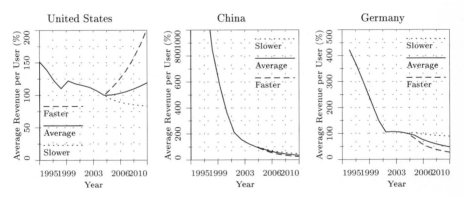

Fig. 4. Corresponding evolution of the average (voice) revenue per user for the three
markets

The corresponding evolution of the average (voice) revenue per user are given
in Fig. 4. The reader is reminded that the currency used is the one defined in Eq.
(14). This means that ARPU is measured relative to GDP such that a constant
ARPU essentially means that it increases at the same rate as the economy as a
whole *etc.*

– For the United States all scenarios but the slow one suggests increasing voice
  ARPU. The reason for this is that a price elasticity of traffic which is greater

than one means that the increases in MoU more than compensates for the decreasing prices except in the slow scenario where the negative impact of new subscribers cannot be fully compensated for.

- For China all scenarios suggest decreasing voice ARPU. Note that the strong decrease of the past partly is an effect of the strong growth of the Chinese economy since our currency is normalised by inflation and GDP. The diagram shows that voice ARPU will drop less the smaller the price cuts and the fundamental reason for this is that the price elasticity less than one cannot compensate for decreasing prices. In addition there is a further negative impact from new subscribers.
- For Germany all scenarios suggest decreasing voice ARPU. Again the drop will be less the smaller the price cuts and the reasons are the same as for China.

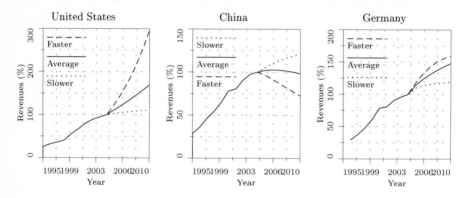

Fig. 5. Corresponding evolution of the total voice revenues for the three markets

The corresponding evolution of the total voice revenues are given in Fig. 4. The reader is again reminded that the currency used is the one defined in Eq. (14). As before, this means that revenues are measured relative to GDP such that a constant revenue essentially means that it increases at the same rate as the economy as a whole *etc.*

- For the United States all scenarios will lead to increasing revenues from voice with the highest increase for the most aggressive price cuts. The main reason for this is the high elasticity of traffic with respect to price.
- For China only the slow scenario will increase revenues from voice. The main reason for this is that decreases in tariffs cannot be compensated for by marginal increases in penetration and traffic.
- For Germany the findings are the same as for the United States except that they are less pronounced. The main reason is, however, the high sensitivity of penetration with respect to price.

8   Conclusions and Further Work

Understanding and forecasting penetration and traffic are essential parts of telecommunications networks management. We have demonstrated that it is possible to identify a few crucial variables that can be used to explain and predict these quantities in cellular mobile networks. Moreover, we have shown that these variables have some universality in that they perform very well for 26 very different markets.

The cost is typically an important factor in explaining and forecasting demand. It is, however, usually difficult to quantify costs of telecommunications services because of the complex price structures on most markets. We show that a correct and easy solution to this problem is to use the operators' revenues instead. Moreover, we show that the state of the economy, an almost mandatory factor when explaining and forecasting demand, can be included in the cost.

Penetration is another important factor and we show that it is possible to model its effect by using Metcalf's model combined with established models in technology diffusion and adoption theory.

The final models show good explanatory and predictive power. They should be useful for analysing scenarios and "what-if" questions and also for understanding differences between different markets.

Current and future work includes a different variable for weighted penetration were costs are logarithmic (as an alternative to Eq. (6)), an additional variable for penetration saturation were population matters (as a compliment to Eq. (7)) and traffic related to different services (in particular SMS).

References

1. SK Telecom Annual Report (2005). Available from www.sktelecom.com.
2. DoCoMo, N.T.T.: Annual Report (2005) Available from www.nttdocomo.com
3. Leijon, H.: Forecasting Theories, PLANITU Doc. 61-E, ITU
4. ITU-T, Forecasting International Traffic, Recommendation E.506, ITU-T (1992)
5. ITU-T, Models for Forecasting International Traffic, Recommendation E.507, ITU-T (1993)
6. ITU-T, Forecasting New Telecommunication Services, Recommendation E.508, ITU-T (1993)
7. Stordahl, K.: Long-Term Telecommunications Forecasting, Ph.D. dissertation, Norwegian University of Science and Technology, Trondheim, Norway (2006)
8. Bass, F.: A New Product Growth Model for Consumer Durables. Management Science 15, 215–227 (1969)
9. Meade, N., et al.: Modelling and Forecasting the Diffusion of Innovation — A 25-Year Review. Journal of Forecasting 22, 519–545 (2006)
10. Zhang, J.: Forecasting the Number of Subscribers of 3G Mobile Services in China Based on Bass Diffusion Model. In: Proc. ITC 19, pp. 1591–1599 (2005)
11. Rouvinen, P.: Diffusion of Digital Mobile Telephony: Are Developing Countries Different? Telecommunications Policy 30, 46–63 (2006)
12. Madden, G., et al.: A Dynamic Model of Mobile Telephony Subscription Incorporating a Network Effect. Telecommunications Policy 28, 133–144 (2004)

13. Arvidsson, Å., et al.: Forecasting Cellular Mobile Traffic: An Econometric Approach. In: Proc. ITC 19, pp. 1581–1590 (2005)
14. Telektronikk, Special Issue on Forecasting, vol. 90(1), ISSN 0085-7130 (1994)
15. Telektronikk, Special Issue on Forecasting, vol. 100(4), ISSN 0085-7130 (2004)
16. Shapiro, C., et al.: Information Rules. Harvard Business School Press, Boston (1998)
17. Rogers, E.: Diffusion of Innovations. Free Press, New York (1995)

Performance Analysis of FEC Recovery Using Finite-Buffer Queueing System with General Renewal and Poisson Inputs

Shun Muraoka, Hiroyuki Masuyama, Shoji Kasahara, and Yutaka Takahashi

Graduate School of Informatics, Kyoto University
Kyoto 606-8501, Japan
{muraoka, masuyama}@sys.i.kyoto-u.ac.jp,
{shoji, takahashi}@i.kyoto-u.ac.jp

Abstract. This paper considers the packet recovery performance of forward error correction (FEC) for a single-server queueing system with a finite buffer fed by two independent input processes; one is a general renewal input process in which the interarrival times of packets are independent and identically distributed according to a general distribution, and the other is a Poisson arrival process. We analyze the packet- and block-level loss probabilities, and investigate the recovery performance of FEC at block level. Numerical examples show that the block-loss probability is significantly improved by FEC when the system can only accommodate a small number of packets. It is also shown that FEC recovery is more effective than packet buffering when the traffic intensity is large.

Keywords: Single-server queue with two input processes, forward error correction, general renewal input, block-loss probability.

1 Introduction

Forward error correction (FEC) is a well-known coding-based error recovery scheme that is expected to guarantee quality of service (QoS) in terms of loss and delay for real-time applications over the Internet, such as voice over IP (VoIP) and Internet TV [2,8]. In FEC, redundant data is generated from the original data, and a sender host transmits both the original and redundant data to a receiver host. If some part of the original data is lost, it can be recovered from the redundant data at the receiver host if the loss is below a prespecified level.

In this paper, we focus on a packet-level FEC scheme [9]. When N redundant data packets are generated from D original data packets, the lost data can be recovered completely if the number of lost packets is less than or equal to N. Because FEC has no retransmission mechanism, it is a suitable packet-loss recovery scheme for real-time applications with stringent delay constraints.

It is well known that the recovery performance of FEC is significantly affected by the packet-loss process. The relation between the recovery performance and

L. Mason, T. Drwiega, and J. Yan (Eds.): ITC 2007, LNCS 4516, pp. 707–718, 2007.
© Springer-Verlag Berlin Heidelberg 2007

the redundancy of FEC has been extensively studied in the literature. Cidon et al. [3] conducted a pioneering study in which the distribution of the number of lost packets within a block of packets is analyzed for an M/M/1/K queue with first-in first-out (FIFO) discipline. Altman and Jean-Marie [1] considered the loss probabilities of a block of packets and investigated the effect of FEC redundancy on these probabilities. Note that the main results of the above works were obtained on the basis that packet arrivals form a single Poissonian source.

Kawahara et al. [7] generalized the packet arrival process by considering a discrete-time finite-buffer queueing model with two arrival processes, and evaluated the recovery capability of FEC. In their study, main packet traffic consisting of original and redundant packets is modelled as an interrupted Bernoulli process, while background traffic is modelled as a Markov modulated Bernoulli process. Hellal et al. [6] considered an M/M/1/K queue in which packets arrive at the system from several independent sources, and the distribution of the number of lost packets within a block of packets was analyzed. These authors also extended the model so that the sequences of service times and interarrival times of packets are ergodic with the system capacity equal to one, and thus were able to discuss the qualitative nature of the redundancy scheme.

Dán et al. [4] considered the effect of the packet-size distribution on the packet-loss process in finite queueing systems with two input processes: one is a Markov-modulated Poisson process (MMPP) while the other is a Poisson process. The FEC performance was numerically evaluated for deterministic and exponential packet-size distributions.

With the recent development of optical networking technology such as wavelength division multiplexing (WDM), the bottleneck of data transmission has shifted from backbone networks to access ones (the last mile bandwidth bottleneck [5]). This suggests that edge routers are likely to be the bottleneck of data transmission for real-time applications. In real-time applications such as VoIP and Internet TV, packets are sent to the network at a constant bit rate. Therefore, it is important to consider the case where interarrival times of packets to a bottleneck edge router are almost the same.

In this paper, we focus on a bottleneck edge router and investigate the packet-loss process observed there. We consider a finite-buffer queueing system with two inputs: main traffic and background traffic. Main traffic consists of original packets along with the corresponding redundant packets, and we assume that interarrival times of packets in main traffic are independent and identically distributed (i.i.d.) according to a general distribution. On the other hand, the packet arrival process of background traffic is modelled as a Poisson process. Note that the general distribution assumption on main traffic enables us to describe various arrival processes including constant interarrival times. Assuming that the packet service times are exponentially distributed, we analyze the packet- and block-level loss probabilities of main traffic, and evaluate the recovery effect of FEC.

The paper is structured as follows: Section 2 specifies the details of the present model, while Section 3 presents an analysis of the model. Numerical examples are presented in Section 4, followed by conclusions in Section 5.

2   Analysis Model

We consider the transmission of a data block consisting of D packets. Suppose N redundant packets are generated from the original D packets, and a set of M ($= D + N$) packets are transmitted as main traffic.

   We assume that packet loss occurs only at the bottleneck edge router. Regarding main traffic, if the number of lost packets among the M packets is smaller than or equal to N, the original data block can be recovered at the destination by FEC decoding. Otherwise not all the original data packets can be recovered, resulting in a block loss.

   We model the bottleneck edge router as a single-server queueing system with a finite buffer. The interarrival times of packets in main traffic are i.i.d. with a general distribution $G(x)$. Background traffic is also multiplexed with main traffic in the system, and we assume that packet arrivals in background traffic form a Poisson process with rate λ. The capacity of the system is K, and hence the buffer size is $K-1$. The service times of packets both in main and background traffic are i.i.d. according to an exponential distribution with parameter μ.

   From the above assumptions, we have a GI+M/M/1/K queue. In the following, we analyze the packet- and block-level loss probabilities of main traffic.

3   Analysis

We first consider the queue length distribution immediately before an arrival from main traffic in the GI+M/M/1/K queue described in the previous section. We then derive the block-loss probability of a block, which consists of D original data packets and $N(= M - D)$ redundant data packets.

3.1   Queue Length Distribution Immediately Before an Arrival from Main Traffic

Let $L(t)$ $(t \geq 0)$ denote the number of packets in the system at time t. We assume that $L(t)$ is right-continuous and is bounded in the limit from below, i.e.,

$$L(t) = \lim_{\Delta \to 0+} L(t + \Delta) , \qquad L(t-) = \lim_{\Delta \to 0+} L(t - \Delta) < \infty , \qquad \forall t > 0 .$$

Let $\widehat{T}_\nu$ $(\nu = 1, 2, \dots)$ denote the arrival epoch of the νth packet in main traffic at the system. We assume $\widehat{T}_1 = 0$ hereafter. We then define $\widehat{L}_\nu^-$ and $\widehat{L}_\nu$ $(\nu = 1, 2, \dots)$ as

$$\widehat{L}_\nu^- = L(\widehat{T}_\nu-) , \qquad \widehat{L}_\nu = L(\widehat{T}_\nu) ,$$

respectively. i.e., the queue lengths immediately before and after the νth arrival from main traffic. Clearly,

$$\widehat{L}_\nu = \min(\widehat{L}_\nu^- + 1, K) , \qquad \nu = 1, 2, \dots . \tag{1}$$

We now define $\boldsymbol{\Lambda}$ as a $(K+1) \times (K+1)$ matrix whose (i,j)-th element $\Lambda_{i,j}$ represents $\Pr[\widehat{L}_\nu = j \mid \widehat{L}_\nu^- = i]$. It follows from (1) that

$$\boldsymbol{\Lambda} = \begin{pmatrix} 0 & 1 & 0 & \cdots & 0 & 0 \\ 0 & 0 & 1 & \ddots & 0 & 0 \\ \vdots & \vdots & \ddots & \ddots & \vdots & \vdots \\ 0 & 0 & 0 & \ddots & 1 & 0 \\ 0 & 0 & 0 & \cdots & 0 & 1 \\ 0 & 0 & 0 & \cdots & 0 & 1 \end{pmatrix} .$$

We also define $\boldsymbol{\Gamma}$ as a $(K+1) \times (K+1)$ matrix whose (i,j)-th element $\Gamma_{i,j}$ represents

$$\Gamma_{i,j} = \Pr[\widehat{L}_{\nu+1}^- = j \mid \widehat{L}_\nu = i] .$$

Recall here that during time interval $(\widehat{T}_\nu, \widehat{T}_{\nu+1})$, there are no arrivals from main traffic, and packets from background traffic arrive at the system according to a Poisson process with rate λ. Recall also that the service times of all packets, both in main and background traffic, are i.i.d. according to an exponential distribution with parameter μ. These facts imply that during time interval $(\widehat{T}_\nu, \widehat{T}_{\nu+1})$, $L(t)$ behaves in the same way as an M/M/1/K queue with arrival rate λ and service rate μ. Furthermore, because interarrival times of main traffic follow a general distribution $G(x)$, we have

$$\boldsymbol{\Gamma} = \int_0^\infty \exp(\boldsymbol{Q}x) dG(x) ,$$

where $\boldsymbol{Q}$ denotes a $(K+1) \times (K+1)$ matrix that is given by

$$\boldsymbol{Q} = \begin{pmatrix} -\lambda & \lambda & 0 & \cdots & 0 & 0 \\ \mu & -(\lambda+\mu) & \lambda & \ddots & \vdots & \vdots \\ 0 & \mu & -(\lambda+\mu) & \ddots & 0 & 0 \\ 0 & 0 & \mu & \ddots & \lambda & 0 \\ \vdots & \vdots & & \ddots & -(\lambda+\mu) & \lambda \\ 0 & 0 & 0 & \ddots & \mu & -\mu \end{pmatrix} .$$

Let T_m $(m = 1, 2, \dots)$ denote $\widehat{T}_{\nu+m}$. In what follows, in examining the limit as $\nu \to \infty$, we assume that the system has already reached the steady state at time T_1. Let $\boldsymbol{\pi}^-$ denote a $1 \times (K+1)$ vector whose jth element π_j^- represents $\Pr[L(T_1-) = j]$, i.e., $\boldsymbol{\pi}^-$ is considered as the queue length distribution immediately before an arrival from main traffic in the steady state. It is easy to see that

$$\boldsymbol{\pi}^-(\boldsymbol{\Lambda}\boldsymbol{\Gamma}) = \boldsymbol{\pi}^- , \qquad \boldsymbol{\pi}^- \boldsymbol{e} = 1 ,$$

where e denotes a column vector of ones with an appropriate dimension. Let P_{loss} denote the packet-loss probability of main traffic. Note here that a packet in main traffic is lost only when the queue length on arrival is equal to K. We then have

$$P_{\text{loss}} = \pi_K^- .$$

3.2   Block-Loss Probability

This subsection derives the block-loss probability $P_{\text{loss}}^{(\text{B})}$, which is defined as the probability that the FEC decoding fails to retrieve data packets of a block, i.e., the probability that the number of lost packets among M packets of a block is greater than that of the redundant packets, N. To do so, we focus on an arbitrary block, which is called a *tagged block* hereafter. We assume that the M packets of the tagged block arrive continuously to the system at times from T_1 through to T_M. We then call the packet arriving at time T_m $(m = 1, 2, \ldots, M)$ packet m.

Let L_m^- and L_m denote the numbers of packets in the system at time T_m- and T_m, respectively, i.e., $L_m^- = L(T_m-)$ and $L_m = L(T_m)$. Let N_m $(m = 1, 2, \ldots, M)$ denote the number of lost packets among packets 1 through m. We define $\boldsymbol{p}_m(k)$ $(m = 1, 2, \ldots, M; k = 0, 1, \ldots, M)$ as a $1 \times K$ vector whose jth element $p_{m,j}(k)$ $(j = 1, 2, \ldots, K)$ is given by

$$p_{m,j}(k) = \Pr[N_m = k, L_m = j] .$$

The block-loss probability $P_{\text{loss}}^{(\text{B})}$ is then given by

$$P_{\text{loss}}^{(\text{B})} = \Pr[N_M > N] = 1 - \sum_{k=0}^{N} \boldsymbol{p}_M(k)\boldsymbol{e} .$$

In the rest of this subsection, we consider recursion for the $\boldsymbol{p}_m(k)$.

We first derive $\boldsymbol{p}_1(k)$: If $L_1^- < K$, packet 1 can enter the system, and so $L_1 = L_1^- + 1$. Thus $p_{1,j}(0)$ $(j = 1, 2, \ldots, K)$ is given by

$$p_{1,j}(0) = \Pr[L_1^- < K, L_1^- = j - 1] = \pi_{j-1}^- . \tag{2}$$

On the other hand, if $L_1^- = K$, packet 1 is lost and $L_1 = K$. Therefore $p_{1,j}(1)$ $(j = 1, 2, \ldots, K)$ is given by

$$p_{1,j}(1) = \begin{cases} 0 , & j = 1, 2, \ldots, K - 1 , \\ \pi_K^- , & j = K . \end{cases} \tag{3}$$

Note here that $N_1 \leq 1$, and hence that, for all $k \geq 2$,

$$p_{1,j}(k) = 0 , \qquad j = 1, 2, \ldots, K . \tag{4}$$

For convenience, we can also express (2), (3) and (4) in vector notation:

$$\boldsymbol{p}_1(0) = (\pi_0^-, \pi_1^-, \ldots, \pi_{K-1}^-) ,$$
$$\boldsymbol{p}_1(1) = (0, 0, \ldots, 0, \pi_K^-) ,$$
$$\boldsymbol{p}_1(k) = (0, 0, \ldots, 0, 0) = \boldsymbol{0} , \qquad k = 2, 3, \ldots, M .$$

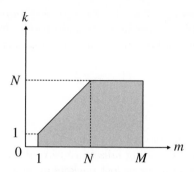

Fig. 1. Domain of indices (m, k) of $\boldsymbol{p}_m(k)$

Next, we discuss $\boldsymbol{p}_m(k)$ $(m = 2, 3, \ldots, M)$. We define $\boldsymbol{A}(\sigma)$ $(\sigma = 0, 1)$ as a $K \times K$ matrix whose (i, j)-th element $A_{i,j}(\sigma)$ $(i, j = 1, 2, \ldots, K)$ is given by

$$A_{i,j}(\sigma) = \Pr[\Theta_m = \sigma, L_m = j \mid L_{m-1} = i] \ ,$$

where $\Theta_m = 1$ if packet m is lost, and otherwise $\Theta_m = 0$. It then follows that for $m = 2, 3, \ldots, M$,

$$\boldsymbol{p}_m(0) = \boldsymbol{p}_{m-1}(0)\boldsymbol{A}(0) \ ,$$
$$\boldsymbol{p}_m(k) = \boldsymbol{p}_{m-1}(k - 1)\boldsymbol{A}(1) + \boldsymbol{p}_{m-1}(k)\boldsymbol{A}(0) \ , \qquad k = 1, 2, \ldots, M \ .$$

Note here that $\boldsymbol{p}_m(k) = \boldsymbol{0}$ for all $k \geq m+1$ because $N_m \leq m$. Figure 1 illustrates the domain of indices (m, k) of $\boldsymbol{p}_m(k)$'s that are needed to calculate $P_{\text{loss}}^{(\text{B})}$.

Finally, we consider $\boldsymbol{A}(\sigma)$ $(\sigma = 0, 1)$. If $L_m^- < K$ then $\Theta_m = 0$ and $L_m = L_m^- + 1$, and so we have

$$A_{i,j}(0) = \Pr[L_m^- = j - 1 \mid L_{m-1} = i] = \Gamma_{i,j-1} \ . \tag{5}$$

Furthermore, since $\{\Theta_m = 1\}$ is equivalent to $\{L_m = L_m^- = K\}$, $A_{i,j}(1)$ $(i, j = 1, 2, \ldots, K)$ is given by

$$A_{i,j}(1) = \begin{cases} 0 \ , & j = 1, 2, \ldots, K - 1 \ , \\ \Gamma_{i,K} \ , & j = K \ , \end{cases} \qquad i = 1, 2, \ldots, K \ . \tag{6}$$

In matrix notation, (5) and (6) can be expressed as

$$\boldsymbol{A}(0) = \begin{pmatrix} \Gamma_{1,0} & \Gamma_{1,1} & \cdots & \Gamma_{1,K-1} \\ \Gamma_{2,0} & \Gamma_{2,1} & \cdots & \Gamma_{2,K-1} \\ \vdots & \vdots & \ddots & \vdots \\ \Gamma_{K,0} & \Gamma_{1,1} & \cdots & \Gamma_{K,K-1} \end{pmatrix} \ ,$$

$$\boldsymbol{A}(1) = \begin{pmatrix} 0 & \cdots & 0 & \Gamma_{1,K} \\ 0 & \cdots & 0 & \Gamma_{2,K} \\ \vdots & \ddots & \vdots & \vdots \\ 0 & \cdots & 0 & \Gamma_{K,K} \end{pmatrix} \ .$$

4   Numerical Examples

We consider video streaming as an example of a real-time application, and evaluate the recovery effect of FEC using the analytical results derived in the previous section. We consider a 10 Mbps transmission rate for the streaming data, and a video frame rate of 30 fps. Assuming that the size of a packet is 1250 bytes, a frame has $D = 34$ original data packets, and a block has the same number of packets as that of a frame. It is also assumed that the interarrival times of packets in main traffic are constant. When the output transmission speed of the bottleneck router is 100 Mbps, the service rate of a 1250 byte-packet μ is 1.0×10^4 [packet/s].

We define FEC redundancy as N/D. Note that when the sender host adds N redundant packets to D original data packets, the resulting packet transmission rate is increased by a factor equal to $(D + N)/D$.

4.1   Impact of Background Traffic

In this subsection, we investigate how the rate of background traffic affects both the block- and packet-level loss probabilities. Figure 2 illustrates the block-loss probability against the rate of background traffic for system capacities $K = 10$ and 100. Note that when the rate of background traffic is γ bps, the corresponding packet arrival rate of background traffic λ is equal to $\gamma \times 10^{-4}$ [packet/s]. For each value of K, we calculated the block-loss probabilities for $N = 0$, 1, 3 and 5. The output transmission speed of the bottleneck router is 100 Mbps ($\mu = 1.0 \times 10^4$ [packet/s]). In Fig. 2, the horizontal axis is the rate of background traffic and the vertical one is the block-loss probability.

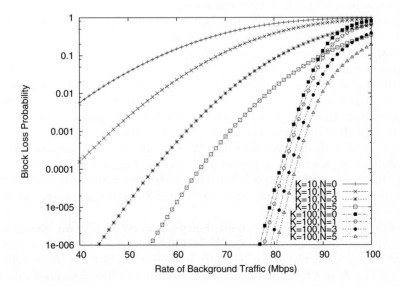

Fig. 2. Block-loss probability vs. rate of background traffic

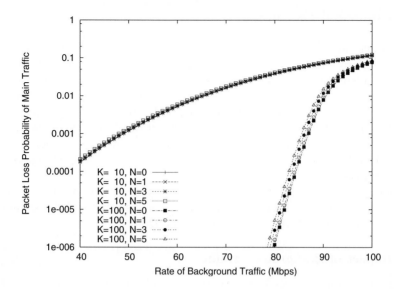

Fig. 3. Packet-loss probability of main traffic vs. rate of background traffic

We observe from Fig. 2 that the block-loss probability increases monotonically as the rate of background traffic increases. This is because a larger rate of background traffic makes the system congested, resulting in a larger block-loss probability. In terms of the FEC recovery effect, the block-loss probability is greatly improved as the number of FEC redundant packets increases. A remarkable point in Fig. 2 is that the improvement in block-loss probability by the application of FEC for $K = 10$ is much stronger than that for $K = 100$.

Figure 3 shows the packet-loss probability of main traffic against the rate of background traffic under the same conditions as in Fig. 2. In Fig. 3, we observe a monotonic increase in the packet-loss probability of main traffic with an increasing rate of background traffic for both $K = 10$ and 100. It is also observed that the packet loss probability is larger the higher the number of redundant packets. This is simply because adding redundant packets increases the traffic intensity. Note that the packet-loss probability of main traffic is not, however, significantly affected by the increase in FEC redundant packets.

Next, we investigate the minimum FEC redundancy such that the block loss probability is smaller than a prespecified value α. Figure 4 illustrates the minimum FEC redundancy against the rate of background traffic for the cases $K = 10$ and 100. In each case, the minimum FEC redundancy was calculated for $\alpha = 10^{-2}$, 10^{-3} and 10^{-4}.

In Fig. 4, the minimum FEC redundancy values for $K = 10$ are always larger than the corresponding values for $K = 100$. This is because the packet-loss probability of main traffic for $K = 10$ is larger than that for $K = 100$, (see Fig. 3.). For $K = 10$, the minimum FEC redundancy increases gradually with

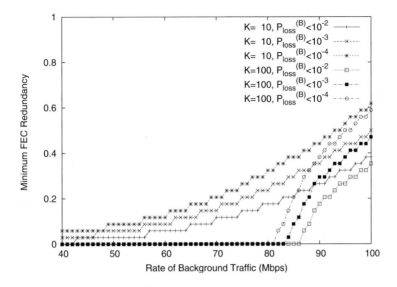

Fig. 4. Minimum FEC redundancy vs. rate of background traffic

an increasing rate of background traffic. For $K = 100$, however, the minimum FEC redundancy remains zero until the rate of background traffic reaches a value around 80 Mbps, and then increases rapidly. Note that in Fig. 3, the packet-loss probability of main traffic for $K = 10$ increases gradually with an increasing rate of background traffic, while, in contrast, for $K = 100$ this probability exhibits a rapid increase.

The above results imply that when the rate of background traffic is small, FEC can significantly improve the recovery of lost packets for a system with small capacity. On the other hand, block-loss events rarely occur in a system with large capacity due to the buffering effect rather than FEC recovery.

4.2   Impact of Service Rate for Bottleneck Router

In this subsection, we investigate how the output transmission speed of the bottleneck router affects the block-loss probability and the minimum FEC redundancy.

Figure 5 plots the block-loss probability against the transmission speed for $K = 10$ and 100. For each value of K, we calculated the block-loss probability for $N = 0$, 1, 3 and 5. Note that when the transmission speed is η bps, the corresponding service rate of a packet at the bottleneck router μ is equal to $\eta \times 10^{-4}$ [packet/s]. The rate of background traffic is set to 50 Mbps.

It is observed from Fig. 5 that the block-loss probability decreases monotonically as the transmission speed increases. For $K = 100$, the block-loss probabilities for the three N vales considered decrease rapidly from around 70 Mbps, and beyond this point there is no significant difference in the results of the three cases. In contrast, for $K = 10$ the block-loss probability decreases gradually with

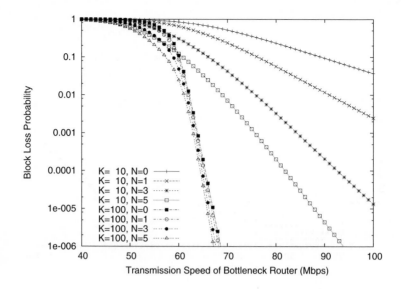

Fig. 5. Block-loss probability vs. transmission speed of bottleneck router

increasing transmission speed. In addition, the block-loss probability for $N = 5$ is the smallest, while that for $N = 0$ is the largest.

These results imply that when the system capacity is large, the transmission speed of the bottleneck router is the dominant influence on the block loss probability and, in particular, its impact is larger than FEC. However, if the system capacity is small, the block-loss probability can be significantly affected by FEC, as well as by the transmission speed.

Next, we investigate how the minimum FEC redundancy is affected by the transmission speed of the bottleneck router. Figure 6 plots the minimum FEC redundancy for $K = 10$ and 100 against the transmission speed of the bottleneck router. The rate of background traffic is set to 50 Mbps. For each value of K, the minimum FEC redundancy was calculated for $\alpha = 10^{-2}$, 10^{-3} and 10^{-4}.

It is observed from Fig. 6 that for each α, the minimum FEC redundancy with $K = 10$ is almost the same as that with $K = 100$ when the transmission speed is smaller than 60 Mbps. For transmission speeds increasing beyond around 60 Mbps, the former decreases gradually while the latter reaches zero. These findings imply that FEC is effective for a wide range of transmission speeds when the system capacity is small. When the system capacity is large, block-level QoS is greatly affected by the transmission speed, and the advantages of applying FEC are reduced.

4.3   Impact of System Capacity

Finally, we investigate how the system capacity affects the minimum FEC redundancy.

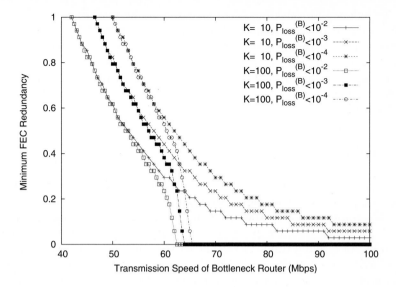

Fig. 6. Minimum FEC redundancy vs. transmission speed of bottleneck router

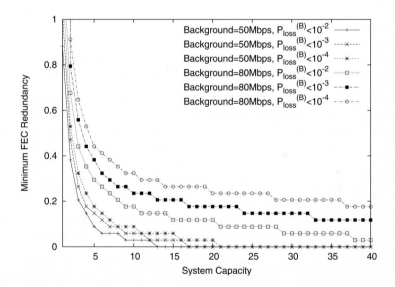

Fig. 7. Minimum FEC redundancy vs. system capacity

Figure 7 illustrates the minimum FEC redundancy plotted against the system capacity for $\alpha = 10^{-2}$, 10^{-3} and 10^{-4}. For each α, the minimum FEC redundancy was calculated for 50 Mbps and 80 Mbps rates of background traffic. The output transmission speed of the bottleneck router is set to 100 Mbps.

In Fig. 7, the minimum FEC redundancy for 50 Mbps of background traffic rapidly decreases as the system capacity increases. In particular, the minimum

FEC redundancy for 50 Mbps of background traffic reaches zero at $K = 20$. On the other hand, the minimum FEC redundancy for 80 Mbps of background traffic decreases gradually with increasing system capacity. A remarkable point of Fig. 7 is that the minimum FEC redundancy is likely to remain constant when the system capacity is large. This result implies that packet buffering is less effective than FEC for improving the block-loss probability when the traffic intensity is large.

5 Conclusions

In this paper, we considered the performance of block recovery with FEC for a GI+M/M/1/K queue. The packet- and block-level loss probabilities were analyzed by a continuous-time Markov chain. It was shown from numerical examples that when the system capacity is large, the block-loss probability is largely affected by background traffic or the output transmission speed, and the advantages of FEC are relatively small. On the contrary, when the traffic intensity is large, the effect on packet recovery by applying FEC is significant in comparison with packet buffering.

References

1. Altman, E., Jean-Marie, A.: Loss Probabilities for Messages with Redundant Packets Feeding a Finite Buffer. IEEE Journal on Selected Areas in Communication 16(5), 778–787 (1998)
2. Carle, G., Biersack, E.W.: Survey on Error Recovery Techniques for IP-Based Audio-Visual Multicast Applications. IEEE Network Magazine 11(6), 24–36 (1997)
3. Cidon, I., Khamisy, A., Sidi, M.: Analysis of Packet Loss Processes in High-Speed Networks. IEEE Transactions on Information Theory 39(1), 98–108 (1993)
4. Dán, G., Fodor, V., Karlsson, G.: On the Effects of the Packet Size Distribution on the Packet Loss Process. Telecommunication Systems 32(1), 31–53 (2006)
5. Green, P.E.: Fiber to the Home: The Next Big Broadband Thing. IEEE Communications Magazine 42(9), 100–106 (2004)
6. Hellal, O.A., Altman, E., Jean-Marie, A., Kurkova, I.A.: On Loss Probabilities in Presence of Redundant Packets and Several Traffic Sources. Performance Evaluation 36-37(1-4), 486–518 (1999)
7. Kawahara, K., Kumazoe, K., Takine, T., Oie, Y.: Forward Error Correction in ATM Networks: An Analysis of Cell Loss Distribution in a Block. In: Proc. IEEE INFOCOM'94, pp. 1150–1159 (1994)
8. Perkins, C., Hodson, O., Hardman, V.: A Survey of Packet Loss Recovery Techniques for Streaming Audio. IEEE Network Magazine 12(5), 40–48 (1998)
9. Shacham, N., Pckenney, P.: Packet Recovery in High-Speed Networks Using Coding and Buffer Management. In: Proc. IEEE INFOCOM, vol. 1, pp. 124–131 (1990)

Queueing Model with Time-Phased Batch Arrivals

Moon Ho Lee[1], Sergey Dudin[2], and Valentina Klimenok[2]

[1] Institute of Information and Communication
Chonbuk National University
Chonju, 561-765, Korea
moonho@chonbuk.ac.kr
[2] Department of Applied Mathematics and Computer Science
Belarusian State University
Minsk 220030, Belarus
dudin@madrid.com, klimenok@bsu.by

Abstract. A novel multi-server queueing model with finite buffer and batch arrival of customers is considered. In contrast to the standard batch arrival when a whole batch arrives into the system at one epoch, we assume that the customers of a batch arrive one by one in exponentially distributed times. Service time is exponentially distributed. Flow of batches is the stationary Poisson arrival process. Batch size distribution is geometric. The number of batches, which can be admitted into the system simultaneously, is subject of control. The problem of maximizing the throughput of the system under the fixed value of the admissible probability of losing the arbitrary customer from admitted batch is considered. Analysis of the joint distribution of the number of batches and customers in the system and sojourn time distribution is implemented by means of the matrix technique and method of catastrophes.

1   Introduction

Queueing systems well describe operation of channels and servers of many communication networks and this is why they have got a lot of attention in probabilistic literature. Important class of queueing systems assumes that customers arrive in batches of random size. It is usually assumed that, at a batch arrival epoch, all customers of this batch arrive into the system simultaneously and if the system capacity is finite they are admitted to the system according to some discipline. Disciplines of partial admission, complete admission and complete rejection are usually applied.

However, it is the typical feature of many nowadays communication networks that customers arrive in batches, but arrival of customers is not instantaneous. The first customer of a batch arrives at the batch arrival epoch while the rest of customers arrive one by one in random intervals. The batch size is random and it may be not known a priori at the batch arrival epoch. Such a situation is typical, e.g., in modeling transmission of video and multimedia information.

L. Mason, T. Drwiega, and J. Yan (Eds.): ITC 2007, LNCS 4516, pp. 719–730, 2007.

This situation is also discussed in [1] with respect to the modeling Scheme of Alternative Packet Overflow Routing in IP networks. In [1], performance measures of this scheme of routing in IP networks is evaluated by means of computer simulation.

In this paper, we construct and investigate analytically the Markovian queueing model that suits for performance evaluation of this routing scheme as well of other real life systems with time distributed arrival of customers in a batch. To the best of our knowledge, such kind of queueing models is not considered and investigated in literature yet.

The rest of the paper is organized as follows. In section 2, the model is described. The steady state joint distribution of the number of batches and customers in the system is analyzed in section 3. Section 4 is devoted to consideration of the customer and batch sojourn time distributions. Section 5 contains formulation of the optimization problem and numerical illustrations and section 6 concludes the paper.

2   Mathematical Model

We consider a queueing system of capacity $N, 1 \leq N < \infty$. The system has R, $1 \leq R \leq N$, servers and a buffer of capacity $N - R \geq 0$. The servers are assumed to be identical and independent of each other. Service time in the server has exponential distribution with parameter μ. The customers arrive to the system in batches. Batches arrive into the system according to a stationary Poisson arrival process with intensity λ. Following to [1], we assume that admission of batches (they are called *flows* in [1]) is restricted by means of *tokens*. The total number of available tokens is assumed to be K, $K \geq 1$. Further we will consider the number K as control parameter and will touch the corresponding optimization problem.

If there is no token available at a batch arrival epoch or the system is full, the batch is rejected. It leaves the system forever. If the number of available tokens at the batch arrival epoch is positive and the system is not already full, this batch is admitted into the system and the number of available tokens decreases by one. We assume that one customer of a batch arrives at the batch arrival epoch and if it meets at least one free server, it occupies this server and is processed. If all servers are busy, the customer moves to a buffer, if it exists, and later it is picked up for the service according to the First Came - First Served discipline.

After admission of the batch, the next customer of this batch can arrive into the system in exponentially distributed with parameter γ time. If the system is not full, the customer is admitted into the system. Otherwise, it is lost. However, it does not mean that the next customers from the batch, to which belongs this customer, will be not admitted to the system. The number of customers in a batch has geometrical distribution with parameter θ, $0 < \theta < 1$, i.e., probability that the batch consists of k customers is equal to $\theta^{k-1}(1 - \theta)$, $k \geq 1$. If the exponentially distributed with parameter γ time since arrival of the previous customer of a batch expires and new customer does not arrive, it means that the arrival of the batch is finished. The token, which was obtained by this batch upon arrival, is returned into the pool of available tokens. The customers of this batch, which stay in the system at the epoch of returning the token, should be processed by the system. When the last customer is served, sojourn time of the batch in the system is considered finished.

It is intuitively clear that this mechanism of arrivals restriction by means of tokens is reasonable. At the expense of rejecting some batches, it allows to decrease probability of the loss of a customer that belongs to admitted batches, batches sojourn time and jitter. It is important in modeling real-life systems because quality of transmission of accepted information units should satisfy imposed requirements of Quality of Service.

Note, that situation when the new batch is rejected while the system is empty is theoretically possible. Probably, this is shortcoming of the considered scheme. But the mechanism of tokens creates better conditions (shorter delay and smaller jitter) for transmission of customers from the accepted batches. Quantitative analysis of advantages and shortcomings of this mechanism and optimal selection of parameter K requires calculation of the main performance measures of the system under any fixed value K of tokens in the system. These measures can be calculated basing on the knowledge of stationary distribution of the random process describing dynamics of the system under study.

3   Joint Distribution of the Number of Batches and Customers in the System

Let i_t be the total number of customers in the system at epoch t, $t \geq 0$, $i_t = \overline{0, N}$, and k_t be the number of batches having token for admission to the system at epoch t, $t \geq 0$, $k_t = \overline{0, K}$.

It is obvious that the two-dimensional process $\xi_t = \{i_t, k_t\}$, $t \geq 0$, is the finite irreducible regular continuous time Markov chain.

Let Q be the generator of the Markov chain ξ_t, $t \geq 0$, with blocks $Q_{i,j}$ consisting of intensities $(Q_{i,j})_{k,k'}$ of the Markov chain ξ_t, $t \geq 0$, transitions from the state (i, k) into the state (j, k'), $k, k' = \overline{0, K}$. The diagonal entries of the matrix $Q_{i,i}$ are negative and the modulus of the entry of $(Q_{i,i})_{k,k}$ defines the total intensity of leaving the state (i, k) of the Markov chain. The block $Q_{i,j}$, $i, j = \overline{0, N}$, has dimension $(K + 1) \times (K + 1)$.

Introduce the following notation:

- $\gamma^- = \gamma(1 - \theta)$, $\gamma^+ = \gamma\theta$;
- $\mu_i = \mu \min\{i, R\}$, $i = \overline{0, N}$;
- $C_K = diag\{0, 1, \ldots, K\}$, i.e., the diagonal matrix with the diagonal entries $\{0, 1, \ldots, K\}$;
- I is identity matrix, $\mathbf{e}$ is column vector consisting of 1's, $\mathbf{0}$ is row vector consisting of 0's. When dimension of the matrix or the vector is not clear from context, it is indicated by suffix, e.g., $\mathbf{e}_{K+1}$ denotes the unit column vector of dimension $K + 1$;

●

$$A = \begin{pmatrix} 0 & 0 & 0 & \dots & 0 & 0 \\ \gamma^- & -\gamma & 0 & \dots & 0 & 0 \\ 0 & 2\gamma^- & -2\gamma & \dots & 0 & 0 \\ \vdots & \vdots & \vdots & \ddots & \vdots & \vdots \\ 0 & 0 & 0 & \dots & K\gamma^- & -K\gamma \end{pmatrix};$$

●

$$A_1 = \begin{pmatrix} -\gamma & 0 & 0 & \dots & 0 & 0 \\ \gamma^- & -2\gamma & 0 & \dots & 0 & 0 \\ 0 & 2\gamma^- & -3\gamma & \dots & 0 & 0 \\ \vdots & \vdots & \vdots & \ddots & \vdots & \vdots \\ 0 & 0 & 0 & \dots & (K-1)\gamma^- & -K\gamma \end{pmatrix}$$

●

$$E^+ = \begin{pmatrix} 0 & 1 & 0 & \dots & 0 \\ 0 & 0 & 1 & \dots & 0 \\ \vdots & \vdots & \vdots & \ddots & \vdots \\ 0 & 0 & 0 & \dots & 1 \\ 0 & 0 & 0 & \dots & 0 \end{pmatrix}; \quad \hat{I} = \begin{pmatrix} 1 & 0 & \dots & 0 & 0 \\ 0 & 1 & \dots & 0 & 0 \\ \vdots & \vdots & \ddots & \vdots & \vdots \\ 0 & 0 & \dots & 1 & 0 \\ 0 & 0 & \dots & 0 & 0 \end{pmatrix};$$

- $\delta_{i,j}$ is Kronecker delta. It is equal to 1 if $i = j$ and equal to 0 otherwise;
- $\otimes$ is the symbol of Kronecker product of matrices;
- $\mathbf{b}^T$ means transposition of the vector $\mathbf{b}$.

Lemma 1. *The generator Q has the three-block-diagonal structure:*

$$Q = \begin{pmatrix} Q_{0,0} & Q_{0,1} & O & \dots & O & O \\ Q_{1,0} & Q_{1,1} & Q_{1,2} & \dots & O & O \\ O & Q_{2,1} & Q_{2,2} & \dots & O & O \\ \vdots & \vdots & \vdots & \ddots & \vdots & \vdots \\ O & O & O & \dots & Q_{N-1,N-1} & Q_{N-1,N} \\ O & O & O & \dots & Q_{N,N-1} & Q_{N,N} \end{pmatrix}$$

where the non-zero blocks $Q_{i,j}$ are computed by

$$Q_{i,i} = A - \lambda\hat{I} - \mu_i I, i = \overline{0, N-1}, \quad Q_{N,N} = A + \gamma^+ C_K - \mu_N I,$$
$$Q_{i-1,i} = \gamma^+ C_K + \lambda E^+, \ i = \overline{1, N}, \quad Q_{i+1,i} = \mu_{i+1} I, \ i = \overline{0, N-1}.$$

Proof of the lemma consists of analysis of the Markov chain $\xi_t, t \geq 0$, transitions during the infinitesimal interval of time and further combining corresponding transition intensities into the matrix blocks. Value γ^- is the intensity of tokens releasing due to the finish of the batch arrival, γ^+ is the intensity of new customers in the batch arrival.

Because the Markov chain $\xi_t = \{i_t, k_t, \}, t \geq 0$, is the irreducible and regular and has the finite state space, the following limits (stationary probabilities) exist:

$$\pi(i, k) = \lim_{t \to \infty} P\{i_t = i, \ k_t = k\}, \ i = \overline{0, N}, \ k = \overline{0, K}.$$

Let us combine these probabilities into the row-vectors

$$\boldsymbol{\pi}_i = (\pi(i,0)\pi(i,1),\dots,\pi(i,K)), \quad i = \overline{0,N}.$$

It is well known that the vector $(\boldsymbol{\pi}_0,\dots,\boldsymbol{\pi}_N)$ is the unique solution to the following system of linear algebraic equations:

$$(\boldsymbol{\pi}_0,\dots,\boldsymbol{\pi}_N)Q = \mathbf{0}, \quad (\boldsymbol{\pi}_0,\dots,\boldsymbol{\pi}_N)\mathbf{e} = 1.$$

In the case if the system capacity N and the number of tokens K are not large, this system can be solved on computer directly. In the case when at least one of values N or K is large, the numerically stable algorithm for solving this system, which was elaborated in [2], can be applied. This algorithm is given by the following assertion.

Theorem 1. *The stationary probability vectors $\boldsymbol{\pi}_i$, $i = \overline{0,N}$, are calculated by*

$$\boldsymbol{\pi}_l = \boldsymbol{\pi}_0 F_l, l = \overline{1,N},$$

where the matrices F_l are calculated recurrently:

$$F_0 = I, \quad F_i = -F_{i-1}Q_{i-1,i}(Q_{i,i}+Q_{i,i+1}G_i)^{-1}, l = \overline{1,N},$$

the matrices $G_i, i = \overline{0,N-1}$, are calculated from the backward recursion:

$$G_i = -(Q_{i+1,i+1} + Q_{i+1,i+2}G_{i+1})^{-1}Q_{i+1,i}, \quad i = N-2, N-3, \dots, 0,$$

with the terminal condition

$$G_{N-1} = -(Q_{N,N})^{-1}Q_{N,N-1},$$

the vector $\boldsymbol{\pi}_0$ is calculated as the unique solution to the following system of linear algebraic equations:

$$\boldsymbol{\pi}_0(Q_{0,0} + Q_{0,1}G_0) = \mathbf{0}, \quad \boldsymbol{\pi}_0 \sum_{l=0}^{N} F_l \mathbf{e} = 1.$$

Note that all inverse matrices in the theorem statement exist because the inverted matrices are sub-generators.

The straightforward algorithmic way for calculating the stationary probability vectors $\boldsymbol{\pi}_i$, $i = \overline{0,N}$, given by this theorem well suits for realization on computer.

Having stationary probability vectors $\boldsymbol{\pi}_i$, $i = \overline{0,N}$, been computed, we can calculate different performance measures of the system. Some of them are given in the following statements.

Corollary 1. *Probability distribution of the number of customers in the system is computed by*

$$\lim_{t\to\infty} P\{i_t = i\} = \boldsymbol{\pi}_i\mathbf{e}, \quad i = \overline{0,N}.$$

Average number L of customers in the system is computed by $L = \sum\limits_{i=0}^{\infty} i\boldsymbol{\pi}_i\mathbf{e}.$

Probability distribution of the number of batches in the system is computed by

$$\lim_{t\to\infty} P\{k_t = k\} = \sum_{i=0}^{N} \pi(i, k), \ k = \overline{0, K}.$$

Average number B of batches in the system is computed by

$$B = \sum_{k=1}^{K}\sum_{i=0}^{N} k\pi(i, k).$$

Mean number T of customers processed by the system at unit of time (throughput) is computed by

$$T = \sum_{i=0}^{N} \mu_i\boldsymbol{\pi}_i\mathbf{e}.$$

Besides the throughput, the main performance measures of this model are probability $P_b^{(loss)}$ of an arbitrary batch rejection and probability $P_c^{(loss)}$ of the rejecting an arbitrary customer from admitted batch.

Theorem 2. *Probability $P_b^{(loss)}$ of an arbitrary batch rejection is computed by*

$$P_b^{(loss)} = \sum_{i=0}^{N-1} \pi(i, K) + \sum_{k=0}^{K} \pi(N, k).$$

Probability $P_c^{(loss)}$ of an arbitrary customer from admitted batch rejection is computed by

$$P_c^{(loss)} = \frac{\sum\limits_{k=1}^{K} k\gamma^+\pi(N, k)}{\sum\limits_{k=1}^{K}\sum\limits_{i=0}^{N} k\gamma^+\pi(i, k)}.$$

4　Distribution of the Sojourn Times

Let $V_b(x)$ and $V_c(x)$ be distribution functions of sojourn time of an arbitrary batch and an arbitrary customer from admitted batch in the system under study and $v_b(s)$ and $v_c(s)$ be their Laplace-Stieltjes Transforms (*LSTs*):

$$v_b(s) = \int\limits_{0}^{\infty} e^{-sx}dV_b(x), \ v_c(s) = \int\limits_{0}^{\infty} e^{-sx}dV_c(x), \ \ Re\ s > 0.$$

The *LST* $v_c(s)$ is computed in a trivial way:

$$v_c(s) = P_c^{(loss)} + \frac{\sum\limits_{k=1}^{K} k\gamma^+ \left[\sum\limits_{i=0}^{R-1} \pi(i, k) + \sum\limits_{i=R}^{N-1} \pi(i, k)\left(\frac{R\mu}{R\mu+s}\right)^{i-R+1} \right]}{\sum\limits_{k=1}^{K}\sum\limits_{i=0}^{N} k\gamma^+\pi(i, k)} \frac{\mu}{\mu+s}.$$

The LST $v_b(s)$ is much more interesting characteristic and is computed not easy. In borders of this paper, we will show how LST $v_b(s)$ is calculated in the case of a single server system $(R = 1)$ with a buffer of capacity $N - 1$. General case can be analyzed by means of a proper generalization of the proposed below constructions.

Recall that sojourn time of an arbitrary batch in the system starts since the epoch of the batch arrival into the system until the moment when all customers, which belong to this batch, leave the system. We will derive expression for the LST $v_b(s)$ by means of the method of collective marks (method of additional event, method of catastrophes). To this end, we interpret the variable s as the intensity of some virtual stationary Poisson flow of catastrophes. So, $v_b(s)$ has meaning of probability that no one catastrophe arrives during the sojourn time of an arbitrary batch.

We will tag an arbitrary batch and will keep track of its staying in the system. Let $v(s, i, l, k)$ be probability that catastrophe will not arrive during the rest of the tagged batch sojourn time in the system conditional that, at the given moment, the number of batches processed in the system is equal to k, $k = \overline{1, K}$, the number of customers is equal to i, $i = \overline{0, N}$, and the last (in the order of arrival) customer of a tagged batch has position number l, $l = \overline{0, i}$, in the system. Position number 0 means that currently there is no one customer of the tagged batch in the system.

It is obvious, that if we compute the LSTs $v(s, i, l, k)$ then the LST $v_b(s)$ is calculated by

$$v_b(s) = P_b^{(loss)} + \sum_{i=0}^{N-1} \sum_{k=0}^{K-1} \pi(i, k) v(s, i+1, i+1, k+1). \tag{1}$$

To derive the system of linear algebraic equations for the LSTs $v(s, i, l, k)$, we use the formula of total probability:

$$v(s, i, l, k) = \frac{1}{s + \lambda(1 - \delta_{k,K})(1 - \delta_{i,N}) + \mu_i + k\gamma} \times \tag{2}$$

$$\times \left[\lambda(1 - \delta_{k,K})(1 - \delta_{i,N}) v(s, i+1, l, k+1) + \mu_i v(s, i-1, l-1, k) + \gamma^+ v(s, i+1, i+1, k) + \right.$$

$$\left. + \gamma^+ (k-1) v(s, i+1, l, k) + \gamma^- \left(\frac{\mu}{\mu+s} \right)^l + \gamma^- (k-1) v(s, i, l, k-1) \right],$$

$$l = \overline{0, i}, \ i = \overline{0, N}, \ k = \overline{1, K}.$$

Let us explain formula (2) in brief. Denominator of the right hand side of (2) is equal to the total intensity of the events which can happen after the arbitrary time moment: catastrophe arrival, new batch arrival, service completion, and expiring the time till the moment of possible customer arrival from batches already admitted into the system. The first term in the square brackets in (2) corresponds to the case when the new batch arrives. The second one corresponds to the case when service completion takes place. The third term corresponds to the case when the new customer of the tagged batch arrives into the system. In this case, the position of the last customer of a tagged batch in the system is reinstalled from l to $i + 1$. The fourth term corresponds to the case when the new customer from another batch, which was already admitted to the system, arrives. The fifth term corresponds to the case when the expected new customer of the tagged batch does not arrive into the system and arrival of customers of the tagged batch is stopped. This batch will not more counted as arriving into the system and the tagged customer finishes its sojourn time when the last customer, who is currently the lth in the system, will leave the system. The sixth term corresponds to the case when some other batch is stopped.

Let us introduce column vectors $\mathbf{v}(s, i, l) = (v(s, i, l, 1), \ldots, v(s, i, l, K))^T$, $l = \overline{0, i}$, $i = \overline{0, N}$, of dimension K.

The system of linear algebraic equations (2) can be rewritten in the following matrix form:

$$-(sI - \hat{Q}_{i,i})\mathbf{v}(s, i, l) + \hat{Q}_{i,i+1}\mathbf{v}(s, i+1, l) + \hat{Q}_{i,i-1}\mathbf{v}(s, i-1, l-1)+$$

$$+\gamma^+\mathbf{v}(s, i+1, i+1) + \gamma^- \left(\frac{\mu}{\mu+s}\right)^l \mathbf{e}_K = \mathbf{0}_K^T, \quad l = \overline{0, i}, \ i = \overline{0, N}. \quad (3)$$

Here it is assumed that $\hat{Q}_{0,-1} = 0$, $\hat{Q}_{N,N+1} = 0$ and

$$\hat{Q}_{i,i} = A_1 - \lambda \hat{I}_K - \mu_i I_K, i = \overline{0, N-1}, \ \hat{Q}_{N,N} = A_1 + (C_{K-1} + I_K)\gamma^+ - \mu_N I_K,$$

$$\hat{Q}_{i,i+1} = \gamma^+ C_{K-1} + \lambda E_K^+, \ i = \overline{0, N-1}, \ \hat{Q}_{i,i-1} = \mu_i I_K, \ i = \overline{1, N}.$$

Now, let us introduce column vectors $\mathbf{v}(s, i) = (\mathbf{v}(s, i, 0), \ldots, \mathbf{v}(s, i, i))^T$ of dimension $K(i+1)$, $i = \overline{0, N}$, and column vector $\mathbf{v}(s) = (\mathbf{v}(s, 0), \ldots, \mathbf{v}(s, N))^T$ and introduce the following notation:

$$\Omega_{i,i}(s) = -I_{i+1} \otimes (sI - \hat{Q}_{i,i}), \ \Omega_{i,i+1} = D_1^{(i)} \otimes \hat{Q}_{i,i+1} + D_2^{(i)} \otimes \gamma^+ I_K,$$

$$\Omega_{i,i-1} = D_3^{(i)} \otimes \hat{Q}_{i,i-1},$$

$$\boldsymbol{\beta}_i(s) = \gamma^- (1, \frac{\mu}{\mu+s}, \ldots, \left(\frac{\mu}{\mu+s}\right)^i)^T \otimes \mathbf{e}_K, \ i = \overline{0, N},$$

The matrix $\boldsymbol{\Omega}(s)$ is the block three-diagonal matrix with non-zero blocks $\Omega_{i,j}(s)$, $j = \max\{0, i-1\}$, i, $\min\{i+1, N\}$, $i = \overline{0, N}$, vector $\boldsymbol{\beta}(s)$ is defined by $\boldsymbol{\beta}(s) = (\boldsymbol{\beta}_0(s), \ldots, \boldsymbol{\beta}_N(s))^T$.

Here, the matrix $D_1^{(i)}$ of dimension $(i+1) \times (i+2)$ is obtained from identity matrix I_{i+1} by means of supplementing it from the right with a column $\mathbf{0}_{i+1}^T$. The matrix $D_2^{(i)}$ of the same dimension has all entries equal to 0 except the entries which are located in the last column and are equal to 1. The matrix $D_3^{(i)}$ of dimension $(i+1) \times i$ is obtained from identity matrix I_i by means of supplementing it from above with a row $\{1, 0, \ldots, 0\}$.

Using this notation, we rewrite the system (3) to the form

$$\Omega_{i,i}(s)\mathbf{v}(s,i)+\Omega_{i,i+1}\mathbf{v}(s,i+1)+\Omega_{i,i-1}\mathbf{v}(s,i-1)+\boldsymbol{\beta}_i(s) = \mathbf{0}^T_{K(i+1)}, \; i = \overline{0,N}, \; (4)$$

and then into the form

$$\boldsymbol{\Omega}(s)\mathbf{v}(s) + \boldsymbol{\beta}(s) = \mathbf{0}^T_{\frac{K(N+1)(N+2)}{2}} . \tag{5}$$

Thus, we have proven the following statement.

Lemma 2. *Vector* $\mathbf{v}(s)$ *consisting of the conditional Laplace-Stieltjes Transforms* $v(s,i,l,k)$, $l = \overline{0,i}$, $i = \overline{0,N}$, $k = \overline{1,K}$, *is computed by*

$$\mathbf{v}(s) = -\boldsymbol{\Omega}^{-1}(s)\boldsymbol{\beta}(s). \tag{6}$$

Inverse matrix in (6) exists for any s, $Re\, s > 0$ and $s = 0$ because the diagonal entries of the matrix $\boldsymbol{\Omega}(s)$ dominate in rows of this matrix, see, e.g., [3].

Theorem 3. *The LST* $v_b(s)$ *of an arbitrary batch sojourn time is computed by formula (1) where the LSTs* $v(s,i,l,k)$, $l = \overline{0,i}$, $i = \overline{0,N}$, $k = \overline{1,K}$, *are defined as components of the vector* $\mathbf{v}(s)$ *having form (6).*

Remark 1. The LST $v_b(s)$ is the LST of an arbitrary batch sojourn time, including the batches which are rejected. The LST $v_b^{(accept)}(s)$ of an arbitrary accepted batch sojourn time is calculated by

$$v_b^{(accept)}(s) = \frac{\displaystyle\sum_{i=0}^{N-1}\sum_{k=0}^{K-1} \pi(i,k)v(s,i+1,i+1,k+1)}{1 - P_b^{(loss)}}.$$

Corollary 2. *The mean sojourn time* V_b *of an arbitrary batch is computed by*

$$V_b = -\sum_{i=0}^{N-1}\sum_{k=0}^{K-1} \pi(i,k)\frac{\partial v(s,i+1,i+1,k+1)}{\partial s}\Big|_{s=0}$$

where the values $\frac{\partial v(s,i+1,i+1,k+1)}{\partial s}\big|_{s=0}$ *are computed as the corresponding entries of the vector* $\frac{d\mathbf{v}(s)}{ds}\big|_{s=0}$, *which is calculated by*

$$\frac{d\mathbf{v}(s)}{ds}\Big|_{s=0} = \boldsymbol{\Omega}^{-1}(0)\left(-\frac{d\boldsymbol{\beta}(s)}{ds}\Big|_{s=0} + \mathbf{e}\right),$$

where

$$\frac{d\boldsymbol{\beta}(s)}{ds}\Big|_{s=0} = -\gamma^-\left(0,0,\frac{1}{\mu},0,\frac{1}{\mu},\frac{2}{\mu},\ldots,0,\frac{1}{\mu},\ldots,\frac{N}{\mu}\right) \otimes \mathbf{e}_K.$$

Corollary 3. *The mean sojourn time* $V_b^{(accept)}$ *of an arbitrary accepted batch is computed by*

$$V_b^{(accept)} = \frac{V_b}{1 - P_b^{(loss)}}.$$

5   Optimization Problem and Numerical Example

It is obvious, that the most important characteristic of the considered model is throughput T of the system because it defines its profit earned by information transmission. However, loss probability $P_c^{(loss)}$ of a customer is also very important. If the number K, that restricts the number of batches, which can be served in the system simultaneously, is increasing, definitely, the throughput T of the system increases and batch loss probability $P_b^{(loss)}$ decreases. But, at the same time loss probability $P_c^{(loss)}$ of a customer increases, what probably implies charge of the system for unsatisfactory Quality of Service. So, the system manager should decide how many batches can be allowed to enter the system simultaneously to fit requirements of Quality of Service and to reach the maximally possible throughput.

Thus, he should solve the following non-trivial optimization problem:

$$T = T(K) \to \max \qquad (7)$$

subject to constraint

$$P_c^{(loss)} = P_c^{(loss)}(K) \leq \varepsilon, \qquad (8)$$

where ε is admissible probability of a customer loss in the system. The value of ε depends on the tolerance of the system to partial packets loss.

In the numerical example, we fix the following parameters of the system:

$$\lambda = 1, \ \mu = 2, \ \gamma = 2, \ R = 3, \ \varepsilon = 0.015.$$

Buffer capacity $N - R$ varies in the interval $[0, 50]$.

We consider the following three values of the parameter θ: 0.4, 0.8, 0.9. Average number of customers in a batch is calculated by $\frac{1}{1-\theta}$ and is equal to 1.5, 5, 10, correspondingly.

Figure 1 gives the dependence of the optimal in the problem (7),(8) value K^* of the threshold K on the buffer capacity N. Figure 2 shows the dependence of the optimal value T^* of throughput T on the buffer capacity N.

Line number 1 corresponds to the value $\theta = 0.4$. Under this value of θ, constraint (8) is not fulfilled if $N = 3$. For $N = 4$, the value of K^* is equal to 1, loss probability $P_c^{(loss)}$ is equal to 0.0023, throughput T^* is equal to 0.908. For $N = 5$, the value of K^* is equal to 2, loss probability $P_c^{(loss)}$ is equal to 0.0051, throughput T^* is equal to 1.397. For $N > 5$, loss probability $P_c^{(loss)}$ is less than the threshold value ε for all values of K. Throughput T^* practically is not changed with increasing the value of K. Load of the system is small and, for $N > 5$, it does not make sense to restrict the number of simultaneously accepted batches.

Thus, line number 1 in figure 1 consists only of two dots surrounded with circle.

Line number 2 corresponds to the value $\theta = 0.8$, and line number 3 corresponds to the value $\theta = 0.9$.

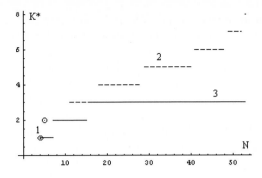

Fig. 1. Dependence of the maximal admissible number K^* of batches in the system on the system capacity N

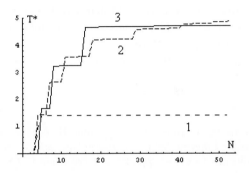

Fig. 2. Dependence of the maximal throughput T^* on the system capacity N

Behavior of loss probability $P_c^{(loss)}$ and throughput T for $\theta = 0.8$, $N = 30$ and different values of K is illustrated by the following numbers:

$K = 1$, $P_c^{(loss)} = 6.65 \times 10^{-18}$, $T = 1.43$; $K = 2$, $P_c^{(loss)} = 1.08 \times 10^{-9}$, $T = 2.64$;

$K = 3$, $P_c^{(loss)} = 1.46 \times 10^{-5}$, $T = 3.59$; $K = 4$, $P_c^{(loss)} = 1.71 \times 10^{-3}$, $T = 4.24$;

$K = 5$, $P_c^{(loss)} = 0.0122$, $T = 4.5841$; $K = 6$, $P_c^{(loss)} = 0.0254$, $T = 4.7037$.

One can see that loss probability is very sensitive with respect to the value of K. So, results of our present paper can be valuable for calculation of the optimal, under the given restriction on QoS, number K of batches (or flows in terminology of [1]) which can be admitted into the system simultaneously.

6   Conclusion

In this paper, novel queueing model with batch arrivals distributed in time is introduced and analyzed. Joint distribution of the number of customers in the system and number of currently admitted batches is computed. Sojourn time distribution of an arbitrary customer is given in terms of the Laplace-Stieltjes Transform. Sojourn time distribution of an arbitrary batch is computed for the single-server system. Problem of the optimal batch admission strategy is discussed. Value of the presented results is illustrated numerically.

Results are planned to be extended to the systems with infinite buffer, more general batch arrival process (e.g., to MAP - Markovian Arrival Process), interarrival times of customers of a batch and service time distributions (e.g., to PH - Phase type distribution), possibility of standard batch customer arrival within an admitted batch, arbitrary distribution of the number of customers in a batch, service intensity depending on the number of customers in the system, etc.

Acknowledgments

This research was supported in part by Ministry of Information and Communication (MIC) Korea, under the IT Foreign Specialist Inviting Program (ITSIP), ITSOC, International Cooperative Research by Ministry of Science and Technology, KOTEF, Chonbuk National University and 2nd stage Brain Korea 21.

References

1. Kist A.A., Lloyd-Smith B., Harris R.J.: A simple IP flow blocking model. Performance Challenges for Efficient Next Generation Networks. In: Proceedings of 19-th International Teletraffic Congress, pp. 355-364 (29 August -2 September 2005) Beijing (2005)
2. Klimenok, V., Kim, C.S., Orlovsky, D., Dudin, A.: Lack of invariant property of Erlang loss model in case of the MAP input. Queueing Systems 49, 187–213 (2005)
3. Gantmakher, F.R.: The Matrix Theory. Moscow, Science (1967)

A Performance Analysis of Tandem Networks with Markovian Sources

X. Song and M. Mehmet Ali

Dept. of Electrical & Comp. Eng., Concordia University
1455 de Maisonneuve Blvd. West, Montreal, Quebec H3G 1M8, Canada
{x_song, mustafa}@ece.concordia.ca

Abstract. In this paper, we present an exact performance analysis of a tandem network with arbitrary number of multiplexers. Each multiplexer is fed by the output of the preceding multiplexer as well as the traffic generated by a number of independent binary Markovian sources. We model the network as a discrete-time queueing system. After determining the unknown boundary function, the probability generating function (PGF) of the distribution of queue length and number of *On* sources for each multiplexer is derived; from these results expressions for the mean and variance of queue length and packet delay have been determined.

Keywords: Tandem networks, multiplexers, Markovian sources, queue length, PGF, packet delay, mean, variance.

1 Introduction

Broadband networks are expected to integrate the transmission of voice, video and data in a single network. The Internet is also expected to transport both real and non real-time traffic; however, various services differ in their QoS requirements. The Internet is based on the packet-switching technology that allows dynamic sharing of the bandwidth among different flows in the network. The packets are stored and forwarded from one node to another with each node making its routing decision until packets reach their destinations. The Internet has been using two transport protocols, TCP and UDP, for information transfer. TCP provides reliable connection-oriented communications with end-to-end control algorithm, while UDP provides a connectionless service between applications. TCP is the choice of transport protocol for loss sensitive and delay tolerant data traffic, while UDP is suitable for delay sensitive and loss tolerant real-time traffic such as voice and video. The network gives priority to UDP over TCP traffic, and the results of this paper will be applicable to the former since it does not have end to end controls.

As a packet is transported in the network, it may be considered as going through a number of multiplexers in tandem. These multiplexers take place at the entry, interior and at the exit of the network. As the packets travel through the network, the statistical properties of the traffic change. For example, the traffic becomes smoother and the long-range dependence of the traffic dissipates due to the statistical

L. Mason, T. Drwiega, and J. Yan (Eds.): ITC 2007, LNCS 4516, pp. 731–742, 2007.

multiplexing gains [1]. Thus the performance analysis of the traffic at the network level is very important.

Due to the lack of exact methods, most of the previous work in the analysis of tandem networks has either focused on simulation experiments or on approximate models. A four node ATM tandem queueing network is considered in [2]. In that work, the output of each node is approximated by a renewal process which is fed to the next node. However, the renewal approximation for the nodal departure process is hard to justify, as correlation is inherent in the output process of each node. The correlation has significant effect on queueing behavior of the downstream nodes.

Next, we mention two works that use exact methods for analysis [3, 4]. In [3], an analysis of a two-node tandem queue network is presented, where the number of external arrivals to the two queues is modulated by a single two-state Markov Chain. An expression for the PGF of the queue length distribution is derived. But the single two-state Markov Chain modeling of external arrivals is restrictive for the network traffic. In [4], an analysis of a two-node tandem network with correlated arrivals has been presented. The expression for mean queue length is given without determining the joint PGF of queue lengths. In the present paper, we consider the performance modeling of n multiplexers in tandem at the entrance of the network. Since these multiplexers are aggregating the traffic, the entire output of each multiplexer is fed to the next one together with the new arrivals. We derive the PGF of the queue length of the n'th multiplexer and determine the mean and variance of the queue length and mean packet delay. The delay result is also applicable to messages with arbitrary number of packets.

The remainder of the paper is organized as follows. The next section presents the analytical model, the section following that is the performance analysis, afterwards the numerical results are given and finally the conclusion of the paper.

2 Analytical Model

We consider a network with n ($n>1$) multiplexers in tandem as shown in Fig. 1 and model the network as a discrete-time queueing system. Each multiplexer is fed by the entire output of the previous multiplexer as well as the traffic generated by a number of Markovian sources. We assume that each multiplexer has an infinite buffer to store packets. The time axis is divided into intervals of equal lengths (slots) and a packet is transmitted at the slot boundaries. It is assumed that a packet cannot be transmitted during the slot that it arrives, and that a packet transmission time is equal to one slot.

The external packet arrivals to the multiplexer i ($1 \le i \le n$) are generated by type i sources. Each type of sources consists of independent binary Markov sources alternating between *On* and *Off* states as shown in Fig. 2. For type i sources, transition from idle to active state occurs with probability $(1-\beta_i)$, while from active to idle state occurs with probability $(1-\alpha_i)$. We assume that an active source generates at least one packet during a slot, while an idle source generates no packets.

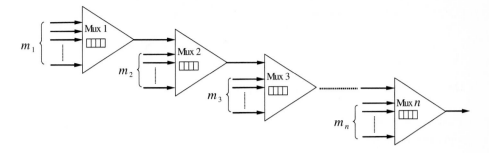

Fig. 1. Tandem Network Model

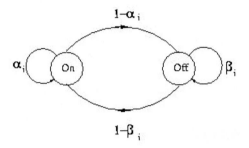

Fig. 2. Source Model for Type-i Sources

Now let us introduce the following notation,

m_i = number of type i sources.

$\ell_{i,k}$ = queue length of multiplexer i at the end of slot k

$a_{i,k}$ = number of type i *On*-sources during slot k.

$f_{ji,k}$ = number of packets generated by the j'th *On*-source of type i during slot k. $f_{ji,k}$ are independent identically distributed (i.i.d.) from slot to slot for type i sources and it has the PGF $f_i(z_i)$.

$b_{i,k}$ = the total number of packets generated by type i sources during slot k.

c_{ji} = a variable that assumes the values of (1, 0) if j'th source from type i is in *On* and *Off* states in the next slot respectively, given that this source is *On* in the present slot.

d_{ji} = a variable that assumes the values of (1, 0) if j'th source from type i is in *On* and *Off* states in the next slot respectively, given that this source is *Off* in the present slot.

The c_{ji}, d_{ji} are i.i.d. Bernoulli random variables with the corresponding PGFs,

$$c_i(z_i) = 1 - \alpha_i + \alpha_i z_i \,, \qquad d_i(z_i) = \beta_i + (1 - \beta_i) z_i \tag{1}$$

From the above definitions, we have

$$b_{i,k} = \sum_{j=1}^{a_{i,k}} f_{ji,k} , \qquad\qquad a_{i,k+1} = \sum_{j=1}^{a_{i,k}} c_{ji} + \sum_{j=1}^{m_i - a_{i,k}} d_{ji} \qquad\qquad (2)$$

The evolution of the first queue length is given by,

$$\ell_{1,k+1} = (\ell_{1,k} - 1)^+ + b_{1,k+1} \qquad\qquad (3)$$

And the evolution of the i'th queue length ($i > 1$) is given by

$$\ell_{i,k+1} = (\ell_{i,k} - 1)^+ + b_{i,k+1} + u_{i,k} \quad , \quad 2 \le i \le n \qquad\qquad (4)$$

where, $u_{i,k}$ is a random variable depending on whether the previous queue is empty or not,

$$u_{i,k} = \begin{cases} 1 & \text{if } \ell_{i-1,k} > 0 \\ 0 & \text{if } \ell_{i-1,k} = 0 \end{cases} \qquad\qquad (5)$$

In the above equations (3, 4) the notation $(x)^+$ denotes $max(x, 0)$.

3 Performance Analysis

The objective of this analysis is to determine the joint PGF of the queue length and the number of *On*-sources for the n'th multiplexer. First, we will derive a functional equation that relates the joint PGF at two consecutive imbedded points. Then, the functional equation will be transformed to a new form that is mathematically more tractable. Afterwards the unknown boundary function will be derived, which will determine the joint PGF for the n'th multiplexer. The effect of the preceding multiplexers on the n'th multiplexer, has been summarized in the output of $(n-1)$'st multiplexer. Thus, in determining the performance of the n'th multiplexer, we only need to consider the joint performance of the $(n-1)$'st and n'th multiplexers. Since the behavior of a multiplexer is not affected by the downstream ones, the following analysis applies to any multiplexer in the tandem network.

The system that consists of $(n-1)$'st and n'th multiplexers can be modeled using a discrete-time four-dimensional Markov chain with the following variables, $(\ell_{n-1,k}, a_{n-1,k}, \ell_{n,k}, a_{n,k})$. Let $Q_k(z_{n-1}, y_{n-1}, z_n, y_n)$ denote the joint PGF of this system, then we have,

$$Q_k(z_{n-1}, y_{n-1}, z_n, y_n) = E\left[z_{n-1}^{\ell_{n-1,k}} y_{n-1}^{a_{n-1,k}} z_n^{\ell_{n,k}} y_n^{a_{n,k}} \right]$$

$$= \sum_{i_{n-1}=0}^{\infty} \sum_{j_{n-1}=0}^{m_{n-1}} \sum_{i_n=0}^{\infty} \sum_{j_n=0}^{m_n} z_{n-1}^{i_{n-1}} y_{n-1}^{j_{n-1}} z_n^{i_n} y_n^{j_n} q_k(i_{n-1}, j_{n-1}, i_n, j_n) \quad , \quad n \ge 2$$

where, $q_k(i_{n-1}, j_{n-1}, i_n, j_n) = \text{Prob}(\ell_{n-1,k} = i_{n-1}, a_{n-1,k} = j_{n-1}, \ell_{n,k} = i_n, a_{n,k} = j_n)$,

Following a derivation similar to that in [5], the joint PGF for two consecutive points may be related as,

$$
Q_{k+1}\left(z_{n-1}, y_{n-1}, z_n, y_n\right)
$$

$$
= B(1)\left\{\frac{1}{z_{n-1}}Q_k\left(z_{n-1}, Y_{n-1}, z_n, Y_n\right) + \frac{z_n - 1}{z_{n-1}}Q_k\left(z_{n-1}, Y_{n-1}, 0_n, 0_n\right)\right.
$$

$$
\left. + \frac{z_{n-1} - z_n}{z_{n-1}z_n}Q_k\left(0_{n-1}, 0_{n-1}, z_n, Y_n\right) + \frac{(z_n - 1)(z_{n-1} - z_n)}{z_{n-1}z_n}Q_k\left(0_{n-1}, 0_{n-1}, 0_n, 0_n\right)\right\}
$$

$$
, \quad k \geq 0, \; n \geq 2 \tag{6}
$$

where, $Y_i = \dfrac{c_i\left(y_i f_i\left(z_i\right)\right)}{d_i\left(y_i f_i\left(z_i\right)\right)}$, $B_i(1) = \left[d_i\left(y_i f_i\left(z_i\right)\right)\right]^{m_i}$, $B(1) = \prod_{i=n-1}^{n} B_i(1)$ and 0_ℓ, 1_ℓ

correspond to $y_\ell = z_\ell = 0$ and 1 respectively.

Defining $Q\left(z_{n-1}, y_{n-1}, z_n, y_n\right) = \lim_{k \to \infty} Q_k\left(z_{n-1}, y_{n-1}, z_n, y_n\right)$, then, the above equation in equilibrium is given by,

$$
Q\left(z_{n-1}, y_{n-1}, z_n, y_n\right)
$$

$$
= B(1)\left\{\frac{1}{z_{n-1}}Q\left(z_{n-1}, Y_{n-1}, z_n, Y_n\right) + \frac{z_n - 1}{z_{n-1}}Q\left(z_{n-1}, Y_{n-1}, 0_n, 0_n\right)\right.
$$

$$
\left. + \frac{z_{n-1} - z_n}{z_{n-1}z_n}Q\left(0_{n-1}, 0_{n-1}, z_n, Y_n\right) + \frac{(z_n - 1)(z_{n-1} - z_n)}{z_{n-1}z_n}Q\left(0_{n-1}, 0_{n-1}, 0_n, 0_n\right)\right\}
$$

$$
, \quad k \geq 0, \; n \geq 2 \tag{7}
$$

Unfortunately, this equation is not immediately solvable due to the presence of $Q(z_{n-1}, Y_{n-1}, z_n, Y_n)$ on the RHS. From (6) the joint PGF of the n'th multiplexer is given by,

$$
Q_{k+1}\left(1_{n-1}, 1_{n-1}, z_n, y_n\right)
$$

$$
= B_n(1)\left\{Q_k\left(1_{n-1}, 1_{n-1}, z_n, Y_n\right) + (z_n - 1)Q_k\left(1_{n-1}, 1_{n-1}, 0_n, 0_n\right)\right. \tag{8}
$$

$$
\left. + \frac{1 - z_n}{z_n}Q_k\left(0_{n-1}, 0_{n-1}, z_n, Y_n\right) - \frac{(z_n - 1)^2}{z_n}Q_k\left(0_{n-1}, 0_{n-1}, 0_n, 0_n\right)\right\}, \quad k \geq 0, n \geq 2
$$

When $n = 1$, the corresponding PGF from [6] is given by,

$$
Q_k(z_1, y_1) = \frac{1}{z_1^{k-1}}B_1(k) + (z_1 - 1)\sum_{h=1}^{k-1}\frac{B_1(k-h)}{z_1^{k-h}}Q_h(0,0), \quad n = 1 \tag{9}
$$

where $Q_h(0,0)$ is the probability that the system is empty at the end of h'th slot. We note that in (8), $Q_k(0_{n-1}, 0_{n-1}, z_n, Y_n)$ is the unknown boundary function.

3.1 Transforming the Functional Equation into a New Form

Next, we will transform the functional equation given in (8) into a mathematically more tractable form. First, we will give a number of preliminary results, let us define,

$$X_i(k+1) = X_i(1)[X_i(k)|_{y_i=Y_i}] \text{ with } X_i(0)=1, \ X_i(1)=\beta_i+(1-\beta_i)y_if_i(z_i) \tag{10}$$

$$U_i(k+1) = X_i(1)[U_i(k)|_{y_i=Y_i}] \text{ with } U_i(0)=y_i, \ U_i(1)=1-\alpha_i+\alpha_iy_if_i(z_i) \tag{11}$$

$X_i(k)$ and $U_i(k)$ defined in (10, 11) have the recurrence relationships given below,

$$X_i(k)=\xi_iX_i(k-1)+\sigma_iX_i(k-2), \ U_i(k)=\xi_iU_i(k-1)+\sigma_iU_i(k-2), \ k\geq 2 \tag{12}$$

where $\xi_i = \beta_i + \alpha_i f_i(z_i)$, $\sigma_i = (1-\alpha_i-\beta_i)f_i(z_i)$
Let us define,

$$\phi_i(k+1)=\phi_i(k)|_{y_i=Y_i} \text{ with } \phi_i(0)=y_i \tag{13}$$

$$B_i(k)=[X_i(k)]^{m_i}, \ B_i^j(k)=B_i(k)|_{y_i=\phi_i(j)}, \ n\geq 0 \tag{14}$$

Then, we have the relation

$$B_i(k+j)=B_i(j)B_i^j(k) \text{ with } B_i^0(k)=B_i(k) \tag{15}$$

The solution of equations in (12) may be given as,

$$X_i(k)=C_{1i}\lambda_{1i}^k+C_{2i}\lambda_{2i}^k, \qquad U_i(k)=D_{1i}\lambda_{1i}^k+D_{2i}\lambda_{2i}^k \tag{16}$$

where,

$$C_{1i,2i}=\frac{1}{2}\mp\frac{2(y_i-y_i\beta_i-\alpha_i)f_i(z_i)+\xi_i}{2\sqrt{\Delta_i}}, \ D_{1i,2i}=\frac{y_i}{2}\mp\frac{2(1-\alpha_i+\alpha_iy_if_i(z_i))-\xi_iy_i}{2\sqrt{\Delta_i}} \tag{17}$$

and $\lambda_{1i,2i}=\frac{1}{2}(\xi_i\mp\sqrt{\Delta_i})$ with $\Delta_i=\xi_i^2+4\sigma_i$ \tag{18}

From the preliminary results, the functional equation in (8) may be expressed through induction as follows,

$$Q_k(1_{n-1},1_{n-1},z_n,y_n)$$

$$= B_n(k)+(z_n-1)\sum_{j=1}^{k-1}B_n(j)Q_{k-j}(1_{n-1},1_{n-1},0,0)$$

$$+\frac{1-z_n}{z_n}\sum_{j=1}^{k-1}B_n(j)Q_{k-j}(0_{n-1},0_{n-1},z_n,\phi_n(j))$$

$$-\frac{(z_n-1)^2}{z_n}\sum_{j=1}^{k-1}B_n(j)Q_{k-j}(0_{n-1},0_{n-1},0_n,0_n), \qquad k\geq 1, n\geq 2 \tag{19}$$

In the above, we assume that if the upper limit of a summation is less than its lower limit, then that summation is empty.

3.2 Solution of the Functional Equation for Two Multiplexers in Tandem

First, we consider the solution of the functional equation for two multiplexers in tandem. Substituting $n = 2$ in (19), it may be seen that the solution contains the unknown boundary function, $Q_k(0,0,z_2,y_2)$. From (6), $Q_k(0,0,z_2,y_2)$ may be expressed in terms of the other boundary function $Q_k(z_1,y_1,0,0)$. The latter has been derived in [4], which is given below,

$$Q_k(z_1,y_1,0,0) = B_1(1)Q_k(1,1,0,0) \quad , \quad k \geq 1$$

Thus from here the unknown boundary function may be determined, which, after some algebra, leads to the determination of the PGF of the second multiplexer. Unfortunately, this solution does not extend to higher number of multiplexers in tandem, because $Q_k(z_{n-1},y_{n-1},0,0)$ cannot be determined for values of $n > 2$.

3.3 Solution of the Functional Equation for the General Case

Next, we present a new method to solve the general case. First, we provide the following motivation for this method through the PGF $Q_k(z_1,y_1)$ given in (9). The terms of the summation on the RHS of (9) may be considered as PGFs conditioned on mutually exclusive events. These events correspond to the last time that the multiplexer queue was empty which may be at the end of any of the k slots. Assuming that the last time this event occurred at the end of h'th slot, then, its probability is given by $Q_h(0,0)$. Since, when the queue length is zero, all the sources must be in the *Off* state; then, $B_1(k-h)$ gives PGF of the number of packet arrivals from the last time the multiplexer queue was empty. z_1^{k-h} in the denominator corresponds to the PGF of the number of packet departures during the $(k-h)$ slots that the multiplexer was busy. We note that the first term on the RHS corresponds to the event that the last time the queue length was zero at the initial state. Finally, this interpretation also applies for the multiplexers fed by multiple type of sources [6].

Now, let us determine the unknown boundary function in (19). Following [5], we assume that a busy period begins and ends with idle slots and that two consecutive idle slots correspond to a busy period with zero duration. Further, consecutive busy periods are separated by an idle slot. Since during an idle slot all the sources are in the *Off* state, the busy periods are independent and identically distributed. Let us define,

$\xi_n(j) = \text{Prob}(n\text{'th multiplexer has a busy period of } j \text{ slots}), j=0, 1, 2,...$

$\varphi_r^{(n)}(\ell) = \text{Prob}(n\text{'th multiplexer has } r \text{ busy periods during an interval of } \ell \text{ slots}).$

Following the interpretation given to (9) in the above, we may write the unknown boundary function as below,

$$Q_k(0_{n-1},0_{n-1},z_n,y_n)$$

$$= \sum_{r=1}^{k} \frac{1}{z_n^{r-1}} \varphi_r^{(n-1)}(k)B_n(k) \tag{20}$$

$$+ (z_n-1)\sum_{h=1}^{k-1}\sum_{r=1}^{k-h+1} \frac{1}{z_n^{r-1}} \varphi_r^{(n-1)}(k-h+1)B_n(k-h)Q_h(1_{n-1},1_{n-1},0_n,0_n), \quad k \geq 1$$

We note that given that the n'th multiplexer is empty at the end of h'th slot, $(n-1)$'st multiplexer would have been empty in the previous slot. Thus in the above, $\varphi_r^{(n-1)}(k-h+1)$ corresponds to the probability that $(n-1)$'st multiplexer will have r complete busy periods during the $(k-h+1)$ slots. Since, busy periods are separated by an idle slot, $(n-1)$'st multiplexer will not generate packets at its output during r of these slots. The PGFs of the packets received and transmitted by the n'th multiplexer during the $(k-h)$ slots are given by $B_n(k-h)z_n^{k-h-r+1}$ and z_n^{k-h} respectively. Later on, we will indicate an indirect proof for the assumed form of the boundary function in (20).

Let us define,

$$\varphi^{(n-1)}(\ell) = \sum_{r=1}^{\ell} \frac{1}{z_n^r} \varphi_r^{(n-1)}(\ell) \qquad \text{where } \varphi^{(n-1)}(0)=0 \tag{21}$$

We note that $\varphi^{(n-1)}(\ell)\big|_{z_n=1}$ corresponds to the probability that $(n-1)$'st multiplexer has integer number of busy periods during an interval of ℓ slots. From this observation, $\varphi^{(n-1)}(\ell)$ may be expressed recursively as follows,

$$\varphi^{(n-1)}(\ell) = \frac{1}{z_{n-1}} \sum_{j=0}^{\ell-1} \xi_{n-1}(\ell-j-1)\varphi^{(n-1)}(j) \tag{22}$$

Next defining the following transform,

$$\Phi^{(n-1)}(\omega) = \sum_{\ell=1}^{\infty} \varphi^{(n-1)}(\ell)\omega^\ell \tag{23}$$

then we obtain,

$$\Phi^{(n-1)}(\omega) = \frac{\omega\Gamma_{n-1}(\omega)}{z_{n-1} - \omega\Gamma_{n-1}(\omega)} \tag{24}$$

where,

$$\Gamma_{n-1}(\omega) = \sum_{j=0}^{\infty} \xi_{n-1}(j)\omega^j \tag{25}$$

is the PGF of the busy period of $(n-1)$'st multiplexer.
Let us define the following transform,

$$Q(1_{n-1},1_{n-1},z_n,y_n,\omega) = \sum_{k=0}^{\infty} Q_k(1_{n-1},1_{n-1},z_n,y_n)\omega^k \tag{26}$$

Substituting (20) in (19) and then taking its transform w.r.t. discrete-time k, after some algebraic manipulations, we have,

$$
\begin{aligned}
&Q(1_{n-1},1_{n-1},z_n,y_n,\omega) \\
&= 1 + \sum_{i=0}^{m_n} \binom{m_n}{i} \frac{(C_{1n}\lambda_{1n})^i (C_{2n}\lambda_{2n})^{m_n-i}\omega}{1-\lambda_{1n}^i\lambda_{2n}^{m_n-i}\omega} \\
&+ (z_n-1)[Q(1_{n-1},1_{n-1},0_n,0_n,\omega)-1]\sum_{i=0}^{m_n} \binom{m_n}{i} \frac{(C_{1n}\lambda_{1n})^i (C_{2n}\lambda_{2n})^{m_n-i}\omega}{1-\lambda_{1n}^i\lambda_{2n}^{m_n-i}\omega} \\
&- (z_n-1)\sum_{i=0}^{m_n} \binom{m_n}{i} \frac{(C_{1n}\lambda_{1n})^i (C_{2n}\lambda_{2n})^{m_n-i}\lambda_{1n}^i\lambda_{2n}^{m_n-i}\omega^2\Gamma_{n-1}(\lambda_{1n}^i\lambda_{2n}^{m_n-i}\omega)}{(1-\lambda_{1n}^i\lambda_{2n}^{m_n-i}\omega)[z_n-\lambda_{1n}^i\lambda_{2n}^{m_n-i}\omega\Gamma_{n-1}(\lambda_{1n}^i\lambda_{2n}^{m_n-i}\omega)]} \\
&- (z_n-1)^2[Q(1_{n-1},1_{n-1},0_n,0_n,\omega)-1] \\
&\quad * \sum_{i=0}^{m_n} \binom{m_n}{i} \frac{(C_{1n}\lambda_{1n})^i (C_{2n}\lambda_{2n})^{m_n-i}\omega\Gamma_{n-1}(\lambda_{1n}^i\lambda_{2n}^{m_n-i}\omega)}{(1-\lambda_{1n}^i\lambda_{2n}^{m_n-i}\omega)[z_n-\lambda_{1n}^i\lambda_{2n}^{m_n-i}\omega\Gamma_{n-1}(\lambda_{1n}^i\lambda_{2n}^{m_n-i}\omega)]}
\end{aligned}
\tag{27}
$$

$Q(1_{n-1},1_{n-1},0_n,0_n,\omega)$ in the above equation may be determined by invoking the analytical property of the function $Q(1_{n-1},1_{n-1},z_n,y_n,\omega)$ inside the poly disk $(|z_n|\le 1; |y_n|\le 1; |\omega|<1)$ and through the application of Rouche's theorem. The result is given below,

$$
Q(1_{n-1},1_{n-1},0_n,0_n,\omega) = 1 + \frac{\lambda_{2n}^{m_n}\omega\big|_{z_n=z_n^*(\omega)}}{1-z_n^*(\omega)}
\tag{28}
$$

where $z_n^*(\omega)$ is the unique root of equation,

$$
z_n^*(\omega) - \lambda_{2n}^{m_n}\omega\Gamma_{n-1}(\lambda_{2n}^{m_n}\omega)\big|_{z_n=z_n^*(\omega)} = 0
$$

Next, application of the final value theorem gives the steady-state PGF of the n'th multiplexer,

$$
\begin{aligned}
&Q(1_{n-1},1_{n-1},z_n,y_n) = \lim_{\omega\to 1}(1-\omega)Q(1_{n-1},1_{n-1},z_n,y_n,\omega) \\
&= (z_n-1)\left(1-\sum_{i=1}^{n}\rho_i\right)\sum_{i=0}^{m_n}\binom{m_n}{i}\frac{(C_{1n}\lambda_{1n})^i (C_{2n}\lambda_{2n})^{m_n-i}}{1-\lambda_{1n}^i\lambda_{2n}^{m_n-i}} \\
&- (z_n-1)^2\left(1-\sum_{i=1}^{n}\rho_i\right)\sum_{i=0}^{m_n}\binom{m_n}{i}\frac{(C_{1n}\lambda_{1n})^i (C_{2n}\lambda_{2n})^{m_n-i}\Gamma_{n-1}(\lambda_{1n}^i\lambda_{2n}^{m_n-i})}{(1-\lambda_{1n}^i\lambda_{2n}^{m_n-i})[z_n-\lambda_{1n}^i\lambda_{2n}^{m_n-i}\Gamma_{n-1}(\lambda_{1n}^i\lambda_{2n}^{m_n-i})]}
\end{aligned}
\tag{29}
$$

We note that this solution for $n = 2$ is same as the one obtained through a different approach in section 3.2. Further, this solution may be shown to satisfy equation (7) that describes the system in equilibrium. Since, the system is modeled as a Markov chain, then, it has a unique solution and therefore (29) must be the correct solution. This also indirectly confirms the correctness of the boundary function given in (20) through observation.

From [5] and (28), the unknown PGF of the busy period of $(n-1)$'st multiplexer, $\Gamma_{n-1}(\omega)$, is given below,

$$\Gamma_{n-1}(\omega) = \frac{1}{\omega}\left(1 - \frac{1}{Q(1_{n-2}, 1_{n-2}, 0_{n-1}, 0_{n-1}, \omega)}\right) = \frac{\left.\lambda_{2(n-1)}^{m_{n-1}}\right|_{z_n = z_n^*(\omega)}}{1 - z_{n-1}^*(\omega) + \left.\lambda_{2(n-1)}^{m_n}\omega\right|_{z_{n-1} = z_{n-1}^*(\omega)}} \tag{30}$$

3.4 Performance Measures of the n'th Multiplexer

In the next, we will determine the mean queue length and packet delay of the n'th multiplexer. Substituting $y_n = 1$ in (29) gives the PGF of the queue length, as follows,

$$P_n(z_n) = (z_n - 1)\left(1 - \sum_{i=1}^{n}\rho_i\right)\left[E_n(z_n) + \frac{G_n(z_n)}{1 - H_n(z_n)}\right]$$
$$- (z_3 - 1)^2\left(1 - \sum_{i=1}^{n}\rho_i\right)\left[F_n(z_n) + \frac{G_n(z_n)\Gamma_{n-1}(H_n(z_n))}{[1 - H_n(z_n)][z_n - \Theta_n(z_n)]}\right] \tag{31}$$

where

$$E_n(z_n) = \sum_{i=1}^{m_n}\binom{m_n}{i}\frac{(\tilde{C}_{1n}\lambda_{1n})^i(\tilde{C}_{2n}\lambda_{2n})^{m_n-i}}{1 - \lambda_{1n}^i\lambda_{2n}^{m_n-i}} \tag{32}$$

$$F_n(z_n) = \sum_{i=1}^{m_n}\binom{m_n}{i}\frac{(\tilde{C}_{1n}\lambda_{1n})^i(\tilde{C}_{2n}\lambda_{2n})^{m_n-i}\Gamma_{n-1}(\lambda_{1n}^i\lambda_{2n}^{m_n-i})}{(1 - \lambda_{1n}^i\lambda_{2n}^{m_n-i})[z_n - \lambda_{1n}^i\lambda_{2n}^{m_n-i}\Gamma_{n-1}(\lambda_{1n}^i\lambda_{2n}^{m_n-i})]} \tag{33}$$

$$H_n(z_n) = \lambda_{2n}^{m_n}, \qquad G_n(z_n) = (\tilde{C}_{2n}\lambda_{2n})^{m_n} \tag{34}$$

$$\Theta_n(z_n) = H_n(z_n)\Gamma_{n-1}(H_n(z_n)) = \lambda_{2n}^{m_n}\Gamma_{n-1}(\lambda_{2n}^{m_n}) \tag{35}$$

$$\tilde{C}_{1n} = \left.C_{1n}\right|_{y_n=1}, \qquad \tilde{C}_{2n} = \left.C_{2n}\right|_{y_n=1} \tag{36}$$

The mean queue length of the n'th multiplexer, $\overline{N}_n$, is given by,

$$\overline{N}_n = P_n'(1) = \frac{\Theta_n''(1)}{2[1 - \Theta_n'(1)]} - \frac{H_n''(1)}{2H_n'(1)} + \frac{1 - \sum_{i=1}^{n}\rho_i}{H_n'(1)[1 - \Theta_n'(1)]}[\Gamma_{n-1}'(1)H_n'(1) + G_n'(1)]$$
$$+ \frac{(1 - \sum_{i=1}^{n}\rho_i)\Theta_n''(1)}{2H_n'(1)[1 - \Theta_n'(1)]} - \frac{1 - \sum_{i=1}^{n}\rho_i}{H_n'(1)}G_n'(1) \tag{37}$$

where ρ_i is the external load for multiplexer i, $\rho_i = H_i'(1)$

Then, application of the Little's result gives the mean delay of a packet at the n'th multiplexer,

$$\overline{D}_n = \overline{N}_n \bigg/ \sum_{j=1}^{n} \rho_j \tag{38}$$

We note that we have also determined the mean message delay and the variance of the queue length but, due to lack of space, these expressions will not be given here and only numerical results will be presented.

4 Numerical Results

In this section, we present some numerical results regarding our analysis. We assume that an *On* source of multiplexer i generates a single packet during a slot, which is $f_i(z_i) = z_i$.

Fig.s 3 and 4 present the mean and standard deviation of queue lengths of Mux-i (i=1, 2, 3) against their total traffic load. In both figures, the curves for Mux-2 and 3 are very close and they overlap with each other. It may be seen that, the mean and standard deviations of queue lengths for Mux-2 and 3 are less than that of Mux-1 except for heavy loading. This is due to smoothing effect of the multiplexers.

We have also simulated the chosen system but in order not to crowd out the figures they are not shown here. The analytical and simulation results match each other very well and curves for Mux-1 show the same crossover behavior with that of Mux-2 and 3. This provides further support for the correctness of the analysis in the paper.

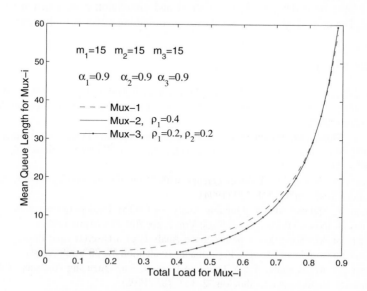

Fig. 3. The mean queue length of Mux-i vs. its total load

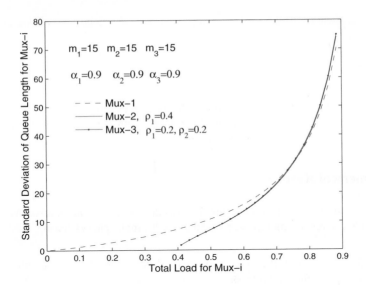

Fig. 4. The standard deviation of queue length of Mux-i vs. its total load

5 Conclusions

In this paper we have presented an exact performance analysis of a discrete-time tandem network with arbitrary number of multiplexers fed by Markovian sources. The PGF of the queue length distribution of each multiplexer has been determined and performance measures have been derived. The analytical and simulation results match each other very well. The numerical results also show the smoothing effect of multiplexers.

References

1. Cao, J., Cleveland, W.S., Lin, D., Sun, D.X.: Internet Traffic Tends Toward Poisson and Independent as the Load Increases. In: Nonlinear Estimation and Classification, Springer, New York (2002)
2. Ohba, Y., Murata, M., Miyahara, H.: Analysis of Interdeparture Processes for Bursty Traffic in ATM Networks. IEEE Journal on Selected Areas in Communications 9(3), 468–476 (1991)
3. Boxma, O.J., Resing, J.: Tandem Queues with Deterministic Service Times. Annals of Operations Research 49, 221–239 (1994)
4. Kamoun, F., Mehmet Ali, M.: Queuing Analysis of ATM Tandem Queues with Correlated Arrivals. In: INFOCOM'95 proceedings, Vol. 2, pp. 709–716 (April 1995)
5. Mehmet Ali, M., Song, X.: A performance Analysis of a Discrete-time Priority Queueing System with Correlated Arrivals. Performance Evaluation 57, 307–339 (2004)
6. Mehmet Ali, M., Kamoun, F.: A Transient Discrete-time Queueing Analysis of the ATM Multiplexer. Performance Evaluation 32, 153–183 (1998)
7. Song, X.: Performance Analysis of Tandem Networks with Markovian Sources. PhD Thesis, Department of Electrical & Computer Engineering, Concordia University, Montreal, Quebec, Canada (2006)

Queues with Message Discard at Non-zero Cost

Tadeusz Drwiega

Nortel, 3500 Carling Avenue, Ottawa, ON, Canada, K2H 8E9
drwiega@nortel.com

Abstract. Some systems discard messages that wait in the queue longer than a certain threshold. Identifying such messages and discarding them takes the processor a fixed amount of time. That time is considerably smaller than an average service time, but greater than zero, therefore the models for queuing systems with reneging do not apply. This paper provides a model for such queues. It describes a method of deriving exact formulas for the distribution of waiting times, and proposes simple, yet accurate approximations. The paper then analyzes the model and discusses the properties of such systems.

Keywords: M/M/1 queue, message discards, level crossing, probability of service, waiting for service.

1 Introduction

Servers that process messages in the session and service control layer, may discard messages that wait in the queue longer than a pre-set threshold. This mechanism is aimed at an efficient use of server resources during periods of overload. Messages that wait too long may be timed out by their senders, and subsequently abandoned or re-sent. The time spent on processing abandoned messages is wasted and in effect decreases the processing capacity of the server. Messages, which wait too long and are timed-out and re-sent by their senders, add to the load even if they are processed. Discarding such messages, in fact, postpones their processing to a later time when, the message arrival rate decreases, and the server clears or reduces the backlog of messages in the queue. Processing only the messages that do not wait too long increases the effectiveness of the server and reduces the overhead traffic of re-sends.

The value of the threshold on waiting time must be set properly to optimize the efficiency of such a system. Large values may not limit the waiting times enough to prevent time-outs. Small values, on the other hand, may result in too hasty discards and, as a consequence in unnecessary reduction of the throughput. An optimal value would be such that maximizes the throughput and is small enough that it does not trigger timeouts.

The decision on what is the maximum delay that does not trigger timeouts must be based on expected delays between the senders of the messages and the server, which

L. Mason, T. Drwiega, and J. Yan (Eds.): ITC 2007, LNCS 4516, pp. 743–753, 2007.

processes them, and on the timing mechanisms and their values. The threshold value that maximizes the throughput should come from a formula for throughput in such queuing system.

Servers which discard messages that wait longer than some limit have been modeled as queuing systems with reneging. Formulas for distribution functions of waiting times in such systems can be found in [1] and [2]. Those models however assume that discarding does not use the processor time. Posner [3] describes a general queuing model, which can be used to systems analyzed here, but solving the model equations is tedious and leads to solutions with complex closed forms. Leung [4] proposes an approximation method for solving a variation of M/G/1 model where service times depend on waiting times. His method calculates approximations of values of the distribution of waiting times without providing the functions in explicit forms.

The objective of this paper is to describe a model for queuing systems with message discards and derive a solution in an explicit form, which offers an insight into the behavior of such systems. The paper provides equations for the density distribution function of queuing time and for two key characteristics, probability that a message will be served and the expected waiting time for the messages that are served. Solutions to the equations for the density distribution function are of closed but quite complex forms. Instead, an approximation is proposed, which has a simple form and very high accuracy, as verified through simulations.

The paper is organized as follows. Section 2 describes the model, its key characteristics and provides the equations for the density function of queuing time. Section 3 describes an approximation of the density function and discusses properties of the obtained solution. In section 4 the accuracy of the approximation is evaluated through simulations. Section 5 shows asymptotic properties of the model. These properties can be used to select thresholds for discarding messages. Section 6 gives key conclusions from the model and the analysis.

2   The Model

The model is a variation of the M/M/1 model, in which the average arrival rate is λ, and the mean service time is μ. However, the service applies only to the messages, which wait for service not more than T. Messages, which wait more than T are discarded and it takes the processor a fixed amount d of time to do a discard.

Equations for the density distribution function of waiting time can be derived by using the level crossing method [5]. A new arriving message increases the amount x of the workload in the system, i.e. the sum of the service times of all the messages waiting in the queue, plus the residual processing time of the message processed (served or discarded), by the time the processor will spend on it. The new workload exceeds a level w $(>x)$, or up-crosses the level w, when the processing time of the new message is greater than $(w\text{-}x)$.

If $w \leq T$, then $x < T$, the new arrival will be served. Its service time is a negative exponential random variable with the mean μ. The probability that the service time is greater than $(w-x)$ is $exp(-\mu (w-x))$. Therefore the long-run proportion of up-crossings that start at a level $x \in I(y) = [y, y+dy]$ is $exp(-\mu(w-x))$. The average rate of crossings that start in $I(y)$ is $\lambda f_S(y)dy$. The value of the density function $f_S(w)$ at w is the rate of down-crossings the level w, so balancing the rates of up- and down-crossings results in the equation

$$f_S(w) = \lambda f_0 e^{-w/\mu} + \lambda \int_0^w e^{-\frac{w-y}{\mu}} f_S(y)dy \qquad for \quad w \leq T \qquad (1)$$

where f_0 is the probability that there is no message in the system, i.e. waiting, processed or being discarded.

Equation (1) can be easily solved to obtain f_S in the explicit form:

$$f_S(w) = f_0 \lambda e^{-\frac{1-\rho}{\mu}w} \qquad \text{for } w \leq T \qquad (2)$$

where $\rho = \lambda \mu$ is the offered load of messages. In M/M/1 ρ is also the average utilization of the processor and it has to be less than 1 for the system to be in an equilibrium state. Here, ρ can take any value. It must be $\lambda d < 1$ for the system with message discard to be in an equilibrium.

Before moving to waiting time levels larger than T, some interesting statements can be made regarding two key characteristics - the probability that a message will be served and the expected waiting time for the messages that are served. The probability P_S that a message will be served is the probability that its waiting time does not exceed T, i.e. $P_S = P(w \leq T)$, and from (2),

$$P_S = P(w \leq T) = f_0 + \int_0^T f_S(y)dy = \begin{cases} f_0(1 - \rho e^{-\frac{1-\rho}{\mu}T})/(1-\rho) \text{ for } \rho \neq 1 \\ f_0(1 + \lambda T) \qquad \text{for } \rho = 1 \end{cases} \qquad (3)$$

It can be seen from (3) that P_S depends on the time d that the processor uses to discard a message, only through the probability of an empty system f_0. In fact, the equation (3) is true for systems, where the time spent on discarding a message is not fixed but a random variable. For example, [3] analyzes another variant of M/M/1, where the service and discard times are exponentially distributed variables with means m for messages with waiting times not exceeding T, and d otherwise. The formulas derived there lead to the equation (3) for the probability waiting time not exceeding T.

The expected mean waiting time of the messages which are served, i.e. which wait in the queue not longer than T, is a conditional expectation $E[w \mid w \leq T]$. It has the following form

$$E[w\,|\,w\le T]=\int_0^T yf_s(y)dy\,/\,P(w\le T)=\begin{cases}\dfrac{\rho\{\mu-[\mu+T(1-\rho)]e^{-\frac{1-\rho}{\mu}T}\}}{(1-\rho)(1-\rho e^{-\frac{1-\rho}{\mu}T})} & \text{for }\rho\ne 1\\[2em]\dfrac{\lambda T^2}{2(1+\lambda T)} & \text{for }\rho=1\end{cases}\qquad(4)$$

The average waiting time of the served messages does not depend on the time d that the processor spends on discarding messages. It is exactly the same as the formula in [1] for M/M/1 systems with reneging. In fact, $E[w\,|\,w\le T\,]$ does not depend on whether d is fixed or is a random variable. Formula (4) is applicable to a wider range of models than the one analyzed here. For example, it applies to the variant of M/M/1 analyzed in [3]. Formulas (3) and (4) lead to the conclusion that the overhead spent on discarding messages impacts the probability that an arriving message will be served, but not the average time it will wait for the service.

The model considered here approaches M/M/1 when $T\to\infty$, and $\rho<1$. Note that under these conditions $E[w\,|\,w\le T\,]$ tends to $\rho\mu/(1-\rho)$, which is the average queuing time in M/M/1.

To derive density distribution function for waiting times greater than T, the following cases are considered. The first case is when $T<w\le(T+d)$. w can be crossed from workload levels $x\le T$ with probability $exp(-\mu(w-x))$, and from levels $x>T$ with probability 1. When a new arrival finds the workload greater than T, it will be discarded, and therefore it increases the workload by d to $(x+d)$, which is greater than $w\le T+d$. The average rate of crossings that start in $T<l(y)<w$ is $\lambda f_d(y)dy$. The equation for the density function $f_d(w)$ is obtained by comparing the up- and down-crossing rates

$$f_d(w)=f_0\lambda e^{-\frac{w}{\mu}}+\lambda\int_0^T e^{-\frac{1-\rho}{\rho}y}f_s(y)dy+\lambda\int_T^w f_d(y)dy\ \text{ for }T<w\le T+d\qquad(5)$$

After calculating the first integral, equation (5) can be simplified to

$$f_d(w)=f_0\lambda e^{\lambda T-\frac{w}{\mu}}+\lambda\int_T^w f_d(y)dy\qquad\text{for }T<w\le T+d\qquad(6)$$

Equation (6) can be solved to obtain $f_d(w)$

$$f_d(w)=f_0\lambda(e^{\lambda T-\frac{w}{\mu}}+\rho e^{\lambda w-\frac{T}{\mu}})/(1+\rho)\qquad\text{for }T<w\le T+d\qquad(7)$$

Any workload level $w>T+d$ can be up-crossed from a lower levels only in the intervals $[0,T]$ or $(w-d,w)$. Messages arriving when the workload x is more than T but less than $(w-d)$ increase it to $(x+d)$, which is less than w. This case is illustrated in Figure 1.

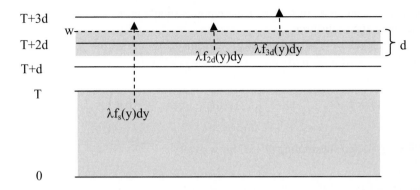

Fig. 1. Up-crossing a level w > T+d

For $T+d<w\leq T+2d$ the rate of up-crossings that start in $(w-d,w)$ is $\lambda f_d(y)$ in the sub-interval $(w-d,T+d]$, and $\lambda f_{2d}(y)$ in the sub-interval $(T+d,w)$, where $f_{2d}(y)$ is the density function for the waiting times in the interval $(T+d,T+2d]$. In a general case, for w such that $T+(i-1)d<w\leq T+id$, $i=2,3,\ldots$, the rate of up-crossings by the value of d, which start in the sub-interval $(w-d,T+(i-1)d]$ is $\lambda f_{(i-1)d}(y)$, and $\lambda f_{id}(y)$ in the sub-interval $(T+(i-1)d,w)$. The level crossing equation for a density function $\lambda f_{id}(y)$ in an interval $(T+(i-1)d,T+id]$ is

$$f_{id}(w) = f_0\lambda e^{-\frac{w}{\mu}} + \lambda\int_0^T e^{-\frac{1-\rho}{\rho}y} f_S(y)dy +$$

$$\lambda\int_{w-d}^{T+(i-1)d} f_{(i-1)d}(y)dy + \lambda\int_{T+(i-1)d}^{w} f_{id}(y)dy = \qquad (8)$$

$$f_0\lambda e^{\lambda T-\frac{w}{\mu}} + \lambda\int_{w-d}^{T+(i-1)d} f_{(i-1)d}(y)dy + \lambda\int_{T+(i-1)d}^{w} f_{id}(y)dy$$

$$\text{for } T+(i-1)d < w \leq T+id; i = 2,3,\ldots$$

Equations (8) can be solved recursively, starting from $i=2$ and using for that case $f_d(y)$ given by (7). However, the forms of the functions become more and more complex as i increases. Solving the equations is tedious and there is no apparent relationship that would allow presenting the results in a concise form.

3   Approximation of Density Function for Queuing Times > T

The approach proposed here is to approximate those values by a function of a simple form, which gives an insight into the behavior of the system. The main idea of the approximation is to solve the equations (8) only for the values of $w = T+2d, T+3d, \ldots$, and find a function whose graph passes through all the points $(id, f_{id}(id))$. Since d is expected be small as compared to μ, these points should provide a basis dense enough for accurate approximation.

Since a level $w = id$ can be crossed from below only from levels in $[0,T]$ or $(T+(i-1)d, T+id)$ equation (8) simplifies for those levels to

$$f_{id}(w) = f_0 \lambda e^{\lambda T - \frac{w}{\mu}} + \lambda \int_{T+(i-1)d}^{w} f_{id}(y)dy \qquad \text{for i} = 2,3,... \qquad (9)$$

Solving (9) and substituting $(T+id)$ for w gives

$$f_{id}(T+id) = \frac{f_0 \lambda}{1+\rho} e^{\frac{1-\rho}{\mu}T} (1+\rho e^{\frac{1+\rho}{\mu}d}) e^{-\frac{id}{\mu}} \qquad \text{for i} = 2,3,... \qquad (10)$$

It should be noted that equation (10) is also true for $f_d(T+d)$ and therefore it is true for $i = 1,2, ... $.

The sought approximation is obtained by extending (10) from the discrete values to the whole interval (T, ∞)

$$f_D(w) = \frac{f_0 \lambda}{1+\rho} e^{-\frac{1-\rho}{\mu}T} (1+\rho e^{\frac{1+\rho}{\mu}d}) e^{-\frac{w-T}{\mu}} \qquad \text{for} \quad w > T \qquad (11)$$

The accuracy of this approximation has been evaluated through simulations and the results are reported in section 4. Now, the approximation is used to calculate f_0. The starting point for the calculations is equation (12),

$$1 = f_0 + \int_0^T f_S(y)dy + \sum_{i=1}^{\infty} \int_{T+(i-1)d}^{T+id} f_{id}(y)dy \approx f_0 + \int_0^T f_S(y)dy + \int_T^{\infty} f_D(y)dy \qquad (12)$$

from which an approximation f_{0app} is obtained

$$f_{0app} = \begin{cases} \dfrac{1-\rho}{1-\rho e^{-\frac{1-\rho}{\mu}T} + (1-\rho)\rho e^{-\frac{1-\rho}{\mu}T+\lambda d}} & \text{for } \rho \neq 1 \\[4mm] \dfrac{1}{1+\lambda T + 0.5(e^{\lambda d} + e^{-\lambda d})} & \text{for } \rho = 1 \end{cases} \qquad (13)$$

It should be noted that the approximation f_{0app} exhibits properties expected from the exact solution. For $T \rightarrow \infty$ and $\rho < 1$, $f_{0app} \rightarrow (1-\rho)$, which is the probability that a corresponding M/M/1 system is empty. For $d=0$ it becomes the probability of an empty M/M/1 system with reneging (comp. [1]).

Now, P_S can be approximated by substituting f_{0app} for f_0 in (3), which gives the following formula for the approximation P_{Sapp}:

$$
P_{Sapp} = \begin{cases} \dfrac{1 - \rho e^{-\frac{1-\rho}{\mu}T}}{1 - \rho e^{-\frac{1-\rho}{\mu}T} + (1-\rho)\rho e^{-\frac{1-\rho}{\mu}T + \lambda d}} & \text{for } \rho \neq 1 \\[4mm] \dfrac{1 + \lambda T}{1 + \lambda T + 0.5(e^{\lambda d} + e^{-\lambda d})} & \text{for } \rho = 1 \end{cases} \qquad (14)
$$

4 Accuracy of the Approximation of P_S

The accuracy of the approximation has been assessed through simulations. In the simulator, a source generated packets as a Poissonian process with the average rate of λ, and sent them to a single-server FIFO queue. The server upon taking a packet from the queue head, compared the queuing time w of the message with a threshold T, and if $w \leq T$ the server entered a busy state for a period of time drawn from a negative exponential distribution with mean μ. If $w > T$, the server entered the busy state for a fixed period d.

Out of the two parameters, probability P_S that a message will be served and the expected mean waiting time $E[w \mid w \leq T]$ of the served messages, only P_S depends on the approximation. The simulator counted all the generated messages and the messages that were served, and at the end of simulations calculated the ratio of the two values as an estimate of P_S.

In all the simulations the value of μ was set to 1. The cost d of discarding a message should be small as compared to μ, so two values were used in simulations, $d=0.01$ and $d=0.1$. They represent the cases when the cost of discarding a message is 1% and 10% of the average time the processor spends on serving a message. The threshold T was set to 10, 50 and 300, which can also be interpreted as multiplicities of μ. For each pair of values of d and T, λ was varied from 0.6 to 4. Because $\mu = 1$, and therefore $\lambda = \rho$, the load of offered messages ranged from moderate 0.6 to a very heavy overload 4.

Table 1 shows the values of P_S calculated from equation (3) and the statistics obtained from the simulations when $d=0.1$ and $T=50$. The differences between the values are very small and are of the range of the statistical error for $\rho \leq 2$. Under high overload the model values overestimate the probability but with an error that is likely to be acceptable for practical purposes.

The results in Table 1 are representative for other values of T when $d=0.1$. When $d=0.01$ the model is even more accurate. This tendency of increasing the accuracy with decreasing d is consistent with the fact that the approximation becomes accurate for $d=0$.

Table 1. P_S values obtained from the model and from simulations ($\mu = 1$, $d=0.1$ and $T = 50$)

λ (and ρ)	model	simulation
0.6	1.000	1.000
0.7	1.000	1.000
0.8	1.000	1.000
0.9	0.999	0.999
0.95	0.995	0.996
1	0.981	0.979
1.1	0.899	0.902
1.2	0.816	0.811
1.3	0.745	0.747
1.5	0.633	0.629
1.7	0.547	0.543
2	0.450	0.447
3	0.270	0.260
4	0.183	0.166

5   Behavior of the System Throughput When $T \to \infty$

The proposed model, including the approximation of the density distribution function for $w>T$, provides an insight into the behavior of systems which discard messages with long queuing times. Some of the properties were discussed in the previous sections. The focus of this section is the asymptotic behavior of the system throughput. As T increases, throughput increases too and tends to a finite limit when $T \to \infty$. The question is how quickly the throughput approaches its limit and what values of T get the throughput close enough to its limit. Since throughput is a product of λ and P_S, it is sufficient to analyze the asymptotic behavior of P_S.

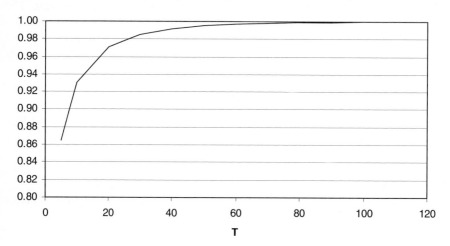

Fig. 2. Values of PSapp for different values of T

It is easy to see from (14) that when the offered load of messages ρ is less than 1, $P_{Sapp} \rightarrow 1$ when $T \rightarrow \infty$, independently of the value of d. The speed of this convergence is illustrated in Figure 2 for $d=0.1$ and $\rho=0.95$. For smaller values of ρ, P_{Sapp} converges to 1 faster.

When $\rho \geq 1$, T must be finite for the system stability. As the values of T increase, initially P_{Sapp} increases too, but eventually it reaches a plateau where its increase is practically unnoticeable. The regions of increase and plateau depend on λ, ρ and d. The larger the value of ρ, the shorter the region of increase of P_{Sapp} and the faster the plateau is reached. Figure 3 shows the regions of increase and the plateau of P_{Sapp} for $d=0.1$ and for two values of ρ, 1.1 and 1.3.

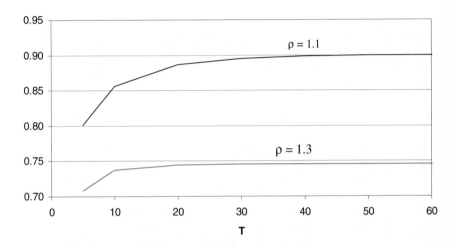

Fig. 3. Growth and plateau of PSapp with increasing T

In both cases, $\rho<1$ and $\rho>1$, the graphs show that initially increasing T results in substantial increases of P_S. As P_S approaches its limit, further increases of T cause insignificant, and even negligible gains of P_S. The slowest convergence is when ρ is close to 1, and therefore the selection of the value for T should be done for ρ closest to 1 in the range expected in practice.

Formula (14) for P_{Sapp} has an interesting asymptotic property. It has a limit when $T \rightarrow \infty$, for $\rho>1$, even though finite T is required for the system stability. That limit is equal to

$$P_L = \frac{1}{1-(1-\rho)e^{\lambda d}} \tag{15}$$

When T is finite, P_S can be interpreted as a probability that a message is served, and $(1 - P_S)$ as a probability that a message is discarded. Consider a queuing system where messages are served or discarded not based on their queuing time being shorter or longer than a threshold, but based on an outcome of a binomial experiment. Each message upon arrival is marked for service with probability q, or for discard with probability *(1-q)*. Service and discard times are the same as in the system considered in this paper.

For λ such that $\lambda\mu=\rho>1$ the fraction q of the arriving messages, which is served, has to be limited to keep the system stable. The upper limit on q can be obtain from the equation

$$1 = \lambda * \mu * q + \lambda * d * (1 - q)$$

(16)

and it is equal to

$$q = \frac{1 - \lambda d}{\rho - \lambda d} .$$

(17)

Values of P_L from (15) are close to the limit (17) on q for a range of values of ρ and d. Table 2 shows values of P_L and q for comparison. This means that the level of plateau reached by P_S is a close approximation of the maximum fraction of messages which can be served.

Table 2. Comparison of P_L and q

ρ	d = 0.1		d = 0.2	
	P_L	q	P_L	q
1.1	0.900	0.899	0.899	0.866
1.2	0.816	0.815	0.797	0.792
1.3	0.745	0.744	0.720	0.712
1.4	0.685	0.683	0.654	0.643
1.5	0.633	0.630	0.597	0.583
2	0.450	0.444	0.401	0.375
3	0.270	0.259	0.215	0.167
4	0.183	0.167	0.130	0.063

6 Conclusions

This paper describes a model for servers which discard messages with queuing times longer than a deterministic threshold, and where discarding a message takes a fixed amount of processor time. The analysis of the model reveals that the expected queuing time for the messages, which are served does not depend on the time spent on discarding a message. It does not even depend on whether that time is deterministic or random.

Another key performance characteristic of such system, probability that a message will be served depends on the discard overhead only through the probability that the system is empty.

The density function of the probability distribution of waiting in the queue can be obtained through the method of level-crossing. However, the solution is very tedious and results in functions of complex forms for waiting times larger than the discard threshold. The approximation proposed in the paper leads to a function of very simple form, which provides an insight into the properties of such systems.

The approximation has been evaluated in two ways. First, it has been shown that in special cases its behavior is that of the exact solution. Second, it has been demonstrated through simulations that its accuracy is excellent for the range of parameters that is required for uses of the model to servers in session or service control layer.

Throughput in such systems converges to fixed values when the discard threshold increases to infinity. The convergence has two phases. Initially the throughput advances towards its limit are very fast, but once it approaches the limit, further gains are negligible.

Acknowledgment

The author is grateful to an anonymous reviewer for his/her insightful comments.

References

1. Doshi, B.T., Heffes, H.: Overload Performance of Several Processor Queueing Disciplines for the M/M/1 Queue. IEEE Transactions on Communications com-34(6), 538–546 (1986)
2. Movaghar, A.: On queueing with customer impatience until the beginning of service. In: Proc. of IEEE Int. Comp. Perf. And Depenability Symposium, pp. 150–157 (1996)
3. Posner, M.J.: Single-server Queues with Service Time Dependent on Waiting Time. Operations Research 21, 610–616 (1973)
4. Leung, K.K.: Load-Dependent Service Queues With Applications to Congestion Control in Broadband Networks. In: Proc. of IEEE Globecom'97, pp. 1674–1679 (1997)
5. Brill, P.H., Posner, M.J.: Level Crossings in Point Processes Applied to Queues: Single-Server Case. Operations Research 25, 662–674 (1977)

Uniform Approximations for Multirate Loss Systems with State Dependent Arrival Rates

Wolfgang Bziuk

Institut for Computer and Communication Network Engineering
w.bziuk@tu-bs.de

Abstract. A multirate loss system with complete sharing is investigated, in which multiple classes of customers arrive as a state dependent Poisson processes. This arrival process includes the Bernoulli-Poisson-Pascal (BPP) and the batched Poisson process with geometric distributed batch sizes. Asymptotic uniform approximations to the blocking probabilities are derived, when the capacity and a parameter of the arrival processes are commensurately large. The results are obtained with the saddle-point method of integration and the approximation uniformly holds across all traffic regimes, where the blocking probabilities may vary by several order of magnitude. Moreover, a numerically stable representation of the approximation is given, which gives accurate results also for the critical traffic region. Numerical results show that while prior asymptotic approximations are quite accurate, the new approximations are very accurate.

Keywords: Multirate Loss System, Complete Sharing, State Dependent Arrival Rates, Uniform Approximation.

1 Introduction

We consider a multirate loss system having capacity C bandwidth units, which is used in a complete sharing mode by S different classes. When there are x_s customers of class s ($1 \leq s \leq S$) in service, type s customers arrive according to a state depending Poisson process with arrival rate $\lambda_s(x_s) = \gamma_s + x_s \alpha_s$, whereas the holding time distribution is exponentially distributed with mean $1/\mu_s$. Each call of class s requires an integer number of bandwidth units b_s and an arriving call is blocked if the available free capacity is less than $C - b_s$, thus the primary performance measure is the call blocking probability. Mixed traffic streams, each having individual peakness above, below or equal to one, can be considered. In the multirate finite sources model (MFS-model) version this system is used by bursty sources, which alternate between random periods in the "on" and "off" states. When a source turns "on" and the free capacity is less than the source requires it is blocked. This can also be considered to be a burst blocking. In the multirate infinite sources model (MIS-model), the arrivals follows a Poisson process with rate λ_s. Further, the arrival rates can be mapped to the

L. Mason, T. Drwiega, and J. Yan (Eds.): ITC 2007, LNCS 4516, pp. 754–766, 2007.

Pascal process, thus the BPP process can be modeled, as well as the batched Poisson arrival process with geometric distributed batch size [1].

Prior work on multiservice, multirate loss systems and networks is extensive. See, for instance, the book by Ross [2] and references therein. For the system under consideration a product form for the stationary distribution of the number of accepted calls of each class exits [3] and one dimensional recursions for the distribution of the number of occupied bandwidth units have been derived in [2] and [3].

For large C the computation time is considerably so several approximations have been derived, a good overview is given in [4]. For the MIS and MFS traffic model the saddle-point method is applied in [4] to the contour integral representation of the normalization constant. This yields a unified asymptotic approximation (UAA) for the loss probability, which is valid for all traffic ranges. This method has been extended to a UAA for the whole distribution [5], which also gives some inside into the nature of the approximate distribution. Also a UAA for the batched Poisson arrival process has been given in [6]. Although the UAA is valid for all traffic ranges, it suffers from some numerical problems in the critical traffic region. To overcome this problem a refined UAA (RUUA), only valid for the MIS traffic model, has been derived in [7].

In this paper a RUAA for state dependent arrival rates is presented, which extends previous results of Morrison, Ramakrishnan and Mitra [7] for constant arrival rate. For this system the probability generating function for the stationary probability of the number of occupied bandwidth units, as well as some contour integral representations are given in Section 2. The analysis in Section 3 focuses on large systems, specifically $C \rightarrow \infty$ and $\gamma_s \rightarrow \infty$, such that γ_s / C is fixed. It should be noted, that the UAA for the batched Poisson arrival process developed in [6] can also be applied to the MIS-traffic model and to the Pascal process, but the approximations developed here include the MFS-traffic model as well and further differs in same important aspects. In contrast to [6] refined approximations are derived, which give more accurate results for a broader range of traffic parameters, e. g. for the BPP-traffic model. And as an important fact, uniform numerical representations of the approximations given in Section 4 overcome numerical problems, which for example arise in the UAA [6] within the critical traffic region. The rest of the paper is organized as follows: Section 5 gives some applications and in Section 6 numerical results are presented.

2 Stationary Distributions

The system state of the multirate loss system with complete sharing is described by the number of accepted calls $\vec{x} = (x_1,...,x_S)$ of each class. For the state dependent arrival rates $\lambda_k(x_k) = \gamma_k + x_k \alpha_k$ and service rates $\mu_k(x_k) = x_k \mu_k$, where without loss of generality μ_k is set to 1, the total number of occupied bandwidth units are obtained from $n = \sum_s x_s b_s$. If we set $\vec{\gamma} = (\gamma_1,...,\gamma_S)$ and $\vec{\alpha} = (\alpha_1,...,\alpha_S)$, their unnormalized probabilities $\tilde{P}(n,\vec{\gamma},\vec{\alpha}) = \tilde{P}(n)$ can be calculated with the recursion [2]

$$n \cdot \tilde{P}(n) = \sum_{s=1}^{S} b_s R_s(n), \quad 1 \leq n \leq C, \tag{1}$$

where $R_s(n) = 0$ and $\tilde{P}(n) = 0$ if $n < 0$, and

$$R_s(n) = \gamma_s \tilde{P}(n - b_s) + \alpha_s R_s(n - b_s). \tag{2}$$

From (1) - (2) it is easy to see that the computational complexity of the recursion scales with $O(S \cdot C)$, thus for large C the computation time can be considerably.

Let us define the probability generating function (PGF) $P_Z(z, s) = \sum_{n=0}^{\infty} P_s(n) \cdot z^n$, than with (1) - (2) a first order differential equation can be derived, which is given by

$$z \frac{d\tilde{P}(z)}{dz} = \tilde{P}(z) \sum_{s=1}^{S} \frac{b_s \gamma_s z^{b_s}}{1 - \alpha_s z^{b_s}} . \tag{3}$$

For $\alpha_s > 0$ it has poles and that pole with the smallest absolute value is denoted as

$$z_P = \min_{s | \alpha_s > 0} \alpha_s^{-1/b_s} . \tag{4}$$

Let us assumed that the complex variable z only can take values out of the domain $D = \{z | z \in C \wedge |z| < z_P\}$. Then the differential equation (3) has the solution

$$\tilde{P}_Z(z, \vec{\gamma}, \vec{\alpha}) = \prod_{s=1}^{S} \left(\frac{1 - \alpha_s}{1 - \alpha_s z^{b_s}} \right)^{\frac{\gamma_s}{\alpha_s}} , \quad z \in D , \tag{5}$$

and (5) is analytical within D. Now, with (5) let us define the function

$$f(z) = \frac{1}{C} Ln(\tilde{P}_Z(z, \vec{\gamma}, \vec{\alpha})) - Ln(z) = \sum_{s=1}^{S} \frac{\gamma_s}{\alpha_s C} Ln\left(\frac{1 - \alpha_s}{1 - \alpha_s z^{b_s}} \right) - Ln(z), \quad z \in D , \tag{6}$$

which is analytic for $z \in D$ and the class dependent function

$$q_k(z) = \frac{1 - \alpha_k}{1 - \alpha_k z^{b_k}} , \quad z \in D . \tag{7}$$

Of primary interest are the probabilities $\tilde{P}_k^{(a)}(n)$ at an arrival instant of a call of class k. It is straightforward to generalize the results for the BPP process given in [3] to show that $\tilde{P}_k^{(a)}(n) = \tilde{P}(n, \vec{\gamma} + \vec{e}_k \alpha_k, \vec{\alpha})$, where $\vec{e}_k$ is the unit vector for class k. Then with (5) - (7) their PGF is given by

$$\tilde{P}_{Z,k}^{(a)}(z) = z^C \cdot q_k(z) \cdot exp(Cf(z)). \tag{8}$$

Observe, that although $f(z)$ is not single-valued, $exp[f(z)]$ is. Now, with the correspondence 13 out of [8, p. 330] the PGF of the normalization at an arrival instant of a call of class k can be derived from (8). With (6), (7) and Cauchy's formula, its inverse PGF (IPGF) has the simple contour integral representation

$$G_k\left(C,\vec{\gamma}+\vec{e}_k\alpha_k,\vec{\alpha}\right)= \sum_{n=0}^{C} \tilde{P}_k^{(a)}(n)=\frac{1}{2\pi i}\oint_{|z|<min(1,z_P)} \frac{q_k(z)}{z(1-z)}\cdot exp(C\cdot f(z))dz, \qquad (9)$$

where the integral is taken in a counter-clockwise direction around a circle of radius less than the minimum of 1 and z_P. Also with (8), the IPGF of the unnormalized call blocking probability of class k is given by

$$\tilde{B}_k = \sum_{n=C-b_k+1}^{C} \tilde{P}_k^{(a)}(n)=\frac{1}{2\pi i}\oint_{|z|<z_P} h_k(z)\cdot exp(C\cdot f(z))dz, \qquad (10)$$

where the function $h_k(z)$ follows to

$$h_k(z)=\frac{1-z^{b_k}}{z(1-z)}q_k(z). \qquad (11)$$

3 Uniform Asymptotic Approximations

The asymptotic scaling in which C and $\vec{\gamma}$ are commensurately large is considered now, thus we assume that $\gamma_s=\eta_s C$, $s=1,...,S$; $C\to\infty$, where η_s is $O(1)$ and bounded away from zero. It is assumed that $z\in D$ and (without loss of generality) that the greatest common divisor (g.c.d.) of $b_1,...,b_S$ is 1. The stationary points of $f(z)$ are given by $f'(z^*)=0$, where the prime denotes derivative, so that

$$z^* f'(z^*)= \sum_{s=1}^{S} \frac{\eta_s b_s\left(z^*\right)^{b_s}}{1-\alpha_s\left(z^*\right)^{b_s}} -1, \quad z^*\in D. \qquad (12)$$

Observe that $zf'(z)\to-1$ as $z\to 0$ and if there is at least one $\alpha_s>0$, we have $zf'(z)\to\infty$ as $z^{b_s}\to 1/\alpha_s$. On the other hand, if $\alpha_s<0,\forall_s$, we have $zf'(z)\to \sum_s\eta_s b_s/|\alpha_s|>0$ as $z\to\infty$. Thus for $z\in D$, $zf'(z)$ is a monotonically increasing function of z and there is a unique positive reel solution z^* of $f'(z)=0$. Also, since $f'(z^*)=0$,

$$v(z^*)=\left(z^*\right)^2 f''(z^*)=\sum_{s=1}^{S}\eta_s \frac{b_s^2\left(z^*\right)^{b_s}}{\left(1-\alpha_s\left(z^*\right)^{b_s}\right)^2}. \qquad (13)$$

Hence $f''(z^*)>0$ for $0<z^*<z_P$, and $f(z)$ has a local minimum on the real axis at $z=z^*$. Having proved these properties the general result derived in [9] can be applied, which states that $|z|=z^*$ is a saddle-point contour.

For a Poisson arrival process (MIS model), we have state independent arrival rates given by $\alpha_k = 0$, and, accordingly, (7) simplifies to $q_k(z) = 1$. For this special case a RUAA to the integrals (9) and (10) has been given in [7]. Thus, if compared to [7], for the model presented here an additional function $q_k(z)$ added to the integrand of the contour integral has to be considered. But the main difference arises from the fact that depending on the traffic parameters the functions (6) and (7) may have a pole. But because $z^* \in D$ is assumed, this pole does not cause an additional mathematical problem and the derivation given in [7] can be generalized to cover the additional function $q_k(z)$. This generalization is proofed in the Appendix. Even though the mathematical derivation can be generalized to deal with the pole, its existence may strongly affect the evaluation of a numerical result. Especially the numerical accuracy suffers if the saddle point is near one or near the pole. This problem will considerably be reduced by a unified numerical representation of the approximation presented in the next section which again is a generalization of the results given in [7].

To summarize the results the notation introduced in [7] is used. Let us define

$$\tau = \left(z^*\right)^3 f^{(3)}\left(z^*\right), \quad y = \left(z^*\right)^4 f^{(4)}\left(z^*\right), \quad w = \left(z^*\right)^5 f^{(5)}\left(z^*\right), \tag{14}$$

where $f^{(j)}$ denotes the j'th derivative, as well as

$$\beta_1 = \frac{z^* \sqrt{2}}{\sqrt{v}}, \quad \beta_2 = \frac{z^* \tau}{3v^2}, \quad \beta_3 = \frac{z^*}{6v\sqrt{2v}}\left(\frac{y}{v} - \frac{5\tau^2}{3v^2}\right). \tag{15}$$

Then, asymptotically, with the abbreviation $G_k = G_k\left(C, \vec{\gamma} + \vec{e}_k \alpha_k, \vec{\alpha}\right)$ we have

$$G_k \approx \hat{G}_k = \frac{q_k(1)}{2}\text{Erfc}\left[\text{sgn}\left(1 - z^*\right)\sqrt{-Cf\left(z^*\right)}\right] + \frac{e^{Cf\left(z^*\right)}}{\sqrt{2\pi Cv}}\left(K_k - \frac{E_k}{Cv}\right), \tag{16}$$

where Erfc is the usual complementary error function and, for $z^* \neq 1$,

$$K_k = \frac{q_k\left(z^*\right)}{1 - z} - \frac{q_k(1)\sqrt{v}\,\text{sgn}\left(1 - z^*\right)}{\sqrt{-2f\left(z^*\right)}}, \tag{17}$$

and

$$E_k = \frac{q_k\left(z^*\right)}{8\left(1 - z^*\right)}\left(\frac{5\tau^2}{3v^2} - \frac{y}{v}\right) + \frac{\tau}{2v}\left(q_k\left(z^*\right)\frac{\left(1 - 2z^*\right)}{\left(1 - z^*\right)^2} - \frac{z^* q_k'\left(z^*\right)}{\left(1 - z^*\right)}\right) + q_k\left(z^*\right)\frac{1 - 3z^* + 3\left(z^*\right)^2}{\left(1 - z^*\right)^3}$$

$$- q_k'\left(z^*\right)\frac{z^*\left(1 - 2z^*\right)}{\left(1 - z^*\right)^2} + \frac{\left(z^*\right)^2 q_k''\left(z^*\right)}{2\left(1 - z^*\right)} - q_k(1)\frac{v\sqrt{v}\,\text{sgn}\left(1 - z^*\right)}{\left(-2f\left(z^*\right)\right)^{3/2}}. \tag{18}$$

Since $f(1) = 0$ and $f'\left(z^*\right) = 0$, the expressions in (17) and (18) remain finite as $z^* \to 1$. Utilizing the series expansion [7, (A8)] their limiting values are given by

$$K_k = 1 + \frac{\tau(1)}{6v(1)} - q'_k(1), \ z^* = 1,$$ (19)

and

$$E_k = \frac{q'_k(1) - 1}{8} \left(\frac{y}{v} - \frac{5\tau^2}{3v^2} \right) + \left(q_k(1) - q'_k(1) + \frac{q''_k(1)}{2} \right) \left(1 + \frac{\tau}{2v} \right)$$

$$+ q_k(1) \left(\frac{35\tau^3}{432v^3} + \frac{w}{40v} - \frac{5y\tau}{48v^2} \right) - \frac{q_k^{(3)}(1)}{6}, \ z^* = 1.$$ (20)

Hence (16) provides a uniform approximation to the normalization (9).

The IPGF (10) for the unnormalized blocking probability $\tilde{B}_k$ depends on (11), which is analytic for all $z^* < z_P$ and especially for $z^* = 1$. Now if we set $h_k(z) \equiv h(z)$ for each class, (10) is exactly of the form given by the proposition 4.2 out of [7], which gives the RUAA $\hat{\tilde{B}}_k \approx \tilde{B}_k$. With (11), (16) and for $z^* < z_P$ we have

$$B_k \approx \frac{\exp(Cf(z^*))}{2\hat{G}_k \sqrt{\pi C}} \left\{ \beta_1 h_k(z^*) + \frac{1}{2C} \left[3\beta_3 h_k(z^*) + 3\beta_1\beta_2 h'_k(z^*) - \frac{1}{2}\beta_1^3 h''_k(z^*) \right] \right\}.$$ (21)

4 Numerical Representation

When α_s is close to zero the numerical evaluation of (6) loses accuracy, although the limit of (6) is well defined. A Laurent series expansion of (6) given by

$$f(z) = \sum_s \frac{\gamma_s}{C} \sum_{n=0}^{\infty} (\alpha_s)^n \frac{z^{(n+1)b_s} - 1}{n+1} - Ln(z), \ |\alpha_s| \le \varepsilon_1,$$ (22)

is used to overcome this problem, whereas accurate numerical results are obtained for $\varepsilon_1 = 0.1$ and truncating the sum at $n = 20$.

Next K_s and E_s are specialized to the case of the MIS-model, that is $q_s(z) = 1$ and $\alpha_s = 0, \forall s$. Thus for each class (17) reduces to

$$K_{MIS} = \frac{1}{1-z} - \frac{\sqrt{v} \, sgn(1-z^*)}{\sqrt{-2f(z^*)}}, \ z^* \ne 1,$$ (23)

and similar, for $z^* \ne 1$ (18) simplifies to

$$E_{MIS} = \frac{1}{8(1-z^*)} \left(\frac{5\tau^2}{3v^2} - \frac{y}{v} \right) + \frac{\tau}{2v} \frac{(1-2z^*)}{(1-z^*)^2} + \frac{1-3z^* + 3(z^*)^2}{(1-z^*)^3} - \frac{v\sqrt{v} \, sgn(1-z^*)}{(-2f(z^*))^{3/2}}.$$ (24)

For the special case $\alpha_s = 0, \forall s$, that is for the MIS-model, an alternative representation for K_{MIS} and E_{MIS} has been derived in ([7], (6.11)-(6.14)), which gives numerically accurate results, whether or not z^* is close to 1. This requires to transform (6) into the form $f(z^*) = (1 - z^*)^2 \hat{f}(z^*)$, at least for $|1 - z^*| \ll 1$. But the method given in [7] can not be used here, because the more general function (6) is assumed. Now, for $|1 - z^*| \ll 1$, $z^* < z_P$ and since $f(1) = 0$ and $f'(z^*) = 0$, the desired transform is achieved by a Taylor series expansion of (6), which is given by

$$f(z^*) = -(1 - z^*)^2 \sum_{j=2}^{n} \frac{f^{(j)}(z^*)}{j!}(1 - z^*)^{j-2} - \frac{(1 - z^*)^{n+1}}{(n+1)!} f^{(n+1)}(z^* + \vartheta(1 - z^*)), \quad 0 < \vartheta < 1. \quad (25)$$

For $n = 6$ the remainder term is $O(|1 - z^*|^7)$ and if $z^* \ll z_P$ or rather $\alpha_s < 0, \forall s$, this term can be neglected. But if z^* is near the pole (4), the accuracy deteriorates. Now, observe that for a specific z^* the Taylor series of the right hand side of (25) must give the same result as (6) and (22) respectively. Thus with a simple line search an optimal value $\hat{\vartheta}$, $0 < \hat{\vartheta} < 1$, for the remainder term can be determined, such that the series (25) is exact for a given z^* and is also in the desired form $f(z^*) = (1 - z^*)^2 \hat{f}(z^*)$. Very accurate results can be obtained for $f(z^*)$ and its derivatives for $|1 - z^*| < 0.05$, also if z^* is near the pole at z_P.

Now let us show how K_{MIS} and E_{MIS} can be used to calculate K_s and E_s. Let

$$H_k(z) = \sum_{m=0}^{b_k - 1} z^m = \begin{cases} (1 - z^{b_k})/(1 - z), & z \neq 1, \\ b_k, & z = 1, \end{cases} \quad (26)$$

and since $q_k(1) = 1$, it is simple to show with (7) and (26) that

$$\Xi_k(z) = \frac{q_k(z) - q_k(1)}{1 - z} = -\frac{\alpha_k H_k(z)}{1 - \alpha_k + \alpha_k (1 - z) H_k(z)}. \quad (27)$$

Then, using (27) a comparison of (17) with (23) yields

$$K_k = q_k(1) K_{MIS} + \frac{q_k(z) - q_k(1)}{1 - z} = q_k(1) K_{MIS} + \Xi_k(z^*). \quad (28)$$

$\Xi_k(z)$ gives accurate numerical results for all $z^* \in D$, as well as K_{MIS} does, thus (28) can be used for all $z^* \in D$. In preparation for the next paragraph, let us define

$$H_k^{(j)}(z) = \frac{d^j}{dz^j} H_k(z), \quad j = 1, 2. \quad (29)$$

The derivation of a numerical stable representation for (18) is a bit cumbersome, but introduces no additional mathematical problems, thus its details are omitted. With (26), (27) and (29) and for all $z^* \in D$, the comparison of (18) with (24) yields

$$
E_k = q_k(1)E_{MIS} + \frac{\Xi_k(z^*)}{8}\left(\frac{5\tau^2}{3v^2} - \frac{y}{v}\right) + \frac{\tau}{2v}\left[q_k(z^*) - q_k(1) + q_k'(z^*) + \frac{\left(\alpha_k z^* H_k(z^*)\right)^2}{\left(1 - \alpha_k(z^*)^{b_k}\right)^2} \right.
$$

$$
\left. + \frac{\alpha_k(1 - \alpha_k)\left(H_k^{(1)}(z^*) - (1 + z^*)H_k(z^*)\right)}{\left(1 - \alpha_k(z^*)^{b_k}\right)^2} + q_k(z^*) - q_k(1) - z^* q_k'(z^*) \right]
$$

$$
+ \frac{\alpha_k}{\left(1 - \alpha_k(z^*)^{b_k}\right)^3}\left[(1 - \alpha_k)(z^*)^2\left(H_k^{(1)}(z^*) - \frac{1}{2}(1 - \alpha_k(z^*)^{b_k+1})H_k^{(2)}(z^*)\right) \right.
$$

$$
- \alpha_k\left(1 - \alpha_k(z^*)^{b_k}\right)(z^*)^3 H_k^{(1)}(z^*)\left(H_k(z^*) + b_k(z^*)^{b_k-1}\right) - \alpha_k^2(z^*)^{2b_k+1}b_k^2 H_k(z^*)
$$

$$
\left. + \frac{\alpha_k}{2}(1 - \alpha_k)b_k(1 + b_k)(z^*)^{2b_k} \right]. \tag{30}
$$

The numerical complexity of this approximation is in principle identical to that given in [7]. To be more precisely an additional cost is introduced if (25) has to be solved for $\hat{\vartheta}$, which is comparable to solve (12) for z^*. To find the root of (12) or (25) typically 20 to 30 iterations are required, thus the complexity is $O(1)$. Also (6), (12), (13) and (22) can be calculated with $O(S)$. If z^* is near one (26) and (29) are solved by summation. This demands for $O(b_k)$, required to solve (28) and (30) too. Thus the total complexity is $O(S \cdot max_s\{b_s\})$ and scales independent of the capacity.

5 Applications

The Bernoulli-Poisson-Pascal (BPP) process is often used for traffic engineering, because each process component has different peakness ς_k [10]. Thus a general traffic process can be approximated by the BPP process by an appropriate mapping of its mean and peakness. The Bernoulli process component is already given by the MFS-model, also called the multirate Engset model, which is commonly used to model at fixed number of sources M_k, where the arrival rate $\lambda_k(x_k) = (M_k - x_k)\varphi_k$ depends on the number of free sources. But in contrast to the MFS-model it is now used to model a process with a peakness below one. Similar the Pascal process is defined by $\lambda_k(x_k) = (M_k + x_k)\varphi_k$ and has peakness above one. If compared to the assumed arrival rates $\lambda_k(x_k) = \gamma_k + x_k\alpha_k$, the following cases can be distinguished:

Bernoulli: $\quad \alpha_k = -\rho_k$, $\gamma_k = M_k\rho_k$, $\rho_k = \varphi_k/\mu_k < 1$, $\varsigma_k = \dfrac{1}{1+\rho_k} < 1$. $\qquad$ (31)

Pascal: $\quad \alpha_k = \rho_k$, $\gamma_k = M_k\rho_k$, $\rho_k = \varphi_k/\mu_k < 1$, $\varsigma_k = \dfrac{1}{1-\rho_k} > 1$. $\qquad$ (32)

Poisson: $\quad \alpha_k = 0$, $\gamma_k = \lambda_k/\mu_k$, $\varsigma_k = 1$. $\qquad\qquad\qquad\qquad\qquad\qquad$ (33)

For the batched Poisson arrival process with geometric batch size distribution $S_j^{(k)} = (1 - \hat{\beta}_k)\hat{\beta}_k^{j-1}$, $j = 1,2,\dots$, it was proven in [1], that the aggregated state distribution of the process is identical to that of a Poisson process with state dependent arrival rates $\lambda_k(x_k) = \lambda_k + x_k\hat{\beta}_k\mu_k$. A comparison gives the parameters

$$\gamma_k = \lambda_k/\mu_k, \ \alpha_k = \hat{\beta}_k \ \text{and} \ \varsigma_k = \frac{1}{1-\hat{\beta}_k} > 1. \qquad (34)$$

Further on for a given blocking probability for the Pascal process, that for the batched Poisson arrival process with geometric batch size distribution follows from [1, (21)]

6   Numerical Results

Exact results for the blocking probabilities are obtained from (1) - (2). The approximation to the blocking probability is called UAA_BPP and is given by (21), where z^* satisfies (12), and v and $\hat{G}_k$ are given by (13) and (16). β_i follows from (15), and τ, y and w are obtained from (14).

For the first example with 2 classes the Pascal traffic model (32) is assumed and the utilization ρ_k of each customer is varied, whereas the offered traffic of each class $a_s = M_s\rho_s/(1-\rho_s) = M_s\rho_s\varsigma_s$ is kept constant. As an alternative to the utilization, the peakness can be calculated (32), which is in the range of 1 to 10. In Fig. 1 the absolute value of the relative approximation errors of the blocking probabilities $(B_k - \hat{B}_k)/\hat{B}_k$ for the UAA_BPP are compared with the errors obtained for the approximation given in [6], which is called UAA_BP. Let us remember that the UAA_BP has been derived for the Batched Poisson process, thus only a peakness of the traffic which is large or equal to 1 is considered first. The UAA_BPP is more accurate if compared to the UAA_BP and particular for high values of the customer utilization, respectively the peakness, this advantage is of practical importance. The UAA_BPP performs better, because the series expansion of the function used to solve the contour integral (9) (see (44) and (45) in the Appendix) takes into account higher order terms than the

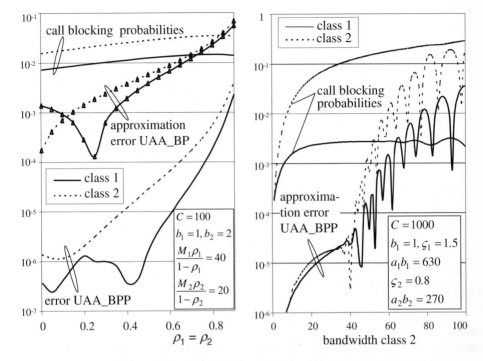

Fig. 1. Exact blocking probabilities and approximation errors for the UAA_BPP and UAA_BP

Fig. 2. Exact blocking probabilities and approximation errors for the UAA_BPP

UAA_BP does. Also observe, that the accuracy of both approximations decreases, if the utilization tends to one. In this case the pole (4) is near the saddle-point, which is the main reason for the decreasing accuracy.

The second example has 2 classes with different peakness. The offered bitrate traffic $a_k b_k$ of each class is kept constant, whereas the bandwidth b_2 is variable. Fig. 2 shows the exact blocking probabilities and the absolute value of the relative approximation error for the UAA_BPP. As expected the error increases with increasing bandwidth of class 2. Up to a ratio of $b_2 / C = 60 / 1000$ the error is below 1% and for ratios up to $b_2 / C = 40 / 1000$ the accuracy is very good for both classes.

Next an example with three classes is considered in Fig. 3. Class 2 has peakness $\varsigma_2 = 1$, whereas for a fixed value of the offered bit rate traffic the ratio $\varsigma_1 / \varsigma_3$ of the peakness of classes 1 and 3 is varied from $10 / 0.55$ to $0.55 / 10$. It can be seen that the accuracy of the approximation UAA_BPP declines if the peakness of class 3 increases, which has the largest bandwidth, even though the peakness of class 1 decreases simultaneously. Thus the accuracy is mainly effected by classes with large peakness and bandwidth. But if the largest value of the peakness is below 2, which is a value of practical interest, again the accuracy is very excellent.

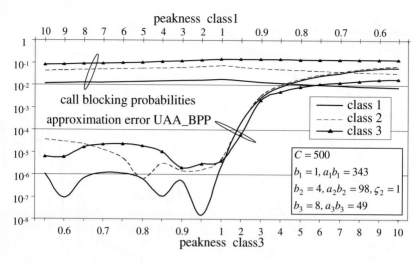

Fig. 3. Exact blocking probabilities and approximation errors for the UAA_BPP for different ratios of the peakness of class 1 and class 3

References

1. Kaufmann, J.S., Rege, K.M.: Blocking in a shared resource environment with batched Poisson arrival processes. Performance Evaluation 24, 249–263 (1996)
2. Ross, K.W.: Multiservice Loss Models for Broadband Telecommunication Networks. Springer, New York (1995)
3. Roberts, J.W.: Teletraffic models for the Telecom 1 integrated services network. In: Proc. 10th. Int. Teletraffic Conference, paper 1.1-2 (1983)
4. Mitra, D., Morrison, J.A.: Erlang Capacity and Uniform Approximations for Shared Unbuffered Resources. IEEE/ACM Transactions Networking 2, 558–570 (1994)
5. Bziuk, W.: Approximate state probabilities in large shared multirate loss systems with an application to trunk reservation. Europ. Trans. on Telecommunications, pp. 205–216 (2005)
6. Morrison, J.A.: Blocking probabilities for multiple class batch Poisson arrivals to a shared resource. Performance Evaluation 25, 131–150 (1996)
7. Morrison, J.A., Ramakrishnan, K.G., Mitra, D.: Refined Asymptotic Approximations to Loss Probabilities and their Sensitivities in Shared Unbuffered Resources. SIAM J. Appl. Math. 59, 494–513 (1998)
8. Kleinrock, L.: Queueing Systems. John Wiley & Sons, New York (1975)
9. Fedoryuk, M.V.: Asymptotic methods in analysis. In: Gamkrelidze, R.V. (ed.) Analysis I, Springer-Verlag, New York (1989)
10. Chung, S., Ross, K.W.: Reduced Load Approximations for Multirate Loss Networks. IEEE Transactions on Communications 41, 1222–1231 (1993)

Appendix: Uniform Approximation

The RUAA to the contour integral (9) derived now is a generalization of the RUAA given in [7], thus only the steps different from [7] are reported. It is assumed for the time that $z^* \neq 1$, $z^* \in D$ and the contour of integration is shifted to $|z| = z^*$, so that

$$G_k = \frac{q_k(1)}{2}\left[1 - \operatorname{sgn}(1 - z^*)\right] + \frac{1}{2\pi i}\oint_{|z|=z^*}\frac{q_k(z)}{z(1-z)} \cdot e^{Cf(z)}dz,$$ (35)

where the first term considers the contribution of the pole at $z = 1$ if $z^* > 1$. We let

$$f(z^*) - f(z) = u^2.$$ (36)

Since $f(1) = 0$, we let $z = 1$ correspond to $u = -i\kappa$, where [7, (4.3)]

$$\kappa = \operatorname{sgn}(1 - z^*)\sqrt{-f(z^*)}.$$ (37)

To approximate the contour integral (35), let us define the function [7]

$$\theta_k(u) = \phi_k + l_k(u)(u + i\kappa),$$ (38)

where $l_k(u)$ is assumed to be analytic at $u = -i\kappa$. Now, for (35) the substitution

$$\frac{q_k(z)}{z(1-z)}\frac{dz}{du} = \frac{\theta_k(u)}{u + i\kappa}$$ (39)

is used. Also it follows from (36) that $dz / du = -2u / f'(z)$ and with (39) we have

$$\theta_k(u) = -q_k(z)\frac{u + i\kappa}{z(1-z)}\frac{2u}{f'(z)}.$$ (40)

From (36) we have $f(z^*) - f(1) = -\kappa^2$ and $f(z) - f(1) = -(u^2 + \kappa^2)$. With a series expansion of $f(z) - f(1)$ and since $f(1) = 0$, it can be shown that [7, (4.5)]

$$f'(1)(z - 1) = 2i\kappa(u + i\kappa) + O\left(|u + i\kappa|^2\right), \quad |z - 1| \ll 1.$$ (41)

Hence, for $u \to -i\kappa$, $z \to 1$, (38) yields $\theta_k(-i\kappa) = \phi_k$ and also from (40) and (41) we have $\theta_k(-i\kappa) = -q_k(1)$, thus $\phi_k = -q_k(1)$. Let us define

$$g_k(z) = \frac{q_k(z)}{z(1-z)},$$ (42)

then from (38) and (39) it follows, that

$$l_k(u) = \frac{q_k(1)}{u + i\kappa} + g_k(z)\frac{dz}{du}.$$ (43)

With a Taylor series of $f(z^*)$ for $|z - z^*| \ll 1$ and (36), u^2 can be expressed as a polynomial function of z, which has been reversed in [7, (4.8)] to get $z - z^* = i\beta_1 u$

$+\beta_2 u^2 + i\beta_3 u^3 + ...,$ for $|u| << 1$. This allows us to derive the derivative $dz/du = i\beta_1 + 2\beta_2 u + i3\beta_3 u^2 + ...,$ as well as a series expansion for $g_k(z)\cdot dz/du$ in terms of the variable u, which is given in [7, (4.14)-(4.17)] for the case $h(z) = g_k(z)$. Now, with

$$\frac{q_k(1)}{(u+i\kappa)} = -i\frac{q_k(1)}{\kappa}\left(1 + \frac{iu}{\kappa} - \frac{u^2}{\kappa^2} + ...\right), \quad |u| << 1,$$

as well as (42) and (43), the series expansion [7, (4.19)] can also be generalized. In particular for $h(z) \equiv g_k(z)$, $z^* \in D$, $z^* \neq 1$, $|z - z^*| << 1$, $|u| << 1$, we have

$$l_k(u) = i\frac{\sqrt{2}}{\sqrt{v}}K_k + \left[2\beta_2 g_k(z^*) - \beta_1^2 g_k'(z^*) + \frac{q_k(1)}{\kappa^2}\right]u - i\frac{2\sqrt{2}}{v\sqrt{v}}E_k u^2 + ..., \quad (44)$$

where

$$K_k = \frac{\sqrt{v}}{\sqrt{2}}\left[\beta_1 g_k(z^*) - \frac{q_k(1)}{\kappa}\right],$$

$$E_k = -\frac{v\sqrt{v}}{2\sqrt{2}}\left[3\beta_3 g_k(z^*) + 3\beta_1\beta_2 g_k'(z^*) - \frac{1}{2}\beta_1^3 g_k''(z^*) + \frac{q_k(1)}{\kappa^3}\right].$$

For $q_k(z) = 1$ (44) reduces to [7, (4.19)]. Now, from (35), (36), (39), (42) and (43),

$$\frac{1}{2\pi i}\oint_{|z|=z^*}\frac{q_k(z)}{z(1-z)}\cdot e^{Cf(z)}dz = \frac{1}{2\pi i}e^{Cf(z^*)}\int_T e^{-Cu^2}\left[l_k(u) - \frac{q_k(1)}{u+i\kappa}\right]du, \quad (45)$$

where the contour T corresponds to $|z| = z^*$. As long as the saddle-point is not too close to the pole z_P, the main contribution to the integral arises from the neighborhood of the saddle-point. For $q_k(z) = 1$ the last integral is of the form [7, (4.20)], which has solution [7, (4.21)-(4.22)]. Utilizing this result and (44) we have

$$\frac{1}{2\pi i}\int_T e^{-Cu^2}l_k(u)du \approx \frac{1}{\sqrt{2\pi vC}}\left(K_k - \frac{E_k}{vC}\right),$$

Similar from [7, (4.23)] it can be concluded, that

$$\frac{e^{Cf(z^*)}}{2\pi i}\int_T e^{-Cu^2}\frac{q_k(1)}{u+i\kappa}du \approx -\frac{q_k(1)}{2}sgn(\kappa)Erfc\left[\sqrt{-Cf(z^*)}\right].$$

Now, with $Erfc[-x] = 2 - Erfc[x]$ and (15) the final relations (16)–(20) are obtained.

Power Control and Rate Adaptation in Multi-hop Access Networks - Is It Worth the Effort?

Carsten Burmeister[1], Ulrich Killat[1], and Kilian Weniger[2]

[1] Hamburg University of Technology, Communication Networks Department
Schwarzenbergstr. 95d, 21073 Hamburg, Germany
[2] Panasonic Research & Development Center Germany GmbH
Monzastr. 4c, 63225 Langen, Germany

Abstract. We investigate the effect that power control and rate adaptation could have on the performance of multi-hop access networks. Therefore, we formulate an optimization problem to find optimum routes in a multi-hop access network. These optimum routes are found using global knowledge and thus have only limited practical relevance. However, they can well be used to calculate theoretical upper limits on the performance, which no routing algorithm could exceed. We calculate the performance limits with and without power control and rate adaptation and compare them with each other to judge whether these mechanisms are worth the effort to be deployed in wireless multi-hop access networks.

1 Introduction

Wireless Local Area Networks (WLAN) see increasing deployment in private and public environments today. In so called Hot Spots Internet access is provided via either isolated access points or small networks consisting of few of these. Even though the users are able to achieve a good quality of service on average, this depends on other terminals connected to the same network. In case of heavily loaded access points, the quality of service per user is degraded and the throughput might decrease to negligible values [2].

To increase the capacity of the networks in multi-rate environments, multi-hop communication is seen as a possible solution. In multi-hop networks terminals could act as routers by forwarding flows for other stations. By doing so, error rates and energy consumption could be reduced and throughput, capacity and coverage could be increased.

Wireless multi-hop networks have been studied for years now in form of isolated ad-hoc networks. These ad-hoc networks are totally self organized networks, without infrastructure support. Multi-hop access networks differ from these by providing access to a fixed network. Still, most lessons learned from ad-hoc networks apply also for wireless multi-hop access networks.

Power control and rate adaptation are two mechanisms of wireless communication techniques which are successfully deployed today. Rate adaptation uses

L. Mason, T. Drwiega, and J. Yan (Eds.): ITC 2007, LNCS 4516, pp. 767–778, 2007.

the existing bandwidth more efficiently, by choosing a modulation scheme that maximizes the throughput for a given signal to noise ratio. Using rate adaptation in multi-hop networks could enhance its performance. As it is already implemented in WLAN cards and access points today, it can be used without additional costs.

Using power control, a station adopts the transmission power. In individual power control every station individually sets the transmission power to the lowest possible value that still enables communication to the intended receiver. Individual power control in WLAN environments according to the IEEE 802.11x standards is problematic. The IEEE 802.11 channel access mechanisms are not designed for power control. Work arounds and solutions to the problem have been proposed, but all of these have some drawbacks and performance issues. Still, it was shown in [1] and [7] that due to the spatial reuse of the transmission frequency, the capacity of the network can be increased.

We consider wireless multi-hop access networks consisting of WLAN stations. Users connect to the Internet via a fixed access point. Possibly other stations of the network forward a flow to the access point. We want to show, to which extend one could theoretically gain from power control and rate adaptation in these networks.

Using global knowledge, we calculate routes from mobile stations to the gateway such, that a certain throughput per flow can be guaranteed. The number of flows that can be routed with these guarantees serves as a theoretical upper performance limit. It is an upper limit, because the route calculation will find a route with these guarantees if one exists. It is a theoretical result, because the routes are calculated using global knowledge, which is usually not available to the routing entity. However, comparing the performance of these routes using power control and/or different modulation schemes lets us see to which extend we can gain from these mechanisms.

As said above, using power control might require to solve some basic problems first. Further, routing algorithms usually do not consider different modulation schemes and thus would need to be enhanced to support rate adaptation. In this work we want to show whether it is worth the effort to do so.

We will first describe the involved technologies in the next section in greater detail. We describe the assumed scenario and a mathematically tractable model of that in Section 3. This lets us formulate an optimization problem in Section 4, which is solved to calculate the optimum routes. The performance gains of using different technologies are presented and evaluated in Section 5. Finally we summarize our results and draw some conclusions.

2 Power Control and Rate Adaptation in Wireless LANs

2.1 Power Control

Power control in ad-hoc networks brings mainly two benefits. On the one hand the energy consumption of the terminals will be reduced, if they transmit with less power while still preserving connectivity. On the other hand the capacity of the network could be increased because of increased spatial reuse of the medium: Stations transmitting with less power generate less interference and possibly enable more stations to transmit simultaneously. In this work we are concerned with the latter benefit only, i.e. we want to investigate the possible effect that power control has on the network capacity.

In general we can divide power control mechanisms in ad-hoc networks into two categories, namely common power control and individual power control. In common power control, all stations of a network agree on the same transmission power level. Usually a power level is selected that assures a certain degree of connectivity between nodes that optimizes some performance parameters. In [6] and [7] it was shown that common power control could increase the throughput in ad-hoc networks.

In individual power control, each station might use a different transmission power level. The transmission power is usually selected as such that the next hop can be reached with a certain error probability or that the station has a certain number of neighbours. Its use in WLAN stations which use the IEEE 802.11 channel access mechanisms is however problematic. The medium access is based on the listen before talk principle, which means that stations have to sense the channel idle for certain periods of time until they can be reasonable sure that no other station is transmitting at the same time. This mechanism is based on symmetric communication, which means that a station is assumed to be able to hear other stations which are disturbed by its own transmission. This assumption is not necessarily valid, if stations transmit with different power levels. Hence, collisions could occur, because some stations might not be aware of ongoing transmissions.

In [8] the authors describe extensions to the WLAN MAC layer that enables individual power control. Thereby the handshake messages, that are exchanged before the data, are transmitted with the maximum power level. This enables stations that overhear these reservation messages to reserve the following medium time for the transmitting station. The data packet is then transmitted with lower power. During the data transmission the power is periodically increased shortly to signal the ongoing transmission to stations that were not able to receive the handshake messages correctly. The authors show that the energy consumption can be reduced significantly, while the stations do not suffer from throughput degradations or an increased number of colliding data transmissions. However, spatial reuse of the medium is not possible with this kind of mechanism. There are other publications that use similar mechanisms, e.g. [1], [9], [10]. All of these have in common that spatial reuse is not possible.

In [11] the authors follow a different approach by introducing busy tones. Busy tones are sent on a different frequency and indicate to other stations ongoing transmissions. While the authors show for this mechanism the possibility to reduce energy consumption and additionally gain from spatial reuse of the medium, the costs are additional frequencies and non standard conform hardware.

Summarizing it can be said that using individual power control in ad-hoc networks to gain from spatial reuse of the channel is possible with additional costs. We will investigate in this paper, whether individual power control in multi-hop access network scenarios is worth the effort.

2.2 Rate Adaptation

Rate adaptive WLAN stations select for unicast data transmissions a modulation scheme that maximizes the throughput for a given signal to noise ratio. The higher the received signal strength, the higher the data rate of the modulation scheme that can be used while still keeping a certain bit or packet error probability. Rate adaptation is widely used in WLAN stations today. In [13] the auto rate fall-back mechanism is described that follows a trial and error approach to select the optimum modulation scheme. The data transmission is first tried with a high bit rate modulation scheme and reduced in cases of transmission errors. In [12] the authors describe a rate adaptation mechanism that measures the signal strength of the handshake messages, which are sent with the low bit rate modulation scheme before the data is transmitted. The signal strength is then used to select the modulation scheme that provides a certain maximum error probability.

In [2] we have described an analytical model of rate adaptive wireless LAN. Thereby, we calculate the MAC layer throughput of a wireless LAN station in dependence of the distance to the receiver and the used modulation scheme. We use this calculation in our model in the next section to select the modulation scheme that maximizes the MAC layer throughput based on the distance to the receiver.

3 Wireless Multi-hop Access Network - Scenario and Model

We assume a certain scenario, where we investigate the effect of power control and rate adaptation. In our scenario an access point is placed in the middle of a circular area with radius $r = 300m$. In this area, N mobile users are placed randomly according to a spatial uniform distribution. From these N users K want to connect to the access point and establish a Voice over IP (VoIP) connection to a host in the Internet. We neglect the influence of the Internet link and assume the connection to terminate at the access point. The VoIP connection is thereby characterized by a fixed packet transmission rate of $R_f = 50$ packets/s, each packet having a packet size of 72 bytes including RTP, UDP and IP header. Each VoIP connection consists of two flows, one from the station to the access point and one from the access point to the station. Hence, we have $F = 2K$ flows in the network. We denote with $s(f)$ the source node of flow f and with $d(f)$ its destination. We assume multi-hop communication capabilities in all nodes. We assume signal strength path loss according to the Walfish Ikegami propagation model [3]. All stations are based on the IEEE 802.11b standard, where four modulation schemes are possible with data rates of 11 Mbps, 5.5 Mbps, 2 Mbps and 1 Mbps. Channel access is according to the Distributed Co-ordination Function (DCF).

A station transmits a packet with one of P different power levels and one of M different modulation schemes. In [2] we have described an analytical model of rate adaptive wireless LAN and have derived the MAC layer throughput in dependence of the received signal strength and the used modulation scheme. The modulation scheme that maximizes the MAC layer throughput is here always used to transmit the packet. Together with the propagation model and the locations of the users, we can for each two users i and j determine the modulation scheme $mmax_{i,j}^p$ that is used when transmitting with transmission power level p. Further, we can calculate from the model described in [2] the expected packet transmission rate $R_{i,j}^{m,p}$ when station i sends packets with modulation scheme m and power level p to station j. As said above, the flows have a fixed packet arrival rate of R_f, where f denotes the considered flow out of F flows in total.

4   Optimization Problem

In this section we describe the optimization problem that lets us find routes for the flows using global knowledge of the network and all flows. A practically relevant routing algorithm would usually not have all this information available and thus might not find these routes. We measure the performance by whether a routing for a certain scenario was found that fulfils the QoS requirements of all flows. We calculate a theoretical upper limit of the performance by means of the optimization. Doing so with and without certain technologies, lets us judge a maximum performance gain, that one could expect from using the corresponding technology.

We first define a routing indicator $r_{i,j}^{f,m,p}$ to be set to one if flow f is routed with modulation scheme m and transmission power p between nodes i and j. Else it is set to zero. Further we define an intereference indicator $a_{i,j}^p$ to be set to one if nodes i and j disturb each others transmission at power level p. Else it is also set to zero.

As communication is always bi-directional in wireless LAN, because acknowledgements are transmitted, we can now calculate the link interference indicator, that describes if communication via link (i,j) with power level p is disturbing communication via link (k,l):

$$b_{(i,j),(k,l)}^p = \max\left(a_{i,k}^p + a_{j,k}^p + a_{i,l}^p + a_{j,l}^p, 1\right) \tag{1}$$

Now, we calculate the link load between nodes i and j, i.e. the load induced by routed transmission over this link and additionally by routed transmission via links in the interference range:

$$L_{i,j} = \sum_{k=1}^{N}\sum_{l=1}^{N}\sum_{p=1}^{P}\sum_{m=1}^{M}\sum_{f=1}^{F} r_{k,l}^{f,m,p} \cdot \frac{R_f}{R_{k,l}^{m,p}} \cdot b_{(k,l),(i,j)}^p \tag{2}$$

With this information we can now formulate the optimization problem as follows:

Minimize

$$\sum_{i=1}^{N}\sum_{j=1}^{N}\sum_{p=1}^{P}\sum_{m=1}^{M}\sum_{f=1}^{F} r_{i,j}^{f,m,p} \cdot \frac{1}{R_{i,j}^{m,p}} \tag{3}$$

subject to

$$\forall f: \quad \sum_{i=1}^{N}\sum_{p=1}^{P}\sum_{m=1}^{M} r_{s(f),i}^{f,m,p} = 1 \tag{4}$$

$$\forall f: \quad \sum_{i=1}^{N}\sum_{p=1}^{P}\sum_{m=1}^{M} r_{i,d(f)}^{f,m,p} = 1 \tag{5}$$

$$\forall(f,i), \text{with } i \neq s \wedge i \neq d: \quad \sum_{j=1}^{N}\sum_{p=1}^{P}\sum_{m=1}^{M} r_{i,j}^{f,m,p} = 0 \tag{6}$$

$$\forall(i,j): \quad L_{i,j} \leq 1 \tag{7}$$

$$\forall(i,j,m,p), \text{with } m > mmax_{i,j}^{p}: \quad r_{i,j}^{f,m,p} = 0 \tag{8}$$

In the objective function (Eqn. 3) we sum up all transmission delays of all routes. By minimizing this objective function, we select routes which have a shorter transmission delay, if that is possible. The objective function is not important for our performance metric: The performance is measured by identifying whether a routing can be found that does not create a link load greater than one on any possible link, which is determined by the constraints only.

In Eqn. 4 we ensure that each node that is a source node for a certain flow has exactly one route of this flow leaving the node. Similarly in Eqn. 5 it is ensured that for a destination node of a certain flow exactly one route of that flow enters the node. Eqn. 6 considers all nodes that are neither source nor destination and restricts the solutions that for each of these nodes the net flow is zero, which means that the same number of routes enter and leave the node for a specific flow.

The physical constraints from the wireless medium are put into the last two constraints. In Eqn. 7 the load of each link is limited to a value smaller or equal to one. The load is defined in Eqn. 2 and considers all routes in the interference range of the link. Another restriction from the wireless medium is that some modulation schemes are not possible between specific nodes. We described above already, that each link has a maximum possible modulation scheme assigned, which depends on the received signal strength. In Eqn. 8 we ensure that routes are used only with the maximum possible modulation scheme.

5   Performance Evaluation

The Mixed-Integer Linear Program (MIP) as described above can be solved for reasonable sized networks with the commercial solver CPLEX [4], which employs the Branch-and-Bound method (see [5] for details). The result is a complete routing of all flows under the constraints given above. If several solutions are possible the routing with the least transmission delay would be chosen. However, it is possible that no solution exists. This could be because too many flows need to be routed via a specific interference region and links in that region would have a link load larger than one. Another reason could be that a node is located too far from any other node and communication is simply not possible. If the optimization problem does not find a feasible solution, the QoS requirements of flows cannot be guaranteed.

We want to evaluate several multi-hop Internet access scenarios regarding the possibility to find a routing guaranteeing the QoS requirements. Therefore we varied different parameters of the scenario. The node density is varied by selecting the number of nodes as $N = 20$ or $N = 40$. The offered load to the networks is varied by selecting the number of flows in the network between $F = 2..30$. Further, we have evaluated the scenarios with power control, where the station can select one out of eight different power levels ($P = \{0.1, 0.5, 1, 5, 20, 50, 100\}$mW), with power control of four different power levels ($P = \{0.5, 5, 20, 100\}$mW) and without power control, where every packet is sent with a transmission power of $P = 100$ mW. The rate adaptation is evaluated by allowing either four modulation schemes with $\{1, 2, 5.5, 11\}$Mbps or only one modulation scheme of 1 Mbps.

For each parameter combination we generate ten random scenarios. Thereby, the users are located randomly across the area according to a uniform distribution. The nodes that use voice communication are selected randomly from all mobile stations. The corresponding node is always the access point. To judge the performance of power control and rate adaptation, we optimize each of these scenarios and evaluate the fraction of scenarios per parameter combination in which a routing was found, where flows achieve the QoS requirements.

5.1   Power Control Performance

First, we evaluate the effect of power control, if rate adaptation is not used. Therefore, we allowed the nodes to use only one modulation scheme with 1 Mbps. Fig. 1 shows the fraction of scenarios where a routing was found for $N = 20$ and $N = 40$ nodes. The number of flows was increased from 2 to 30.

The first thing to note is that the number of used power levels has hardly any effect on the performance. In the scenarios with $N = 20$ nodes, we are able to calculate routes for all scenarios with up to six flows. In higher loaded networks, a routing was found for only a small fraction of scenarios. With one power level only, it was not possible to route 14 flows or more. With four or eight power levels we could eventually find a correct routing for some higher load scenarios, but the gain is not significant.

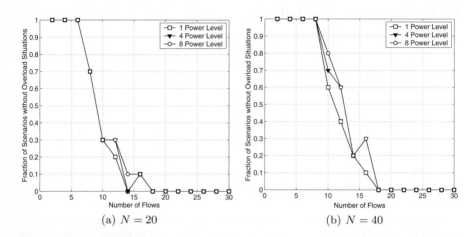

Fig. 1. Fraction of scenarios in which routing was found for one modulation scheme and different power levels

The overall performance is increased if we increase the number of nodes. For $N = 40$ nodes, we could find a feasible routing for roughly two more flows in the networks. The small performance gain for eight power levels is due to the higher node density. Routes are found with small distance hops, where a reduced transmission power leads to less interference. However, the relative performance between the different power level configurations is not significant.

Now, we wanted to see the effect of power control in rate adaptive scenarios. Therefore, the nodes were allowed to use any of the four modulation schemes. The results are shown in Fig. 2.

Basically, the curves follow the same trend as in the previous scenarios in that the difference between different numbers of power levels is hardly visible. For $N = 40$ nodes power control could lead to small performance benefits.

The result, that power control has nearly no influence on the number of flows that can be routed is surprising. With power control the interference can be reduced and an increased spatial reuse of the wireless channel should be possible. However, in the considered multi-hop Internet access scenario it is not possible to benefit from spatial reuse.

Now, we calculate the mean link utilization of the network as the average link load (as defined in Eqn. 2) calculated over all links, where communication is possible. We expect that the utilization is lower when using power control, because sending with lower transmission power interferes less links. The result is depicted in Fig. 3: Here, we show the mean link utilization in the case of $N = 40$ nodes using all modulation schemes. The three curves show the mean link utilization for different numbers of power control level and the confidence intervals (95%) of these. We see that the differences are not that high; even

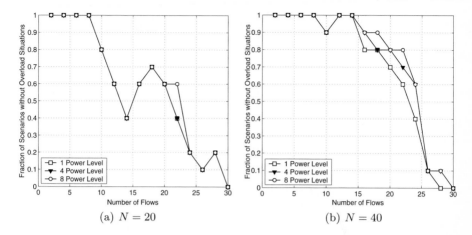

Fig. 2. Fraction of scenarios in which a routing was found for all modulation scheme and different power levels

if the mean link utilizations over all scenarios is lower with power control, the confidence intervals overlap in most of the cases.

5.2 Rate Adaptation Performance

In this section we show the results of the evaluation of rate adaptation in multi-hop Internet access scenarios. Basically the results are already included in Fig. 1 and Fig. 2, where we show the power control results with and without rate adaptation. In order to see the effect of rate adaptation more clearly we summarize these figures in Fig. 4. We compare the performance of using rate adaptation to the performance of using only one modulation scheme. In the figure the results are shown using all eight power levels. The results for one power level are similar as can be seen from Fig. 1 and Fig. 2.

A strong performance gain by using rate adaptation can be seen. Already in the $N = 20$ node scenario, in many scenarios a routing can be found using rate adaptation, where it is not possible without. In the $N = 40$ nodes scenarios the effect is even more significant. Here, roughly double the amount of flows can be routed with rate adaptation. The strong effect can be reasoned by the significantly reduced transmission times if higher modulation schemes are used. In this way the interference is reduced as it is done using power control, but on a temporal basis in contrary to the spatial basis. We see that in contrast to the interference reduction on a spatial basis, the temporal reduction has a great effect on the performance.

Looking at the link utilization in Fig. 5, this trend can be verified. Still, some of the confidence intervals overlap, but the difference between the two curves is higher than for the power control case. Further the utilization values of up to ten flows are not necessarily the least possible utilizations in these scenarios. Because

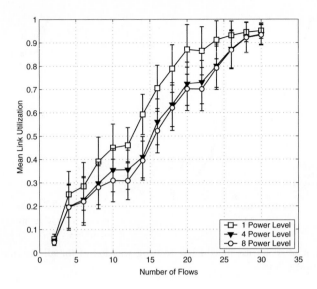

Fig. 3. Mean link utilization for different power control configurations

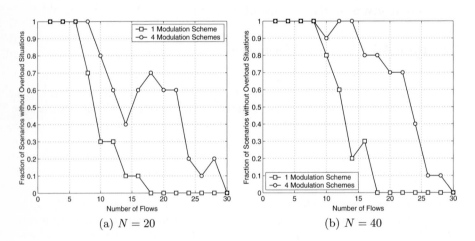

Fig. 4. Fraction of correctly routed scenarios with and without rate adaptation

of the chosen objective function in the optimization problem, low delay routes are chosen in case several solutions are found. Low delay routes are usually not routes that optimize the link utilization. In higher load scenarios, less solutions are found (if any). The routes that fulfil the constraints given in the optimization problem do minimize the link utilization. This is what we see in Fig. 5 in the higher load scenarios. From ten flows onwards, the difference between the curves

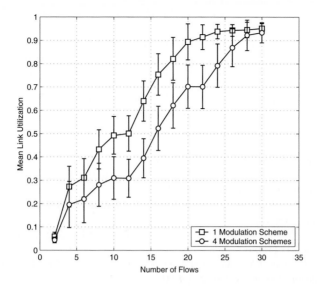

Fig. 5. Mean link utilization with and without rate adaptation

is more significant and confidence intervals do not overlap in most cases. These are the scenarios, where we could optimize the performance significantly using multiple modulation schemes.

6   Conclusion

We have evaluated the performance of rate adaptation and power control in a special wireless multi-hop access network scenario, consisting of one access point and a number of mobile stations. The performance was thereby measured as the probability that for a specificly loaded network a routing could be found as such no overload situations occur.

The most striking result is that power control has nearly no influence in the considered access scenarios. The mean link utilization is reduced marginally by using power control, which shows that spatial reuse of the radio channel is facilitated, but the resuls show that in the considered scenarios, this can hardly be exploited at all.

Rate adaptation however, which reduces the interference in the network on a temporal basis by reducing the transmission time when using higher modulation schemes, has a great effect on the performance. In some scenarios about double the number of flows could be routed by adding three higher modulation schemes to the basic one.

Considering the deployment issues of power control and further the existing implementations of rate adapatation in WLAN cards today, it can be said that in our scenario power control is not worth the effort, while rate adaptation surely is.

References

1. Agarwal, S., Katz, R.H., Krishnamurthy, S.V., Dao, S.K.: Distributed Power Control in Ad-hoc Wireless Networks. In: 12th IEEE International Symposium on Personal, Indoor and Mobile Radio Communications (PIMRC 2001), San Diego, CA, USA (2001)
2. Burmeister, C., Killat, U., Bachmann, J.: An Analytical Model of Wireless-LAN and its Simulative Verification. In: Third ACM International Workshop on Wireless Mobile Applications and Services on WLAN Hotspots, WMASH 2005, Cologne, Germany (2005)
3. Damasso, E., Correia, L.M. (eds.): Digital mobile radio: COST 231 view on the evolution towards 3rd generation systems. European Commission - COST Telecommunications, Brussels, Belgium (1998)
4. ILOG: ILOG CPLEX 8.1 Reference Manual (2002) Available from http://www.ilog.com
5. Lawler, E.L., Wood, D.E.: Branch-and-bound method: A survey. Operations Research, vol.14(4) (1966)
6. Gupta, P., Kumar, P.R.: The Capacity of Wireless Networks. IEEE Transactions on Information Theory, vol.46(2) (2000)
7. Narayanaswamy, S., Kawadia, V., Sreenivas, R.S., Kumar, P.R.: Power control in ad-hoc networks: Theory, architecture, algorithm and implementation of the COMPOW protocol. European Wireless 2002, Florence, Italy (2002)
8. Jung, E.-S., Vaidya, N.H.: A Power Control MAC Protocol for Ad Hoc Networks. In: Wireless Networks, vol. 11(1-2), Springer, Netherlands (2005)
9. Gomez, J., Campbell, A.T., Naghshineh, M., Bisdikian, C.: Conserving transmission power in wireless ad hoc networks. In: 9th International Conference on Network Protocols (ICNP 2001), Riverside, CA, USA (2001)
10. Pursley, M.B., Russell, H.B., Wysocarski, J.S.: Energy-efficient transmission and routing protocols for wireless multiple-hop networks and spread-spectrum radios. EuroComm 2000, Munich, Germany (2000)
11. Monks, J.P., Bharghavan, V., Hwu, W.M.W.: A power controlled multiple access protocol for wireless packet networks. In: 20th Annual Joint Conference of the IEEE Computer and Communications Societies (INFOCOM 2001), Anchorage, AL, USA (2001)
12. Holland, G., Vaidya, N., Bahl, P.: A Rate-Adaptive MAC Protocol for Multi-Hop Wireless Networks. In: 7th Annual International Conference on Mobile Computing and Networking (MobiCom 2001), Rome, Italy (2001)
13. Kamerman, A., Monteban, L.: WaveLAN-II: A high-performance wireless LAN for the unlicensed band. Bell Labs Technical Journal (1997)

Random Multi-access Algorithms in Networks with Partial Interaction: A Mean Field Analysis*

Charles Bordenave[1], David McDonald[2], and Alexandre Proutiere[3],**

[1] University of California, Berkeley, USA
[2] University of Ottawa, Canada
[3] KTH, the Royal Institute of Technology, Sweden

Abstract. We consider a network with a fixed number of links whose transmitters are saturated and access a channel using a random back-off algorithm. Some of the links may be hidden in the sense that they do not interfere with all other links but rather with a subset of the links. Using mean field techniques, we analyze a variety of random back-off algorithms by explicit calculating the throughput of the links in such networks. We apply our results to analyze the performance of the exponential back-off algorithm in networks with partial interaction. The results are striking and confirm experimental results. Hidden transmitters that fail to sense collisions with other links unfairly grab too much bandwidth at the expense of transmitters that comply with the back-off rules. We believe the model can be used to develop new algorithms realizing an adequate trade-off between fairness and efficiency.

1   Introduction

Random multi-access algorithms, from the first version of the Aloha algorithm [2] to more recent protocols used in wired and wireless LANs [1], have played an increasingly important role in telecommunication networks for more than 30 years. They have therefore generated a lot of research interest, but their performance remains extremely complicated to analyze, particularly in the case of adaptive random multi-access algorithms (such as the exponential back-off algorithm) where the processes representing the state of transmitters are inherently correlated.

In this paper, we consider networks where the simultaneous transmissions of two transmitters result in a collision when the corresponding links are close to each other, i.e., interfere. When modeling an access point in wireless LAN for example, all links are assumed to interfere with each other; i.e. there is a full interaction. In such a scenario Bianchi [4] proposed a heuristic leading to a simple yet precise analysis of the performance of the exponential back-off algorithm. His heuristic is based on a decoupling approach; i.e. assuming that the back-off

---

* This work was done during C. Bordenave's and A. Proutiere's visit to University of Ottawa. We wish to thank Prof. McDonald for his kind hospitality.
** A. Proutiere is currently on sabbatical leave from France Telecom R&D.

L. Mason, T. Drwiega, and J. Yan (Eds.): ITC 2007, LNCS 4516, pp. 779–790, 2007.

processes of the various links are mutually independent. This assumption was recently theoretically justified [6,12] using mean field asymptotics, and has been extensively used to predict the performance of networks with full interaction under various traffic scenarios.

In many situations however, the interaction between links is partial, which means that some links do not interfere with all the other links, but rather with a restricted subset of links. This is the case of AdHoc and mesh networks, but also of certain networks with access points (cellular-type networks). In these networks some links have more interfering neighbors than others, leading to inevitable throughput unfairness when the random algorithms are not adapted to the network topology. Due to the system complexity, the performance of networks with partial interaction has only been studied in a very limited set of cases. The performance of non-adaptive Aloha-type algorithms can be approximately derived using loss networks [8,9]. More practical adaptive algorithms, such the exponential back-off algorithm, are much more challenging to analyze.

In this paper we provide a general mathematical model, based on the mean field analysis of systems of interacting particles, leading to exact performance results for a large class of random back-off algorithms in networks with full or partial interaction (see Theorem 4 in Section 5). Our model is used to predict the performance of the classical exponential back-off algorithm, and in particular to quantify the throughput unfairness in case of networks with partial interaction, (see Corollary 1 in Section 5). The results are striking and confirm experimental results [10]. Hidden nodes that fail to sense collisions with other links unfairly grab too much bandwidth at the expense of nodes that comply with the back-off rules. The model can be also useful to design new algorithms, that for example, could realize a better trade-off between fairness and efficiency.

2   System Models and MAC Protocols

2.1   Network and Traffic Models

We consider a network consisting of a set $\mathcal{N}$ of N saturated wireless links. A link is saturated in the sense that there are always packets to be transmitted in the corresponding buffer. This assumption is relevant when considering data networks: in such networks, TCP tends to saturate buffers. The N links interact through interference depending on the geographical locations of the corresponding transmitters and receivers. The interference model is kept simple here: we assume that no transmission on link i can be successful if one of the interfering links is transmitting at the same time. We restrict the analysis to symmetric interference scenarios, where when link i interferes link j, the j interferes i (the model can be generalized to account for asymmetric interference). Links are classified according to their interference properties, i.e., two links belong to the same class if they interfere with the same set of links. Denote by $\mathcal{C}$ the set of link classes, and by μ_c the proportion of links of class c. $i \in c$ denotes the fact that link i is of class c. Interference between classes is characterized by the symmetric matrix A such that $A_{cd} = 1$ if class-c links interfere class-d links, and $A_{cd} = 0$

otherwise. We denote by $\mathcal{V}_c = \{d \in \mathcal{C} : A_{cd} = 1\}$ the set of classes of links interfering with class-c links. A typical example where the above model applies is as follows.

Example (Networks with access points and overlapping cells). Consider the uplink of a wireless LANs with two access points (APs) sharing the same channel (which means that users attached to different access points may interfere). This model was considered in [11]. The network is divided into 3 geographical zones: Zones 1 and 3 where users can only transmit to a single AP and do not interfere with each other; Zone 2 where users may transmit to both APs and interfere any other transmission. The network is illustrated in Figure 1. Links are from users to the APs. The class of a link is just the zone where the corresponding user is located.

The performance of the network is measured in terms of the long-term throughputs realized by the various links. The aim of our analysis is to quantify these throughputs. Since links of the same class experience the same performance, we will only compute the global throughput of links of the different classes. We denote by γ_c the throughput of class-c links. It is by definition the proportion of time links of class c transmit successfully.

The networks considered here may have natural spatial heterogeneity as some links may suffer interference from more links than others. This may create substantial unfairness in terms of link throughput. We will then be interested in quantifying the basic trade-off between the total network throughput and fairness realized by the random access algorithms considered.

2.2   Random Multi-access Algorithms

Transmitters access the channel in a distributed manner using some random multi-access algorithm. We aim at developing a very broad model, able of capturing the characteristics of a large class of such algorithms in order to compare existing algorithms and to design new algorithms with improved performance.

We consider a time slotted system. The duration of a slot will be taken as the time unit. For simplicity, the transmissions of all packets have the same average duration, equal to L slots. This assumption means for example that if all packets are of identical size then so must be the rates of all links. Note that it is not crucial, but simplifies the exposition of our theoretical model. The average duration of collisions is denoted by L_c and unless otherwise specified, $L_c = L$.

Each transmitter is assumed to run a random back-off algorithm. When it has a packet to transmit, it randomly chooses a back-off timer expressed in slots. The back-off timer is then decremented at the end of each slot when the transmitter observes an idle channel, until the back-off timer reaches 0, in which case, the transmitter attempts to send the packet. To simplify the notation, we assume that all timers are geometrically distributed, so that we can represent the state of a transmitter by the probability it attempts to use the channel at the end of an idle slot. This probability p takes values in an at most countable set $\mathcal{B}$, and evolves as follows. In case of successful transmission, it becomes $S(p)$, and in case of collision $C(p)$, where $S : \mathcal{B} \to \mathcal{B}$ (resp. $C : \mathcal{B} \to \mathcal{B}$) is an increasing (resp. decreasing) function. Random multi-access algorithms are usually categorized according to their ability to adapt to the number of sources competing for the use of the channel.

- *Non-adaptive algorithms.* The first random multi-access algorithm, Aloha, was introduced by Abramson [2]. It is a non-adaptive algorithm where basically, each source transmits with a constant probability (S and C are identical and constant functions). The problem with such an algorithm is that as the population of sources grows large, the network throughput decreases and ultimately reaches 0.
- *Adaptive algorithms.* Adaptive random algorithms are more suited to the case where the environment of a link may change, which is basically always the case in practical networks. These algorithms are designed so that a transmitter can *discover* its environment, i.e., the number of active links competing for the use of its channel. In fact, all implemented multi-access algorithms are adaptive. They are often based on the so-called *exponential back-off algorithm*, where the transmission probability becomes p_0 after a successful transmission and divided by a factor 2 in case of collision: $S(p) = p_0$ and $C(p) = p/2$. In that case, $\mathcal{B} = \{p_0 2^{-n}, n \in \mathbb{N}\}$. The Ethernet (in wired LANs) and the IEEE802.11 DCF (in wireless LANs) protocols implement this algorithm, usually with $p_0 = 1/16$.

2.3   Results

The main contribution of this paper is to derive explicit analytical expressions for the throughput of any link. We do so in Theorem 4 where we characterize the stationary proportion Q_c^n of links that are of class c and that we will attempt to use the channel at the end of an idle slot with probability $p_0 2^{-n}$. From these proportions we deduce the stationary throughputs of links in class c, see (10)-(11)-(12).

3   Performance Analysis of Non-adaptive Algorithms

In this section, we analyze the performance of networks under non-adaptive multi-access algorithms. The transmitter of link i attempts to use the channel at the end of an empty slot with constant probability p_i.

To partially capture the network dynamics, we define a process $Z = \{Z(k), k \geq 0\}$ representing the classes for which there exists at least one active link during slot k: $Z(k) \in \mathcal{Z} = \{0, 1\}^{|\mathcal{C}|}$, where $Z_c(k) = 0$ if and only if there is no active link of class c. Note that if $d \in \mathcal{V}_c$ and $Z_c(k) = 1 = Z_d(k)$, there is necessarily a collision between links of classes c, d. Given that $Z_c(k) = 1$, and that for all $d \in \mathcal{V}_c$, $Z_d(k) = 0$, there is either a successful transmission or a collision involving links of class c only. We also introduce the *clear-to-send* functions C_c as follows. If $Z(k) = z$, a class-c link is clear to send at the end of slot k if $C_c(z) = 1$ where

$C_c(z) = 1$ if $z_d = 0$ for $d \in \mathcal{V}_c$, otherwise $C_c(z) = 0$. We say that two classes c and d are consecutive if $A_{cd} = 1$. Then for all state z, the active classes c are classified into maximal connected components. c and d belong to the same component if one can find of sequence of transmitting, consecutive classes containing both c and d. Let $N(z)$ denote the set of maximal connected components in state z. Z is a Markov chain with transition kernel K defined as follows: for all $z, z' \in \mathcal{Z}$,

$$K(z, z') = K_A(z, z')K_D(z, z')K_0(z, z'), \tag{1}$$

where $K_A(z, z')$ (resp. $K_D(z, z')$) corresponds to the set of links that become active (resp. inactive) from z to z', and $K_0(z, z')$ corresponds to the links that do not change state in a transition from z to z'. Let $A(z, z')$, respectively $D(z, z')$, denote the maximal connected components created, respectively removed, in a transition from z to z'. Then, we have:

$$K_A(z, z') = \prod_{E \in A(z,z')} \prod_{c \in E} C_c(z)(1 - \prod_{i \in c}(1 - p_i)), \quad K_D(z, z') = \left(\frac{1}{L}\right)^{|D(z,z')|},$$

$$K_0(z, z') = \left(1 - \frac{1}{L}\right)^{|N(z) \setminus D(z,z')|} \prod_{c:z_c=0=z'_c} (C_c(z) \prod_{i \in c}(1 - p_i) + 1 - C_c(z)).$$

Note that the Markov chain Z evolves in a finite state space and is provably irreducible. Z is then stationary ergodic, and we denote by π its stationary distribution. π can be easily numerically evaluated using K, but in general it has no explicit expression. Also note that the evolution of Z can be interpreted as the evolution of the number of clients in a discrete time loss network, where any maximal connected component corresponds to a route (refer to [8] for details). The arrival rates on routes depend on the system state, which explains why π can not explicitly evaluated. Fortunately, when the collision probability is negligible, i.e., when $p_i \ll 1$, the arrival rates become independent of the state, and π can be approximated by [8]: $\pi(z) \approx \pi(0)K_A(0, z)L^{|N(z)|}$.

From the stationary distribution π, we can deduce the mean throughput γ_c obtained by links of class c. Consider the point process of returns to the set $\mathcal{A} = \{z : C_c(z) = 1\}$. Let T_1 denote the first return time after time zero. By the cycle formula (see (1.3.2) in [3]) we may express the steady state probability of a link in c successfully transmitting a packet by the mean time spent in the transmission state per cycle divided by the mean cycle length. The expectation is calculated with respect to the Palm measure of the point process of returns to $\mathcal{A}$ but in this Markovian case this just means starting on $\mathcal{A}$ with probability $\pi^{\mathcal{A}}$ which is π renormalized to be a probability on $\mathcal{A}$.

A link in c can only go into a successful transmission state once per cycle; i.e. no other link in c transmits and other links in $\mathcal{V}_c$ are either blocked or remain silent. Hence the mean time per cycle spent in a transmission state is

$\sum_{z \in \mathcal{A}} \pi^{\mathcal{A}}(z) L g(z)$ where $g(z) = \sum_{i \in c} p_i \prod_{j \in c, j \neq i} (1 - p_j) \prod_{d \in \mathcal{V}_c, d \neq c} (C_d(z) (\prod_{j \in d}$
$(1 - p_j) - 1) + 1)$. Moreover, $\sum_{z \in \mathcal{A}} \pi^{\mathcal{A}}(z) E_z[T_1] = \frac{1}{\pi(A)}$; i.e. the intensity of the point process of visits to $\mathcal{A}$. Finally the throughput is given by:

$$\gamma_c = \sum_{z: C_c(z)=1} \pi(z) L \left(\sum_{i \in c} \frac{p_i}{1 - p_i} \prod_{d \in \mathcal{V}_c} \left(C_d(z) (\prod_{j \in d} (1 - p_j) - 1) + 1 \right) \right). \quad (2)$$

4   Mean-Field Limits of a Particle System Interacting with a Rapidly Varying Environment

The analysis of adaptive multi-access algorithms is much more challenging than that of non-adaptive algorithms. This is due to the complexity of the underlying Markov chain representing for example the states of the various back-off processes at the transmitters. To circumvent this difficulty we study the system when the number of links grows large. To do so, we use mathematical tools, usually developed in statistical physics to capture the behavior of a system of particles. In this section, we present a general model of interacting particles in a Markovian environment, and its mean field analysis. All the detailed assumptions and proofs of the results can be found in our companion paper [7], and have been removed due to space limitations.

Notation. Let $\mathcal{S}$ be a separable, complete metric space, $\mathcal{P}(\mathcal{S})$ denotes the space of probability measures on $\mathcal{S}$. $\mathcal{L}(X)$ is the law of the $\mathcal{S}$-valued random variable X. $D(\mathbb{R}^+, \mathcal{S})$ the space of right-continuous functions with left-handed limits, with the Skorohod topology [5]. We extend a discrete time trajectory $(X(k), k \in \mathbb{N})$, in $D(\mathbb{N}, \mathcal{S})$ in a continuous time trajectory in $D(\mathbb{R}^+, \mathcal{S})$ by setting for $t \in \mathbb{R}^+$, $X(t) = X([t])$, where $[\cdot]$ denotes the integer part.

4.1   Model Description

The particles. We consider N particles evolving in a countable state space $\mathcal{X}$ at discrete time slots $k \in \mathbb{N}$. For simplicity we assume the particles are exchangeable. At time k, the state of the ith particle is $X_i^N(k) \in \mathcal{X}$. The state of the system at time k is described by the empirical measure $\nu^N(k) \in \mathcal{P}(\mathcal{X})$ while the entire history of the process is described by the empirical measure ν^N on path space $\mathcal{P}(D(\mathbb{N}, \mathcal{X}))$: $\nu^N(k) = \frac{1}{N} \sum_{i=1}^N \delta_{X_i^N(k)}$ and $\nu^N = \frac{1}{N} \sum_{i=1}^N \delta_{X_i^N}$.

The Markovian Environment. In the system we consider, the evolution of the particles depends not only on the state of the particle system but also on a background Markovian process $Z^N \in D(\mathbb{N}, \mathcal{Z})$, where $\mathcal{Z}$ is a countable state space. This process evolves as follows: $P(Z^N(k+1) = z | \mathcal{F}_k) = K_{\nu^N(k)}^N(Z^N(k), z)$, where K_μ^N is a transition kernel on $\mathcal{Z}$ depending on a probability measure μ on $\mathcal{P}(\mathcal{X})$, and where $\mathcal{F}_k = \sigma((\nu^N(0), Z^N(0)), \cdots, (\nu^N(k), Z^N(k)))$. Z^N is a Markov chain whose transition kernel evolves with the empirical measure of the state of the particle system.

Evolution of the Particles. We represent the possible transitions for a particle by a countable set $\mathcal{S}$ of mappings from $\mathcal{X}$ to $\mathcal{X}$. A s-transition for a particle in state x leads this particle to the state $s(x)$. In each time slot the state of a particle has a transition with probability $1/N$ independently of everything else. If a transition occurs, this transition is a s-transition with probability $F_s^N(x, \nu, z)$, where x, ν, and z are the state of the particle, the empirical measure, and the state of the background process before the transition respectively. Hence, in this state, a s-transition occurs with probability: $\frac{1}{N}F_s^N(x, \nu, z)$, with $\sum_s F_s^N(x, \alpha, z) = 1$ for all $(x, \alpha, z) \in \mathcal{X} \times \mathcal{P}(\mathcal{X}) \times \mathcal{Z}$. Note that the process Z^N evolves quickly while the empirical measure $\nu^N(k)$ evolves slowly. Also note that the s-transitions of the various particles may be correlated, and that the process Z^N may depend on the transitions of the particles. The particle system is thus in interaction with its environment.

4.2   Mean Field Asymptotics

We now characterize the evolution of the system when the number of particles grows. We assume that when $N \to \infty$, $F_s^N \to F_s$, $K_\nu^N \to K_\nu$ (see [7] for precise convergence definitions). As $N \to \infty$, the chains $X_i^N(t)$ slow down hence to derive a limiting behavior we define $q_i^N(t) = X_i^N([Nt])$ and $\mu^N = \frac{1}{N}\sum_{i=1}^N \delta_{q_i^N} \in \mathcal{P}(D(\mathbb{R}_+, \mathcal{X}))$.

Theorem 1. *Assume that the initial values $q_i^N(0)$, $i = 1, \ldots, N$, are exchangeable and chaotic: i.e. the empirical measure μ_0^N of the $q_i^N(0)$ converges weakly to a deterministic limit Q_0. There exists a probability measure Q on $D(\mathbb{R}^+, \mathcal{X})$ such that for all subsets $I \subset \mathbb{N}$ of finite cardinal $|I|$,*

$$\lim_{N\to\infty} \mathcal{L}\left((q_i^N(.))_{i\in I}\right) = Q^{\otimes|I|} \quad \text{weakly in } \mathcal{P}(D(\mathbb{R}^+, \mathcal{X})^{|I|}). \tag{3}$$

This theorem, whose proof can be found in [7], states that the particles become asymptotically independent. This phenomenon is known as the propagation of chaos [13].

The independence allows us to derive explicit expression for the system state evolution. As explained earlier, intuitively, when N is large, the evolution of the background process is very fast compared to that of the particle system. The particles then see an average of the background process. The following theorem formalizes this observation. For $\alpha \in \mathcal{P}(\mathcal{X})$, let π_α denote the stationary distribution of the Markov chain with transition kernel K_α. We define the average transition rates for a particle in state x by $\overline{F}_s(x, \alpha) = \sum_{z\in\mathcal{Z}} F_s(x, \alpha, z)\pi_\alpha(z)$. Define $Q^n(t) = Q(t)(\{x_n\})$ where $\mathcal{X} = \{x_n, n \in \mathbb{N}\}$. $Q^n(t)$ is the limiting (when $N \to \infty$) proportion of particles in state x_n at time t. We have:

Theorem 2

$$\forall n \in \mathbb{N}, \quad \frac{dQ^n}{dt} = \sum_{s\in\mathcal{S}} \sum_{m:s(x_m)=x_n} Q^m(t)\overline{F}_s(x_m, Q(t)) - Q^n(t)\sum_{s\in\mathcal{S}} \overline{F}_s(x_n, Q(t)). \tag{4}$$

The differential equations (4) have the following interpretation: if $s(x_m) = x_n$ then $Q^m(t)\overline{F}_s(x_m, Q(t))$ is a mean flow of particles from state x_m to x_n. Hence, $\sum_{s \in \mathcal{S}} \sum_{m:s(x_m)=x_n} Q^m(t)\overline{F}_s(x_m, Q(t))$, is the total mean incoming flow of particle to x_n and $\sum_{s \in \mathcal{S}} Q^n(t)\overline{F}_s(x_n, Q(t))$ is the mean outgoing flow from x_n.

Now, assume that the equations (4) describe the evolution of a globally stable dynamical system, and define by $Q^{st} = (Q^n_{st}) \in \mathcal{P}(\mathcal{X})$ its stable point: for all n, $\lim_{t \to +\infty} Q^n(t) = Q^n_{st}$. Then the asymptotic independence of the particles also holds in the stationary regime:

Theorem 3. *For all subsets $I \subset \mathbb{N}$ of finite cardinal $|I|$,*

$$\lim_{N \to \infty} \mathcal{L}_{st}\left((q^N_i(.))_{i \in I}\right) = Q^{\otimes|I|}_{st} \quad weakly \ in \ \mathcal{P}(D(\mathbb{R}^+, \mathcal{X})^{|I|}).$$

5   Performance Analysis of Adaptive Algorithms

In this section we apply the results of Section 4 to analyze the performance of networks under random multiple-access algorithms.

5.1   Complete Interaction

In [6], we have studied networks with complete interaction, i.e., networks where each link interferes any other link (the matrix A is such that $A_{cd} = 1$ for all c, d), and where the collision and successful transmissions may have different durations ($L_c \neq L$). The main result of [6] can be derived using the results in Section 4. It states that in the limit, the users behave independently and in case of exponential backoff algorithm, the proportion of users with renormalized transmission probability $p_0 2^{-n}$ at time t tends to a deterministic proportion $Q^n(t)$. For $n \geq 1$,

$$\frac{dQ^n}{dt}(t) = 2^{1-n}p_0 \left(Q^{n-1}(t)\left(1 - \exp(-\sum_{i=0}^{\infty} 2^{-i}p_0 Q^i(t))\right) - \frac{Q^n(t)}{2}\right), \quad (5)$$

$$\frac{dQ^0}{dt}(t) = \sum_{n=0}^{\infty} 2^{-n}p_0 Q^n(t)\exp(-\sum_{i=0}^{\infty} 2^{-i}p_0 Q^i(t)) - p_0 Q^0(t). \quad (6)$$

It can be easily proved that the dynamic system described by differential equations (5)-(6) admits a unique equilibrium point (Q^n) defined by:

$$\forall n \geq 1, \quad Q^n = (2(1 - e^{-\rho}))^n Q^0, \quad and \quad Q^0 = \rho e^{-\rho}/p_0,$$

where ρ satisfies $p_0 e^\rho + \rho - 2p_0 = 0$ or $p_0 = \rho/(2 - e^\rho)$. Note that the steady state always exists since, for any $p_0 > 0$, there exists a unique $\rho > 0$ such that $p_0 = \rho/(2 - e^\rho)$ with $2 - e^\rho > 0$. We can also verify that $\rho = \sum_n 2^{-n}p_0 Q^n$, and ρ may be interpreted as the probability that one transmitter accesses the channel. Similarly $e^{-\rho}$ may be seen as the probability that no other transmitter accesses the channel. From these observations, we can readily deduce the system throughput γ in steady state:

$$\gamma = \frac{L}{L + \frac{1}{\rho} + L_c(e^\rho/\rho - 1/\rho - 1)}.$$

5.2 Partial Interaction

Consider now the case of partial interaction. We analyze the system at the beginning of each slot. Denote by $p_i^N(k)/N$ the probability link i becomes active at the end of the k-th slot, if idle. We first show how to model the network as a set of interacting particles as described in Section 4.

- The particles: the i-th link corresponds to the i-th particle with state describing the class of the link and the transmission probability at the end of the next idle slot $X_i^N(k) = (c_i, p_i^N(k)) \in \mathcal{X} = \mathcal{C} \times \mathcal{B}$.
- The background process: it represents the classes for which there exists at least one active link: $Z^N(k) \in \mathcal{Z} = \{0,1\}^{|\mathcal{C}|}$, where $Z_c^N(k) = 0$ if and only if there is no active link of class c.

Particle Transitions. We first compute the transition probabilities for the various particles. The set $\mathcal{S}$ of possible transitions is composed by two functions, the first one representing a successful transmission $p \mapsto S(p)$ and the other one collisions $p \mapsto C(p)$. Note that the class of a particle / link does not change. Let $\nu_c^N(k) = \frac{1}{N}\sum_{i=1}^N \delta_{p_i^N(k)} \mathbf{1}_{c(i)=c}$ and $\nu^N(k) = (\nu_c^N(k))_{c \in \mathcal{C}}$.

Assume that at some slot k, the system is in state

$$((c_i^N(k), p_i^N(k))_{i=1,\ldots N}, \nu^N(k), Z^N(k))) = ((c_i, p_i)_{i=1,\ldots,N}, \alpha, z).$$

A class-c link i may have a transition at the end of slot k only if $C_c(z) = 1$. In this case it can either initiate a successful transmission or experience a collision. If $C_c(z) = 1$, the event that none of the users in c transmits at the end of slot k is given by $D_c^N = \prod_{i \in \mathcal{C}_c} \mathbf{1}_{(NU_i > p_i)}$, where the U_i's are i.i.d. r.v. uniformly distributed on $[0, 1]$. The event that user $i \in \mathcal{C}_c$ accesses the channel with success at the end of slot k is given by the indicator:

$$\mathbf{1}_{NU_i \leq p_i} C_c(z) \prod_{j \in \mathcal{C}_c, j \neq i} \left(\mathbf{1}_{NU_j > p_j}\right) \prod_{d \in \mathcal{V}_c, d \neq c} \left(\mathbf{1}_{C_d(z)=1} D_d^N + \mathbf{1}_{C_d(z)=0} \right).$$

Averaging the above quantity gives the transition probability $F_S^N((c, p_i), \alpha, z)/N$ corresponding to a successful transmission. For all $\alpha \in \mathcal{P}(\mathcal{B})$ and all f $\mathcal{B}$-valued function, define $\langle f, \alpha \rangle = \sum_p f(p)\alpha(p)$. Moreover let α_c denote the restriction of α to class-c links. Let I denote the identity function. One can readily see that:

$$F_S^N((c, p_i), \alpha, z) = \frac{p_i}{1 - p_i/N} C_c(z) \prod_{d \in \mathcal{V}_c} \left(C_d(z)(e^{\langle N \log(1 - \frac{I}{N}), \alpha_d \rangle} - 1) + 1 \right) \quad (7)$$

$$F_S((c, p_i), \alpha, z) = p_i C_c(z) \prod_{d \in \mathcal{V}_c} \left(C_d(z)(e^{-\langle I, \alpha_d \rangle} - 1) + 1 \right), \tag{8}$$

A similar derivation leads to the following asymptotic collision rates:

$$F_C((c, p_i), \alpha, z) = p_i C_c(z) \left(1 - \prod_{d \in \mathcal{V}_c} \left(C_d(z)(e^{-\langle I, \alpha_d \rangle} - 1) + 1 \right) \right). \tag{9}$$

Transitions of the Background Process. Assume that the system is in state $((c_i, p_i)_{i=1,\dots N}, \alpha, z)$. The transition kernel K_α^N is obtained replacing p_i by p_i/N in (1). When $N \to \infty$, the asymptotic kernel is $K_\alpha = K_{\alpha,A} K_{\alpha,D} K_{\alpha,0}$ with

$$K_{\alpha,A}(z, z') = \prod_{E \in A(z,z')} \prod_{c \in E} C_c(z)(1 - e^{-\langle I, \alpha_c \rangle}), \quad K_{\alpha,D}(z, z') = \left(\frac{1}{L} \right)^{|D(z,z')|},$$

$$K_{\alpha,0}(z, z') = \left(1 - \frac{1}{L} \right)^{|N(z) \backslash D(z,z')|} \prod_{c : z_c = 0 = z'_c} (C_c(z)e^{-\langle I, \alpha_c \rangle} + 1 - C_c(z)).$$

5.2.1   Stationary Throughputs

We are interested in deriving the stationary throughputs achieved by links of various classes. To do so, we apply the results of Section 4 to derive the stationary distribution Q_{st} and $\pi_{Q_{st}}$ of the particles and the background process. To simplify the notation we write $Q_{st} = Q$ and $\pi_Q = \pi$. Also denote $Q_c^p = Q(\{c, p\})$ the stationary proportion of links of class c transmitting with probability p.

Following the same reasoning leading to (2) in Section 3, the total throughput of the links of class c is

$$\gamma_c = \sum_{z : C_c(z) = 1} \pi(z) L \rho_c \prod_{d \in \mathcal{V}_c} \left(C_d(z)(e^{-\rho_d} - 1) + 1 \right), \tag{10}$$

where

$$\rho_c = \sum_{p \in B} p Q_c^p, \tag{11}$$

which can be interpreted as the probability that a link of class c attempts to use the channel at the end of an empty slot. We now evaluate Q and π. Note that π depends on Q through the ρ_c's only, and can be easily computed. As in Section 3 we can approximate π:

$$\pi(z) \approx \pi(0) K_{\alpha,A}(0, z) L^{|N(z)|}. \tag{12}$$

Now define G_c, H_c as follows:

$$G_c = \sum_z \pi(z) C_c(z) \prod_{d \in \mathcal{V}_c} \left(C_d(z)(e^{-\rho_d} - 1) + 1 \right), \tag{13}$$

$$H_c = \sum_z \pi(z) C_c(z) \left(1 - \prod_{d \in \mathcal{V}_c} \left(C_d(z)(e^{-\rho_d} - 1) + 1 \right) \right), . \tag{14}$$

G_c, H_C depend on Q through the ρ_c's only. We have for all c, p: $pG_c = \overline{F}_S((c,p), Q)$, $pH_c = \overline{F}_C((p,c), Q)$. The marginals Q_c^p satisfy for all c, p,

$$G_c \left(\sum_{p' \in \mathcal{B}: S(p')=p} p' Q_c^{p'} - p Q_c^p \right) + H_c \left(\sum_{p' \in \mathcal{B}: C(p')=p} p' Q_c^{p'} - p Q_c^p \right) = 0. \quad (15)$$

They also satisfy:

$$\forall c \in \mathcal{C}, \quad \sum_{p \in \mathcal{B}} Q_c^p = \mu_c. \quad (16)$$

Summarizing the above analysis, we have:

Theorem 4. *The stationary distribution Q is characterized by the set of equations (11), (12), (13), (14), (15), (16).*

The exponential Back-Off Algorithm. We now examine the specific case of the exponential back-off algorithm. Here we denote $Q_c^n = Q(\{c, p_0 2^{-n}\})$ for all $n \in \mathbb{N}$. One can then show that:

$$\rho_c = p_0 \mu_c \frac{G_c - H_c}{G_c}, \quad (17)$$

and that the system behavior in steady state is given by the following corollary.

Corollary 1. *The stationary distribution Q is given by: for all $c \in \mathcal{C}$,*

$$Q_c^n = \mu_c \frac{G_c - H_c}{G_c + H_c} \left(\frac{2H_c}{G_c + H_c} \right)^n,$$

where the $G_c's$, H_s's, and ρ_c's are the unique solutions of the system of equations (11), (12), (13), (14), (17).

We now illustrate the above results on the simple network of Figure 1. Here the state space of the background process is $\mathcal{Z} = \{0, 1\}^3$ and its stationary distribution can be easily evaluated. Applying Corollary 1, we may compute the throughputs γ_1, γ_2, and γ_3. In Figure 1, the throughputs of the various link classes are presented as a function of the proportion of the number of class-2 links. The packet transmission and collision durations are fixed to $L = 100$ slots, which roughly corresponds to the case of packets of size 1000 bytes transmitted at rate 54 Mbit/s in a IEEE802.11g-based network. The total network throughput decreases when the proportion of class-2 links increases, which illustrates the loss of efficiency due to the network spatial heterogeneity. Note also that if the links are equally distributed in the 3 zones, i.e., $\mu_1 = \mu_2 = \mu_3$, the network is highly unfair: the throughput of a link of class 1 or 3 is almost 5 times greater than that of a link of class 2. To use the above mean field limit to approximate the throughput of classes in a system with a finite population N and $p_0 = 1/16$ say, one must redefine p_0 so $p_0/N = 1/16$.

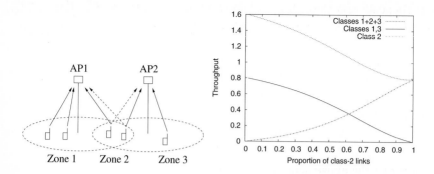

Fig. 1. A cellular network with 2 overlapping cells and its throughput performance

References

1. IEEE Standard for Wireless LAN Medium Access Protocol and Physical Layer Specifications, IEEE std. (August 1999)
2. Abramson, N.: The ALOHA system - Another alternative for computer communications. In: Proc. of AFIPS press (1970)
3. Baccelli, F., Brémaud, P.: Elements of Queueing Theory: Palm-Martingale Calculus and Stochastic Recurrences. Springer, Heidelberg (1994)
4. Bianchi, G.: Performance Analysis of the IEEE 802.11 Distributed Coordination Function. IEEE Journal on Selected Areas in Communications 18(3), 535–547 (2000)
5. Billingsley, P.: Convergence of Probability Measures. Wiley, New York (1968)
6. Bordenave, C., McDonald, D., Proutiere, A.: A Random Multi-access Algorithms: A Mean Field analysis. In: Proceedings of the Allerton Conference on Communication, Control, and Computing (2005)
7. Bordenave, C., McDonald, D., Proutiere, A.: Mean-field limits of a particle system interacting with a rapidly varying environment. Submitted (2006) Available at http://www.cos.ict.kth.se/alepro/mathpap.pdf
8. Durvy, M., Thiran, P.: A Packing Approach to Compare Slotted and Non-slotted Medium Access Control. In: Proc. of IEEE Infocom (2006)
9. Kelly, F.P.: Loss Networks. The Annals of Applied Probability 1(3), 319–378 (1991)
10. Li, J., Blake, C., De Couto, D.S.J., Lee, H.I., Morris, R.: Capacity of Ad Hoc wireless networks. In: Proc. of ACM Mobicom (2001)
11. Nguyen, G.D., Wieselthier, J., Ephremides, A.: Random Access in Overlapping Cells. In: Proc. of Eurpean Wireless (2005)
12. Sharma, G., Ganesh, A., Key, P.: Performance Analysis of Contention Based Medium Access Control Protocols. In: Proc. IEEE Infocom (2006)
13. Sznitman, A.S.: Propagation of chaos. In: Ecole d'été de probabilités de Saint-Flour XIX. lecture notes in Maths, vol. 1464, Springer, Berlin (1991)

A Phase-Type Based Markov Chain Model for IEEE 802.16e Sleep Mode and Its Performance Analysis

Zhisheng Niu[1], Yanfeng Zhu[1], and Vilius Benetis[2]

[1] Department of Electronic Engineering, Tsinghua University Beijing P.R. China
[2] Research Center COM Technical University of Denmark

Abstract. To support battery powered mobile broadband wireless access devices efficiently, IEEE 802.16e defines a sleep mode operation for conserving the power of mobile terminals. In this paper we propose a theoretical Phase-type (PH) based Markov chain model to analyze the performance of IEEE 802.16e sleep mode operation. The model describes the behavior of the mobile stations working in sleep mode. In particular, the service process is designed as a discrete PH model. By means of the mathematic tools of PH theory we then derive the closed-form expressions of the mean sojourn time of packets and power consumption. Comparison with simulation results shows that the model provides an accurate prediction of the system performance. Furthermore, we propose a simple utility function to quantify the efficiency of sleep mode operation which takes the joint effect of sojourn time and power saving into account. This function allows mobile stations to decide when to enable sleep mode operation for power saving.

1 Introduction

IEEE 802.16e [1] provides enhancement to IEEE Std 802.16-2004 [2] to support subscriber stations moving at vehicular speeds and thereby specifies a system for combined fixed and mobile Broadband Wireless Access (BWA). IEEE 802.16e will increase the market for BWA solutions by taking advantage of the inherent mobility of wireless media. In mobile BWA, mobile stations (MS) are usually powered by battery, and therefore power saving is a critical concern in designing the medium access control. The basic principle of power saving mechanism is to employ sleep mode operation to minimize MS power usage. In power saving scenarios an MS has two modes: a sleep mode and an awake mode. As defined in [1], the sleep mode is a state in which an MS conducts pre-negotiated periods of absence from the serving base station (BS) air interface. These periods are characterized by the non-availability of the MS, as observed from the serving BS, to downlink and uplink traffic. The MS can decrease power consumption by increasing the interval working in sleep mode.

When the MS works in sleep mode, the BS buffers all packets addressed to the MS until it wakes up. From the queueing theory, it is well known that the

L. Mason, T. Drwiega, and J. Yan (Eds.): ITC 2007, LNCS 4516, pp. 791–802, 2007.

queueing length in the buffer increases with the sleep interval, and so that packets' sojourn time and packet loss increases with the sleep interval. Therefore, the parameters of sleep mechanism are the keys of tradeoff between power consumption and Quality-of-Service (QoS), which is generally measured by the mean sojourn time. To balance the tradeoff in practical system, an analytic model is required to clarify the relationship among these performance metrics.

In the literature, a similar sleep mode operation employed in a cellular digital packet data systems have been investigated in [5] by simulations. For the same system, the authors in [8] developed a queueing model to analyze the packet delay. In [3], the authors extended the work to IEEE 802.16e sleep mode operation and proposed an M/GI/1/N queueing system with multiple vacations, by constructing Hessenberg matrix representing an M/G/1 queueing system [6][9]. However, to construct a vacation model, the authors ignore the relationship between consecutive sleep intervals (in fact, any sleep interval depends on the preceding one) and treat all sleep intervals as the same as the interval averaged over all precalculated sleep intervals. In [7][4], the authors proposed some Markov models to investigate the variation of sleep intervals, with which the sojourn time and power consumption due to the sleep interval can be computed. However, the effect of queueing delay is not taken in account precisely. Moreover, due to the complexity of these models, the tradeoff between the sojourn time and the power consumption is not discussed.

In this paper,we develop a phase type (PH) based Markov chain model that accounts for all details of the sleep mode operation and allows the closed-form expressions of the sojourn time and the power consumption to be calculated. In contrast to the existing works, we employ a PH model to describe the service process, so that complicated queueing model analysis is avoided. Furthermore, based on the performance measures, we propose a utility function which considers the joint effect of sojourn time and power saving to quantify the efficiency of sleep mode operation.

The rest of this paper is organized as follows. In Section 2, we provide a simple description of the considered IEEE 802.16e sleep mode operation. Then, Section 3 proposes an analytic model to describe the behavior of a single station working in sleep mode operation. In Section 4 we deduce the closed-form expressions of the sojourn time and the power consumption, and then the utility function is introduced. The performance of the analytic model is validated by extensive simulations in Section 5. Finally, the conclusions are drawn in Section 6.

2   System Description

The standard of IEEE 802.16e [1] indicates that the MS is capable of waking up at any time when (from MS to BS) uplink traffic arrives, i.e., the delay of uplink traffic is independent of the sleep mechanism. Thus, we focus on the downlink (from BS to MS) traffic only. The sleep mode operation is illustrated in Fig. 1.

As described in the draft [1], before entering sleep mode, the MS sends a MOB-SLP-REQ message to the BS and waits for the approval of the BS. The

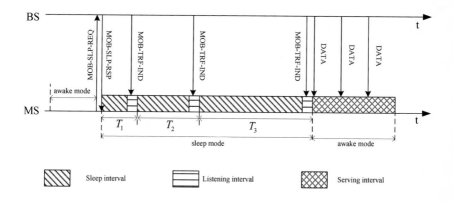

Fig. 1. Sleep mode operation in IEEE 802.16e

MS enters the sleep-mode for an interval defined by initial sleep window T_b after receiving the approving MOB-SLP-RSP message, which carries the initial sleep window, listening window, final sleep window exponent, and traffic triggered wakening flag. When the sleep interval finishes, the MS wakes up and, during the following listen interval, checks whether there are any packets for it. If there are packets addressed to the MS in the buffer of the BS, the MS goes to the awake mode. Otherwise, the MS returns to the sleep mode, and the sleep interval is doubled, up to a maximum value $2^M T_b$ where M is the final sleep window exponent defined in [1].

In the listening interval the BS ends the power saving state of the MS by sending a MOB-TRF-IND message in broadcast mode to alert the MS of presence of downlink traffic. When the MS senses a positive MOB-TRF-IND message indication, it stays awake until all packets addressed to it (in the buffer of the BS) are received. Herein, an exhaustive access policy [10] is employed, which means that the BS transmits all packets addressed to the MS in its buffer as well as packets arriving at the BS whilst the MS is receiving queued packets.

3   A PH Based Markov Chain Model

The system of interest is assumed to comprise a BS with a separate per-MS-limited buffer. The listening interval is generally much shorter than the sleep interval. Therefore, during any listening interval the MS is also considered to be in the sleep mode even if it has been physically already awake. Let T_b denote the initial sleep window and T_l the listening interval. Then, the length of the ith sleep interval, T_i, which is the sum of the ith sleep cycle and the following listening interval, is given by

$$T_i = 2^i T_b + T_l, \qquad 0 \le i \le M. \tag{1}$$

In addition, we assume that the packet arrival process from network to the BS follows a Poisson process with arrival rate λ. The service time of a packet

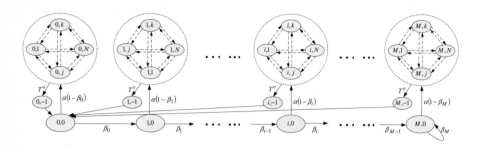

Fig. 2. Phase-type based Markov chain model for the sleep mechanism of IEEE 802.16e

is assumed to be generally distributed with probability density function (pdf) $v(t)$.

Let $s(t)$ $(0 \leq s(t) \leq M)$ be a stochastic process representing the sleep stage of the MS at time t, and then $s(t) = i$ means the MS stays in the ith sleep interval at time t. Let $b(t)$ be the stochastic process representing the number of packets for the MS buffered in the BS at time t. Then, the bi-dimensional random process $\{s(t), b(t)\}$ is a discrete-time Markov chain. The state transition diagram of the proposed model is shown in Fig.2. Herein, N represents the buffer size (in packets) of the BS allocated to store downlink packets of the considered MS. In addition, state $(i, -1)$ is a supplementary state named "absorption state" in PH model [11] and is used to describe the end of a service process. Once all packets in the buffer are served, the MS enters the absorption state and then transits to state $(0, 0)$ immediately. Note that the absorption state is introduced only for constructing the PH model, and no time is costed in such an absorption state. In this way, the service process after each sleep state is in fact a typical discrete PH model.

In Fig.2, β_i represents the transition probability from state $(i, 0)$ to state $(i + 1, 0)$ (or from $(M, 0)$ to $(M, 0)$ when $i = M$), which is equal to the probability that no packets arrive in the ith sleep state. Then, we have

$$\beta_i = e^{-\lambda T_i}, \qquad 0 \leq i \leq M. \tag{2}$$

The vector $\alpha_i = (\alpha_{i1}, \alpha_{i2}, ..., \alpha_{iN})$ represents the conditional transition probabilities from the ith sleep state $(i, 0)$ to the child states of the ith service state, i.e. $\alpha_{ij} = P\{j$ packets arrive $\mid$ at least a packet arrive during state (i,0)$\}$. The vector α_i is the initial probability of each child state in the ith service state. Due to the Poisson arrival, we have

$$\alpha_{ij} = \begin{cases} \dfrac{(\lambda T_i)^j e^{-\lambda T_i}}{j!(1 - e^{-\lambda T_i})}, & 1 \leq j \leq N - 1 \\[3mm] \dfrac{e^{-\lambda T_i}}{(1 - e^{-\lambda T_i})} \displaystyle\sum_{k=N}^{\infty} \dfrac{(\lambda T_i)^k}{k!}, & j = N. \end{cases} \tag{3}$$

Herein, $\sum_{j=1}^{N} \alpha_{ij} = 1$.

To simplify the presentation, the transition probability $P\{s(t+1) = i, b(t+1) = k | s(t) = i, b(t) = j\}$ is denoted by $P(i, k | i, j)$. In this model, the only non-zero one-step transition probabilities are:

$$\begin{cases} P(i+1, 0 | i, 0) = \beta_i, & 0 \leq i \leq M - 1; \\ P(M, 0 | M, 0) = \beta_M; \\ P(i, j | i, 0) = \alpha_{ij}(1 - \beta_i), & 0 \leq i \leq M; \\ P(0, 0 | i, -1) = 1, & 0 \leq i \leq M. \end{cases} \tag{4}$$

Let $p_{i,j} = \lim_{t \to \infty} P\{s(t) = i, b(t) = j\}$ be the stationary distribution of the chain. Then, from (4), we have

$$\begin{cases} p_{i+1,0} = \beta_i p_{i,0}, & 0 \leq i \leq M - 1; \\ p_{M,0} = \dfrac{\beta_{M-1}}{1 - \beta_M} p_{M-1,0}; \\ p_{i,j} = (1 - \beta_i) p_{i,0} \alpha_{ij}, & 0 \leq i \leq M. \end{cases} \tag{5}$$

Then, from $1 = \sum_{i=0}^{M} \sum_{j=1}^{N} p_{i,j}$, we compute $p_{0,0}$:

$$p_{0,0} = \frac{1}{1 + \sum_{i=1}^{M-1}(2 - \beta_i) \left(\prod_{m=0}^{i-1} \beta_m \right) + \frac{2 - \beta_M}{1 - \beta_M} \prod_{m=0}^{M-1} \beta_m}. \tag{6}$$

In addition, the transition process inside the service state is a typical embedded Markov process, and the embedded point is the departure moment of each packet. Let matrix $T = (T_{jk})_{1 \leq j, k \leq N}$ denote the transition probability from state $\{i, j\}$ to $\{i, k\}$. Then T_{jk} is equal to the probability that $k - j + 1$ packets arrive during one packet serving in the state (i, j). Then T_{jk} is given by

$$T_{jk} = \begin{cases} \displaystyle\int_0^\infty \frac{(\lambda t)^{k-j+1}}{(k-j+1)!} e^{-\lambda t} v(t) \, dt, & j \leq k + 1 \,\&\, k < N; \\ \displaystyle\sum_{k=N}^{\infty} \int_0^\infty \frac{(\lambda t)^{k-j+1}}{(k-j+1)!} e^{-\lambda t} v(t) \, dt, & k = N; \\ 0, & \text{otherwise.} \end{cases} \tag{7}$$

Let vector $T^o = (T_1^o, T_2^o, ..., T_N^o)$ denote the transition probabilities from the service state to the related absorption state, where T_j^o denote the probability that there are no packets addressed to the MS in the buffer of the BS after serving in state (i, j). Then,

$$T_j^o = \begin{cases} \int_0^\infty e^{-\lambda t} v(t) \, dt, & j = 1 \\ 0, & j > 1. \end{cases} \tag{8}$$

So far we have obtained the triple elements of the PH model $\{\alpha_i, T, T^o\}$. Given the packet arrival rate λ, β_i $(0 \leq i \leq M)$ can be computed with (2), and then the closed-form expressions for all stable probabilities can be calculated with (3)-(8).

4   Performance Measures

By means of the mathematic tools of PH theory [11], we deduce the closed-form expressions of the mean sojourn time and power consumption.

4.1   Mean Sojourn Time

Packets' sojourn time is defined as the time interval from the moment when a packet arrives at the BS till the moment when the packet is successfully received by the related MS. Since packet arrivals follow Poisson distribution, the arrival events are random observers to the sleep intervals. Then, if there are j packets addressed to the MS buffered in the BS after the ith sleep state, the sojourn time averaged over these j packets is given by $\frac{T_i + (j+1)\mathrm{E}[v]}{2}$. During these j packets being served, some new packets may arrive, and during the new arrived packets being served some more packets may arrive, and a such iterative process terminates till no packets arrive during the preceding step. In this way, calculating the mean sojourn time is a complex recursion.

From [12], the energy consumption for the startup procedure (from sleep to awake) is non negligible. Therefore, if the MS works in high traffic load, frequent startups will cost so much energy that such a sleep mode operation is not power efficient[1]. From this standpoint, we only consider the scenario with light traffic load ($\lambda/\mu < 50\%$) where the probability that there are packets available after the first order recursion is very low. To simplify the calculation, we only consider the first order recursion, i.e., all packets except the initial j packets are assumed to arrive during the service time of initial j packets.

Define vector $E_j = (e_{j,1}, e_{j,2}, ..., e_{j,N})$, where $e_{j,j} = 1$ and $e_{j,k} = 0$ for $j \neq k$. Let c_j denote the mean transition times before entering the absorption state given that there are j packets in the BS. Following the characteristics of PH model, we have

$$c_j = \sum_{k=1}^{\infty} k E_j T^{k-1} T^o \tag{9}$$

Then, we obtain the sojourn time averaged over the packets served in ith service state with j initial packets as

$$d_{i,j} = \frac{1}{c_j} \left[j \frac{T_i + (j+1)\mathrm{E}[v]}{2} + (c_j - j) \frac{(c_j - j + 1)\mathrm{E}[v]}{2} \right] \tag{10}$$

Consequently, the sojourn time averaged over all packets is given by

$$D = \frac{\sum_{i=0}^{M} p_{i,0}(1 - \beta_i) \sum_{j=1}^{N} \alpha_{i,j} \, c_j \, d_{i,j}}{\sum_{i=0}^{M} p_{i,0}(1 - \beta_i) \sum_{j=1}^{N} \alpha_{i,j} \, c_j} \tag{11}$$

Substituting (3), (9), and (10) into (11), we get the mean sojourn time.

---

[1] The issue on efficiency of sleep mode operation will be discussed in detail in the next section.

4.2   Efficiency of Sleep Mode Operation

As mentioned in the introduction, the key purpose of employing sleep mode in IEEE 802.16e is to conserve the power consumption of MS. The power consumption of the MS can be divided into two parts: one is for receiving the expected data packets, and the other is for sleep operation, which consists of sleep, idle listening and switch (from sleep to awake or from awake to sleep). It is worth noting that the first part is decided by the traffic load and independent of the sleep operation. Let P_s and P_r denote the power consumption for sleep and receiving[2], respectively. Let E_{switch} denote the energy consumption for a pair of switches, which includes a sleep-awake and an awake-sleep. The power consumption for receiving the expected packets is given by ρP_r, where $\rho = \lambda E[v]$ is the traffic load. In addition, the second part of the power consumption can be computed by the probability sum of the power consumption in each state in each state. Based on the analysis above, the total power consumption P_{ps} is given by

$$P_{ps} = \rho P_r + \frac{\sum_{i=0}^{M} p_{i,0}[E_{switch} + P_r T_l + P_s T_i]}{\sum_{i=0}^{M} T_i p_{i,0}} \tag{12}$$

To quantify the power saving of the sleep mode operation, we introduce a power saving factor R_P defined as the power consumption ratio of a power saving MS and a non-power saving MS. For a non-power saving MS, the transceiver is always working in receiving state. Therefore, R_P is given by

$$R_P = \frac{P_{ps}}{P_r} \tag{13}$$

Clearly, the smaller the power saving factor, the more efficient the sleep mode operation.

Intuitively, the power saving factor increases with the traffic load, which implies that the sleep mode operation is much more efficient in light traffic load. However, from (11) we derive that the mean sojourn time decreases with the traffic load. Therefore, there exists a tradeoff between the power saving and sojourn time. To quantify the efficiency of the sleep mode operation, the joint effect of power saving and sojourn time should be considered.

The same to the introduction of the power saving factor, we define a sojourn time factor R_D as the ratio of the mean sojourn time between a power saving MS and a non-power saving MS. For a Markovian arrival system, the mean sojourn time of the packets of a non-power saving MS can be computed by means of a M/G/1 queueing model. Let D_n denote the mean sojourn time of the packets of a non-power saving MS. From the Pollaczek-Khintchine formula, we have

$$D_n = \frac{1}{\mu} + \frac{\rho(1 + c_b^2)/\mu}{2(1 - \rho)} \tag{14}$$

---

[2] From [12], the power consumptions for idle listening and receiving are nearly the same, and thus we omit the definition of the power consumption for idle listening.

where c_b^2 is the variance coefficient of the serving time of a packet. Then, the sojourn time factor is given by $R_D = D/D_n$. Substituting (11) and (14) into $R_D = D/D_n$, we obtain that R_D is a monotone decreasing function with respect to ρ.

In practical implementation, the efficiency of sleep mode operation is usually identified by some utility formulas. In this paper, we propose a simple utility formula which is just the weighted-sum of the power saving factor and the sojourn time factor.

$$R = c_d \, \mathrm{Norm}(R_D) + R_P \qquad (15)$$

where $\mathrm{Norm}(R_D) = \frac{R_D}{\mu T_M/2+1}$ is a normalizing procedure. From [12], E_{switch} is much smaller than the energy consumption for receiving, and thus $R_P \in (0,1)$ for $\rho \in (0,1)$. We introduce $\mathrm{Norm}(\cdot)$ to bound the effect of sojourn time factor in value region $(0,1)$. From (11), we have $\lim_{\rho \to 0} R_D = \mu T_M/2 + 1$, and thus $\mathrm{Norm}(R_D) \in (0,1)$ for $\rho \in (0,1)$. In addition, coefficient c_d named delay component coefficient models the importance of sojourn time factor in the efficiency of sleep mode operation.

5   Performance Evaluation

To validate the proposed analytic model and demonstrate the performance of IEEE 802.16e sleep mode operation, we show numerical results and simulation results with the parameters shown in Table.1. Herein, the service time of a packet is fixed to $1/\mu$. The simulation is achieved via discrete event simulation tool NS2 [13]. The simulation time for each scenario is 50 hours, and the results are obtained via averaging values from 10 different runs with different seeds.

Table 1. Parameters for Numerical calculation and Simulation

Parameter	value
Listenin interval T_l (ms)	5
Final sleep window exponent M	5
service rate μ (packets/second)	25
Buffer size N (packets)	5, 10, 15

5.1   Mean sojourn time

In this subsection, we concentrate on the mean sojourn time. The effects of the arrival rate and the initial sleep window are plotted in Figs.3 and 4, respectively. For the purpose of performance comparison, the analytical results from [7] are also plotted. From both figures it is observed that our PH based Markov chain model significantly outperforms the one from [7], which ignores the queueing time. Due to the absence of queueing analysis, the results of the model proposed in [7] show that the sojourn time monotonically decreases with the traffic load, which is acceptable in light traffic load region but not in heavy traffic load

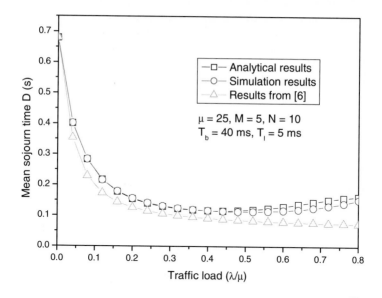

Fig. 3. The mean sojourn time D versus arrival rate λ

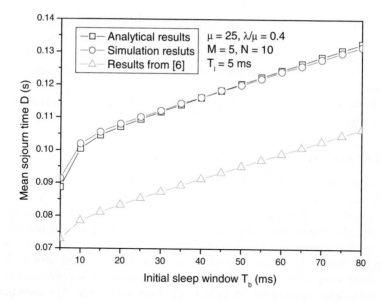

Fig. 4. The mean sojourn time D versus initial sleep window T_0

region. As shown in our analytical and simulation results, the mean sojourn time decreases with the arrival rate in the light load region, and then increases slowly with the arrival rate in the heavy load region.

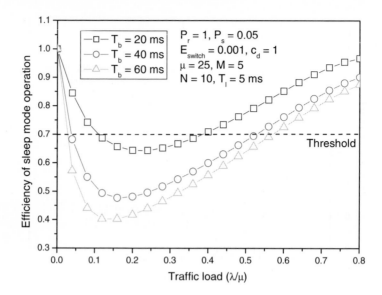

Fig. 5. The efficiency of sleep mode operation versus traffic load for various c_d

In the light load region, the probability that new packets arrive during the serving period is very small, and thus the sojourn time mainly depends on the sleep interval. Clearly, the mean sleep time decreases with the arrival rate, and thus the mean sojourn time decreases with the arrival rate in light load region. In heavy load region, the queueing time increases quickly with the arrival rate and gradually dominates the sojourn time, and thus the mean sojourn time increases with the arrival rate in heavy load region.

5.2 Efficiency of Sleep Mode Operation

In this subsection, the efficiency of sleep mode operation defined as (15) is evaluated by numerical results.

In Fig.5, the dashed horizontal line is an arbitrary pre-set threshold and it splits operation zones into sleep-zone efficient below the line and not efficient - above the line. Lower part of curve indicates the traffic load values where sleep mode is beneficent. The curve part above the threshold line indicates traffic load values where sleep mode is disadvantageous. Sleep mode looses efficiency because of too much delay introduced (at light traffic load) or because MS does not save much power anymore (at high load).

Delay component coefficient c_d allows to adjust how important is altered delay compared to altered power consumption (15). From Fig.5 it is clear that by reducing importance of delay ($c_d = 0.5$, meaning that mobile prefers to save more power and allows longer delays), sleep mode is beneficent under loads up to 60%. Otherwise, when delay bears higher importance ($c_d = 1.5$, meaning that mobile prefers lower delays), sleep mode brings benefit around traffic load of

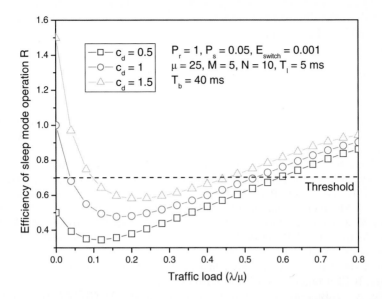

Fig. 6. The efficiency of sleep mode operation versus traffic load with initial sleep window $T_b = 20$ ms, 40 ms, and 60 ms

$10 - 50\%$. For practical applications the threshold value should be adjusted from empirical studies. Flexibility in changing c_d allows to set different power saving profiles in MS.

Fig.6 shows the influence of initial sleep window towards the efficiency of sleep mode. Larger sleep window values allows to be more power efficient, as then the number of listen periods is reduced and packets are served in bigger batches.

The procedure described in this paper provides a framework how to design and evaluate the efficiency of sleep mode. MS is able to calculate should it use sleep mode or not depending on the power saving profile and observed traffic load on the interface from the parameters discussed above.

6   Conclusions

In this paper we have presented a PH based Markov chain model to investigate the performance of IEEE 802.16e sleep mode operation. This model properly describes the behavior of the MS working in sleep-mode. By means of the mathematic tools of PH theory we have derived the closed-form expressions of various performance measures. A comparison with simulated results shows that the model is very accurate in predicting system performance. The results also show that the mean serving time and the mean sojourn time are strongly dependent on the traffic load and the initial sleep window. Furthermore, we quantified the efficiency of sleep mode operation with a utility function combining the effect of sojourn time and power saving. Then by presetting an efficiency threshold the

MS can dynamically decide whether to enable sleep mode operation according to the traffic load it observes.

References

1. IEEE802.16e-2006, IEEE Standard for Local and metropolitan area networks, Part 16: Air Interface for Broadband Wireless Access Systems C Amendment for Physical and Medium Access Control Layers for Combined Fixed and Mobile Operation in Licensed Bands (February 2006)
2. IEEE802.16-2004, IEEE Standard for Local and metropolitan area networks, Part 16: Air Interface for Broadband Wireless Access Systems (October 1, 2004)
3. Seo, J.-B., Lee, S.-Q., Park, N.-H., Lee, H.-W., Cho, C.-H.: Performance Analysis of Sleep Mode Operation in IEEE 802.16e. VTC2004-Fall. 2, 1169–1173 (2004)
4. Han, K., Choi, S.: Performance Analysis of Sleep Mode Operation in IEEE 802.16e Mobile Broadband Wireless Access Systems. In: Proc. IEEE VTC'2006-Spring, Melbourne, Australia (May 7-10, 2006)
5. Lin, Y.-B., Chuang, Y.-M.: Modeling the Sleep Mode for Cellular Digital Packet Data. IEEE Comm. Letters 3(3), 63–65 (1999)
6. Frey, A., Takahashi, Y.: Explicit Solutions for the M/GI/1/N Finite Capacity Queues With and Without Vacation Time. In: Proc. 15th International Teletraffic Congress, pp. 507–516, (June 1997)
7. Xiao, Y.: Energy Saving Mechanism in the IEEE 802.16e Wireless MAN. IEEE Communications Letters 9(7), 595–597 (2005)
8. Kwon, S.-J., Chung, Y.W., Sung, D.K.: Queueing model of sleep-mode operation in cellular digital packet data. IEEE Trans. Veh. Tech. 52(4), 1158–1162 (2003)
9. Takagi, H.: Analysis of a Finite Capacity M/G/1 Queue with Resume Level. Performance Evaluation 5(3), 197–203 (1985)
10. Fantacci, R., Zoppi, L.: Performance evaluation of polling systems for wireless local communication networks. IEEE Trans. Veh. Tech 49(6), 2148–2157 (2000)
11. Neuts, M.F.: Matrix-Geometric Solution in Stochastic Models: An Algorithmic Approach. The John Hopkins University Press, Baltimore, MD (1981)
12. Langendoen, K., Halkes, G.: Energy-Efficient Medium Access Control. Technical report, http://www.isa.ewi.tudelft.nl/koen/papers/MAC-chapter.pdf
13. Fall, K., Varadhan, K. (eds.): ns notes and documentation. The VINT Project, UC BERKELY, LBL, USC/ISI, and Xerox PARC (November 1997)

The Impact of Interference on Optimal Multi-path Routing in Ad Hoc Networks

Roland de Haan, Richard J. Boucherie, and Jan-Kees van Ommeren

Stochastic Operations Research, University of Twente, The Netherlands

Abstract. We develop a queueing model characterizing explicitly the impact of interference on end-to-end performance measures such as throughput in ad hoc networks, emphasizing the performance trade-off between single-path and multi-path routing. It may seem attractive to employ multi-path routing, but as all nodes share a single channel, efficiency may drop due to increased interference levels thus yielding single-path performance for some topologies. We formulate a nonlinear programming problem to optimize network performance. Next, we focus on network capacity and show that for this objective the optimum could be found by solving an exponential number of linear programmes. We propose a greedy algorithm that efficiently searches these programmes to approximate the optimal solution. Numerical results for small topologies provide structural insight in optimal path selection and demonstrate the excellent performance of the proposed algorithm. Besides, larger networks and more advanced scenarios with multiple source-destination pairs and different radio ranges are analyzed.

Keywords: Ad hoc networks, Interference, Capacity, Multi-path routing, Network optimization.

1 Introduction

Ad hoc networks have received considerable attention in the recent literature. The main focus has been on the development of routing protocols, since protocols for wired networks cannot be employed efficiently in a wireless let alone an ad hoc environment. The main body of routing protocol proposals regards single-path routing, i.e., for each source-destination pair a single (shortest) path is discovered and used for data transmission (see e.g., [1,2,3]). An alternative is multi-path routing (see [4]) in which multiple paths are used, thereby offering more opportunities for regulating the traffic over the network.

Multi-path routing enhances single-path routing mainly in two directions: (i) to have backup paths available in case of path failures (see e.g., [5,6]), and (ii) to spread traffic to increase the effective bandwidth (see e.g. [7,8]). However, there are also a number of drawbacks to employing multiple paths. In single-path routing the shortest path is normally selected; hence, any additional path will typically be longer. This may not matter when considering capacity questions, but it definitely does when considering the transfer time of a packet. Another,

L. Mason, T. Drwiega, and J. Yan (Eds.): ITC 2007, LNCS 4516, pp. 803–815, 2007.

frequently underexposed, drawback, which occurs typically in wireless environments, is that nodes situated on nearby paths may interfere. As a result, resources need to be shared (to prevent packet collisions) and throughput may decrease drastically. Therefore, the actual performance gain (in terms of bandwidth) of using multiple paths over using a single path is uncertain.

Recently, this bandwidth gain has received more attention, with the emphasis on the development of routing protocols. These protocols often aim at finding link- or node-independent paths and do not explicitly take signal interference between paths into account (see e.g. [9]). An explanation for this is perhaps that the notion of interference is hard to quantify. In fact, it is still an open question how one can accurately measure the level of interference in a network, despite the several metrics that have been suggested (see e.g., [10,11]).

Following a graph-theoretical approach for a given (finite) network instance, [12] and [13] analytically assess the network capacity by taking interference into account. In [12], a multi-commodity flow problem is formulated and extended by interference-related constraints in order to find lower and upper bounds for the capacity. These additional constraints follow by regarding cliques and independent sets in the so-called conflict graph of the network. The disadvantages of this approach are that a central scheduling entity is assumed and that extensive computations are required, even for small networks. The approach of the authors in [13] is similar but aims at a more distributed way of controlling the traffic streams in the network. They develop a low-complexity algorithm to find approximate cliques and use this to calculate lower and upper bounds for the capacity.

In this paper, we study optimization of network performance while explicitly considering the interference between nodes (and thus also paths). More specifically, the trade-off between the bandwidth gain using multiple paths on the one hand and the loss of bandwidth due to the additional interference involved when using those paths on the one hand is investigated. In contrast to most of the work that has appeared in this area, which focuses primarily on the capacity value itself, our interest is mainly in the underlying fundamental aspects so as to gain structural insights into the network performance under interference. First, we discuss the ad hoc network model and then present a general stochastic framework in which many performance metrics of interest can be investigated. The analysis is partly based on the MAC layer IEEE 802.11 protocol for which it has been shown that single-hop ad hoc networks can successfully be modelled by a Processor Sharing queue [14]. We formulate a nonlinear programming problem to optimize network performance. This program is then customized to meet the objective of maximal network capacity as might be achieved by a central scheduler. To solve the resulting quadratic programming problem, an approximate greedy algorithm is proposed. Finally, example topologies are analyzed both to assess the quality of the algorithm and to uncover structural relations in the network. These insights may be applied in the development of routing protocols. However, the design of such protocols is outside the scope of this paper.

The paper is organized as follows. Section 2 describes the ad hoc network model, our mathematical framework and the network optimization problem.

Section 3 presents techniques towards solving the optimization problem, and Sect. 4 discusses the impact of interference on network capacity via illustrative example topologies. A discussion and conclusions are presented in Sect. 5.

2 Model

2.1 Ad Hoc Network Model

Consider an ad hoc network in which all stations (or nodes) are equipped with an identical packet radio (with omnidirectional antenna) operating in half-duplex mode, and transmit over a common channel at identical (maximum) power. A node may transmit data to nodes that are within its *transmission range*. During data reception at a node, all nodes within its *interference range* must be silent for the reception to be successful. Typically, the interference range exceeds the transmission range. Links (or connections) between nodes are assumed error free. However, link errors may easily be incorporated in our model as will be discussed in Sect. 5. We assume a distributed transmission scheduling mechanism that mimics the IEEE 802.11 MAC protocol, which aims to preventing packet collisions. Data transmission in the network is between source-destination pairs (SD-pairs).

2.2 Mathematical Framework

We consider a network consisting of a set of nodes $\mathcal{N} = \{1, \ldots, N\}$. This set comprises a collection of source nodes $\mathcal{S} = \{s_1, \ldots, s_F\}$ and destination nodes $\mathcal{D} = \{d_1, \ldots, d_F\}$, where F denotes the total number of SD-pairs. The remaining nodes can be seen as (pure) relay nodes, but we note that source and destination nodes can also relay traffic. For $j \in \mathcal{N}$, let $\mathcal{N}_T(j) \subset \mathcal{N}$ denote the transmission neighborhood of node j, that is, the set of nodes (possibly sources or destinations) that node j can successfully transmit packets to, and let $\mathcal{N}_I(j) \subset \mathcal{N}$ denote the interference neighborhood of node j, that is, the set of nodes that must be quiet for successful packet reception at node j. The neighborhood relation is not necessarily a symmetrical relation, e.g., $n \in \mathcal{N}_T(n')$ does not imply $n' \in \mathcal{N}_T(n)$. When $n' \in \mathcal{N}_T(n)$, we say that the network contains a *link* from node n to node n'. Paths in the network consists of a number of links starting at a source node and ending at a destination node. On a path from source s to destination d consisting of $\ell + 2$ links via nodes $n_1, \ldots, n_\ell$, the nodes must be such that $n_1 \in \mathcal{N}_T(s)$, $n_j \in \mathcal{N}_T(n_{j-1})$, $j = 2, \ldots, \ell$, $d \in \mathcal{N}_T(n_\ell)$.

Source s_f generates data flows according to a Poisson process at rate $\lambda^{(f)}$ for SD-pair f. A flow consists of a series of data packets. Let $\beta^{(f)}$ denote the mean number of packets per flow between SD-pair f. Then $\alpha_j^{(f)} = \lambda^{(f)}\beta^{(f)}$, for $j = s_f$, is the mean rate at which source s_f generates packets for SD-pair f; moreover, $\alpha_j^{(f)} = 0$, $\forall_{j \in \mathcal{N}: j \neq s_f, d_f}$, and we set $\alpha_j^{(f)} = -\lambda^{(f)}\beta^{(f)}$, for $j = d_f$. We assume that ordering of packets at the destination can be handled without loss of information. Thus, a flow may be split and its packets may be transferred over different paths selected according to a suitable network optimization mechanism.

Let node j forward a fraction $p_{jk}^{(f)}$ of its incoming packets for SD-pair f to node k. Denote by $\rho_{t,j}^{(f)}$ and $\rho_{r,j}^{(f)}$ the average number of packets transmitted and received, respectively, per time unit by node j for SD-pair f. It is assumed here that for any node transmitting a packet takes exactly one unit of time. We may thus also interpret $\rho_{t,j}^{(f)}$ as the average fraction of time node j is transmitting packets for SD-pair f. Then, for the network to sustain all these packet transmissions, the following relations must hold:

$$\rho_{t,j}^{(f)} = \alpha_j^{(f)} + \rho_{r,j}^{(f)}, \ \forall_{j\in\mathcal{N}}, \ \forall_{f\in\mathcal{F}}, \tag{1}$$

$$\rho_{r,j}^{(f)} = \sum_{k\in\mathcal{N}} \rho_{t,k}^{(f)} p_{kj}^{(f)}, \ \forall_{j\in\mathcal{N}}, \ \forall_{f\in\mathcal{F}}. \tag{2}$$

Further, we define for node j, $\rho_{t,j} := \sum_f \rho_{t,j}^{(f)}$ and $\rho_{r,j} := \sum_f \rho_{r,j}^{(f)}$. Notice that for nodes j which are pure relay nodes $\rho_{t,j}^{(f)} = \rho_{r,j}^{(f)}, \ \forall_{f\in\mathcal{F}}$. Optimal network design then corresponds to optimal selection of the fractions $p_{kj}^{(f)}$ in (2), so as to maximize a network performance criterion. These fractions determine the optimal path selection for flows in the network.

The number of packets arriving at a node determines the workload of the node. As a consequence, from the flow perspective, flows share the transmission capacity of the nodes. A single node can handle packets originating from different flows and these packets will be transferred in order of arrival. As nodes are identical, each node transmits packets at a normalized unit rate in the absence of signal interference. Clearly, interference among neighboring nodes reduces the amount of time a node is allowed to transmit packets. In a coordinated network, when neighboring nodes each have packets to transmit, these nodes will share the resources. This is achieved, for example, by the MAC layer protocol of IEEE 802.11, where each node uses its fair share of the medium. It is demonstrated in [14] that the PS queue is an adequate model for MAC layer sharing among multiple nodes. The overhead of the MAC layer protocol results in a reduced data rate: for IEEE 802.11 under RTS/CTS the MAC layer operates at roughly 85% efficiency. As a consequence, we may model all nodes in the interference neighborhood (which include the nodes that transmit packets to j) of node j as a single PS queue. We normalize this maximum transmission rate of the nodes at 1. The resulting interference restriction under which node j can still receive data successfully is

$$\sum_{m\in\mathcal{N}_I(j)} \rho_{t,m} + \rho_{t,j} \le 1, \ \forall_{j\in\mathcal{N}}. \tag{3}$$

Notice that this restriction is conservative: when none of the multiple transmissions overheard at a node are directed to this node, then those are unnecessarily prohibited. Moreover, the capacity restriction need not be imposed when node j is not a recipient of any data in the optimal design. This is incorporated via the following modification:

$$\rho_{r,j}\left(\sum_{m\in\mathcal{N}_I(j)} \rho_{t,m} + \rho_{t,j}\right) \le \rho_{r,j}, \ \forall_{j\in\mathcal{N}}. \tag{4}$$

Our aim is to investigate the performance trade-off between single-path and multi-path routing. In particular, we investigate the maximum data rate (i.e., capacity) that can be sustained by this optimal path selection under interference. Therefore, we consider capacity optimization of the network in equilibrium. However, our framework also allows us to consider alternative performance measures such as the maximal delay for a given traffic load.

2.3 Network Optimization Formulation

Optimal design of paths in the ad hoc network comes down to determining the $p_{jk}^{(f)}$ as the fraction of SD-flow f routed from node j to node k, or equivalently as the probability of routing traffic of SD-flow f from node j to node k. Define the matrices $\boldsymbol{\rho_t} = (\rho_{t,j}^{(f)})_{j \in \mathcal{N}, f \in \mathcal{F}}$ and $\boldsymbol{p} = (p_{jk}^{(f)})_{j,k \in \mathcal{N}, f \in \mathcal{F}}$. Network optimization can then be formulated as a nonlinear programming problem:

$$\max \qquad h(\boldsymbol{\rho_t}, \boldsymbol{p}) \qquad\qquad (5)$$

$$\text{s.t.} \qquad \rho_{t,j}^{(f)} - \rho_{r,j}^{(f)} = \alpha_j^{(f)}, \ \forall_{j \in \mathcal{N}}, \ \forall_{f \in \mathcal{F}} , \qquad (6)$$

$$\rho_{r,j}^{(f)} - \textstyle\sum_{k:j \in \mathcal{N}_T(k)} \rho_{t,k}^{(f)} p_{kj}^{(f)} = 0, \ \forall_{j \in \mathcal{N}}, \ \forall_{f \in \mathcal{F}} , \qquad (7)$$

$$\rho_{r,j} \cdot (\textstyle\sum_{m \in \mathcal{N}_I(j)} \rho_{t,m} + \rho_{t,j}) - \rho_{r,j} \leq 0, \ \forall_{j \in \mathcal{N}} , \qquad (8)$$

$$1 - \textstyle\sum_{k \in \mathcal{N}_T(j)} p_{jk}^{(f)} = 0, \ \forall_{j \in \mathcal{N}:j \neq d_f}, \forall_{f \in \mathcal{F}} , \qquad (9)$$

$$\rho_{t,j} - \textstyle\sum_f \rho_{t,j}^{(f)} = 0, \ \forall_{j \in \mathcal{N}} , \qquad (10)$$

$$\rho_{r,j} - \textstyle\sum_f \rho_{r,j}^{(f)} = 0, \ \forall_{j \in \mathcal{N}} , \qquad (11)$$

$$\rho_{t,j}^{(f)}, \ \rho_{r,j}^{(f)} \geq 0, \ \forall_{j \in \mathcal{N}}, \ \forall_{f \in \mathcal{F}} , \qquad (12)$$

$$p_{jk}^{(f)} \geq 0, \ \forall_{j,k \in \mathcal{N}}, \forall_{f \in \mathcal{F}} , \qquad (13)$$

where $\mathbf{1}$ is the indicator function and $h(\boldsymbol{\rho_t}, \boldsymbol{p})$ is a general objective function. For instance, for $h(\boldsymbol{\rho_t}, \boldsymbol{p}) = \sum_{i=1}^{F} w_i \cdot \rho_{t,s_i}^{(i)}$ with weights $w_i \geq 0$, $i = 1, \ldots, F$, the total weighted capacity in the network is optimized, and for $h(\boldsymbol{\rho_t}, \boldsymbol{p}) = - \sum_{j=1}^{N} \frac{\rho_{t,j}}{1 - \rho_{t,j}}$ the mean number of packets in the network can be minimized for a given set of flows $a_1, \ldots, a_F$. The first set of constraints (6)–(7) refers to the traffic equations (1) and (2), and describes flow conservation; Eq. (8) is the interference constraint; Eq. (9) indicates that the fractions $p_{jk}^{(f)}$ must sum to one.

The feasible region (6)–(13) shows that we are dealing with a nonlinear programming problem in the unknowns $\rho_{t,j}^{(f)}$ and $p_{jk}^{(f)}$. Our interference assumptions yield constraints that are quadratic in $\rho_{t,j}^{(f)}$ (see Eqs. (7), (8) and (11)) and moreover interacting terms of $p_{jk}^{(f)}$ and $\rho_{t,j}^{(f)}$ show up in Eq. (7). These latter interactions terms can, however, conveniently be eliminated by introducing new variables referring to *link* flows. To this end, introduce $\lambda_{jk}^{(f)} :=$ $\rho_{t,j}^{(f)} \cdot p_{jk}^{(f)}$, $\forall_{j,k \in \mathcal{N}}$, $\forall_{f \in F}$, which represent the amount of traffic going from node

j to node k per time unit for SD-pair f. The constraints (6)–(8) then become linear in $\lambda_{jk}^{(f)}$.

In the rest of this paper, we will focus on the linear objective of capacity optimization. We note that this optimization problem for a single SD-pair resembles the well-known max-flow problem in discrete optimization. For multiple SD-pairs, it resembles the multi-commodity flow problem [15]. Unfortunately, due to the nonlinear interference constraints, our optimization problem cannot be recast in the framework of these problems.

3   Solution Techniques

We approach the program of (5) by exact and approximative solution techniques which are based on the following important observation. Let the problem be formulated in the link flows $\lambda_{jk}^{(f)}$ $(= \rho_{t,j}^{(f)} \cdot p_{jk}^{(f)})$, so that the program of (5) becomes linear, except for the interference constraints. We observe that these quadratic constraints can be replaced by linear ones if the nodes which do not receive any data packets in the optimal traffic distribution (i.e., the nodes with $\rho_{r,j} = 0$) are known. More explicitly, when $\rho_{r,j} = 0$, the constraint is always satisfied, and when $\rho_{r,j} > 0$, we can divide by $\rho_{r,j}$ to get the equivalent constraint:

$$\sum_{m \in \mathcal{N}_\mathcal{I}(j)} \rho_{t,m} + \rho_{t,j} \leq 1, \ \forall_{j \in \mathcal{N}} . \tag{14}$$

3.1   Exact Approach

The above observation shows that any feasible solution of the nonlinear programming problem characterizes a linear programming problem. The global optimum is also a feasible solution and, therefore, we could solve our nonlinear programming problem by consecutively solving 2^N linear problems. However, in practice such an approach is not computationally feasible.

Common techniques would then define the interference constraints by means of functions which indicate whether a node receives data or not. Instead of Eq. (8), we can write:

$$\sum_{m \in \mathcal{N}_I(j)} \rho_{t,m} + \rho_{t,j} \leq 1 + N(1 - r_j), \ \forall_{j \in \mathcal{N}} ,$$
$$\rho_{r,j} \leq r_j, \ \forall_{j \in \mathcal{N}} , \tag{15}$$
$$r_j \in \{0,1\}, \ \forall_{j \in \mathcal{N}} .$$

The introduction of the indicator function r_j transforms the quadratic programming problem into a mixed integer programming problem that is linear in the $\lambda_{jk}^{(f)}$. Although such a mixed problem is NP-hard, the advantage of this formulation is that standard solvers for this class of programming problems are widely available. Such standard solvers often embed branch-and-bound techniques to reduce the number of LP problems to be solved. Although such branch-and-bound techniques provide the optimal solution, unfortunately no guarantees as to the number of programs to be solved can be given.

3.2   Greedy Approximation Approach

For the evaluation of large networks, a more efficient technique than inspecting all LP problems or applying a branch-and-bound technique will be required. To this end, we introduce an approximate greedy algorithm which works linearly in N, the number of nodes. Our greedy algorithm is defined as follows.

Initially, assume that all nodes in the network receive data (i.e., $\rho_{r,j} > 0, \forall_{j \in \mathcal{N}}$) and then solve the linear program (i.e., with all interference constraints included) and its corresponding dual. In each following step, a node j^* is eliminated from the network and the resulting (linear) program is then analyzed again. That means, the program with ρ_{r,j^*} set to zero and thus with one interference constraint removed. This iteration process is continued until the optimum decreases after a node elimination.

The key element of the algorithm is the elimination step. This elimination takes place based on the values of the dual variables related to the interference constraints, since these dual values indicate the importance of the primal constraints (which have a one-to-one correspondence to the nodes). Therefore, in each step we eliminate the node with the greatest dual value (which is strictly positive for a connected network), which means that its constraint is definitely tight by appealing to the complementary slackness conditions. Notice that removing a node corresponding to a primal constraint that is not tight (i.e., its dual value is zero) would never lead to an improvement of the objective function. In the case of a tie, i.e., multiple nodes attain the highest dual value, then we select one of the nodes (by solving the primal problem) with the lowest fraction of transmitted packets per unit time. The rationale behind this is that a limited fraction can more easily be diverted via other paths.

The great advantage of this greedy approach is that since the set of removable nodes is of size N, if the greedy algorithm is used only order N of the 2^N LP problems have to be solved.

4   Numerical Results

We evaluate the impact of interference on the network capacity for various scenarios. We discuss a number of basic topologies that provide structural insights. Next, we move to more general topologies to assess the performance of our proposed greedy algorithm. Finally, capacity results for multiple SD-pairs and for extended interference and transmission ranges will be discussed. Recall that the network capacity is defined in terms of the fractions of time that the sources are transmitting and is thus a dimensionless metric.

4.1   Single Source-Destination Pair: Basic Topologies

In the topology figures presented in this section, a node can transmit data to (and also interfere with) another node (i.e., there exists a link between them), if and only if these nodes are connected by an arrow. Additional (symmetric) interference

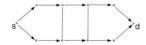

Fig. 1. Bridges between paths

Fig. 2. Triangular structure

Fig. 3. Claw structure

Fig. 4. Example of scenario for 3 node-independent paths of 5 nodes each

relations are indicated by solid lines. The interference and transmission neighborhoods follow accordingly. The paths are assumed to be given and can be seen as provided by a multi-path routing protocol which aims to select node-independent paths. Our interest here is to evaluate the impact of certain interference structures between paths and to find out in which situations using multiple paths is attractive. All the numerical values for the network capacity are computed by solving the program (5) via an exact approach. We note that the greedy approximate algorithm also provides the optimal solution in the cases presented.

Consider a single-path of at least three links from a source s to a destination d. The only interference is the self-interference within the path. Clearly, for capacity optimization all nodes on the path must be utilized in an identical fashion. The network capacity then equals $1/3$ (i.e., the source transmits data $1/3$ of the time), since nodes are not allowed to transmit simultaneously with their neighbors. If the network is shortened to a chain with only one or two links, the capacities would become 1 and $1/2$, respectively.

Next, let us define *independent paths* as paths from source s to destination d that do not interfere with each other except that they share the resources at s and d. The capacity of the SD-flow equals the sum of the flows over all individual paths from s to d. Using only one path, the network capacity clearly equals $1/3$. However, when using both paths, it follows that a capacity of $1/2$ can be attained. Hence, in a situation with interference only at the endpoints of the paths, it is favorable to split the traffic at the source and distribute it over multiple paths. When paths mildly interfere at some points, as in the situation with so-called "bridges" between the paths shown in Fig. 1, a capacity of $1/2$ may still be attained. However, increasing the interference between the paths further may lead to a drastic reduction of the capacity. If a single node is interfering with two nodes on the other path leading to a triangular structure between the paths (see Fig. 2), the capacity becomes $3/7$. Ultimately, when at least one node interferes with three (or even more) nodes (see Fig. 3) on the adjacent path, a claw structure arises and the capacity decreases to $1/3$, the single-path capacity. Thus, in a situation of two paths with at least one

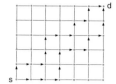

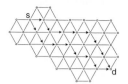

Fig. 5. Grid structure

Fig. 6. Honeycomb structure

heavily interfering node, employing an extra path does not affect the capacity in any meaningful way.

As an example, we consider networks with a more general structure: a grid (or mesh) network (see Fig. 5) and a honeycomb network (see Fig. 6). Each of the intersections corresponds to a node and the edges indicate that node pairs can communicate. Typically, in such networks there exist many equal-length paths between source s and destination d. Employing two independent shortest paths would yield the optimal capacity of $1/2$. One could argue that paths that more closely mimic the straight line between s and d should be preferred as those for example can be found more easily using geographical information, or alternatively, such paths restrict interference to a small corridor between s and d, thus causing less interference with other SD-pairs. We observe that several basic structures, namely triangles, bridges and even claws (see Fig. 6), will then appear between such paths and thus determine the network capacity.

4.2   Single Source-Destination Pair: General Topologies

To assess the quality of our greedy algorithm, we have studied several general topologies. More precisely, we have constructed topologies comprising equal-length, parallel, node-independent paths and randomly generated links between nodes on adjacent paths according to independent Bernoulli experiments with success probability p. A sample scenario for such a topology is provided in Fig. 4 where the solid lines indicate the links on the node-independent paths and the dashed lines the randomly generated links between these paths. The interference and transmission range are set to be equal, i.e., any link in the network induces interference, but can also be used for transmission. For all $j \in \mathcal{N}$, the neighborhoods $\mathcal{N}_{\mathcal{T}}(j)$ and $\mathcal{N}_{\mathcal{I}}(j)$ follow directly from this link set and we have $\mathcal{N}_{\mathcal{T}}(j) = \mathcal{N}_{\mathcal{I}}(j)$. For each scenario, we study the performance of the greedy algorithm by carrying out 50 runs of different random link configurations between the paths.

For moderate-size topologies (i.e., fewer than 20 nodes), we are still able to attain the optimal capacity by means of standard solvers. In Table 1, the average capacity values are presented for several scenarios and compared with the outcomes achieved by the greedy algorithm. Also included are the percentages of runs for which the greedy algorithm deviates more than 5% or 10% from the optimum. In the final scenario (5 paths of 3 nodes, $p = 0.80$), the greedy algorithm deviates more from the optimum, because it only succeeds in finding two independent paths while three such paths are present. However, the results show that, on average, the

Table 1. Capacity results for moderate-size topologies

Scenario		Optimal	Greedy	Err.>5%	Err.>10%	#LP solved
3 paths, 5 nodes	p=0.40	0.518	0.516	0%	0%	4.50
"	p=0.80	0.500	0.500	0%	0%	7.70
4 paths, 4 nodes	p=0.40	0.568	0.566	0%	0%	4.66
"	p=0.80	0.512	0.512	2%	0%	6.38
5 paths, 3 nodes	p=0.40	0.630	0.630	0%	0%	4.80
"	p=0.80	0.602	0.588	18%	10%	6.84

Table 2. Capacity results for large topologies

Scenario	Total nr. of nodes	p=0.4		p=0.8	
		All used	Greedy	All used	Greedy
4 paths, 6 nodes	26	0.564	0.568	0.502	0.512
5 paths, 6 nodes	32	0.616	0.624	0.522	0.590
6 paths, 6 nodes	38	0.654	0.660	0.568	0.612
7 paths, 6 nodes	44	0.684	0.692	0.612	0.646

greedy algorithm performs close to optimal, that its solution rarely deviates far from the optimal capacity, and that it requires limited computational effort (see the last column which indicates the average number of LPs solved per run).

For larger topologies (i.e., greater or equal than 20 nodes), exact solution approaches are no longer computationally feasible. On the contrary, we can rely on the greedy algorithm to find a fast approximation for the capacity. In Table 2, we compare the greedy approximation with the solution for the case that all interference constraints (8) are taken into account and only a single LP problem is to be solved; this corresponds to replacing Eq. (8) by Eq. (14). The comparison shows that for $p = 0.8$ our algorithm yields a much better approximation for the capacity, while for $p = 0.4$ the approximations are similar. We conclude that application of the greedy algorithm is especially valuable in situations in which there is much interference. Another observation from our experiments is that for high values of p, the source forwards traffic via fewer nodes than for low values of p. This suggests that in the case of heavy interference only a few paths need to be used.

4.3   Advanced Scenarios

Many networks are employed in situations where typically more than a single subject wants to transmit data. To this end, we study the network capacity in the situation of two (intersecting) SD-pairs. Regarding the objective function $h(\rho_t, p)$, we do not merely optimize the total capacity (as one pair may consume all the capacity, while the other pair may starve), but we impose additional restrictions on the individual flows. More specifically, we let $h(\rho_t, p) = \rho_{t,s_1}^{(1)} + \rho_{t,s_2}^{(2)}$, and impose $\rho_{t,s_2}^{(2)} = \frac{1-c}{c} \cdot \rho_{t,s_1}^{(1)}$, $0 \leq c \leq 1$. That means that, e.g., for $c = 1/2$, both flows are equally important, while for $c > 1/2$, SD-pair 1 is prioritized.

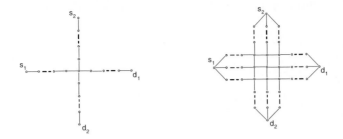

Fig. 7. a. Two single-paths (left), and b. Two multi-paths (right)

We construct two scenarios (along the approach in Sect. 4.2) with paths chosen to be long enough to avoid boundary effects. The capacity is determined using the approximate greedy algorithm. In the first scenario (see Fig. 7a), we consider two intersecting single-path flows. Hence, the central node can be active for at most $1/3$ of the time (as it must share resources with its neighbors), so that the capacity equals $1/3$ independent of the value for c. The second scenario (see Fig. 7b) comprises two intersecting multi-paths which each consist of three node-independent paths. Notice that the value of c influences the capacity in a symmetrical fashion. We discuss here the extreme cases $p = 0$ (independent paths) and $p = 1$ (heavily interfering paths). For $p = 1$, the impact of c is limited as for $c = 1/2$ the capacity equals 0.48, while for $c \to 0$ it approaches 0.50. Contrary, for $p = 0$, the capacity for $c = 1/2$ is 0.74, while for $c \to 0$ it is only 0.60. Thus, prioritizing one flow at the expense of the other can eventually be undesirable from a capacity viewpoint in the latter case. Again, we observe that the "multi-path" flow tends to use more paths for $p = 0$ than for $p = 1$.

As a final illustration, in accordance with more realistic applications, we analyze different interference and transmission neighborhoods. We construct scenarios which conform to the approach in Sect. 4.2 with link probabilities $p = 0$ and $p = 1$ (for a single SD-pair). For these choices of p, the neighborhoods $\mathcal{N}_T(j)$ and $\mathcal{N}_I(j)$ are fixed and identical for a given $j \in \mathcal{N}$. Let us denote these neighborhoods by $\mathcal{N}_0(j)$ and $\mathcal{N}_1(j)$, respectively. Next, we define extended neighborhoods as follows. For $p = 0$, let $\mathcal{N}_0^*(j)$ extend $\mathcal{N}_0(j)$ with the three parallel nodes on the two nearest paths and with the two adjacent nodes on the same path. For $p = 1$, let $\mathcal{N}_1^*(j)$ extend $\mathcal{N}_0^*(j)$ with the three parallel nodes of the paths located next to the adjacent paths (whenever present). Thus, we have that $\mathcal{N}_0^*(j) \supseteq \mathcal{N}_0(j)$ and $\mathcal{N}_1^*(j) \supseteq \mathcal{N}_1(j)$ We stress that the exact structure of the neighborhoods is not especially important here. Our intention is to gain some understanding of what might happen in realistic cases. In practice, the exact neighborhoods could be constructed based on network measurements.

We have evaluated scenarios with 1, 3 and 5 node-independent paths consisting of six "single-links" from source to destination. In Table 3, the capacity results (obtained via our greedy algorithm) for the different neighborhoods and link probabilities are presented. There are two main conclusions that can be drawn from our experiments. First, increasing the interference range for a fixed transmission

Table 3. Capacity results for different neighborhoods

Scenario	Neighborhoods	1 path	3 paths		5 paths	
		p=0/1	p=0	p=1	p=0	p=1
Basic scenario	$\mathcal{N}_T(j) = \mathcal{N}_I(j) = \mathcal{N}_{0/1}(j)$	0.333	0.600	0.500	0.714	0.583
Ext. interference	$\mathcal{N}_T(j) = \mathcal{N}_{0/1}(j)$, $\mathcal{N}_I(j) = \mathcal{N}_{0/1}^*(j)$	0.200	0.333	0.250	0.429	0.382
Ext. power	$\mathcal{N}_T(j) = \mathcal{N}_I(j) = \mathcal{N}_{0/1}^*(j)$	0.333	0.500	0.400	0.600	0.529

range has a clear negative effect on the capacity, and this is shown to hold for all scenarios. Second, also increasing both ranges (in an equal fashion) has a (minor) negative effect on the capacity, except for the case of a single path. Further, we observe that the solutions found comprise fewer paths when the interference range is extended; this indicates again that the impact of interference on path selection cannot be ignored.

5   Concluding Remarks

We have first presented a general stochastic framework for the analysis of network performance under interference. We have then developed a generic mathematical model to assess the network capacity in a multi-path environment. A nonlinear programming problem formulation incorporating interference has been introduced, and it has been shown that the optimal network capacity could be obtained by solving an exponential number of linear programming problems. We have presented a greedy algorithm that can closely approximate this optimum by solving only a small number of these linear programmes. Our examples further show that paths do not need to be independent to attain the optimal capacity under interference. Moreover, using multiple paths that moderately interfere appears to be better than using a single path, whereas in situations with heavily interfering nodes, use of multiple paths does not improve the network performance. Finally, more general scenarios have shown the significant impact of interference on the optimal route selection.

An important element of our modelling approach is the construction of interference and transmission neighborhoods. Our model facilitates consideration of any neighborhood of interest; in particular, the neighborhoods do not need to depend on the distance between nodes. This means that by performing certain network measurements before the actual network operation starts neighborhoods can accurately be defined and the capacity of the constructed network can be determined using the techniques presented in this paper.

We have assumed that links in the network are error-free. This is a valid assumption if lossy links are excluded in advance (cf. the approach in e.g. [16]). Otherwise, non error-free links may be incorporated by reducing the transmission rate so as to account for retransmission overhead. The validation of our results for more general networks in which nodes are not identical is among our aims for further research. It is further relevant to study the impact of interference in the context of delay optimization, because blindly employing multiple independent paths may yield undesirable delays due to the divergence in path lengths.

References

1. Clausen, T. (eds.): P.J.: Optimized Link State Routing protocol (OLSR) (2003)
2. Johnson, D., Maltz, D.: Dynamic source routing in ad hoc wireless networks. Mobile Computing, vol. 353 (1996)
3. Perkins, C., Royer, E.: Ad-hoc on-demand distance vector routing. In: Proc. of the 2nd IEEE Workshop on Mobile Computing System and Applications, pp. 90–100 New Orleans, LA, USA (1999)
4. Mueller, S., Tsang, R., Ghosal, D.: Multipath routing in mobile ad hoc networks: Issues and challenges. Lecture Notes in Computer Science (2004)
5. Lee, S.J., Gerla, M.: AODV-BR: Backup routing in ad hoc networks. In: Proc. of WCNC, Chicago, Illinois, USA (2000)
6. Nasipuri, A., Castañeda, R., Das, S.: Peformance of multipath routing for on-demand protocols in mobile ad hoc networks. ACM/Baltzer Mobile Networks and Applications (MONET) Journal 6, 339–349 (2001)
7. Lee, S.J., Gerla, M.: Split multipath routing with maximally disjoint paths in ad hoc networks. In: Proc. of IEEE ICC, pp. 3201–3205. Helsinki, Finland (2001)
8. Pham, P., Perreau, S.: Increasing the network performance using multi-path routing mechanism with load balance. Ad. Hoc. Networks 2, 433–459 (2004)
9. Tsirigos, A., Haas, Z.: Multipath routing in the presence of frequent topological changes. IEEE Communications Magazine 39(11), 132–138 (2001)
10. Wu, K., Harms, J.: Multipath routing for mobile ad hoc networks. Journal of Communication and Networks, vol. 4 (2002)
11. Pearlman, M.R., Haas, Z.J., Sholander, P., Tabrizi, S.S.: On the impact of alternate path routing for load balancing in mobile ad hoc networks. In: Proc. of MobiHoc, Boston, Massachusetts, USA (2000)
12. Jain, K., Padhye, J., Padmanabhan, V., Qiu, L.: Impact of interference on multi-hop wireless network performance. In: Proc. of Mobicom, San Diego, California, USA (2003)
13. Gupta, R., Musacchio, J., Walrand, J.: Sufficient rate constraints for QoS flows in ad-hoc networks (UCB/ERL Technical Memorandum M04/42, 2004)
14. Litjens, R., van den Berg, H., Boucherie, R.J., Roijers, F., Fleuren, M.: Performance analysis of wireless LANs: an integrated packet/flow level approach. In: Proceedings of ITC-18, Berlin, Germany (2003)
15. Schrijver, A.: Combinatorial Optimization: Polyhedra and Efficiency. Springer, Heidelberg (2006)
16. Padhye, J., Agarwal, S., Padmanabhan, V., Qiu, L., Rao, A., Zill, B.: Estimation of link interference in static multi-hop wireless networks. In: Proc. of IMC, Berkeley, CA, United States (2005)

Modeling IEEE 802.11 in Mesh Networks

Ali Ibrahim and James Roberts

France Telecom R&D

Abstract. This paper present an analytical model for the DCF function in 802.11 Mesh Networks where nodes are not all able to hear each other. The aim is to obtain insight into behavior rather than numerical accuracy. The model is quite general and accounts for the fundamental parameters that affect performance, namely network and flow topologies, channel reservation, exponential backoff and hidden nodes. Performances of the most elementary mesh networks (linear, tree, star,...) are analysed using this model. NS simulations are also provided to compare with the analytic results.

1 Introduction and Related Works

In the past few years, we have witnessed the widespread deployment of 802.11 WLAN networks. Two modes of operation have been specified for these networks in [1]: ad hoc (i.e., without any infrastructure equipment) and infrastructure mode which is the most common mode of deployment. In infrastructure mode, users generally connect to a fixed network through an Access Point which has limited transmission range. To achieve larger coverage areas, many access points are deployed as in cellular systems, and networks operators are obliged to deploy a high cost wired network to interconnect these access points. When there is no convincing reason to not interconnect these access points through wireless links, Wireless Mesh Networks tend to avoid this drawback by allowing access points to be interconnected by wireless links. Mesh networks thus increase the coverage area and reduce installation costs. From a performance evaluation point of view, in the first type of network deployment, nodes are relatively close to each other. Consequently, each node of the network is able to hear the transmission of all other nodes. We call this type a single collision domain or single cell network. The performances of these networks are well known analytically and by simulations [3], [7]. The Bianchi model [3] is amongst the mostly accurate models for these networks. The second type of network (i.e., Mesh) where, by definition, network nodes are not all able to hear each other is called a multiple collision domain (or multiple cell) network. The performance of this type of network, especially those based on the 802.11 protocol, is poorly understood at least analytically. The aim of this paper is to present an analytic model to analyse 802.11-based Mesh networks. Obtaining insight into behavior rather than numerical accuracy is our main objective.

The fundamental access method for IEEE 802.11 WLAN is called DCF. DCF is a CSMA/CA based channel access mechanism with binary slotted exponential

L. Mason, T. Drwiega, and J. Yan (Eds.): ITC 2007, LNCS 4516, pp. 816–828, 2007.

backoff and channel reservation mechanism. A station operating under DCF may transmit directly its frame when it senses that the medium has been idle for a period greater than or equal to a predefined period DIFS. Otherwise, the station enters the backoff mode. In backoff mode, the station generates a random integer distributed uniformly between 0 and $CW - 1$. This random integer is used to initialise a backoff counter. CW is called the current contention window size. This backoff counter measures the number of idle physical time slots that the station must wait before accessing the channel. The physical time slot time is an elementary predefined period which depends only on the physical layer. Hidden nodes constitute an inherent problem for accessing the shared wireless medium. To cope with this problem, DCF uses a channel reservation mechanism. This mechanism consists in exchanging two short frames called RTS/CTS between the sender and the receiver prior to any transmission. Each of these frames contains a field which indicates the duration of the intended transmission. Upon receiving an RTS or CTS frame, a node in the neighbourhood of the sender or the receiver is informed about the intended transmission and therefore defers its channel access.

Let us now try to analyse the operation of nodes in 802.11 networks. In the single cell case, this can be summarised in the following way: after a non-idle period, all stations (in backoff mode) wait for a DIFS, then they decrement their backoff counters as long as the channel is idle. When the backoff counter of a station reaches zero, an RTS or DATA frame is transmitted (the decision depends on the frame size). Only two cases must be considered as a result of this transmission: (a) Collision: when at least one other station of the network has accessed the channel during the same slot, (b) Successful transmission: when there is no other transmitting station. Thus collisions occur only as a result of equalities between backoff counters. Networks with multiple collision domains or meshes exhibit an extremely complex behaviour due partly to the fact that collisions may occur between any type of frame (RTS with DATA, RTS with ACK, etc...). The main reason for this is the fact that nodes are not able to hear each other. In fact, in CSMA based MAC protocols, node behaviour depends on what they see on the channel. In the single cell case, all nodes have the same global vision of the channel, node behaviour is then based on this global information. In the multi-cell case, nodes are not able to hear each other, consequently each node has only a local vision of the network, and its operation is based on partial information. Moreover, 802.11 provides a channel reservation mechanism and frame acknowledgement that include exchanging some information between the nodes (such as channel reservation time). The 4-way handshake, information exchange combined with decisions based only on local information are the main reasons for this complex interaction. Several previous works [5], [4] have pointed out to this complex interaction .

This paper is organised as follows. In section 2 we describe our model, model analysis is provided in sections 3 and 4. Numerical results are provided in section 5. We conclude in section 6.

2   The Model

The protocol: We consider a modified version of the 802.11 protocol. In order to specify this new protocol, we will describe only its differences with the original 802.11. In this new MAC, time is divided into slots of equal duration σ. Network nodes are synchronised in the sense that all stations see the same time axis, but not necessarily all transmissions, and a transmission may occur only at the beginning of a time slot. We also suppose that all nodes use the RTS/CTS mechanism whatever the packet length. Moreover the transmission time of a control frame RTS, CTS, ACK and the physical slot time are all equal to σ and the transmission time of any DATA frame is an integral multiple of σ. Finally, the inter-frame spaces SIFS and DIFS are assumed to be null. All other mechanisms such as exponential backoff, freezing the backoff counter when the channel is not idle, channel reservation by mean of RTS/CTS and frame acknowledgement are exactly the same as in the original 802.11 MAC.

The network: We consider a set of nodes using the previously specified protocol as MAC. The topology of the network is defined by an undirected graph $G = (V, E)$. V is the set of nodes and G is the set of links. A link $e = \{u, v\}$ between stations u and v means that these stations are able to hear each other.

The flows: In single cell networks, it has been common to consider saturated stations [3], [7], i.e., at any instant of time, each station has a least one packet to transmit. This assumption can be easily justified because maximal throughput occurs when nodes are saturated. In addition to providing maximal performance, it has been pointed out that, under this assumption, nodes become synchronised and this greatly simplifies the analysis. Multi-cell or Mesh networks differ significantly from their single cell counterparts. Mesh networks are multi-hop, with frames belonging to the same flow transmitted hop-by-hop along a predefined path. These paths must be completely specified. Moreover, saturating the sources of these flows does not mean maximal throughput, nor does it simplify the analysis as nodes cannot be synchronised even if they were all saturated. To make the analysis tractable we assume the following: we consider only single hop flows with saturated sources. Two reasons justify our choice: 1) this flow model is the first step before studying a more general flow models, 2) this model is sufficient to analyse the most elementary network scenarios that are likely to occur in practical mesh deployment. Flow topology is thus defined using a directed sub-graph $H = (V, F)$ of G. Another important parameter that directly affects throughput is the packet length. In our flow model, all packets have the same fixed size. Choosing a fixed packet length model as in [7] is sufficient to gain insight into performance in most situations, although it remains a special case of the general random packet length model adopted in [3].

3   Model Analysis

Under all the above assumptions, we can now analyse the operation of a station v using the new MAC protocol. Without loss of generality, consider a slot n

and suppose that at the beginning of this slot the station is in backoff stage i ($0 \leq i \leq m$) having j ($1 \leq j \leq W_i - 1$) as a value for the backoff counter where m is the number of backoff stages and W_i is the backoff window in the ith backoff stage. We refer to this state by $S_{Back}(i,j)$, The letter S in the notation stands for Silent. During slot n, 4 events can occur (see Figure 1):

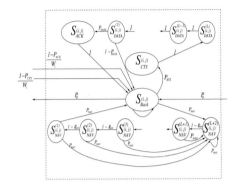 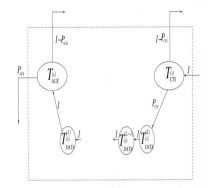

Fig. 1. Silent mode   **Fig. 2.** Transmitting DATA

a. Station v receives an RTS frame, it moves to state $S_{CTS}(i,j)$ in slot $n+1$. In this state, the station sends a CTS frame and then waits L slots for the reception of the DATA frame, we refer to these waiting states by $S_{DATA}^{(k)}(i,j)$ $1 \leq k \leq L$. If, at the end of these L slots, the DATA frame was received successfully, the station acknowledges this frame by sending an ACK frame to the sender in state $S_{ACK}(i,j)$ and then goes back to its previous backoff state $S_{Back}(i,j)$. Conversely, if the reception of the DATA frame was not correct, the station simply returns to state $S_{Back}(i,j)$ without sending any acknowledgement.

b. The station receives a control frame causing its NAV vector to be set. More precisely, station v receives an RTS or CTS frame whose destination address is different from its own address, we refer to such frame by NAV frame. In this case the station updates its NAV, i.e. it waits for $L+2$ slots before resuming its backoff operation. We denote these waiting states by $S_{NAV}^{(k)}(i,j)$ $1 \leq k \leq L+2$. The transition from any state $S_{NAV}^{(k)}(i,j)$ to the first state $S_{NAV}^{(L+2)}(i,j)$ simply indicates that station v may receive at any instant an other NAV frame.

c. The station senses that the medium was not idle during slot n, due for example to a collision or a neighbouring transmission, the station remains in its current state.

d. The channel is idle, the station decrements its backoff counter and moves to state $S_{Back}(i,j-1)$.

When the backoff counter reaches zero in state $T_{RTS}(i)$ (see Figure 3), the station transmits an RTS frame and then waits for the CTS in state $T_{CTS}(i)$ (Figure 2) during the next slot. Note that in state $T_{RTS}(i)$ the station cannot receive an RTS frame nor it can update its NAV as in the previous backoff states. If no

CTS frame is received in response to the transmitted RTS then, according to 802.11, the station must double the size of its contention window and choose a random integer in the new contention window. If the CTS response is received, the station transmits its DATA frame during the next L slots and waits for the acknowledgement. We denote these states by $T^{(k)}_{DATA}(i,j)$ $1 \leq k \leq L$. Similarly for the CTS frame, two cases must be considered: an ACK reception in state $T_{ACK}(i)$ causes the station to move back to its first backoff stage, while absence of acknowledgement causes the station to move to the next backoff stage.

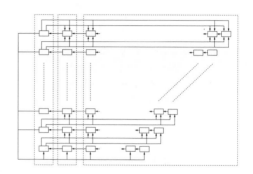

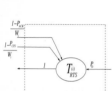

Fig. 3. Transmit RTS   **Fig. 4.** Chain of station v

Taking into account the binary exponential backoff algorithm as specified in [1], and arranging the states in the correct order, we obtain the chain of Figure 4 which is similar to that obtained in [3]. However, each box now represents one of the sets of states defined by the finite state machine of Figures 1 to 3.

Notations: We use the letter S which stand for "Silent" to denote the states of Figure 1, because the station is not accessing the channel in these states. The letter T used in Figures 2 and 3 stands for "Transmission". An omitted sub/superscript from state notations means that a state aggregation has been performed on that sub/superscript, for example

$$S_{DATA}(i,j) = \bigcup_{1 \leq k \leq L} S^{(k)}_{DATA}(i,j) \qquad S_{DATA} = \bigcup_{\substack{0 \leq i \leq m \\ 0 < j < W_i}} S_{DATA}(i,j)$$

$$S^k_{NAV} = \bigcup_{\substack{0 \leq i \leq m \\ 0 < j < W_i}} S^{(k)}_{NAV}(i,j) \ .$$

And let $S = S_{DATA} \bigcup S_{ACK} \bigcup S_{CTS} \bigcup S_{Back} \bigcup S_{NAV}$.

For a flow $f \in H$ and $u, v, w \in V$, we denote by $s(f)$ (resp. $r(f)$) the sender (resp. receiver) station of flow f, $N(u) = \{v \in V \ / \ \{u,v\} \in E\}$ the neighbours of u, $N(u,v) = (N(u) \cup N(v)) \backslash \{u,v\}$, $N(u,v,w) = (N(u) \cup N(v) \cup N(w)) \backslash \{u,v,w\}$. When $u = s(f)$ and $v = r(f)$, $N(s(f), r(f))$ is called the first neighbourhood of flow f. $H(f) = N(r(f)) \backslash (N(s(f)) \cup \{s(f)\})$ denotes the set of hidden nodes relative to flow f, $N^2(f) = \cup_{u \in H(f)} N(u) \backslash (N(s(f)) \cup N(r(f)))$ the second neighbourhood of flow f, i.e. the neighbours of the hidden nodes which are not in the first neighbourhood of f. Finally, we denote by $NAV_{RTS}(v) = \{ f \in F \ / \ v \in N(s(f))\backslash\{r(f)\} \}$ the set of flows which may cause the NAV of station v to be set by an RTS frame and by $NAV_{CTS}(v) = \{ f \in F \ / \ v \in N(r(f))\backslash(\{s(f)\} \cup N(s(f)))\ \}$ the set of flows which are able to update the NAV of station v only by CTS frames.

3.1   Networks with Multiple Collision Domains: Key Observations

Having a well defined state space and state transitions, we can now describe the operation of each station by a stochastic process $(X_n^v)_{n\in\mathbb{Z}}$. We denote by $(\Omega, \mathcal{A}, \mu)$ its sample space, and by ω the elements of Ω. The overall network operation is described by the stochastic process $(\prod_{v\in V} X_n^v)_{n\in\mathbb{N}}$ which will be analysed in the current section. The analysis is based on establishing what we call minimal relations of dependence. The basic idea of these relations is: given some station v, a slot n and a subset E of the state space, what is, in the general case, the maximal amount of information that we can deduce regarding the states of neighbouring nodes ? To establish these relations, we will begin by analysing the transmission of a frame belonging to a flow f. Without loss of generality, we suppose that station $s(f)$ which is the flow source, transmits an RTS frame at time $t = 0$. According to the 802.11 protocol, the station must have sensed an idle channel for a period of time. This means that all the neighbours of $s(f)$ are in the silent mode. As illustrated on Figure 1, silent mode means one of the following: (i) virtual reservation, (ii) reception or (iii) backoff. We deduce the following observation and the corresponding dependence relation:

Observation 1: a necessary and sufficient condition for a station $s(f)$ to access the channel by sending a RTS frame during slot n is the following: (i) The backoff counter of the station is null at slot n. (ii) All the neighbours $N(s(f))$ of the station are in silent mode.

$$\left\{\omega/X_n^{s(f)} \in T_{RTS}\right\} = \left\{\omega/X_n^{s(f)} \in T_{RTS}, \forall u \in N(s(f))\ X_n^u \in S\right\} . \qquad (1)$$

To correctly receive the RTS frame, the channel must also be idle in the neighbour of $r(f)$. According to *observation 1*, receiver $r(f)$ is in one of the following modes: (i.) Virtual reservation, in this case, the station must discard the received RTS frame.(ii.) Reception, $r(f)$ cannot interpret the RTS frame due to the previous ongoing reception. (iii.) Backoff, only in this case, the station will send a CTS response to $s(f)$. We deduce:

Observation 2: A necessary and sufficient condition to have a successful RTS/CTS exchange sequence between sender $s(f)$ and receiver $r(f)$ during slots n and $n+1$ is: (i.) The backoff counter of the sender is null during slot n. (ii.) All the neighbours of the sender and the receiver are in silent mode during slots n and $n + 1$. (iii.) The receiver $r(f)$ is in the backoff mode during slot n.

$$\left\{\omega/X_{n+1}^{r(f)} \in S_{CTS}\right\} = \bigcup_{f^*\in F \ / \ r(f^*)=r(f)} \left\{\omega/X_n^{s(f^*)} \in T_{RTS},\ X_n^{r(f^*)} \in S_{Back},\right.$$
$$\left.\forall u \in N(s(f^*), r(f^*))\ X_n^u \in S\right\} \quad (2)$$

$$\{\omega/X_{n+1}^{s(f)} \in T_{DATA}^L\}=\{\omega/X_{n-1}^{s(f)} \in T_{RTS},\ X_{n-1}^{r(f)} \in S_{Back}, \forall u \in N(s(f),r(f))$$
$$X_{n-1}^u \in S\} \ . \quad (3)$$

Until now, the RTS/CTS exchange sequence has been established between the sender and the receiver, but this does not imply that all neighbouring nodes are able to correctly receive one of these frames in order to update their NAV and therefore refrain from channel access. In other words, the success of the current transmission is not guaranteed. Depending on the position of the node that has missed the RTS/CTS sequence, we distinguish two cases: (i) the node is in the sender neighbourhood, then it will hear the transmission of the DATA frame and therefore the station will refrain from channel access because the medium will not be idle, (in summary, it has no effect on the success of the current transmission). (ii) if the station that has missed the RTS/CTS sequence is only in the receiver neighbourhood, i. e., a hidden node, then not setting the NAV increases significantly the probability of collision, because in this case the station is completely unaware of the current transmission. Therefore we can deduce the following observation:

Observation 3: Suppose that we have a successful RTS/CTS exchange sequence between the sender and the receiver, then the transmission of the DATA frame will be successful iff all the hidden nodes set their NAV.

Finally, it remains to analyse the necessary and sufficient condition for a station v to update its NAV vector. According the IEEE standard, a station must set its NAV upon receiving an RTS or a CTS frame having a destination address different from its own address, we call such a frame a NAV frame. In order to update its NAV, the station also must be in Backoff or virtual reservation mode and it must receive correctly the control frame, i.e., all the neighbours of the considered station must be in the silent mode. We, thus have the following observation :

Observation 4: Suppose that a station v receives a NAV frame during slot n, then it will set its NAV iff: (i.) The station v is in backoff or virtual reservation mode. (ii.) All the neighbours of station v are in a silent state. From all the previous observations we deduce the following relations:

$$\{\omega/X_{n+1}^v \in S_{NAV}^{(L+2)}\} = \bigcup_{f \in NAV_{RTS}(v)} \{\omega/X_n^{s(f)} \in T_{RTS}, X_n^v \in (S_{Back} \cup S_{NAV}),$$
$$\forall u \in N(s(f),v)\ X_n^u \in S\} \bigcup_{f \in NAV_{CTS}(v)} \{\omega/X_{n-1}^{s(f)} \in T_{RTS}, X_{n-1}^{r(f)} \in S_{Back},$$
$$X_n^v \in (S_{Back} \cup S_{NAV}),\ \forall u \in N(s(f),r(f),v)\ X_{n-1}^u \in S\} \quad (4)$$

$$\{\omega/\ X_{n+1}^v \in S_{ACK}\} = \bigcup_{f \in F\ /\ r(f)=v} \{\omega/\ X_{n-L-1}^{s(f)} \in T_{RTS},\ X_{n-L-1}^{r(f)} \in S_{Back},$$
$$\forall u \in N(s(f)) \setminus (\{v\})\ X_{n-L-1}^u \in S,\ \forall u \in H(r(f))\ X_{n-L-1}^u \in (S_{NAV} \cup S_{Back}),$$
$$\forall t \in N^2(r(f))\ X_{n-L-1}^t \in S\} \quad (5)$$

$$\{\omega/X^v_{n+1} \in S_{Back}(0;.)\cup T_{RTS}(0)\} = \{\omega/X^{s(f)}_{n-L-1} \in T_{RTS},\ X^{r(f)}_{n-L-1} \in S_{Back},$$

$$\forall u \in N(s(f),r(f))\ X^u_{n-L-1} \in S, \forall u \in H(r(f))\ X^u_{n-L} \in S_{Back} \cup S_{NAV},$$

$$\forall t \in N^2(v)\ X^t_{n-L} \in S\} \ . \quad (6)$$

4 Stochastic Analysis: Markovian Approach

Network operation is described by the stochastic process $(\prod_{v\in V} X^v_n)_{n\in\mathbb{N}}$. In general, this process is not Markovian and the component processes (i.e., the processes describing the operation of each individual node) are not independent. Such processes occur frequently when modelling communication networks. Surprisingly, the classical approach for deriving analytic results for these processes is to assume the inverse, i.e., to decouple the Markovian component processes. Adopting this approach, 3 assumptions A_1, A_2 and A_3 will be made. Assumption A_1 enables us to assign transition probabilities, assumption A_2 defines these transition probabilities, and assumption A_3 decouples the component processes.

A_1: $\forall v \in V$ the process $(X^v_n)_{n\in\mathbb{Z}}$ is Markovien.
A_2: The transition probabilities of the process $(X^v_n)_{n\in\mathbb{Z}}$ are as given in Figures 1, 2 and 3.
A_3: The processes $(X^v_n)_{n\in\mathbb{Z}}$ are independent.

The key idea of this model is to represent the operation of a station by a Markov chain. This Markov chain is defined using 6 probabilities: P_{NAV}, P_{RTS}, P_{DATA}, P_D, P_{CTS} et P_{ACK}. Conversely, dependence relations established in the previous section and the topology of the flows and the network allow us to compute these transition probabilities as functions of the stationary probabilities of the neighbouring nodes. We obtain a fixed point equation which determines the operating point of the network.

4.1 Stationary Probabilities

The state transition diagram and transition probabilities are depicted on Figures 1, 2, 3 and 4. This chain can be solved easily for the stationary probabilities. Due to lack of space, the details of the calculation are omitted. Let

$$p = 1 - P_{ACK}P_{CTS} \qquad\qquad k_1 = P_{RTS}(1 + L + P_{DATA}) + \frac{1}{(1-P_{NAV})^{L+2}}$$

$$k_2 = \frac{(W_0-1)(1-2p)+pW_0(1-(2p)^m)}{2P_D P_{CTS}(1-p)(1-2p)} \qquad\qquad k_3 = (L+1+\frac{2}{P_{CTS}})\frac{1}{1-p}$$

Then the stationary probabilities are given by:

$$\pi_{ACK}(0) = \frac{1}{k_1 k_2 + k_3} \qquad\qquad\qquad \pi(S_{Back}) = k_2 \pi_{ACK}(0)$$

$$\pi(S) = k_1 k_2 \pi_{ACK}(0) \qquad\qquad \pi(S_{Back} \cup S_{NAV}) = \frac{1}{(1-P_{NAV})^{L+2}} k_2 \pi_{ACK}(0)$$

$$\pi(T_{RTS}) = \frac{\pi_{ACK}(0)}{P_{CTS}(1-p)} \ .$$

4.2 Transition Probabilities

Due to lack of space, we will only illustrate the method used to derive the transition probability P_{NAV}^v for station v. The other transition probabilities are derived in a similar way. Considering the one time step evolution for the chain $(X_n^v)_{n \in \mathbb{Z}}$ at state $S_{NAV}^{(L+2)}(i,j)$ we obtain the following equation

$$P[X_{n+1}^v = S_{NAV}^{(L+2)}(i,j)] = P_{NAV}^v \left(\sum_{1 \leq k \leq L+2} P[X_n^v = S_{NAV}^{(k)}(i,j)] + P[X_n^v = S_{Back}(i,j)] \right) .$$

(7)

Summing both sides of equation (7) over i and j and using the notation of Section 3 we obtain

$$P[X_{n+1}^v \in S_{NAV}^{(L+2)}] = P_{NAV}^v P[X_n^v \in S_{NAV} \bigcup S_{Back}] .$$

(8)

The above equation can be rewritten in terms of probability measure in the sample space $(\Omega, \mathcal{A}, \mu)$

$$\mu\{\omega \, / \, X_{n+1}^v(\omega) \in S_{NAV}^{(L+2)}\} = P_{NAV}^v \mu\{\omega \, / \, X_n^v(\omega) \in S_{NAV} \bigcup S_{Back}\} .$$

(9)

Using equation (4) and assumption $\mathbf{A_3}$, then taking the limits when $n \longrightarrow +\infty$, we obtain the following equation

$$P_{NAV}^v = \sum_{f \in NAV_{RTS}(v)} \pi^{s(f)}(T_{RTS}) \prod_{u \in N(s(f),v)} \pi^u(S)$$

$$+ \sum_{f \in NAV_{CTS}(v)} \pi^{s(f)}(T_{RTS}) \pi^{r(f)}(S_{Back}) \prod_{u \in N(s(f),r(f),v)} \pi^u(S) \quad (10)$$

where $\pi^u(E) = \lim_{n \longrightarrow +\infty} \mu\{\omega \, / \, X_n^v(\omega) \in E\}$. For the other transition probabilities we obtain the following equations

$$P_{RTS}^v = \sum_{f \in F, r(f)=v} \pi^{s(f)}(T_{RTS}) \prod_{u \in N(s(f),r(f))} \pi^u(S) \quad (11)$$

$$P_D^v = \prod_{u \in N(v)} \pi^u(S) \qquad P_{CTS}^v = \frac{\pi^{r(f)}(S_{back})}{\pi^{r(f)}(S)} \prod_{u \in H(r(f))} \pi^u(S) \quad (12)$$

$$P_{ACK}^v = \prod_{u \in H(r(f))} \frac{\pi^u(S_{Back} \cup S_{NAV})}{\pi^u(S)} \prod_{u \in N^2(r(f))} \pi^t(S) \quad (13)$$

$$P_{DATA}^v = \frac{\sum_{f \in F \, / \, r(f)=v} \pi^{s(f)}(T_{RTS}) \pi^{r(f)}(S_{Back}) \prod_{u \in N(s(f))} \pi^u(S)}{\sum_{f \in F \, / \, r(f)=v} \pi^{s(f)}(T_{RTS}) \pi^{r(f)}(S_{Back})}$$

$$\frac{\prod_{u \in H(r(f))} \pi^u(S_{Back} \cup S_{NAV}) \prod_{u \in N^2(r(f))} \pi^u(S)}{\prod_{u \in N(s(f),r(f))} \pi^u(S)} . \quad (14)$$

5   Numerical Results

In this section we will apply the above described analytical model to calculate the throughput in some elementary 802.11b mesh networks which are likely to occur in real deployments. In the 802.11 standard, any frame (DATA or Control i.e., RTS, CTS or ACK) consists of PHY header, MAC header and a payload. In the particular case of 802.11b, headers lengths and the transmission times of RTS, CTS and ACK frames are given in Table 1. The physical header is always transmitted at 1 Mbit/s while the MAC header and the payload are transmitted at one of the following rates : 1, 2, 5.5 and 11 Mbit/s depending on signal quality. In this model, we assume that all stations use the same maximal transmission rate of 11 Mbits/s, and the packet payload is 1000 bytes. Simulation results are obtained using ns2.29.

Throughput: In Section 2, we have assumed that the transmission time of an RTS, CTS or ACK frame and the physical slot time ($\sigma_{PHY} = 20\mu s$) all have the same duration σ and that inter-frame spaces DIFS ($50\mu s$) and SIFS ($10\mu s$) are null. In this case the throughput of a station v will be

$$R^v = \frac{(\sum_{0 \leq i \leq m} \pi_{ACK}(i))^v P^v_{ACK}}{\sigma} \times \text{Frame length} . \tag{15}$$

$\sum_{0 \leq i \leq m} \pi_{ACK}(i))^v P^v_{ACK}$ is the probability in stationary regime that station v receives an ACK frame in response to its transmitted DATA frame. Equation (15) indicate that a transmission will be considered as successful only when the ACK frame is received by the sender.

Table 1. PHY and MAC headers

	RTS	CTS	DATA	ACK
PHY header (bits)	192	192	192	192
MAC header (bits)	160	112	272	112
Payload	0	0	variable	0
Transmission time (μs)	209	202	variable	202

So a reasonable choice of σ is $\sigma_{CON} = 215\mu s$ which is an approximation of the transmission time of a control frame (RTS, CTS or ACK)+SIFS. The parameter L of the Markov chain is then calculated for the relative transmission time of the DATA frame:

$$L = \frac{\text{transmission time of DATA}}{\sigma_{CON}} . \tag{16}$$

Finally, to compare simulations and analytical results we must estimate σ taking into account the real value of the physical slot. We propose the following equation to estimate the value the mean slot time seen by station v then:

$$\sigma^v = \sigma_{PHY} \pi^v(S_{Back})P_D^v + \sigma_{CON}(1 - \pi^v(S_{Back})P_D^v) \ . \tag{17}$$

Then the Throughput R^v is then

$$R^v = \frac{(\sum_{0 \le i \le m} \pi_{ACK}(i))^v P_{ACK}^v}{\sigma^v} \times \text{Frame length} \ . \tag{18}$$

The global throughput $\sum R^v$ is the sum of the individual throughputs. For each scenario we provide two sets of numerical results. In the first set, we apply the model using the real parameters of 802.11b as provided in Table 1, i.e. $\sigma_{PHY} = 20\mu s$ and $\sigma_{CON} = 215\mu s$. In the second set, the model is applied using $\sigma_{PHY} = \sigma_{CON} = 200\mu s$. The objective of providing this second set of numerical values is to understand the impact of the assumptions we have made on the accuracy of the model, notably the assumption that physical slot is equal to the transmission of a control frame.

Three hop linear networks: We consider the 3 hop linear network shown in Figure 5. Simulation and analytic results are illustrated in Table 2. These results show the poor performances of station A and unfairness in bandwidth allocation. This unfairness in bandwidth allocation can be interpreted as follows: transmissions from C to D are not affected by any other transmission, therefore the backoff counter for station C will remain in the first stage. Station B collides only when transmitting at the same time as station C. For station A, the situation is completely different. We observe the following phenomenon: during transmissions from C to D, station B is blocked by the virtual channel reservation mechanism and therefore will not respond to the RTS requests of station A. Consequently station A will continue to double its contention window as long as it does not receive the CTS frame from B, thus reducing its probability of accessing to win the channel. We observe also that for some flows, the model is not quite accurate. The differences between simulation and analytical results are explained in the conclusion.

Fig. 5. 3 hops linear networks

Table 2. Numerical results for the linear network

| | $\sigma_{PHY} = 20\mu s$ | | | | | | $\sigma_{PHY} = 200\mu s$ | | | | | |
| | Simulation | | | Model | | | Simulation | | | Model | | |
	f_{AB}	f_{BC}	f_{CB}	f_{AB}	f_{BC}	f_{CB}	f_{AB}	f_{BC}	f_{CB}	f_{AB}	f_{BC}	f_{CB}
R^v	205	1 880	2 281	647	1 217	2 411	547	1 038	1 204	514	995	1 230
$\sum R^v$	4 366			4 275			2 789			2 739		

Tree network: In this scenario, we consider a tree network of three nodes as shown in Figure 6. This topology is most likely to occur in wireless access networks. Numerical results are illustrated in Table 5.Unfairness regarding stations A and C can be explained as follows: after a successful transmission by station B, stations A and C are in backoff mode since they are saturated, since these stations do not hear each other, an RTS transmission by one of these stations will not be detected by other stations until the reception of the CTS response of B. Collision occurs between these two station occurs when transmitting instants are separated by a period less than RTS which is approximately 10 physical time slots. Therefore, collisions between stations C and D are more less likely to occur as if these two stations are within the hearing range of each other.

Fig. 6. Tree network

Table 3. Numerical results for the tree network

	$\sigma_{PHY} = 20\mu s$						$\sigma_{PHY} = 200\mu s$					
	NS Simulation			Model			NS Simulation			Model		
	f_{AB}	f_{BD}	f_{CB}	f_{AB}	f_{BD}	f_{CB}	f_{AB}	f_{BD}	f_{CB}	f_{AB}	f_{BD}	f_{CB}
R^v	1 007	2 192	1 017	993	2 058	993	829	1 152	837	757	1 128	757
$\sum R^v$	2 818			4 044			2 818			2 642		

Star Network: In the star network of Figure 7, unfairness is not present due to network symmetry but global throughput is much more degraded than all previous cases. The interpretation is similar to the case of tree network.

Table 4. Numerical results for the star network

	$\sigma_{PHY} = 20\mu s$		$\sigma_{PHY} = 200\mu s$	
	Sim.	Model	Sim.	Model
	f_{AB}	f_{AB}	f_{AB}	f_{AB}
R^v	1190	1 438	872	797
$\sum R^v$	3 570	4 314	2 616	2 391

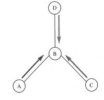

Fig. 7. Star network

6   Conclusion

In this paper we have presented an analytical model for the DCF function in multiple collision domain networks. The model is quite general and accounts for the fundamental parameters that affect performance, namely network and flow topologies, channel reservation, exponential backoff and hidden nodes. Differences between the analytic results and simulations are due essentially to the assumptions that we have made in order to make the model tractable: (i) assuming that nodes are synchronised, (ii) the duration of the physical slot is equal to the transmission time of a control frame, (iii) the inter-frame spaces SIFS and DIFS are null (iv) and Finally, we have neglected the loss phenomena due to the finite number of retransmissions since in our model, the frames are retransmitted until success. The model is somewhat complex and is not intended for operational use. It is meant as an aid to understanding the fundamental behaviour of multiple collision domain networks. We would ideally like to identify simpler models that build on the intuition gained in the present work. This is the object of our current research.

References

1. IEEE standard for Wireless LAN Medium Access Protocol and Physical Layer Specifications, IEEE Std. (1999)
2. Medepalli, K., Tobagi, F.: Towards Performance Modeling of IEEE 802.11 based Wireless Netwroks: A Unified Framework and its applications. In: Proceedings of IEEE INFOCOM (2006)
3. Bianchi, G.: Performance Analysis of the IEEE 802.11 Distributed Coordination Function. IEEE Journal on Selected Areas in Communications 18, 535–547 (2000)
4. Tsertou, A., Laurenson, D.I., Thompson, J.S.: A New Approach for the Throughput Analysis of IEEE 802.11 in Networks with Hidden terminals, IWWAN (2005)
5. Ray, S., Carruthers, J.B., Starobinski, D.: Evaluation of the Masked Node Problem in Ad-Hoc Wireless LANs. IEEE Transactions on Mobile Computing, Vol. 4(5), (September 2005)
6. Hegde, N., Proutiere, A.: Packet and Flow level Performance of Wireless Multihop Data Networks, Globcom (2006)
7. Kumar, A., Altman, E., Miorandi, D., Goyal, M.: New insights from a Fixed Point Analysis of Single Cell IEEE 802.11 WLANs. In: Proc. INFOCOM (2005)

Self-tuning Optimal PI Rate Controller for End-to-End Congestion With LQR Approach

Yang Hong and Oliver W.W. Yang

[1] CCNR Lab, SITE, University of Ottawa,
Ottawa, Ontario, Canada K1N 6N5
{yhong, yang}@site.uottawa.ca

Abstract. This paper presents a self-tuning optimal PI (Proportional-Integral) rate controller for end-to-end congestion in the IP-based Internet. We employ LQR (Linear Quadratic Regulator) approach from modern control theory in the optimal controller design that would allow the user to achieve good transient performance of IP networks by selecting proper positive definite matrices Q and R. Self-tuning PI rate controller self-tunes only when the changes in network traffic have drifted the network monitoring parameter outside its specified interval. The PI rate controller is located in each router and can calculate a desirable advertised source window size (i.e. source sending rate) based on the instantaneous queue length of the buffer. Our OPNET simulations demonstrate that our network traffic control algorithm can provide the network with a good transient behavior in AQM and flow throughput.

Keywords: Rate-Based Control, Active Queue Management, Best-Effort Traffic, LQR Approach, Streaming Media Traffic.

1 Introduction

Congestion control has been a critical factor in the robustness of the Internet [1] and its photonic infrastructure e.g., [2]. As a dominant end-to-end congestion control algorithm, TCP (Transmission Control Protocol) is widely implemented in current IP (Internet Protocol) networks [1]. In TCP/IP networks, the router applies AQM (Active Queue Management) algorithm (e.g., RED [3], PI-RED [4, 5] etc.) to implicitly inform the source the network congestion by dropping the packets. In the mean time, the source uses the AIMD (Additive Increase and Multiplicative Decrease) algorithm [6] to adjust its window size to prevent the congestion. They are all *window-based control*.

The AIMD algorithm is very effective in the conventional IP networks where the propagation delays are short compared to the source sending rates. However, under the network environment with the long round trip delay and dynamically changing available bandwidth (e.g., the edge switches of the Agile All-Photonic Networks [2]), the AIMD control with AQM often yields severe fluctuations in the source sending rates and oscillation of queuing sizes in the routers [7, 8]. Such behavior greatly degrades the network performance and makes the AIMD control unsuitable for streaming media transmission [9, 10].

L. Mason, T. Drwiega, and J. Yan (Eds.): ITC 2007, LNCS 4516, pp. 829–840, 2007.

Many attempts have been made to achieve a high QoS (Quality of Service) for streaming media transmission in the Internet. One of the challenges in designing congestion controller is to support streaming media traffic consisting of a variety of traffic classes with a different quality of service requirement. *Rate-based control* allows the sources to adjust their sending rates to support best-effort service traffic and makes the optimum use of network resources. Most of rate-based control schemes were originally used for high-speed ATM networks, e.g., [11, 12]. Some rate-based control methods have been proposed recently for the AQM router to support best effort service traffic (e.g., audio and video stream) in the IP networks. For example, the TFRC (TCP Friendly Rate Control) scheme proposed an equation in the source to calculate the source sending rate for unicast traffic based on the packet loss event [10, 13]. The VCP (Variable-Structure Congestion Control) Protocol [14] employed an improved AIMD scheme in the sources by using the existing two ECN bits for network congestion feedback, but the fluctuations of the source sending rates still remain. A rate-based control scheme was proposed for transparently augmenting the end-to-end performance by controlling the source sending rate in [15]. A two-state adaptive rate control mechanism was proposed for streaming media [9]. A rate-based window control method was presented in co-operation with the RED scheme to control best-effort traffic in an environment with packet loss and varying round-trip delays [16]. However, most of these rate-based feedback control algorithms for IP networks are heuristics that lack a control theoretical analysis, so they cannot always guarantee the closed-loop stability in different traffic conditions. The XCP (Explicit Congestion Control) protocol [7] did apply the simple control theoretical analysis to calculate the advertised source sending rates in the routers based on the spare network bandwidth and the queue length, but it cannot cancel the steady state error due to the estimation error of the link capacity [17]. Furthermore, the choices of the XCP parameters cannot guarantee its desirable network performance in different network environment [17, 18]. In summary, all these discussed rate-based algorithms did not consider the optimal solution for the network performance of AQM control system.

The main contributions of this paper are: 1) Unlike all other papers that propose heuristic rate-based algorithms for the AQM control system (either applying frequency domain or time domain analysis from the classical control theory), we use instead the state space approach from the modern control theory to describe the network model for a typical AQM router in the IP-based Internet (shown in Fig. 2 later on). 2) We employ the LQR (Linear Quadratic Regulator) approach to obtain our optimal PI rate controller that can achieve good transient performance. We have derived for the first time Equations (24) and (25) to calculate the PI controller parameters, where PI controller is used to regulate the source sending rates. 3) Allowing our controller to self-tune the controller parameters upon the network traffic changes so that the closed-loop stability of rate-based AQM control is always guaranteed. To the best of our knowledge, no other papers on *rate-based control* algorithms applied the optimal control and adaptive control jointly in the controller design.

We have verified all the claims above by OPNET simulations to show that our control scheme serves both guaranteed service traffic and best-effort service traffic well, while achieving maximum bandwidth utilization of the Internet.

2 Network Model

We consider a data communication network consisting of a number of geographically distributed source/destination nodes. IP packets generated at a source node are delivered to their destination through a sequence of AQM routers in the Internet. The PI rate controller is located in the router and can calculate a desirable source window size (i.e. source sending rate $\lambda_i'(t)$) based on the instantaneous queue length of the buffer in the router, and advertise it to the source through the IP packet and the ACK packet so that the source can regulate its current sending rate $\lambda_i(t)$ to provide the best-effort service traffic. There is also an uncontrolled guaranteed service traffic flows (both local and from the upstream router) with input rate $v(t)$ into the router.

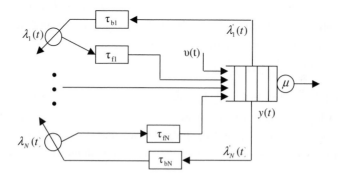

Fig. 1. System Model of the AQM Router

The AQM router under consideration accumulates multiple best-effort service traffic streams and one guaranteed service traffic stream, as shown in Fig. 1. There are N controlled source nodes that transmit the packets routed through the AQM router. The AQM router has a finite buffer space K to store the incoming packets and an output link to serve them at a constant data rate of μ. It is assumed that the service for the queue is FIFO (First-Input First-Output) and the best-effort service traffic and the guaranteed service traffic enter the same queue and share the common bandwidth of the output link. The guaranteed service traffic will not be influenced by the congestion situation of the router, i.e. their source sending rates are guaranteed. There are two kinds of time delays: (1) the varying forward time delay τ_{fi} from the controlled source node i to the router; which includes propagation delay, queuing delay and processing delay; (2) the varying feedback time delay τ_{bi} of both from the router to the controlled source node i via the destination, which also includes propagation delay, queuing delay and processing delay. We can let $\tau_i = \tau_{fi} + \tau_{bi}$ be the varying round trip time of the controlled source node i.

From the AQM router model in Fig. 1, the change of the queue length in the buffer of the router is the sum of the input rates of both the N best-effort service traffic flows and the guaranteed service traffic flows minus the service rate. It can be written as

$$\dot{y}(t) = \sum_{i=1}^{N} \lambda_i'(t - \tau_{fi}) + \upsilon(t) - \mu \; . \tag{1}$$

3   Self-tuning Optimal PI Rate Controller Design

We propose a self-tuning optimal PI rate controller that can achieve zero queue deviation in the router and prevent the Internet from congestion. Every controlled source node is allowed to send the packets into the network at its maximum allowed sending rate in order to utilize the spare bandwidth unused by the guarantee service traffic. Let y_0 be the target buffer occupancy of the buffer in the AQM router. Let the parameters K_P and K_I be the proportional gain and integral gain of PI rate controller at the router, which are two important parameters in our controller design. Based on the instantaneous queue length $y(t)$ of the buffer, the advertised sending rate $\lambda_i'(t)$ for the controlled source node i can be obtained by the following PI rate control structure:

$$\lambda_i'(t) = \lambda'(t) = K_P e(t) + K_I \int_0^t e(\tau) d\tau = K_P(y_0 - y(t)) + K_I \int_0^t (y_0 - y(\tau)) d\tau \; . \tag{2}$$

where $e(t)$ indicates the difference between $y(t)$ and y_0. It can be seen from the control structure in Equation (2) that the advertised source rate $\lambda_i'(t)$ is calculated based on the instantaneous queue length of the buffer. Since the advertised source sending rate $\lambda_i'(t)$ becomes the source rate $\lambda_i(t)$ after a feedback time delay of τ_{bi}, we can write

$$\lambda_i(t) = \lambda_i'(t - \tau_{bi}) \; . \tag{3}$$

By substituting Eq. (3) into Eq. (1), we can capture the dynamics of the router as

$$\dot{y}(t) = \sum_{i=1}^{N} \lambda_i'(t - \tau_{bi} - \tau_{fi}) + \upsilon(t) - \mu = \sum_{i=1}^{N} \lambda_i'(t - \tau_i) + \upsilon(t) - \mu \; . \tag{4}$$

To simplify the controller design, we can approximate the standardized round trip time by the average value τ of τ_i, i.e. $\tau = \sum_{i=1}^{N} \tau_i / N$. Such control plant approximation has been widely adopted in industrial process control system design, e.g., [19-23], which has found its application in network traffic control recently, e.g., [4, 7]. The validity of this approximation has been verified by our performance evaluation later on. Therefore, under this approximation, we can update Equation (4) as

$$\dot{y}(t) = N\lambda'(t - \tau) + \upsilon(t) - \mu \; . \tag{5}$$

Note that we only design our optimal controller based on the plant approximation. We implement our control algorithm in the real network environment by OPNET simulations without any plant approximation in this paper.

3.1   Controller Design with LQR Approach

To design our optimal PI rate controller via LQR approach [19], we need to translate the AQM control system described by time domain method in Fig. 1 into a state feedback AQM control system. We will summarize the key steps for the optimal controller design in the following due to space limitation.

To start, let x=[x_1 x_2]^T such that $x_1 = \int_0^t e(\tau)d\tau$ and x_2=e(t) represent/capture the queue deviation. Also let $u = \lambda(t)$ such that $u = Kx = [K_I \quad K_P][x_1 \quad x_2]^T$. $\qquad$ (6)

Equations (2) and (5) can now be represented by the state space equations as follows:

$$\dot{x} = \begin{bmatrix} 0 & 1 \\ 0 & 0 \end{bmatrix} x + \begin{bmatrix} 0 \\ -N \end{bmatrix} u(t-\tau) + \begin{bmatrix} 0 \\ -1 \end{bmatrix} \upsilon + \begin{bmatrix} 0 \\ 1 \end{bmatrix} \mu .$$ $\qquad$ (7)

Equation (7) can now be represented by Fig. 2. Our LQR problem is to find the optimal control u(t) such that J is minimized when

$$J = \int_0^\infty (x^T(t)Qx(t) + u^T(t)Ru(t))dt .$$ $\qquad$ (8)

where Q and R are given positive definite matrices with proper dimensions (u(t)=0 when t<0, Q≥0 and R>0) [19, 24].

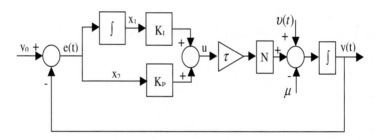

Fig. 2. State Feedback AQM Control System

Considering that μ and υ are uncontrolled variables, we can now decompose the control system in Equation (7) into two different time periods [19, 20]:

$$\dot{x}(t) = Ax(t) + Bu(t-\tau) = Ax(t) \text{ when } 0 \leq t < \tau,$$ $\qquad$ (9)

$$\dot{x}(t) = Ax(t) + Bu(t-\tau) = Ax(t) + B\hat{u}(t) \text{ when } t \geq \tau.$$ $\qquad$ (10)

The state matrices $A = \begin{bmatrix} 0 & 1 \\ 0 & 0 \end{bmatrix}$ and $B = \begin{bmatrix} 0 \\ -N \end{bmatrix}$ in Equation (7) are the important consideration in the optimal controller design in order to make use of LQR approach. According to the LQR optimal control law [19], we need

$$\hat{u}(t) = -R^{-1}B^T Px(t) \text{ with } t \geq \tau,$$ $\qquad$ (11)

where $P = \begin{bmatrix} p_{11} & p_{12} \\ p_{12} & p_{22} \end{bmatrix}$ is a positive definite matrix that can be solved by the Riccati equation such that

$$A^T P + PA - PBR^{-1}B^T P + Q = 0 . \tag{12}$$

Then we can obtain $u(t) = \hat{u}(t + \tau) = -R^{-1}B^T Px(t + \tau)$ with t≥0. \hfill (13)

However, $x(t + \tau)$ is not directly available at time t. Based on the theorem in [25], Equation (13) can be decomposed as

$$u(t) = \begin{cases} -R^{-1}B^T Pe^{(A-BR^{-1}B^T P)t} e^{A(\tau-t)} x(t) & 0 \le t < \tau \\ -R^{-1}B^T Pe^{(A-BR^{-1}B^T P)\tau} x(t) & t \ge \tau \end{cases} . \tag{14}$$

Let $Q = \begin{bmatrix} q_1 & 0 \\ 0 & q_2 \end{bmatrix}$. Then substituting P into Riccati Equation (12) yields positive definite analytical solution [19]:

$$\begin{cases} p_{12} = \sqrt{Rq_1} / N \\ p_{22} = \sqrt{R(2p_{12} + q_2)} / N \\ p_{11} = \sqrt{q_1(2p_{12} + q_2)} \end{cases} . \tag{15}$$

Next, let $K' = R^{-1}B^T P = R^{-1}\begin{bmatrix} 0 & -N \end{bmatrix}\begin{bmatrix} p_{11} & p_{12} \\ p_{12} & p_{22} \end{bmatrix} = -R^{-1}N\begin{bmatrix} p_{12} & p_{22} \end{bmatrix}$ \hfill (16)

and
$$\begin{aligned} A_c = A - BK' &= \begin{bmatrix} 0 & 1 \\ 0 & 0 \end{bmatrix} + \begin{bmatrix} 0 \\ -N \end{bmatrix} R^{-1}N\begin{bmatrix} p_{12} & p_{22} \end{bmatrix} \\ &= \begin{bmatrix} 0 & 1 \\ -R^{-1}N^2 p_{12} & -N\sqrt{R^{-1}(2p_{12} + q_2)} \end{bmatrix} \end{aligned} . \tag{17}$$

Hence, the optimal controller in Equation (14) can be simplified as [24]

$$u(t) = \begin{cases} -K' e^{A_c t} e^{A(\tau-t)} x(t) & 0 \le t < \tau \\ -K' e^{A_c \tau} x(t) & t \ge \tau \end{cases} . \tag{18}$$

Considering the slow-start mechanism of TCP window-based control protocol, the parameters of the controller are unimportant during the first average round trip time τ, so the optimal controller can be further simplified as

$$u(t) = -K' e^{A_c \tau} x(t) = Kx(t) \quad t \ge 0 . \tag{19}$$

Such simplification has little impact on the network performance, as shown in our OPNET simulations later on. To obtain the state feedback gains (i.e. PI controller parameters K_P and K_I) explicitly, we need to calculate $e^{A_c \tau}$ by Laplace transform approach [19]. Since

$$sI - A_c = \begin{bmatrix} s & -1 \\ R^{-1}N^2 p_{12} & s + N\sqrt{R^{-1}(2p_{12} + q_2)} \end{bmatrix} = \begin{bmatrix} s & -1 \\ \alpha_2 & s + \alpha_1 \end{bmatrix} , \tag{20}$$

the inverse Laplace transform of $(sI-A_c)^{-1}$ gives

$$e^{A_c t} = L^{-1}\left[(sI - A_c)^{-1}\right] = \begin{bmatrix} m_{11}(t) & m_{12}(t) \\ m_{21}(t) & m_{22}(t) \end{bmatrix} . \tag{21}$$

To simplify the expression of the elements m_{ij}, we let β_1 and β_2 be the roots of the characteristic equation $s^2 + \alpha_1 s + \alpha_2 = s^2 + 2\zeta\omega_n s + \omega_n^2 = 0$, (22)

i.e. $\beta_1 = (-\alpha_1 + \sqrt{\alpha_1^2 - 4\alpha_2})/2$ and $\beta_2 = (-\alpha_1 - \sqrt{\alpha_1^2 - 4\alpha_2})/2$ [24]. Hence, we have

$$\begin{cases} m_{11}(t) = \left[(\alpha_1 + \beta_1)e^{\beta_1 t} - (\alpha_1 + \beta_2)e^{\beta_2 t}\right]/(\beta_1 - \beta_2) \\ m_{12}(t) = (e^{\beta_1 t} - e^{\beta_2 t})/(\beta_1 - \beta_2) \\ m_{21}(t) = -\alpha_2(e^{\beta_1 t} - e^{\beta_2 t})/(\beta_1 - \beta_2) \\ m_{22}(t) = (\beta_1 e^{\beta_1 t} - \beta_2 e^{\beta_2 t})/(\beta_1 - \beta_2) \end{cases} . \tag{23}$$

Combining and solving Equations (6), (15), (16), (19), (21) and (23), we can obtain the parameters of optimal PI rate controller as

$$K_p = R^{-1}N\left[p_{12}m_{12}(\tau) + p_{22}m_{22}(\tau)\right] , \tag{24}$$

$$K_I = R^{-1}N\left[p_{12}m_{11}(\tau) + p_{22}m_{21}(\tau)\right] . \tag{25}$$

Now K_P and K_I are the functions of Q (i.e. q_1 and q_2) and R. The choice of Q and R matrix can affect the network performance a lot. Experience from control practice suggest that R=1 is a good choice [24]. Then we can select Q based on the damping ratio ζ and the natural frequency ω_n of the closed-loop AQM control system. Solving from Equations (15), (20) and (22) (proof omitted due to page limit), we can obtain

$$q_1 = \omega_n^4 R / N^2 , \tag{26}$$

$$q_2 = (4\zeta^2 - 2)\omega_n^2 R / N^2 . \tag{27}$$

Experience from the control practice [19, 20] and many of our OPNET simulations (see [26]) suggest that $0.6 \le \zeta \le 1$ and $0.6/\tau \le \omega_n \le 1/\tau$ would give a satisfactory transient network performance in AQM and source sending rates.

When the network parameters (N and τ) change, the PI rate controller parameters K_P and K_I do not change, which will force the ζ and ω_n to change. When the network parameters (i.e., the number N of active controlled flows and the average round-trip time τ) change so significantly that the system ζ and ω_n has fallen outside their specified interval, the PI rate controller needs to self-tune to give new values for ζ and ω_n. Therefore, it would be practical to introduce an adaptive control. Adaptive control has been widely applied in industrial process control e.g., [20, 27, 28, 29], and also found some application in Internet traffic control (e.g., [5, 30]).

Based on our practical experience from many simulation results (some of which shown in [26, 30]), we may want to choose a parameter $\alpha = N\tau$ to monitor N (the number of the active flows from the controlled sources) and τ (the average round trip time). When the network parameters N and τ change so greatly that causes the monitoring   parameter $|\alpha_{new}-\alpha_{old}|>0.5\min\{\alpha_{new},\alpha_{old}\}$ (that   is $\alpha_{new}>1.5\alpha_{old}$ or $\alpha_{new}<0.6\alpha_{old}$), our adaptive PI rate controller will self-tune based on the Equations (24) and (25) and then update $\alpha_{old}=\alpha_{new}$.

To be compatible for the current dominant TCP/IP protocol in the Internet, we convert the advertised source sending rate into the advertised window $win_i^{'}(t)$ as $win_i^{'}(t)=\tau_i \cdot \lambda_i^{'}(t)$. More details on rate-based control protocol can be referred to [30].

4   Simulations and Performance Evaluation

We have verified self-tuning our optimal PI rate controller using OPNET® Modeler [31] based on the network topology shown in Fig. 3. We consider two types of TCP Reno sources in the simulations – the long lived sources (i.e., controlled FTP sources for best-effort service traffic), which always have IP packets to send as long as their congestion windows allow; and the short lived sources (i.e. HTTP sources), which enters the network after an undetermined think time, sends a file of an undetermined length, and then waits for another think time period. The window size of the FTP source that transmits uncontrolled guaranteed service traffic will not be influenced by the network congestion information, which can also be treated as UDP sources. The destination's advertised window size is set sufficiently large so that TCP connections are not constrained at the destination.

To validate our PI rate control algorithm, we have made performance comparison with AIMD control with RED [3] and AIMD control with PI-RED [4] under the comparable network environment in [4]. During the simulation, all the IP packets have the same size of 1024 bytes. We have specified the $\zeta=0.8$ in the range of $0.6\leq\zeta\leq1$, $\omega_h=0.8/\tau$ in the range of $0.6/\tau\leq\omega_h\leq1/\tau$. The maximum window size allowed for all the TCP sources is specified as 2000 packets. The link bandwidth μ (i.e., service rate of AQM Router 1) is T3 (*44,736,000bps*, i.e., *5,461packets/sec*). The sampling time for PI rate controller and PI-RED controller [4] is chosen to be per packet interval (i.e., $1/\mu$) [3, 4]. The buffer size B of the AQM Router 1 is *1000* packets, while the buffer size of the Router 2 is big enough so that no packets will be lost [4]. The target buffer occupancy is 400 packets. The window size of the uncontrolled FTP source starts at 100 packets at time t=0s, increases to 500 packets at time t=80s and then decreases to 200 packets at time t=240s.

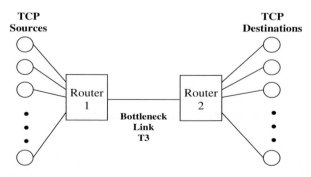

Fig. 3. Network Simulation Topology for Performance Evaluation

Table 1 provides the round trip propagation delays and the active periods (i.e. simulation time) for 100 controlled FTP sources, 1 uncontrolled FTP source and 100 http sources. To allow further comparison, we adopt comparable RED parameters from [4]. That is, the maximum value for drop probability p_{max} is 0.1; the maximum queue threshold max_{th} is 700; the minimum queue threshold min_{th} is 150; the queue weight w_q is 1.33×10^{-6}. We adopt Equations (6) and (7) in [4] to calculate the PI-RED controller based on the estimated parameters ($N=60$, $R_0=0.25s$ and $C=5461packets/sec$) of the current network. We can then obtain our optimal PI rate controller: $K_P=0.0339$ and $K_I=0.0244$.

Table 1. Round Trip Propagation Delay (RTPD) Configuration for Experiment

Source ID	RTPD	Time	Source ID	RTPD	Time
FTP 1~20	80ms	0<t≤320	HTTP 1~20	80ms	0<t≤320
FTP 21~40	160ms	0<t≤320	HTTP 21~40	160ms	0<t≤320
FTP 41~60	240ms	0<t≤320	HTTP 41~60	240ms	0<t≤320
FTP 61~80	280ms	160<t≤320	HTTP 61~80	280ms	160<t≤320
FTP 81~100	340ms	160<t≤320	HTTP 81~100	340ms	160<t≤320
Uncontrolled	160ms	0<t≤320			

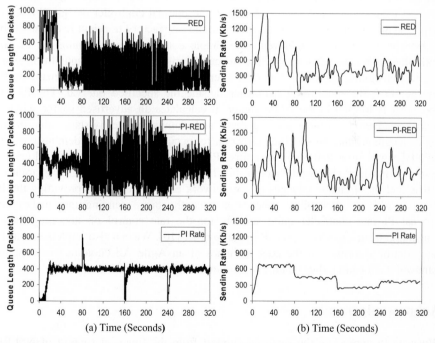

(a) Time (Seconds)   (b) Time (Seconds)

Fig. 4. (a) Instantaneous Queue Length Comparison (b) Instantaneous Sending rate Comparison for Controlled FTP Source among RED Controller, PI-RED Controller and Optimal PI Rate Controller

Fig. 4(a) shows the instantaneous queue length of the RED controller, the PI-RED controller and self-tuning optimal PI rate controller upon the change of the network traffic. It can be seen that the queue length in the RED router fluctuates with large amplitude. The PI-RED controller can clamp the queue length around the target buffer occupancy (400 packets), but its queue length oscillation is still noticeable. With our optimal PI rate controller applied, the queue length remains around the target buffer occupancy quite stably upon the change of the network traffic. Fig. 4(b) shows the instantaneous sending rates (i.e., flow throughput) of the controlled FTP source node 21 (whose RTPD=160ms) for the RED controller, the PI-RED controller and the optimal PI rate controller upon changes in the network traffic. Both sending rates of RED and PI-RED exhibit large fluctuations. Our optimal PI rate controller provides a relatively smooth source sending rate. Other controlled FTP source nodes have comparable sending rates. Due to the page limit, we only show one typical source node here. We will also make comparison with PI rate controller in [30] in our future work.

5 Conclusions

We have developed a self-tuning optimal PI rate controller for an AQM router supporting best-effort service traffic in the Internet. We employ LQR approach method from the modern control theory in the optimal controller design. We select proper Q and R based on ζ and ω_n such that the satisfactory transient response of the AQM control system is achieved. The simple equations for obtaining the PI rate controller have been derived. We also select a heuristic parameter to monitor network traffic. Self-tuning optimal PI rate controller would self-tune once the dramatic traffic changes drift the monitoring parameter outside the specified interval. The PI rate controller is located in the AQM router. It calculates an advertised source window size (i.e. advertised source sending rate), and clamps the steady-state value of the queue length around the specified target buffer occupancy. We have demonstrated by OPNET simulations that with our optimal PI rate controller applied, the AQM control system can adapt very well to both the fluctuation of the uncontrolled guaranteed traffic and the dramatic changes of the network environment, thus providing the IP network with good stability robustness. Smooth source sending rate makes our optimal PI rate controller quite suitable for the rate control of streaming media transmission such as video over IP in the Internet [10]. We can also implement our PI rate control algorithm in the edge switches of an Agile All-Photonic Network for bandwidth allocation, thus making maximum use of the network resources.

Acknowledgement

We want to appreciate the financial support from the financial support of NSERC Research Discovery Grant (#RGPIN42878) and the partial support from the industrial partners through the AAPN Research Network, NSERC.

References

[1] Floyd, S., Fall, K.: Promoting the Use of End-to-End Congestion Control in the Internet. IEEE/ACM Transactions on Networking 7(4), 458–472 (1999)

[2] Rahbar, A.G.P., Yang, O.: Integrated TDM Single-Hop Slotted All-Optical OPS Networks. In: Proceedings of IEEE 21st International Conference on Advanced Information Networking and Applications, Niagara Falls, Canada (2007)

[3] Floyd, S., Jacobson, V.: Random Early Detection Gateways for Congestion Avoidance. IEEE/ACM Transactions on Networking 1(4), 397–413 (1993)

[4] Hollot, C.V., Misra, V., Towsley, D., Gong, W.-B.: On Designing Improved Controllers for AQM Routers Supporting TCP Flows. In: Proceedings of IEEE INFOCOM, pp. 1726–1734 (2001)

[5] Hong, Y., Yang, O.W.W., Huang, C.: Self-Tuning PI TCP Flow Controller for AQM Routers With Interval Gain and Phase Margin Assignment. In: Proceedings of IEEE Globecom, Dallas, TX, U.S.A , pp. 1324–1328 (2004)

[6] Stevens, W.R.: TCP/IP Illustrated, vol. 1. Addison-Wesley, Boston (1994)

[7] Katabi, D., Handley, M., Rohrs, C.: Congestion Control for High Bandwidth Delay Product Networks. In: Proceedings of ACM SIGCOMM (2002)

[8] Low, S., Paganini, F., Wang, J., Adlakha, S., Doyle, J.C.: Dynamics of TCP/RED and a Scalable Control. In: Proceedings of IEEE INFOCOM, pp. 239–248 (2002)

[9] Huang, C., Xu, L.H.: SRC: Stable Rate Control for Streaming Media. In: Proceedings of IEEE Globecom, San Francisco, U.S.A (2003)

[10] Hanley, M., Floyd, S., Padhye, J., Widmer, J.: TCP Friendly Rate Control (TFRC): Protocol Specification, RFC 3448, Proposed Standard (2003)

[11] Zhang, H.Y., Yang, O.W.W., Mouftah, H.: Design of robust congestion controllers for ATM networks. In: Proceedings of IEEE INFOCOM, pp. 302–309 (1997)

[12] Blanchini, F., Lo Cigno, R., Tempo, R.: Robust Rate Control for Integrated Services Packet Networks. IEEE/ACM Transactions on Networking 10(5), 644–652 (2002)

[13] Floyd, S., Hanley, M., Padhye, J., Widmer, J.: Equation-Based Congestion Control for Unicast Applications. In: Proceedings of ACM SIGCOMM, pp. 43–56 (2000)

[14] Xia, Y., Subramanian, L., Stoica, I., Kalyanaraman, S.: One more bit is enough. In: Proceedings of ACM SIGCOMM (2005)

[15] Karandikar, S., Kalyanaraman, S., Bagal, P., Packer, B.: TCP Rate Control. ACM Computer Communications Review 30(1), 45–58 (2000)

[16] Zhang, H.Y., Cong, J., Yang, O.W.W.: Rate Control Over RED With Data Loss and Varying Delays. In: Proceedings of IEEE Globecom, pp. 3035–3040 (2003)

[17] Zhang, Y., Ahmed, M.: A control theoretic analysis of XCP. In: Proceedings of IEEE INFOCOM, pp. 2831–2835 (2005)

[18] Low, S.H., Andrew, L.L.H., Wydrowski, B.P.: Understanding XCP: equilibrium and fairness. In: Proceedings of IEEE INFOCOM (2005)

[19] Ogata, K.: Modern Control Engineering, 4th edn. Prentice Hall, New Jersey (2002)

[20] Astrom, K.J., Wittenmark, B.: Adaptive Control, 2nd edn. Addison-Wesley Publication Co, Boston (1995)

[21] Astrom, K.J., Hagglund, T.: PID Controllers: Theory, Design, and Tuning, Instrument Society of America, 2nd edn., North Carolina (1995)

[22] Astrom, K.J., Hang, C.C., Persson, P., Ho, W.K.: Towards intelligent PID control. Automatica 28(1), 1–9 (1992)

[23] Ho, W.K., Hong, Y., Hansson, A., Hjalmarsson, H., Deng, J.W.: Relay Auto-Tuning of PID Controllers using Iterative Feedback Tuning. Automatica 39(1), 149–157 (2003)

[24] He, J.B., Wang, Q.G., Lee, T.H.: PI/PID Controller Tuning Via LQR Approach. Chemical Engineering Science 55(13), 2429–2439 (2000)

[25] Marshall, J.E.: Control of time-delay systems, Peter Peregrinus, London (1979)

[26] Hong, Y., Yang, O.W.W.: Self-Tuning Optimal PI Rate Controller for End-to-End Congestion With LQR Approach, CCNR Technical Report 06-0601, University of Ottawa, Canada, Available upon request (2006)

[27] Astrom, K.J., Goodwin, G.C., Kumar, P.R. (eds.): Adaptive Control, Filtering, and Signal Processing, vol. 74, The IMA Volumes in Mathematics and its Applications. Springer-Verlag, Heidelberg (1995)

[28] Astrom, K.J., Hagglund, T., Hang, C.C., Ho, W.K.: Automatic Tuning and Adaptation for PID Controllers-A Survey, IFAC Journal of Control Engineering Practice, vol. 1(4) (1993)

[29] Ho, W.K., Lee, T.H., Han, H.P., Hong, Y.: Self-Tuning IMC-PID Control with Interval Gain and Phase Margin Assignment. IEEE Transactions on Control Systems Technology 9(3), 535–541 (2001)

[30] Hong, Y., Yang, O.W.W.: Design of Adaptive PI Rate Controller for Best-Effort Traffic in the Internet Based on Phase Margin. IEEE Transactions on Parallel and Distributed Systems 18(4), 550–561 (2007)

[31] OPNET Technologies Inc, OPNET Modeler Manuals, OPNET Version 7.0 (2000)

Quality-of-Service Provisioning for Multi-service TDMA Mesh Networks

Petar Djukic and Shahrokh Valaee*

The Edward S. Rogers Sr. Department of Electrical and Computer Engineering
University of Toronto
10 King's College Road, Toronto, ON, M5S 3G4, Canada
{djukic,valaee}@comm.utoronto.ca

Abstract. Multi-service mesh networks allow existence of guaranteed delay Quality-of-Service (QoS) traffic streams such as Voice over IP and best effort QoS traffic streams such as file transfer. We present an optimization that performs a linear search for the minimum number of TDMA slots required to support the guaranteed QoS flows. At each stage of the search a linear integer program is solved to find if there is a feasible schedule supporting the required end-to-end bandwidth and delay. Our optimization results in a relative order of transmissions in the frame that guarantees a maximum end-to-end delay in the network. The ordering of the transmissions can be used later to find feasible schedules with the Bellman-Ford algorithm on the conflict graph for the network. We use the optimization in numerical simulations showing the efficiency of 802.16 mesh networks with VoIP traffic.

1   Introduction

Wireless mesh networks interconnect access points (APs) spread out over a large geographical area. Wireless terminals (WTs) connect to the access points on their first hop and their traffic is carried by the wireless mesh to the Point-of-Presence (POP) where it can go to the Internet. The POP is the only node in the network connected to the Internet and can also act as a base station (mesh coordinator). Current mesh networks use 802.11 technology to interconnect the mesh backbone [1, 2]. However, 802.11 technology is a decade old and was not designed for mesh networks. In particular, 802.11 lacks the extensions to provide Quality–of–Service (QoS) in multihop wireless environments [3].

New mesh network technologies such as 802.16 (WiMax) and 802.11s are designed to provide QoS with Time Division Multiple Access (TDMA) [4,5]. In TDMA, end-to-end QoS is negotiated in terms of end-to-end bandwidth reserved for each AP on the links connecting it to the POP. QoS is enforced at each link with scheduled access to the wireless channel. A schedule assigns slots from each TDMA frame to links so that a number of non-conflicting links can transmit simultaneously in each slot. Link bandwidth is given by the number of slots assigned to it in each frame and the modulation used in the slots.

---

* This work was sponsored in part by the LG Electronics Corporation.

L. Mason, T. Drwiega, and J. Yan (Eds.): ITC 2007, LNCS 4516, pp. 841–852, 2007.
© Springer-Verlag Berlin Heidelberg 2007

If all slots in a frame are reserved, TDMA mesh networks would not allow statistical multiplexing of best effort data streams at the MAC layer. So, both 802.16 and 802.11s divide the slots in every frame between guaranteed service traffic streams and best effort traffic streams. In 802.16, slots reserved for guaranteed QoS traffic are assigned with the centralized scheduling protocol, while other slots are assigned with the decentralized scheduling protocol. In the centralized scheduling protocol, the mesh coordinator assigns bandwidth to all links in the network based on traffic demands from the APs. On the other hand, in the decentralized scheduling protocol, mesh routers are free to negotiate pairwise TDMA assignments, with no QoS guarantees on the bandwidth. In 802.11s networks, slots dedicated to TDMA access are negotiated in a pairwise fashion, while the rest of the frame is dedicated to best effort service with 802.11 EDCF.

In this paper we answer the following QoS provisioning question: *What is the minimum number of TDMA slots required to support a required guaranteed QoS in the network?* The QoS is specified both in terms of bandwidth and delay, for traffic streams such as Voice over Internet Protocol (VoIP). The end-to-end bandwidth of each AP is guaranteed with a TDMA schedule that assigns the appropriate bandwidth to links connecting the AP to the POP. The delay in the network consists of queueing delay due to traffic variations and TDMA propagation delay. TDMA propagation delay occurs when an outgoing link on a mesh node is scheduled to transmit before an incoming link in the path of a packet [6]. In this paper, we assume that the queueing delay is minimized in the network layer with the assignment of link bandwidths and concentrate on scheduling algorithms that guarantee a bound on the TDMA delay. We have shown in [6] that end-to-end TDMA propagation delay accumulates at each hop and can be very large – multiples of TDMA frame duration. In 802.16 frame duration can be as large as 20ms, so TDMA delay can be relatively large compared to the 150ms delay budget required for VoIP quality [7].

We formulate an optimization that minimizes the number of TDMA slots allocated for guaranteed QoS traffic, subject to the constraint that the delay introduced with TDMA scheduling is bellow a given threshold. The bound on TDMA propagation delay can be found by delay budgeting the network across the mesh and the wired backbones. The maximum allowed TDMA delay is found by subtracting the delay due to voice processing and the jitter buffer delay from the overall delay budget [7]. The optimization performs a linear search for the minimum number of TDMA slots. At each step of the search, the optimization increases the number of slots required for guaranteed QoS and solves an integer program that finds a transmission order that supports the required banwidth at each hop, subject to the TDMA delay. The optimization stops as soon as the number of guaranteed slots with a feasible transmission schedule is found. We have shown that with a known transmission order TDMA schedules can be found with the Bellman-Ford algorithm run on the conflict graph for the wireless network [6,8]. Since end-to-end TDMA delay depends on the transmission order alone, it can be distributed to the nodes as a part of their QoS provisioning information, thus making sure that the resulting TDMA schedules have a fixed maximum TDMA delay. Using this method, schedules can be changed dynamically when the bandwidth changes, but still maintain the maximum end-to-end delay.

Our optimization is suitable for mesh network planning. During the planning process, locations of mesh nodes are chosen based on the expected interference from neighbouring nodes [4]. With the location known, the expected interference is also known, making it possible to predict maximum modulation at each link, as well as the mesh topology. Given the network topology and the maximum bit-rate on each link, it is possible to plan end-to-end bandwidth to support a specified number of VoIP connections. Our optimization adds an additional level of QoS planning for mesh networks – the scheduling delay through the mesh.

We study the planing for 802.16 mesh networks with numerical simulations. We examine the effect of 802.16 frame size on efficiency of carrying VoIP traffic. Efficiency is defined as the number of slots required by guaranteed QoS traffic divided by the total number of TDMA slots in the frame. We show that increasing the frame size increases the efficiency in the network almost 50% for a low number of VoIP calls, however for a high number of VoIP calls the increase in the efficiency is less than 5%. The efficiency increases with the frame size because as the frame size increases, more transmission orders can produce TDMA schedules with the required bandwidth requirement.

2   Network and Transmission Model

The mesh network is using a time division multiple access (TDMA) MAC protocol [4]. In TDMA MAC protocols, the time is divided into slots of fixed duration, which are then grouped into frames. A fixed portion of each frame is dedicated to control traffic, while the rest of the slots are used for data traffic. Each frame consists of N_f slots, where N_c of the slots are allocated for control traffic, N_g are reserved for guaranteed service data traffic and N_b are reserved for best effort data traffic (Fig. 1). Frames have duration of $T_f = N_f T_S$ seconds, where T_S is the slot duration. In this paper, we minimize the number of slots reserved for guaranteed service traffic, N_g.

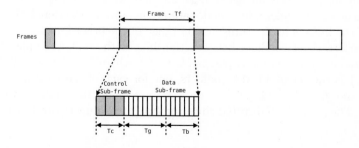

Fig. 1. Multi-service TDMA frames

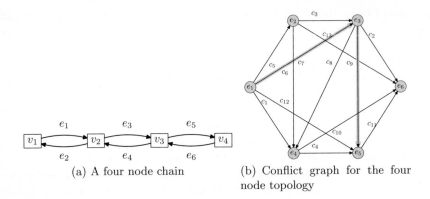

(a) A four node chain

(b) Conflict graph for the four node topology

Fig. 2. Chain topology and its conflict graph

We model the mesh with a topology graph connecting wireless routers in the range of each other. We assume that if two routers are in the range of each other, they establish links in the MAC layer, so the TDMA network can be represented with a connectivity graph $G(V, E, f_t)$, where $V = \{v_1, \ldots, v_n\}$ is the set of nodes[1], $E = \{e_1, \ldots, e_m\}$ are directional links between neighbouring nodes, and $f_t : E \rightarrow V \times V$ assigns links to pairs of nodes. Links are directional, so for a link $e_k \in E$, $f_t(e_k) = (v_i, v_j)$ means that traffic on the link is transmitted from v_i to v_j. The links operate at different bitrates, which depend on the signal-to-noise ratio. Signal-to-noise ratio is divided into several discrete levels and each is associated with its corresponding bitrate. We define the link bitrate as the number of bits transmitted in a TDMA slot, represented with the mapping $b : E \rightarrow \{M_1, M_2, \ldots, M_{\max}\}$, where M_1 is the number of data bits carried in a slot with the minimum modulation and $M_{\max}$ is the number of bits carried in a slot with the maximum modulation and coding.

We assume that the signal-to-noise ratio of each link depends on the wireless channel alone and not other links in the network, meaning that competing links do not transmit at the same time. Under this model of transmission, in TDMA networks, a receiver can only have one active link at any given time. In a single hop neighbourhood, this means all links interfere with each other. In a two hop neighbourhood, two links, whose nodes are two hops away, interfere if the receiver of one of the links is in the transmission range of the other link.

We keep track of conflicts between the links with conflict graphs. Conflict graphs can be defined with a triplet $G_c(E, C, f_c)$, where E is the set of links, $C = \{c_1, \ldots c_r\}$ is the set of TDMA conflicts, one for each of the r conflicting pairs of links, and $f_c : C \rightarrow \{\{e_i, e_j\}, e_i, e_j \in E\}$ associates the conflicts with pairs of links.[2] The graph is undirected since conflicts are symmetrical. In this paper,

---

[1] We use the convention that v_n is the POP.

[2] We use the notation $\{\cdot\}$ for unordered sets and $(\cdot)$ for ordered sets, so f_c defines an undirected graph.

we use a conflict graph with an arbitrary assignment of directions to the arcs, $\overrightarrow{G}_c(E, C, \overrightarrow{f}_c)$, where $\overrightarrow{f}_c : C \rightarrow E \times E$. The directed conflict graph simplifies the derivation of formulas, however, as we have shown in [6], the arbitrary orientation of arcs does not cause any loss of generality. We use the four node example from Fig. 2a to demonstrate how the arcs in the conflict graph are created. The vertices in the conflict graph are the six links from the topology graph. All of the links conflict with each other, except for pairs e_1 and e_6 and e_2 and e_5, so they are not connected in the conflict graph (Fig. 2b). The orientation of the conflicts was chosen randomly, since it does not change the resulting delay or wireless capacity [6].

Link bandwidths are assigned so that a certain number of VoIP connections can be carried between each AP and the POP. The assignment of bandwidths is performed during mesh network planning. The assignment of link bandwidths is provided as the mapping $R : E \rightarrow \mathbb{R}_{[0,\infty)}$. The scheduling algorithm assigns link bandwith through the number of slots a link can use in a frame $d : E \rightarrow \mathbb{Z}_{[0,T]}$. The number of slots required to achieve bandwidth R_i on link e_i can be found with:

$$d_i = \left\lceil \frac{R_i T_f}{b_i} \right\rceil = \left\lceil \frac{R_i N_f T_S}{b_i} \right\rceil, \tag{1}$$

where $\lceil \cdot \rceil$ denotes the ceiling of a real number, T_f is the duration of the frame, N_f is the number of slots in the frame, T_s is the duration of mini-slot in seconds and b_i is the number of bits in each slot.

We assume that after the link bandwidths have been assigned, there are $2q$ one-way paths terminating or originating at the POP. The paths connect the POP with $q < n - 1$ APs acting as VoIP cells for WTs. We denote a path from the POP to node v_l with $\mathcal{P}_l$ and the path from the node v_l to the POP with $\mathcal{P}_{q+l}$. The set of all paths is denoted with $P = \{\mathcal{P}_1, \ldots, \mathcal{P}_{2q}\}$. We use a mapping function $\overrightarrow{f}_P : P \rightarrow E \times E$ to associate a path with its starting and ending links, so $\overrightarrow{f}_P(\mathcal{P}_l) = (e_i, e_j)$ means the link for the first hop is e_i and the link for the last hop is e_j.

3   TDMA Scheduling

We present a set of conditions that guarantees that the transmission schedule for guaranteed QoS slots is both *valid* and *conflict-free*. A valid transmission schedule assigns the number of slots allocated to the links due to QoS require-ments. A conflict-free schedule ensures that transmissions of conflicting links do not overlap. These conditions are used in the minimization of N_g as constraints, to ensure that a given N_g results in a feasible schedule for the guaranteed QoS slots. A transmission schedule assigns slots from each TDMA frame to links so that a number of non-conflicting links can transmit simultaneously during each slot.

We define the TDMA schedule, used for guaranteed QoS service slots, with a pair of vectors $\boldsymbol{\pi}$ and $\boldsymbol{d}$, where $\boldsymbol{\pi} = [\pi_1, \ldots, \pi_m]^T$ is the vector of link starting

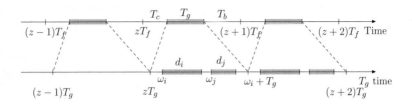

Fig. 3. Conflict-free Conditions

times in the part of the frame allocated for guaranteed QoS and $\boldsymbol{d} = [d_1, \ldots, d_m]^T$ is the vector of the number of slots each link transmits in the frame. The activation times need to be limited to $\pi_i \in [0, N_g)$, $\forall e_i \in E$, so that each link transmits in every frame. If we assume that the slots in the frame are numbered $[0, N_f - 1]$, the transmission takes place in slot $N_c + \pi_i$ with the duration of d_i slots. We note that the schedule is valid by definition, since every link will be scheduled to transmit for the number of slots required with its bandwidth assignment. We have defined conditions for conflict-free scheduling in [6]; we briefly summarize those results here.

The guaranteed TDMA schedule repeats in every frame, until a new set of link bandwidths is assigned. The slots allocated for guaranteed QoS are always sandwiched between the control slots and the best effort TDMA slots. If we ignore the non-guaranteed QoS slots, we can view the uninterrupted sequence of guaranteed QoS slots on its own axis (Fig. 3). On this axis, the activation times, $\boldsymbol{\pi}$, are periodic. Periodicity of the schedule means that the start time π_i for link e_i actually represents a series of activation times, which can be derived from π_i by adding multiples of N_g slots (Fig. 3). We denote with $\Pi_i = \{\pi_i + z_i N_g, z_i \in \mathbb{Z}\}$ the series of activation times for link e_i, generated with π_i. The actual activation time in the frame π_i can be found from any activation time $\omega_i \in \Pi_i$ with the modulo operator: $\pi_i = \omega_i \pmod{N_g}$.

The number of times a link transmits in the frame depends on its starting time and the duration of its transmission. If for some link e_i, $\pi_i + d_i \leq N_g$, the link will transmit once per frame. On the other hand, if $\pi_i + d_i > N_g$, the link will be scheduled twice for transmission in the frame. The first transmission starts in slot N_c, with the duration $\pi_i + d_i - N_g$ slots and the second transmission starts in slot $N_c + \pi_i$ with the duration of $N_g - \pi_i$ slots. So, our scheme limits the number of transmissions by any link to at most two in a frame. This is good for protocols such as 802.16 where the overhead of each transmission can be as much as 324 bytes at the highest modulation [4]. In [6], we also show how this method can be extended to find schedules for multiple activation times in the guaranteed QoS part of the frame.

The conflict-free conditions for a schedule can be expressed in terms of points in the sequences Π_i, $\forall e_i \in E$. We have shown in [6] that a schedule is conflict-free, if for any two conflicting links e_i and e_j whose conflict is $c_k \in C : \overrightarrow{f}_c(c_k) = (e_i, e_j)$:

$$d_i \leq \omega_j - \omega_i + p_k N_g \leq N_g - d_j, \tag{2}$$

where $\omega_i \in \Pi_i$ and $\omega_j \in \Pi_j$ and $p_k = 0$ if $\omega_j - \omega_i > 0$ and $p_k = 1$ if $\omega_j - \omega_i < 0$. Variable p_k specifies a relative order of transmissions, which prompts us to refer to it as the "transmission order" in the rest of the paper. A schedule is conflict free if (2) is true for all conflicts in the network. Fig. 3 shows why (2) is necessary for the schedule to be conflict-free. In the figure, $p_k = 0$, so we are comparing the timing of e_i's transmission to the first transmission of e_j that follows it. Clearly, it is necessary that $\omega_j \geq \omega_i + d_i$ since e_j cannot start its transmission before e_i finishes. Also, the next transmission of e_i should be after e_j has finished its transmission, so $\omega_i + n_g \geq \omega_j + d_j$. Full proof of necessity and sufficiency of (2) can be found in [6].

We show next that the TDMA delay depends on the transmission order and a feasible $\boldsymbol{\omega} = [\omega_1, \ldots, \omega_m]^T$. However, we also show that the feasible $\boldsymbol{\omega}$ can be compressed into a single parameter, leading us to a two step procedure to optimize TDMA delay. First, TDMA propagation delay is minimized subject to an existence of a feasible schedule. Second, the transmission order and the feasibility parameter are distributed among the mesh nodes, so they can find the transmission schedule using the Bellman-Ford algorithm.

4 TDMA Delay

We show how to calculate and minimize return trip TDMA propagation delay in the mesh in [6]. While that approach is appropriate for TCP flows, it is not appropriate for VoIP connections, since perceived voice quality depends on the *one-way* delay between a sender and its receiver [7]. In this section, we find the expression for one way TDMA propagation delay on a path. We first find TDMA propagation delay at single router on the path and then add up the delay at every router on the path to find an expression for the one-way end-to-end TDMA delay on the path.

TDMA propagation delay occurs if an ingress link is scheduled to transmit after an egress link on the router. So, on a single mesh router it is measured as the time between the transmission of an ingress link, to the time when the egress link transmits, excluding the queueing delay. We note that the TDMA propagation delay experienced by a packet traversing a mesh router from an ingress link e_i to an egress link e_j, in slots, is given by:

$$\Delta_k = \begin{cases} \omega_j - \omega_i + p_k N_f & \text{if } \overrightarrow{f}_c(c_k) = (e_i, e_j) \\ \omega_j - \omega_i + (1 - p_k) N_f & \text{if } \overrightarrow{f}_c(c_k) = (e_j, e_i), \end{cases} \tag{3}$$

where $c_k \in C$ is the conflict connecting the two links in the conflict graph and ω_i, ω_j and p_k correspond to a fixed feasible schedule $S(\boldsymbol{\pi}, \boldsymbol{d})$. The delay in seconds can be found by multiplying Δ_k with the slot duration T_S. For example, if $\overrightarrow{f}_c(c_k) = (e_i, e_j)$ and $p_k = 0$ it is easy to see that $\Delta_k = \omega_j - \omega_i$ since the packet can be transmitted in the same frame on both links. However, if $p_k = 1$,

$\Delta_k = \omega_j - \omega_i + N_f$ since the packet has to wait for new frame to be transmitted by e_j.

The total TDMA delay on a path is found by adding up the delay at each router on the path in the topology graph. We now show that each path in the topology graph corresponds to a path in the conflict graph, which lead us to a simpler formulation of the TDMA delay. The path in the conflict graph, corresponding to a path in the topology graph, can be obtained by traversing the conflicts in $\overrightarrow{G}_c(V, E, \overrightarrow{f}_c)$ corresponding to conflicts between ingress and egress links at each router in the path. For example, path $e_1 \rightsquigarrow e_3 \rightsquigarrow e_5$ in the four node topology shown in Fig. 2a, corresponds to the path $c_6 \rightsquigarrow c_9$ in Fig. 2b. We represent the paths in the conflict graph with r-sized vectors in the $\{-1, 0, 1\}^r$ path space of the conflict graph [9]. The meaning of the entries of $\boldsymbol{\theta}_l = [\theta_1, \ldots, \theta_r]^T$, corresponding to path $\mathcal{P}_l$ in the conflict graph, is:

$$\forall c_k \in C, \quad \theta_k = \begin{cases} 1, & \text{if } c_k \in \boldsymbol{\theta}_l^+ \\ -1, & \text{if } c_k \in \boldsymbol{\theta}_l^+ \\ 0, & \text{otherwise,} \end{cases} \tag{4}$$

where $\boldsymbol{\theta}_l^+$ is the set of arcs in the positive direction of $\boldsymbol{\theta}_l$ and $\boldsymbol{\theta}_l^+$ is the set of arcs in the negative direction of $\boldsymbol{\theta}_l^+$. For example, the path emphasized in Fig. 2 corresponds to the vector $\boldsymbol{\theta} = [0, 0, 0, 0, 0, 1, 0, 0, 1, 0, 0, 0, 0]^T$.

The total delay on path $\mathcal{P}_l$ is found by adding up the single hop delay incurred for the conflicts between ingress and egress links at each router in the path. The delay on path $\mathcal{P}_l$ is given by:

$$D(\mathcal{P}_l) = \sum_{k=1}^{r} \theta_k \left(\tau_k + p_k N_f \right) I(\theta_k > 0) + \sum_{k=1}^{r} \theta_k \left(\tau_k + p_k N_f - N_f \right) I(\theta_k < 0) \tag{5}$$

where $\tau_k = \omega_j - \omega_i$ is the tension for the conflict c_k, $\overrightarrow{f}_c(c_k) = (e_i, e_j)$, and $I(\cdot)$ is the indicator function, that is 0 when its argument is false and 1 when its argument is true. A well known property of tensions is that the sum of tensions along a path is equal to the tension between end vertices [9]. This property allows us to express the delay on the path with:

$$D(\mathcal{P}_l) = \omega_j - \omega_i + \boldsymbol{\theta}_l^T \boldsymbol{p} N_f + \sum_{k=1}^{r} N_f I(\theta_k < 0), \tag{6}$$

where e_i and e_j are the first and the last link on the path, $\overrightarrow{f}_{\mathcal{P}}(\mathcal{P}_l) = (e_i, e_j)$, and we have used vector product to express the summation of $\theta_k p_k$ on the path. Since the last term in the delay is a constant depending only on the orientation in the conflict graph, we will refer to it with $D_l = \sum_{k=1}^{r} N_f I(\theta_k < 0)$ for path $\mathcal{P}_l$ in the rest of the paper.[3]

---

[3] D_l depends on the orientation of the conflict graph. However, we show in [6] that since $\boldsymbol{p}$ also depends on the orientation of the conflict graph, the total TDMA propagation delay does not change if the orientation changes.

5   QoS Provisioning for Minimum Delay

In this section, we present an algorithm that can be used to find the minimum number of guaranteed QoS slots, required to support a given bandwidth subject to maximum TDMA delay. The maximum TDMA delay in slots is found with delay budgeting and is denoted with $N_{\max} = D_{\max}/T_S$, where $D_{\max}$ is the maximum allowed delay and T_S is the slot duration. We present the algorithm first and then show how to compress a feasible schedule associated with the minimum N_g into two provisioning parameters that should be distributed to all mesh routers.

The minimum N_g can be found with a non-linear $\{0,1\}$-integer program. However, we simplify this optimization by finding the minimum N_g with a search algorithm. The algorithms starts with $N_g = 1$, and increments N_g in every iteration. At each step of the search, the algorithm solves a $\{0,1\}$-integer program, which is a linear program for a fixed N_g. The search for the number of required slots stops when a schedule with the required QoS properties is found.

At each step, the algorithm solves the following $\{0,1\}$-integer linear program:

$$\text{Find} \quad \boldsymbol{\omega}, \boldsymbol{p} \tag{7a}$$

$$\text{s.t.} \quad \omega_j - \omega_i + \boldsymbol{\theta}_l^T \boldsymbol{p} N_f \le N_{\max} - D_l, \quad \forall \mathcal{P}_l \in P, \ \overrightarrow{f}_p(\mathcal{P}_l) = (e_i, e_j) \tag{7b}$$

$$d_i \le \omega_j - \omega_i + p_k N_g \le N_g - d_j, \quad \forall c_k \in C : \overrightarrow{f}_c(c_k) = (e_i, e_j) \tag{7c}$$

$$\boldsymbol{\omega} \in \mathbb{Z}^m, \boldsymbol{p} \in \{0,1\}^r. \tag{7d}$$

The linear program finds a feasible $\boldsymbol{\omega}$ and a feasible $\boldsymbol{p}$, ensuring a feasible schedule exists for a given N_g in the iteration. The first $2q$ constraints, (7b), ensure that the total delay on all paths is less then $D_{\max}$. The next r constraints, (7c), ensure that there is a feasible schedule satisfying the delay constraints. The algorithm either runs until a feasible set of $\boldsymbol{\omega}$ and $\boldsymbol{p}$ is found or until N_g reaches $N_f - N_c$. Since we perform a linear search of all possible values of N_g, we are guaranteed to find the minimum N_g for which there is a feasible schedule with a TDMA delay less than $D_{\max}$ on every path.

In order to allow the mesh routers to schedule links without the knowledge of a specific feasible $\boldsymbol{\omega}$, we introduce a new variable into the optimization. The new variable represents the maximum allowed difference between the activation time of the last link on a path and the first link on the path. We substitute t instead of the first two terms in (6), so delay on the path becomes:

$$D(\mathcal{P}_l) = t + \boldsymbol{\theta}_l^T \boldsymbol{p} N_f + D_l, \quad \forall \mathcal{P}_l \in P. \tag{8}$$

The required constraint on TDMA delay, (7b), is still true if:

$$\omega_j - \omega_i \le t, \quad \forall \mathcal{P}_l \in P, \ \overrightarrow{f}_p(\mathcal{P}_l) = (e_i, e_j). \tag{9}$$

This leads us to the following $\{0,1\}$-integer program, to be run for each N_g, instead of (7):

$$\text{Find} \quad \boldsymbol{\omega}, \boldsymbol{p}, t \tag{10a}$$

$$\text{s.t.} \quad t + \boldsymbol{\theta}_l^T \boldsymbol{p} N_f \leq N_{\max} - D_l, \qquad\qquad\qquad \forall \mathcal{P}_l \in P \tag{10b}$$

$$d_i \leq \omega_j - \omega_i + p_k N_g \leq N_g - d_j, \quad \forall c_k \in C : \overrightarrow{f}_c(c_k) = (e_i, e_j) \tag{10c}$$

$$\omega_j - \omega_i \leq t, \qquad\qquad \forall \mathcal{P}_l \in P, \overrightarrow{f}_p(\mathcal{P}_l) = (e_i, e_j) \tag{10d}$$

$$\boldsymbol{\omega} \in \mathbb{Z}^m, \boldsymbol{p} \in \{0,1\}^r, t \in \mathbb{R}, \tag{10e}$$

where the combination of (10b) and (10d) replaces (7b).

Using the symmetry between the paths we can see that (10d) is equivalent to half as many double sided constraints:

$$-t \leq \omega_j - \omega_i \leq t, \quad \forall \mathcal{P}_l, l = 1, \ldots q, \overrightarrow{f}_p(\mathcal{P}_l) = (e_i, e_j). \tag{11}$$

So when $\boldsymbol{p}$ and t are fixed a feasible schedule can be found using the Bellman-Ford algorithm on a modified conflict graph. We create a new scheduling graph, $G_S(E, C_S, \overrightarrow{f}_S)$, from the conflict graph by adding arcs between the start link and the end link of the first for every path originating at the POP. This adds q additional arcs to the conflict graph to create $C_S = C \cap \{c_{r+1}, \ldots, c_{r+q}\}$ arcs for the scheduling graph. The function connecting the arcs of the scheduling graph to the links $\overrightarrow{f}_S$ by combining $\overrightarrow{f}_c$ and $\overrightarrow{f}_p$:

$$\forall c_l \in C_s, \quad \overrightarrow{f}_s(c_l) = \begin{cases} \overrightarrow{f}_c(c_l), & \text{if } l \leq r \\ \overrightarrow{f}_p(\mathcal{P}_l), & \text{if } r < l \leq q + r. \end{cases} \tag{12}$$

Since the scheduling also has a set of inequalities associated with every arc, the schedules can be found from the scheduling graph the same way they are found from the conflict graph [6, 8].

6   Numerical Results

In this section, we present numerical results for the application of VoIP traffic in 802.16 mesh networks. In 802.16 mesh networks, N_g is specified as the network parameter MSH-CSCH-DATA-FRACTION [4, p. 86]. This parameter specifies the percentage of each frame that should be used for centralized TDMA scheduling. Here, we find the percentage of the frame that should be scheduled with the 802.16 centralized scheduling protocol, so that VoIP QoS is met. The results from this section can also be used to decide the frame sizes for 802.16 mesh networks.

We assume that WTs are using the G.729 codec to encode voice. With the G.729 codec, the bandwidth of each VoIP call is 8.0kbps [7], so we assume that the end-to-end bandwidth required by each VoIP call is 8.0kbps. We use the delay budgeting presented in [7] to derive the bound on TDMA propagation delay required in the network. The delay budgeting assumes that the voice quality requires an end-to-end delay of 150ms. The delay components, not associated to voice processing, consist of the jitter buffer delay of 60ms and the Public Switched Telephone Network (PSTN) of 30ms. We assume that the PSTN delay is fixed and examine how much jitter delay can be allowed in the Internet. We use the values of $D_{max} = 40$ms and $D_{max} = 60$ms, corresponding to the jitter buffer delay of 20ms and 0ms, respectively.

We have generated 100 random mesh network topologies, and performed mesh network planning for each of them. Each topology was generated by placing the POP in the center of a square area of 500m × 500m and then randomly placing 29 mesh nodes in the square area. The topology graph for the network is created from the transmit power of the nodes and signal path loss. Each mesh node is given transmit power of 40dbm. We use the sample calculation given in [4] and the ECC-33 path loss model for medium city environments [10] to calculate the path loss due to the distance between the nodes. The modulation on each link is chosen based on received signal strength, as specified in [4, p. 765]. We assume that the network is using OFDM with 10Mhz bandwidth, so the OFDM symbol size is $25\mu s$ [4, p. 812].

The area where the mesh is located is partitioned into 25 cells, each with the radius of 50m. The purpose of the cells is to simulate short range 802.11 APs, which allow WTs to connect to the network. Each cell is assigned the mesh router closest to it as the AP. We use the minimum spanning tree algorithm to find a tree topology connecting all the mesh routers to the POP. Each router is assigned an end-to-end bandwidth to support a certain number of VoIP calls, and the end-to-end bandwidths are used to calculate link bandwidths required on every link in the network. The number of guaranteed service slots required on every link is calculated from the modulation used on the link and the symbol size.

Table 1. Percentage of Slots Required for VoIP Traffic ($D_{max} = 40$ms)

Calls	802.16 Frame Size			
	2.5ms	5.0ms	10.0ms	20.0ms
4	47%	29%	27%	27%
8	55%	53%	52%	50%

Table 1 summarizes the results of our numerical simulations for $D_{max} = 40$ms. We have used the GNU Linear Programming Kit (GLPK) [11] to perform the main $\{0, 1\}$-integer optimization in the search problem. The table represents the percentage of the slots required for VoIP traffic for 4 and 8 VoIP calls and different frame size. As observed, it is advantageous to increase the frame size since it decreases the percentage of slots needed to carry VoIP traffic. The results for $D_{max} = 60$ms are within 2% of the values reported in Table 1. The number of slots required for guaranteed traffic does not decrease if the delay is allowed to increase up to 60ms.

7   Conclusion

We have presented a method to minimize the number of TDMA slots required to support a given end-to-end QoS in mesh networks. Our optimization works by performing a linear search over the number of slots required to support the given end-to-end bandwidth. At each iteration of the search, the optimization solves a $\{0, 1\}$-integer program that finds an order of transmissions in the frame, so that the maximum TDMA propagation delay is kept bellow a given QoS level and end-to-end bandwidths can be scheduled. It is important to limit the TDMA propagation delay for traffic streams such as VoIP calls, requiring a guaranteed end-to-end delay. The optimization method in this paper is appropriate for mesh network planning, since the order of transmissions can later be distributed to the nodes to create schedules. The schedules will have the same maximum TDMA propagation delay, since the delay depends on transmission ordering in the frame. We have also used numerical simulations to show the efficiency of 802.16 network in carrying VoIP traffic.

References

1. Camp, J., Robinson, J., Steger, C., Knightly, E.: Measurement driven deployment of a two-tier urban mesh access network. Technical Report TREE0505, Rice University (2005)
2. Nortel Networks: Wireless mesh network - extending the reach of wireless LAN, securely and cost-effectively. http://www.nortelnetworks.com/solutions/wlan/ (2003)
3. Xu, S., Saadawi, T.: Does the IEEE 802.11 MAC protocol work well in multihop wireless ad hoc networks. **39**(6) (2001) 130–137
4. IEEE: Standard for local and metropolitan area networks part 16: Air interface for fixed broadband wireless access systems (2004)
5. IEEE: 802.11 TGs MAC enhacement proposal. Protocol Proposal IEEE 802.11-05/0575r3, IEEE (2005)
6. Djukic, P., Valaee, S.: Link scheduling for minimum delay in spatial re-use TDMA. In: Proceedings of INFOCOM. (2007)
7. Goode, B.: Voice over internet protocol VoIP. Proceedings of the IEEE **90**(9) (2002) 1495–1517
8. Djukic, P., Valaee, S.: Distributed link scheduling for TDMA mesh networks. In: Proceedings of ICC. (2007)
9. Rockafellar, R.T.: Network Flows and Monotropic Optimization. John Wiley & Sons (1984)
10. Electronic Communication Committee (ECC) within the European Conferenceof Postal and Telecommunications Administration (CEPT): The analysis of the coexistance of FWA cells in the 3.4-3.8 GHz band. ECC Report 33 (2003)
11. Makhorin, A.: GNU linear programming kit. Technical Report Version 4.8 (2005)

Provisioning Dedicated Class of Service for Reliable Transfer of Signaling Traffic

Jordi Mongay Batalla and Robert Janowski

Warsaw University of Technology, Institute of Telecommunications, Nowowiejska 15/19
00-665 Warsaw, Poland
{jordim,robert}@tele.pw.edu.pl

Abstract. In IP QoS (*Quality of Service*) networks, classes of service for different kinds of data traffic have been proposed, analyzed and currently implemented. On the contrary, signaling traffic has been mostly considered as negligible. However in the last time, it has become more critical due to the necessity of controlling setup latencies. As a consequence, a new class of service (*CoS*) dedicated to carry the user signaling traffic was proposed by IETF. This paper deeps into the signaling CoS and proposes a model for its provisioning. It also considers the impact of timers (for reliable data transfer) on the network load and the setup latencies. It concludes aspects about the transport protocols used for the signaling class of service.

Keywords: Signaling, setup latency, SIP, Class of Service, QoS.

1 Introduction

On the tracks of circuit switched networks, packet switched networks are lastly facing the challenge to effectively transfer the signaling traffic. Solutions proposed so far have been mainly concerned with the application level for which new signaling protocols were developed and implemented. These protocols generally use TCP or UDP as the transport. However, the main weakness persists at the network level, as the transfer of the signaling packets over the IP network definitively impacts on the performance of signaling procedures at the application level, e.g. setup latency.

IETF has lastly released a new document (RFC 4594) [1], which proposes classes of service as a solution for providing QoS in IP networks. Three of them have been dedicated for transferring signaling traffic over IP network and they are: Network Control, Signaling and OAM (Operation, Administration and Management).

This paper addresses the problem of assuring target values of setup latencies in IP QoS networks by implementing the signaling class of service. In further part, we refer to this class as S-CoS. Our studies are placed in the mark of the EuQoS IST project [4], [5], which implements the concept of classes of service to ensure end-to-end quality of service. Following the signaling scheme defined in EuQoS system we present a model to dimension the network resources for S-CoS necessary to assure the target values of setup latencies including the issue of retransmitted messages.

The remaining part of the paper is organized as follows: in section 2 we discuss some important aspects about the setup latency, which is the starting point for our

L. Mason, T. Drwiega, and J. Yan (Eds.): ITC 2007, LNCS 4516, pp. 853–864, 2007.

model. Afterwards, in section 3, the models for signaling traffic and the underlying network are presented. In section 4 we present a method for dimensioning the resources of S-CoS when the retransmissions are not considered (section 4.1) and when they are (section 4.2). In the latter case, we dig into the important problem of timeouts, which provoke retransmissions and prolong the setup procedure. In section 5, we provide some conclusions about design and implementation of S-CoS.

2   The Setup Latency

The setup latency is the time counted from the moment of sending the caller's request until the moment of receiving the corresponding response indicating the readiness of the network and the callee to start the actual connection. The maximum setup latencies for ISDN have been well defined by ITU in the document E-721 [6]. Since at this moment, no recommendation exists concerning the requirement for the target values of setup latencies in the IP networks (e.g. for VoIP applications), the values suggested therein might be also taken as a referent.

In the document E-721 [6] the acceptable setup latency for international calls is 11 seconds (assuming normal load conditions). This value is the sum of processing times of signaling messages and the time of transferring the packets, which carry them over the network. In the next part of the text, the latter one will be referred as "*transfer packets setup latency*" or briefly "*tp setup latency*". Note that by implementing the signaling Class of Service, we are only able to control the *tp setup latency* and not the whole setup latency. Supported by the measurements from the prototype deployment of the EuQoS system, which revealed that the contribution of processing times is near 8 seconds, we fix the target value of maximum *tp setup latency* to be 3 seconds in order to meet the requirement of total setup latency not greater than 11 seconds.

Document E-721 also recommends that the maximum blocking probability of an international connection is 5% assuming normal load conditions. The reason for call blocking might be two-fold: first, a lack of network resources for payload traffic and secondly, a lack of network resources for signaling traffic. In our case, the latter reason would result in the setup latency beyond maximum value. However, if the resources of S-CoS are well-provisioned, this reason should be negligible (e.g. at least ten times smaller: 0.5%). Following this guideline, we demand that the probability of violating the maximum tp setup latency is not greater than 0.5%.

3   Definition of S-CoS

Assuming the values of maximum tp setup latency and the probability of its violation, we will try to determine the acceptable values of QoS parameters relevant at the network level. As proposed in RFC 4594, the objective of S-CoS is to assure the target values (low!) for maximum IP Packet Transfer Delay (IPTD) and IP Packet Loss Ratio (IPLR) parameters while the value of the IP Packet Delay Variation (IPDV) is not crucial. The signaling traffic is generated/received by the system signaling entities using point-to-point connections. Therefore, signaling entities must set the appropriate Differentiated Services Code Point (DSCP) into the signaling packets submitted to S-CoS. We assume that in each part of the network supporting classes of service, S-CoS will constitute a separate class with its own buffer size ($B_{S\text{-}CoS}$) and the allocated bit-rate on the link ($C_{S\text{-}CoS}$).

3.1 Network Model

We assume the network topology as depicted in Fig. 1, where each terminal is connected to the server with a separate link. Furthermore, we assume that S-CoS is implemented only on the links between the following pairs of elements: each terminal and server, and between routers. The links between servers and routers are highly over-provisioned. These assumptions imply that the bottlenecks in transferring signaling traffic may occur only at the links with S-CoS. We model the Transit Network 2 in a simplest way using a single bottleneck. For more complicated network models the results must be re-calculated, but our method is still valid. The main aim of this paper is to present an approach for provisioning S-CoS, which is illustrated by the example. On the other hand, the example is taken from representative scenario and the results provide useful information for designing S-CoS.

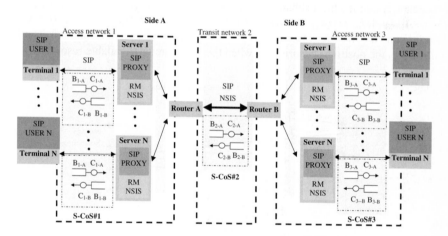

Fig. 1. Network model

In such a network model, S-CoS in represented in the following way. For access network 1, S-CoS is modeled between the terminal and the server by two separate single queues, one serving the traffic from the terminal to the server (the buffer size and link rate are denoted as $B_{1\text{-}A}$ and $C_{1\text{-}A}$ respectively) and one serving the traffic from the server to the terminal (the buffer size and link rates are denoted as $B_{1\text{-}B}$ and $C_{1\text{-}B}$ respectively). The letter "1" denotes the access network 1. The letter "A" denotes the queue for direction from side A to side B, and the letter "B" denotes the queue for direction from side B to side A. Let us remark that since the traffic is the same between each pair Terminal K – server, S-CoS#1 is provisioned with the same resources in all these links. In the same way S-CoS in transit network 2 is modeled with two single queues. The packet transmission and propagation times on the links between servers and routers are considered negligible i.e., the link rate is several magnitudes higher than the rates of other links.

3.2 Setup Procedure

The sequence of messages exchanged in order to establish connection with a callee user is depicted in Fig 2 and it is drawn based on the packet traces collected from the prototype of EuQoS system. The numbers shown in the Fig. 2 denote sizes of signaling messages without headers of lower layers protocols (no IP, UDP nor TCP headers). Although this particular sequence is EuQoS specific, the setup procedures in other IP QoS networks are quite similar, e.g. [12], [13]. That's why we believe it is representative example for studying the problem of designing and dimensioning S-CoS.

In general, the purpose of the setup procedure in EuQoS is two-fold. First, a caller user has to contact the remote side (callee) and agree the codecs used for voice/video communication using SIP [11] protocol. Secondly, the availability of the required network resources must be checked in the domains along the path from the caller to the callee. For the first aim, each Access Network provides a SIP Proxy which participates in the SIP message exchange and is responsible for locating the callee user. The second aim is accomplished by implementing Resource Managers (RMs) which communicate between themselves with NSIS [14] protocol. Thus the setup procedure combines the exchange of messages from SIP (i.e. *Invite, Auth Request, Invite Auth, Ack, Trying, Progress, Update, Prack, 200 OK Prack, 200 OK* and *Ringing*) and the messages from NSIS protocols (i.e. *Reserve* and *Response*). The Fig. 2 illustrates the positive scenario, i.e. when there are enough available resources.

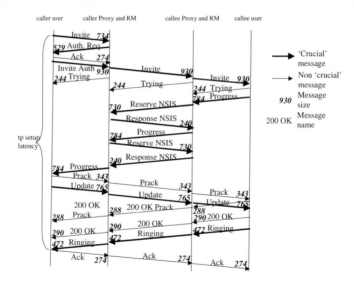

Fig. 2. Messages sequence for setup procedure

The setup procedure is started when a caller user sends *Invite* message and is finished upon receiving corresponding *ACK* by the callee user. However, *Ringing* message which reaches the caller user already informs him about successful connection establishment. Accordingly we define the tp setup latency as the time elapsed between the moment of sending *Invite* and receiving *Ringing* message.

Since UDP does not provide reliable message transfer, it must be assured at the application level (SIP and NSIS) by implementing the mechanism of timers and retransmissions. Timers are activated when sending the following SIP messages: *Invite, Invite Auth, Progress, Update, Ringing* and NSIS message: *Reserve*. In case of TCP the reliability is assured by the transport protocol.

4 Provisioning of S-CoS

As explained in RFC 4594 [1], signaling for S-CoS is not applicable, i.e. it is not possible to apply Admission Control function. Instead an adequate provisioning is required.

To efficiently manage the problem of dimensioning the resources for S-CoS, which carries the aggregated traffic from different signaling procedures e.g. setup, release, others, we propose to use the decomposition approach. In this approach we consider the traffic generated by each procedure "in isolation" and determine the required resources for these sub-streams in a separate way. Then, S-CoS carrying the aggregated traffic related to different procedures requires, for the worst case, the resources equal to the sum of resources calculated separately. The summation is possible when for each type of separately analyzed signaling traffic we assume the same QoS objectives (IPTD and IPLR values).

The method proposed below determines the required capacity and buffer size for a single setup procedure. The resources required for N such procedures are the sum of the resources calculated for a single one. The simple summing operation performed on the calculated resources neglects the multiplexing gain and this is the reason of overestimating the resources.

Let us remark that S-CoS#1, S-CoS#2 and S-CoS#3 (see Fig. 1) will require different amount of network resources, since the volume of signaling traffic submitted to them is different (see Fig. 2). Although the setup procedures might be initiated in both directions we evaluate them running only in one direction. This let us for performing studies of the worst-case scenario when the procedures start at the same time and simultaneously submit their packets to S-CoS#2.

A critical point in transferring the signaling traffic is the retransmission of packets as they provoke more traffic in the network and prolong tp setup latency. The retransmissions are a consequence of the reliable data transfer, which might be accomplished at the application or transport level. The three possible transport protocols are: UDP, TCP or SCTP [2]. TCP and SCTP have no differences when serving only one connection. Performance differences, in favor of SCTP [9], appear only when multiplexing more connections. As our model is based on the study of a single isolated connection, we consider only TCP.

For the reliable transfer, SIP activates timers for its own messages as explained in [11]. The default value of this timer is 500 ms but in [11] –section 17.1.1.2- it is recommended to use larger values that fit RTT if it is known in advance. The value of timer is set in the signalling application (e.g. SIP) and the same value is used for local and long distance connections. That's why it is very difficult to approximate the optimal value for this timer [15]. We assume the value of 500 ms for SIP and NSIS over UDP. On the contrary when using TCP (or SCTP), the timers for retransmissions are adjusted by the transport protocol, and are determined by the values of RTT.

4.1 Provisioning S-CoS Assuming No Retransmissions

In this section we treat only the case of no-retransmissions. Generally, the retransmissions are provoked by two events: packet losses or excessive delay of packets. To avoid retransmissions, it is necessary to assure no packet losses (IPLR=0) and either to control the maximum packet delays (maximum IPTD) or, in case of UDP, set long timers at the application level.

The analysis of Fig. 2 indicates that the maximum number of packets, which can be potentially stored together in a buffer of S-CoS is 2 for A to B direction and 3 for B to A. Since we demanded IPLR equal 0 we must provide the following buffer sizes for each S-CoS: $B_{1-A}= B_{2-A}= B_{3-A}=2$ packets and $B_{1-B}= B_{2-B}= B_{3-B}=3$ packets.

To determine the required capacity, we propose to use a conservative dimensioning method based on the deterministic QoS guarantees [7], [8] i.e. maximum IPTD. For single sent packet, e.g. *Invite*, IPTD is the ratio of the packet size and the capacity of S-CoSs. Whereas for packets sent in a batch, e.g. *Ack* and *Invite Auth*, the IPTD is increased by the transmission times of all preceding packets from the batch. We impose the same maximum IPTD in all S-CoSs, which implies the following set of equalities (considering UDP and IP headers):

$$maximum\ IPTD = \frac{8\times(302+958)}{C_{1-A}} = \frac{8\times812}{C_{1-B}} = \frac{8\times958}{C_{2-A}} = \frac{8\times812}{C_{2-B}} = \frac{8\times958}{C_{3-A}} = \frac{8\times(272+812)}{C_{3-B}} \tag{1}$$

Additional constraint related to maximum tp setup latency of 3 seconds constitutes a new equality. This equality states that tp setup latency is the sum of IPTDs related to the transfer of all crucial messages (marked with bold line in Fig. 2).

$$tp\ setup\ latency = 3secs = \frac{8\times(762+302+958+371+793)}{C_{1-A}} + \frac{8\times(557+812+500)}{C_{1-B}} + $$
$$+\frac{8\times(958+268+758+793)}{C_{2-A}} + \frac{8\times(758+812+268+500)}{C_{2-B}} + \frac{8\times(958+793)}{C_{3-A}} + \frac{8\times(272+812+318+500)}{C_{3-B}} \tag{2}$$

Formulas 1 and 2 altogether let for calculating the following capacities: $C_{1-A}= 47.7$ Kbps, $C_{2-A}=36.3$ kbps, $C_{3-A}=36.3$ kbps, $C_{1-B}= 30.7$ Kbps, $C_{2-B}=30.7$ kbps, $C_{3-B}=41.0$ kbps. The value of maximum IPTD is 220.96 ms for every S-CoS.

To verify the correctness of the conservative approach for dimensioning the resources of S-CoS, we performed the simulation studies assuming the worst-case traffic condition. We check whether the target values of IPLR, maximum IPTD and maximum tp setup latencies are assured when we have N concurrent setup procedures

For simulations, we implemented our own modules for the applications (SIP and NSIS) in ns-2 [3]. In the simulation studies we have N (N=1,5,10,...100) concurrent setup procedures originating from side A. Let us recall that for S-CoS#2 the resources in the network must be N times the resources of single setup procedure. Since the simulation are performed assuming the worst-case traffic conditions it is enough to perform a single run.

The simulated maximum IPTD is 220.96 ms and IPLR is 0, confirming the validity of the proposed model. The values of tp setup latency are presented in Table 1. We can observe that, the more setup procedures are simultaneously transferred in transit network 2, the shorter tp setup latency we obtain. This is due to the multiplexing gain.

Table 1. Simulation results of tp setup latencies (for UDP)

Number of connections	max. tp setup latency [s]	Mean tp setup latency [s]	Number of connections	max. tp setup latency [s]	Mean tp setup latency [s]
1	3.000	3.000	20	2.327	2.176
5	2.407	2.303	50	2.311	2.150
10	2.354	2.218	100	2.306	2.141

We can conclude that this model is valid to dimension the signaling traffic and provision S-CoS when we limit the number of the setup procedures handled by the signaling system in parallel.

4.2 Provisioning S-CoS Taking into Account Retransmissions

In this section we dig into the details of the retransmissions. Remind that they are provoked by two events: excessive delay of packets or packet losses.

4.2.1 Excessive Delay of Packets
The excessive delay of packets does not impact on the amount of provisioned resources because for our request/response communication pattern the interference of the retransmitted packets with the original sequence is negligible. Despite, they increment the network load.

When there are no packet losses it is unlikely that TCP retransmits a packet since its timer continuously adjusts to the changes in RTT. On the contrary, for UDP, timers are set to constant values so they may timeout before the awaited response arrives.

Thus, for a single setup procedure and TCP as the transport protocol, the sequence of messages exchanged between network elements (Fig. 1) is exactly as depicted in Fig. 2. Its total volume, assuming TCP with Forward Acknowledgment [10] is equal to 18007 bytes. On the other hand, considering UDP as the transport protocol with default value of timers and the capacities (C_{1-A}, C_{2-A}, C_{3-A}, C_{1-B}, C_{2-B}, C_{3-B}) calculated in section 4.1, the total of 26687 bytes are exchanged due to the retransmissions of delayed packets (48.2% more traffic volume!). UDP unnecessarily increases the network load when no packets are lost.

4.2.2 Packet Losses
Analysis of the message sequence from Fig. 2, shows that loosing some of the packets affects the tp setup latency while loosing others, does not. This is because some packets are crucial from the point of view of setup and their loss ceases the whole procedure until successful retransmission. As a result, the setup procedure is delayed as much as the value of retransmission timer. For example, the loss of *Invite* provokes its retransmission after the timeout, and the tp setup latency is extended by this time;

but the loss of *Trying* forces its retransmission although this does not delay the setup procedure. There are 18 crucial packets (see Fig. 2), which can delay the setup procedure if they are lost. In this case, in order to assure that the tp setup latency is the same as in case of no packet losses, it is necessary to speed up the transmission of packets by providing higher capacity. Since the packet losses are due to the limited buffer size there is a trade off between these two quantities: the smaller buffer size in S-CoS, the more capacity we need, and opposite.

In this section we want to determine the relation between buffer size and capacity. In our model for provisioning, the capacity and the maximum IPTD are strongly related (higher values of maximum IPTD mean less capacity). Therefore, we calculate the relation between IPLR and maximum IPTD. We proceed as follows:

- For TCP we assume the worst-case timeout value which is 2 x RTT_{max} = 4 x maximum IPTD. We search the pairs IPLR-maximum IPTD, which induce k losses. Increasing k from 0 to a given value N, we perform the following steps:

First step: determine the probability P_k that no more than k losses related to crucial packets occur. This probability is calculated using combinatorial approach:

$$P_k = \sum_{j=0}^{k} IPLR^j \times (1 - IPLR)^{18} \times CR_{18,j} \quad where \quad CR_{n,m} = \frac{(n+m-1)!}{m!(n-1)!} \tag{3}$$

Second step: Since the probability of violating the maximum tp setup latency is fixed at 0.5%, determine such a value of IPLR that meets the condition: $P_k = 0.995$

Third step: Calculate such a value of maximum IPTD and such C_{1-A}, C_{2-A}, C_{3-A}, C_{1-B}, C_{2-B}, C_{3-B} (as in the model in section 4.1), that assure the sequence is completed within the time t equal:

$$t = maximum\ tp\ setup\ latency - k \times 4 \times maximum\ IPTD \tag{4}$$

Fourth step: Represent the result on the graph as a single point with coordinates equal to the obtained values of IPLR and maximum IPTD.

Let us recall that each obtained value of IPLR corresponds to 0, 1, 2... and so on packet losses delaying the setup procedure by 0, 1, 2... timeouts respectively. For example, IPLR=0.01273 induces the maximum IPTD=134.41ms corresponding to the capacities, which assure that sequences with k=2 losses, finish in 3 seconds (see Table 3). And IPLR=0.026299 induces the maximum IPTD=113.98 ms corresponding to the capacities, which assure that sequences with k=3 losses, finish in 3 seconds. For the values of IPLR between 0.01273 and 0.026299, the number of sequences with 0, 1, 2 or 3 losses is higher than 99.5% (acceptable) but the percentile of sequences with 0, 1 or 2 losses is lower than 99.5% (unacceptable). Therefore, for these values of IPLR, we must provision S-CoS with the same amount of resources as for the case of IPLR=0.026299. The effect is a stepwise curve as shown in Fig. 3.

- For UDP we consider the adaptive retransmission timer of SIP, i.e. the mechanism which doubles the value of timer after each timeout as defined in [11].

By the loss pattern LP_k we define all of the possible occurrences of packet losses that prolong the tp setup latency by the time equal to k times default timer value (500 ms). Since we are only interested in such LP_ks that does not extend tp setup latency beyond the maximum value, then k takes the values in the range from 0 to:

$$k_{max} = \left\lfloor \frac{maximum\ tp\ setup\ latency}{default\ timer\ value} \right\rfloor \tag{5}$$

Below we give the algorithm accompanied by the example when maximum tp setup latency is 3 seconds and the tp setup latency is delayed not more than 1.5 seconds due to timeouts (i.e. $k=3$). For each LP_k:

First step: Taking into account IPLR value, calculate the probability $Prob_k$ that the loss pattern LP_k occurs; e.g. LP_3 occurs when there are two losses of packets related to the same timer (i.e. delay=0.5s+1s) or when there are 3 losses of packets related to different timers (i.e. delay=0.5s+0.5s+0.5s). The first case is referred in Formula 6 and the second one in Formula 7. Formulas 6 and 7 can be explained by the fact that some packets are related to the same timer (we name them a group). If two of them are lost, then the setup procedure is delayed 0.5s for the first loss, and 1s for the second loss. We can distinguish 4 groups of such packets, composed of 2 (*Invite* and *Auth Req*), 7 (*Progress, Reserve, ..., Progress*), 3 (*Update, ..., Update*) or 3 (*Ringing, ..., Ringing*) crucial packets (see Fig. 2).

$$IPLR^2 \times (1-IPLR)^{18} \times \{CR_{2,2} + CR_{7,2} + 2xCR_{3,2} + (18-2-7-2\times3) \times CR_{1,2}\} \tag{6}$$

$$\begin{aligned} IPLR^3 \times (1-IPLR)^{18} \times \{CR_{18,3} - (CR_{2,3} + CR_{7,3} + 2\times CR_{3,3} + \\ + (18-2-7-2\times3) \times CR_{1,3}) - (CR_{2,2}(18-2) + CR_{7,2}(18-7) + \\ + 2\times CR_{3,2}(18-3) + (18-2-7-2\times3) \times CR_{1,2}(18-1))\} \end{aligned} \tag{7}$$

Second step: Calculate the probability P_k that the timeout $\leq k$ x *default timer value*, as the sum of probabilities of occurrence of all loss patterns LP_j such that $j \leq k$, e.g.:

$$\begin{aligned} P_3 = \sum_{j=1}^{3} Prob_j = (1-IPLR)^{18} + IPLR \times (1-IPLR)^{18} \times CR_{18,1} + IPLR^2 \times (1-IPLR)^{18} \times CR_{18,2} + \\ + IPLR^3 \times (1-IPLR)^{18} \{CR_{18,3} - [CR_{2,3} + CR_{7,3} + 2\times CR_{3,3} + (18-2-7-2\times3) \times CR_{1,3} + \\ + CR_{2,2}(18-2) + CR_{7,2}(18-7) + 2\times CR_{3,2}(18-3) + (18-2-7-2\times3) \times CR_{1,2}(18-1)]\} \end{aligned} \tag{8}$$

Determine such a IPLR value that meets the condition: $P_k=0.995$

E.g.: $\quad P_3 = 0.995 \quad \Rightarrow \quad IPLR = 0.01489 \tag{9}$

Third step: Calculate such a maximum IPTD and such $C_{1-A}, C_{2-A}, C_{3-A}, C_{1-B}, C_{2-B}, C_{3-B}$ (as in section 4.1.), that assure the sequence is completed within the time t equal:

$$t = maximum\ tp\ setup\ latency - k \times default\ timeout \tag{10}$$

E.g.: $\quad t = 3\ s - 3\times0.5\ s \quad \Rightarrow \quad maximum\ IPTD = 110.479\ ms \tag{11}$

Fourth step: Represent the result on the graph as a single point with coordinates equal to the obtained values of IPLR and maximum IPTD.

Fig. 3 provides the information about the values of IPLR and maximum IPTD that S-CoS should be designed for, in order to assure maximum tp setup latencies. The points located below the curves define such values of IPLR and maximum IPTD pairs, which assure appropriate provisioning of S-CoS. It means that for any point in this area, its coordinates determine the values of IPLR and maximum IPTD that guarantee maximum tp setup latency. The optimal points are located on the curve since for the given IPLR value they imply the most relaxed constraints on the maximum IPTD, which lets to provision minimum needed capacity.

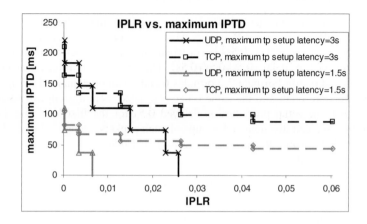

Fig. 3. IPLR vs. maximum IPTD assuming the probability of violating maximum tp setup latency equal 0.5%

Fig. 3 reveals the main difference between the cases when UDP or TCP is used to transport signaling messages. First of all, the working area for UDP is bounded while for TCP it is not. It means that, for UDP, it is not possible to guarantee maximum tp setup latencies when IPLR values are too high. The reason of this fact is that the values of timer are constant (500 ms) and they are not adapted to the RTT, whereas for TCP they are.

The other important observation is that almost in all cases, the results of TCP are better than the results of UDP, i.e. TCP allows higher maximum IPTD for the same value of IPLR and, because of this, it needs less capacity to ensure the same probability of violating the maximum tp setup latency. UDP is better only for low values of IPLR (less than 6.5×10^{-3} in the case of maximum tp setup latency equal 3 seconds and 1.1×10^{-4} in the case of maximum tp setup latency equal 1.5 seconds).

If the values of timers in SIP (over UDP) were, at least, as long as the maximum tp setup latency, any loss of crucial packet would cause the violation of maximum tp setup latency. If the SIP timers are set to this value, the IPLR in S-CoS cannot be greater than 0.0111% to ensure the probability of violation equal 0.5%.

4.2.3 Simulations

The analytical method presented in section 4.2.2 does not consider some effects of the retransmissions, e.g. retransmissions of non-crucial packets. These effects are considered negligible because do not interfere in the tp setup latency. Now we demonstrate it with simulations. We proceed as follows: in all the links we force the IPLR and the capacities obtained from the model shown in section 4.2.2, and we measure tp setup latency. Let us note that TCP connections are already established when we start each setup procedure. In Table 2 and Table 3 we report the time in which 99.5% of setup procedures is completed. The values of k indicate the number of timeout events as referred in the analytical studies.

Table 2. Simulation results of tp setup latency (UDP) with 95% confidence intervals

k	Maximum tp setup latency equal 3 secs.			Maximum tp setup latency equal 1.5 seconds		
	IPLR	maximum IPTD [ms]	tp setup latency [s] for 99.5% of setup procedures	IPLR	maximum IPTD [ms]	tp setup latency [s] for 99.5% of setup procedures
1	0,003488	184.13	2.9 ± 0.01	0.003488	73.65	1.41 ± 0.01
2	0,00647	147.31	2.7 ± 0.02	0,00647	36.83	1.43 ± 0.02
3	0,01489	110.48	2.76 ± 0.04	-	-	-
4	0,02289	73.65	2.53 ± 0.02	-	-	-
5	0,02579	36.83	2.42 ± 0.03	-	-	-

Table 3. Simulation results of tp setup latency (TCP) with 95% confidence intervals

k	Maximum tp setup latency equal 3 seconds			Maximum tp setup latency equal 1.5 seconds		
	IPLR	maximum IPTD [ms]	tp setup latency [s] for 99.5% of setup procedures	IPLR	maximum IPTD [ms]	tp setup latency [s] for 99.5% of setup procedures
1	0,003488	163.76	2.85 ± 0.05	0,003488	81.88	1.44 ± 0.02
2	0,01273	134.41	2.79 ± 0.01	0,01273	67.21	1.38 ± 0.02
3	0,026299	113.98	2.73 ± 0.01	0,026299	56.99	1.36 ± 0.02
4	0,042554	98.95	2.76 ± 0.06	0,042554	49.47	1.41 ± 0.04
5	0,06038	87.41	2.95 ± 0.10	0,06038	43.71	1.48 ± 0.02

For very low values of IPLR, the simulations should be very long to achieve credible results. Because of this, we only simulate the cases when k>0.

The simulation results indicate that, for each value of IPLR, the calculated capacity is enough to ensure that the 99.5% of setup procedures finishes within the time not longer than the maximum tp setup latency. We can conclude that the analytical model, as well as the comparison between TCP and UDP, presented in 4.2.2 are valid.

5 Conclusions

In this paper we discussed the performance of S-CoS. Its deployment seems to be essential in order to control the signaling traffic at the packet level (the values of IPLR and IPTD) and at the application level (tp setup latency).

In section 4.1 we proposed a model based on the decomposition approach, deterministic QoS guaranties and the worst-case traffic conditions. With this model, it is possible to provision S-CoS if we control the maximum number of setup procedures in the system. So, it is desirable to implement in S-CoS a congestion avoidance mechanism as it is done in telephone exchanges.

In section 4.2, we addressed the problem of retransmissions and the choice of transport protocol to transfer signaling messages. The advantages of UDP for transferring signaling traffic come from its simplicity. Despite this, IP QoS networks aim at implementing stable QoS solutions and the simplicity is not a definitive criterion. On the other hand, in this paper we demonstrated the advantages of TCP as compared to UDP. They are especially pronounced when there are many packet losses or when packets are excessively delayed. When considering message processing times, the advantages of TCP increase because TCP waits for local acknowledgements (between signaling entities: terminal, proxy, RM), while UDP waits for end-to-end acknowledgments (between terminals). Concluding, we suggest transferring signaling traffic by the couple TCP and S-CoS.

References

1. Babiarz, J., Chan, K., Baker, F.: Request For Comments 4594: Configuration Guidelines for DiffServ Service Classes (August 2006)
2. Stewart, R., et al.: Request For Comments 2960: Stream Control Transmission Protocol (2000)
3. The Network Simulator ns-2: http://www.isi.edu/nsnam/ns/
4. Dugeon, O., Morris, D., Monteiro, E., Burakowski, W., Diaz, M.: End to End Quality of Service over Heterogeneous Networks (EuQoS), Lannion, France (November 14-18, 2005)
5. EuQoS project Homepage, http://www.euqos.org
6. ITU-T E.721. Network grade of service parameters and target values for circuit-switched services in the evolving ISDN (05/99)
7. Le Boudec, J.Y.: A proven delay bound for a network with Aggregate Scheduling, EPFL-DCS Technical Report DSC,2000/002 (2000), http://www.epfl.ch/PS_files/ds2.pdf
8. Listanti, M., Ricciato, F., Salsano, S., Ventri, L.: Worst case analysis for deterministic allocation in the Differentiated Services network. In: Proc.of IEEE Globecom'00, San Francisco (November 27-30, 2000)
9. Camarillo, G. et al.: Evaluation of Transport Protocols for the Session Initiation Protocol. IEEE Network (October 2003)
10. Mathis, M., Mahdavi, J.: Forward Acknowledgment: Refining TCP congestion protocol. ACMSIGCOMM (August 1996)
11. Rosenberget, J., et al.: Request For Comments 3261: SIP: Session Initiation Protocol (June 2002)
12. Sargento, S., et al.: End-to-end QoS Architecture for 4G Scenarios (2005)
13. Camarillo, G. et al.: Request For Comments 3312: Integration of Resource Management and Session Initiation Protocol (SIP) (October 2002)
14. Hancock, R., et al.: Request For Comments 4080: Next Steps in Signaling (NSIS): Framework (June 2005)
15. Cisco SIP IP Phone 7960 Administrator Guide, Version 2.0. http://www.cisco.com/univercd/cc/td/doc/product/voice/c_ipphon/sip7960/sipadm2/install.htm

An Algorithmic Framework for Discrete-Time Flow-Level Simulation of Data Networks

Lasse Jansen[1,2], Ivan Gojmerac[1], Michael Menth[2],
Peter Reichl[1], and Phuoc Tran-Gia[2]

[1] Telecommunications Research Center Vienna (ftw.)
Donau-City-Str. 1, 1220 Vienna, Austria
[2] University of Würzburg, Institute of Computer Science
Am Hubland, 97074 Würzburg, Germany
{jansen,gojmerac,reichl}@ftw.at,
{menth,trangia}@informatik.uni-wuerzburg.de

Abstract. In this paper, we present a comprehensive algorithmic framework for discrete-time flow-level simulation of data networks. We first provide a simple algorithm based upon iterative equations useful for the simulation of networks with static traffic demands, and we show how to determine packet loss and throughput rates using a simple example network. We then extend these basic equations to a simulation method capable of handling queue and link delays in dynamic traffic scenarios and compare results from flow-level simulation to those obtained by packet-level simulation. Finally, we illustrate the tradeoff between computational complexity and simulation accuracy which is controlled by the duration of a single iteration interval Δ.

1 Introduction

Simulation has traditionally been an important tool for performance evaluation of data networks, mostly in the form of packet-level simulation by employing discrete-event simulation techniques [1]. Every packet arrival and departure at each link is modeled as a separate event. Although packet-level simulation still represents the most widely used approach, the simulation of today's networks with very high packet rates is often not feasible, as too many simulation events must be generated even for small intervals of simulated time.

However, in many cases the overhead of packet-level simulations is not necessary at all in order to achieve a realistic estimation of network statistics like throughput rates, queue sizes, or loss probabilities. In those cases, an efficient alternative to packet-level simulation is the simulation of networks at the level of individual flows, for which there exists a multitude of different techniques, commonly summarized under the terms *fluid simulation* or *flow-level simulation*.

In this paper, we concentrate upon *discrete-time flow-level simulation*. Traffic is not modeled in terms of discrete packets but rather in terms of a continuous amount of data. The data is shifted in fixed intervals Δ on predefined routes through the network. However, to our best knowledge, literature in this field of research lacks a discrete, easy-to-implement formulation of discrete-time flow-level simulation which is able to model end-to-end connections. Addressing this issue, in this paper, we provide such a

L. Mason, T. Drwiega, and J. Yan (Eds.): ITC 2007, LNCS 4516, pp. 865–877, 2007.

formulation, which additionally allows the network to be simulated at different levels of detail. We develop the fundamental flow-level simulation techniques step by step, first presenting a simple algorithm for throughput calculation, and then extending this algorithm to capture network dynamics like link and queueing delay.

The paper is structured as follows: Section 2 gives an overview of related work. In Section 3 we describe the basic equations for calculating the time-dependent aggregate throughput rates and the loss probabilities on the links, and we present a method for the calculation of the stationary network state in the presence of static traffic demands. Subsequently, in Section 4 we extend these basic equations to scenarios with dynamic traffic patterns by introducing queue and link delay modeling. Section 5 provides a comparison of flow-level and packet-level simulation results and demonstrates the influence of different durations of iteration interval Δ. Finally, Section 6 concludes the paper with summarizing remarks.

2   Related Work

This section provides a brief description of previous work and relevant applications of flow-level simulation. There are two main variants of flow-level simulation. The foundation for the *continuous-time* variant was given in [2] and [3], and has since been further developed and widely applied by other authors [4,5,6,7]. The basic principle of this approach is to model flow rates and rate changes without considering discrete data packets. Each flow is assigned a certain transmission rate, and rate reductions due to bottleneck links are tracked as events in the event chain of the simulator. Although widely used, under particular circumstances this approach has been shown to suffer from the so called *ripple effect* which can cause severe performance degradations concerning computation time. It occurs in networks with circularly overlapping end-to-end connections and overloaded links, causing rate change events to reproduce themselves in a circular manner [8,9].

An alternative to the continuous-time variant is *time-stepped fluid simulation* which is the foundation of the simulation method presented in this paper. Here, the data is modeled in terms of quantities of a continuous amount of data which are shifted through the network at a fixed time step. This model was first proposed in [10], but the routing model applied therein is hardly applicable to realistic networks. In particular, a traffic demand matrix cannot be represented, as end-to-end connections have not been modeled at all.

Another approach which defines the evolution of network data streams in terms of differential equations is used in [11]. End-to-end connections are modeled as traffic aggregates which enable the simulation of realistic network and routing scenarios. The equations are numerically solved using the Runge-Kutta algorithm. In this paper, we use the basic idea from [10], and additionally enable end-to-end connections like in [11]. Furthermore, we provide a clean formulation of discrete-time flow-level simulation which is well suited for practical implementation.

The possibility of dynamic traffic rate adjustments and delay modeling in the presented simulation approach allows for multipath routing simulations in which traffic aggregates between individual pairs of nodes are carried via multiple parallel paths. Implementing the simulation approach described in this paper, we have already evaluated the Adaptive Multi-Path algorithm (AMP) [12,13] using our flow-level simulation environment (cf. [14,15]).

Furthermore, the throughput and delay approximation capabilities of this approach enable more complex elastic traffic models, like e.g. TCP, to be integrated into such a flow-level simulation framework, as the sending rate of individual sources can be dynamically adjusted based upon the information about flow round-trip time (RTT) and loss probability on the path, employing differential equation models of TCP as presented in [11,16,17].

3   Basic Iterative Algorithm

In this section we first clarify the notation, after which we introduce the iterative algorithm that calculates throughput rates and loss probabilities. Finally, we show the application of this algorithm for the calculation of stationary throughput in a network with static traffic demands.

3.1   Definitions

The simulation process is based upon shifting data through the network in fixed time intervals Δ. In our model, a data rate λ denotes the total amount of data in a single iteration interval, normalized by Δ as the time between two successive iteration steps. In other words, a data rate of λ in interval Δ corresponds to an amount of data, $\lambda \cdot \Delta$, which may represent e.g. bits, bytes, or equally sized packets.

In the following we list some definitions for reference. Expressions indexed by $[n]$ are time-dependent and calculated in each iteration.

L	Set of all links.
A	Set of all aggregates.
$L^a \subseteq L$	Set of links crossed by aggregate $a \in A$.
$A_l \subseteq A$	Set of aggregates crossing link $l \in L$.
$first(a) \in L^a$	First link on the route of aggregate $a \in A$.
$next(a,l) \in L^a$	Successor link of $l \in L^a$ on the route of aggregate $a \in A_l$.
$sink(a) \in L^a$	Virtual link representing the sink of aggregate $a \in A$ and succeeding the last link of a.
c_l	Link capacity of link $l \in L$, i.e. data that can be served during a single iteration interval Δ.
$p_l[n]$	Loss probability at link $l \in L$.
$\lambda_l^a[n]$	Arrival rate of aggregate $a \in A$ at link $l \in L^a$.
$\lambda^a[n]$	Sending rate of aggregate $a \in A$, equal to $\lambda_{first(a)}^a[n]$.
$\lambda_l[n]$	Sum of all arrival rates at link $l \in L$.
$\theta^a[n]$	Throughput of aggregate $a \in A$.
Δ	Iteration duration interval, i.e. the time between two iteration steps.

3.2   Iterative Equations for Discrete-Time Flow-Level Simulation

We assume the network to be empty at initialization time ($n = 0$). Each traffic aggregate $a \in A$ has a specific route consisting of a sequence of links $l \in L^a$ along which the arrival rates $\lambda_l^a[n]$ of this aggregate are shifted in each iteration step. The overall arrival rate at link l in iteration step n is then:

$$\lambda_l[n] = \sum_{a \in A_l} \lambda_l^a[n].$$
(1)

Based on this sum and the capacity c_l of this link, its loss probability $p_l[n]$ can be calculated for the corresponding iteration step n. The capacity c_l of a link l is assumed to be proportionally partitioned among the competing aggregates $a \in A_l$.

$$p_l[n] = \max\left\{1 - \frac{c_l}{\lambda_l[n]}, \ 0\right\}.$$
(2)

The arrival rate $\lambda_l^a[n]$ of aggregate a and the loss probability at link l in iteration step n determine the arrival rate $\lambda_{next(a,l)}^a[n + 1]$ of a at its next link at the next iteration step $n + 1$.

$$\lambda_{next(a,l)}^a[n + 1] = \lambda_l^a[n] \cdot (1 - p_l[n]).$$
(3)

The throughput of aggregate a is given by

$$\theta^a[n] = \lambda_{sink(a)}^a[n].$$
(4)

3.3   Calculation of Stationary Throughput for Static Traffic Demands

We now use the equations from the previous section to calculate the stationary throughput rates in settings with static traffic demands. By first calculating the throughput rates of a small example network analytically, we show that this can be a non-trivial task and propose an algorithm based upon the iterative equations as a simple alternative.

Figure 1(a) shows a triangle network with all link capacities set to $c = 1$ and three overlapping aggregates with static, non-adaptive sending rates $\lambda = 1$, as well. Intuition might lead to the belief that each aggregate should achieve a throughput of $\theta = 0.5$, as this corresponds to the most reasonable bandwidth distribution with respect to both fairness and throughput. However, the analytical calculation of the throughput yields significantly different results. Due to the symmetrical nature of the example, we assume that the loss probability p is the same on all links. The throughput, which equals the arrival rate at the second link of each aggregate, can then be expressed by $\theta = \lambda \cdot (1-p)^2$. The arrival rate of each aggregate at its respective first link equals $\lambda \cdot (1 - p)$. As each link is crossed by two aggregates, i.e. one at its initial hop and the other at its second hop, the following equation must be solved: $c = \lambda(1 - p) + \lambda(1 - p)^2$. With $\lambda = 1$ and $c = 1$, and presuming that the loss probability is greater than 0, we can calculate the throughput resulting in $\theta \approx 0.38$.

While in this simple example we are able to calculate the throughput quite easily, by solving just one quadratic equation, we would have to solve much more complex

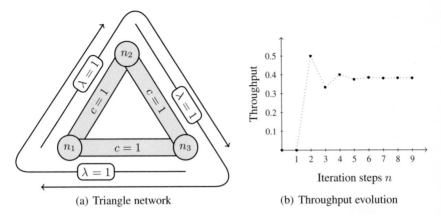

(a) Triangle network   (b) Throughput evolution

Fig. 1. Throughput calculation for an triangle network

sets of nonlinear equations in the case of larger networks. When instead applying the proposed iterative equations from the last section to the example network, we observe in Figure 1(b) that the stationary throughput of all aggregates converges at the same point as the analytical solution, i.e. at the value of about 0.38, after only a few iterations of the algorithm.

The exact solution for throughput rates in arbitrary networks is achieved after a number of iterations which corresponds to the hop count of the longest possible path composable of overlapping aggregates in the network. As in this example we can construct a path of infinite length due to circularly overlapping aggregates, an infinite number of iterations is required to obtain an exact solution. Therefore, in order to achieve reasonably precise results, we iteratively calculate the throughput of all aggregates until the differences between two consecutive arrival rates of all aggregates at each link drop below a predefined small value of ε. A formal description is provided in Algorithm 1 which may serve as an efficient and conceptually simple alternative to analytical throughput calculation.

4   Capturing Network Dynamics

In this section, we extend the flow-level simulation to dynamic scenarios, in which link and queue delay is modeled and the sending rates of aggregates can change in each iteration step independently of the link delay.

4.1   Definitions

In addition to the definitions from Section 3 we list the notations needed for the calculation of delay and queue size.

$d_l \cdot \Delta$ Link propagation delay of link l, $d_l \in \mathbb{N}$.
$q_l[n]$ Queue length at link l, initialized with $q_l[0] = 0$.

Algorithm 1. Throughput calculation for static traffic demands

---

1: $n \Leftarrow 0$
2: initialize $\lambda_l^a[n]$ with 0 $\forall a \in A, \forall l \in L^a \setminus \{first_a\}$
3: **repeat**
4:      $stop \Leftarrow$ **true**
5:      **for all** $l \in L$ **do**
6:          calculate $\lambda_l[n]$ according to Equation (1)
7:          calculate $p_l[n]$ according to Equation (2)
8:          **for all** $a \in A_l$ **do**
9:              **if** $sink(a) \neq l$ **then**
10:                 calculate $\lambda_{next(a,l)}^a[n+1]$ according to Equation (3)
11:                 **if** $\left|\lambda_{next(a,l)}^a[n+1] - \lambda_{next(a,l)}^a[n]\right| > \varepsilon$ **then**
12:                     $stop \Leftarrow$ **false**
13:                 **end if**
14:             **end if**
15:         **end for**
16:     **end for**
17:     $n \Leftarrow n+1$
18: **until** $stop = true$
19: **for all** $a \in A_l$ **do**
20:     $\theta^a \Leftarrow \lambda_{sink(a)}^a[n]$
21: **end for**

---

q_l^{max}	Maximum queue length at link l.
$\delta_l^a[n]$	Present delay of the data of aggregate a arriving at link l, initialized with $\delta_{first(a)}^a[0] = 0$.
$\delta^a[n]$	End-to-end delay of the data of aggregate a arriving at its destination in iteration interval n.

4.2   Modeling Link Propagation Delay

The propagation delay $(d_l \cdot \Delta)$ is expressed by an integer multiple d_l of the iteration interval Δ. Due to this delay, the arrival rates $\lambda_l^a[n]$ at link l at iteration step n are propagated to their respective next link only at iteration step $n + d_l$:

$$\lambda_{next(a,l)}^a[n + d_l] = \lambda_l^a[n] \cdot \left(1 - p_l[n]\right). \tag{5}$$

In concrete implementations of our algorithm, the formulation of this equation implies that for each aggregate a, at every link l, a number of d_l values must be stored for future arrival at the respective next link. This can be viewed as dividing each link in d_l slots, and then shifting the data amount $(\lambda_l^a[n] \cdot \Delta)$ one slot further on its route in each iteration, as illustrated in Figure 2. This method can be implemented quite efficiently by using an array of fixed size d_l for each aggregate a at each link $l \in L^a$ and using $(n$ modulo $d_l)$ as index operator, as e.g. done in [18].

Each iteration maps the arrival rates $\lambda_l^a[n]$ for each link l and aggregate a to arrival rates $\lambda_{next(a,l)}^a[n + d_l]$ for the successor link $next(a, l)$ of the respective aggregate in

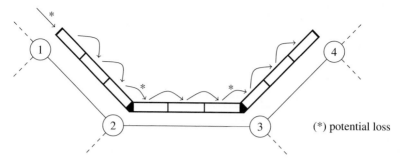

Fig. 2. Network with 4 nodes and a traffic aggregate that crosses 3 links, each with a link delay of $d_l = 3$

the future. Thus, the traffic faces a delay of d_l iteration intervals until then. We take advantage of this fact for tracking the end-to-end delay, by using the following data structure. In addition to each $\lambda_l^a[n]$ we define a $\delta_l^a[n]$ and add the following operation to each iteration step to capture the delay up to this point.

$$\delta_{next(a,l)}^a[n+d_l] = \delta_l^a[n] + d_l. \tag{6}$$

The end-to-end delay is then given by

$$\delta^a[n] = \delta_{sink(a)}^a[n]. \tag{7}$$

For now, this does not yield insightful results because all end-to-end delays are constant. But we will use this data structure as a basis for keeping track of queueing delay in the next section.

4.3 Modeling Queuing Delay

So far, we have considered networks without queues, meaning that traffic exceeding the link bandwidth, i.e. $\lambda_l[n] > c_l$, is dropped. When using queues, the excess traffic that fits in the queue is buffered and only the carryover is dropped.

During one iteration interval, $(\lambda_l \cdot \Delta)$ new data arrives while $(c_l \cdot \Delta)$ data can be transmitted by link l. As the queue size can be at most q_l^{max}, the queue size for the next iteration step $(n+1)$ can be calculated by:

$$q_l[n+1] = \min\left\{q_l^{max}, \quad \max\left\{q_l[n] - c_l \cdot \Delta + \lambda_l[n] \cdot \Delta, \quad 0\right\}\right\}. \tag{8}$$

At most $(q_l^{max} + c_l \cdot \Delta - q_l[n])$ can be buffered in the queue, while the exceeding traffic is dropped. Therefore, we can calculate the loss probability by:

$$p_l[n] = \max\left\{1 - \frac{q_l^{max} + c_l \cdot \Delta - q_l[n]}{\lambda_l[n] \cdot \Delta}, \quad 0\right\}. \tag{9}$$

The flow-level simulation works so far by calculating the arrival rate at a link l based on the current arrival rate at its predecessor for d_l iteration steps in the future. While

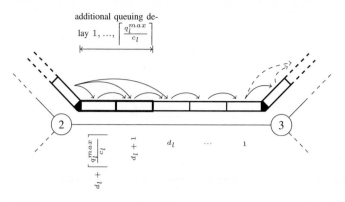

Fig. 3. Array data structure for traffic aggregates with queueing delays

the link delay d_l is constant, a queue introduces additional delay which can vary in each iteration step. We therefore substitute Equation (5) with the following equations.

Within iteration step n, $(\lambda_l[n] \cdot \Delta)$ data arrives which can be sent at the earliest in $i = \left\lfloor \frac{q_l[n]}{c_l \cdot \Delta} \right\rfloor$ iteration steps. Thus, i is the minimum queuing delay for the traffic of all aggregates arrived during interval n. The maximum queuing delay for the arriving data corresponds to $k = \left\lceil \frac{q_l[n+1]}{c_l \cdot \Delta} \right\rceil$. We now calculate how the arriving data is distributed between the delays from i to k.

In order to compute the amount of data which can be dequeued in i iteration steps, we need to consider that it is possible that there is already data stored for this iteration step, which does not use the full link capacity. Therefore, we calculate the remaining capacity for this iteration step by $((i+1) \cdot c_l \cdot \Delta - q_l[n])$. This free capacity is shared proportionally by all aggregates competing for the link. As a consequence, the amount of data that will arrive at interval $(n + d_l + i)$ at the next link, $\lambda^a_{next(a,l)}[n + d_l + i] \cdot \Delta$, is increased. We flag $\lambda^a_{next(a,l)}[n + d_l + i]$ before the increase with a '−', and after the increase with a '+'.

$$\lambda^a_{next(a,l)}[n + d_l + i]^+ = \lambda^a_{next(a,l)}[n + d_l + i]^-$$
$$+ \min \left\{ \frac{(i+1) \cdot c_l \cdot \Delta - q_l[n]}{\Delta} \cdot \frac{\lambda^a_l[n]}{\lambda_l[n]}, \quad \lambda^a_l[n] \cdot (1 - p_l[n]) \right\}. \tag{10}$$

The arrival rates $\lambda^a_{next(a,l)}[n + d_l + j]$ with queueing delay j, $i < j < k$, make proportional use of the full link capacity c_l.

$$\lambda^a_{next(a,l)}[n + d_l + j] = c_l \cdot \frac{\lambda^a_l[n]}{\lambda_l[n]}, \qquad \forall j, \ i < j < k. \tag{11}$$

If $k > i$, the arrival rate $\lambda^a_{next(a,l)}[n + d_l + k]$ with queueing delay k corresponds to the proportional fraction of the latest buffered data in the queue $(q_l[n+1] - (k-1) \cdot c_l \cdot \Delta)$.

$$\lambda^a_{next(a,l)}[n + d_l + k] = \frac{q_l[n+1] - (k-1) \cdot c_l \cdot \Delta}{\Delta} \cdot \frac{\lambda^a_l[n]}{\lambda_l[n]}. \tag{12}$$

We observe that in the case of queues we calculate the arrival rates for several future iteration steps at once if $k > i$, which is illustrated in Figure 3. The data structure for $\lambda_l^a[n]$ in a concrete implementation can essentially stay as in Equation (5) with the exception that the size of the respective modulo array must now be $d_l + \left\lceil \frac{q_l^{max}}{c_l} \right\rceil$.

To keep track of the end-to-end delay of aggregates we use the same data structure as in Equation (6) but add the additional queuing delay. For j with $i < j \leq k$ we can simply write

$$\delta^a_{next(a,l)}[n + d_l + j] = \delta_l^a[n] + d_l + j, \qquad \forall j, \ i < j \leq k. \tag{13}$$

However, for the minimum delay i, it is possible that there is already data stored for arrival at interval $(n + d_l + i)$. Therefore, we have to calculate the weighted average of the delay of the data rate that was stored earlier $(\lambda^a_{next(a,l)}[n + d_l + i]^-)$ and the delay of the data arrived in the last interval $(\lambda^a_{next(a,l)}[n + d_l + i]^+ - \lambda^a_{next(a,l)}[n + d_l + i]^-)$. Again, we flag $\delta^a_{next(a,l)}$ before modifying with a '$-$' and after modifying with a '$+$'.

$$\begin{aligned}
\delta^a_{next(a,l)}[n + d_l + i]^+ = {} & \delta^a_{next(a,l)}[n + d_l + i]^- \cdot \left(\frac{\lambda^a_{next(a,l)}[n + d_l + i]^-}{\lambda^a_{next(a,l)}[n + d_l + i]^+} \right) \\
& + (\delta_l^a[n] + d_l + i) \cdot \left(1 - \frac{\lambda^a_{next(a,l)}[n + d_l + i]^-}{\lambda^a_{next(a,l)}[n + d_l + i]^+} \right).
\end{aligned} \tag{14}$$

5   Analysis of Simulation Accuracy

In this section we demonstrate the application of the proposed flow-level simulation model using a small example network in order to provide the reader with an insight about the character of results which may be expected.

The network in Figure 4(a) consists of six nodes and three traffic aggregates. The traffic is *Poisson*, meaning that the packet interarrival times are exponentially distributed. In addition, in order to generate a dynamic network-wide traffic matrix, the mean rate of each traffic source varies in an oscillatory manner with different frequencies.

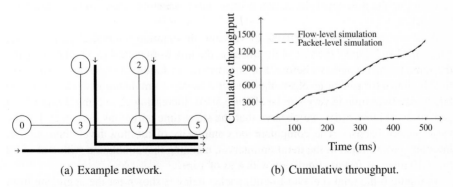

(a) Example network.   (b) Cumulative throughput.

Fig. 4. Simple example network with cumulative throughput for the aggregate from 0 to 5

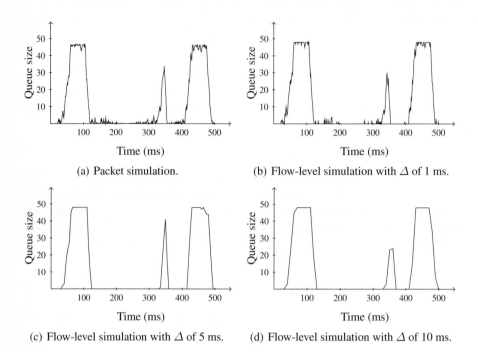

(a) Packet simulation.   (b) Flow-level simulation with Δ of 1 ms.

(c) Flow-level simulation with Δ of 5 ms.   (d) Flow-level simulation with Δ of 10 ms.

Fig. 5. Queue sizes for packet-level simulation and flow-level simulation with different iteration intervals

The presented simulation setting results in temporary overload at both the link from 3 to 4 and from 4 to 5, allowing us to observe the throughput of aggregates which experience queueing and loss at more than one link. We performed the packet-level simulations in the *ns-2* network simulator [19] and the the flow-level simulations in our own simulator.

We concentrate on the aggregate from 0 to 5. In Figure 4(b) the cumulative through-put (the total amount of data arrived until iteration interval n) is shown for this aggregate for both packet-level simulation and flow-level simulation (with Δ of 1 ms), demonstrating that the flow-level simulation results closely resemble those from packet-level simulation.

Next we demonstrate the effects of varying the iteration interval Δ in flow-level simulation. In Figure 5 the size of the queue at the link from node 4 to 5 is shown, using the same traffic scenario as before. For comparison, in Figure 5(a) the queue size over time is shown for packet-level simulation. For flow-level simulation with $\Delta = 1$ ms the queue size over time is very similar (Figure 5(b)). Increasing Δ to 5 ms (Figure 5(c)) and 10 ms (Figure 5(d)), we observe that the short-time variations disappear, but the basic pattern persists. The computational complexity of the flow-level simulation is inversely proprotional to the iteration interval, meaning that simulation with a Δ of 10 ms is 10 times as fast as simulation with a Δ of 1 ms.

In Figure 6 the total (i.e. end-to-end) packet delay is shown for the aggregate from 0 to 5 for packet simulation and flow-level simulation. In packet-level simulation, the

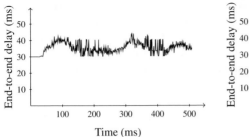

(a) End-to-end delay in packet simulation.

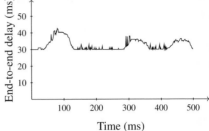

(b) End-to-end delay in flow-level simulation with Δ of 1 ms.

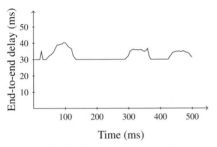

(c) End-to-end delay in flow-level simulation with Δ of 5 ms.

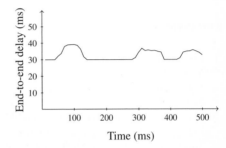

(d) End-to-end delay in flow-level simulation with Δ of 10 ms.

Fig. 6. End-to-end delay statistics for the aggregate from 0 to 5

end-to-end delay was determined by calculating the average delay for all packets that arrived at their destination within an interval of 1 ms. In flow-level simulation we apply Equations (13) and (14). We observe that the calculated end-to-end delays for packet and flow-level simulation are quite similar when using a Δ of 1 ms, with the exception that the short-time variations are higher in packet-level simulation (Figures 6(a)–6(b)). In Figures 6(c)–6(d) we notice that these variations disappear completely, but the simulation still captures the basic end-to-end delay characteristics quite well.

6   Conclusion

In this paper we have provided a comprehensive formulation of discrete-time flow-level simulation at different levels of detail. We have first introduced a basic algorithm capable of calculating the loss probabilities and throughput rates of end-to-end traffic aggregates in a setting with static traffic demands, and have demonstrated this algorithm to be a conceptually very simple and efficient alternative to complex analytical throughput calculation.

We have then extended the algorithm to handle time-dependent behavior of networks by introducing link and queueing delay and a technique for measurement of end-to-end delay. We have compared results from flow-level simulation examples to results from packet-level simulation, and have shown that the they are very similar, especially if a sufficiently small iteration interval Δ is chosen for flow-level simulation. A smaller interval leads to more accurate results while a larger interval speeds up the simulation. However, we have demonstrated that even for larger intervals the basic network behavior is captured. Future work may include the analysis of realistic ISP networks, both in terms of network size and traffic patterns observed in today's Internet.

Acknowledgements

This work has been performed partially in the framework of the Austrian Kplus Competence Center Program.

References

1. Law, A.M., Kelton, W.D.: Simulation Modeling and Analysis, 3rd edn. McGraw-Hill, New York (2000)
2. Mitra, D., Anick, D., Sondhi, M.M.: Stochastic Theory of a Data Handling System with Multiple Sources. Bell. Systems Technical Journal 61, 1871–1894 (1982)
3. Mitra, D.: Stochastic Theory of a Fluid Model of Producers and Consumers Coupled by a Buffer. Advances in Applied Probability 20, 646–676 (1988)
4. Tucker, R.C.F.: Accurate Method for Analysis of a Packet-Speech Multiplexer with Limited Delay. IEEE Transactions on Communications 36, 479–483 (1988)
5. Pitts, J.M.: Cell-Rate Modelling for Accelerated Simulation of ATM at the Burst Level, in Communications, IEE Proceedings (1995)
6. Kesidis, G., Singh, A., Cheung, D., Kwok, W.: Feasibility of Fluid Event-Driven Simulation for ATM Networks. In: IEEE Globecom, London, UK (November 1996)
7. Bahr, M., Butenweg, S.: On Rate-Based Simulation of Communication Networks, in Design, Analysis, and Simulation of Distributed Systems (DASD), (April 2003)
8. Liu, B., Figueiredo, D.R., Guo, Y., Kurose, J.F., Towsley, D.F.: A Study of Networks Simulation Efficiency: Fluid Simulation vs. Packet-level Simulation, in IEEE Infocom, pp. 1244–1253 (2001)
9. Liu, B., Guo, Y., Kurose, J., Towsley, D., Gong, W.: Fluid Simulation of Large Scale Networks: Issues and Tradeoffs, Tech. Rep. UM-CS-1999-038 (1999)
10. Yan, A., Gong, W.-B.: Time-Driven Fluid Simulation for High-Speed Networks. IEEE Transactions on Information Theory 45, 1588–1599 (1999)
11. Liu, Y., Presti, F.L., Misra, V., Towsley, D., Gu, Y.: Fluid Models and Solutions for Large-Scale IP Networks, in ACM SIGMETRICS, pp. 91–101 (2003)
12. Gojmerac, I., Ziegler, T., Ricciato, F., Reichl, P.: Adaptive Multipath Routing for Dynamic Traffic Engineering. In: Proc. IEEE Globecom, San Francisco, USA, 2003, pp. 3058–3062 (2003)
13. Gojmerac, I., Ziegler, T., Reichl, P.: Adaptive Multipath Routing Based on Local Distribution of Link Load Information. In: International Workshop on Quality of future Internet Services (QofIS), pp. 122–131 (2003)
14. Gojmerac, I., Jansen, L., Ziegler, T., Reichl, P.: Feasibility Aspects of AMP Performance Evaluation in a Fluid Simulation Environment. In: MMBnet Workshop, Hamburg, Germany, 2005.
15. Gojmerac, I., Jansen, L., Ziegler, T., Reichl, P.: A Simulation Study of Microscopic AMP Behavior. In: Polish-German Teletraffic Symposium (PGTS) (2006)

16. Misra, V., Gong, W.-B., Towsley, D.F.: Fluid-Based Analysis of a Network of AQM Routers Supporting TCP Flows with an Application to RED. In: ACM SIGCOMM, pp. 151–160 (2000)
17. Marsan, M.A., Garetto, M., Giaccone, P., Leonardi, E., Schiattarella, E., Tarello, A.: Using Partial Differential Equations to Model TCP Mice and Elephants in Large IP Networks, In: IEEE Infocom (2004)
18. Fluid Flow Model in NS, Source code (2003), available at `ftp://gaia.cs.umass.edu/pub/ffm_in_ns.tar.gz`
19. The Network Simulator (Version 2), Source code and documentation (1995), available at `http://www.isi.edu/nsnam/ns/`

On the Application of the Asymmetric Convolution Algorithm in Modeling of Full-Availability Group with Bandwidth Reservation

Mariusz Głąbowski, Adam Kaliszan, and Maciej Stasiak

Chair of Communication and Computer Networks, Poznań University of Technology
ul. Piotrowo 3A, 60-965 Poznań
mariusz.glabowski@et.put.poznan.pl

Abstract. The paper proposes a new method to calculate the occupancy distribution and the blocking probability in the full-availability group with multi-rate traffic and bandwidth reservation. In the presented method, the occupancy distribution in a group is calculated with the application of the proposed asymmetric convolution operation that takes into consideration the influence of the reservation mechanism upon the probability of passing between the states. The results of the analytical calculations of the blocking probability in the system with bandwidth reservation obtained on the basis of the proposed analytical method are compared to the data obtained on the basis of other known analytical methods and to the results of the simulation.

1 Introduction

The full-availability group (FAG) is a single link model with unlimited access to resources that services multi-rate traffic streams. Recently, the FAG model is used more and more for modeling wireless systems, and UTMS systems in particular [1, 2]. The FAG belongs to the class of systems known as multi-rate systems [3, 4, 5, 6, 7, 8]. In multi-rate models, the system services call demands having an integer number of the so-called BBU (Basic Bandwidth Unit). When constructing multirate models for broadband systems, it is assumed that BBU is the greatest common divisor of the equivalent bandwidths of all call streams offered to the system [4, 9, 10].

The traffic characteristics in the FAG may be obtained on the basis of the multi-dimensional Markov process occurring in the system [5, 8, 11, 12, 13, 14, 15, 16, 17]. However, the calculations based on the multi-dimensional Markov process are not very convenient because of an excessive number of states in which the process can be found. Therefore, two other groups of methods for the occupancy distribution calculation in the multi-rate systems are known in the literature. The first one consists in approximate transformation of the multi-dimensional process into the one-dimensional Markov chain [18, 19, 20, 21, 22, 23, 24]. Such a reduction is the base for the determination of the occupancy distribution in the group by means of the recurrent Fortet-Grandjean formula [22] which is generally known as the Kaufman-Roberts recursion [23,24]. In the other group the methods based on the convolution algorithm [5, 8] are considered, first introduced in [25]. The advantage of the first approach is simplicity and possibility of

L. Mason, T. Drwiega, and J. Yan (Eds.): ITC 2007, LNCS 4516, pp. 878–889, 2007.

state–dependent systems calculation. However, this technique can only be applied when the offered traffic is considered as Poisson stream. The convolution algorithm is more complicated but it allows us to calculate the *state–independent systems* with arbitrary traffic streams (e.g. BPP - Bernoulli-Poisson-Pascal). Unfortunately, only the FAG without bandwidth reservation can be treated as the state-independent system. The introduction of the reservation mechanism causes the full-availability group to become a system dependable on the state, i.e. the system in which the admission of a new call becomes is conditioned by not only the sufficient number of free BBU's to service the call of a given class, but also by the current state of the occupancy distribution of the system[1].

The reservation mechanism has been considered in many works [4, 26, 27, 28, 29, 30, 31, 32, 33, 34, 35, 36, 37]. In [26], for a full-availability group with bandwidth reservation (FAG-BR), a modified Kaufman-Roberts algorithm, called the Roberts algorithm, is proposed. Due to the limitations the Kaufman-Roberts imposes as far as the type of the offered traffic is concerned (only Poisson traffic), [38, 39, 40] present the attempt at the application of the convolution algorithm used for modeling the FAG-BR. The first solution, proposed in [38], is computationally too complex for practical calculations, while the methods presented in [39] and [40] estimate the lower and upper bound of blocking probability in FAG-BR. In the present paper, a new simple analytical method based on the convolution algorithm that makes it possible to determine the blocking probability in the full-availability group with bandwidth reservation with any distribution of the offered traffic is proposed.

Further on in the article, the content of this paper is arranged in the following way. Section 2 describes the basic version of the Iversen algorithm for a group without reservation. Section 3 discusses the proposed modification of the convolution algorithm for a system with bandwidth reservation. The next section presents the results of the analytical calculations that are then compared to the simulation results. Section 5 sums up the most important results of the study.

2   The Full-Availability Group Without Reservation – FAG

Let us consider now the full-availability group [4] with the capacity of V BBUs servicing m independent traffic classes of the M set. Let us assume that the class i demands t_i BBUs to set up a connection and that the order of indexes indicates the growing number of BBUs demanded by particular classes. With thus adopted order, the so-called "oldest" m class demands the maximum number of BBUs. The intensities of the streams of the offered call to particular classes are respectively $\lambda_1, \lambda_2, \ldots, \lambda_m$, while the holding times of particular classes have an exponential distribution with the parameters $\mu_1, \mu_2, \ldots, \mu_m$.

The occupancy distribution and the blocking probability for calls of particular classes in the FAG with integrated traffic may be determined on the basis of the convolution algorithm [25, 41]. The Iversen convolution algorithm calculates the occupancy

---

[1] In the FAG with reservation, a threshold Q is introduced which determines such a limiting state of a group, in which servicing all calls is still possible. All states higher than Q belong to the so-called *reservation space* R, in which calls of given classes will be blocked.

distribution of the system through calls of all traffic classes (the so-called aggregated occupancy distribution) on the basis of occupancy distributions of individual classes.

The calculation of the occupancy distribution of each single traffic class, with the initial assumption that only calls of this class are offered to the group, forms then the basis of the Iversen algorithm. Therefore, for each $i \in M$ class we obtain:

$$[p]_V^{\{i\}} = \left\{ [p_0]_V^{\{i\}}, [p_1]_V^{\{i\}}, \ldots, [p_V]_V^{\{i\}} \right\} , \tag{1}$$

where $[p]_V^{\{i\}}$ is the occupancy distribution for class i and $[p_n]_V^{\{i\}}$ is the probability of n BBUs being occupied by class i calls in the group with the capacity equal to V BBUs ($i = 1, 2, \ldots, m$, where m is the number of elements in the M set). Due to the fact that the occupancy distributions $[p]_V^{\{i\}}$ are determined individually for each traffic class, they may be determined on the basis of classical (single-rate) models elaborated in traffic theory.

Having the occupancy distributions $[p]_V^{\{i\}}$, we can determine the aggregated occupancy distribution $[P]_{mV}^M$ on the basis of the convolution operation performed successively for all traffic classes:

$$[P]_{mV}^M = [p]_V^{\{1\}} * [p]_V^{\{2\}} * \ldots * [p]_V^{\{m\}} , \tag{2}$$

where "$*$" designates the the convolution operation. A single element $[P_n]_{mV}^M$ of (2) is equal to:

$$[P_n]_{mV}^M = \sum_{l=0}^{n} [p_l]_V^{\{i\}} [P_{n-l}]_{(m-1)V}^{M \setminus \{i\}} , \tag{3}$$

where $[P_n]_{mV}^M$ is the probability of n BBUs being occupied by calls of all classes, $[p_l]_V^{\{i\}}$ is the probability of occupying l BBUs by class i calls, and $[P_{n-l}]_{(m-1)V}^{M \setminus \{i\}}$ is the probability of occupying $n - l$ BBUs by calls of other classes (except for class i).

It should be stressed here that the number of elements (mV) in the aggregated distribution (2) (the lower index of the expression $[P]_{mV}^M$) is greater than the capacity of the considered group which is equal to V. This phenomenon is the result of the convolution operation. Because the investigated group has V common BBUs it is necessary to truncate the distribution $[P]_{mV}$ to the distribution $[P]_V$. Consequently, $\sum_{n=0}^{V} [P_n]_{mV}^M \leqslant 1$. Therefore, in order to obtain a sum of probabilities within a range of 0 to V BBUs, calculated for the set of M aggregated classes, equal to 1, it is necessary to introduce the normalization coefficient k:

$$k = \left(\sum_{n=0}^{V} [P_n]_{mV}^M \right)^{-1} . \tag{4}$$

The probabilities of states within a range of 0 to V, determined on the basis of the distribution with mV elements, may be then presented as follows:

$$[P]_V^M = \{k[P_0]_{mV}^M, k[P_1]_{mV}^M, \ldots, k[P_V]_{mV}^M\} = k \left([p]_V^{\{1\}} * [p]_V^{\{2\}} * \ldots * [p]_V^{\{m\}} \right) . \tag{5}$$

After determining the aggregated occupancy distribution we can calculate the blocking probability E_i for class i traffic stream:

$$E_i = \sum_{n=V-t_i+1}^{V} [P_n]_V^M , \tag{6}$$

where $[P_n]_V^M$ is a single element of (5).

3   Full-Availability Group with Bandwidth Reservation – FAG-BR

3.1   Reservation Mechanism

Bandwidth reservation is one of the possible strategies of Call Admission Control function, considered in numerous works, e.g. in [3, 4, 6, 26, 35, 42]. The reservation mechanism makes access to network resources more proper for all traffic classes. In particular cases, the reservation leads to equalisation of blocking probabilities for all traffic classes serviced in the system. In the systems with reservation, the *reservation threshold* Q_i for each traffic class is designated. The parameter Q_i determines the borderline state of a system, in which servicing calls of class i is still possible. All states higher than Q_i belong to the *reservation space* $R_i = V - Q_i$, in which class i calls are blocked. According to the equalization rule [4, 26, 35], the blocking probability in the FAG with reservation will be the same for all calls if the threshold Q_i for all traffic classes is identical and equal to the difference between the group capacity and the number of BBUs required by a call with maximum demands: $Q_i = Q = V - t_{\max}$.

3.2   Asymmetric Convolution Algorithm

In the paper we have proposed the calculation method of the occupancy distribution and the blocking probability in the FAG-BR. The proposed method is based on the convolution algorithm. The original convolution operation had been modified in order to exclude states combinations which are forbidden in the FAG due to the reservation mechanism. To determine the "forbidden" combinations of states, it is necessary to include the class of the last call admittance. The proposed modified convolution operation, the so-called asymmetric convolution, assumes that a call of this class, which corresponds to the second argument of the convolution operation, is the last call accepted by the system[2]. The distribution obtained as the result of the convolution operation is then a conditional distribution that takes into account the class of the accepted call,

In order to explain the basic assumptions of the proposed calculation method let us consider FAG with the capacity $V = 6$ BBUs which is offered three traffic classes with the following bandwidth requirements: $t_1 = 1$ BBU, $t_2 = 2$ BBUs, $t_3 = 3$ BBUs. According to the equalisation rule, the reservation threshold $Q = 3$.

Figure 1 shows the allowed (by the reservation mechanism) combinations of states during the distribution aggregation, respectively: $[P]_V^{\{2,3\}}$ and $[p]_V^{\{1\}}$ (Fig.1(a)), $[P]_V^{\{1,3\}}$ and $[p]_V^{\{2\}}$ (Fig.1(b)), $[P]_V^{\{1,2\}}$ and $[p]_V^{\{3\}}$ (Fig. 1(c)). A given combination is available if an element with the number $l_{\{x,y\}}$, on the left side of each of Figs. 1(a), 1(b), 1(c), is combined with the element $l_{\{z\}}$, on the right side of Fig. ($l_{\{x,y\}}$ – is the number of the total number of busy BBUs by a class x and y calls, $l_{\{z\}}$ – is the number of busy BBUs with class z calls, n – is the number of busy BBUs in all x, y, and z classes ($n = l_{\{x,y\}} + l_{\{z\}}$)). The sense of the arrow linking the elements indicates the class whose call was the last one accepted to be serviced by the system. A lack of a link between the elements means (indicates) that a given combination is forbidden. Additionally, for

---

[2] The algorithm considerations include m number of different events in which class $1, 2, \ldots, m$ is the class whose call was the last one accepted by the system.

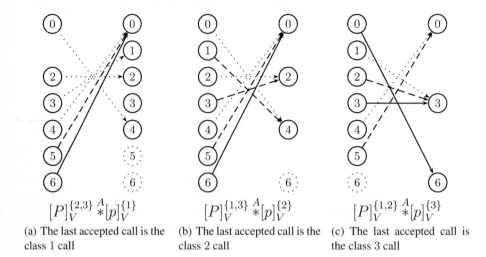

(a) The last accepted call is the class 1 call

(b) The last accepted call is the class 2 call

(c) The last accepted call is the class 3 call

Fig. 1. Conditionally allowed states in a FAG-BR ($V = 6$, $Q = 3$, $t_1 = 1$, $t_2 = 2$, $t_3 = 3$)

each of the occupancy distributions shown in the figure, those states that cannot be achieved as the result of the applied reservation mechanism are indicated by a dotted line. For better clarity, the occupancy distributions n lower than 4 BBUs are omitted in the figure. The combinations whose results is the state $n = 6$ is indicated in Fig. 1 by a continuous line, the state $n = 5$ by a dashed line, and the state $n = 4$ by a dotted line.

Let us consider more deeply Fig. 1(a) which shows the allowed combinations of states with the assumption that the last admitted call was a class 1 call that demanded $t_1 = 1$ BBU. Let $(l_{\{2,3\}}, l_{\{1\}})$ denotes such a combination in which the class 2 and 3 calls occupy together $l_{\{2,3\}}$ BBUs, while class 1 occupies $l_{\{1\}}$ BBUs. Figure 1(a) does not show, in the distribution $[P]_V^{\{2,3\}}$, the state $l_{\{2,3\}} = 1$ as it is impossible for the state to occur with the calls demanding respectively 2 and 3 BBUs.

It is noticeable that in the considered case, for the state $n = l_{\{2,3\}} + l_{\{1\}} = 6$, the only allowed combination is (6,0), which can be interpreted in the following way: in the occupancy state 6 BBU with class 2 and 3 calls, no class 1 call appeared. The combinations: (5,1), (4,2), (3,3), (2,4), (0,6) are, in turn, forbidden as their occurrences (in accordance with the assumption that the last accepted call was the class 1 call) would require the acceptance of the class 1 call in the occupancy state $n = 5$, i.e. in a state that already belongs to the reservation space.

For the state $n = l_{\{2,3\}} + l_{\{1\}} = 5$, the only possible combination is (5,0), while the combinations (4,1), (3,2), (0,5) are impossible because for all these combinations the penultimate state was the state $n = 4$, which also belongs to the reservation space.

In the state $n = l_{\{2,3\}} + l_{\{1\}} = 4$, the possible combinations of the states are: (4,0), (3,1), (2,2) and (0,4). The first one indicates that in the occupancy state $n = 4$, no class 1 call appeared. In the following three, the penultimate state was the state $n = l_{\{2,3\}} + l_{\{1\}} = 3$ in which it is possible – according to the reservation mechanism

– to accept a new class 1 call. Similar considerations can be applied to Fig. 1(b) (the last admitted call is a class 2 call), and Fig. 1(c) (the last admitted call is the class 3 call).

The above considerations indicate that the number of allowed states combinations in the reservation space depends on the class of the last accepted call. Therefore, the aggregated occupancy distribution $[W]_{mV}^{M|\{i\}}$ in the group under consideration, obtained as the result of the application of the convolution operation $[P]_{(m-1)V}^{M\backslash\{i\}}$ and $[p]_V^{\{i\}}$, is the conditional distribution in which it was assumed that the last accepted call to the system was the class i call:

$$[W]_{mV}^{M|\{i\}} = [P]_{(m-1)V}^{M\backslash\{i\}} \overset{A}{*} [p]_V^{\{i\}} \; , \tag{7}$$

where:

– $\overset{A}{*}$ is the asymmetric convolution operation whose n-th element of the convoluted distributions is defined in the following way ($n \leq V$):

$$[W_n]_{mV}^{M|\{i\}} = \sum_{l=0}^{n} \gamma_{\{i\}}^{M\backslash\{i\}}(n-l,l) \cdot [P_{n-l}]_{(m-1)V}^{M\backslash\{i\}} \cdot [p_l]_V^{\{i\}} \; . \tag{8}$$

– $[P]_{(m-1)V}^{M\backslash\{i\}}$ is not normalized occupancy distribution of all classes, except class i, defined on the basis of (2):

$$[P]_{(m-1)V}^{M\backslash i} = [p]_V^{\{1\}} * [p]_V^{\{2\}} * \ldots * [p]_V^{\{i-1\}} * [p]_V^{\{i+1\}} * \ldots * [p]_V^{\{m\}} \; . \tag{9}$$

In Equation (8) the coefficient $\gamma_{\{i\}}^{M\backslash\{i\}}(n-l,l)$ determines whether a given occupancy state n, in which l BBUs would be busy by class i calls, and $n-l$ BBUs by calls of the remaining classes ($M\backslash\{i\}$), is allowed. According to the reservation mechanism, fo the FAG-BR, we get:

$$\gamma_{\{i\}}^{M\backslash\{i\}}(n-l,l) = \begin{cases} 0 & \text{for } (n-l > Q + \max(t_j) : j \in \{M\backslash\{i\}\}) \vee (n > Q + t_i) \; , \\ 1 & \text{for the other cases} \; . \end{cases} \tag{10}$$

Unconditional, normalized occupancy distribution $[P]_{V,Q}^{M}$ for all m classes, serviced in the FAG-BR, with the capacity V and the reservation threshold Q, can be calculated (determined) on the basis of the following formula:

$$[P]_{V,Q}^{M} = k \sum_{i=0}^{m} P(i)[W]_{(mV)}^{M|\{i\}} \; , \tag{11}$$

where $P(i)$ determines the probability of an event in which the last accepted call for servicing was a class i call. The assumption taken in the article is that the probability of such an event is proportional to the class i traffic, expressed in BBUs:

$$P(i) = a_i t_i \bigg/ \sum_{j=0}^{m} a_j t_j \; . \tag{12}$$

To determine the blocking probability for i class calls in the FAG-BR it is necessary to sum up all the states of the unconditional normalized distribution with reservation mechanism in which admission of a class i call (Equation (11)) is not possible.

$$B_i(t) = \sum_{n=Q+1}^{V} [P_n]_{V,Q}^{M} \; . \tag{13}$$

Concluding, the asymmetric convolution algorithm consists of three steps presented in Fig 2. Each column, separated by a dotted frame, presents consecutive steps to be taken within the algorithm.

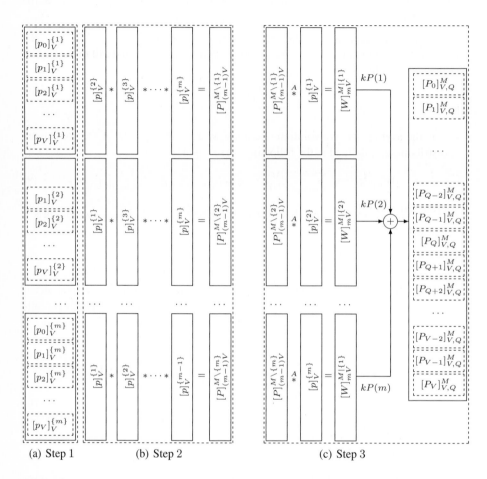

(a) Step 1   (b) Step 2   (c) Step 3

Fig. 2. The illustration of individual steps to be taken in the proposed asymmetric convolution algorithm for the calculation of the occupancy distribution and the blocking probability in the FAG-BR

1. Calculation of the occupancy distribution $[p]_V^{\{i\}}$ for each of the class i, with the assumption that only this class is serviced by the system (Fig. 2(a)).
2. Calculation of the not normalized occupancy distribution $[P]_{(m-1)V}^{M\setminus\{i\}}$ of all classes except class i (Equation (9), Fig. 2(b)) for each of the classes.
3. Calculation of the occupancy distribution on the basis of (11) (Fig. 2(c)) and the blocking probability on the basis of (13).

4 Numerical Results

In order to confirm the theoretical assumptions used in the proposed asymmetric convolution algorithm, the results of analytical calculations carried out for the FAG-BR were compared with the simulation results. The results, presented in the paper, have been obtained in the FAG with a capacity of $V = 30$ BBUs. The groups were offerred different multi-rate Bernoulli, Poisson and Pascal traffic streams. The full specification of traffic streams are given together with numerical results.

The analytical and simulation results are presented in Tab. 1 and in Figs. 3, 4 and 5 in relation to the value of traffic $a = \sum_{i=1}^{M} \frac{a_i t_i}{V}$ offered to a single BBU. The results of the simulation are given with 95% confidence intervals that were calculated after the t–Student distribution for five series with 100000 calls of this traffic class that generates the lowest number of calls. The calculation results in Tab. 1 (for infinite source population) are presented in three columns, where the first two columns contain the results obtained according to the method proposed in [26] (column "Roberts") and in [43] (column "Stasiak & Glabowski"), respectively. The third column contains the results obtained according to the proposed asymmetric convolution algorithm. The proposed method may be also applied to systems with finite source population. The results obtained in the FAG with bandwidth reservation and finite source population are presented in Figs. 4 and 5, for Bernoulli and Pascal traffic streams, respectively.

The results presented indicate that the proposed method assures high accuracy which is comparable to this obtained with the use of the methods proposed in [26] and [43]. Numerous simulation experiments carried out so far indicate that similar accuracy can be obtained for greater link capacity and greater number of traffic classes.

Table 1. Blocking probability in the FAG; infinite source population; Poisson traffic streams, $V = 30$, $M = 2$, $t_1 = 1$, $t_2 = 4$, $Q = 26$, $a_1 t_1 : a_2 t_2 = 1 : 2$

a	Calculation			Simulation
	Roberts [26]	Stasiak & Glabowski [43]	Proposed algorithm	
0.5	0.0315	0.0316	0.0287	0.0329±0.0022
0.6	0.0638	0.0641	0.0594	0.0668±0.0026
0.7	0.1055	0.1060	0.1004	0.1083±0.0043
0.8	0.1523	0.1532	0.1481	0.1580±0.0051
0.9	0.2005	0.2018	0.1989	0.2064±0.0061
1.0	0.2476	0.2493	0.2501	0.2557±0.0082
1.1	0.2922	0.2943	0.2999	0.2991±0.0038
1.2	0.3336	0.3360	0.3473	0.3410±0.0077
1.3	0.3716	0.3744	0.3918	0.3805±0.0053
1.4	0.4063	0.4094	0.4332	0.4145±0.0065
1.5	0.4380	0.4414	0.4716	0.4471±0.0071

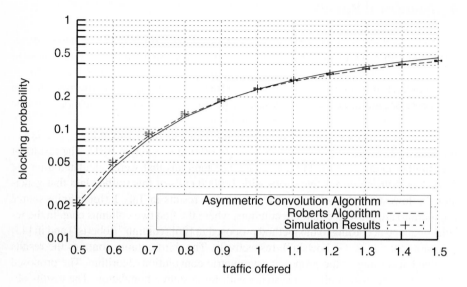

Fig. 3. Blocking probability in the FAG-BR, infinite source population, Poisson traffic streams, $V = 30, Q = 26, m = 3, a_1 t_1 : a_2 t_2 : a_3 t_3 = 1 : 1 : 1, t_1 = 1, t_2 = 2, t_3 = 4$

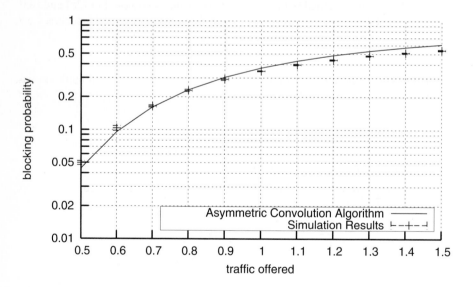

Fig. 4. Blocking probability in the FAG-BR, finite source population, Pascal traffic streams, $V = 30, Q = 26, m = 3, a_1 t_1 : a_2 t_2 : a_3 t_3 = 1 : 1 : 1, t_1 = 1, t_2 = 2, t_3 = 4, S_1 = 30, S_2 = 15, S_3 = 9$

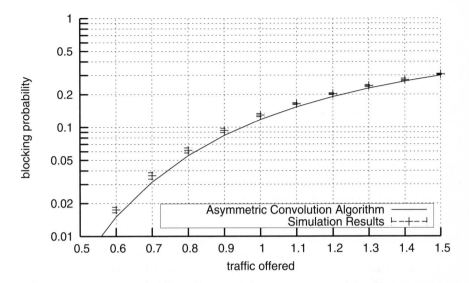

Fig. 5. Blocking probability in the FAG-BR, infinite source population, Bernoulli traffic streams, $V = 30$, $Q = 26$, $m = 3$, $a_1 t_1 : a_2 t_2 : a_3 t_3 = 1 : 1 : 1$, $t_1 = 1$, $t_2 = 2$, $t_3 = 4$, $S_1 = 30$, $S_2 = 15$, $S_3 = 9$

5   Conclusions

The article presents a new approximate method for blocking probability calculations in the FAG-BR. The proposed method is based on the Iversen convolution algorithm. The basic element of the calculation algorithm is the introduction of the asymmetric convolution operation to determine occupancy states forbidden by reservation mechanism applied. The method conceived and presented in the article is characterized by lower complexity than the convolution algorithm for the FAG-BR devised earlier [38]. It needs to be stressed that the proposed solution can be used for dimensioning systems that are offered traffic with any distribution. The Roberts algorithm, which is also an approximate algorithm, dimensions only systems with traffic characterized by Poison traffic stream.

References

1. Staehle, D., Mäder, A.: An analytic approximation of the uplink capacity in a UMTS network with heterogeneous traffic. In: 18th International Teletraffic Congress (ITC18), Berlin, pp. 81–91 (2003)
2. Głąbowski, M., Stasiak, M., Zwierzykowski, P.: Uplink blocking probability calculation for cellular systems with WCDMA radio interface, finite source population and differently loaded neighbouring cells. In: Proceedings of Asia-Pacific Conference on Communications, Perth, Australia, pp. 138–142 (2005)
3. Akimuru, H., Kawashima, K.: Teletraffic: Theory and Application. Springer Verlag, Berlin-Heidelberg-New York (1999)

4. Roberts, J., Mocci, V., Virtamo, I. (eds.): Broadband Network Teletraffic, Final Report of Action COST 242. Commission of the European Communities. Springer, Berlin Heidelberg (1996)
5. Ross, K.: Multiservice Loss Models for Broadband Telecommunication Network, London. Springer Verlag, Heidelberg (1995)
6. Bziuk, W.: Approximate state probabilities in large shared multirate loss systems with an application to trunk reservation. In: Proceedings of 2nd Polish-German Teletraffic Symposium (9th Polish Teletraffic Symposium), Gdańsk, Poland, pp. 145–152 (2002)
7. Moscholios, I., Logothetis, M., Kokkinakis, G.: Connection-dependent threshold model: a generalization of the Erlang multiple rate loss model. Journal of Performance Evaluation 48, 177–200 (2002)
8. Iversen, V. (ed.): Teletraffic Engineering Handbook. ITU-D, Study Group 2, Question 16/2, Geneva (2003)
9. Kelly, F.: Loss networks. The. Annals of Applied Probability 1(3), 319–378 (1991)
10. Roberts, J. (ed.): Performance Evaluation and Design of Multiservice Networks, Final Report COST 224. Commission of the European Communities, Brussels, Holland (1992)
11. Kogan, Y., Shenfild, M.: Asymptotic solution of generalized multiclass Engset model. In: Labetoulle, J., Roberts, J. (eds.) Proceedings of 14th International Teletraffic Congress, Antibes Juan-les-Pins, France, vol. 1b, pp. 1239–1249. Elsevier, North-Holland, Amsterdam (1994)
12. Choudhury, G., Leung, K., Whitt, W.: An inversion algorithm to compute blocking probabilities in loss networks with state-dependent rates. IEEE/ACM Trans. on Networking 3, 585–601 (1995)
13. Berezner, S., Krzesinski, A.: An efficient stable recursion to compute multiservice blocking probabilities. Journal of Performance Evaluation 43(2–3), 151–164 (2001)
14. Nilson, A., Perry, M.: Multi-rate blocking probabilities: numerically stable computation. In: Ramaswami, V., Wirth, P. (eds.) Proceedings of 15th International Teletraffic Congress, Washington D.C., USA, pp. 1359–1368. Elsevier, North-Holland, Amsterdam (1997)
15. Stamatelos, G., Hayes, J.: Admission-control technics with application to broadband networks. Computer Communication 17(9), 663–673 (1994)
16. Aein, J.M.: A multi-user-class, blocked-calls-cleared, demand access model. IEEE Transactions on Communications COM-26(3), 378–385 (1978)
17. Gimpelson, L.: Analysis of mixtures of wide and narrow-band traffic. IEEE Trans. Comm. Technology 13(3), 258–266 (1953)
18. Beshai, M., Manfield, D.: Multichannel services performance of switching networks. In: Proceedings of 12th International Teletraffic Congress, Torino, Italy, pp. 857–864. Elsevier, Amsterdam (1988)
19. Stasiak, M.: Blocking probability in a limited-availability group carrying mixture of different multichannel traffic streams. Annales des Télécommunications 48(1-2), 71–76 (1993)
20. Stasiak, M.: An approximate model of a switching network carrying mixture of different multichannel traffic streams. IEEE Transactions on Communications 41(6), 836–840 (1993)
21. Delbrouck, L.: On the steady-state distribution in a service facility carrying mixtures of traffic with different peakedness factors and capacity requirements. IEEE Transactions on Communications 31(11), 1209–1211 (1983)
22. Fortet, R., Grandjean, C.: Congestion in a loss system when some calls want several devices simultaneously. Electrical Communication 39(4), 513–526 (1964)
23. Kaufman, J.: Blocking in a shared resource environment. IEEE Transactions on Communications 29(10), 1474–1481 (1981)
24. Roberts, J.: A service system with heterogeneous user requirements — application to multiservice telecommunications systems. In: Pujolle, G. (ed.) Proceedings of Performance of Data Communications Systems and their Applications, pp. 423–431. North Holland, Amsterdam (1981)

25. Iversen, V.: The exact evaluation of multi-service loss systems with access control. In: Seventh Nordic Teletraffic Seminar (NTS-7), Lund,Sweden, pp. 56–61 (1987)

26. Roberts, J.: Teletraffic models for the Telcom 1 integrated services network. In: Proceedings of 10th International Teletraffic Congress, Montreal, Canada, 1.1.2 (1983)

27. Lindberger, K.: Simple approximations of overflow system quantities for additional demands in the optimisation. In: Proceedings of 10th International Teletraffic Congress, Montreal, Canada, paper 5.3.3 (1983)

28. Whitt, W.: Blocking when service is required from several facilities simultaneously. Bell. System Technical Journal 64(8), 1807–1856 (1985)

29. Kawashima, K.: Trunk reservation models in telecommunication systems. Volume Teletraffic Analysis and Computer Performance Evaluation of Studies in Telecommunication. North Holland, Amsterdam (1986)

30. Lindberger, K.: Blocking for multislot heterogeneous traffic streams offered to a trunk group with reservation. In: Proceeding of the 5th ITC Seminar, Lake Como (1987)

31. Takagi, K., Sakita, Y.: Analysis of loss probability equalised by trunk reservation for mixtures of several bandwidth traffic (5.1.A.1). In: Proceedings of 12th International Teletraffic Congress, Torino, Italy, Elsevier, Amsterdam 5.1.A.1 (1988)

32. Gersht, A., Lee, K.: Virtual-circuit load control in fast packet-switched broadband networks. In: Proceedings of IEEE Global Telecommunications Conference (GLOBECOM 1989), Dallas, TX, USA, pp. 214–220 (1989)

33. Korner, U., Lubacz, J., Pióro, M.: Traffic engineering problems in multiservice circuit switched networks. In: Proceedings of International Teletraffic Congress Specialist Seminar - ITC–12, Adelaide (1989)

34. Roberts, J.: Traffic control in the B-ISDN. Computer Networks and ISDN Systems 25(10), 1055–1064 (1993)

35. Tran-Gia, P., Hubner, F.: An analysis of trunk reservation and grade of service balancing mechanisms in multiservice broadband networks. In: IFIP Workshop TC6. Volume Modelling and Performance Evaluation of ATM Technology, La Martinique, paper 2.1 (1993)

36. Wajda, K.: Estimation of call blocking probabilities in telecommunication networks. Technical Report 93007, Kyoto University, Univeristy of Mining and Metallurgy, Kyoto, Kraków (1993)

37. Ritter, M., Tran-Gia, P.: Multi-rate models for dimesioning and performance evaluation of ATM networks. Technical report, Institue of Computer Science, University of Würzburg (1994)

38. Głąbowski, M., Kaliszan, A., Stasiak, M.: Blocking probability calculation in a fag with bandwidth reservation. In: Proceedings of the Advanced International Conference on Telecommunications, Guadeloupe (2006)

39. Głąbowski, M., Kaliszan, A., Stasiak, M.: A new convolution algorithm of blocking probability calculation in full-availability group with bandwidth reservation. In: Proceedings of the Fifth International Symposium on Communication Systems, Networks and Digital Signal Processing, Patras (2006)

40. Głąbowski, M., Kaliszan, A., Stasiak, M.: Asymmetric convolution algorithm for full-availability group with bandwidth reservation. In: Proceedings of the Asia-Pacific Conference on Communications, Busan, Korea (2006)

41. Iversen, V.: Teletraffic Engineering Handbook. ITU-D SG 2/16 and ITC Draft (2001)

42. Stasiak, M., Głąbowski, M., Zwierzykowski, P.: Equalisation of blocking probability in switching systems with limited availability. In: Kouvatsos, D. (ed.) Performance Analysis of ATM Networks, pp. 358–376. Kluwer Academic Publishers, Dordrecht (2000)

43. Stasiak, M., Głąbowski, M.: A simple approximation of the link model with reservation by a one-dimensional Markov chain. Journal of Performance Evaluation 41(2–3), 195–208 (2000)

Analysis and Provisioning of a Circuit-Switched Link with Variable-Demand Customers

Wenhong Tian and Harry Perros

Department of Computer Science, North Carolina State University, NC 27606, USA

Abstract. We consider a single circuit-switched communication link, depicted by a Erlang multi-class loss queue, where a customer may vary its required bandwidth during its service. We obtain approximately the steady-state blocking probability of each class of customer. Comparisons with simulation results show that the approximation solution has a good accuracy. For the proposed model, we also provide an efficient capacity provisioning algorithm.

1 Introduction

In circuit-switched communication systems and connection-oriented packet switched networks, a connection is typically allocated a fixed bandwidth which does not vary over the life of the connection. However, in today's dynamically changing communication networks, the bandwidth allocated to a connection may have to vary in order to accommodate load fluctuations. In this paper, we consider the case where the customer's bandwidth requirements change during its service time. This case has been motivated by the Link Capacity Adjustment Scheme (LCAS) in the Data over SONET (Dos) architecture.

Traditional SONET/SDH was optimized to carry voice traffic. It was also defined to carry ATM traffic and IP packets (PoS). Changes in the capacity allocated to a connection are done manually. Recently, a novel architecture has been proposed, referred to as data over SONET/SDH (DoS) which provides a mechanism for the efficient transport of integrated data services.

It utilizes three schemes, namely, the Generic Frame Procedure (GFP), Virtual Concatenation (VCAT), and Link Capacity Adjustment Scheme (LCAS) [12]. GFP is a simple adaptation scheme that extends the ability of SONET/SDH to carrying different types of traffic. Specifically, it permits the transport of frame-oriented traffic, such as Ethernet and IP over PPP. It also permits continuous-bit-rate block-coded data from Storage Area Networks (SAN) transported by networks, such as Fiber Channel, Fiber Connection (FICON), and Enterprise System Connect (ECON).

Virtual concatenation maps an incoming traffic stream into a number of individual subrate payloads. The subrate payloads are switched through the SONET/SDH network independently of each other (see for example, Perros [7]).

Virtual concatenation is only required to be implemented at the originating node where the incoming traffic is demultiplexed into subrate payloads and at the terminating node, where the payloads are multiplexed back to the original stream. The individual payloads might not necessarily be contiguous within the same OC-N payload. Finally,

L. Mason, T. Drwiega, and J. Yan (Eds.): ITC 2007, LNCS 4516, pp. 890–900, 2007.

the number of subrate payloads allocated to an application is typically determined in advance. However, the transmission rate of the application may vary over time. In view of this, it can be useful to dynamically vary the number of subrate payloads allocated to an application. This can be done using the link capacity adjustment scheme (LCAS). In LCAS, signaling messages are exchanged between the originating and terminating SONET/SDH node to determine and adjust the number of required subrate payloads. LCAS makes sure that the adjustment process is done without losing any data.

The calculation of call blocking probabilities in circuit-switched networks has been extensively analyzed. However, this has been done under the assumption that the bandwidth allocated to a customer does not change throughout the customer's service. For instance, Kaufman [4], Roberts [8], and Nilsson et al.[6] developed efficient algorithms for calculating the blocking probabilities of a multi-rate loss queue. In this case, customers belong to different classes and each class is associated with a class-dependent arrival rate, class dependent service rate, and a class-dependent bandwidth requirement expressed in number of servers. However, a class r customer cannot switch classes during its service time, and as a result, it cannot change the number of servers allocated to it. Call blocking probabilities over an entire circuit-switched network have been computed under a variety of assumptions, see for instance Kelly [5], Ross [9], Alnowibet and Perros [1], Washington and Perros [11], under assumptions similar to the above case of a single loss queue.

In this paper, we consider the multirate single loss queue depicting a circuit-switched communication link, this link may be an optical link, or a wired or wireless TDM link. Each server represents a time slot in a TDM link or a subrate stream in an optical link. Class r calls arrive in a Poisson fashion at the rate of λ_r, and require initially b_r servers. During the service time of the call, the number of servers required may change. The call is not blocked if fewer servers than currently allocated to it are required. However, the call will get blocked if additional servers are required and these servers are not available at that instance. We describe an approximation algorithm for the calculation of the call blocking probability of each class. To the best of our knowledge this queueing system has not been analyzed before.

We also use a provisioning method based on Hampshire et al. [3] to determine the minimum number of required servers.

This paper is organized as follows. In section 2, we describe in detail the multi-class loss queue under study and how it can be used to model various cases where a customer may change its bandwidth requirements during its service. In section 3, we describe the approximation algorithm and in section 4, we describe how to calculate the minimum number of servers of the loss system so that the blocking probability of any class is less than a pre-specified value. Numerical examples are given in section 5, and finally the conclusions are given in section 6.

2 The Multi-class Loss Queue with Variable-Demand Customers

Let us consider a multi-class loss queue. There are R classes of traffic, and class i customers (i=1,2,...,R) arrive at the loss queue in a Poisson fashion with a class-dependent arrival rate λ_i requiring b_i servers. Class i customers receive an exponentially distributed service time with mean $1/\mu_i$. The required number of servers is ordered for convenience, that is, $0 < b_1 \leq b_2 \ldots \leq b_R$. Upon arrival at the loss queue, a customer is blocked if the required number of servers is not available. After an exponentially distributed service with rate μ_i, a class i customer may depart from the system with probability p_{i0} or it may change its class to class k with probability p_{ik} where $\sum p_{ij} = 1$. A class change implies that the customer's bandwidth requirements change from b_i to b_k. If $b_k < b_i$, then the class change is successful and the $b_i - b_k$ remaining unused servers join in the pool of available servers. However, if $b_k > b_i$, then $b_k - b_i$ additional servers are required. The customer is blocked (i.e. lost) if these $b_k - b_i$ additional servers are not all available at that moment.

Define P1(i,j)=p_{ij} which is a matrix with size $R \times (R + 1)$. Let P be submatrix of P1 and P has dimension $R \times R$. We have to assure that $(I - P)$ is invertible so that a customer entering the system eventually exists.

As will be seen below, the analysis of this system permits a large number of servers which allows us to model high-bandwidth circuit-switched links. For instance, an OC-768 link will be modeled by a loss queue with 768 servers where a server represents an OC-1 subrate stream. The analysis of this model also permits a large number of classes. This feature gives the model the required flexibility to depict users with a given bandwidth profile. For instance, let us consider an example where one group of users requires initially 10 servers. This bandwidth requirement is changed to 20 servers and after that to 15 servers. Then, we say that the bandwidth profile of this group of users is: $\{10, 20, 15\}$. This will be modeled using three classes, say 1, 2, and 3, as follows. A customers with this profile arrives at the loss queue with an arrival rate λ_1 (arrival rates of λ_2 and λ_3 are equal to zero) requiring $b_1 = 10$ servers. After an exponential service time with a mean of $1/\mu_1$, a customer changes to class 2, thus requiring $b_2 = 20$ servers with probability $p_{12}=1$. Following an exponential service time with a mean of of $1/\mu_2$, the customer changes to class 3 with probability $p_{23}=1$. Finally, after an exponential service time with a mean of $1/\mu_3$, the customer departs, i.e. $p_{30}=1$. A customer changing from class 1 to 2, may get blocked if the additional 5 servers are not available. However, a customer going from class 2 to 3 will never get blocked since it requires fewer servers than those it held. More complex bandwidth profiles can be constructed by selecting a set of unused classes, and associating each class i with a set of values for $1/\mu_i$, b_i, p_{ij}. Each set of classes associated with a specific bandwidth profile can be seen as forming a closed super-class within which class changes are allowed in a pre-specified manner. The case where a customer can change bandwidth requirements in a random manner can be readily accommodated.

2.1   A Random Walk Look at the System

It can be seen as a randwom walk for the dynamic adjustment problem. Howerver it may not capture the nature of capacity increasing.

3   Calculation of Call Blocking Probabilities

For the classical multi-class loss system without bandwidth adjustments, there are well-known results.

Let us assume that the system has a total of C identical servers (channels or units of bandwidths), and each can provide service to any class of arrivals. Let $n=(n_1, n_2, ..., n_R)$ where n_r is the number of class r customers in the system, and let $b=(b_1, b_2, ..., b_R)$. The total number of busy servers in state n is

$$bn^T = b_1 n_1 + b_2 n_2 + ... + b_R n_R. \tag{1}$$

The set of all possible states of the system can be described as

$$S^b = \{n : bn^T \leq C\}. \tag{2}$$

It is well known that the multi-class loss system has a product-form solution given by:

$$P(n) = \prod_{i=1}^{R} \frac{\rho_i^{n_i}}{n_i!} G^{-1}(\Omega), \forall n \in \Omega \tag{3}$$

where

$$G(\Omega) = \sum_{n \in \Omega} \prod_{i=1}^{R} \frac{\rho_i^{n_i}}{n_i!} \tag{4}$$

and $\rho_i = \lambda_i / \mu_i$. The challenge is to obtain the blocking probability for each class. Computing the blocking probabilities by directly enumerating all possible states of the system requires an $O(C^R)$ amount of time. The direct method is computationally cumbersome and grows exponentially fast even for relatively small systems. Several methods have been presented in the literature to avoid the exponential complexity of the computations. One of the most powerful methods for obtaining the blocking probabilities was published independently by Kaufman (1981) [4] and Roberts (1981) [8]. The Kaufman-Roberts method is a fast recursive algorithm that has a linear complexity of $O(CR)$. The recursive formula is as follows:

$$w(k) = \frac{1}{k} \sum_{r=1}^{R} \rho_r b_r w(k - b_r), k = 1, 2, ..., C. \tag{5}$$

where $w(0)=1$ and $\rho_r = \lambda_r / \mu_r$. Then, the blocking probability of class r arrivals is given by:

$$B_r = \frac{\sum_{j=C-(b_r-1)}^{C} w(j)}{\sum_{j=0}^{C} w(j)}, r = 1, 2, ..., R. \tag{6}$$

It is interesting to know that this formula can be applied to the single class model, as a fast way of obtaining the blocking probability. Given the blocking probabilities, the average number of class r customers in the system is

$$E[Q_r] = \rho_r(1 - B_r), r = 1, 2, ..., R. \tag{7}$$

The multi-rate loss model with variable-demand customers described in the previous section can be analyzed numerically by setting up the underlying rate matrix and subsequently solving it in order to obtain the stationary probability vector. However, this numerical approach is limited to small size problems due to the complexity involved in setting up the rate matrix. It is also difficult to obtain a closed-form expression because of the variable-demand customers.

In view of these considerations, we solve this loss system approximately as follows.

We assume that when a customer changes its class from i to another class, say class j, it simply departs from the loss queue and it re-joins it as a new class j customer. Its departure from the loss queue and its arrival to the loss queue are not synchronized. That is, we simply calculate a new arrival rate for class i customers based on the external arrival rate and all the possible feedbacks due to customers changing their class to class i. Specifically, we have that the departure rate of class i customers from the loss model is:

$$\mu_i E[Q_i] = \mu_i \rho_i (1 - B_i) = \lambda_i (1 - B_i) \tag{8}$$

Then, the total class i arrival rate due to feedback from other classes is:

$$\lambda_{hi} = \sum_{k=1}^{R} \bar{\lambda}_k (1 - B_k) p_{ki} \tag{9}$$

where p_{ki} is the probability that a class k customer will change to class i and $\bar{\lambda}_k$ is the total class k effective arrival rate (i.e., external arrival rate plus feedbacks from the other classes).

Thus the total effective arrival rate $\bar{\lambda}_i$ of class-i to the loss model is:

$$\bar{\lambda}_i = \lambda_i + \lambda_{hi} = \lambda_i + \sum_{k=1}^{R} \bar{\lambda}_k (1 - B_k) p_{ki} \tag{10}$$

where λ_i is the class i external arrival rate to the loss queue.

This equation is often called the traffic equation. The effective arrival rate and the blocking probability of each class are unknown and have to be decided iteratively.

The total offered load for each class is given by the following nonlinear matrix equations obtained from (10):

$$\bar{\rho} = (I - P^T \bar{B})^{-1} \rho' \tag{11}$$

where I is the identity matrix, $\bar{B}$=diag($[1 - B_1, 1 - B_2, \ldots, 1 - B_R]$) is a diagonal matrix, P is the class-changing probability matrix with its elements $P(i, j)=p_{ij}$, $\bar{\rho}=[\bar{\rho}_1, \bar{\rho}_2, \ldots, \bar{\rho}_R]$ where $\bar{\rho}_i=\bar{\lambda}_i/\mu_i$ is the effective offered load of class i and $\rho=[\rho_1, \rho_2,\ldots, \rho_R]$. We can now use (11) in expression (5) in order to calculate the class blocking probabilities.

The resulting system of equations is solved by a fixed-point procedure summarized below.

Summary of algorithm
set small value for degree of accuracy ϵ
do (the following steps)
    Step 1: Set initially all blocking probabilities and λ_{hi} to be zero
    Step 2: compute values for total offered load per $\rho_i=\lambda_i/\mu_i$
    Step 3: compute values for blocking probabilities B_i per (6)

Step 4: update values for total offered load per (11)
Step 5: update values for blocking probabilities per (6)
while (relative error of two successive blocking prob.$> \epsilon$)
end while

The above algorithm for the calculation of call blocking probabilities has a time complexity of $O(log2(CR/\epsilon)CR^2)$. This algorithm is scalable in the number of classes. A proof of convergence and complexity for a single-class traffic using the bisection algorithm has been sketched out in [10]. For multi-class case considered in this paper, we do not provide a proof of convergence. Through numerous examples, however, we observed that this algorithm converges very fast, often in a few tens of steps.

As is known, the Kaufman-Roberts algorithm is numerically unstable, i.e., it causes overflows when the offered load and/or the total number of servers is very large. This can be avoided by using a dynamic factoring technique. We use a small number α as a scaling factor to avoid potential overflows. If upon inspection, it is found that an overflow would occur in the computation of $w(k)$ in the Kaufman-Roberts formula, all $w(i)$ are scaled, i.e., w(i)=w(i)α for i=0, 1, ..k, so that each w(i) is small enough. The process of dynamic scaling increases the computational costs, but the order of the overall complexity remains unchanged.

We observed that a simpler algorithm can be used in the following two cases:

1) When the total number of servers is very large comparing to the offered loads so that the blocking probability for each class is very small (for example, less than 0.001), equation (11) can be approximated by $\bar{\rho} = (I - P^T)^{-1}\rho$, i.e., we can set all the blocking probabilities equal to zero. In this case, we can calculate the effective offered load from (11) without iterating on the blocking probabilities.

2) If total capacity C is very large, then the feedback rate from class i to any class j can be simply expressed as $\lambda_i p_{ij}$. In this case, the solution is simplified as in case (1) above.

The above two cases provide an upper bound on the effective offered load.

4   Capacity Provisioning

Provisioning optimal total capacity is one of practical ways to meet the blocking probability and other QoS requirements. In this section, we describe how to calculate the minimum number of servers C of the loss model so that the maximum blocking probability of any class is less than a pre-specified value ϵ for a given load. This permits the blocking probabilities of the remaining classes to also be less than ϵ.

This minimum value of C can be calculated iteratively using the fixed-point algorithm described in the previous section. However, when the required capacity C is very large, this iterative approach becomes CPU intensive since its time complexity is $O(log2(CR/\epsilon)CR^2)$.

It is a long-standing conjecture that the optimal number of servers is of the form $\rho+K\sqrt{\rho}$ for single class traffic where K is a constant depending on the offered load and blocking probability. This approximation yields very accurate results. Indeed, based on extensive sensitivity tests, the actual optimum and approximate values rarely deviate by more than one server, or by more than one percent, whichever is greater (see Grassman [2]). Hampshire et al. [3] obtained the following asymptotic expression for the optimum value of C in the multiclass case:

$$C = \sum_{i=1}^{R} b_i \bar{\rho}_i + \psi\left(\min_{1 \le i \le R} \frac{\epsilon_i}{b_i} \sqrt{\sum_{i=1}^{R} b_i^2 \bar{\rho}_i}\right) \sqrt{\sum_{i=1}^{R} b_i^2 \bar{\rho}_i} \qquad (12)$$

where ϵ_i is the blocking probability requirement for class i and $\psi(x)$ is the unique solution of the following differential equation

$$\psi'(x) = \frac{-1}{(\psi(x) + x)x}, \psi(\sqrt{2/\pi}) = 0 \qquad (13)$$

In [3], the authors suggest to use a lookup table for values of $\psi(x)$ by using a second order Runge-Kutta method to compute $\psi(x)$. However, a lookup table may not be practical if the step size is very small and we do not know the starting point x. In this paper, we have solved equation (13) to obtain

$$x^{-1}e^{-0.5\psi(x)^2} - \sqrt{2\pi}erf(0.5^{1/2}\psi(x)) - x\sqrt{0.5\pi} = 0 \qquad (14)$$

where $erf(.)$ function is defined as follow:

$$erf(x) = \frac{2}{\sqrt{\pi}} \int_0^x exp(-t^2)dt \qquad (15)$$

Given x, equation (14) can be easily solved numerically for $\psi(x)$. Applied to equation (12), we obtain the requested total capacity. Because of the asymptotic rule [3], satisfying the requirements provides more than enough capacity for all the other classes. Through many numerical examples, we observed that the minimum capacity C obtained using equation (12) is very closed to the exact solution.

5 Numerical Examples

In this section, we validate the accuracy of our approximation and provide some insights into the multi-rate loss queue with variable-demand customers. We also provide a capacity provisioning example.

The approximation results were compared against simulation data. 95% confidence intervals were also calculated, but since they are extremely small, they are not given in the results below. In Table 1, we give the approximate and simulation results of call blocking probabilities for three classes customers with the number of servers C varying from 20 to 50. The following parameters were used: $\rho=[1,2,3],b=[1,2,3], p_{i0}=0.5, i=1, 2, 3$. Any class i customer can change to any other class j customer, including its own, with probability $p_{ij}=0.5/3, j=1, 2, 3$.

Table 2 gives similar results for a large problem with 100 classes and 1000 servers. The following traffic parameters were used: $\rho_i=i/1000, b_i=i, p_{i0}=0.5$ and $p_{ij}=0.5/100, i=1,2,..100$. Table 3 gives similar results as Table 2. The assumptions are the same, with the exception that $\rho_i = i/300$, i=1,2,..100. We observe that the approximation model match the simulation results quite well. Some deviations were observed when the blocking probabilities are high (see for instance, Table 1, C=20, approximate and simulation results for class 3).

As mentioned above, our algorithm runs very fast. For instance, the approximation results given in Table 2 were obtained in 0.363709 seconds in Matlab 7.0.4. However, the simulation needs much longer time. The simulation results in Table 2 and 3 required around 30000 seconds. (The simulation was implemented in C program on a Pentium(R) 4 CPU 3GHz PC).

Table 1. Approximation and simulation results for 3 classes customers

	Approximation	Approximation	Approximation	Simulation	Simulation	Simulation
Capacity	class-1	class-2	class-3	class-1	class-2	class-3
C=20	0.1129	0.2252	0.3347	0.1176	0.2345	0.3485
C=25	0.0703	0.1446	0.2221	0.0722	0.1487	0.2285
C=30	0.0405	0.0856	0.1352	0.0413	0.0874	0.1380
C=35	0.0219	0.0474	0.0764	0.0225	0.0486	0.0779
C=40	0.0098	0.0217	0.0359	0.0100	0.0221	0.0366
C=45	0.0036	0.0081	0.0138	0.0037	0.0083	0.0139
C=50	0.0010	0.0025	0.0043	0.0011	0.0026	0.0045

Table 2. Approximation (Appr.) and simulation (Sim) results for 100 classes customers (1)

$class_i$	5	10	15	20	25
Appr.	0.00017	0.00035	0.00054	0.00074	0.00096
Sim	0.00016	0.00034	0.00053	0.00073	0.00094
$class_i$	30	35	40	45	50
Appr.	0.00119	0.00143	0.00169	0.00196	0.00225
Sim	0.00117	0.00140	0.00166	0.00192	0.00220
$class_i$	55	60	65	70	75
Appr.	0.00255	0.00287	0.00321	0.00357	0.00395
Sim	0.00250	0.00281	0.00315	0.00350	0.00387
$class_i$	80	85	90	95	100
Appr.	0.00435	0.00477	0.00522	0.00569	0.00618
Sim	0.00426	0.00468	0.00512	0.00558	0.00606

Next we consider the case where all customers arriving at the loss queue have the same bandwidth profile. Specifically, new customers arrive at the loss queue as class 1 and require 1 server. After an exponentially distributed service time with mean $1/\mu$, a class-1 customer changes to a class-2 customer with a bandwidth requirement of 2 servers with probability $p_{12}=1$. After another exponentially distributed service time with mean $1/\mu$, the class-2 customer changes to a class-3 customer with a bandwidth

Table 3. Approximation (Appr.) and simulation (Sim) results for 100 classes customers (2)

$class_i$	5	10	15	20	25
Appr.	0.02635	0.05229	0.07780	0.10290	0.12758
Sim	0.02891	0.05751	0.08573	0.11291	0.14021
$class_i$	30	35	40	45	50
Appr.	0.15184	0.17568	0.19909	0.22208	0.24465
Sim	0.17061	0.19782	0.22293	0.24923	0.27341
$class_i$	55	60	65	70	75
Appr.	0.26679	0.28852	0.30982	0.33069	0.35115
Sim	0.30208	0.32619	0.35060	0.37237	0.39684
$class_i$	80	85	90	95	100
Appr.	0.37118	0.39081	0.41001	0.42880	0.44718
Sim	0.42342	0.44494	0.46745	0.49198	0.50767

requirement of 3 servers with probability $p_{23}=1$. Finally, the class-3 customer departs with probability $p_{30}=1$ after an exponentially distributed service time with mean $1/\mu$. The offered loads are $\rho_1 > 0$, $\rho_2=0$, $\rho_3=0$, i.e., no external class-2 and class-3 arrivals occur. Given this load profile, we compare the following three bandwidth allocation strategies.

Case 1 (variable-demand policy): bandwidth is allocated on demand whenever a customer changes a class. In this case, a customer may be blocked upon arrival to the loss queue as class-1 customer and each time it changes a class. The class-dependent mean service time is $1/\mu$.

Case 2 (maximum service policy): A class 1 customer is allocated the maximum number of servers, i.e., 3 servers, upon arrival as class 1 customer to the loss queue. The mean service time is : (a). $2/\mu$ in order to keep the product of bandwidth and service-time the same as case 1; or (b). $3/\mu$ so that the arrival will use the same mean service time as case 1. The implication in case (a) is that the customer will take full advantage of the 3 servers allocated to it. In case (b) on the other hand, we assume that the customer follows its bandwidth profile and it uses only the required number of servers. A class 1 customer is blocked if these servers are not available upon arrival.

Case 3 (minimum service policy): A class 1 customer is not allowed to change bandwidth requirements. It is allocated the minimum number of customers, i.e., 1 server for a service time $6/\mu$ in order to keep the product of bandwidth and service-time the same as cases 1 and 2. A customer is blocked if no server is available upon arrival.

Case 1 was analyzed using our approximation algorithm, whereas cases 2 and 3 were analyzed using the Erlang loss formula for single class traffic. In order to facilitate the comparison among these three cases, we calculate the average call blocking probability for the case 1 as follows:

$$B_{avg} = \sum_{r=1}^{R} \bar{\rho}_r b_r B_r / \sum_{r=1}^{R} \bar{\rho}_r b_r \tag{16}$$

Table 4. Call blocking comparison among variable-demand service, Max and Min service

ρ_1	15	16	17	18	19	20	21	22
Case1	0.0495	0.0691	0.0915	0.1126	0.1336	0.1541	0.1740	0.1932
Case2a	0.0805	0.1109	0.1430	0.1755	0.2075	0.2384	0.2680	0.2959
Case2b	0.3093	0.3472	0.3817	0.4131	0.4417	0.4678	0.4917	0.5136
Case3	0.0270	0.0539	0.0874	0.1238	0.1606	0.1963	0.2302	0.2620

We show the results in Table 4 for various values of the class-1 offered load ρ_1 for a total capacity C=100.

We note that the variable demand policy outperforms the maximum and minimum service policies when the offered load is medium or large. This observation holds for many other similar examples (not reported here).

Finally, in Table 5 we show the minimum required number of servers for a 3-class Erlang loss queue, so that the blocking probability is less than 0.01 for all three classes. The required bandwidth for the three classes is b=[1,2,3], p_{i0}=0.5 for i=1,2,3 and p_{ij}=0.5/3, j=1,2, 3 and i=1,2,3. The external offered load $\rho=[\rho_1, \rho_2, \rho_3]$ was varied. For each set of value of ρ, we computed the minimum required servers using equation (12) (labelled as 'Asmp') and also using our algorithm in an iterative manner as explained at the beginning of section 4 (labelled as 'Appr.').

Table 5. The optimized capacity vs. offered load (ρ) for 3-classes traffic

Offered load	Method	Capacity	Offered load	Method	Capacity
[0.14 0.29 0.43]	Appr.	15	[10,20,30]	Appr.	301
[0.14 0.29 0.43]	Asmp	13	[10,20,30]	Asmp	300
[1,2,3]	Appr.	46	[40,80,120]	Appr.	1088
[1,2,3]	Asmp	44	[40,80,120]	Asmp	1087
[3,6,9]	Appr.	107	[100,200,300]	Appr.	2629
[3,6,9]	Asmp	105	[100,200,300]	Asmp	2629

6   Conclusion

In this paper, we described a model for calculation of call blocking probabilities in a multi-rate Erlang loss queue where the customers are allowed to change their bandwidth requirements during their service. Comparisons against simulation data showed that the algorithm has a good accuracy. The model was also used to evaluate different allocation policies and capacity provisioning, and we hope to expend this work in an upcoming paper. Also we are currently extending the algorithm to the case where the bandwidth requirements of each customer in service are modified by the network manager as a function of the congestion level.

References

1. Alnowibet, K.: Ph.D Dissertaion, North Carolina State University Nonstationary Erlang Loss Queues and Networks (2004)
2. Grassmann, W.K.: Is the Fact That the Emperor Wears No Clothes a Subject Worthy of Publication? Interfaces 16, 43–46 (1986)
3. Hampshire, R.C., Massey, W.A., Mitra, D., Wang, Q.: Provisioning For Bandwidth Sharing and Exchange. In: Hampshire, R.C., Massey, W.A., Mitra, D., Wang, Q. (eds.) Oper. Res./Comput. Sci. Interfaces Ser., vol. 23, Kluwer Academic Publishers, Boston, MA (2003)
4. Kaufman, J.: Blocking in a Shared Resource Environment. IEEE Transactions On. Communications vol. COM-29(10), (1981)
5. Kelly, F.: Blocking probabilities in large circuit-switched networks. Adv. Appl. Prob. 18, 473–505 (1986)
6. Nilsson, A.A., Perry, M.J., Gersht, A., Iversen, V.B.: On multi-rate Erlang-B computations. In: Proceedings of ITC 16, pp. 1051–1060 (1999)
7. Perros, H.: Connection-Oriented Networks: SONET/SDH, ATM, MPLS and Optical Networks. John Wiley & Sons, England (2005)
8. Roberts, J.W.: A serverice system with heterogeneous user requirements. In: Performance of Data Communications Systems and Their Applications, pp. 423–431 (1981)
9. Ross, K.W.: Multiservice Loss Models for Broadband Telecommunication Networks. Springer-Verlag, London (1995)
10. Sarkar, D., Jewell, T., Ramakrishnan, S.: Convergence in the Calculation of the Handoff Arrival Rate: A Log-Time Iterative Algorithm, EURASIP Journal on Wireless (2006)
11. Washington, A.N., Perros, H.G.: Call blocking probabilities in a traffic groomed tandem optical network", Special issue dedicated to the memory of Professor Olga Casals, Blondia and Stavrakakis (eds.) Journal of Computer Networks, Vol. 45 (2004)
12. ITU-T G.7042/Y. 1035, Link Capacity Adjustment Scheme (LCAS) for Virtual Concatenated Signals (January 2006)

An Approximation Method for Multiservice Loss Performance in Hierarchical Networks[*]

Qian Huang[1], King-Tim Ko[1], and Villy Bæk Iversen[2]

[1] Department of Electronic Engineering, City University of Hong Kong,
Hong Kong SAR, China
{eeqhuang, eektko}@cityu.edu.hk
[2] COM·DTU, Technical University of Denmark,
DK-2800 Kgs. Lyngby, Denmark
vbi@com.dtu.dk

Abstract. This paper presents an approximation method — Multiservice Overflow Approximation (MOA), to compute traffic loss in multitier hierarchical networks with multiservice overflows. With the MOA method, we first obtain the variances of multiservice non-Poisson overflows in each tier by a blocking probability matching approach, and then compute the call blocking probability of multiservice overflow traffic by a modified Fredericks & Hayward's approximation. The results obtained by the MOA method are verified by simulations. Compared with an existing approximation method based on multi-dimensional Markov-Modulated Poisson Process analysis, the MOA method achieves an accurate estimation of overflow traffic loss at a much lower computational cost, particularly for high-load multi-tier hierarchical networks.

Keywords: Hierarchical network, overlay network, overflow.

1 Introduction

Hierarchical network structures with traffic overflow have been studied in circuit-switched networks for capacity improvement [1]. It has led to various techniques for loss analysis of single-service overflow traffic in hierarchical networks, such as Equivalent Random Theory (ERT) [2] and Fredericks & Hayward's method [3,4]. For hierarchical architecture supporting multimedia applications, multiservice traffic overflows are expected not only in homogenous networks, such as Third Generation (3G) cellular networks [5], but also among heterogeneous wireless networks which support internetworking overflow, such as traffic overflow from WiFi to WiMAX [6]. In such cases, multiple types of services share the common link capacity, and the resultant overflow traffic streams with different service types are statistically correlated. The loss analysis of multiservice overflow traffic poses significant challenges in terms of accuracy and computational complexity for multiservice hierarchical networks designs. An alternative

---

[*] The work described in this paper was fully supported by a grant from City University of Hong Kong (Project No. 7001848).

L. Mason, T. Drwiega, and J. Yan (Eds.): ITC 2007, LNCS 4516, pp. 901–912, 2007.

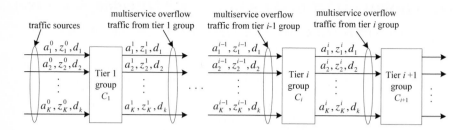

Fig. 1. A multi-tier multiservice overflow system

loss analysis technique for multiservice overflows is based on Markov-modulated Poisson process (MMPP). As a simplified case of MMPP, Interrupted Poisson Process (IPP) has been widely used to model overflow traffic in overlay cellular networks [7]–[9]. With the IPP model, the overflowed calls arrive according to a Poisson process during ON periods, whereas no calls arrive during OFF periods. Both ON and OFF periods of an IPP are assumed to be exponentially distributed. A more complicated MMPP model is used in [10] to evaluate the performance of correlated multiservice overflow traffic in a two-tier hierarchical cellular network. The overflow traffic of each service type is modeled as an MMPP with two overflow states, representing *high* and *low* congestion periods. During different congestion periods, the overflowed call arrivals of each service type are approximated by a different Poisson process. The average call arrival rates during *high* and *low* congestion periods and the transition rates between these two periods are determined by moments matching [10]. However, the MMPP methods are usually computationally intensive for the calculation of multiservice traffic loss or the characterization of overflow traffic, particularly the variance of overflow traffic for each service type. For example, in [9,10], the variance of correlated overflow traffic for each service type is calculated by solving a large number of equilibrium state equations, in a system which admits the overflowed calls to a group with "infinite" number of servers. For an overflow system with K types of services over a shared link with capacity C, the order of computational complexity required by the MMPP method is $O(C^K)$, and the loss analysis becomes intractable as the values of K or/and C increase.

In this paper, we present an efficient approximation method, referred to as Multiservice Overflow Approximation (MOA), for loss analysis in multiservice hierarchical networks. The details of the MOA method are described in Section 2, followed by the performance evaluation in Section 3. Finally, conclusions are provided in Section 4.

2   The Proposed Approximation Method

We consider a multi-tier multiservice hierarchical overflow system as shown in Fig. 1. The system supports K types of services, which differ in call arrival rates, mean holding times and bandwidth requirements. Tier i, $i \geq 1$, is overlaid by tier $i + 1$ and share the capacity of tier $i + 1$. The capacity of tier i is denoted as C_i. The tiers can be micro-cells/macro-cells in hierarchical cellular networks. Hereafter, we refer to them as groups. The traffic blocked by a tier i group is overflowed to a tier $i + 1$ group for service.

The traffic sources to the considered system model are offered to tier 1. Without loss of generality, for type k $(1 \leq k \leq K)$ traffic source, its call arrivals follow a Poisson process with a mean arrival rate λ_k^0; its call holding time is exponentially distributed with a mean holding time $1/\mu_k^0$; the offered traffic load is denoted as $a_k^0 = \lambda_k^0 / \mu_k^0$; the bandwidth required by each arriving call is denoted as d_k. For type k overflow traffic from tier i to tier $i + 1$, it is described by the traffic load a_k^i, the peakedness z_k^i and the required bandwidth d_k for each overflowed call. For simplicity, we assume that the group capacity of each tier and the bandwidth required by each type of calls are measured by the number of Bandwidth Units (BUs). The BU can be a basic data rate carried by a channel, e.g., a time slot or a carrier frequency. We also assume that the link capacity of a group is completely shared by all types of traffic.

2.1 Equilibrium Equations for Variance of Overflow Traffic

As an alternative solution, the variances of overflow traffic with multiple services sharing the common link capacity can be calculated by solving the related equilibrium equations. Let tier i be the reference group. The calls blocked by tier i are assumed to overflow to a "virtual" group with infinite number of servers. Let n_k and o_k be the number of type k calls in tier i and the "virtual" group, respectively. Such a queueing system can be specified by a $2K$-dimensional Markov chain with the state vector denoted $\boldsymbol{s} = (n_1, n_2, ..., n_k, ..., n_K, o_1, o_2, ..., o_k, ..., o_K)$. The space of $\boldsymbol{s}$, denoted Ω_i, is determined by

$$\Omega_i = \{ \boldsymbol{s} : \sum_{k=1}^{K} n_k d_k \leq C_i, n_k \geq 0, o_k \geq 0, 1 \leq k \leq K \}. \tag{1}$$

Let $\pi(\boldsymbol{s})$ denote the equilibrium distribution of $\boldsymbol{s} \in \Omega_i$. For systems with small link capacity and small number of service types (e.g. two), $\pi(\boldsymbol{s})$ can be obtained by solving the associated equilibrium state equations of the $2K$-dimensional Markov chain. Based on the resultant $\pi(\boldsymbol{s})$, the mean and the variance of type k overflow traffic from tier i, denoted as m_k^i and v_k^i, are calculated by,

$$m_k^i = \sum_{\boldsymbol{s} \in \Omega_i} o_k \pi(\boldsymbol{s}) \tag{2}$$

$$v_k^i = \sum_{\boldsymbol{s} \in \Omega_i} (o_k)^2 \pi(\boldsymbol{s}) - (m_k^i)^2. \tag{3}$$

However, this approach becomes intractable for systems with large capacity or large number of service types, due to high computational cost required by solving large number of equilibrium equations for $\pi(\boldsymbol{s})$.

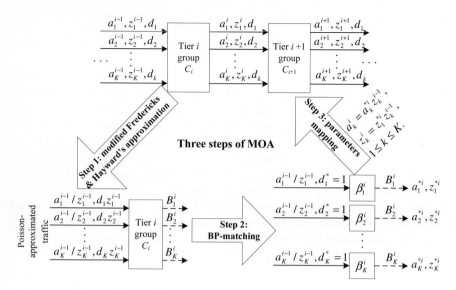

Fig. 2. The MOA method for overflow loss analysis in tier i

2.2   The MOA Method

To combat the computational complexity problem, we propose the MOA method to calculate the variances and blocking probabilities of multiservice overflow traffic in large systems. The MOA method includes three steps as shown in Fig. 2, where a_k^{i-1} and z_k^{i-1} denote the offered load and the peakedness for type k $(1 \leq k \leq K)$ traffic to tier i $(i \geq 1)$.

Step 1: Modification of Fredericks & Hayward's approximation. The first step of the MOA method is to obtain the call blocking probabilities for type k traffic in tier i, denoted as B_k^i. Once the values of a_k^{i-1} and z_k^{i-1} are obtained, B_k^i can be calculated by a modification of the Fredericks & Hayward's approximation method [3]. The original Fredericks & Hayward's method only considers single-service overflow. A non-Poisson traffic stream with intensity A and peakedness Z direct to a circuit group with X servers can be approximated by a Poisson traffic stream with intensity A/Z and peakedness $Z = 1$ direct to a circuit group with X/Z servers, that is $E(X, A, Z) \sim E(X/Z, A/Z, 1)$. Here, $E()$ denotes the loss probability from Erlang loss function. The Fredericks & Hayward's method can be enhanced to handle multi-rate traffic by the approximation of $E(X, A, Z, d) \sim E(X/Z, A/Z, 1, d)$ [4]. Here d represents the bandwidth required by each arriving call. In our considered scenario, the capacity of tier i group is completely shared by multiple types of overflow traffic differing in a_k^{i-1}, z_k^{i-1} and d_k. To keep fixed group capacity being seen by any type of traffic in tier i, we further modify the Fredericks & Hayward's approximation as follows,

$$E(C_i, a_k^i, z_k^{i-1}, d_k) \sim E(C_i/z_k^{i-1}, a_k^{i-1}/z_k^{i-1}, 1, d_k)$$
$$= E(C_i, a_k^{i-1}/z_k^{i-1}, 1, d_k \cdot z_k^{i-1}). \tag{4}$$

In terms of (4), type k overflow traffic to tier i can be approximated by Poisson traffic with load a_k^{i-1}/z_k^{i-1} and required bandwidth $d_k \cdot z_k^{i-1}$; and the call blocking probability for each type of overflow traffic in tier i is equivalent to the call blocking probability for each type of Poisson-approximated traffic. We denote the latter as B_k^{*i}, which can be calculated by existing loss analysis methods for multiservice Poisson traffic, such as Kaufman-Roberts' recursion [11,12] or the convolution algorithm [13,14]. However, Kaufman-Roberts' recursion cannot achieve accurate results when the mean service times of different traffic largely differ from each other [14]. In the MOA method, we use the convolution algorithm [13] to calculate the call blocking probability. Based on the modified Fredericks & Hayward's approximation (4), we have $B_k^i = B_k^{*i}$.

Step 2: Blocking probability matching (BP matching). The second step is to calculate the mean and the peakedness of type k overflow traffic from tier i to tier $i+1$. These two parameters are used as the offered load and the peakedness of the overflow traffic to tier $i+1$ for calculating the call blocking probabilities in tier $i+1$ by the approach described in Step 1.

The mean of the overflow traffic from tier i to tier $i+1$ is given by the average traffic loss in tier i. For type k overflow traffic to tier $i+1$, it is calculated by:

$$a_k^i = a_k^{i-1} B_k^i. \tag{5}$$

For the peakedness of the overflow traffic from tier i, we first calculate the variance by the following BP matching approach. We map tier i into an "imaginary" overflow model, where the Poisson-approximated traffic streams which are originally offered to tier i are assumed to direct to K "imaginary" circuit groups classified by types of traffic, as show in Fig. 2. Here, β_k^i denotes the number of "imaginary" circuits for type k traffic, a_k^{i-1}/z_k^{i-1} denotes the offered load of type k traffic to the "imaginary" β_k^i circuits, $1 \leq k \leq K$. Each call in the "imaginary" circuits is assumed to have the same mean holding time as in tier i but require unit bandwidth denoted as $d_k^* = 1$ for type k traffic. Such an assumption does not change the resultant peakedness of the overflow traffic from the "imaginary" circuits, because mean and variance of overflow traffic are derived from the statistical distribution of the number of overflowed calls in an infinite overflow group. We solve for β_k^i so as to make the call blocking probability of Poisson-approximated traffic in the "imaginary" model equal the call blocking probability in tier i, that is finding β_k^i which satisfies $E(\beta_k^i, a_k^{i-1}/z_k^{i-1}) = B_k^i$. Denoting z_k^{*i} the peakedness of type k overflow traffic from the "imaginary" β_k^i circuits, it can be calculated by Riordan's equation [2,4] as follows,

$$z_k^{*i} = 1 - a_k^{*i} + \frac{a_k^{i-1}/z_k^{i-1}}{1 + \beta_k^i - a_k^{i-1}/z_k^{i-1} + a_k^{*i}}, \tag{6}$$

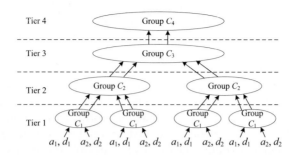

Fig. 3. A four-tier hierarchical overflow loss system

where a_k^{*i} is the mean of type k overflow traffic from the "imaginary" β_k^i circuits; it is given by

$$a_k^{*i} = \frac{a_k^{i-1}}{z_k^{i-1}} \cdot B_k^i. \tag{7}$$

Step 3: Parameters mapping. We have obtained a_k^{*i} and z_k^{*i} for type k overflow traffic from the "imaginary" overflow model, corresponding to the offered Poisson-approximated traffic with load a_k^{i-1}/z_k^{i-1}. For the original group of tier i offered by non-Poisson traffic with load a_k^{i-1} for type k, the mean of type k overflow traffic from tier i is finally obtained by

$$a_k^i = a_k^{*i} z_k^{i-1} = a_k^{i-1} B_k^i, \tag{8}$$

which matches the derivation of Eq. (5); and the peakedness of type k overflow traffic from tier i is obtained by [3,15]

$$z_k^i = z_k^{*i} z_k^{i-1}. \tag{9}$$

3   Validation of the MOA Method

We evaluate the MOA method in a multi-tier hierarchical overflow network shown in Fig. 3. Two types of Poisson traffic are offered to tier 1, and the call arrivals of the same service type are statistically identical in each tier 1 group. For type k ($k = 1, 2$) traffic, a_k and d_k denote the offered load and the number of BUs required by each call. The group capacity of tier i is denoted as C_i in number of BUs. Both types of traffic blocked by the lower-tier are overflowed to the higher-tier and completely share the capacity of the higher-tier. The results obtained by the MOA method are verified by simulation in OPNET [16].

3.1   Two-Tier Hierarchical Networks

The approximation methods for two-tier hierarchical overflow networks have been investigated in both [10] and our previous work [17]. For comparison, in

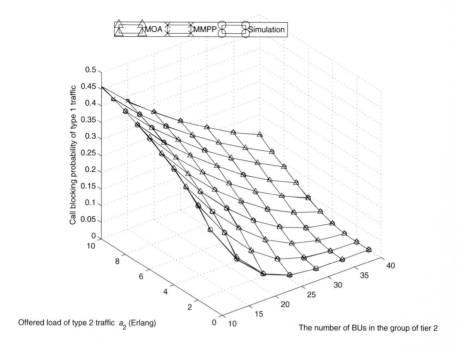

Fig. 4. Call blocking probability of type 1 traffic in the 2-tier overflow system

Figs. 4 and 5 we present the results of the call blocking probability for the two types of traffic obtained by our method MOA, the two-state MMPP method of [10] and simulations, under $a_1 = 10$ Erlangs, $d_1 = 1$ BU; $a_2 = 1 \sim 10$ Erlangs, $d_2 = 2$ BUs; $C_1 = 5$ BUs, $C_2 = 10 \sim 40$ BUs. The confidence intervals of the simulations are kept within 0.2% of the results with a 95% confidence level based on Student's t-distribution. We observe that for both types of traffic, the curves of MOA, MMPP and simulation overlap each other in most areas. It demonstrates that the results obtained by MOA match well with the simulations, except for the cases of light loading for type 2 traffic, e.g., $a_2 \leq 3$ in Fig. 5.

3.2   Three-Tier and Four-Tier Hierarchical Networks

Now we verify the validity of the MOA method in three and four tiers systems as shown in Fig. 3. Here, tier 3 and tier 4 groups are set to $C_3 = C_4 = 40$ BUs; the others are set to $C_1 = 6$ BUs and $C_2 = 20$ BUs. Besides, we set $a_1 = 10$ Erlangs, $d_1 = 1$ BU; $a_2 = 1 \sim 20$ Erlangs, $d_2 = 2$ BUs. Figs. 6 and 7 present the call blocking probabilities obtained in two, three and four tiers overflow systems, respectively for two types of traffic. The results obtained by the MOA method are verified by simulation with a confidence level of 95%.

We have the following observations from Figs. 6 and 7. First, the results obtained by MOA match well with the simulations in the two-tier system. Second,

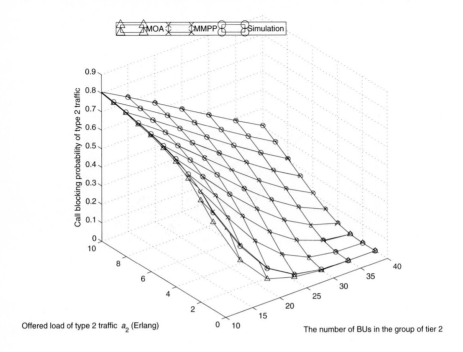

Fig. 5. Call blocking probability of type 2 traffic in the 2-tier overflow system

the discrepancy between MOA and simulation becomes large in the systems with three and four tiers. This demonstrates the accumulation of errors introduced by MOA due to the repeated use of the modified Hayward's approximation and the BP matching. Third, the errors are obvious in the cases where traffic load is low. In contrast, for the useful cases of high traffic loading, MOA gives accurate blocking probabilities for three and four-tier hierarchical networks.

To evaluate the approximation errors for multi-tier overflows, we compare the mean and the peakedness of overflow traffic obtained by MOA and simulation in each tier for both types of traffic. The results demonstrate that the means of overflow traffic for both service types obtained by MOA match well with the simulations. Figs. 8 and 9 plot the peakedness obtained by MOA, respectively for the two types of overflow traffic from the groups of tier 1, tier 2 and tier 3. The discrepancy between MOA and simulation in the peakedness for the two types of overflow traffic are plotted in Figs. 10 and 11. It demonstrates that the peakedness obtained by MOA is close to the simulations for the overflows from tier 1 to tier 2. However, due to the accumulated errors introduced by the approximations in MOA, we observe increased discrepancy in the peakedness of the overflows from tier 2 to tier 3 and from tier 3 to tier 4. It therefore leads to the aforementioned discrepancy between the results of the MOA and the simulations in call blocking probabilities under low traffic loading.

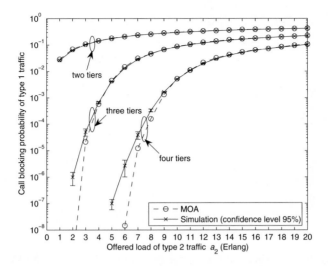

Fig. 6. Call blocking probabilities of type 1 traffic in multi-tier overflow systems

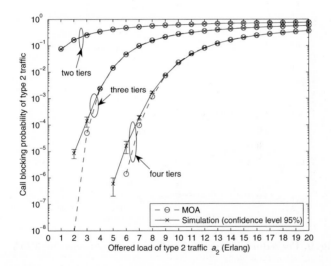

Fig. 7. Call blocking probabilities of type 2 traffic in multi-tier overflow systems

An additional interest is the computational complexity of the MOA method. Compared with the approximation based on the two-state MMPP model [10], the MOA method requires lower computational cost. For example, K types of traffic are offered to a multi-tier hierarchical overflow network, with C_i BUs for a tier i group. The two-state MMPP method of [10] requires intensive computation to determine the parameters of the MMPP model, by solving a

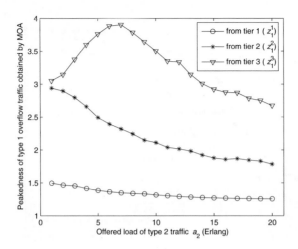

Fig. 8. Peakedness of type 1 overflow traffic obtained by MOA

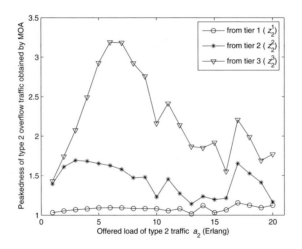

Fig. 9. Peakedness of type 2 overflow traffic obtained by MOA

$(K + 1)$-dimensional Markov chain with the order of complexity $O(C_i^K)$. As the number of service types or/and the group capacity of each tier increase, the computational complexity of the MMPP method grows exponentially. In contrast, the main computational effort required by the MOA method is the convolution operators for calculating the call blocking probability of each service type. For tier i, the order of complexity involved in the MOA method for K types of traffic is $O(K^2 C_i^2)$, much lower than that of MMPP.

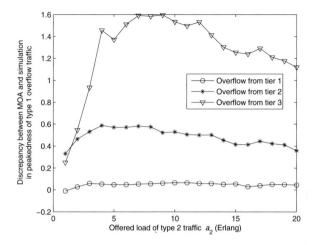

Fig. 10. Difference between MOA and simulation in peakedness of type 1 overflow traffic

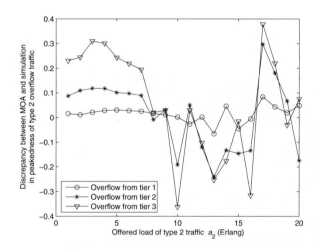

Fig. 11. Difference between MOA and simulation in peakedness of type 2 overflow traffic

4   Conclusions

We have proposed an efficient approximation method for loss calculation in hierarchical multiservice overflow networks. Compared with the existing solution such as the MMPP method, the proposed MOA method achieves accurate blocking probabilities in two-tier multiservice hierarchical networks at a much lower computational cost. Moreover, verified by simulations, the results obtained by the MOA method in three and four-tier multiservice hierarchical networks are accurate for the useful cases of high traffic loading. The inaccuracy for low traffic loading is due to high percentage error from repeated approximations.

References

1. Ghanbari, M., Hughes, C.J., Sinclair, M.C., Eade, J.P.: Principles of Performance Engineering for Telecommunication and Information Systems. Institution of Electrical Engineering, London (1997)
2. Wilkinson, R.I.: Theories for toll traffic engineering in the U.S.A. Bell Syst. Tech. J. 35, 421–514 (1956)
3. Fredericks, A.A.: Congestion in blocking systems — a simple approximation technique. Bell Syst. Tech. J. 59, 805–827 (1980)
4. Iversen, V. B.: Teletraffic Engineering. ITU-D Study Group 2 Question 16/2, Geneva, Chapter 9 (January 2005)
   http://www.com.dtu.dk/teletraffic/handbook/telenook.pdf
5. Wu, X., Mukherjee, B., Ghosal, D.: Hierarchical architectures in the third-generation cellular network. IEEE Wireless Commun. 11, 62–71 (2004)
6. Ghosh, A., Wolter, D.R., Andrews, J.G., Chen, R.: Broadband wireless access with WiMax/802.16: current performance benchmarks and future potential. IEEE Commun. Mag. 43, 129–136 (2005)
7. Kuczura, A.: The interrupted Poisson process as an overflow process. Bell Syst. Tech. J. 52, 437–448 (1973)
8. Matsumoto, J., Watanabe, Y.: Individual traffic characteristics of queueing systems with multiple Poisson and overflow inputs. IEEE Trans. Commun. 33, 1–9 (1985)
9. Hu, L-R., Rappaport, S.S.: Personal communication systems using multiple hierarchical cellular overlays. IEEE J. Select. Areas Commun. 13, 406–415 (1995)
10. Chung, S-P., Lee, J-C.: Performance analysis and overflowed traffic characterization in multiservice hierarchical wireless networks. IEEE Trans. Wireless Commun. 4, 904–918 (2005)
11. Kaufman, J.S.: Blocking in a shared resource environment. IEEE Trans. Commun. 29, 1474–1481 (1981)
12. Roberts, J.W.: A service system with heterogeneous user requirements — Application to multi-service telecommunications systems. In: Pujolle, G. (ed.) Performance of Data Communications Systems and Their Applications, pp. 423–431. North Holland, New York (1981)
13. Iversen, V.B.: The exact evaluation of multi-service loss system with access control. Teleteknik (English Edition) 31, 56–61 (1987)
14. Ross, K.W.: Multiservice Loss Models for Broadband Telecommunication Networks, London, pp. 80–89. Springer, Heidelberg (1995)
15. Jagerman, D.L., Melamed, B., Willinger, W.: Stochastic Modeling of Traffic Processes. In: Jagerman, D.L., Melamed, B., Willinger, W. (eds.) Frontiers in queueing: models and applications in science and engineering, pp. 271–320. CRC Press, Inc, Boca Raton, FL (1998)
16. OPNET University Program: http://www.opnet.com/services/university/
17. Huang, Q., Ko, K.T., Iversen, V.B.: Approximation of loss calculation for hierarchical networks with multiservice overflows. Accepted for publication by IEEE Transactions on Communications

Analysis of Losses in a Bufferless Transmission Link*

Valeriy A. Naumov[1,2] and Peder J. Emstad[2]

[1] Department of Telecommunication Systems, Peoples' Friendship University of Russia
(PFUR), Moscow, Russia
[2] Centre for Quantifiable Quality of Service in Communication Systems,
Norwegian University of Science and Technology (NTNU),
Trondheim, Norway
vnaumov@sci.pfu.edu.ru, peder@q2s.ntnu.no

Abstract. In this paper we use a multivariate Markov Modulated Fluid Flow model to study the loss process for a bufferless transmission link. We propose a method for the analysis of moments and correlations of congestion periods and cumulative amount of lost bits and lost packets from different sources. Such knowledge is useful for the encoding of voice and video. Then we demonstrate how the proposed method can be used for the analysis of capacity sharing policies.

Keywords: Bufferless system, fluid flow, congestion, Markov additive process.

1 Introduction

Several authors have modeled communication systems as bufferless system, even if they have buffers ([1] – [4]). The justifications for that are several. First, short queueing delays are generally wanted and especially for interactive audio and multimedia systems; consequently the buffers are made small. Second, with low transmission costs the systems are often dimensioned such that queueing delays become small anyway. Third, it is a pessimistic assumption for assessing losses.

From a modeling point of view the bufferless assumption leads to simpler models and therefore allows for the study of more system properties. In this paper we propose a method for the calculation of moments of congestion periods, their spacing and lengths, and amount of bits and packets lost during congestion periods. Such knowledge is useful for the encoding of voice and video.

The paper is organized as follows. In the next section we interrelate between bit and packet losses in a bufferless transmission link. In section 3 we consider a particular case using a multivariate Markov Modulated Fluid Flow (MMFF) model to describe the cumulative amount of arrived bits and packets from a single source. A method for the analysis of general multivariate MMFF with partitioned space state is presented in section 4. In section 5 we demonstrate how the proposed method can be applied for the analysis of capacity sharing policies in a bufferless system.

---

* Centre for Quantifiable Quality of Service in Communication Systems is a Centre of Excellence appointed by the Research Council of Norway and funded by the Research Council, NTNU and UNINETT, http://www.q2s.ntnu.no.

L. Mason, T. Drwiega, and J. Yan (Eds.): ITC 2007, LNCS 4516, pp. 913–924, 2007.

2 Loss Probabilities

Consider a traffic flow processed by a bufferless transmission link of capacity C. The flow is described by two stationary ergodic processes. The process $a(t)$ represents the packet arrival rate and the process $L(t)$ represents the length of a packet arrived at time t. Thus the bit rate of the flow is given by $b(t) = a(t)L(t)$. We assume that the processes $a(t)$, $b(t)$ and $L(t)$ have finite means

$$\overline{a} = \int_0^\infty x dF_a(x), \quad \overline{L} = \int_0^\infty x dF_L(x), \quad \overline{b} = \int_0^\infty x dF_b(x),$$

where $F_a(x)$, $F_b(x)$, $F_L(x)$ are corresponding probability distribution functions.

2.1 Bit Loss Probability

The bit traffic suffers losses if the bit arrival rate $b(t)$ exceeds the link capacity C (see Fig.1).

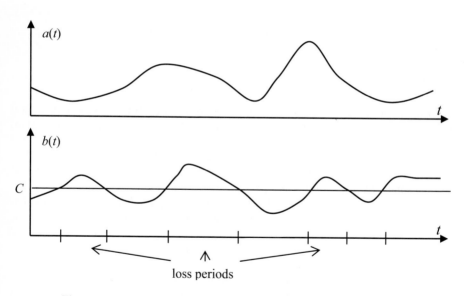

Fig. 1. The packet rate process $a(t)$, bit rate process $b(t)$, and loss periods

Then the long-run average bit loss probability is given by

$$\pi_{\text{bit}} = \lim_{T \to \infty} \frac{\int_0^T (b(t) - C)_+ dt}{\int_0^T b(t) dt} = \frac{\Lambda}{\overline{b}}, \tag{1}$$

where $\Lambda = \mathbb{E}[(b(t) - C)_+]$ is the mean bit loss rate, $x_+ = x$ if $x > 0$ and $x_+ = 0$ otherwise. The expression for the mean bit loss rate can also be written as

$$\Lambda = \mathbb{P}[b(t) > C]\left(\mathbb{E}[b(t)|b(t) > C] - C\right) = P_{loss}(\overline{b}_{loss} - C). \tag{2}$$

Here the loss period probability P_{loss} and the mean bit arrival rate in a loss period $\overline{b}_{loss}$ are defined by

$$P_{loss} = \mathbb{P}[b(t) > C] = 1 - F_b(C), \tag{3}$$

$$\overline{b}_{loss} = \mathbb{E}[b(t)|b(t) > C] = \frac{1}{P_{loss}}\int_C^{\infty} x\,dF_b(x). \tag{4}$$

Therefore the mean bit loss rate Λ can be calculated as

$$\Lambda = \int_C^{\infty}(x - C)dF_b(x) = \int_C^{\infty}(1 - F_b(x))dx, \tag{5}$$

which leads to the following formula for the bit loss probability

$$\pi_{bit} = 1 - \widehat{F}_b(C), \tag{6}$$

where

$$\widehat{F}_b(x) = \frac{1}{\overline{b}}\int_0^x(1 - F_b(z))dz. \tag{7}$$

2.2 Packet Loss Probability

If the bit arrival rate $b(t) = a(t)L(t)$ exceeds the link capacity C then the packet loss rate $r(t)$ satisfies $(a(t) - r(t))L(t) = C$, i.e. $r(t) = \dfrac{(b(t) - C)_+}{L(t)}$. Therefore the long-run average packet loss probability is given by

$$\pi_{packet} = \lim_{T \to \infty} \frac{\displaystyle\int_0^T \frac{(b(t) - C)_+}{L(t)}\,dt}{\displaystyle\int_0^T a(t)dt} = \frac{\lambda}{a}, \tag{8}$$

Here $\lambda = \mathbb{E}\left[\dfrac{(b(t) - C)_+}{L(t)}\right]$ is the mean packet loss rate, which can be presented in the following form

$$\lambda = \mathbb{P}[b(t) > C]\left(\mathbb{E}[a(t)|b(t) > C] - \mathbb{E}\left[\frac{C}{L(t)}\middle|b(t) > C\right]\right) =$$

$$= P_{loss}\left(\overline{a}_{loss} - \frac{C}{\widetilde{L}_{loss}}\right), \tag{9}$$

where

$$\bar{a}_{loss} = \mathbb{E}\big[a(t)\big|b(t) > C\big]$$

is the mean packet arrival rate in a loss period, and

$$\tilde{L}_{loss} = \left(\mathbb{E}\left[\frac{1}{|L(t)|}\bigg|b(t) > C\right]\right)^{-1} \tag{10}$$

is the harmonic mean of the packet lengths in a loss period. The mean length of packets arrived in a loss period $\bar{L}_{loss} = \mathbb{E}\big[L(t)\big|b(t) > C\big]$ can be used to estimate the mean packet loss rate. It follows from Jensen inequality that $\tilde{L}_{loss} \leq \bar{L}_{loss}$, and hence the following upper bound for the packet loss rate is valid:

$$\lambda \leq P_{loss}\left(\bar{a}_{loss} - \frac{C}{\bar{L}_{loss}}\right). \tag{11}$$

Let $F_{a,L}(x, y)$ be the joint probability distribution function of the packet arrival rate $a(t)$ and packet length $L(t)$. Then the mean packet loss rate λ can be calculated as

$$\lambda = \int_{xy>C,y>0} \left(x - \frac{C}{y}\right) F_{a,L}(dx, dy). \tag{12}$$

Particularly, when the packet rate process $a(t)$ and the packet length process $L(t)$ are independent, we have

$$\lambda = \int_0^\infty \int_{C/y}^\infty \left(x - \frac{C}{y}\right) dF_a(x) dF_L(y) = \int_0^\infty \int_{C/y}^\infty (1 - F_a(x)) dx \, dF_L(y). \tag{13}$$

In this case the packet loss probability can be expressed in the following form

$$\pi_{packet} = \int_0^\infty \left(1 - \widehat{F}_a\left(\frac{C}{y}\right)\right) dF_L(y) \tag{14}$$

where the probability distribution function $\widehat{F}_a(x)$ is given by

$$\widehat{F}_a(x) = \frac{1}{\bar{a}} \int_0^x (1 - F_a(z)) dz. \tag{15}$$

Let the bit and packet rates be driven by an irreducible Markov Chain $J(t)$. Characteristics of busy periods and cumulative amount of bits and packets lost during a busy period can be derived using matrix analytic methods. In the next sections we develop tools for such analysis.

3 Arrival Flows Driven by a Markov Chain

Assume that the bit and packet rates are driven by an irreducible Markov Chain $J(t)$ with a finite state space $\mathfrak{J}$. Let α_i be the packet arrival rate and d_i be the length of arriving packets when $J(t)$ is in state i. Then $\beta_i = \alpha_i d_i$ is the bit arrival rate in state i. Random processes represent the cumulative amount of packets and

$$A(T) = \int_0^T \alpha_{J(t)} dt, \ B(T) = \int_0^T \beta_{J(t)} dt$$

bits arrived in $(0, T)$. Observe that the set $S = \{i \in \mathfrak{J} \,|\, \beta_i > C\}$ specifies the phases of $J(t)$ where the bit rate exceeds the link capacity, and successive sojourn times in S form a sequence of loss periods.

Loss period probability, as well as packet loss probability and bit loss probability, can be easily calculated as

$$P_{\text{loss}} = \sum_{i \in S} q_i, \quad \pi_{\text{packet}} = \frac{\sum\limits_{i \in S}(\alpha_i - \frac{C}{d_i})q_i}{\sum\limits_{i \in \mathfrak{J}} \alpha_i q_i}, \quad \pi_{\text{bit}} = \frac{\sum\limits_{i \in S}(\beta_i - C)q_i}{\sum\limits_{i \in \mathfrak{J}} \beta_i q_i},$$

where q_i, $i \in \mathfrak{J}$, is the stationary distribution of Markov Chain $J(t)$.

A phase-type (PH-type) distribution $F(x)$ of the order N is a function expressed as $F(x) = 1 - \boldsymbol{\tau} \exp(\mathbf{T}x)\mathbf{1}$ for $x > 0$, where $\boldsymbol{\tau}$ is a stochastic row vector and $\mathbf{T}$ is a subgenerator matrix of the order N [9]. PH-type distributions are widely used in stochastic modeling due to their tractability by matrix-analytical methods. If a random variable has the mean m and the coefficient of variation c then its probability distribution function can be approximated by a PH-type distribution $F(x)$ as follows.

a) For $c > 1$ the vector $\boldsymbol{\tau}$ and matrix $\mathbf{T}$ of the order $N = 2$ are given by

$$\boldsymbol{\tau} = (t_1, t_2), \qquad \mathbf{T} = \frac{2}{m}\begin{bmatrix} -t_1 & 0 \\ 0 & -t_2 \end{bmatrix},$$

$$t_1 = \frac{1}{2}\left(1 + \sqrt{\frac{c^2-1}{c^2+1}}\right), \quad t_1 = \frac{1}{2}\left(1 - \sqrt{\frac{c^2-1}{c^2+1}}\right);$$

b) For $c \le 1$ the order $N = \lceil c^{-2} \rceil$ of the PH-distribution is equal to the minimum integer greater or equal than c^{-2}, and the vector $\boldsymbol{\tau}$ and matrix $\mathbf{T}$ are given by

$$\boldsymbol{\tau} = (0, \ldots, 0, 1), \qquad \mathbf{T} = \frac{1}{m}\begin{bmatrix} -t_1 & & & \mathbf{O} \\ t_2 & -t_2 & & \\ & \ddots & \ddots & \\ \mathbf{O} & & t_2 & -t_2 \end{bmatrix},$$

$$t_1 = \frac{N}{1 + \sqrt{(N-1)(c^2N-1)}} \, , \qquad t_2 = \frac{N}{1 - \sqrt{\dfrac{c^2N-1}{N-1}}} \, .$$

The third and higher moments can also be used to get better PH-type approximation of probability distributions as shown in [10] and [11].

Consider a traffic source with alternating on- and off-periods. The lengths of the on and off-periods are independent random variables, the on-periods of the source j have a common irreducible PH-type distribution $F_{\mathrm{on}}(x) = 1 - \boldsymbol{\tau}_{\mathrm{on}} \exp(\mathbf{S}_{\mathrm{on}}x)\mathbf{1}$ of an order n_{on} and the off-periods have another irreducible PH-type distribution $F_{\mathrm{off}}(x) = 1 - \boldsymbol{\tau}_{\mathrm{off}} \exp(\mathbf{S}_{\mathrm{off}}x)\mathbf{1}$ of an order n_{off}. Then the phases of on and off-periods of the traffic source can be described by the Markov chain $J(t)$ with $n = n_{\mathrm{on}} + n_{\mathrm{off}}$ states and the generator matrix $\mathbf{Q}$ given by

$$\mathbf{Q} = \begin{bmatrix} \mathbf{S}_{\mathrm{on}} & \mathbf{s}_{\mathrm{on}}\boldsymbol{\tau}_{\mathrm{off}} \\ \mathbf{s}_{\mathrm{off}}\boldsymbol{\tau}_{\mathrm{on}} & \mathbf{S}_{\mathrm{off}} \end{bmatrix}, \quad \mathbf{s}_{\mathrm{on}} = -\mathbf{S}_{\mathrm{on}}\mathbf{1}, \quad \mathbf{s}_{\mathrm{off}} = -\mathbf{S}_{\mathrm{off}}\mathbf{1}.$$

Note that by definition bit arrival rates α_i and packet arrival rates β_i are equal to 0 for all phases i of off period i.e. when $n_{\mathrm{on}} < i \leq n$.

4 Multivariate Markov-Modulated Fluid Flow

Multivariate Markov-modulated fluid flow (MMFF) is a particular case of Markov additive process [5]. It is a two-component Markov process $(\mathbf{F}(t), J(t))$, $t \geq 0$, where $J(t)$, called phase, is a Markov chain with a finite state space $\mathfrak{J}$, and $\mathbf{F}(t) = (F^{(1)}(t), F^{(2)}(t), \ldots, F^{(m)}(t)) \in \mathbb{R}^m_+$ represents arriving flows. Here $F^{(k)}(t)$ is the cumulative amount of the type k fluid arrived in $(0, t)$, which increases at a constant rate $r_i^{(k)} \geq 0$ if $J(t) = i$, i.e.

$$F^{(k)}(t) = \int\limits_0^t r_{J(x)}^{(k)} dx \, .$$

4.1 Transition Probabilities

MMFF $(\mathbf{F}(t), J(t))$ for all $i, j \in \mathfrak{J}$, $\mathbf{y}, \mathbf{z} \in \mathbb{R}^m_+$ and nonnegative real numbers h, t satisfies the following property

$$\begin{aligned} \mathbb{P}\big[\mathbf{F}(h+t) - \mathbf{F}(h) < \mathbf{z}, J(h+t) = j \mid \mathbf{F}(h) = \mathbf{y}, J(h) = i\big] = \\ = \mathbb{P}\big[\mathbf{F}(t) < \mathbf{z}, J(t) = j \mid \mathbf{F}(0) = \mathbf{0}, J(0) = i\big], \end{aligned} \quad (16)$$

i.e. the process $(J(t), \mathbf{F}(t))$ is homogeneous in the second components ([6]), for which many results are available. Let $I(\cdot)$ be an indicator function. For

$\mathrm{Re}(x_k) \geq 0$, $k = 1, 2, \ldots, m$, and $\mathbf{x} = (x_1, x_2, \ldots x_m)$, the matrix $\mathbf{\Pi}(t, \mathbf{x})$ with elements

$$\pi_{i,j}(t, \mathbf{x}) = \mathbb{E}\left[\exp\left(-\sum_{k=1}^{m} x_k F_k(t)\right) \cdot I(J(t) = j)\,\Big|\,\mathbf{F}(0) = \mathbf{0}, J(0) = i\right],$$

is given by

$$\mathbf{\Pi}(t, \mathbf{x}) = e^{\mathbf{Q}(\mathbf{x})t}, \tag{17}$$

where

$$\mathbf{Q}(\mathbf{x}) = \mathbf{Q} - \sum_{k=1}^{m} x_k \mathbf{R}^{(k)},$$

$\mathbf{Q}$ is the infinitesimal generator of $J(t)$ and $\mathbf{R}^{(k)} = \mathrm{diag}(r_i^{(k)})$ is the diagonal matrix ([6]). It follows from (16)-(17) that the transition probabilities of MMFF $(\mathbf{F}(t), J(t))$ are uniquely characterized by the generator matrix $\mathbf{Q}$ of the Markov chain $J(t)$ and m diagonal matrices $\mathbf{R}^{(1)}, \mathbf{R}^{(2)}, \ldots, \mathbf{R}^{(m)}$ specifying the fluid rates.

We assume that $\mathbf{Q}$ is an irreducible infinitesimal generator. Then the stationary probability vector $\mathbf{q}$ of $J(t)$ is the unique solution of the linear system

$$\mathbf{qQ} = 0, \quad \mathbf{q1} = 1,$$

where $\mathbf{1}$ is the all-ones column vector of the appropriate size.

Basic stationary characteristics of the rate processes $r_{J(t)}^{(k)}$ are as follows.

The expected rate of the type k flow:

$$r^{(k)} = \lim_{T \to \infty} \frac{1}{T} \mathbb{E}[r_{J(T)}^{(k)}] = \mathbf{qR}^{(k)}\mathbf{1}. \tag{18}$$

The cross-correlation functions:

$$\rho^{(k,l)}(t) = \lim_{T \to \infty} \mathbb{E}\left[r_{J(t)}^{(k)} r_{J(T+t)}^{(l)}\right] = \mathbf{qR}^{(k)}\mathbf{\Pi}(t)\mathbf{R}^{(l)}\mathbf{1}, \quad t \geq 0, \tag{19}$$

where $\mathbf{\Pi}(t) = \mathbf{\Pi}(t, \mathbf{0}) = e^{\mathbf{Q}t}$ is the matrix of transition probabilities of the phase process $J(t)$.

4.2 Multivariate MMFF with Partitioned State Space

Let $\mathcal{Z}$ be a nonempty subset of the phase space $\mathfrak{J}$ with nonempty complement $\bar{\mathcal{Z}} = \mathfrak{J} \setminus \mathcal{Z}$. We partition matrices and vectors to conform with the partition $\mathfrak{J} = \mathcal{Z} \cup \bar{\mathcal{Z}}$, thus we write $\mathbf{q} = [\mathbf{q}_\mathcal{Z}, \mathbf{q}_{\bar{\mathcal{Z}}}]$ and

$$\mathbf{Q} = \begin{bmatrix} \mathbf{Q}_{\mathcal{Z}, \mathcal{Z}} & \mathbf{Q}_{\mathcal{Z}, \bar{\mathcal{Z}}} \\ \mathbf{Q}_{\bar{\mathcal{Z}}, \mathcal{Z}} & \mathbf{Q}_{\bar{\mathcal{Z}}, \bar{\mathcal{Z}}} \end{bmatrix}.$$

Consider the sequence of passage times to subsets $\mathcal{Z}$ and $\bar{\mathcal{Z}}$ defined by

$$t_{\bar{\mathcal{Z}}}(0) = 0, \quad t_{\mathcal{Z}}(n) = \inf\{t > t_{\bar{\mathcal{Z}}}(n-1) \mid J(t) \in \mathcal{Z}\}, \quad n \geq 1.$$

Then for $n \geq 1$,

$\tau_Z(n) = t_{\bar{Z}}(n) - t_Z(n)$ is the n-th sojourn time in Z,

$\varphi_Z^{(k)}(n) = F^{(k)}(t_{\bar{Z}}(n)) - F^{(k)}(t_Z(n))$ is the cumulative amount of the type k fluid generated during the n-th sojourn time in Z,

$\tau_{\bar{Z}}(n) = t_Z(n+1) - t_{\bar{Z}}(n)$ is the n-th sojourn time in $\bar{Z}$,

$\varphi_{\bar{Z}}^{(k)}(n) = F_k(t_Z(n+1)) - F_k(t_{\bar{Z}}(n))$ is the cumulative amount of the type k fluid generated during n-th sojourn time in $\bar{Z}$.

It follows from (16) and (17) that the matrix $\mathbf{G}_{Z,\bar{Z}}(s,\mathbf{x})$ with elements

$$\mathbb{E}\left[\exp\left(-s\tau_Z(n) - \sum_{k=1}^{m} x_k \varphi_Z^{(k)}(n)\right) \cdot I(J(t_{\bar{Z}}(n)) = j)\bigg| J(t_Z(n)) = i\right], \quad i \in Z, j \in \bar{Z}$$

does not dependent on n and for $\mathrm{Re}(s) \geq 0$ is given by

$$\mathbf{G}_{Z,\bar{Z}}(s,\mathbf{x}) = (s\mathbf{I} - \mathbf{Q}_{Z,Z}(\mathbf{x}))^{-1}\mathbf{Q}_{Z,\bar{Z}}, \tag{20}$$

where $\mathbf{I}$ is the identity matrix of the appropriate size.

The matrix (20) can be used for calculation of different characteristics of multivariate MMFF with partitioned state space. We use it for analysis of first and second order characteristics of amounts of fluid generated during sojourn times in a subset of the phase space.

It follows from (20) (see also [8]) that the embedded Markov Chain $J(t_Z(n))$, $n \geq 1$, has the matrix of transition probabilities $\mathbf{P}_Z$ and the stationary vector $\mathbf{p}_Z$ given by

$$\mathbf{P}_Z = \mathbf{P}_{Z,\bar{Z}}\mathbf{P}_{\bar{Z},Z}, \qquad \mathbf{p}_Z = \theta_Z \mathbf{q}_{\bar{Z}}\mathbf{Q}_{\bar{Z},Z}, \tag{21}$$

where

$$\mathbf{P}_{Z,\bar{Z}} = -\mathbf{Q}_{Z,Z}^{-1}\mathbf{Q}_{Z,\bar{Z}}, \quad \theta_Z = (\mathbf{q}_Z \mathbf{Q}_{Z,\bar{Z}}\mathbf{1})^{-1}.$$

Note that $\theta_Z = \theta_{\bar{Z}}$ and probability vector $\mathbf{p}_Z$ satisfies the following property:

$$\mathbf{p}_Z\mathbf{P}_{Z,\bar{Z}} = \mathbf{p}_{\bar{Z}}. \tag{22}$$

The Laplace transform of the stationary distribution of the sojourn time $\tau_Z(n)$ and the cumulative amount of fluid $\varphi_Z^{(k)}(n)$ generated during this sojourn time is given by

$$g_Z(s,\mathbf{x}) = \lim_{n\to\infty} \mathbb{E}\left[\exp\left(-\left(s\tau_Z(n) + \sum_{k=1}^{m} x_k \varphi_Z^{(k)}(n)\right)\right)\right] = \mathbf{p}_Z\mathbf{G}_{Z,\bar{Z}}(s,\mathbf{x})\mathbf{1}. \tag{23}$$

The Laplace transform of the stationary joint distribution of the sojourn times $\tau_Z(n+1), \tau_{\bar{Z}}(n+1), \tau_Z(n+2), \tau_{\bar{Z}}(n+2),\ldots,\tau_Z(n+N), \tau_{\bar{Z}}(n+N)$ and corresponding cumulative amounts of generated fluid is given by

$$g_{Z,n}\left(s_{Z,1}, \mathbf{x}_{Z,1}, s_{\bar{Z},1}, \mathbf{x}_{\bar{Z},1}, \ldots, s_{Z,N}, \mathbf{x}_{Z,N}, s_{\bar{Z},N}, \mathbf{x}_{\bar{Z},N}\right) =$$

$$\lim_{n\to\infty} \mathbb{E}\left[\exp\left(-\sum_{r=1}^{N}\left(s_{Z,r}\tau_Z(n+r) + s_{\bar{Z},r}\tau_{\bar{Z}}(n+r)\right.\right.\right.$$

$$\left.\left.\left. + \sum_{k=1}^{m} x_{Z,r,k}\,\varphi_Z^{(k)}(n+r) + \sum_{k=1}^{m} x_{\bar{Z},r,k}\,\varphi_{\bar{Z}}^{(k)}(n+r)\right)\right)\right] =$$

$$= \mathbf{p}_Z \mathbf{G}_Z\left(s_{Z,1}, \mathbf{x}_{Z,1}, s_{\bar{Z},1}, \mathbf{x}_{\bar{Z},1}\right) \cdots \mathbf{G}_Z\left(s_{Z,N}, \mathbf{x}_{Z,N}, s_{\bar{Z},N}, \mathbf{x}_{\bar{Z},N}\right)\mathbf{1}, \qquad (24)$$

where

$$\mathbf{G}_Z\left(s_Z, \mathbf{x}_Z, s_{\bar{Z}}, \mathbf{x}_{\bar{Z}}\right) = \mathbf{G}_{Z,\bar{Z}}\left(s_Z, \mathbf{x}_Z\right)\mathbf{G}_{\bar{Z},Z}\left(s_{\bar{Z}}, \mathbf{x}_{\bar{Z}}\right).$$

By differentiating (23) successively with respect to s and setting $s = x_1 = x_2 = \cdots x_m = 0$, we obtain the stationary r th moment of the sojourn time in $\mathcal{Z}$ as

$$u_{Z,r} = \lim_{n\to\infty} \mathbb{E}\left[\left(\tau_Z(n)\right)^r\right] = r!\,\mathbf{p}_Z\mathbf{U}_Z^r\mathbf{1}. \qquad (25)$$

where $\mathbf{U}_Z = -\mathbf{Q}_{Z,Z}^{-1}$, the matrix of conditional expected sojourn times in the states of $\mathcal{Z}$, satisfies

$$\mathbf{p}_Z\mathbf{U}_Z = \theta_Z\mathbf{q}_Z.$$

Particularly, the stationary mean $u_Z = u_{Z,1}$ and the variance $\sigma_Z^2 = u_{Z,2} - u_{Z,1}^2$ of the sojourn time in $\mathcal{Z}$ are given by

$$u_Z = \theta_Z\mathbf{q}_Z\mathbf{1}, \qquad \sigma_Z^2 = 2\,\theta_Z\mathbf{q}_Z\mathbf{U}_Z\mathbf{1} - \left(\theta_Z\mathbf{q}_Z\mathbf{1}\right)^2. \qquad (26)$$

By differentiating (23) with respect to x_k and setting $s = x_1 = x_2 = \cdots x_m = 0$, we obtain the stationary mean $m_Z^{(k)}$ and variance $v_Z^{(k)}$ of the amount of the type k fluid $\varphi_Z^{(k)}(n)$ generated during the sojourn time $\tau_Z(n)$ as

$$m_Z^{(k)} = \theta_Z\mathbf{q}_Z\mathbf{R}_{Z,Z}^{(k)}\mathbf{1}, v_Z^{(k)} = 2\theta_Z\mathbf{q}_Z\mathbf{R}_{Z,Z}^{(k)}\mathbf{U}_Z\mathbf{R}_{Z,Z}^{(k)}\mathbf{1} - \left(\theta_Z\mathbf{q}_Z\mathbf{R}_{Z,Z}^{(k)}\mathbf{1}\right)^2. \qquad (27)$$

Similarly using (23) we obtain the stationary covariance of the amount of the type k_1 fluid $\varphi_Z^{(k_1)}(n)$ and amount of the type k_2 fluid $\varphi_Z^{(k_2)}(n)$ generated during the sojourn time $\tau_Z(n)$ as

$$\gamma_Z^{(k_1,k_2)} = \theta_Z\left(\mathbf{q}_Z\mathbf{R}_{Z,Z}^{(k_1)}\mathbf{U}_Z\mathbf{R}_{Z,Z}^{(k_2)}\mathbf{1} + \mathbf{q}_Z\mathbf{R}_{Z,Z}^{(k_2)}\mathbf{U}_Z\mathbf{R}_{Z,Z}^{(k_1)}\mathbf{1}\right) - m_Z^{(k_1)}m_Z^{(k_2)}, \qquad (28)$$

and the covariance between the amount of type k_1 fluid $\varphi_Z^{(k_1)}(n)$ generated during sojourn time $\tau_Z(n)$ and the amount of type k_2 fluid $\varphi_{\bar{Z}}^{(k_2)}(n)$ generated during next sojourn time $\tau_{\bar{Z}}(n)$ as

$$\gamma_{Z,\bar{Z}}^{(k_1,k_2)} = \theta_Z\,\mathbf{q}_Z\mathbf{R}_Z^{(k_1)}\mathbf{U}_Z\mathbf{Q}_{Z,\bar{Z}}\mathbf{U}_{\bar{Z}}\mathbf{R}_{\bar{Z}}^{(k_2)}\mathbf{1} - m_Z^{(k_1)}m_Z^{(k_2)}. \qquad (29)$$

The complexity of the calculation of the stationary moments and covariance functions in formulas (25) – (29) is determined by the complexity of the calculation of the matrices $\mathbf{U}_z = -\mathbf{Q}_{z,z}^{-1}$ and $\mathbf{U}_{\bar{z}} = -\mathbf{Q}_{\bar{z},\bar{z}}^{-1}$. The nonsingular matrices $-\mathbf{Q}_{z,z}$ and $-\mathbf{Q}_{\bar{z},\bar{z}}$ are diagonally dominant with entries positive on the diagonal and nonpositive elsewhere, consequently, it is a M-matrix ([12]). Stable inversion of $n \times n$ M-matrix may be achieved using LU-decomposition without pivoting with complexity $O(n^3)$.

5 Bufferless Markov-Modulated Fluid Flow System

Now we consider a bufferless transmission link with capacity C that carries several classes of traffic flows described by a multivariate Markov-modulated fluid flow $(\mathbf{A}(t), \mathbf{B}(t), J(t))$. Here $\mathbf{A}(t) = (A^{(1)}(t), A^{(2)}(t), ..., A^{(m)}(t))$ represents cumulative amount of arriving packets, and $\mathbf{B}(t) = (B^{(1)}(t), B^{(2)}(t), ..., B^{(m)}(t))$ represents cumulative amount of arriving bits. The process $A^{(k)}(t)$ is characterized by packet arrival rates $\alpha_i^{(k)}$ and process $B^{(k)}(t)$ is characterized by bit arrival rates $\beta_i^{(k)}$.

The rates $\mu_i^{(k)}$ of the fluid flow $\mathbf{D}(t) = (D^{(1)}(t), D^{(2)}(t), ..., D^{(m)}(t))$, which represents the cumulative amount of processed fluid, depend of on the policy which specify how available capacity is shared among arrival flows.

Capacity sharing policy can be identified with a probability vector $\boldsymbol{\pi} = (\pi^{(1)}, \pi^{(2)}, ..., \pi^{(m)})$. The capacity offered to a source l will be at least $\pi^{(l)}\Theta$. When the arrival bit rate $\beta_i^{(l)}$ does not exceed the threshold $\pi^{(l)}C$, the bits from the source l are processed with the rate $\mu_i^{(l)} = \beta_i^{(l)}$. The capacity $\pi^{(l)}C - \beta_i^{(l)}$ that is not used by the source l is consumed by the remaining sources. Thus the total capacity $\mu_i^{(k)}$ allocated to a source with arrival rate $\lambda_i^{(k)} > \pi^{(k)}C$ is given by

$$\mu_i^{(k)} = \frac{\pi^{(k)}}{\sum\limits_{l:\,\lambda_i^{(l)} > \pi^{(l)}C} \pi^{(l)}} \left(C - \sum\limits_{l:\,\lambda_i^{(l)} \le \pi^{(l)}C} \lambda_i^{(l)} \right).$$

In *priority policy*, each flow is assigned a threshold $\Theta^{(k)}$, where $0 < \Theta^{(m)} \le \Theta^{(m-1)} \le \cdots \le \Theta^{(1)} = C$. Flow 1 has the highest priority, flow 2 the second highest priority and so on. Lower priority flow fluid can only be processed when there is no demand from higher priority flows. The processing rates of the type k fluid are calculated as:

$$\mu_i^{(1)} = \min\{\lambda_i^{(1)}, \Theta^{(1)}\}, \quad \mu_i^{(k)} = \min\left\{\beta_i^{(k)}, \left(\Theta^{(k)} - \sum\limits_{l=1}^{k-1}\mu_i^{(l)}\right)_+\right\},$$

$$k = 2, 3, ..., m,$$

where $x_+ = x$ if $x \ge 0$ and $x_+ = 0$ if $x < 0$.

Fluid flow $R(t) = (R^{(1)}(t), R^{(2)}(t), ..., R^{(m)}(t))$ with $R^{(k)}(t) = B^{(k)}(t) - D^{(k)}(t)$ represents the amount of lost bits and its rates $\eta_i^{(k)}$ are given by

$$\eta_i^{(k)} = \beta_i^{(k)} - \mu_i^{(k)}, \ k = 1, 2, ..., m.$$

The rates $\phi_i^{(k)}$ of fluid flow representing the amount of lost packets are given by

$$\phi_i^{(k)} = \alpha_i^{(k)} \left(1 - \frac{\mu_i^{(k)}}{\beta_i^{(k)}} \right), \ k = 1, 2, ..., m.$$

For each traffic flow we can define a subset $\mathcal{S}_k$ of the phase space $\mathfrak{J}$ by $\mathcal{S}_k = \{ i \in \mathfrak{X} \mid \beta_i^{(k)} > \mu_i^{(k)} \}$. This is the set of all phases, in which type k traffic suffers losses. Applying results from the previous section we can obtain various characteristics of congestion periods and amount of lost bits and packets of different types.

6 Conclusion

The use of bufferless models allows analysis of different characteristics of interactive audio and multimedia systems. In this paper we consider a bufferless system processing several fluid flows driven by a Markov Chain. We propose a method for the calculation of moments of congestion periods, their spacing and lengths, and amount of bits and packets lost during congestion periods. We also demonstrate how the proposed method can be used for efficiency analysis of policies that specify how available transmission capacity is shared among arrival flows in a bufferless system.

Acknowledgment. Sincere thanks are given to the anonymous referees, who provided the authors with helpful comments.

References

1. Knightly, E.W., Shroff, N.B.: Admission Control for Statistical QoS: Theory and Practice. IEEE Network 13(2), 20–29 (1999)
2. Haßlinger, G., Hartleb, F., Fiedler, M.: The Relevance of the Bufferless Analysis for Traffic Management in Telecommunication Networks, Proc. IEEE European Conference on Universal Multiservice Networks, Colmar, France, pp. 823–827 (2000)
3. Haßlinger, G., Fiedler, M.: Why buffers in routing and switching systems do not essentially improve QoS: An analytical case study for aggregated On-Off traffic, Internet performance & Control of Network Systems, SPIE, Vol. 4865, pp. 47–58 (2002)
4. Haßlinger, G., Takes, P.: Real Time Video Traffic Characteristics and Dimensioning Regarding QOS Demands, Proc. of ITC-18, pp. 1211–1220 (2003)
5. Cinlar, E.: Introduction to Stochastic Processes. Prentice-Hall, Englewood Cliffs, NJ (1975)

6. Ezhov, I.I., Skorokhod, A.V.: Markov processes with homogeneous second component, Theory of Probability and Its Applications. Part I, vol. 14(1), pp. 1–13, Part II, vol. 14(4), pp. 652–667 (1969)
7. Ferrandiz, J.M.: Analysis of Fluid Buffer Models with Markov Modulated Rates, Hewlett-Packard Laboratories, Technical report HPLB-NSMG-92-8 (1992)
8. Latouche, G., Ramaswami, V.: Introduction to Matrix Analytic Methods in Stochastic Modeling, SIAM (1999)
9. Neuts, M.F.: Matrix-Geometric Solutions in Stochastic Models—An Algorithmic Approach. The Johns Hopkins University Press, Baltimore (1981)
10. Osagami, T., Harchol-Balter, M.: A Closed-Form Solution for Mapping General Distributions to Minimal PH Distribution. In: Kemper, P., Sanders, W.H. (eds.) TOOLS 2003. LNCS, vol. 2794, pp. 200–217. Springer, Heidelberg (2003)
11. Bobbio, A., Horváth, A., Telek, M.: Matching three moments with minimal acyclic phase type distributions. Stochastic Models 21, 303–326 (2005)
12. Berman, A., Plemmons, R.J.: Nonnegative Matrices in the Mathematical Sciences. Academic Press, NY (1979)

Adaptive Admission Control in Mobile Cellular Networks with Streaming and Elastic Traffic

David Garcia-Roger, M.ª Jose Domenech-Benlloch, Jorge Martinez-Bauset, and Vicent Pla

Departamento de Comunicaciones, Universidad Politecnica de Valencia
Camino de Vera s/n, 46022, Valencia, Spain
Phone: +34 963879733; Fax: +34 963877309
{dagarro,mdoben}@doctor.upv.es,{jmartinez,vpla}@dcom.upv.es

Abstract. We propose a novel adaptive reservation scheme that handles, in an integrated way, streaming and elastic traffic. The scheme continuously adjusts the quality of service perceived by users, adapting to any mix of traffic and enforcing a differentiated treatment among services, both in fixed and variable capacity systems. The performance evaluation carried out verifies that the QoS objective is met with an excellent precision and that it converges rapidly to new operating conditions. Other key features of our scheme are its simplicity and its oscillation-free behavior.

1 Introduction

Applications expected to produce the bulk of traffic in the future multiservice Internet can be broadly categorized as streaming or elastic [1]. Streaming traffic requires a minimum transfer rate in order to work properly as well as some time related requirements such as bounded delay and jitter. Elastic traffic has loose time requirements and can adapt to the available resources. In the light of the above it seems natural to give priority to streaming traffic and leave elastic traffic use the remaining capacity (a small amount of resources might be reserved for the elastic traffic to prevent starvation in case of overload of the streaming traffic). Elastic flows are generally transported over TCP which takes care of rate adaptation and bandwidth sharing among the different flows. If the total traffic demand of elastic flows exceeds the available capacity some flows might be aborted due to impatience. Flow impatience due to a very low throughput can arise from human user impatience or because TCP or higher layer protocols interpret that the connection is broken. Abandonments are useful to cope with overload and serve to stabilize the system but, on the other hand, this phenomenon will negatively impact on the efficiency because capacity is wasted by non-completed flows [1]. This drop of efficiency led the authors of [1] to claim that session admission control (SAC) should be enforced for elastic traffic.

In this paper we propose an adaptive SAC scheme for mobile wireless cellular networks that handles in an integrated way both streaming and elastic traffic

L. Mason, T. Drwiega, and J. Yan (Eds.): ITC 2007, LNCS 4516, pp. 925–937, 2007.

and tries to maximize the carried traffic while meeting certain quality of service (QoS) objectives. The QoS objective for streaming traffic is expressed as upper bounds for the blocking probabilities of new and handover requests, while for elastic traffic it is defined as a bound for the abandonment probability. The proposed scheme is adaptive in the sense that if the offered load is above the system capacity, or the number of resource units decreases, or both simultaneously, the SAC system will react trying to meet the QoS objective for as many services as possible. Therefore the proposed scheme might be deployed in both fixed capacity systems (e.g. FDMA/TDMA) and systems limited by interference where capacity is variable (e.g. CDMA).

Our work is motivated in part by the fact that previous adaptive proposals like [2,3,4,5,6] deploy long measurement windows to estimate system parameters, which make the convergence period too long to cope with real operating conditions, or do not provide explicit indication of how the time window must be configured [7,8,9]. Another motivation is the fact that most of the studies devoted to adaptive schemes only consider the stationary regime and no evidence is provided about their behavior in the transient regime. Therefore, we consider that a fundamental characteristic of an adaptive scheme like its convergence speed to new operating conditions has not been sufficiently explored.

Our scheme does not rely on measurement intervals to estimate the value of system parameters. It generalizes the novel SAC adaptive strategy introduced in [10], which operates in coordination with two well known trunk reservation policies named Multiple Guard Channel (MGC) and Multiple Fractional Guard Channel, although only its operation with the MGC is described here. It has been shown that deploying trunk reservation instead of complete-sharing policies in mobile networks allows the operator to achieve higher system capacity, i.e. to carry more traffic while meeting certain QoS objectives [11].

Our new scheme has four key features that enhance the scheme in [10]. First it handles in an integrated way both streaming and elastic traffic. Second, it allows to enforce a differentiated treatment among different streaming services during under load and overload episodes. In the latter case, this differentiated treatment guarantees that higher priority services will be able to meet their QoS objective, possibly at the expense of lower priority ones. Third, the prioritization order of the streaming services can be fully specified by the operator. And fourth, the operator has the possibility of identifying one of the streaming services as best-effort, being it useful to concentrate on it the penalty that has to be unavoidably paid during overloads.

Adaptive SAC mechanisms have also been studied, for example in [4,5,6,12], both in single service and multiservice scenarios, but in a context which is somewhat different to the one of this paper. There, the adjustment of the SAC policy configuration is based on estimates of both the mobility pattern and the handover arrival rates derived from the current number of ongoing calls in neighboring cells. It is expected that the performance of our scheme would improve when provided with such predictive information but this is left for further study.

An extension of our scheme to operate with rate-adaptive multimedia applications [9] is also left for further study.

Given that the operation of the SAC scheme when handling streaming traffic is independent of the elastic traffic because the former has higher priority than the latter, we describe first the operation of the SAC scheme and evaluate its performance with streaming traffic. Section 2 describes the model of the system and defines the relevant SAC policies for our study. Section 3 describes the fundamentals of the adaptive scheme, introducing the policy adjustment strategy and how multiple streaming services are handled. In Section 4 we present the performance evaluation of the proposed adaptive scheme when handling streaming traffic in different scenarios, both under stationary and non-stationary traffic conditions. Section 5 describes the operation of the scheme when handling elastic traffic and evaluates its performance. Finally, Section 6 concludes the paper.

2 System Model and SAC Policies

We consider the homogeneous case where all cells are statistically identical and independent. Consequently the global performance of the system can be analyzed focusing on a single cell. Nevertheless, given that the proposed scheme is adaptive it could also be deployed in non-homogeneous scenarios.

In each cell a set of R different streaming services contend for C resource units, where the meaning of a unit of resource depends on the specific implementation of the radio interface. For each streaming service, new and handover arrival requests are distinguished, which defines $2R$ arrival classes. For convenience, we denote by s_i the arrival class i, $1 \leq i \leq 2R$. Additionally we denote by s_r^n (s_r^h), the arrival class associated to new (handover) requests of streaming service r, $1 \leq r \leq R$, being $s_r^n = s_r$ and $s_r^h = s_{r+R}$, $1 \leq r \leq R$. For brevity, when we refer to a service or to a class, we mean a streaming service or a streaming arrival class respectively. Elastic traffic is discussed in Section 5.

For any service r, new (handover) requests arrive according to an inhomogeneous Poisson process with time-varying rate $\lambda_r^n(t)$ ($\lambda_r^h(t)$). For mathematical tractability we make the common assumption to model the inter-arrival time of handover requests as an exponential distribution, which is considered a good approximation [13]. Besides, although our scheme does not require any relationship between $\lambda_r^h(t)$ and $\lambda_r^n(t)$, for simplicity we suppose that $\lambda_r^h(t)$ is a constant fraction of $\lambda_r^n(t)$ [14,15]. Service r requests require d_r resource units per session. As each service has two associated arrival classes, if we denote by c_i the amount of resource units that an arrival class requires for each session, then $d_r = c_r = c_{r+R}$, $1 \leq r \leq R$. For variable bit rate sources d_r resource units denotes the effective bandwidth of the session [15,16].

For a service r session, both its duration and its cell residence (dwell) time are also assumed to be exponentially distributed with rates μ_r^s and μ_r^d. Hence, the resource holding time for a service r session in a cell is also exponentially distributed with rate $\mu_r = \mu_r^s + \mu_r^d$. Note that the proposed scheme can easily take into account terminals moving at different speeds by defining additional arrival classes for any service. Note also that the exponential assumption also represents a good approximation for the cell dwell time (essentially, only its average matters), when the performance of the system is evaluated by computing blocking probabilities [17]. It should be highlighted that the operation of our scheme is based on the simple balance equations described in Section 3, which hold for any arrival process and holding time distribution. Hence the basis of the adaptive scheme holds beyond the assumptions made for modeling purposes.

We denote by P_i, $1 \leq i \leq 2R$, the blocking probability perceived by class i requests and by $P_r^n = P_r$ ($P_r^h = P_{R+r}$) the blocking probability perceived by new (handover) requests of service r. The QoS objective is expressed as upper bounds for the blocking probabilities of each arrival class. Thus, we denote by B_r^n (B_r^h) the bound for new (handover) blocking probabilities. Let the ongoing sessions vector be $\boldsymbol{n} := (n_1, \ldots, n_R)$, where n_r is the number of sessions in progress of service r in the cell initiated as new or handover requests. We denote by $c(\boldsymbol{n}) = \sum_{r=1}^{R} n_r d_r$ the number of busy resource units in state $\boldsymbol{n}$.

Finally, we denote by λ_{max} the system capacity, i.e. the maximum λ that can be offered to the system while meeting the QoS objectives, where λ is the aggregated arrival rate of new requests $\lambda = \sum_{r=1}^{R} \lambda_r^n$, $\lambda_r^n = f_r \lambda$ and $\sum_{r=1}^{R} f_r = 1$. Defining service penetrations (f_r) is a common approach when studying these systems [15].

The definition of the MGC SAC policy is as follows: one configuration parameter is associated with each arrival class i, $l_i \in \mathbb{N}$. An arrival of class i in state $\boldsymbol{n}$ is accepted if $c(\boldsymbol{n}) + c_i \leq l_i$ and blocked otherwise. Therefore, l_i is the amount of resources that class i has access to and increasing (decreasing) it reduces (augments) P_i. Number based SAC, that is a common technique in systems whose capacity is limited by blocking, has also been considered a good approach for those systems whose capacity is limited by interference, see for example [18] and references therein.

3   Operation of the SAC Adaptive Scheme

Most of the adaptive schemes proposed for single service scenarios deploy a reservation strategy based on *guard channels*, increasing its number when the QoS objective of the handover arrival class is not met. The extension of this heuristic to a scenario with multiple services is much more difficult to manage because the adjustment of the configuration parameter l_i has an impact not only on the QoS perceived by class i but also on the QoS perceived by the rest of classes. Our scheme has been designed to handle this difficulty.

As a first step to handle this difficulty, we classify the different arrival classes into two generic categories: i) several *protected* classes, for which specific QoS objectives must be met; ii) one *Best-Effort Class* (BEC), with no specific QoS objective. Additionally, in a multiservice scenario the operator can define priorities for the protected classes at its convenience in order to give greater protection to the most important classes. Note that BEC arrival requests perceive an

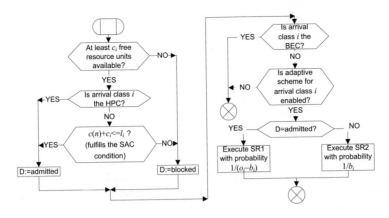

Fig. 1. Operation of the SAC adaptive scheme

unpredictable blocking probability but those sessions accepted are allocated a constant amount of resources during its lifetime.

Let $\mathbf{s} = (s_1, \ldots, s_{2R})$ be the set of arrival classes, $\Pi := \{(\pi_1, \ldots, \pi_{2R}) : \pi_i \in \mathbb{N}, 1 \leq \pi_i \leq 2R, \pi_i \neq \pi_j \text{ if } i \neq j\}$ the set of all possible permutations of $\{1, 2, \ldots, 2R\}$ and $\pi^* \in \Pi$ the order defined by the operator, then $\mathbf{s}^* = (s_{\pi_1}, \ldots, s_{\pi_{2R}})$ is called the *prioritization order*, being s_{π_1} the *Highest-Priority Class* (HPC) and $s_{\pi_{2R}}$ the *Lowest-Priority Class* (LPC). If there is a BEC, this class will be the LPC in the prioritization order. We study two implementations, one in which the LPC is treated as a protected class and one in which the LPC is the BEC.

For the sake of clarity, the operation of our scheme is described assuming that the arrival processes are stationary and the system is in steady state. In practice, we can assume without loss of generality that the QoS objective for s_i can be expressed as $B_i = b_i/o_i$, where $b_i, o_i \in \mathbb{N}$. Then it is expected that if $P_i = B_i$ the class i will experience, in average, b_i rejected requests and $o_i - b_i$ admitted requests, out of o_i offered requests. For example, a QoS objective for s_i of $B_i = 1/100$ implies that $b_i = 1$ and $o_i = 100$. It seems intuitive to think that the adaptive scheme should not change the configuration parameters of those arrival classes meeting their QoS objective and, on the contrary, adjust them on the required direction if the perceived QoS is different from its target. Therefore, assuming integer values for the configuration parameters, like those of the MGC policy, we propose to perform a probabilistic adjustment each time a request is processed in the following way: i) if accepted, do $\{l_i \leftarrow (l_i - 1)\}$ with probability $1/(o_i - b_i)$; ii) if rejected, do $\{l_i \leftarrow (l_i + 1)\}$ with probability $1/b_i$. Therefore, when deploying this adjustment scheme under stationary traffic, if $P_i = B_i$, then, in average, l_i is increased by 1 and decreased by 1 every o_i offered new requests, i.e. its mean value is kept constant. Finally, note that when the traffic is non-stationary, the adaptive scheme will continuously adjust the QoS perceived by each class in order to meet its objective if possible. Our approach, although simple, is innovative because to the best of our knowledge, something based on the same or a similar idea has not been proposed before.

Figure 1 shows the operation of the SAC policy and the adaptive scheme in more detail. As shown, to admit a class i request it is first checked that at least c_i free resource units are available. Note that once this is verified, HPC requests are always admitted, while the rest of classes must also fulfill the admission condition imposed by the SAC policy. Note also that the $l_{\pi 1}$ is always updated to detect when the HPS becomes congested. Due to paper length limitations, the subroutines SR1 and SR2 mentioned in Fig. 1 are not explained in detail. In general, the adaptive scheme associated to each class is always operating (except for the BEC), but to be able to guarantee that the QoS objective is met for as many classes as possible, particularly during overload episodes or changes in the traffic mix, the adjustment algorithm described before requires additional mechanisms which might include the disabling of the adaptive scheme associated to other low priority classes.

When the QoS objective for class i is not met, the MGC policy configuration will be adjusted using two different mechanisms. The *direct* way is to increase the configuration parameter l_i, but its maximum value is C, i.e. when $l_i = C$, full access to resources is provided to class i and setting $l_i > C$ does not provide additional benefits. In these cases, an *indirect* way to help class i is to limit the access to resources of lower priority classes by reducing their associated configuration parameters. It is clear that when a higher priority class s_i needs to adjust the configuration parameter of a lower priority class s_j, the adaptive scheme must adjust l_j only when arrivals from s_i occur, while no adjustments must be carried out when arrivals from s_j occur. To operate in this way the adaptive scheme associated to s_j is disabled.

4   Performance Evaluation

We evaluate the performance of the proposed adaptive SAC scheme by solving the continuous-time Markov chain (CTMC) that describes its operation, both in the stationary and transient regimes. In both regimes P_i is determined as the percentage of time an arrival request from s_i would be rejected.

In general, we have a multidimensional CTMC which state space is given by $(n_1, \ldots, n_R, l_1, \ldots, l_{2R})$. We allow l_i to take positive and negative values as a means to remember past adjustments and to identify the adjustment type the scheme uses (direct or indirect). Given that the general multidimensional diagram is difficult to draw, we show a bidimensional CTMC in Fig. 2 as an example. This system has only one service and therefore two classes, s^h and s^n, with $d = 1$ and C resource units. It is assumed that s^h is the HPC and therefore their requests are always accepted (if free resources are available), while s^n is a BEC. The system state vector is defined as (n, l^h), where n is the number of resource units occupied. In this system, l^h is adjusted following the probabilistic adjustment rule described previously and $l^n = C - \max\{0, (l^h - C)\}$. Note that during under load episodes $l^n = C$, but during overload episodes s^h might have

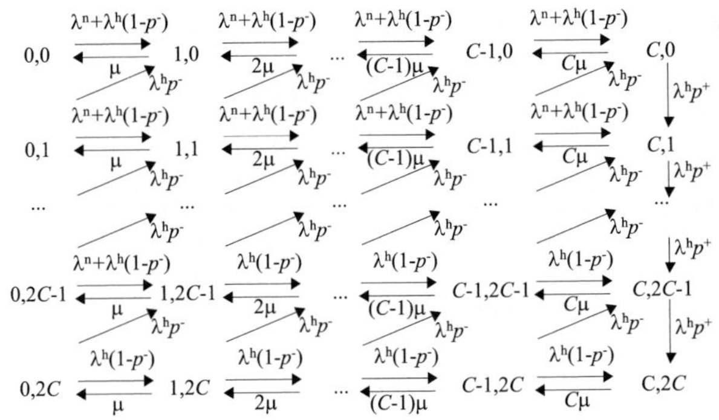

Transition labels (top rows): $\lambda^n + \lambda^h(1-p^-)$ $\lambda^n + \lambda^h(1-p^-)$ $\lambda^n + \lambda^h(1-p^-)$ $\lambda^n + \lambda^h(1-p^-)$

$0,0 \;\rightleftarrows\; 1,0 \;\rightleftarrows\; \cdots \;\rightleftarrows\; C\text{-}1,0 \;\rightleftarrows\; C,0$ (forward rates $\mu,\,2\mu,\,(C\text{-}1)\mu,\,C\mu$; diagonal $\lambda^h p^-$; down $\lambda^h p^+$)

$0,1 \;\rightleftarrows\; 1,1 \;\rightleftarrows\; \cdots \;\rightleftarrows\; C\text{-}1,1 \;\rightleftarrows\; C,1$

Row labels for lower rows: $\lambda^n + \lambda^h(1-p^-)$, then $\lambda^h(1-p^-)$

$0,2C\text{-}1 \;\rightleftarrows\; 1,2C\text{-}1 \;\rightleftarrows\; \cdots \;\rightleftarrows\; C\text{-}1,2C\text{-}1 \;\rightleftarrows\; C,2C\text{-}1$

Bottom row labels: $\lambda^h(1-p^-)$

$0,2C \;\rightleftarrows\; 1,2C \;\rightleftarrows\; \cdots \;\rightleftarrows\; C\text{-}1,2C \;\rightleftarrows\; C,2C$ (forward rates $\mu,\,2\mu,\,(C\text{-}1)\mu,\,C\mu$)

Fig. 2. State diagram of the CTMC in a scenario with two classes

Table 1. Definition of the scenarios under study

	d_1	d_2	f_1	f_2	$B_1^n(\%)$	$B_2^n(\%)$	$B_r^h(\%)$	λ_r^n	λ_r^h	μ_1	μ_2
A	1	2	0.8	0.2	5		1				
B	1	4	0.8	0.2	5		1				
C	1	2	0.2	0.8	5		1	$0.1B_r^n$	$f_r\lambda$	$0.5\lambda_r^n$	1	3
D	1	2	0.8	0.2	1		2				
E	1	2	0.8	0.2	1		1				

to resort to the indirect adjustment in which case l^n is decreased accordingly. If the QoS objective for s^h is expressed as $B = b/o$, then $p^- = 1/(o-b)$ and $p^+ = 1/b$.

The performance evaluation is carried out for five different scenarios (A, B, C, D and E) that are defined in Table 1, being the QoS parameters B_i expressed as percentage values. The parameters in Table 1 have been selected to explore possible trends in the numerical results, i.e., taking scenario A as a reference, scenario B represents the case where the ratio c_1/c_2 is smaller, scenario C where f_1/f_2 is smaller, scenario D where B_1/B_2 is smaller and scenario E where B_1 and B_2 are equal.

The system capacities when deploying the MGC policy without the adaptive scheme for the five scenarios defined in Table 1, $\{A, B, C, D, E\}$, with $C = 10$ are $\lambda_{max} = \{1.89, 0.40, 1.52, 1.97, 1.74\}$, respectively. Refer to [11] for details on how to determine the system capacity. For all scenarios defined in Table 1 we assume the following prioritization order $\mathbf{s}^* = (s_2^h, s_1^h, s_2^n, s_1^n)$. We evaluate two implementations that differ in the treatment of the LPC, (s_1^n), one in which it is a protected class and one in which it is the BEC.

4.1 Performance Under Stationary Traffic

For the two implementations of the adaptive scheme, Table 2 shows the ratio P_i/B_i for the four arrival classes in the five scenarios considered. In all cases, an

Table 2. P_i/B_i when deploying the MGC policy and a stationary load equal to λ_{max}

a) LPC is a protected class.						b) LPC is the best-effort class.					
P_i/B_i	Scenario					P_i/B_i	Scenario				
	A	B	C	D	E		A	B	C	D	E
Class 1N	1.004	1.030	1.036	1.841	1.223	Class 1N	0.938	1.404	0.007	2.348	1.857
Class 2N	0.998	0.992	1.001	0.998	1.007	Class 2N	1.003	1.065	1.000	1.004	0.999
Class 1H	1.006	0.992	1.002	1.007	0.999	Class 1H	1.007	1.001	1.007	0.999	0.999
Class 2H	0.848	0.899	0.803	0.988	0.985	Class 2H	0.993	1.006	0.988	0.989	0.999

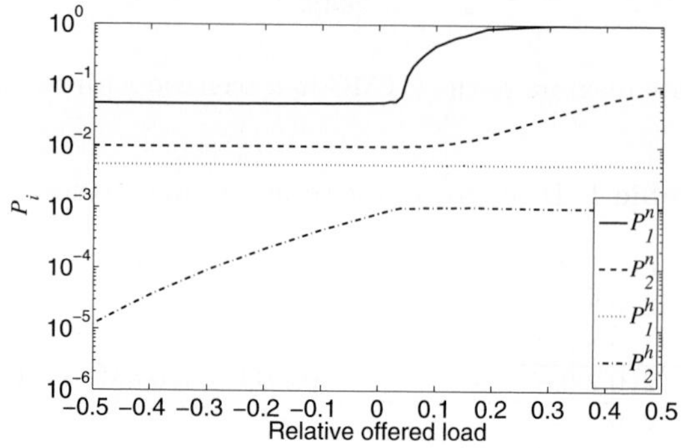

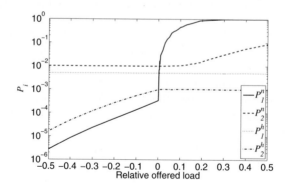

Fig. 3. Variation of P_i with the relative offered load in stationary conditions when the LPC is a protected class

Fig. 4. Variation of P_i with the relative offered load in stationary conditions when the LPC is the BEC

aggregated load equal to the system capacity (λ_{max}) is offered. Note that the adjustment is much more precise when the LPC is the BEC.

Figures 3 and 4 show the variation of P_i with the relative offered load ($(\lambda - \lambda_{max})/\lambda_{max}$) in scenario C with $C = 10$ resource units. Note that the adaptive scheme tries to enforce $P_i = B_i$ when possible for the protected classes,

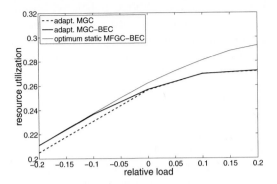

Fig. 5. Resource utilization factor

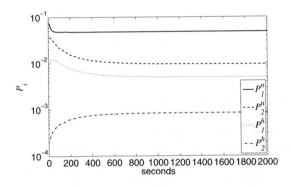

Fig. 6. Transient behavior of blocking probabilities

and therefore during under load episodes the system is rejecting more requests than strictly required. Nevertheless, some classes (BEC and/or HPC) benefit from this extra capacity. When the LPC is a protected class (Fig. 3), it does not benefit from the capacity surplus during under load episodes and it is the first to be penalized during overload episodes. On the other hand, when the LPC is the BEC (Fig. 4), it benefits during under load episodes and, as before, it is the first to be penalized during overload episodes. In both implementations, note that s_2^n is also penalized when keeping on reducing l_1^n (below zero) would be ineffective to meet the QoS objective of higher priority classes.

In Fig. 5 the resource utilization factor $E[c(\boldsymbol{n})]/C$ of the adaptive scheme in scenario A is compared to the one of an optimum static Multiple Fractional Guard Channel (MFGC) policy, which performance is close to the performance of an optimal policy [11]. The configuration parameters of the MFGC policy have been determined by formulating the problem as a non-linear programming algorithm in which for each λ we search for the values of the configuration parameters that maximize the carried traffic while still meeting the QoS objective. Therefore we refer to this policy as the *optimum* MFGC policy. We also refer to it as *static* because for each arrival rate studied the optimum configuration parameters are determined and they are unique. On the other hand, the adaptive scheme does not know the arrival rates a priori and therefore it continuously changes the configuration parameters of the MGC policy to meet the QoS objective.

The term "adapt. MGC" refers to the adaptive scheme when the LPC is a protected class, while "adapt. MGC-BEC" refers to the adaptive scheme when the LPC is the BEC. Note that for $\lambda = \lambda_{max}$ the utilization achieved by the MFGC policy is only 2% higher than the utilization achieved by the adaptive scheme. The system capacity achieved by the MFGC policy is 8.6% higher than the one achieved by the MGC and therefore when $\lambda > \lambda_{max}$, it achieves a slight better the resource utilization. Our adaptive scheme can also operate with the MFGC instead of the MGC policy but its operation in not discussed in this paper. Note that both implementations of the adaptive scheme behave identically during overload ($\lambda > \lambda_{max}$).

4.2   Performance Under Non-stationary Traffic

In this section we study the transient regime after a step-type traffic increase from $0.66\lambda_{max}$ to λ_{max} is applied to the system in scenario A when the LPC is a protected class. Before the step increase is applied the system is in the steady state regime.

Figure 6 shows the transient behavior of the blocking probabilities. The convergence period is lower than 1000 s. A comparative performance evaluation of our scheme and the schemes proposed in [2,3] in a single service scenario shows that the convergence period obtained is 10 to 100 times lower than in both previous proposals [19]. Besides, our scheme shows a non-oscillating behavior unlike [2,3]. Note that the convergence period will be even shorter when the offered load is above the system capacity thanks to the increase of the rate of probabilistic-adjustment actions, which is an additional advantage of the scheme.

5   Adaptive Scheme for Elastic Flows

Like in other studies of the same nature [1] we focus on the flow level and ignore the detailed mechanisms operating at the packet level. Since our focus is the radio interface at the access network, we assume that each elastic flow is rate limited either by terminal capabilities or because it is bottlenecked at the radio link, i.e. it will receive its fair share of the radio link bandwidth up to a maximum which has a common value for all terminals. For the sake of mathematical tractability we assume that the flow size (given in bytes) is exponentially distributed. While it is commonly accepted that the statistical distribution of Internet document sizes shows a greater variability than the exponential distribution, in the light of the results in [1] the numerical results obtained by using an exponential document size can be considered as a lower bound of performance.

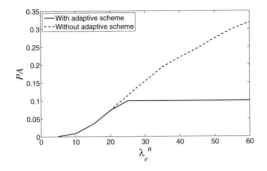

Fig. 7. Abandonment probability of elastic flows with and without adaptive scheme

We consider the same system model described in Section 2, adding a service with elastic demands, as follows. We denote by s_e^n (s_e^t), the arrival class associated to new (handover) requests of elastic flows. Their requests arrive according to Poisson processes with time-varying rate $\lambda_e^n(t)$ ($\lambda_e^h(t)$). For an elastic session, its cell residence (dwell) time is exponentially distributed with rate μ_e^d. If we denote by d_e the maximum number of resource units an elastic flow uses, and by n_e then number of elastic flows in the system, then we define the flow service rate as μ_e^s when $n_e d_e \leq (C - c(\boldsymbol{n}))$ and $\mu_e^s(C - c(\boldsymbol{n}))/(n_e d_e)$ when $n_e d_e > (C - c(\boldsymbol{n}))$, where $c(\boldsymbol{n})$ is the number of resource units occupied by streaming sessions.

To model the behavior of users we consider the impatience time as an independent exponentially distributed random variable. The impatience rate μ_I is assumed to be inversely proportional to the share of resources allocated to each elastic flow, thus we define $\mu_I = 0$ when $n_e d_e \leq (C - c(\boldsymbol{n}))$ and $\mu_I = K(n_e d_e/(C - c(\boldsymbol{n})))$ when $n_e d_e > (C - c(\boldsymbol{n}))$, where K is a constant. We denote by BA the QoS objective expressed as an upper bound for the abandonment probability, i.e. the ratio between unsuccessfully completed flows and accepted flows, and by PA the actual perceived abandonment probability.

The SAC policy for elastic service is as follows: one configuration parameter is associated with s_e^n, $l_e \in \mathbb{N}$. When there are n_e ongoing elastic flows, a new request is accepted if $n_e + 1 \leq l_e$ and blocked otherwise. Handover requests of elastic flows are always accepted. The adaptive scheme for elastic flows follows a similar approach to that described in Section 3. When the QoS objective for elastic flows can be expressed as $BA = a/b$, where $a, b \in \mathbb{N}$, then we propose to perform a probabilistic adjustment in the following way: i) $\{l_e \leftarrow (l_e - 1)\}$ with probability $1/a$ each time an elastic flow abandons due to impatience; ii) $\{l_e \leftarrow (l_e + 1)\}$ with probability $1/(b - a)$ each time an elastic flows completes its service successfully, i.e. either it finishes or it hands over to another cell. A methodology to infer TCP flow interruption has been proposed in [20].

We evaluate by simulation the performance of the scheme with streaming traffic and elastic flows in scenario A with $C = 10$, considering the LPC as the best-effort class. The streaming traffic offers a constant load equal to the system capacity ($\lambda = \lambda_{max} = 1.89$). To avoid starvation the system reserves 1 resource unit for elastic traffic. The rest of the parameters that model the elastic traffic are: $\mu_e^s = 2.0$, $\mu_e^d = 2.0$, $K = 0.4$, $\lambda_e^h = 0.5\lambda_e^n$, with a QoS objective of $BA = 0.1$. Figure 7 shows that the adaptive scheme assures the QoS objective. Without the adaptive scheme the abandonment probability increases as the elastic arrival rate increases. This is due to the fact that less resources are available per elastic flow as more elastic flows are accepted in the system, which consequently increases the abandonment rate. Finally, note that high abandonment probabilities bring as a consequence an inefficient use of system resources because resources assigned to flows that are not completed are totally wasted.

6   Conclusions

We developed a novel adaptive reservation scheme that can adapt to non-stationary traffic both in fixed and variable capacity systems. The operation of our scheme is based on simple balance equations which hold for any arrival process and holding time distribution.

Three relevant features of our proposal are: its capability to handle streaming and elastic traffic, its ability to continuously track and adjust the QoS perceived by users and the simplicity of its implementation.

We evaluated the performance of the scheme when handling multiple streaming services and showed that the QoS objective is met with an excellent precision while achieving an oscillation-free convergence period. This confirms that our scheme can handle satisfactorily the non-stationarity of a real traffic. We also evaluated the performance of the scheme when handling elastic flows in a scenario with streaming background traffic. We showed that the scheme guarantees an upper bound for the abandonment probability of elastic flows.

Acknowledgments

This work has been supported by the Spanish Ministry of Education and Science (30%) and by the European Union (FEDER 70%) under projects TSI2005-07520-C03-03, TEC2004-06437-C05-01 and under contract AP-2004-3332, and by the Generalitat of Valencia under contract CTB/PRB/2002/267.

References

1. Bonald, T., Roberts, J.: Congestion at flow level and the impact of user behaviour. Comp. Net. 42, 521–536 (2003)
2. Zhang, Y., Liu, D.: An adaptive algorithm for call admission control in wireless networks. In: Proc. IEEE Glob. Comm. Conf (GLOBECOM'01), San Antonio, TX (USA), pp. 3628–3632 (2001)
3. Wang, X.P., Zheng, J.L., Zeng, W., Zhang, G.D.: A probability-based adaptive algorithm for call admission control in wireless network. In: Proc. Int. Conf. on Comp. Netw. & Mob. Comp (ICCNMC'03), Shanghai, China, 197–204 (2003)

4. Yu, O.T.W., Leung, V.C.M.: Adaptive resource allocation for prioritized call admission over an ATM-based wireless PCN. IEEE Jour. on Sel. Areas in Comm. 15(7), 1208–1225 (1997)
5. Ramanathan, P., Sivalingam, K.M., Agrawal, P., Kishore, S.: Dynamic resource allocation schemes during handoff for mobile multimedia wireless networks. IEEE Jour. on Sel. Areas in Comm. 17(7), 1270–1283 (1999)
6. Yu, O., Khanvilkar, S.: Dynamic adaptive QoS provisioning over GPRS wireless mobile links. In: Proc., I.E.E.E. (ed.) Proc. IEEE Int. Conf. on Comm (ICC 2002), New York, NY (USA), vol. 2, pp. 1100–1104 (2002)
7. Jeon, W.S., Jeong, D.G.: Call admission control for mobile multimedia communications with traffic asymmetry between uplink and downlink. IEEE Trans. Veh. Tech. 50(1), 59–66 (2001)
8. Wei, Y., Lina, C., Rena, F., Raad, R., Dutkiewiczb, E.: Dynamic handoff scheme in differentiated QoS wireless multimedia networks. Comp. Comm. 27(10), 1001–1011 (2004)
9. Huang, L., Kumar, S., Kuo, C.C.J.: Adaptive resource allocation for multimedia QoS management in wireless networks. IEEE Trans. Veh. Tech. 53(2), 547–558 (2004)
10. Garcia-Roger, D., Domenech-Benlloch, M.J., Martinez-Bauset, J., Pla, V.: Adaptive admission control scheme for multiservice mobile cellular networks. In: Proc. 1st EuroNGI Conf. on Next Gen. Internet Net. Traf. Eng (NGI), Roma, Italy, 288–295 (2005)
11. García, D., Martínez, J., Pla, V.: Admission control policies in multiservice cellular networks: Optimum configuration and sensitivity. In: Kotsis, G., Spaniol, O. (eds.) Wireless Systems and Mobility in Next Generation Internet. LNCS, vol. 3427, pp. 121–135. Springer, Heidelberg (2005)
12. Soh, W.S., Kim, H.S.: Dynamic bandwidth reservation in cellular networks using road topology based mobility prediction. In: Proc. 23rd Ann. Joint Conf. IEEE Comp. & Comm. Soc (INFOCOM)., Hong Kong, China, vol. 4, pp. 2766–2777 (2004)
13. Orlik, P.V., Rappaport, S.S.: On the handoff arrival process in cellular communications. Wi. Net. Journal 7(2), 147–157 (2001)
14. Jabbari, B.: Teletraffic aspects of evolving and next-generation networks. IEEE Pers. Comm. 3(6), 4–9 (1996)
15. Biswas, S., Sengupta, B.: Call admissibility for multirate traffic in wireless ATM networks. In: Proc. 16th Ann. Joint Conf. IEEE Comp. & Comm. Soc (INFOCOM)., vol. 2, pp. 649–657 (1997)
16. Evans, J.S., Everitt, D.: Effective bandwidth-based admission control for multiservice CDMA cellular networks. IEEE Trans. Veh. Tech. 48(1), 36–46 (1999)
17. Khan, F., Zeghlache, D.: Effect of cell residence time distribution on the performance of cellular mobile networks. In: Proc. IEEE 47th Veh. Tech. Conf (VTC'97-Spring), Phoenix (USA), pp. 949–953 (1997)
18. Koo, I., Furuskar, A., Zander, J., Kim, K.: Erlang capacity of multiaccess systems with service-based access selection. IEEE Comm. Lett. 8(11), 662–664 (2004)
19. Garcia-Roger, D., Domenech-Benlloch, M.J., Martinez-Bauset, J., Pla, V.: Comparative evaluation of adaptive trunk reservation schemes for mobile cellular networks. In: Proc. 3rd Int. W. Conf. Perf. Mod. & Eval. Het. Net (HET-NETs), Ilkley, UK (2005)
20. Rossi, D., Mellia, M., Casetti, C.: User patience and the web: a hands-on investigation. In: Proc. IEEE Glob. Tel. Conf (GLOBECOM), vol. 7, pp. 4163–4168 (2003)

Queueing Analysis of Mobile Cellular Networks Considering Wireless Channel Unreliability and Resource Insufficiency*

Andrés Rico-Páez[1], Carmen B. Rodríguez-Estrello[1], Felipe A. Cruz-Pérez[1], and Genaro Hernández-Valdez[2]

[1] Electrical Engineering Department, CINVESTAV-IPN, Av. IPN 2508, Col. San Pedro Zacatenco, CP 07360, Mexico City, Mexico
arico@cinvestav.mx, crodriguez@cinvestav.mx, facruz@cinvestav.mx
[2] Electronics Department, UAM-A, Av. San Pablo 180, Col. Reynosa Tamaulipas, CP 02200, Mexico City, Mexico
ghv@correo.azc.uam.mx

Abstract. A queueing model for the system level performance evaluation of mobile cellular networks considering both resource insufficiency and wireless channel unreliability is proposed and mathematically analyzed. The proposed mathematical approach is based on the use of simple call interruption processes to model the effect of wireless channel unreliability. More importantly, this paper develops a system level teletraffic model for the performance evaluation considering that channel holding times for new and handed off calls are general (but not necessary identically) distributed random variables. Additionally, an approximated one-dimensional recursive approach based on the well known Kauffman-Roberts formula is proposed for the case when the channel holding time variables can be adequately characterized by Hyper-Exponential distributions. Also, the case when the channel holding time variables are Mixed-Erlang distributed, a distribution having universal approximation capability, is analyzed. Thus, our teletraffic model allows obtaining more general, realistic, and easily computable analytical results.

Keywords: Mobile cellular networks, queueing analysis, link unreliability, channel holding time, and phase-type probability distributions.

1 Introduction

The limiting resource in mobile cellular systems is radio channels. The channels of a given cell are loaded by a mix of calls: new calls initiated inside the cell and handed off calls from neighboring cells. Therefore, the teletraffic model developed to evaluate the performance of these networks needs to differentiate between channel holding time, unencumbered service time, and cell residence time. Due to the negative exponential probability distribution assumptions, many of the classic teletraffic models are insensitive to the probability distribution of these time variables; only the

---

* This research was performed under the support of CONACYT project 50434.

L. Mason, T. Drwiega, and J. Yan (Eds.): ITC 2007, LNCS 4516, pp. 938–949, 2007.

mean value being relevant for the performance metrics [1]-[2]. Additionally, it is typically considered that the random variables used to model the channel holding time of new and handed off calls have the same mean and/or probability distribution [3]-[4]. Recent papers [1], [5], have concluded that in order to capture the overall effects of the cellular shape and the user's mobility patterns, both the channel holding time and cell dwell time need to be modeled as random variables with more general distribution. Furthermore, it has been shown that, in general, the channel holding time for new and handed off calls has different probability distribution [6]. On the other hand, system level performance evaluation of mobile cellular networks has been traditionally addressed by considering only resource insufficiency, while link unreliability has received not enough attention because of the complexity involved in its study.

In this paper, we propose a queueing model for the system level performance evaluation of mobile cellular networks taking into account the inherent wireless channel unreliability and considering that the channel holding time variables for new calls and handed off calls are general (but not necessary identically) distributed random variables. Due to space limitations, we could not give a detailed discussion on the previously published related work. In its place, the main difference relative to previous work and contribution of this paper are pointed out through the text. The primary difference between the present work and most of the approaches reported in the literature (see [7], and the references therein) is that instead of basing the analysis on link-level statistics, a general system level analytical model for the performance evaluation of mobile wireless communication networks considering both resource insufficiency and link unreliability is proposed. In a previous work [7], a simple call interruption processes to model the effect of wireless channel unreliability was introduced. However, both cell dwell time and call interruption time due to link unreliability were considered to be only exponentially distributed. In this paper, channel holding time random variables for new calls and handed off calls are assumed general and not identically distributed random variables. We think that the model here proposed represents a step toward the development of a general and analytical tractable modeling tool for the design of mobile wireless communication networks under more realistic considerations.

The remainder of this article is structured as follows. In Section 2, the interruption process and its associated interruption time are modeled and described. The different pdf's used to model channel holding time variables are studied in Section 3. Channel holding time is expressed as a function of the involved time variables in Section 4. Section 5 presents the teletraffic analysis when the channel holding time for both new and handed off calls is considered to have either Hyper-Exponencial or Mixed-Erlang distributions. Finally, the conclusions are given in Section 6.

2 Modeling Wireless Channel Unreliability

In this section, the proposed model in [7] to include call dropping due to the wireless channel unreliability is described. Let us define some time variables involved in the process of a call considering interruption *due* to the wireless link unreliability:

Channel holding time $\mathbf{X}_c^{(j)}$ in the j-th (for $j=0,1,..$) handed off cell, that is, the time that a user, which is accepted in the j-th handed off cell and is assigned resources, will use these resources before either its call being completed, handed off to another cell, or interrupted due to the wireless channel unreliability. *Cell dwell time* $\mathbf{X}_d^{(j)}$ is defined as the time that a mobile station spends in the j-th (for $j = 0, 1,..$) handed off cell irrespective of whether it is engaged in a session or not. On the other hand, the *residual cell dwell time* is define as the time between the instant a new call is initiated at and the instant the user is handed off to another cell. The random variable used to represent this time is $\mathbf{X}_r$. *Unencumbered service time* per call $\mathbf{X}_s$ represents the duration of the requested call connection when the network has both unlimited resources and link reliability. It has been widely accepted in the literature that the unencumbered service time can be adequately modeled by a negative exponentially distributed random variable [8]. Considering the different physical processes in which a call is involved, let us explain the call interruption process: After a call is originated and accepted by a base station (BS) (as a new call or as a handed off call), the physical link between BS and mobile station (MS) may become degraded during the call. Thus, in this case the call is forced terminated due to the wireless link unreliability. The proposed model in [7] considers an interruption process and an associated time to this process which is called "*unencumbered interruption time*" and it is denoted by $\mathbf{X}_i^{(j)}$. Thus, the *unencumbered interruption time* $\mathbf{X}_i^{(j)}$ is defined as the period of time from the instant the MS establish a link with the j-th handed-off cell until the time the call would be interrupted due to the wireless link unreliability assuming that the network has unlimited resources and the service time is of infinite duration.

3 System Model

3.1 Description

In this section, the general guidelines for the mathematical analysis are given. A homogeneous multi-cellular system with omni-directional antennas located at the centre of each cell is assumed. Each cell has a maximum number s of radio channels and it is considered that N channels are reserved for hand off prioritization. As it has been widely accepted in the related literature [7], [9]-[10], both the new call arrivals and handoff attempts are assumed to follow independent Poisson processes with mean arrival rate λ_n and $\lambda_h{}^1$, respectively, per cell.

3.2 Phase-Type Probability Distributions

Similar to [1] and [5], in this paper, the channel holding times for both new and handed off calls are considered to be phase type distributed. This provides a model that is not based on the classical exponential assumptions but still analytically

---

[1] λ_h depends on the rest of the system parameters (i.e., new call arrival rate; cell dwell, service, and interruption times, etc.) and on the admission policy (i.e., call blocking and dropping probabilities). In a homogeneous system, it can be calculated iteratively as shown in [11].

tractable and which easily can handle mixtures of different kinds of traffic. A phase-type distribution is a random variable describing the time until absorption of a Markov Chain with some transient states and one absorbing state [1]. The importance of the phase-type probability distributions relies on the fact that some of them can be used to approximate the behavior of any positive random variable [10], [12]. Additionally, its use allows one to consider a broad class of pdf's while retaining the Markovian properties that are required by the multidimensional birth–death (MDBD) framework. If m identically-distributed negative exponential random variables with parameter φ are added (series combination), then an m-th order Erlang distribution is obtained. The elements of the sum are called *phases*. On the other hand, if m negative exponential random variables with parameter φ_i are combined in parallel, where the probability of choosing the i-th negative exponential random variable is given by α_i (i=1, 2,..., m, and $\alpha_1 + \alpha_2 + \ldots + \alpha_m = 1$); then, it is obtained a Hyper-Exponential distribution of order m. Each one of the m possible options is called *stage*. If m Erlang random variables with parameter φ_{ij} are combined in parallel, where the probability of choosing the i-th Erlang random variable is given by α_i (i=1,2,..., m, and $\alpha_1 + \alpha_2 + \ldots + \alpha_m = 1$) and j denotes phase of stage i; then, it is obtained a Mixed-Erlang distribution (or called Hyper-Erlang distribution) of order m.

In this paper, the involved time random variables are considered to have either Hyper-Exponencial or Mixed-Erlang distributions. This is because these types of distributions allow a very simple mathematical analysis and, more importantly, the Mixed-Erlang distribution has the universal approximation capability [13]. Additionally, the Mixed-Erlang includes as particular cases[2] other phase type distributions that are widely used in literature (i.e., Negative Exponential, Hyper-Exponencial, and Erlang) [3], [7]. Table 1 shows probability density function, expected value, and variance of both the Hyper-Exponential and Mixed-Erlang distributions. For the Mixed-Erlang distribution shown in Table 1, variable i denotes selected stage and variable j denotes phase of stage i; each stage i has u_i phases.

Table 1. Probability density function, expected value, and variance of the Hyper-Exponential and Mixed-Erlang distributions

Probability density function $f(t)$	Mean E[X]	Variance σ^2
m-th order Hyper-Exponential $$\sum_{i=1}^{m}\alpha_i\varphi_i e^{-\varphi_i t},\ \varphi_i > 0, t \geq 0, 0 \leq \alpha_i \leq 1, \sum_{i=1}^{m}\alpha_i = 1$$	$\displaystyle\sum_{i=1}^{m}\frac{\alpha_i}{\varphi_i}$	$\displaystyle\sum_{i=1}^{m}\frac{2\alpha_i}{\varphi_i^2} - \left(\sum_{i=1}^{m}\frac{\alpha_i}{\varphi_i}\right)^2$
m-th order Mixed-Erlang $$\sum_{i=1}^{m}\alpha_i\frac{\varphi_{ij}^{u_i} t^{u_i-1}}{(u_i-1)!}e^{-\varphi_{ij}t},\ \varphi_{ij} > 0, t \geq 0, 0 \leq \alpha_i \leq 1, \sum_{i=1}^{m}\alpha_i = 1,$$ $\varphi_{ik} = \varphi_{il}$ for $k = 1,2,\ldots u_i$ and $l = 1,2,\ldots u_i$	$\displaystyle\sum_{i=1}^{m}\frac{\alpha_i u_i}{\varphi_{ij}}$	$\displaystyle\sum_{i=1}^{m}\frac{\alpha_i u_i (u_i+1)}{\varphi_{ij}^2} - \left(\sum_{i=1}^{m}\frac{\alpha_i u_i}{\varphi_{ij}}\right)^2$

---

[2] Negative Exponential distribution is obtained from Mixed-Erlang distribution if the number of stages and phases is one ($i = 1$ and $j = 1$). Erlang distribution is obtained if the number of stages is one ($i = 1$). Hyper-Exponential distribution is obtained if the number of phases of each stage is one ($j = 1$ for any i).

4   Channel Holding Time

In order to model channel holding time, it is necessary to express it as a function of the unencumbered service, cell dwell, and unencumbered call interruption time variables. The random variable $\mathbf{X}_c^{(j)}$, which models channel holding time in the j-th handed off cell, is defined as:

$$
\mathbf{X}_c^{(j)} = \begin{cases} \min\left(\mathbf{X}_i^{(0)}, \mathbf{X}_s, \mathbf{X}_r\right) & ;\quad j = 0 \\ \min\left(\mathbf{X}_i^{(j)}, \mathbf{X}_s, \mathbf{X}_d^{(j)}\right) & ;\quad j > 0 \end{cases}
\tag{1}
$$

Thus, call holding time should be defined for the following type of calls: new calls [$j=0$ in (1)] and handed off calls [$j>0$ in (1)]. Assuming that the different involved time variables are independent, the cumulative distribution probability (CDF) of the channel holding time (i.e., the CDF of the minimum of the cell dwell time, service time, and call interruption time) is given by:

$$
F_{\mathbf{X}_c^{(j)}}(t) = \begin{cases} 1 - \left[1 - F_{\mathbf{X}_i^{(0)}}(t)\right]\left[1 - F_{\mathbf{X}_s}(t)\right]\left[1 - F_{\mathbf{X}_r}(t)\right]; & j = 0 \\ 1 - \left[1 - F_{\mathbf{X}_i^{(j)}}(t)\right]\left[1 - F_{\mathbf{X}_s}(t)\right]\left[1 - F_{\mathbf{X}_d^{(j)}}(t)\right]; & j > 0 \end{cases}
\tag{2}
$$

Where $\mathbf{X}_c^{(0)}$ represents channel holding time of new calls ($\mathbf{X}_c^{(n)}$) and $\mathbf{X}_c^{(j)}$ for $j>0$ represents channel holding time of handed off calls ($\mathbf{X}_c^{(h)}$).

5   Teletraffic Analysis

Most of the work published in the literature that consider general probability distributions for the cell dwell and/or unencumbered service times to evaluate the system performance of mobile cellular networks do not consider birth and death processes. Exception of these studies are [10], [14]-[15]. In [15], the authors propose a queueing model to directly consider probability distributions of the cell dwell and service times by means of a multidimensional birth and death process. However, this queueing model is incorrect, as it is explained next. Firstly, it is important to notice that a time variable follows a given phase type probability distribution if and only if it passes through all the phases of the given distribution. Thus, a phase-type random distributed time cannot expire in an intermediate phase (i.e., a phase different to the last one). This is not considered by the model proposed in [15]. The authors of [15] consider that calls can be completed in an intermediate phase of the cell dwell time. Also, they consider that users can leave cells in a given intermediate phase of the unencumbered service time, but do not consider that users must continue their service time in that same given phase of the unencumbered service time in the cell to which they are handed off. Orlik and Rappaport partially realize this mistake [16]. In [16], Orlik and Rappaport corrected equations (42) and (43) of [15] to consider the fact that handoff departures occur only when platforms complete the final phase of the cell dwell time. However, the rest of the analysis was not revised. On the other hand, in

[14] also general probability distributions for the cell dwell and unencumbered service times are considered. However, the mathematical model developed in [14] is only applicable to a particular call admission control strategy (i.e., new call bounding) whose performance depends on the distribution of the channel holding time only through its average (that is to say, it has the insensitivity property) [17]. It is important to mention that when the complete sharing strategy is considered, the steady state probability distribution has product form and, therefore, it is insensitive to the probability distribution of the channel holding time [17], [18].

Contrary to the just above described work, in this paper, a queueing analysis considering both channel reservation for handoff prioritization and that the channel holding times for new calls and handoff calls are general (but not necessary identically) distributed random variables. Also, our queueing analysis is applicable to a wide range of call admission control strategies whose performance depends on the distribution of the channel holding time. For the analysis presented below, the mathematical tools of multidimensional birth and death processes, the method of the phases, and the formula of Kauffman-Roberts are used [17].

The method of the phases is an approach procedure by which a random variable with arbitrary distribution is replaced by a sum, or mixes, or sum-mixes combined of independent phase-type random variables (each one constituting one phase of the original random variable). This method has been widely used in the performance evaluation of wireless cellular networks, for example, see [1], [3]-[5], [10], [12] and [19]-[20]. However, those studies have not taken into account the inherent wireless channel unreliability. For the teletraffic evaluation of the mobile cellular systems when the cell dwell time and/or call interruption time are generally distributed, the well known phases' method [21] works as follows.

- First, the distribution of the channel holding time under the condition that both the unencumbered call interruption time (due to link unreliability) and the cell dwell time are modeled as general phase-type distributed random variables is derived. Alternatively, it can be obtained directly from real system statistics.
- Then, the pdf of the channel holding time is approximated by a suitable phase-type probability distribution.
- Finally, with this distribution characterizing the channel holding time, the system is modeled by a multidimensional birth and death process.

In [4], the cumulative distribution function and the moment generating function of the channel holding time, considering that the unencumbered service time is exponentially distributed and both the cell dwell time and the unencumbered call interruption time are modeled by a phase-type distribution are derived. From the moment generating function, the moments of the channel holding time can be obtained easily and can be used as parameters to fit a particular probability distribution to a phase-type distribution. The Mixed-Erlang distribution is of arbitrary order or degree (i.e., phases and stages), therefore, can approximate any probability distribution with an arbitrary precision. Next, we present the teletraffic analysis when the channel holding time for both new and handed off calls is assumed to be a RV

Hyper-Exponentially distributed. As a homogenous multi-cellular system is considered in this paper, it is sufficient to analyze a single cell.

5.1 Case I: The Channel Holding Time for Both New and Handed Off Calls Is Assumed Hyper-Exponentially Distributed

Let us assume $\mathbf{X}_c^{(y)}$ to be a $m^{(y)}$-th order Hyper-Exponential random variable (where $y=\{n, h\}$ for new and handoff calls, respectively). The diagram of stages of these random variables is shown in Fig. 1.

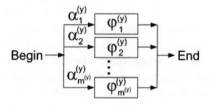

Fig. 1. Diagram of the stages of the probability distribution of $\mathbf{X}_c^{(y)}$, where $y=\{n\}$ for new calls and $y=\{h\}$ for handoff calls

In Fig. 1, $1/\varphi_i^{(y)}$ represents the mean time duration of stage i (for $i = 1, 2,..., m^{(y)}$) and $\alpha_i^{(y)}$ represents the probability of choosing the stage i. In order to model this system by using a multidimensional birth and death process, $m^{(n)} + m^{(h)}$ state variables are needed. Let us define $k_i^{(y)}$ as the number of users in stage i of the channel holding time. To simplify mathematical notation, the following vectors are defined: $\mathbf{K}^{(y)} = [k_1^{(y)}, k_2^{(y)},..., k_{m^{(y)}}^{(y)}]$ and $\mathbf{e}_i^{(y)} =$ unit vector of dimension $m^{(y)}$ whose all entries are 0 except the i-th one which is 1 (for $i = 1, 2, ..., m^{(y)}$). Let us define the current state of the analyzed cell as the vector $[\mathbf{K}^{(n)}, \mathbf{K}^{(h)}]$. Table 2 provides the rules that determine transition rates to the different successor states (shown in the second column).

Table 2. Transition rules and rates for the case when the channel holding time for both new and handed off calls is Hyper-Exponentially distributed

Event	Successor State	Rate
New call enters stage i ($i=1,2,...,m^{(n)}$)	$\left[\mathbf{K}^{(n)} + \mathbf{e}_i^{(n)}, \mathbf{K}^{(h)}\right]$	$a_i^{(n)}\left(\left[\mathbf{K}^{(n)}, \mathbf{K}^{(h)}\right]\right)$
New call leaves stage i ($i=1,2,...,m^{(n)}$)	$\left[\mathbf{K}^{(n)} - \mathbf{e}_i^{(n)}, \mathbf{K}^{(h)}\right]$	$b_i^{(n)}\left(\left[\mathbf{K}^{(n)}, \mathbf{K}^{(h)}\right]\right)$
Handed off call enters stage i ($i = 1, 2,..., m^{(h)}$)	$\left[\mathbf{K}^{(n)}, \mathbf{K}^{(h)} + \mathbf{e}_i^{(h)}\right]$	$a_i^{(h)}\left(\left[\mathbf{K}^{(n)}, \mathbf{K}^{(h)}\right]\right)$
Handed off call leaves stage i ($i = 1, 2,...,m^{(h)}$)	$\left[\mathbf{K}^{(n)}, \mathbf{K}^{(h)} - \mathbf{e}_i^{(h)}\right]$	$b_i^{(h)}\left(\left[\mathbf{K}^{(n)}, \mathbf{K}^{(h)}\right]\right)$

Let us assume that the channels reserved for handoff prioritization are given by $N^{(n)}=N$ and $N^{(h)}=0$. Transition rates shown in Table 2 are given by (for $y = \{n, h\}$):

$$a_i^{(y)}\left(\left[\mathbf{K}^{(n)},\mathbf{K}^{(h)}\right]\right) = \begin{cases} \alpha_i^{(y)}\lambda^{(y)}; & \sum_{i=1}^{m^{(n)}} k_i^{(n)} + \sum_{i=1}^{m^{(h)}} k_i^{(h)} < s - N^{(y)} \ \cap \ k_i^{(y)} \geq 0 \\ 0 & ; \quad \text{otherwise} \end{cases} \tag{3}$$

$$b_i^{(y)}\left(\left[\mathbf{K}^{(n)},\mathbf{K}^{(h)}\right]\right) = \begin{cases} k_i^{(y)}\varphi_i^{(y)}; k_i^{(n)} > 0 \cap \sum_{i=1}^{m^{(n)}} k_i^{(n)} <= s - N^{(y)} \cap \sum_{i=1}^{m^{(n)}} k_i^{(n)} + \sum_{i=1}^{m^{(h)}} k_i^{(h)} <= s \\ 0 & ;\text{otherwise} \end{cases} \tag{4}$$

The new call blocking ($P^{(n)}$) and handoff failure ($P^{(h)}$) probabilities are given by

$$P^{(y)} = \sum_{k_1^{(n)}=0}^{s} \cdots \sum_{k_{m^{(n)}}^{(n)}=0}^{s} \sum_{k_1^{(h)}=0}^{s} \cdots \sum_{k_{m^{(h)}}^{(h)}=0}^{s} P\left(\left[\mathbf{K}^{(n)},\mathbf{K}^{(h)}\right]\right) \ ; \quad \text{for } y = \{n,h\}$$

$$\left\{ s - N^{(y)} \leq \sum_{i=1}^{m^{(n)}} k_i^{(n)} + \sum_{i=1}^{m^{(h)}} k_i^{(h)} \leq s \right\} \tag{5}$$

It is important to realize that the performance of this system can be approximately evaluated by using the well known Kauffman-Roberts formula [17]. The Kauffman-Roberts' formula uses the technique of macrostates to reduce a multidimensional problem to a one-dimensional recursive formula that is much simpler to evaluate. Let $q(s)$ denotes the equilibrium channel occupancy probability when exactly s channels are occupied in the analyzed cell. Then, this probability distribution can be calculated by employing [22]:

$$q(j) = \frac{1}{j}\left[\sum_{i=1}^{m^{(n)}} \frac{\alpha_i^{(n)}\theta^{(n)}(j-1,i)}{\varphi_i^{(n)}} + \sum_{i=1}^{m^{(h)}} \frac{\alpha_i^{(h)}\theta^{(h)}(j-1,i)}{\varphi_i^{(h)}}\right] q(j-1) \ ; \quad \text{for } j = 1,2,...,s \tag{6}$$

$$\theta^{(y)}(j,i) = \begin{cases} \lambda^{(y)} ; & j < s - N^{(y)} \\ 0 ; & \text{otherwise} \end{cases} \ ; \quad \text{for } y = \{n,h\} \tag{7}$$

Thus, the new call blocking ($P^{(n)}$) and handoff failure ($P^{(h)}$) probabilities can be computed as follows

$$P^{(y)} = \sum_{j=s-N^{(y)}}^{s} q(j) \ ; \quad \text{for } y = \{n,h\} \tag{8}$$

5.2 Case II: The Channel Holding Time for Both New and Handed Off Calls Is Assumed Mixed-Erlang Distributed

Let us assume $\mathbf{X}_c^{(y)}$ to be a $m^{(y)}$-th order Mixed-Erlang random variable (where $y = \{n, h\}$ for new and handed off calls, respectively). The diagram of phases and stages of these random variables is shown in Fig. 2.

In Fig. 2, $1/\varphi_{\sum_{x=1}^{i-1} u_x^{(y)}+j}^{(y)}$ represents the mean time duration of stage i (for $i = 1, 2,...,$ $m^{(y)}$) and phase j (for $j = 1, 2,..., u_i^{(y)}$) and $\alpha_i^{(y)}$ represents the probability of choosing the stage i. For modeling this system through a multidimensional birth and death

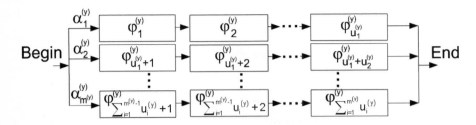

Fig. 2. Diagram of phases and stages of the probability distribution of $\mathbf{X}_c^{(y)}$, where $y=\{n\}$ for new calls and $y=\{h\}$ for handoff calls

Table 3. Transition rules and rates for the case when the channel holding time for both new and handed off calls is Mixed-Exponentially distributed

Event	Successor State	Rate
New call enters first phase of stage i $(i=1,2,...,m^{(n)})$	$\left[\mathbf{K}^{(n)} + \mathbf{e}^{(n)}_{\sum\limits_{x=1}^{i-1} u_x^{(n)}+1}, \mathbf{K}^{(h)} \right]$	$a_i^{(n)}\left(\left[\mathbf{K}^{(n)}, \mathbf{K}^{(h)} \right] \right)$
New call leaves phase j of stage i and enters phase $j+1$ of stage i $(i = 1, 2, ..., m^{(n)})$ $(j = 1, 2, ..., u_i^{(n)}-1)$	$\left[\mathbf{K}^{(n)} - \mathbf{e}^{(n)}_{\sum\limits_{x=1}^{i-1} u_x^{(n)}+j} + \mathbf{e}^{(n)}_{\sum\limits_{x=1}^{i-1} u_x^{(n)}+j+1}, \mathbf{K}^{(h)} \right]$	$b^{(n)}_{\sum\limits_{x=1}^{i-1}\left(u_x^{(n)}-1\right)+j}\left(\left[\mathbf{K}^{(n)}, \mathbf{K}^{(h)} \right] \right)$
New call leaves last phase of stage i $(i=1,2,...,m^{(n)})$	$\left[\mathbf{K}^{(n)} - \mathbf{e}^{(n)}_{\sum\limits_{x=1}^{i} u_x^{(n)}}, \mathbf{K}^{(h)} \right]$	$c_i^{(n)}\left(\left[\mathbf{K}^{(n)}, \mathbf{K}^{(h)} \right] \right)$
Handed off call enters first phase of stage i $(i=1,2,...,m^{(h)})$	$\left[\mathbf{K}^{(n)}, \mathbf{K}^{(h)} + \mathbf{e}^{(h)}_{\sum\limits_{x=1}^{i-1} u_x^{(h)}+1} \right]$	$a_i^{(h)}\left(\left[\mathbf{K}^{(n)}, \mathbf{K}^{(h)} \right] \right)$
Handed off call leaves phase j of stage i and enters phase $j+1$ of stage i $(i = 1, 2, ..., m^{(h)})$ $(j = 1, 2, ..., u_i^{(h)}-1)$	$\left[\mathbf{K}^{(n)}, \mathbf{K}^{(h)} - \mathbf{e}^{(h)}_{\sum\limits_{x=1}^{i-1} u_x^{(h)}+j} + \mathbf{e}^{(h)}_{\sum\limits_{x=1}^{i-1} u_x^{(h)}+j+1} \right]$	$b^{(h)}_{\sum\limits_{x=1}^{i-1}\left(u_x^{(h)}-1\right)+j}\left(\left[\mathbf{K}^{(n)}, \mathbf{K}^{(h)} \right] \right)$
Handed off call leaves last phase of stage i $(i=1,2,...,m^{(h)})$	$\left[\mathbf{K}^{(n)}, \mathbf{K}^{(h)} - \mathbf{e}^{(h)}_{\sum\limits_{x=1}^{i} u_x^{(h)}} \right]$	$c_i^{(h)}\left(\left[\mathbf{K}^{(n)}, \mathbf{K}^{(h)} \right] \right)$

process, $\sum_{x=1}^{m^{(n)}} u_x^{(n)} + \sum_{x=1}^{m^{(h)}} u_x^{(h)}$ state variables are needed. Let us define $k^{(y)}_{\sum_{x=1}^{i-1} u_x^{(y)}+j}$ as the number of users in stage i and phase j of the channel holding time. To simplify mathematical notations, the following vectors are defined:

$$\mathbf{K}^{(y)} = \left[k_1^{(y)}, k_2^{(y)}, ..., k^{(y)}_{\sum_{i=1}^{m^{(y)}} u_i^{(y)}} \right]; \qquad \mathbf{\Phi}^{(y)} = \left[\varphi_1^{(y)}, \varphi_2^{(y)}, ..., \varphi^{(y)}_{\sum_{i=1}^{m^{(y)}} u_i^{(y)}} \right];$$

$\mathbf{e}_i^{(y)}$ = unit vector of dimension $m^{(y)}$ whose all entries are 0 except the i-th one which is 1 (for $i = 1, 2, \ldots, \Sigma_{i=1}^{m^{(y)}} u_i^{(y)}$).

Let us define the current state of the analyzed cell as the vector $[\mathbf{K}^{(n)}, \mathbf{K}^{(h)}]$. Table 3 provides the rules that determine transition rates to the different successor states (shown in the second column).

Let us assume that the channels reserved for handoff prioritization are given by $N^{(n)}=N$ and $N^{(h)}=0$. Transition rates shown in Table 3 are given by (for $y = \{n, h\}$):

$$a_i^{(y)}\left(\left[\mathbf{K}^{(n)},\mathbf{K}^{(h)}\right]\right)=\begin{cases}\alpha_i^{(y)}\lambda^{(y)} \;;\; \displaystyle\sum_{i=1}^{\sum_{x=1}^{m^{(n)}}u_x^{(n)}}k_i^{(n)}+\sum_{i=1}^{\sum_{x=1}^{m^{(h)}}u_x^{(h)}}k_i^{(h)}<s-N^{(y)} \;\cap\; k_{\sum_{x=1}^{(y)}u_x^{(y)}+1}^{(y)} \geq 0 \\ 0 \qquad ;\;\; \text{otherwise}\end{cases} \tag{9}$$

$$b_{\sum_{x=1}^{i-1}\left(u_x^{(y)}-1\right)+j}^{(y)}\left(\left[\mathbf{K}^{(n)},\mathbf{K}^{(h)}\right]\right)=\begin{cases}k_{\sum_{x=1}^{i-1}u_x^{(y)}+j}^{(y)}\dfrac{\varphi_i^{(y)}}{\sum_{x=1}^{i-1}u_x^{(y)}+j}\;;k_{\sum_{x=1}^{i-1}u_x^{(y)}+j+1}^{(y)}\geq 0 \;\cap\; \displaystyle\sum_{i=1}^{\sum_{x=1}^{m^{(n)}}u_x^{(n)}}k_i^{(n)}<=s-N^{(y)} \\[4mm] \qquad \cap\, k_{\sum_{x=1}^{i-1}u_x^{(y)}+j}^{(y)}>0 \;\cap\; \displaystyle\sum_{i=1}^{\sum_{x=1}^{m^{(n)}}u_x^{(n)}}k_i^{(n)}+\sum_{i=1}^{\sum_{x=1}^{m^{(h)}}u_x^{(h)}}k_i^{(h)}<=s \\[4mm] 0 \qquad\qquad ;\text{otherwise}\end{cases} \tag{10}$$

$$c_i^{(y)}\left(\left[\mathbf{K}^{(n)},\mathbf{K}^{(h)}\right]\right)=\begin{cases}k_{\sum_{x=1}^{i}u_x^{(y)}}^{(y)}\dfrac{\varphi_i^{(y)}}{\sum_{x=1}^{i}u_x^{(y)}}\;;\; k_{\sum_{x=1}^{i}u_x^{(y)}}^{(y)}>0\cap \displaystyle\sum_{i=1}^{\sum_{x=1}^{m^{(n)}}u_x^{(n)}}k_i^{(n)}<=s-N^{(y)} \\[4mm] \qquad \cap \displaystyle\sum_{i=1}^{\sum_{x=1}^{m^{(n)}}u_x^{(n)}}k_i^{(n)}+\sum_{i=1}^{\sum_{x=1}^{m^{(h)}}u_x^{(h)}}k_i^{(h)}<=s \\[4mm] 0 \qquad ;\;\; \text{otherwise}\end{cases} \tag{11}$$

The new call blocking ($P^{(n)}$) and handoff failure ($P^{(h)}$) probabilities are given by

$$P^{(y)}=\sum_{k_1^{(n)}=0}^{s}\cdots\sum_{k_{\sum_{x=1}^{m^{(n)}}u_x^{(n)}}^{(n)}=0}^{s}\sum_{k_1^{(h)}=0}^{s}\cdots\sum_{k_{\sum_{x=1}^{m^{(h)}}u_x^{(h)}}^{(h)}=0}^{s}P\left(\left[\mathbf{K}^{(n)},\mathbf{K}^{(h)}\right]\right) \quad;\quad \text{for } y=\{n,h\}$$

$$\left\{s-N^{(y)}\leq\sum_{i=1}^{\sum_{x=1}^{m^{(n)}}u_x^{(n)}}k_i^{(n)}+\sum_{i=1}^{\sum_{x=1}^{m^{(h)}}u_x^{(h)}}k_i^{(h)}\leq s\right\} \tag{12}$$

6 Conclusions

Performance evaluation under more realistic assumption is an important part in modeling and designing effective schemes for an efficient use of radio resources in wireless cellular networks. In this paper, we propose a novel queueing model for the system level performance evaluation of mobile cellular networks taking into account the inherent wireless channel unreliability. The effect of link unreliability is captured through the appropriate characterization of the probability distribution of the unencumbered call interruption time in the proposed mathematical model. More importantly, the queueing model is developed assuming that the channel holding times for new calls and handoff calls are general and non identically distributed random variables. In particular, the cases when the channel holding time variables are either Mixed-Erlang or Hyper-Exponentially distributed are analyzed. These types of distributions allow a very simple mathematical analysis and, more importantly, the Mixed Erlang distribution is able to arbitrary closely approximate to the distribution of any positive random variable as well as measured data. Additionally, when the channel holding time variables are Hyper-Exponentially distributed, an approximated one-dimensional recursive approach based on the well known Kauffman-Roberts' formula is proposed. Thus, our proposed queueing model allows developing general and easily computable mathematical expressions for many useful performance metrics in the evaluation of wireless cellular networks under more realistic considerations.

References

1. Christensen, T.K., Nielsen, B.F., Iversen, V.B.: Phase-type models of channel-holding times in cellular communication systems. IEEE Trans. Veh. Technol. 3, 725–733 (2004)
2. Guérin, R.: Channel occupancy time distribution in a cellular radio system. IEEE Trans. Veh. Technol. VT-35, 89–99 (1987)
3. Fang, Y., Chlamtac, I., Lin, Y.-B.: Modeling PCS networks under general call holding time and cell residence time distributions. IEEE/ACM Trans. on Networking 5(6), 893–906 (1997)
4. Soong, B.H., Barria, J.A.: A Coaxian model for channel holding time distribution for teletraffic mobility modeling. IEEE Commun. Letters 4(12), 402–404 (2000)
5. Alfa, A.S., Li, W.: A homogeneous PCS network with Markov call arrival process and phase type cell residence time. Wireless Networks 8, 597–605 (2002)
6. Xie, H., Goodman, D.J.: Mobility models and biased sampling problem. In: Proc. IEEE International Conference on Universal Personal Communications (ICUPC'93), Ottawa, Canada, pp. 803–807 (1993)
7. Rodríguez-Estrello, C.B., Cruz-Pérez, F.A., Hernández-Valdez, G.: System level model for the wireless channel unreliability in mobile cellular networks. In: Proc. IEEE Mobile Computing and Wireless Communications International Conference (MCWC'06), Amman, Jordan (September 2006)
8. Zhang, Y., Soong, B.: Performance of mobile network with wireless channel unreliability and resource insufficiency. IEEE Trans. Wirel. Commun. vol. 5(5), pp. 990–995
9. Hong, D., Rappaport, S.S.: Traffic model and performance analysis for cellular mobile radio telephone systems with prioritized and nonprioritized handoff procedures. IEEE Trans. Veh. Technol. 35(3), 77–92 (1986)

10. Orlik, P.V., Rappaport, S.S.: A model for teletraffic performance and channel holding time characterization in wireless cellular communication with general session and dwell time distributions. IEEE J. Select. Areas Commun. 16(5), 788–803 (1998)

11. Lin, Y.-B., Mohan, S., Noerpel, A.: Queuing priority channel assignment strategies for PCS and handoff initial access. IEEE Trans. Veh. Technol. 43(3), 704–712 (1994)

12. Fang, Y., Chlamtac, I.: Teletraffic analysis and mobility modeling of PCS networks. IEEE Trans. Commun. 47(7), 1062–1072 (1999)

13. Kelly, F.P.: Reversibility and Stochastic Networks. Wiley, New York (1979)

14. Yeo, K., Jun, C.-H.: Teletraffic analysis of cellular communication systems with general mobility based on hyper-Erlang characterization. Computers & Industrial Engineering 42, 507–520 (2002)

15. Orlik, P.V., Rappaport, S.S.: Traffic performance and mobility modeling of cellular communications with mixed platforms and highly variable mobilities. Proc. IEEE 86(7), 1464–1479 (1998)

16. Orlik, P.V., Rappaport, S.S.: Corrections to Traffic performance and mobility modeling of cellular communications with mixed platforms and highly variable mobilities. Proc. IEEE 86(10), 2111 (1998)

17. Kaufman, J.: Blocking in a shared resource environment. IEEE Trans. Commun. 29, 1474–1481 (1981)

18. Document 8F/434-E, Refined method of multi-dimensional Erlang-B formula and its sensitivity analysis calculation for mixed circuit-switched traffic, ITU, Radiocommunication Study Groups, April 12, 2005 (2005)

19. Bolotin, A.A.: Modeling call holding time distributions for CCS network design and performance analysis. IEEE J. Select. Areas Commun. 12(3), 433–438 (1994)

20. Orlik, P.V., Rappaport, S.S.: On the hand-off arrival process in cellular communications. In: Proc. IEEE WCNC'99, New Orleans, LA, pp. 545-549 (September 1999)

21. Cooper, R.B.: Introduction to Queuing Theory, 2nd edn. Elsevier North-Holland, New York (1981)

22. Logothetis, M., Shioda, S.: Centralized virtual path bandwidth allocation scheme for ATM networks. IEICE Trans. Commun. E75-B(10), 1071–1080 (1992)

A Novel Performance Model for the HSDPA with Adaptive Resource Allocation

Andreas Mäder[1], Dirk Staehle[1], and Hans Barth[2]

[1] University of Würzburg, Department of Distributed Systems
Am Hubland, D-97074 Würzburg
{maeder,staehle}@informatik.uni-wuerzburg.de
[2] T-Systems Enterprise Services GmbH
hans.barth@t-systems.com

Abstract. We propose a novel performance model for the HSDPA in presence of circuit-switched dedicated channels. The model consists of two parts: An HSDPA bandwidth model which considers the SIR distribution according to the multi-path model and the number of available channelization codes, and an analytical capacity model which integrates HSDPA and dedicated channels under assumption of adaptive resource allocation for the HSDPA. Additionally, the model considers the impact of location dependent bandwidths on the spatial user distribution. The accuracy of the model is demonstrated for an example scenario.

1 Introduction

Mobile network operators continue to deploy the High Speed Downlink Packet Access (HSDPA) service in their existing UMTS networks. From the users perspective, the HSDPA offers high bit rates (promised are up to 14.4 Mbps) and low latency. From operators perspective, the HSDPA is hoped to play a key role for the much longed for break through of high quality mobile data services. From a technical perspective, the HSDPA brings a new paradigm to UMTS: Instead of adapting transmit power to the radio channel condition in order to ensure constant link quality, HSDPA adapts the link quality to the radio channel conditions. This enables a more efficient use of scarce resources like transmit power, code resources and also hardware resources.

The basic principle of the HSDPA is to adapt the link to the radio channel condition with help of adaptive modulation and coding (AMC). For this reason, it employs a shared channel, the HS-DSCH, which is used by all HSDPA users. By using a shared channel, radio resources are occupied only if a transmission occurs which enables a more efficient transport of bursty traffic. In each transport time interval (TTI), the scheduler located in the NodeB decides which users will be scheduled at which rate. The scheduling decision can either be on behalf of channel quality indicator (CQI) reports from the UEs to enable opportunistic scheduling schemes which use the air interface more efficiently, or simple non-opportunistic schemes like Round-Robin can be used, which distributes the resources time-fair between the users. The rate selection in the NodeB is done

L. Mason, T. Drwiega, and J. Yan (Eds.): ITC 2007, LNCS 4516, pp. 950–961, 2007.

with a direct relation between CQI and the transport format resource combination (TFRC), which describes the number of information bits per TTI, the modulation scheme and following from this the code rate.

In the literature, a wide range of publications on several aspects of the HSDPA exists. The capacity of the HSDPA, mostly in terms of throughput, is the focus of many works which use simulations to obtain their results. The models in early publications like [1] and [2] concentrate on aspects of scheduling, HARQ and physical layer techniques. In [3], link-layer simulations have been performed which are used to fit the signal-to-noise ratio to CQIs. All these models do not consider the impact of coexistent dedicated channels on the HSDPA. This is done in [4], which assumes a fixed number of codes reserved for the HS-DSCH in their extensive simulation. The impact of the HSDPA on network planning is the focus of [5], [6], [7] and [8]. All these works use simulations for their results. The impact of code restraints is considered in [5] and [6], while [7] and [8] concentrate on the influence of the multi-path model and scheduling. In [9], a method for the estimation of the interference for the HSPDA is proposed.

In the next section we give a short overview of the HSDPA and relevant parts of the UMTS system. In Sec. 3, we introduce the model for the DCH users, for the HSDPA bandwidth and for the stochastic capacity model. In Sec. 5, we provide some numerical results and discuss the key impact factors on the performance of the HSDPA. Finally, in Sec. 6, we give a conclusion and point out some further topics of research.

2 System Description

The HSDPA is part of the UMTS Rel. 5. The core of the HSDPA is a new transport channel, the HS-DSCH (high speed downlink shared channel), which is a channel which is shared between all UEs in a sector. The HS-DSCH enables two types of multiplexing: Time multiplex by scheduling the subframes to different users, and code multiplex by assigning each user a non-overlapping subset of the available codes. Figure 1 shows a schematic view of the HS-DSCH over a short time period. The time axis is divided in subframes of 2 ms. In each subframe, one of three users is assigned a number of codes with SF 16 between 0 and 15. Throughout this work we only consider time multiplex.

In contrast to DCH, where the transmit power is adapted to the propagation loss with fast power control and thus enabling a more or less constant bit rate, the HS-DSCH adapts the channel to the propagation loss with adaptive modulation and coding (AMC). This means that depending on the received SIR, the scheduler in the NodeB chooses a transport format combination (TFC) with a pre-defined target FER (frame error rate), which is often chosen as 10%. The TFC contains information about the modulation (QPSK or 16QAM), the number of used codes (from 1 to 15), and the coding rate resulting in a certain transport block size (TBS) that defines the information bits transmitted during a TTI. Which TFC to choose is determined by a number of tables in [10] which map the channel quality indicator (CQI) to TFC. The channel quality indicator

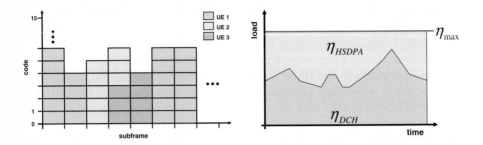

Fig. 1. Schematic view of the HS-DSCH   **Fig. 2.** Fast power adaptation

is a discretization of the received SIR at the UE and ranges from 0 to 30. The mapping from CQI to TFC depends on the technical capabilities of the UE and there exist five different UE classes.

Since DCH connections are power controlled and have a fixed bit rate, the load η_D for DCH connections depends on the other-cell interference, on the bit rates and on the number of connections only. The unused resource (i.e. transmit power) on the downlink is available for the HSDPA. We consider fast power adaptation for the HSDPA, which means that the NodeB adapts the HSDPA transmit power instantaneously on the load situation and therefore always meets a operator-defined target load $\eta_{\max}$, cf. Fig. 2.

3   System Model

3.1   HSDPA Bandwidth Model

In the following we propose a simple model for computing the long-term bandwidth for a system state that provides us the number of codes and the transmit power available for the HSDPA as well as the number of users that share these resources. We assume the adaptive HSDPA configuration which means that the HSDPA uses or may use all transmit power and all codes remaining from the dedicated channels. In particular, this means that all cells are transmitting with maximum/desired power T_{max} if there is at least one HSDPA user, and in the following we make the worst-case assumption that the surrounding cells always serve an HSDPA user and consequently produce interference according to their maximum power. Furthermore, we assume the round robin scheduling discipline which gives us a direct relationship of long-term bandwidth R and mean TBS $E[TBS]$ for a certain number N_{hs} of HSDPA users:

$$R = \frac{1s}{N_{hs} \cdot 2ms \ TTI} \cdot E[TBS]. \tag{1}$$

The standard specifies a set of TFCs according to the technical capabilities of the UE. A TFC relates the CQI to the TBS and the required number of parallel HSDPA codes. The CQI is a random variable that depends on the instantaneous channel conditions. Let $p_{CQI}(q)$ be the probability for a CQI of q corresponding to the TFC with TBS $TBS(q)$. For a number C_{hs} of usable HSDPA codes, we obtain the mean TBS as

$$E[TBS] = \sum_{q=0}^{30} p_{CQI}(q) \cdot \min\left(TBS^*, TBS(q)\right), \tag{2}$$

where TBS^* is the maximum TBS that is possible with C_{hs} codes. The CQI relates to the SIR in decibels according to the formula given in [3]

$$CQI = \max\left(0, \min\left(30, \left\lfloor \frac{SIR[dB]}{1.02} + 16.62 \right\rfloor\right)\right). \tag{3}$$

The SIR depends on the HSDPA transmit power T_{hs}, the location f of the mobile that defines the average propagation gain $d_{y,f}$ from a NodeB y to the mobile, and finally the multi-path propagation. The multi-path profile for the channel between NodeB y and location f corresponds to a set of paths $\mathcal{P}_{y,f}$. Every single path p corresponds to a Rayleigh fading channel with an average relative power β_p. For numerical results we used the ITU Vehicular A model:

path p	1	2	3	4	5	6
average power β_p [dB]	0	-1	-9	-10	-15	-20

Assuming optimal maximum ratio combining by the RAKE receiver, the SIR γ_f at a certain position f and for a certain ratio $\Delta_T = T_{hs}/T_{max}$ of HSDPA power to total cell power is given by

$$\gamma_f(\Delta_T) = \Delta_T \cdot \sum_{p \in \mathcal{P}_{x,f}} \frac{\xi_p}{\frac{I_f^{other}}{T_{max} \cdot d_{x,f}} + \sum_{r \in \mathcal{P}_{x,f} \backslash p} \xi_r} \tag{4}$$

$$\text{with} \quad I_f^{other} = \sum_y d_{y,f} \cdot T_{max} \cdot \sum_{r \in \mathcal{P}_{y,f}} \xi_r, \tag{5}$$

where ξ_p is an exponential random variable with mean β_p that describes the instantaneous propagation gain on path p. Note, that this equation neglects the thermal noise as it has almost no impact on the SIR for reasonably sized cells. Let us now introduce the variable $\Gamma_f = \gamma_f(1)$ for the SIR with $\Delta_T = 1$ that corresponds to the main sum in Eq. 4, and the variable $\Sigma_f = I_f^{other}/(T_{max} \cdot d_{x,f})$ for the ratio of average other-cell received power to average own-cell received power. In the following we refer to the random variable Γ_f as the normalized SIR at location f. Now, we make the assumption that the distribution of the normalized SIR is a function of Σ_f. Simulations have shown that for $\Sigma_f < 0.1$ the distribution of Γ_f in decibel scale is well-approximated by an inverse Gaussian distribution and for $\Sigma_f >= 0.1$ the distribution of Γ_f in linear scale is also well-approximated by an inverse Gaussian distribution. Based on the assumption that the distribution of Γ_f is a function of Σ_f we identified the following functions for the mean and standard deviation of Γ_f:

$$\begin{aligned}
E[\Gamma_f] &= 0.1424 + 4.0627 \cdot exp\left(-2.0073 \cdot \Sigma_f^{0.4220}\right), \text{ for } \Sigma_f \geq 0.1 \\
STD[\Gamma_f] &= 0.1283 + 5.7680 \cdot exp\left(-3.0712 \cdot \Sigma_f^{0.2819}\right), \text{ for } \Sigma_f \geq 0.1 \\
E[\Gamma_f[dB]] &= 0.0807 + 3.9085 \cdot exp\left(-4.3624 \cdot \Sigma_f^{0.9728}\right), \text{ for } \Sigma_f < 0.1 \\
STD[\Gamma_f[dB]] &= 1.1039 + 1.2365 \cdot exp\left(-0.7480 \cdot \Sigma_f^{0.4109}\right), \text{ for } \Sigma_f < 0.1
\end{aligned} \tag{6}$$

The probability density function (PDF) $a(x)$ of an inverse Gaussian distributed random variable X is given by

$$a(x) = \sqrt{\frac{\lambda}{2\pi x^3}} \cdot e^{\frac{-\lambda(x-\mu)^2}{(2x\mu^2)}} \text{ with } \mu = E[X] \text{ and } \lambda = \frac{E[X]^3}{VAR[X]}. \tag{7}$$

Consequently, we are now able to determine the PDF $a_{\Gamma_f}(x)$ of the normalized SIR at a certain location f that corresponds to the ratio of average other- and own-cell received powers. Applying the formula in Eq. (3) that relates SIR to CQI we obtain the following distribution for the CQI:

$$p_{CQI}(q,f) = \begin{cases} A_{\Gamma_f}(\phi_{max}(q)), \text{ for } q = 0 \\ A_{\Gamma_f}(\phi_{max}(q)) - A_{\Gamma_f}(\phi_{min}(q)), \text{ for } q = 1, ..., 29 \\ 1 - A_{\Gamma_f}(\phi_{min}(q)), \text{ for } q = 30 \end{cases} \tag{8}$$

where A_{Γ_f} denotes the CDF of the normalized SIR and the functions $\phi_{max}(q)$ and $\phi_{min}(q)$ relate to the maximum and minimum normalized SIR for a certain HSDPA power that lead to CQI q. The functions are given as

$$\begin{aligned} \phi_{max}(q) &= (q - 15.62) \cdot 1.02 + \Delta_T[dB] \\ \phi_{min}(q) &= (q - 16.62) \cdot 1.02 + \Delta_T[dB]. \end{aligned} \tag{9}$$

Now, Eq. (8) yields the distribution of the CQI to be used in Eq. (2) and Eq. (1) for computing the mean TBS and the long-term bandwidth $R_f(C_{hs}, \Delta_T)$ of an HSDPA user at a certain position f within the cell area, a ratio Δ_T of HSDPA power to total cell power, and C_{hs} available HSDPA codes.

In the following, we demonstrate the accuracy of the model. Therefore, we consider a network with 19 hexagonal cells arranged in two tiers around a central cell. The distance between two NodeBs is 1.2 km. For this network we produce 1000 random situations that means a random mobile position within the central cell and independent total transmit powers for the 19 NodeBs which are uniformly distributed between 2 W and 10 W. For every situation we generate 10000 independent instances of the multi-path profile and compute the resulting normalized SIR according to Eq. (4). Thus, we obtain the mean, the standard deviation, and the distribution of the normalized SIR, and also the associated ratio of average other- to own-cell received power for every situation. Figure 3 demonstrates the accuracy of the functions for estimating the mean and the standard deviation of the normalized SIR as defined in Eq. (6). The left figure shows the functions for $\Sigma < 0.1$ and the right figure for $\Sigma \geq 0.1$. The functions displayed as solid lines match the center of the simulated points which mark the 1000 different situations quite well. The deviation of the simulated points from the estimated function results from the fact that the distribution of the normalized SIR is precisely a function of all received powers and the aggregation of these values to the single value Σ is an approximation leading to a certain inaccuracy.

Next, we demonstrate the final accuracy of the long-term bandwidth estimation for a more specific scenario. The values of Σ correspond to points on a grid with a resolution of 50m covering the central cell. Furthermore, we apply now

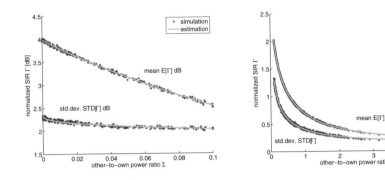

Fig. 3. Accuracy of the functions defined in Eq. (6)

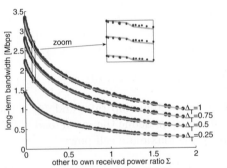

Fig. 4. Accuracy of long-term bandwidth estimation

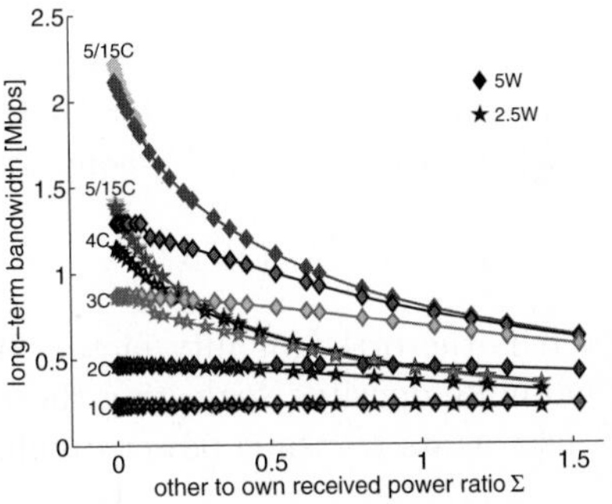

Fig. 5. Impact of codes and power on long-term bandwidth. (xC means x HSDPA codex)

the adaptive HSDPA configuration, which means that all cells transmit with equal power. The ratio Δ_T of HSDPA power to total cell power is set to $1/4$, $1/2, 3/4$, and 1. The maximum of 15 parallel HSDPA codes is available. Figure 4 compares the estimated bandwidth displayed as solid lines with the bandwidth resulting from the simulations presented by dot markers. We can observe a very good match for all points. The main inaccuracy occurs for values of Σ around 0.1 which is specifically illustrated in the small box. For Σ just below 0.1 we underestimate the bandwidth and for values just above 0.1 we overestimate the bandwidth. The switching point is exactly where we change estimating the distribution of Γ from linear scale to decibel scale. Figure 5 shows the impact of available power and available codes on the HSDPA bandwidth for a single user. The power is set to 2.5 W and 5 W, the available codes are varied from only a single code up to 15 codes. The first point to observe is that in the considered scenario, i.e. full other-cell power, multi-path profile, and UE class four, five parallel codes are enough to achieve the maximum bandwidth if the HSDPA transmits with 5 W which is half of the total cell power assumed here with 10 W. On the other hand, if we have only one or two codes available, there is only a small difference in the resulting bandwidth whether the HSDPA uses 2.5 W or 5 W. For three and four codes there is a distinct gap between HSDPA powers of 2.5 W and 5 W, and up to a maximum of five, the availability of more codes also leads to an improved performance.

3.2   Sharing Resources Between DCH and HSDPA

The main difference between a DCH and an HSDPA users is that the former receives a certain QoS expressed as the data rate R_s and the target bit-energy-to-noise ratio (E_b/N_0) ε_s while the latter utilizes the remaining resources. The data rate and the E_b/N_0 value correspond to a certain spreading factor SF_s and a certain transmit power that depends on the location of the user. Applying the adaptive HSDPA configuration, i.e. the HSDPA consumes all power remaining from the DCH users and accordingly all NodeBs transmit with target power $T_{\max}$, the power requirement for DCH users is given by the following equation, which follows from the E_b/N_0-equation for power controlled CDMA systems:

$$\hat{T}_k = \frac{\varepsilon_s R_s}{W} \cdot \left(W N_0 \cdot \frac{1}{d_{x,k}} + \sum_{y \neq x} T_{\max} \cdot \frac{d_{y,k}}{d_{x,k}} + \alpha \cdot T_{\max} \right), \qquad (10)$$

where α is the orthogonality factor, W is the system chip rate, and N_0 is the thermal noise density. Neglecting thermal noise, we define the mean load of one DCH user as its transmit power divided by the maximum cell power:

$$\omega_s = \nu_s \cdot \frac{\varepsilon_s R_s}{W} \cdot \left(\sum_{y \neq x} E\left[\frac{d_{y,k}}{d_{x,k}} \right] + \alpha \right), \qquad (11)$$

where ν_s is the activity factor. The total DCH load η_d and the mean HSDPA transmit power are then given by

$$\eta_D = \sum_{s \in \mathcal{S}} n_s \omega_s, \quad \text{and} \quad T_H = T_{\max} - T_c - \eta_D \cdot T_{\max}, \qquad (12)$$

where T_c is the power required for common channels.

The UMTS downlink uses orthogonal variable spreading factor (OSVF) codes with spreading factors (SF) between 4 and 512. If we use the code c_{512} with SF 512 as the basic code unit, all other codes can be expressed in terms of a multiple of c_{512}, such that $c_s = \frac{512}{SF_s}$. With the assumption of perfect reordering the system has a total code capacity C of $512 - c_c$, where $c_c = 32$ is reserved for common channels. The HSDPA is able to use multiple codes with spreading factor 16 in parallel. Accordingly, $C_{hs} = \left\lfloor \frac{480 - \sum_s n_s \cdot c_s}{32} \right\rfloor$ HSDPA codes are available.

4   Capacity Model

We consider a UMTS cell with users arriving according to a Poisson process with rate λ_s for DCH users and λ_H for HSDPA users. DCH users depart with rate μ_s

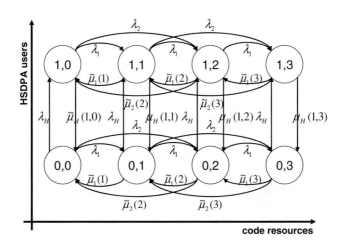

Fig. 6. Structure of a two-dimensional state space

and have exponentially distributed service times whereas HSDPA users transmit an exponentially distributed data volume with mean $E[V_H]$. The user location is uniformly distributed over the cell area which is divided into a number q of square area units. We develop a common state space for DCH and HSDPA traffic in order to estimate the performance of the HSDPA.

We consider the number of occupied code units as state space description. Each DCH service connection requires a number c_s of SF 512 code equivalents, so the code resources form a shared resource which can be used in the Kaufman-Roberts recursion [11] to form an one-dimensional state space. However, since we also want to include the HSDPA users in the state space, we construct an 2-dimensional state space with the number of occupied code resources by DCH users on the first dimension, and the number of active HSDPA flows on the second dimension. A state j in the DCH dimension corresponds to the number of occupied code units $c_u = min_s(c_s)$, i.e. in state j the DCH users occupy $j \cdot c_u$ codes with SF 512. The size of the state space is $C/c_u \times n_{H,max}$, where $n_{H,max}$ is the maximum number of HSDPA users. Note that we assume here a simple count-based admission control for the HSDPA. Figure 6 shows a fraction of the state space. The departure rates for the DCH dimension are calculated according to the following equation (see [11]):

$$\tilde{\mu}_s(j) = \mu_s \cdot E[n_s|j], \tag{13}$$

where $E[n_s|j]$ is the mean number of service class s connections in state j:

$$E[n_s|j] = \frac{\lambda_s}{\mu_s} \cdot \frac{\tilde{p}_{kr}(j - \frac{c_s}{g_c})}{\tilde{p}_{kr}j}. \tag{14}$$

The variable $\tilde{p}_{kr}(j)$ denotes the un-normalized probability for the DCH state $j = \frac{C}{g_c}$. It is calculated recursively as

$$\tilde{p}_{kr}(j) = \frac{g_c}{C} \sum_{s \in \mathcal{S}} \frac{\lambda_s}{\mu_s} c_s \tilde{p}_{kr}\left(j - \frac{c_s}{g_c}\right). \tag{15}$$

Note that this model does not allow an explicit computation of the DCH load in a state but only its mean value. This prohibits the computation of soft blocking probabilities, i.e. blocking due to transmit power limitation, for DCH users. However, in [12] it has been shown that the code capacity is the dominating factor for the system capacity except for nearly full DCH activity and concurrent bad orthogonality conditions, i.e. high orthogonality factors. Nevertheless, incorporating soft blocking and investigating the impact of the DCH load distribution within a code-based system state will be an item for future work.

The HSDPA user throughput depends on both the current DCH load and on the number of active HSDPA flows. While the code resources are directly available through the state, the current transmit power for the HSDPA requires knowledge of the DCH downlink load, which is calculated as in (12) using the mean number of DCH users $E[n_s|j]$. Consequently, the number of usable HSDPA codes is $C_{hs}(j, n_H) = \lfloor (480 - j * c_u)/32 \rfloor$ and the mean ratio of HSDPA power to total cell power is

$$\Delta_T(j, n_H) = 1 - \frac{T_c}{T_{\max}} - \sum_s E[n_s|j] \cdot \omega_s. \tag{16}$$

The HSDPA bandwidth for a certain location f follows according to the model defined in Section 3.1

$$R_f(j, n_H) = R_f(C_{hs}(j, n_h), \Delta_T(j, n_H)). \tag{17}$$

A straightforward method for computing the departure rate would be to determine the mean bandwidth for an HSDPA user by averaging over the cell area. We will later refer to this approach as the "naïve" approach. However, we have to consider the following: With a volume-based user model, the life-time of an HSDPA user depends on its data volume V_H and its bandwidth. Even with round-robin scheduling, users at the cell border receive a smaller bandwidth than users in the cell center, and accordingly, they stay in the system for a longer time. Consequently, the probability p_f to meet a user at position f when looking into the system at a random instance of time is larger for a location f close to the cell border than for one close to the cell center. More precisely, the probability p_f is proportional to the reciprocal bandwidth available at this position: $p_f \sim \frac{1}{R_f(j, n_H)}$. This effect is also mentioned by Litjens et al [8] regarding Monte Carlo simulations. We approximate the mean time $E[T|j]$ by summing over all positions in the cell and, after some algebraic operations, obtain the following formulation:

$$E[T|j] = E[V_H] \cdot E\left[\frac{1}{R_f(j, n_H)^2}\right] \cdot E\left[\frac{1}{R_f(j, n_H)}\right]^{-1} \tag{18}$$

In the following, we will refer to this method as the "location-aware" approach. In order to calculate the stead-state distribution, we arrange the transition rate matrix Q with help of an index function $\phi(j, n_H) \to \mathbb{N}$ according to the following rules for all valid states:

$$
\begin{aligned}
Q(\phi(j, n_H), \phi(j + \tfrac{c_s}{g_c}, n_H)) &= \lambda_s \\
Q(\phi(j, n_H), \phi(j - \tfrac{c_s}{g_c}, n_H)) &= \tilde{\mu}_s(j) \\
Q(\phi(j, n_H), \phi(j, n_H + 1)) &= \lambda_H \\
Q(\phi(j, n_H), \phi(j, n_H - 1)) &= \tfrac{1}{E[T|j]}
\end{aligned}
\tag{19}
$$

In all other cases $Q(i, j)$ is set to zero and $Q(i, i)$ is set to the negative row-sum of all entries to keep the state equations balanced. The steady-state distribution is then obtained by solving $Q \cdot \bar{\pi} = 0$ s.t. $\sum \pi = 1$ for the state vector $\bar{\pi}$. Performance measures like blocking probabilities or moments of the user throughput are then calculated with help of the steady-state distribution. For example, the mean HSDPA user throughput at a random time instance is

$$
E[R_U] = \sum_{(j, n_H)|n_H > 0} R_U(j, n_H) \cdot \frac{n_H \cdot \pi(\phi(j, n_H))}{\sum_{(j', n'_H)|n'_H > 0} n'_H \cdot \pi(\phi(j', n'_H))}.
\tag{20}
$$

5 Numerical Example

Let us now define an example scenario with the following parameters: We consider two DCH service classes with 128 kbps and 384 kbps. The service mix is 0.6 to 0.4. The activity factor is 0.55 for both service classes. HSDPA flows arrive with rate $\lambda_H = 1$. The orthogonality factor α for the DCH part of the air interface model is set to 0.35 corresponding to the HSDPA bandwidth model. The distance between the NodeBs is 1.2 km and the COST 231 Hata Model is used as propagation loss model. We validate our analytical results with an event-based simulation. The simulation places new users on random locations and calculates the exact air interface load for DCH users as well as the mean bandwidth for the HSDPA users according to the method specified in Section 3.1. The users keep their positions during their life time.

Figure 7 shows the mean user throughput versus the offered DCH code load which is defined as $a_c = \sum_{s \in \mathcal{S}} \lambda_s / \mu_s \cdot c_s / C$. The figure shows two scenarios, one with a mean HSDPA data volume of 50 kbyte and one with 100 kbyte. The influence of the spatial user distribution can be clearly seen on the large difference between the naïve approach (square marker) and the location-aware approach (circle marker). Especially for a low DCH load the difference is nearly 50%. With location-awareness, the analytical and the simulation results match well. The curves for 50 kbyte and 100 kbyte converge with a higher offered load for the DCH users since in this case the HSDPA is in an overload situation due to insufficient power and code resources.

In Fig. 8 we show the impact of the HSDPA admission control on the trade-off between user throughput and HSDPA blocking probabilities. The data volume

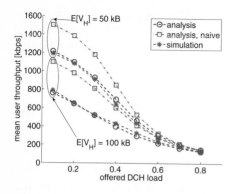
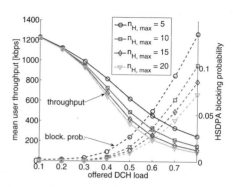

Fig. 7. Mean HSDPA user throughput vs. offered DCH load

Fig. 8. Impact of $n_{H,max}$ on throughput and blocking probabilities

is $E[V_H] = 50$ kbyte. The solid lines indicate the user throughput, while the dashed lines indicate the HSDPA blocking probabilities. The maximum number of allowed HSDPA users, $n_{H,max}$ is set to 5, 10, 15 and 20, respectively. We see that with higher DCH loads the increasing blocking probabilities for low values of $n_{H,max}$ leads to a significant improvement of the user throughput. The difference between the curves become smaller with higher values of $n_{H,max}$ which indicates that they will converge if $n_{H,max}$ would be increased further.

6   Conclusion

We presented an analytical capacity model for the HSDPA. It includes a novel bandwidth model for the HSDPA which estimates the location dependent instantaneous user throughput with help of the SIR distribution. The model also considers DCH users, which have a crucial impact on the performance of the HSDPA both due to the transmit powers and due to the available code resources. The common state space integrates all DCH service classes into one dimension to reduce the state space dimensions and computational complexity. Furthermore, we approximated the effect of a shifted active user distribution due to location-dependent user bandwidths. The analytical results have been validated with help of a full featured simulation. The results from the simulation and the analytical results match well. Further research topics are the impact of radio resource management strategies, different multi-path profiles as well as other scheduling schemes than round robin.

Acknowledgments

The authors want to thank Prof. Phuoc Tran-Gia, Markus Spahn and Tobias Hoßfeld, University of Wuerzburg, as well as Bernhard Liesenfeld, T-Systems Germany, for the helpful discussions.

References

1. Kolding, T.E., Frederiksen, F., Mogensen, P.E.: Performance Aspects of WCDMA Systems with High Speed Downlink Packet Access (HSDPA). In: Proc. of IEEE VTC Fall '02, Vancouver, Canada, vol. 1, pp. 477–481 (2002)
2. Assaad, M., Zeghlache, D.: On the Capacity of HSDPA. In: Proc. of GLOBECOM '03, San Francisco, USA, vol. 1, pp. 60–64 (2003)
3. Brouwer, F., de Bruin, I., de Bruin, J., Souto, N., Cercas, F., Correia, A.: Useage of Link-Level Performance Indicators for HSDPA Network-Level Simulations in E-UMTS. In: Proc. of IEEE ISSSTA '04, Sidney, Australia, pp. 844–848 (2004)
4. Pedersen, K.I., Lootsma, T.F., Støttrup, M., Frederiksen, F., Kolding, T.E., Mogensen, P.E.: Network Performance of Mixed Traffic on High Speed Downlink Packet Access and Dedicated Channels in WCDMA. In: Proc. of IEEE VTC Fall '04. Vol. 6., Milan, Italy, pp. 2296–4500 (2004)
5. Voigt, J., Deissner, J., Hübner, J., Hunold, D., Möbius, S.: Optimizing HSDPA Performance in the UMTS Network Planning Process. In: Proc. of IEEE VTC Spring '05. Vol. 4., Stockholm, Sweden, pp. 2384–2388 (2005)
6. Türke, U., Koonert, M., Schelb, R., Görg, C.: HSDPA performance analysis in UMTS radio network planning simulations. In: Proc. of IEEE VTC Spring '04, Milan, Italy, vol. 5, pp. 2555–2559 (2004)
7. van den Berg, J.L., Litjens, R., Laverman, J.: HSDPA flow level performance: the impact of key system and traffic aspects. In: Proc. of MSWiM '04, Venice, Italy, pp. 283–292 (2004)
8. Litjens, R., van den Berg, J.L., Fleuren, M.: Spatial Traffic Heterogeneity in HSDPA Networks and its Impact on Network Planning. In: Proc. of ITC 19, Bejing, China, pp. 653–666 (2005)
9. Mäder, A., Staehle, D.: Interference estimation for the HSDPA service in heterogeneous UMTS networks. In: Cesana, M., Fratta, L. (eds.) Wireless Systems and Network Architectures in Next Generation Internet. LNCS, vol. 3883, Springer, Heidelberg, Como, Italy (2006)
10. 3GPP: 3GPP TS 25.306 V6.5.0 UE Radio Access capabilities (Release 6). Technical report, 3GPP (2004)
11. Kaufman, J.: Blocking in a Shared Resource Environment. IEEE Transactions on Communications 29(10), 1474–1481 (1981)
12. Staehle, D.: On the Code and Soft Capacity of the UMTS FDD Downlink and the Capacity Increase by using a Secondary Scrambling Code. In: Proc. of IEEE International Symposium on Personal Indoor and Mobile Radio Communications (PIMRC), Berlin, Germany (2005)

On the Interaction Between Internet Applications and TCP*

M. Siekkinen[1,**], G. Urvoy-Keller[2], and E.W. Biersack[2]

[1] University of Oslo, Dept. of Informatics, Postbox 1080 Blindern, 0316 Oslo, Norway
siekkine@ifi.uio.no
[2] Institut Eurecom, 2229, route des crêtes, 06904 Sophia-Antipolis, France
{urvoy,erbi}@eurecom.fr

Abstract. We focus in this paper on passive traffic measurement techniques that collect traces of TCP packets and analyze them to derive, for example, round-trip times or aggregate metrics such as average throughput. The seminal work of Zhang [1] has shown that for more than 50% of the TCP connections observed, it is not the network bandwidth that limits the throughput but rather the application or mechanisms such as TCP slow start or too small a receiver window. Certain types of analysis of the network characteristics are meaningful only when performed on TCP traffic that experiences minimal interference by the application. To eliminate such interference, we propose a generic method that partitions the packets of a TCP connection in bulk data transfer and in application limited periods: The packets of a bulk data transfer period (BTP) experience minimal interference from the application, while the packets of an application limited period (ALP) experience interference from the application that prevents TCP from fully utilizing the network resources because the application does not produce data fast enough. As a proof of concept, we apply our algorithm to public Internet traffic traces and show that unless the effects of the application are filtered out, studying the end-to-end path and traffic characteristics from a network point of view can produce biased results.

1 Introduction

While the majority of Internet applications today use TCP as a transport protocol, they differ very much in the way they use TCP. Consider, for instance, a P2P application such as BitTorrent that may establish 20-30 connections of which only a subset is actively used at any time for transmitting data. On the other hand, FTP establishes one control and one data connection. FTP transfers the data at once", whereas BitTorrent connections alternate between busy (unchoked) and idle (choked) periods.

Much research has been done to detect anomalies and to characterize TCP traffic in the Internet through passive measurements. This work usually focuses

* This work has been partly supported by France Telecom, project CRE-46126878.
** Done while at Institut Eurecom.

L. Mason, T. Drwiega, and J. Yan (Eds.): ITC 2007, LNCS 4516, pp. 962–973, 2007.
© Springer-Verlag Berlin Heidelberg 2007

on the TCP and IP layers, but often ignores the effects of the application on top. When seeking to explain certain characteristics, e.g. burstiness of TCP traffic [2], it is crucial to account for the effects of the application. In addition, it is very important to understand to what extent the applications themselves are responsible for the transmission rates observed in today's Internet, and not limited by the available resources in the network [1].

The contributions of this paper are two fold: First, we present an algorithm that allows to isolate *bulk data transfer periods* (BTP) and *application limited periods* (ALP) within a connection for an application using TCP as transport protocol. We define a BTP as a period where the TCP sender never needs to wait for the application on top to provide data to transfer. Other time periods are defined as ALPs. Our algorithm to identify BTPs within a TCP connection is *generic* in the sense that it works regardless of the type of application on top of TCP. In this way, it can be applied to traces containing traffic from an unknown mixture of applications. Second, as a proof of concept, we show through examples that unless the effects of the application are filtered out, studying the end-to-end path and traffic characteristics from a network point of view can produce biased results.

Pioneering research work on TCP root cause analysis was done by Zhang et al. in [1] where they identify application limitation as one of the possible causes for achieving a given throughput. Our work differs from theirs by providing a method to isolate the bulk data transfers for further analysis and to evaluate the effect of the application in a quantitative way. Allman [3] recommends to carefully choose the application when evaluating TCP performance. We propose to minimize the effect of the application in the case when it is impossible to make such a choice, e.g. when analyzing traffic traces recorded by third parties.

2   Applications and TCP

Figure 1 describes the way data flows from sender to receiver application using a single TCP connection. The interaction happens through buffers: at the sender side the application stores data to be transmitted by TCP in buffer b1, while at the receiver side TCP stores correctly received and ordered data in buffer b2 that is consequently read by the receiving application. We focus only on the behavior of the sending application since it projects directly the application protocol behavior while the receiving application should always read the buffer b2 whenever it contains data[1]. When the application sends data constantly, buffer b1 in Figure 1 always contains data waiting to be transferred. We refer to such a period as Bulk Transfer Period (BTP). In other cases, when the application limits the throughput achieved, TCP is unable to fully utilize the network resources due to lack of data to send. We call such a period Application Limited Period (ALP). The interaction between the sending application and TCP manifests itself in the traffic in diverse ways depending on the type of application.

---

[1] When the receiving application can not read the buffer b2 fast enough, the receiving TCP will notify the sending TCP by lowering the receiver advertised window.

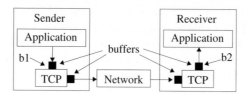

Fig. 1. Data flow from the sender to the receiver application through a single TCP connection

Table 1. Summary of Different Application Types

Type	Main Characteristics	Example Applications
1	constant application limited transmission rate, consists of a single ALP	Skype and other live streaming, and client rate limited eDonkey
2	user dependent transmission rate, typically a single ALP	telnet and instant messaging applications
3	transmission bursts separated by idle periods, applications using persistent connections, BTPs interspersed with ALPs	Web w/ persistent HTTP connections, BitTorrent (choked and unchoked periods)
4	transmit all data at once, single BTP	FTP
5	mixture of 1 and 3	BitTorrent with rate limit imposed by client application

We depict the diverse classes of applications operating on top of TCP in Table 1. This classification illustrates the multiple ways in which the application can influence the TCP traffic. Given such a diversity, it is challenging to design a generic algorithm that separates BTPs from ALPs since the application may interfere on very different time scales.

3   The Isolate & Merge (IM) Algorithm

3.1   Context

We call our algorithm for identifying BTPs and ALPs the Isolate & Merge (IM) Algorithm due to the way it proceeds. The algorithm is generic in that it can be applied without any calibration to traffic from any application. Additionally, it does not depend on the version of TCP used. Instead, the algorithm relies only on observing generic behavior common to all TCP versions. The algorithm processes bidirectional TCP/IP traffic passively collected at a single measurement point[2].

---

[2] It may not always be possible to capture the traffic in both directions, e.g. in the backbone where connections may have asymmetric upstream and downstream paths [4]. Nevertheless, we argue that unidirectional traces are often not sufficient for in-depth analysis, as is the case for round-trip time (RTT) estimation.

We define a TCP *connection* as a sequence of packets having the same source-destination or destination-source pairs of IP addresses and TCP port numbers. The IM algorithm processes only connections consisting of at least 130 data packets: Connections with fewer than 130 packets are very likely to be dominated by the TCP slow start algorithm and therefore convey little information about the TCP/IP data path for future analysis. We have chosen this threshold since a TCP sender that starts in slow start needs to transmit approximately 130 data packets (assuming a MSS of 1460 bytes) in order to reach a congestion window size equal to 64 Kbytes, which is a common size for the receiver advertised window [5]. For the same reasons, we define the minimum required size of a BTP to be 130 data packets. We also define as *short transfer period* (STP) a sequence of packets that contains fewer than 130 data packets and whose rate of transfer is not application limited. We use the term *transfer period* (TP) to refer to either a BTP or STP. The IM algorithm identifies BTPs for a single direction of a connection at a time.

3.2   Procedures

The IM algorithm consists of two phases: First, the *Isolate* phase partitions the connection into TPs separated by ALPs. In the second, *Merge* phase, the algorithm attempts to merge two consecutive TPs including the ALP that separates these two TPs in order to create a new BTP.

The key insight of the Isolate phase is to use small-size packets and idle times between packets as indications that the application does not provide data fast enough to TCP. Therefore, a large fraction of transmitted packets that are smaller than MSS or an idle time longer than a RTT following a packet smaller than MSS trigger the beginning of an ALP. Similarly, many consecutive MSS packets mark the beginning of a BTP.

In order to understand the reasoning behind the Merge phase, let us consider how the ALPs differ from the TPs. ALPs achieve by definition a lower throughput than TPs, since the application prevents TCP from fully using the network resources. Thus, the application interference is visible as a reduced throughput. The merging phase is needed because, after the isolate phase, a connection may be divided into many BTPs and STPs separated by very short ALPs. It would be often desirable to combine these periods into one long BTP for subsequent analysis if the effect of these short ALPs on the overall throughput achieved is small. For the above reasons, the procedure of merging periods is based on comparing the throughput of the periods involved in the merger.

The mergers are controlled with the threshold parameter $drop \in [0,1]$. Figures 2 and 3 demonstrate successful and failed mergers, respectively. Periods can be merged if and only if the throughput of the resulting merged BTP (total bytes divided by total duration) is higher than the *drop* value times the throughput of the TPs combined together excluding the ALP in the middle (sum of bytes of TPs divided by sum of durations of TPs). In this way, the *drop* parameter value limits the maximum amount of application interference, i.e. throughput decrease, within the resulting merged periods. Hence, by selecting

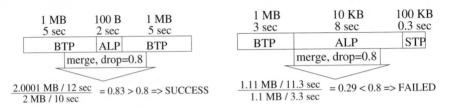

Fig. 2. Successful merger   **Fig. 3.** Failed merger

a specific value of the *drop* parameter, the user can choose the desired maximum amount of application interference allowed to be present in the resulting BTPs that can then be used for further analysis. The algorithm for merging periods proceeds in an iterative manner, which ensures that eventually all possible mergers, and only those allowed, are performed.

Experimenting with different values of the *drop* parameter allows for a quantitative analysis of the application impact on the throughput achieved, which, as we show in the extended version of this paper [6], can provide interesting input for characterizing the application behavior. Due to space limitations, we refer the reader for an algorithmic description of the IM procedures to our technical report [6]. The procedures are more complex that one would think at first sight because they are designed to work correctly regardless of the location of the measurement point on the TCP/IP path (i.e. at sender/receiver or middle of path). Because of that, the measurement point location needs first to be determined, as it is not always known a priori, and furthermore, the location has implications to RTT estimation and calculations of the inter-arrival times (IAT) between packets.

We validated the IM algorithm with the help of the Web100 software [7] that allows querying the TCP state information of active TCP connections. We cross-checked the results of applying IM on a large trace of Internet traffic with the Web100 data captured at the sender and observed that the results matched in more than 95% of the cases. For more details please consult [6].

4   Data Sets

We applied the IM algorithm to eight application-specific traffic traces in order to investigate the impact of different applications in our studies in Section 5. Except for the SSH data set, all of the application specific traces were extracted from the same original public ADSL access network traces (the first 19 days from Location 4 traces in the M2C Measurement Data Repository at http://m2c-a.cs.utwente.nl/repository/, two traces per day, 15 minutes each) by filtering on the well-known TCP port numbers of these applications. This method gives in most of the cases solely the traffic from the expected application except for some cases where well-known TCP ports, such as port 80, are used by P2P applications to bypass firewalls, for example. The SSH traffic data set consists of scp downloads from various locations all over the world to a single destination.

Table 2. Trace characteristics

traffic type	BitTorrent	eDonkey	FTP data	SSH	Gnutella	HTTP(S)	FastTrack	WinMX
port numbers	6881-6889	4661,4662	20	22	6346,6347	80,443	1214	6699
duration	4d 22h	4d 22h	18d 22h	7d	18d 22h	4d 22h	18d 22h	18d 22h
packets	31M	44M	9M	3.6M	8M	14M	20M	13M
bytes	19GB	20GB	7GB	2.9GB	2GB	9GB	14GB	5GB
cnxs	150K	1.6M	5.9K	48K	410K	590K	360K	6.3K
cnxs carrying >10KB	30K	23K	1.1K	670	8.0K	53K	11K	3.3K
cnxs with BTPs	10K	5.5K	390	442	940	3.2K	5.6K	480
bytes in BTPs ($drop=1$)	2.9GB	690MB	4.3GB	2.7GB	560MB	4.5GB	7.0GB	150MB
bytes in BTPs ($drop=0.9$)	7.4GB	3.0GB	5.2GB	2.8GB	1.0GB	4.8GB	11GB	1.2GB
avg BTP size ($drop=1$)	640KB	550KB	2.9MB	590KB	590KB	1.6MB	770KB	290KB
avg BTP size ($drop=0.9$)	850KB	780KB	13.4MB	5.9MB	1.2MB	1.9MB	1.9MB	3.7MB
avg BTP dur. ($drop=1$)	38s	2m 23s	55s	14s	51s	35s	1m 47s	27s
avg BTP dur. ($drop=0.9$)	1m 45s	4m 14s	4m 31s	17s	2m 26s	51s	5m 13s	6m 7s

Table 2 summarizes the characteristics of the traces. Regardless of the application type, BTPs were found only in a small fraction of the connections, which is mostly explained by the large number of small connections and the fact that BTPs are required to contain at least 130 packets. BTPs generally carry the majority of the bytes. However, BitTorrent and eDonkey traffic are exceptions with BTPs containing a smaller fraction of the bytes, which can be explained by the fact that these applications often throttle their transmission rates, hence, generating only ALPs. The average size of the connections including no BTPs was below 30KB for all applications except for FTP which had an average of 220KB. Oddly enough, the largest ones of these FTP connections, carrying up to 90MB, appeared clearly to be rate limited by the application sending constantly small packets. These unexpected examples clearly emphasize the need to identify the BTPs even for "bulk transfer applications" such as FTP.

5   Distortion Due to ALPs on End-to-end Path Studies

BTPs can capture the TCP/IP path properties in a different way than do the entire connections. If TCP sends at full rate, the effects for the data path (e.g. congestion) and, thus, the behavior observed (e.g. retransmissions), are different from the situation when the application limits the transmission rate. In many cases (network health monitoring, network aware applications etc.), it will be necessary to capture only the effects of the TCP/IP data path excluding the application impact.

In this section, we attempt to quantify what we call distortion in the TCP/IP data path analysis due to the presence of ALPs. This distortion is the biasing effects due to the application protocol in measures/estimations of metrics that typically measure the end-to-end TCP/IP path properties. RTT and throughput are generally the two principal ones among such metrics. That is why they are the focus of our two case studies. In the first case study, we study the impact of the application protocol on the characteristics of TCP transfer rates, i.e. the throughput achieved on a given data path. In the second case study, we investigate the impact of the application protocol on the accuracy of RTT estimation of TCP connections. Our goal is not to demonstrate that connection-level measurements (vs. BTP-level measurements) yield necessarily wrong results (especially in the first case study). Instead, we want to underline the fact that one should carefully consider what is the role and impact of the application running on top of TCP on the measurements when drawing conclusions from them. While there are cases when connection-level measurements are desirable, there are also other cases where they can be misleading.

5.1 Studying Characteristics of Rates

Throughput is a key performance metrics for many Internet application. It can be seen as a manifestation of the underlying TCP/IP data path characteristics at a given time instant. However, if the application controls the transmission rate, such an interpretation is false. In order to demonstrate the difference in measuring the mean throughput directly at the connection level vs. first filtering out the application impact, we compared the mean rates of BTPs within a connection to the mean rates of entire connections. Throughout this study, we used $drop = 0.9$ when identifying the BTPs used in the analysis, which means that we allowed a maximum of 10% of reduction in the throughput of the merged TPs. We computed the ratio $\frac{\left(\frac{connection\_bytes}{connection\_duration}\right)}{\left(\frac{\sum BTP\_bytes}{\sum BTP\_duration}\right)}$, which is the throughput computed for the entire connection divided by the throughput obtained when including only the bytes and durations of the BTPs of the connection. The mean values of the ratios for each application data set are in Table 3. These values show that the results can differ a lot depending on the application. The interpretation depends also on the application: For example, while the average download throughput of a BitTorrent BTP might express the average achievable throughput of that specific TCP/IP path, the average download throughput of an entire BitTorrent connection could be interpreted as the average rate a specific peer is providing to another peer. On the other hand, in the case of web browsing using persistent HTTP connections, a difference between these two throughput values could be interpreted as a sign of particular user behavior. For example, a large difference means that the user spends a long time reading the current page before clicking on a new link.

The authors found in [1] that the rates and sizes of transfers were highly correlated (coefficients of correlation consistently over 0.8) which they considered as an indication of specific user behavior: the users choose what they download based on the available bandwidth. Table 4 contains the coefficients of correlation between the rates (throughput) and sizes (number of bytes transferred) computed for entire connections and BTPs of our data sets. In the case of BTPs, the average throughput and the sum of bytes transferred was computed for each connection. As in [1], we compared the logarithms of the rates and sizes because of the large range of values.

While we observe correlation throughout our data sets, the amount of correlation varies a lot depending on the application. Furthermore, when we compare correlation at connection- and BTP-level, the difference in the degree of correla-

Table 3. Mean Values of the Throughput Ratio

traffic type	BitTorrent	eDonkey	FTP data	SSH
avg tput ratio	0.36	0.86	0.96	0.73
traffic type	Gnutella	HTTP(S)	FastTrack	WinMX
avg tput ratio	0.74	0.64	0.94	0.87

Table 4. Coefficients of correlation between log of throughput and log of number of bytes transferred. Only connections transferring at least 100KB were included and $drop = 0.9$ was used when determining the BTPs.

traffic type	BitTorrent	eDonkey	FTP data	SSH
connections	0.92	0.66	0.41	0.83
BTPs	0.37	0.42	0.32	0.16
traffic type	Gnutella	HTTP(S)	FastTrack	WinMX
connections	0.63	0.19	0.56	0.91
BTPs	0.48	0.13	0.52	0.77

tion varies from negligible (HTTP and FTP) to very large (BitTorrent and SSH). The large difference for BitTorrent traffic is due to two characteristics specific to the application protocol. First, BitTorrent favors fast peers. Fast peers, i.e. the peers having large available bandwidth, are less likely to be choked and, hence, manage to exchange more bytes than slower peers. Slow peers are more likely to be choked more often and, thus, exchange less bytes. While this effect is also visible in the correlations when looking at the BTPs, it is amplified when the throughput is computed for the entire connection. The first reason is that the choked periods, during which the peer is idle, are identified as ALPs, which, therefore, decrease the connection-level throughput but do not affect the throughput of the BTPs. In this way, connections of slow peers mainly contribute to the difference of the correlations between connection and BTP level for BitTorrent traffic in Table 4. The second reason is the BitTorrent download connections that are not used simultaneously to upload data. In this case, the upstream data traffic of these connections consists only of periodically sent very small packets containing requests for new chunks and other control messages. This type of traffic generates very low-rate ($< 5Kbit/s$) small-size connections that are identified as ALPs and, therefore, excluded when studying the BTPs.

For SSH traffic, the large difference in the degree of correlation in connection-level and BTP-level is explained by the parameter negotiation in the beginning of the connection. This negotiation takes a relatively long time (even up to a few seconds) during which few bytes are transferred. Since the very low throughput during this negotiation phase is controlled by the application, this period is identified as an ALP. Therefore, it only decreases the connection-level throughput. Moreover, the fewer are the bytes transmitted in total, the larger is the impact on the rate of the connection.

Overall, the degrees of correlation seem to be slightly higher for the P2P applications (BitTorrent, eDonkey, Gnutella, FastTrack, and WinMX). One explanation is their ability to download a given file in pieces from several sources simultaneously: the faster the peer, the more it contributes by transferring more pieces, i.e. a larger portion of the file. This behavior may be reflected in both, connection-level and BTP-level results, depending on whether a transfer of multiple pieces is identified as a single or multiple BTPs. In the case that the application waits for a download of a piece to finish before requesting a new piece, which can cause an ALP, a transfer of a piece is likely to be identified as a separate BTP. If the application "pipelines" the requests, the transfer of multiple pieces is likely to be continuous and be identified as a single BTP.

The relatively low correlation for FTP and HTTP contradicts with the results in [1] and suggests that users download content regardless of the available bandwidth. In other words, the content is what matters most. The data sets used in [1] date back to 2001 and 2002, which could partially explain this change in behavior. Five years ago, many users were still accessing the Internet using standard modem and ISDN lines with relatively low access capacities and paying for each minute of connection. In such a case, a cost-aware user may not want to wait a long time for a large download to finish and, thus, aborts it or does not start it at all if the available bandwidth is low. An impatient user may do the same. Today, the standard is broadband access (e.g. DSL and cable modems) which is typically flat rate, as is the case for our data sets originating from an ADSL access network. In addition, the typical access link speeds have multiplied. Therefore, there are fewer reasons for choosing the content size based on the available bandwidth today.

5.2   Case Study on RTT Estimation

The case of RTT estimation is particularly interesting, since the ALPs may distort the estimates in yet another way when the measurement point is in the middle of the TCP/IP path, a common situation in traffic monitoring [8]. In this case, when estimating the RTT from passively collected packet headers and using techniques based on observing bidirectional traffic, the RTT needs to be computed in two parts (see Figure 4): (i) delay between observing a data packet and the corresponding ack packet and (ii) delay between observing the ack packet and the data packet the transmission of which the ack packet triggered. The main challenge in such a technique is to be able to compute the second part (d2 in Figure 4), i.e. to associate a data packet to an ack packet that triggered its transmission. However, as Figure 4 shows, there is an additional error d3 added to the RTT estimate whenever the application delays giving more data to TCP for transmission. This error affects every RTT estimation technique that is based on bidirectional traffic observations, such as the technique relying on TCP timestamps to do the ack-to-data association (the authors of [9] acknowledge the problem in Section 4.1) or the technique described in [4] which reconstructs the TCP state machine to track the sender's congestion window, which in turn enables the ack-to-data association.

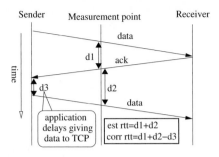

Fig. 4. Problem with RTT estimation during an ALP

In the case that data is constantly transferred in both directions of a connection, an RTT estimate can be obtained by simply computing d1 in Figure 4 for both directions and then summing them up. Unfortunately, it is rare to have constant data transfers in both directions within a connection.

Other RTT estimation techniques based only on unidirectional traffic observations have been proposed in [9] and in [1]. These techniques base the estimation on observing the self clocking behavior of TCP in the traffic pattern. However, when the application dominates the transfer rate and, hence, controls the pace at which new data packets are sent, this pattern will cease to exist and the estimations will be distorted.

To compute the running RTT samples we used primarily the technique from [9] relying on TCP timestamps and secondarily the technique from [4] when TCP timestamps were not available. For each connection, we computed the average for two sets of estimated RTT samples: first, including only the BTPs and second, including only the ALPs. We used $drop = 0.9$. The CDF plot in Figure 5 shows how the ratio of the average RTTs ($\frac{\overline{RTT}_{ALP}}{\overline{RTT}_{BTP}}$) is distributed.

We can first observe that the differences between RTTs during BTPs and ALPs are striking: For instance, approximately 18% of the eDonkey connections have ten times longer and approximately 10% of them ten times shorter average RTT during ALPs than during BTPs. Second, the results vary significantly from one application to another. Many of the inflated RTT values during the ALPs can be due to large values of d3 (see Figure 4).

Figure 6 shows an example of the error that can be introduced by the application. The figure contains estimated RTT samples from a short piece of an HTTP connection that incorporates several BTPs interleaved with ALPs. Similar sawtooth pattern of the RTTs persists throughout the lifetime of the connection. It could be a persistent HTTP connection transferring several objects of a web page. The first RTT sample of each BTP after an ALP is clearly longer than the other samples and corresponds to the situation depicted in Figure 4. These RTTs are erroneously inflated by the delay due to the application (d3 in Figure 4), which makes them on the average much larger than the actual RTT. In this example, this delay d3 due to the application inflates the first RTT sample of each BTP (right after the dotted vertical line in Figure 6) because no packets

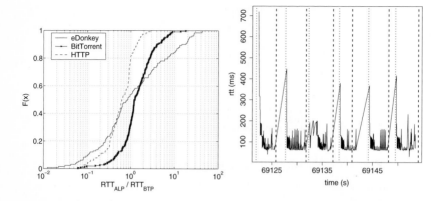

Fig. 5. CDFs of ratio of the mean RTTs: $\frac{\overline{RTT}_{ALP}}{\overline{RTT}_{BTP}}$ ($drop = 0.90$)

Fig. 6. Piece of an HTTP connection. Dashed and dotted vertical lines start an ALP and BTP ($drop = 0.95$), respectively

are transmitted and, thus, no samples obtained during the ALPs. As a consequence, to correct the error introduced by the application, the first sample of each BTP should be filtered out. This example may partially explain why in Figure 5 we also observe larger RTTs during the BTPs than during the ALPs of a connection. In order to fully explain all of these particular findings (e.g. why we observe many shorter RTTs during ALPs especially in eDonkey traffic) requires further research including experimentations with these applications in a controlled environment. This work is beyond the scope of this paper whose goal is to raise awareness on the potential impact of the application on the analysis results when studying the TCP/IP path and to propose an application agnostic solution for these studies.

6   Conclusions

TCP is the dominant transport protocol carrying a large fraction of the total traffic in the Internet. It is therefore quite natural to use TCP packet traces for network tomography purposes. However, as we have shown in this paper, the application can interfere in various ways with the flow of packets injected into the network. The IM algorithm is able to isolate the "contribution" of the application for TCP packet traces containing traffic of any kind of application. We applied the IM algorithm to a variety of different application traffic and demonstrated the impact of the application when studying the TCP/IP path properties in the case of throughput achieved and RTT estimation.

As a continuation of this work, we have applied the IM algorithm on traffic traces in order to study the role of applications as origins of the rates and throughput performance observed by today's Internet applications. In specific, we have evaluated to what extent clients of ADSL access network experience throughput limitation by the application [10]. As future work, we would also like to investigate some of the TCP traffic and network path properties, such as RTT and burstiness [2], using publicly available traffic traces (such as the ones used in [2]) in order to quantify the impact of applications on these properties.

References

1. Zhang, Y., Breslau, L., Paxson, V., Shenker, S.: On the characteristics and origins of internet flow rates. In: Proceedings of ACM SIGCOMM 2002 Conference, Pittsburgh, PA, USA (2002)
2. Jiang, H., Dovrolis, C.: Source-level IP packet bursts: causes and effects. In: IMC '03: Proceedings of the 3rd ACM SIGCOMM conference on Internet measurement, New York, NY, USA, pp. 301–306. ACM Press, New York (2003)
3. Allman, M., Falk, A.: On the effective evaluation of TCP. Comput. Commun. Rev. 29, 59–70 (1999)
4. Jaiswal, S., Iannaccone, G., Diot, C., Kurose, J., Towsley, D.: Inferring tcp connections characteristics from passive measurements. In: Proc. Infocom 2004 (2004)
5. Medina, A., Allman, M., Floyd, S.: Measuring the evolution of transport protocols in the internet. Comput. Commun. Rev. 35, 37–52 (2005)
6. Siekkinen, M., Biersack, E.W., Urvoy-Keller, G.: On the interaction between internet applications and tcp. Technical report, Institut Eurecom (2006) http://www.eurecom.fr/s̃iekkine/pub.hml
7. Mathis, M., Heffner, J., Reddy, R.: Web100: extended TCP instrumentation for research, education and diagnosis. Comput. Commun. Rev. 33, 69–79 (2003)
8. Jaiswal, S.: Measurements-in-the-middle: Inferring end-end path properties and characteristics of TCP connections through passive measurements. PhD thesis, Univ. of Massachusetts, Amherst (2005)
9. Veal, B., Li, K., Lowenthal, D.: New methods for passive estimation of TCP round-trip times. In: Proceedings of Passive and Active Measurements (PAM) (2005)
10. Siekkinen, M., Collange, D., Urvoy-Keller, G., Biersack, E.W.: Performance limitations of ADSL users: A case study. In: Proceedings of the Eighth Passive and Active Measurement Conference (PAM) (2007)

Measurement Based Modeling of eDonkey Peer-to-Peer File Sharing System

Walid Saddi and Fabrice Guillemin

France Telecom, Division R&D, 2 Avenue Pierre Marzin, 22300 Lannion, France
{Fabrice.Guillemin,Walid.Saddi}@orange-ftgroup.com
http://perso.rd.francetelecom.fr/guillemin

Abstract. We analyze in this paper eDonkey traffic on the basis of measurements from the France Telecom IP collect network. Specifically, we investigate the structure of the eDonkey network by analyzing addresses and port numbers involved in eDonkey transactions. This analysis allows us to understand the dynamics of eDonkey traffic in a subnetwork of the whole eDonkey network. In particular, we stress the fact that eDonkey gives rise to a huge amount of signaling, identified as small TCP data transfers. To qualitatively explain the observed phenomena, we develop a very simple qualitative model to understand the dynamics of an eDonkey community sharing a given object (a chunk or a file). This model is intended to capture the transient behavior of a network relying on eDonkey principles when sharing an object. This simple model allows us to identify two regimes depending on the congestion level of servers (expansion and collapse).

1 Introduction

The emergence of peer-to-peer (p2p) applications is certainly the event, which has the most significantly impacted in the past few years the structure of IP traffic in commercial networks. With legal actions against copyrighted material sharing, however, one may have thought that p2p traffic would rapidly decrease. In fact, a careful analysis of traffic shows that p2p file sharing is still very active and generates the prevalent part of traffic in today's IP platforms. With the generalization of the use of dynamic port numbers [1] and encryption technologies, it simply becomes more and more difficult to identify p2p traffic.

In this paper, we study the p2p protocol, which gives rise to the prevalent part of traffic in some European countries, namely eDonkey; see for instance [2,3] for traffic characterization in France and Germany. To be more specific, we consider a traffic trace from the France Telecom backbone network carrying ADSL traffic. This kind of commercial traffic is mainly composed of eDonkey traffic generated by residential customers. In the traffic trace considered in this paper, about 3,090 users are connected to the observed ADSL area. The objective of the present study is to outline salient characteristics of this kind of traffic. In particular, we point out that a significant part of eDonkey traffic is transmitted on standard port numbers. When analyzing the activity based on IP addresses,

L. Mason, T. Drwiega, and J. Yan (Eds.): ITC 2007, LNCS 4516, pp. 974–985, 2007.

we come up with the conclusion that a large proportion of IP addresses transmit at least one large file on a standard port number.

From the analysis of a real traffic trace, we can observe that eDonkey gives rise to a huge amount of signaling and that a small number of peers seem to be permanently active. To qualitatively understand these phenomena, we propose a very simple mathematical model for describing how an object (a file or a chunk) is shared by a community of eDonkey peers. This simple model allows us to highlight the importance of congestion and to point out two regimes of the eDonkey network, namely expansion and collapse. In the first case, we can see that the eDonkey network is rapidly growing so that all peer retrieve the desired object with minimum latency. In the latter regime, we observe the formation of a core of congested nodes so that the majority of peers get trapped at congested servers and suffer from high latency when downloading the object.

The goal of the mathematical model developed in this paper is to understand the formation of an eDonkey community, when an object begin to be shared among different peers. We are specifically interested in transient phenomena. In this respect, the present model is different from those developed in [4,5], which study different p2p networks in equilibrium with many peers joining and leaving the system. In the present paper, we consider a community of N peers sharing an object according to the principles of eDonkey and we are interested in the structure of the eDonkey network, when the N peers have joined the system.

This paper is organized as follows: In Section 2, we investigate on the basis of a traffic trace from the France Telecom IP collect network carrying ADSL traffic some basic facts concerning the behavior of eDonkey peers. In Section 3, we develop a very simple mathematical model for explaining the observations made from the traffic trace analysis. Concluding remarks are presented in Section 4.

2   Traffic Trace Analysis

2.1   Experimental Setting

To study the behavior of eDonkey peers, we consider in this paper traffic measurements from a high speed link of the France Telecom IP collect network carrying ADSL traffic. We monitored traffic in both directions of an OC3 link during the time period between 6:00 pm and 9:00 pm on September 11, 2006. The downstream direction goes from the backbone to the ADSL area. The opposite direction is referred to as upstream direction.

The IP addresses of those customers connected to the ADSL area are called internal addresses. Conversely, IP addresses of those customers not connected to the ADSL area under consideration are called external addresses. The present analysis is limited to TCP traffic even though information on UDP traffic is available. The reason for this limitation is that TCP traffic represents more than 96 % of the total amount of traffic and that almost all p2p traffic (in particular data transfers) is transmitted via TCP connections. In addition, since the goal of this paper is to qualitatively explain the behavior of eDonkey peers, all the events of interest take place on TCP connections.

We have classified packets into applications via a payload analysis. We have also added the proportion of traffic on "standard" port numbers. Specifically, if a source or a destination port number of a packet matches one of the well-known port numbers [6], the packet is considered as belonging to that specified application. For example, packets are marked as eDonkey if the destination or source port number is equal to 4661 or the 4662 for TCP and 4665 or the 4672 for UDP.

A preliminary analysis of port numbers based on traffic traces captured in 2004 and 2005 have shown that a significant amount of eDonkey traffic is transferred on port numbers 5662, 40662 and 14662. These port numbers are not standard but close to the standard ones. Traffic on these port numbers will be aggregated in the following with "standard" traffic.

2.2   Global Statistics

Table 1 gives the percentage of volume per application; only TCP traffic is considered. For port analysis, packets are labeled as "eDonkey" not only if the destination or source port numbers are equal to 4661 or 4662 but also to 5662, 40662 or 14662. Table 1(b) gives the contribution of these different port numbers. Note that traffic on port numbers 5662, 40662 and 14662 is now very small.

Table 1. Composition of ADSL traffic per application (Downstream and Upstream)

(a) Application breakthrough

Appl.	Payload analysis		Port analysis	
	% up	% down	% up	% down
eDonkey	64.36%	42.53%	42.66%	23.94%
Gnutella	3.58%	4.87%	2.49%	2.23%
Bittorrent	19.79%	14.38%	3.37%	7.74%
http	4.45%	23.68%	4.48%	23.30%
nntp	< 1%	3.02%	≪ 1%	3.02%
Unknown	3.93%	4.25%	46.69%	40.59%

(b) Contribution per port

Port	% up	% down
4662	98.80%	97.70%
4661	≪ 1%	≪ 1%
5662, 14662, 40662	1.1%	2.3%

From Table 1, we see that a significant part of traffic generated by eDonkey and Gnutella are on standard port numbers, while BitTorrent seems to more systematically use dynamic port numbers. By using application layer information, it is possible to check that the amount of traffic transferred on conventional and assimilated ports represents 56% (resp. 66%) of whole eDonkey traffic in the downstream (resp. upstream) direction. In addition, as observed in many European operational networks (see [3] for instance), eDonkey is still the dominating p2p protocol, even though BitTorrent now represents a significant part of p2p traffic.

In Table 2, we present address statistics. The first column of this table gives the number of internal and external IP addresses, which have transmitted TCP packets. A TCP connection gives rise to a couple (i.e., the source and destination IP addresses of the connection). The second column gives the number of external and internal IP addresses, which have transmitted an eDonkey packet. The last column gives similar statistics for HTTP transactions.

Table 2. Number of IP addresses (in thousands) involved in TCP, eDonkey and HTTP transactions

	TCP down	eDonkey down		http down
		Application analysis	Port analysis	
Ext.	584,610	409,900 (70.12%)	268,530 (45.93%)	31,701 (5.4%)
Int.	3,090	353 (10.70%)	341 (11.03%)	2,668 (86.34%)
Couples	932,920	570,140 (61.11%)	359,000 (38.48%)	102,270 (10.96%)
	TCP up	eDonkey up		http up
		Application analysis	Port analysis	
Ext.	917,700	410,070 (44.68%)	268,620 (29.27%)	33,170 (3.6%)
Int.	2,700	346 (12.81%)	341 (12.62%)	2,508 (92.88%)
Couples	1256,320	570,570 (45.42%)	359,23 (28.59%)	112,899 (8.98%)

From this table, it clearly appears that a huge number of peers with external IP addresses have sent eDonkey packets with "standard" port numbers. This indicates that even if large file transfers take place on dynamic port numbers, eDonkey is characterized by a high activity (in terms of packets and not in terms of volume) on standard or assimilated port numbers, certainly due to signaling. This experimental observation may be used for p2p identification. Even if p2p protocols have been designed with many features to hide p2p transactions, many real customers run the most basic versions of protocols.

To refine the above observation, we introduce the mouse and elephant dichotomy in order to distinguish between signaling and data transfers. Let us first define a flow as a group of those TCP packets with the same source and destination IP addresses together with the same source and destination port numbers. We say that a flow is small if it comprises less than 20 packets; such a flow is referred to as a mouse. Moreover, a flow is said to be large if it comprises more than 20 packets; we call such a flow an elephant.

In addition, we are interested in those elephants carrying actual data and not only acknowledgment TCP segments. We shall call regular elephant an elephant with a mean packet size greater than 80 bytes; such a regular elephant is supposed to carry a file (or at least a chunk) from a peer to another. An elephant, which is not a regular elephant, is referred to as an ACK elephant.

Mice comprise a small number of packets (less than 8) and we can reasonably assume that they are associated with signaling (communications with index servers, maintenance, file requests, etc.). The above definitions may appear very crude but they are sufficient to develop mathematical models for representing with a reasonable accuracy traffic on a Gigabit Ethernet link (see [2] for details).

The internal and external addresses in the traffic trace are classified according to the above mouse/elephant dichotomy and the results are given in Table 3. The address statistics show that signaling traffic (mice) involves much more addresses than file transfers (regular elephants). This phenomenon is particularly evident for external addresses. In the same table, we have indicated address statistics for eDonkey traffic (identified through payload and port analysis and encompassing together mice, ACK and regular elephants). From these figures, we can deduce that the signaling activity in the global eDonkey network is huge. The part of the eDonkey network formed by those customers connected to the observed ADSL area is a very small proportion of the global network but is highly connected to the rest of the eDonkey network.

Table 3. Number of (external and internal) IP addresses involved in mice and regular elephants

		mice		Reg. elephants		ACK. elephants	
		Payload	Port	Payload	Port	Payload	Port
up	Int.	346	341	316	308	346	341
	Ext.	407,129	266,347	10,457	7,190	410,072	268,620
down	Int.	353	341	298	270	353	341
	Ext.	406,976	266,316	12,171	6,910	409,907	268,533

From the above analysis, we can make the following points:

– eDonkey gives rise to a huge amount of signaling activity (observed by tracking mice),
– actual data transfers take place only on a small number of TCP connections.

These two observations may indicate that the eDonkey network is in fact saturated. Indeed, a large number of peers search for objects, but object transfers are not so numerous. In the next section, we develop a very simple mathematical model to represent the behavior of an "ideal" eDonkey community sharing an object. This community is ideal in the sense that we get rid of all technological constraints and we consider the community in isolation, i.e., we ignore the interactions between communities sharing different objects. These restrictive assumptions nevertheless allow us to point out two regimes of an eDonkey network: expansion and collapse. The transition between these two modes depends upon a congestion factor, which is related to the size of objects and the uplink transmission capacities of peers.

3   Simplified Model for the Expansion of an eDonkey Network

In the previous section, we have observed that the eDonkey protocol gives rises to a huge amount of signaling and that peers are highly connected one to each other, at least in terms of signaling, even if file transfers involve only a small fraction of the total eDonkey community. In this section, we propose a mathematical model to explain how the eDonkey network expands when an object (a file or a chunk) begins to be shared between a given population of peers. We first consider the case when there are no free riders, that is, each peer downloading the object accepts to share it with other peers. Note that this point is a characteristic feature of eDoneky. BitTorrent, which implements enforcement of the "tit-for-tat policy", should be described according to different principles. In a second step, we analyze the impact of free riders.

The objective of this model is to capture the dynamics of the formation of the eDonkey network built up when sharing an object. We are specifically interested in the transient behavior of the system in order to distinguish between two regimes (collapse and expansion). In this respect, the proposed model is different from the studies [4,5], since these latter are interested in the steady state analysis of a p2p network (namely BitTorrent in [5] and different centralized and decentralized p2p systems in [4]). In [5], a large p2p network is considered and peers join and leave so that a steady state is reached. Subsequently, oscillations around the steady state are modeled by means of a multidimensional Ornstein-Ulhenbeck process. In [4], the system is described as a closed queueing network with different degrees of freedom in order to account for different features of the p2p protocols. The steady state is studied via a bottleneck analysis technique.

3.1   The Expansion of the Network in the Absence of Free Riders

We consider a given object, which is to be shared among a population of N peers, and we make the following assumptions:

1. Once a peer has downloaded the object, he immediately registers on the list of peers ready to share the object, this list being maintained by the indexation server of the eDonkey network; this peer becomes a potential server from which the object can be downloaded by any other peer.
2. If n peers are in the system (either as servers or as peers downloading the object), requests for the object by other peers occur as a Poisson process with intensity $(N - n)\lambda$, where λ is a positive real number.
3. The time to download the object from a peer is exponentially distributed with parameter μ, where μ is a positive real number; the parameter μ depends upon the distribution of uplink transmission capacities of peers: the object is of a given size and the download time depends upon many factors such as the uplink transmission capacity, the CPU of the host, etc.
4. When a peer uploads the object to other peers, his transmission capacity is shared according to the processor sharing discipline, ensuring a certain fairness among the different competing peers.
5. When a peer wants to download the object, he picks up at random *one and only one* server having the object.

Under the above assumptions, we investigate the formation of the network composed of servers. We assume that all the transmission capacity of a server is dedicated to the upload of the object considered. In other words, we ignore the interactions with other networks formed by sharing other objects; we analyze the community in isolation.

We describe the system by means of three variables:

N_s: the number of servers (i.e., the number of peers sharing the object),
N_q: the number of customers queued at server, i.e., peers downloading simultaneously the object from a server,
N_a: the number of servers with queued customers.

Note that the quantity $N_s + N_q - 1$ describes the number of peers in the system (the server initiating the system is not taken into account).

The system is initialized when a peer registers at the indexation server. Other peers can then know which server has the object and initiate a download. We describe the system by means of the variables (N_s, N_a, N_q) as follows: we say that an event occurs in the system when a new peer joins the system or when a download is completed. Under these assumptions, an event corresponds to a peer arrival with probability p given by

$$p = \frac{(N - (N_s + N_q - 1))\lambda}{(N - (N_s + N_q - 1))\lambda + N_a\mu}$$

and to a download completion with probability $1 - p$. Note that a service completion gives birth to a new server. An arriving peer chooses a server at random (a server is chosen with probability $1/N_s$).

When observing the system at event instants, we have a discrete time system. Let $N_s(t)$, $N_q(t)$, $N_a(t)$ be the respective values of the variables N_s, N_a and N_q at instant $t = 1, 2, 3, \ldots$. The process $(N_s(t), N_q(t), N_a(t))$ evolves as follows: the process jumps from state (i, j, k) at time t to

- state $(i+1, j-1, k)$ with probability $(1 - p(t))$ multiplied by the probability that the peer completing his download leaves queued peers behind him at the corresponding server,
- $(i + 1, j - 1, k - 1)$ with probability $(1 - p(t))$ multiplied by the probability that the peer completing his download leaves *no* queued peers behind him at the server,
- $(i, j + 1, k)$ with probability $p(t)k/i$ (the arriving peer joins a server with queued peers),
- $(i, j + 1, k + 1)$ with probability $p(t)(1 - k/i)$ (the arriving peer joins a server with no queued peers),

at time $t + 1$, where

$$p(t) = \frac{(N - (N_s(t) + N_q(t) - 1))\rho}{(N - (N_s(t) + N_q(t) - 1))\rho + N_a(t)}$$

with $\rho = \lambda/\mu$ denoting the load of a processor sharing queue with arrival rate λ and service rate μ. Note that the process $(N_s(t), N_q(t), N_a(t), t = 1, \ldots, N)$ is not a Markov process since the transitions depend upon the number of queued customers at servers. Indeed, the transition probabilities depend on the fact that customers leave servers empty or not after download completion, which information is not captured in the process $(N_s(t), N_q(t), N_a(t), t = 1, \ldots, N)$.

To evaluate the performance of the system, we analyze the final values of the quantities N_a, N_s and N_q when all peers have joined the network. We specifically estimate the efficiency of the system as follows: If N_s is of the same order of magnitude as N and N_a is very small, this means that the system is very efficient. The object is disseminated among the peers very rapidly and only a few peers are queued at servers. On the contrary, if the number of servers is much smaller than N and N_q is of the same order of magnitude as N, the system is not efficient in the sense that only a small fraction of peers share the object and many peers are queued at the servers. This means that the object is slowly disseminated among peers, which wait for a long time to download the object. In some sense, the system is congested. Of course, this congestion does not last for ever since all peers eventually download the desired object. Nevertheless, in the formation of the network, peers accumulate at a few servers and we can say that the system collapses. In the first case, when $N_a \ll N$, the network expands. Note that this concept of congestion is not evident, since the system is not congested in the usual sense but only during the transient regime.

Figure 1(a) represents the fraction of queued peers and clearly exhibits the critical role played by the load ρ. When the parameter ρ becomes close to 1, the fraction of queued peers rapidly grows to 1, indicating a poor performance of the system. This performance degradation is confirmed by Figure 1(b) showing the evolution of the fraction N_a/N_s. A ratio N_a/N_s close to one indicates that almost all servers have queued peers, which then have to wait for a long time in order to complete their download. In addition, the number of servers becomes very small when compared to the total number of peers, as illustrated in Figure 1(c).

Finally, we examine the evolution of the number of servers with queued customers. In some sense, these servers form the "core" of the eDonkey network. Indeed, during the formation of the network, we can note that the first peers completing their download become servers and the subsequent peers accumulate at these primal servers. Figure 1(d) represents the evolution with ρ of the ratio of the number of servers with queued customers to the square root of the number of peers. We can observe that for moderate values of ρ, the number of servers with queued customers grows as the square root of the population size N before rapidly decreasing when ρ becomes close to 1. This indicates that the size of the "core" of the network is much smaller than the total population size in non overloaded regimes (network expansion). In case of collapse, data transfers take place from a small number of peers (the congested nodes) and the signaling activity on the status of queues in servers is huge.

The above study shows the critical role of the load ρ with regard to the expansion of the network. When ρ is less than 1, the network expands and the system is very efficient for disseminating a object. In terms of traffic, this entails that there are a large number of servers for a given object and certainly that

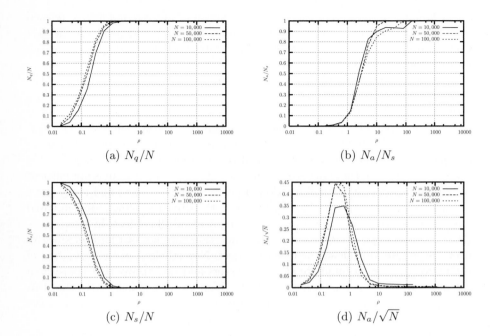

(a) N_q/N

(b) N_a/N_s

(c) N_s/N

(d) $N_a/\sqrt{N}$

Fig. 1. Number of servers and active servers peers in the systems for different values of N

there is a huge signaling activity by peers for choosing a server. When the load is close to or greater than 1, the performance of the system rapidly degrades. The network builds up around a "core" of congested nodes with limited size (corresponding to a collapse of the network). The majority of peers get trapped at servers and have to wait for long time for completing their download. The network then offers a bad performance for disseminating an object. Note that even in the case of congestion, all peers get served because the total population size N is finite. In the present context, congestion means that peers accumulate at congested nodes but eventually complete their downloads.

The way the network collapses depends upon the transient behavior of the processor sharing queues representing the servers. In the present paper, we have assumed that service times (the time needed to retrieve a file from a server when there is only one customer) is exponentially distributed. This assumption is in fact a little bit optimistic. It is well known in the literature that the behavior of a processor sharing queue in overload depends upon the distribution of service times, in particular on the squared coefficient of variation (see for instance [7]). Hence, the collapse phenomenon could be more severe with more variable service time distribution.

Besides the impact of the load ρ, we can note that when the total population size N becomes very large, the different quantities N_a, N_s and N_q

properly rescaled seem to tend to limits. Let $(X_1^{[N]}, X_2^{[N]}, \ldots)$ denote the random vector describing the number of queued customers at servers $1, 2, \ldots$ when the load is greater than or equal 1. Let $M^{[N]}$ denote the distribution of the vector $(X_1^{[N]}, X_2^{[N]}, \ldots)/N$. The marginals of $M^{[N]}$ for $j = 10$, $j = 20$ and $j = 30$ (i.e., the distribution of $X_j^{[N]}/N$ for $j = 10, 20, 30$) are displayed in Figures 2(a), 2(b) and 2(c), respectively for $\rho = 1.2$. It clearly appears that these marginals tend to limits. This supports the fact that the complete distribution $M^{[N]}$ tends to a deterministic measure μ on $\mathbb{N}^{\mathbb{N}}$ and that an asymptotic regime exists for a congested system with a very large number of customers. This point will be investigated in further studies.

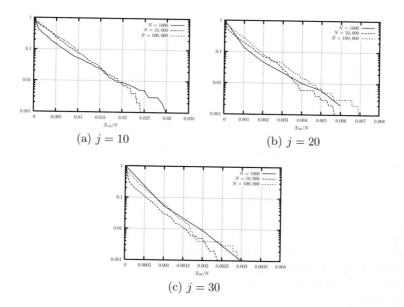

(a) $j = 10$ (b) $j = 20$

(c) $j = 30$

Fig. 2. Convergence of some marginals of the measure $M^{[N]}$ when N tends to infinity (load $\rho = 1.2$)

3.2   Impact of Free Riders

So far, we have ignored the presence of free riders by assuming that each peer downloading an object immediately accepts to share it with other peers. Even in this simple case, we have highlighted the negative impact, which may be due to congestion of servers (a load ρ greater than or equal to unity). Because of the sharing principles implemented by the eDonkey protocol, congestion limits the expansion of the network (formation of a core of congested nodes) and as a consequence slows down the dissemination of an object, even if the object is eventually shared among all peers. In this section, we investigate the impact of free riders.

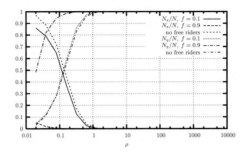

Fig. 3. Ratios of the number of queued peers and servers in the presence of free riders

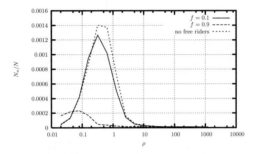

Fig. 4. Ratio of the number of servers with queued peers in the presence of free riders

We specifically assume that a peer having completed his download leaves the network with a probability f. In other words, a peer accepts with probability $1-f$ to share the downloaded object and is a free rider with probability f. Figure 3 displays the ratios N_s/N and N_q/N as functions of the load ρ for different probabilities f and for $N = 100,000$. Contrary to the analysis performed in [4], which shows that free riders can benefit from the system without degrading the global performance, we clearly see in the present model that free riders have a very negative impact on the performance of the network in the sense that the ratio of queued customers rapidly increases for large probabilities f. This difference is due to the fact that in [4], the analysis is performed for the steady regime and does not account for transient overload as in the present paper. Here, a core of congested nodes appears for loads much smaller than 1, namely at a load equal to $1 - f$. The number of servers rapidly becomes negligible when compared to the total population size N.

The negative impact of free riders is confirmed by examining the ratio of servers with queued peers as a function of the load ρ for different probabilities f. We see, as in the absence of free riders, that this ratio is increasing with the load for moderate load values before reaching a maximum and decreasing for larger load values. The presence of free riders entails that the maximum is reached for smaller values of the load (namely $1 - f$) and the value of the supremum is rapidly decreasing when the proportion f of free riders is increasing. This means that the core of congested nodes is forming for smaller load values with a smaller number of servers.

4   Conclusion

By analyzing a traffic trace from a high speed link from the France Telecom IP collect network, we have outlined some salient features of eDonkey traffic, representing a significant part of global traffic. First, we can observe that eDonkey peers massively use "standard" port numbers, even if large data transfers can take place with dynamic port numbers (about 60% of elephants are on standard port numbers).

Furthermore, it is empirically observed that eDonkey gives rise to a huge amount of signaling. In order to qualitatively explain the observed phenomena, we have developed a simple mathematical model, which transient behavior represents how an eDonkey community shares a given object. This analysis has allowed us to exhibit two regimes: expansion and collapse of the network. The transition between these two regimes depends upon the load of servers and the fraction of free riders. In the case of collapse, it has been checked that some marginals of the vector describing the number of queued customers at the different servers converge, when properly rescaled, to limits. This observation supports the intuition that an asymptotic regime for a congested network should hold. The detailed description of these phenomena requires the investigation of the transient behavior of a queueing network composed of processor sharing queues and will be addressed in further studies.

References

1. Karagiannis, T., Broido, A., Faloutsos, M., Claffy, K.: Transport layer identification of p2p traffic. In: IMC'04, Taormina, Italy (October 2004)
2. Azzouna, N.B., Clérot, F., Fricker, C., Guillemin, F.: A flow-based approach to modeling ADSL traffic on an IP backbone link. Annals of Telecommunications 59(11-12), 1260–1299 (2004)
3. Hasslinger, G.: ISP platforms under a heavy peer-to-peer workload. In: Peer-to-Peer Systems and Applications, pp. 369–381 (2005)
4. Ge, Z., Figueiredo, D., Jaiswal, S., Kurose, J., Towsley, D.: Modeling peer-to-peer file sharing systems. In: Proc. Infocom 2003 (2003)
5. Qiu, D., Srikant, R.: Modeling and performance analysis of BitTorrent-like peer-to-peer networks. In: Proc. Sigcomm 2004 (2004)
6. The Internet Assigned Network Authority website. [Online]. Available: http://www.iana.org/
7. Jean-Marie, A., Robert, P.: On the transient behavior of the processor sharing queue. Queueing Systems, Theory and Applications 17, 129–136 (1994)

Internet Data Flow Characterization and Bandwidth Sharing Modelling

Philippe Olivier

France Telecom, R&D Division
CORE/CPN Laboratory, 38-40 rue du Général Leclerc
92794 Issy-les-Moulineaux Cedex 9, France
phil.olivier@orange-ftgroup.com

Abstract. The purpose of the paper is two-fold: first, characterize the Internet data traffic at high levels, i.e., flows and sessions, with the principal aim of comparing the salient features shown by different types of applications, namely P2P on one hand and classical Internet applications in the other. Second, among the observed traffic characteristics, we particularly focus on the way bandwidth is actually shared by flows and users by developing an empirical bandwidth sharing model. We infer from it some trends shown by the different classes of traffic in terms of elasticity, i.e., their potential reaction against congestion.

Keywords: Internet, traffic characterization, data flows, bandwidth sharing.

1 Introduction

Many papers have studied the characteristics of IP traffic at flow level, see, e.g., [5], [7] and [8]. They were mainly motivated by the well known complexity of data traffic at packet level, where features like Long Range Dependence (LRD), self-similarity or multi-fractality typically lead to intractable performance models. Moreover, flow level performance models are best suited for the analysis of TCP traffic: the flow transfer time constitutes the proper performance parameter associated with user-perceived Quality of Service (QoS). Most of these works have sought to characterize flows in terms of statistical parameters such as the arrival process and flow length distribution. For instance, it was recognized that the flow arrival process is not Poisson in most cases and that flow length, both in time and volume, is highly variable and statistically distributed according to heavy-tailed laws.

One of the primary purposes of the present work is to check the relevance of these previous flow-level representation models on more recent data sets. To go a bit further, we will seek a better understanding of the interaction between user sessions and flows that are generated within them. We first expect that the great increase of Internet traffic volumes generated in the past few years, due especially to the large take up of residential broadband access services (ADSL), will have some impact on the models due notably to volume aggregation. Secondly, the appearance of Peer-to-Peer (P2P) applications has led to the most noticeable change in data traffic composition on the last few years [1], [10]. It is therefore of considerable interest to

L. Mason, T. Drwiega, and J. Yan (Eds.): ITC 2007, LNCS 4516, pp. 986–997, 2007.

identify what are the specific characteristics of P2P traffic, compared to that generated by classical client/server applications such as HTTP (Web), FTP, mail, etc.

Our second main objective is to describe how link capacity is actually shared by users and flows, and how this share evolves in time as the number of flows dynamically varies according to successive flow arrivals and departures. Little effort in this direction has been devoted so far in the literature. However, it seems to be a topic of some importance in performance evaluation, at least for elastic traffic, since the amount of bandwidth allocated to a flow directly determines the time it takes to complete the transfer, i.e., its throughput. Some theoretical works analyze how the available capacity is allocated to flows in a network, or how it should be in order to preserve some interesting properties such as fairness or performance insensitivity [3]. Such analyses are generally supported by simulations, but much more rarely by experimental results based on real traces.

The purpose here is thus to try to fill the gap between theory and practice by providing some results on empirical bandwidth sharing. Section 2 provides some definitions and the theoretical framework for the proposed model; a summary of the data traces used and of the traffic composition is then given in Section 3; in Section 4 we perform a comparative analysis of flow and session characteristics; the experimental results on bandwidth sharing are presented in Section 5; finally, Section 6 concludes the paper by summarizing the main results of the study.

2   High-Level Traffic Modelling

2.1   Flow and Session Definitions

We consider a hierarchical traffic model distinguishing entities at three levels: packets, flows and sessions. Loosely speaking, a session corresponds to a user activity period during which he may run different applications simultaneously. In generic terms, a flow can be thought of as a continuous packet train associated with a single instance of a given application: a document transfer (for data traffic) or a real-time call (voice, video, etc.). Note that we only consider data traffic in this paper. Up to now, there is no commonly agreed or standardized definition of a flow. Moreover, whatever the chosen definition, the traffic intensity on a given resource can always be written as the product *"flow arrival rate × average flow size"*. Thus, for the purposes of our study, we are led to define several classes of flows.

According to [4], a *flow* can be defined as a set of packets corresponding to a given *flow specification* and which follow each other by time intervals no greater than a given threshold called *Time Out* (TO). To implement flow processing in routers and maintain flow state tables, the value of TO should be optimized in order to economize memory and CPU resources. For our purpose of traffic modelling and performance evaluation, it should provide a natural threshold to distinguish between intra-flow and inter-flow intervals. An empirical method to evaluate an appropriate Time Out value, assumed constant for simplicity, was given in [8] based on the analysis of cumulative histograms of the packet inter-arrival time within packet streams.

We first define the *micro-flow specification* as determined by the standard 5-tuple *{IP source address, IP destination address, source port number, destination port* *number, transport protocol}*. (Note that the port numbers only exist if TCP or UDP is the transport protocol used.) Following earlier investigations [8], we chose a TO value of 20 s for the micro-flow definition. Note that micro-flows defined this way do not necessarily correspond to TCP connections. It is sometimes suggested to use the SYN/FIN information to identify flows [5]. However, this seems too restrictive: it limits flow analysis to TCP traffic and does not fit the concept of "time-consistent" flows presented above (for instance, Web connections with HTTP 1.1 protocol may be composed of several well-separated transfers using the same TCP connection).

At a higher level of aggregation, we define the *user specification*. On ADSL aggregation links supported by an ATM layer-2 infrastructure, the user specification is simply defined as a set *{VP, VC}*, the Virtual Path and Virtual Channel identifiers of the ATM connection. On a backbone link, or whenever we can use only the IP layer, it is proposed to consider one of the two IP packet addresses as the user specification. In the backbone link considered below, we chose the *{destination address}* singleton as the user specification since there is a natural asymmetry in this case. *User-flows* will be determined by a TO value of 20 s, the same as for micro-flows. Then we introduce the entity of *user-sessions*, or simply *sessions*, which is defined by the same user specification and a TO value of 100 s.

What are the reasons to distinguish between two kinds of flows, user-flows and micro-flows? User-flows are well-adapted to a hierarchical traffic model and to its use for performance evaluation since they are generated sequentially within sessions. It was reported in [9] that, under the assumptions of Poisson session arrivals and perfect bandwidth sharing, the performance of flows is insensitive to the detailed characteristics of traffic: the mean throughput realized by flows only depends on the mean values of flow inter-arrival time and flow length. However, there is no reason *a priori* that user-flows should share bandwidth one way or another. On the other hand, micro-flows are closer to TCP connections and then more able to realize some kind of bandwidth sharing, although the share is unfair in reality due to different Round Trip Times, TCP windows and mechanisms. But the fact that micro-flows are frequently generated in parallel, within sessions, precludes simple stochastic modelling.

2.2 Model for Bandwidth Sharing

Performance of IP links carrying data traffic (so-called elastic traffic) is conveniently evaluated by means of Processor Sharing (PS) queuing systems where flows are served according to a fair sharing policy, possibly generalized, of the link bandwidth [9]. To analyse such queuing systems, a good knowledge of three kinds of parameters is fundamental: the client arrival process, the required service (or workload) distribution, and the service discipline (here the bandwidth sharing mechanism). Let us begin by introducing some theoretical considerations on bandwidth sharing that will be useful in interpreting our experimental results in section 5.

We consider a population of users competing for a given resource capacity, i. e., the bandwidth (bit-rate) of a packet network link. We focus on some portion of that population which shows similar service properties: for example, same access rate, same application, same flow-granularity level, or any combination of these. In this section, we call a user an entity generically defined as a flow without a Time Out. A user session is composed of a succession of active (On) and inactive (Off) periods corresponding to flow transfers and "think times", respectively. Let us denote by N the size of the considered population, T the mean transfer time, S the mean silence or think time, and L the mean amount of data to be transferred during an On period.

Consider the stochastic process of the number $n(t)$ of simultaneously active flows at a given time t. We seek to write down the equilibrium equations which govern the evolution of this process in a stationary regime. Assuming for the moment that think time duration has an exponential distribution (this will be generalized below), the flow arrival rate when there are n active flows is $\lambda(n) = (N - n) / S$.

Suppose a very general bandwidth allocation $\phi(n)$ is applied to the considered population, depending only on the number n of active flows, so that a bandwidth equal to $\phi(n)/n$ is available for each flow of the considered class. Then, denoting by $\pi(n)$ the stationary state probability that n flows are present, the system evolves according to the local balance equation [3]: $\pi(n - 1)\lambda(n - 1) = \pi(n)\phi(n) / L$.

By simple recurrence, we obtain the state probabilities $\pi(n)$:

$$\pi(n) = \pi(0) \frac{N!}{(N - n)!} \frac{d^n}{\prod_{i=1}^{n} \phi(i)}, \quad 0 \le n \le N, \tag{1}$$

where the rate $d = L/S$ represents what the average user demand would be without any network limitation (no congestion and infinite bandwidth). The probability $\pi(0)$, as usual, is given by the normalization condition that the probabilities sum to 1.

The general performance properties of data networks derived in [3] show that expression in equ. (1) is actually insensitive: the state probabilities, and all system parameters that could be derived from them, do not depend on detailed traffic characteristics but only on mean demand through the rate parameter d. Particularly, the exponential distribution assumption for think time duration can be removed.

We seek a generic form of the bandwidth allocation function which includes, as special cases, a wide spectrum of known models of traffic variability. The choice of power law functions seems the most natural one; so we restrain the allocation functions to the following:

$$\phi(n) = \frac{c}{n^{b-1}}, \quad c > 0, 0 \le b \le 1 \cdot \tag{2}$$

The parameters b and c will be called the "exponent of the power law" and the "virtual allocated bandwidth", respectively. Such an allocation function leads to a share of $\phi(n)/n = c/n^b$ for each active flow. The state probabilities may be written:

$$\pi(n) = \pi(0) \, n!^b \binom{N}{n} \left(\frac{d}{c}\right)^n. \tag{3}$$

If $b = 1$, the allocation function $\phi(n)$ is a constant independent of the number of concurrent flows, and the bandwidth sharing discipline is pure Processor Sharing. In this case, equ. (3) reduces to the state probabilities for a PS queue with finite source population size [2]. We obtain the well-known geometric distribution for an infinite population by letting N tend towards infinity and d towards 0 in such a way that the link load, $\rho = Nd/C$, remains constant and less than 1.

When $b = 0$, $\phi(n)$ is a linear function of the number of flows: $\phi(n) = nc$. We thus deal with a circuit-like allocation where each active flow has a constant and fixed rate at its disposal. The system capacity is supposed to be high enough to handle the maximum number of flows, so that there is no blocking. In this case, the state probabilities follow a Binomial distribution that we might derive from (1), and asymptotically a Poisson distribution for an infinite population.

2.3 Fitting the Model to Experimental Data

To fit this model, called "Power-PS" or simply P-PS in the following, to experimental results, the following methodology is adopted. First, to characterize the bandwidth sharing mechanism for a considered class of flows, we fit the power law allocation function to the estimated average of the "instantaneous" flow bit-rate as a function of the number of active flows. With a time resolution of 1 s, flows that are actually active at a given time are detected, and then their instantaneous bit-rates are estimated. This procedure is performed along successive 1 s spaced samples over a time-limited period where the process of the number of flows appears stationary, typically a few tens of minutes at most (10 mn was chosen in Section 5). The optimal power law parameters, c and b, are quite simply obtained by a linear regression on the experimental points in logarithmic scales. Second, focusing now on the parameters which govern the arrival process and traffic demand, we estimate the optimal N and d parameters that provide the best fit to the number of flows distribution according to equ. (3). The latter is performed by means of a MLSE (Minimum Least Square Estimation) method using a search algorithm provided by the MATLAB software.

The P-PS model describes the behaviour of one given class of flows independently of the other classes which actually share the same resource. Nevertheless, the obtained model parameters implicitly incorporate the effects of this co-existence in a given situation. Besides, the model captures the bandwidth sharing mechanisms very roughly, i.e., to first order, since only the mean of the instantaneous bit-rate is computed at each time sample. This very simple approach is thus designed to provide a kind of "equivalent" representation model, the main purpose of which is to qualitatively infer the general trends shown by each considered class of flows, in terms of elasticity in reaction against possible congestion.

3 Data Traces and Traffic Composition

The data traces we used in this paper were collected on two sites of the France Telecom IP network in 2003 and 2005, using PCs equipped with DAG cards capable of collecting IP packet headers from any of several link layer protocols such as ATM or GBitEthernet. The first trace was recorded in August 2003 on the Fontenay ADSL aggregation area (near Paris) and consisted in a 3-hour measurement in both traffic directions. The other trace, of nearly 2 hours, was obtained in September 2005 on a backbone link connecting a POP (Point of Presence) in Paris to an overseas POP located in La Reunion Island, actually deserving ADSL subscribers in majority, too. Here, the downstream traffic alone was captured on two link interfaces at 155 Mbps.

Table 1. Data traces and traffic breakdown by transport protocol and application

Experiment	Total		UDP		TCP		HTTP		eDonkey		Kazaa		Gnutella		BitTorr.		Napster	
	MPkt	GB	%P	%B	%P	%B	%P	%B	%P	%B	%P	%B	%P	%B	%P	%B	%P	%B
Fontenay Up	109.9	35.7	11.3	2.3	87.6	97.2	6.7	2.7	59.1	63.7	2.8	6.5	0.9	0.6	1.2	1.7	1.6	2.4
Fontenay Dn	108.2	57.2	9.4	2.4	89.4	97.4	8.7	17.4	58.2	47.1	2.7	3.7	0.8	0.3	1.3	2.1	1.5	1.1
La Réunion L0	163.9	97.4	15.3	4.5	82.1	92.7	18.2	26.6	32.2	24.3	-	-	8.0	7.9	6.3	6.2	-	-
La Réunion L1	168.1	99.6	18.0	7.4	79.5	90.2	18.2	26.2	32.1	23.5	-	-	7.9	7.7	6.3	5.9	-	-

Table 1 provides a summary of the data traces used and of traffic composition. When the ADSL trace was recorded, TCP was still by far the most prominent transport protocol used in the Internet. The presence of other transport protocols than UDP or TCP is almost negligible (ICMP or others). Nowadays, real-time streaming applications are more significant than before due to the emergence of audio- and video services such as VoIP, Visiophony, VoD, TV on ADSL, etc. Nevertheless, in the overseas link experiment, we hardly see any noticeable amount of streaming traffic. Moreover, most of the UDP traffic in our trace is composed of P2P service signalling and of L2TP (Layer 2 Tunneling Protocol) transport of other ISPs traffic. Thus our concerns will be on TCP traffic only in the rest of the paper.

Concerning the application breakdown, eDonkey, by far, and HTTP appear to be the most popular applications (proportions given in Table 1 are computed relative to TCP). They will therefore be considered in the following as representatives of P2P and classical Internet applications, respectively. In the present study, it was assumed that identifying these applications by the well-known port numbers was sufficient.

4   Flow and Session Characterization

In this section, we highlight some distinctive features displayed by eDonkey and HTTP applications in the flow and session-related processes. We mainly deal with the arrival processes and the statistical distributions of the size of these entities. When applicable, we make use of the following standard probability laws to fit the experimentally obtained statistical distributions: Exponential, Gamma and Weibull for arrival processes, Log-normal and Pareto for size distributions. All are commonly employed to model positive random variable distributions in teletraffic studies.

4.1   Flow and Session Arrival Processes

So far, the flow arrival process is generally known to be non-Poisson [5], [7], [8], mainly because flows may be correlatively emitted within user sessions and generated in bursts by some applications (opening several parallel connections to mirror sites in HTTP and FTP, looking for a file in flooding mode in P2P services, etc.).

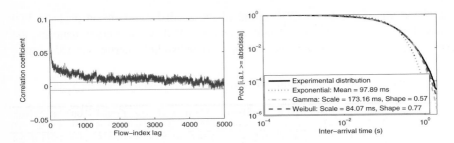

Fig. 1. Autocorrelation function and ccdf of the HTTP micro-flow i.a.t – Fontenay Down

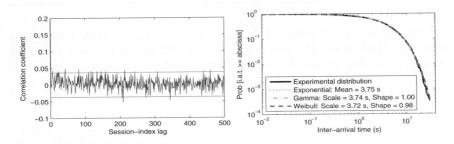

Fig. 2. Autocorrelation function and ccdf of the HTTP session i.a.t – Fontenay Down

Figs. 1 and 2 show the principal characteristics of the arrival process for HTTP micro-flows and HTTP sessions, respectively, as obtained from a sub-set of the ADSL traces (a full three hour measurement in the downstream for a single VP). We characterize it by the autocorrelation function and the complementary cumulative distribution function (ccdf), plotted in logarithmic scales, of the flow inter-arrival time (i.a.t.). These two presented figures are typical, among all our observations, of the arrival process of "classical Internet" applications micro-flows, for Fig. 1, and that of P2P micro-flows as well as of all applications user-flows or sessions, for Fig. 2.

Slight correlation is observable on the HTTP micro-flow i.a.t. autocorrelation function. This was consistently observed on other data sets considered in this and previous analyses [8]. Regarding the distribution of flow i.a.t., the Gamma and Weibull distributions provide an excellent fit to the experimental data while the Exponential fails to correctly match the tail of the distribution. (Note that in [5], the Weibull distribution was also found to provide good fit to the TCP connection inter-arrivals.) The micro-flow arrival process for HTTP traffic is thus clearly not Poisson.

For all other types of flows or sessions, including P2P micro-flows, the arrival process perfectly matches a Poisson process, at least up to second statistical order. Fig. 2 left plot shows no noticeable correlation between successive micro-flow inter-arrival times: the autocorrelation curve mostly lies within the zero-correlation 95% confidence interval. On Fig. 2 right plot, the simple exponential law appears to perfectly represent the i.a.t. distribution, which is quantitatively confirmed by the obtained shape parameters, close to 1, of the Gamma and Weibull distributions. Such results provide an appreciable clue to differentiate P2P micro-flows from those of non-P2P classical applications. Besides, that the session arrival process is Poisson is no surprise: this is generally recognized as one of the rare traffic invariants of the Internet. It naturally occurs when the user population is very large and when all consumers express small and uncorrelated demands. Experimental evidence of this property has already been shown [7]; our results provide some further elements to confirm it in the form of distribution fitting and autocorrelation function estimation.

The behaviours discussed so far were consistently observed on all data traces we have at our disposal. Comparing with previously published results, the flow arrival process shows a general tendency to be "more Poisson" than some years ago, partly due to the ever-increasing aggregation level of traffic on high-speed packet networks [6]. However, we have just seen that this is not true for all traffic classes. For classical Internet applications, the generation of a sequence of flows within a session reflects human activity, e.g., a user navigating on the Web generates a succession of transfer and think times, while bursts of flows occur due to the various objects included in a Web page. For P2P applications, the organization and segmentation of user demands in several chunks is mainly under control of the application [10]. This fact and the naturally distributed character of P2P overlay networks tend, in our opinion, to randomize the flow generation process so that it becomes closer to a Poisson process.

4.2 Flow and Session Length Distributions

Fig. 3 shows the size distributions in packets for eDonkey and HTTP micro-flows (left plot) and sessions (right plot). The corresponding figures for user-flow length distributions are very similar to those of sessions, at least qualitatively. All display a heavy-tail but, generally speaking, HTTP flow length distributions appear smoother, with no specific mode. We might model them by mixtures of Log-normal and Pareto distributions for the body and tail parts, respectively. Size distributions for eDonkey exhibit two principal modes, one corresponding to short flows mainly composed of P2P service signalling traffic (also present in UDP) [10], and the other to long flows. Note that the tail of the eDonkey session length distribution is highly biased by the necessary time-limited observation period, so that it cannot be correctly estimated.

The power parameters of the fitting Pareto distributions for HTTP (no reliable fit could be performed for eDonkey) are close to or less than 1, i.e., at the border line or in the area where even the mean is theoretically infinite (the variance alone is infinite when this power coefficient is between 1 and 2). This reflects a very high degree of variability of the distributions, the so-called "mice/elephants" phenomenon [1].

We now provide some new insights on the internal structure of user sessions. For that purpose, Fig. 4 gives the ccdf of the number of user flows per session for the eDonkey and HTTP applications. Both applications show very similar distributions in terms of the number of user-flows, except that the HTTP distribution has a slightly wider dynamic range. In both cases, though the dynamic range is rather limited, Pareto distributions provide very good fits to the overall empirical distributions. The corresponding distributions for the number of micro-flows per session are not presented since they pretty much look like those of the session length distributions in packets shown above. This may be explained by the fact that most of the micro-flows are only composed of a few packets.

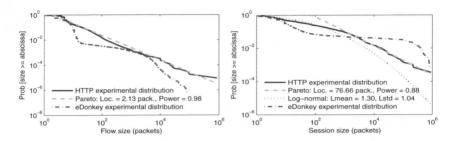

Fig. 3. Ccdf of the eDonkey and HTTP micro-flow (left) and session (right) length in packets

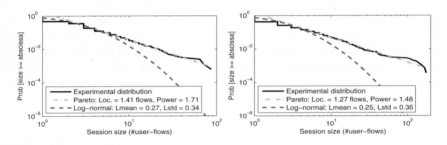

Fig. 4. Ccdf of the eDonkey (left) and HTTP (right) per-session number of user-flows

5 Bandwidth Sharing Experimental Results

The bandwidth sharing model introduced in Section 2 is now applied to the experimental data from the backbone link, "La Réunion L0" trace according to the names given in Table 1, focussing successively on micro-flows and on user-flows.

5.1 Bandwidth Sharing by Micro-Flows

For the considered applications, eDonkey and HTTP, Figs. 5 and 6 both show that the overall behaviour of micro-flows lies mid-way between pure PS and circuit-like bandwidth sharing, since the exponent parameter of the best fit power law (left plots) stands between 0.4 and 0.7 in both cases. The average bit-rate of eDonkey micro-flows, showing a higher exponent parameter, seems to react a little bit more to the number of active flows than that of HTTP micro-flows. Obviously, this kind of observation must be taken as a rough general tendency (as it is the very purpose of the approach indeed), given the first order aspect of the model. Accounting for flow rate variability appears as a fairly difficult task; thus, possible improvements of the model in that direction should be reserved for future work.

A feature jointly shared by both applications is the good quality of fit provided by the Gaussian law to model the distribution of the number of active micro-flows. Besides, it is interesting to compare with the Poisson distribution, since this is the natural outcome of an M/G/∞ queue where the service rate is not constrained. The number of flows distribution shows a statistical dispersion (variance-to-mean ratio) less than 1 in the case of eDonkey flows and greater than 1 in the case of HTTP flows, while it should be 1 for Poisson. As for the representation by the P-PS model, the size N of the hypothetical source population is quite high for both applications, and the average individual rate demand d is correlatively low at around a few kbps. This seems rather logical because the probability is not so great that the same micro-flow specification would be re-used after a Time Out delay or a think time period.

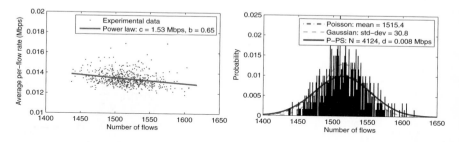

Fig. 5. Average instantaneous bit-rate per active micro-flow (left) and distribution of the number of active micro-flows (right), for eDonkey application

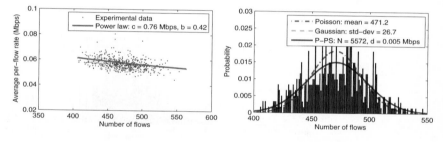

Fig. 6. Average instantaneous bit-rate per active micro-flow (left) and distribution of the number of active micro-flows (right), for HTTP application

5.2 Bandwidth Sharing by User-Flows

Concerning bandwidth sharing at user-flow level, Fig. 7 shows that eDonkey shares bandwidth in a reasonably fair way since the power exponent b is closer to 1 than for micro-flows. The small statistical dispersion of the number of active users explains the very low quality obtained by Poisson distribution fitting. These characteristics of the active user distribution, together with the N and d parameters provided by the P-PS representation model, would indicate a stable population of users being in active phase for quite a long time on the average. Note in particular the obtained source population size N which is barely greater than the mean number of active flows.

HTTP user-flows behave differently. The exponent of the power law is even lower than for micro-flows. Nevertheless, all three considered theoretical laws provide a good fit to the active user-flow distribution. The fact that the distribution is not far from Poisson confirms the lack of reactivity of HTTP user-flows. Moreover, the Poisson character is compatible with the high size N of the potential population source, compared to the mean number of active flows. Finally, the robustness of the model was checked by tracing the evolution of obtained parameters along successive 10 mn windows within the whole observation period: it was observed that the P-PS model parameters for user-flows are reasonably stable during the period, while those for micro-flows are more variable. This may be due to the high statistical dispersion shown by the distribution of active micro-flows (see Fig. 6 right plot) which does not facilitate algorithm convergence during the parameter optimization step.

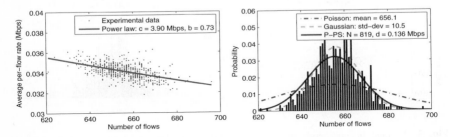

Fig. 7. Average instantaneous bit-rate per active user (left) and distribution of the number of active users (right), for eDonkey application

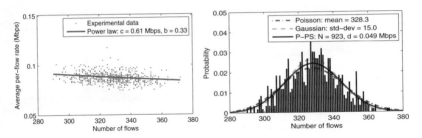

Fig. 8. Average instantaneous bit-rate per active user (left) and distribution of the number of active users (right), for HTTP application

6   Conclusion

In this paper we first updated some representation models of high-level traffic entities, pointing out the discrepancies shown between Peer-to-Peer services and classical Internet applications. The generation process of P2P flows is more widely distributed and randomized than for other applications, so that the arrival process of this type of flows is close to Poisson even at micro-flow level. The arrival process is Poisson, too, for all applications at user level, for both user-flows and sessions. The distributions of flow or session size are clearly distinguishable between these two types of traffic, the P2P distributions showing a bi-modal feature related to the short/long flows dichotomy. The internal structure of sessions shows an interesting property in that they generate a "heavy-tailed"-distributed number of user-flows in sequence.

We then introduced a simple bandwidth sharing model describing the average instantaneous bit-rate and number of active flows behaviours within different traffic classes. We mainly learnt from it that P2P (eDonkey) user-flows show more "elasticity" than classical Internet (HTTP) flows and exhibit a very narrow distribution typical of a somewhat reduced number of users with a very high activity ratio. Some of these observations illustrate a tendency for P2P applications to regularize to some extent the bit-rate processes of data traffic. This may be due partly to the "high-activity/reduced size" population effect pointed out above, and partly to the existence of a bottleneck in the upward access rates for this type of traffic.

It is hoped that the presented approach might have some interest for application in the fields of traffic classification and performance evaluation (if not for tractable mathematical analysis, at least for more realistic simulations of different classes of traffic). As a very first step in this direction, we suggest that the granularity of the user level seems more adequate than the micro-flow level to undertake performance or traffic control investigations: traffic is more regular at user level; differentiation between P2P and classical applications is easier; theoretical models are more tractable (flows generated in sequence within sessions), etc.

Acknowledgments. The author expresses many thanks to S. Oueslati, J. Augé and J. Roberts for many fruitful discussions, and to the France Telecom R&D teams, especially J. François, T. Houdoin, J. L. Simon and R. Cicutto, who set up the experiments, collected the data and provided some necessary pre-processing tools.

References

1. Ben Azzouna, N., Clérot, F., Fricker, C., Guillemin, F.: A flow-based approach to modeling ADSL traffic on an IP backbone link. Annals of Telecommunication 59(11-12) (2004)
2. Bonald, T., Olivier, P., Roberts, J.W.: Dimensioning IP access links carrying data traffic. Annals of Telecommunication, vol. 59(11-12) (2004)
3. Bonald, T., Proutière, A.: On performance bounds for balanced fairness. Performance Evaluation 55, 25–50 (2004)
4. Claffy, K., Braun, H.-W., Polyzos, G.: A parametrizable methodology for Internet traffic flow profiling. IEEE JSAC, vol. 13(8) (1995)
5. Feldmann, A.: Characteristics of TCP connection arrivals. In: Park, K., Willinger, W. (eds.) Self-similar network traffic and performance evaluation, J. Wiley & Sons, New York (2000)
6. Karagiannis, T., Molle, M., Faloutsos, M., Broido, A.: A nonstationary Poisson view of Internet traffic. In: Proceedings of IEEE Infocom 2004, Hong-Kong (2004)
7. Nuzman, C.J., Saniee, I., Sweldens, W., Weiss, A.: A compound model for TCP connection arrivals. In: Proceedings of 13th ITC Specialist Seminar on IP Traffic Modeling, Measurement and Management, Monterey, USA (2000)
8. Olivier, P., Benameur, N.: Flow level IP traffic characterization. In: de Souza, M., de Souza e Silva, M. (eds.) Proceedings of ITC'17, Elsevier, Amsterdam (2001)
9. Roberts, J.W.: A survey on statistical bandwidth sharing. Computer Networks 45, 319–332 (2004)
10. Tutschku, K., Tran-Gia, Ph.: Traffic characteristics and performance evaluation of Peer-to-Peer systems. In: Steinmetz, R., Wehrle, K. (eds.) Peer-to-Peer Systems and Applications. LNCS, pp. 383–397. Springer, Heidelberg (2005)

Measurement and Characteristics of Aggregated Traffic in Broadband Access Networks

Gerhard Haßlinger[1], Joachim Mende[1], Rüdiger Geib[1], Thomas Beckhaus[2], and Franz Hartleb[1]

[1] T-Systems, Deutsche-Telekom-Allee 7
[2] T-Com, Deutsche-Telekom-Allee 1, D-64295 Darmstadt, Germany
{gerhard.hasslinger, joachim.mende, ruediger.geib,
franz.hartleb}@t-systems.com, thomas.beckhaus@t-com.net

Abstract. We investigate statistical properties of the traffic especially for ADSL broadband access platforms, which have been widely deployed in recent years. Measurement traces of aggregated traffic are evaluated on multiple time scales and show an unexpected smooth profile with less relevance of long range correlation than experienced for traffic from Ethernet LANs.

A reason for the different characteristics lies in the shift to an increasing population of residential users generating most traffic on IP platforms via ADSL access. In addition, data transfer protocols of peer-to-peer networks strengthen the smoothing effect observed in current IP traffic profiles.

Keywords: traffic measurement, ADSL access networks, variability on multiple time scales.

1 Traffic Variability on Different Time Scales

Standard measurement in IP networks includes 5- or 15-minute mean values of the traffic rate or the load on the links. In IP platforms with underlying multiprotocol label switching (MPLS) the traffic matrix of flow intensities between all edge routers of a (sub-)network is usually available again for 5- or 15-minute intervals. The data forms a basis for network planning and the process of network resource upgrades to cope with the steadily increasing Internet traffic volume. The measurement can be collected from IP and MPLS routers without stressing the performance of the routing equipment when taken at intervals of several minutes length. The daily traffic profiles can be revealed in this way, showing the peak rates during busy hours, which are most relevant for network dimensioning.

The standard statistics does not include all relevant time scales to ensure quality of service, which is affected by congestion even on small time scales of less than a second. Short term overload is often invisible on longer time scales due to compensation by alternating phases of low load. Buffers can bridge temporary overload on account of delay for the buffered data, but only to a limited extend until buffer overflows occur. Real time applications with strict delay bounds of e.g. less than 0.2s for conversational services impose restrictions on waiting times and corresponding buffer sizes.

L. Mason, T. Drwiega, and J. Yan (Eds.): ITC 2007, LNCS 4516, pp. 998–1010, 2007.

Since more than a decade, many evaluations of IP traffic measurement revealed long range dependencies and self similar patterns over the relevant time scales [1][5][11]. While most of this measurement was conducted on Ethernet LANs, ADSL broadband access presently connects a population of more than 170 million residential users worldwide to the Internet, still with increasing tendency [6]. Different traffic profiles are experienced for Ethernet and ADSL using measurement at the digital subscriber line access modules (DSLAMs) [3].

We investigate comparable traffic measurement at the interconnection of the ADSL access network and the IP backbone. In Section 2, we analyse the variability of samples on several time scales starting from 1ms. In addition to the analysis of the complete traffic on a link, we filter HTTP, UDP and a part of the peer-to-peer traffic to investigate their influence on variability in section 3. Section 4 studies the implications of traffic profiles for waiting times as the main QoS indicator in order to estimate load thresholds on transmission links indicating critical QoS conditions.

2 IP Traffic Measurement and Evaluation

For measurement purposes, we consider the amount of arriving data in a time slotted system [12], where the time is subdivided into subsequent intervals of arbitrary but constant length Δ. In order to capture the process of arriving traffic in detail, each arriving IP or MPLS packet can be registered with a time stamp as well as its packet size. The storage demand for measurement traces in this representation is increasing with the line speed, where millions of packets are counted per second on high speed links in the Gbit/s range. In a time slotted view, the data volume V of all packets arriving during a slot Δ is computed and traffic is represented as a series $V_1, V_2, V_3, \ldots$ of data volumes per slot. The slot length Δ determines the accuracy of the representation.

Measurement equipment is capable to capture the amount of data arriving e.g. per millisecond, from which corresponding buffer occupation and waiting times are derived as main QoS indicators at the same precision of milliseconds. An advantage of a time slotted approach lies in a limited storage demand for $M = S/\Delta$ integers to represent a traffic trace over S seconds, which can be controlled by appropriate choice of Δ independent of the transmission speed. The traffic rate in each time slot is given by $R_j^{(\Delta)} = V_j/\Delta$ for $j = 1, \ldots, M$.

Figure 1 represents a corresponding traffic trace from a 2.5Gbit/s link at the border of the ADSL aggregation and the IP backbone network with intervals starting at the time scale $\Delta = 1$ms. Typical examples have been extracted from measurements running over about a month in December 2005 and January 2006. Traffic rates $R_m^{(K\Delta)}$ for longer time frames of multiples $K \cdot \Delta$ ($K = 2, 3, \ldots$) of a slot are simply computed by the mean over K subsequent Δ–intervals

$$R_m^{(K\Delta)} = \frac{1}{K} \sum_{j=m(K-1)+1}^{mK} R_j^{(\Delta)}.$$

Figure 1 includes 4 time scales for $\Delta = 1$ms, $\ldots$, 1s with multiples $K = 10, 100, 1000$. It is apparent, that traffic becomes smoother when observed on longer time

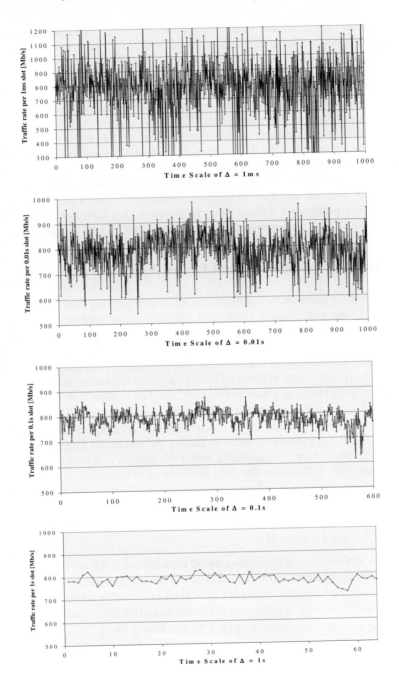

Fig. 1. Traffic variability on different time scales

scales. The coefficient of variation $\sigma^{(\Delta)} / \mu^{(\Delta)}$ is a usual measure of variability computed from the mean $\mu^{(\Delta)}$ and standard deviation $\sigma^{(\Delta)}$:

$$\mu^{(\Delta)} = \frac{1}{M} \sum_{j=1}^{M} R_j^{(\Delta)} \quad \text{and} \quad \sigma^{(\Delta)} = \sqrt{\frac{\sum_{j=1}^{M} (R_j^{(\Delta)} - \mu^{(\Delta)})^2}{M}}.$$

Considering longer time scales $K\Delta$, the mean value is always preserved $\mu^{(K\Delta)} = \mu^{(\Delta)}$. The standard deviation and the coefficient of variation $\sigma^{(\Delta)} / \mu^{(\Delta)}$ are preserved if and only if $R_{nK+1}^{(\Delta)} = R_{nK+2}^{(\Delta)} = \cdots = R_{n(K+1)}^{(\Delta)} = R_n^{(K\Delta)}$ for $n = 0, 1, \cdots, (M/K) - 1,$ such that the traffic rate is constant in each subsequence of K intervals which is comprised on time scale $K\Delta$. Otherwise, the coefficient of variation is smaller $\sigma^{(K\Delta)} / \mu^{(K\Delta)} \leq \sigma^{(\Delta)} / \mu^{(\Delta)}$.

Traffic measurement has been studied at different time scales by [3] starting from 1s intervals. In addition, this work compares modeling approaches including M/G/∞ for the arrival process. The analysis is carried out with different assumptions on the distribution of flow sizes and derives their impact on the autocorrelation of $R_m^{(\Delta)}$. Wavelet-based approaches provide an alternative analysis method on multiple time scales [1].

Another comparative study in [3] reveals higher variability of Ethernet traffic, while measurements taken from the DSLAMs in an ADSL access platforms show smooth pattern, characterized by the fact that about 99% of the mean rates $R_m^{(1s)}$ over 1s intervals stay below $\mu(R) + \sqrt{\mu(R)}$, where a 5-minute mean value is taken for $\mu(R)$:

$$\Pr\left\{ R_m^{(1s)} \leq \mu(R) + \sqrt{\mu(R)} \right\} \approx 99\% \quad \text{for rates measured in Mbit/s.} \tag{1}$$

Therefore $C \geq \mu(R) + \sqrt{\mu(R)}$ is proposed as a minimum threshold for bandwidth provisioning on a link. Note that this bound is derived from traffic variability on short

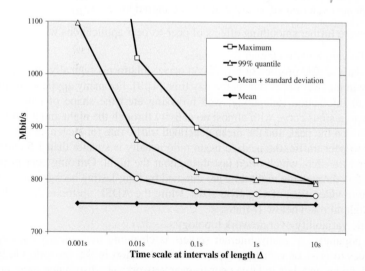

Fig. 2. Statistical traffic parameters on multiple time scales

to medium time scales, while further aspects have to be involved in network planning including long term traffic fluctuation, link upgrade processes for growing traffic as well as failure resilience. Those factors may lead to provisioning of capacity far beyond the criterion (1).

We consider comparable measurement taken at broadband access routers of Deutsche Telekom's IP platform, which connects regions for ADSL access to the IP backbone. Figure 2 summarizes measurement statistics over multiple time scales, including the maximum, 99%-quantiles $\gamma_{99\%}$ and the standard deviation of a 30-minute trace with mean rate $\mu^{(1800s)} \approx 753$ Mbit/s, which confirms essential reduction of the variability on longer time scales.

Internet traffic measurement on the contrary revealed long range dependencies over many time scales as a motivation to introduce self-similar traffic modeling [1][5][11]. They show only a minor smoothing effect of the variability on longer time scales. There are at least two reasons for a different behaviour observed in our measurement:

➢ Most measurements showing self-similar pattern have been conducted on LANs and aggregation platforms with prevalent Ethernet access. Ethernet is equipped with at least 10Mbit/s to the terminals, ranging up to Gigabit Ethernet nowadays, whereas most ADSL lines are still limited to a few Mbit/s. On the other hand, the residential user population on ADSL platforms counts in millions [6] leading to a high multiplexing level of small flows in aggregation stages.
➢ Most of the traffic volume is generated by peer-to-peer file sharing protocols, which subdivide downloads of a large file into fixed size data units to be transmitted in parallel TCP connections from different sources [1][9][15]. Those multi source downloads lead to many small flows per user and keep the throughput at an almost constant rate even when some sources are going offline and have to be replaced. Usually the uplinks are the bottleneck of the P2P network throughput, which are even limited in speed to a few hundred kbit/s [14].

There are further smoothing effects of peer-to-peer applications with regard to

➢ Traffic variability on longer time scales:
The daily traffic profiles in broadband access platforms typically show peak activity in the evening or during the day time [4][9]. For many applications involving human interaction (telephony, web browsing etc.) the shape of daily activity is close to a sinus curve with almost no activity through the night and a ratio of about 2 between the peak and the mean generated traffic rate for such a daily profile. For peer-to-peer traffic, the peak to mean ratio usually is smaller than 1.5 due to background transfers which often last throughout the night. Ongoing peer-to-peer data transfers through the night time are initiated by long-lasting background downloads of large video files of Gigabyte sizes, while the ADSL upstream speed limits the throughput often below 1Mbit/s.
➢ Traffic variability over network topology:
The popularity of many Internet servers is changing dynamically causing traffic sources to arise or vanish at one or another location in the network. On the other hand, nodes and data in large peer-to-peer networks are distributed more uniformly over the access areas. During search phases the peer-to-peer protocols often in-

volve supernodes which are comparable to servers, but the P2P downloads, which transport most of the data volume, are running distributed among the peer nodes. While spontaneously increasing popularity of a server can lead to access bottle-necks, frequently referenced data is soon replicated and then available from many nodes in a peer-to-peer system. Hence, peer-to-peer applications strengthen a uni-form distribution of traffic sources over the network independent of sudden changes in the popularity of content.

Regarding the QoS criterion (1), we determine the quantiles of the sequence of traffic rates $R_m^{(1s)}$ of a typical aggregation stream of rate 753Mb/s in Figure 1, which results in

$$\Pr\left\{ R_m^{(1s)} \leq \mu(R) + k \sqrt{\mu(R)} \right\} = 99\% \quad \text{for } k \approx 1.5 \tag{2}$$

with mean rates again measured in Mbit/s. For a set of 11 included MPLS flows with mean rates from $10 - 40$ Mbit/s the same evaluation results in factors in the range $1.7 < k < 2.5$. Thus our measurement traces confirm the form of equation (2) as pro-posed by [3] with $k = 1$, but the factor k is experienced to be larger.

3   Variability for Different Application Types

In addition to the characteristics of the total aggregated traffic, some applications can be filtered out to study their traffic profiles. Table 1 and Figure 3 show corresponding results for transport layer differentiation of HTTP, UDP and peer-to-peer traffic, where HTTP is identified by TCP port 80 and P2P by the set of ports 4661, 4662, 6881, 6882, 6346, 6348, 6883, 6884 and 52525, which are known to be used by the popular P2P protocols eDonkey, BitTorrent and Gnutella. The port lists are neither complete, nor can the application types be clearly distinguished based on ports. More P2P traffic is being disguised over the HTTP or other ports.

Table 1. Parameters for variability of different traffic types

Mbit/s	Mean Rate μ	Standard deviation $\sigma^{(\Delta)}$ over multiple time scales Δ				
		0.001s	0.01s	0.1s	1s	10s
HTTP port 80	278.7	88.1	35.3	20.5	17.2	14.4
P2P ports	165.6	40.1	12.5	4.1	1.7	1.3
Other ports	387.2	83.0	30.0	12.4	9.3	7.6
UDP	50.8	16.3	6.2	2.7	2.2	2.0
Total traffic	753.9	127.3	47.9	24.3	19.5	16.3

About 22% of the total traffic is observed on the set of P2P ports, whereas applica-tion layer analysis reveals that P2P represents the major portion of the traffic [14]. On the other hand, we expect only a negligible portion of non-P2P applications on the considered P2P ports as false positives. Therefore the port filtering covers only a part of the P2P traffic, but serves as a simple online method extracting almost pure P2P traffic, whereas HTTP and UDP are composed of a mixture of several application types including P2P.

When we compare the first three columns of Table 1 with the statistics of the total traffic, a perfect confirmation of the statistical multiplexing effect is observed, i.e. $\sigma^2_{HTTP} + \sigma^2_{P2P} + \sigma^2_{Other} = \sigma^2_{Total}$ holds on all time scales with deviations less than 1%. Figure 3 shows that the identified P2P traffic portion has a smaller ratio σ/μ than the other traffic filtered by HTTP, UDP, which becomes most apparent on the longer time scales 1s and 10s. The statistics of Table 1 is given for the downstream direction. The upstream traffic has a similar profile with no essential deviations. Moreover, the total traffic volume is almost symmetrical in both directions, which again is a typical P2P characteristics.

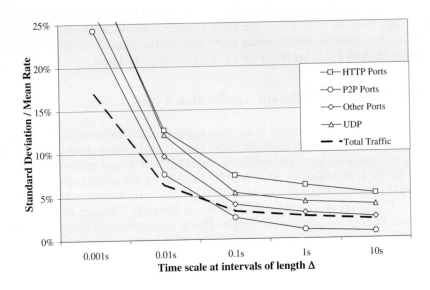

Fig. 3. Variability of traffic types distinguished by transport layer filtering

4 Load Dependent Waiting Times Computed from Traces

The traces of the amount of arriving traffic $R_m^{(\Delta)}$ can be used to determine the development of the waiting time at the accuracy of a slot time Δ. Therefore we presume

➢ a constant available bandwidth C (Mbit/s) and
➢ a buffer size B (Mbit) at the router interface.

Then the waiting time after the 1., 2., 3. ... slot of a considered trace can be iteratively computed, where $C\Delta$ is the amount of data which can be served per slot. We assume that the capacity is available at the end of each slot for all data that arrives during the slot and for buffered data. The latter assumption is optimistic, since the data may arrive non-uniformly over a slot time, while forwarding can be assumed as a continuous constant rate process, since the considered time scales start from 1ms, whereas packets are interspaced in the order of microseconds on Gigabit links. The difference to the pessimistic assumption that forwarding of all data arriving in a slot has to be deferred until the next slot, means an increase in waiting time by no more than one slot time Δ. Let

➤ A_k denote the amount of data arriving in the k-th slot and

➤ W_k denote the waiting time after the k-th time slot.

The waiting time W_{k+1} after the next slot can be calculated from the previous W_k

$$W_{k+1} = \text{Max}(\text{Min}(W_k + A_k / C - \Delta, B/C), 0).$$

This form of Lindely's equation accounts for a per slot difference $A_k/C - \Delta$ in the workload and a limitation B/C of the waiting time according to a finite buffer size B. Considering a traffic trace over M intervals, a corresponding series of waiting times $W_1, W_2, W_3, ..., W_M$ after each time slot is determined starting from $W_0 = 0$. In addition, we obtain the statistical parameters including the mean, the maximum and the quantiles of the waiting time. The analysis can be executed for arbitrary capacities C and corresponding utilization $\mu(R)/C$.

We applied the evaluation to the complete traffic on the link and to the largest involved MPLS traffic flows with mean rates from $10 - 40$ Mbit/s. The waiting time is computed at the 1ms time scale $W_{k+1} = \text{Max}(W_k + A_k/C - 0.001, 0)$ and for an infinite buffer ($B \rightarrow \infty$), with QoS degradation corresponding to long waiting times.

Evaluations for two MPLS traffic traces of about 15 minute length are shown in Figure 4. Each MPLS flow represents a source to destination traffic demand between edge routers of the backbone. Both examples have a mean rate of 18.0Mbit/s and have been filtered out of the total traffic with mean 753Mbit/s on a 2.5Gbit/s link.

The measurement trace shows a moderate link utilization of about 30% and does not lead to overload neither for total traffic nor for any included MPLS flows. We analyzed the same measurement trace for smaller capacity C corresponding to higher utilization. Then overload is occurring at some level and the waiting time at each time slot in the trace is increasing with the utilization. In this way, we can determine a critical utilization level, when the mean or maximum waiting time exceeds a predefined threshold. The analysis does not include the TCP congestion control mechanism [7]. Nevertheless, the analysis for the original source traffic is relevant especially for the non-congested regime, where overload phases are seldom and most of them are short, such that TCP has no essential influence. In addition, there is an increasing portion of inelastic and non-responsive applications including UDP traffic, which caused the IETF standardization to consider extension of rate control mechanisms.

In both MPLS flow examples, a maximum waiting time of 50ms is exceeded at different utilization of about 90% for $C = 20$ Mbit/s in the first case and at about 40% or $C = 45$ Mbit/s in the second case. The representation in the figures does not give a resolution on the 1ms level, but instead includes the maximum waiting time value during each second. The smooth traffic in the first example exhibits a similar behaviour as for the total traffic on the link, where sufficient QoS properties can be met even up to 90% load. Beyond this threshold, QoS is sharply decreasing and long lasting overload phases become visible, e.g. over the last minute of the first trace in

Figure 4 at 95% utilization. In this case, TCP surely would have an influence on the long lasting overload phase, but the example is by far the smoothest case for a set of considered MPLS examples.

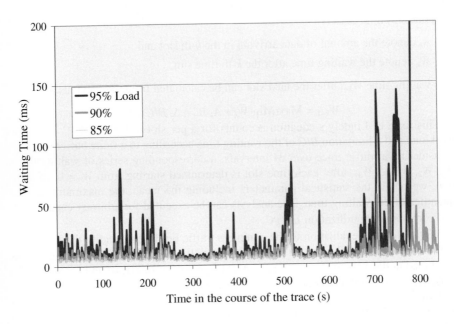

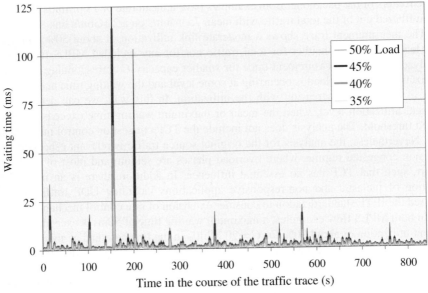

Fig. 4. Waiting time development during a traffic trace

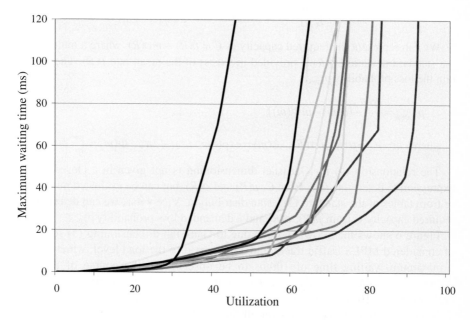

Fig. 5. Load and maximum waiting time in 11 examples of MPLS traffic traces

The other MPLS traffic flows show higher variability $\sigma^{(\Delta)} / \mu^{(\Delta)}$ than the entire traffic on a link, due to a smaller rate and multiplexing level. Consequently, the QoS-critical thresholds of the utilization shift to lower loads for smaller traffic aggregates.

While Figure 4 gives an impression of the distribution of load peaks during a trace for several levels of the long term utilization, Figure 5 indicates the increase of the maximum waiting time with growing utilization. Each curve corresponds to an MPLS traffic flow, where the examples of Figure 4 represent the extreme cases at both sides.

Next, we compare the observed relationship between maximum waiting times and long term utilization obtained from the traffic traces with Gaussian dimensioning approaches. The latter only include the mean and standard deviation or the 99%-quantile as parameters. The statistical multiplexing effect suggests that the distribution of the traffic rate on relevant time scales approximately shows a Gaussian shape [8][10][13]. This is confirmed by histograms for the traffic rates per time slot, despite of increasing deviations in the outer ranges $x < \mu - 2\sigma$ and $x > \mu + 2\sigma$, which occur more frequently than expected for a Gaussian distribution.

Based on Gaussian distributions, dimensioning rules are available with and without including buffers [8][7]. The zero-buffer analysis determines the loss rate r_{Loss} of a switching system with constant forwarding capacity C in a continuous flow model. It is a simplified worst case approach, since buffers may prevent losses and thus improve the QoS. It accounts for data lost during overload phases when the arrival rate exceeds the forwarding capacity. In general, the loss rate is determined with regard to the rate distribution function $F_R(x)$. The loss probability p_{Loss} is given by the ratio of the mean loss rate to the mean traffic rate:

$$r_{\text{Loss}} = \int_{x > C}(x - C)\,dF_R(x) \quad \text{and} \quad p_{\text{Loss}} = r_{\text{Loss}}\,/\,\mu(R). \tag{3}$$

We can represent the required capacity as $C = \mu(R) + m\sigma(R)$, where a multiple m of the standard deviation $\sigma(R)$ is provided in excess of the mean rate $\mu(R)$. Then we obtain the loss probability p_{Loss}:

$$p_{\text{Loss}} = \frac{\sigma(R)}{\mu(R)}(\phi(m) - m\Phi(m)); \tag{4}$$

where $m = (C - \mu(R))/\sigma(R); \quad \phi(m) = \exp(-m^2/2)/\sqrt{2\pi}; \quad \Phi(m) = \int_m^\infty \phi(x)\,dx.$

The relationship (4) for Gaussian dimensioning is not given in a closed formula expression to determine p_{Loss} from C, $\mu(R)$ and $\sigma(R)$, but can be evaluated numerically or from tables of the standard Gaussian distribution. Vice versa, we can determine the required capacity C from $\mu(R)$, $\sigma(R)$ and a demanded loss probability p_{Loss}.

Figure 6 shows the allowable load due to Gaussian dimensioning (3) for the set of considered MPLS traffic traces and compares it to the load level, which leads to a maximum waiting time of 100ms to be observed in the course of the trace. Again, the capacity $C = \mu(R) + m\sigma(R)$ and utilization $\mu(R)/C$ directly correspond to each other.

While $\mu(R)$ remains constant over all time scales, $\sigma(R)$ is reducing on larger time scales. Thus the question arises, which time scale is appropriate to determine $\sigma(R)$? The relevant time scale should be of about the same order as the delay introduced by buffers, since fluctuations on smaller time scales can be compensated by the buffer whereas overload phases on longer time scales lead to buffer overflows. The maximum delay in buffers of usual size is in the range of 0.1s – 1s. Thus we focus on those time scales.

As the main results of the comparison, Gaussian dimensioning for $\sigma(R)$ taken from the 1s time scale establishes an optimistic upper bound, resulting in the uppermost curve in Figure 6. When $\sigma(R)$ is taken instead from the 0.1s time scale, the allowable load in the Gaussian model is reducing by 3-10%. The corresponding curve is alternating with the load levels for 100ms maximum waiting time computed from the traces, with 5 cases lying above and below and thus establishes the most reasonable estimate due to Gaussian modeling. As an alternative, we determined $\tilde{\sigma}(R)$ from the 99%-quantile $\gamma_{99\%}$ of the 0.1s time scale, using the fact that $\gamma_{99\%} \approx \mu(R) + 2.33\,\tilde{\sigma}(R)$ $\Rightarrow \tilde{\sigma}(R) \approx (\gamma_{99\%} - \mu(R))\,/\,2.33$ for a Gaussian distribution. This leads to a third curve in Figure 6, which is again below the Gaussian model with $\sigma(R)$ taken directly from the 0.1s time scale.

The second example of Figure 4 is also included as the tenth case in Figure 6, which has a 43% load limit computed from the trace, largely deviating from each of the Gaussian models. The maximum waiting time is encountered in two peaks visible in Figure 4, both of which are shorter than 2s and cannot be predicted from $\mu(R)$, $\sigma(R)$ and $\gamma_{99\%}$. Although the maximum waiting time is exceptional for this case, the mean waiting time is smaller than for most other MPLS flows for the same load.

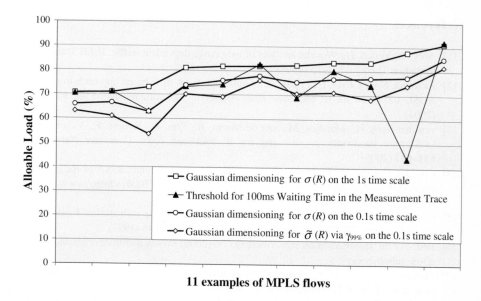

11 examples of MPLS flows

Fig. 6. Allowable load for Gaussian distribution of the traffic rate compared to the load level which exceeds 100ms as maximum waiting time in the trace

5 Summary and Conclusions

The variability of traffic generated on ADSL broadband access platforms essentially reduces on longer time scales and differs from classical IP traffic measurement suggesting a self-similar structure [11], although long range dependency is still visible. The statistical multiplexing effect is strengthened by a large number of transport layer flows at small rates being generated by residential users whose ADSL access speed is still limited to a few Mbit/s, while the variability of traffic via Ethernet access networks remains higher [3]. Peer-to-peer protocols presently contribute most of the traffic using multi source connections for each download. They increase the number of flows and stabilize download rates at an almost constant level for each user. Therefore P2P traffic is observed to be much smoother than classical applications in the Internet.

As indicators of QoS properties, the maximum and the quantiles of the waiting time are determined based on traffic traces and as a function of the utilization. Variability is experienced to be much higher for smaller aggregates of flows, which suggests lower allowable load thresholds with regard to QoS. A comparison to dimensioning for Gaussian traffic based only on the mean rate and standard deviation leads to reasonable estimates in most cases, although with exceptions. Further study is required to determine the most significant time scales with regard to QoS aspects and whether the evaluation can be based on only a few measurement parameters.

References

[1] Abry, P., Veitch, D.: Wavelet analysis of long range dependent traffic. IEEE Trans. on Information Theory 44, 2–15 (1998)

[2] Azzouna, N.B., Clérot, F., Fricker, C., Guillemin, F.: A flow-based approach to modeling ADSL traffic on an IP backbone link. Annals of Telecommunication 59, 1252–1255 (2004)

[3] van den Berg, H., Mandjes, M., van de Meent, R., Pras, A., Roijers, F., Venemans, P.: QoS-aware bandwidth provisioning for IP backbone links. Computer Networks 50, 631–647 (2006)

[4] Cho, K., Fukuda, K., Esaki, H., Kato, A.: The impact and implications of the growth in residential user-to-user traffic, ACM Sigcomm Conf., Pisa (2006) <http://www.acm.org/sigs/sigcomm/sigcomm2006>

[5] Crovella, M.E., Bestavros, A.: Self-similarity in world wide web traffic: Evidence and possible causes. IEEE/ACM Trans. on Networking 5, 835–846 (1997)

[6] DSL Forum, Information on subscribers Q3'06 (2006) <www.dslforum.org/dslnews/pdfs/q306_subscribers.pdf>

[7] Hartleb, F., Haßlinger, G.: Comparison of Link Dimensioning Methods including TCP Behaviour. Proc. IEEE Globecom Conf, San Antonio, USA, pp. 2240–2247 (2001)

[8] Haßlinger, G.: QoS analysis for statistical multiplexing with Gaussian and autoregressive input. Telecommunication Systems 16, 315–334 (2001)

[9] Haßlinger, G.: ISP Platforms under a heavy peer-to-peer workload. In: Steinmetz, R., Wehrle, K. (eds.) Peer-to-Peer Systems and Applications. LNCS, vol. 3485, pp. 369–382. Springer, Heidelberg (2005)

[10] Kilpi, J., Norros, I.: Testing the Gaussian approximation of aggregate traffic. Proc. Internet Measurement Workshop, Marseille, France (2002)

[11] Leland, W., Taqqu, M., Willinger, W., Wilson, D.: On the self-similar nature of Ethernet traffic. IEEE Trans. on Networking 2, 1–15 (1994)

[12] Li, S.-Q.: A general solution technique for discrete queueing analysis of multi-media traffic on ATM. IEEE Trans. on Communication, pp. 1115–1132 (1991)

[13] Norros, I., Pruthi, P.: On the applicability of Gaussian traffic models. In: Emstad, P.J., et al. (eds.) 13. Nordic Teletraffic Seminar, Trondheim, pp. 37–50 (1996)

[14] Siekkinen, M., Collange, D., Urvoy-Keller, G., Biersack, E.W.: Performance limitations of ADSL users: a case study. PAM 2007, 8th Passive and Active Measurement conf, Louvain-la-neuve, Belgium (2007)

[15] Tutschku, K., Tran-Gia, P.: Traffic characteristics and performance evaluation of peer-to-peer systems. In: Steinmetz, R., Wehrle, K. (eds.) Peer-to-Peer Systems and Applications. LNCS, vol. 3485, pp. 383–398. Springer, Heidelberg (2005)

Revisiting the Optimal Partitioning of Zones in Next Generation Cellular Networks: A Network Capacity Impact Perspective

Samik Ghosh[1], Huan Li[2], Hee Lee[2], Prabir Das[2], Kalyan Basu[1], and Sajal K. Das[1]

[1] Center For Research In Wireless Mobility and Networking (CreWMaN)
Department of Computer Science & Engineering, The Univeristy of texas at Arlington,
Arlington TX- 76010
{sghosh,basu,das}@cse.uta.edu
[2] Wireless Core Systems Engineering, Converged Multimedia Networks, Nortel Networks,
Richardson, TC 75081
{huanli,heele,daszprab}@nortel.com

Abstract. While the problem of optimal cell-site partitioning has been primarily studied from the perspective of scarce radio resources in the access network, recent field measurements have shown significant impact of paging load on the core signaling subsystem capacity. In this paper, we revisit the problem of optimal zone partitioning, identifying factors affecting partitioning cost at the core switch. We develop a general integer programming formulation for joint optimization of two key issues: (i) assignment of cells to a base station controller and controllers to the core signaling subsystem, (ii) optimal partitioning of zones at the access and network levels. Given the exponential nature of the location planning problem, we develop a genetic algorithm based approach for solving the general zone partitioning and configuration problem, both for incumbent and greenfield networks. Our results demonstrate significant cost and performance benefits at the network level for next generation converged services.

Keywords: converged networks, location area, genetic algorithms.

1 Introduction

The key differentiator for wide area cellular services has been their ability to provide ubiquitous coverage over large areas and mobility support for end users. One of the key components for mobility support is the mobility management component of the network, responsible for tracking users as they move through the service area maintaining ongoing calls and changing their network attachment. In order to track a user for an incoming call, the network needs to page the cells in the access network. A simple technique for determining the current user location employs paging all the cells in the network, termed system wide paging, as shown in Fig.1.

L. Mason, T. Drwiega, and J. Yan (Eds.): ITC 2007, LNCS 4516, pp. 1011–1023, 2007.
© Springer-Verlag Berlin Heidelberg 2007

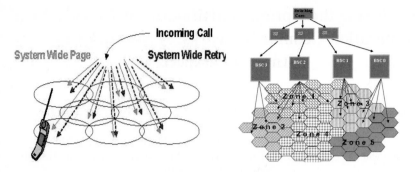

Fig. 1. System Wide Paging and retry   **Fig. 2.** Zone partitions of system

However, this system-wide paging approach incurs significant cost to the network, both in terms of radio resources as well network capacity. The location management scheme which has been deployed in GSM/UMTS and CDMA standards for efficiently tracking users involves dividing the coverage area of the system into groups of cells forming a continuous geographic area, variably called *location areas* (LA) or *zones*, as shown in Fig.2. In this scheme, when a user crosses zone boundaries, the system sends location update or zone registration message and the system updates the user location information in the database.

As is evident from the above discussion, the zone or LA based paging scheme introduces a new cost factor: the cost of zone registration or location update. While zone partitioning reduces the paging cost, it increases the registration cost, as more zones imply more frequent location updates due to user mobility, presenting a tradeoff. Thus, the problem of partitioning a cellular service area into zones essentially involves achieving an optimal partition, taking into account the cost of paging and zone registration. The problem increases a significant order of magnitude in a green-field deployment scenario where the physical configuration of the base stations, controller (BSCs) and the switch have to be considered in the light of their impact on the zone partitioning cost.

Most of the work on optimal zone partitioning has focused on considering the cost on the access network due to the scarcity of radio bandwidth. Saraydar et al. investigated into zone-based schemes focusing on optimizing the location area (LA) borders [1]. In [2], the authors have looked into zone-based scheme with selective using a tree-location area algorithm (TrLA). Distance-based schemes are outlined in [3] while a hybrid of zone-based and distance-based scheme is studied in [4]. A common thread in all existing work is the focus on the access network impact on the paging and registration costs assuming sufficient capacity at the core network segment. While all these methods provide strong theoretical results for optimally designing zones, the real-life application of these techniques are limited by the network characteristics. Moreover, data from real-life service providers have shown that the impact of zone partition is also severely felt in the core network segment. The problem becomes further significant in next generation converged networks, where the optimal planning of zones would play a key role in network performance.

In this paper, we revisit the zone partitioning problem in this light. We identify the key factors affecting the core network capacity and their impact on zone partitioning parameters. We develop a general integer programming formulation for joint optimization of two key issues: (i) assignment of cells to a BSC as well as assigning BSCs' to the signaling peripheral cards (SS) in a core switch, and, (ii) optimal partitioning of zones at the access level (base stations) as well as network level (BSC). The zone partitioning problem has been proven to be NP-hard for even simple cases [2], [3] and involves exponential time computation. We develop a genetic algorithm (GA) based heuristic solution to the generic problem to obtain near optimal solutions which give significant benefits in terms of system cost.

In Section 2, we revisit the zone partitioning problem in terms of network capacity impact. We formulate the new zone partitioning problem in Section 3, with the solution technique outlined in Section 4 and summarizing the results in Section 5. The paper concludes with a brief outline of future work in Section 6.

2   Revisiting the Zone Partitioning Problem

As mentioned in the previous section, zone optimization techniques have been primarily focused on the access network radio resources. Current analysis, based on network service provider data, have however, shown that even with system-wide paging, the main bottleneck is the signaling part of the switch instead of the access network. Depending on the physical configurations of the signaling sub-system (SS) shelves on the switch, the BSCs can be connected to one or more SS cards which can have a crucial impact on their performance in terms of paging. Fig.3 shows how the number of paging messages increases exponentially as the number of SS shelves on the switch increases. Fig.4 shows the impact of system wide paging on the SS capacity under a scenario where all SS cards have same configuration, (i.e. the best case) and when the configuration is skewed (worst case). The plots show that system wide paging causes nearly 50% degradation in the performance of the SS.

The above results motivate us to revisit the zone partitioning problem in the perspective of the impact of paging and registration of the network, specifically the signaling subsystem capacity. We observe that the key factor which causes the capacity degradation is the configuration of the SS shelves on the switch and assignment of SS resources to the BSCs depending on their load and other factors.

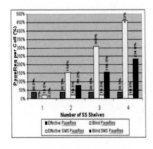

Fig. 3. Exponential increase in paging messages

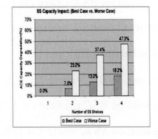

Fig. 4. SS capacity impact due to system wide paging

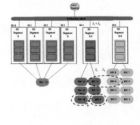

Fig. 5. Impact of switch configuration on zone partitioning

Fig.5 provides an example showing the impact of the switch configuration on the paging cost to the system. Assume the paging loads on BSC 13 and 6 are λ_1 and λ_2 respectively and the handover traffic between BSC 10 and 9 is h. If the zone partitioning scheme does not take the current configuration into account, it can possibly group BSC 13 and 6 in one zone, which causes the load on the network to double, since BSC 13 and 6 are on different signaling sub-systems and in order to page that zone, the load of ($\lambda_1 + \lambda_2$) needs to be sent to SS 4 and SS 5. Similarly, if only the network cost is considered, then grouping BSC 10 and 9 into different zones does not affect the cost from the network perspective as they belong to same SS. However, if the handover traffic h between them is very high, the network cost saving is negated by the increasing in zone registration costs between them. Thus, ideally forming zones around BSC boundaries would be optimal from a network perspective, though it might affect other factors in the total cost.

3   The Generic Configuration and Zone Partitioning Problem

In this section, we formally define the problem of optimally partitioning the cells of a service area together with assigning base stations (BS) to BSCs and BSCs to SS cards, for a generic configuration and optimization problem. We also consider two special cases: (i) Optimizing zone partitioning in an incumbent service provider scenario for a given configuration, (ii) Optimizing zone partitioning for a given configuration at the network or BSC level only.

Before formulating the non-linear integer program for the optimization problem, we define some notations and parameters used in the analysis.

N_{BSC} = Number of BSCs in the system

N_{BS} = Number of cells or base stations in the system

N_{zone} = Maximum number of zones in the system,

N_{SS} = Number of SS cards for a switch in the system

f = Re-page rate in the system (assuming system wide re-paging strategy)

λ_l = Original paging load on BS l, where $l = 1, 2, 3... N_{BS}$

c_l = Traffic load (BHCA) on BS l, where $l = 1, 2, 3... N_{BS}$

$h_{l,m}$ = Handover traffic between BS l and m.

α_l = Actual paging load on BS l, where $l = 1, 2, 3... N_{BS}$

α_i^{BSC} = Actual paging load on BSC i, where $i = 1, 2, 3... N_{BSC}$

α_k^{SS} = Actual paging load on SS k, where $k = 1, 2, 3... N_{SS}$

β_l = Registration load on BS l, where $l = 1, 2, 3... N_{BS}$

β_i^{BSC} = Registration load on BSC i, where $i = 1, 2, 3... N_{BSC}$

β_k^{SS} = Registration load on SS k, where $k = 1, 2, 3... \ N_{SS}$

P_i^{BSC} = Paging capacity of BSC i

P_l^{BS} = Paging capacity of BS l

C_i^{BSC} = Busy Hour Call Attempt (BHCA) capacity of BSC i

C_k^{LPP} = Busy Hour Call Attempt (BHCA) capacity of SS k

a = idle registration traffic load factor

C_r = Cost factor due to registration

C_p = Cost factor due to paging

$X_{i,j}$ = 1, if BSC i and BSC j belong to different zone, 0, otherwise.

$Y_{l,n}$ = 1, if BS l is assigned to zone n, $n = 1, 2, 3..... N_{zone}$, 0, otherwise.

$Z_{l,m}$ = 1 if BS l and BS m belong to different zones, 0, otherwise.

$U_{i,k}$ = 1, if BSC i is assigned to SS k, 0, otherwise.

$V_{l,j}$ = 1 if BS l is assigned to BSC j, 0, otherwise.

3.1   Estimation of System Parameters

An important step in formulating the optimization problem is the estimation of system parameters which play an important role in the cost function. Most of the system parameters can obtained from measurements made on the switch, e.g. determining the mobility pattern between cells in the network (i.e., matrix H($h_{l,m}$)). The idle registration load factor is assumed to be equivalent to the out-of-VLR registration to Inter-System Handoffs. The repaging rate, f, with system-wide initial paging can be obtained from the field data. C_p and C_r represents the impact of page and registration messages to the SS CPU capacity and are obtained from the lab measurements together with the paging capacities of BSC and BS.

3.2   Generic Integer Programming Formulation

In this section, we focus on the joint optimization problem: namely, optimizing the assignment of cells to BSCs and BSCs to SS on a switch in a green-field deployment scenario together with zone partitioning of the cells. As mentioned earlier, the zone partitioning takes into account the cost on the access as well as the core network due to the system configuration. Based on the notations defined at the beginning of this section, we can define the actual paging load on a BS l, α_l, as follows:

$$\alpha_l = \sum_{m=1}^{N_{BS}} \lambda_m (1 - Z_{l,m}) \tag{1}$$

The repaging cost, R, assuming system-wide repaging is given by,

$$R = f \sum_{m=1}^{N_{BS}} \lambda_m \tag{2}$$

The zone registration cost, is difficult to trace based on operational measurements (OMs). While the handover traffic is a good measure of registration load, it does not include idle registration when mobiles change zones in idle state. Based on estimation of this load factor a, the total registration cost can be computed as.

$$\beta_l = \sum_{m=1}^{N_{BSC}} (1+a) h_{l,m} Z_{l,m} \tag{3}$$

Once the total paging and registration loads are computed at the cell or access level, we need to compute the same at the level of BSCs. It may be observed here that the paging and registration costs at the BSC level depend on:

(i)   The physical assignment of cells to BSC, i.e. the values of the variables $\mathbf{V}$ ($V_{l,j}$) .

(ii) The assignment of BSCs to zones, i.e. the values of the variable $\mathbf{X}$ ($X_{i,j}$).

Thus, the total paging load on a BSC i is given by,

$$\alpha_i^{BSC} = \sum_{j=1}^{N_{BSC}} \sum_{l=1}^{N_{BS}} ((\sum_{l=1}^{} \alpha_l V_{l,j}) \times (1 - X_{i,j})) \tag{4}$$

The zone registration cost at the BSC level is given as,

$$\beta_i^{BSC} = \sum_{l=1}^{N_{BS}} \beta_l V_{l,i} \tag{5}$$

Moving on to the level of SS, the paging and registration cost views at this level of the network are essentially a function of the assignment of BSCs to SS, i.e. the values of the variable $\mathbf{U}$ ($U_{i,k}$) and are computed as follows,

$$\alpha_k^{SS} = \sum_{i=1}^{N_{BSC}} \alpha_i^{BSC} U_{i,k} \tag{6}$$

$$\beta_k^{SS} = \sum_{i=1}^{N_{BSC}} \beta_i^{BSC} U_{i,k} \tag{7}$$

At the SS level, the total cost of system-wide repaging is given as,

$$R_{LPP} = C_p f \sum_{m=1}^{N_{BS}} \lambda_m \tag{8}$$

Based on Eqn. (6), (7) and (8), the total cost to the system can be defined as,

$$C_{SYS} = \sum_{k=1}^{N_{SS}} (C_p \times \alpha_k^{SS} + C_r \times \beta_k^{SS}) + N_{SS} C_p f \sum_{m=1}^{N_{BS}} \lambda_m \tag{9}$$

Thus, the optimization problem can be defined as an integer programming problem of finding an optimal assignment of the variables **X, U, V** and **Z**, such as to minimize the total system cost i.e.

Min (C_{SYS}), subject to the following constraints,

$$\sum_{j=1}^{N_{BSC}} V_{l,j} = 1 \quad \text{for all } l = 1, 2, 3..., N_{BS} \tag{10}$$

which ensures that each cell is assigned to one BSC only.

$$\sum_{k=1}^{N_{SS}} U_{i,k} = 1 \quad \text{for all } i = 1, 2, 3..., N_{BSC} \tag{11}$$

which ensures that each BSC is assigned to one SS only.

$$\sum_{n=1}^{N_{zone}} Y_{l,n} = 1 \quad \text{for all } l = 1, 2, 3..., N_{BS} \tag{12}$$

which ensures that each cell is assigned to one zone only.

$$1 - Z_{l,m} = \sum_{n=1}^{N_{zone}} Y_{l,n} \times Y_{m,n} \quad \text{for all } l, m = 1, 2, 3..., N_{BS} \tag{13}$$

which maintains consistency between the matrices **Z** and **Y**.

$$1 - X_{i,j} \leq \sum_{k=1}^{N_{SS}} U_{i,k} \times U_{j,k} \tag{14}$$

which constraints that BSCs on different SS should not belong to the same zone.
The paging capacity of a cell should not be exceeded, which is represented as,

$$\alpha_l \leq P_l^{BS} \quad \text{for all } l = 1, 2, 3..., N_{BS} \tag{15}$$

The paging capacity of a BSC should not be exceeded, which is represented as,

$$\alpha_i^{BSC} \leq P_i^{BSC} \quad \text{for all } i = 1, 2, 3..., N_{BSC} \tag{16}$$

The BHCA call capacity of a BSC should not be exceeded in the assignment of cells to a BSC, which is represented as,

$$\sum_{l=1}^{N_{BS}} c_l \times V_{l,i} \leq C_i^{BSC} \quad \text{for all } i = 1, 2, 3..., N_{BSC} \tag{17}$$

The BHCA call capacity of a SS should not be exceeded in assignment of BSCs to an SS, which is represented as,

$$\sum_{i=1}^{N_{BSC}} (\sum_{l=1}^{N_{BS}} c_l \times V_{l,i}) \times U_{i,k} \leq C_k^{SS} \quad \text{for all } k = 1, 2, 3..., N_{SS} \tag{18}$$

Based on the above formulation, it is possible to obtain a solution for the configuration and zone partitioning problem which minimizes the total cost to the system. An important observation in this regard is the computational complexity of the configuration as well as the zone partitioning problem is non-polynomial $O(n^m + n^n)$ with n number of base stations and m number of BSCs.

It may be noted here that for the special case of incumbent networks, the assignments of BSC to SS and BS to BSCs are fixed, the above formulation can be applied for finding the optimal zone partitioning of base stations.

3.3 Special Case of Optimization at Network (BSC) Level Only

Now, we consider a special case, where the focus of the zone partitioning scheme is at the network level, i.e. assignment of BSCs to optimal zones. This particular case is important under certain network characteristics, especially in existing scenarios where the focus is on optimally partitioning the BSCs into zones, which fixes the assignment of BS to zones for a given configuration of the network.

In this case, since we focus only at the level of BSCs, we define some new notations which simply the problem definition:

$h_{i,j}^{BSC}$ = Handover traffic between BSC i and j, which is computed in terms of $h_{l,m}$

$Y_{i,n}^{BSC}$ = 1, if BSC i is assigned to zone n, $n = 1, 2, 3....., N_{Zone}$, 0, otherwise.

Since, we are interested in the BSC level cost, the effective paging load on BSC is defined as,

$$\alpha_i^{BSC} = \sum_{j=1}^{N_{BSC}} \alpha_j^{BSC} \times (1 - X_{i,j})) \tag{19}$$

The zone registration cost at the BSC level is given as,

$$\beta_i^{BSC} = \sum_{j=1}^{N_{BSC}} (1+a) h_{i,j}^{BSC} \times Y_{i,j}^{BSC} \tag{20}$$

The cost at the SS level are given by Eqn. 6 and 7, while the re-paging cost is given as,

$$R_{LPP} = C_P f \sum_{i=1}^{N_{BSC}} \alpha_i^{BSC} \tag{21}$$

Now, the total cost to the system can be computed as earlier,

$$C_{SYS}^{BSC} = \sum_{k=1}^{N_{SS}} (C_p \times \alpha_k^{SS} + C_r \times \beta_k^{SS} + C_p f \sum_{i=1}^{N_{BSC}} \alpha_i^{BSC}) \tag{22}$$

Now, replacing from Eqn. 21 and 22, we have,

$$C_{SYS}^{BSC} = \sum_{k=1}^{N_{SS}} (C_p \times \sum_{i=1}^{N_{BSC}} (\sum_{j=1}^{N_{BSC}} \alpha_j^{BSC} \times (1 - X_{i,j}))) U_{i,k} + C_r \times \sum_{i=1}^{N_{BSC}} (\sum_{j=1}^{N_{BSC}} (1+a) h_{i,j}^{BSC} \times Y_{i,j}^{BSC}) U_{i,k} + C_p f \sum_{i=1}^{N_{BSC}} \alpha_i^{BSC})$$

The optimization problem goal is to minimize C_{SYS}^{BSC} subject to constraints which re a subject of the general formulation with two modifications,

$$\sum_{n=1}^{N_{zone}} Y_{i,n}^{BSC} = 1 \quad \text{for all } i = 1, 2, 3..., N_{BSC} \tag{23}$$

which ensures that each BSC is assigned to one zone only.

$$1 - X_{i,j} = \sum_{n=1}^{N_{zone}} Y_{i,n}^{BSC} \times Y_{j,n}^{BSC} \quad \text{for all } i, j = 1, 2, 3..., N_{BSC} \tag{24}$$

which maintains consistency between the matrices $\mathbf{X}$ and Y^{BSC}.

It may be noted here that the computational complexity of the BSC level optimization problem is a function of the number of BSCs, N_{BSC}, which typically is low for a given switch.

4   An Evolutionary Algorithmic Approach

As observed in the problem formulation section, the joint optimization of physical configuration and zone partitioning becomes a computational prohibitive problem. Thus, different heuristic techniques need to be applied in order to obtain a near optimal solution to the problem space. Various techniques, based on graph partitioning, simulated annealing etc. can be used in such scenarios [7]. In this section, we outline the details of an evolutionary algorithm based heuristic approach, employing genetic algorithm (GA), to obtain a cost-minimized zone partitioning.

4.1   The Genetic Algorithm Formulation

The central theme of genetic algorithms (GA's) revolves around the process of utilizing natural evolution techniques to search optimal solutions for complex problems [6]. The algorithm begins with a *set of initial solutions*, represented by *chromosomes*, which forms the *population*. Each chromosome consists of a set of strings, called *genes*, which encode a possible solution to the problem. Solutions from one population are taken and reproduced to form a new population, following specific *crossover* and *mutation* operators, to yield new solutions (*off springs*) which have better *fitness* criteria as defined by the GA. We outline below the specification of our genetic algorithm employed for the zone partitioning and configuration problem.

A. Chromosome: The entire population is denoted by a set c of chromosomes, where c_i denotes chromosome i. The chromosome consists of *genes* which represent an encoding of a solution. In our case, there are four solution spaces, each of which is represented by a gene and they together form a chromosome c_i in the population:

(i) Gene 1: The assignment of cells to zones, represented by the vector $\mathbf{Y}$ of size N_{BS} where each entry represents the zone assigned to that cell, i.e. each entry can take values $= 1, 2, 3... N_{zones}$.

(ii) Gene 2: The assignment of the BSCs to zones, which is captured in the vector $\mathbf{X}$, represented by a vector of N_{BSC} elements.

(iii) **Gene 3:** The assignment of cells to BSCs, represented by a matrix of size $N_{BS} \times N_{BSC}$, giving the corresponding gene $N_{BS} \times N_{BSC}$ elements.

(iv) **Gene 4:** The assignment of BSCs to SS, which is represented by the matrix **U** with size $N_{BSC} \times N_{SS}$, giving the corresponding gene $N_{BSC} \times N_{SS}$ elements

Based on the above defined genes, the complete solution can be encoded in a chromosome formed by a linear combination of these genes, giving the size of the chromosome as $N_{BS} + N_{BSC} + N_{BS} \times N_{BSC} + N_{BSC} \times N_{SS}$, which is of $O(n^2)$ where $n = max$ (N_{BS}, N_{BSC}, N_{SS}). Fig. 6 gives the chromosome of one possible configuration for $N_{BS} = 5$, $N_{BSC} = 3$ and $N_{SS} = 2$.

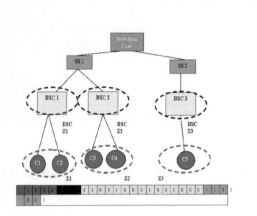 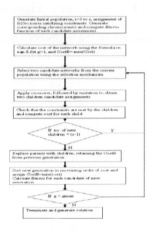

Fig. 6. Sample zone assignment and configuration

Fig. 7. GA flowchart for zone partitioning

B. Initial population: The initial population consists of an assignment of physical configuration and zone partitioning as shown in the previous example, ensuring that the initial population satisfies all the constraints of the optimization problem

C. Selection: Two parents are selected to breed using a *rank based quadratic procedure* [6]. All the parents in the population are ordered according to their level of *fitness* with 1 being the highest level of *fitness*. A randomized technique is then employed to select two parents for breeding.

D. Crossover: Breeding is done using *uniform crossover*. This type of breeding is accomplished by selecting two parent solutions and randomly taking a component from one parent to form the corresponding component of the child. For example, suppose parent c1 and parent c2 are chosen to breed.

c1= {11223123100100010010001101001},
c2={11223213100100010010001100101}, Child = {11223213100010010001101001}

In the child above, the first component of the chromosome comes from parent c1. The second component of the child has to be a 1 since both parents have a 1 in this position of the chromosome. The third component of the child comes from parent $c2$, and so on.

E. Mutation: Mutation (m) is a way to add new genetic material to a population. This is done to help the GA avoid getting stuck at a local optimum. A child undergoes mutation according to the percentage of population mutated, m. For example, if m = 20% and s = 20, then four children in the new population are randomly chosen for mutation. Once a child is chosen to be mutated, then the probability of mutation for each vector component is equal to the mutation rate, r_m. If $r_m = 0.3$, then each component of a selected child will be mutated with a probability of 0.3. Thus, in our case for a mutation, change occurs in one of the four genes. As an example, a child and its possible mutation is given below,

Child= {11223123100010010001101001}
Mutated child = {11223213100010010001101001}

F. Fitness: The genetic algorithm attempts to find the minimum cost zone partitioning scheme along with the physical configuration of the cell sites and BSCs. The GA is constructed so that it may consider infeasible network zone partitions which are penalized. Infeasible solutions may contain beneficial parts. Consequently, breeding two infeasible solutions or an infeasible with a feasible solution can yield a good feasible solution. The optimal design will lie on the boundary between feasible and infeasible designs In GA, the fitness function $F(c_i)$, traditionally improves with an improved objective function, creating a maximization problem. Thus, the fitness in this paper is defined as $F(c_i) = \dfrac{1}{C_p(c_i)}$ where $C_p(c_i)$ is the penalized network cost,

G. Termination Condition: The criterion is the total number of generations, *gmax*, where *gmax* varies according to the size of the network, N, under study.

The steps of the GA based zone partitioning algorithm are outlined in the flowchart illustrated in Fig. 7, which yield an optimal solution to the problem which can be decoded from the chromosome.

5 Implementation and Performance

In this section, we outline the implementation of the zone partitioning tool developed to implement the GA algorithm defined in the previous section along with results on sample network scenarios. The algorithm was implemented using the Java based JGAP package for GA solutions (http://jgap.sourceforge.net/). A simple front-end user interface for inputting the parameters of the problem was developed. Behind the UI is the core engine based on JGAP which implements the algorithm of the Zone Optimizer tool. We depict a sample example network configuration in Fig. 8 with the various input parameters outlined in Table 1. In a GA based tool, the quality of the solution, which is measured by average best fitness and the performance of the algorithm in terms of speed are significant factors. In Table 2, we tabulate the network configurations and system parameters, along with the saving of cost in terms of using zone partitioning against a system-wide paging scheme.

As can be seen from Table 2 and Fig. 9, the zone partitioning scheme gives significant savings in terms of percentage cost against a system-wide zone paging scheme. The cost savings become significant for larger systems as is expected from previous discussions. Also, we find from Table 2 that the Zone Optimizer algorithm gives better solutions for larger network sizes while the cost in terms of cpu time is not significantly high compared to the savings in systems cost (around 70% saving).

Genetic Algorithm Parameters	Values
Termination criteria	1000
Population size	100
Percentage of population mutated, m	20%
Mutation rate, rm	0.3
Penalty rate, rp	6

Table 1. GA parameters

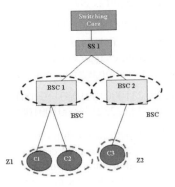

Fig. 8. Sample network configuration

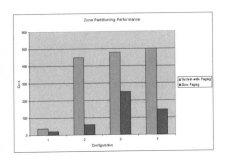

Fig. 9. Zone Partitioning performance

Configuration	System Parameters	Avg. Fitness	System-wide Paging Cost	Zone Paging Cost	Cost Saving (%)	Run time(ms)
Cell :3, BSC: 2SS :1	f=a=Cr=Cp=1	3.5521E-4	36.0	21.0	41.66	140804.0
Cell :3, BSC: 2SS :1	f=a=1 Cr=2, Cp=5	0.001116	450.0	60.0	86.6	148265.0
Cell :5, BSC: 3SS :2	f=a=Cr=Cp=1	0.001	480.0	250.0	47.9	269590.0
Cell :5, BSC: 3SS :2	f=a=1 Cr=2, Cp=5	0.005	500.0	146.0	70.8	283027.0

Table 2. Performance of Zone Optimizer

6 Conclusion

In this paper, we have revisited the problem of optimal zone partitioning from a network perspective, identifying key network characteristics which affect the zone partitioning problem. A generic integer program has been devised for the optimal zone partitioning, together with physical configuration of a switch. We have developed a randomized heuristic algorithm based on evolutionary principles of the genetic algorithms, for finding solutions to the zone partitioning which gives significant cost benefits. As an extension, we are implementing the devised algorithm into a custom zone partitioning tool which can be eventually deployed to customers for zone planning in the field.

References

1. Lei, Z., Saraydar, C.U., Mandayam, N.B.: Paging area optimization based on interval estimation in wireless personal communication networks. Mobile Networks and Applications 5(1), 85–99 (2000)
2. Tabbane, S.: An alternative strategy for location tracking. IEEE Journal on Selected Areas in Communications, JSAC 13(5), 880–892 (1995)
3. Bar-Noy, A., Kessler, I., Sidi, M.: Mobile users: to update or not to update? Wireless Networks 1(2), 175–185 (1995)
4. Liang, B., Haas, Z.J.: Predictive distance-based mobility management for pcs networks. In: Proceedings of IEEE INFOCOM'01, New York, NY, USA, March 1999, vol. 3, pp. 1377–1384 (1999)
5. Goldberg, D.E: Genetic Algorithms in Search, Optimization and Machine Learning. Kluwer Academic Publishers, Boston, MA (1989)

Distributed Dynamic Load Balancing
in Wireless Networks

Sem Borst, Iraj Saniee, and Phil Whiting

Bell Labs, Alcatel-Lucent, 600 Mountain Avenue, Murray Hill, NJ 07974, USA

Abstract. Spatial and temporal load variations, e.g. flash overloads and traffic hot spots that persist for minutes to hours, are intrinsic features of wireless networks, and give rise to potentially huge performance repercussions. Dynamic load balancing strategies provide a natural mechanism for dealing with load fluctuations and alleviating the performance impact. In the present paper we propose a distributed shadow-price-based approach to dynamic load balancing in wireless data networks. We examine two related problem versions: (i) minimizing a convex function of the transmitter loads for given user throughput requirements; and (ii) maximizing a concave function of the user throughputs subject to constraints on the transmitter loads. As conceptual counterparts, these two formulations turn out to be amenable to a common primal-dual decomposition framework. Numerical experiments show that dynamic load balancing yields significant performance gains in terms of user throughputs and delays, even in scenarios where the long-term loads are perfectly balanced.

1   Introduction

Even more so than other communication networks, wireless data networks are characterized by the occurrence of large spatial and temporal load variations. The spatial variations manifest themselves in flash overloads in persistent hot spots in dense urban areas due to mobility, transportation busy hours, accidents, and other unpredictable events. Temporal fluctuations occur in time scales from milliseconds and minutes to hours and days. On the latter time scales, the traffic load varies according to predictable day-of-week and hour-of-day aggregate patterns. On the former time scales, the load fluctuates not only because of the intrinsic randomness in user arrivals and session durations, but also due to variations in the transmission rates that rapidly change due to fast fading. Clearly, the spatial and temporal variation and uncertainty in the traffic will only tend to be more pronounced in ad-hoc deployment environments compared to carefully planned commercial cellular networks.

Third-generation cellular systems aim to provide high-speed data services despite these spatio-temporal variations. In fact, the fast temporal fluctuations are taken advantage of by the base station to allocate resources based on channel feedback and backlog through careful scheduling [1,5,9,15,17]. In all existing systems, however, including CDMA, UMTS and even IEEE 802.11, each base station, or access point, independently arbitrates among users in its coverage

L. Mason, T. Drwiega, and J. Yan (Eds.): ITC 2007, LNCS 4516, pp. 1024–1037, 2007.

area. Users simply select the strongest received base station, and each base station allocates resources without any coordination with other base stations in its vicinity. As a result, one base station or access point may experience severe overload, while resources might be abundant at surrounding base stations, thus providing scope for performance gains through some form of coordination.

Coordinated resource allocation has recently been considered in several studies, see for instance [7,8,18]. The work in [8] shows the gains due to coordination to be significant. Despite these gains, however, coordinated resource allocation among cells remains a challenging task. Centralized coordination requires huge processing capability as well as exchange of vast amounts of information among all users in a geographical area and the coordinating entity.

A possible remedy comes from decentralized or self-organizing schemes where a sufficient degree of coordination is achieved with minimal exchange of state. In previous work [6], the authors considered distributed load balancing as a mechanism for achieving such functionalities for power-controlled services, such as voice connections. In that work it was shown that shadow prices for carefully selected critical resources provide the additional means to dynamically allocate users to cells based on load considerations, in addition to the standard notion of proximity, and thus achieve a high degree of optimization, without the need for centralization. Dynamic association of users with access points has also been considered in the context of IEEE 802.11 networks, see for instance [2].

In the present paper we combine the utility maximization framework that has been successfully leveraged for scheduling and resource allocation in the uncoordinated case [20,21] with the distributed optimization approach developed in [6]. Even though the aim is not to enable full-fledge network-wide scheduling, we show that significant gains result from suitable assignment of users to cells and efficient allocation of resources. Additional per-cell scheduling will obviously further improve the performance. The proposed mechanism relies on distributed shadow prices for dynamic load balancing. We examine two related problem versions: (i) minimizing a convex function of the transmitter loads for given user throughput requirements; and (ii) maximizing a concave function of the user throughputs subject to constraints on the transmitter loads. As conceptual counterparts, these two formulations turn out to be amenable to a common primal-dual decomposition framework. Numerical experiments indicate that dynamic load balancing yields significant performance gains in terms of user throughputs and delays, even in scenarios where the long-term loads are perfectly balanced.

The remainder of the paper is organized as follows. In Section 2 we examine the problem of minimizing a convex function of the transmitter loads for given throughput requirements. In Section 3 we turn attention to the problem of maximizing a concave function of the user throughputs subject to constraints on the transmitter loads. We describe how the merits of load balancing schemes can be evaluated in terms of transfer delays and user throughputs in Section 4. In Section 5 we present the numerical experiments that we conducted to benchmark the performance gains from dynamic load balancing. We make some concluding remarks in Section 6.

2 Load Minimization

2.1 Model Description

We consider a wireless data network with C transmitters. For now we will focus on a static scenario with M users, and address the problem of determining which users should be allocated resources from which transmitters.

Denote by r_{mc} the feasible transmission rate of user m when served by transmitter c. By feasible rate, we mean the long-term rate that the user would receive if it were allocated all the transmission resources (time slots, power, frequencies) of the transmitter. Let x_{mc} be the actual amount of resources of transmitter c allocated to user m. We assume the transmissions to be (roughly) orthogonal, so that user m receives a total rate of (approximately) $T_m := \sum_{c=1}^{C} r_{mc} x_{mc}$. Let $L_c := \sum_{m=1}^{M} x_{mc}$ be the total load (resource utilization) at transmitter c. Denote by τ_m the rate requirement of user m, and by σ_c the maximum sustainable load on transmitter c, when applicable.

The coefficients r_{mc} only serve as a parsimonious representation of the rate statistics, and likewise the parameters τ_m are only meant to provide a coarse characterization of the traffic demands. We abstract from the specific details of the air-interface structure, and also ignore the burstiness in the traffic processes and the fact that actual transmission rates vary over time because of fast fading. While the latter aspects are clearly crucial for the scheduling at each of the transmitters on a fast time scale, they are less relevant in deciding which users should be served by which transmitters. Also, the quantities x_{mc} will only play the role of decision variables in coordinating the assignment of users to transmitters, with the actual allocation of resources governed by local schedulers residing at the individual transmitters.

2.2 Problem Formulation

We first examine the problem of minimizing a convex function $F(L_1, \ldots, L_C)$ of the transmitter loads for given throughput requirements $\tau_1, \ldots, \tau_M$. This formulation is particularly natural when the users have intrinsic rate requirements and conservation of transmission resources (e.g. battery life) is of vital importance, or when the level of congestion is a critical performance measure.

$$\min \ F(L_1, \ldots, L_C) \tag{1}$$

$$\text{sub } L_c = \sum_{m=1}^{M} x_{mc} \qquad c = 1, \ldots, C \tag{2}$$

$$T_m = \sum_{c=1}^{C} r_{mc} x_{mc} \geq \tau_m \qquad m = 1, \ldots, M \tag{3}$$

$$x_{mc} \geq 0 \qquad m = 1, \ldots, M, c = 1, \ldots, C.$$

Convex duality implies that the optimal solution to the above problem may be found from the Lagrangian formulation $\max_{\mu \in \mathbb{R}_+^M} F^*(\mu)$, with $F^*(\mu) = \min_{x \in \mathbb{R}_+^{M \times C}} \mathcal{L}(x, \mu)$,

$$\mathcal{L}(x, \mu) := F(L_1, \ldots, L_C) + \sum_{m=1}^{M} \mu_m (\tau_m - \sum_{c=1}^{C} r_{mc} x_{mc}),$$

and $\mu_1, \ldots, \mu_M$ Lagrangian multipliers.

The optimality conditions for the latter formulation read $\frac{\partial F}{\partial L_c} \geq r_{mc} \mu_m^*$, with the complementary slackness conditions $x_{mc}^* \left(\frac{\partial F}{\partial L_c} - r_{mc} \mu_m^* \right) = 0$ for all $m = 1, \ldots, M$, $c = 1, \ldots, C$, and $\mu_m^* (\tau_m - \sum_{c=1}^{C} r_{mc} x_{mc}^*) = 0$ for all $m = 1, \ldots, M$.

Note that the problem (1)–(3) will have a feasible solution (and the Lagrangian will have a finite solution) as long as $\min_{c=1,\ldots,C} r_{mc} > 0$ for all $m = 1, \ldots, M$. Also, there exists an optimal solution with at most $M + C - 1$ non-zero variables x_{mc}^*, which means that there will be at most $C - 1$ additional 'legs' beyond the minimum number that is necessary to connect all the M users.

We now focus on the case where the objective function is of the form $F(L_1, \ldots, L_C) = \sum_{c=1}^{C} K_c(L_c)$, with $K_c(\cdot)$ some strictly convex differentiable function. In that case, x_{mc}^* satisfies $K_c'(L_c^*) = r_{mc} \mu_m^*$ for all $m \in \mathcal{M}_c$, $\mathcal{M}_c := \arg \max_{m=1,\ldots,M} r_{mc} \mu_m^*$, and $x_{mc}^* = 0$ for all $m \notin \mathcal{M}_c$. A particular example is $K_c(L_c) = L_c^{1+\beta}/(1+\beta)$ for some parameter $\beta > 0$, which governs the trade-off between minimizing the total load and the maximum load across all transmitters. As $\beta \downarrow 0$, the objective function becomes $\sum_{c=1}^{C} L_c$, which is minimized by simply assigning each individual user m to the strongest transmitter $c_m := \arg \max_{c=1,\ldots,C} r_{mc}$. In contrast, when $\beta \to \infty$, the problem amounts to minimizing $\max_{c=1,\ldots,C} L_c$, which deserves special treatment and will be examined in further detail below.

The Lagrangian formulation may be interpreted as follows. Each of the users can be allocated resources from each of the transmitters. The cost associated with the load imposed on the transmitter c is specified by the function $K_c(\cdot)$, while each unit of throughput obtained by user m carries a reward μ_m. There are two opposing players. Player 1 aims to allocate the resources to users so as to minimize the net cost (or maximize the net revenue) for given rewards μ_m. Player 2 aims to set rewards μ_m, so as to maximize the net cost incurred (minimize the net revenue earned) by the first player.

In principle, the above problems may be readily solved using standard routines. However, these algorithms generally involve centralized computation and require global knowledge of all parameters. To circumvent these issues, an Arrow-Hurwicz type dual-ascent scheme [11] may be used which, while slower to converge, is mostly distributed in nature, and only involves a limited exchange of information among transmitters and users. Such a scheme may be interpreted as a repeated game between the two opposing players as described above. Because of page constraints a detailed description of the algorithm and the convergence proof is omitted.

We now investigate the case where the objective function is of the form $F(L_1, \ldots, L_C) = \max\limits_{c=1,\ldots,C} w_c L_c$, i.e., the maximum weighted load across all transmitters. In that case, problem (1)–(3) reduces to the following linear program:

$$\min \; L \tag{4}$$

$$\text{sub } L \geq w_c L_c = w_c \sum_{m=1}^{M} x_{mc} \qquad c = 1, \ldots, C \tag{5}$$

$$T_m = \sum_{c=1}^{C} r_{mc} x_{mc} \geq \tau_m \qquad m = 1, \ldots, M \tag{6}$$

$$x_{mc} \geq 0 \qquad m = 1, \ldots, M, c = 1, \ldots, C.$$

The dual version of the above linear program reads:

$$\max \; \sum_{m=1}^{M} \tau_m \mu_m$$

$$\text{sub } \sum_{c=1}^{C} \lambda_c \leq 1$$

$$r_{mc} \mu_m \leq w_c \lambda_c \qquad m = 1, \ldots, M, c = 1, \ldots, C$$

$$\lambda_c, \mu_m \geq 0 \qquad m = 1, \ldots, M, c = 1, \ldots, C,$$

with λ_c and μ_m representing the dual variables or shadow prices associated with the constraints (5) and (6), respectively.

Since optimality demands $\sum_{c=1}^{C} \lambda_c^* = 1$ and $\mu_m^* = \min\limits_{c=1,\ldots,C} w_c \lambda_c^*/r_{mc}$, the latter variables may be eliminated, and the dual problem may be more succinctly cast as maximizing $V(\lambda_1, \ldots, \lambda_C)$ subject to $\sum_{c=1}^{C} \lambda_c = 1$ and $\lambda_c \geq 0$, $c = 1, \ldots, C$, with $V(\lambda_1, \ldots, \lambda_C) := \sum_{m=1}^{M} \tau_m \min\limits_{c=1,\ldots,C} w_c \lambda_c/r_{mc}$. The latter problem may be interpreted in a similar fashion as above. Each of the users can be allocated resources from each of the transmitters. User m needs to receive throughput τ_m, while each unit of resource allocated by transmitter costs $w_c \lambda_c$, so the cost when transmitter m provides the entire througput required by user m is $w_c \lambda_c \tau_m$. There are two 'opposing' players. Player 1 aims to allocate transmission resources to the users so as to minimize the total cost for given prices λ_c while satisfying the throughput requirements. Player 2 aims to set prices λ_c so as to maximize the total cost incurred by the first player.

The problem (4)–(6) may be solved by a dual-ascent scheme similar to the one that will be described in the next section.

3   Throughput Maximization

We now turn attention to the problem of maximizing a concave function $G(T_1, \ldots, T_M)$ of the user throughputs for given load (resource utilization) constraints $\sigma_1, \ldots, \sigma_C$. This formulation is appropriate when users have elastic traffic demands and the consumption of transmission resources (e.g. power) is constrained by hard limits, but not a crucial criterion otherwise.

$$\max \; G(T_1, \ldots, T_M) \tag{7}$$

$$\text{sub} \; T_m = \sum_{c=1}^{C} r_{mc} x_{mc} \qquad m = 1, \ldots, M \tag{8}$$

$$L_c = \sum_{m=1}^{M} x_{mc} \leq \sigma_c \qquad c = 1, \ldots, C \tag{9}$$

$$x_{mc} \geq 0 \qquad m = 1, \ldots, M, c = 1, \ldots, C.$$

The above formulation is conceptually similar to the multi-path extension of the basic utility maximization problem in [12], i.e., joint routing and rate control, see also for instance [13,14,19,22].

Convex duality implies that the optimal solution to the above problem may be found from the Lagrangian formulation $\min_{\lambda \in \mathbb{R}_+^C} G^*(\lambda)$, with $G^*(\lambda) = \max_{x \in \mathbb{R}_+^{M \times C}} \mathcal{L}(x, \lambda)$,

$$\mathcal{L}(x, \lambda) := G(T_1, \ldots, T_M) + \sum_{c=1}^{C} \lambda_c (\sigma_c - \sum_{m=1}^{M} x_{mc}),$$

and $\lambda_1, \ldots, \lambda_C$ Lagrangian multipliers.

The optimality conditions for the latter formulation read $\frac{\partial G}{\partial T_m} \leq \lambda_c^*/r_{mc}$, with the complementary slackness conditions $x_{mc}^* \left(\frac{\partial G}{\partial T_m} - \lambda_c^*/r_{mc} \right) = 0$ for all $m = 1, \ldots, M$, $c = 1, \ldots, C$, and $\lambda_c^* (\sigma_c - \sum_{m=1}^{M} x_{mc}^*) = 0$ for all $c = 1, \ldots, C$.

It may be checked that there exists an optimal solution of the problem (7)–(9) with at most $M + C - 1$ non-zero variables x_{mc}^*, which means that there will be at most $C - 1$ additional 'legs' beyond the minimum number that is necessary to connect all the M users.

We now focus on the case where the objective function is of the form $G(T_1, \ldots, T_M) = \sum_{m=1}^{M} U_m(T_m)$, with $U_m(\cdot)$ some strictly concave differentiable function. In that case, x_{mc}^* satisfies $U_m'(T_m^*) = r_{mc} \mu_m^*$ for all $c \in \mathcal{C}_m$, $\mathcal{C}_m := \arg \min_{c=1,\ldots,C} \lambda_c^*/r_{mc}$, and $x_{mc}^* = 0$ for all $c \notin \mathcal{C}_m$.

We will specifically consider the family of α-fair utility functions defined by $U_m(T_m) = U^\alpha(T_m) = \frac{T_m^{1-\alpha}}{1-\alpha}$ for some $\alpha > 0$. The parameter α represents a fairness coefficient which characterizes the trade-off between the total throughput and the minimum throughput across all users [16]. In particular, the cases $\alpha \downarrow 0$, $\alpha \to 1$ and $\alpha \to \infty$ correspond to maximum throughput, Propertional Fairness and max-min fairness, respectively. As $\alpha \downarrow 0$, optimality is achieved by simply allocating all the resources of each individual transmitter c to the strongest received user $m_c := \arg\max_{c=1,\ldots,C} r_{mc}$. In contrast, when $\alpha \to \infty$, the problem merits special treatment and will be revisited below.

Algorithm description for problem (7)–(9)

1. Initialize $\lambda = (\lambda_1, \ldots, \lambda_C)$, e.g., $\lambda_c^{(0)} = M/C$ for all $c = 1, \ldots, C$.
2. For given $\lambda = (\lambda_1, \ldots, \lambda_C)$, find resource allocations x_{mc} that maximize $\mathcal{L}(x, \lambda) :=$ $\sum_{m=1}^{M} U_m(T_m) + \sum_{c=1}^{C} \lambda_c(\sigma_c - \sum_{m=1}^{M} x_{mc})$. This amounts to allocating each individual user m resources from the most attractive transmitter $c_m := \arg\min_{c=1,\ldots,C} \lambda_c/r_{mc}$; x_{mc_m} satisfies $U'_m(x_{mc_m} r_{mc_m}) = \lambda_{c_m}/r_{mc_m}$, and $x_{mc} = 0$ for all $c \neq c_m$. In the special case where $U_m(x) = U^\alpha(x) = x^{1-\alpha}/(1-\alpha)$ for some $\alpha > 0$, we obtain $x_{mc_m} = (\lambda_{mc_m}/r_{cm})^{-1/\alpha}/r_{mc_m} = r_{mc_m}^{1/\alpha-1} \lambda_{cm}^{-1/\alpha}$.
3. Let $L_c(\lambda^{(i)})$ be the optimal load at transmitter c for given $\lambda^{(i)} = (\lambda_1^{(i)}, \ldots, \lambda_C^{(i)})$ as determined in step 2. Update $\lambda_c^{(i)}$ as

$$\lambda_c^{(i+1)} := \lambda_c^{(i)} + \varrho_i(L_c(\lambda^{(i)}) - \sigma_c).$$

To guarantee convergence, it is required that $\lim_{i \to \infty} \varrho_i = 0$ and $\sum_{i=0}^{\infty} \varrho_i = \infty$. For example, one may take $\varrho_i = \varrho i^{-1/2+\epsilon}$ for positive constants ϵ, ϱ. To ensure that $\lambda_c^{(i+1)} > 0$ for all $c = 1, \ldots, C$, truncate the update step if needed.
4. Let $x_{mc}(\lambda^{(i)})$ be the optimal amount of resources allocated by transmitter c to user m for given $\lambda^{(i)} = (\lambda_1^{(i)}, \ldots, \lambda_C^{(i)})$ as determined in step 2. Update $x_{mc}^{(i)}$ as

$$x_{mc}^{(i+1)} = (1 - \varsigma_i)x_{mc}^{(i)} + \varsigma_i x_{mc}(\lambda^{(i)}),$$

with $\varsigma_i := \varrho_i / \sum_{j=0}^{i} \varrho_j$.
5. Repeat the above steps until some convergence/stopping criterion is satisfied.

The convergence proof is skipped because of page limitations.

Observe that, in view of the complementary slackness conditions, the optimal shadow price vector $\lambda^* \equiv (\lambda_1^*, \ldots, \lambda_C^*)$ suffices to determine which users should be served by which transmitters. The exact amount of resources allocated to the various users will in practice be governed by local schedulers at each of the transmitters. In that sense step 4 is optional, as it only serves to obtain the optimal resource allocations x_{mc}^*, and plays no role in finding the optimal shadow price vector.

In the case where the objective function is of the form $G(T_1, \ldots, T_M) := \min_{m=1,\ldots,M} T_m$, i.e., the minimum throughput across all users, problem (7)–(9) is equivalent to problem (4)–(6) with $v_m = 1/\tau_m$ and $\sigma_c = 1/w_c$ in the sense that the optimal solutions are related.

4   Dynamic Setting

In the previous section we addressed the problem of maximizing a through-put utility function for a given static user population. While utility maximization provides a useful guiding principle for fair and efficient resource sharing among competing users, the utility function does not necessarily have any physical meaning in terms of actual perceived performance. In particular, the exact numerical value of the utility function or the fact that the aggregate system utility has been maximized may not be of any direct relevance to a data user. What a data user does perceive, is the performance experienced in terms of delays or actual received throughputs for example, and hence we will evaluate the merits of the load balancing schemes in terms of these metrics. In order to do so, we will consider a dynamic setting where users generate random finite-size data transfers over time. For convenience, we assume that users belong to one of K classes, with transmission rates taking values in a discrete set of values, but the results easily extend to scenarios with a continuum of rates. Class-k users arrive as a stationary ergodic process of rate ν_k, and have generally distributed service requirements with mean β_k (bits). Define $\rho :=$ $(\rho_1, \ldots, \rho_K)$, with $\rho_k := \nu_k \beta_k$ the traffic intensity associated with class-k users. Denote by R_{kc} the feasible transmission rate of class-k users when served by transmitter c.

In order for delays and user throughputs to be meaningful, a first prerequisite is that the system is stable. Define the rate region of the system by

$$\mathcal{R} := \{r \in \mathbb{R}_+^K : \exists x \in \mathcal{X} : r_k \leq \sum_{c=1}^{C} x_{kc} R_{kc} \text{ for all } k = 1, \ldots, K\}, \text{ with}$$

$$\mathcal{X} := \{x \in \mathbb{R}_+^{K \times C} : \sum_{k=1}^{K} x_{kc} \leq \sigma_c \text{ for all } c = 1, \ldots, C\} \text{ representing the set of fea-}$$

sible resource allocations. Clearly, $\rho \in \mathcal{R}$ is a necessary condition for the existence of a resource allocation strategy that achieves stability, while $\rho \in \text{interior}(\mathcal{R})$ is a sufficient condition.

We will consider four different scenarios.

(i) The 'greedy' scheme simply assigns users to the strongest received transmitter. Thus, class-k users are statically assigned to transmitter $c_k :=$ $\arg \max_{c=1,\ldots,C} R_{kc}$. Let $\mathcal{K}_c := \{k : c_k = c\}$ be the set of user classes assigned to transmitter c, and define $\bar{\sigma}_c := \sum_{k \in \mathcal{K}_c} \rho_k / R_{kc}$ as the resulting load on transmitter c. It is easily seen that the greedy assignment achieves stability if and only if $\bar{\sigma}_c < \sigma_c$ for all $c = 1, \ldots, C$.

(ii) The 'fractional' α-fair strategy assigns users to transmitters so as to maximize the aggregate α-fair utility. Specifically, suppose that there are N_k class-k users at some point in time. Then the fractional α-fair strategy solves the following optimization problem:

$$\max \sum_{k=1}^{K} \sum_{n=1}^{N_k} U^\alpha(T_{kn})$$

$$\text{sub } T_{kn} = \sum_{c=1}^{C} R_{kc} x_{knc} \qquad n = 1, \ldots, N_k, k = 1, \ldots, K \qquad (10)$$

$$L_c = \sum_{k=1}^{K} \sum_{n=1}^{N_k} x_{knc} \leq \sigma_c \qquad c = 1, \ldots, C \qquad (11)$$

$$x_{knc} \geq 0.$$

It is easily verified that all the class-k users will receive the same throughput, and thus the above problem can alternatively be phrased as maximizing $\sum_{k=1}^{K} U^\alpha(T_k/N_k)$ subject to $(T_1, \ldots, T_K) \in \mathcal{R}$. Since $\mathcal{R}$ is a convex set, it then follows that the fractional α-fair strategy achieves stability for any $\rho \in \text{interior}(\mathcal{R})$, provided $\alpha > 0$ [3].

(iii) The 'integral' α-fair strategy also assigns users to transmitters so as to maximize the aggregate utility, but subject to the additional constraint that users can only be assigned to a single transmitter. It is readily checked that in this case all the class-k users assigned to the same transmitter will receive the same throughput. Also, the above problem can be equivalently stated as maximizing $\sum_{k=1}^{K} U^\alpha(T_k/N_k)$ subject to $(T_1, \ldots, T_K) \in \mathcal{R}(N_1, \ldots, N_K)$. Here $\mathcal{R}(N_1, \ldots, N_K) \subseteq \mathcal{R}$ is some subset with the property that $\lim_{N_1, \ldots, N_K \to \infty} \mathcal{R}(N_1, \ldots, N_K) = \mathcal{R}$. This implies that the integral α-fair strategy also achieves stability for any $\rho \in \text{interior}(\mathcal{R})$, provided $\alpha > 0$ [4].

(iv) The 'ideal' scenario is where the resources of all the transmitters can be pooled into a single transmitter which offers a transmission rate $R_k^{\max} := \max_{c=1,\ldots,C} R_{kc}$ to class-k users. This is a hypothetical scenario in typical propagation conditions, and is only meant to provide an absolute bound on the achievable performance. It is easily seen that stability occurs in the ideal scenario if and only if $\sum_{c=1}^{C} \bar{\sigma}_c < \sum_{c=1}^{C} \sigma_c$.

In conclusion, both the fractional and the integral α-fair strategies achieve stability whenever feasible. The greedy assignment may generally fail to do so, while the stability region for the ideal scenario will typically be strictly larger than $\mathcal{R}$, except in the rather special circumstance that $\bar{\sigma}_c \equiv \bar{\sigma}$ for all $c = 1, \ldots, C$. In that case the stability regions for both the greedy assignment and the ideal scenario coincide with interior$(\mathcal{R})$.

We now compare the performance in terms of delays and perceived throughputs in the various scenarios. Clearly, if $\rho \in \text{interior}(\mathcal{R})$, but $\bar{\sigma}_c > \sigma_c$ for some $c = 1, \ldots, C$, i.e., the loads are imbalanced, but the total load is sustainable, then the delay under the greedy assignment will be infinite, whereas it is finite under both the fractional and integral α-fair schemes. Now consider the case $\bar{\sigma}_c < \sigma_c \equiv 1$ for all $c = 1, \ldots, C$. If we assume Poisson arrivals and fair sharing of resources among competing users, then the mean number of active users under the greedy assignment is $EN^{greedy} = \sum_{c=1}^{C} \frac{\bar{\sigma}_c}{1-\bar{\sigma}_c}$, whereas the mean number of users in the ideal scenario is $EN^{ideal} = \left(\sum_{c=1}^{C} \bar{\sigma}_c \right) / \left(C - \sum_{c=1}^{C} \bar{\sigma}_c \right)$. It is easily verified that for a given value of $\sum_{c=1}^{C} \bar{\sigma}_c$, EN^{greedy} is minimal if $\bar{\sigma}_c \equiv \bar{\sigma} = \sum_{c=1}^{C} \bar{\sigma}_c / C$, and then equal to $C\bar{\sigma}/(1-\bar{\sigma}) = \sum_{c=1}^{C} \bar{\sigma}_c / (1 - \sum_{c=1}^{C} \bar{\sigma}_c/C) = CEN^{ideal}$.

Thus, even in the best-case scenario where the loads are perfectly balanced, the mean number of active users under the greedy assignment is C times as large as in the ideal scenario. Because of Little's law, this implies that the mean delays will be C times as large as well, and thus the throughputs (defined as the ratio of service requirement and delay) will be C times lower. Although the delays for the fractional and integral α-fair strategies are expected to be "somewhere in between", this appears difficult to prove, let alone quantify where exactly they fall relative to the greedy scheme and the ideal scenario. Hence, we will examine the delay performance in the various scenarios in the next section through numerical means.

5   Numerical Experiments

We now discuss the numerical experiments that we conducted to benchmark the performance gains from dynamic load balancing. We consider a dynamic setting where users generate random finite-size data transfers as described in the previous section, and evaluate the performance in terms of transfer delays and blocking rates.

We first examine a linear network with just two transmitters. While admittedly simple, a two-transmitter scenario is likely to provide conservative estimates for the potential gains, since the scope for load balancing increases with the number of neighboring transmitters, as will in fact be confirmed later.

The two transmitters cover an interval $[0, D]$ and are located at positions $D/6$ and $5D/6$, respectively. The path loss q behaves as a function $q = d^{-\gamma}$ of distance d, with $\gamma = 3.5$. The feasible transmission rate r at transmitter c (in bits/second behaves as a function $r_c = \zeta \log(1 + snr_c)$ of the Signal-to-Noise Ratio (SNR), with $snr_c = q_c/(\eta + \theta q_c + q_{3-c})$, $c = 1, 2$, with ζ, θ, η system-specific parameters. Throughout we take $\eta = 0.01$, $\theta = 0.1$, $\zeta = 800$.

Users arrive as a Poisson process of rate ν (per second), and have service requirements with mean β (in bits). Throughout we take $\beta = 250$ Kbits (31.25 Kbytes). At most 40 users are admitted into the system simultaneously. Users that generate a transfer request when there are already 40 transfers in progress, are blocked and lost. Let (R_1, R_2) be the rate pair of an arbitrary user. The nominal average load on transmitter c under the greedy assignment may then be derived as

$$\nu\beta\mathbb{E}\{\frac{1}{R_c}\mathbb{I}_{\{R_c>R_{3-c}\}}\} = \nu\beta\mathbb{E}\{\frac{1}{\max\{R_1,R_2\}}\}\mathbb{P}\{R_c > R_{3-c}\}.$$

We compare the four scenarios described in the previous section as well as the dual-ascent scheme. We take $U(x) = \log(x)$, i.e., $\alpha \to 1$, which corresponds to Proportional Fair (PF) scheduling. In the dual-ascent scheme, we only executed 30 iterations for every change in the user population, with $\varrho_i = 0.5/\sqrt{i}$.

We first consider a scenario where the user locations are uniformly distributed across the coverage area. In this case, the nominal load on each of the two transmitters is $\rho/2$, with $\rho := \nu\beta\mathbb{E}\{\frac{1}{\max\{R_1,R_2\}}\}$. Since the long-term loads are perfectly balanced, this provides a lower bound for the potential gains from load balancing.

Figure 1 shows the mean transfer delay as function of the arrival rate. Note that the delay in the ideal scenario is roughly half of that under the greedy assignment, as indicated by the analysis in the previous section. At high load, the relative difference reduces though. This may be explained from the fact that a significant fraction of the users are blocked under the greedy assignment (not shown in the figure), effectively reducing the load on the system, while the blocking in the ideal scenario remains negligible throughout. Also recall here that the latter scenario is entirely hypothetical and only serves to provide an absolute performance bound, as it assumes that users can be offered the same transmission rate by all transmitters, which will be far from the case under power-law propagation conditions. Further observe that the performance of the dual-ascent scheme is virtually indistinguishable from that of the globally optimal integral or fractional PF allocation. Given the small number of iterations, this indirectly demonstrates that the dual-ascent scheme converges rapidly enough to achieve similar performance as a globally optimal allocation. It also suggests that the dynamic load balancing is hardly hindered by refraining from soft hand-off and assigning users to just a single transmitter. The mean transfer delay in each of these three cases is approximately 15–25% lower than under the greedy assignment, even though the long-term loads are perfectly balanced.

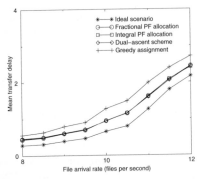
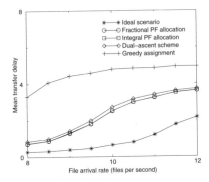

Fig. 1. Mean transfer delay as function of arrival rate; uniform traffic

Fig. 2. Mean transfer delay as function of arrival rate; non-uniform traffic

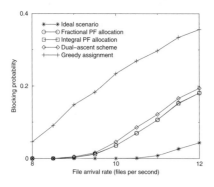
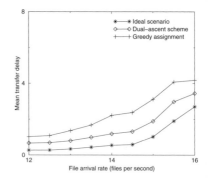

Fig. 3. Blocking probability as function of arrival rate; non-uniform traffic

Fig. 4. Mean transfer delay as function of arrival rate; uniform traffic

We now look at a scenario where the density of users is three times higher in one half of the coverage area than the other. In this case, the nominal loads on the two transmitters are $\rho/4$ and $3\rho/4$, respectively.

Figure 2 plots the mean transfer delay as function of the arrival rate. As before, the performance of the dual-ascent scheme is practically identical to that of the globally optimal integral or fractional PF allocation. The reduction in mean transfer delay in each of these three cases varies from 30% to 80%, which means that the improvement in perceived user throughput can be as large as 500%. Further note that at high load the relative improvement diminishes. This reflects the fact that a substantial fraction of the users are blocked under the greedy assignment, as shown in Figure 3, essentially lowering the load on the system, while the blocking with load balancing remains moderate. Observe that even with load balancing the system will become overloaded at some point, but load balancing helps to move that point significantly further out.

We now investigate a network with a square coverage area $[0, D] \times [0, D]$, with four transmitters located at positions $(D/6, D/6)$, $(D/6, 5D/6)$, $(5D/6, D/6)$, $(5D/6, 5D/6)$. In this case, the SNR at transmitter c is determined by $snr_c = q_c/(\eta + \sum_{d=1}^{4} q_d - (1 - \theta)q_c)$, with q_d the path loss to transmitter d governed by a negative power-law as function of distance with exponent $\gamma = 3.5$ as specified earlier. At most 80 users are admitted into the system simultaneously. We only present the results for the dual-ascent scheme, and not for the globally optimal integral and fractional PF allocation.

We first consider a scenario where the user locations are uniformly distributed across the coverage area. In this case, the nominal load on each of the four transmitters is $\rho/4$, with $\rho := \nu\beta\mathbb{E}\{\frac{1}{\max\{R_1, R_2, R_3, R_4\}}\}$.

Figure 4 shows the mean transfer delay as function of the arrival rate. Note that the delay in the ideal scenario is now roughly a quarter of that under the greedy assignment, as predicted by the analysis in the previous section. At high load, the relative difference decreases again because a significant fraction of the users are blocked under the greedy assignment, effectively shedding load from the system, while the blocking in the ideal scenario remains negligible throughout.

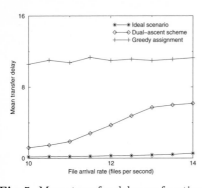 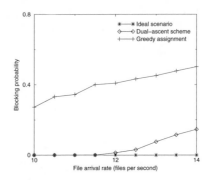

Fig. 5. Mean transfer delay as function of arrival rate; non-uniform traffic

Fig. 6. Blocking probability as function of arrival rate; non-uniform traffic

The mean transfer delay for the dual-ascent scheme is now 30–50% lower than under the greedy assignment, which corroborates the earlier assertion that the gains from load balancing tend to increase with the number of transmitters.

We now look at a scenario where the density of users is three times higher in one square corner of the coverage area than elsewhere. In this case, the nominal load on one transmitter is $\rho/2$ and $\rho/6$ on the other three.

Figure 5 plots the mean transfer delay as function of the arrival rate. The reduction in mean transfer delay achieved by the dual-ascent scheme ranges from 50% to 90%, which means that the improvement in perceived user throughput can be as large as 1000%. Further note that at high load the relative improvement diminishes again because a substantial fraction of the users are blocked under the greedy assignment, as shown in Figure 6.

6 Conclusion

In the current generation of wireless networks, base stations or access points either statically or independently allocate resources such as frequencies or time slots to users within their coverage areas. Because wireless networks typically cater to mobiles, it is common for these networks to experience significant spatio-temporal load fluctuations. Coordinated action during these typical load surges, it has now been shown, significantly improves throughput and performance. Co-ordinated action via centralized resource controllers is not scalable for emerging micro- and pico-cellular structures, however. In this paper we constructed a utility optimization framework for decentralized network-wide resource control. The basic approach involves a decentralized mechanism consisting of two steps: 1) propagation of candidate "shadow prices" by the base stations (e.g. via pilot signals) and 2) asynchronous updating of these prices via bids from mobiles (e.g. tentative association of the mobiles with their "least-cost suppliers") through power-control type of iterative updates. We show through numerical experiments that the proposed schemes yield significant performance gains in terms of throughput and/or delays, even in scenarios where the long-term loads in the cell's coverage areas are completely balanced.

References

1. Andrews, D.M., Qian, L., Stolyar, A.L.: Optimal utility-based multi-user throughput allocation subject to throughput constraints. In: Proc. Infocom 2005 (2005)
2. Bejerano, Y., Han, S.-J., Li, L.: Fairness and load balancing in wireless LAN's using association control. In: Proc. ACM MobiCom 2004, pp. 315–329 (2004)
3. Bonald, T., Massoulié, L., Proutière, A., Virtamo, J.: A queueing analysis of max-min fairness, proportional fairness and balanced fairness. Queueing Systems 53, 65–84 (2006)
4. Borst, S.C., Jonckheere, M.: Flow-level stability of channel-aware scheduling algorithms. In: Proc. WiOpt '06 Conf (2006)
5. Chaponniere, E.F., Black, P.J., Holtzman, J.M., Tse, D.N.C.: Transmitter directed code division multiple access system using path diversity to equitably maximize throughput. US Patent 6, 449–490 (2002)
6. Borst, S.C., Hampel, G., Saniee, I., Whiting, P.A.: Load balancing in cellular wireless networks. In: Resende, M.G.C., Pardalos, P.M. (eds.) Handbook of Optimization in Telecommunication, Springer, Heidelberg (2006)
7. Bu, T., Li, L., Ramjee, R.: Generalized Proportional Fair scheduling in third-generation wireless networks. In: Proc. Infocom 2006 (2006)
8. Das, S., Viswanathan, H., Rittenhouse, G.: Dynamic load balancing through coordinated scheduling in packet data systems. In: Proc. Infocom 2003 (2003)
9. Eryilmaz, E., Srikant, R.: Fair resource allocation in wireless networks using queue length based scheduling and congestion control. In: Proc. Infocom 2005 (2005)
10. Han, H., Shakkottai, S., Hollot, C.V., Srikant, R., Towsley, D.: Multi-path TCP: a joint congestion control and routing scheme to exploit path diversity in the Internet. IEEE/ACM Trans. Netw. 14, 1260–1271 (2006)
11. Hurwicz, L., Arrow, K., Uzawa, H.: Studies in Linear and Non-Linear Programming. Stanford University Press (1958)
12. Kelly, F.P., Maulloo, A., Tan, D.: Rate control in communication networks: shadow prices, proportional fairness, and stability. J. Oper. Res. Soc. 49, 237–252 (1998)
13. Kelly, F.P., Voice, T.: Stability of end-to-end algorithms for joint routing and rate control. Comp. Commun. Rev. 35, 5–12 (2005)
14. Lin, X., Shroff, N.B.: Utility maximization for communication networks with multi-path routing. IEEE Trans. Aut. Control 51, 766–781 (2003)
15. Liu, X., Chong, E.K.P., Shroff, N.B.: A framework for opportunistic scheduling in wireless networks. Comp. Netw. 41, 451–474 (2003)
16. Mo, J., Walrand, J.: Fair end-to-end window-based congestion control. IEEE/ACM Trans. Netw. 8, 556–567 (2000)
17. Neely, M.J., Modiano, E., Li, C.: Fairness and optimal stochastic control for heterogeneous networks. In: Proc. Infocom 2005 (2005)
18. Sang, A., Wang, X., Madihian, M., Gitlin, R.D.: Coordinated load balancing / cell-site selection and scheduling in multi-cell packet data systems. In: Proc. ACM Mobicom 2004 (2004)
19. Srinivasan, V., Chiasserini, C., Nuggehalli, P., Rao, R.: Optimal rate allocation for energy-efficient multi-path routing in wireless ad hoc networks. IEEE Trans. Wireless Commun. 3, 891–899 (2005)
20. Stolyar, A.L.: Maximizing queueing network utility subject to stability: greedy primal-dual algorithm. Queueing Systems 50, 401–457 (2005)
21. Stolyar, A.L.: On the asymptotic optimality of the gradient scheduling algorithm for multi-user throughput allocation. Oper. Res. 53, 12–25 (2005)
22. Wang, W.H., Palaniswami, M., Low, S.H.: Optimal flow control and routing in multi-path networks. Perf. Eval. 52, 119–132 (2002)

Algorithms for Designing WDM Networks and Grooming the Traffic

Abderraouf Bahri and Steven Chamberland

Department of Computer Engineering, École Polytechnique de Montréal,
C.P. 6079, Succ. Centre-Ville, Montréal (Québec), Canada H3C 3A7
{abderraouf.bahri,steven.chamberland}@polymtl.ca

Abstract. In this paper, we propose a model for the global design problem of wavelength division multiplexing (WDM) networks including the traffic grooming. This problem consists in finding the number of fibers between each pair of node, selecting the configuration of each node, choosing the set of lightpaths (i.e., the virtual topology), routing these lightpaths over the physical topology and finally, grooming and routing the traffic over the lightpaths. Since this problem is NP-hard, we propose heuristic algorithms and a tabu search metaheuristic algorithm to find good solutions for real-size instances rapidly.

Keywords: Wavelength division multiplexing (WDM) networks, network design, traffic grooming, virtual and physical topologies, tabu search.

1 Introduction

The wavelength division multiplexing (WDM) is widely deployed over different optical networks (e.g., synchronous optical network (SONET)) to exploit the important optical fiber bandwidth. It allows the use of multiple wavelengths over a single fiber. This leads to a huge bandwidth available in each fiber but, nowadays, a connection between two network ports has rarely more than the gigabit per second. Therefore, this fiber bandwidth is outsized compare to the requests. This under exploitation may lead to a premature saturation of the network and to a loss of the provider incomes.

A solution is the traffic grooming which is a traffic engineering technique that allows grouping several requests within a lightpath. A lightpath is a path that transits traffic form a source to a destination node, passing possibly by several intermediate nodes, while remaining in the optical domain.

In this paper, we address the global problem of designing WDM networks including the traffic grooming. It consists in finding the number of fibers between each pair of node, selecting the configuration of each node (in terms of the number of transponders), choosing the set of lightpaths (i.e., the set of lightpaths), routing these lightpaths over the physical topology and finally, grooming and routing the traffic over the lightpaths.

The literature contains many articles relating to traffic grooming [1, 4, 7, 8, 9, 10, 11] but the global problem defined here has not been considered before.

L. Mason, T. Drwiega, and J. Yan (Eds.): ITC 2007, LNCS 4516, pp. 1038–1047, 2007.
© Springer-Verlag Berlin Heidelberg 2007

Moreover, in this paper we provide a general framework considering the use of many optical fibers per link, many wavelengths per fiber and many grooming granularities.

This paper is organized as follows. Section 2 presents a mathematical model for the design problem and Section 3 presents two heuristics and a tabu search metaheurstic algorithm. Numerical results are presented in Section 4. Conclusions follow in Section 5.

2   The Global WDM Network Design Problem

2.1   Problem Formulation

For the global WDM network design problem, the following information is considered known: (I1) the location of the optical nodes; (I2) the origin-destination connection demand (i.e., the number of OC-3, OC-12, etc.) between each pair of nodes; (I3) the maximum number of optical fibers that can be installed between each pair of nodes; (I4) the maximum number of wavelengths that can be used into a fiber; (I5) the cost of the links between each pair of nodes as a function of the number of fibers including the installation cost (for the patch panels, the patch cords, the labor, etc.); (I6) the cost the transponders including the installation cost.

We also make the following assumptions concerning the organization of the network: (A1) a link can be installed between each pair of nodes; (A2) a link is composed of one or more fibers; (A3) the number of fibers installed between two nodes cannot exceed the maximum allowed; (A4) an optical fiber can contain multiple wavelengths; (A5) the number of wavelengths used into a fiber cannot exceed the maximum allowed; (A6) each node have full grooming facilities; (A7) an optical cross-connect is installed in each node; (A8) there is no wavelength conversion in the network; (A9) a connection can use more than one lightpath (multi-hop routing); (A10) a connection stay packed during its transport.

The global WDM network design problem consists in finding the number of fibers between each pair of nodes, the number of transponders at each node, the virtual topology as well as routing the lightpaths over the physical topology, grooming the traffic and finally, routing the traffic over the virtual topology. The objective is to minimize the cost of the network.

There is a strong interaction between these subproblems and several of them are NP-hard (e.g., the traffic grooming [6] and the wavelength assignment subproblems [2]).

2.2   Notation

Sets: Let N be the set of nodes, T, the set of connection types (e.g., OC-3 and OC-12), and finally, Ω, the set of wavelengths that can be used into each fiber.

Decision Variables: Let w_{mn} be the number of lightpaths between nodes $m \in N$ and $n \in N$, x_{ij}, a 0–1 variable such that $x_{ij} = 1$ if and only if node $i \in N$ is connected to node $j \in N$, x_{ij}^f, a 0–1 variable such that $x_{ij}^f = 1$ if and only if the fiber f is used from node $i \in N$ to node $j \in N$, $x_{ij}^{f\omega}$, a 0–1 variable such that $x_{ij}^{f\omega} = 1$ if and only if wavelength $\omega \in \Omega$ is used into fiber f from node $i \in N$ to node $j \in N$, and $x_{ijmn}^{f\omega}$, a 0–1 variable such that $x_{ijmn}^{f\omega} = 1$ if and only if wavelength $\omega \in \Omega$ is used on fiber f from node $i \in N$ to node $j \in N$ for a lightpath from node $m \in N$ to $n \in N$. Let y_{mn}^{odkt} be a 0–1 variable such that $y_{mn}^{odkt} = 1$ if and only if the connection k of type $t \in T$ from the origin $o \in N$ to the destination $d \in N$ uses a lightpath between the nodes $m \in N$ and $n \in N$, and finally, z^{odkt}, a 0–1 variable such that $z^{odkt} = 1$ if and only if the connection k of type $t \in T$ from the origin $o \in N$ to the destination $d \in N$ is established.

Constants: Let α^{odt} be the number of connections of type $t \in T$ requested from $o \in N$ to $d \in N$, β_{ij}, the maximum of fibers that can be installed between nodes $i \in N$ and $j \in N$, δ^t, the capacity needed for a connection of type $t \in T$ (i.e., the number of OC-1 for a connection of type $t \in T$), and finally, ϵ, the capacity of a wavelength (i.e., the number of OC-1 that a wavelength can carry).

Cost Parameters: Let $a_{ij}(n)$ be the link cost for n fibers between nodes $i \in N$ and $j \in N$ including the installation cost and b_i, the transponder cost including the installation cost at node $i \in N$.

2.3   The Mathematical Model

The model for the global WDM network design problem, denoted GWNDP, is presented in Appendix A.

3   Heuristics and Tabu Search

In this section, we propose two heuristics and a tabu search metaheuristic algorithm for GWNDP.

3.1   The Heuristics H1 and H2

The main steps of the heuristics (denoted H1 and H2) are: (1) construct a physical topology; (2) construct a virtual topology; (3) route the requests in single hop over the virtual topology and (4) route the remaining requests in multi-hop.

The heuristic H1, partially inspired by [10], takes into account the traffic between the pair of nodes to built the physical and the logical topologies. The heuristic H2 builds a lightpath on each wavelength into each fiber.

Let *floor* be the minimum value of a traffic to be exchanged to set up a fiber between a pair of nodes, *volume*(k), the volume of the requests of the node pair k and *LP_list*, the list of the lightpaths containing the remaining capacity.

In the heuristics H1 and H2, we also use the function *sort()* used for sorting the node pairs k in the decreasing order of the volume of remaining requests.

Heuristic H1

Step 1: (Physical topology construction)

For all node pairs k from i_k to j_k do

1.1 For $n := \beta_{i_k j_k}$ to 1 do

If $volume(k) \geq n \times floor$, build n fibers between the nodes i_k and j_k and go to 2.1.

Step 2: (Virtual topology construction)

2.1 Set $LP\_list := \emptyset$, call the function $sort()$ and set $k := 0$.

2.2 While $volume(k) > 0$ and $k < |N|(|N| - 1)$ do

2.2.1 If the links on shortest path found using Dijkstra (at the physical level) from i_k and j_k have enough resources to create a lightpath do

2.2.1.1 Set up a lightpath from i_k to j_k, update $volume(k) := (volume(k) - \epsilon)^+$, set $k := 0$, call the function $sort()$ and update the LP_list.

Otherwise set $k := k + 1$.

Step 3: (Traffic routing over the virtual topology — single hop)

3.1 Call the function $sort()$.

3.2 For all the lightpaths

3.2.1 For $t := |T|$ to 1 do

If a request of type t exists between the source and the destination of the lightpath and its remaining capacity is sufficient, route this request over the lightpath, update LP_list, call the function $sort()$ and set $t := |T|$.

Step 4: (Traffic routing over the virtual topology — multi-hop)

4.1 Set $k := 0$.

4.2 While $k < |N|(|N| - 1)$ do

4.2.1 If $volume(k) > 0$ and no shortest path is found using Dijkstra (at the lightpath level), set $k := k + 1$ and go to 4.2.

4.2.2 If $volume(k) > 0$ and a shortest path is found with Dijkstra (at the lightpath level) do

4.2.2.1 For $t := |T|$ to 1 do

If a request of type t exists between this node pair and the remaining capacity is sufficient, route this request over the virtual path, update LP_list and set $k := 0$ and $t := 0$.

4.2.2.2 If no request is routed between this node pair, set $k := k + 1$ and go to 4.2.

Note that a shortest path found in Step 4 is defined as a set of lightpaths between the source and the destination. If more than one lightpath exits between a source and a destination, only the lightpath containing the largest remaining capacity is used in the adjacency table of Dijkstra.

For the heuristic H2, we replace the Step 2 of H1 by the Step 2' presented below.

Heuristic H2

Step 2': (Virtual topology construction)

2.1 For all the fibers (i, j) and wavelengths on this fiber, create one lightpath from i to j or from j to i.

3.2   The Tabu Search Algorithm

In this section, we propose a tabu search (TS) algorithm for GWNDP, denoted TS. The basic principle of tabu search is to define a set of possible solutions, and starting from the current solution, to find a better one in its neighbourhood. A neighbourhood is a set of solutions that are found by applying an appropriate transformation of the current solution. In order for the algorithm to move away from a local minimum, the search allows moves resulting in a degradation of the objective function value, thus avoiding the trap of local optimality. To prevent the search from cycling, solutions obtained recently and moves that reverse the effect of recent moves are considered tabu (for more details, see [3]).

In this paper, a solution is obtained with the heuristic H1 or H2 for a fixed physical topology. The neighbourhood of a current solution is a set of solutions for the physical topologies obtained by adding or removing a single fiber. Thus, a movement is a single fiber add or drop. The initial solution is found using the heuristic H1 or H2 and the search stops after a fixed number of iterations.

The cost of a solution is evaluated using the equation (A.1) plus a penalty factor for the blocked requests. Thus, the cost function is

$$\sum_{\substack{i \in N}} \sum_{\substack{j \in N \\ i \neq j}} a_{ij} \left(\sum_{f=1}^{\beta_{ij}} x_{ij}^f \right) + \sum_{m \in N} \sum_{\substack{n \in N \\ m \neq n}} (b_m + b_n) \, w_{mn} + \gamma \sum_{o \in N} \sum_{\substack{d \in N \\ o \neq d}} \frac{a_{od}(1) \theta^{odt} \delta^t}{\epsilon |F_{od}|} \quad (1)$$

where θ^{odt} is the number of blocked requests of type t from node o to d and γ a penalty factor.

4   Numerical Results

All algorithms are programmed in the C language on a Sun Java workstation under Linux with a AMD Opteron 150 CPU and 2 GB of RAM.

Table 1 presents the cost of the optical network components. Note that for each lightpath, two transponders are needed. Table 2 presents the (x, y) coordinates of the nodes and Table 3 presents the technological parameters. Full grooming is considered and the capacity of each wavelength was set to OC-48.

In this paper, the notation 4/2/0.1 is related to the traffic profile and it means that between two nodes there is a connection demand of up to four OC-3, up to two OC-12 and a probability of 0.1 to have a demand for an OC-48. Four traffic profile scenarios are considered and presented in Table 4.

For the heuristics H1 and H2, the parameter γ was set to 100 and *floor* to zero. It means that the tabu search begin with a complete graph as a physical topology. For the tabu search algorithm, the best move at each iteration of the search is considered tabu for a number of iterations found randomly in the interval [1,5]. The number of iterations of the search was set to 100.

Table 1. Cost of the network components

Component	Cost
One-fiber link	3 750$/km
Two-fiber link	6 250$/km
Multi-rate transponder	20 000$

Table 2. Coordinates of the optical nodes

Node	x (km)	y (km)
1	0	600
2	740	840
3	2080	620
4	3500	0
5	3800	180
6	3960	200

Table 3. Technological parameters

Parameter	Value
$|F_{ij}|$	2
$|\Omega|$	2

Table 4. Demand scenarios

Scenario	Traffic profile
1	2/2/0.1
2	4/2/0.1
3	4/4/0.1
4	4/4/0.2

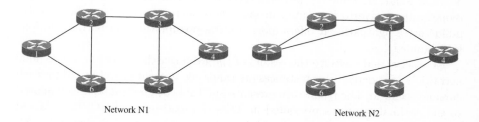

Network N1   Network N2

Fig. 1. Networks of the tests

First, the performance of heuristics H1 and H2 is tested with a fixed physical topology. Figure 1 presents the network tests N1 and N2. For each link, two fibers are installed. The results of the heuristics are presented in Table 5.

Table 5. Performance of the heuristics

Network	Scenario	Blocked requests with H1	Blocked requests with H2
N1 (8 links)	1 (138 OC-3)	0.0% (18 LPs)	11.6% (32 LPs)
	2 (189 OC-3)	9.0% (20 LPs)	11.6% (32 LPs)
	3 (231 OC-3)	10.8% (20 LPs)	14.7% (32 LPs)
	4 (321 OC-3)	25.8% (19 LPs)	21.2% (32 LPs)
N2 (7 links)	1 (138 OC-3)	4.3% (16 LPs)	14.5% (28 LPs)
	2 (189 OC-3)	9.0% (17 LPs)	11.6% (28 LPs)
	3 (231 OC-3)	14.3% (18 LPs)	13.4% (28 LPs)
	4 (321 OC-3)	35.5% (15 LPs)	27.4% (28 LPs)

Table 6. Average results from the tabu search combined with H1 and H2

Scenario	TS-H1 cost (k$)	TS-H2 cost (k$)	TS-H1 GAP	TS-H1 GAP
1	40 416	35 774	33.7%	18.3%
2	40 930	38 415	20.0%	12.7%
3	52 775	54 410	4.2%	7.4%
4	65 898	60 602	13.8%	4.7%

Table 7. Average results from CPLEX

Scenario	Cost(k$)
1	30 228
2	34 089
3	50 643
4	57 886

It can be observed that heuristic H2 provides better results when the volume of traffic is important. This can be explained by the fact that when the volume of traffic is larger, while the physical topology remains fixed, the number of the requests that have to be routed in multi-hop, increases. However, because H2 builds more lightpaths, it offers more routing possibilities and better solutions are obtained.

To observe and measure the results of the tabu search algorithm, a five-node network is considered. Five instances are randomly generated and tested for each demand scenario. The results are presented in Table 6. The results are compared to the optimal solutions presented in Table 7 found using the CPLEX Mixed Integer Optimizer 9.0 [5] to solve the mathematical model (see Appendix A). Note that the resolution time with CPLEX is large (the resolution time is 178 901 seconds for the last scenario). That is why we considered only a five-node test network.

As can be gathered from the tables, the tabu search combined with H2 performs in general better than when it is combined with H1. Also, a tendency can be observed concerning the relation between the GAP and the traffic growth for

TS-H1 and TS-H2. In fact, with CPLEX, the average proportion of multi-hop routed requests is 49.5%, 48.4%, 29.2% and 23.0% for respectively scenario 1, 2, 3 and 4. H1 and H2 perform better in the latest scenarios since both algorithms advantage single-hop routed requests.

5   Conclusions

In this paper, we have studied the global problem of designing WDM networks including the traffic grooming. We have presented a mathematical model, two heuristics, a local search and tabu search algorithms for this problem. The objective is to find the minimum cost WDM network for real size instances. The results have shown the importance of the physical topology selection.

Several research avenues are open at this point. First, we currently explore the use of alternate paths to improve the quality of the heuristics. Another research avenue is to exploit several path selections to route a request with different metrics and search different rules to set the processing order of the requests.

Acknowledgements

The completion of this research was made possible thanks to Bell Canada's support through its Bell University Laboratories R&D program.

References

1. Bouabdallah, N., Pujolle, G., Dotaro, E., Le Sauze, N., Ciavaglia, L.: Distributed Aggregation in All-Optical Wavelength Routed Networks. IEEE Conference on Communications, pp. 1806–1810 (2004)
2. Chamberland, S., Khyda, D.O., Pierre, S.: Joint Routing and Wavelength Assignment in WDM Networks for Permanent and Reliable Paths. Computers and Operations Research 32(5), 1073–1087 (2005)
3. Glover, F., Laguna, M.: Tabu Search. Kluwer Academic Publishers, Dordrecht (1997)
4. Hu, J.Q., Leida, B.: Traffic Grooming, Routing and Wavelength assignment in Optical WDM Mesh Network. IEEE International Conference on Computer Communications, pp. 495–501 (2004)
5. ILOG, Inc., Using the CPLEX Callable Library and CPLEX Mixed Integer Library, ILOG, Inc., (2000)
6. Modiano, E., Lin, P.J.: Traffic Grooming in WDM Networks. IEEE Communications Magazine 36(7), 127–129 (2001)
7. Prathombutr, P., Stach, J., Park, E.K.: An Algorithm for traffic grooming in WDM optical mesh networks with multiple objectives. IEEE Conference on Computer Communications and Networks, pp. 405–411 (2003)
8. Xiang, B., Yu, H., Wang, S., Li, L.: QoS-specified Traffic Grooming Algorithm in WDM Mesh Networks. IEEE Conference on Communications, Circuits and systems, pp. 633–637 (2004)

9. Zhu, K., Mukherjee, B.: A Review of Traffic Grooming in WDM Optical Networks: Architectures and Challenges. Optical Networks Magazine 4(2), 122–133 (2002)
10. Zhu, K., Mukherjee, B.: Traffic Grooming in an Optical WDM Mesh Network. IEEE Communications Magazine 20(1), 55–64 (2003)
11. Zhu, K., Mukherjee, B.: A Novel Generic Graph Model for Traffic Grooming in heterogeneous WDM Mesh Networks. IEEE/ACM Transactions on Networking 11(2), 285–299 (2003)

A   Model Formulation

The model for the global WDM network design problem, is given below.

$$\min \sum_{i \in N} \sum_{\substack{j \in N \\ i \neq j}} a_{ij} \left(\sum_{f=1}^{\beta_{ij}} x_{ij}^f \right) + \sum_{m \in N} \sum_{\substack{n \in N \\ m \neq n}} (b_m + b_n) \, w_{mn} \tag{A.1}$$

subject to
Physical topological constraints

$$\sum_{m \in N} \sum_{\substack{n \in N \\ m \neq n}} x_{ijmn}^{f\omega} \leq x_{ij}^{f\omega} \quad \forall_{i,j \in N (i \neq j),\ f \in F_{ij},\ \omega \in \Omega} \tag{A.2}$$

where $F_{ij} = \{1, \ldots, \beta_{ij}\}$.

$$x_{ij}^{f\omega} \leq x_{ij}^f \quad \forall_{i,j \in N (i \neq j),\ f \in F_{ij},\ \omega \in \Omega} \tag{A.3}$$

$$x_{ij}^f \leq x_{ij} \quad \forall_{i,j \in N (i \neq j),\ f \in F_{ij}} \tag{A.4}$$

$$\sum_{m \in N} \sum_{\substack{n \in N \\ m \neq n}} \sum_{\omega \in \Omega} x_{ijmn}^{f\omega} \leq |\Omega| \quad \forall_{i,j \in N (i \neq j),\ f \in F_{ij}} \tag{A.5}$$

$$\boxed{\text{Additional topological constraints}} \tag{A.6}$$

Virtual topology over physical topology constraints

$$\sum_{\ell \in N \setminus \{m\}} \sum_{f \in F_{m\ell}} \sum_{\omega \in \Omega} x_{m\ell mn}^{f\omega} = w_{mn} \quad \forall_{m,n \in N (m \neq n)} \tag{A.7}$$

$$\sum_{\ell \in N \setminus \{n\}} \sum_{f \in F_{\ell n}} \sum_{\omega \in \Omega} x_{\ell nmn}^{f\omega} = w_{mn} \quad \forall_{m,n \in N (m \neq n)} \tag{A.8}$$

$$\sum_{\substack{i \in N \\ i \neq j}} \sum_{f \in F_{ij}} x_{ijmn}^{f\omega} = \sum_{\substack{\ell \in N \\ j \neq \ell}} \sum_{f \in F_{j\ell}} x_{j\ell mn}^{f\omega} \quad \forall_{j \in N,\ m,n \in N \setminus \{j\} (m \neq n),\ \omega \in \Omega} \tag{A.9}$$

$$\sum_{\ell \in N \setminus \{m\}} \sum_{f \in F_{\ell m}} \sum_{\omega \in \Omega} x_{\ell mmn}^{f\omega} = 0 \quad \forall_{m,n \in N (m \neq n)} \tag{A.10}$$

$$\sum_{\ell \in N \setminus \{n\}} \sum_{f \in F_{n\ell}} \sum_{\omega \in \Omega} x_{n\ell mn}^{f\omega} = 0 \quad \forall m,n \in N(m \neq n) \tag{A.11}$$

Traffic constraints

$$\sum_{m \in N \setminus \{d\}} y_{md}^{odkt} = z^{odkt} \quad \forall o,d \in N(o \neq d), t \in T, k \in K^{odt} \tag{A.12}$$

where $k \in K^{odt} = \{1,\dots,\alpha^{odt}\}$.

$$\sum_{n \in N \setminus \{o\}} y_{on}^{odkt} = z^{odkt} \quad \forall o,d \in N(o \neq d), t \in T, k \in K^{odt} \tag{A.13}$$

$$\sum_{m \in N \setminus \{\ell\}} y_{m\ell}^{odkt} = \sum_{n \in N \setminus \{\ell\}} y_{\ell n}^{odkt} \quad \forall \ell \in N, o,d \in N \setminus \{\ell\}(o \neq d), t \in T, k \in K^{odt} \tag{A.14}$$

$$\sum_{m \in N} y_{mo}^{odkt} = 0 \quad \forall o,d \in N(o \neq d), t \in T, k \in K^{odt} \tag{A.15}$$

$$\sum_{n \in N} y_{dn}^{odkt} = 0 \quad \forall o,d \in N(o \neq d), t \in T, k \in K^{odt} \tag{A.16}$$

$$\sum_{k=1}^{\alpha^{odt}} z^{odkt} = \alpha^{odt} \quad \forall o,d \in N(o \neq d), t \in T \tag{A.17}$$

$$\sum_{\substack{o \in N \\ }} \sum_{\substack{d \in N \\ o \neq d}} \sum_{k=1}^{\alpha^{odt}} \sum_{t \in T} \delta^t y_{mn}^{odkt} \leq \epsilon w_{mn} \quad \forall m,n \in N(m \neq n) \tag{A.18}$$

$$w_{mn} \in \mathbb{N},\ x_{ij} \in \mathsf{B},\ x_{ij}^f \in \mathsf{B},\ x_{ij}^{f\omega} \in \mathsf{B},\ x_{ijmn}^{f\omega} \in \mathsf{B}, y_{mn}^{odkt} \in \mathsf{B},\ z^{odkt} \in \mathsf{B}. \tag{A.19}$$

The objective function (A.1) represents the cost of the links and of the transponders installed in the network. Constraints (A.2) require that a wavelength is used only if a lightpath uses this wavelength and constraints (A.3) impose to use the fibre $f \in F_{ij}$ from i to j only if a wavelength is used into this fibre. Constraints (A.4) impose to install a link between nodes i and j only if a fibre is used between these nodes and constraints (A.5) ensure the number of wavelengths used into a fibre be less or equal to the maximum number of wavelength allowed. Constraints (A.6) include additional topological constraints defined by the network planner.

Constraints (A.7) necessitate that the number of lightpaths starting from a node be equal to the number of lightpaths having this node as a source and constraints (A.8) require the number of lightpaths terminating at a node be equal to the number of lightpaths having this node as a destination. Constraints (A.9) dictate that a lightpath passing through an intermediate node uses only one wavelength, constraints (A.10) impose that a lightpath cannot enter to the source node of this lightpath and constraints (A.11) require that a lightpath cannot leave the destination node of this lightpath.

Constraints (A.12) to (A.18) are traffic constraints and constraints (A.19) are integrality constraints where $\mathsf{B} = \{0,1\}$.

Performance Analysis of Traffic-Groomed Optical Networks Employing Alternate Routing Techniques

Nicki Washington[1] and Harry Perros[2]

[1] Systems and Computer Science Department Howard University
Washington, DC
a_n_washington@howard.edu
[2] Computer Science Department North Carolina State University
Raleigh, NC
hp@csc.ncsu.edu

Abstract. Recent advances in telecommunication networks have allowed WDM to emerge as a viable solution to the ever-increasing demands of the Internet. In a wavelength-routed optical network, traffic is transported over lightpaths, which exclusively occupy an entire wavelength on each hop of the source-destination path. Because these networks carry large amounts of traffic, alternate routing methods are designed in order to allow traffic to be properly re-routed from source to destination in the event of certain events, such as link blocking or failure. In this paper, we consider a tandem traffic-groomed optical network, modeled as a multi-level overflow system, where each level represents a wavelength between adjacent nodes. The queueing network is analyzed using a combination of methods. As will be shown, the decomposition method provides a good approximate analysis of large overflow systems supporting traffic from multiple sources.

1 Introduction

In a wavelength-routed optical network, traffic is transported over lightpaths, which exclusively occupy an entire wavelength on each hop of the source-destination path. Traffic grooming divides the bandwidth of a lightpath into lower sub-rate units, so that it can carry traffic streams transmitted at lower rates. A traffic stream uses one or a multiple of these sub-rate units. The lowest sub-rate stream carried, hereafter referred to as a sub-wavelength unit, defines a lightpath's granularity.

Because these networks carry large amounts of traffic, alternate routing methods are designed in order to allow traffic to be properly re-routed from source to destination in the event of certain events, such as link blocking, or failure. A call arriving at a network is assigned a primary path, upon which it initially attempts to be routed. If the call cannot be routed on this path, then it is re-routed to an alternate path. If there are multiple alternate paths available, then they are attempted in either a random or specific order until the call is established. The call is blocked from the network if the primary and alternate paths cannot carry it.

In this paper, we consider a tandem traffic-groomed optical network composed of N nodes linked in series. Each optical node is an add/drop multiplexer (ADM). The

L. Mason, T. Drwiega, and J. Yan (Eds.): ITC 2007, LNCS 4516, pp. 1048–1059, 2007.

network supports traffic from multiple sources, each with a resource requirement measured in sub-wavelength units. We assume that a number of lightpaths have been established. A path between two nodes may be served by a single direct lightpath or a sequence of lightpaths. Traffic bifurcation is permitted, meaning calls are allowed to use sub-rate traffic streams belonging to different wavelengths.

We model the optical network as a multi-level overflow system, where each level represents a single wavelength between adjacent nodes. Within each level is a tandem queueing network of multi-rate Erlang loss nodes, where a customer occupies a number of servers on one or more adjacent nodes.

The analysis of overflow systems, where blocked calls arriving to a primary Erlang loss system overflow to a secondary system, has been extensively studied. The majority of this work has focused on systems where all calls require a single resource[1],[2],[5],[6],[7],[9],[11],[12],[13],[15].

In this work, we decompose the queuing network into subsystems, where each subsystem is analyzed using a combination of methods developed by Nilsson et. al [14], Frederick and Hayward, and Basharin and Kurenkov [3]. As will be shown, the decomposition method provides a good approximate analysis of large overflow systems supporting traffic from multiple sources.

Washington et. al [19] and [20] proposed a decomposition algorithm for the analysis of tandem optical networks, whereby the queueing network was decomposed into subsystems, each consisting of two adjacent nodes. In this work, we extend the work of [19] to allow for overflow analysis.

The remainder of the paper is organized as follows. Section 2 describes the queuing network model used to calculate call blocking probabilities in the traffic-groomed optical network. Section 3 describes the decomposition algorithm. Section 4 presents numerical comparisons between the algorithm and simulation model. Finally, section 5 concludes the paper.

2 The Queueing Network Model

We consider a traffic-groomed optical network composed of N nodes linked in series. Each link between adjacent nodes is composed of W wavelengths. Each wavelength contains s sub-wavelength units, where s is assumed to be large.

Figure 1 illustrates a four-node optical network with $W=3$ wavelengths on each link. The following lightpaths have been established: *0-3* on wavelength *1*, *0-2* and *2-3* on wavelength *2*, and *0-1*, *1-2*, and *2-3* on wavelength *3*. Due to the traffic bifurcation assumption, each of the three wavelengths in Figure 1 logically represents the combining of multiple wavelengths between the same source destination pair. In other words, all wavelengths where calls have a direct path from node 0 to node 2, for example, are grouped to form a single pool of resources from which sub-rate streams are allocated to each call on the direct path. Calls arrive at any node i, $i=0$, 1, 2 and are destined for any node j, $j=1$, 2, 3, where $j>i$. All calls, irrespective of their source-destination path, are grouped into R classes. Each class r call, $r=1$, 2,..., R, has an associated demand, d_r, measured in sub-wavelength units.

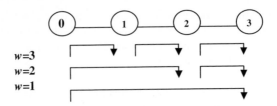

Fig. 1. A four-node, three-wavelength optical network and all source-destination paths

A class r call arriving at node i with destination j, $j>i$, is initially offered to the direct lightpath between nodes i and j, if there is any. The call will be accepted on the direct lightpath, which we will refer to as the primary path, if d_r sub-wavelength units are available on the same wavelength along links $i+1$, $i+2$, ...j. These sub-wavelength units are simultaneously allocated on all links upon call arrival. Upon the call departure, all d_r sub-wavelength units are simultaneously released along each link of the source-destination path. If a call cannot be serviced on its primary path, the call is offered to its pre-determined alternate paths in a sequential fashion. The alternate paths may traverse one or more lightpaths. This process repeats until the call is either accepted on one of its alternate paths or is blocked.

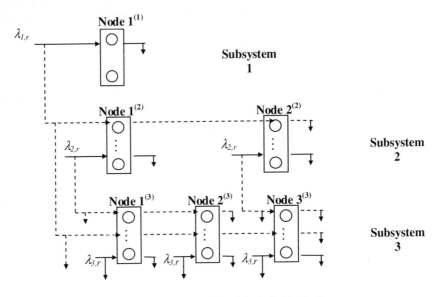

Fig. 2. The queueing network and the individual subsystems

We model the optical network of Figure 1 as a W-level overflow system, where each level represents the lightpaths existing on a wavelength and each lightpath is represented by a multi-rate loss node with $n=s$ servers. Figure 2 illustrates the queueing network model. The solid lines represent random Poisson traffic arriving to the node. The dashed lines represent overflow traffic arriving to the node from a previous wavelength. A server represents a sub-wavelength unit. Each customer in the queueing network represents a call. Customers arrive at any node on any wavelength w, $w=1,2,...W$, with Poisson rate $\lambda_{w,r}$ and exponential service rate $\mu_{w,r}$. In addition to the local arriving traffic, traffic overflowing from wavelength w-1 may also arrive to the node.

A class r local customer arriving to a node at wavelength w simultaneously occupies d_r servers and simultaneously releases them at the time the customer leaves the network. If there are not enough free servers available, the customer overflows to the pre-determined alternate node(s) on wavelength w+1. If the customer is accepted on the alternate path, then it simultaneously occupies and releases d_r servers on each node of the alternate path. If there are not enough free servers available at wavelength w+1, the customer overflows to the alternate node(s) on wavelength w+2, and so forth. Upon reaching wavelength W, if the class r customer finds an insufficient number of servers available, the customer is blocked, i.e. lost. We note that, at wavelength W, random Poisson customers arriving to any node for service are blocked, i.e. do not overflow, if they cannot be serviced at the node.

3   The Decomposition Algorithm

In this section, we present the decomposition algorithm developed for the analysis of the above-described queueing network. The algorithm first decomposes the network into W subsystems. Each subsystem is then analyzed as a single-wavelength network composed of the nodes that correspond to the lightpaths established over the wavelength. Below, we describe the approximation developed for the analysis of overflow systems by Frederick and Hayward, the extension by Basharin and Kurenkov, as well as the modification of the iterative scheme in [19]. Following this, we present our approximation method.

3.1   Frederick and Hayward's Approximation

Consider a single class of customers arriving to a loss system, referred to as the primary system, consisting of N servers. All customers arrive to the node with Poisson rate λ, are serviced with rate μ, and require a single server. Customers blocked in the primary system overflow to a secondary system of N servers where, if the customers cannot be serviced, they are blocked, i.e. lost. We define the offered traffic of an arriving customer as $A=\lambda/\mu$. The blocking probability, b_1, of an arriving customer in the primary system is determined using the Erlang-B equation

$$b_1 = E(A, N) . \tag{1}$$

where $E(A,N) = AE(A, N-1)/(N+AE(A, N-1))$ and $E(A,0)=1$. The mean and variance, M and V respectively, of the traffic offered to the secondary system are given by the expressions:

$$M = A*b_1 . \tag{2}$$

$$V = 1-M+(A/(N+1-A+M)) . \tag{3}$$

We note that, while the traffic offered to the primary system is Poisson, the traffic offered to the secondary system is bursty, i.e. non-Poisson. By assuming that this traffic is Poisson, the blocking probability of the secondary system is underestimated. Frederick and Hayward [6] developed the following approximation to address this issue.

We define the peakedness, $Z=V/M$. For Poisson traffic, $Z=1$. Frederick and Hayward determined that a non-Poisson traffic stream with mean M and $Z \neq 1$, arriving to a secondary system of N servers experiences the same blocking as an equivalent system of N/Z servers offered traffic with mean M/Z and peakedness $Z=1$. The analysis of the secondary system is performed using the equivalent Poisson flow and equivalent maximum servers. The blocking probability of the customers in the secondary system, b_2, is approximated using

$$b_2 = E_r(A/Z, N/Z) . \qquad (4)$$

We note that $E_r(A, N) = \lambda^N e^{-\lambda} (\Gamma(N+1, \lambda))^{-1}$ is the generalized Erlang-B equation that allows for non-integer values. $\Gamma(N+1, \lambda)$ is the incomplete Gamma function.

Basharin and Kurenkov [3] extended the work of Frederick and Hayward to the analysis of networks supporting multi-rate calls with simultaneous resource possession. Assume that customers arrive to a primary system requesting d servers, which are simultaneously allocated and released upon call arrival and departure, respectively. The blocking of calls in the primary system is determined using the product-form solution

$$p(n) = G^{-1}(A^n/n!) . \qquad (5)$$

where n is the number of customers present in the network and G is the normalizing constant, defined as

$$G = \Sigma A^n/n! . \qquad (6)$$

The authors noted that the peakedness, Z, is d times bigger as a result of the simultaneous resource possession, and subsequent increased mean and variance. The secondary system is analyzed in one of two manners. First, the demand is fixed and number of servers is decreased by Z. That is, an equivalent system of N/Z servers is offered traffic with mean M/Z and demand d. Second, the maximum number of servers is fixed and the demand is increased by Z. That is, an equivalent system of N servers is offered traffic with mean M/Z and demand $d*Z$. We note that, the second case results in non-integer values for the call demand, which must be appropriately addressed.

3.2 The Approximation Method

As previously stated, the queueing network consists of levels of nodes, representing lightpaths on specific wavelengths in the original network. There are two types of nodes that exist in the network. The analysis of each is described below.

3.2.1 Nodes Supporting Poisson Traffic.
First, we consider the case where the node supports Poisson traffic only. That is, there is no overflow traffic arriving to the node. This is the case for traffic arriving to node *1* on wavelength *1*, in Figure 2. The

node supports traffic from R classes, each with a demand of d_r servers, $r=1,2,...,R$. The analysis of the node is performed using the recursion developed by Nilsson et. al. Let m denote the maximum number of servers in the node. In addition, let k denote the number of free servers in the node. The function $\beta(m,k)$, defined as the probability that m-k servers are busy in the node, is determined as

$$\beta(m,k)= \beta(m\text{-}1,k\text{-}1)/[1+(1/m)\sum_r A_r d_r \beta(m-1,d_r-1)\,]\,. \tag{7}$$

$$\beta(m,0)= (1/m) \sum_r A_r d_r \beta(m-1,d_r-1)\,[/1+(1/m)\sum_r A_r d_r \beta(m-1,d_r-1)\,]\,. \tag{8}$$

$A_r=A_{w,r}=\lambda_{w,r}/\mu_{w,r}$ is the offered traffic of a class r customer. The blocking probability is determined as

$$b_r= \sum_{j=0}^{d_r-1} \beta(m,\,j)\;. \tag{9}$$

Nodes Supporting Overflow and Poisson Traffic. The second type of node supports both local arriving Poisson traffic and traffic overflowing from a previous subsystem, such as those on wavelength 2 of Figure 2. If either traffic cannot be serviced on the node, it overflows to the subsequent wavelength until it is blocked. We note that there may be overflow traffic from more than one node.

Consider, for example, the traffic arriving to node $1^{(2)}$ in Figure 2. The node supports R classes of traffic overflowing from wavelength 1, as well as R classes of local Poisson traffic. As noted before, the overflow traffic is not Poisson. In order to analyze the node, we must determine an equivalent Poisson flow to represent the overflow traffic. We propose the following approximation.

The mean of the offered traffic of each class r overflow customer, M_r, is determined using (2). However, the variance, V_r, cannot be determined using Expression (3). This is due to the fact that the overflow of each class r customer at node 1 on wavelength 1 is not independent of the overflow from the other classes. To determine V_r, we assume that each class r customer arrives to an independent set of primary servers, N_r, on wavelength 1. Each customer, regardless of class and original demand, requires one server. The blocking experienced by the class r customer is equal to the blocking experienced by a class r customer in the original node. This analysis allows us to analyze each class r customer using the Erlang-B equation, such that we can determine an equivalent number of servers that would produce the same blocking probability, assuming single-server demands, as in the original system.

The value of N_r is determined as follows. For each class r, N_r is initialized to zero. We define b_{test} as the current value of the blocking probability of the single-flow equivalent system, such that

$$b_{test}=E_r(A_r,\,N_r)\,. \tag{10}$$

Beginning with $N_r=0$, b_{test} is determined using (10). The difference between b_{test} and b_r is calculated. If the difference is less than a pre-specified value, then the current value of N_r is a suitable approximation for the single-flow system. Otherwise, if b_{test} is greater than b_r, the value of N_r is too low, and it is increased by a specific amount, referred to as *delta*. If b_{test} is lower than b_r, N_r is too high, and it is decreased by *delta*. The value of *delta* is initialized to 1 and successively halved, depending on the difference between b_{test} and b_r. While the difference is higher, the value of *delta* remains unchanged. As the difference decreases, the value of *delta* is halved. This process is repeated until the difference between b_{test} and b_r is less than the pre-specified value.

At this point, we have determined the equivalent number of servers, N_r, for each class r customer, that would produce the blocking probability, b_r, experienced by the node of the previous subsystem, assuming each class r customer arrived to an independent group of servers with rate $A_{w,r}$ and requesting a single server. N_r is used to approximate the variance of each overflowing traffic stream as

$$V_r = 1 - M_r + (A_r/(N_r + 1 - A_r + M_r)) . \tag{11}$$

The peakedness of each overflowing customer, Z_r, is defined as

$$Z_r = V_r/M_r . \tag{12}$$

In this work, we attempted to analyze the blocking of the network using three approaches. The first was to scale the offered traffic and number of servers using Basharin and Kurenkov's approximation. The blocking was then determined using (7), (8), and (9). The last two approaches were to determine the blocking of each call individually using (10), the generalized Erlang-B equation. One used the equivalent Poisson flow, A_r/Z_r, and equivalent number of servers, N_r/Z_r, to determine the blocking probability of each individual call. The other used the offered traffic of a class i call calculated using (2) and the maximum number of servers on the original overflow link. We compared these variations of analysis to see if the analyzing each individual class I call's blocking using the generalized Erlang-B function would produce better results than (7), (8), and (9).

3.3 The Decomposition Algorithm

We describe the decomposition algorithm using the four-node optical network modeled by the queueing network shown in Figure 2. The queueing network is decomposed into three subsystems, corresponding to the three wavelengths shown in Figure 1. Subsystem *1* represents the lightpath on wavelength *1*, subsystem *2* represents the lightpaths on wavelength *2*, and subsystem *3* represents the lightpaths on wavelength *3*. $A_r^{(x)} = \lambda_r^{(x)}/\mu_r^{(x)}$ is the offered traffic of a class r customer on subsystem x, where $x=1,2,3$. $M_{y,r}^{(x)}$ and $Z_{y,r}^{(x)}$ denote the mean and peakedness, respectively, of the offered traffic of a class r customer overflowing from node y of subsystem x. $\underline{A}_{y,r}^{(x)}$ and $\underline{d}_{y,r}^{(x)}$ denote the equivalent offered traffic and equivalent demand, respectively, of a class r customer overflowing to node y on subsystem x. Finally, $\overline{b}_{y,r}^{(x)}$ is the blocking probability of a class r customer at node y of subsystem x. All blocking probabilities are initialized to zero.

Beginning with subsystem *1*, each subsystem is analyzed in sequence. Within a subsystem, each node is analyzed iteratively using the single-node decomposition method assuming link independence. For subsystem *1*, the only traffic arriving to the node is Poisson traffic. The node is analyzed using (7), (8), and (9). The blocking probabilities, $b_{1,r}^{(1)}$, are stored and the algorithm shifts to subsystem *2*.

At subsystem *2*, the algorithm begins with node *1*. The equivalent Poisson traffic for each class r customer overflowing from subsystem *1* is first determined, using (2), (11), and (12), and the most recently calculated values for $b_{1,r}^{(1)}$. We note that the overflow traffic arriving to this node continues on to node *2*, if accepted. As a result, the equivalent Poisson traffic is

$$\underline{A}_{1,r}^{(2)} = M_{1,r}^{(1)} \ (1 - b_{2,r}^{(2)}) / Z_{1,r}^{(1)} \ . \tag{13}$$

Expression (13) assumes that the overflow traffic is not blocked at the downstream nodes of the alternate path, in this case, node *2*. This allows for the calculation of the blocking experienced by a customer, due to an insufficient number of servers on node *1* of the alternate path. The equivalent demands of the overflow traffic are determined as

$$\underline{d}_{1,r}^{(2)} = d_r{}^* \ Z_{1,r}^{(1)} \ . \tag{14}$$

Because there is both Poisson and overflow traffic arriving to the node, the maximum number of servers per node is fixed and the demands of the overflow traffic are scaled by their corresponding peakedness.

At this point, we have a total of $2R$ classes arriving to node *1* on wavelength *2*, R classes of overflow traffic with equivalent offered traffic $\underline{A}_{1,r}^{(2)}$ and equivalent demands $\underline{d}_{1,r}^{(2)}$, and R classes of local Poisson traffic with offered traffic $A_r^{(x)}$ and demands d_r. The nodes are analyzed using (7), (8), and (9) assuming $2R$ classes. The algorithm then shifts to node *2* on wavelength *2*, which is analyzed in similar fashion.

At this point, a single iteration has completed within subsystem *2*. The algorithm executes successive iterations until the blocking probabilities of each node within the subsystem have converged within a tolerance, ε, which is set to 10^{-4}. Once the results have converged, the blocking probabilities of all source-destination paths within the subsystem are calculated by appropriately combining the results from each node. For example, the blocking probability of the class r renewal traffic on path *(0,3)* of subsystem *2* is $1 - (1 - b_{1,r}^{(2)})(1 - b_{2,r}^{(2)})$. The algorithm then shifts to subsystem *3*.

At subsystem *3*, each node is analyzed in iteration, similar to subsystem *2*, using the most recently calculated blocking probabilities from each node in subsystem *2*. Beginning with node *1* on wavelength *3*, the equivalent Poisson traffic for each class r customer overflowing from node *1* on wavelength *2* is determined. We note there now are a total of $2R$ classes of overflow traffic offered to node *1* on wavelength *3*, R classes from the overflow path traversing both nodes *1* and *2* on wavelength *2*, and R classes from the local traffic traversing node *1*.

The algorithm executes successive iterations in subsystem 3 until the blocking probabilities of each node within the subsystem have converged within ε. Once the results have converged, the blocking probabilities of all source-destination paths within the subsystem are calculated by appropriately combining the results from each node. Because this is the last subsystem, the blocking is determined as follows. For local traffic on node 1 of wavelength 3, the blocking probability is simply $b_{1,r}^{(3)}$, $r=2R+1,2R+2,...,3R$. The blocking experienced by a call in a previous subsystem, such as subsystem 1, is $M_{1,r}^{(2)}$ $[1-(1-b_{1,r}^{(3)})(1-b_{2,r}^{(3)})(1-b_{3,r}^{(3)})]/A_r^{(1)}$, $r=1,2,...,R$. We note that the blocking for this path uses the offered traffic arriving from subsystem 2, the original offered traffic to subsystem 1, and the blocking probability of the call in subsystem 3.

4 Numerical Results

In this section, we discuss the accuracy of the decomposition algorithm by comparing the results obtained to simulation data. For each case, we consider the four-ADM optical network of Figure 1, modeled by the W-level queueing network shown in Figure 2. Each link between two adjacent ADMs is composed of 3 wavelengths, each wavelength containing s servers.

All possible source-destination paths are considered and, for each path, the call is first offered to the direct lightpath between the source and destination node. If the call cannot be serviced on the direct lightpath, it overflows to its alternate paths sequentially, until the call is either accepted or blocked. We assume four classes of traffic, i.e. $R=4$, with demands of $d_1=2$, $d_2=4$, $d_3=6$, and $d_4=8$ servers.

The accuracy of the decomposition algorithm was evaluated under different cases, where the arrival rates and the number of servers vary. We assume, for presentation purposes, that the arrival rate of all class r calls, regardless of source-destination path, is λ_r. For each case, λ_1 was assigned the highest arrival rate, and the arrival rate of each successive class r call was decreased by a small amount from the previous as follows. We assume $s=128, 256$ servers. For $s=128$ servers, cases 1, 2, and 3, set λ_1 to 5000, 8000, and 10000, respectively. For $s=256$, cases 1 and 2 set λ_1 to 8000 and 10000, respectively. For both cases, the arrival rate of each successive class r call was decreased by 500, and the service rate of all calls was set to 500. Each case was considered for the three approaches discussed. It was determined that the first approach, discussed above, produced the best results of the three.

Figures 3 and 4 show the results for wavelengths 1 and 2, and wavelengths 3 when $s=256$ servers. Each figure plots the blocking probability of each source-destination path, labeled as (source, destination), as a function of the call class. In addition, the simulation results are plotted with a 95% confidence interval. We note that the simulation results for each class are plotted as lines. This was done for presentation purposes, to help illustrate the accuracy of the approximation. It is to be noted, however, that the actual points represent the simulated data.

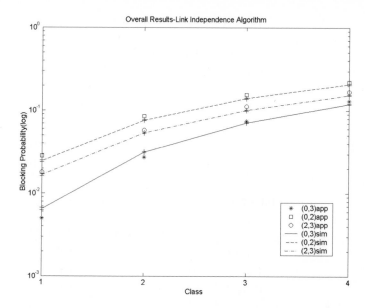

Fig. 3. Wavelength 1 and 2 results for s=256 servers

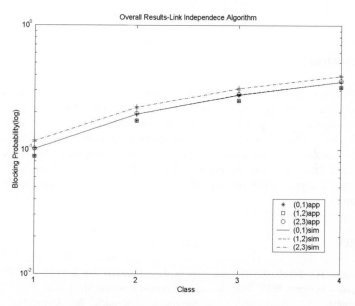

Fig. 4. Wavelength 3 results for s=256 servers

We observe that overall, the decomposition provides good accuracy for all cases. Tables 1 and 2 show the per-class and per wavelength relative errors for each case. As is shown, the average relative errors are fairly low, meaning that the decomposition algorithm, using the single-node decomposition algorithm assuming link independence has good accuracy.

Table 1. Per-class relative errors

		Class 1	Class 2	Class 3	Class 4
128	**Case 1**	0.029	0.022	0.015	0.013
	Case 2	0.039	0.014	0.013	0.014
	Case 3	0.038	0.013	0.014	0.013
256	**Case 1**	0.041	0.031	0.027	0.025
	Case 2	0.046	0.027	0.019	0.015

Table 2. Per-wavelength relative errors

		Wave 1	Wave 2	Wave 3
128	**Case 1**	0.044	0.048	0.092
	Case 2	0.047	0.070	0.076
	Case 3	0.054	0.081	0.065
256	**Case 1**	0.125	0.099	0.108
	Case 2	0.058	0.090	0.109

5 Conclusion

In this paper, we consider a tandem traffic-groomed optical network composed of N nodes linked in series, supporting alternate routing. The network is modeled as a multi-level overflow system, where each level represents a single wavelength between adjacent nodes. The queueing network is decomposed into subsystems and each subsystem is analyzed using a combination of methods developed by Nilsson et. al, Frederick and Hayward, and Basharin and Kurenkov. This method was shown to provide a good approximate analysis of large overflow systems supporting multiple traffic sources.

References

[1] Akimaru, H., Takahashi, H.: An Approximate Formula for Individual Call Losses in Overflow Systems. IEEE Transactions on Communications 31(6), 808–811 (1983)
[2] Alyatama, A.: Wavelength Decomposition Approach for Computing Blocking Probabilities in WDM Optical Networks Without Wavelength Conversion, Journal of Computer Networks (To appear)
[3] Basharin, G.P., Kurenkov, B.E.: Analysis of Overflow in Circuit Switching Networks. Problems of Information Transmission 23(3), 216–223 (1987)
[4] Birman, A.: Computing Approximate Blocking Probabilities for a Class of All-Optical Networks. IEEE Journal on Selected Areas in Communications 14(5), 852–857 (1996)

[5] Brochin, F., Pradel, E.: A Call Traffic Model for Integrated Services Digital Networks. IEE Global Telecommunications Conference (5), 1508–1512 (1992)

[6] Frederick, A.A.: Congestion in Blocking Systems-A Simple Approximation Technique. Bell. System Technical Journal 59(6), 805–827 (1980)

[7] Overflow Approximations for Non-Random Inputs, ITU-T Recommendation E.524 (1999)

[8] Iversen, V., Stepanov, S.N., Kostrov, V.O.: The Derivation of a Stable Recursion for Multi-Service Models, International Conference on Next Generation Teletraffic and WiredWireless Advanced Networking (2004)

[9] Iversen, V.: Teletraffic Engineering and Network Planning, COM, DTU (2005)

[10] Kaufman, J.S.: Blocking in a Shared Resource Environment. IEEE Transactions on Communications 29(10), 1474–1481 (1981)

[11] Kuczura, A.: The Interrupted Poisson Process as an Overflow Process. Bell. System Technical Journal 52, 437–448 (1973)

[12] Listanti, M., Parisi, D., Sabella, R.: Optical Data Network Based on the Switchless Concept: Analysis and Dimensioning. Photonic Network Communications 3(4), 363–375 (2001)

[13] Liu, Y., Iversen, V., Dickmeiss, A., Larsen, M.: Individual Overflow Characteristics for Loss System with Multi-Slot Traffic and State-Dependent Poisson Input Processes, 11th Nordic Teletraffic Seminar (1993)

[14] Nilsson, A., Perry, M., Iversen, V., Gershtc, A.: On Multi-Rate Erlang-B Computations. In: 16th ITC (1999)

[15] Schehrer, R.: On the Calculation of Overflow Systems with a Finite Number of Sources and Full Available Groups. IEEE Transactions on Communications 26(1), 75–82 (1978)

[16] Srinivasan, R., Somani, A.K.: Analysis of Multi-Rate Traffic in WDM Grooming Networks. In: 11th International Conference on Computer Communications and Networks, pp. 295–301 (2002)

[17] Thiagarajan, S., Somani, A.K.: Capacity Correlation Model for WDM Networks with Constrained Grooming Capabilities. IEEE International Conference on Communications 5, 1592–1596 (2001)

[18] Thiagarajan, S., Somani, A.K.: Performance Analysis of WDM Optical Networks with Grooming Capabilities, SPIE Technical Conference on Terabit Optical Networking: Architecture, Control, and Management Issues (2000)

[19] Washington, A.N., Hsu, C., Perros, H.G., Devetsikiotis, M.: Approximation Techniques for the Analysis of Large Traffic-Groomed Tandem Optical Networks. 8th Annual Simulation Symposium (ANSS-38 2005), San Diego, California, April 2-8, 2005 (2005)

[20] Washington, A.N., Perros, H.G.: Call Blocking Probabilities in a Traffic Groomed Tandem Optical Network. Journal of Computer Networks 45 (2004)

[21] Xin, C., Qiao, C., Dixit, S.: Traffic Grooming in Mesh WDM Optical Networks-Performance Analysis, OPTICOMM (2003)

Approximating the Admission Region for Multi-resource CAC in Wireless Networks

Taka Sakurai, Milosh Ivanovich, and Paul Fitzpatrick

Telstra, 13/242 Exhibition St Melbourne, Australia
{taka.sakurai, milosh.ivanovich,
paul.g.fitzpatrick}@team.telstra.com

Abstract. This paper studies the problem of dimensioning a wireless base station that operates with CAC. It models the base station as a multi-rate, multi-class, multi-resource loss model and proposes two methods for fast estimation of the admission region. The results show that our methods are typically more accurate than other more computationally intensive approximations, such as the reduced-load approximation, and provide an excellent basis for the dimensioning of wireless networks with CAC.

Keywords: Loss networks, wireless, CAC.

1 Introduction

The trend in wireless network development is towards supporting a wide range of real-time and non real-time services mapped to appropriate quality of service (QoS) traffic classes to guarantee adequate performance. An integral part of achieving performance targets is radio resource management (RRM), encompassing call admission control (CAC) [1] and congestion control. Wireless networks utilise a range of resources on the air interface depending on the exact nature of the system implementation. For example, within the 3GPP standards, the WCDMA 3G cellular systems air interface supports a range of bearer types operating at different data rates, each of which must share system resources such as codes and power. Other physical constraints such as the level of interference (in both the uplink and the downlink) also limit the number of services that can be simultaneously supported by WCDMA. CDMA systems generally exhibit soft capacity (i.e. the ability to trade-off quality with capacity), therefore CAC is essential to maintaining quality at the expense of capacity by avoiding overload [2], [3]. This paper considers the overload avoidance function of CAC via call blocking.

The assignment of multiple resources per traffic class subject to the additional constraint of interference in both transmission directions presents a multi-dimensional problem when deciding whether to admit or block new connections. Importantly, our modelling approach does not distinguish between system resources and constraints. Rather, it treats them equally as "resources", none of which must be fully depleted in order for successful call admission. In this paper, we model a wireless base station

L. Mason, T. Drwiega, and J. Yan (Eds.): ITC 2007, LNCS 4516, pp. 1060–1071, 2007.
© Springer-Verlag Berlin Heidelberg 2007

with CAC as a multi-rate, multi-class, multi-resource loss model [4]. There have been a few previous papers that model CAC in a wireless base station as a loss model, such as [5] and [6]. However, we are not aware of any other work that uses a multi-resource loss model to capture the multi-resource nature of CAC in a single base station. The emphasis in [5] is on the use of effective bandwidths to approximate the resource usage on a single resource, namely the interference level. In [6], a multi-class, multi-resource loss model is used to represent a multi-cell scenario, where each cell has a single resource (bandwidth).

Loss models have been extensively studied in the literature (refer to Section 2), however for dimensioning purposes it is necessary to solve the inverse of the blocking probability problem to determine the admission region for pre-defined blocking probability limits. Determining the admission region for a multi-resource loss model can be difficult and computationally expensive. The number of resources in current and emerging base station technologies is of the order of ten, which is extremely challenging for known exact computational methods. Furthermore, in the context of dimensioning a network of base stations, typically numbering in the thousands, the computational cost will be prohibitive because base stations are typically configured differently, resulting in many variants of the loss model that require solution. The loss models may also need to be recomputed in response to network re-configurations. The problem then is to solve the loss model to determine the blocking probabilities of the individual traffic classes in a realistic timeframe. The contributions of this paper are fast approximate methods for accurately estimating the admission region of the multi-rate, multi-class, multi resource loss system model of the wireless base station. A broad range of examples are studied in order to demonstrate that the admission regions derived using our approximations are typically more accurate than those obtained using the well-known reduced load approximation, and are obtained considerably faster. This result provides an excellent basis for the use of these approximations for the engineering level dimensioning of wireless systems.

The remainder of this paper is organised along the following lines. Section 2 contains a description of the system model and its formulation as a loss model. The merits and applicability of existing work on loss models to solving the problems outlined above is also discussed. In Section 3 we present our approximations. Results are presented in Section 4 comparing the approximations from Section 3 with other exact and approximate solutions for loss systems for a range of examples. Finally, conclusions, discussion and ideas for future work are presented in Section 5.

2 System Model

Consider a wireless base station with J resources, where resource j has capacity K_j. Let there be R traffic classes, where connections from traffic class r arrive with Poisson rate λ_r and require A_{jr} units of resource from resource j. For a typical wireless technology, J is of the order of 10 or less and all A_{jr} are non-zero, implying that each traffic class requires resource units from every resource of the base station.

The system described above can be modelled by a multi-rate, multi-class, multi-resource loss model using the complete sharing admission policy. Alternatively, if we impose thresholds t_{jr} such that a call of class r is only accepted if the resulting free capacity on resource j is at least t_{jr}, then we will have a loss model with trunk reservation. Trunk reservation could be used to give precedence for handover traffic classes over non-handover traffic classes, or real-time traffic classes over non-real-time traffic classes. We assume that connections represent user sessions rather than flows, and arrive according to a Poisson process. A connection may be a single web browsing session encompassing multiple TCP flows for object retrievals and reading time. This follows from [7].

Let us denote the offered traffic intensity of traffic class r, for $1 \leq r \leq R$, by ρ_r, and the vector of traffic intensities by $\rho \equiv (\rho_1, \rho_2, ..., \rho_R)$. Further, denote the blocking probability experienced by traffic class r by $B_r(\rho)$, the blocking limit for traffic class r by δ_r, and the vector of blocking limits by $\delta \equiv (\delta_1, \delta_2, ..., \delta_R)$. Our specific interest in this work is to find an efficient method to determine the admission region of offered traffic intensities. We define the admission region as the set of traffic vectors ρ for which the respective blocking probabilities satisfy the blocking limits; we denote the admission region by $S(\delta)$ and formally define

$$S(\delta) = \left\{ \rho : B_r(\rho) \leq \delta_r, \ 1 \leq r \leq R \right\}. \tag{1}$$

For given δ_r, $1 \leq r \leq R$, the boundary of the admission region can be determined by combining a method to compute the $B_r(\rho)$ with an overarching iterative search algorithm to find the values of ρ on the boundary. The iteration ends when we hit the admission region limit, where $B_r(\rho) \leq \delta_r$, with equality for at least one r.

Blocking probabilities in multi-resource, complete-sharing loss models can be computed exactly using analytical approaches [4]. However, the computational complexity grows exponentially in J, making the calculation practical only for small systems. One of the most efficient exact approaches is to use numerical transform inversion (NTI) [8]. Assuming for simplicity that the capacity of each resource is K, it is shown in [8] that the NTI algorithm to compute $B_r(\rho)$ for all r has computational complexity $O(RK^J)$. For systems of practical interest, computation times can be excessive for more than two resources; moreover, machine precision dictates that the NTI method is limited to about 6 resources. For larger J, blocking probabilities can be approximated using a reduced load approximation [4], [11]. As is well-known, a reduced load approximation avoids the state space explosion of the multi-resource problem by decomposing it into multiple single resource problems that are connected through a fixed-point approximation.

The classical reduced load approximation is the *Erlang fixed-point approximation*, which was originally developed for loss models with single-rate traffic but was subsequently generalized to multi-rate traffic by incorporating an approximation that multi-rate requirements can be modelled as collections of single-rate requirements [11]. Another type of reduced load approximation is the *knapsack approximation* [9], in which the single resource problems are solved exactly but still connected through a fixed-point approximation. It has been shown that the knapsack approximation tends to perform better than the Erlang fixed-point approximation [9]. The knapsack approximation can be summarized as follows. We denote by $b_{jr}(\rho)$ the blocking probability of class r for load ρ when resource j is considered in isolation. The knapsack approximation for a J resource problem with load ρ involves the blocking probabilities

$$L_{jr} = b_{jr}(\tilde{\rho}_j), \tag{2}$$

where $\tilde{\rho}_j = (\tilde{\rho}_{j1}, \ldots, \tilde{\rho}_{jR})$, and

$$\tilde{\rho}_{jr} = \rho_r \prod_{k \neq j} (1 - L_{kr}). \tag{3}$$

Equations (2) and (3) define a set of fixed-point equations that can be solved by repeated substitution. After a solution to (2) is obtained, the knapsack approximation to the blocking probability for class r, $B_r^N(\rho)$, is found from

$$B_r^N(\rho) = 1 - \prod_{j=1}^{J} (1 - L_{jr}),$$

and the associated approximation to the admission region is

$$N(\delta) = \left\{ \rho : B_r^N(\rho) \leq \delta_r, \ 1 \leq r \leq R \right\}.$$

Reduced load approximations involve making the assumption that blocking events on different resources are independent. The independence assumption tends to hold true when the system is large in some sense, such as when the capacities and arrival rates are large or when the number of resources and the number of classes using diverse subsets of resources is large [11]. Conversely, reduced load approximations tend to be inaccurate for loss systems in which there is strong dependence of blocking between resources. In our wireless base station application, strong dependence can be expected, since the number of resources is usually not large, and all traffic classes require units from each resource.

Bebbington et. al. [12] propose an improvement to the EFP that provides better estimates of the blocking probability. This is achieved by including dependency between blocking for a traffic class on different resources, but with additional computational expense. For this reason we do not consider it in this paper.

3   Proposed Approximations for the Admission Region

3.1   The Single Resource Intersection (SRI) Method

For calculating the admission region, we propose a simpler approximation than the reduced load approximation, that does not require computation of a fixed-point solution. Our approach is to consider each of the J resources in isolation and to compute the associated single resource admission regions, and then to use the *intersection* of these J admission regions as an approximation to the true (joint) admission region. To define the SRI Method admission region in a formal way, we first define the J single resource admission regions by

$$s^j(\delta) = \left\{ \rho : b_{jr}(\rho) \le \delta_r, \ 1 \le r \le R \right\}, \qquad j = 1, \ 2 \ldots, J .$$

Then the SRI Method admission region is $I(\delta) = \bigcap_{j=1}^{J} s^j(\delta)$.

In addition to the above graphical interpretation, $I(\delta)$ can be defined in a similar way to (1) with an appropriate definition of the blocking function. To see this, let $B_r^I(\rho)$ be equal to the maximum value of $b_{jr}(\rho)$ over all resources:

$$B_r^I(\rho) = \max(b_{1r}(\rho), \ldots, b_{Jr}(\rho)) .$$

Then $I(\delta)$ is obtained by using $B_r^I(\rho)$ as an approximation to $B_r(\rho)$ in (1):

$$I(\delta) = \left\{ \rho : B_r^I(\rho) \le \delta_r, \ 1 \le r \le R \right\}.$$

The approximation will be good if ρ is such that the maximum $B_r^I(\rho)$ is much larger than the other blocking probabilities, since then, resources other than the resource that yields the maximum blocking will not have a significant influence in the joint problem. If all resources are of size K, the computational complexity of the SRI Method in computing $B_r^I(\rho)$ for all r is $O(RJK)$.

3.2 The One-Step Load Reduction (OSLR) Method

The SRI Method considers each resource in isolation and ignores the impact of the other resources. One way in which other resources can affect the blocking of class r on resource j is by significant blocking of other traffic classes. The impact on class r will be beneficial, in that more resources will be available for class r traffic on resource j. To account for this effect, we propose a variant of the SRI Method, which we call the One-Step Load Reduction Method, in which the load of all traffic classes on resource j other than class r is reduced by the maximum blocking suffered on the other resources. We provide a precise definition of the OSLR Method below.

In the OSLR Method, we consider each class r in turn and first identify the resource for which $b_{jr}(\rho)$ is greatest. Denoting the set of all resources by ψ, we find $k = \arg\max_{j \in \psi}(b_{jr}(\rho))$ and approximate $B_r(\rho)$ by $B_r^L(\rho)$, where $B_r^L(\rho)$ is

$$B_r^L(\rho) = \max(b_{1r}(\overline{\rho}^{(r)}), \ldots, b_{Jr}(\overline{\rho}^{(r)})), \tag{4}$$

where $\bar{\rho}^{(r)}$, is the *reduced load vector*. The reduced load vector is comprised of *reduced load elements*, except in the rth location, namely

$$\bar{\rho}^{(r)} = (\bar{\rho}_1^{(r)}, \ldots, \bar{\rho}_{r-1}^{(r)}, \rho_r, \bar{\rho}_{r+1}^{(r)}, \ldots, \bar{\rho}_R^{(r)}). \tag{5}$$

Let ψ_k denote the set of all resources except resource k. The reduced load element $\bar{\rho}_q^{(r)}$, is obtained by reducing ρ_q by the maximum blocking experienced by class q on resources in ψ_k, namely

$$\bar{\rho}_q^{(r)} = \rho_q (1 - b_{mq}(\boldsymbol{\rho})), \tag{6}$$

$$m = \arg \max_{j \in \psi_k} (b_{jq}(\boldsymbol{\rho})). \tag{7}$$

Finally, the Load Reduction Method admission region $L(\boldsymbol{\delta})$ is defined as

$$L(\boldsymbol{\delta}) = \left\{ \boldsymbol{\rho} : B_r^L(\boldsymbol{\rho}) \le \delta_r, \ 1 \le r \le R \right\}.$$

If all resources are of size K, the computational complexity of the OSLR Method in computing $B_r^L(\boldsymbol{\rho})$ for all r is $O(2RJK + (R-1)(J-1)K)$.

4   Results

In this section results are presented for a range of examples using only two traffic classes. This has been done so that the presentation of results can be kept to two dimensions. This model is easily extendible to $J>2$. The examples have been selected with significant asymmetry in the resource requirements of the two classes on the resources in an attempt to generate results exhibiting the greatest difference between the approximate and true admission regions. This characteristic results in quite different gradients for the boundaries of the single resource admission regions. It should be pointed out that these extreme asymmetries are less likely in real wireless systems, where there is a high correlation of resource usage within each traffic class. That is, a traffic class that requires a large/small number of resource units on one resource, tends to require a large/small number of resource units on all resources. A range of resource capacities (K_j) and target blocking probabilities $(\boldsymbol{\delta})$ are also used so that behaviour of the admission region with varying parameters can be identified and studied.

4.1   Two Resource Examples

The results in Fig. 1 and Fig. 2 below are for systems with equal capacities of $K=[30 \ 30]$ resources, $A = \begin{bmatrix} 1 & 4 \\ 4 & 1 \end{bmatrix}$ and blocking targets of $\delta=[0.02 \ 0.02]$ (referred to as low blocking) and $\delta = [0.2\ 0.2]$ (referred to as high blocking) respectively. We have plotted the boundaries of $S(\delta)$, $s^1(\delta)$, $s^2(\delta)$, $N(\delta)$ and $L(\delta)$; the $I(\delta)$ region is easy to visualize as the intersection of the $s^1(\delta)$ and $s^2(\delta)$ regions. It can be seen for both examples that the reduced load (RL) approximation under-estimates the exact admission region. The SRI method over-estimates in the low blocking case and under-estimates in the high blocking case and OSLR always over estimates. All approximations perform worst around the region where the single resource lines intersect. Away from these regions, where the blocking of one resource dominates, all approximations tend to converge on the exact admission region. This is a trend that is always evident throughout the results for SRI and OSLR, but not necessarily for RL.

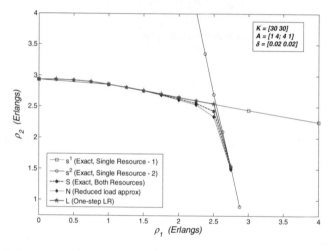

Fig. 1. System capacity $K = [30\ 30]$, $\delta = [0.02\ 0.02]$. The low blocking example.

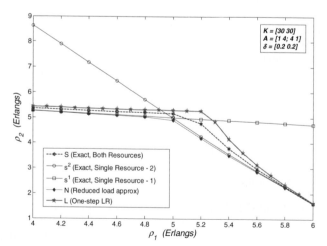

Fig. 2. System capacity $K = [30\ 30]$, $\delta = [0.2\ 0.2]$. The high blocking example.

Results for systems with asymmetrical capacity of K =[20 10] and K =[200 100] are shown in Fig. 4. These have the resource requirements $A = \begin{bmatrix} 5 & 3 \\ 3 & 1 \end{bmatrix}$ for the two traffic classes and the same δ =[0.02 0.02]. This allows a comparison of different system sizes with equal target blocking probabilities.

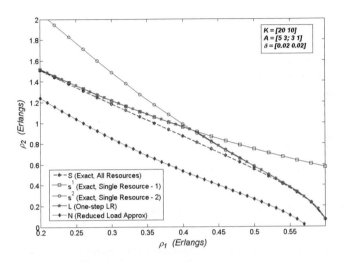

Fig. 3. System capacity K =[20 10], δ = [0.02 0.02]. Low blocking example.

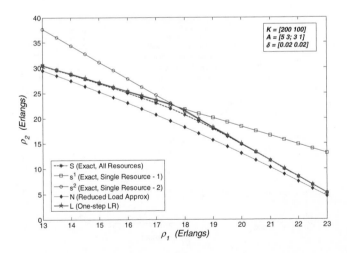

Fig. 4. System capacity K =[200 100], δ = [0.02 0.02]. Low blocking example.

All approximations improve as the system size grows. This is expected for the RL approximation due to RL/EFP being asymptotically exact with system size going to infinity. Both the SRI and the OSLR approximations give a superior result compared to RL. Of the example studies, $K = [200\ 100]$ with low blocking is the most representative of a CDMA/OFDM cellular system, since such systems are typically operated with maximum blocking probabilities of the order $2 - 5\ \%$. The accuracy of SRI and OSLR methods is particularly encouraging for such practical parameter ranges. A key factor for these large systems is the speed advantage of the SRI/OSLR methods. The order of magnitude for the admission region computation time for SIR/OSLR is $\sim 10^0$ sec, compared to $\sim 10^1$s for RL, and $\sim 10^4$ s for NTI. For larger systems (both in number of resources and capacity per resource) the speed advantages are significantly greater, as expected, given the relative computational complexities summarized in Sections 2 and 3. In practice, for systems with $J \geq 4$ resources of moderate size ($K > 30$), computation of the exact admission regions using NTI was either time prohibitive or failed due to numerical rounding errors. While the per-class blocking probabilities $B_r(\rho)$ could be computed in a feasible period of time for $J=4$, this was no longer possible for $J > 4$.

4.2 Three Resource Examples

This section presents results for systems of three resources. First in Fig. 5 is a system of capacity of $K = [5\ 5\ 5]$, resource requirements of $A = \begin{bmatrix} 1 & 4 \\ 4 & 1 \\ 3 & 3 \end{bmatrix}$ and $\delta = [0.2\ 0.2]$.

Here it can be seen that one resource dominates the capacity bound (i.e. provides the highest blocking probability) and so the exact and the SRI admission regions are almost identical, while the OSLR is still close to the exact but RL is poor.

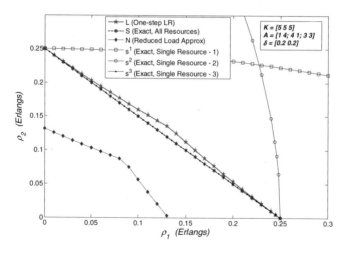

Fig. 5. System capacity $K=[5\ 5\ 5]$, $\delta = [0.2\ 0.2]$. High blocking probability target.

The results in Fig. 6 and Fig. 7 are for two systems with the same resource requirements as above, but larger capacity $K = [10\ 10\ 10]$ and low and high blocking probability targets of $\delta = [0.02\ 0.02]$ and $\delta = [0.2\ 0.2]$ respectively. While in Fig. 8 and

Fig. 9 the resource capacities are further increased to K =[20 20 20] and the other values remain the same. In these examples we see the continuing trend of all approximations converging to the exact result at the edges, where a single resource dominates the admission region and the maximum deviation occurring near the intersection of the single resource admission regions.

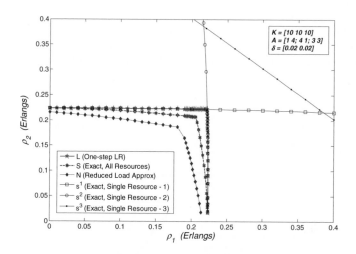

Fig. 6. System capacity $K = [10\ 10\ 10]$, $\delta = [0.02\ 0.02]$. Low blocking probability target.

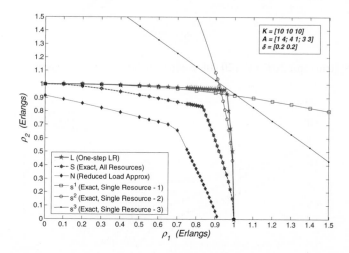

Fig. 7. System capacity $K = [10\ 10\ 10]$, $\delta = [0.2\ 0.2]$. High blocking probability target.

The impact of increasing system capacity for the high blocking cases can be seen by comparing Figures 5, 7 and 9. In Fig. 5 there is one constraining resource, which explains the excellent level of agreement between the exact region and both SRI and OSLR. As the system capacity increases the number of resources in the constraining set increases from two to three. Thus the SRI and OSLR approximations perform slightly worse, rather than improve with system size, as observed in the two dimensional case. Interestingly, SRI and OSLR tend to give the worst fit only near the single resource admission region intersections. Away from these, they converge rapidly to the exact region, unlike RL which can sometimes perform very poorly over the entire range of offered loads.

Similar results for the low blocking cases in Figures 6 and 8 also show how the RL approximation always improves with system size, but is generally worse than either the SRI or OSLR methods. Also the SRI and OSLR tend to converge to the exact admission region more rapidly than RL and are significantly faster to compute.

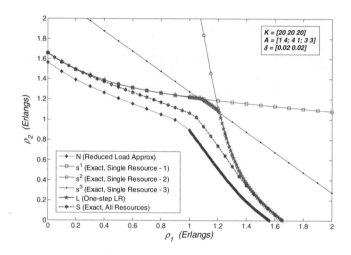

Fig. 8. System capacity $K = [20\ 20\ 20]$, $\delta = [0.02\ 0.02]$. Low blocking probability target.

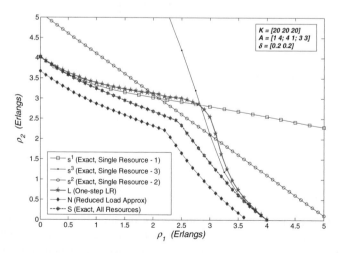

Fig. 9. System capacity $K = [20\ 20\ 20]$, $\delta = [0.2\ 0.2]$. High blocking probability target.

5 Conclusions

In this paper we propose two methods (SRI and OSLR) for approximating the admission region of a multi-rate, multi-class, multi-resource blocking system. In the examples studied, both SRI and OSLR perform very well when compared to the exact admission region, consistently out-perform RL, and achieve the main stated goal of significantly improved speed. The major deviation from the exact results appears to occur when the blocking probability targets are high (beyond that of typical wireless system performance). While these methods are applicable for dimensioning purposes, further work is required to understand their behaviour more generally for a broader range of system sizes and blocking targets. Some initial work to assess their accuracy with trunk reservation has given promising results, and both of these areas remain for future work.

Acknowledgements. The authors acknowledge the permission of Telstra to publish the results herein.

References

[1] Ahmed, M.H.: Call Admission Control in Wireless Networks: A Comprehensive Survey. IEEE Communications Surveys & Tutorials, vol. 7(1) (2005)

[2] 3GPP, 3rd Generation Partnership Project; Technical Specification Group Services and System Aspects; Quality of Service (QoS) concept and architecture (Release 6), 3GPP TS 23.107 V6.2.0 (2004-12)

[3] 3GPP, 3rd Generation Partnership Project; Technical Specification Group Radio Access Network; Radio resource management strategies (Release 6), 3GPP TR 25.922 V6.2.0 (2005-09)

[4] Ross, K.: Multiservice Loss Models for Broadband Communication Networks. Springer, New York (1995)

[5] Evans, J., Everitt, D.: Effective bandwidth-based admission control for multiservice CDMA cellular networks. IEEE Trans. Veh. Tech. 48, 36–46 (1999)

[6] Mitchell, K., Sohraby, K.: An analysis of the effects of mobility on bandwidth allocation strategies in multi-class cellular wireless networks. Proc. IEEE INFOCOM, pp. 1005–1011 (2001)

[7] Bonald, T., Proutiere, A., Regnie, G., Roberts, J.W.: Insensitivity Results in Statistical Bandwidth Sharing. In: Proc.17th International Teletraffic Congress (2001)

[8] Choudhury, G., Leung, K.K., Whitt, W.: An Algorithm to Calculate Blocking Probabilities in Multi-Rate Multi-Class Multi-Resource Loss Models. Advances in Applied Probability 27, 1104–1143 (1995)

[9] Chung, S-P., Ross, K.W.: Reduced load approximations for multirate loss networks. IEEE Trans. Commun. 41, 1222–1231 (1993)

[10] Roberts, J.W.: Teletraffic models for the Telecom 1 integrated services network. In: Proc.10th International Teletraffic Congress (1983)

[11] Kelly, F.: Loss Networks. The. Annals of Applied Probability 1, 319–378 (1991)

[12] Bebington, M., Pollett, P., Zieddins, I.: Improved Fixed Point Methods for Loss Networks with Linear Structure. Proc. 4th ICT, Melbourne, vol. 3, pp. 1411–1416. (1997)

Co-MAC: Energy Efficient MAC Protocol for Large-Scale Sensor Networks

Jinhun Bae, Kitae Kim, and Keonwook Kim*

High-performance Computing and Simulation Research Lab.
Dept. of Electronics Engineering, Dongguk Univ.,
Seoul, Korea
{coreahun, kkta4, kwkim}@dongguk.edu

Abstract. Sensor nodes in large-scale sensor networks autonomously and proactively report diverse information obtained from extensive target area to a base station, which is energy unconstrained node. To efficiently handle the variable traffic, we present Co-MAC, an energy efficient medium access control protocol for large-scale sensor networks. In Co-MAC, an overall network is divided into independent *subnets*, and each subnet orthogonally operates on time line in a temporal fashion. The novelty in this protocol is to evenly distribute sensor nodes *in a certain geographic area* to subnets at the association process with probability *p*. In our simulation, it was observed that energy efficiency of Co-MAC outperforms S-MAC by 6 times and T-MAC by 2 times under identical conditions. The reason for this phenomenon is that the overhearing frequency between sensor nodes is relatively lower in Co-MAC. The results of these analyses demonstrate that Co-MAC obtains significant energy savings.

Keywords: medium access control, large-scale sensor networks, wireless sensor networks, energy efficient protocol.

1 Introduction

With recent technological advances, wireless sensor networks have a wide range of applications. Among the applications, monitoring, e.g., civil environment [14] and natural habitat [15], is an area of research devoted to systems that can autonomously and proactively access to diverse information of real-world. Such applications consist of a large number of distributed sensor nodes in order to cover the target area successfully, where sensor nodes are battery operated and normally are impossible to recharge the energy on the fly. Under the circumstances, the design of a medium access control (MAC) protocol is critical factor to efficiently use the shared channel as maintaining low power consumption.

In large-scale sensor networks, energy conservation is a dominant constraint to design MAC protocol. The major sources of energy dissipation are collision, overhearing, control packet overhead, and idle listening [2]. Among the energy

---

* Corresponding author.

L. Mason, T. Drwiega, and J. Yan (Eds.): ITC 2007, LNCS 4516, pp. 1072–1083, 2007.

dissipations, idle listening is a major factor of energy consumption. For example, simulation outcomes from Stemm and Katz [4] and Digital Wireless LAN Module (IEEE 802.11) specification [5] report that the energy consumption ratios between idle mode, receive mode and send mode are 1:1.05:1.4 and 1:2:2.5, respectively. Thus, most of MAC protocols for wireless sensor networks compel sensor nodes to operate on low duty cycle scheme in order to reduce idle listening.

Sensor nodes in large-scale sensor networks usually encounter *spatially and temporally correlated contention*, which is the result that numerous sensor nodes in a certain geographic area simultaneously transmit data. The traffic concentrates into short communication period due to the low duty cycle scheme, and contention rate and latency rapidly increases. These restricted conditions result in degraded throughput, which is fraction of the channel capacity used for data transmission.

In conventional MAC protocols like S-MAC [2] and T-MAC [3], most of sensor nodes operate on identical schedule. Through in-network data processing, e.g., data aggregation, sensor nodes reduce energy consumption by fusing or aggregating sensing data collected from a group of sensor nodes before forwarding the aggregated data on to a base station. However, as sensing technology develops rapidly, variable kinds of sensing data coexists in large-scale sensor networks. The correlation between the sensing data diminishes, and energy efficiency through in-network processing relatively decreases. Thus, with low duty cycle scheme, short communication bandwidth results in high contention between sensor nodes. In sporadic traffic, asynchronous protocols such as B-MAC [8] and WiseMAC [9], which rely on low power listening called preamble sampling, have a significant effect on performance. However, if the traffic is increased, the low duty cycle scheme is always not guaranteed due to the overhearing.

This paper presents *Coexistence-MAC (Co-MAC)* designed for large-scale sensor networks. Our protocol divides an overall network into independent *subnets*, and each subnet orthogonally operates on time line in a temporal fashion. Our goal is evenly distributed sensor nodes in a certain geographical area into subnets at the association process in order to handle diverse traffic.

The remainder of this paper is organized as follows. Section 2 describes the Co-MAC protocol design, and section 3 analyzes its performance, considering energy efficiency which is a significant factor in designing wireless sensor networks. The related work, conclusion, and future work are presented in section 4 and section 5, respectively.

2 Co-MAC Protocol Design

Co-MAC is a slotted protocol like S-MAC [2] and T-MAC [3], which divides time into discrete intervals (*slots*) and schedule whether the radio is in active mode or sleep mode in terms of these slots. In such protocol, the data transmission and reception are enabled if sensor nodes share identical slots for activation. In Co-MAC, we define N slots as a *frame*, and split up a frame into M *segments* organized by $L=N/M$ slots each. For activation, each sensor node chooses one among M segments at association

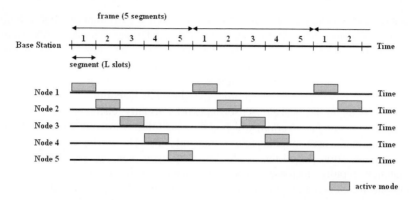

Fig. 1. A time-line of five sensor nodes running the Co-MAC protocol

process with probability *p*, and only operates on the selected segment periodically. Sensor nodes, which operate on an identical segment, construct a *subnet* in an overall network. The fundamental idea of Co-MAC design is to evenly distribute sensor nodes *in a certain geographical area* to subnets to diminish contention between the sensor nodes. Therefore, the duty cycle of sensor nodes should be on the order of *1/M*, assuming that all *L* slots of the segment are in active mode. We can expect that increasing the factor *M* reduces the duty cycle.

Figure 1 shows a simple case of Co-MAC protocol. The five sensor nodes are within the communication range from base station. Note that this case is only considered for single hop within the reachable communication range from base station, but Co-MAC is inherently designed for multi-hop networks. Therefore, the shown example is interpreted as a subset among the overall network.

In the rest of this section, the policy of SYNC packet transmission for synchronization and Co-MAC probability model for association are described in detail.

2.1 SYNC Packet Transmission Scheme for Synchronization

We define *f* frames as a *round*. The base station broadcasts SYNC packet at the beginning of one among M segments in the first frame of the round. As above described, a subnet is constructed with sensor nodes, which periodically operate on an identical one among M segments. In the round, synchronization is only achieved in the subnet which periodically operates on the segment broadcasted by the base station. At the second frame in the round, sensor nodes, which receive the SYNC packet from the base station, attempts to broadcast SYNC packet at the beginning of the segment.

To avoid collision between SYNC packets, the sensor nodes use a *small and fixed contention window (CW)*. For picking a contention slot, we employ the *Sift, the truncated geometrically-increasing probability distribution*, developed by K. Jamieson and H. Balakrishnan [12]. The Sift probability distribution is given by

$$p(r) = \frac{(1-\alpha) \cdot \alpha^{CW}}{1-\alpha^{CW}} \cdot \alpha^{-r} \quad for \ \ r = 1,...,CW \tag{1}$$

where $0 < \alpha < 1$ is a parameter for probability distribution. In the work [16], a near-optimal choice of α for a wide range of population sizes is

$$\alpha = N_1^{-\frac{1}{CW-1}} \tag{2}$$

where N_1 is a fixed parameter that defines the maximum population size. In Co-MAC, the N_1 means that the maximum number of sensor nodes which attempt to broadcast SYNC packet each frame. It is difficult to evaluate the number of sensor nodes which preciously attempt to broadcast the SYNC packet. In this situation, by using the Sift probability distribution, only an extremely small portion of sensor nodes choose low slot numbers, and attempt to broadcast SYNC packet.

A sensor node, which selects the least slot number, broadcasts the SYNC packet at the beginning of third frame in the round. If sensor nodes, which simultaneously attempt to broadcast SYNC packet in the same frame, receive the SYNC packet, they abandon the attempt. Note that sensor nodes, which receive the SYNC packet, continuously neglect SYNC packets received in the round. On the other hand, sensor nodes, which newly receive the SYNC packet in the round, maintain synchronization and reserve the contention window for SYNC packet transmission by using the Sift probability distribution with $p(r)$.

The use of a small and fixed contention window causes time delay which results in time gap between sensor nodes for synchronization. To compensate for loss by the time delay, sensor nodes broadcast the SYNC packet which includes the number of slots, r used for contention window. Sensor nodes, which receive the SYNC packet, apply r to calculation for correct synchronization.

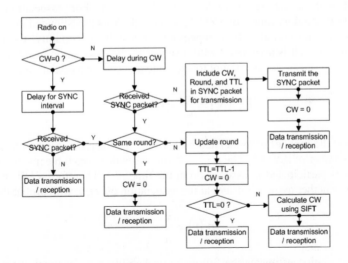

Fig. 2. Block diagram for SYNC packet transmission scheme

Because a round is constructed with total f frames, a SYNC packet transmission, which is begun from the base station, is able to generate until maximum f times. By contrast, at the last segment in a round, every SYNC packet transmission should be autonomously expired for another subnet. To achieve the autonomous expiration of SYNC packet transmission, we employ *Time-To-Live (TTL)*, which is a limit on number of SYNC packet transmission. We configure the initial value of TTL as f. When a certain sensor node transmits SYNC packet, TTL in the SYNC packet decreases by 1. Once TTL in SYNC packet decreases to zero, the SYNC packet will be dropped and no longer forward.

In the next round, the base station broadcasts the SYNC packet at the beginning of next segment in round robin fashion, and the above work is repeated. Total M rounds based on M segments in a frame periodically operate on the time line.

Figure 2 shows the block diagram of SYNC packet transmission scheme planned on sensor nodes. Each sensor node maintains three variables, CW, Round, and TTL for SYNC packet transmission scheme. The CW is a fixed and small contention window for SYNC packet transmission. If the value of CW is 0, SYNC packet is not transmitted in the frame. The sequence number of round managed by the base station is included in SYNC packet. TTL is a limit on the number of SYNC packet transmission.

2.2 Co-MAC Probability Model for Association

For association, Co-MAC proposes a probability model inspired by LEACH [13]. When a SYNC packet is generated by a certain sensor node, each pending sensor node, i, which received the SYNC packet, determines itself whether associate or not with probability $p_i(t)$. We define this situation as *association event* at the time of t. The fundamental idea and implementation are to evenly distribute sensor nodes in a certain geographic area by restricting the number of sensor nodes, k, which is associated with a subnet at each association event. For association, lopsided opportunity by random routes of SYNC packet and errors by estimated probability are reasons for the uncertainty. To compensate for loss by uncertainty, the number of sensor nodes, which is associated with a subnet at the association event, is restricted as k. Thus, if n sensor nodes attempt to associate at a certain association event

$$E[AN] = \sum_{i=1}^{n} P_i(t) = k \tag{3}$$

where AN is the number of associated sensor node at a certain association event. The expected number of AN is k. The n is an estimated value which means the number of sensor nodes participated in association at the time of t. Therefore, each sensor node determines whether associate or not at the association event, t, with probability

$$p_i(t) = \frac{k}{n} \tag{4}$$

The base station estimates the following probability $p_i(t)$, contains the probability $p_i(t)$ into SYNC packet, and broadcast the SYNC packet at the first frame of each round. According to the above described SYNC packet transmission scheme, the SYNC packets are transmitted from the base station or associated sensor nodes. If pending sensor nodes receive the SYNC packet, they determine whether associate or not with the probability $p_i(t)$ in the SYNC packet.

The k in the probability $p_i(t)$ is deterministic, but the n is an estimated value. The base station updates n each round shown below:

$$n = \frac{N_{pending}}{N_{total}} \cdot S \tag{5}$$

where N_{total} is the number of total sensor nodes, and $N_{pending}$ is the number of pending sensor nodes in an overall network. The S is the number of sensor nodes, which is within communication range of a certain sensor node, and we assume that this value is fixed.

Due to unidirectional SYNC packet transmission scheme, sensor nodes in an overall network spatially have a fair opportunity for association comparatively. Also, by configuring k as a low value, we expect that the number of associated sensor nodes is regularly maintained at each round. Therefore, every sensor node is evenly distributed to subnets at the association process.

3 Performance Evaluation

In our simulations, we consider a deployment scenario for large-scale sensor networks on a unit square region R. The disk is configured as an independent region like a room, or the inside of building. We randomly generate N disks that each radius is l to simulate deployments. Next, N disks are independently and randomly distributed in the unit square region R with uniform distribution, and n sensor nodes are distributed in each disk with randomly uniform distribution as well.

Energy efficiency is evaluated on the deployment scenario between S-MAC, T-MAC and Co-MAC.

3.1 Deployment of Sensor Nodes

Senor nodes are deployed on the unit square region R. As described above, we independently and randomly generate N distributed disks, are distributed in the square region R with 500×500 size. 555 sensor nodes are deployed in the each disk with uniform distribution.

Figure 3 shows the deployment result. In this case, each sensor node has an equal likelihood of locating at a certain geographical area. Figure 4 illustrates the simulation result based on the deployment of sensor nodes in figure 3, and sensor nodes in the deployment are performed on total 5 segments (subnets). Sensor nodes marked with a given symbol belong to the identical subnet, and total 5 subnets exist in the network.

For sift probability distribution, we configure α and CW as 0.691 and 16, respectively. For the probability $p_i(t)$ for association, parameter k is defined as 1. The area A represents the reachable communication range within a sensor node located at (200, 200). In a geographical area A, we can verify that the sensor nodes were evenly distributed into each subnet based on intuitional analysis. Precisely, each subnet includes 24, 23, 20, 20 and 19 number of sensor nodes, respectively. If sensor nodes in the area A operate on an identical segment, the sensor node located at (200, 200)

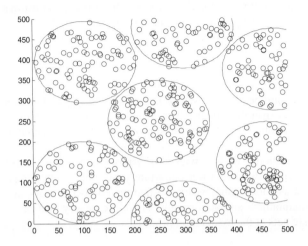

Fig. 3. Deployment of sensor nodes. In 500×500 size, 7 disks are randomly deployed with each radius 100, and 555 sensor nodes is randomly deployed in the each disk with uniform probability.

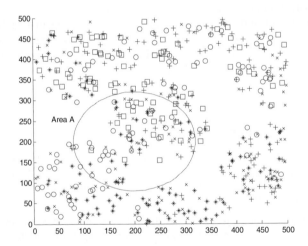

Fig. 4. Subnet formation based on deployment of sensor nodes in Figure 3

contends with 106 sensor nodes in the worst case. In the overall network, each subnet contains 101, 130, 140, 98 and 86 number of sensor nodes, correspondingly. For that reason, our network is constructed into 5 independent networks in the simulation. By administering 5 small-scale networks than a vast network, we expect that the tremendous burden of sensor nodes is reduced to some extent for contention between sensor nodes, memory usage, and energy usage, and so forth.

The disparity of the geographical deployment in terms of subnets is caused by uncertainty with various other aspects, such as SYNC packet transmission route, probability for association, and etc. Sensor nodes around the base station are fairly distributed into subnets by relatively even opportunities of SYNC packet reception and updated probability for association. On the other hand, in the edge of the network, the geographical deployment in terms of subnets is shown as a trend that sensor nodes are relatively concentrated into a small number of a certain minority subnet due to previously mentioned uncertainty.

3.2 Energy Consumption

In this section, we compare S-MAC, T-MAC, and Co-MAC by evaluating the energy consumption under the above described deployment scenario. In our simulation, we have referred EYE nodes [17] for data transmission and reception. Some of the parameters for simulation are listed in Table 1 and Table 2.

Table 1. Power breakdown EYE nodes

Parameter	Value
CPU	
Active	2.1 mA
Sleep	1.6 μA
Radio	
Transmit	10 mA
Receive	4 mA
Sleep	20 μA

Table 2. Parameters for Analysis

Parameter	Value
Bandwidth	115 Kbps
Data packet (T_d sec)	150 Bytes
SYNC packet	12 Bytes
Co-MAC	
Segment time (T_{seg})	0.4 s
# of Segment (M)	5
S-MAC / T-MAC	
Frame time (T_f)	2 s
Active period (T_a)	0.4 s
Activity time-out (TA)	15 ms
Contend time (T_c)	3.05-9.15ms

Our simulation employs constant-bit-rate traffic. Sensor nodes for data transmission are randomly selected during the uniform time. For simplicity, RTS/CTS mechanism and link level acknowledgement is ignored, and the traffic is only considered as data transmission. The maximum duty cycle is equally configured as 20% in S-MAC, T-MAC, and Co-MAC. By contrast, the duty cycle of T-MAC and Co-MAC in the simulation may be applied according to the traffic below 20% by using timeout (TA) scheme which adaptively ends the active time.

In S-MAC protocol, total energy consumption of a certain sensor node in a frame is shown as below

$$E_{smac} = n_t(T_c + T_d) \cdot P_t + n_r(T_c + T_d) \cdot P_r$$
$$+ [T_a - n_t(T_c + T_d) - n_r(T_c + T_d)] \cdot P_r + [T_f - T_a] \cdot P_s \tag{6}$$

where n_t is the number of data transmission, and n_r is the number of data reception in the sensor node. P_t is transmission power, P_r is receiving power.

Total energy consumption of T-MAC each frame in a certain sensor node is given by

$$E_{tmac} = n_t\left(T_c + T_d\right)\cdot P_t + n_r\left(T_c + T_d\right)\cdot p_r + TA\cdot P_r$$
$$+\left[T_f - n_t\left(T_c + T_d\right) - n_r\left(T_c + T_d\right) - TA\right]\cdot P_s \tag{7}$$

In Co-MAC with timeout, total energy consumption is

$$E_{comac} = n_t\left(T_c + T_d\right)\cdot P_t + n_r\left(T_c + T_d\right)\cdot p_r + TA\cdot P_r$$
$$+\left[T_{seg}\cdot M - n_t\left(T_c + T_d\right) - n_r\left(T_c + T_d\right) - TA\right]\cdot P_s \tag{8}$$

Figure 5 shows average energy consumption between S-MAC, T-MAC and Co-MAC with constant-bit-rate traffic under the proposed deployment scenario for both small-scale network (50 sensor nodes) and large-scale network (500 sensor nodes). Co-MAC outperforms S-MAC by 6 times and T-MAC by 2 times in small-scale sensor network. In large-scale sensor network, Co-MAC outperforms S-MAC 8 times and T-MAC by 2 times. The reason for this phenomenon is that the overhearing frequency between sensor nodes is relatively lower in Co-MAC.

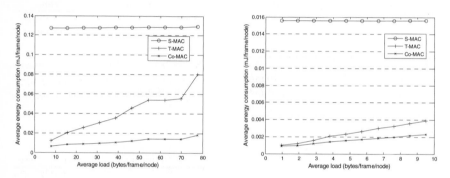

Fig. 5. Average current consumption using constant-bit-rate traffic between S-MAC, T-MAC, and Co-MAC for small-scale network (left) and large-scale network (right)

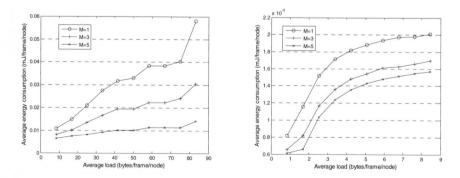

Fig. 6. Average current consumption using constant-bit-rate traffic according to number of segments in a frame for small-scale network (left) and large-scale network (right)

Figure 6 illustrates average energy consumption according to the number of segments in a frame. The fixed duty cycle of sensor nodes should be on the order of $1/M$, where M is the number of segments in a frame. In our simulation, increasing the factor M reduces the duty cycle, and relatively decreases overhearing frequency as well. In large-scale sensor network, the simulation results show that Co-MAC ($M=3$) consumes 1.2 times more than Co-MAC ($M=5$), and preserve 1.5 times more than Co-MAC ($M=1$). In small-scale sensor network, Co-MAC ($M=3$) consumes 2 times more than Co-MAC ($M=5$), and conserve 1.8 times more than Co-MAC ($M=1$).

4 Related Work

In wireless sensor networks, a number of MAC protocols have proposed over recent years for reason of the novel constraints, particularly energy conservation. The choice of an appropriate MAC protocol is important factor for the performance of wireless sensor networks.

These MAC protocols fall into two basic classes, slotted protocols and sampling protocols [1]. Slotted protocols divide time into discrete intervals (slots) and schedule whether the radio is in active mode and sleep mode in terms of slots. S-MAC [2] and T-MAC [3] is representative slotted protocols as benchmarks for the area of MAC protocols of wireless sensor networks.

S-MAC is an energy efficient MAC protocol for wireless sensor network inspired by PAMAS [6] and IEEE 802.11 [7]. In S-MAC, sensor nodes maintain static duty cycle to listen for channel activity and transmit data. For communications with neighboring nodes, S-MAC schedules periodic listen-sleep scheme based on locally managed synchronization throughout SYNC packet exchange. T-MAC improves energy efficiency of S-MAC by dynamically adapting the listen interval according to the load variations in time and location. In other words, T-MAC employs an adaptive duty cycle. In these slotted protocols like S-MAC and T-MAC, as communication periods of sensor nodes decrease for the reason of low duty cycle, contention rate between sensor nodes increase. Also, synchronization maintenance costs both power and bandwidth.

The second class, sampling protocols, take a different approach in comparison with slotted protocols. These protocols consist in regularly sampling the medium to check for activity. By sampling the channel, sensor nodes listen to the radio channel for a short duration. If the channel is busy, sensor nodes decode the medium until a data is received or until the medium becomes idle again. The transmitter uses preamble sampling technique [10] that a preamble to wake up neighboring nodes is inserted in front of the data packet to guarantee that the receiver will be awake before the data packet arrives.

B-MAC [8], which is used as the default MAC for Mica2, employs an adaptive preamble sampling which is called Low Power Listening (LPL) for low power consumption, clear channel assessment (CCA) and packet backoffs for channel arbitration, and link layer acknowledgments for reliability. However, sampling protocols like B-MAC must transmit expensive messages to wake up neighboring nodes.

WiseMAC [9] uses non-persistent CSMA with preamble sampling as in [10] to reduce idle listening. The novel idea of WiseMAC is a scheme that learns the sampling schedule of direct neighbors and exploits this knowledge to minimize the wake-up preamble length.

With the classification, diverse MAC designs have proposed as research topics relative to higher level protocols or special applications. Among these protocols, CSMA/p^* [11] and Sift [12] are MAC protocols proposed for event-driven environments. Both protocols are non-persistent CSMA with a chosen probability distribution p^* that sensor nodes utilize to randomly select contention slots. CSMA/p^* is used for a network where the number of senders of N relative to the event is known. On the other hand, Sift adjusts CSMA/p^* for a network where N is unknown. However, the optimal probability distribution operates only when sensor nodes simultaneously participate in contention for data transmission generated by event.

5   Conclusion and Future Work

In this paper, we have proposed an energy-efficient MAC protocol for large-scale sensor networks. The Co-MAC is a complementary version of S-MAC and T-MAC to efficiently handle variable traffic in low duty cycle scheme. The key idea of Co-MAC protocol described in the paper is to evenly distribute sensor nodes in a certain geographical area into independent subnets divided in the temporal fashion. Our preliminary performance results show that the proposed approach improves energy efficiency with low duty cycle scheme in both large-scale sensor networks and small-scale sensor networks.

In Co-MAC, network configuration is necessarily achieved by the base station, and managing the network is accomplished by the base station as well. Future work will diminish the dependency of the base station for robustness of the network. Furthermore, the solution for disparity of the geographical deployment in terms of subnets will be studied.

Acknowledgments

This work was supported by the SRC/ERC program of MOST/KOSEF (grant R11-1999-058-01007-0).

References

1. Polastre, J., Hui, J., Levis, P., Zhao, J., Culler, D., Shenker, S., Stoica, I.: A Unifying Link Abstraction for Wireless Sensor Networks. In: Proc. the ACM SenSys Conf., San Diego, California (2005)
2. Ye, W., Heidemann, J., Estrin, D.: Medium Access Control With Coordinated Adaptive Sleeping for Wireless Sensor Networks. IEEE/ACM Trans. on Networking, vol. 493–506 (2004)
3. Van Dam, T., Langendoen, K.: An Adaptive Energy-Efficient MAC Protocol for Wireless Sensor Networks. In: Proc. the ACM SenSys Conf, Los Angeles, California (2003)

4. Stemm, M., Katz, r.H: Measuring and reducing energy consumption of network interfaces in hand-held devices. IEICE Trans. on communications, pp. 1125–1131 (1997)
5. Kasten, O.: Energy Consumption, Eldgenossische Technishche Hochschule Zurich. [online] http://www.Inf.ethz.ch/kasten/researh/bathtub/energy_ consumption.html
6. Singh, S., Raghavendra, C.S.: PAMAS: Power Aware Multi-Access protocol with Signaling for Ad Hoc Networks. ACM Computer Communication Review, pp. 5–26 (1998)
7. LAN MAN Standards Committee of the IEEE Computer Society.: IEEE Std 802.11 1999, wireless LAN Medium Access Control (MAC) and Physical layer (PHY) specification. IEEE (1999)
8. Polastre, J., Hill, J., Culler, D.: Versatile Low Power Media Access for Wireless Sensor Networks. In: Proc. The ACM SenSys Conf., Baltimore, Maryland (2004)
9. Enz, C.C., El-Hoiydi, A., Decotignie, J.-D., Peiris, V.: WiseNET : An ultralow-Power Wireless Sensor Network solution. IEEE Computer Society Press, Los Alamitos (2004)
10. El-Hoiydi, A.: Spatial TDMA and CSMA with Preamble Sampling for Low Power Ad Hoc Wireless Sensor Networks. In: Proc ISCC, pp. 685–692 (2002)
11. Tay, Y.C., Jamieson, K., Balakrishnan, H.: Collision-Minimizing CSMA and Its Applications to Wireless Sensor Networks. IEEE Journal on selected areas in Communications, pp. 1048–1057 (2004)
12. Jamieson, K., Balakrishnan, H., Tay, Y.C.: Sift: A MAC Protocol for Event-Driven Wireless Sensor Networks. In: Römer, K., Karl, H., Mattern, F. (eds.) EWSN 2006. LNCS, vol. 3868, pp. 260–275. Springer, Heidelberg (2006)
13. Heinzelman, W.B., Chandrakasan, A.P., Balakrishnan, H.: An Application-Specific Protocol Architecture for Wireless Microsensor Networks. IEEE Transaction on Wireless Communications, pp. 660–670 (2002)
14. Chintalapudi, K., Fu, T., Paek, J., Kothari, N., Rangwala, S., Caffrey, J., Govindan, R., Johnson, E., Masri, S.: Monitoring Civil Structures with a Wireless Sensor Network. IEEE Internet Computing, pp. 26–34 (2006)
15. Mainwaring, A., Polastre, J., Szewczyk, R., Culler, D., Anderson, J.: Wireless Sensor Networks for Habitat Monitoring. WSNA conf., Atlanta, Georgia (2002)
16. Jamieson, K., Balakrishnan, H., Tay, Y.C.: Sift: A MAC Protocol for Event-Driven Wireless Sensor networks. Technical Report MIT-LCS-TR-894, Massachusetts Institute of Technology (2003)
17. Halkes, G.P., Van Dam, T., Langgendoen, K.G.: Comparing Energy-Saving MAC protocols for Wireless Sensor Networks. Mobile Networks and Applications, vol. 783–791 (2005)

Performance Evaluation of IEEE 802.11 DCF Networks

Krzysztof Szczypiorski and Józef Lubacz

Warsaw University of Technology, Institute of Telecommunications,
ul. Nowowiejska 15/19, 00-665 Warsaw, Poland
{ksz, jl}@tele.pw.edu.pl

Abstract. The paper presents a new analytical saturation throughput model of IEEE 802.11 DCF (*Distributed Coordination Function*) with basic access in ad-hoc mode. The model takes into account *freezing of the backoff timer* when a station senses a busy channel. It is shown that taking into account this feature of DCF is important in modeling saturation throughput – yields more accurate results than models known from literature. The proposed analytical model also takes into account the effect of transmission errors.

Keywords: WLAN, IEEE 802.11, DCF, CSMA/CA, modeling.

1 Introduction

The paper concerns IEEE 802.11 DCF (*Distributed Coordination Function*), also referred to as CSMA/CA (*Carrier Sense Multiple Access with Collision Avoidance*), with *basic access in ad-hoc mode* [7]. For IEEE 802.11 DCF networks a new analytical model for throughput evaluation is proposed, assuming saturated conditions, i.e. when all stations involved in transmission have no empty queues. Saturation throughput is an efficiency measure of maximum load in saturated conditions. According to the DCF protocol, when a station senses a busy channel the backoff is suspended in effect of *freezing of the station backoff timer*. It is shown that taking into account this feature of DCF is important in modeling saturation throughput – yields more accurate results than models known from literature. Moreover, the influence of transmission errors is taken into account. The proposed model is based on a Markov chain.

According the authors' knowledge the first analytical model of DCF was proposed by G. Bianchi [2]. Bianchi proposed a Markov chain based model to evaluate saturation throughput, assuming a finite number of stations and ideal channel conditions (no errors).

H. Wu *et al.* [18] modified Bianchi's model through introducing a limit on the number of retransmissions (maximum number of backoff stages) and a maximum size of the contention window. E. Ziouva and T. Antonakopoulos [19], and probably independently M. Ergen and P. Varaiya [5], extended Bianchi's model through taking into account freezing of the backoff timer during a busy channel occurrences. In [19] it is assumed that, after successful transmission, a station can access the medium without backoff; this assumption does not comply with the IEEE 802.11 standard [7]. In [5] the presented analytical solution of the introduced Markov chain is erroneous.

L. Mason, T. Drwiega, and J. Yan (Eds.): ITC 2007, LNCS 4516, pp. 1084–1095, 2007.

The above mentioned models assume ideal channel conditions, i.e. no transmission errors. P. Chatzimisios *et al.* [4] and Q. Ni *et al.* [14] extended Wu's model [18] to take account of transmission failure. In [14] ACK frames loss due to errors is taken into account; in [4] ACK frames loss is not considered.

In [2], [5] and [18] RTS/CTS (*Request to Send/Clear to Send*) is considered, but without taking into account the two independent retransmission counters: SLRC – *Station Long Transmission Retry* and SSRC – *Station Short Transmission Retry*. In effect these models cannot be extended to take account of transmission errors. In [4] and [14] transmission errors are considered, however only for the case of basic access (i.e. only with the account of the SLRC counter).

It should be noted that in [2], [4] and [18] the authors have mistakenly taken DIFS (*DCF InterFrame Space*) for EIFS (*Extended InterFrame Space*). This mistake does not however have a very important impact on the evaluation of saturation throughput.

All the aforementioned analyses are based on Markov chains. Also other approaches were presented, e.g. in [1], [3] and [15]. These approaches make several simplifying assumptions and thus do not take into account important features of DCF.

The model presented in this paper is, generally speaking, in line with the extensions of the basic Bianchi's model [2] which were proposed in [18] and [14]. The essential difference of the presented model with respect to the latter two is in that it takes into account the effect of freezing of the stations' backoff timer along with the limitation of the number of retransmissions, maximum size of the contention window and the impact of transmission errors.

2   The Model

2.1 Assumptions

1. Saturated conditions are considered; stations have no empty queues – there is always a frame to be sent.
2. n stations compete for medium access (for $n=1$ only one station sends frames to other station which can only reply with ACK).
3. Errors in the transmission medium are randomly distributed; this is the worst case for the *frame error rate* – FER. All stations have the same *bit error rate* (BER).
4. All stations are in transmission range and there are no hidden terminals.
5. Stations communicate in ad hoc mode (BSS – *Basic Service Set*) with basic access method.
6. All stations use the same physical layer (PHY).
7. The transmission data rate R is the same and constant for all stations.
8. All frames are of constant length L.
9. Only data frames and ACK frames are exchanged.
10. Collided frames are discarded – the capture effect [12] is not considered.

2.2 Saturation Throughput S Expressed Through Characteristics of the Physical Channel

The saturation throughput S is defined as in [2]:

$$S = \frac{E[DATA]}{E[T]} \tag{1}$$

where E[*DATA*] is the mean value of the successfully transmitted payload, and E[*T*] is the mean value of the duration of the following *channel states*:

T_I – idle slot,
T_S – successful transmission,
T_C – transmission with collision,
T_{E_DATA} – unsuccessful transmission with data frame error,
T_{E_ACK} – unsuccessful transmission with ACK error.

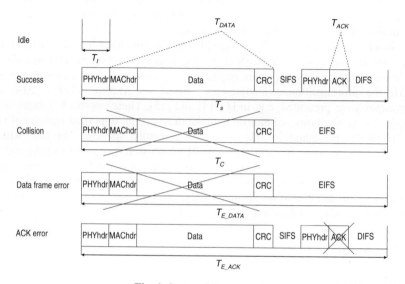

Fig. 1. States of the channel

Fig. 1 illustrates the dependence of the above channel states on:

T_{PHYhdr} – duration of a PLCP (*PHY Layer Convergence Procedure*) preamble and a PLCP header,
T_{DATA} – duration to transmit a data frame,
T_{ACK} – duration to transmit an ACK frame,
T_{SIFS} – duration of SIFS (*Short InterFrame Space*),
T_{DIFS} – duration of DIFS,
T_{EIFS} – duration of EIFS.

The relation of the saturation throughput to physical channel characteristics is calculated similarly as in [14]:

$$\begin{cases} T_I = \sigma \\ T_S = 2T_{PHYhdr} + T_{DATA} + 2\delta + T_{SIFS} + T_{ACK} + T_{DIFS} \\ T_C = T_{PHYhdr} + T_{DATA} + \delta + T_{EIFS} \\ T_{E\_DATA} = T_{PHYhdr} + \delta + T_{DATA} + T_{EIFS} \\ T_{E\_ACK} = T_S \end{cases} \quad (2)$$

where σ is the duration of idle slot (*aSlotTime* [7]) and δ is the propagation delay.

For OFDM (*Orthogonal Frequency Division Multiplexing*) PHY, i.e. 802.11a [8] and 802.11g [11]:

$$T_{ACK} = T_{symbol} \left\lceil \frac{L_{SER} + L_{TAIL} + L_{ACK}}{N_{BpS}} \right\rceil \tag{3}$$

$$T_{DATA} = T_{symbol} \left\lceil \frac{L_{SER} + L_{TAIL} + L_{DATA}}{N_{BpS}} \right\rceil \tag{4}$$

where:
T_{symbol} – duration of a transmission symbol,
L_{SER} – ODFM PHY layer SERVICE field size,
L_{TAIL} – OFDM PHY layer TAIL fields size,
N_{BpS} – number of encoded bits per one symbol,
L_{ACK} – size of an ACK frame,
L_{DATA} – size of a data frame.

For DSSS (*Direct Sequence Spread Spectrum*) PHY (i.e. 802.11 1 and 2 Mbps [7], 802.11b [9] with long preamble) formulas (3) and (4) may be applied with $L_{SER}=L_{TAIL}=0$ (there are no such fields). Values of σ, T_{PHYhdr}, T_{SIFS}, T_{DIFS}, T_{EIFS}, T_{symbol}, N_{BpS}, L_{SER} and L_{TAIL} are defined in accordance with 802.11 standard ([7], [8], [9], or [11]).

Probabilities corresponding to states of the channel are denoted as follows:
P_I – probability of idle slot,
P_S – probability of successful transmission,
P_C – probability of collision,
P_{E_DATA} – probability of unsuccessful transmission due to data frame error,
P_{E_ACK} – probability of unsuccessful transmission due to ACK error.

Let τ be the probability of frame transmission, p_{e_data} the probability of data frame error and p_{e_ACK} the probability of ACK error. These are related to channel state probabilities as follows:

$$\begin{cases} P_I = (1-\tau)^n \\ P_S = n\tau(1-\tau)^{n-1}(1-p_{e\_data})(1-p_{e\_ACK}) \\ P_C = 1-(1-\tau)^n - n\tau(1-\tau)^{n-1} \\ P_{E\_DATA} = n\tau(1-\tau)^{n-1}p_{e\_data} \\ P_{E\_ACK} = n\tau(1-\tau)^{n-1}(1-p_{e\_data})p_{e\_ACK} \end{cases} \tag{5}$$

The saturation throughput S equals

$$S = \frac{P_S L_{pld}}{T_I P_I + T_S P_S + T_C P_C + T_{E\_DATA}P_{E\_DATA} + T_{E\_ACK}P_{E\_ACK}} \tag{6}$$

where L_{pld} is MAC (*Medium Access Control*) payload size and $L_{pld} = L - L_{MAChdr}$, where L_{MAChdr} is the size of the MAC header plus the size of FCS (*Frame Checksum Sequence*).

S can be normalized to data rate R:

$$\overline{S} = \frac{S}{R} \qquad (7)$$

where

$$R = \frac{N_{BpS}}{T_{symbol}} \qquad (8)$$

As a result, saturation throughput S is expressed as a function of τ, p_{e_data} and p_{e_ACK}. In the following sections these probabilities are evaluated.

2.3   Probability of Frame Transmission τ

Let $s(t)$ be a random variable describing DCF backoff stage at time t, with values from set $\{0, 1, 2,...,m\}$. Let $b(t)$ be a random variable describing the value of the backoff timer at time t, with values from set $\{0, 1, 2,..., W_i\text{-}1\}$. These random variables are dependent because the maximum value of the backoff timer depends on backoff stage:

$$W_i = \begin{cases} 2^i W_0, & i \le m' \\ 2^{m'} W_0 = W_m, & i > m' \end{cases} \qquad (9)$$

where W_0 is an initial size of contention window and m' is a maximum number by which the contention window can be doubled; m' can be both greater and smaller than m and also equal to m. W_0 and $W_{m'}$ depend on CW_{min} and CW_{max} [7]:

$$W_0 = CW_{min} + 1 \qquad (10)$$

$$W_{m'} = CW_{max} + 1 = 2^{m'} W_0 \qquad (11)$$

The two-dimensional process $(s(t), b(t))$ will be analyzed with an embedded Markov chain (in steady state) at time instants at which the channel state changes. Let (i,k) denote the state of this process. The one-step conditional state transition probabilities will be denoted by $P = (\cdot,\cdot\,|\,\cdot,\cdot)$.

Let p_f be the probability of transmission failure and p_{coll} the probability of collision. The non-null transition probabilities are determined as follows (comp. Fig. 2):

$$\begin{aligned}
&(a)\, P(i,k\,|\,i,k+1) = 1 - p_{coll}, && 0 \le i \le m, 0 \le k \le W_i - 2 \\
&(b)\, P(i,k\,|\,i,k) = p_{coll}, && 0 \le i \le m, 1 \le k \le W_i - 1 \\
&(c)\, P(0,k\,|\,i,0) = (1 - p_f)/W_0, && 0 \le i \le m-1, 0 \le k \le W_0 - 1 \\
&(d)\, P(i,k\,|\,i-1,0) = p_f/W_i, && 1 \le i \le m, 0 \le k \le W_i - 1 \\
&(e)\, P(0,k\,|\,m,0) = 1/W_0, && 0 \le k \le W_0 - 1
\end{aligned} \qquad (12)$$

Ad (a): The station's backoff timer is decremented from $k+1$ to k at fixed i backoff stage, i.e. the station has detected an idle slot, so the channel is idle. The probability of this event $Pr\{channel\ is\ idle\} = 1 - Pr\{one\ or\ more\ station\ is\ transmitting\}$. We consider saturated conditions, so $Pr\{one\ or\ more\ station\ is\ transmitting\}$ equals p_{coll}.

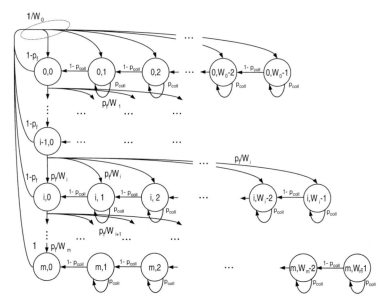

Fig. 2. Markov chain transitions

Ad (b): The station's backoff timer is frozen at fixed i backoff stage, i.e. the channel is busy. *Pr{channel is busy} = Pr{one or more station is transmitting} = p_{coll}.*

Ad (c): The station's backoff timer is changed from 0 to k and the backoff stage is changed from i to 0. The probability of this event equals: *Pr{transmission is successful and number k was randomly chosen to initiate the backoff timer at stage 0} = Pr{transmission is successful}· Pr{number k was randomly chosen to initiate the backoff timer at stage 0}.* The probability of successful transmission is equal to $1 - p_f$ and the probability that number k was randomly chosen to initiate the backoff timer at stage 0 equals $1/W_0$.

Ad (d): The station's backoff timer is changed from 0 to k and the backoff stage is changed from $i-1$ to i. The probability of this event equals: *Pr{transmission is unsuccessful and number k was randomly chosen to initiate the backoff timer at stage i}= Pr{transmission is unsuccessful}· Pr{number k was randomly chosen to initiate the backoff timer at stage i}.* The probability of unsuccessful transmission equals p_f and the probability that number k was randomly chosen to initiate the backoff timer at stage i equals $1/W_i$.

Ad (e): The station's backoff timer is changed from 0 to k and the backoff stage is changed from m to 0, i.e. the station has reached maximum retransmission count. The probability of this event equals the probability that number k was randomly chosen to initiate the backoff timer at stage 0, i.e. $1/W_0$.

Let $b_{i,k}$ be the probability of state (i,k). It can be shown that:

$$b_{i,0} = p_f \cdot b_{i-1,0} \tag{13}$$

$$b_{i,0} = p_f^{\ i} \cdot b_{0,0} \tag{14}$$

and

$$b_{i,k} = \begin{cases} \dfrac{W_i - k}{W_i(1 - p_{coll})} p_f^{\,i} \cdot b_{0,0}, & 0 < k \le W_i - 1 \\ p_f^{\,i} \cdot b_{0,0}, & k = 0 \end{cases} \tag{15}$$

From:

$$\sum_{i=0}^{m} \sum_{k=0}^{W_i-1} b_{i,k} = 1 \tag{16}$$

and

$$\sum_{i=0}^{m} b_{i,0} = b_{0,0} \frac{1 - p_f^{\,m+1}}{1 - p_f} \tag{17}$$

we get

$$b_{0,0}^{\,-1} = \begin{cases} \dfrac{(1-p_f)W_0(1-(2p_f)^{m+1}) - (1-2p_f)(1-p_f^{\,m+1})}{2(1-2p_f)(1-p_f)(1-p_{coll})} + \dfrac{1-p_f^{\,m+1}}{1-p_f}, & m \le m' \\[3mm] \dfrac{\Psi}{2(1-2p_f)(1-p_f)(1-p_{coll})} + \dfrac{1-p_f^{\,m+1}}{1-p_f}, & m > m' \end{cases} \tag{18}$$

where

$$\Psi = (1-p_f)W_0(1-(2p_f)^{m'+1}) - (1-2p_f)(1-p_f^{\,m+1}) + W_0 2^{m'} p_f^{\,m'+1}(1-2p_f)(1-p_f^{\,m-m'}) \tag{19}$$

The probability of frame transmission τ is equal to *Pr{backoff timer equals 0}* and thus:

$$\tau = \sum_{i=0}^{m} b_{i,0} =$$

$$= \begin{cases} \left(\dfrac{(1-p_f)W_0(1-(2p_f)^{m+1}) - (1-2p_f)(1-p_f^{\,m+1})}{2(1-2p_f)(1-p_f)(1-p_{coll})} + \dfrac{1-p_f^{\,m+1}}{1-p_f} \right)^{-1} \dfrac{1-p_f^{\,m+1}}{1-p_f}, & m \le m' \\[4mm] \left(\dfrac{\Psi}{2(1-2p_f)(1-p_f)(1-p_{coll})} + \dfrac{1-p_f^{\,m+1}}{1-p_f} \right)^{-1} \dfrac{1-p_f^{\,m+1}}{1-p_f}, & m > m' \end{cases} \tag{20}$$

For $p_{coll} = 0$ the above solution is the same as presented in [14].

2.4 Probability of Transmission Failure p_f and Probability of Collision p_{coll}

The probability of transmission failure

$$p_f = 1 - (1 - p_{coll})(1 - p_e) \tag{21}$$

where p_e is the frame error probability:

$$p_e = 1 - (1 - p_{e\_data})(1 - p_{e\_ACK}) \tag{22}$$

where p_{e_data} is FER for data frames and p_{e_ACK} is FER for ACK frames. p_{e_data} and p_{e_ACK} can be calculated from bit error probability (i.e. BER) p_b:

$$p_{e\_data} = 1 - (1 - p_b)^{L_{data}} \tag{23}$$

$$p_{e\_ACK} = 1 - (1 - p_b)^{L_{ACK}} \tag{24}$$

The probability of collision

$$p_{coll} = 1 - (1 - \tau)^{n-1}$$ (25)

Finally

$$p_f = 1 - (1 - p_{coll})(1 - p_e) = 1 - (1 - \tau)^{n-1}(1 - p_e)$$ (26)

Equations (20) and (26) form a non-linear system with two unknown variables τ and p_f which may be solved numerically.

3 Validation

The presented model was validated with the use of simulation in two steps. The aim of Step 1 was to compare the proposed model with (i) models presented in [2] and [18] in which channel errors are not taken into account, with (ii) the special case of model [14] for which BER is assumed zero, and with (iii) simulations (also presented in [6]). In Step 2, channel errors are taken into account; the accuracy of the presented model is evaluated with simulations presented in [13] and compared to the model presented in [14].

Step 1
The *ns-2* simulator version 2.29 [17] was used. The IEEE 802.11 DSSS 1 Mbps PHY was simulated (OFDM PHY is not implemented in the standard version of *ns-2*). The simulation was performed for saturated conditions with static routing and for 1000 bytes MAC frames UDP traffic. The results are presented in Table 1 and Fig. 3.

The proposed model was also compared with simulation results presented in [6], which were obtained with the simulation tool created at *Universitat Politècnica de Catalunya* in Barcelona [15]. In Table 2 the condition of simulation and simulation results are presented. Note that non-aggregated values of saturation throughput (S/n) are presented.

Step 2
Although the *ns-2* simulator enables to simulate channel errors, the mechanism of errors occurrence is based on physical features and is thus different from the one

Table 1. Normalized values of saturation throughput for IEEE 802.11 DSSS 1 Mbps with $L=1000$ bytes and $BER = 0$

n	Bianchi model [2]	Wu *et al.* model [18]	Ni *et al.* model [14]	Proposed model	Simulation (average)	Standard deviation
1	0.8769	0.8769	0.8769	0.8769	0.8780	0.000
2	0.8666	0.8666	0.8657	0.8661	0.8635	0.000
4	0.8329	0.8329	0.8306	0.8367	0.8354	0.003
10	0.7602	0.7586	0.7540	0.7779	0.7625	0.005
20	0.6929	0.6846	0.6783	0.7238	0.7200	0.002
30	0.6497	0.6330	0.6258	0.6891	0.6872	0.004
50	0.5904	0.5558	0.5477	0.6421	0.6303	0.003
80	0.5297	0.4684	0.4599	0.5955	0.5633	0.004

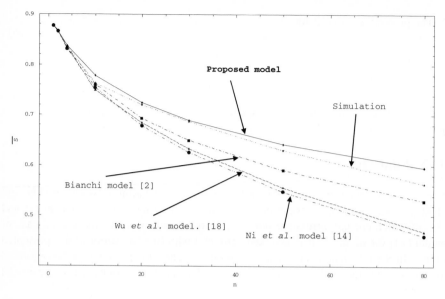

Fig. 3. Normalized saturation throughput: analytical and simulation results for IEEE 802.11 DSSS 1 Mbps with L=1000 bytes and BER=0

Table 2. Non-aggregated values (in Mbps) of saturation throughput for IEEE 802.11g 54 Mbps ERP-OFDM with L=1500 bytes and BER=0

n	Bianchi model [2]	Wu *et al.* model [18]	Ni *et al.* model [14]	Proposed model	Simulation
1	31.36	31.36	31.36	31.36	31.79
2	16.24	16.24	16.15	16.05	16.18
4	7.90	7.90	7.79	7.86	7.85
10	2.87	2.86	2.79	2.93	2.92
15	1.82	1.78	1.72	1.88	1.87
20	1.30	1.26	1.21	1.36	1.36
25	1.00	0.95	0.91	1.06	1.06
50	0.43	0.37	0,35	0.47	0.49
100	0.17	0.11	0.10	0.21	0.22

assumed in the proposed analytical model (randomly distributed bit errors). For this reason the *ns-2* simulator was not used.

The accuracy of the proposed model was compared with results obtained by solving the model presented in [14] for assumptions concerning the physical layer and its parameters which are presented in Table 3 and Fig. 4. The table also presents simulation results which were obtained by the authors of [13] (according to [15], the authors used the same simulation tool as authors of [6], i.e. the simulator mentioned above). In these simulations a random pattern of bit-error occurrence was assumed (i.e. as assumed in the model presented in this paper).

Table 3. Normalized values of saturation throughput for IEEE 802.11g 54 Mbps with L=1500 bytes and $BER=10^{-5}$ and $BER=10^{-4}$

n	Ni *et al.* model [14]	Proposed model	Simulation	Ni *et al.* model [14]	Proposed model	Simulation
	BER=10^{-5}			BER=10^{-4}		
2	0.5246	0.5207	0.5080	0.1425	0.1412	0.0711
4	0.5148	0.5167	0.4989	0.1640	0.1619	0.0993
10	0.4672	0.4880	0.4739	0.1699	0.1705	0.1365
15	0.4354	0.4693	0.4593	0.1640	0.1682	0.1461
20	0.4081	0.4541	0.4476	0.1564	0.1648	0.1517
25	0.3843	0.4413	0.4383	0.1486	0.1612	0.1537
50	0.2906	0.3965	0.4052	0.1128	0.1459	0.1572
100	0.1656	0.3448	0.3607	0.0619	0.1260	0.1504

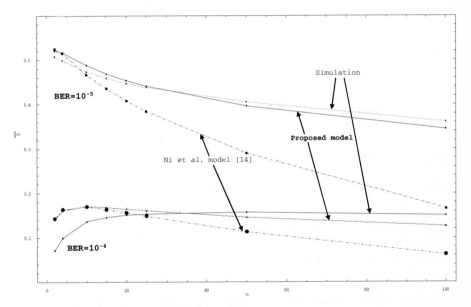

Fig. 4. Normalized saturation throughput: analytical and simulation results for IEEE 802.11g 54 Mbps ERP-OFDM with L=1500 bytes and $BER=10^{-5}$ and $BER=10^{-4}$

To conclude: the results presented in the tables and figures above show that the proposed model has good accuracy both in the case of error-free and error-prone channels. For error-free conditions the model yields some overestimation while other models known from literature tend to underestimate the saturation throughput. For both error-free and error-prone cases the proposed model shows better accuracy than the literature models with which it was compared, especially for large number of stations. The latter is the consequence of the fact that the proposed model takes into account freezing of the backoff timer; the impact of the freezing of the backoff timer on throughput evaluation increases with the increase of the number of station competing for access to the transmission medium.

Future work could be focused on taking into account such features of the IEEE 802.11 protocol as the RTS/CTS and EDCA (*Enhanced Distributed Channel Access*) [10].

References

1. Bianchi, G., Tinnirello, I.: Remarks on IEEE 802.11 DCF Performance Analysis. IEEE Communications Letters 9, 765–767 (2005)
2. Bianchi, G.: Performance Analysis of the IEEE 802.11 Distributed Coordination Function. IEEE Journal on Selected Areas in Communications 18(3), 535–547 (2000)
3. Cali, F., Conti, M., Gregori, E.: Dynamic Tuning of the IEEE 802.11 Protocol to Achieve a Theoretical Throughput Limit. IEEE/ACM Trans. Networking 8(6), 785–799 (2000)
4. Chatzimisios, P., Boucouvalas, A., Vitsas, V.: Influence of Channel BER on IEEE 802.11 DCF. IEE Electronics Letters, vol. 39(23) (2003)
5. Ergen, M., Varaiya, P.: Throughput Analysis and Admission Control in IEEE 802.11a. Springer Mobile Networks and Applications 10(5), 705–706 (2005)
6. Heusse, M., Rousseau, F., Guillier, R., Duda, A.: Idle Sense: An Optimal Access Method for High Throughput and Fairness in Rate Diverse Wireless LANs. In: SIGCOMM'05 Conference on Applications, Technologies, Architectures and Protocols for Computer Communications, Philadelphia, pp. 121–132 (2005)
7. IEEE 802.11, 1999 Edition (ISO/IEC 8802-11: 1999) IEEE Standards for Information Technology – Telecommunications and Information Exchange between Systems – Local and Metropolitan Area Network – Specific Requirements – Part 11: Wireless LAN Medium Access Control (MAC) and Physical Layer (PHY) Specifications (1999)
8. IEEE 802.11a-1999 (8802-11:1999/Amd 1:2000(E)), IEEE Standard for Information technology – Telecommunications and information exchange between systems – Local and metropolitan area networks – Specific requirements – Part 11: Wireless LAN Medium Access Control (MAC) and Physical Layer (PHY) specifications – Amendment 1: High-speed Physical Layer in the 5 GHz band (1999)
9. IEEE 802.11b-1999 Supplement to 802.11-1999, Wireless LAN MAC and PHY specifications: Higher speed Physical Layer (PHY) extension in the 2.4 GHz band (1999)
10. IEEE 802.11e-2005, IEEE Standard for Information technology – Telecommunications and information exchange between systems – Local and metropolitan area networks – Specific requirements Part 11: Wireless LAN Medium Access Control (MAC) and Physical Layer (PHY) specifications: Amendment 8: Medium Access Control (MAC) Quality of Service Enhancements (2005)
11. IEEE 802.11g-2003 IEEE Standard for Information technology – Telecommunications and information exchange between systems – Local and metropolitan area networks – Specific requirements – Part 11: Wireless LAN Medium Access Control (MAC) and Physical Layer (PHY) specifications – Amendment 4: Further Higher-Speed Physical Layer Extension in the 2.4 GHz Band (2003)
12. Kochut, A., Vasan, A., Shankar, A., Agrawala, A.: Sniffing Out the Correct Physical Layer Capture Model in 802. In: 12th IEEE International Conference on Network Protocols (ICNP 2004), Berlin (2004)
13. Lopez-Aguilera, E., Heusse, M., Rousseau, F., Duda, A., Casademont, J.: Evaluating Wireless LAN Access Methods in Presence of Transmission Errors. In: IEEE INFOCOM, Barcelona (2006)

14. Ni, Q., Li, T., Turletti, T., Xiao, Y.: Saturation Throughput Analysis of Error-Prone 802.11 Wireless Networks. Wiley Journal of Wireless Communications and Mobile Computing (JWCMC) 5(8), 945–956 (2005)
15. Private communication with Prof. Andrzej Duda
16. Tay, Y., Chua, K.: A Capacity Analysis for the IEEE 802.11 MAC Protocol. Wireless Networks 7(2), 159–171 (2001)
17. The Network Simulator – ns-2. URL: http://nsnam.isi.edu/nsnam/index.php/Main_Page
18. Wu, H., Peng, Y., Long, K., Cheng, S., Ma, J.: Performance of Reliable Transport Protocol over IEEE 802.11 Wireless LAN: Analysis and Enhancement. In: IEEE INFOCOM'02 (2002)
19. Ziouva, E., Antonakopoulos, T.: CSMA/CA Performance under High Traffic Conditions: Throughput and Delay Analysis. Computer Communications 25, 313–321 (2002)

Impact of Size-Based Scheduling on Flow Level Performance in Wireless Downlink Data Channels

Samuli Aalto and Pasi Lassila

TKK Helsinki University of Technology, Networking Laboratory, P.O. Box 3000,
FI-02015 TKK, Finland
{samuli.aalto,pasi.lassila}@tkk.fi

Abstract. We analyze the impact of size-based scheduling on the flow level performance of elastic traffic in wireless downlink data channels. The impact is assessed by comparing the flow level delay of the simple RR scheduler to two optimized non-anticipating schedulers (FB and FB°) and SRPT. The optimized distance-aware scheduler FB° is derived by applying the Gittins index approach. Our results show that for Pareto-type file size distributions, the size-based information is more important than the location information. Additionally, FB not only decreases the overall mean delay, but it can also decrease considerably the mean delay of all users independently of their location.

Keywords: Scheduling, cellular system, HSDPA, HDR, mean delay, elastic traffic, Pareto distribution, Gittins index.

1 Introduction

We consider the optimal scheduling problem for downlink data traffic in a single cell of a cellular system. We assume that the traffic consists of elastic flows, such as file transfers using TCP, with each flow characterized by its size, *i.e.*, the total amount of bits to be transferred. An important performance measure for such elastic flows is the total time needed for transferring all the bits, which we call flow level delay, or just briefly, *delay*.

Standardized systems like HSDPA and HDR minimize the intra-cell interference by having a time-slotted system where the base station transmits at full power to only one terminal in each slot. Typically the time slot is very short (milliseconds) compared to the flow level delay (ranging from seconds to minutes). The "air time" is shared in a fair way among the receiving terminals by the simple Round Robin (RR) scheduling discipline. For short time slots this is well approximated by the Processor Sharing (PS) discipline.

The cell capacity may be increased by applying *channel-aware* scheduling disciplines that utilize the fast fading effect. An example is the Proportionally Fair (PF) discipline implemented in HSDPA/HDR systems, where the time slot is scheduled to the terminal with the highest momentary receiving rate proportionally to its average receiving rate. This will lead to the PS discipline in the

L. Mason, T. Drwiega, and J. Yan (Eds.): ITC 2007, LNCS 4516, pp. 1096–1107, 2007.

limit as the time slot shrinks down to zero, however, with a higher cell capacity compared to the limit of the RR discipline [8,2,4]. The improvement in the cell capacity has been studied in [2] with the number of flows fixed and in [3] assuming a randomly varying number of flows. The latter concludes that the improvement in a dynamic setting, however, is not that large.

A fundamental result related to single server queueing systems says that the number of jobs is minimized pathwise by applying the Shortest Remaining Processing Time (SRPT) discipline [13]. This implies, by Little's result, that SRPT minimizes the mean delay as well. The benefit is achieved by utilizing the *size-dependent information*. However, applying SRPT as such in the downlink data traffic problem is intractable due to unpredictable factors such as fading effects. Hu *et al.* [9] developed heuristic algorithms that combine channel-aware and size-dependent scheduling.

In this paper we focus on non-anticipating disciplines, for which the remaining service times are *not* known. If the service times are of type Decreasing Hazard Rate (DHR), the Foreground Background (FB) discipline is optimal among the non-anticipating disciplines in an M/G/1 queue [14,11,15]. FB is a size-based discipline giving full priority to the job with least amount of attained service, see [10]. If there are multiple jobs with the same least amount of attained service, then the service is shared evenly between these jobs. DHR distributions, such as Pareto, have been used to model flow sizes in the Internet [5,6].

As compared to an ordinary M/G/1 queue, the difference in the downlink data traffic problem is the *location information*. Due to the path loss effect, terminals far away from the base station have a lower receiving rate than the near-by terminals. Thus, an optimal scheduler that utilizes this information achieves even better delay performance than FB. As the main theoretical contribution, we determine the optimal non-anticipating and distance-aware discipline for DHR flow sizes by applying the so called Gittins index approach. For Pareto distributions, the optimal discipline proves to be a simple modification of FB. We demonstrate that the improvement is, however, marginal when compared to the ordinary FB discipline for Pareto flow size distributions. This implies that utilizing the size-dependent information is more important than utilizing the location information when scheduling the time slots for downlink data traffic. On the other hand, we also demonstrate that there is a clear improvement in the delay performance when the simple RR discipline is replaced by FB.

Bonald and Proutière [3] state that FB would exacerbate the discrimination against far terminals. This is indeed the case: the reduction in the mean delay is greater for the near terminals. Our observation, however, is that there can be *some reduction* in the mean delay *for all terminals*, even for the farthermost ones. The losers are the huge flows independent of their location.

The rest of the paper is organized as follows. The model is explained in Section 2. In Section 3 we give the mean delay formulas for the reference disciplines, while in Section 4 the optimal non-anticipating and distance-aware discipline is determined. Different disciplines are compared in Section 5 based on simulations and numerical evaluations. Section 6 concludes the paper.

2   Model

Consider downlink data traffic in a single cell of a cellular system. The traffic consists of elastic flows. We assume that the flows arrive at the base station according to a Poisson process with rate λ. The flow sizes X are assumed to be independent and identically distributed (positive real-valued) random variables with the cumulative distribution function denoted by $F_X(x) = P\{X \leq x\} = \int_0^x f_X(y)\,dy$, where $f_X(x)$ refers to the corresponding density function. The hazard rate $h_X(x)$ is defined by $h_X(x) = f_X(x)/\bar{F}_X(x)$, where $\bar{F}_X(x) = 1 - F_X(x)$. We assume that the flow size distribution belongs to the class DHR with a differentiable density function $f_X(x)$. Thus, $h'_X(x) \leq 0$ for all $x \geq 0$.

Each flow is associated with a receiving terminal. We assume that the terminals are independently and uniformly distributed in the cell area, which is a circular disk with radius r_1. Thus, the distance R from the base station to the receiving terminal for a flow has the following distribution:

$$P\{R \leq r\} = \frac{1}{\pi r_1^2} \int_0^r 2\pi s\,ds = \left(\frac{r}{r_1}\right)^2, \quad r \leq r_1.$$

Furthermore, we assume that the transmission rate (bits per time unit) from the base station to the receiving terminal depends on the distance r between them as follows (see also [3]):

$$c(r) = \begin{cases} c_0, & r \leq r_0, \\ c_0 \left(\frac{r_0}{r}\right)^\alpha, & r > r_0, \end{cases} \tag{1}$$

where c_0 is the maximum data rate of the system and α, called the attenuation factor, typically takes values in the range from 2 to 4. The received power attenuates due to path loss and the rate is linear in the received power. Basically, we consider an ideal case with no fast fading nor intercell interference, and where the set of achievable rates is continuous omitting possible coding constraints.

Given the location of the receiving terminal, the service time for a flow is just the flow size divided by the constant rate. Let $F_r(t)$ denote the cumulative distribution function of the service time $S_r = X/c(r)$ for a flow with the terminal located at distance r from the base station. Now $E[S_r] = E[X]/c(r)$ and

$$F_r(t) = P\{S_r \leq t\} = P\{X \leq c(r)\,t\} = F_X(c(r)\,t).$$

The corresponding density function is clearly $f_r(t) = c(r)f_X(c(r)\,t)$ and the corresponding hazard rate $h_r(t) = c(r)h_X(c(r)\,t)$. Since the flow sizes X are DHR, also these conditional service times S_r belong to the class DHR.

On the other hand, if the location of the receiving terminal is *not* known, then the service time for a flow is the ratio between two independent random variables, $S = X/c(R)$, with mean

$$E[S] = E[X] \int_0^{r_1} \frac{1}{c(r)} P\{R \in dr\} = \frac{E[X]}{c_0} \left(\frac{\alpha}{\alpha+2} \left(\frac{r_0}{r_1}\right)^2 + \frac{2}{\alpha+2} \left(\frac{r_1}{r_0}\right)^\alpha\right).$$

The cumulative distribution function of the service time S is denoted by $F(t) = P\{S \leq t\} = \int_0^x f(s)\,ds$ with $f(t)$ referring to the corresponding density function. It is easy to see that

$$F(t) = F_X\left(c_0 t \left(\frac{r_0}{r_1}\right)^\alpha\right) + \left(\frac{r_0}{r_1}\right)^2 \int_{c_0 t (r_0/r_1)^\alpha}^{c_0 t} f_X(x) \left(\frac{c_0 t}{x}\right)^{\frac{2}{\alpha}} dx,$$

from which the density can be derived with the following result:

$$f(t) = f_X(c_0 t) c_0 \left(\frac{r_0}{r_1}\right)^2 + \frac{2}{\alpha t}\left(\frac{r_0}{r_1}\right)^2 \int_{c_0 t(r_0/r_1)^\alpha}^{c_0 t} f_X(x) \left(\frac{c_0 t}{x}\right)^{\frac{2}{\alpha}} dx.$$

The hazard rate of S is denoted by $h(t)$.

Proposition 1. *If the flow size distribution $F_X(x)$ belongs to the class DHR, so does the service time distribution $F(t)$.*

Proof. Since $F(t) = \int_0^{r_1} F_r(t)\,P\{R \in dr\}$, this is a special case of the result given in [1, Theorem 3.4]. □

Throughout the paper we assume that the time slot used for scheduling is negligible compared to the flow sizes, cf. [4]. So we have an M/G/1 queue with arrival rate λ, service times S, and load $\rho = \lambda E[S]$. For stability, we assume that $\rho < 1$.

3   Reference Schedulers

In this section we give the mean delay formulas for the disciplines PS, FB, SRPT, and GR. The PS discipline (representing the limiting case of RR) is our main reference scheduler. FB is the optimal non-anticipating discipline that does not utilize location information. SRPT is the optimal (hypothetical) scheduler which gives the lower bound for the delay performance. The last one, GR, refers to the greedy distance-aware discipline that gives the full priority to the receiving terminal with the shortest distance from the base station.

PS. The conditional mean delay of a flow with service time t is for the PS discipline as follows [10]:

$$E[T^{\mathrm{PS}}(t)] = \frac{t}{1-\rho}. \tag{2}$$

Thus, the mean delay of a flow is

$$E[T^{\mathrm{PS}}] = \int_0^\infty E[T^{\mathrm{PS}}(t)]f(t)\,dt = \frac{E[S]}{1-\rho}. \tag{3}$$

In addition, the mean delay of a flow with the receiving terminal located at distance r from the base station becomes

$$E[T_r^{\mathrm{PS}}] = \int_0^\infty E[T^{\mathrm{PS}}(t)]f_r(t)\,dt = \frac{E[S_r]}{1-\rho}. \tag{4}$$

FB. The conditional mean delay for FB reads as follows [10]:

$$E[T^{\mathrm{FB}}(t)] = \frac{\lambda E[(S \wedge t)^2]}{2(1 - \rho_t)^2} + \frac{t}{1 - \rho_t}, \tag{5}$$

where $S \wedge t = \min\{S, t\}$ and ρ_t refers to the truncated load, $\rho_t = \lambda E[(S \wedge t)]$. The mean delays $E[T^{\mathrm{FB}}]$ and $E[T_r^{\mathrm{FB}}]$ are calculated from the conditional mean delay $E[T^{\mathrm{FB}}(t)]$ similarly as for the PS discipline, see (3) and (4).

SRPT. The conditional mean delay formula for SRPT originates from [12]:

$$E[T^{\mathrm{SRPT}}(t)] = \frac{\lambda E[(S \wedge t)^2]}{2(1 - \rho(t))^2} + \int_0^t \frac{1}{1 - \rho(s)}\, ds. \tag{6}$$

Here $\rho(t)$ refers to $\rho(t) = \lambda \int_0^t s f(s)\, ds$. Again, the mean delays $E[T^{\mathrm{SRPT}}]$ and $E[T_r^{\mathrm{SRPT}}]$ are calculated from the conditional mean delay $E[T^{\mathrm{SRPT}}(t)]$ similarly as for the PS discipline, see (3) and (4).

GR. Discipline GR results in a pre-emptive priority M/G/1 queue with a continuum of priority classes. By applying the well known results for priority queues [10], we conclude that the mean delay for a flow with the receiving terminal located at distance r from the base station is as follows:

$$E[T_r^{\mathrm{GR}}] = \frac{E[S_r^*]}{(1 - \sigma_r)^2} + \frac{E[S_r]}{1 - \sigma_r}. \tag{7}$$

Here, σ_r is the expected load up to distance r defined by

$$\sigma_r = \lambda \int_0^r E[S_z]\, P\{R \in dz\} = \lambda E[X] \int_0^r \frac{1}{c(z)} P\{R \in dz\}$$

$$= \begin{cases} \lambda \dfrac{E[X]}{c_0} \left(\dfrac{r}{r_1}\right)^2, & r \le r_0, \\[3mm] \lambda \dfrac{E[X]}{c_0} \left(\dfrac{\alpha}{\alpha + 2} \left(\dfrac{r_0}{r_1}\right)^2 + \dfrac{2}{\alpha + 2} \left(\dfrac{r}{r_0}\right)^\alpha \left(\dfrac{r}{r_1}\right)^2\right), & r > r_0, \end{cases}$$

and S_r^* refers to the so called remaining service time for terminals up to distance r with mean

$$E[S_r^*] = \frac{\lambda}{2} \int_0^r E[S_z^2]\, P\{R \in dz\} = \frac{\lambda}{2} E[X^2] \int_0^r \frac{1}{c^2(z)} P\{R \in dz\}$$

$$= \begin{cases} \dfrac{\lambda}{2} \dfrac{E[X^2]}{c_0^2} \left(\dfrac{r}{r_1}\right)^2, & r \le r_0, \\[3mm] \dfrac{\lambda}{2} \dfrac{E[X^2]}{c_0^2} \left(\dfrac{\alpha}{\alpha + 1} \left(\dfrac{r_0}{r_1}\right)^2 + \dfrac{1}{\alpha + 1} \left(\dfrac{r}{r_0}\right)^{2\alpha} \left(\dfrac{r}{r_1}\right)^2\right), & r > r_0. \end{cases}$$

The mean delay is given by

$$E[T^{\mathrm{GR}}] = \int_0^{r_1} E[T_r^{\mathrm{GR}}]\, P\{R \in dr\}. \tag{8}$$

4   Optimal Non-anticipating Distance-Aware Scheduler

In this section we determine the optimal non-anticipating and distance-aware discipline for DHR flow sizes. Let Π denote the family of non-anticipating scheduling disciplines that do not utilize location information. In addition, let Π° denote the whole family of non-anticipating scheduling disciplines including also those that utilize the location information. Thus, our purpose is to find the optimal discipline in Π°.

4.1   Gittins Index

In this subsection we first recall results related to the so called Gittins index, see [7,15]. Then we apply the results to the case where the location information is not utilized.

For any $a, \Delta \geq 0$, let

$$J(a, \Delta) = \frac{\int_0^\Delta f(a + t)\, dt}{\int_0^\Delta \overline{F}(a + t)\, dt}. \tag{9}$$

Note that $J(a, 0) = h(a)$ for any a. Function $J(a, \Delta)$ is clearly continuous with respect to both arguments. In addition, the one-sided partial derivatives with respect to Δ are defined for any pair (a, Δ),

$$\frac{\partial}{\partial \Delta} J(a, \Delta) = \frac{f(a + \Delta) \int_0^\Delta \overline{F}(a + t)\, dt - \overline{F}(a + \Delta) \int_0^\Delta f(a + t)\, dt}{(\int_0^\Delta \overline{F}(a + t)\, dt)^2} \tag{10}$$

For a flow with attained service a, the *Gittins index* is defined as follows [7,15]:

$$G(a) = \sup_{\Delta \geq 0} J(a, \Delta), \tag{11}$$

Consider now a non-anticipating discipline π^* which always gives service to the job with the highest Gittins index. We call this discipline the *Gittins discipline*. It is known that the Gittins discipline is optimal with respect to the mean delay for an M/G/1 queue, see [7, Theorem 3.28], [15, Theorem 4.7].

The optimality result can be used to prove the optimality of the FB discipline for DHR flow sizes. Recall from Proposition 1 that in this case also the service times belong to the class DHR so that the hazard rate $h(t)$ is a decreasing function.

Proposition 2. *Function $J(a, \Delta)$ is decreasing with respect to Δ for any a.*

Proof. Let $a, \Delta \geq 0$. Since $h(t)$ is decreasing, we have $h(a + t) \geq h(a + \Delta)$ for all $0 \leq t \leq \Delta$, which is equivalent with

$$\frac{f(a + t)}{f(a + \Delta)} \geq \frac{\overline{F}(a + t)}{\overline{F}(a + \Delta)}. \tag{12}$$

By (10), we have

$$\frac{\partial}{\partial \Delta} J(a, \Delta) \leq 0 \quad \Longleftrightarrow \quad \frac{1}{\int_0^\Delta \frac{f(a+t)}{f(a+\Delta)}\, dt} \leq \frac{1}{\int_0^\Delta \frac{\overline{F}(a+t)}{\overline{F}(a+\Delta)}\, dt}.$$

The claim follows from this by (12). □

Proposition 3. $G(a) = h(a)$ for all a.

Proof. Let $a \geq 0$. By Proposition 2, $G(a) = J(a, 0) = h(a)$. □

Theorem 1. $E[T^{\mathrm{FB}}] \leq E[T^\pi]$ for any $\pi \in \Pi$.

Proof. By Proposition 3 and the fact that $h(a)$ is decreasing, the flow with least amount of service has the highest Gittins index. Thus, in this case the Gittins discipline corresponds to FB. The claim follows now from the optimality of the Gittins discipline. □

4.2 Utilizing Location Information

Consider now how a non-anticipating discipline can be improved if the scheduler is aware of the distances between the base station and the receiving terminals.

Recall that $h_r(t)$ refers to the hazard rate related to the cumulative distribution function of the service time S_r for a flow with the terminal located at distance r from the base station. As mentioned in Section 2, the service times S_r belong to the class DHR. Thus, the Gittins index for such a flow with attained service a is $G_r(a) = h_r(a) = c(r)h_X(c(r)a)$.

Let R_i and $A_i(t)$, respectively, denote the distance from the base station and the attained service time related to flow i at time t. Furthermore, let $\gamma_i(t)$ denote the proportion of time that is scheduled for flow i at time t. It follows that $A_i(t) = \int_0^t \gamma_i(s)\, ds$. According to the Gittins discipline, the optimal scheduler transmits to terminal j such that

$$h_{R_j}(A_j(t)) = \max_i h_{R_i}(A_i(t)) = \max_i c(R_i)\, h_X\left(c(R_i)A_i(t)\right). \tag{13}$$

If this maximum is not unique, the service capacity shall be shared between the maximizing terminals $j \in \mathcal{J}(t)$ in such a way that

$$\frac{d}{dt} h_{R_j}(A_j(t)) = c(R_j)^2\, h_X'\left(c(R_j)A_j(t)\right)\, \gamma_j(t) \tag{14}$$

is the same for all $j \in \mathcal{J}(t)$. The optimal shares $\gamma_j(t)$ can be determined from this condition together with the constraint

$$\sum_{j \in \mathcal{J}(t)} \gamma_j(t) = 1. \tag{15}$$

We denote this optimal non-anticipating distance-aware discipline by FB°.

Theorem 2. $E[T^{\mathrm{FB}°}] \leq E[T^\pi]$ for any $\pi \in \Pi°$.

Proof. The claim follows from the optimality of the Gittins discipline together with the derivations made above. □

4.3   Pareto Flow Sizes

Assume now that the flow size distribution is Pareto with shape parameter $\beta > 1$ and scale parameter $b > 0$ such that, for all $x \geq 0$,

$$\bar{F}_X(x) = \left(\frac{1}{1+bx}\right)^{\beta}, \quad h_X(x) = \frac{\beta b}{1+bx}, \quad h'_X(x) = \frac{-\beta b^2}{(1+bx)^2}.$$

By (13), the optimal distance-aware discipline FB° transmits to terminal j such that

$$\frac{1}{b\,c(R_j)} + A_j(t) = \min_i \left(\frac{1}{b\,c(R_i)} + A_i(t)\right). \tag{16}$$

Interestingly, the optimal rule is independent of the shape parameter β. In addition, the inverse of the Gittins index grows linearly with the amount of attained service. Within the constant transmission rate area ($r \leq r_0$), the flow with the least amount of attained service is the preferred one. In particular, if $r_0 = r_1$, then FB° reduces back to FB.

If the maximum is not unique, then

$$\frac{1}{c(R_j)} + b\,A_j(t) \tag{17}$$

is the same for all maximizing terminals $j \in \mathcal{J}(t)$. By (14), if the maximum is not unique, the service capacity shall be shared between the maximizing terminals $j \in \mathcal{J}(t)$ so that

$$c(R_j)^2\,h'_X\left(c(R_j)\,A_j(t)\right)\gamma_j(t) = \frac{-\beta\,\gamma_j(t)}{\left(\frac{1}{c(R_j)} + b\,A_j(t)\right)^2} \tag{18}$$

is the same for all $j \in \mathcal{J}(t)$. But now by (17) we conclude that, in fact, the optimal shares $\gamma_j(t)$ are the same for all $j \in \mathcal{J}(t)$. Thus, by (15),

$$\gamma_j(t) = \frac{1}{|\mathcal{J}(t)|}, \tag{19}$$

where $|\mathcal{J}(t)|$ refers to the magnitude of the set $\mathcal{J}(t)$. Thus, for the Pareto distribution, FB° applies PS among the maximizing terminals.

4.4   Exponential Flow Sizes

Assume now that the flow size distribution is exponential with parameter μ so that, for all $x \geq 0$,

$$\bar{F}_X(x) = e^{-\mu x}, \quad h_X(x) = \mu, \quad h'_X(x) = 0.$$

By (13), FB° transmits to terminal j such that

$$c(R_j) = \max_i c(R_i). \tag{20}$$

It is easy to see that, in this case, it does not matter how the service is shared among the flows with the same transmission rate. Thus, for exponential flow sizes, FB° is equivalent with the greedy discipline GR.

5   Numerical Results

In the following we give numerical evidence of the amount of performance gains achievable with the various scheduling disciplines. As the baseline scheduler we use the PS discipline (corresponding to the limit of the RR scheduler). The idea is to compare the performance of the other disciplines with respect to PS, *i.e.*, we are only interested in the relative performance of the policies. Thus, in our numerical examples we have scaled the parameters such that flow sizes have unit length, $E[X] = 1$, the maximum transmission rate $c_0 = 1$, and the cell radius $r_1 = 1$ (unit circle). We evaluate the relative performance for exponentially distributed and Pareto distributed flow sizes. The parameters affecting the performance are the attenuation factor α and the radius of the constant rate area r_0, which jointly define the variability of the transmission rates. Additionally, for the Pareto distribution we have the shape parameter β.

In practice, the variability in the rates (or the range of the possible rates) is determined by the capabilities of the technology and the amount of transmission power the base station has available. To have a realistic scenario for the variability in the rates, we use the parameters from Table 1 in [3], where it is given that in current HSDPA/HDR systems for $\alpha = 2$ the ratio of the cell radius to the constant rate area's radius $r_1/r_0 = 7.94$.

5.1   Overall Mean Delay

We first compare the optimal non-anticipating distance-aware scheduler FB° with the plain FB. We have simulated the system at a fixed load, $\rho = 0.9$, with Pareto distributed flow sizes under both policies using the same stochastic input to minimize the variance. In the simulations we vary the loss exponent $\alpha = 2$ and 4, and the Pareto shape parameter $\beta = 2$ and 3. With exponentially distributed flow sizes, the FB° discipline corresponds to the GR discipline, and the results can be obtained analytically.

The results in Table 1 depict the relative mean delay difference $\Delta = (E[T^{\text{FB}}] - E[T^{\text{FB}^\circ}])/E[T^{\text{FB}}]$, together with the confidence intervals for simulation results. As seen from the results, FB° performs better than FB, especially for exponential flow sizes. However, for Pareto flow sizes the difference is rather small so that one does not benefit much from the location information.

Next we study the mean overall delays as a function of the load ρ. Since the delay difference between FB° and FB is so small, results for FB° are not shown in the Pareto cases. The idea is to compare how much better/worse the

Table 1. Mean delay comparison between FB° and FB. For simulation results (Pareto distributions), the 95% confidence intervals are given

	Exponential		Pareto, $\beta = 3$		Pareto, $\beta = 2$	
	$\alpha = 2$	$\alpha = 4$	$\alpha = 2$	$\alpha = 4$	$\alpha = 2$	$\alpha = 4$
Δ	15.7%	17.3%	2.0% ± 0.1%	2.6% ± 0.1%	0.7% ± 0.1%	1.0% ± 0.1%

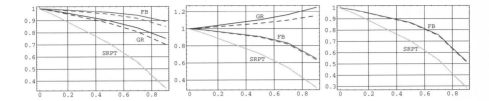

Fig. 1. Mean delay ratio as a function of ρ for exponential distribution (left), Pareto with $\beta = 3$ (middle) and Pareto with $\beta = 2$ (right)

disciplines GR, FB and SRPT perform than PS by considering the normalized results $E[T^\pi]/E[T^{\mathrm{PS}}]$ for $\pi = \mathrm{GR, FB, SRPT}$.

The results are shown in Figure 1. In the graphs, solid lines correspond to $\alpha = 2$ and the dashed lines to $\alpha = 4$. The scheduling disciplines are indicated next to each curve. The left figure corresponds to exponentially distributed flow sizes. The middle and right figures correspond to the Pareto case with shape parameters $\beta = 3$ and 2, respectively. Under the exponential distribution (left figure), all disciplines achieve a better performance than PS. In the Pareto cases, for $\beta = 3$ (middle figure), the mean delay for GR is finite, but the delay is greater than under PS for both $\alpha = 2$ and 4. For $\beta = 2$ (right figure), $E[X^2] \to \infty$, and correspondingly the mean delay for GR becomes infinite. For FB and SRPT, the delays are always smaller than under PS and the benefit increases with load. Notably, the benefit does not seem to be affected by the value of α.

In conclusion, it is clear that the size-based scheduling mechanisms FB, FB° and SRPT yield substantial performance gains in terms of the overall mean delay. In addition, due to the marginal difference between FB and FB° for Pareto flow sizes, we conclude that utilizing the size-dependent information is more important than utilizing the location information.

5.2   Near-Far Unfairness

Near-far unfairness refers to the inherent property of the HSDPA/HDR systems that users far away from the base station experience much worse performance than users near the base station. This is also addressed in [3], where it is remarked that size-based schedulers only exacerbate the near-far unfairness property. Thus, next we examine in more detail the near-far unfairness issue in terms of the conditional mean delays of the different disciplines when the receiving terminal is at a given distance from the base station.

The results as a function of the distance r are given in Figure 2 for the exponential distribution (left) and Pareto distributions with $\beta = 3$ (middle) and $\beta = 2$ (right). The results show the normalized conditional delay of the other disciplines with respect to PS, i.e., we plot $E[T_r^\pi]/E[T_r^{\mathrm{PS}}]$ for $\pi = \mathrm{GR, FB}$. Results for SRPT are not shown to keep the figures clear. However, under SRPT the conditional delays are always smaller than under FB by a similar margin as in the overall delays in Figure 1. From the figures we can see that GR results

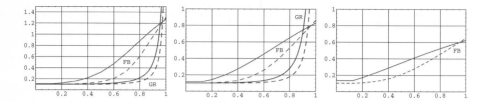

Fig. 2. Conditional mean delay ratio as a function of distance r for exponential distribution (left) and Pareto distributions with $\beta = 3$ (middle) and $\beta = 2$ (right)

in extreme near-far unfairness, *i.e.*, the performance benefits are achieved at the expense of the far users (again, for $\beta = 2$ the GR discipline does not even have a finite mean). On the other hand, the FB discipline makes the performance slightly worse for the users on the border in the exponential case. However, in the Pareto cases the conditional delays are uniformly in the cell better than under PS. Thus, roughly speaking, everybody "wins".

6   Conclusions

We have analyzed the impact of size-based scheduling on the flow level performance of elastic traffic in the downlink data channels in a single cell. In such systems the service time of a flow is determined both by its size and transmission rate (as determined by the random location of a user in the cell). Both information can be used to minimize the flow level delay. The baseline scheduling discipline was provided by the simple RR policy (corresponding to a PS system in the limit), which does not employ any information about the rates nor the sizes. The other studied schedulers were the GR scheduler (only rate information), plain FB scheduler (distribution of rates and sizes), FB° (distribution of sizes and exact knowledge of rate) and SRPT (exact knowledge of size and rate). Notably, the FB° policy was derived by applying the Gittins index approach, and it yields the optimal non-anticipating scheduling discipline that utilizes exact knowledge of a user's rate. In the case of Pareto distributions, which have been used to model flow sizes in the Internet, the optimal discipline FB° proves to be a simple modification of FB.

The results showed that the GR discipline can sometimes (*e.g.*, for exponentially distributed flow sizes) decrease the overall mean delay compared with RR. However, it is always at the expense of the users that are on the border of the cell. Comparing plain FB and FB°, it appears that for Pareto distributed flow sizes, knowledge of the location is not that important. Additionally, for Pareto distributed files we demonstrated that considerable gains can be achieved by applying size-based scheduling (FB and SRPT), both in terms of the overall mean delay, as well as the conditional mean delay at a given distance from the base station. The latter property implies that age-based scheduling can increase the performance of all users, even those on the border of the cell.

In this paper the baseline policy was provided by RR. In practical systems, channel-aware schedulers, such as PF, are used, and they are known to achieve a better performance under fast fading conditions. However, as shown in [3], the performance gain in a dynamic traffic situation may not be that great. Thus, it can be argued that the significant gains achievable with FB may be higher than those achievable with channel-aware schedulers. Nevertheless, a detailed performance comparison with channel-aware scheduling remains as future work. Additionally, multi-cell scenarios offer another area of extensions, as well as the inclusion of coding constraints in the rate function defined in (1).

References

1. Barlow, R.E., Marshall, A.W., Proschan, F.: Properties of probability distributions with monotone hazard rate. The. Annals of Mathematical Statistics 34, 375–389 (1963)
2. Berggren, F., Jäntti, R.: Asymptotically fair scheduling on fading channels. In: Proceedings of IEEE VTC 2002-Fall, Vancouver, Canada, pp. 1934–1938 (2002)
3. Bonald, T., Proutière, A.: Wireless downlink data channels: user performance and cell dimensioning. In: Proceedings of ACM MobiCom 2003, San Diego, CA, pp. 339–352 (2003)
4. Borst, S.: User-level performance of channel-aware scheduling algorithms in wireless data networks. In: Proceedings of IEEE Infocom 2003, San Fransisco, CA, pp. 321–331 (2003)
5. Crovella, M., Bestavros, A.: Self-similarity in world wide web traffic: evidence and possible causes. In: Proceedings of ACM SIGMETRICS 1996, Philadelphia, PA, pp. 160–169 (1996)
6. Feldmann, A., Whitt, W.: Fitting mixtures of exponentials to long-tail distributions to analyze network performance models. In: Proceedings of IEEE Infocom 1997, Kobe, Japan, pp. 1096–1104 (1997)
7. Gittins, J.C.: Multi-armed Bandit Allocation Indices. Wiley, Chichester (1989)
8. Holzman, J.M.: CDMA forward link waterfilling power control. In: Proceedings of IEEE VTC 2000-Spring, Tokyo, Japan, pp. 1663–1667 (2000)
9. Hu, M., Zhang, J., Sadowsky, J.: Traffic aided opportunistic scheduling for downlink transmissions: algorithms and performance bounds. In: Proceedings of IEEE Infocom 2004, Hong Kong, pp. 1652–1661 (2004)
10. Kleinrock, L.: Queueing Systems, Volume II: Computer Applications. Wiley, New York (1976)
11. Righter, R., Shanthikumar, J.G.: Scheduling multiclass single server queueing systems to stochastically maximize the number of successful departures. Probability in the Engineering and Informational Sciences 3, 323–333 (1989)
12. Schrage, L.E., Miller, L.W.: The queue M/G/1 with the shortest remaining processing time discipline. Operations Research 14, 670–683 (1966)
13. Schrage, L.E.: A proof of the optimality of the shortest remaining processing time discipline. Operations Research 16, 687–690 (1968)
14. Yashkov, S.F.: Processor-sharing queues: Some progress in analysis. Queueing Systems 2, 1–17 (1987)
15. Yashkov, S.F.: Mathematical problems in the theory of shared-processor systems. Journal of Mathematical Sciences 58, 101–147 (1992)

A Resource Scheduling Scheme for Radio Access in Portable Internet

Bongkyo Moon

Dongguk University, 3-26 Pil-dong, Chung-gu, Seoul 100-715, Korea
bkmoon@dongguk.edu

Abstract. Wireless mobile networks generally provide radio access bearer services over wireless links between a terminal equipment (UE) and a base station (BS). In this paper, a radio resource scheduling scheme is proposed for providing a merged service of both realtime and non-realtime flow connections over the wireless link in the Portable Internet (e.g., IEEE 802.16e) with frequent host mobility. This scheme can efficiently improve the connection blocking probability, connection dropping probability, and bandwidth utilization.

1 Introduction

Wireless networks generally provide radio access bearer services to support the layer 2 connection between a user terminal equipment (UE) and a base station (BS). Recently IEEE standard 802.16 was designed to evolve as a set of air interfaces based on a common MAC protocol and has now been updated and extended to the 802.16e (Portable Internet) for mobile access. This could provide better data handling than that of the third-generation (3G) cellular system before fourth-generation (4G) systems arrive.

In wireless mobile Internet, IP is not only independent of the actual radio access technology but also tolerates a variety of radio protocols. IEEE 802.16e is essentially designed to provide all-IP and packet services, such as streaming video, music on demand, online gaming, and broadcasting under wireless mobile scenario [1].

In the Portable Internet (PI) like IEEE 802.16e, voice or multimedia flow request may allow the radio access system (RAS) to exercise admission control over wireless link. Hereby low priority flow can typically be transported amongst high priority flow using spare resources [2]. Typically, efficient radio resource sharing mechanism is required in order to react to rapidly changing traffic flow conditions due to host mobility.

Cheng *et al.* [3] proposed the resource allocation scheme for fast handover in an IP-based wireless mobile Internet. This scheme assumes that wireless links are bottlenecks in the domain. This scheme assumes that the complete partitioning (CP) mechanism for radio resource scheduling among the service classes is used, so the admission control of each class can be considered separately. That is, the radio resource can be partitioned into distinct channel groups, and then they are

L. Mason, T. Drwiega, and J. Yan (Eds.): ITC 2007, LNCS 4516, pp. 1108–1119, 2007.

assigned exclusively to each traffic. However, such CP-based strategy is difficult to adapt to the varying traffic flow loads due to the difficulty of dynamically changing the partitioning of the channels which are allocated exclusively to each type of flow.

In this paper, a radio resource scheduling scheme is proposed for providing a merged service of both realtime flow and non-realtime flow in a PI. In section 3, the proposed scheme is described, and analytic model is derived. In section 4, performance measures are presented. Examples and results are illustrated and discussed respectively in section 5.

2   Portable Internet for Broadband Wireless Access

The Fig. 1 presents the network architecture of Wireless Broadband (WiBro), one of the IEEE802.16e implementations. There are four main components in the WiBro architecture: portable subscriber station (PSS), radio access system (RAS), access control router (ACR), and core network. PSS communicates with RAS using wireless access technology [4]. The PSS also provides the functions of MAC processing, mobile IP, authentication, packet retransmission, and handover. The RAS provides wireless interfaces for the PSS and takes care of wireless resource management, QoS support, and handover control. The ACR plays a key-role in IP-based data services including IP packet routing, security, QoS and handover control, and foreign agent (FA) in the mobile IP. To provide mobility for PSS, the ACR supports handover between the RASs while the mobile IP provides handover between the ACRs.

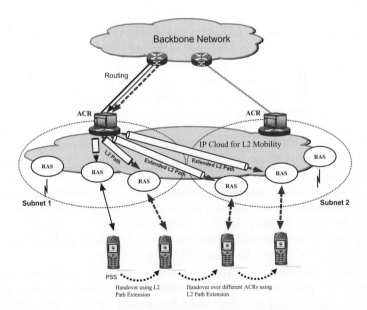

Fig. 1. Network Architecture of WiBro (IEEE802.16e)

The 802.16 MAC provides a wide range of service types analogous to the classic asynchronous transfer mode (ATM) service categories. Thus the variety of services includes legacy time-division multiplex (TDM) voice and data, Internet Protocol (IP) connectivity, and packetized voice over IP (VoIP). The MAC includes service-specific convergence sublayers that interface to higher layers, above the core MAC common part sublayer that carries out the key MAC functions. Below the common part sublayer is the privacy sublayer [1][6].

Convergence sublayers are used to map the transport-layer-specific traffic to a MAC that is flexible enough to efficiently carry and traffic type. Through such features as payload header suppression, packing, and fragmentation, the convergence sublayers and MAC work together to carry traffic in a form that is often more efficient than the original transport mechanism.

The 802.16 MAC is connection-oriented. All services, including inherently connectionless services, are mapped to a connection. The common part sublayer provides a mechanism for requesting bandwidth, associating QoS and traffic parameters, transporting and routing data to the appropriate convergence sublayer.

While extensive bandwidth allocation and QoS mechanisms are provided, the details of scheduling and reservation management are left unstandardized and provide an important mechanism for vendors to differentiate their equipment [1][6]. Along with the fundamental task of allocating bandwidth and transporting data, the MAC includes a privacy sublayer that provides authentication of network access and connection establishment to avoid theft of service, and it provides key exchange and encryption for data privacy.

3   Proposed Scheme

IEEE 802.16 uses the concept of service flow connections to define unidirectional transport of packets on either downlink or uplink. Service flow connections are characterized by a set of QoS parameters such as latency and jitter [6]. To most efficiently utilize network resources such as bandwidth and memory, 802.16 adopt a two-phase activation model in which resources assigned to a particular admitted service flow may not be actually committed until the service flow is activated. Each admitted or active service flow is mapped to a MAC connection.

In this section, a radio resource scheduling scheme is proposed for providing a merged service of both realtime and non-realtime flow connections over the wireless link in the Portable Internet (e.g., IEEE 802.16e) with frequent host mobility. This scheme allows non-realtime (NRT) flow connections to overflow into the region reserved for realtime (RT) flow connections with the risk of being preempted by newly arriving RT flow connections. The remaining bandwidth, except for the regions reserved for realtime flow connections, is dedicated to both NRT streaming flow connections and NRT flow connections.

This scheme has infinite waiting queue for both NRT streaming and NRT handover flow connections and, guard channels for RT handover flow connections (see Fig. 2). The NRT flow connections preempted by the newly arriving RT flow connections are queued, and NRT streaming flow handover requests have preemptive priority over NRT flow connections existing in the dedicated region. The pseudo code for simulation of the proposed scheme is described in Fig. 3.

For analytic simplicity, however, it is assumed that a NRT flow occupies a basic unit bandwidth unit (BBU) and thus each flow is considered at call level. As a result, the congestion of NRT flow connections can be estimated with the mean system time including queue waiting time.

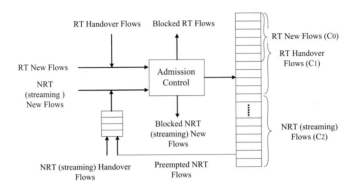

Fig. 2. Diagram of Proposed Radio Resource Scheduling Scheme

3.1   Analytic Model

For analytic model, it is assumed that service classes are classified into RT flow and NRT flow for simplicity of analysis. It is also assumed that a total capacity of cell is divided by m Basic Bandwidth Units (BBUs), where BBU corresponds to a logical channel. It is necessary to assume that a NRT flow has one BBU for the convenience of the comparison, an RT flow requires nBBUs $(n < m)$, and all BBUs assigned to an RT flow are occupied and released together.

In Fig. 2, c_0 is the number of BBUs reserved for originating RT flow connections and $(c_1 - c_0)$ is the number of BBUs reserved exclusively for RT handover flow connections. It is assumed that the RT service and the NRT service flow connections arrive according to a Poisson process with mean arrival rate λ_k for a type k flow $(k = 1, 2)$, and that service time is exponentially distributed with mean service time of μ_k^{-1} for a type k $(k = 1, 2)$ flow. Then, the system can be modelled as a two-dimensional Markov process, characterized by $\{n_2(t), n_1(t)\}$, where $n_2(t)$ and $n_1(t)$ are the numbers of NRT and RT flow connections in the system at time t, respectively, and the state space is represented by the set $\{s(n_2, n_1) \mid 0 \leq n_2, 0 \leq n_1 \leq \lfloor \frac{c_1}{n} \rfloor\}$. Also, the steady-state probability that the system is in state $s(n_2, n_1)$ be $p(n_2, n_1)$. The steady-state probability vector P is then partitioned as $p = (p_0, p_1, \cdots)$ where for $i = 0, 1, 2, \cdots$, $p_i = (p(i, 0), p(i, 1), \cdots, p(i, \lfloor \frac{c_1}{n} \rfloor))$. The vector P is the solution of equations

$$pQ = 0$$
$$pe = 1$$

(1)

```
Pseudo Code: Radio Resource Scheduling Scheme

C1: Bandwidth Resource Reserved for Overall RT Flows
C0: Bandwidth Resource Reserved for RT New Flows and Handover Flows
C1 - C0: Guard Bandwidth Resource Reserved Exclusively for RT Handover Flows
C2: Bandwidth Resource Reserved for both NRT streaming and NRT Flows
Queue: Waiting Queue for NRT Handover Flows, NRT streaming Handover Flows and Preempted NRT Flows
// NRT streaming flows have priority over NRT flows in Queue//

If (Requested Flow == RT New Flow)
{
        if (Bandwidth (BW) Available in C0 >= Bandwidth Required for the RT New Flow)
                Allocate BW for the RT New Flow;
        else
                Reject the RT New Flow;
}
else if (Requested Flow == RT Handover Flow)
{
        if (BW Available in C0 >= BW Required for the RT Handover Flow)
                Allocate BW for the RT Handover Flow;
        else if (BW Available in C1 - C0 >= BW Required for RT Handover Flow)
                Allocate BW for the RT Handover Flow;
        else if (Total BW of NRT Flows in C1 >= BW Required for RT Handover Flow)
                Preempt and Queue Some of Them to Make   BW Available for RT Handover Flow;
                Allocate BW for the RT Handover Flow;
        else
                Reject the RT Handover Flow;
}
else if (Requested Flow == NRT streaming New Flow)
{
        if (There is Any Flow Waiting for Service in Queue)
                Reject the NRT streaming New Flow;
        else if (BW Available in C2 >= BW Required for NRT streaming New Flow)
                Allocate BW for the NRT streaming New Flow;
        else
                Reject the NRT streaming New Flow;
}
else if (Requested Flow == NRT streaming Handover Flow)
{
        if (BW Available in C2 >= BW Required for NRT streaming Handover Flow)
                Allocate BW for the NRT streaming Handover Flow;
        else if (Total BW of NRT Flows in C2 >= BW Required for NRT streaming Handover Flow)
                Preempt and Queue Some of Them to Make   BW Available for the NRT streaming Handover Flow;
                Allocate BW for the NRT streaming Handover Flow;
        else
                Put the NRT streaming Handover Flow in Queue;
}
else if (Requested Flow == NRT New Flow)
{
        if (There is Any Flow Waiting for Service in Queue)
                Reject the NRT New Flow;
        else if (BW Available in C2 >= BW Required for NRT New Flow)
                Allocate BW for the NRT New Flow;
        else if (Any Available BW in C1)
                Allow NRT New Flow to Overflow into the C1 Region, and thus Allocate BW;
        else
                Reject the NRT New Flow;
}
else if (Requested Flow == NRT Handover Flow)
{
        if (BW Available in C2 >= BW Required for NRT Handover Flow)
                Allocate BW for the NRT Handover Flow;
        else if (Any Available BW in C1)
                Allow NRT Handover Flow to Overflow into the C1 Region, and thus Allocate BW;
        else
                Put the NRT Handover Flow in Queue;
}
```

Fig. 3. Pseudo Code of Proposed Radio Resource Scheduling Scheme

where e and 0 are vectors of all ones and zeros, respectively, and Q is the transition rate matrix of the Markov process. The state probabilities can be obtained by using state-transition equations since the system possesses a quasi-birth-death process [9]. The analytic method in this section heavily relies on previous works [2][8]. Here an example of the state transition diagram for the proposed scheme is abbreviated. Instead, the transition rate matrix Q of the Markov process can be presented: the transition matrix is of the block-partitioned form can be formulated as quasi-birth-and-death processes (QBDs) [9].

$$
Q = \begin{bmatrix}
A_0 & D_0 & & & & \\
B_1 & A_1 & D_1 & & & \\
& \ddots & \ddots & \ddots & & \\
& & B_r & A_r & D_r & \\
& & & B_r & A_r & D_r \\
& & & & \ddots & \ddots
\end{bmatrix} \tag{2}
$$

where the sub-matrices are defined for $i, j = 0, 1, \cdots, \lfloor \frac{c_1}{n} \rfloor$ and $l = 0, 1, \cdots, r$ ($n \cdot \lfloor \frac{c_1}{n} \rfloor + r \geq m$) by

$$
A_l(i, j) = \begin{cases}
\lambda_1 & \text{if } i = j - 1 \text{ and } j \leq \lfloor \frac{c_0}{n} \rfloor \\
\lambda_{1h} & \text{if } i = j - 1 \text{ and } \lfloor \frac{c_0}{n} \rfloor < j \leq \lfloor \frac{c_1}{n} \rfloor \\
(j + 1)\mu_1 & \text{if } i = j + 1 \\
a_l(i) & \text{if } i = j \\
0 & \text{otherwise}
\end{cases} \tag{3}
$$

$$
B_l(i, j) = \begin{cases}
min(l, m - n \cdot j)\mu_2 & \text{if } i = j \\
0 & \text{otherwise}
\end{cases} \tag{4}
$$

$$
D_l(i, j) = \begin{cases}
\lambda_2 & \text{if } i = j \text{ and } l < m - n \cdot j \\
\lambda_{2h} & \text{if } i = j \text{ and } l \geq m - n \cdot j \\
0 & \text{otherwise}
\end{cases} \tag{5}
$$

where $a_l(i)$ is the value that makes the sum of the row elements of Q equal to zero, $\lfloor x \rfloor$ denotes the greatest integer smaller than or equal to x. To solve the equation (1) with this transition rate matrix Q, the Neuts' theorem [9] is applied to the matrix Q. Next, the minimal nonnegative matrix R of the matrix equation is determined with

$$
R^2 B_r + R A_r + D_r = 0. \tag{6}
$$

by iteration. The steady-state probability vector p_k $(k \geq r)$ is given by

$$
p_k = p_{r-1} R^{k-r+1}, \quad k \geq r. \tag{7}
$$

Next, the probability vector $\tilde{p} = (p_0, \cdots, p_{r-1})$ is solved from the equations

$$
\tilde{p}T = 0
$$
$$
\tilde{p}e + p_{r-1}R(I - R)^{-1}e = 1 \tag{8}
$$

where the matrix T is a generator $(Te = 0)$ given by

$$
T = \begin{bmatrix}
A_0 & D_0 & & & & \\
B_1 & A_1 & D_1 & & & \\
& \ddots & \ddots & \ddots & & \\
& & B_{r-2} & A_{r-2} & D_{r-2} & \\
& & & B_{r-1} & A_{r-1} + R B_r &
\end{bmatrix} \tag{9}
$$

Once the matrix R and the boundary probabilities have been computed, the performance measures are easily obtained.

4   Performance Measures

The probability that a radio resource allocation request for an originating RT flow fails is given by

$$P_{B_{RT}} = \sum_{n_2=0}^{\infty} \sum_{n_1=k(c_0)}^{k(c_1)} p(n_2, n_1) \tag{10}$$

which, using (7), is easily reduced to

$$P_{B_{RT}} = \sum_{n_2=0}^{r-1} \sum_{n_1=k(c_0)}^{k(c_1)} p(n_2, n_1) + \left[p_{r-1} R(I-R)^{-1} \right]_{(k(c_0)..k(c_1))} \tag{11}$$

where $k(l)$ denotes the function of $\lfloor l/n \rfloor$, and $[\ ](a..b)$ denotes the sum to the bth element from the ath element of the vector in the bracket.

The handover failure probability of an RT flow is the probability that an arriving RT handover request finds that all the bandwidth channels assigned for RT handover flow connections in a cell are busy, and thereby the base station (BS) fails to allocate an amount of radio resource required to keep going with a multimedia flow. This metric should be minimized to give statistic QoS guarantee. It is given by

$$P_{D_{RT}} = \sum_{n_2=0}^{\infty} p(n_2, k(c_1))$$

$$= \sum_{n_2=0}^{r-1} p(n_2, k(c_1)) + \left[p_{r-1} R(I-R)^{-1} \right]_{(k(c_1))} \tag{12}$$

The forced termination probability of an RT flow, $P_{F_{RT}}$, is the probability that an initially accepted RT flow is interrupted due to bandwidth allocation failure over the wireless link at handover during its lifetime. Typically, a new RT flow is initially accommodated by a cell and then managed with priority so as not to be terminated due to bandwidth allocation failure over the wireless link during handover. Let α_1 be the probability that an RT flow currently served by a cell requires another handover request which is accommodated by neighboring cells. Similarly, let α_2 denote the probability that RT flow requires another handover request which fails. Then, using the Markovian properties of the model, it is determined that

$$\alpha_1 = (1 - P_{D_{RT}}) \frac{U_1}{\mu_1 + U_1} \tag{13}$$

$$\alpha_2 = P_{D_{RT}} \frac{U_1}{\mu_1 + U_1} \tag{14}$$

where μ_1 is average duration time of the RT flow and U_1 is average channel holding time of the RT flow in a cell. Now let us focus on a new RT flow that is initially accommodated by a cell. When this flow leaves the service of a cell, it can be forced into termination by bandwidth allocation failure over wireless link. Thus, the probability that an RT flow currently accommodated by a cell is forced into termination is given by

$$P_{F_{RT}} = \sum_{i=0}^{\infty} \alpha_1^i \alpha_2 \tag{15}$$

Now the average system time (i.e., sum of queuing time and service time) of NRT flow connections can be considered. The mean number of a NRT flow in a cell is

$$N_{NRT} = \sum_{n_1=0}^{k(c_1)} \sum_{n_2=0}^{\infty} n_2 p(n_2, n_1) \tag{16}$$

which, also using (7), reduces to

$$N_{NRT} = \sum_{n_1=0}^{k(c_1)} \sum_{n_2=0}^{r-1} n_2 p(n_2, n_1) + P_{r-1} R^2 (I - R)^{-2} e$$
$$+ r P_{r-1} R (I - R)^{-1} e. \tag{17}$$

The blocking probability of NRT flow connections is given by

$$P_{B_{NRT}} = \sum_{n_1=0}^{k(c_1)} \sum_{n_2=(m-n_1 \cdot n)}^{\infty} p(n_2, n_1)$$
$$= \sum_{n_1=0}^{k(c_1)} \sum_{n_2=(m-n_1 \cdot n)}^{r-1} p(n_2, n_1) + P_{r-1} R (I - R)^{-1} e. \tag{18}$$

Therefore, the mean system time for NRT flow connections from Little's formula is given by

$$W_{NRT} = \frac{N_{NRT}}{\lambda_{2h} + \lambda_{2o}(1 - P_{B_{NRT}})} \tag{19}$$

Another important system performance measure is carried traffic. For a given number of channels, a large carried traffic value implies efficient use of bandwidth. Carried traffic per RT flow in a cell, E_{RT}, and carried traffic per NRT flow, E_{NRT}, can be calculated easily once the state probabilities are determined. They are simply the average number of occupied channels per flow, which can be given by

$$E_{RT} = \sum_{n_1=0}^{k(c_1)} \sum_{n_2=0}^{\infty} n_1 p(n_2, n_1) = \sum_{n_1=0}^{k(c_1)} \sum_{n_2=0}^{r-1} n_1 p(n_2, n_1) +$$
$$2^{-1} k(c_1)(k(c_1) + 1) P_{r-1} R (I - R)^{-1} e \tag{20}$$

$$E_{NRT} = \sum_{n_1=0}^{k(c_1)} \sum_{n_2=0}^{m-n_1 \cdot n} n_2 p(n_2, n_1) \tag{21}$$

Because an RT flow requires nBBUs, the total carried traffic in a cell is given by

$$E = n E_{RT} + E_{NRT} \tag{22}$$

5   Examples and Discussions

For simplicity of analysis and simulation, the number of total bandwidth units (m) of each cell is given as 6, and the number (n) of $BBUs$ which occupies or releases simultaneously for an RT flow is given as 1 or 2. The number of bandwidth units assigned for originating RT flow connections is basically 2. In addition, RT handover flow connections can use 2 extra bandwidth units exclusively. That is, $n = 2$, $m = 6$, $c_0 = 2$ and $c_1 = 4$. The average channel holding time of RT flow connections in each cell is 200 s. When the arrival rate of new RT flow connections in a cell reaches to the limit c_0, new RT flow connections start to be blocked for reducing the input traffic load.

Fig. 4 shows the forced termination probability (η) for RT flow connections when the traffic load (ρ) of NRT flow connections is 11 Er, n is 1 or 2, and handover rate (h) is 0.25 or 0.5. In the case when the RT flow arrival rate (λ_1) is less than 10^{-3}, all the η's stay below 10^{-2}, and in the case when λ_1 is more than 10^{-2}, the η's exceed 10^{-1}. In this figure, we can know that the forced termination probability of RT handover flow connections can be kept at a reasonable level (e.g. 10^{-2}) by the use of exclusive guard channels. That is, the statistical guarantee of handover success of RT flow connections can be maintained by adjusting well the size of guard channels for handover flow connections.

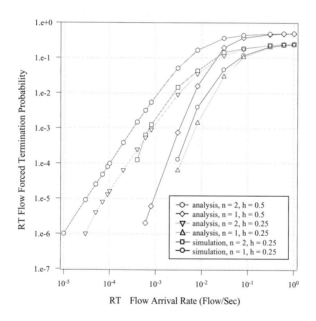

Fig. 4. Forced Termination Probability for RT Flows as a Function of λ_1 When $\rho = 11$ Er

Figure 5 shows the RT (NRT) flow blocking probability as a function of RT flow arrival rate when n is 2 and h is 0.25. In this figure, as the RT flow arrival rate (λ_1) increases, the RT flow blocking probability also goes up significantly until it reaches 1.0. However, under a heavy constant load of NRT flows (3 Er), NRT flow blocking probability is kept at the same value (0.3) until λ_1 reaches 3.0×10^{-3}. As λ_1 becomes greater, NRT flows are much often preempted by newly arriving RT flows, and hereby, the mean number of server channels for NRT flows is reduced

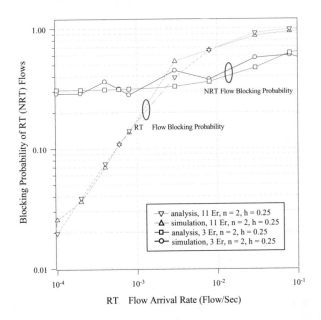

Fig. 5. RT (NRT) Flow Blocking Probability as a Function of λ_1 When $n = 2$ and $h = 0.25$

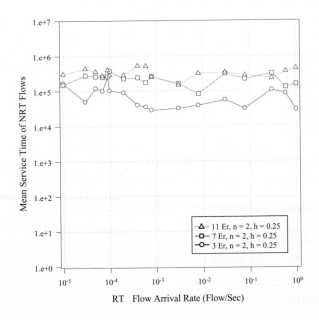

Fig. 6. Mean Service Time of NRT Flows as a Function of λ_1 with Different NRT Flow Loads When $n = 2$ and $h = 0.25$

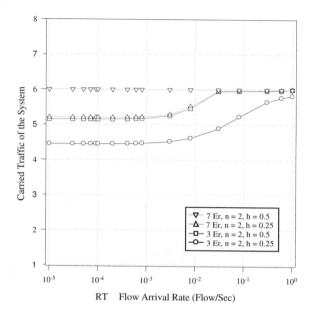

Fig. 7. Carried Traffic of the System as a Function of λ_1 with Different NRT Flow Loads and Different Handover Rate h When $n = 2$

until it becomes 2. Thus, originating NRT flows start to be blocked earlier because this makes the mean queue size of NRT flows become longer.

Fig. 6 shows the mean service time of NRT flow connections as a function of RT flow arrival rate with different NRT flow loads when n is 2 and h is 0.25. In this example, as the load of NRT flow (ρ) becomes higher, the mean service time (τ) also becomes greater. That is, the τ in the case of 11 Er load has much greater than that in the case of 3 Erlang load. On the other hand, when the RT flow arrival rate (λ_1) is low, NRT flow connections might overflow into the channels reserved for RT flow connections. As λ_1 increases, RT flow connections might make NRT flow connections preempted quickly and queued.

Fig. 7 shows the carried traffic as a function of RT flow arrival rate with different NRT flow loads and different handover rates when n is 2. In this example, the number of total bandwidth units (m) of each radio cell is given as 6. In the case of $\rho = 7$ Er, the carried traffic is more than that in the case of $\rho = 3$ Er since heavy NRT flow load makes NRT flow connections overflow into the RT flow region. In addition, as RT flow arrival rate (λ_1) increases, NRT flow connections in the channels reserved for RT flow connections are preempted and queued. At this point, the length of waiting queue becomes much longer.

6   Conclusions

In this paper, a radio resource scheduling scheme was proposed for providing a merged service of both realtime and non-realtime flow connections over the wireless link in the Portable Internet (e.g., IEEE 802.16e) with frequent host mobility. In addition to basic simulation, a simple analytic model is developed for the proposed scheme. The forced termination probability of RT flow, the blocking probability RT and NRT flow connections, the mean service time of NRT flow connections and carried traffic are derived and evaluated with several examples.

References

1. Kwon, T., Lee, H., Choi, S., Kim, J., Cho, D., Cho, S., Yun, S., Park, W., Kim, K.: Design and Implementation of a Simulator Based on a Cross-Layer Protocol between MAC and PHY Layers in a WiBro Compatible IEEE 802.16e OFDMA System. In: IEEE Communications, December 2005, pp. 136–146 (2005)
2. Moon, B.-K., Aghvami, A.H.: Quality of Service Mechanisms in All-IP Wireless Access Networks. IEEE Journal on Selected Areas in Communications 22(5), 873–888 (2004)
3. Cheng, Y., Zhuang, W.: DiffServ Resource Allocation for Fast Handoff in Wireless Mobile Internet, IEEE Communications (May 2002)
4. Chang, H., Chang, Y., Cho, J.: A Tightly-coupled Integration Scheme between WiBro and cdma2000 mobile networks. International Transaction on Computer Science and Engineering 6(1), 195–205 (2005)
5. ITU Internet Reports, The Portable Internet, International Telecommunication Union (September 2004)
6. Eklund, C., Marks, R.B., Stanwood, K.L., Wang, S.: IEEE Standard 802.16: A technical Overview of the WirelessMAN Air Interface for Broadband Wireless Access, IEEE Communications, pp. 98–107 (June 2002)
7. Andrews, M., Kumaran, K., Ramanan, K., Stolyar, A., Whiting, P., Vijayakumar, R.: Providing Quality of Service over a Shared Wireless Link, IEEE Communications, pp. 150–154 (February 2001)
8. Serres, Y.D., Mason, L.G.: A multiserver queue with narrow and wide band customers and wide band restricted access. IEEE Trans. Commun. vol. COM–36 (1988)
9. Neuts, M.F.: Matrix-Geometric Solutions in Stochastic Models: An Algorithmic Approach. Dover Publications, Inc., Mineola (1994)
10. MacDougall, M.H.: Simulating Computer Systems: Techniques and Tools. The MIT Press, Cambridge (1987)

Opportunistic Scheduling of Wireless Links

Vinod Sharma[1], D.K. Prasad[1], and Eitan Altman[2,*]

[1] Dept Elect. Comm. Engg., Indian Istitute of Science,
Bangalore, 560012, India
{vinod, dkp}@ece.iisc.ernet.in
[2] INRIA B.O.93, 2004 Route des Lucioles, 06902
Sophia-Antipolis Cedex, France
Eitman.Altman@sophia.inria.fr

Abstract. We consider the problem of scheduling of a wireless channel (server) to several queues. Each queue has its own link (transmission) rate. The link rate of a queue can vary randomly from slot to slot. The queue lengths and channel states of all users are known at the beginning of each slot. We show the existence of an optimal policy that minimizes the long term (discounted) average sum of queue lengths. The optimal policy, in general needs to be computed numerically. Then we identify a greedy (one step optimal) policy, MAX-TRANS which is easy to implement and does not require the channel and traffic statistics. The cost of this policy is close to optimal and better than other well-known policies (when stable) although it is not throughput optimal for asymmetric systems. We (approximately) identify its stability region and obtain approximations for its mean queue lengths and mean delays. We also modify this policy to make it throughput optimal while retaining good performance.

Keywords: Wireless channel, opportunistic scheduling, greedy scheduling, throughput optimal scheduling, performance analysis.

1 Introduction

We consider the problem of scheduling of several users on a wireless link shared by them. This problem is of relevance in scheduling in uplink and downlink of cellular systems as well as multihop wireless networks. Such problems have been addressed in wireline networks also (e.g., in a router, LAN, switches etc) but in the wireless scenario an added complexity is that the link rates seen by different queues can be different and usually vary randomly in time. Thus the wireline solutions, TDMA, weighted round-robin (when queue lengths and packet sizes may not be available) or strict priority rules such as the $c\mu$ rule ([21]) do not provide reasonable performance in the wireless systems. For good performance one needs to exploit the multiuser rate diversity and use opportunistic scheduling.

---

* The work of this author was supported by a contract with France Telecom.

L. Mason, T. Drwiega, and J. Yan (Eds.): ITC 2007, LNCS 4516, pp. 1120–1134, 2007.

In principle one may allow simultaneous transmission by different users. However, in this paper we will limit ourselves to the case where at a time only one user can transmit.

In [10] it is shown that in a slotted wireless system allocating the slot to the user with the best channel will provide maximum throughput if all the users have always enough data to send. But, by this scheme the throughput received by different users can be very unfair and the mean delays obtained can be quite large ([4]). Thus, if all users always have data to transmit then different approaches have been followed in [3], [5], [12]. When we remove the constraint that all users always have data to transmit, then [2], [6], [17], provide policies which are throughput optimal, i.e., these policies will stabilize the system if any feasible policy will. Throughput optimal policies when the link rates are 0 or 1 were earlier obtained in [19]. Generalization of these results when the arrival processes and the channel availability processes satisfy some burstiness constraints is provided in [20]. Two recent excellent survey-cum-tutorial papers on this topic are [8] and [13]. When we do not consider a slotted system then optimal scheduling policies are also obtained in [7] and [21] Chapter X. When the link rates are constant then the $c\mu$ rule is known to be optimal in many different scenarios ([21], [14]).

In this paper we consider a slotted single hop wireless system. The schedular knows the queue lengths and the channel states of each of the queues. The packets arrive at each queue as sequences of independent, identically distributed (*iid*) random variables. The channel rates of each queue also form *iid* sequences. In this scenario, as mentioned above, [2], [6], [17] provide scheduling policies which are throughput optimal. However, often mean delays are of concern. Although the policies provided in [17] also minimize the mean delays under heavy traffic, the policies in [6] provide less mean delays than those provided in [17] when it is not a heavy traffic scenario. Optimal policies which minimize mean delays or queue lengths under heavy traffic have also been studied in [1] when the link rates can be 0 or 1. In this paper we look for policies which minimize mean delay at different operating points. We start with considering the problem of minimizing mean weighted delay. We first show the existence of an optimal policy. This policy is also throughput optimal. However it is obtained numerically and requires the knowledge of the statistics of the arrival traffic and the link rates. Next we look for good sub-optimal policies. MAX-TRANS, also considered in [18], is a one-step optimal greedy policy. This does not require knowledge of channel and traffic statistics and is easy to implement. We compare its performance (discounted average and mean queuelength) to the optimal policy and the policies in [6], [17] and a generalization of the policy in [19]. The MAX-TRANS performs better than the policies in [6], [17] and [19] and is also close to the optimal. However, we will show that it is not throughput optimal for asymmetric traffic (thus there will be traffic rates when the mean delays for MAX-TRANS will be infinite but for the policies in [6] and [17] finite. Even for such cases we have observed that MAX-TRANS provides better performance for the discounted cost problem). We will obtain its (approximate) stability region. We will also provide formulae for approximate mean delay and mean queue length of this policy. Formulae for mean delay and mean queue lengths of policies in [6], [17] and [19] are not available. We will also modify this policy to make it throughput optimal while retaining its performance.

Greedy policies, like MAX-TRANS, have also been considered before (see [8] for comments) and have *not* been recommended because they are not throughput optimal. However, our study indicates that such policies can indeed be useful from performance (e.g., mean delay) point of view and can often be modified to obtain throughput optimal policies.

The paper is organized as follows. In Section 2 we formulate the problem as a Markov Decision Problem and show the existence of an optimal policy. In Section 3 we identify a one-step optimal policy and compare this policy to the optimal policy and other policies available in literature. We find that its performance is quite good as compared to other policies and is also close to the optimal. Thus in Section 4 we study its performance theoretically: we find approximately its stability region and its mean queue length and delays. We verify the accuracy of our approximations with simulations in Section 5. Section 6 concludes the paper.

2   Problem Formulation and Existence of an Optimal Policy

We consider the problem of scheduling transmission of $N \geq 2$ data users (flows) sharing the same wireless channel (server). The system is slotted and multiple transmissions in a slot are allowed. Each user has an infinite buffer to store the data. At the time of transmission, the packets can be arbitrarily fragmented for efficient transmission. We ignore the fragmentation overhead. These assumptions have also been made in [6], [8], [12]. Thus the buffer contents can be considered at the bit level (i.e., queue lengths will be the number of bits in the queue). One of the links is to be scheduled in a slot depending on the current queue lengths and link rates of different users. We denote the queue size of the ith queue at the beginning of the time slot k by $q_k(i)$, the number of arrivals (bits) to queue i in slot k by $X_k(i)$, and the amount of service offered to queue i in slot k by $r_k(i)$. We assume that these parameters can only take non-negative integer values (not really needed). The evolution of the size of the ith queue is given by

$$q_{k+1}(i) = (q_k(i) + X_k(i) - y_k(i)r_k(i))^+, \quad i = 1, ..., N \qquad (1)$$

where $(y)^+ = \max(0, y)$ and $y_k(i) = 1$ if i^{th} queue is scheduled in k^{th} slot; otherwise 0.

We assume that the channels of different users can be in any one of the M states in a given slot, where $M < \infty$. The channel state is assumed to be fixed within a slot, but may vary from slot to slot and hence the model captures the time-varying characteristics of a fading channel. The channel rate processes $\{r_k(i), k \geq 1\}$ and the arrival processes $\{X_k(i), k \geq 1\}$ are assumed to be *iid* sequences. Also, these sequences are assumed to be independent of each other. Some of these assumptions may not be true in practical systems but we think this setup captures the essential elements of the general problem and the solutions proposed and the conditions presented should be relevant in the general setup.

We consider the problem of scheduling the channel such that

$$\limsup_{n \to \infty} \sum_{k=1}^{n} \sum_{i=1}^{N} w(i) q_k(i) \alpha^n \qquad (2)$$

is minimized where $w(i) > 0$ are weights which reflect the priorities of different users and $0 < \alpha < 1$. We also consider the optimization of the average cost

$$\limsup_{n \to \infty} \frac{1}{n} \sum_{k=1}^{n} \sum_{i=1}^{N} w(i) q_k(i). \qquad (3)$$

A policy optimizing (2) will be called an α-discounted optimal policy and a policy optimizing (3) will be called an average cost optimal. By Little's law a policy optimizing (3) also optimizes weighted sums of mean stationary delays. In addition it is throughput optimal. If the Quality of Service (QoS) requirement demands more than mean delays, i.e., has soft or hard delay constraints then (2) can be relevant.

Next we prove the existence of optimal policies for (2) and (3). Obtaining the optimal policies requires information about the statistics of the traffic and link rates and needs to be numerically computed. Therefore, in Section 3 we identify a policy which has performance close to the optimal cost and is better than other well known policies. In Section 4 we study the performance of this policy.

Existence of Optimal Policy

We assume that the rate process $r_k = (r_k(1), ..., r_k(N))$ is component-wise upper bounded by $\bar{r}$, i.e., $\bar{r}(i)$ is the largest value $r_k(i)$ can take, $i = 1, ..., N$. We use the notation $q_k = (q_k(1), ..., q_k(N))$ and $X_k = (X_k(1), ..., X_k(N))$. Also, $J(q, r)$ and $J_\alpha(q, r)$ denote the optimal average cost and α-discounted optimal cost where (q, r) denotes the queue and link rates at time $k = 1$. In addition $J_{\alpha,n}(q, r)$ and $J_n(q, r)$ will denote the n-step optimal costs.

Proof of the following theorem is available in [16](see also [8]).

Theorem 1. Under our assumptions, there exists an α-discounted optimal policy for any α, $0 < \alpha < 1$ and $J_{\alpha,n}(q, r) \to J_\alpha(q, r)$ as $n \to \infty$. If there exists a stationary policy under which the system can reach state $(0, \bar{r})$ in a finite mean time starting from any initial state (q, r) then there also exists an optimal average cost policy.

Under the above conditions, value iteration and policy iteration algorithms for the discounted problem also converge ([9]). Furthermore, for any $\alpha_k \to 1$, $J^* = J(q, r) = \lim_{k \to \infty} J_{\alpha_k}(q, r)$ ([9]).

The condition in the above theorem that there is a stationary policy under which the system can reach state $(0, \bar{r})$ in a finite mean time is implied by the condition that there is a stationary policy under which the process $\{q_k, r_k\}$ is an ergodic Markov chain. This is obviously a necessary condition to have a stable stationary ergodic optimal policy.

3   MAX-TRANS: A Good Suboptimal Policy

Theorem 1 provides the existence of α-discount and average cost optimal policies. We have also seen above that the optimal policy can be obtained numerically via value or policy iteration. However, numerical computations can be cumbersome and may not be feasible in real time for a large number of queues (possible in practical scenarios). Neither does it provide any insight into the problem nor into the optimal policies. Thus in the following we consider suboptimal policies which may be easier to implement and still provide good performance. These policies have been taken from the literature and have been found to have many desirable features. In particular they do not require the statistics of the input traffic and link rates. Also, several of them are throughput optimal ie., they will stabilize a system if it is possible to do so by any feasible policy. We compare the performance of these policies to identify a good policy.

In the following z_k will denote the index of the queue selected by a policy in slot k. Often more than one index will be picked by the criterion used by a policy. Then one can pick one of the selected queues probabilistically.

1. *Maximum Transmission Scheme (MAX-TRANS)*:

$$z_k \overset{\triangle}{=} \arg\max w(i)\left(\min(r_k(i), q_k(i) + X_k(i))\right).$$

This policy was used in [18] and compared with several other policies. It was found to provide good performance even when used for a whole frame and in a multihop environment. This is a 'greedy' policy and can be shown to be 1-step optimal for (2) and (3). However we will show that it is not throughput optimal.

2. *Modified Longer Queue First*:

$$z_k = \arg\max\left(w(i)q_k(i).1\{r_k(i) > 0\}\right).$$

For $r(i) \in \{0, 1\}$ this policy is known to be throughput optimal ([19]). For a symmetric system it also minimizes the mean delay for such $r(i)$.

3. *Eryilmaz, Srikant and Perkins Policy [6]*:

$$z_k \overset{\triangle}{=} \arg\max w(i)r_k(i)q_k(i).$$

This policy is throughput optimal.

4. *Shakkottai and Stolyar Policy [17]* : Define $\bar{Q}_k = \frac{1}{N}\sum_{i=1}^{N} a_i q_k(i)$ with $a_i = 2$. Then

$$z_k = \arg\max\left\{w(i)\ r_k(i)\ \exp\left(\frac{a_i q_k(i) - \bar{Q}_k}{1 + \sqrt{\bar{Q}_k}}\right)\right\}.$$

This policy is also throughput optimal and provides minimum delays under heavy traffic.

We compare the performance of these policies for the discounted cost (2) and the average cost (3) for various queueing systems. In the following we provide one example. We also obtain the optimal policies for (2) and (3) via Policy and Value iteration. In addition the costs of the above four policies are obtained via value iteration as well as via simulations.

We consider three queues and a single server. First we consider the discounted (discount factor $= 0.94$) case and then the undiscounted case. All $w(i)$ in (2) and (3) are taken as 1. Arrival and service distributions for different queues are given in Table 1. We have considered a buffer size of 4 (to keep the complexity of computations small), for numerical computation of the optimal policies and for computing the costs of the above suboptimal policies via value iteration. Thus there are 3375 possible states. We refer to state $(0, 0, 0, 0, 0, 0)$ as state number 1 and state $(4, 4, 4, 2, 3, 4)$ as 3375. The discounted costs are provided in Table 2 for a few initial states. Table 3 has the discounted costs obtained via simulations. Now we consider the queues with infinite buffer. We ran the simulations for 150 slots, two thousand times and obtained their average.

From Tables 2 and 3 we conclude that MAX-TRANS (Policy 1) is the best among all suboptimal policies and is quite close to the optimal for all initial values (although we have shown in these tables the cost for a few initial states due to lack of space, the conclusions hold for other states also). Next is Policy 3. The other two policies can have much worse values. Policy 2 is the worst.

Table 4 contains the results for the average cost problem: via numerical computations and simulations for a buffer size of 4 and via simulations for a buffer size of infinite. Each simulation was run for 2 million slots. Here also we observe the same trend as for the discounted case.

Table 1. Parameters for an example to compare

Q	Link distributions	Input distributions
	2, w.p. $= 0.6$	2, w.p. $= 0.250$
1	1, w.p. $= 0.3$	1, w.p. $= 0.168164$
	0, w.p. $= 0.1$	0, w.p. $= 0.581836$
	3, w.p. $= 0.2$	3, w.p. $= 0.15$
2	1, w.p. $= 0.5$	1, w.p. $= 0.151524$
	0, w.p. $= 0.3$	0, w.p. $= 0.698476$
	4, w.p. $= 0.05$	3, w.p. $= 0.15$
3	2, w.p. $= 0.6$	2, w.p. $= 0.20228$
	0, w.p. $= 0.35$	0, w.p. $= 0.64772$

Table 2. Comparison of policies for discount 0.94; analytical results for buffer size 4

state no.	Optimal	Policy 1	Policy 2	Policy 3	Policy 4
500	79.061	82.275	92.608	83.012	83.965
750	66.469	69.574	76.331	70.076	70.326
1000	82.021	85.221	94.659	85.906	87.013
1250	76.846	80.195	87.415	80.598	81.301
1500	71.546	74.551	83.096	76.078	76.731
2000	93.088	96.026	107.080	97.430	99.213
2250	78.951	81.984	90.004	82.748	83.571
2500	82.918	85.948	94.107	86.597	87.744
3000	82.977	86.120	95.151	86.874	88.171
3250	88.761	91.673	101.870	92.615	93.841

We simulated a large number of different queueing systems and observed the same trend. Actually we observed that MAX-TRANS was always the best among the four suboptimal policies for the discounted cost problem but sometimes the policy 3 was slightly better than policy 1 for the average cost problem (of course there were cases when MAX-TRANS was unstable but Policy 3 was stable). The other two policies were generally worse than policies 1 and 3.

We present one more example to show that the improvement provided by MAX-TRANS can be significant even with respect to the policy 3 which in the above example is always quite close in performance to MAX-TRANS. Let $N = 2$,

Table 3. Comparison for discount 0.94 (simulation with buffer = infinity)

Table 4. Comparison for undiscounted case

state no.	Policy 1	Policy 2	Policy 3	Policy 4
500	142.73	179.25	145.83	157.90
750	117.60	146.09	120.09	130.20
1000	133.35	167.83	136.57	148.19
1250	149.83	185.79	152.43	164.98
1500	119.76	149.99	122.54	133.12
2000	179.89	228.12	184.98	202.10
2250	154.21	192.40	157.12	170.40
2500	158.88	205.27	163.12	177.17
3000	152.84	191.26	156.11	169.92
3250	160.72	207.31	165.65	180.26

Policy	Theoretical cost	Sim cost	Sim with infinite buffer
Optimal	4.4489	4.0587	—
Policy1	4.6770	4.1764	106.5219
Policy2	5.3521	5.2444	$3.3814 * 10^5$
Policy3	4.7452	4.3857	108.9936
Policy4	4.8190	4.6256	$3.5326 * 10^3$

$r_k(1) \equiv 1$, $r_k(2) \equiv 2$ and $P[X_k(1) = 1] = 0.58$ and 0 otherwise, $P[X_k(2) = 2] = 0.4$ and 0 otherwise. Then via simulations we found the average cost of the four policies is 10.820, 14.246, 12.691, 13.863. Thus MAX-TRANS provides 17.3% lower cost than policy 3.

As mentioned above one drawback of MAX-TRANS is that it may not be throughput optimal when the system is not symmetric. We show it by an example. Consider an example with $N = 2$. Let $P[r_k(1) = 1] = 0.01 = 1 - P[r(1) = 0]$ and $P[r_k(2) = 2] = 1$. Also let $P[X_k(1) = 1] = 0.009 = 1 - P[X_k(1) = 0]$ and $P[X_k(2) = 2] = 0.9 = 1 - P[X_k(2) = 0]$. For this example, MAX-TRANS will always serve queue 2 whenever it is non-empty. This happens with probability 0.9. Thus queue 1 gets service at most for probability 0.1. This will make queue 1 unstable. As against this consider a policy that serves queue 1 whenever $r_k(1)=1$. Otherwise serve queue 2. This makes both the queues stable. Thus MAX-TRANS is not throughput optimal.

We will show in Section 4 that for symmetric traffic and link statistics MAX-TRANS is throughput optimal. For asymmetric statistics, as the above example shows, a user with bad channel can get starved for unusually long times making it an unstable queue. Throughput optimal policies do not starve any queue for very long time. Thus we can make MAX-TRANS into a throughput optimal policy, by modifying it appropriately. For example, if we make the following policy: select an appropriately large constant $L > 0$. If all queue lengths are less than L then use MAX-TRANS for scheduling. If one or more queues is larger than L then use policy 3 on those queues to select a queue to serve. If L is very large, the performance of this algorithm will be close to MAX-TRANS. If L is taken small, its performance will be close to policy 3. But for any $L > 0$, the policy will be throughput optimal.

We apply this modified MAX-TRANS on the above example where the MAX-TRANS is unstable with $L = 5$. Then the average cost of this policy and that of policies 2, 3, 4 is 10.70, 8.61, 9.66, 8.81. Thus one sees that the modified MAX-TRANS is stable.

We propose that if the system statistics are symmetric then use MAX-TRANS. Otherwise, if the system is well within the stability region of the MAX-TRANS, then just use this policy. If it is close to the stability region of

MAX-TRANS or outside it but within the stability region of the throughput optimal policies then we can use the modified policy. If the traffic statistics are not known but are asymmetric, then one can use the modified policy with somewhat large L.

Of course to be able to use the above scheme, we need to know the stability region of MAX-TRANS policy. We provide this in the next section. We show that for symmetric conditions MAX-TRANS is throughput optimal. We will also study the performance (mean delays and queue lengths) of this policy. When L is large, then the performance of the modified policy will be close to this. This encourages us to study the MAX-TRANS scheme in more detail.

4   Performance Analysis of MAX-TRANS Policy

As we observed in the last section MAX-TRANS is a promising policy. It is a greedy and one-step optimal policy and compares well with other well-known policies. Thus, it is useful to study its performance in more detail. In this section we obtain the stability region of this policy. We first show that this policy may have a smaller stability region for asymmetric traffic than policies 3 and 4 which are throughput optimal. For simplicity we will first consider this policy for two users and then generalize to multiple users. After studying the stability region of this policy, we obtain approximations for mean queue lengths and delays. We will show via simulations that our approximations work well atleast under heavy traffic. Of course from practical point of view this is the most important region of operation because QoS violation may happen in this region only.

First we provide sufficient conditions for stability for $N = 2$. We will show later on, that for symmetric inter-arrival and link rate distributions these conditions indeed are also necessary. The stability region provided by this theorem is suboptimal for asymmetric traffic. Later we will improve upon these conditions and provide approximate stability region for the asymmetric scenario.

In MAX-TRANS policy if $\min(q_k(1)+X_k(1), r_k(1)) = \min(q_k(2)+X_k(2), r_k(2))$ then we will assign k^{th} slot to queue 1 with probability p, $0 \le p \le 1$ and to queue 2 with probability $(1-p)$. Since $X(i), r(i)$ will often be discrete valued, this event can happen with nonzero probability.

Let $\bar{r} < \infty$ be the largest value $r_k(1)$ and $r_k(2)$ can take.

Theorem 2. If

$$\mathbb{E}[X(1)] < \mathbb{E}[r(1)1_{\{r(1)>r(2)\}}] + p\mathbb{E}[r(1)1_{\{r(1)=r(2)\}}], \qquad (4)$$

$$\mathbb{E}[X(2)] < \mathbb{E}[r(2)1_{\{r(2)>r(1)\}}] + (1-p)\mathbb{E}[r(2)1_{\{r(2)=r(1)\}}] \qquad (5)$$

then $\{(q_k(1), q_k(2))\}$ stays bounded with probability 1 for any initial condition. Also the inter visit time to set $A = \{x : x \le \bar{r}\}$ by $\{q_k(i)\}$ has finite mean, for $i = 1, 2$.

Proof. We consider another system $\{(\bar{q}_k(1), \bar{q}_k(2))\}$ where the $\bar{q}_{k+1} \triangleq (\bar{q}_{k+1}(1), \bar{q}_{k+1}(2))$ behaves as q_k as long as $\bar{q}_k(1) > \bar{r}$ and $\bar{q}_k(2) > \bar{r}$. If $\bar{q}_k(i) \leq \bar{r}$ then the ith queue is not served in the kth slot. Thus

$$\bar{q}_{k+1}(1) = (\bar{q}_k(1) + X_k(1) - z_k(1))^+$$

where $z_k(1) = 0$, if $\bar{q}_k(1) \leq \bar{r}$ and if $\bar{q}_k(1) > \bar{r}$, then

$$
\begin{aligned}
z_k(1) &= r_k(1), && \text{if } r_k(1) > r_k(2), \\
&= r_k(1), && \text{w. p. } p \text{ if } r_k(1) = r_k(2), \qquad (6)\\
&= 0, && \text{otherwise.}
\end{aligned}
$$

The dynamics of $\bar{q}_k(2)$ behave in the same way. Observe that if $q_k(i) > \bar{r}$, it gets the service as in (6) but in addition it can get service even when $z_k(i) = 0$. Unlike $\{q_k(i), k \geq 0\}$, $\{\bar{q}_k(i), k \geq 0\}$, $i = 1$, 2 is a Markov chain.

Consider the Markov chain $\{(\bar{q}_k(1), \bar{q}_k(2)), k \geq 0\}$. Let $B = \{(x, y) : x \geq x', y \geq y'\}$ for some (x', y'). Then one can easily show that

$$P[(\bar{q}_1(1), \bar{q}_1(2)) \in B | (\bar{q}_0(1), \bar{q}_0(2)) = (q_1, q_2)]$$
$$\geq P[(q_1(1), q_1(2)) \in B | (q_0(1), q_0(2)) = (q_1, q_2)]$$

for any (x', y') and (q_1, q_2). We can also construct $\{(\bar{q}_k(1), \bar{q}_k(2), q_k(1), q_k(2)), k \geq 1\}$ on a common probability space ([15]) such that starting from any initial conditions (q_1, q_2), for all $k \geq 1$

$$P[(\bar{q}_k(1), \bar{q}_k(2)) \geq (q_k(1), q_k(2)) | \bar{q}_0(1) = q_0(1) = q_1, \bar{q}_0(2) = q_0(2) = q_2] = 1.(7)$$

Next we show that under (4) and (5), $\{(\bar{q}_k(1), \bar{q}_k(2))\}$ stays bounded with probability 1 for any initial conditions. But considering $\{\bar{q}_k(i)\}$, $i = 1$, 2, we realize it behaves as a GI/GI/1 queue with service times $X_k(i)$ and inter-arrival times $z_k(i)$ whenever $\bar{q}_k(i) > \bar{r}$. Also, with (4) and (5), its traffic intensity $\rho(i) < 1$. Therefore, $\{\bar{q}_k(i)\}$ stays bounded with probability 1. Furthermore, from results on GI/GI/1 queues, the inter-visit times of $q_k(i)$ to the set $\{x : x \leq \bar{r}\}$ have finite mean (this is busy period for the corresponding GI/GI/1 queue).

The above results for $(\bar{q}_k(1), \bar{q}_k(2))$ along with (7) provide the corresponding results for the process $\{(q_k(1), q_k(2))\}$. $\square$

The results of the above theorem do not guarantee existence of a stationary distribution for the process $\{(q_k(1), q_k(2))\}$. However, if we also assume that

$$P[X_k(i) = 0] > 0, \ i = 1, 2 \qquad (8)$$

then with some additional work, we can show that the inter-visit time of the process $\{(\bar{q}_k(1), \bar{q}_k(2))\}$ to the set $\{(x, y) : x \leq \bar{r}, y \leq \bar{r}\}$ has a finite mean time. Then from (7) the same is true for the process $\{(q_k(1), q_k(2))\}$. Furthermore, the process $\{(q_k(1), q_k(2))\}$ can indeed visit the state $(0, 0)$ with finite mean inter-visit times and the process is an irreducible and aperiodic Markov chain. This guarantees that the Markov chain $\{q_k\}$ has a unique stationary distribution and starting from any initial conditions, it converges to the stationary distribution.

Although (4) and (5) have been shown to be sufficient conditions for stability, for symmetric rate and traffic conditions they are actually also necessary (take $p = \frac{1}{2}$ in this case). To see this, one only has to observe that summing (4) and (5) provides the maximum possible link rates one can obtain for this system. For symmetric case, taking half of the sum provides the largest possible throughput each queue can get. This argument holds for more than two queues (considered below) also.

The conditions (4) and (5) can be immediately generalized to $N \geq 2$ user case: for all i,

$$\mathbb{E}[X(i)] < \mathbb{E}[r(i)1_{\{r(i)>r(j),\forall j\neq i\}}] + \sum_{(j_1,j_2,\ldots j_k)} p(i,j_1,j_2,\ldots,j_k).$$
$$\mathbb{E}[r(i)1_{\{r(i)=r(j),j=j_1,j_2,\ldots,j_k,r(i)>r(j),j\neq i,j_1,\ldots,j_k\}}] \tag{9}$$

where $p(i,j_1,\ldots,j_k)$ are the probabilities used for breaking ties and the summation is over all possible subsets of $\{1,\ldots,N\} - \{i\}$.

Next let us consider the asymmetric case. For simplicity we restrict ourselves to two queues. See the generalization in the next section.

If (say) $\mathbb{E}[X(2)]$ is much smaller than the RHS of (5), then (4) is too conservative for $\mathbb{E}[X(1)]$. At least when $q_k(2) + X_k(2) = 0$, then MAX-TRANS will allow Q_1 to transmit as long as $q_k(1) + X_k(1) > 0$. Thus (4) can be relaxed to

$$\mathbb{E}[X_k(1)] < \mathbb{E}[r_k(1)1_{\{r_k(1)>r_k(2)\}}] + p\mathbb{E}[r_k(1)1_{\{r_k(1)=r_k(2)\}}]$$
$$+ (1-p)\mathbb{E}[r_k(1)1_{\{r_k(1)=r_k(2)\}}1_{\{q_k(2)+X_k(2)=0\}}]$$
$$+ \mathbb{E}[r_k(1)1_{\{r_k(1)<r_k(2)\}}1_{\{q_k(2)+X_k(2)=0\}}]. \tag{10}$$

Since $q_k(i) + X_k(i)$ is independent of $r_k(1)$ and $r_k(2)$ the last two expressions in the RHS of the above inequality can be simplified. For example, the last expression becomes $\mathbb{E}[r_k(1)1_{\{r_k(1)<r_k(2)\}}]P(q_k(2) + X_k(2) = 0)$ where $P(q_k(2) + X_k(2) = 0)$ is the probability under stationarity. But we do not know $P(q_k(2) + X_k(2) = 0)$. Motivated by GI/GI/1 queues, we approximate it by $1 - \rho(2)$ where

$$\rho(2) = \frac{\mathbb{E}[X(2)]}{\mathbb{E}[r(2)1_{\{r(2)>r(1)\}}] + (1-p)\mathbb{E}[r(2)1_{\{r(2)=r(1)\}}]}. \tag{11}$$

Finally we provide approximations for mean queue lengths and delays. For simplicity we limit to two queue case. The general case is considered in the next section. First consider $\mathbb{E}[q(i)]$, the mean queue length of the ith queue under steady state. We use the approximation formula for a GI/GI/I queue available in [11]:

$$\mathbb{E}[q] \approx \frac{\rho \; g \; \mathbb{E}[X](C_X^2 + C_r^2)}{2(1-\rho)} \tag{12}$$

where g, C_X^2, C_r^2 are

$$\rho = \mathbb{E}[X]/\mathbb{E}[r], \quad C_X^2 = \frac{\mathrm{var}(X)}{(\mathbb{E}[X])^2}, \quad C_r^2 = \frac{\mathrm{var}(r)}{(\mathbb{E}[r])^2},$$

$$g = \exp\left[-2\,\frac{1-\rho}{3\rho}\,\frac{(1-C_r^2)^2}{C_r^2+C_X^2}\right], \quad if \quad C_r^2 < 1,$$

$$= \exp\left[-(1-\rho)\,\frac{C_r^2-1}{C_r^2+4C_X^2}\right], \quad if \quad C_r^2 \geq 1.$$

We use statistics for $X(i)$ and $r(i)$ in (12) but the $\rho(i)$ will be as follows. Queue1 at least under heavy traffic receives rate $\mathbb{E}[z_k(1)]$ where

$$\mathbb{E}[z_k(1)] = \mathbb{E}[r_k(1)1\{r_k(1) > r_k(2)\}] + p\mathbb{E}[r_k(1)1\{r_k(1) = r_k(2)\}]. \quad (13)$$

But it will also receive (as argued above for stability) an extra service

$$(1-p)(1-\rho(2))\mathbb{E}[r(1)1_{\{r(1)=r(2)\}}] + (1-\rho(2))\mathbb{E}[r(1)1_{\{r(1)<r(2)\}}]. \quad (14)$$

Thus we define $\rho(1)$ for queue1 as $\mathbb{E}[X(1)]$ divided by the sum of (13) and (14). Similarly we define $\mathbb{E}[z_k(2)]$ and $\rho(2)$.

Once $\mathbb{E}[q(i)]$ has been approximated we obtain approximations for $\mathbb{E}[D(i)]$, the mean delay of the first bit arriving at a slot under heavy traffic as (see details in [16])

$$\mathbb{E}[D(i)] = \frac{\mathbb{E}[q(i)]}{\mathbb{E}[z(i)]} + \frac{\mathbb{E}[z(i)^2]}{2(\mathbb{E}[z(i)])^2} + \frac{d}{2\mathbb{E}[z(i)]} \quad (15)$$

where d is the lattice span of the arithmetic distribution of r (and equals 0 if it is non-arithmetic). We will see in the next section that (15) in fact provides a good approximation even under low traffic intensities (we will in fact show it for more than two queues).

Similarly one can get approximation of the mean delay of the last bit arriving at a slot and the mean delay of any bit (average delay of all bits).

5   Simulations

Now we compare the stability results and the approximate mean delays and queue lengths obtained in Section 4 with simulations and verify our claims.

First we consider a system with 10 queues and symmetric statistics. We take $r_k(i) = 25, 15, 0$ with probabilities 0.3, 0.4 and 0.3 respectively. We break the ties assigning a slot with equal probabilities. Then stability boundary from (9) is 2.47. Let $X_k(i)$ take values 5, 2 and 1. The distributions of $X_k(i)$ are the same and we change them so that $\mathbb{E}[X_k(i)]$ is increased from 2.2 to 2.8. For each case simulations were run for 2 million slots. The mean queue lengths of the first five queues are plotted in Figure 1. We see that all the queues start becoming unstable around 2.47 verifying our claim. The same is true for the other five queues also.

Next we consider an asymmetric case with 10 queues. Now the $r_k(i)$ are 50 with probability 0.65 and 0 otherwise. The $X_k(i)$ take values 0 and 10. For $i = 1, 2, 3$, $P[X_k(i) = 10] = 0.2$.

The stability region for the queues $i = 4, ..., 10$ is (approximately) obtained by adding to (9) the extra terms

$$\frac{3(1-\rho)}{2}\mathbb{E}[r(i)1_{\{r(i)=r(1),r(i)>r(j),\text{for all }j\neq1,i\}}]$$

$$+\frac{3.2.(1-\rho)}{3}\mathbb{E}[r(i)1_{\{r(i)=r(1)=r(2),r(i)>r(j),j\neq,1,2,i\}}]$$

$$+\frac{3}{4}(1-\rho)\mathbb{E}[r(i)1_{\{r(i)=r(1)=r(2)=r(3),r(i)>r(j),j\neq1,2,3,i\}}].$$

The justification for adding these extra terms is that queues 1, 2, 3 are lightly loaded and hence the other queues can get their share of slots if these are empty. The probability that any of the queues 1, 2, 3 is empty is approximated by $(1-\rho)$ where $\rho = \mathbb{E}[X(1)]/\mathbb{E}[r'(1)]$ and $\mathbb{E}[r'(1)]$ is obtained from (9). Thus the stability region for queues 4, ..., 10 is 6.162.

The $P[X_k(i) = 10]$, $i = 4, ..., 10$ is the same and increased such that $\mathbb{E}[X(i)]$ varies from 5.80 to 6.80.

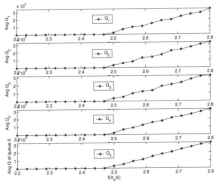

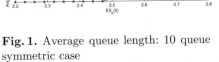

Fig. 1. Average queue length: 10 queue symmetric case

Fig. 2. Average queue length: 10 queue asymmetric case

The plots for $\mathbb{E}[q(i)]$ $i = 1, 2, ..., 5$ are provided in Figure 2. We see that queues 1, 2, 3 stay stable but queues 4 and 5 become unstable at 6.3. The same happens with the queues 6, 7, 8, 9, 10.

Finally we verify the accuracy of the approximate formulae (12) and (15) for mean queue lengths and mean delays. We consider a 2 queue symmetric case. The $r(i)$ take values 10 and 0 with probabilities 0.75 and 0.25. The $X(i)$ take values 15 and 0 and their probabilities are changed to vary the traffic intensity.

The theoretical and simulated values of the mean delays and the mean queue lengths are provided in Table 5. We see that our approximations match the simulated values, quite well.

Table 5. Comparison of theoretical and simulated values of $\mathbb{E}[Q_1]$ and $\mathbb{E}[D_1]$ for 2 queue case

traffic	simulated $\mathbb{E}[Q_1]$	theoretical $\mathbb{E}[Q_1]$	simulated $\mathbb{E}[D_1]$	theoretical $\mathbb{E}[D_1]$ with correction
0.30	1.60	1.68	2.54	2.49
0.50	3.76	3.86	2.90	2.96
0.70	8.82	8.87	3.86	4.03
0.90	33.95	33.87	9.11	9.31
0.95	71.72	70. 78	17.07	17.23
0.98	187.52	182.11	41.76	40.98

Table 6. queue, symmetric case

traffic	simulated $\mathbb{E}[Q_1]$	theoretical $\mathbb{E}[Q_1]$	simulated $\mathbb{E}[D_1]$	theoretical $\mathbb{E}[D_1]$ with correction
0.30	3.86	3.77	3.02	2.94
0.50	8.76	5.88	3.63	3.15
0.70	15.09	11.66	3.99	3.73
0.80	19.76	19.57	4.22	4.53
0.85	23.18	27.84	4.42	5.36

Table 7. queue, asymmetric case

traffic	simulated $\mathbb{E}[Q_1]$	theoretical $\mathbb{E}[Q_1]$	simulated $\mathbb{E}[D_1]$	theoretical $\mathbb{E}[D_1]$ with correction
0.30	6.53	7.38	2.41	1.76
0.50	15.00	16.48	2.73	2.25
0.70	33.87	36.76	3.49	3.33
0.80	59.65	61.58	4.64	4.66
0.85	86.59	86.23	5.89	5.98

Next we verify these formulae for a five queue example. First consider the symmetric case. The $r(i)$ take values 50 and 1 with probabilities 0.65 and 0.35 respectively. The theoretical and simulated $\mathbb{E}[Q_1]$ and $\mathbb{E}[D_1]$ are shown in Table 6 for different $\mathbb{E}[X(i)]$. We see that the approximations are quite reasonable, particularly for $\mathbb{E}[D_1]$.

We also consider a five queue asymmetric case. The $r(i)$ have the same distributions as for the symmetric case. The $X(i)$ take values 40 and 0. We take $P[X(i) = 40] = 0.05$, for $i = 1, 2, 3$, and $P[X(i) = 40]$, $i = 4, 5$ are varied to have different traffic intensities. The values of $\mathbb{E}[Q_4]$ and $\mathbb{E}[D_4]$ are provided in Table 7. We see a good match between theory and simulations.

6   Conclusions

In this paper we considered the problem of optimal scheduling of wireless links. We proved the existence of optimal and average cost optimal policies. Then we identified a greedy policy, called MAX-TRANS which is easy to implement, performs close to optimal and does not require the channel and link rate statistics. Its performance (mean queue lengths and delays) is better than the other well known policies. However, it is not throughput optimal (for asymmetric systems). Thus, we obtained stability region of this policy and approximate mean queue lengths and delays. We also modified this policy to make it throughput optimal.

References

1. Altman, E., Kushner, H.J.: Control of polling in presence of vacations in heavy traffic with applications to satellite and mobile radio systems. SIAM J. Control and Optimization 41, 217–252 (2000)
2. Andrews, M., Kumaran, K., Ramanan, K., Stolyer, A., Vijaykumar, R., Whiting, P.: Scheduling in a queueing system with asynchronously varying service rates. Probability in the Engineering & Informational Sciences 18, 191–217 (2004)
3. Berggren, F., Jantti, R.: Asymptotically fair transmission scheduling over fading channels. IEEE Trans. Wireless Communication 3, 326–336 (2004)
4. Berggren, F., Jantti, R.: Multiuser scheduling over Rayleigh fading channels. Global Telecommunications Conference, IEEE GLOBECOM, PP. 158–162 (2003)
5. Borst, S.: User level performance of channel aware scheduling algorithms in wireless data networks. IEEE/ACM Trans. Networking (TON) 13, 636–647 (2005)
6. Eryilamz, A., Srikant, R., Perkins, J.R.: Stable scheduling policies for fading wireless channels. IEEE/ACM Trans. Networking 13, 411–424 (2005)
7. Goyal, M., Kumar, A., Sharma, V.: A stochastic control approach for scheduling multimedia transmissions over a polled multiaccess fading channel. ACM/SPRINGER Journal on Wireless Networks, to appear (2006)
8. Georgiadis, L., Neely, M.J., Tassiulas, L.: Resource allocation and cross-layer control in Wireless networks. In: Foundations and Trends in Networking, vol. 1, pp. 1–144. New Publishers, Inc, Hanover, MA (2006)
9. Hernandez-Lerma, O., Lassere, J.B.: Discrete time Markov Control Processes. Springer, Heidelberg (1996)
10. Knopp, R., Humblet, P.: Information capacity and power control in single-cell multiuser communication. Proc. IEEE ICC 1995, pp. 331–335 (1995)
11. Krämer, W., Langenbach-Bellz, M.: Approximate formulae for the delay in the queueing system GI/G/1. 8th International Teletrafic Congress, pp. 235–1–8 (1976)
12. Lin, X., Chong, E.K.P., Shroff, N.B.: Opportunistic transmission schdeuling with resource-sharing constraints in Wireless networks. IEEE Journal on Selected Areas in Communications 19, 2053–2064 (2001)
13. Lin, X., Shroff, N.B., Srikant, R.: A tutorial on cross-layer optimization in wireless networks. IEEE Journal on Selected Areas in Communications 24, 1452–1463 (2006)
14. Lin, P., Berry, R., Honig, M.: A Fluid Analysis of a Utility-based Wireless Scheduling Policy, to appear in IEEE Transaction on Information Theory
15. Muller, A., Stoyan, D.: Comparison methods for stochastic Models and Risks. Wiley, Chichester, England (2002)
16. Prasad, D.K.: Scheduling and performance analysis of queues in wireless systems M.E. Thesis, Dept. of Electrical Communication Engg., Indian Institute of Science, Bangalore (July 2006)

17. Shakkottai, S., Stolyar, A.: Scheduling algorithms for a mixture of real-time and non-real time data in HDR. "citeseer.ist.edu/466123.html"
18. Shetiya, H., Sharma, V.: Algorithms for routing and centralized scheduling in IEEE 802.16 mesh networks. In: Proc IEEE WCNC (2006)
19. Tassiulas, L., Ephremides, A.: Dynamic server allocation to parallel queues with randomly varying connectivity. IEEE Trans., Information Theory 39, 466–478 (1993)
20. Tsibonis, V., Georgiadis, L., Tassiulas, L.: Exploiting wireless channel state information for throughput maximization. IEEE Trans. Inf. Theory 50, 2566–2582 (2004)
21. Walrand, J.: An Introduction to Queueing Networks. Printice Hall, Englewood Cliffs (1988)

Reversible Fair Scheduling: The Teletraffic Theory Revisited

Villy Bæk Iversen

COM-DTU, Bld. 343, Technical University of Denmark
DK-2800 Lyngby, Denmark
Fax +45 45936581
vbi@com.dtu.dk

Abstract. This paper presents a unified model of classical teletraffic models covering both circuit-switched and packet-switched networks. The models of Erlang, Engset, Palm, and Delbrouck are generalized to loss and delay systems with multi-rate BPP (Binomial, Poisson, Pascal) traffic streams which all are insensitive to the holding time distributions. The Engset and Pascal traffic streams are also insensitive to the idle time distribution. Engset traffic corresponds to a finite number of users and Pascal traffic results in same state probabilities as Pareto-distributed inter-arrival times. The model is similar to balanced fair scheduling by Bonald and Proutière, but the approach is much simpler and allows for explicit results for finite source traffic and performance measures of each individual service. The model is evaluated by a simple stable algorithm with a minimal memory requirement and a time requirement which is linear in number of streams and number of servers plus queueing positions.

Keywords: reversible scheduling, fair balanced scheduling, multi-service, multi-rate traffic, generalized processor sharing.

1 Introduction

In classical teletraffic theory we have general models for loss systems and delay systems. *For loss systems* the multi-rate loss model with BPP–traffic is very general and insensitive to service time distribution, i.e. the state probabilities of the system depend only upon the holding time distribution through its mean value. Algorithms for evaluating multi-rate loss systems can be classified into two groups. One group is based on the aggregation of traffic streams. Most important algorithm is the convolution algorithm (Iversen, 1987 [5]) which allows for minimum and maximum capacity allocation to each individual stream. The other group is based on aggregation of the state space and most important algorithm for Poisson arrivals is the Fortet & Grandjean model, known as Kaufman & Roberts algorithm. The more general algorithm by Delbrouck [3] allows for BPP–traffic. This group of algorithms only calculates the time-congestion. Recently a new effective state-based algorithm has been published which also allows

L. Mason, T. Drwiega, and J. Yan (Eds.): ITC 2007, LNCS 4516, pp. 1135–1148, 2007.

for individual performance measures for each stream as time, call, traffic congestion. The algorithm is very fast (linear in number of channels and services), accurate, and requires a minimum of memory.

For *delay systems* the Generalized Processor Sharing (*GPS*) model is insensitive to the service time distribution. Each *GPS* connection occupies only one channel even if there is idle capacity. If there are more connections than channels, then processor sharing is applied. So if the number of channels is n and the number of connections is $x \geq n$, then each connection gets the bandwidth n/x. The service rate is thus reduced and the time to transmit a certain traffic volume (bytes) is increased. The increase in service time as compared to a system with infinite capacity is denoted the virtual waiting time. For one Poisson arrival process the system corresponds to an $M/G/n$ *PS*–system (*PS* = Processor Sharing) and has the same state probabilities as Erlang's delay system $M/M/n$. The processor sharing insensitivity property is not related to the arrival process, but to the service process and is also valid for a finite source system with n servers. For delay systems with multi-rate traffic the solution becomes much more complex. Holmberg & Iversen (2006 [4]) present a model where the above down-scaling during overload is applied to each global state. This model looks insensitive when investigated by simulations, but theoretical investigations show that it is sensitive.

It would be desirable to combine the loss model valid for underload region with the processor sharing model valid for overload regime as we then would have a combined model which is insensitive for both under- and overload, and allows connections to have individual bandwidth when the system is in underload region. In the following we combine these two models by accepting the multi-rate loss model for underload region and forcing reversibility in the overload region. It turns out to be a model which in heavy overload region in the limit operates as the GPS–model, but which in a intermediate region allocate more resources to broad-band calls than to narrow-band calls.

In section 2 we describe the traffic model and the performance measures of importance. Section 3 introduces the Kolmogorov-cycle conditions for reversibility. The main theoretical part is section 4 where we construct a state transition diagram which is reversible and also insensitive. The general recursion formula is given in section 5. Section 6 describes a new simple algorithm for calculating the state global state probabilities made up of contributions from each individual service. Detailed performance measures are derived in section 7, and the complexity – or rather simplicity – is described in section 8. Section 9 outlines generalizations of the above model. Finally, conclusions are presented in section 10.

2   Multi-rate Traffic Model

We will consider BPP traffic, i.e. the arrival processes are Poisson processes with linearly state-dependent arrival rates. The restriction to linearity is not necessary for the reversibility, but only for the compact effective algorithm presented in section 6. The arrival rate of service j ($j = 1, 2, \ldots, N$) is supposed to depend on the number of connections of this type. So the arrival rate of service j is $\lambda_j(x)$ when we have x connections of this type. The bandwidth demand of service j is d_j, so that each connection requires d_j channels (= Basic Bandwidth Units = BBU). The *service time* of a connection of type j is exponentially distributed with rate $d_j \mu_j$ (the factor d_j is only for convenience, the connection terminates all channels at the same time). In the mode considered (reversible scheduling), this model will be insensitive to the service time distribution, so that results obtained for exponential distributions are valid for more general distributions.

Finite source traffic is characterized by number of sources S and offered traffic per idle source β. Alternatively, we in practice often use the offered traffic A and the peakedness Z. We have the following relations between the two representations:

$$A = S \cdot \frac{\beta}{1+\beta}, \qquad Z = \frac{1}{1+\beta}, \qquad (1)$$

$$\beta = \frac{\gamma}{d\,\mu} = \frac{1-Z}{Z}, \qquad S = \frac{A}{1-Z}. \qquad (2)$$

3   Reversibility and Insensitivity

As a basic requirement to the combined process it must be reversible in order to be insensitive. Our approach is based on local balance in the state-transition diagram approach and application of the Kolmogorov cycle conditions as described in the textbook [6]. This is a simple and stringent approach which allows for generalizations of the model to finite number of source and to include minimum and maximum bandwidth allocations for individual streams (services) by truncating the state space. It may be shown that the system considered is insensitive to Cox distributions, i.e. to distributions having rational Laplace transforms [4].

4   Construction of State Transition Diagram

In this section we consider a system with $n + k$ channels. The n channels are servers, and the k channels are buffers or queueing positions. We define the state of the system $\{x_1, x_2, \ldots, x_N\}$ as the number of channels occupied by each service (traffic stream). As an alternative we might choose the state equal to number of connections of each type. The global state x is the total number of channels occupied by all services:

$$x = \sum_{j=1}^{N} x_j, \qquad 0 \leq x \leq n+k.$$

If we in states above the global state $x = n$ want to keep the processor sharing model with bandwidth allocation proportional with the bandwidth demand, then it is not possible to obtain the multi-rate model below the global state n as there is no feasible transition between the two models which maintain the reversibility for the total system. The model obtained would not exploit the full capacity when possible. This is only possible for the generalized processor sharing (GPS) model. So the only feasible approach is to accept the multi-rate state transition diagram up to global state n and then maintain the reversibility in a reasonable way. This will give us a unique solution which is the only feasible solution if we want to keep the maximum utilization all the time.

As compared with the infinite server model, we have to modify the service rates for states above n to obtain a reversible state transition diagram. We choose the service rate of stream j equal to $d_j \mu_j$. We only include the factor d_j in the service rate for stream j in order to make the following formulæ more simple. Note that all channels of a connection are released simultaneously as in the multi-rate loss model. As an example we first consider a system with only two services, $N = 2$. In the two-dimensional state-transition diagram we denote stream one by index 1 and stream two by index 2. In state (x_1, x_2) the service rates in the infinite server system will be

$$\mu_1(x_1, x_2) = x_1 \cdot \mu_1, \quad \text{respectively} \quad \mu_2(x_1, x_2) = x_2 \cdot \mu_2.$$

The actual service rates in the finite capacity system will be reduced by a reduction factors $g_1(x_1, x_2)$, respectively $g_2(x_1, x_2)$, so the actual service rates in state (x_1, x_2) becomes:

$$\mu_1(x_1, x_2) = g_1(x_1, x_2) \cdot x_1 \mu_1, \tag{3}$$
$$\mu_2(x_1, x_2) = g_2(x_1, x_2) \cdot x_2 \mu_2, \tag{4}$$

The task is to find the reduction factors $g_1(x_1, x_2)$ and $g_2(x_1, x_2)$. For states $(x_1 < 0)$ and/or $(x_2 < 0)$ we of course have service rates zero, and the reduction factors may be put equal to any value appropriate for the recursion formula.

4.1   States with Demand Less Than or Equal to the Capacity

In underload states, i.e. states $(x_1 + x_2 \leq n)$, there is no reduction in service rates. All connections, including multi-rate connections, get the full service, so the reduction factors are one:

$$g_1(x_1, x_2) = 1 \qquad 0 \leq x_1 + x_2 \leq n, \tag{5}$$
$$g_2(x_1, x_2) = 1 \qquad 0 \leq x_1 + x_2 \leq n. \tag{6}$$

This part of the state transition diagram is similar to the diagram of a multi-rate loss system.

4.2   States with Demand Bigger Than the Capacity

– When we are in state $\{(x_1, 0), x_1 > n\}$, the demand is bigger than the capacity. A natural requirement is to choose the service rate equal to $n \cdot \mu_i$ when we are in an overload state with only type one connections present. To fully exploit the capacity of n servers the service rate should be $n \mu_1$. This

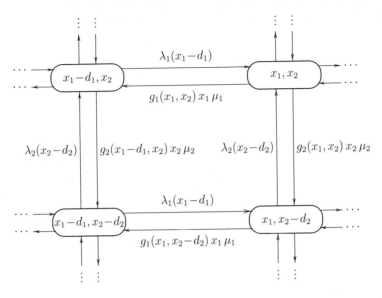

Fig. 1. *State-transition diagram for four neighbouring states in a system with two traffic streams*

corresponds to processor sharing mode for one service, and thus we have the reduction factors:

$$g_1(x_1, 0) = \frac{n}{x_1} \qquad x_1 \geq n, \tag{7}$$

$$g_2(0, x_2) = \frac{n}{x_2} \qquad x_2 \geq n, \tag{8}$$

as we may use the same argument when service two is alone.

– When neither x_1 nor x_2 is zero we have the general case. Let us consider a state (x_1, x_2) and assume that we know the reduction factors $g_1(x_1, x_2 - d_2)$ and $g_2(x_1 - d_1, x_2)$, then we want to find the reduction factors $g_1(x_1, x_2)$ and $g_2(x_1, x_2)$. Considering the four states in a square (Fig.1)

$$\{(x_1 - d_1, x_2 - d_2), \ (x_1, x_2 - d_2), \ (x_1, x_2), \ (x_1 - d_1, x_2)\}$$

we apply
1. the Kolmogorov cycle requirement for reversibility:

    clockwise: $g_2(x_1, x_2) \cdot x_2\, \mu_2 \ \cdot \ g_1(x_1, x_2 - d_2) \cdot x_1\, \mu_1$

    counter-clockwise: $g_1(x_1, x_2) \cdot x_1\, \mu_1 \ \cdot \ g_2(x_1 - d_1, x_2) \cdot x_2\, \mu_2$

We assume that the arrival rate of stream one only depends on the state of this stream so that the arrival rates $\lambda_1(x_1 - d_1) \cdot \lambda_2(x_2 - d_2)$ always balance out. This is e.g. fulfilled for BPP–traffic.
2. the requirement of a total capacity equal to n:

$$x_1 \cdot g_1(x_1, x_2) + x_2 \cdot g_2(x_1, x_2) = n \tag{9}$$

This yields two equations with the two unknown reduction factors. Solving the equations we get the unique solution:

$$g_1(x_1, x_2) = \frac{n \cdot g_1(x_1, x_2 - d_2)}{x_1 \cdot g_1(x_1, x_2 - d_2) + x_2 \cdot g_2(x_1 - d_1, x_2)}, \tag{10}$$

$$g_2(x_1, x_2) = \frac{n \cdot g_2(x_1 - d_1, x_2)}{x_1 \cdot g_1(x_1, x_2 - d_2) + x_2 \cdot g_2(x_1 - d_1, x_2)}. \tag{11}$$

Thus it is easy to construct an algorithm for calculation of reduction factors. We may also write the formulæ using the number of connections as variables, but many aspects become simpler by using the basic bandwidth unit.

As we know the reduction factors on the x_1 and x_2 axes, we can recursively calculate all reduction factors and construct the total state transition diagram.

5   General Case

It is easy to extend the above derivations to more than two services as we for each service in a given state get one additional equation and one unknown reduction factor more, which can be obtained from the additional equation. Due to reversibility (local balance) it is easy to calculate the reduction factors. For N traffic streams we get the reduction factor for stream j ($j = 1, 2, \ldots, N$):

$$g_j(x_1, x_2, \ldots, x_j, \ldots, x_N) = \tag{12}$$
$$\frac{n \cdot g_j(x_1 - d_1, x_2 - d_2, \ldots, x_{j-1} - d_{j-1}, \; x_j \;, x_{j+1} - d_{j+1}, \ldots, x_N - d_N)}{\sum_{i=1}^{N} x_i \cdot g_i(x_1 - d_1, x_2 - d_2, \ldots, x_{i-1} - d_{i-1}, \; x_i \;, x_{i+1} - d_{i+1}, \ldots, x_N - d_N)}.$$

As mentioned above the reduction factors are independent of the traffic processes, and the approach includes Engset and Pascal traffic and any state-dependent Poisson arrival process.

It can be shown by studying the reduction factors for increasing load that if the system becomes highly overloaded, then in the limit all connections will be allocated the same capacity independent of the slot-sizes d_j corresponding to GPS.

6   Generalized Recursion Formula

The above system is reversible, but has no product form. The general procedure for obtaining state probabilities is to express all states probabilities by probability of state zero and then find state zero by normalizing the total probability mass. From the state probabilities we can find performance measures of interest. In this case we are able to derive a simple recursion formula as shown in the following.

The detailed state $\{x_1, x_2, \ldots, x_N\}$ is defined as the state where x_j channels are occupied by stream j. As we for stream j only occupies an integer multiple of d_j channels, many states of course get the probability zero. But for the following derivations it is an advantage to consider a channel (= basic bandwidth unit) as the bandwidth unit, not connections. Each state (detailed or global) is in the following split up into disjoint sub-states, one for each stream indicated by an index. The splitting is done according to the reduction factors in the following way (here exemplified by 2-stream case):

$$p_1(x_1, x_2) = \frac{x_1}{n} \cdot g_1(x_1, x_2) \cdot p(x_1, x_2) \tag{13}$$

$$p_2(x_1, x_2) = \frac{x_2}{n} \cdot g_2(x_1, x_2) \cdot p(x_1, x_2) \tag{14}$$

so that we get by (10) and (11):

$$p(x_1, x_2) = p_1(x_1, x_2) + p_2(x_1, x_2). \tag{15}$$

Associating probabilities with each state, we have:

$$p(x_1, x_2, \dots, x_N) = \sum_{j=1}^{N} p_j(x_1, x_2, \dots, x_N). \tag{16}$$

The global state x is the summation of all states with a total number of x channels being busy:

$$p(x) = \sum_{\{x \mid x = \sum_{j=1}^{N} x_j\}} p(x_1, x_2, \dots, x_N) = \sum_{j=1}^{N} p_j(x) \tag{17}$$

Here $p_j(x)$ is the contribution of stream j to the global state x, i.e. the traffic carried by type j in state x is $x \cdot p_j(x)$, and the total traffic carried in state x is $x \cdot p(x)$.

We consider a state above n knowing that for traffic stream one the number of changes per time unit from state $(x_1 - d_1, x_2)$ to state (x_1, x_2) is equal to the number of changes per time unit from state (x_1, x_2) to state $(x_1 - d_1, x_2)$ (local balance = reversibility). First we consider Poisson arrival processes. Considering the reversible state transition diagram in Fig. 1 we get the following flow balance equation between the two neighbouring states $(x_1 - d_1, x_2)$ and (x_1, x_2)):

$$p(x_1 - d_1, x_2) \cdot \lambda_1 = g_1(x_1, x_2) \cdot x_1 \mu_1 \cdot p(x_1, x_2). \tag{18}$$

Exploiting (13) and summing over all j traffic streams we get, remembering that $A_j = \lambda_j / (\mu_j d_j)$:

$$p(x_1 - d_1, x_2) \cdot \lambda_1 = n \cdot \mu_1 \cdot p_1(x_1, x_2) \tag{19}$$

$$p_1(x_1, x_2) = \frac{d_1 \cdot A_1}{n} \cdot p(x_1 - d_1, x_2), \tag{20}$$

$$p_1(x) = \sum_{i=0}^{x} p_1(x - i, i) \tag{21}$$

$$p_1(x) = \frac{d_1 \cdot A_1}{n} p(x - d_1). \tag{22}$$

We should notice that on the left hand side we have the state probability for stream one and on right-hand side we have the global state probability. The term $p_1(x)$ denotes the contribution of stream one to the global state probability $p(x)$. We assume that in state x all x channels are occupied by stream j during the proportion of time $p_j(x)$ so that we in state x split the time axes into periods where the system works with only one type at a time. This is only correct if the arrival rate is a linear function of the state of this traffic stream. Then the number of call attempts will be the same if we for example have y idle sources during one time unit or one source idle during y time unit. In a similar way we get for Engset and Pascal traffic processes (Fig. 1):

$$p(x_1-d_1, x_2) \cdot S_1\, \gamma_1 - p_1(x_1-d_1, x_2) \cdot \frac{x_1 - d_1}{d_1}\, \gamma_1$$
$$= g_1(x_1, x_2) \cdot x_1 \cdot \mu_1 \cdot p(x_1, x_2)$$
$$= n \cdot \mu_1 \cdot p_1(x_1, x_2)\,.$$

The first term on the left hand assumes that all S_1 sources are idle. The second term reduces the number of idle sources with the average number of busy sources. Rearranging the terms and using (2) we get:

$$p_1(x_1, x_2) = \frac{S_1\, \gamma_1}{n\, \mu_1} \cdot p(x_1-d_1, x_2) - \frac{x_1-d_1}{d_1}\, \gamma_1 \cdot \frac{1}{n\, \mu_1} \cdot p_1(x_1-d_1, x_2)$$
$$= \frac{d_1}{n} \cdot S_1\, \beta_1 \cdot p(x_1-d_1, x_2) - \frac{x_1-d_1}{n} \cdot \beta_1 \cdot p_1(x_1-d_1, x_2)\,.$$

Summing over all states $x_1 + x_2 = x$ we get:

$$p_1(x) = \frac{d_1}{n} \cdot S_1\, \beta_1 \cdot p(x-d_1) - \frac{i-d_1}{n} \cdot \beta_1 \cdot p_1(x-d_1)\,. \tag{23}$$

We now consider a system with n servers, k buffers and N BPP traffic streams. The generalized recursion formula for multi-rate loss systems by Iversen [7] can be combined with the above model and we find the general formula:

$$p(x) = \begin{cases} 0 & x < 0 \\ 1 & x = 0 \\ \displaystyle\sum_{i=j}^{N} p_j(x) & x = 1, 2, \ldots, k+n \end{cases} \tag{24}$$

where

$$p_j(x) = \max\left\{\frac{x}{n}, 1\right\} \cdot \left\{ \frac{d_j}{x} \cdot S_j\, \beta_j \cdot p(x - d_j) - \frac{x - d_j}{x} \cdot \beta_j \cdot p_j(x - d_j) \right\} \tag{25}$$

or by replacing the parameters (S_j, β_j) by (A_j, Z_j) (1):

$$p_j(x) = \max\left\{\frac{x}{n}, 1\right\} \cdot \left\{ \frac{d_j}{x} \cdot \frac{A_j}{Z_j} \cdot p(x-d_j) - \frac{x-d_j}{x} \cdot \frac{1-Z_j}{Z_j} \cdot p_j(x-d_j) \right\} \tag{26}$$

The initialization values of $p_j(x)$ are $\{p_j(x) = 0, x < d_j\}$. This is a simple general recursion formula covering all classical Markovian models. In general the model is equivalent to the Balanced Fair Scheduling model of Bonald & Proutière [1]. For $k = 0$ we get a generalized Delbrouck formula. The approach in this paper is mathematically much simpler, but more general, and it allows for simple numerical evaluation, not only for Poisson arrival processes as the complex algorithm in Bonald & Virtamo [2], but also for finite source traffic. The properties of the algorithm are analysed in section 8.

Remark: For infinite number of buffers we require $A < n$ to attain statistical equilibrium. For Pascal arrival processes there are more strict requirements. For a system with buffers the Pascal case result in a carried traffic which is bigger than the offered traffic, because the arrival rate increases linearly with the number of customer being served or waiting.

7 Performance Measures

We consider a system with n channels and k buffers, both given in basic bandwidth units. The model includes loss systems (blocked calls cleared and blocked calls held), classical delay systems, processor sharing systems etc. The performance measures become rather diversified, and we only derive the basic performance measures.

7.1 Time Average Performance Measures

The time congestion Eb_j of stream j is defined as the proportion of time new connections are blocked:

$$Eb_j = \sum_{x=n+k-d_j+1}^{n+k} p(x), \quad j = 1, 2, \ldots, N. \tag{27}$$

In a similar way the proportion of time new calls are delayed becomes:

$$Ed_j = \sum_{x=n-d_j+1}^{n+k-d_j} p(x), \quad j = 1, 2, \ldots, N. \tag{28}$$

The proportion of time new calls get full service at the time of arrival becomes:

$$Es_j = \sum_{x=0}^{n-d_j} p(x), \quad j = 1, 2, \ldots, N. \tag{29}$$

Of course, we have $Eb_j + Ed_j + Es_j = 1$. When the system operates as a classical single-slot delay or loss system these measures are simple to understand. However, we should remember that in case of processor sharing systems, calls arriving later may influence the service of existing calls. The above performance measures are time averages. The more useful call averages are derived below.

7.2   Traffic Average Performance Measures

It should be noticed that these mean values are much more important than time and call average values. The carried traffic for stream j measured in channels is given by:

$$Y_j = \sum_{x=0}^{n} x \cdot p_j(x) + n \cdot \sum_{x=n+1}^{n+k} p_j(x) \tag{30}$$

As the offered traffic of type j measured in channels is $d_j \cdot A_j$, the lost traffic is $(d_j \cdot A_j - Y_j)$. The traffic congestion C_j for stream j, which is the proportion of offered traffic blocked, becomes:

$$C_j = \frac{d_j \cdot A_j - Y_j}{d_j \cdot A_j}, \quad j = 1, 2, \ldots, N. \tag{31}$$

The total traffic congestion is:

$$C = \frac{A - Y}{A} \tag{32}$$

where

$$A = \sum_{j=1}^{N} d_j \cdot A_j \quad \text{and} \quad Y = \sum_{j=1}^{N} Y_j = \sum_{x=0}^{n} x \cdot p(x) + \sum_{x=n+1}^{n+k} n \cdot p(x).$$

7.3   Call Average Performance Measures

The call congestion can always be obtained from the traffic congestion as follows. The carried traffic Y corresponds to $(Y \cdot \mu)$ accepted call attempts per time unit. The average number of idle sources is $(S - Y)$, so the average number of call attempts per time unit is $(S - Y) \cdot \gamma$. The call congestion is the ratio between the number of rejected call attempts and the total number of call attempts, both per time unit:

$$B = \frac{(S - Y) \gamma - Y \cdot \mu}{(S - Y) \gamma}$$
$$= \frac{(S - Y) \beta - Y}{(S - Y) \beta}.$$

By definition $Y = A(1 - C)$ and from (1) we have $S = A(1 + \beta)/\beta$. Inserting this we get:

$$B = \frac{A(1 + \beta) - A(1 - C)\beta - A(1 - C)}{A(1 + \beta) - A(1 - C)\beta},$$
$$B = \frac{(1 + \beta) C}{1 + \beta C}. \tag{33}$$

This relation between B and C is always valid. For systems with trunk reservation we may for Engset traffic find cases where for example the call congestion is bigger than the time congestion.

7.4 Mean Waiting Times and Mean Queue Lengths

The mean queue length of stream j (stream j traffic carried by the queueing positions) measured in unit of channels becomes:

$$L_j = \sum_{x=n+1}^{n+k} (x-n)\,p_j(x)\,, \quad j = 1, 2, \ldots, N\,. \tag{34}$$

As the same calls are waiting (including no waiting time) and served, the mean waiting time for all accepted customers of type j becomes:

$$W_j = s_j \cdot \frac{L_j}{Y_j}\,, \quad j = 1, 2, \ldots, N\,, \tag{35}$$

where s_j is the mean service time of type j calls, and both L_j and Y_j (30) are measured in channels. The total mean queue length measured in channels is:

$$L = \sum_{x=n+1}^{n+k} (x-n) \cdot p(x) = \sum_{j=1}^{N} L_j\,. \tag{36}$$

The overall mean waiting time for all accepted connections is given by:

$$W = s \cdot \frac{L}{Y}\,. \tag{37}$$

The mean waiting time of delayed calls type j excluding blocked calls is obtained from (35):

$$w_j = \frac{W_j}{D_j} \cdot (X_j + D_j)\,. \tag{38}$$

We notice that mean waiting times are measured in mean service times. Let us denote the mean service time by s for transfer of a fixed amount of data (bytes) using bandwidth $d = 1$. Then the mean service time when using a bandwidth d_j will be s/d_j, and also the mean waiting time will be reduced. By choosing a bigger bandwidth we may give priority to a traffic stream (reduce transfer time) and/or increase the amount of data transferred. Only when the system is heavily overloaded all connections will be allocated the same capacity (generalized processor sharing).

8 Properties of the Algorithm

The above theory results in a simple algorithm with the following basic features:

1. Initialization of variables
2. Let $x := x + 1$
3. Calculate $p_i(x)$ using (25) or (26) and $p(x)$ using (24)
4. Normalize all states by dividing by $\{1 + p(x)\}$

5. Go to step 2 if $x < n + k$

6. Calculate performance measures

If we know the normalized state probabilities $p_j(x-d_j)$, then we can calculate $p_j(x)$ for x. As $p_j(x) = 0$ for $x < d_j$ we are able to calculate the state probabilities by recursion. The implementation is described elsewhere. To find the performance measures we only need to know the d_j previous states of traffic stream j, as we may accumulate the necessary information on carried traffic and other statistics in a few variables. The memory requirements m_m of the algorithm, respectively the number of operations m_c, are of the order of size:

$$m_m = O\left\{\sum_{j=1}^{N} d_j\right\} , \quad \text{respectively} \quad m_c = O\{(n+k) \cdot (m_m + N)\} \qquad (39)$$

as we for a given number of channels need to calculate N new terms from $\max\{d_j\}$ previous global states and normalize m_m terms. Thus the algorithm requires very little memory, and it is linear in number of channels (servers + buffers) and number of services. The accuracy is optimal as we always operate with normalized values and always normalize (divide) with constants greater than one. It may be mentioned that the famous recursion formula for Erlang-B formula is a special case with one single-slot Poisson traffic stream. Also the recursion formula for Engset is a special case, as well as Delbrouck's formula.

9   Generalizations

The approach in this paper is mathematically very simple and allows for many generalizations. Some are outlined below.

Batched Poisson Arrival Processes
Above we have used the Pascal model to describe bursty traffic. This model have some drawbacks. In particular for delay systems there are several problems in applying it. A better model to describe traffic more bursty than the Poisson process is to use batched Poisson arrival processes, where events occur according to a Poisson process, but each event comprises of several calls. This model is easy to include in the above algorithm.

State-Dependent Blocking
In systems with limited accessibility (e.g. CDMA systems) a call may be blocked even if there is idle capacity. The blocking probability of a d-slot call in state x is denoted by $b_{x,d}$. The acceptance probability of the call is then $1 - b_{x,d}$, which is called the *passage factor*. As mentioned earlier the reversibility is ensured when the flow among the four neighbor states is the same clockwise and counter-clockwise (Kolmogorov cycles). Let us denote the passage factor for a single slot call in state x as $1-b_x = 1-b_{x,1}$, letting $b_{x,1} = b_x$. When b_x is greater than zero, then we may choose b_x so that the diagram is still reversible, but the product

form will be lost. For a d–slot call we have to choose the passage factor in state x as:

$$1 - b_{x,d} = \prod_{j=x}^{x+d-1} (1 - b_j) = (1 - b_x)(1 - b_{x+1}) \dots (1 - b_{x+d-1}).$$

This corresponds to that a d-slot call chooses one single channel d times, and the call is only accepted if all d channels are successfully obtained. This is a natural requirement, and it can be shown to be a necessary and sufficient condition for maintaining the reversibility of the process [7].

Trunk Reservation

Above we considered a system where the state-dependent blocking probability is similar for all traffic streams. We now consider systems where each stream may have individual state-dependent blocking probabilities. In wireless systems we may introduce trunk reservation (guard channels) so that call attempts from traffic stream i observe a system with r_i channels. Thus a call attempt of type j will be blocked in the states $r_j - d_j + 1, r_j - d_j + 2, \dots, n$. If a stream has full accessibility, then $r_j = n$. An obvious modification of the above algorithm is to let $p_j(x) = 0$, $x > r_j$, because we do not enter these states due to an arrival of type j. Introducing trunk reservation implies that the process becomes non-reversible, and the system becomes sensitive to the holding time distribution. For single-rate calls all having the same mean holding time the above modification will give the exact state probabilities when all arrival processes are Poisson processes. For multi-rate traffic and individual mean holding times the above modification is an approximation to the exact solution. The exact solution is only obtainable by solving the node balance equations. In [8] we study the accuracy of this approximation.

Maximum number of simultaneous connections

By a simple truncation of the state space we are able to put an upper limit to the number of connections of a given service (number of servers and buffer positions occupied). When we truncate the state space for individual services we can no longer use the simple recursive formula, but have to modify it or to calculate the individual state probabilities exploiting the reversibility. It is in principle easy to express all state probabilities by state zero, and then normalize [7].

Loss Networks and Queueing Networks

For circuit switched networks (loss networks) with several links the above model cannot be used, because we do not keep account of the number of connections of given type on a given link. In this case we have to use the convolution algorithm [5]. For store-and-forward (queueing) networks composed of nodes of the above-mentioned type we have a product form between the nodes. For Poisson traffic the nodes behave as if they were independent, and we are able to calculate mean delay for each service independently in each node. This corresponds to open queueing networks. For Engset/Pascal traffic streams we still have a product form between nodes and we may apply the convolution algorithm between nodes. But as the total number of customers is fixed we need the detailed state probability for each node as for ordinary closed queueing networks, and thus the state space becomes large. In this way we get a theory of queueing networks with multi-rate traffic.

10   Conclusions

This paper has presented a generalization of all the classical theory of loss and delay systems with multi-rate BPP–traffic. Furthermore, a simple recursive formula to evaluate the individual performance measures of each stream is presented. A computer-program to evaluate the model is available from the author. Due to the limited space, emphasis has been put on presentation of the simple basic ideas behind the model.

References

1. Bonald, T., Proutière, A.: Insensitive bandwidth sharing in data networks. Queueing Systems 44, 69–100 (2003)
2. Bonald, T.: A recursive formula for multirate systems with elastic traffic. IEEE Communications Letters 9(8), 753–755 (2005)
3. Delbrouck, L.E.N.: On the steady–state distribution in a service facility carrying mixtures of traffic with different peakedness factor and capacity requirements. IEEE Transactions on Communications COM–31(11), 1209–1211 (1983)
4. Holmberg, T., Iversen, V.B.: Resource sharing models for Quality-of-Service with multi-rate traffic. ITC19, Performance Challenges for Efficient Next Generation Networks. Ninteenth International Teletraffic Congress, Beijing, August 29 – September 2, Proceedings, pp. 1275–1284 (2005)
5. Iversen, V.B.: The exact evaluation of multi–service loss system with access control. Teleteknik, English ed., vol. 31(2), pp. 56–61. NTS–7, Seventh Nordic Teletraffic Seminar, Lund, Sweden, August 25–27, 1987, p. 22 (1987)
6. Iversen, V.B.: Teletraffic Theory and Network Planning. COM, Technical University of Denmark, 2006. p. 354. (2006) Textbook. Available at http://com.dtu.dk/educaton/34340/material/telenookpdf.pdf
7. Iversen, V.B.: Modelling restricted accessibility for wireless multi-service systems. In: Cesana, M. (ed.) Proceedings of Second Euro-NGI Workshop on Wireless and Mobility. LNCS, vol. 3883, pp. 93–102. Springer, Heidelberg (2006)
8. Zheng, H., Zhang, Q., Iversen, V.B.: Trunk reservation in multiservice networks with BPP traffic. In: García-Vidal, J., Cerdà-Alabern, L. (eds.) Euro-NGI 2007. LNCS, vol. 4396, pp. 200–212. Springer, Heidelberg (2007)

Is ALOHA Causing Power Law Delays? *

Predrag R. Jelenković and Jian Tan

Department of Electrical Engineering,
Columbia University
New York, NY 10027
{predrag,jiantan}@ee.columbia.edu

Abstract. Renewed interest in ALOHA-based Medium Access Control (MAC) protocols stems from their proposed applications to wireless ad hoc and sensor networks that require distributed and low complexity channel access algorithms. In this paper, unlike in the traditional work that focused on mean value (throughput) and stability analysis, we study the distributional properties of packet transmission delays over an ALOHA channel. We discover a new phenomenon showing that a basic finite population ALOHA model with variable size (exponential) packets is characterized by power law transmission delays, possibly even resulting in zero throughput. This power law effect might be diminished, or perhaps eliminated, by reducing the variability of packets. However, we show that even a slotted (synchronized) ALOHA with packets of constant size can exhibit power law delays when the number of active users is random. From an engineering perspective, our results imply that the variability of packet sizes and number of active users need to be taken into consideration when designing robust MAC protocols, especially for ad-hoc/sensor networks where other factors, such as link failures and mobility, might further compound the problem.

Keywords: ALOHA, medium access control, power laws, heavy-tailed distributions, light-tailed distributions, ad-hoc/sensor networks.

1 Introduction

ALOHA represents one of the first and most basic distributed Medium Access Control (MAC) protocols [1]. It is easy to implement since it does not require any user coordination or complicated controls and, thus, represents a basis for many modern MAC protocols, e.g., Carrier Sense Multiple Access (CSMA). Basically, ALOHA enables multiple users to share a common communication medium (channel) in a completely uncoordinated manner. Namely, a user attempts to send a packet over the common channel and, if there are no other user (packet) transmissions during the same time, the packet is considered successfully transmitted. Otherwise, if the transmissions of more than one packet

---

* This work is supported by the National Science Foundation under grant number CNS-0435168.

L. Mason, T. Drwiega, and J. Yan (Eds.): ITC 2007, LNCS 4516, pp. 1149–1160, 2007.

(user) overlap, we say that there is a collision and the colliding packets need to be retransmitted. Each user retransmits a packet after waiting for an independent (usually exponential/geometric) period of time, making ALOHA entirely decentralized and asynchronous. The desirable properties of ALOHA, including its low complexity and distributed/asynchronous nature, make it especially beneficial for wireless sensor networks with limited resources as well as for wireless ad hoc networks that have difficulty in carrier sensing due to hidden terminal problems and mobility. This explains the recent renewed interests in ALOHA type protocols.

Traditionally, the performance evaluation of ALOHA has focused on mean value (throughput) and stability analysis, the examples of which can be found in every standard textbook on networking, e.g., see [3,9,8]; for more recent references see [7] and the references therein (due to space limitations, we do not provide comprehensive literature review on ALOHA in this paper). However, it appears that there are no explicit and general studies (more than two users) of the distributional properties of ALOHA, e.g., delay distributions. In this regard, in Subsection 2.1, we consider a standard finite population ALOHA model with variable length packets [4,2] that have an asymptotically exponential tail. Surprisingly, we discover a new phenomenon that the distribution of the number of retransmissions (collisions) and time between two successful transmissions follow power law distributions, as stated in Theorem 1. Informally, our theorem shows that when the exponential decay rate of the packet distribution is smaller than the parameter of the exponential backoff distribution, even the finite population ALOHA has zero throughput. This is contrary to the common belief that the finite population ALOHA system always has a positive, albeit possibly small, throughput. Furthermore, even when the long term throughput is positive, the high variability of power laws (infinite variance when the power law exponent is less than 2) may cause long periods of very high congestion/low throughput. It also may appear counterintuitive that the system is characterized by power laws even though the distributions of all the variables (arrivals, backoffs and packets) of the system are of exponential type. However, this is in line with the very recent results in [5,10,6], which show that job completion times in systems with failures where jobs restart from the very beginning exhibit similar power law behavior. Our study in [6] was done in the communication context where job completion times are represented by document/packet transmission delays. It may also be worth noting that [6] reveals the existence of power law delays regardless of how light or heavy the packet/document and link failure distributions may be (e.g., Gaussian), as long as they have proportional hazard functions. Furthermore, from a mathematical perspective, our Theorem 1 analyzes a more complex setting than the one in [6,10] and, thus, requires a novel proof. Hence, when compared with [6,10], this paper both discovers a new related phenomenon in a communication MAC layer application area and provides a novel analysis of it.

As already stated in the abstract, the preceding power law phenomenon is a result of combined effects of packet variability and collisions. Hence, one can see easily that the power law delays can be eliminated by reducing the variability of packets. Indeed, for slotted ALOHA with constant size packets the delays are geometrically distributed. However, we show in Section 3 that, when the number of users sharing the channel is geometrically distributed, the slotted ALOHA exhibits power law delays as well.

In Section 4, we illustrate our results with simulation experiments, which show that the asymptotic power law regime is valid for relatively small delays and reasonably large probability values. Furthermore, the distribution of packets in practice might have a bounded support. To this end, we show by a simulation experiment that this situation results in distributions that have power law main body with an exponentiated (stretched) support in relation to the support of the packet size/number of active users. Hence, although exponentially bounded, the delays may be prohibitively long.

In practical applications, we may have combined effects of both variable packets and a random number of users, implying that the delay and congestion is likely to be even worse than predicted by our results. Thus, from an engineering perspective, one has to pay special attention to the packet variability and the number of users when designing robust MAC protocols, especially for ad-hoc/sensor networks where link failures [6], mobility and many other factors might further worsen the performance.

2   Power Laws in the Finite Population ALOHA with Variable Size Packets

In this section we show that the variability of packet sizes, when coupled with the contention nature of ALOHA, is a cause of power law delays. This study is motivated by the well-known fact that packets in todays Internet have variable sizes. To further emphasize that packet variability is a sole cause of power laws, we assume a finite population ALOHA model where each user can hold (queue) up to one packet at the time since the increased queueing only further exacerbates the problem. In addition, in Section 3 we show that the user variability in an infinite population model may be a cause of power law delays as well. In the remainder of this section, we describe the model and introduce the necessary notation in Subsection 2.1 and then in Subsection 2.2 we formulate and prove our main result on the logarithmic asymptotics of the transmission delay in Theorem 1.

2.1   Model Description

Consider $M \geq 2$ users sharing a common communication link (channel) of unit capacity. Each user can hold at most one packet in its queue and, when the queue is empty, a new packet is generated after an independent (from all other variables) exponential time with mean $1/\lambda$. Each packet has an independent length that is equal in distribution to a generic random variable L. A user with a newly generated packet attempts its transmission immediately and, if there are no other users transmitting during the same time, the packet is considered successfully transmitted. Otherwise, if the transmissions of more than one packet overlap, we say that there is a collision and the colliding packets need to be retransmitted; for a visual representation of the system see Figure 1. After a collision, each participating user waits (backoffs) for an independent exponential period of time with mean $1/\nu$ and then attempts to retransmit its packet. Each user continues this procedure until its packet is successfully transmitted and then it generates a new packet after an independent exponential time of mean $1/\lambda$.

Without loss of generality, assume that there is a successful transmission at time $t = 0$ and let $\{T_i\}_{i \geq 1}$ be an increasing sequence of positive time points when either a collision or successful transmission occurs. Let N be the smallest index i such that at time $T \equiv T_N$ there was a successful transmission. We will study the asymptotic properties of the distributions of N and T, representing the total number of transmission attempts per one successful transmission and the time between the two consecutive successful transmissions, respectively.

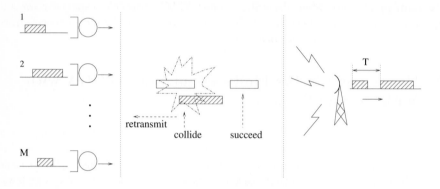

Fig. 1. Finite population ALOHA model with variable packet sizes

2.2   Power Law Asymptotics

The following theorem on the logarithmic asymptotics of the number of transmission attempts N per successful transmission and delay T between two successful transmissions is our main result of this section.

Theorem 1. *If*

$$\lim_{x \to \infty} \frac{\log \mathbb{P}[L > x]}{x} = -\mu, \quad \mu > 0, \tag{1}$$

then, we have

$$\lim_{n \to \infty} \frac{\log \mathbb{P}[N > n]}{\log n} = -\frac{M\mu}{(M-1)\nu} \tag{2}$$

and

$$\lim_{t \to \infty} \frac{\log \mathbb{P}[T > t]}{\log t} = -\frac{M\mu}{(M-1)\nu}. \tag{3}$$

Remark 1. The proof of this result reveals that if the transmission delay is long, then the shortest packet from all M users will be the most likely one that is successfully transmitted, i.e., ALOHA is unfair to longer packets.

Remark 2. This theorem indicates that the distribution tails of N and T are essentially power laws when the packet distribution is approximately exponential ($\approx e^{-\mu x}$). Thus, the finite population ALOHA may exhibit high variations and possible zero throughput. More precisely, by the strong law of large numbers for stationary and ergodic point processes, the system has zero throughput when $0 < M\mu/(M-1)\nu < 1$; and when $1 < M\mu/(M-1)\nu < 2$, the transmission time has finite mean but infinite variance. Furthermore, for large M, $M\mu/(M-1) \approx \mu/\nu$ and thus, the system has zero throughput if the backoff parameter $\nu \gtrsim \mu$. It might be worth noting that this may even occur when the expected packet length is much smaller than the expected backoff time $\mathbb{E}L \ll 1/\nu$.

Proof. Let us first prove equation (2) assuming that at time $t = 0$ a collision happens and all users have a packet waiting to be send, i.e., $U(0) = M$. Since each user i has an equal probability $1/M$ of being the first one to attempt a transmission, and $e^{-L_i(M-1)\nu}$ is the conditional probability, given L_i, that such an attempt is successful, we obtain, for $x_\epsilon > 0$,

$$\mathbb{P}[N > n] = \mathbb{E}\left[\left(1 - \frac{1}{M}\left(\sum_{i=1}^{M} e^{-L_i(M-1)\nu}\right)\right)^n\right]$$

$$= \mathbb{E}\left[\left(1 - \frac{1}{M}\left(\sum_{i=1}^{M} e^{-L_i(M-1)\nu}\right)\right)^n \mathbf{1}\left(\bigcap_{i=1}^{M}\{L_i > x_\epsilon\}\right)\right]$$

$$+ \mathbb{E}\left[\left(1 - \frac{1}{M}\left(\sum_{i=1}^{M} e^{-L_i(M-1)\nu}\right)\right)^n \mathbf{1}\left(\bigcup_{i=1}^{M}\{L_i \le x_\epsilon\}\right)\right].$$

Next, by using $1 - x \le e^{-x}$ and the independence of L_i, we derive

$$\mathbb{P}[N > n] \le \left(\mathbb{E}\left[e^{-\frac{n}{M}e^{-L(M-1)\nu}}\mathbf{1}(L > x_\epsilon)\right]\right)^M + \left(1 - \frac{1}{M}e^{-x_\epsilon(M-1)\nu}\right)^n$$

$$\le \left(\mathbb{E}\left[e^{-\frac{n}{M}e^{-L\mathbf{1}(L>x_\epsilon)(M-1)\nu}}\right]\right)^M + \eta^n, \tag{4}$$

where $\eta \triangleq 1 - e^{-x_\epsilon(M-1)\nu}/M < 1$. Then, by assumption (1), for any $0 < \epsilon < \mu$, we can choose x_ϵ such that $\mathbb{P}[L > x] \le e^{-(\mu-\epsilon)x}$ for all $x \ge x_\epsilon$, which, by defining an exponential random variable L_ϵ with $\mathbb{P}[L_\epsilon > x] = e^{-(\mu-\epsilon)x}, x \ge 0$, implies $L\mathbf{1}(L > x_\epsilon) \overset{d}{\le} L_\epsilon$, where "$\overset{d}{\le}$" denotes inequality in distribution. Therefore, (4) implies

$$\mathbb{P}[N > n] \le \left(\mathbb{E}\left[e^{-\frac{n}{M}e^{-L_\epsilon(M-1)\nu}}\right]\right)^M + \eta^n. \tag{5}$$

Now, for any $0 < x < 1$,

$$\mathbb{P}\left[e^{-(\mu-\epsilon)L_\epsilon} < x\right] = \mathbb{P}[(\mu-\epsilon)L_\epsilon > -\log x] = x,$$

implying that $e^{-(\mu-\epsilon)L_\epsilon} \overset{d}{=} U$, where " $\overset{d}{=}$ " denotes equality in distribution and U is a uniform random variable between 0 and 1. Thus,

$$\mathbb{P}[N > n] \leq \left(\mathbb{E}\left[e^{-\frac{n}{M}U^{(M-1)\nu/(\mu-\epsilon)}}\right]\right)^M + \eta^n.$$

By using the identity $\mathbb{E}[e^{-\theta U^{1/\alpha}}] = \Gamma(\alpha+1)/\theta^\alpha$, one can easily obtain

$$\varlimsup_{n\to\infty} \frac{\log \mathbb{P}[N > n]}{\log n} \leq -\frac{M(\mu - \epsilon)}{(M-1)\nu}, \tag{6}$$

which, by passing $\epsilon \to 0$, completes the proof of the upper bound.

For the lower bound, define $L_o \triangleq \min\{L_1, L_2, \cdots, L_M\}$, and observe that

$$\mathbb{P}[N > n] = \mathbb{E}\left[\left(1 - \frac{1}{M}\left(\sum_{i=1}^M e^{-L_i(M-1)\nu}\right)\right)^n\right]$$

$$\geq \mathbb{E}\left[\left(1 - e^{-L_o(M-1)\nu}\right)^n\right]. \tag{7}$$

The complementary cumulative distribution function $\bar{F}_o(x) \triangleq \mathbb{P}[L_o \geq x]$ satisfies

$$\lim_{x\to\infty} \frac{\log \bar{F}_o(x)}{x} = -M\mu,$$

implying that, for any $\epsilon > 0$, there exists x_ϵ such that $\mathbb{P}[L_o > x] \geq e^{-(M\mu+\epsilon)x}$ for all $x \geq x_\epsilon$. Next, if we define random variable L_o^ϵ such that $\mathbb{P}[L_o^\epsilon > x] = e^{-(M\mu+\epsilon)x}, x \geq 0$, then,

$$L_o \overset{d}{\geq} L_o^\epsilon \mathbf{1}(L_o^\epsilon > x_\epsilon),$$

which, by (7), implies

$$\mathbb{P}[N > n] \geq \mathbb{E}\left[\left(1 - e^{-L_o^\epsilon \mathbf{1}(L_o^\epsilon > x_\epsilon)(M-1)\nu}\right)^n\right]$$

$$\geq \mathbb{E}\left[\left(1 - e^{-L_o^\epsilon(M-1)\nu}\right)^n \mathbf{1}(L_o^\epsilon > x_\epsilon)\right].$$

Noticing that for any $0 < \delta < 1$, there exists $x_\delta > 0$ such that $1 - x \geq e^{(1-\delta)x}$ for all $0 < x < x_\delta$, we can choose x_ϵ large enough, such that

$$\mathbb{P}[N > n] \geq \mathbb{E}\left[e^{-(1-\epsilon)ne^{-L_o^\epsilon(M-1)\nu}} \mathbf{1}(L_o^\epsilon > x_\epsilon)\right]$$

$$\geq \mathbb{E}\left[e^{-(1-\epsilon)ne^{-L_o^\epsilon(M-1)\nu}}\right] - \mathbb{E}\left[e^{-(1-\epsilon)ne^{-L_o^\epsilon(M-1)\nu}} \mathbf{1}(L_o^\epsilon \leq x_\epsilon)\right]$$

$$\geq \mathbb{E}\left[e^{-(1-\epsilon)ne^{-L_o^\epsilon(M-1)\nu}}\right] - \zeta^n, \tag{8}$$

where $\zeta = e^{-(1-\epsilon)e^{-x_\epsilon(M-1)\nu}} < 1$. Similarly as in the proof of the upper bound, it is easy to check that $e^{-(M\mu+\epsilon)L_o^\epsilon} \overset{d}{=} U$, and therefore,

$$\mathbb{P}[N > n] \geq \mathbb{E}\left[e^{-(1-\epsilon)nU^{(M-1)\nu/(M\mu+\epsilon)}}\right] - \zeta^n,$$

which, by recalling the identity $\mathbb{E}[e^{-\theta U^{1/\alpha}}] = \Gamma(\alpha + 1)/\theta^{\alpha}$, yields

$$\lim_{n\to\infty} \frac{\log \mathbb{P}[N > n]}{\log n} \geq -\frac{M\mu + \epsilon}{(M-1)\nu}. \tag{9}$$

Finally, passing $\epsilon \to 0$ in (9) completes the proof of the lower bound. Combining the lower and upper bound, we finish the proof of (2) for the case $U(0) = M$.

Next, define $N_s \triangleq \min\{n \geq 0 : U(T_n) = M\}$, $N_l \triangleq \min\{N, N_s\}$ and $N_e \triangleq N - N_l$. It can be shown that

$$\mathbb{P}[N_l > n] \leq \mathbb{P}[N_s > n] = o(e^{-\theta n}) \tag{10}$$

for some $\theta > 0$; due to space limits, the details of this proof will be presented in the full version of this paper. Assuming that the preceding bound holds, by the memoryless property of exponential distributions, we obtain

$$\begin{aligned}
\mathbb{P}[N_e > n] &= \mathbb{P}[N - N_s > n, N > N_s] \\
&= \mathbb{P}[N > N_s]\mathbb{P}[N - N_s > n \mid N > N_s] \\
&= \mathbb{P}[N > N_s]\mathbb{P}[N > n \mid N_s = 0].
\end{aligned}$$

Noting that $\mathbb{P}[N > n \mid N_s = 0]$ is the case of $U(0) = M$ that has already been proved, we conclude

$$\lim_{n\to\infty} \frac{\log \mathbb{P}[N_e > n]}{\log n} = -\frac{M\mu}{(M-1)\nu},$$

which, combined with (10) and the union bound, yields

$$\lim_{n\to\infty} \frac{\log \mathbb{P}[N > n]}{\log n} = \lim_{n\to\infty} \frac{\log \mathbb{P}[N_l + N_e > n]}{\log n} = -\frac{M\mu}{(M-1)\nu}. \tag{11}$$

The proof of (3) is presented in Section 4. $\square$

3   Power Laws in Slotted ALOHA with Random Number of Users

It is clear from the preceding section that the power law delays arise due to the combination of collisions and packet variability. Hence, it is reasonable to expect an improved performance when this variability is reduced. Indeed, it is easy to see that the delays are geometrically bounded in a slotted ALOHA with constant size packets and a finite number of users. However, in this section we will show that, when the number of users sharing the channel has asymptotically an exponential distribution, the slotted ALOHA exhibits power law delays as well. Situations with random number of users are essentially predominant in practice, e.g., in sensor networks, the number of active sensors in a neighborhood is a random variable since sensors may switch between sleep/active modes, as shown

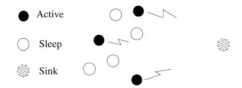

Fig. 2. Random number of active neighbors in a sensor network

in Figure 2; similarly in ad hoc wireless networks the variability of users may arise due to mobility, new users joining the network, etc.

More formally, consider a slotted ALOHA model (e.g., see Section 4.2.2 of [3]) with packets/slots of unit size and a random number of users $M \geq 1$ that are fixed over time. Similarly as in Section 2, each user holds at most one packet at a time and after a successful transmission a new packet is generated according to an independent Bernoulli process with success probability $0 < \lambda \leq 1$. In case of a collision, each colliding user backs off according to an independent geometric random variable with parameter $e^{-\nu}, \nu > 0$. Denote the number of slots where transmissions are attempted but failed and the total time between two successful packet transmissions as N and T, respectively.

Theorem 2. *If there exists $\alpha > 0$, such that*

$$\lim_{x \to \infty} \frac{\log \mathbb{P}[M > x]}{x} = -\alpha,$$

then, we have

$$\lim_{n \to \infty} \frac{\log \mathbb{P}[N > n]}{\log n} = \lim_{t \to \infty} \frac{\log \mathbb{P}[T > t]}{\log t} = -\frac{\alpha}{\nu}. \tag{12}$$

Remark 3. Similarly as in Theorem 1, this result shows that the distributions of N and T are essentially power laws, i.e., $\mathbb{P}[T > t] \approx t^{-\alpha/\nu}$ and, clearly, if $\alpha < \nu$, then $\mathbb{E}N = \mathbb{E}T = \infty$.

Proof. First consider a situation where all the users are backlogged, i.e., have a packet to send. In this case the total number of collisions between two successful transmissions is geometrically distributed given M,

$$\mathbb{P}[N > n \mid M] = \left(1 - \frac{Me^{-(M-1)\nu}(1 - e^{-\nu})}{1 - e^{-M\nu}}\right)^n, \quad n \in \mathbb{N},$$

since, given M, $1 - e^{-M\nu}$ is the conditional probability that there is an attempt to transmit a packet, and $1 - e^{-M\nu} - Me^{-(M-1)\nu}(1 - e^{-\nu})$ is the conditional probability that there is a collision. Therefore,

$$\mathbb{P}[N > n] = \mathbb{E}\left[\left(1 - \frac{Me^{-(M-1)\nu}(1 - e^{-\nu})}{1 - e^{-M\nu}}\right)^n\right]. \tag{13}$$

On the other hand, we have

$$\mathbb{P}[T > t] = \mathbb{E}\left[\left(1 - Me^{-(M-1)\nu}(1 - e^{-\nu})\right)^t\right], \; t \in \mathbb{N}. \qquad (14)$$

Now, following the same arguments as in the proof of Theorem 1, we can prove (12). Similarly, one can show that the same asymptotic results hold if the initial number of backlogged users is less than M. Due to space limitations, a complete proof of this theorem will be presented in the extended version of this paper. $\square$

Actually, using the technique developed in [6] with some modifications, we can compute the exact asymptotics under a bit more restrictive conditions. Again, due to space limitations, the proof of the following theorem is deferred to the full version of the paper.

Theorem 3. *If $\lambda = \nu$ and $\bar{F}(x) \triangleq \mathbb{P}[M > x]$ satisfies $H\left(-\log \bar{F}(x)\right) \bar{F}(x)^{1/\beta} \sim xe^{-\nu x}$ with $H(x)$ being continuous and regularly varying, then, as $t \to \infty$,*

$$\mathbb{P}[T > t] \sim \frac{\Gamma(\beta+1)(e^\nu - 1)^\beta}{t^\beta H(\beta \log t)^\beta}.$$

4   Simulation Examples

In this section, we illustrate our theoretical results with simulation experiments. In particular, we emphasize the characteristics of the studied ALOHA protocol that may not be immediately apparent from our theorems. For example, in practice, the distributions of packets and number of random users might have a bounded supports. We show that this situation may result in truncated power law distributions for T. To this end, it is also important to note that the distributions of N and T will have a power law main body with a stretched support in relation to the support of L and M and, thus, may result in very long, although, exponentially bounded delays. We will study the case where M has a bounded support in our second experiment.

Example 1 (Finite population model). For the finite population model described in Subsection 2.1, we study the distribution of time T between two consecutive successful transmissions. In this regard, we conduct four experiments for $M = 2, 4, 10, 20$ users, respectively. The packets are assumed i.i.d. exponential with mean 1 and the arrival intervals and backoffs follow exponential distribution with mean $2/3$. Simulation result with 10^5 samples are shown in Figure 3, which indicates a power law transmission delay. We can see from the figure that, as M gets large ($M = 10, 20$), the slopes of the distributions that represent the power law exponents on the log / log plot are essentially the same, as predicted by our Theorem 1.

Example 2 (Random number of users). As stated in Section 3, the situation when the number of users M is random may cause heavy-tailed transmission delays even for slotted ALOHA. However, in many practical applications the

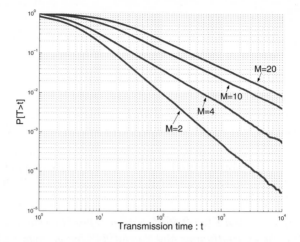

Fig. 3. Interval distribution between successfully transmitted packets for finite population ALOHA with variable size packets

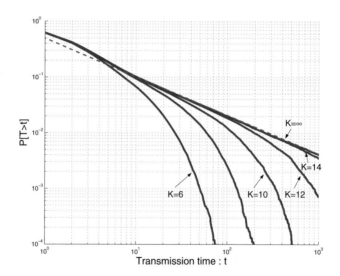

Fig. 4. Illustration of the stretched support of the power law main body when the number of users is $\min(M, K)$, where M is geometrically distributed

number of active users M may be bounded, i.e., the distribution $\mathbb{P}[M > x]$ has a bounded support. Thus, from equation (14) it is easy to see that the distribution of T is exponentially bounded. However, this exponential behavior may happen for very small probabilities, while the delays of interest can fall inside the region of the distribution (main body) that behaves as the power law. This example is aimed to illustrate this important phenomenon. Assume that initially $M \geq 1$ users have unit size packets ready to send and M follows geometric distribution with mean 3. The backoff times of colliding users are independent and geometrically distributed with mean 2. We take the number of users to have finite support $[1, K]$ and show how this results in a truncated power law distribution for T in the main body, even though the tails are exponentially bounded. This example is parameterized by K where K ranges from 6 to 14 and for each K we set the number of users to be equal to $M_K = \min(M, K)$.

We plot the distribution of $\mathbb{P}[T > t]$, parameterized by K, in Figure 4. From the figure we can see that, when we increase the support of the distributions from $K = 6$ to $K = 14$, the main (power law) body of the distribution of T increases from less than 5 to almost 700. This effect is what we call the stretched support of the main body of $\mathbb{P}[T > t]$ in relation to the support K of M. In fact, it can be rigorously shown that the support of the main body of $\mathbb{P}[T > t]$ grows exponentially fast. Furthermore, it is important to note that, if $K = 14$ and the probabilities of interest for $\mathbb{P}[T > t]$ are bigger than $1/500$, then the result of this experiment is basically the same as for $K = \infty$; see Figure 4.

References

1. Abramson, N.: The Aloha system - another alternative for computer communications. In: Proc. of the Fall Joint Computer Conference, pp. 281–285 (1970)
2. Bellini, S., Borgonovo, F.: On the throughput of an ALOHA channel with variable length packets. IEEE Transactions on Communications 28, 1932–1935 (1980)
3. Bertsekas, D.P., Gallager, R.: Data Networks, 2nd edn. Prentice-Hall, Englewood Cliffs (1992)
4. Ferguson, M.J.: An approximate analysis of delay for fixed and variable length packets in an unslotted ALOHA channel. IEEE Transactions on Communications 25, 644–654 (1977)
5. Fiorini, P.M., Sheahan, R., Lipsky, L.: On unreliable computing systems when heavy-tails appear as a result of the recovery procedure. ACM SIGMETRICS Performance Evaluation Review 33(2), 15–17 (2005)
6. Jelenković, P.R., Tan, J.: Can retransmissions of superexponential documents cause subexponential delays? In: Proceedings of IEEE INFOCOM'07, Anchorage, Alaska, USA (May 2007) URL: http://www.comet.columbia.edu/~predrag/pub_list.html
7. Naware, V., Mergen, G., Tong, L.: Stability and delay of finite user slotted. ALOHA with multipacket reception 51(7), 2636–2656 (2005)
8. Rom, R., Sidi, M.: Multiple access protocols: performance and analysis, New York, NY, USA. Springer-Verlag, Heidelberg (1990)
9. Schwartz, M.: Telecommunication networks: protocols, modeling and analysis. Addison-Wesley Longman Publishing Co., Inc, Boston, MA, USA (1986)
10. Sheahan, R., Lipsky, L., Fiorini, P., Asmussen, S.: On the completion time distribution for tasks that must restart from the beginning if a failure occurs. MAMA 2006 Workshop, Saint-Malo, France (June 2006)

Appendix

Proof (of equation (3)). Similarly as in proving (2), we first assume that at time $t = 0$ there is a collision and that $U(0) = M$. Next, let $\{Y_i\}_{i \geq 1}$ be an i.i.d. sequence of exponential random variables with parameter $M\mu$ and define $L^* = \max_{1 \leq i \leq M}\{L_i\}$ where L_i is the packet size. Then,

$$\sum_{i=1}^{N} Y_i \overset{d}{\leq} T \overset{d}{\leq} NL^* + \sum_{i=1}^{N} Y_i. \tag{15}$$

Now, we establish the upper bound. For $H > 0$,

$$\mathbb{P}[T > t] \leq \mathbb{P}\left[N > \frac{t}{H\log t}\right] + \mathbb{P}\left[\sum_{1 \leq i \leq t/H\log t} Y_i > \frac{t}{2}\right] + \mathbb{P}\left[\frac{tL^*}{H\log t} > \frac{t}{2}\right]$$

$$\triangleq I_1 + I_2 + I_3. \tag{16}$$

Since L^* is exponentially bounded, we can choose H large enough, such that for any fixed $\alpha > 0$,

$$I_3 = o\left(\frac{1}{t^\alpha}\right). \tag{17}$$

For the second term, by applying union bound, we obtain

$$I_2 \leq \left(\frac{t}{H\log t} + 1\right)\mathbb{P}\left[Y_i > \frac{H\log t}{2}\right] = o\left(\frac{1}{t^\alpha}\right), \tag{18}$$

for any fixed $\alpha > 0$ and H large enough. Next, by (2), the asymptotics of I_1 is equal to

$$\lim_{t \to \infty} \frac{\log \mathbb{P}\left[N > \frac{t}{H\log t}\right]}{\log t} = -\frac{M\mu}{(M-1)\nu},$$

which, combined with (16), (17) and (18), completes the proof of the upper bound. Next, we prove the the lower bound. By the left inequality of (15), for $1 > \delta > 0$,

$$\mathbb{P}[T > t] \geq \mathbb{P}\left[N \geq \frac{t(1+\delta)}{\mathbb{E}[Y_1]} + 1\right] - \mathbb{P}\left[\sum_{i=1}^{N} Y_i \leq t, N \geq \frac{t(1+\delta)}{\mathbb{E}[Y_1]} + 1\right]. \tag{19}$$

Now, by using the standard large deviation result (Chernoff bound), it immediately follows that, for some $\eta > 0$, the second term on the right-hand side of (19) is bounded by

$$\mathbb{P}\left[\sum_{i \leq t(1+\delta)/\mathbb{E}[Y_1]+1} Y_i \leq t\right] = \mathbb{P}\left[\sum_{i \leq t(1+\delta)/\mathbb{E}[Y_1]+1} (\mathbb{E}[Y_1] - Y_i) \geq \delta t\right] = o\left(e^{-\eta t}\right).$$

Again, by (2), the first term on the right-hand side of (19) gives the right asymptotics, which proves the lower bound. The proof of the case $U(0) < M$ is more involved and, therefore, due to space limitation, we defer it to the full version of this paper. $\square$

Peer-to-Peer vs. Client/Server: Reliability and Efficiency of a Content Distribution Service

Kenji Leibnitz[1], Tobias Hoßfeld[2], Naoki Wakamiya[1], and Masayuki Murata[1]

[1] Graduate School of Information Science and Technology
Osaka University, 1-5 Yamadaoka, Suita, Osaka 565-0871, Japan
[leibnitz,wakamiya,murata]@ist.osaka-u.ac.jp
[2] University of Würzburg, Institute of Computer Science
Am Hubland, 97074 Würzburg, Germany
hossfeld@informatik.uni-wuerzburg.de

Abstract. In this paper we evaluate the performance of a content distribution service with respect to reliability and efficiency. The considered technology for realizing such a service can either be a traditional client/server (CS) architecture or a peer-to-peer (P2P) network. In CS, the capacity of the server is the bottleneck and has to be dimensioned in such a way that all requests can be accommodated at any time, while a P2P system does not burden a single server since the content is distributed in the network among sharing peers. However, corrupted or fake files may diminish the reliability of the P2P service due to downloading of useless contents. We compare a CS system to P2P and evaluate the downloading time, success ratio, and fairness while considering flash crowd arrivals and corrupted contents.

1 Introduction

The volume of traffic transported over the Internet has drastically increased over the last few years. The download of multimedia contents or software packages may consist of large files imposing high requirements on the bandwidth of the file servers. In conventional systems this means that the servers must be properly dimensioned with sufficient capacity in order to service all incoming file requests from clients. On the other hand, *peer-to-peer* (P2P) technology offers a simple and cost-effective way for sharing content. Providers offering large volume distributions (e.g. Linux) have recognized the potential of P2P and increasingly offer downloads via eDonkey or BitTorrent.

In P2P, all participating peers act simultaneously as clients and as servers, and the file is not offered at a single server location, but by multiple sharing peers. Since the load is distributed among all sharing peers, the risk of overloading servers with requests is reduced, especially in the presence of flash crowd arrivals. However, this flexibility comes at a slight risk. Since the shared file is no longer at a single trusted server location, peers may offer a corrupted version of a file or parts of it. This is referred to as *poisoning* or *pollution* [1] depending on whether the decoy was offered deliberately or not. When the number of fake peers is large, the dissemination of the file may be severely disrupted. All of this leads to a trade-off consideration between high reliability at the risk of overloaded servers and good scalability where the received data may

L. Mason, T. Drwiega, and J. Yan (Eds.): ITC 2007, LNCS 4516, pp. 1161–1172, 2007.

be corrupt. In this context, we define reliability as the availability of a single file over time in a disruptive environment. This is expressed by the success ratio of downloads. While in some structured P2P network types, the disconnection or segmentation of the network topology due to node failure may influence the availability of content, we will only focus on unstructured P2P systems and assume that each peer can contact any other peer with the same probability (epidemic model). This is valid e.g. in eDonkey networks. In this case, the availability of a file is expressed by the number of sharing peers. Hence, the availability in the P2P networks we consider is predominantly influenced by the user behavior [2], like churn, willingness to share a file, or impatience during downloading.

In this paper, we investigate the trade-off between client/server (CS) and P2P file sharing using simple models. While we assume the file structure and download mechanism to be operating like in eDonkey, the model can be easily extended to any other P2P network. Our focus of interest lies hereby on the downloading time until successful completion of the file and the number of aborted downloads due to the impatience of users. With these performance metrics we can justify under which conditions a P2P network outperforms CS. In addition, fairness of the CDS is considered as well.

The paper is organized as follows. In Section 2 we briefly summarize existing work related to evaluating content distribution systems and comparing P2P with CS. Section 3 provides the models and assumptions that we impose. In Section 4, we provide numerical results for the comparison of the performance of P2P with CS in terms of success ratio, download duration, and fairness. Finally, this paper is concluded with a summary and an outlook on future work.

2   Related Work

Most studies on the performance of P2P systems as content distribution network rely on measurements or simulations of existing P2P networks. For example, Saroiu et al. [3] conducted measurement studies of content delivery systems that were accessed by the University of Washington. The authors distinguished traffic from P2P, WWW, and the Akamai content distribution network and they found that the majority of volume was transported over P2P. A comprehensive survey of different P2P-based content distribution technologies is given in [4]. In [5] a simulation study of P2P file dissemination using multicast agents is performed and the propagation under different conditions is studied. Hoßfeld et al. [6] provide a simulation study of the well-known eDonkey network and investigate the file diffusion properties under constant and flash crowd arrivals. However, most work on P2P file diffusion as those mentioned above usually do not assume any fake files from pollution or poisoning.

Han et al. [7] study the distribution of content over P2P and consider rewarding strategies as incentives to improve the diffusion. They show that the network structure in terms of hierarchy and clustering improve the diffusion over flat structures and that compensating referrers improves the speed of diffusion and an optimal referral payment can be derived. The user behavior and an analysis of the rationale in file sharing is studied in [8] using game theory. The focus lies on free riding in the network and the authors offer suggestions on how to improve the willingness of peers to share. Qiu et al. [9] model a BitTorrent network using a fluid model and investigate the performance in steady state. They study the effectiveness of the incentive mechanism in BitTorrent and prove the existence of a Nash equilibrium. Rubenstein and Sahu [10] provide a mathematical model of unstructured P2P and show that P2P networks show good scalability and are well suited to cope with flash crowd arrivals. Another fluid-diffusive P2P model from statistical physics is presented by Carofiglio et al. [11]. Both, the user and the content dynamics are included, but this is only done on file level and without pollution. These studies show that by providing incentives to the peers for sharing a file, the diffusion properties are improved. We include appropriate paramters in our model which capture this effect, while also considering pollution.

Christin et al. [1] measured content availability of popular P2P file sharing networks and used this measurement data for simulating different pollution and poisoning strategies. They showed that only a small number of fake peers can seriously impact the user's perception of content availability. In [12] a diffusion model for modeling eDonkey-like P2P networks was presented based on an epidemic SIR [13] model. This model includes pollution and a peer patience threshold at which the peer aborts its download attempt and retries later again. It was shown that an evaluation of the diffusion process is not accurate enough when steady state is assumed or the model only considers the transmission of the complete file, especially in the presence of flash crowd arrivals.

3   Modeling the Content Distribution Service

In the following we will consider two alternative architectures for content distribution: P2P and a traditional client/server structure (e.g. HTTP or FTP server). We include pollution from malicious peers in the P2P model offering fake content. On the other hand, the client/server system is limited by the server bandwidth. In both systems we assume that the user is willing to wait only for a limited time until the download completes. If the downloading process exceeds a patience threshold, the user will abort his attempt. We will use these models to later analyze the benefits and drawbacks of each architecture.

3.1   Peer-to-Peer Network

In the P2P model we assume that the file sharing process of a file with size f_{size} operates similar to the eDonkey network. The sharing itself is performed in units of 9.5MB, so-called *chunks*, and the data of each chunk is transfered in *blocks* of 180kB. In order to make the model more tractable, we simply consider that each file consists of M download data units. After each chunk is downloaded, it is checked using MD5 hashes and in case an error is detected e.g. due to transmission errors, the chunk is discarded and downloaded again. After all chunks of a file have been successfully downloaded, it is up to the peer if the file is kept as a *seeder* for other peers to download or if it is removed from share (*leecher* or *free rider*). In this study, we only consider a file that consists of a single chunk with $M = 53$ download units which corresponds to the number of blocks in a chunk. Thus, the terms block and download/data unit will be used interchangeably.

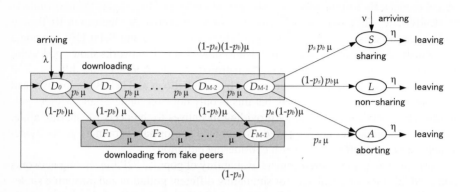

Fig. 1. Flow diagram of P2P file sharing model

Description of the Flow Model. As mentioned above, errors may occur during the download process of a chunk rendering it useless. This mechanism is used by malicious peers deliberately introducing erroneous chunks to the file sharing network. In order to characterize the dynamic behavior of the P2P network with K malicious peers, referred to as *fake peers*, we extend the model in [12]. This model is based on the epidemic diffusion of diseases [13] and is characterized by a differential equation system describing the transitions between each of the states a peer traverses. Initially, there are only S_0 peers in the system sharing a correct version of the file and K fake peers. Requests for downloading the file arrive with rate λ. A peer downloads M units of the file where it has the possibility of reaching a correct version of the data block with probability p_b. Since we assume an equal probability for reaching a sharing or fake peer, p_b can be given as in Eqn. (1) at time t.

$$p_b(t) = \frac{S(t)}{S(t) + K} \tag{1}$$

The population of peers with successful downloads of i units is defined as D_i. After having successfully downloaded M data units, an error check is performed and the chunk is discarded in case of an error. If the download of the entire chunk was successful, the peer either shares the file and enters population S with the *sharing probability* p_s or enters L of non-sharing peers with the complementary probability $1 - p_s$. On the other hand, if the download attempt of the chunk failed because of downloading at least one block from a fake peer, the peer aborts with probability p_a and retries the download attempt with $1 - p_a$.

The download of i data units of which at least one is corrupt is represented by state F_i. The number of fake peers K is assumed to remain constant throughout the observation period. The system model with all populations and their transitions is shown in Fig. 1. The system of differential equations describing the dynamic behavior of each population is given in Eqns. (2-8).

$$\dot{D}_0 = \lambda + \mu \left(1 - p_a\right) \left[F_{M-1} + \left(1 - p_b\right) D_{M-1}\right] - \mu D_0 \tag{2}$$

$$\dot{D}_i = \mu p_b D_{i-1} - \mu D_i \qquad\qquad i = 1, \ldots, M - 1 \tag{3}$$

$$\dot{F}_1 = \mu \left(1 - p_b\right) D_0 - \mu F_1 \tag{4}$$

$$\dot{F}_i = \mu \left(1 - p_b\right) D_{i-1} + \mu F_{i-1} - \mu F_i \qquad\qquad i = 2, \ldots, M-1 \tag{5}$$

$$\dot{S} = \nu + \mu p_s \, p_b \, D_{M-1} - \eta \, S \tag{6}$$

$$\dot{L} = \mu \left(1 - p_s\right) p_b \, D_{M-1} - \eta \, L \tag{7}$$

$$\dot{A} = \mu p_a \left[F_{M-1} + \left(1 - p_b\right) D_{M-1}\right] - \eta \, A \tag{8}$$

The other variables that have not yet been discussed are the file request rate λ and the rates for leaving the system η. Furthermore, ν is the rate of arrivals of peers that share the file which they obtained from another source than from this network. For peers in the network, we will assume *flash crowd* arrivals as $\dot{\lambda} = -\alpha \lambda$ with initial value of $\lambda(0) = \lambda_0$. Hence, the flash crowd scenario corresponds to an exponentially decreasing arrival rate with parameter α.

$$\lambda(t) = \lambda_0 \, e^{-\alpha t} \tag{9}$$

For the sake of simplicity we assume that a peer decides to leave only if he either has successfully completed the download (S and L) or when he aborts the download attempt (A). In F_{M-1}, the peer may enter the population A with *abort probability* p_a or else retries the attempt. Perhaps the most important variable in the model is the download rate per data unit $\mu(t)$. We use the same approximation as in [12] which assumes that if there are enough sharers, the download bandwidth r_{dn} of a peer will be the limitation, otherwise all requesting peers fairly share the upload bandwidth r_{up} of all sharing peers, see Eqn. (10).

$$\mu(t) = \frac{M}{f_{size}} \min \left\{ \frac{r_{up} \left(S(t) + K\right)}{\sum_{i=0}^{M-1} D_i(t) + \sum_{i=1}^{M-1} F_i(t)}, r_{dn} \right\} \tag{10}$$

Note that all variables in the equation system are in fact functions of time resulting in a highly non-stationary behavior. Finally, it should be remarked that the continuous transition rates lead to a slight inaccuracy from non-integer population sizes which do not appear in reality, but reflect the average values.

Evaluation of the Download Duration. From the solution of the dynamic system in Eqns. (2-8), we can indirectly derive the transmission durations until reaching an absorbing population S, L, or A. The states S and D_i allow from Eqn. (10) the computation of the download rates per data unit $\mu(t)$. For the computation of the download duration $\delta(t)$, let us consider the start of the download attempt of a chunk at time t_0 and a series of time instants $t_1, \ldots, t_M$. Each t_i indicates the time at which the downloading of one data unit is completed. Since the transmission rates are with respect to the transmission of a block, the t_i values can be computed by numerical solution of Eqn. (11) for a given t_0.

$$\int_{t_{i-1}}^{t_i} \mu(t) \, dt = 1 \qquad\qquad 1 \leq i \leq M \tag{11}$$

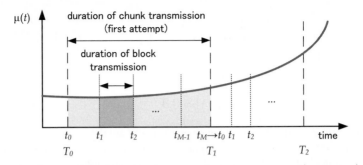

Fig. 2. Computation of block and chunk transmission durations from $\mu(t)$

Once the whole chunk is downloaded, we also define this time instant as $T_j, j > 1$ indicating with j the number of attempts a download attempt was made starting at T_0. Thus, t_0 is always set to the starting time of a new chunk download and is considered only within the context of a chunk. The relationship between $\mu(t)$, t_i, and T_j is illustrated in Fig. 2.

At time instants T_j we compute the probability that the chunk was correctly received by considering the possibilities of encountering a fake source at all t_i. The probability for a correct block $p_b(t_i)$ at the start of each block download interval $[t_i, t_{i+1}]$ and the probability of the chunk being correctly received is the product over each of the correct block probabilities beginning at t_0.

$$p_c(t_0) = \prod_{i=0}^{M-1} p_b(t_i) \tag{12}$$

If the chunk was not successfully downloaded, the peer chooses to retry its attempt with probability $1 - p_a$. The average successful download duration $\delta(t)$ is then computed considering $p_c(t)$ and p_a. If we define the random variable of trials $X_s(T_0)$ needed for successfully completing the download which started at T_0 after the j-th download attempt, we obtain the probabilities in Eqn. (13).

$$P\left(X_s(T_0) = 1\right) = p_c(T_0)$$
$$P\left(X_s(T_0) = j\right) = (1 - p_a)^{j-1} p_c(T_{j-1}) \prod_{k=0}^{j-2} (1 - p_c(T_k)) \qquad j \geq 2 \tag{13}$$

The average time until successfully completing the chunk download which the peer started at time T_0 follows then as shown in Eqn. (14). The probabilities for $X_s(T_0)$ must be normalized by all possible realizations in order to only take the successful download completions into account.

$$\delta(T_0) = \sum_{j=1}^{\infty} (T_j - T_0) \frac{P\left(X_s(T_0) = j\right)}{\sum_{k=1}^{\infty} P\left(X_s(T_0) = k\right)} \tag{14}$$

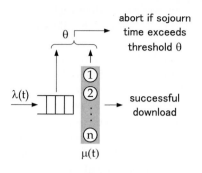

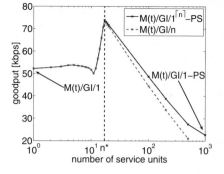

Fig. 3. Model of server-based content distribution system

Fig. 4. Influence of the number of servers on the download bandwidth of the clients

3.2 Client/Server System

In order to compare the performance of P2P and a server-based system, we need to match the conditions like the available capacity of the system and aborted downloads. A server in the Internet, e.g. HTTP or FTP server, transfers the complete file and does not split it into chunks. Hence, the client behavior must be modeled in a different way from P2P w.r.t. aborting the download. The server model in this paper needs to consider impatient users which cancel their downloading attempt if the total sojourn time in the system exceeds an impatience threshold θ. For comparing the CS system with a P2P system, θ can be obtained from the P2P system for a given abort probability p_a using the sojourn time of peers until they abort the download. With an abuse of the Kendall-notation, we will denote the server system as $M(t)/GI/1^{\lceil n \rceil} - PS$, see Fig. 3. The queue length at the server is assumed to be infinite.

$M(t)$ means that requests arrive at the server with a non-stationary Poisson process using the flash crowd arrival rate $\lambda(t)$ described in Eqn. (9). The system itself has a total constant capacity $C = S_0 \, r_{up}$ which corresponds to the total bandwidth available in the P2P system at time $t = 0$.

We assume that the complete bandwidth C is split equally among all downloading clients with the processor sharing discipline. However, the number $D(t)$ of simultaneously served clients is restricted to a maximum n. Each client is served by one virtual service unit and is guaranteed a minimal offered download bandwidth of C/n. If less than the maximum number of n service units (or parallel download slots) are actually occupied, a client receives $C/D(t)$. Thus, the average service rate of the system is then either limited by the bandwidth that each simultaneously downloading client gets or the maximum client download bandwidth r_{dn} as given in Eqn. (15). We use $1^{\lceil n \rceil} - PS$ in the notation for the server model to describe this service behavior.

$$\mu(t) = \min \left\{ r_{dn}, \frac{C}{D(t)} \right\} \tag{15}$$

The service requirement follows a general distribution GI and describes the sizes of the files to be downloaded. As we consider only a single file with a fixed size, the corresponding system is $M(t)/D/1^{\lceil n \rceil} - PS$.

The impact of the number of service units n on the average goodput each user experiences is illustrated in Fig. 4. The goodput is the ratio between the file size and the sojourn time of a user, where latter is the sum of the service time and the waiting time. This figure also shows the equivalent curve for an $M(t)/GI/n$ system with rate $\mu = C/n$. In the processor sharing model, if $n < C/r_{dn}$, the downlink of the client is the bottleneck in the system and $M(t)/GI/1^{\lceil n \rceil} - PS$ is equivalent to the $M(t)/GI/n$ system. For $n > C/r_{dn}$, both systems show a different behavior. The processor sharing discipline utilizes the entire capacity C and can therefore be seen as the best case scenario in terms of bandwidth efficiency. From Fig. 4, we can also recognize the existence of a maximum value at $n^* = \lceil C/r_{dn} \rceil$ where the highest efficiency can be found. While for $n < n^*$ the average bandwidth is limited by the client download bandwidth, for $n > n^*$ the capacity of the server is the limiting factor. Note that for $n = \frac{\lambda_0}{\alpha}$, the system results in a pure $M(t)/GI/1 - PS$ queue, as the total number of arriving users in the system is limited in the considered flash-crowd scenario: $\lim_{t \to \infty} \int_0^t \lambda(t)dt = \frac{\lambda_0}{\alpha}$.

4   Numerical Results

We will now show numerical results and compare the P2P and CS system in performance. Unless stated otherwise, we will make the following assumptions as summarized in Tab. 1. Note that with $\eta = 0$, $\nu = 0$, and the limited number of arrivals ($\lim_{t \to \infty} \lambda(t) = 0$), all peers remain in the system after their either successful or unsuccessful download attempt. Therefore, the populations S, L, and A increase monotonically. The capacity of CS is $C = 12.8$Mbps. Due to the complexity of the CS system and since we focus on the performance of P2P, we will provide numerical results for the client/server system by simulation.

We investigate the influence of the maximum number n of parallel downloads at the server, the number K of fake peers, and the user's patience θ on the number of successful downloads and the expected download time. The download has to be finished within the time θ the user is willing to wait. In addition, fairness of the CDS has to be considered as well. In a fair system each user experiences a similar download duration like others.

4.1   Evaluation of the P2P Flow Model

First, we validate the analytical P2P flow model with simulation. Fig. 5(a) shows the final average population sizes of sharing peers and aborting peers over the number of

Table 1. Default parameters for evaluation of P2P and CS system

general parameters			P2P parameters		
file size	f_{size}	9.5MB	initial sharing peers	S_0	100
upload bandwidth	r_{up}	128kbps	seeder arrival rate	ν	0
download bandwidth	r_{dn}	768kbps	departure rate	η	0
initial arrival rate	λ_0	1	sharing probability	p_s	0.8
flash crowd decay	α	10^{-3}	abort probability	p_a	0.2

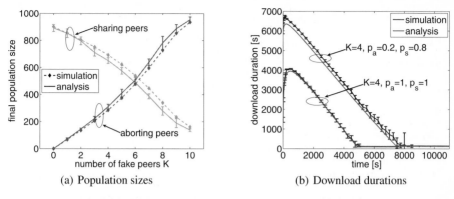

(a) Population sizes

(b) Download durations

Fig. 5. Comparison of simulation results with analytic flow model

fake peers K, when the whole population is in the absorbing states, S, L, A. The values are obtained from 20 simulation runs and error bars represent the 99% confidence intervals. The analysis matches the simulations well with only slight differences due to the underlying Markovian assumption at state transitions. The accuracy can be increased by inserting additional intermediate states at the cost of a higher computational complexity for solving the equations. Fig. 5(a) shows that a small number of $K = 10$ fake peers is almost sufficient to prevent any peer from completing the download.

Since we consider a non-stationary system, the download duration varies over time according to the current system state. Fig. 5(b) shows the average duration of a peer as function of the starting time of the download for $K = 4$. The analytical result is computed directly from Eqn. (14) and compared to values obtained from 20 simulation experiments. In both scenarios with different abort and sharing probabilities, the curves show a good match. The flash crowd arrival causes in both cases a strong increase with a linear decrease and in the case of no retrials and altruistic users ($p_a = 1, p_s = 1$), the duration is significantly smaller since peers only attempt to download the file once. On the other hand, when $p_a < 1$ the number of trials has an average greater than one resulting in longer download durations, see Eqn. (13).

4.2   Success Ratio

The performance of P2P and CS is now compared regarding the success ratio, i.e., the ratio of successful downloads to the sum of successful and aborted downloads. In order to make a fair comparison, we now use a deterministic patience threshold $\theta = 50, 100, 150, 200$ minutes after which a user in both systems cancels the download. The success ratio in P2P is 100% for $\theta > 50$ and small K, see Fig. 6(a). However, when K increases from 6 to 7, the success ratio with $\theta = 200$ reduces to about 50% and for even larger K no peer completes the download.

Fig. 6(b) shows the equivalent results for CS as function of the number of service units n. Except when n is too small, the success ratio lies above that of P2P for each θ, especially when the optimal value n^* is chosen.

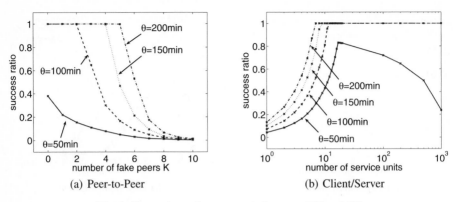

Fig. 6. Comparison of success ratio between P2P and CS

We conclude that the client/server system has at least the success ratio of P2P, if the client bandwidths are known a priori for dimensioning the optimal number of service units. The P2P system strongly suffers from the presence of too many fake peers.

4.3 Download Duration

The key performance indicator from the user's viewpoint is the overall download duration, i.e., the interval from the request of a file until its successful download. In Fig. 7(a), the time for successful downloads and the sojourn time of aborted downloads is depicted. Since the patience time is deterministic, the abort time is given as straight lines for each θ. The lines begin at values of K where the success ratios become less than 1. The successful download duration increases with K until impatience manifests itself in increased canceled downloads. Peers beginning their download later benefit from this effect. As a result the mean download time stays constant or even decreases again with K and the 99%-confidence intervals from the simulation runs increase due to the decreasing number of successful downloads which can be used to compute the averages.

The results in Fig. 7(b) show that well dimensioned systems show the best download performance. However, if the optimal capacity is a priori unknown, the P2P system

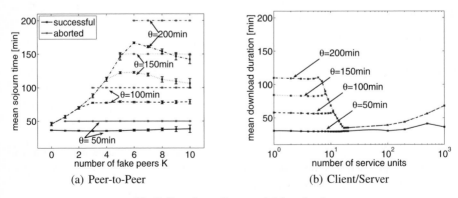

Fig. 7. Durations of successful downloads

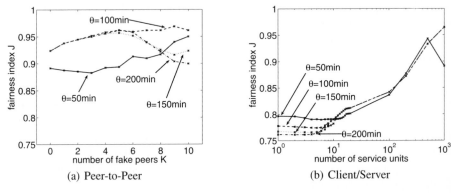

Fig. 8. Fairness index of successful downloads w.r.t. download duration

outperforms the server as the capacity of P2P increases with the number of sharers. If the peers behave altruistic, the P2P system has its advantages and might cope with even more extreme flash crowds, which will crash a server with fixed capacity. The P2P system mainly benefits from incentives and its multiple source technique when sharing already received chunks to other peers, thus fostering the cooperation among peers [2].

4.4 Fairness

We choose the fairness indicator $J = \left(1 + c_x^2\right)^{-1}$ given in [14] which returns values between 0 and 1. Low values of the fairness index indicate an unfair system, while a fairness index of one describes a completely fair system, where all users experience exactly the same download time. The term c_x^2 is the coefficient of variance of the considered performance measure x, which is the download time a user experiences. Independent of the number of fake peers K or the patience time θ, the P2P system is a more fair system with higher fairness index above 0.9, cf. Fig. 8(a). On the other hand, CS reaches such fairness only for very large n in Fig. 8(b). In that case, the average download time, however, is larger than in the P2P system (for a small number of fake peers).

We can conclude that a well dimensioned CS with a priori knowledge of the clients' bandwidths outperforms P2P at the cost of fairness. Furthermore, we could see that the influence from only few fake peers is sufficient to severely cut down the performance of the P2P system.

5 Conclusion

In this paper we presented a flow model for a P2P file sharing network and compared its performance to a client/server system. While in general it is not easy to compare both types of networks due to their inherently different structures, we could qualitatively investigate both architectures under comparable situations.

Basically, when it comes to the reliability, servers are the better choice, as manipulated data is not being injected into the network. However, from the view of the end user, the same effect may be experienced when downloading from a trusted server as with P2P networks with pollution or poisoning. Especially, when the request arrival rate is high, the waiting time until the download can be processed or its duration may become too long. The problems in CS performance can be overcome by adding further server capacity.

On the other hand, P2P systems can be easily made inoperable when many fake sources exist. If the initial number of sources is small there is a risk of these peers leaving the system which would make the network lose content due to churn. For this reason, it is important that incentives are being provided to peers to increase the willingness to share the data. Enhanced error detection mechanisms must be provided to reduce the number of retransmission in case of errors. This could be done in combination with a caching peer which acts like a server but whose content is being determined by the requests of the peers.

References

1. Christin, N., Weigend, A.S., Chuang, J.: Content availability, pollution and poisoning in file sharing peer-to-peer networks. In: Proc. of ACM EC'05, Vancouver, BC (2005)
2. Schlosser, D., Hoßfeld, T., Tutschku, K.: Comparison of robust cooperation strategies for P2P content distribution networks with multiple source download. In: Proc. of IEEE P2P2006, Cambridge, UK (2006)
3. Saroiu, S., Gummadi, K.P., Dunn, R.J., Gribble, S.D., Levy, H.M.: An analysis of internet content delivery systems. In: Proc. of USENIX OSDI 2002, Boston, MA (2002)
4. Androutsellis-Theotokis, S., Spinellis, D.: A survey of peer-to-peer content distribution technologies. ACM Computing Surveys 36, 335–371 (2004)
5. Zerfiridis, K.G., Karatza, H.D.: File distribution using a peer-to-peer network–a simulation study. Journal of Systems and Software 73, 31–44 (2004)
6. Hoßfeld, T., Leibnitz, K., Pries, R., Tutschku, K., Tran-Gia, P., Pawlikowski, K.: Information diffusion in eDonkey-like P2P networks. In: ATNAC 2004, Bondi Beach, Australia (2004)
7. Han, P., Hosanagar, K., Tan, Y.: Diffusion of digital products in peer-to-peer networks. In: 25th Intern. Conf. on Information Systems, Washington, DC (2004)
8. Becker, J.U., Clement, M.: The economic rationale of offering media files in peer-to-peer networks. In: Proc. of HICSS-37, Hawaii, HI (2004)
9. Qiu, D., Srikant, R.: Modeling and performance analysis of BitTorrent-like peer-to-peer networks. In: ACM SIGCOMM'04, Portland, OR (2004)
10. Rubenstein, D., Sahu, S.: Can unstructured P2P protocols survive flash crowds? IEEE/ACM Trans Netw. 13, 501–512 (2005)
11. Carofiglio, G., Gaeta, R., Garetto, M., Giaccone, P., Leonardi, E., Sereno, M.: A statistical physics approach for modelling P2P systems. In: ACM MAMA'05, Banff, Canada (2005)
12. Leibnitz, K., Hoßfeld, T., Wakamiya, N., Murata, M.: On pollution in eDonkey-like peer-to-peer file-sharing networks. In: Proc. of GI/ITG MMB 2006, Nuremberg, Germany (2006)
13. Murray, J.: Mathematical Biology, I: An Introduction. Springer, Heidelberg (2002)
14. Jain, R., Chiu, D., Hawe, W.: A Quantitative Measure of Fairness and Discrimination for Resource Allocation in Shared Systems. Technical Report DEC TR- 301, Digital Equipment Corporation (1984)

Globally Optimal User-Network Association in an 802.11 WLAN & 3G UMTS Hybrid Cell[*]

Dinesh Kumar[1], Eitan Altman[1], and Jean-Marc Kelif[2]

[1] INRIA, Sophia Antipolis, France
{dkumar, altman}@sophia.inria.fr
[2] France Telecom R&D, Issy les Moulineaux, France
jeanmarc.kelif@orange-ft.com

Abstract. We study globally optimal user-network association in an integrated WLAN and UMTS *hybrid cell*. The association problem is formulated as a generic MDP (Markov Decision Process) connection routing decision problem. In the formulation, mobile arrivals are assumed to follow Poisson process and a *uniformization* technique is applied in order to transform the otherwise state-dependent mobile departures into an i.i.d. process. We solve the MDP problem using a particular network model for WLAN and UMTS networks and with rewards comprising financial gain and throughput components. The corresponding Dynamic Programming equation is solved using Value Iteration and a stationary optimal policy with neither convex nor concave type switching curve structure is obtained. Threshold type and symmetric switching curves are observed for the analogous homogenous network cases.

1 Introduction

Consider a hybrid network comprising two independent 802.11 WLAN and 3G UMTS networks, that offers connectivity to mobile users arriving in the combined coverage area of these two networks. By independent we mean that transmission activity in one network does not create interference in the other. Our goal in this paper is to study the dynamics of optimal user-network association in such a WLAN-UMTS hybrid network. We concentrate only on streaming and interactive (HTTP like) data transfers. Note that we do not propose a full fledged cell-load or interference based connection admission control (CAC) policy in this paper. We instead assume that a CAC precedes the association decision control. Thereafter, an association decision only ensures an optimal performance of the hybrid cell and it is not proposed as an alternative to the CAC decision.

Study of WLAN-UMTS hybrid networks is an emerging area of research and not much related work is available. Authors in some related papers [1,2,3,4] have studied some common issues but questions related to load balancing or optimal user-network association have not been explored much. Premkumar et al. in [5]

---

[*] This work was sponsored through a research contract between INRIA and France Telecom R&D.

L. Mason, T. Drwiega, and J. Yan (Eds.): ITC 2007, LNCS 4516, pp. 1173–1187, 2007.
© Springer-Verlag Berlin Heidelberg 2007

propose a *near optimal* solution for a hybrid network within a combinatorial optimization framework, which is different from our approach. To the best of our knowledge, ours is the first attempt to present a generic formulation of the user-network association problem under an MDP decision control framework.

2   Framework for the Decision Control Problem

Our focus is only on a single pair of an 802.11 WLAN Access Point (AP) and 3G UMTS Base Station (NodeB) that are located sufficiently close to each other so that mobile users arriving in the combined coverage area of this AP-NodeB pair have a choice to connect to either of the two networks. We call the combined coverage area of a single AP cell and a single NodeB micro-cell [9] as a *hybrid cell*. The cell coverage radius of a UMTS micro-cell is usually around $400m$ to $1000m$ whereas that of a WLAN cell varies from a few tens to a few hundreds of meters. Therefore, some mobiles arriving in the hybrid cell may only be able to connect to the NodeB, either because they fall outside the transmission range of the AP or they are equipped with only 3G technology electronics. While other mobiles that are equipped with only 802.11 technology can connect exclusively to the WLAN AP. Apart from these two categories, mobiles equipped with both 802.11 WLAN and 3G UMTS technologies can connect to any one of the two networks. The decision to connect to either of the two networks may involve a utility criteria that could comprise the total packet throughput of the hybrid network. Moreover, the connection or association decision involves two different possible decision makers, the mobile user and the network operator. We focus only on the globally optimal control problem in which the network operator dictates the decision of mobile users to connect to one of the two networks, so as to optimize a certain global cell utility. In Section 3, we model this global optimality problem under an MDP (Markov Decision Process) control framework. Our MDP control formulation is a *generic* formulation of the user-network association problem in a WLAN-UMTS hybrid network and is independent of the network model assumed for WLAN and UMTS networks. Thereafter in Section 5, we solve the MDP problem assuming a particular network model (described in Section 4) which is based on some reasonable simplifying assumptions.

2.1   Mobile Arrivals

We model the hybrid cell of an 802.11 WLAN AP and a 3G UMTS NodeB as a two-server processing system (Figure 1) with each server having a separate finite capacity of M_A and M_N mobiles, respectively. Mobiles are considered as candidates to connect to the hybrid cell only after being admitted by a CAC such as the one described in [6]. Some of the mobiles (after they have been admitted by the CAC) can connect only to the WLAN AP and some others only to the UMTS NodeB. These two set of mobiles (or sessions) are each assumed to constitute two separate dedicated arrival streams with Poisson rates λ_A and λ_N, respectively. The remaining set of mobiles which can connect to both networks

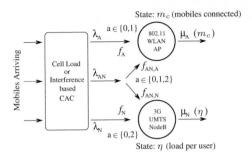

Fig. 1. Hybrid cell scenario

form a common arrival stream with Poisson rate λ_{AN}. The mobiles of the two dedicated streams can either directly join their respective AP or NodeB network without any connection decision choice involved, or they can be rejected. For mobiles of common stream, either a rejection or a connection routing decision has to be taken as to which of the two networks will the arriving mobiles join while optimizing a certain utility.

2.2   Service Requirements and Departure Rates

It is assumed that all arriving mobiles have a downlink data service requirement which is exponentially distributed with parameter ζ. In other words, every arriving mobile seeks to download a data file of average size $1/\zeta$ bits on the downlink. Let $\theta_A(m_c)$ denote the downlink packet (or file) throughput of each mobile in the AP network when m_c mobiles are connected to it at any given instant. If η_{DL} denotes the *total cell load* in downlink of the NodeB cell, then assuming N active mobiles to be connected to the NodeB, $\eta \triangleq \frac{\eta_{DL}}{N}$ denotes the average *load per user* in the cell (Chapter 8 in [9]). Let $\theta_N(\eta)$ denote the downlink packet (or file) throughput of each mobile in the NodeB network when its average load per user is η. With these notations, the effective departure rates of mobiles (or sessions) in each network or server can be denoted by,

$$\mu_A(m_c) = \zeta \cdot \theta_A(m_c) \quad \text{and} \quad \mu_N(\eta) = \zeta \cdot \theta_N(\eta). \tag{1}$$

3   MDP Control Formulation

For a globally optimal decision control it is the network operator that takes the decision for each mobile as to which of the AP or NodeB networks the mobile will connect to, after it has been admitted into the hybrid cell by the CAC controller (Figure 1). Since decisions have to be made in continuous time, this gives a *continuous time* MDP structure [14] to the decision problem and we state the equivalent MDP problem as follows:

- *States:* The state of a hybrid cell system is denoted by the tuple (m_c, η) where m_c ($m_c \in \mathbb{Z}, 0 \le m_c \le M_A$) denotes the number of mobiles connected to the AP and η ($\eta \in \mathbb{R}, 0.05 \leq \eta \leq 0.9$) is the load per user of the NodeB cell (see Sections 4.2 & 4.3 for details on bounds for m_c and η).

- *Events:* We consider two distinguishable events: (i) arrival of a new mobile after admission by CAC, (ii) departure of a mobile after service completion.
- *Decisions:* For mobiles arriving in the common stream a decision action $a \in \{0, 1, 2\}$ must be taken. $a = 0$ represents rejecting the mobile, $a = 1$ represents routing the mobile connection to AP network and $a = 2$ represents routing it to NodeB network. For the dedicated arrival streams to AP and NodeB, a decision action $a \in \{0, 1\}$ & $a \in \{0, 2\}$, respectively, must be taken.
- *Rewards:* Whenever a new incoming mobile is either rejected or routed to one of the two networks, it generates an instantaneous *state dependent* reward. $R_A(m_c, \eta; a)$ and $R_N(m_c, \eta; a)$ denote the rewards generated at dedicated arrival streams for AP and NodeB, respectively, when action 'a' is taken and the state of the system is (m_c, η). Similarly, $R_{AN}(m_c, \eta; a)$ denotes the reward generated at the common stream.

The criterion is to maximize the total expected discounted reward over an infinite horizon and obtain a *deterministic* and *stationary* optimal policy. Note that in the MDP problem statement above, state transition probabilities have not been mentioned because depending on the action taken, the system moves into a unique new state deterministically, i.e., w.p. 1. For instance when action $a = 1$ is taken, the state evolves from (m_c, η) to the unique new state (m_c+1, η).

Now, it is important to see that though events corresponding to mobile arrivals follow the Poisson process and are i.i.d., events corresponding to mobile departures are not i.i.d.. This is so because the departure rate of mobiles in both AP and NodeB depends on the state of the system (Equations 1) and is not fixed. However, applying the well-known *uniformization* technique from [13] we can introduce *virtual* (or dummy) departure events in the departure process. We can thus say that events (i.e., both arrival and departure) occur at the jump times of the combined Poisson process of all types of events with rate $\Lambda := \lambda_A + \lambda_N + \lambda_{AN} + \breve{\mu}_A + \breve{\mu}_N$, where $\breve{\mu}_A := \max_{m_c} \mu_A(m_c)$ and $\breve{\mu}_N := \max_{\eta} \mu_N(\eta)$. The departure of a mobile is now considered as either a real departure or a virtual departure. Then, any event occurring corresponds to an arrival on the dedicated streams with probability λ_A/Λ and λ_N/Λ, an arrival on the common stream with probability λ_{AN}/Λ, a real departure with probability $\mu_A(m_c)/\Lambda$ or $\mu_N(\eta)/\Lambda$ and a virtual departure with probability $1 - (\lambda_A + \lambda_N + \lambda_{AN} + \mu_A(m_c) + \mu_N(\eta))/\Lambda$. As a result, the time *periods* between consecutive events (including virtual departures) are i.i.d. and we can consider an n-stage *discrete time* MDP decision problem [14]. Let $V_n(m_c, \eta)$ denote the maximum expected n-stage discounted reward for the hybrid cell when the system is in state (m_c, η). The stationary optimal policy that achieves the maximum total expected discounted reward over an infinite horizon can then be obtained as a solution of the n-stage problem as $n \to \infty$ [14].

The discount factor is denoted by γ ($\gamma \in \mathbb{R}, 0 < \gamma < 1$) and determines the relative worth of present reward v/s future rewards. State (m_c, η) of the system is observed right after the occurrence of an event, for example, right after a newly arrived mobile in the common stream has been routed to one of the two networks, or right after the departure of a mobile. Given that an arrival event has occurred and that action 'a' will be taken for this newly arrived mobile, let $U_n(m_c, \eta; a)$ denote the maximum expected $n-$stage discounted reward for the hybrid cell when the system is in state (m_c, η). We can then write down the following recursive Dynamic Programming (DP) [14] equation to solve our MDP decision problem,

$\forall n \geq 0$ and $0 \leq m_c \leq M_A$, $0.05 \leq \eta \leq 0.9$,

$$
\begin{aligned}
V_{n+1}(m_c, \eta) = {} & \frac{\lambda_A}{\Lambda} \max_{a \in \{0,1\}} \{R_A(m_c, \eta; a) + \gamma\, U_n(m_c, \eta; a)\} \\
& + \frac{\lambda_N}{\Lambda} \max_{a \in \{0,2\}} \{R_N(m_c, \eta; a) + \gamma\, U_n(m_c, \eta; a)\} \\
& + \frac{\lambda_{AN}}{\Lambda} \max_{a \in \{0,1,2\}} \{R_{AN}(m_c, \eta; a) + \gamma\, U_n(m_c, \eta; a)\} \qquad (2) \\
& + \frac{\mu_A(m_c)}{\Lambda} \gamma\, V_n(m_c', \eta) \quad + \quad \frac{\mu_N(\eta)}{\Lambda} \gamma\, V_n(m_c, \eta') \\
& + \frac{\Lambda - (\lambda_A + \lambda_N + \lambda_{AN} + \mu_A(m_c) + \mu_N(\eta))}{\Lambda} \gamma\, V_n(m_c, \eta),
\end{aligned}
$$

where, states (m_c', η) and (m_c, η') are the new states that the system evolves into when a departure occurs at AP and NodeB, respectively. The fact that dedicated stream mobiles can only join one network or the other has been incorporated in the first two terms in R.H.S. Equation 2 is a very *generic* formulation of our user-network association decision problem and it can be solved using any particular definition for the rewards and the new states (m_c', η) and (m_c, η'). In Section 5, we will solve the DP formulation of Equation 2 assuming a specific definition for the rewards based on throughput expressions obtained from a specific network model for the WLAN and UMTS networks. We first present this network model in the following section along with some simplifying assumptions.

4   WLAN and UMTS Network Models

Since the bulk of data transfer for a mobile engaged in streaming or interactive data transmission is carried over downlink (AP to mobile or NodeB to mobile) and since TCP is the most commonly used transport protocol (streaming protocols based on TCP also exist, e.g., Real Time Streaming Protocol), we are interested here in network models for computing TCP throughput on downlink.

4.1   Simplifying Assumptions

Assumption on QoS and TCP: We assume a single QoS class of arriving mobiles so that each mobile has an identical minimum downlink throughput requirement of θ_{min}, i.e., each arriving mobile must achieve a downlink packet throughput of at least θ_{min} bps in either of the two networks. Several versions of TCP have been proposed in literature for wireless environments. For our purposes we assume that the wireless TCP algorithm operates in *split mode* [15]. In brief, the split mode divides the TCP connection into wireless and wired portions, and acks are generated for both portions separately. Therefore, in our hybrid cell scenario TCP acks are generated separately for the single hop between mobiles and AP or NodeB. It is further assumed that each mobile's or receiver's advertised window W^* is set to 1 in the wireless portion of TCP protocol. This is in fact known to provide the best performance of TCP in a single hop case (see [7,8] and references therein).

Resource allocation in AP: We assume *saturated resource allocation* in the downlink of AP and NodeB networks. Specifically, this assumption for the AP network means the following. The AP is *saturated* and has infinitely many packets backlogged in its transmission buffer, i.e., there is always a packet in the AP's transmission buffer waiting to be transmitted to each of the connected mobiles. Now, in a WLAN resource allocation to the AP on downlink is carried out through the contention based DCF (Distributed Coordination Function) protocol. If the AP is saturated for a particular mobile's connection and W^* is set to 1, then this particular mobile can benefit from higher number of transmission opportunities (*TxOPs*) won by the AP for downlink transmission to this mobile (hence higher downlink throughput), than if the AP was not saturated or W^* was not set to 1. Thus, mobiles can be allocated downlink packet throughputs greater than their QoS requirements of θ_{min} and cell resources in terms of *TxOPs* on the downlink will be maximally utilized.

Resource allocation in NodeB: For the NodeB network it is assumed that at any given instant, the NodeB cell resources on downlink are fully utilized resulting in a constant maximum cell load of η_{DL}^{max}. This is analogous to the maximal utilization of *TxOPs* in the AP network discussed above. With this maximum cell load assumption even if a mobile has a minimum packet throughput requirement of only θ_{min} bps, it can actually be allocated a higher throughput if additional unutilized cell resources are available, so that the cell load is always at its maximum of η_{DL}^{max}. If say a new mobile j arrives and if it is possible to accommodate its connection while maintaining the QoS requirements of the presently connected mobiles (this will be decided by the CAC), then the NodeB will initiate a *renegotiation* of QoS attributes (or bearer attributes) procedure with all the presently connected mobiles. All presently connected mobiles will then be allocated a lower throughput than the one prior to the set-up of mobile j's connection. However, this new lower throughput will still be higher than each mobile's QoS requirement. This kind of a renegotiation of QoS attributes is a special feature in UMTS ([9], Chapter 7). Also note a key point here that the average load per user, η (Section 2.2), decreases with increasing number of mobiles connected to the NodeB. Though the total cell load is always at its maximum of η_{DL}^{max}, contribution to this total load from a single mobile (i.e., load per user, η) decreases as more mobiles connect to the NodeB cell. We define $\Delta_d(\eta)$ and $\Delta_i(\eta)$ as the average change in η caused by a new mobile's connection and an already connected mobile's disconnection, respectively. Therefore, when a new mobile connects the load per user drops from η to $\eta - \Delta_d(\eta)$ and when a mobile disconnects the load per user increases from η to $\eta + \Delta_i(\eta)$.

Power control & location of mobiles in NodeB: In downlink, the inter-cell to intra-cell interference ratio denoted by i_j and the orthogonality factor denoted by α_j are different for each mobile j depending on its location in the NodeB cell. Moreover, the throughput achieved by each mobile is interference limited and depends on the signal to interference plus noise ratio (SINR) received at that mobile. Thus, in the absence of any power control the throughput also depends on the location of mobile in the NodeB cell. We however assume a uniform SINR scenario where closed-loop fast power control is applied in the NodeB cell so that each mobile receives approximately the same SINR ([9], Section 3.5). We therefore assume that all mobiles in the NodeB cell are allocated equal throughputs. This kind of a power control will allocate more power to users far away from the NodeB that are subject to higher path-loss, fading and neighboring cell interference. Users closer to the NodeB will be allocated relatively less power since they are susceptible to weaker signal attenuation. In fact, such a fair throughput allocation can also be achieved by adopting a fair and power-efficient channel dependent scheduling scheme as described in [10]. Now since all mobiles are allocated equal throughputs, it can be said that mobiles arrive at an *average* location in the NodeB cell. Therefore all mobiles are assumed to have an identical average inter-cell to intra-cell interference ratio $\bar{i}$ and an identical average orthogonality factor $\bar{\alpha}$ ([9], Section 8.2.2.2).

Justification: The assumption on saturated resource allocation is a standard assumption, usually adopted to simplify modeling of complex network frameworks like those of WLAN and UMTS [9,11]. Mobiles in NodeB cell are assumed to be allocated equal throughputs in order to have a comparable scenario to that of an AP cell in which mobiles are also known to achieve fair and equal throughput allocation (Section 4.2). The assumption of mobiles arriving at an average location in the NodeB cell is essential in order to simplify our MDP formulation. For instance, without this assumption the hybrid network system state will have to include the location of each mobile. This will result in an MDP problem with higher dimensional state space which is known to be analytically intractable and not have an exact solution [14]. We therefore assume mobiles arriving at an average location and seek to compute the optimal association policy more from a network planning and dimensioning point of view.

4.2 Downlink Throughput in 802.11 WLAN AP

We reuse the downlink TCP throughput formula for a mobile in a WLAN from [12]. Here we briefly mention the network model that has been extensively studied in [12] and then simply restate the throughput expression without going into much details. Each mobile connected to the AP uses the Distributed Coordination Function (DCF) protocol with an RTS/CTS frame exchange before any

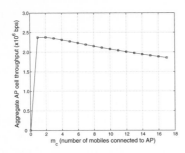

Fig. 2. Total throughput in AP cell   **Fig. 3.** Total throughput in NodeB cell

data-ack frame exchange and each mobile (including the AP) has an equal probability of the channel being allocated to it. The AP does not employ any rate control algorithm and transmits at a fixed PHY data rate of R_{data} bps to all mobiles. If the AP is always saturated then with the assumption of W^* being set to 1 (Section 4.1), the average number of backlogged mobiles contending for the channel is given by $m_b = 1 + \frac{m_c}{2}$ [12]. Based on this assumption the downlink TCP throughput of a single mobile is given by Section 3.2 in [12] as,

$$\theta_A(m_c) = \frac{L_{TCP}}{m_c(T_{TCPdata} + T_{TCPack} + 2T_{tbo} + 2T_w)}, \qquad (3)$$

where L_{TCP} is the size of TCP packets and $T_{TCPdata}$ and T_{TCPack} are the raw transmission times of a TCP data and a TCP ack packet, respectively. T_{tbo} and T_w denote the mean total time spent in *back-off* and the average total time wasted in collisions for any successful packet transmission and are computed assuming m_b backlogged mobiles. The explicit expressions for $T_{TCPdata}, T_{TCPack}, T_{tbo}$ and T_w can be referred to in [12]. However, we mention here that they depend on certain quantities whose numerical values have been provided in Section 5.2. Note that all mobiles connected to the AP achieve equal downlink TCP throughputs (given by Equation 3) in a fair manner [12]. Figure 2 shows a plot of *total* cell throughput in an AP cell for an example scenario. Since the total throughput monotonically decreases with increasing number of mobiles, the capacity of an AP cell, M_A, is limited by the QoS requirement θ_{min} bps of each mobile.

4.3   Downlink Throughput in 3G UMTS NodeB

Consider a standard model for the NodeB cell. Let W be the WCDMA modulation bandwidth and if $SINR$ denotes the signal to interference plus noise ratio received at a mobile then its energy per bit to noise density ratio is given by,

$$\frac{E_b}{N_o} = \frac{W}{\theta_N} \times SINR. \qquad (4)$$

Now, under the assumptions of identical throughput allocation to each mobile arriving at an average location and application of power control so that each

Table 1.

η	$log(\eta)$	$N(\eta)$	$SINR$ (dB)	θ_N (kbps)	$\frac{E_b}{N_o}$ (dB)
0.9	−0.10536	1	0.8423	572	9.0612
0.45	−0.79851	2	−2.1804	465	6.9503
0.3	−1.204	3	−3.7341	405	5.7894
0.225	−1.4917	4	−5.1034	360	5.0515
0.18	−1.7148	5	−6.0327	322	4.5669
0.15	−1.8971	6	−6.5093	285	4.3052
0.1286	−2.0513	7	−7.2075	242	4.3460
0.1125	−2.1848	8	−8.8312	191	4.7939
0.1	−2.3026	9	−8.9641	144	5.5091
0.09	−2.4079	10	−9.1832	115	6.0281
0.0818	−2.5033	11	−9.9324	96	6.3985
0.0750	−2.5903	12	−10.1847	83	6.6525
0.0692	−2.6703	13	−10.7294	73	6.8625
0.0643	−2.7444	14	−10.9023	65	7.0447
0.06	−2.8134	15	−10.9983	60	7.0927
0.0563	−2.8779	16	−11.1832	55	7.1903
0.0529	−2.9386	17	−11.3802	51	7.2549
0.05	−2.9957	18	−11.9231	47	7.3614

mobile receives the same SINR (Section 4.1), we deduce from Equation 4 that each mobile requires the same E_b/N_o ratio in order to be able to successfully decode NodeB's transmission. From Chapter 8 in [9] we can thus say that the downlink TCP throughput θ_N of any mobile, in a NodeB cell with saturated resource allocation, as a function of load per user η is given by,

$$\theta_N(\eta) = \frac{\eta W}{(E_b/N_o)(1 - \bar{\alpha} + \bar{i})}, \tag{5}$$

where $\bar{\alpha}$ and $\bar{i}$ have been defined earlier in Section 4.1. For an example scenario, Figure 3 shows a plot of *total* cell throughput of all mobiles (against $log(\eta)$) in a UMTS NodeB cell. The load per user η has been stretched to a logarithmic scale for better presentation. Also note that throughput values have been plotted in the second quadrant. As we go away from origin on the horizontal axis, $log(\eta)$ (and η) decreases or equivalently number of connected mobiles increase. The equivalence between η and $log(\eta)$ scales and number of mobiles $N(\eta)$ can be referred to in Table 1.

It is to be noted here that the required E_b/N_o ratio by each mobile is a function of its throughput. Also, if the NodeB cell is fully loaded with $\eta_{DL} = \eta_{DL}^{max}$ and if each mobile operates at its minimum throughput requirement of θ_{min} then we can easily compute the capacity, M_N, of the cell as,

$$M_{3G} = \frac{\eta_{DL}^{max} W}{\theta_{min}(E_b/N_o)(1 - \bar{\alpha} + \bar{i})}. \tag{6}$$

For $\eta_{DL}^{max} = 0.9$, $\theta_{min} = 46$ kbps and a typical NodeB cell scenario that employs the closed-loop fast power control mechanism mentioned previously in Section 4.1, Table 1 shows the SINR (fourth column) received at each mobile as a function of the average load per user (first column). Note that we consider a maximum cell load of 0.9 and not 1 in order to avoid instability conditions in the cell. These values of SINR have been obtained at France Telecom R&D from radio layer simulations of a NodeB cell. The fifth column shows the downlink packet throughput with a block error rate (BLER) of 10^{-2} that can be achieved by each mobile as a function of the SINR observed at that mobile. And the sixth column lists the corresponding values of E_b/N_o ratio (obtained from Equation 4) that are required at each mobile to successfully decode NodeB's transmission.

5   Solving the MDP Control Problem

As mentioned earlier, the MDP formulation can be solved for any given definition of rewards. Here we will motivate the choice of a particular definition based on aggregate throughput of WLAN and UMTS networks.

5.1   Defining the Rewards and State Evolution

If we consider the global performance of hybrid cell in terms of throughput and financial revenue earned by the network operator, it is natural from the network operator's point of view to maximize both aggregate network throughput and financial revenue. Except for a certain band of values of η (or $log(\eta)$), generally the aggregate throughput of an AP or NodeB cell drops when an additional new mobile connects to it (Figures 2 & 3). However, the network operator gains some financial revenue from the mobile user at the same time. There is thus a trade-off between revenue gain and the aggregate network throughput which motivates us to formulate an instantaneous, *state dependent*, linear (non-linear can also be considered) reward as follows. The reward consists of the sum of a fixed financial revenue price component and β times an aggregate network throughput component which is state dependent. Here β is an appropriate proportionality constant. When a mobile of the dedicated arrival streams is routed to the corresponding AP or NodeB, it generates a financial revenue of f_A and f_N, respectively. A mobile of the common stream generates a financial revenue of $f_{AN,A}$ on being routed to the AP and $f_{AN,N}$ on being routed to the NodeB. Any mobile that is rejected does not generate any financial revenue. The throughput component of the reward is represented by the aggregate network throughput of the corresponding AP or NodeB network to which a newly arrived mobile connects, taking into account the *change* in the state of the system caused by this new mobile's connection. Where as, if a newly arrived mobile in a dedicated stream is rejected then the throughput component represents the same, but taking into account the *unchanged* state of the system. For a rejected mobile belonging to the common stream, it is the maximum of the aggregate throughputs of the two networks that is considered.

With the foregoing discussion in mind, we may define the instantaneous reward functions R_A, R_N and R_{AN} introduced earlier in Section 3 as,

$$
R_A(m_c, \eta; a) = \begin{cases} \beta\, m_c\, \theta_A(m_c) & : a = 0 \\ f_A + \beta\,(m_c + 1)\,\theta_A(m_c + 1) & : a = 1, m_c < M_A \\ \beta\, m_c\, \theta_A(m_c) & : a = 1, m_c = M_A \end{cases} \tag{7}
$$

$$R_N(m_c, \eta; a) = \begin{cases} \beta \, N(\eta) \, \theta_N(\eta) & : a = 0 \\ f_N + \beta \, N(\eta - \Delta_d(\eta)) \, \theta_N(\eta - \Delta_d(\eta)) & : a = 2, N(\eta) < M_N \\ \beta \, N(\eta) \, \theta_N(\eta) & : a = 2, N(\eta) = M_N \end{cases} \tag{8}$$

$$R_{AN}(m_c, \eta; a) = \begin{cases} \max\{\beta \, m_c \, \theta_A(m_c), \beta \, N(\eta) \, \theta_N(\eta)\} & : a = 0 \\ f_{AN,A} + \beta \, (m_c + 1) \, \theta_A(m_c + 1) & : a = 1, m_c < M_A \\ \beta \, m_c \, \theta_A(m_c) & : a = 1, m_c = M_A \\ f_{AN,N} + \beta \, N(\eta - \Delta_d(\eta)) \, \theta_N(\eta - \Delta_d(\eta)) & : a = 2, N(\eta) < M_N \\ \beta \, N(\eta) \, \theta_N(\eta) & : a = 2, N(\eta) = M_N \end{cases} \tag{9}$$

where, $\theta_A(\cdot)$ and $\theta_N(\cdot)$ have been defined earlier in Equations 3 & 5 and $N(\cdot)$ can be obtained from Table 1. Note that the discount factor, γ, has already been incorporated in Equation 2. Also, based on the discussion in Section 4.1 we may define the new states at departure events as,

$$(m'_c, \eta) = ((m_c - 1) \vee 0, \eta) \quad \text{and} \quad (m_c, \eta') = (m_c, (\eta + \Delta_i(\eta)) \wedge 0.9), \tag{10}$$

for departures at AP and NodeB, respectively. Additionally, the following entities that were introduced in Section 3 may be defined as, $U_n(m_c, \eta; 0) := V_n(m_c, \eta)$, $U_n(m_c, \eta; 1) := V_n((m_c + 1) \wedge M_A, \eta)$ and $U_n(m_c, \eta; 2) := V_n(m_c, (\eta - \Delta_d(\eta)) \vee 0.05)$ for $\theta_{min} = 46$ kbps (Table 1).

5.2   Numerical Analysis

The focus of our numerical analysis is to study the optimal association policy under an ordinary network scenario. We do not investigate in detail the effects of specific TCP parameters and it is outside the scope of this paper. Plugging Equations 7, 8, 9 & 10 in the Dynamic Programming Equation 2, we solve it for an ordinary scenario using the Value Iteration method [14]. The scenario that we consider is as follows: $L_{TCP} = 8000$ bits (size of TCP packets), $L_{MAC} = 272$ bits, $L_{IPH} = 320$ bits (size of MAC and TCP/IP headers), $L_{ACK} = 112$ bits (size of MAC layer ACK), $L_{RTS} = 180$ bits, $L_{CTS} = 112$ bits (size of RTS and CTS frames), $R_{data} = 11$ Mbits/s, $R_{control} = 2$ Mbits/s (802.11 PHY data transmission and control rates), $CW_{min} = 32$ (minimum 802.11 contention window), $T_P = 144\mu s$, $T_{PHY} = 48\mu s$ (times to transmit the PLCP preamble and PHY layer header), $T_{DIFS} = 50\mu s$, $T_{SIFS} = 10\mu s$ (distributed inter-frame spacing time and short inter-frame spacing time), $T_{slot} = 20\mu s$ (slot size time), $K = 7$ (*retry limit* in 802.11 standard), $b_0 = 16$ (initial mean back-off), $p = 2$ (exponential back-off multiplier), $\gamma = 0.8$, $\lambda_A = 0.03$, $\lambda_N = 0.03$, $\lambda_{AN} = 0.01$, $1/\zeta = 10^6$ bits, $\beta = 10^{-6}$, $M_A = 18$ and $M_N = 18$ for $\theta_{min} = 46$ kbps, $\bar{\alpha} = 0.9$ for ITU Pedestrian A channel, $\bar{i} = 0.7$, $W = 3.84$ Mcps and other values as illustrated in Table 1.

The DP equation has been solved for three different kinds of network setups. We first study the simple *homogenous* network case where both networks are AP

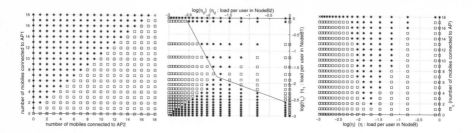

Fig. 4. Common flow policy. *First*: AP1, *Second*: AP2.   **Fig. 5.** Common flow policy. *First*: NodeB1, *Second*: NodeB2.   **Fig. 6.** Common flow policy. *First*: AP, *Second*: NodeB.

and hence an incoming mobile belonging to the common stream is offered a connection choice between two identical AP networks. Next, we study an analogous case where both networks are NodeB terminals. We study these two cases in order to gain some insight into connection routing dynamics in simple homogenous network setups before studying the third more complex, hybrid AP-NodeB scenario. Figures 4-8 show the optimal connection routing policy for the three network setups. Note that the plot in Figure 5 is in 3rd quadrant and the plots in Figures 6-8 are in 2nd quadrant. In all these figures a square box symbol ($\square$) denotes routing a mobile's connection to the *first* network, a star symbol ($*$) denotes routing to the *second* network and a cross symbol ($\times$) denotes rejecting a mobile all together.

AP-AP homogenous case: In Figure 4, optimal policy for the common stream in an AP-AP homogenous network setup is shown with $f_{A1A2,A1} = f_{A1A2,A2} = 5$ (with some abuse of notation). The optimal policy routes mobiles of common stream to the network which has lesser number of mobiles than the other one. We refer to this behavior as *mobile-balancing* network phenomenon. This happens because the total throughput of an AP network decreases with increasing number of mobiles (Figure 2). Therefore, an AP network with higher number of mobiles offers lesser reward in terms of network throughput and a mobile generates greater incentive by joining the network with fewer mobiles. Also note that the optimal routing policy in this case is *symmetric* and of *threshold type* with the threshold switching curve being the coordinate line $y = x$.

NodeB-NodeB homogenous case: Figure 5 shows optimal routing policy for the common stream in a NodeB-NodeB homogenous network setup. With equal financial incentives for the mobiles, i.e., $f_{N1N2,N1} = f_{N1N2,N2} = 5$ (with some abuse of notation), we observe a very interesting switching curve structure. The state space in Figure 5 is divided into an *L-shaped* region (at bottom-left) and a *quadrilateral shaped* region (at top-right) under the optimal policy. Each region separately, is *symmetric* around the coordinate diagonal line $y = x$. Consider the point $(log(\eta_1), log(\eta_2)) = (-0.79851, -1.4917)$ on logarithmic scale in the upper triangle of the quadrilateral region. From Table 1 this corresponds to the network state when load per user in the first NodeB network is 0.45 which is

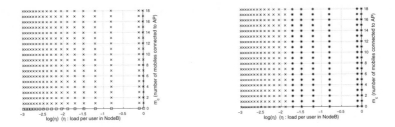

Fig. 7. Policy for AP dedicated flow in AP-NodeB hybrid cell   **Fig. 8.** Policy for NodeB dedicated flow in AP-NodeB hybrid cell

more than the load per user of 0.225 in the second NodeB network. Equivalently, there are less mobiles connected to the first network as compared to the second network. Ideally, one would expect new mobiles to be routed to the first network rather than the second network. However, according to Figure 5 in this state the optimal policy is to route to the second network even though the number of mobiles connected to it is more than those in the first. We refer to this behavior as *mobile-greedy* network phenomenon and explain the intuition behind it in the following paragraph. The routing policies on boundary coordinate lines are clearly comprehensible. On $y = -2.9957$ line when the first network is full (i.e., with least possible load per user), incoming mobiles are routed to second network (if possible) and vice-versa for the line $x = -2.9957$. When both networks are full, incoming mobiles are rejected which is indicated by the cross at coordinate point $(x, y) = (-2.9957, -2.9957)$.

The reason behind the mobile-greedy phenomenon in Figure 5 can be attributed to the fact that in a NodeB network, the total throughput increases with decreasing average load per user up to a particular threshold (say η_{thres}) and then decreases thereafter (Figure 3). Therefore, routing new mobiles to a network with lesser (but greater than η_{thres}) load per user (greater number of mobiles) results in a higher reward in terms of total network throughput, than routing new mobiles to the other network with greater load per user (lesser number of mobiles). However, the mobile-greedy phenomenon is only limited to the quadrilateral shaped region. In the L-shaped region, the throughput of a NodeB network decreases with decreasing load per user, contrary to the quadrilateral region where the throughput increases with decreasing load per user. Hence, in the L-shaped region higher reward is obtained by routing to the network having higher load per user (lesser number of mobiles) than by routing to the network with lesser load per user (greater number of mobiles). In this sense the L-shaped region shows similar characteristics to mobile-balancing phenomenon observed in AP-AP network setup (Figure 4).

AP-NodeB hybrid cell: We finally discuss now the hybrid AP-NodeB network setup. Here we consider financial revenue gains of $f_{AN,A} = 5$ and $f_{AN,N} = 6$ motivated by the fact that a network operator can charge more for a UMTS connection since it offers a larger coverage area. Moreover, UMTS equipment is more expensive to install and maintain than WLAN equipment. In Figure 6, we observe that the state space is divided into two regions by the optimal policy switching curve which is *neither convex nor concave*. Besides, in some regions of state space the mobile-balancing network phenomenon is observed, where as in some other regions the mobile-greedy network phenomenon is observed. In some sense, this can be attributed to the symmetric threshold type switching curve and the symmetric L-shaped and quadrilateral shaped regions in the corresponding AP-AP and NodeB-NodeB homogenous network setups, respectively. Figures 7 and 8 show the optimal policies for dedicated streams in an AP-NodeB hybrid cell with $f_A = f_N = 0$. The optimal policy accepts new mobiles in the AP network only when there are none already connected. This happens because initially the network throughput of an AP is zero when there are no mobiles connected and a non-zero reward is obtained by accepting a mobile. Thereafter, since $f_A = 0$ the policy rejects all incoming mobiles due to decrease in network throughput with increasing number of mobiles and hence decrease in corresponding reward. Similarly, for the dedicated mobiles to the NodeB network, the optimal policy accepts new mobiles until the network throughput increases (Figure 3) and rejects them thereafter due to absence of any financial reward component and decrease in the network throughput. Note that we have considered zero financial gains here ($f_A = f_N = 0$) to be able to exhibit existence of these *threshold type* policies for the dedicated streams.

6   Conclusion

In this paper, we have considered globally optimal user-network association (load balancing) in an AP-NodeB hybrid cell. To the best of our knowledge this study is the first of its kind. Since it is infeasible to solve an MDP formulation for an exhaustive set of network scenarios, we have considered an ordinary network scenario and computed the optimal association policy. Even though the characteristics of the solution to our particular scenario are not depictive of the complete solution space, they can certainly be helpful in acquiring an intuition about the underlying dynamics of user-network association in a hybrid cell.

References

1. Ma, et al.: A New Method to support UMTS/WLAN Vertical Handover using SCTP. IEEE Wireless Commun. 11(4), 44–51 (2004)
2. Song, et al.: Hybrid Coupling Scheme for UMTS and Wireless LAN Interworking. VTC-Fall 4, 2247–2251 (2003)
3. Vulic, et al.: Common Radio Resource Management for WLAN-UMTS Integration Radio Access Level, IST Mobile & Wireless Communications, Germany (2005)
4. Falowo, et al.: AAA and Mobility Management in UMTS-WLAN Interworking 12th International Conference on Telecommications (ICT), Cape Town (May 2005)
5. Premkumar, et al.: Optimal Association of Mobile Wireless Devices with a WLAN-3G Access Network, IEEE ICC (June 2006)

6. Yu, et al.: Efficient Radio Resource Management in Integrated WLAN/CDMA Mobile Networks. Telecommunication Systems Journal 30(1-3), 177–192 (2005)
7. Fu, et al.: The Impact of Multihop Wireless Channel on TCP Throughput and Loss, IEEE INFOCOM 2003 (2003)
8. Lebeugle, et al.: User-level performance in WLAN hotspots, ITC, Beijing (2005)
9. Holma, H., Toskala, A.: WCDMA for UMTS (Revised Edition). Wiley, Chichester (2001)
10. Zan, et al.: Fair & Power-Efficient Channel-Dependent Scheduling for CDMA Packet Nets, ICWN, USA (2003)
11. Kumar, et al.: New insights from a fixed point analysis of single cell IEEE 802.11 WLANs, IEEE Infocom, USA (March 2005)
12. Miorandi, et al.: A Queueing Model for HTTP Traffic over IEEE 802.11 WLANs. Computer Networks 50(1), 63–79 (2006)
13. Lippman, S.A.: Applying a new device in the optimization of exponential queueing systems. Operations Research, pp. 687–710 (1975)
14. Puterman, M.L.: Markov Decision Processes: Discrete Stochastic Dynamic Programming. Wiley, Chichester (1994)
15. Tian, et al.: TCP in Wireless Environments: Problems and Solutions. IEEE (Radio) Communications Magazine 43(3), S27–S32 (2005)

Author Index

Printing: Mercedes-Druck, Berlin
Binding: Stein+Lehmann, Berlin

Lecture Notes in Computer Science

For information about Vols. 1–4429

please contact your bookseller or Springer